The Handbook of Mites of Stored Products

The Handbook of Mites of Stored Products

Vincenzo Vacante and Carlos H. W. Flechtmann

CABI

CABI is a trading name of CAB International

CABI
Nosworthy Way
Wallingford
Oxfordshire OX10 8DE
UK

Tel: +44 (0)1491 832111
E-mail: info@cabi.org
Website: www.cabi.org

CABI
200 Portland Street
Boston
MA 02114
USA

Tel: +1 (617)682-9015
E-mail: cabi-nao@cabi.org

A catalogue record for this book is available from the British Library, London, UK.

ISBN-13: 9781786390899 (hardback)
9781786390905 (ePDF)
9781786390912 (ePub)

DOI: 10.1079/9781786390912.0000

Commissioning Editor: Jamie Lee
Editorial Assistant: Helen Elliott
Production Editor: James Bishop

Typeset by Straive, Pondicherry, India
Printed and bound in the UK by CPI Group (UK) Ltd, Croydon, CR4 0YY

Contents

Preface xi

Acknowledgements xiii

PART I INTRODUCTION 1

1 **Introduction to the Acari** 3
Systematic position and ecological role 3
Origins and phylogenesis 4

2 **External morphology** 7
Integument 7
Body division 7
Gnathosoma 9
Chelicerae 11
Palpi 14
Idiosoma 15
Legs 18
Respiratory organs 21
Reproductive organs 23
Sensory receptors 26
Notation of mechanosensory and chemosensory setae 28
Secretory organs 28
Osmoregulatory organs 28

3 **Internal morphology** 29
Circulatory system 29
Nervous system 29
Supraoesophageal ganglion 30
Suboesophageal ganglion 31
Reproductive system 31
Female system 31
Male system 33
Glandular system 33
Salivary glands 33
Coxal glands 33
Podocephalic glands 34
Silk glands 34
Digestive system 35
Excretory system 37
Malpighian tubules 37
Other structures 37
Water balance 37

4 **Biology** 40
Biological Instars 40
The egg 40
The prelarva 40
The larva 41
The protonymph 41
The deutonymph and the hypopus 41
The tritonymph 42
The adult 42
The reproduction 43

5 **Ecology** 45
Feeding Habits and Habitats 45
Free living species 45
Parasitic species 46
Host associations 46
Evolution of parasitism 47
Habitats 48
Habitats created by plants and fungi 49
Habitats created by animals through non-interactive processes 50
Habitats created by interaction of hosts with the environment 51
Habitats created by insects 51
Mammal hosts 53
Bird hosts 53
Habitat shifts in nest-inhabiting Astigmatid mites 53
Chemical Communication 54
Pheromones 54
Allomones 55
Kairomones 56
Synomones 56
Dispersal 56
Phoresy 57
Phoretic dispersal by deutonymphs 58
Non-phoretic dispersal by deutonymphs 59
Loss of the dispersal (deutonymphal) morph 59
Dispersal of permanent parasites 59
Dispersal and adverse changes in the environment 60
Factors Influencing Population Dynamics 60
Temperature and Relative Humidity 63
Food Resource 66
Natural Enemies 68

6 **Collecting, Detecting, Preserving, Preparing and Rearing** 71
Collecting, and Detecting 71
Sieves 71
Hand-operated vacuum apparatus 72
Tullgren-Berlese funnels 72
Mite traps 72
Flotation test 74
Immuno-diagnostic method 75
Near-infrared reflectance spectroscopy 76
Hyperspectral imaging system 76

Electronic nose 77
Packaging tests 77
Monitoring and Threshold 77
Preserving, Preparing and Rearing 78
Identification of Storage Mites 78

7 **Damage** **80**
Direct damage to stored food 80
Mites as Vectors of Micromycetes in Stored Products 84
Stored Mites and Bacteria 87
Storage Mites Allergy 88
Pulmonary acariasis 92
Enteric acariasis 92
Urinary acariasis 92
Dermatitis and other disorders 92
Systemic anaphylaxis 94
Stored Mites and Animal Health 95
Forensic Importance of Storage Mites 96

8 **Control** **98**
Food Preservation Techniques 98
Prevention 99
Physical Methods 99
Aeration 100
Low air pressure 100
Cold 100
Heat 100
Washing 101
Inert dusts 101
Screening, Sorting and Sanitation 102
Chemical Methods 102
Organophosphorus Compounds and other Chemicals 102
Fumigants 103
Methyl bromide 103
Phosphine 103
Sulfuryl fluoride 104
Ozone 104
Carbon dioxide 105
Carbon sulfide 105
Nitric oxide 105
Biological control 105
Food Grade Oils and Preservatives 106
Plant Extracts and Oils 106
Genetic Control 109
Management of Stored Mites 109

PART II **113**

9 **Classification and Key for the Identification of Major Taxa** **115**
Superorder Acariformes 115
Superorder Parasitiformes 116
Key for the Identification of the Higher Categories and Families (adults) of Mites of Stored Food or sometime occurring on Stored Food 117

Superorder Acariformes, Order Trombidiformes, Suborder Prostigmata (predators) 130
10 The Family Acarophenacidae Cross 130

11 The Family Bdellidae Dugès 132

12 The Family Caligonellidae Grandjean 134

13 The Family Cheyletidae Leach 136

14 The Family Cunaxidae Thor 146

15 The Family Eupodidae Koch 148

16 The Family Camerobiidae Southcott 149

17 The Family Pterygosomatidae Oudemans 150

18 The Family Raphignatidae Kramer 151

19 The Family Smarididae Vitzthum 154

20 The Family Erythraeidae Robineau-Desvoidy 155

21 The Family Stigmaeidae Oudemans 156

Superorder Acariformes, Order Sarcoptiformes, Suborder Endeostigmata 159
22 The Family Terpnacaridae Grandjean 159

Superorder Acariformes, Order Trombidiformes, Suborder Prostigmata (injurious) 161
23 The Family Pyemotidae Amerling 161
Morphological Characteristics, Systematics and Bio-ecology 161
The Genus *Pyemotes* Amerling 163

24 The Family Tarsonemidae Canestrini *et* Fanzago 168
Morphological Characteristics, Systematics and Bio-ecology 168
The Genus *Tarsonemus* Canestrini *et* Fanzago 171

25 The Family Tydeidae Kramer 175
Morphological characteristics, systematic and bio-ecology 175

26 Family Ereynetidae Oudemans 178

Superorder Acariformes, Order Sarcoptiformes, Suborder Oribatida, Cohort Astigmatina 179
27 The Family Lardoglyphidae Oudemans 179
Morphological Characteristics, Systematics and Bio-ecology 179
The Genus *Lardoglyphus* Oudemans 179

28 The Family Acaridae Latreille 184
Morphological Characteristics, Systematics and Bio-ecology 184
Key for the identification of Genera 186
The Genus *Acarus* Linnaeus 187
The Genus *Tyrolichus* Oudemans 206
The Genus *Tyrophagus* Oudemans 208
The Genus *Tyroborus* Oudemans 242
The Genus *Aleuroglyphus* Zachvatkin 245
The Genus *Mycetoglyphus* Oudemans 250
The Genus *Rhizoglyphus* Claparède 252
The Genus *Sancassania* Oudemans 265

The Genus Thyreophagus Rondani 271
Other genera 275

29 The Family Carpoglyphidae Oudemans **276**
Morphological characteristics, systematic and bio-ecology 276
The Genus *Carpoglyphus* Robin 277

30 The Family Chortoglyphidae Berlese **287**
Morphological characteristics, systematic and bio-ecology 287
The Genus *Chortoglyphus* Berlese 287

31 The Family Aëroglyphidae Zachvatkin **292**
Morphological characteristics, systematic and bio-ecology 292
The Genus *Aëroglyphus* Zachvatkin 292
The Genus *Glycycometus* Pillai 297

32 The Family Echimyopodidae Fain **300**
Morphological characteristics, systematic and bio-ecology 300
The Genus *Blomia* Oudemans 300

33 The Family Ctenoglyphidae Zachvatkin **304**
Morphological characteristics, systematic and bio-ecology 304
The Genus *Ctenoglyphus* Berlese 304

34 The Family Labidophoridae Zachvatkin **308**
Morphological characteristics, systematic and bio-ecology 308
The Genus *Gohieria* Oudemans 308

35 The Family Glycyphagidae Berlese **313**
Morphological characteristics, systematic and bio-ecology 313
The Genus *Glycyphagus* Hering 314
The Genus *Lepidoglyphus* Zachvatkin 334

36 The Family Hemisarcoptidae Oudemans **336**
Morphological characteristics, systematic and bio-ecology 336

37 The Family Histiostomatidae Berlese **338**
Morphological characteristics, systematic and bio-ecology 338
The Genus *Histiostoma* Kramer 339

38 The Family Pyroglyphidae Cunliffe **348**
Morphological characteristics, systematic and bio-ecology 348
The Genus *Dermatophagoides* Bogdanov 351
The Genus *Euroglyphus* Fain 363

39 The Family Suidasiidae Hughes **367**
Morphological characteristics, systematic and bio-ecology 367
The Genus *Suidasia* Oudemans 367

40 The Family Winterschmidtiidae Oudemans (= Saproglyphidae Oudemans) **375**

41 The Family Achipteriidae Thor **377**

42 The Family Camisiidae Oudemans **378**

43 The Family Carabodidae Koch **379**

44 The Family Ceratozetidae Jacot **380**

45 The Family Chamobatidae Thor 381
46 The Family Cosmochthoniidae Grandjean 382
47 The Family Cymbaeremaeidae Sellnick 383
48 The Family Euphthiracaridae Jacot 384
49 The Family Haplochthoniidae Van der Hammen 385
50 The Family Oppiidae Sellnick 386
51 The Family Oribatulidae Thor 387
52 The Family Pediculochelidae Lavoipierre 389
53 The Family Protoplophoridae Ewing 390
Other Oribatid species 390

Order Parasitiformes, Suborder Mesostigmata 391
54 The Family Ameroseiidae Evans 391
55 The Family Ascidae Voigts *et* Oudemans 395
56 The Family Melicharidae Hirschmann 399
57 The Family Blattisociidae Garman 404
58 The Family Dermanyssidae Kolenati 416
59 The Family Digamasellidae Evans 437
60 The Family Ologamasidae Ryke 439
61 The Family Rhodacaridae Oudemans 440
62 The Family Oplitidae Johnston 441
63 The Family Trematuridae Berlese 443
64 The Family Eviphididae Berlese 447
65 The Family Laelapidae Berlese 450
66 The Family Macrochelidae Vitzthum 465
67 The Family Parholaspididae Evans 477
68 The Family Parasitidae Oudemans 478
69 The Family Phytoseidae Berlese 480
70 The Family Uropodellidae Camin 488
71 The Family Uropodidae Kramer 490

References 495
Taxonomic Index 641
Subject Index 659

Preface

Mites are arthropods characterized by a broad adaptive radiation, and, for the most significant bio-ecological aspects, they can be connoted in agreement with the great Antonio Berlese (1927), as 'similar to insects'. In this sense, they occupy almost all ecosystems, including the marine one. Some taxa play a fundamental ecological role in the processes of recovery and recycling of organic matter in agricultural and forest soils, others actively participate in the maintenance of biological equilibrium in agricultural and natural ecosystems, and still others relate more directly with the human economy, both as consumers of cultivated plants and their products (stored food) and as agents responsible for the transmission of serious diseases to humans and domestic animals. The size of the impact is considerable, and its economic quantification is not simple. More or less, reliable estimates are reported in the literature for the more economically developed countries, a little less for developing countries. Regardless of the value of this last point, it is certain that the study of these organisms and their impact on the human economy identify a branch of applied sciences (agricultural, veterinary, medical, etc.) identifiable as 'economic acarology'.

This last aspect is testified by a rich and qualified bibliographic source, which, over the decades, has specialized in dealing with the impact of mites on cultivated and non-cultivated plants, with foodstuffs, and with human and animal health. From a technical point of view arises the need for a continuous acquisition of knowledge. This observation prompted an editorial project, welcomed with availability by CABI Publishing, which we sincerely thank, articulated in the first publication of a specialized volume *The Handbook of Mites of Economic Plants* (Vacante, 2016), followed by this second volume *The Handbook of Mites of Stored Products*. What has been reported suggests the hypothesis of a third contribution dedicated to Acari of medical and veterinary importance to complete the trilogy on the 'economic acarology'.

This volume has largely drawn on authoritative contributions published by different authors over time, without prejudice to the need for its publication, based above all on an accurate updating of the knowledge acquired in recent decades. The treatment of the different species has been related with a brief discussion on the systematic categories they belong to and on the main morphological characters that allow their identification. According to the technical spirit of the editorial proposal, a broader discussion was reserved for the bio-ecology of the various entities, with particular reference to the nature and typology of the damages caused by them and to the means of control realistically usable for their containment. In no case did the work have the purpose of intervening on topics of a systematic nature, in order to avoid the risk of generating confusion with the most accredited bibliographic information, even if in some cases, needs worthy of modifications and adjustments. In this sense, we ask our colleagues to excuse any approximations and invite them to fill any gaps with their welcome contributions.

Finally, it must be specified that the drafting of this volume was handled equally by both authors, who, exchanging opinions on pertinent topics, spontaneously agreed to develop the project together.

The authors

Acknowledgements

The authors thank several colleagues from international acarology journals for allowing the use of various figures in the text. We list below their names and institutions in geographical order. We cordially thank Prof. Jerry L. Cook, Director of Sam Houston Natural History Collections, Department of Biological Sciences, Sam Houston State University Huntsville, USA. A special thanks to Prof. Serge Kreiter of SUPAGRO of Montpellier, France, Editor-in-Chief of the historic journal *Acarologia* and to Dr Isabelle Coppée, Coordinator of the Société royale belge d'Entomologie, Belgium. Furthermore, we sincerely thank the Turkish colleagues Hüsein Özdikmen of the Department of Biology of Gazi University, Editor-in-Chief of *MEZ*, Salih Dogan of Erzincan Binali Yıldırım University, Sultan Cobanoglu of Ankara University, and Prof. Halit Apaydin, Editor-in-Chief of the *Journal of Agricultural Sciences*. A cordial thanks also to Prof. Badamdorj Bayartogtokh of the Department of Biology, School of Arts and Sciences of the National University of Mongolia, and to Dr Andrei Tolstikov, Editor-in-Chief of the journal *Acarina*, Russia. We also extend thanks to Prof. Dariusz Gwiazdowicz of the University of Life Sciences, Poznan, and Prof. Dariusz Iwan of the Muzeum i Instytut Zoologii of the Polska Akademia Nauk, Warsaw, Poland. We sincerely thank Prof. Alireza Saboori, Editor-in-Chief of the *Persian Journal of Acarology*, Department of Plant Protection, Faculty of Agriculture, University of Tehran, Iran. We are also grateful to Prof. Long-sho Li of the Department of Plant Protection, People's Republic of China. We address a special thanks to Dr Bruce Halliday of the Australian National Insect Collection, CSIRO, Australia, for their help in providing several reprints.

Last but not least, we sincerely thank Miss Lucy Pritchard, Managing Editor at CABI, for her competence, courtesy and active availability in the various steps in printing this book.

PART I
Introduction

The first part of this book covers, in eight chapters, the mite systematics, their external morphology, the fundamental aspects of mite bio-ecology, mite collection and laboratory study, the damage caused by mites to stored food, the means and the strategies of mite control.

1 Introduction to the Acari

Systematic Position and Ecological Role

The Acari, or mites, are small animals; the adults range from 300 to 500 μm in body length, except for some eriophyoids which are approximately 100 μm long and certain ticks whose engorged females can grow up to 30,000 μm long. They belong to the phylum Arthropoda, a monophyletic taxon of terrestrial and marine invertebrates (Weygoldt, 1998), which are characterized by jointed legs and a chitinous exoskeleton. Two groups of Arthropoda, the Pantopoda and the Chelicerata, lack antennae and mandibles, and molecular analyses have shown that the Chelicerata is a sister taxon of a clade including the Crustacea and the Insecta (Telford and Thomas, 1998). From a systematic point of view, the Chelicerata is a subphylum that includes the class Arachnida, a primarily terrestrial group that includes the subclass Acari along with several orders (Amblypygi, Araneae, Opiliones, Palpigradi, Pseudoscorpionida, Ricinulei, Schizomida, Scorpiones, Solifugae and Uropygi) (Weygoldt and Paulus, 1979a, b). Members of the subclass Acari lack opisthosomatic spinnerets and conspicuous primary segmentations (or tergites arranged segmentally). The Acari includes the superorders, Acariformes and Parasitiformes, and is morphologically differentiated from other relatively simple arachnids (Krantz, 2009a). Adults of the Acari usually have four pairs of legs, with the exception of eriophyoids, which have two pairs, and a few Tenuipalpidae (*Larvacarus*) and a few Podapolipidae, which have less than four pairs of legs. Their biological cycle progresses through the stages of egg, larva and nymph. The larva is six-legged and the nymph eight-legged. Another characteristic that sets the Acari apart from the other Arachnida is that the phase which hatches from the egg has fewer legs than subsequent phases, while the eggs of spiders, scorpions, pseudoscorpions and opilionids all hatch into eight-legged organisms. Up to 1999, approximately 55,000 species of Acari have been described, but the group is likely to grow to number over a million species (Walter and Proctor, 1999). By 2011, a total of 54,617 mite and tick species had been reported (Zhang, 2011). This high number of species can be explained by assuming that if in each of the 400,000 plants currently known there lives almost one species-specific mite, then the number of plant-feeding mites could be as many as 400,000 species. Similarly, if each of the 60,000 known species of vertebrate animals harbours almost two species-specific mites, the number of species-specific mites might be at least 120,000 species (Hoy, 2011). That there could be such a high number of species is not surprising when one considers that at present our knowledge of the Acari is small and fragmentary. A bibliographic analysis of the acarological papers published in *Zootaxa* from 2001 to 2005 gives a good representation of the strong development in this subject. In this period, the list of new taxa described and published in this journal only includes 16 genera and 185 species (Fan, 2006). Krantz (2009b) reported that the large number of species that have been described has led to the designation of about 5500 genera and 1200 subgenera in 540 families. The small size and the strong evolutionary plasticity of the Acari have permitted these singular organisms to colonize most of the available aquatic and terrestrial habitats, where they exhibit a great ability to exploit the available trophic resources owing to their different and varied dietary regimes – phytophagous, parasitic, predacious, mycophagous, saprophagous, coprophagous and necrophagous. In short, the Acari may be found worldwide in any environment capable of supporting life. As such, they can be collected from regimes as diverse as windswept arctic tundra, hot desert sands, icy depths or the hair follicles of eyebrows (Krantz, 2009b). For humans, the Acari form a significant component of the fauna of any region of the world, either as pests injurious to economic plants (Jeppson *et al.*, 1975; Vacante, 2016), to stored food (Zachvatkin, 1959;

DOI: 10.1079/9781786390912.0001

Hughes, 1976; Campbell *et al.*, 2004; Henszel *et al.*, 2011; Hagstrum *et al.*, 2013), to their health and to that of other animals or as beneficial organisms (Huffaker *et al.*, 1970; McMurtry *et al.*, 1970; Gerson *et al.*, 2003), etc. Furthermore, these little animals play a fundamental role in the litter and humus layers of forests, grasslands and agricultural soils, where they recycle minerals (Balogh, 1972). Bader (1989), in an interesting booklet written in German, discussed the fundamental features of the Acari. Walter and Proctor (1999) very effectively summarized the ecological importance and the economic role of these small and interesting animals. They wrote that: 'The average mite is minuscule, barely perceptible to even the sharpest eyes. When enough are present, however, mites can exert efforts disproportionate to their size. When a crop disappears under a blanket of silk or stored food turns into a seething mass of hairy motes, then the numbers of mites will be legion. If the mite is an itch mite burrowing into one's skin or a tick injecting virulent rickettsiae into one's blood, then "enough" can be a single mite'.

Origins and Phylogenesis

Incomplete conservation, inadequate sampling and taxonomic problems make it very difficult to assess the exact evolution of the Acari over a long period of time. In addition, species such as the Acari are very small in size and have very little tough exoskeleton, making them unsuitable for preservation. So to better uncover the essential features of the origins and phylogenesis of the Acari, one must make some assumptions. In particular, the first fossil records of marine invertebrates date back about 600 million years ago, and it is estimated that these evolved in a few million years. In the previous 2500 million years, life on earth appeared to be exclusively represented by algae and bacteria. The abundant availability of a monoculture of algae and eukaryotic cells formed in that time could have led to the largest explosion of evolutionary activity so far known (Gould, 1981). During the Permian period, there was a major decrease in the number of families of animals living at the bottom of the sea (Valentine, 1970), and Schopf (1974) reported that the later Permian was the time when the earth's continents came together to form the (only) supercontinent of Pangaea. This union produced a decrease in the surface of the earth covered by shallow seas and, consequently, a decrease in the habitat available to shallow-water invertebrates. Likewise, the increase in the abundance of fauna recorded after the Permian can be explained by the subsequent increase in shallow seas produced by the separation of Pangea into smaller continents. An interesting analysis of the history of marine animal diversity has recently been published by Stanley (2007). In addition to the above-mentioned fossil records of marine invertebrates 600 mya, early fossil records of vascular plants appeared in the early Silurian, followed by records of pteridophytes in the Carboniferous, and records of gymnosperms to angiosperms in the Cretaceous and Tertiary (Niklas *et al.*, 1983). The earliest phytophagous Insecta date from the Carboniferous (Strong *et al.*, 1984), and it is well known that there is reciprocal evolution between plants and insects, which represents an important mechanism for increasing the wealth of wildlife (Begon *et al.*, 1986). It is commonly believed that the first fossil records of the Arachnida (and the Acari) originate in the late Silurian to early Devonian periods (*c.* 425 mya) in the form of a pulmonate trigonotarbid (Arachnida, Trigonotarbida), which is now extinct. Dunlop and Selden (2009) report that the oldest mite fossil is from the Devonian (410 mya), while the age of scorpion and spider fossils is 428 and 312 mya, respectively. The advent of the Acari probably also relates to the latter part of the evolution of the arthropods, with diversification having occurred about 200 million years earlier in the Cambrian period (Gould, 1989; Walter and Proctor, 1999). Fossil records show that the terrestrial Arachnida acquired respiratory organs of different types at different times during the transition from the marine environment to land, as in the scorpions, trigonotarbids, etc. Formally, this pattern of developmental behaviour contrasts with the hypothesis of the monophyletic origin of the Arachnida (Bergstrom, 1979). The fossil records now available seem to indicate that the Acari had achieved a certain amount of diversity by the early to mid-Devonian (Hirst, 1923a; Woolley, 1961; Norton *et al.*, 1988; Evans, 1992; Subías and Arillo, 2002), which suggests that they colonized terrestrial environments as early as the late Silurian. Moreover, in the early Devonian, all fossil Acari now known belonged to the superorder Acariformes (Walter and Proctor, 1999), and the fossil records of the Parasitiformes dating to the late Cretaceous (Witalinski, 2000; Poinar and Brown, 2003); the parasitiformes Trigynaspida may date to as early as the upper Triassic (*c.*220 mya) (Kim, 2004). Studies of mitochondrial phylogeny have shown that the orders and classes of spiders, scorpions, mites and ticks diversified in the late Palaeozoic (Jeyaprakash and Hoy, 2009). Recently, Sidorchuk and Bertrand (2013) described *Sellnickiella balticae* Sidorchuk *et* Bertrand (Labidostomatidae), thus highlighting the presence in Eocene Baltic amber (more than 40 mya) of the genus *Sellnickiella* Feider *et* Vasiliu and a second species of the genus *Labidostoma* Kramer (*Labidostoma* (*Pseudocornutella*) *electri* Sidorchuk *et* Bertrand), after the congeneric species previously described by Dunlop and Bertrand (2011). The report pointed out that notwithstanding the few fossil mite descriptions, the discovery of these mite carnivores in Baltic amber suggests that Palaeogene mite biodiversity was comparable to current mite diversity (Sidorchuk and Bertrand, 2013). The fossil records now available seem to show that after the extinction events registered in the Permo-Triassic and after the statement of modern lineages, a significant adaptive process may have occurred in the Acari during the late Mesozoic and early Cenozoic periods (60–100 mya). At this time, there was a strong morphological and ecological diversity among the Acari that permitted significant exploitation of the many co-evolving plants and animals, and there was a consistent diversification of angiosperm plants and Insecta (Lindquist, 1975a; Krantz, 2009c). Specialized plant-parasitic mites, such as the species *Triasacarus fedelei* Lindquist *et* Grimaldi and *Ampezzoa triassica* Lindquist *et* Grimaldi, existed as early as the Triassic (Schmidt *et al.*, 2012). According to Sidorchuk *et al.* (2014), both species are ascribed to Triasacaroidea Lindquist *et* Sidorchuk superfam. nov. and Triasacaridae Lindquist *et* Sidorchuk fam. nov. (having the same composition and inclusiveness as the superfamily), with *Triasacarus* Lindquist *et* Grimaldi (Schmidt *et al.*, 2012) designated as type genus, to accommodate *Triasacarus*, *Minyacarus*, *Cheirolepidoptus* and *Ampezzoa*, distinct from the superfamily Eriophyoidea, in the lineage Tetrapodili. Acari are often associated with Insecta as phoretic or parasitic mites in fossil records dating from the Cretaceous about 20–40 million years earlier (Poinar, 1992; Poinar *et al.*, 1993, 1997). The same phenomenon has been observed between parasitic Acari and vertebrate hosts in the mid-Tertiary (Poinar, 1995). Peñalver *et al.* (2017) have referred that the feeding habits and hosts of the ticks, retained today among the most prevalent blood-feeding ectoparasites, in deep time have long remained speculative. These authors reported that direct and indirect evidence in 99-million-year-old Cretaceous amber, showing that hard ticks and ticks of the extinct new family Deinocrotonidae fed on blood from feathered dinosaurs, non-avialan or avialan excluding crown-group birds. A †*Cornupalpatum burmanicum* Poinar *et* Brown hard tick is entangled in a pennaceous feather. Two deinocrotonids described as †*Deinocroton draculin* gen. *et* sp. nov. Peñalver *et al.* possess specialized setae from dermestid beetle larvae (hastisetae) attached to their bodies, likely highlighting a cohabitation in a feathered dinosaur nest. A third conspecific specimen is blood-engorged, its anatomical features suggesting that deinocrotonids fed rapidly to engorgement and had multiple gonotrophic cycles. These findings provide insight into early tick evolution and ecology and shed light on poorly known arthropod-vertebrate interactions and potential disease transmission during the Mesozoic. Mites of the suborder Oribatidae have been found in fossils of the late Carboniferous (Labandeira *et al.*, 1997). The oldest oribatid fossils are from the Devonian (Norton *et al.*, 1988), and some of them belong to modern families (Subías and Arillo, 2002). All mites described from the Carboniferous, Jurassic and Cretaceous periods are included in modern families, with one exception (Subías and Arillo, 2002; Arillo *et al.*, 2012). Regardless of the above information, in the last half-century, a need has developed to understand whether mites are characterized by a condition of dichotomy between acariforms and non-acariforms, or not. In short, we need to know whether in the course of the evolution of the Arachnida, an event has resulted in a monophylogenous origin of the Acari or whether they result from the convergence of two or more groups that have evolved independently of one other and are now similar. Woolley (1961) notes that 'no known connections with other arachnids can be ascertained from the fossil record, but the chelicerae and other morphological features indicate that the Acari probably had an arachnid origin in opilionid-like ancestors, but developed with inherent peculiarities, principally the gnathosoma. They are at least diphyletic, probably polyphyletic, in origin, with two main groups, the older Anactinochitinosi (Notostigmata, Mesostigmata, Ixodides) and the younger Actinochitinosi (Trombidiformes, Sarcoptiformes)'. Following Jeyaprakash and Hoy (2008), an estimate of the evolutionary age of mites places their chelicerate ancestors in the Precambrian and their divergence as mites to at least 424 mya. Studies of the morphological, developmental and behavioural aspects of mites from various research support different positions on their evolutionary origin (Zakhvatkin, 1952; Lindquist, 1984; Van der Hammen, 1989; Walter and Proctor, 1999). Examination of 40 characters of 12 extant orders of arachnids (including the Acari), and in particular, the examination of the common characters – possession of a movable gnathosoma and the existence of exapod larva and of the three octopod nymphal instars – suggests that the mites are derived from a common ancestral stock (Lindquist, 1984) and that monophyletic Acari and Ricinulei are sister groups (Kaestner, 1980). However, spermatological evidence from the Chelicerata does not encourage support for a common origin of acariforms and non-acariforms (Alberti, 2000).

2 External morphology

Integument (Fig. 2.1)

In the Acarina, the developing integument appears first as undifferentiated tissue, covered by a thin layer of 'cuticulin' and separated from the underlying 'epidermis layer' by a 'Schmidt layer'. As development proceeds, the anatomical organization of the integument undergoes further transformation and shows from the outside inwards an 'epicuticular layer' (the epicuticle), a 'procuticular layer' (the procuticle), which consists of an outer 'exocuticular layer' and an inner 'endocuticular layer', an underlying Schmidt layer, an epidermal layer and, finally, a basal lamina set below the epidermis. This anatomical organization underlies the external 'cerotegument', which consists of an outer 'cement layer' and an underlying 'wax layer' that are set on the outer epicuticle and the underlying inner epicuticle. Together, the epicuticular complex – the cerotegument and the outer and inner epicuticle – provides multilayered protection. The cerotegument is a secreted layer that is formed by the epidermis, from which 'pore canals' travel towards the inner epicuticle, which includes an underlying cuticulin layer and a covering cuticulin layer. Overall, the cerotegument may have various consistencies, but its cement layer is dense and frequently sculptured. Together, the wax and cement layers of the cerotegument are involved in the regulation of water loss and/or water intake (Alberti *et al.*, 1981; Norton *et al.*, 1997). In the different regions of the body, tanning by orthoquinones can induce differential sclerotization of the two layers of the procuticle. The secretion of granules, which coalesce with 'microfibres', occurs before the secretion of new cuticle below the endocuticular layer, and the microfibres consolidate and form laminations (Brody, 1970). The width of these laminations progressively decreases towards the cuticle surface. The pore canals in the endocuticular layers are first organized in a helical pattern, but as they move towards the surface, they gradually become linear, sometimes with branching, and finish at the micropores of the cuticulin layer below the epicuticle (Norton *et al.*, 1997). In addition to having micropores, the body surface of mites shows a variety of 'macropores' that are involved in different functions (secretory and sensory – including proprioception), and belong to three fundamental categories of cuticular apertures: poroidotactic (lyrifissures or cupules), setal (openings for the insertion of tactile, olfactory and chemosensory hairs and pegs) and adenotactic (openings of secretory structures, such as solenostomes or structures with a calyx) (Athias Henriot, 1969a, 1969b). Further information on these topics may be found in subsequent paragraphs and other information can be found in Krantz (2009a) and Vacante (2016).

The Astigmatina have a pallid and soft cuticle variable in colour from pearly white to yellow brown, and may become pigmented in some species, such as *Gohieria* or many hypopial stages. However, it never becomes hard or densely sclerotized. The surface of the cuticle may be smooth, rough or finely grooved; such grooving characterizes the parasitic forms but is unknown among the free-living species (Hughes, 1976).

Body Division (Figs 2.2 and 2.3)

Primary segmentation is largely absent in the Acari, and instead, a segmental fusion or 'tagmosis' has become established. Some aspects of mite ontogenesis assist in a better understanding of this phenomenon. Several factors influence the establishment of somatic derivatives for most Acari, among these the posteroventral idiosomatic curvature or 'caudal bent' (Evans, 1992). Several interpretations of somatic origins and structural homologies in the Acari have been based on setae, pores or gland patterns. According to some specialists (Zakhvatkin, 1952; Van der Hammen, 1970; Sitnikova, 1978; Jesionowska, 1991), primitive segmentation in the

DOI: 10.1079/9781786390912.0002

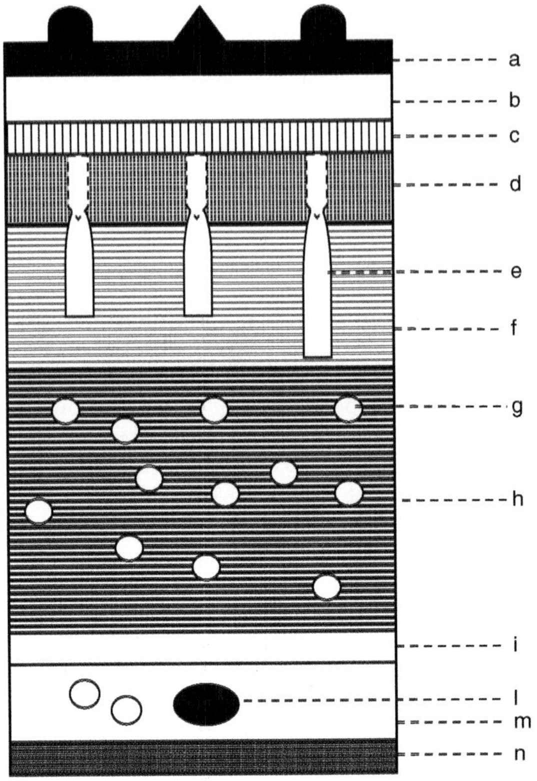

Fig. 2.1. Schematic drawing of acarine cuticle: a, cement layer; b, wax layer; c, outer epicuticle; d, inner epicuticle; e, pore canal; f, exocuticle; g, evolving pore canal; h, endocuticle; i, Shimidt layer; l, nucleus; m, epidermis; n, basal lamina (according to Norton *et al.*, 1997).

Acari ranged from 12 to 22 segments, but observations carried out on caeculid mites by Coineau (1974a, b), supported by the analysis of Grandjean (1954), have led to a mapping in which the archetypal adult acariform is characterized by 16 somites plus a cheliceral segment. The larval idiosoma (the part of the body that follows on from the first region, or gnathosoma) includes 11 somites, and one segment is added posteriorly at each postlarval moult, according to the phenomenon of 'anamorphosis'. The larva itself has six idiosomal segments (C, D, E, F, H and Ps), and one terminal segment (AD) is added to its idiosoma when it moults into a protonymph, another segment (AN) is added to the protonymph idiosoma when it moults into a tritonymph and a third segment (PA) is added to the tritonymph idiosoma when it moults into an adult. These changes can affect the numbers of genital papillae, the setae of the genital and anal fields, and the cupules; the latter number four pairs in the larva, five in the protonymph and six in the other biological stages. Fusion of the primitive body segments just described has led to the formation of the two major body regions or 'tagmata' of mites mentioned above: the anterior 'gnathosoma' and the posterior 'idiosoma'. The gnathosoma is derived from the first two anterior segments and has the primitive somatic primary organs for feeding. The idiosoma is derived from the remaining segments. The two body regions are separated by a 'circumcapitular furrow', which is clear in some groups, but is absent in others owing

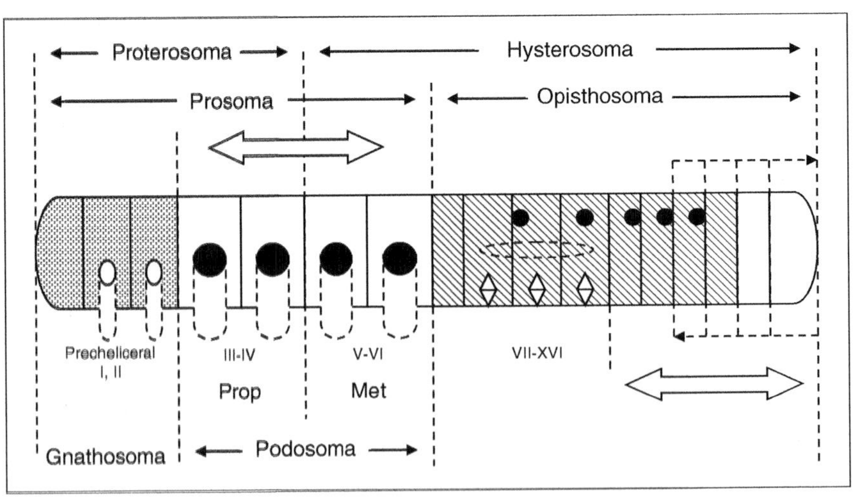

Fig. 2.2. Diagram of somatic adult archetypical acariform (according to Coineau, 1974b). The segment XII-XVI correspond to H, PS, AD, AN, and PA respectively.

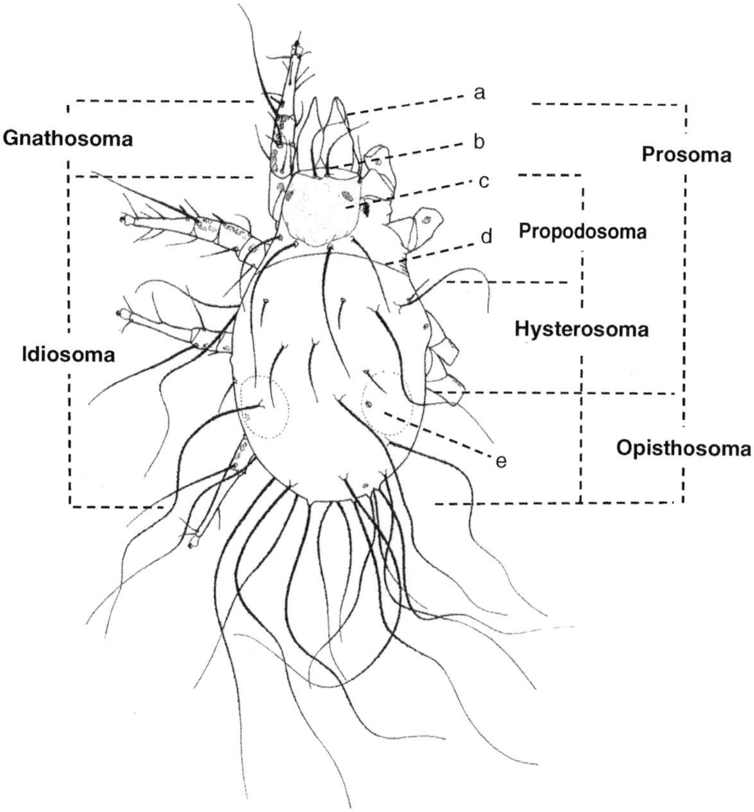

Fig. 2.3. Different body areas of *Tyrophagus neiswanderi* Johnston *et* Bruce. Female: a, chelicerae; b, circumcapitular furrow; c, prodorsal shield; d, dorsosejugal furrow; e, opisthonotal gland (from Vacante, 1988).

to secondary somatic development. Other furrows can occur in several groups of Acari, such as a 'dorsosejugal furrow' which is set between the second and third pairs of legs of acariform mites, and a 'postpedal furrow' that may be set behind the pair of fourth legs. In some groups, such as the Prostigmata, a certain somatic plasticity can complicate this scheme with the appearance of other furrows (the 'abjugal furrow', 'disjugal furrow', etc.).

Gnathosoma (Figs 2.4 and 2.5)

The gnathosoma, or 'capitulum', is the first region of the body; it is very complex and specialized, and is characterized by important adaptations (sensory and feeding organs, etc.) (Alberti and Coons, 1999). It includes the chelicerae and the palpi, which are the fundamental mite-feeding organs and are variously adapted to apprehension, chewing, piercing, tearing and sucking. Ontogenetically, the gnathosoma derives from the cheliceral segment and the paired biramous appendages of the second segment; the cheliceral segment is also homologous with the first antennal segment of insects (Manton, 1977; Telford and Thomas, 1998). The palpal coxae are derived from primitive paired protopods and contribute to the formation of the 'subcapitulum' and 'hypostome', and the gnathosomatic floor and the walls, respectively; moreover, they form a channel for the paired chelicerae that are set above it and an outlet for the buccal cavity that opens below the chelicerae. The malar processes ('corniculi' or 'rutella') are sometimes set alongside the lateral lips of the hypostome. An 'epistome' or 'gnathotectum' can be extended over the gnathosomatic dorsum to partially protect the chelicerae. A 'subcheliceral plate' constitutes a base for the chelicerae and provides attachment for the muscles of the preoral 'labrum', an underlying expansion of the dorsal pharyngeal wall (Evans, 1992; Alberti and Coons, 1999). Depending on the taxonomic group, the labrum can develop the function of a prepharyngeal valve,

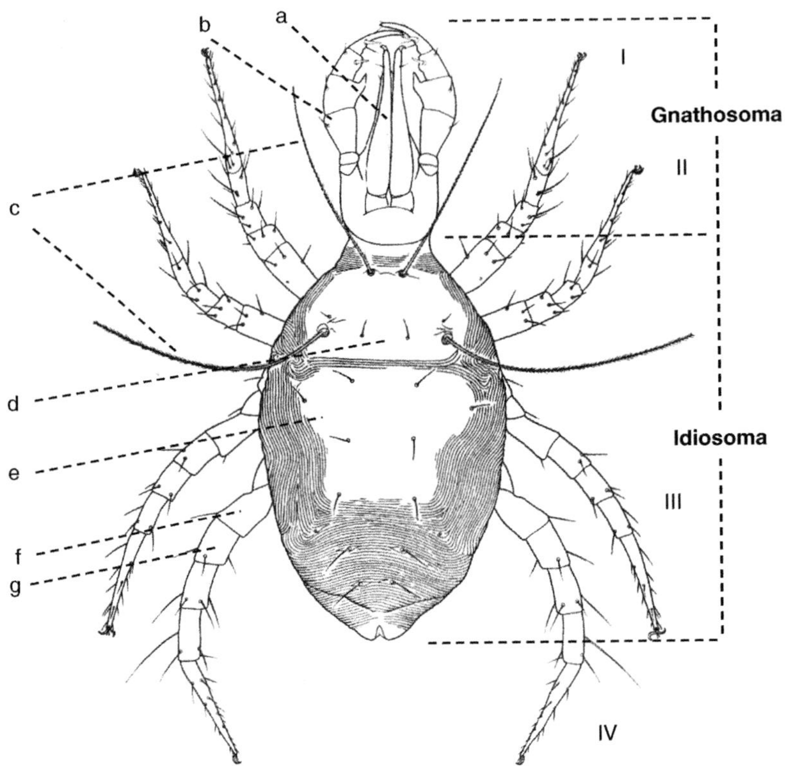

Fig. 2.4. *Cunaxa capreolus* (Berlese). Female: a, chelicerae; b, palpus; c, bothridial setae; d, prodorsal shield; e, opisthosomal shield; f, basifemur; g, telofemur; I-IV, legs (from Vacante and Nucifora, 1985).

which is an adaptation to prevent food loss, as in the Ixodida (Sonenshine and Gregson, 1970) and other acarine groups; it can also help to prevent the ingestion of solid food particles (Evans and Loots, 1975). The buccal cavity opens into the 'pharynx', a suction pump for food whose action is related to the constrictor and dilatator muscles. The shape of the pharynx varies in different orders – it is triradiate in cross-section in the Mesostigmata and reniform in the acariform mites (Alberti and Coons, 1999); the form of the pharynx can be taken into account in the taxonomy of families and genera. The buccal cavity may show paired ducts of idiosomatic 'salivary glands' below the subcheliceral plate (Sonenshine and Gregson, 1970; Evans, 1992), or other adaptations. In many acarine species, the salivary glands produce enzymes for the preoral digestion of food (Moss, 1962), while other glandular systems are involved in silk production or osmoregulation (Alberti, 1973; Brody *et al.*, 1976).

The Astigmatina have a gnathosoma very mobile and partially retractile within the body. In life, it is commonly held at an angle to the idiosoma, and the tips of the chelicerae come in contact with the food. It may become concealed by a rostral prolongation of the dorsal surface of the propodosoma. The dorsal surface of the gnathosoma is reduced to form a small lobe between the bases of the chelicerae which are thus exposed dorsally (Hughes, 1976).

In the Astigmatina, the labrum, or true epistome, lies below the chelicerae, a hollow structure forming the roof of the mouth. The labrum continues backwards into the body as a plate, the lateral walls of which are continuous with the ventral part of the gnathosoma, and from it originate the dilator muscles of the pharynx (Hughes T., 1953). The ventral part of gnathosoma, the hypostome or infracapitulum, is formed mainly by the fused coxae of the pedipalps, produced forward as a pair of internal lobes or malae, and a pair of external two-segmented palps. From the basal portion arise two pairs of setae. The palp is a flattened structure whose proximal segment bears a pair of setae and the distal, a single seta and an eccentric rod-shaped structure,

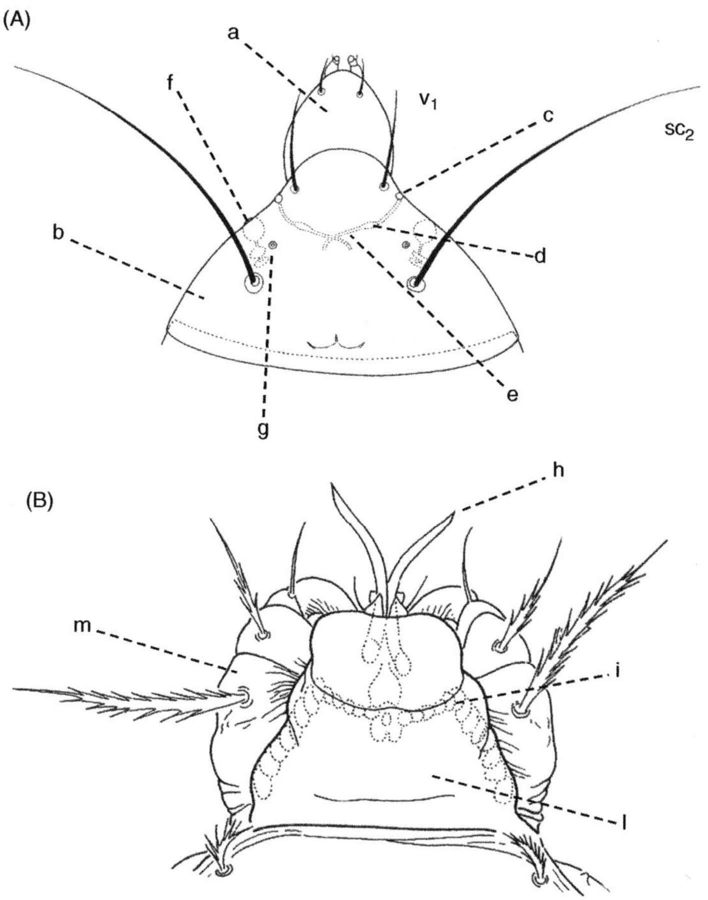

Fig. 2.5. Gnathosoma and prodorsum of female of *Tarsonemus waitei* (Banks) (from Vacante and Nucifora, 1985, partially modified) (A). Gnathosoma of female of *Cheyletiella blakei* Smiley (from Patti and Vacante, 1985, Vacante del.) (B); a, gnathosomal capsule; b, prodorsum; c, stigma; d, atrium; e, tracheal trunk; f, bothridial sensillum or pseudostigmatic organ; g, pit; h, chelicera; i, peritreme; l, gnathosoma; m, palpus.

possibly the vestige of a third segment or a sense organ. The mouth parts of some species may become modified in connection with some particular way of life, such as in the Histiostomidae whose mouth parts are adapted for the concentration of particles of food from a fluid medium (Hughes T., 1953), or those of Listrophoridae adapted to clinging to hair and scraping its surface (Hughes T., 1954). The Cytodytidae inhabit the lungs of birds and possess reduced mouth parts (Hughes, 1976).

Chelicerae (Figs 2.6 and 2.7)

The chelicerae are placed dorsally in relation to the mouth opening and commonly consist of three segments called the 'cheliceral base', 'digitus fixus' and 'digitus mobilis'. The first segment is basal and bears the digitus fixus articulated with the distal digitus mobilis dorsally. From an ontogenetic point of view, the chelicerae derive from the endopodal appendages of the first segment. These organs evolve from a leg-like to a mandible-like appendage, resulting from a gradual series of fusions and setal modifications. The basal cheliceral segment derives from the endopodal trochanter of the first somite, the attached digitus fixus from the fusion of the femur, genu, tibia and tarsus, and the terminal digit from the pretarsus (Grandjean, 1947). Cheliceral morphology shows a strong variation within the subclass Acari, as a result of the large number of evolutionary adaptations that have occurred and the variable feeding habits (Akimov *et al.*, 2002).

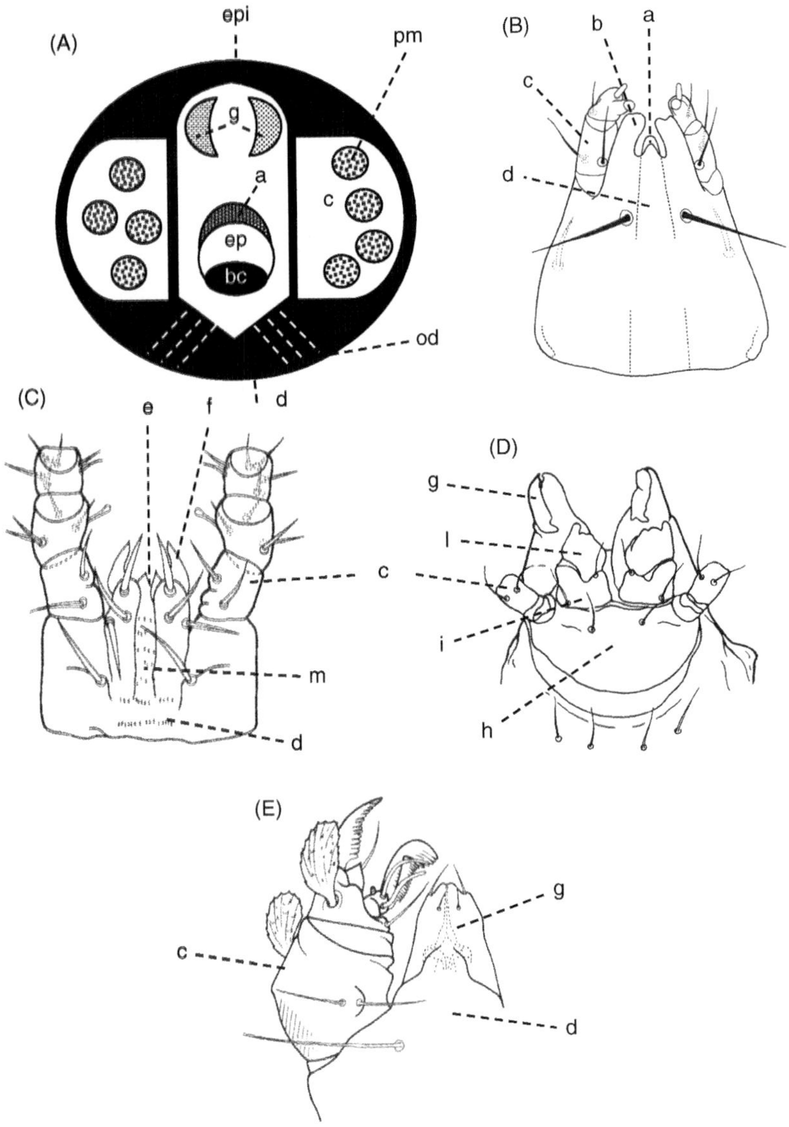

Fig. 2.6. Diagram of acarine gnathosoma in cross section (A) and ventral view of gnathosoma of acarine species (B-E). *Tyrophagus* sp. Female (B) (from Vacante, 1989a). *Proctolaelaps pigmaeus* (Müller). Female (C). *Micreremus gracilior* Willmann. Female (D). *Cheletogenes ornatus* (Canestrini *et* Fanzago). Female (E) (drawings C-E from Vacante and Nucifora, 1985) (from Vacante, 2016); a, labrum; b, mala; bc, buccal cavity; c, palpus; d, hypostome; e, internal mala; ep, epipharinx; epi, epistoma; f, corniculus; g, chelicerae; h, mentum; i, gena; l, rutellum; m, deutosternum; od, oral dilators; pm, palpal muscle.

The mandibulate form occurs in many parasitiform and acariform mites, while in phytophagous and parasitic species, there are often cheliceral modifications that result in styliform, hook-like or finely toothed organs adapted to pierce plant or animal tissues or attack bacterial films (Krantz, 2009a). The bases of the chelicerae may be partially or completely fused to form the 'stylophore', as in the Tetranychoidea, while the digitus mobilis is modified into a stylet that can pierce the (plant) cell wall. The chelicerae are characterized by extrusion and retraction movements. Extrusion movements depend

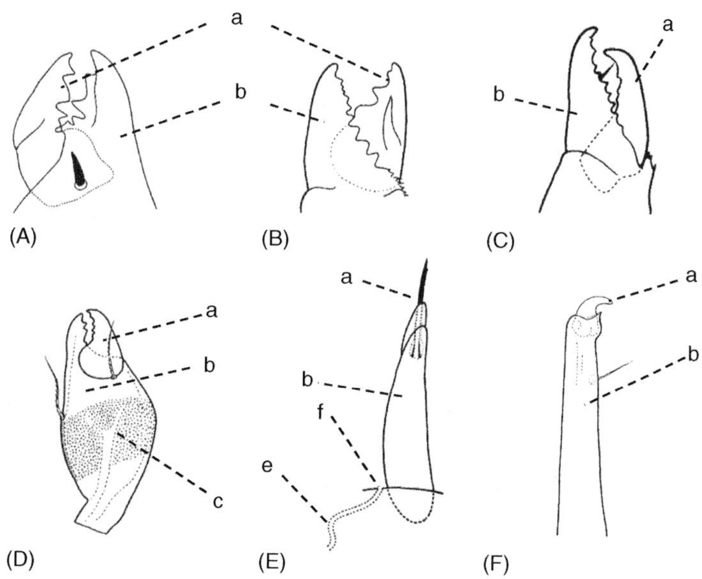

Fig. 2.7. Various kinds of chelicera (from Vacante, 2016). *Tyrophagus* sp. Chelicerae of female (A) (from Vacante, 1989a). *Proctolaelaps pigmaeus* (Müller). Chelicera of female (B). *Neoseiulus californicus* (McGregor). Chelicera of female (C). *Siculobata sicula* (Berlese). Chelicera of female (D). *Ledermuelleriopsis plumosus* Willman. Chelicera and tracheal duct with stigmata (E). *Cunaxa capreolus* (Berlese). Chelicera of female (partially drawn) (F). Drawings B-G from Vacante and Nucifora, 1985; a, digitus mobilis; b, digitus fixus; c, area porosa; d, palpus; e, tracheal duct; f, stigma; g, stylophore.

fundamentally on hydrostatic pressure produced by the contraction of the dorsoventral idiosomatic muscles, while retraction involves the action of retractor muscles arising from the dorsal gnathosoma or from the 'idionotum'. In some groups, the chelicerae are also involved in sperm transfer. Krantz (2009a) reports that the cheliceral structures involved in the transfer of sperm have evolved in a least four lineages of Mesostigmata. Male Parasitiae possesses a median foramen in the movable cheliceral digit, called 'spermatotreme', derived through the terminal fusion of the digit with a process arising at its base. The 'spermatophore' casing is extruded through the spermatotreme in its collapsed state and then filled with sperm and accessory gland products so that the spermatotreme ends up holding the balloon-like spermatophore by a narrow neck (Hyatt, 1980). Some groups of Mesostigmata (Dermanyssiae, Heterozerconina and the Celaenopsoidea) possess chelicerae with semi-detached or fused process modified for sperm transfer and a male genital opening that has migrated anteriorly to the base of the tritosternum. Males of Dermanyssiae show an often elaborate, species-specific outgrowth on the movable cheliceral digit called a 'spermatodactyl', employed to transfer sperm material to the female. In Rhodacaridae, the males produce spermatophores with distinct casings (Lee, 1974), and others, such as the Macrochelidae, extrude seminal droplets from the genital openings (Krantz and Wernz, 1979).

Other mites, such as ticks and early derivative Mesostigmata, deposit the spermatophores directly on or just inside the female primary genital opening. Athias Henriot (1969c) called this mechanism 'tocospermy', employing either the chelicerae or direct venter-to-venter contact. However, the males of Dermanyssiae transfer the sperm to a pore-like secondary genital opening, called the 'sperm induction pore', commonly near the base of the legs of the female. Athias Henriot (1969c) called this mechanism 'podospermy', also referred to as porospermy.

In the Astigmatina, each chelicera is bilaterally compressed and consists of a large basal region prolonged anteriorly as the fixed digit and articulating with a movable one. The shears work in a vertical plane and their edges are often toothed or serrated. On the internal side of the fixed digit is a conical spur and beyond it, the mandibular spine. The muscles acting on the chelicera originate from the dorsal propodosomal plate and the two chelicerae can move independently of one another (Hughes T., 1953).

Palpi (Fig. 2.8)

The palpi are postoral appendices that consist of a variable number of segments – the trochanter, femur, genu, tibia, tarsus and apotele – which may be from one or two to six. The two coxae merge and contribute to form the subcapitulum, with the mouth opening anteriorly, bounded dorsally by the labrum, and at the sides by various structures (the lateral labra, rutella, malae). Ontogenetically, the palpi derive from the leg-like endopodal appendages of the second somatic segments. Their function is primarily related to searching for and handling food, with the distal segments usually leading to chemosensory and thigmotactic sensory receptors, which act as a support

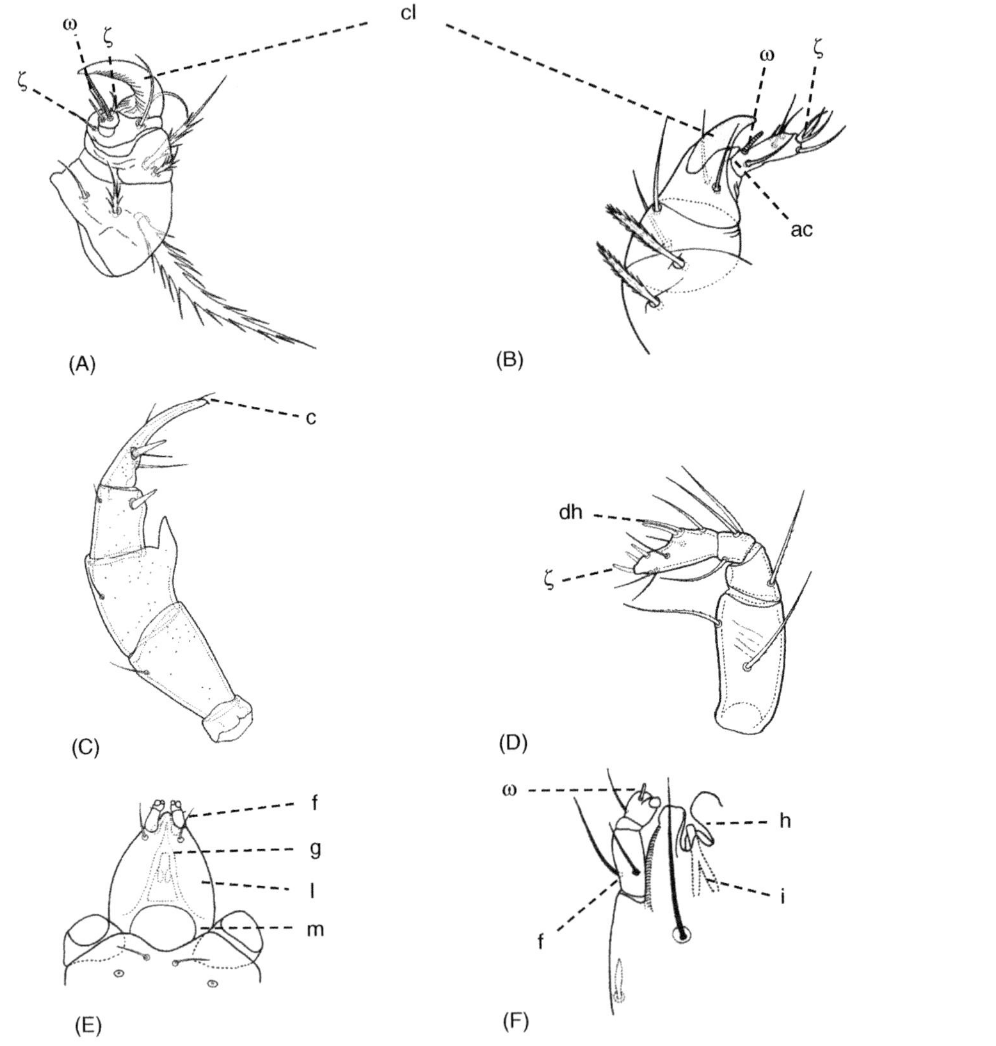

Fig. 2.8. Palpi of acarine species (from Vacante, 2016). *Cheyletiella blakei* Smiley. Female (A) (from Patti and Vacante, 1985, Vacante del.). *Ledermuelleriopsis plumosus* Willmann. Female (B). *Cunaxa capreolus* (Berlese). Female (C). *Humerobates rostrolamellatus* Grandjean. Female (D). *Tarsonemus waitei* Banks. Female (E) (drawings B-G from Vacante and Nucifora, 1985). *Tyrophagus neiswanderi* Johnston *et* Bruce. Male (F) (from Vacante, 1988); a, b, c, simple setae; ac, accessory claw; cl, claw; dh, fused solenidion and eupathidium (double horn); e, supracoxal seta; ω, solenidion; ζ, eupathidium; f, palpus; g, chelicera; h, mala; i, labrum; l, gnathosomal capsule; m, ventral apodemes.

for the feeding activity, with various modifications related to various evolutionary adaptations. Modifications can occur in the number of free segments, which are one or two in some Astigmatina and Prostigmata, three or four in the Ixodida, and five in many Mesostigmata and Oribatida (Krantz, 2009a). In the Tetranychidae, the palpi contain part of the silk gland, whose secretions reach a distal hollow eupathidium (also eupathidion) or 'spinneret' (Alberti and Crooker, 1985). Palp chaetotaxy (the use of setal position and form in taxonomy) is briefly discussed below under the family Tetranychidae. Incremental addition of palpal setae through ontogeny provides a useful means for identifying immature stages in many Mesostigmata (Evans, 1963a), and variations on the general ontogenetic pattern may occasionally be helpful in defining taxa, such as in the genus *Blattisocius* of the family Ascidae (Lindquist and Evans, 1965).

Idiosoma (Figs 2.2, 2.3, 2.4, 2.9, 2.10 and 2.11)

A simple pattern divides the idiosoma into the anterior 'propodosoma' and the posterior 'hysterosoma', distinct or not by a sejugal furrow, as in the acariform mites. The region bearing the legs derives from four embryonal somites and is also called the 'podosoma'; the region of the body deriving from the first two somites includes the first two pairs of legs and represents the 'propodosoma', while the last part, which is derived from two other somites, bears the other two pairs of legs and is referred as the 'metapodosoma' (Coineau, 1974a, b; Krantz, 2009a; Alberti and Coons, 1999). The part of the hysterosoma excluding legs III and IV constitutes the opisthosoma and derives, according to systematic groupings, from 6 to 13 embryonal somites; it bears the genital and anal openings (Coineau, 1974a, b). Its shape is ovoid to fusiform and presents a cuticle that is variously striate, with furrows or covered by 'shields' or plates of a shape and size that is variable in the different systematic groups. Male and female genital openings open both anteriorly or posteriorly. In addition to the terminology used above, other old terms can be found in the literature, such as 'prosoma', which refers to the gnathosoma plus podosoma, and 'proterosoma' which refers to the gnathosoma plus the propodosoma. Moreover, the idiosoma can bear one or more sclerotized shields or plates that are weakly tanned or strongly developed and heavily tanned, according to group. Retention of larval shield characteristics by sexually mature adults (neoteny) is common in the Astigmatina and some Oribatida; however, the degree of sclerotization in most Acari normally increases through the ontogenetic development. Idiosomatic shields provide a degree of protection from predation and environmental perturbation and serve as sites of muscle attachment. In various mite groups, cheliceral extension with retractile chelicerae is hydrostatically driven, and gnathosomatic extension in all Acari is controlled to some extent by hydrostatic pressure, strong dorsoventral muscles are involved to produce necessary movement. Consequently, anterodorsal and anteroventral portions of the idiosoma may develop shields or shield complexes to support the attachment of these muscles. The dorsal shield may cover a small portion of the propodosoma or it may extend laterally or posteriorly to cover all of it. Extension to include the entire podonotum is common in the Mesostigmata, which possess a 'podonotal shield' abutting an 'opisthonotal shield'. The opisthonotum may bear a reduced 'mesonotal shield', or two or more 'mesonotal scutellae' and a 'pygidial shield' from which cheliceral retractor muscles may arise in uropodine Mesostigmata (Krantz, 2009a). The proterosoma of some Oribatida is covered by a 'proterosomatic shield', obscuring completely the gnathosoma dorsally, and abuts the anterodorsally extended opisthosoma posteriorly. The opisthosomatic shield forms a covering dorsally and often extends lateroventrally to produce a shell-like, circumgastric 'notogaster'. The Mesostigmata commonly show a fusion of podonotal/hysteronotal shield elements into a 'holonotal shield' (Krantz, 2009a).

Dorsal and ventral shields may be lacking in many Prostigmata. The pattern of ornamentation of the dorsal shield is often used in taxonomic studies.

The ventral idiosoma may or may not be sclerotized. Variation in the ventral shield characterizes the morphology of different taxa and species. In the Acariformes, the shield remnants, commonly called 'epimera', mark the attachment for the legs, and in several groups, they are modified to form the 'epimeral plates'. Epimera can be distinguished by 'epimeric furrows' that extend internally to form typical 'apodemes' (Evans, 1992). Leg articulations and genital and anal openings have a surrounding sclerite (hardened part), that can be variously sclerotized and of different shapes and sizes. Ventral shield expansion, reduction or fusion into

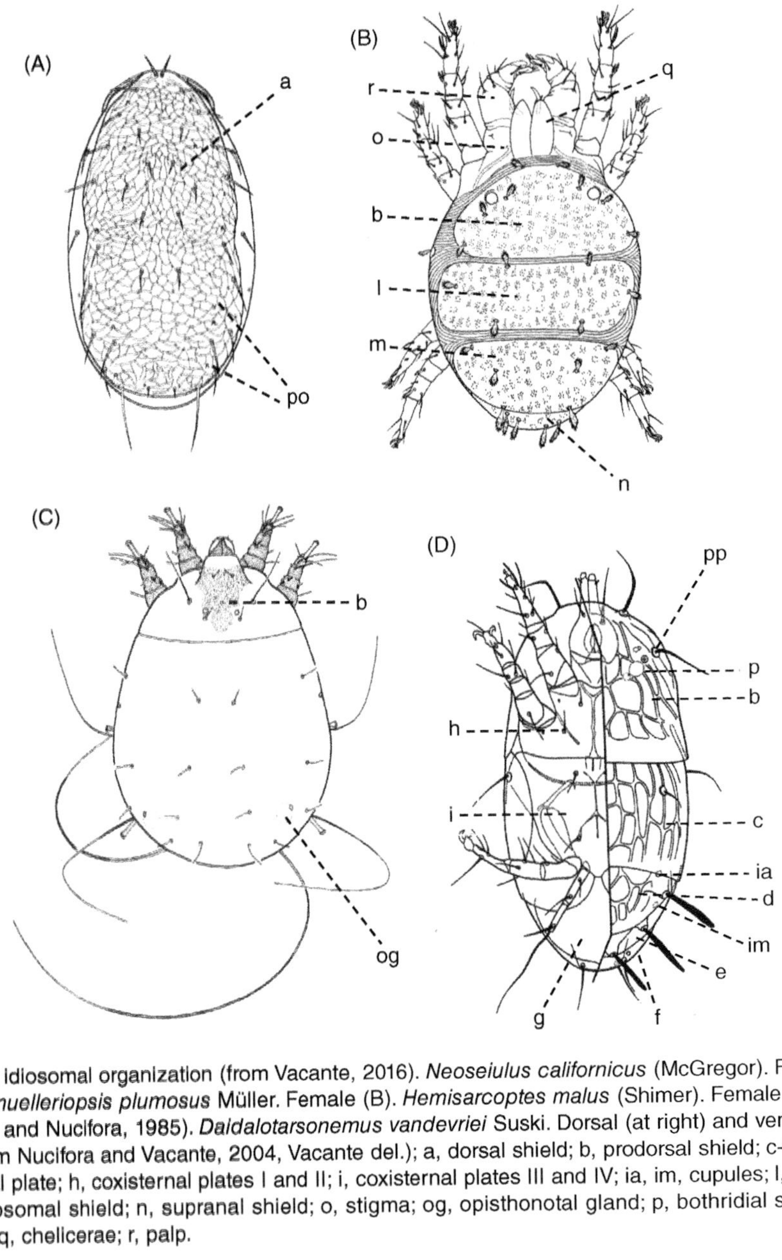

Fig. 2.9. Different idiosomal organization (from Vacante, 2016). *Neoseiulus californicus* (McGregor). Female partially drawn (A). *Ledermuelleriopsis plumosus* Müller. Female (B). *Hemisarcoptes malus* (Shimer). Female (C) (drawings A-C from Vacante and Nucifora, 1985). *Daidalotarsonemus vandevriei* Suski. Dorsal (at right) and ventral (at left) view of female (D) (from Nucifora and Vacante, 2004, Vacante del.); a, dorsal shield; b, prodorsal shield; c-f, tergites C, D, EF, H; g, aggenital plate; h, coxisternal plates I and II; i, coxisternal plates III and IV; ia, im, cupules; l, metapodosomal shield; m, opisthosomal shield; n, supranal shield; o, stigma; og, opisthonotal gland; p, bothridial seta; po, pores; pp, prodorsal pit; q, chelicerae; r, palp.

more or less fixed patterns are very common in Mesostigmata and Oribatida and are used as taxonomic indicators.

The primary external structures of the idiosoma relate to locomotion, respiration, sperm transfer, sensory reception, secretion and osmoregulation (Krantz, 2009a).

The idiosoma of Astigmatina is commonly ovoid, characterized by little or no traces of segmentation, with the exception of transverse constriction, dividing it into propodosoma and hysterosoma. The posterior edge of the body is sometimes lobed, such as in the males of *Thyreophagus* or *Histiogaster*, or more often in ectoparasites families, such as

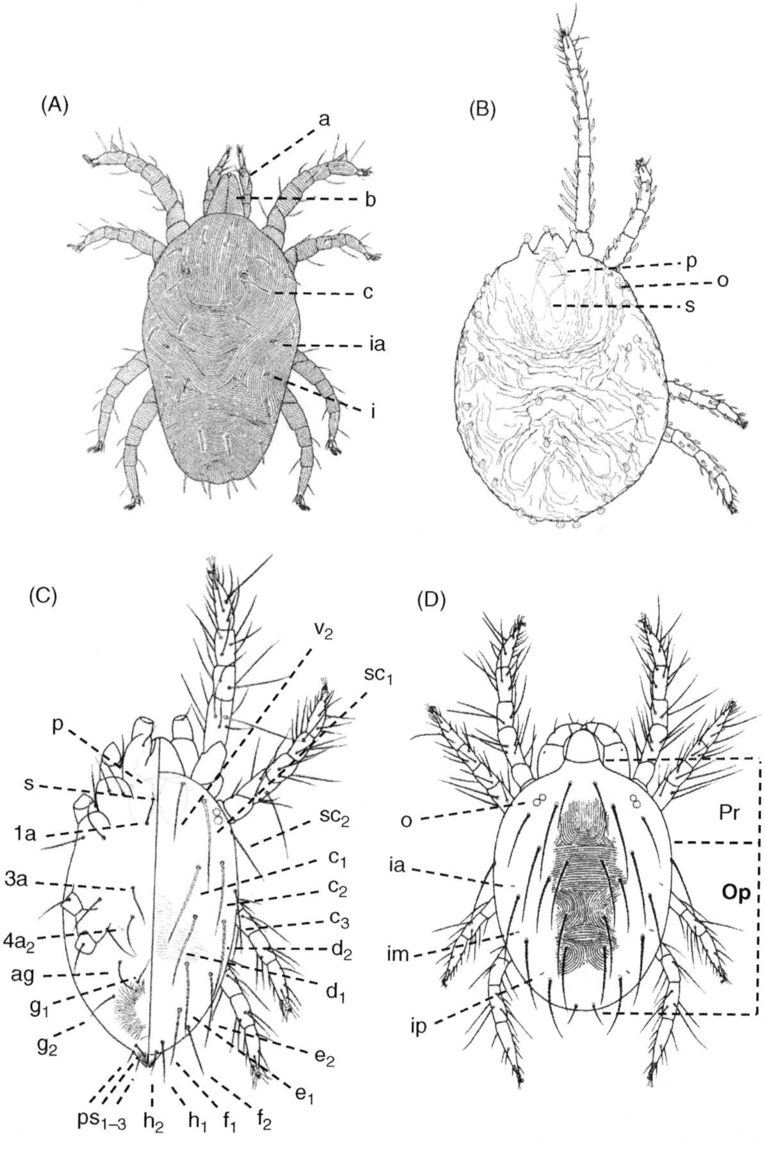

Fig. 2.10. *Tydeus australensis* Baker. Female (A) (from Vacante and Nucifora, 1985). *Bryobia cavalloroi* Vacante. Female (B) (from Vacante and van Eyndhoven, 1989). *Eotetranychus rubiphilus* (Reck). Dorsal (at right) and ventral (at left) view of female (C) (from Vacante, 1985a). *Tetranychus urticae* Koch. Female (D) (from Vacante, 2010); a, palpus; b, chelicerae; c, bothridial seta; ia, im, ip, lyrifessures; o, ocelli; op, opisthosoma; p, peritreme; pr, prodorsum; s, stigma. The setal notation is explicated in the text.

Proctophyllodidae or Psoptidae. A pair of longitudinal grooves extending the length of the idiosoma marks the position of the dorsoventral muscles and, since the cuticle is inextensible, these grooves are particularly conspicuous in specimens that have been kept under dry conditions (Hughes, 1976). The idiosoma possesses a number of setae, characterized by various lengths and shapes, arranged typically in each systematic group. These are discussed in more detail in each group.

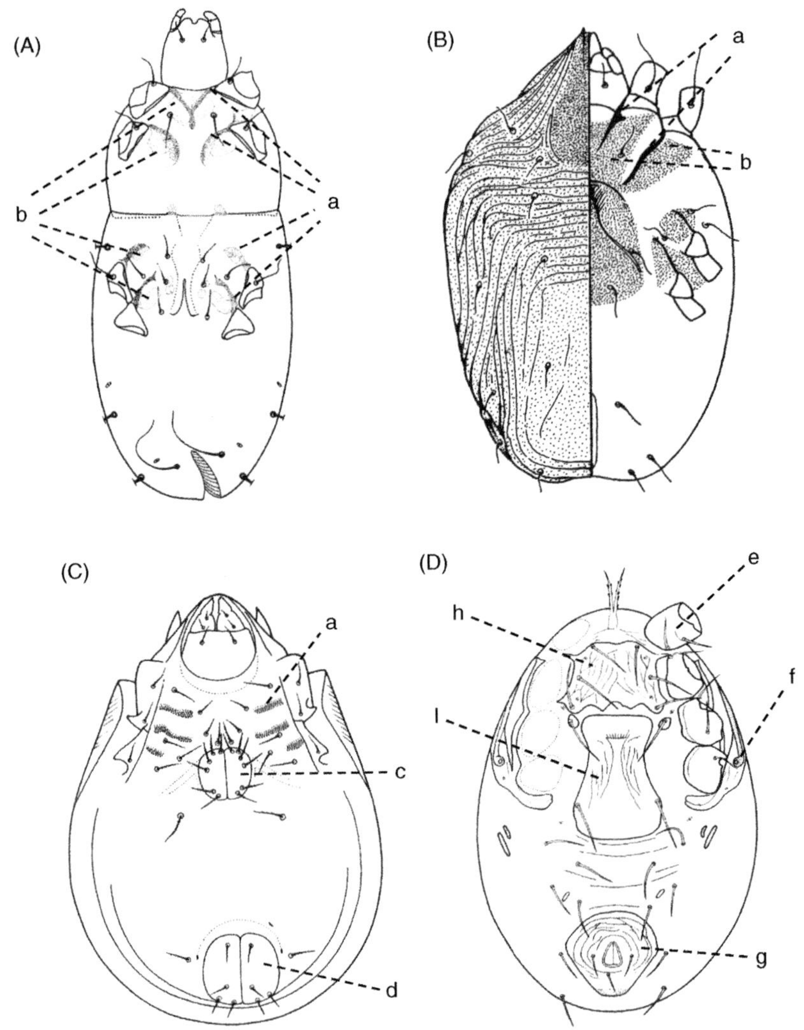

Fig. 2.11. *Tyreophagus entomophagus* (Laboulbène). Ventral view of female (partially drawn) (A) (from Vacante, 1989a). *Euroglyphus maynei* (Cooreman). Dorsal (at left) and ventral view (at right) of female (partially drawn) (B) (from Vacante, 1997a). *Humerobates rostrolamellatus* Grandjean. Ventral view of female (partially drawn) (C). *Phytoseiulus persimilis* Athias Henriot (D). Ventral view of female (partially drawn) (drawings C and D from Vacante and Nucifora, 1985); a, epimera; b, apodemes; c, genital plate; d, anal plate; e, coxa; f, stigma; g, ventrianal shield; h, sternal shield; l, epigynial shield.

Legs (Figs 2.12–2.17)

Adult and nymphs of Acari, except in some Prostigmata and Astigmatina, have four pairs of jointed legs, and the larval instar has three pairs of legs. The legs have seven primary segments called, in proximal to distal order, the coxa, trochanter, femur, genu, tibia, tarsus and pretarsus. As the order is always given as proximal–distal, the last segment (the most distal) could formally be called a 'posttarsus' instead of a pretarsus; however, in almost every text, the term used is pretarsus. The tarsus I in some derivative Mesostigmata may have a separate distal podomeric 'acrotarsus' or 'apicotarsus' on which the pretarsus is inserted. Deutonymphs of some Parasitidae possess acrotarsi (Evans, 1992). The family Anistidae is characterized by tarsi with multiple podomeric divisions (Otto, 1999). Larval and protonymphal stages of early derivative acariform taxa lack subdivision of

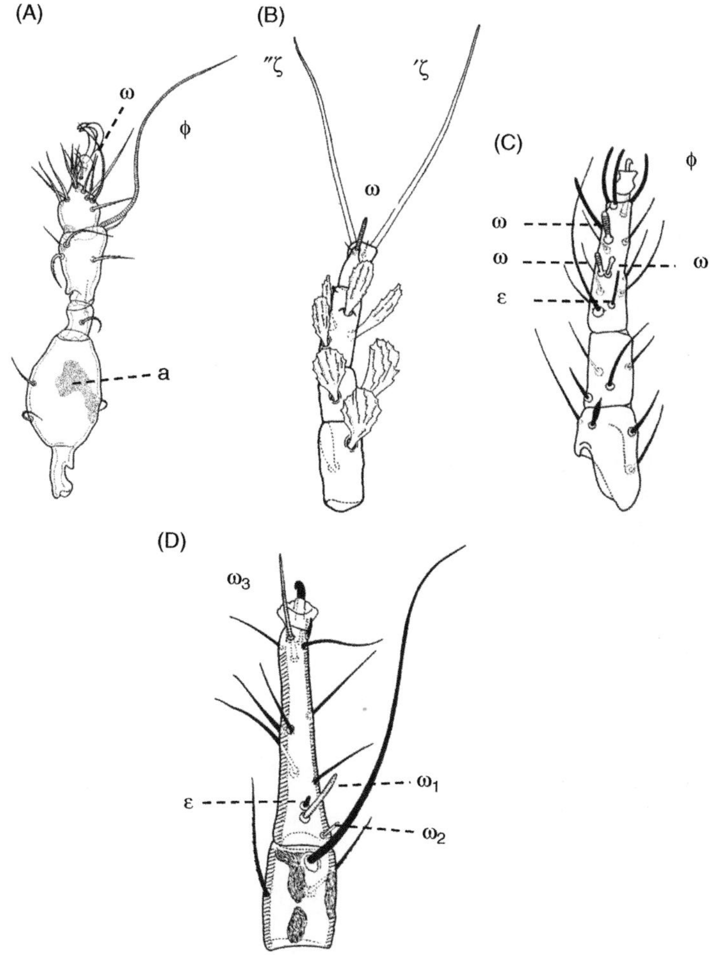

Fig. 2.12. Legs I of acarine species. *Micreremus gracilior* Willmann. Female (femur, genu, tibia and tarsus) (A). *Cheletogenes ornatus* (Canestrini *et* Fanzago). Female (femur, genu, tibia, tarsus) (B) (drawings A-C from Vacante and Nucifora, 1985). *Tarsonemus parawaitei* Kim. Qin *et* Lindquist. Female (femur, genu, tibiotarsus) (C) (from Nucifora and Vacante, 2004, Vacante del.). *Tyrophagus neiswanderi* Johnston *et* Bruce. Female (tibia and tarsus) (D) (from Vacante, 1988); a, area porosa; db, trichobothridial seta; ω, tarsal solenidia; φ, tibial solenidion; ε, famulus; ζ, eupathidia.

femora, but the division as a vestigial suture may be observed between basi- and telofemora III–IV in the Heterostigmata (Lindquist, 1984). A podomeric fusion of two or more postcoxal primary segments may occur in the Acariformes (Hughes, 1959; Evans, 1992). Femur and genu or tibia and tarsus may be amalgamated and form a 'femurogenu' or a 'tibiotarsus'.

The leg segments are variously organized in different taxa. Segment movement is led by intersegmental monocondylar or dicondylar articulations, and leg flexion and extension are supported by two different physiological mechanisms. Extension exploits the hydrostatic pressure resulting from the contraction of the dorsoventral idiosomatic muscles, while flexion results from the action of intrinsic and extrinsic intersegmental muscles (Krantz, 2009a). According to the systematic group, the coxae may be free or fused with the ventral podosoma, and the femur may be divided into the basifemur and the telofemur. The various leg segments bear a number of setae, tactile or specialized (solenidia, eupathidia, famuli, microsetae, trichobothria), commonly corresponding to the pattern of ontogenetic appearance in the taxon, and the tactile setae can be organized in whorls or 'verticils', and set in dorsal,

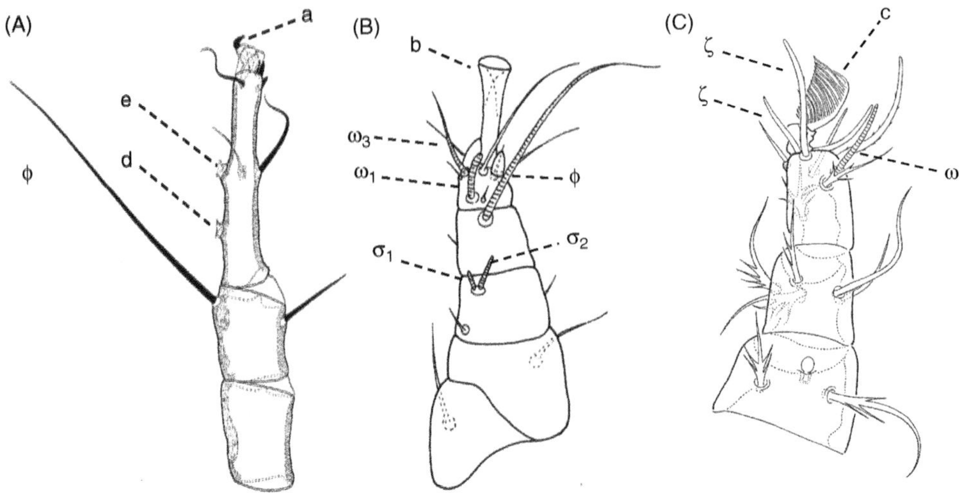

Fig. 2.13. Legs of acarine species (from Vacante, 2016). *Tyrophagus neiswanderi* Johnston *et* Bruce. Leg IV (genu, tibia and tarsus) of male (A) (from Vacante, 1988). *Hemisarcoptes malus* (Shimer). Leg I (trocanther, femus, genu, tibia, tarsus) of female (B) (from Vacante and Nucifora, 1985). *Cheyletiella blakei* Smiley. Leg I (genu, tibia, tarsus) of female (C) (from Patti and Vacante, 1985, Vacante del.); a, claw; b, sucker-like ambulacrum; c, empodium; d, e, tarsal suckers; ω, φ, σ, solenidia; ζ, eupathidia.

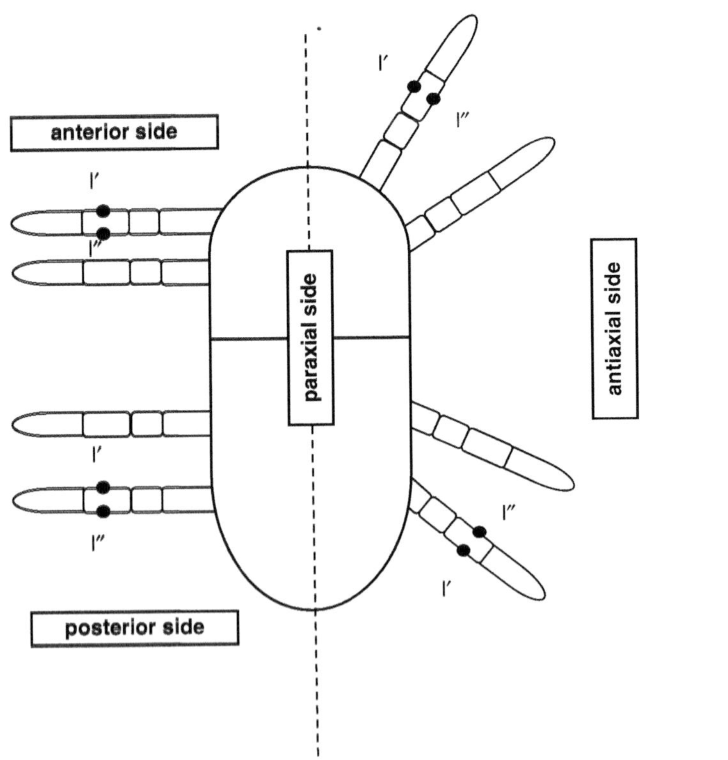

Fig. 2.14. Schematic drawing of Acariform mite showing the paraxial and antiaxial position of leg setae (according to Norton, 1977, from Vacante, 2016).

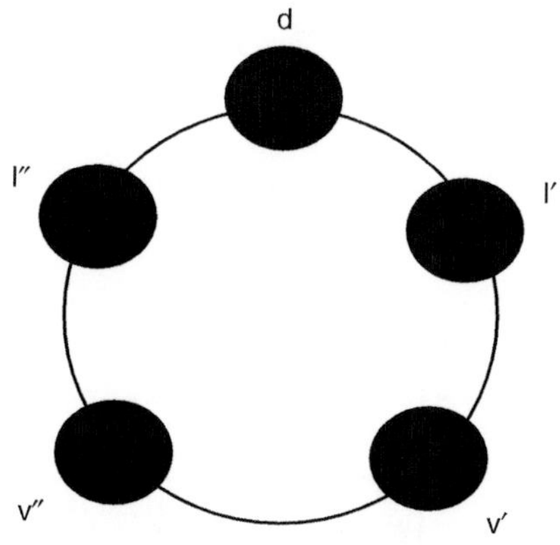

Fig. 2.15. Shematic drawing of verticill of 5 setae (following Norton, 1977, from Vacante, 2016); d, dorsal; l, lateral; v, ventral.

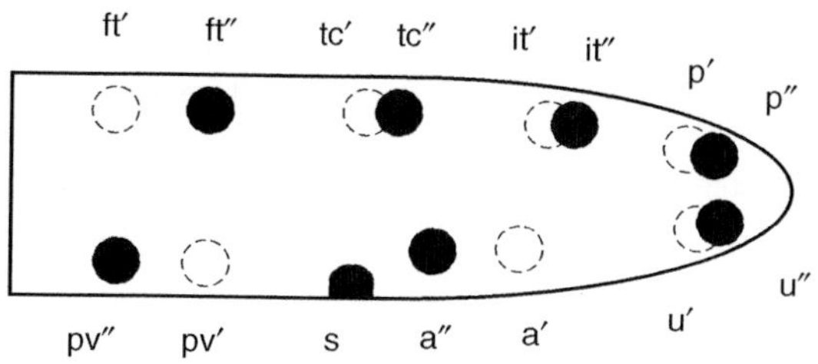

Fig. 2.16. Schematic drawing showing nomenclature and notation of tarsal setae (according to Norton, 1977, from Vacante, 2016).

ventral or lateral positions. In the Mesostigmata (and other groups), the distribution of setae on each leg segment can be examined by a formula that is useful in systematic and phylogenetic analysis (Evans, 1963b). The system was applied by Grandjean (1935a, 1940, 1941, 1942, 1946a, 1958) to the leg chaetotaxy of the Oribatida and successively reviewed by Norton (1977), and variously applied to other acarine groups. In this system, the leg faces with respect to the central body axis are identified as a 'paraxial face' and an 'antiaxial face', and the schematic verticils of setae of a leg segment are represented by one dorsal (*d*), two lateral (*l'*, *l"*) and two ventral setae (*v'*, *v"*), respectively; taking into account that the paired setae are both paraxial (indicated by ') and antiaxial (indicated by "). The setae *l'* on leg I are paraxial, whereas the setae on leg IV are antiaxial. The setal notation of the tarsi includes a greater number of elements, and from the proximal to the distal part of the segment, there may be four pairs of dorsal setae, called fastigial (*ft'*, *ft"*), tectal (*tc*), iteral (*it*) and proral (*p*); the ventrolateral setae are the primiventral pair (*pv'*, *pv"*), the unpaired subunguinal seta (*s*), the anterolateral pair (*a'*, *a"*) and the unguinal pair (*u*) (Norton, 1977). Usually, the tarsi and pretarsi of legs I and II have various sensory structures (see section on sensory receptors). The pretarsus includes a 'basilar sclerite' and other structures, such as a median 'empodium' (an unpaired structure arising between the tarsal claws), a pad-like or sucker-like 'pulvillus' (a membranous, pad-like structure associated with the claws) or paired 'claws'. When the pretarsus bears median elements, it is called an 'ambulacrum'. The basilar sclerite acts as a fulcrum articulated with the tarsus by condylophores and is related to the extensor and depressor muscles arising in the tibia or in the tarsus (Krantz, 2009a). There can be strong diversity in the structure of the ambulacrum; it may be absent, it may have a claw-like or pad-like empodium or the empodium may be absent, paired claws may be absent, there may be a sucker-like pulvillus, etc.

The Astigmatina have an arrangement and number of setae and solenidia characterized by the same basic pattern. In some species setae and solenidia may be lost or may occupy a different position, as well as the two sexes may show differences (Hughes, 1976).

Respiratory Organs (Fig. 2.18)

Respiration in the Acari can be 'cuticular' but is more commonly based on an internal branched transport system that includes spiracular tegumental openings called 'stigmata' or 'stigma'. The number, presence and position of the stigmata vary in different systematic groups, and these variations are used in taxonomy for the identification of acarine orders and suborders. In all orders of the superorder Parasitiformes, the stigmata are set in the median idiosoma, while in the suborder Prostigmata of the superorder Acariformes, paired stigmata are located anteriorly, behind or between the cheliceral bases, or in relation to the lateral humeral angles of the propodosoma. Parasitiformes and Trombidiformes possess stigmatic openings that can be connected to tubular 'peritremes', which are sometimes extended throughout most of the propodosoma, or are arranged simply with ducts connecting to a tracheal system. Moreover, in the different acarine groups, there may be morphological and physiological variation of the respiratory system

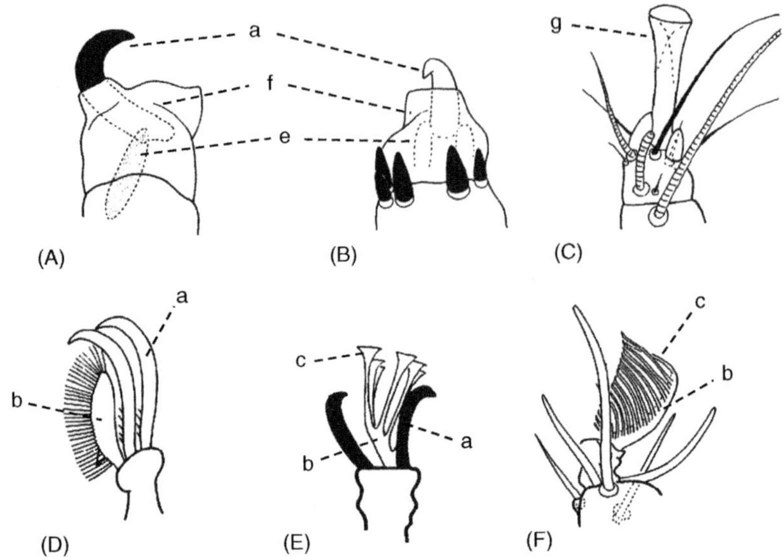

Fig. 2.17 Pretarsi of acarine species (from Vacante, 2016). *Tyrophagus tropicus* Robertson. Pretarsus I of female (A). *Michaelopus corticalis* (Michael). Distal end of tarsus and pretarsus III of female (B). *Hemisarcoptes malus* (Shimer). Distal end of tibia and tarsus I of female (C) (drawings B-D from Vacante, 1989a). *Tydeus australensis* Baker. Pretarsus I of female (D) (from Vacante and Nucifora, 1985). *Eryngiopus siculus* Vacante *et* Gerson. Pretarsus I of female (E) (from Vacante and Gerson, 1987). *Cheyletiella blakei* Smiley. Distal end of tarsus and pretarsus I of female (F) (from Patti and Vacante, 1985, Vacante del.); a, claw; b, empodium; c, tenent hairs; d, basilar sclerite; e, condylophore; f, ambulacrum; g, sucker-like ambulacrum.

that involves different structures (encircling plates, microvilli, areae porosae, brachytracheae, ostia, maculae, aeropyles) (Krantz, 2009a). In the Ixodida, the peritremes take the form of paired 'encircling plates' around the stigmata, and those of Mesostigmata, generally, extend anteriorly or/and posteriorly from the stigmatic opening as an external groove of varying length and complexity. The peritrematic groove or plate may be lined with cuticular 'microvilli', reminiscent of the filtering devices found in the spiracular atria of insects (Chapman, 1982). They serve as filters for preventing debris from clogging the groove (Alberti and Coons, 1999), and provide a means for conserving an air layer for plastron respiration when submersion in ground or sea water may limit normal respiration (Krantz, 1974). In the Oribatida, the tracheal ducts often open via obscure stigmata, set in the acetabular invaginations between the trochanters I–IV and the adjacent body wall, in the sejugal groove, on the appendages. Or in the prodorsal bothridia, where they appear as short 'brachytracheae' (Norton *et al.*, 1997). The tracheae may be derived from internalized 'areae porosae', which develop from surface pores as a correlate of increased sclerotization and concomitant loss of gas permeability (Norton and Alberti, 1997). The paired stigmata of ixodid ticks are slit-like 'ostia', each surrounded by a heavily sclerotized peritrematic plate, comprising a pigmented portion called 'macula', and a sieve plate containing a number of fine pores called 'aeropyles'. The function of these aeropyles appears unclear (Krantz, 2009a).

Stigmata are commonly connected to a developed tracheal system, but exceptions may occur as in Trombiculidae and in water mites, in which tracheae either do not occur, or serve particular organs in which they form a subcuticular network with independent tracheal strands serving particular organs (Mitchell, 1972a). However, the lack of stigmata does not correspond to a lack of a tracheal system, as in the glycyphagid *Gohieria fusca* (Oudemans), in which, the paired trachea-like structures open in the genital atrium (Hughes, 1959). Evans (1992) reported that other glycyphagid mites have similar structures often ending in the genital or anal region. However, it is possible that they represent glandular, and not respiratory structures. Lindquist (1984) reported that early derivative Prostigmata and Oribatida possess genital tracheae retained as possible remnants of an earlier, better-developed respiratory system.

In the Astigmatina, the main respiratory surface is the cuticle and must also play a part in whatever control mechanism there is, regulating the loss and uptake of water (Hughes, 1976).

Reproductive Organs (Figs 2.19, 2.20 and 2.21)

The reproductive organs of mites can be discussed in terms of their external and internal morphology, and their physiology and ecology, mainly in relation to strategies of sperm transfer (Alberti and Coons, 1999). In this section, the idiosomal external features of the reproductive organs are briefly discussed. Sperm transfer may take place either directly

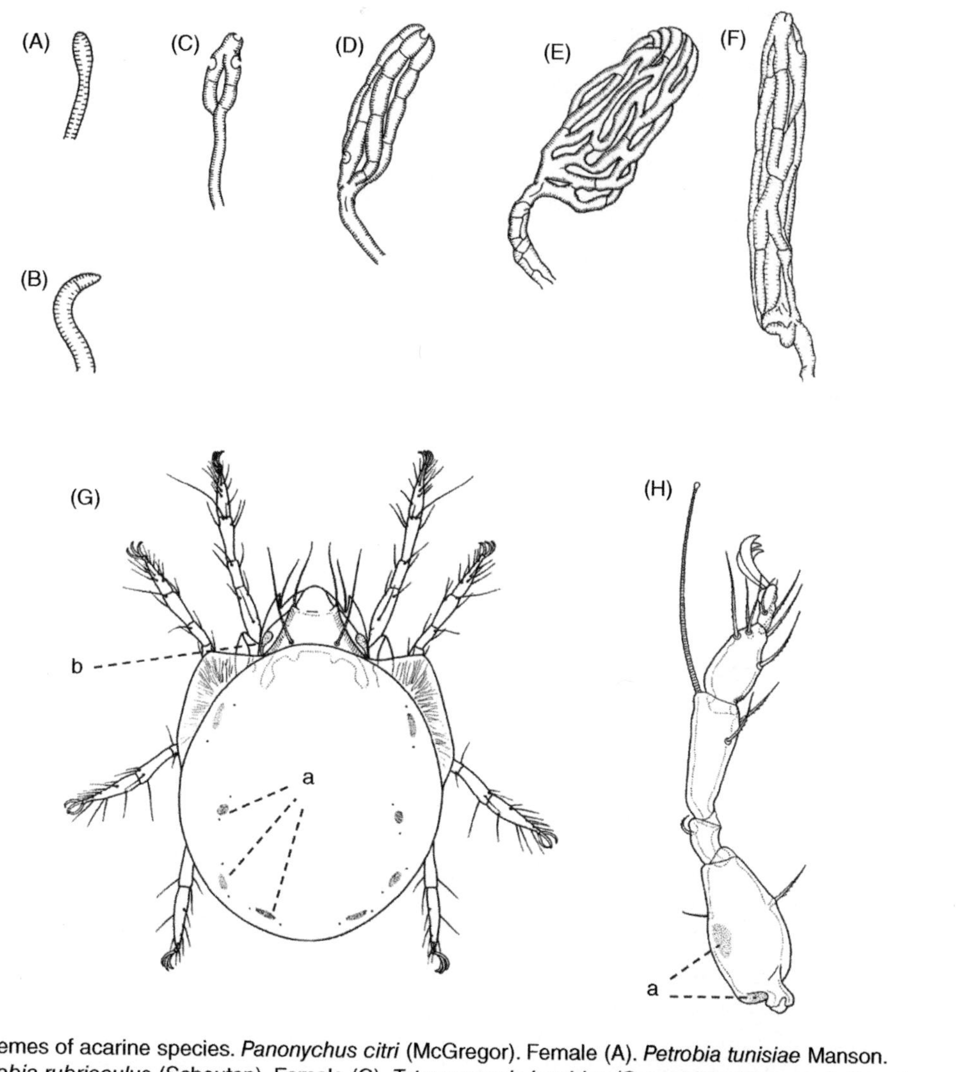

Fig. 2.18. Peritremes of acarine species. *Panonychus citri* (McGregor). Female (A). *Petrobia tunisiae* Manson. Female (B). *Bryobia rubrioculus* (Scheuten). Female (C). *Tetranycopsis horridus* (Canestrini *et* Fanzago). Female (D). *Aplonobia histricina* (Berlese). Female (E) (Drawings A-E are taken from Vacante, 1985a). *Bryobia cavalloroi* Vacante *et* Van Eyndhoven. Female (F) (from Vacante and Van Eyndhoven, 1989). Areae porosae. *Humerobates rostrolamellatus* Grandjean. Female (G). *Micreremus gracilior* Willmann. Female (H) (drawings G and H from Vacante and Nucifora, 1985); a, area porosa; b, bothridial seta.

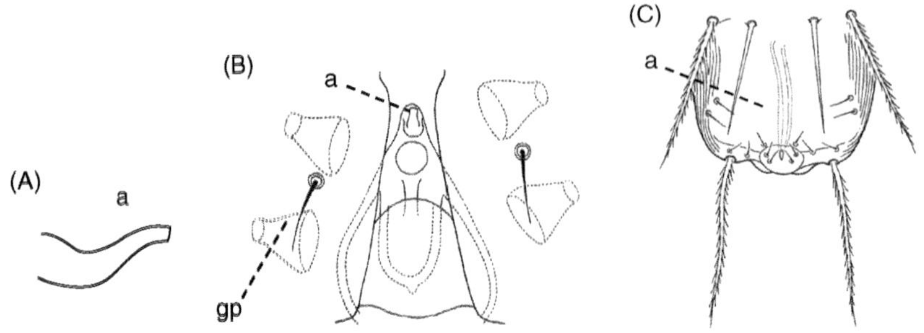

Fig. 2.19. Aedeagi of acarine species (from Vacante, 2016). *Tyrophagus neiswanderi* Johnston *et* Bruce (A) (from Vacante, 1988). *Tyrophagus palmarum* Oudemans (B) (from Vacante, 1989a). *Cheyletiella blakei* Smiley. Male opisthosoma (C) (from Patti and Vacante, 1985, Vacante del., partially modified); a, aedeagus; gp, genital papilla.

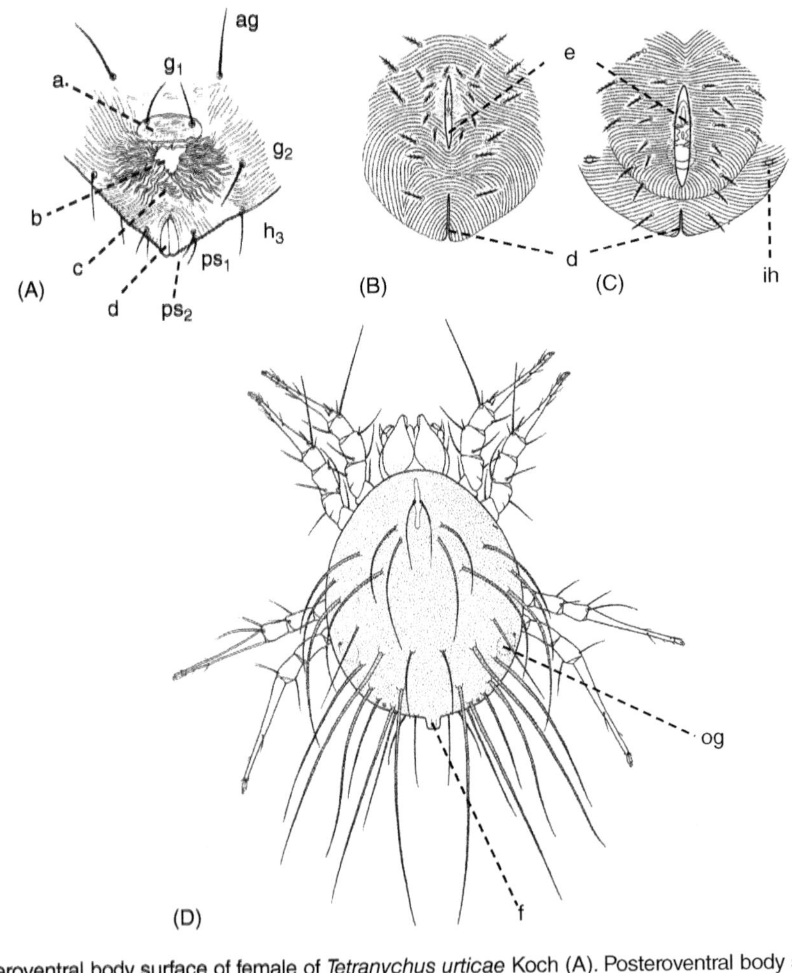

Fig. 2.20. Posteroventral body surface of female of *Tetranychus urticae* Koch (A). Posteroventral body surface of female (B) and male (C) of *Triophtydeus triophthalmus* (Oudemans). *Glycyphagus domesticus* (De Geer). Female (D) (from Vacante and Nucifora, 1985); a, genital flap; b, ovipositor; c, copulatory orifice; d, anal opening; e, progenital aperture; f, bursa copulatrix; ih, lyrifissure; og, opisthonotal gland. The setal notation is explicated in the text.

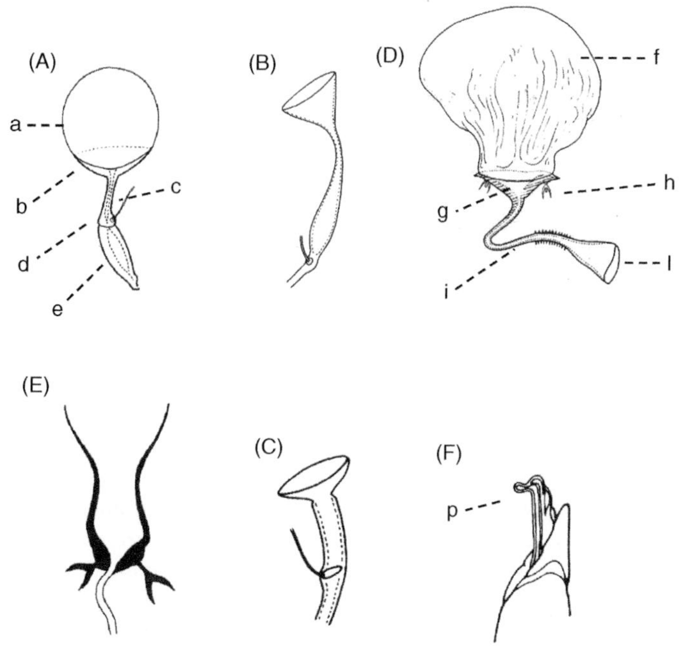

Fig. 2.21. Receptacula seminis of acarine species (from Vacante, 2016). *Phytoseius finitimus* Ribaga (A). *Phytoseiulus persimilis* Ahtias Henriot (B). *Euseius stipulatus* (Athias Henriot) (C) (from Vacante and Nucifora, 1987). *Tyrophagus neiswanderi* Johnston *et* Bruce (D) (from Vacante, 1988). *Thyreophagus entomophagus* (Laboulbène) (E) (from Vacante, 1989a). Chelicera of male of *Amblyseius californicus* (McGregor) with spermatodactyl (F) (from Vacante and Nucifora, 1985); a, vesicle; b, cervix; c, minor duct; d, atrium; e, major duct; f, spermathecal sac; g, base of spermathecal sac; h, sclerite of oviduct; i, spermathecal duct; l, copulatory opening; m, spermathecal tube; n, spermatheca; o, anterior apodeme; p, spermatodactyl.

(direct sperm transfer) or indirectly (indirect sperm transfer). Direct sperm transfer involves the entry of the male copulatory organ or 'aedeagus' into the female copulatory orifice, which can be located either between the ovipositor and the anal opening, as in several Prostigmata (Alberti and Crooker, 1985), or in a special extragenital structure, the 'bursa copulatrix', at the rear of the idiosoma, as in the Astigmatina and some Prostigmata. The bursa is connected to the internal seminal receptacle, or 'receptaculum seminis', from which two ducts lead into the ovaries (Evans, 1992). The function of the genital opening is fundamentally that of oviposition, and for this reason it is called an 'oviporus'. The morphology of the receptaculum seminis varies according to the systematic group. In several Prostigmata, such as the Tetranychidae (Pritchard and Baker, 1955; Gutierrez and Helle, 1985; Vacante, 1985; Van Eyndhoven and Vacante, 1985) or the Eriophyidae (Amrine *et al.*, 2003), the morphology of the aedeagus and the receptaculum seminis (spermatheca) is used in specific or generic taxonomy. Non-copulatory sperm transfer uses various strategies, ranging from a transfer of sperm during male/female pairing (Walter and Proctor, 1999) to the production by the male of 'spermatophores' that are deposited on a support surface. The morphology of spermatophores varies from elongate to sculpted, and from small to very small, and is distinctive by family and genus (Fernandez *et al.*, 1991). The deposition of spermatophores can be 'proximal', when the females are close to or paired with the males, or 'dissociative' if it occurs in the absence of females, as in many Eriophyidae (Oldfield *et al.*, 1970, 1972; Oldfield and Michalska, 1996). Spermatophores can also be deposited in variable numbers, singly or at high numbers in masses or organized in rows. In general, the presence of the male assists the possibility of contact between females and spermatophores by various means (physical, chemical or behavioural). Indirect sperm transfer can occur by 'tocospermy', in which the male uses one or both chelicerae to transfer its sperm into the female genital opening, as in the family Parasitidae of Mesostigmata (Evans and Till, 1979), or by 'podospermy', which uses a cheliceral sperm transfer organ or 'spermatodactyl' to transfer sperm into solenostomes or sperm induction pores (Athias Henriot, 1969c), variously located in the idiosoma, and analogous to the bursa copulatrix of the Astigmatina (Krantz, 2009a). The reproductive behaviour of Acari, especially sperm transfer and mating, can be influenced by pheromones that are emitted by receptive biological stages (adult and immature females). Three types of these sex pheromones are known: 'arrestants', 'attractants' and 'contact pheromones' (Sonenshine, 1985). The arrestant pheromones are emitted by immature females, and act by stimulating precopulatory guarding by adult males. The attractant pheromones are produced by adult females, and act by attracting the males. The contact pheromones are emitted by adult females and stimulate copulatory behaviour in the males. The phenomenon has been investigated in several groups of Acari (Krantz, 2009a).

Sensory Receptors (Fig. 2.22)

The idiosoma has different types of sensory receptors (chemoreceptors, mechanoreceptors and photoreceptors), commonly with a setal structure (Evans, 1992; Alberti and Coons, 1999). According to Grandjean (1935b), tactile and chemosensory setae may or may not contain an optically active iodophilic material, 'actinopilin', which is birefringent in polarized light and constitutes a layer surrounding part of the basal setal nerve cell. The Acariformes have 'anisotropic' setae that are optically active, and for this reason, they also have the name Actinotrichida, while the Parasitiformes have 'isotropic' setae that are optically inactive, so they also have the name Anactinotrichida (Van der Hammen, 1961; Evans, 1992). The cuticular surface also has different pore-like openings, some of which have a sensory function (Krantz, 2009a). Both cuticular and intracuticular pore-like organs may receive inputs from changes in physical tension within the body (Alberti and Crooker, 1985). These proprioceptors have various shapes, but they are commonly small membrane-covered clefts that are referred to as 'slit organs', 'poroids', 'cupules' or 'lyrifissures' (Alberti, 1998). Commonly, chemosensory receptors have a setal appearance; they are usually anisotropic and located on distal segments of legs I and II, and on the palpi. Inside, they have a sheath of actinopilin, and externally, their structure (and morphology) varies significantly. Among the variations found are the 'eupathidia', which are spine-like, apically rounded or divided and set on the distal segment of tarsi I and II, and at the end of the palptarsus. They probably have a single terminal pore and serve as contact or gustatory receptors or chemoreceptors (Alberti, 1998; Alberti and Coons, 1999). Their presence on the legs and palpi has been associated with a gustatory-mechanosensory function (Evans, 1992). Another receptor, whose function is not clear, is the famulus, which is a simple or peg-like seta, sometimes falcate, branched, etc. The eupathidia have a terminal pore, which is usually small, and set on tarsus I in the Acariformes, and sometimes on tarsus II. In many cases, it is invaginated into pits (depressions) (Alberti and Coons, 1999). Another type of setal receptor is the solenidia; these are common on the tarsi, and on tibiae I and II (sometimes on femur III and IV) and on the palptarsus of Acariformes. They are rounded, pointed or peglike, and lack the actinopilin that is present in the eupathidia and famuli; they are also birefringent and sometimes striated, as in the Tetranychidae. Their function is probably olfactory (Evans, 1992). Other probable olfactory receptors are the 'sensilla basiconica' of various Mesostigmata, and the 'Haller's organ' of the Ixodida.

Most mechanoreceptors are simple tactile setae, with an isotropic or anisotropic shaft, a membranous socket and basal receptor cells that send signals from the setae to the central nervous system. Morphologically, the tactile setae are usually simple and spinose, but they may also have different shapes, such as expanded or variously ornate, etc. On the base of leg I of the Astigmatina, there is a supracoxal seta that is variable in form – from a short spine-like seta, in those species inhabiting wetter habitats to a many-branched seta in those species inhabiting drier environments; the latter type of seta seems to cover gland openings and reduce the loss of water. The shape, number and position of the tactile setae can form a pattern that is used in the taxonomy of various groups. A particular type of tactile setae is the 'trichobotria', which are specialized sensilla, variously present on the body, legs and appendages of some acarine groups. Prodorsal trichobothria, also called the 'bothridial setae' or 'bothridial organs' or 'pseudostigmatic organs', are located in cup-like cavities or 'bothridia', and consist of a long and specialized seta of various shapes (simple, clavate, pectinate, etc.) (Evans, 1992).

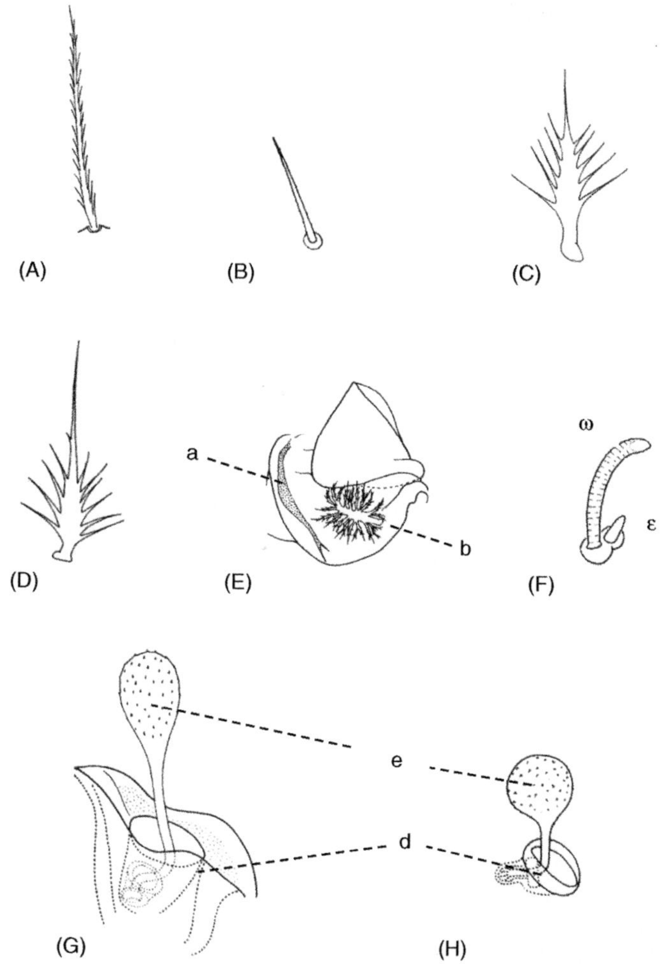

Fig. 2.22. Various setae of acarine species (from Vacante, 2016). Opisthosomal seta of male of *Cheyletiella blakei* Smiley (A). Seta *f* of tergite EF of female of *Tarsonemus waitei* Banks (B) (from Vacante and Nucifora, 1985). Supracoxal seta of *Tyrophagus palmarum* Oudemans (C). Supracoxal seta of *Tyrophagus tropicus* Robertson (D). Grandjean's organ and supracoxal seta of *Glycyphagus domesticus* (De Geer) (E). Solenidion and famulus of tarsus I of female of *Thyreophagus cooremani* Fain (F) (from Vacante, 1989a). Bothridial setae of *Humerobates rostrolamellatus* Grandjean (G) and *Micreremus gracilior* Willmann (H) (from Vacante and Nucifora, 1985); a, Grandjean's organ; b, supracoxal seta; c, tactile seta; d, bothridium; e, bothridial seta; ω, solenidium; ε, famulus.

Despite the variable form of the bothridial setae, they are similar to the acariform tactile setae, and have a solid core of actinopilin. Their main function is anemotactic (Evans, 1992), and they may also be an acoustic organ (Coineau *et al.*, 1997), act as vibroreceptors and anemoreceptors (Alberti *et al.*, 1994), and have a geotaxic function (Aoki, 1973). Moreover, trichobothria could be involved as capacitors in deflecting electromagnetic signals produced by their host or food substrate (Wergin *et al.*, 2000), or in maintaining balance (Evans, 1992).

Several acarines have eyes, eyespots and photoreceptive organs. One or two pairs of simple 'photoreceptors' or 'ocelli' are set on the anteromedian or anterolateral prodorsum (the dorsal surface of the propodosoma) in various groups (Evans, 1992; Alberti and Coons, 1999; Coons and Alberti, 1999). Each ocellus consists of a cuticular lens, biconvex or convex-concave, and an underlying photoreceptor, with axons forming the optic nerve (Evans, 1992). Tetranychid mites have lateral prodorsal ocelli divided into two parts, each functioning differently from the other; the anterior part represents a biconvex eye and may act as a scanning-point detector for green and UV light, while the posterior part receives signals in the near-UV that are probably non-directional (Alberti and Crooker, 1985). Other variations may occur in different groups.

Notation of Mechanosensory and Chemosensory Setae

There has been a certain confusion around the designation of mechanosensory and chemosensory setae, and their derivative structures on the appendages of actinotrichid mites. This problem was investigated by Grandjean (1935a, 1939, 1958), who indicated a solution for identifying the homologies between the structures in the various taxa. In the 'Grandjean system', the setae are indicated by Greek symbols, with the solenidia of the femur, genu, tibia and tarsus designated by θ (theta), σ (sigma), φ (phi) and ω (omega), respectively. Famuli and eupathidia are designated by ε (epsilon) and ζ (zeta), respectively.

Secretory Organs

The secretory organs discussed in this section concern external secretions. It is very difficult to summarize the amount of information that is available on this topic, and at the same time to present an account that is consistent in character. Pore-like openings located in the acarine cuticle are common among the secretory organs and may be connected by ducts to subcuticular gland cells (Alberti and Seeman, 2005). The nature of the products that are secreted is heterogeneous and varies from cement-like to waxy (Evans, 1992; Coons and Alberti, 1999) to sex pheromones (Hoy and Smilanick, 1979) or to defensive secretions, as a means of protection from predators (Kerfoot, 1982). Some oribatid mites have lateral 'opisthonotal glands', which are easily seen as variously coloured spots below the transparent integument (Evans, 1992; Alberti and Coons, 1999). These oil glands have a duct, with a pore and a cuticular flap. The secretions of these glands are of varying chemical nature (they may be monoterpenes, hydrocarbons, esters, aromatics, etc.) and may have various functions, such as acting as pheromones (alarm, aggregation, sexual, defence, etc.). Neryl formate, a compound secreted by the opisthonotal glands of some Oribatida, has been identified as a key alarm pheromone, and is also found in the injurious acarid mite *Rhizoglyphus robini* Claparède (Mizoguchi *et al.*, 2003), as well as the S-isorobinal as the female sexual pheromone from an alarm pheromone emitting by the acarid *Rhizoglyphus setosus* Manson (Mizoguchi *et al.*, 2005). Raspotnig *et al.* (2011) found that the secretions from the opisthonotal glands of four species of oribatid mites contain the iridoid monoterpenes chrysomelidial and epi-chrysomelidial and the diterpene b-springene, and discuss the chemosystematic discrimination of morphologically similar, syntopic or cryptic species based on their oil gland secretion profiles. Other secretory organs, such as the 'sensilla sagittiformia', are known for ixodid mites. A detailed treatment of this topic can be found in Alberti and Coons (1999).

Osmoregulatory Organs

Acarine osmoregulation relies on a pair of 'urstigmata' or 'organs of Claparède'. Serially homologous genital papillae are present in nymphs and adults, whose larvae have Claparède's organ (Oudeman's Rule), and tydeid mites may retain the urstigmata beyond the larval stage. In the larval stage, the organs of Claparède are located ventrally on coxae I or II or between them and have been retained as homologous to the genital papillae. According to Vercammen-Grandjean (1975), in representative Prostigmata, the urstigmata and the genital papillae play a respiratory function, but an alternative interpretation concludes that these organs play a role in osmoregulation (Alberti, 1979; Alberti and Bader, 1990). There is also a diversity of other structures that might be related to the function of osmoregulation.

3 Internal Morphology

Like other Arthropoda, the internal morphology of mites primarily involves the idiosoma, which includes the 'haemocoel', a cavity that also extends to the legs and surrounds the internal organs that constitute the different systems of the body (circulatory, nervous, reproductive, glandular, digestive and excretory).

Circulatory System (Figs 3.1 and 3.2)

The haemocoel is filled with 'haemolymph', a free-flowing liquid whose circulation is facilitated by body movement and the dorsoventral contraction of the idiosomal muscles. The haemolymph contains various substances, such as amino acids found in the Ixodoidea (Boctor, 1972), lipids in the Uropodidae (Woodring and Galbraith, 1976), sugars in the Argasidae (Aboul-Nasr and Bassal, 1971), etc. and several cellular elements, the 'haemocytes', which are adapted to different functions, such as clotting, macrophagy and tissue dissolution, the development of which may involve germinal cells (Evans, 1992; Alberti and Coons, 1999; Coons and Alberti, 1999). The Ixodida and some of the larger sized Gamasida may also have a 'heart', which is set in the anterior opisthosoma, and is hexagonal or rhomboid in shape and flattened dorsoventrally. This has a single chamber with two dorsal ostia and is related anteriorly to the long aorta.

Pritchard *et al.* (2015) reported that the circulatory system of the Acari hosts an immune system composed of phagocytizing haemocytes and antimicrobial peptides, such as defensins and lysozymes. The midgut is the primary site for the destruction of bacterial and viral pathogens that are ingested with a blood meal, but if these microbes successfully traverse the midgut epithelium, then defensins and lysozymes are secreted into the haemolymph and fat body (Ceraul *et al.*, 2003; Simser *et al.*, 2004; Taylor, 2006). Lysozymes in astigmatid mites can function in both defence and digestion when microbes are used as a secondary food source (Childs and Bowman, 1981; Erban and Hubert, 2008). Greater understanding of *D. gallinae* (De Geer) lysozymes and the cells that contain them could contribute to novel controls against the mites by affecting the ability of the mite to process ingested pathogens that may affect or be transmitted by PRM, as demonstrated for ticks (Simser *et al.*, 2004).

Smrž (2006) studied the microanatomy of several oribatid, and one acarid mite was studied to determine the role of free cells (haemocytes) in mites. Mites from the field as well as laboratory cultures were observed and analysed histologically using Masson triple stain. The mites were offered various foods and kept in fluctuating moisture conditions. The presence of haemocytes was significantly correlated with the transport between internal organs of various substances. Three types of transport were recorded: (i) enzymes into the alimentary tract, including the incorporation of haemocytes into the gut walls. This process seemed to be correlated with the amount and type of food, and frequently with the presence of internal extraintestinal bacteria associated with mesenchyma; (ii) metabolites, like guanine from mesenchyma into the alimentary tract followed by expulsion from the body via the gut. This process is correlated with food of high nitrogen content or dry conditions; (iii) resorption of nutrients from eggs during an induced quiescent state under unfavourable conditions by small haemocytes.

Nervous System

The nervous system consists primarily of a 'synganglion', which is a nervous mass extended around the oesophagus, and basically corresponds to the central nervous system. It is characterized by a protocerebral 'supraoesophageal ganglion' and a tritocerebral 'suboesophageal ganglion' (Alberti and Crooker, 1985). The synganglion has an outer

DOI: 10.1079/9781786390912.0003

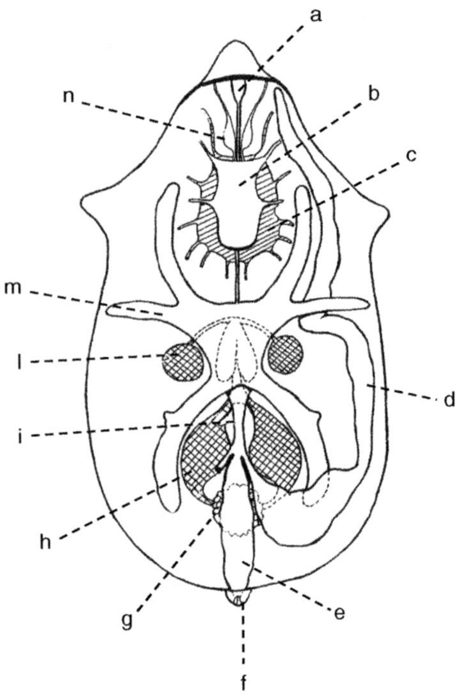

Fig. 3.1. *Caminella peraphora* Krantz *et* Ainscough. Internal aspect of female (from Evans, 1992); a, esophagous; b, supraesophageal ganglion; c, subesophageal ganglion; d, Malpighian tubule; e, anal atrium; f, anus; g, ovary; h, oviduct; i, colon; l, accessory gland; m, gastric caeca; n, salivary gland.

'cortex' and an inner 'neuropil' but lacks a deutocerebral portion owing to the absence of antennae, the lack of which is the cause of this secondary loss (Evans, 1992). The nervous system of mites does not have separate dorsal or ventral portions, but there are dorsal and ventral ganglia related by commissures (circumoesophageal, etc.) that are located in the prosoma. In some groups, such as the Ixodoidea, the cortex has a 'perineurium', 'glial cells' and 'neuronal cell bodies' that are involved in the nerve impulse and in the secretion of neuromediators (Coons *et al.*, 1974). Studies carried out on ixodids have shown that the central nervous system includes neurosecretory cells that produce different substances that mediate various physiological processes (nymphal cuticular apolysis, moulting, oviposition, salivation, etc.) (Hughes, 1954; Ioffe, 1965; Binnington and Obenchain, 1982; Hänel, 1986). In parasitiform mites, a degree of sexual dimorphism may depend on neurosecretory development (Krantz, 2009a). The cholinergic system (acetylcholine, choline acetylase, cholinesterase) and synaptic action are basic to nerve function.

Supraoesophageal Ganglion

Much of the knowledge on the supraoesophageal ganglion is from studies that have been carried out on the Ixodoidea, Astigmatina and some Prostigmata. This ganglion may be variously connected with

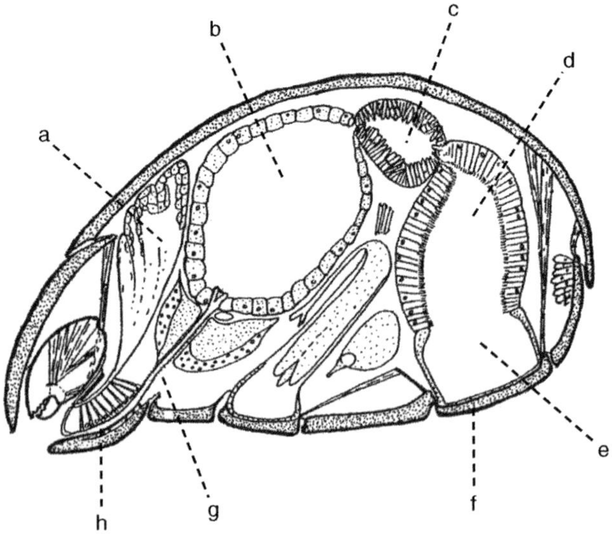

Fig. 3.2. Schematic midsagittal section of *Euzetes* sp. (from Evans, 1992); a, salivary gland; b, ventriculus; c, colon; d, postcolon; e, anal atrium; f, anal plate; g, esophagus; h, pharynx.

other nervous organs. Severino *et al.* (1984) observed in the dermanyssid *D. gallinae* (De Geer) four pairs of pedal ganglia extending distally from the supra-oesophageal mass, each ganglion connecting to each of the eight legs of the mite. The suboesophageal mass is bisected longitudinally by the oesophagus, and surrounded by fat tissue (Pritchard *et al.*, 2015). In *Dermacentor variabilis* (Say), for example, an 'optic lobe' and 'cheliceral lobes' have been found, with a posterior 'stomodaeal bridge' and 'pedipalpal ganglia' (Ivanov and Leonovich, 1983). In the Astigmatina, these pedipalpal ganglia serve the palpi and the pharyngeal muscles (Hughes, 1959). In the two-spotted spider mite, *Tetranychus urticae* Koch, the nervous structure is different from that observed in ticks. In *T. urticae*, an anterior unpaired 'rostral nerve' reaches the capitulum, and the pedipalpal ganglia originate in the ventral suboesophageal part of the synganglion (Alberti and Crooker, 1985).

Pritchard *et al.* (2015) reported that chemical acaricides against *D. gallinae* predominantly target neurotransmitters and synapses between neurons within the synganglion tissue. These substances classically target the voltage-gated Na^+ channels of pre-synaptic axons, propagating a continually depolarized membrane, leading to loss of action potential and eventually mite paralysis. Mites that cannot move to find food or escape environmental factors eventually die. Sprayed acaricides are most likely taken up via sites of gaseous exchange in the *D. gallinae* principally through the stigmata, located adjacent and dorsally to coxae II and III, through the peritreme branching network, into the haemolymph, and finally through the synganglion tissue. Mite synganglion tissue is reported in several mite species to be covered in an acellular sheath of neural lamellae, which allows access to nutrients and other compounds (Coons and Axtell, 1971; Woodring and Galbraith, 1976). A rind of perikaryon (neural somata) cells further surrounds a central neuropile of axons and dendrites (Severino *et al.*, 1984), where it is likely that PRM neurological controls are mostly active.

Suboesophageal Ganglion

The pedal ganglia and an unpaired posterior 'opisthosmatic ganglion' arise from the suboesophageal synganglion; from the latter, an unpaired abdominal nerve may originate that extends posteroventrally beyond the podosoma. Numerous peripheral nerves, which are variable in the different suborders and in their nomenclature (e.g. genital, scutal, stomodaeal, splanchnic, tritosternal, rectal), are connected to the synganglion (Hughes, 1959; Woodring and Galbraith, 1976; Woolley, 1988).

Reproductive System

The female and male reproductive systems of mites are composed of different organs that are structured in different ways (paired, fused or multiple) and are variable in the higher taxa.

Female System (Fig. 3.3)

In the Astigmatina, the genital opening in both male and female is on the ventral side of the body, between the bases of the legs. It is often concealed by a pair of diverging genital folds on the internal side of which is a pair of blunt tubular structures, called 'genital suckers' or 'sense organs'. In some species, such as *Blomia* sp. an additional unpaired fold may also be present, closing the genital opening from behind. In the female, the genital opening is large, to allow the passage of eggs, and it may be longitudinal or transverse. Sometimes, the anterior edge of the female opening is strengthened by a small crescentic epigynum, as in *Glycyphagus* sp., and this may be united with the sternum, as in *Carpoglyphus* sp. or with a circumgenital ring surrounding the opening of the oviduct, as in *Gohieria* sp. (Hughes, 1976).

In various Acariformes, sperm transfer occurs via an extrudable aedeagus, that transfers sperm directly through the female genital aperture (Alberti and Crooker, 1985), while in the Astigmatina, the sperm may be transferred to the female through a 'bursa copulatrix', which is related to a 'seminal receptacle' or 'receptaculum seminis' and the ovary, the genital aperture functions as an 'oviporus', an opening for the discharge of eggs (Hughes, 1959). The bursa copulatrix is an additional opening set at the posterior end of the body. It is commonly a circular pore, more rarely set at the end of a tubular structure, as in *Blomia* sp. or *Glycyphagus* sp., and internally, it may lead into a receptaculum seminis which communicate with the ovary.

In some Mesostigmata, sperm transfer by extragenital insemination occurs through two 'induction pores' or 'solenostomes' (Athias Henriot, 1968) that are connected to a 'sacculus vestibulus' and to a 'tubulus annulatus', which leads the sperm into

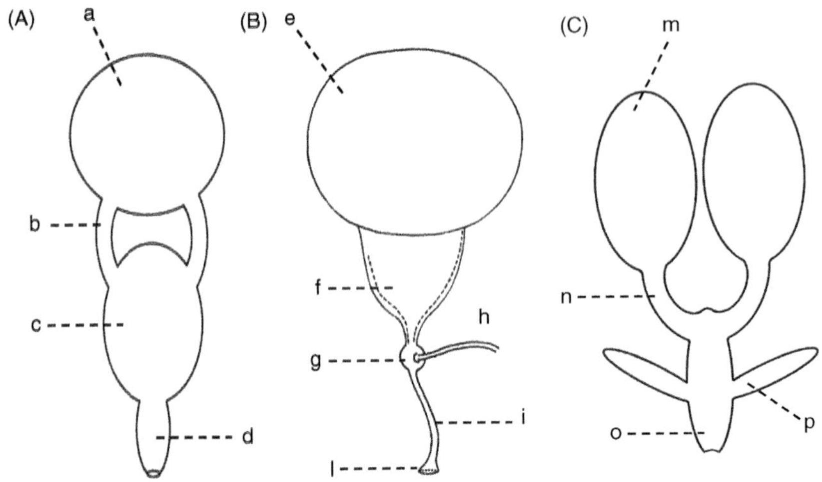

Fig. 3.3. Schematic drawing of female reproductive system (A), type B complex of a phytoseiid mite (B) and male reproductive system (C); a, ovary; b, oviduct; c, uterus; d, vagina; e, vesicle; f, calyx; g, atrium; h, minor duct; i, major duct; l, solenostome; m, testes; n, vas deferens; o, ejaculatory duct; p, accessory gland (from Vacante, 2016).

the organ complex. In podospermous Mesostigmata, the organ complex may have two different forms, which are designated the type A and type B complex. The phytoseiid mites have a type B complex, structured in a solenostome, a major duct, an atrium, a calyx and a vesicle; a minor duct originates from an enlargement of the major duct and ends in a lyrate organ that is related to an ovary (Alberti and Di Palma, 2002). The number of ovaries varies between acarine groups. The Parasitiformes, Trombidiformes and several acariform Oribatida have an unpaired ovary, while the Astigmatina may have two ovaries. The oviducts are commonly paired, but a single oviduct occurs in some Mesostigmata and Trombidiformes (Evans, 1992). The Eriophyidae and Tarsonemidae have an unpaired oviduct, probably as a result of an adaptation that is related to its small size. The oviducts may secrete substances involved in eggshell production (Woodring and Galbraith, 1976). The vagina is next to the oviducts and may develop in an 'ovipositor' that is variously structured in different acarine groups; it may be a tube-like eversion of the vaginal wall, as in the Tetranychidae, or it may be more complex.

Witaliński *et al.* (1990) summarized the chronology of studies on this topic, and they reported that the first studies on the reproductive systems of this group were carried out by light-microscopy by different authors (Nevin, 1935; Oboussier, 1939; Perron, 1954). Most important studies concerned the reproductive system of *Caloglyphus mycophagus* (Megin) (Kuo and Nesbitt, 1970), *Caloglyphus boharti* deutonymph (Woodring and Carter, 1974), *Lardoglyphus konoi* (Vijayambika and John, 1975) and *Rhizoglyphus robini* Claparède (Baker and Krantz, 1985). The general structure of the female reproductive system in Acaridida was also discussed by Griffiths and Boczek (1977). A detailed light-microscopy study on oogenesis and spermatogenesis in *Caloglyphus berlesei* Michael and *Caloglyphus michaeli* Oudemans was published by Prasse (1968). Successively, spermatogenesis in Acaridida was investigated by electron microscopy in *Caloglyphus anomalus* Nesbitt by Reger (1977), and in *Acarus siro* Linnaeus and *Tyrophagus putrescentiae* (Schrank) by Witaliński *et al.* (1986). Some years later, Witaliński *et al.* (1990) studied the anatomy and ultrastructure of the female and male reproductive system in *A. siro* using light and electron microscopy. The female system consists of paired ovaries of nutrimentary type in which oogonia and oocytes are connected by bridges with a large central cell. The oviducts empty into the uterus, which passes into preoviporal duct lined by cuticle, and opening as a longitudinal slit (oviporus). An elongated accessory gland, composed of one type of secretory cell, is located along each oviduct. The copulatory opening occurs at the posterior margin of the body and leads, via the inseminatory canal, to the receptaculum seminis, consisting of the basal and saccular part. Both inseminatory canal and basal part of receptaculum seminis are lined by cuticle, whereas the wall of the sac is formed by cells covered only by long, numerous microvilli. The basal part of the receptaculum seminis joins the ovaries via two lumenless transitory cones (Lustgraaf, 1977).

Male System (Fig. 3.3)

The male reproductive system consists of paired or fused 'testes' and 'vasa deferentia', a variable number of 'accessory glands' and an 'ejaculatory duct'. A 'seminal vesicle' may occur in some groups. Some of the Mesostigmata, Astigmatina and Trombidiformes have paired testes, while other groups, such as small Prostigmata, may have unpaired testes or still others elongate, acinous or multi-lobed testes. In direct insemination, a muscular sperm pump interposed between the vasa deferentia and the ejaculatory duct may push the sperm into the aedeagus and from this into the female seminal receptacle (Alberti and Crooker, 1985; Evans, 1992). The accessory glands may play different roles, such as the production of the seminal fluid matrix (Woodring and Galbraith, 1976), or biochemicals involved in spermiogenesis or in the prevention of desiccation of the spermatophore stalk (Evans, 1992). All acarine spermatozoa are flagellated, but they are variable in form – from ribbon-like or vacuolate to small or simple; their ultrastructure tends to be taxon-specific at subcohort levels (Alberti, 2000), and it may be a potential tool for the examination of systematic relations (Alberti, 1984).

In the Astigmatina, the anterior end of the genital opening may be supported by an epiandrium. Copulatory organs may take the form of large adhesive suckers on either side of the anal slit and a pair of smaller suckers on the fourth tarsus, such as in *Acarus* and *Tyrophagus*, or one only on tarsi I and II, such as in *Calvolia*. The modification of the shape of the legs has a similar function, such as of legs I and II in *Glycyphagus ornatus* (Kramer) or legs III and IV in different parasites species (Hughes, 1976).

Witaliński *et al.* (1990) reported that the male reproductive system of *A. siro* contains paired testes, in which spermatogonia tightly surround the central cell. The proximal part of the paired vasa deferentia serves as a sperm reservoir, while the distal one has a glandular character. An unpaired, cuticle-lined ejaculatory duct opens into the apex of the aedeagus. The single accessory gland is located asymmetrically at the level of, or slightly posterior to, coxae IV. The structure of the genital papillae, which are topographically related to the genital opening in both sexes, is also briefly described.

Glandular System

The 'glandular system' discussed here is a non-integumental system – the complex of glands that are variously located in the acarine body, but not in the integument. The main glands are the 'salivary glands', the 'coxal glands' and the 'podocephalic glands'. Some acarine groups also have 'tracheal glands' (Tetranychidae) (Mills, 1973; Mothes and Seitz, 1981a; Alberti and Crooker, 1985) and 'silk glands'.

Salivary Glands

The structure and role of the salivary glands vary in the different acarine taxa. The Parasitiformes have a simpler gland type, consisting of paired acinous glands set in the prosoma, while the Acariformes have a slightly more complex organization. The eriophyoid mites have 'prosomal glands', which are glandular structures associated with the mouthparts; these consist of three glands that pass their secretions through ducts between the stylets of the mouthparts at the base of the labrum. One pair of these glands is retained as the podocephalic glands, while the unpaired gland represents the 'tracheal gland' (see further on). The function of the secretions of these glands is unclear, but they are probably involved in the production of saliva or of a lubricant substance, facilitating the action of the stylet (Nuzzaci and Alberti, 1996). The Tetranychidae have a tracheal gland set in the dorsal anterior part of the propodosoma, and its duct crosses the stylophore medioventrally, and carries the secretion to between the cheliceral stylets or into the infracapitulum (Alberti and Crooker, 1985).

The dermanyssid *D. gallinae* (De Geer) possesses salivary glands involved in the blood suction. A recent research on sequencing the PRM transcriptome identified 24 potential salivary proteins likely to be involved in blood digestion (Schicht *et al.*, 2013), some of which have hypothesized antibacterial functions.

Coxal Glands

The coxal glands are probably involved in osmoregulation or the ion/water balance and excretion, and no solid excretory products have been found in them (Alberti and Crooker, 1985). Mills (1973) notes that in the female, these glands may secrete a pheromone additive that attracts the males to female webbing (see section on water balance). Regardless of their function, the coxal glands have a nephridial origin, and consist of a 'coelomic sacculus' and a 'coiled duct' or 'labyrinth'; they may be present in different body regions but are commonly found in the coxal regions. In the Tetranychidae, the sacculus seems to be absent, and only the tubulus has been found. The tubulus is divided into three parts (proximal, middle and distal parts); the proximal and middle parts double back on one another, and the distal part leads through a duct to the podocephalic canal (Alberti and Crooker, 1985). Several species bear coxal glands in the prosoma, near the legs; in the Mesostigmata, they may be set in the opisthosoma near a long duct opening in coxa I (Alberti *et al.*, 1996). In the Oribatida and Astigmatina, the typical structure consists of the 'supracoxal complex', located at the base of leg I; this covers a pair of coxal glands which open by a small slit at the base of the 'supracoxal seta'. In many species of Astigmatina, the anterior lateral edge of the propodosoma is produced forward as a thin cuticular plate, called 'Grandjean's organ', embracing the base of the gnathosoma. In *Otodectes*, it is very small, while in *Acarus gracilis*, Hughes is produced into flame-like extensions. From the base of the Grandjean's organ, a lateral sclerite extends posteriorly to curve around the base of leg I. At the posterior end of this sclerite is the supracoxal fossa, from which may arise the supracoxal seta, which may be a simple rod, such as in *C. berlesei* (Michael), or richly branched, such as in *Glycyphagus*. In many acarids, at the posterior end of the lateral sclerite, adjacent to the supracoxal seta, is a slit pore, the opening of coxal glands lying beneath the cuticle. Similar slits, the openings of the latero-abdominal or 'fat glands', leading into a sac lined with chitin, are also present, on the sides of the body between the anterior and posterior lateral setae. The latero-abdominal glands are present in all stages of the life cycle and contain a highly refractile liquid, colourless, sometimes yellow or brown and red in some species, such as *Austroglycyphagus geniculatus* (Vitzthum). The lining of the sac is shed at ecdysis (Michael, 1901), and the opening in some mites can be partially closed by a cuticular flap (Brody and Wharton, 1970). The hysterosoma possesses four pairs of circular rings, one pair set near the setae c_2 and c_3, one near the d_2, another near the posterior end of the body and the remaining pair on either side of the anus, respectively. These rings are well developed in *Histiostoma laboratorium* R. Hughes, leading to a semiaquatic adaptation, and for this reason, it has been suggested that they act as osmoregulating organs (Perron, 1954).

Podocephalic Glands

All Acariformes have a pair of lateral 'podocephalic canals' (Van der Hammen, 1980), which are connected to the propodosomatic glands, collect the product from the glands, and carry it to the gnathosomatic region (Evans, 1992; Alberti and Coons, 1999). The structure of these canals may vary; sometimes they are like an open external gutter, while in other cases they are clearly internal, as in the Prostigmata. In the Astigmatina, they are short, and in the Oribatida, they are long. The two podocephalic ducts collect the products secreted by the coxal glands and by a variable number (from one to three) of additional gland pairs. The functions of these secretions are variable and include endocrine substances for moulting, products of the salivary glands (Woodring, 1973) and products for silk production (Alberti, 1973). In the Tetranychidae, each podocephalic canal receives the secretion of three glands. The two canals meet at the basis of the gnathosoma to form a single canal, the 'median salivary duct', which travels along the infracapitulum to the gnathosoma. The dorsal and anterior podocephalic glands contain numerous granules, probably of a protein nature. The dorsal glands consist of two cells and the anterior glands of four cells, as in the tetranychid *T. urticae* Koch (Alberti and Crooker, 1985). These glands are retained salivary glands (Mothes and Seitz, 1981a), but their function is unclear, and they may be fat bodies (Summers *et al.*, 1973).

Silk Glands

Different mite families produce silk from salivary gland secretions (Gerson, 1985), such as the Tydeidae (Schuster, 1972), Cunaxidae (Walter and Kaplan, 1991) and Cheyletidae (Furman and Sousa, 1969). Silk production assumes considerable importance in several tetranychid species. In these mites, the silk glands consist of a unicellular gland located in each palpus, which extends back to the central nervous mass (Alberti and Storch, 1974; Mothes and Seitz, 1981a). The gland cell is very large and contains many vacuoles filled with proteinaceous secretions. In the distal palpal segments, the droplets are contained in the reservoir-like area of the cell apex, which narrows, and is connected to a terminal outlet at the tip of the spinneret. Mills (1973) reported that the podocephalic complex of the two-spotted spider mite, *T. urticae* was involved entirely in silk production, while Alberti and Crooker (1985) reported that the secretions of other glands may also be involved in silk production.

Digestive System

The digestive system functions both as a 'preoral digestive system' and a 'postoral digestive system'. The preoral system is based on the secretion of salivary enzymes onto or into the food, in order to predigest food and facilitate its ingestion as a liquid (Evans, 1992; Alberti and Coons, 1999; Coons and Alberti, 1999), while postoral digestion is the typical digestive process. The fundamental features of the postoral digestive system are discussed in this section. The postoral digestive system of mites consists of various organs. Anteriorly, the 'buccal cavity' or 'mouth' opens into a muscular 'pharynx', which is connected posteriorly by the 'oesophagus' to the 'stomach' or 'ventriculus'. The oesophagus is narrow and, in some groups, such as the Tetranychidae, passes through a part of the anterior salivary or silk gland (Alberti and Crooker, 1985) and the synganglion, from whence it continues into the ventriculus. The ventriculus may be differently shaped and sized, ranging from simple to sac-like, or equipped with two or more gastric 'diverticula' or 'caeca' (Evans, 1992) of complex function. In general, the Trombidiformes have a larger ventriculus and two large caeca (Woodring and Galbraith, 1976). Any feeding adaptation is generally correlated with the type and size of caeca, such as phytophagous mites exhibit expanded large caeca during feeding. Intracellular digestion may occur in the anterior portion of the ventriculus, and this involves vacuolated epithelial digestive cells that absorb nutrients and secrete digestive enzymes. The ventriculus opens directly into the 'colon' or 'postventricular midgut', which is sometimes divided by a 'valvular intercolon', into an anterior part and a 'postcolon' or 'anterior hindgut'. In the Tetranychidae, the midgut has a type of cell that functions in different ways at different sites: the cells absorb nutrients in the midgut and excrete them between the ventriculus and the colon (Alberti, 1973; Mothes and Seitz, 1981b).

Pritchard *et al.* (2015) reported that the mite digestive tract of the Acaridida mites has a comparatively well-studied part of the anatomy of several species including the storage mite *Lepidoglyphus destructor* (Schrank) (Erban and Hubert, 2011), the house-dust mite *Dermatophagoides farinae* Hughes (Dumez *et al.*, 2014) and a range of synanthropic species (Erban and Hubert, 2010). In combination, these studies provide an outline of the general anatomy of mites (Mehlhorn, 2001). Hubert *et al.* (2004a) published an extensive review of the digestive system of Astigmatina mites. They reported that the mite gut is divided into three compartments; the first includes the foregut, pharynx and the oesophagus; the second, the midgut-ventriculus, paired caeca, colon, intercolon and postcolon; and the third, the anal atrium, respectively (Hughes, 1950; Brody *et al.*, 1972; Alberti and Coons 1999; Alberti *et al.*, 2003). The pharynx is formed by sclerotized shields, with attached pharyngeal muscles. The oesophagus is a tube formed by simple epithelial cells covered by a cuticle. The oesophageal lumen contains food particles and/or regurgitated digestive enzymes. Epithelial cells possess microvilli in all parts of the midgut. The microvilli are larger and denser than in the ventriculus, caeca and colon. The ventricular and caecal lumen contain fine particles, probably of mucoid nature, and digestive enzymes. Endogenous enzymes are produced by secretory cells in the ventriculus and caeca (Hughes, 1950; Kuo and Nesbitt, 1970). The ingested food is concentrated in the middle of the ventriculus, and then packed into a peritrophic membrane (Wharton and Brody, 1972; Wurst, 1993). In *D. farinae*, a 'peritrophic membrane' may provide a protective wall that envelops the food bolus as it moves in the postcolon (Wharton and Brody, 1972). The gut and the fat body contain associated microflora-producing exogenous enzymes (Smřz, 2003). Members of the intestinal microbial community have the ability to decompose cellulose, chitin and lignin (Stefaniak and Seniczak, 1976; Smřz, 2003). The evidence for microbial activity is based on cultivation and ultrastructural observation. The enzymatic activity of bacteria (bacterial exoenzymes) and yeasts results in a digestion of the 'substrate' around their cells. This 'substrate' is formed by ingested food in the gut, and the digestion is visible as empty space around microbial cells. The pH of the gut environment is important, as it affects the activity of DE (Terra and Ferreira, 1994). Due to the small size of mites, pH in their guts is estimated by the colour changes of ingested colour shift indicators (Akimov, 1985). The ventriculus and caeca showed the same low acid pH in the range from 5.4 to 6. The colon had low acid or neutral pH conditions (5.9–7.4), while the pH of the postcolon was more alkaline (6.5–8). The low acid conditions in the ventriculus correspond to the pH optima of α-amylases and glucosidases. The shift in the pH of food contents to alkaline conditions in the colon indicates the start of protein hydrolysis in this gut compartment (Akimov, 1973, 1985). The biochemical and physiological processes that result in digestion and utilization of nutrients in the midgut are arranged both spatially and temporally. The spatial compartmentalization is based on the peritrophic membrane (PM) (Terra *et al.*,1996) that covers the food bolus. The PM separates the gut into endoperitrophic spaces inside the food boli, and an ectoperitrophic space outside. The relative quantities of DE differ in these compartments (Terra, 2001). In insects, the initial digestion occurs in the endoperitrophic space, the intermediate digestion in the luminal contents, and the final digestion is accomplished by enzymes bound to cell surfaces (Terra, 2001). It is well-documented that mite faeces are the most important source of mite allergens (Tovey *et al.*,1981; van Hage-Hamsten and Johanson, 1998). The ability to produce laboratory cultures of the stored product and house-dust mites, and to collect their faeces in their 'spent growth medium' (SGM), has allowed identifying enzymatic activities associated with the faeces (Stewart *et al.*, 1991). The presence of enzymes in faeces suggests their role in digestion (Stewart *et al.*, 1998). These enzymes are produced by secretory cells and released into the gut lumen where they penetrate into the endoperitrophic space.

Enzymes of lower and medium molecular weight, like α-amylases and proteases, probably regularly penetrate the endoperitrophic space. Although a certain proportion of these enzymes are remobilized, the rest of the enzymes are passed through the gut together with undigested food, into faeces. Cysteine and serine proteases, α-amylases and lysozyme-like enzymes are allergenic enzymes isolated from the faeces of stored products and house-dust mites (Yasueda *et al.*, 1986; Chua *et al.*, 1988; Lake *et al.*, 1991; Stewart *et al.*, 1992a, 1992b, 1998; King *et al.*, 1996; Olsson and Hage-Hamsten, 2000).

Wu *et al.* (2009a) reported that the domestic mite species *Blomia tropicalis* Bronswijk, Cock *et* Oshima is an important indoor allergen source related to asthma and other allergic diseases in tropical and subtropical regions. Here, we describe the alimentary canal of *B. tropicalis* with the particular application of three-dimensional reconstruction technology. The alimentary canal of *B. tropicalis* resembles the typical acarid form, consisting of the cuticle-lined foregut and hindgut separated by a cuticle-free midgut. The foregut is divided into a muscular pharynx and an oesophagus. The midgut is composed of a central ventriculus, two lateral caeca, a globular colon and a postcolon with two tubiform postcolonic diverticula. The most common cells forming the epithelium of ventriculus and caeca are squamous and cuboidal. The globular cells contain a big central vacuole in the posterior region of the caeca. The epithelium of the colon and postcolon has significantly longer microvilli. The anal atrium is a simple tube with flattened epithelial cells. The spatial measurements of the three-dimensional model suggest that the paired caeca and central ventriculus occupy 55.1% and 34.6%, respectively, of the total volume of the alimentary canal and may play a key role in food digestion.

Šobotník *et al.* (2008a) studied the ultrastructure of the digestive tract in *A. siro* Linnaeus. They observed that the colon leads to an 'anal atrium' or 'posterior hindgut', which leads to the 'anal opening'. The gut of this mite is characterized on the ultrastructural level. It consists of the foregut (pharynx, oesophagus), midgut (ventriculus, caeca, colon, intercolon, postcolonic diverticula, postcolon), and hindgut (anal atrium). The gut wall is formed by a single-layered epithelium; only regenerative cells are located basally and these have no contact with the lumen. Eight cell types form the whole gut; they are the simple epithelial cells forming fore and hindgut, the cells that probably produce the peritrophic membrane, the regenerative cells occurring in the ventriculus, caeca, colon and intercolon, the spherite cells and digestive cells forming the ventriculus and caeca, the colonic cells and intercolonic cells, and the cells forming the walls of postcolonic diverticula and postcolon. Spherite and digestive cells change in structure during secretory cycles. The cycle of spherite, colonic, and intercolonic cells ends by apoptosis. Ingested food is packed into a food bolus, surrounded by a single homogeneous peritrophic membrane formed by the addition of lamellae that subsequently fuse together. The diverticula serve as a shelter for filamentous bacteria, which also are abundant in the intercolon.

Šobotník *et al.* (2008b) reported that the peritrophic membrane in *A. siro* Linnaeus is produced by distinct cells located in the ventriculus. This study detected the chitin inside the peritrophic membrane using wheat germ-lectin conjugated with colloidal gold (10 nm). The chitin of the peritrophic membrane was a target for chitin effectors, including (i) chitinase, which hydrolyses chitin inside the peritrophic membrane; (ii) calcofluor, which binds to chitin and destroys the peritrophic membrane mesh structure; and (iii) diflubenzuron, which inhibits chitin synthesis. In addition, soybean trypsin protease inhibitor (STI) and cocktails of chitinase/calcoßuor, dißubenzuron/calcoßuor and chitinase/STI were tested. These compounds were supplemented in diets, and an increase of population initiated from 50 individuals was observed after 21 days of cultivation. Final *A. siro* densities on experimental and control diets were compared. The chitin in the peritrophic membrane was determined to be a suitable target for novel acaricidal compounds, for suppressing the population growth of *A. siro*. The most effective compounds were calcofluor and diflubenzuron, whereas the suppressive effects of chitinase and STI were low. The failure of chitinase could be due to its degradation by endogenous proteases. The combination of chitinase and STI suppressed *A. siro* population growth more effectively than when they were tested in oral admission separately. The combinations of calcofluor/chitinase or calcofluor/difluorbenzuron showed no additive effects on final *A. siro* density. The presence of chitin in the peritrophic membrane provides a target for novel acaricidal compounds, which disrupt the peritrophic membrane structure. The suitability of chitin effectors and their practical application in the management of stored-product mites is discussed.

Excretory System

Mites excrete liquid and solid waste products through the 'Malpighian tubules' and other structures. They can also both lose water and regulate its balance. The latter capacity makes it difficult to separate the two functions.

Malpighian Tubules

In the Parasitiformes and some Astigmatina, the posterior midgut, postcolon or intercolon may bear one or two pairs of Malpighian tubules (Woodring and Galbraith, 1976) (Fig. 3.1), which collect and store guanine or uric acid (insoluble nitrogenous waste); these substances can reabsorb water and become paste-like and are slowly excreted (Hughes, 1959).

Other Structures

The discontinuity of the alimentary duct in some acarine species may complicate the digestive process. The complication primarily involves indigestible solid products, such as chromolipids, which are stored in the caeca, until the accumulated faecal material becomes overabundant, leading to the critical situation of 'schizeckenosy', in which there is integument tearing and extrusion, and caecal lobes filled with faeces are pinched off, but without leaving any strong indications of the event (Mitchell and Nadchatram, 1969).

Water Balance

The small size of mites means that there is a high surface-to-volume ratio, which affects their water balance. Thus, fluctuations in temperature and relative humidity (RH) oblige the mites to regulate their water loss by means of various adaptations; these include simple diffusion from the body surface or mechanisms such as defaecation, excretion, the secretion of digestive fluids and the production of various other fluids or reproductive product, or moving to more favourable sites.

Knülle and Wharton (1964) reported that above a critical equilibrium point, uptake of water vapour can be compared with water uptake in non-living hygroscopic substances. In *A. siro* Linnaeus this point lies at an R.H. of 71%, and above this value, the mite freely gains or loses moisture, according to whether it is being transferred to a higher or lower RH. Below this value, a shortening of life coincides with an increase in water loss. An RH of 71% is likely to be encountered by *A. siro* under natural conditions and would be the condition prevailing in stored cereals with a moisture content (MC) of 14%. Solomon (1966) reported that *A. siro* can live at humidities down to 62%, probably below the critical equilibrium point, the cuticle ceases to behave like a non-living hygroscopic system. When a mite is killed, there is a great increase in the rate of cuticular water loss. This shows the existence of some energy-consuming mechanism that tends to reduce water loss at relatively low humidities (Hughes, 1976).

Larson (1969) reported that below a certain critical activity, living arthropods have been found incapable of effecting a net gain of water. This activity has been called the 'critical equilibrium activity' (Wharton and Devine, 1968), and after being designated as the CEA, a water gain occurs whenever an organism is transferred from a lower to a higher activity; but at activities below the CEA, transferring organisms from lower to higher activities, only results in a continuation of water loss until death by desiccation results and until static equilibrium with the environmental activity is reached. The grain mite, *A. siro* L., has a CEA of 0.71 (Knülle, 1965).

Two different methods have been employed to determine the CEA of arthropods (Wharton, 1963, in Larson, 1969). In one method, an organism or group of organisms of a particular instar and/or conditions is enclosed in a relatively small air space, in terms of the amount of water in the air compared to the amount of water in the organism. The water in the air and in the organism will come into equilibrium at approximately the CEA, for the organism and instar exposed.

Another method has been to determine the lowest activity at which desiccated organisms maintain or gain weight by exposing them to a graded series of different activities, where a change in weight reflects a change in the water mass in the organism. Starved desiccated animals are usually used in such determinations, in order to prevent excretion, reproduction or excess water in the digestive tract from influencing the water mass in the organism.

Arlian and Veselica (1979) reported that insects and mites must maintain a body water content and water concentration within tolerable limits in order to maintain physiological integrity. Viewed thermodynamically, this body water pool responds as an open system in which there may be simultaneous, independent influx and efflux of water. As such, water balance maintenance requires that the quantity of water entering and leaving the system approximately balance during some increment of time so that body water content does not fall below or rise above critical lower or upper limits. The instantaneous body water content (m) is equal to the instantaneous difference between the gain (g) and the loss (l) of water in the individual. That is:

$$dm/dt = dg/dt - dl/dt$$

Over measurable time intervals, the change in body water content (Δm) is more conveniently expressed as:

$$\Delta m = g - l$$

where $\Delta m > 0$ when $g > l$, $\Delta m < 0$ when $g < l$, and $\Delta m = 0$ when $g = l$.

Avenues of water gain are imbibition through the ingestion of hydrated foods or drinking of free water, the production of metabolic water, and the active and passive uptake from the ambient air or aqueous environment. Water is lost by simple diffusion from the respiratory and general body surfaces, by secretion of digestive fluids, defensive fluids, pheromones, reproductive products, and during oviposition, defecation and excretion. The significance of each of these avenues of gain and loss varies with the species and the situation (Arlian and Veselica, 1979).

Arlian (1992) reported that the house-dust mites, *D. farinae* Hughes, *Dermatophagoides pteronyssinus* (Trouessart) and *Euroglyphus maynéi* (Cooreman) are prevalent in homes in humid geographical areas throughout the world and thrive in humid environments in human dwellings where there is no liquid water to drink. However, their bodies contain 70–75% water by weight, which must be maintained in order to reproduce. Their primary source of water is water vapour, which is actively extracted from unsaturated air. At relative humidities above 65–70%, adequate amounts of water can be extracted from unsaturated air to compensate for that lost by all avenues. Active uptake is associated with ingestion of a hyperosmotic solution that is secreted by the supracoxal glands. Active mites do not survive longer than 6–11 days at RHs ≤50%. They survive extended dry periods, by forming a desiccation-resistant protonymphal stage, which can survive for months at RHs below the critical humidity for active stages. Feeding rate and allergen production are directly influenced by RH. Mites feed, multiply and produce more faecal matter at higher RHs than at lower ones.

Schmidt (2009) studied the atmospheric water vapour and critical equilibrium humidity of the predatory mite *Cheyletus malaccensis* Oudemans living in stored products. The aim of this investigation was to know the effect of RH on the body water budget and on the longevity of the cheyletid mite. Metabolic production of water contributes very little to the body water budget of *C. malaccensis*. If only fats were oxidized (from which more water is produced per unit mass oxidized than from other energy reserve substrates), the production of water would be at most 0.07 µg/day. *C. malaccensis* can physiologically absorb water vapour from unsaturated air, to balance the net losses of water in other processes, or to recover a deficit of body water. This capability has exceptional importance for its survival. The lowest or 'critical' RH in which *C. malaccensis* can absorb water vapour at a rate sufficient to recover a body water deficit, and maintain a steady-state body water mass during long periods of time, lies between 76% RH and 85% RH. Both the steady-state mass of body water and the rate of vapour absorption from the atmosphere in individuals having a deficit of body water increase with the relative humidity of the environment. In ambient relative humidities less than the critical value, body water is lost at such a great rate as to reduce the longevity, to less than half the value pertaining to higher humidities.

An example of mite-water relations is given by research on the net water exchange of the predatory phytoseiid mite *Phytoseiulus persimilis* Athias Henriot. Water loss by transpiration increases with environmental temperature. The transpiration rate is directly related to the saturation deficit of the air and at constant temperature is linearly dependent on the water vapour activity. Above the critical equilibrium activity (CEA), the mite can absorb water vapour from the unsaturated atmosphere. The amount of water vapour within the laminar layer of the leaf surface has periods during which it is sufficient for the phytoseiid to employ its water vapour sorption capability. *P. persimilis* also exhibits positive hygrotactic behaviour and can absorb water vapour that can at least double the survival time of starving individuals. The supply of a prey mite (water content about 5.5 mg) every 3 days, and a water vapour activity (av) of 0.76 (at 20°C), can allow a survival time of about 120 days (Gaede, 1991).

4 Biology

'Biological instars' and 'reproduction' are two basic factors of mite life. They affect the ecology of mite populations, and knowledge of their characteristics is of strategic importance in the control of injurious species.

Biological instars

The life cycle of Acari develops through the egg and six (sometimes more) biological instars that follow: the prelarva, larva, protonymph, deutonymph, tritonymph and adult. There are usually two or three nymphal instars between the larval instar and the adult. In no more than a few cases is there only a single nymphal instar, as in the ticks. Protonymphal, deutonymphal and tritonymphal instars are found in the Prostigmata, Astigmatina and some other acarine groups. The adult represents the final stage after a series of three or four moultings, although in some groups there may be post-imaginal moulting during the adult stages. The life cycle of eriophyoid pest mites develops through the stages of egg, larva, nymph and adult, with two 'quiescent' or 'resting stages' that occur between the larva and the nymph, and between the nymph and the adult (Manson and Oldfield, 1996), which are called the 'nymphochrysalis' and the 'imagochrysalis', respectively (Sternlicht and Goldenberg, 1971). The life cycle of the Kanzawa spider mite, *Tetranychus kanzawai* Kishida, has three active biological stages (larva, protonymph and deutonymph) and four quiescent stages (egg, protochrysalis, deutochrysalis and teleiochrysalis) (Oku *et al.*, 2003). André and Van Impe (2012) reported that the two-spotted spider mite, *T. urticae* Koch lacks an imago, the name that is proposed for the last 'stase', and that its post-embryonic development is characterized by the 'stases' of prelarva (calyptostase), larva (six-legged stase), protonymph (eight-legged stase), deutonymph (eight-legged stase) and tritonymph. This last stase shows a precocious development of sexual maturity (paedogenesis). According to the approach emphasized by Grandjean (1938, 1970) and by André (1988, 1989), André and Van Impe (2012) proposed the use of the word 'stase' instead of the words 'instar', 'stade' or 'stage'. In their view, the use of the word stase (and its derivatives) is based on 'idionymy' (the existence of structures or stages that differ in discontinuous surface characters), and it is thus possible to homologize the different stases. The same words are used to define ontogenetic levels, larvae and adults.

The Egg

The egg is deposited on the substrate through the genital aperture or the ovipositor. According to the group, this instar exhibits a typical morphology, size and colour. It is oval, long, flattened, or some other shape, with the external surface smooth, striate, reticulate, etc., and variously coloured, such as white, red or orange. During the process of deposition, it usually has strong elasticity. In the Astigmatina, the eggs are relatively large, owing to the presence of yolk, and covered with a smooth or sculptured shell (Hughes, 1976).

The Prelarva

The prelarva is also called the 'prolarva' or 'deutovum' and represents a non-feeding form, which is quiescent and closed within the egg chorion, and sometimes without legs and mouthparts, as in some Prostigmata (Walter and Proctor, 1999). In other groups, the prelarva bears three pairs of legs, mouthparts and setae. The first condition is referred to as 'calyptostasis' (strong inhibition) and the second as 'elattostasis' (incomplete inhibition) (Coineau, 1974a, b). A typical deutovum occurs in the Penthaleidae. A calyptostase is reported for *T. urticae* (André and Van Impe, 2012).

DOI: 10.1079/9781786390912.0004

The Larva

The larva is a typical six-legged instar, commonly lacking external genitalia. Many larval instars do not feed, but others may be active predators or parasites. The primordia of legs IV may occur in the larva of the Opilioacaridida (Walter and Proctor, 1999), while in the Eriophyoidea, legs III and IV are suppressed in all stages (Lindquist, 1996). In this last group of mites, the larva develops into the nymph through a quiescent or resting stage (Manson and Oldfield, 1996), also called the nymphochrysalis (Sternlicht and Goldenberg, 1971). The larva of the tetranychid *T. kanzawai* develops into the protonymph through the quiescent stage of the protochrysalis (Oku *et al.*, 2003). In the Acaridida, the larva lacks the fourth pair of legs and sometimes possesses the 'coxal rods', also called 'Bruststiele' or 'Claparèdè's organ', set in the area of coxae I; these are similar to solenidia, as they are distally enlarged hollow rods, but their function is unknown (Hughes, 1976).

Similar coxal rods are present in the larvae of Bdellidae (Prostigmata), and Grandjean (1938) suggested that they are homologous with the genital sense organs, the 'suckers', observed in the nymphs and adults. The larva has the hysterosoma not fully developed, and some setae or solenidia of the body and legs are lacking. In the genus *Glycyphagus*, some setae may be very long and may be involved in dispersal by wind. External genitalia are not visible (Hughes, 1976).

The Protonymph

The protonymph represents the first eight-legged nymphal instar; it is free living and active. It may feed or not and develops in the same habitat as the later instars. The protonymph develops into the deutonymph through a quiescent stage. The protonymph of Acaridida lost its coxal rods and possesses the rudiments of a genital opening, the rudiments of a pair of genital sense organs, one pair of genital setae and adanal setae when they exist in the adult. It has a simplification of setae on legs IV, and the chaetotaxy of legs I–III is the same as that of the adult (Hughes, 1976).

The Deutonymph and the Hypopus

This second nymphal instar is eight-legged and greatly resembles the adult but does not have its external sexual characters or size. Some groups have non-feeding deutonymphs but, in general, this stage feeds actively. In the Astigmatina, there is a heteromorphic deutonymph, also called the 'hypopus', whose appearance is influenced by various factors (environmental, genetic, trophic, etc.). It is a heteromorphic stage, and this type of life cycle is probably unique in the animal kingdom (Hughes, 1976). The hypopus is characterized by a typical morphology, lacks mouthparts, has a closed gut and is strongly resistant to environmental stresses; it is adapted to phoretic dispersal, and attaches to animal carriers by ventral suckers (Fain, 1971a).

In *Tyrophagus* spp., the hypopus is absent, while in the Hemisarcoptidae family, it absorbs nutrients from the haemolymph of its carrier coccinellid, the ladybird, through the ventral plate suckers (Houck, 1994). The behaviour of hypopodes of some Glycyphagidae is very unusual in that they confine themselves to the subdermal tissues of their carriers to absorb nutrients (Fain *et al.*, 1973). When the tritonymph instar is absent, the deutonymph develops into the adult through a teleiochrysalis.

Two kinds of hypopi are known – the first is the active one, which moves freely about and is adapted for clinging to other Arthropoda or to Mammalia, and the other is inert almost or completely devoid of movement, and often remains within the pronymphal cuticle to await passively for more favourable living conditions; in some cases, they are resistant to desiccation. The distinction between the two types is not a rigid one, and they tend structurally to grade into one another (Hughes, 1976).

The active hypopus is yellow or brown in colour and has a hard and rigid cuticle. The body is round or oval and dorsoventrally flattened. The ventral surface being concave and the dorsal convex; the curvature enables the hypopus to fit more closely to the body of another arthopod. The dorsal surface is completely covered by propodosomal and hysterosomal shields. Two corneae may be present on the rostral region of the propodosoma, such as in *Calvolia*. The mouth lacks and the gnathosoma is reduced to an unpaired plate with the anterior bilobed edge; on each lobe is set a flagelliform setae. On the ventral surface of the body, the coxal plates are well-defined and bounded anteriorly and posteriorly by apodemes and epimerites, respectively. In the opisthosomal region, there is a well-defined 'sucker plate' forming part of the adhesive apparatus. The hypopi adapted to clinging to

Arthropoda have the sucker plate perforated and the spaces occupied by suckers which can be projected owing to an increase in turgor (Perron, 1954). The two central suckers are the most prominent ones and are important in attaching the hypopus to its host. Anteriorly are two smaller ones, often radially striated; in *Histiostoma laboratorium* R. Hughes, they can be protruded at a distance of 15 µm. Perron (1954) hypothesized that these smaller suckers make the first contact with the host and later the other suckers coming into action. Posterior to the central suckers are four smaller ones, and between these are clear areas which may represent vestigial suckers. Anterior to the sucker plate is a rudimentary genital opening flanked by a pair of suckers and also by a pair of genital setae. Beneath the cuticle are two pairs of genital sense organs, similar to those found in the tritonymph. Between the large central suckers is the anus which in living specimens can be seen to open and close.

The Labidophorinae have a sucker plate replaced by an apparatus for clinging to mammalian hair. This apparatus consists of a pair of movable folds, with a ridged internal surface, which covers two pairs of transversely grooved claspers. The whole organ acts as a vice that grips the hair securely (Hughes, 1976).

Legs I and II are more developed than the posterior ones, which are almost completely tucked away beneath the body and, in Histiostomatidae, are bent sharply towards the head. Some species have legs I and II make a questing movement in the air, the body being supported by the remaining legs and the sucker plate. The leg chaetotaxy varies from that of the other stages; the differences involve a change in shape of setae, and in the expression and suppression of some setae and solenidia. Tarsi I and II may bear setae widened and leaf-like or expanded distally as small suckers and employed as a part of the adhesive apparatus. Tarsus IV may end in one or two long setae and is used for clinging to insects (Hughes, 1976).

A small number of genera (*Acarus*, *Glycyphagus*, *Chaetodactylus*, etc.) possess an inert hypopus, and in an extreme form, such as in *Glycyphagus domesticus* (De Geer), the hypopus has the form of an oval cyst enclosed within the dry protonymphal cuticle. The internal body has muscles and the alimentary canal degenerated into a structure-less mass, and only the nervous system retaining its form (Hughes and Hughes, 1939). Boczek *et al.* (1969) reported that the motile hypopus of *Acarus farris* (Oudemans) has an internal structure similar to that of *G. domesticus*, except that a muscular system is formed which operates the sucker plate and legs. In some species of Hypodectidae, parasites of birds, the hypopus has become the dominant stage of the life cycle. The hypopus of *Hypodectes propus* (Nitzch) (Hypodectidae) has been found beneath the skin of pigeons and obtains its food from the tissue fluid of the host. Larva, protonymph and tritonymph are suppressed and the adult stage, which lives freely in the bird's nest, is only found during the nesting period. It is probable that the adults do not feed, since the chelicerae of the female are rudimentary (Fain, 1967a), and it is not known how the hypopus manages to penetrate the host's skin (Hughes, 1976).

The Tritonymph

This third nymphal instar is eight-legged and uncommon. When present, it is active. It is absent in the Mesostigmata and in many Prostigmata. *T. urticae* lacks both the adult and the tritonymph and is characterized by paedogenesis (André and Van Impe, 2012).

The Adult

Formally, the adult instar concludes the biological cycle. It is eight-legged, with the exception of the Eriophyoidea, which are secondary tetrapods, and the neotenic genus *Larvacarus* of the Tenuipalpidae which are hexapod, and according to species may belong to one or both sexes, which can either be morphologically very similar except in the genital region or show strong diversity. Body growth usually ends with the last moult and sexual maturity. However, in few cases, post-imaginal moulting has been observed, as in some Prostigmata and Astigmatina. Some groups may exhibit different 'morphofunctional' adaptations to selected food regimes. In *T. urticae*, as already noted, the adult instar is missing and is replaced by the tritonymph instar, which is able to reproduce by paedogenesis (André and Van Impe, 2012).

Some species of the Astigmatina are characterized by polymorphism. The adaptation was first observed by Canestrini (1888), who reported that *Rhizoglyphus* sp. can form two types of males: one morphologically similar to the female and one with the third pair of legs enlarged, and the tarsi of these legs are replaced by a slightly curved claw (Hughes,

1976). These legs are not employed in walking (Michael, 1901) but are held away from the ground, and in *Caloglyphus anomalus* Nesbitt can be used as a weapon to kill young homomorphic males or even to fight among themselves (Woodring, 1969a). Abnormal specimens are sometimes observed where the hypertrophied leg and its associated heavily sclerotized apodemes are found on one side of the body only. The most elaborate forms of these males are found among the bird parasites such as *Analges*, which possess a third pair of legs almost as large as the rest of the body. In *Lardoglyphus*, the heteromorphic male is present, and the homomorphic lacks (Hughes, 1976).

Some species belonging to the genera *Caloglyphus*, *Rhizoglyphus* and *Schwiebea* possess four types of males. Türk and Türk (1957) stated that the 'homomorphic' male has the body shape and length of dorsal setae, as the non-gravid female; the 'bimorphic' male has a larger body and longer dorsal setae; the 'heteromorphic' male resembles the homomorphic male but has a modified third pair of legs; and the 'pleomorphic' male has the modified third pair of legs and a bimorphic body. Woodring (1969a) arranged the four types of males into two groups, according to body shape and length of dorsal setae. These groups are named as 'homotype' and 'bimotype' male; the first has the female-type body and includes the homomorphic and the heteromorphic male, with modified legs III, while the second group, characterized by a longer body, includes the bimorphic and the pleomorphic male with modified legs III. *C. anomalus* possesses the four types of males; other species of the same genus, such as *C. berlesei* (Michael) have the bimorphic and the pleomorphic male (Hughes, 1976).

Woodring (1969a) recognized two different types of polymorphism. The first type includes either of homotypes and one of the bimotypes, in a fixed ratio, under a given set of environmental conditions. Thus, *C. berlesei* is characterized by a definite pleomorphic-homomorphic male ratio. Changes in temperature or alterations of the food modify this ratio. In the second type of polymorphism, any one of the four male types can be the dominant form, irrespective of external conditions, and the other male type of that particular body shape, homotype or bimotype, rarely occurs. In *Rhizoglyphus echinopus* (Fumouze *et* Robin) (*sensu* Woodring), the homomorphic male is the dominant one and the heteromorphic rarely occurs. In *C. boharti* (Cross), the bimorphic male is predominant, and the pleomorphic male occurs occasionally. Both types of polymorphism may be observed in the same species, and in *C. anomalus*, the pleiomorphic and homomorphic males occur in a regular ratio, and heteromorphic and bimorphic males are found rarely. Woodring (1969a) observed that in *C. boharti*, pleomorphic male formation was inhibited by the presence of other mites which produced a volatile pheromone, the threshold pheromone level depending on the degree of separation of the developing immature stages (Hughes, 1976).

The Reproduction

Mites reproduce sexually and exhibit a common gonochoristic condition, which may or may not be anphigonic. Parthenogenesis is also a common mode of reproduction, and paedogenesis has been observed in various families of the Acariformes (Vacante, 2016).

Aspects related to reproduction are the sex ratio and the genetic system. The sex ratio is the proportion of adult males to females in a natural field population at a certain moment. Knowledge of the sex ratio helps to determine the mode of reproduction of a species, and the management of pest mite populations. A method that is commonly used to evaluate the proportion of the two sexes is to measure the 'apparent sex ratio' – the observed proportion of both sexes at each biological stage of the species in the field. However, various factors (differential mortality of the sexes and instars, sex dispersion, etc.) may distort the evaluation of the apparent sex ratio, and it has been suggested that instead, the 'primary sex ratio' – the proportion of each sex in the offspring of a brood at fertilization – should be considered. Theoretically, an estimate of the proportion of the sexes at the adult stage (the time when oviposition first occurs), which is designated the 'secondary sex ratio', could be a better measurement, but this is very difficult to realize. Therefore, in order to obtain an estimate of the primary sex ratio, it is considered preferable to calculate the 'tertiary sex ratio', which is the sex ratio at the time of the adult moult (Walter, 2009a). Among extrinsic factors affecting the sex ratio, the temperature (Hazan *et al.*, 1973), the population density (Wrensch and Young, 1978) and the quality of leaf on which the mite is feeding (Wrensch and Young, 1983) are very important. Intrinsic factors affecting the sex ratio include the amount of sperm transferred to the female, which varies according to the length of mating and sperm availability (Overmeer, 1972). Age also influences the sex ratio and old females generate more males or all males (Shih, 1979; Wrensch, 1979). Females fertilize a certain number of eggs daily during the oviposition period (Kongchuensin, 1983; Young *et al.*, 1986), and the first egg laid by a mated female is unfertilized and male-determined (Helle, 1967), as sperm cannot reach it in time for internal fertilization to occur. Movement of the sperm towards the oocytes does not appear to be controlled by the female. The ventral part of the ovary receives less sperm than the dorsal region, and in the medioventral region, the oocytes probably escape sperm penetration (Feiertag-Koppen and Pijnacker, 1982). The sex ratio has a hereditary basis, and a maternal effect has also been reported (Overmeer and Harrison, 1969; Mitchell, 1972b).

Reproduction in mites is characterized by a haploid-diploid genetic system that is based on haploid and diploid chromosomal complements. The main genetic systems reported for mites are 'diplodiploidy', 'haplodiploidy' and 'thelytoky'. All these systems are sexual.

The diplodiploidy is an ancestral genetic system, in which both sexes have two sets of chromosomes. The adaptation is known in various acarine groups, such as the Mesostigmata. The sex ratio of the species involved could be 1:1.

In the haplodiploidy, the males have half the number of chromosomes of the females, and they are produced parthenogenetically from unfertilized and haploid eggs, while the females are produced from fertilized and diploid eggs. This genetic system is known in several acarine groups, including the Prostigmata and the Astigmatina. The formation of haploid males from unfertilized eggs is also called 'arrhenotoky'.

Haplodiploidy has been demonstrated by karyotype analysis of over 100 bisexual tetranychid species. Only seven species out of 126 species investigated exhibited thelytoky (obligate parthenogenesis producing only females), two species exhibited both thelytoky and arrhenotoky, and 117 only arrhenotoky (Helle and Pijnacker, 1985). The haploid chromosomal complement may be obtained also by elimination or deactivation of the paternal genome, as occurs in the Phytoseiidae, where the male chromosomes are either totally or in part eliminated, and the zygote acquires a haploid condition; this is called 'parahaploidy' (Helle *et al.*, 1978). The thelytoky involves obligate parthenogenesis, and the females reproduce only parthenogenetically. It is common in many acarine groups (Walter and Proctor, 1999), but only rarely does it reach the condition of 'amphitoky' or 'deuterotoky', in which both sexes are produced parthenogenetically. Thelytoky occurs in many Prostigmata, and it is well known in the Tetranychidae family (Bryobiinae) (Helle, 1985).

5 Ecology

Feeding Habits and Habitats

The treatment of the fundamental aspects of ecology is largely drawn from Vacante (2016) for Prostigmata and from O'Connor (1994) for the Astigmatina mites, respectively. Feeding habits of mites involve various aspects of their morphology and lifestyle. Their morphological features (mouthparts, postoral digestive systems, etc.) usually reflect the ecological approach that these animals exhibit with respect to various food resources and habitats. From this viewpoint, morphology, habitats and behavioural patterns are interdependent and may markedly influence systematic features. Mites can be free living, symbiotic or parasitic, and occupy various terrestrial and aquatic habitats. Terrestrial species may be arboreal or nidicolous or live in niches related to soil and/ or litter. Aquatic species may be littoral, marine or freshwater species. Mites are common in all ecosystems and are rarely lacking in any habitat. This latter characteristic is very important and has been described as an 'evolutionary revolution' in this subclass of organisms (Krantz, 2009d). Mites can also feed on a large variety of organic food or other materials, including both living and dead parts of plants and animals. Several species are adapted to ectoparasitism or endoparasitism and feed on organic substrates from vertebrates or invertebrates, such as blood, tissue fluids or other parts. Many species are fungivorous, others feed on decaying leaves or other vegetable tissues, while others feed on living plant tissues or stored food. The various species may be either general or specialized feeders. Most of the mites that live and feed on plants belong to the Prostigmata, and among these, there are phytophagous mites that have characteristic structures of their mouthparts, such as the Tetranychoidea and Eriophyoidea. Other mite groups, such as some of the phytophagous Penthaleidae and Acaridae, have mouthparts that are not strictly adapted to phytophagy. The last group includes many species adapted to live and/ or feed on stored food.

Free-Living Species

Free-living species occupy soil habitats, so-called 'soil mites', and aboveground (aerial) habitats, so-called 'aerial mites'. Soil mites commonly inhabit the underlying rhizosphere, which hosts a rich microfauna and microflora. They are variously associated with plant roots and rhizomes, which represent a considerable food resource for organisms with different feeding habits, such as fungivores, microbivores, detritivores, herbivores and predators. Soil mites have a fundamental role in the transformation of litter into humus. According to the environment, their ecological characteristics and other variables (extraction method, depth of layer, water content, etc.), mites can be collected in numbers ranging from 50,000 to 250,000 or more from each 1 m^2 area of soil surface (Petersen, 1982); they can number up to 400,000 in winter in forest soil (Wallwork, 1959). Microarthropods in general can reach a density greater than 1,000,000 individuals m^{-2} (André *et al.*, 1994). Among the soil fauna, the Oribatida play a key role in the dispersal of ectomycorrhizal fungal spores and in the breakdown of detritus. Other mites feed on algae, lichens and mosses. Soil mites also include predators belonging to several families. A restricted group of phytophages inhabits the soil, where the mites feed on root tissue, corms and bulbs in the rhizosphere, sometimes causing severe damage to ornamentals and vegetables; examples include *Rhizoglyphus* spp., acarids, etc. Aerial mites typically include predators and phytophages. Often associated with these important groups are other mites that feed on fungi and algae. Predators are very common in the aerial parts of plants, where they feed on different instars of phytophagous or

(V. Vacante and C.H.W. Flechtmann)
DOI: 10.1079/9781786390912.0005

mycophagous mites and small insects. Some dozens of families of predatory mites are known (Gerson *et al.*, 2003), and among these are the Phytoseiidae, Stigmaeidae, Cheyletidae, Hemisarcoptidae and Cunaxidae. Species of these families are reported also on stored food. The most important predatory mites belong to the Phytoseiidae family, which numbers thousands of species (2250 in 2004) (Moraes *et al.*, 2004). Phytophages include species adapted to phytophagy, pollenophagy and nectarivory. They mostly belong to the Trombidiformes, a group with over 4000 species and 300 genera (Bolland *et al.*, 1998; Amrine *et al.*, 2003; Migeon and Dorkeld, 2006–2013). The most important families are the Tetranychidae, Eriophyidae, Phytoptidae, Diptilomiopidae, Tenuipalpidae and Tarsonemidae. Other families may exhibit a partial adaptation to phytophagy, such as the Penthaleidae, Acaridae, etc.

Hughes (1976) reported that free-living Astigmata show the same wide range of habitat as other groups of mites. Some species, such as *Rhizoglyphus* sp., *Tyrophagus palmarum* Oudemans, etc. are found in soil and surface litter and feed on decaying plant remains. Some Saproglyphidae occur on the leaves of trees and others such as *T. palmarum*, *Acarus farris* (Oudemans), *Tyrophagus longior* (Gervais), *Mycetoglyphus fungivorus* Oudemans may inhabit grassland and pastures (Sheals, 1956; Griffiths, 1960), and may be found in stacks of grain and hay in open fields. A reduced number of species (Hyadesidae) are aquatic, and others live in the sap exudated from trees (*Hericia* sp.) or in the surface film of sewage beds.

Parasitic Species

According to O'Connor (1994), parasitism may be defined as 'the ability of a species to live permanently on a host or feeding directly from it'. Parasitic species are ectoparasites and endoparasites of vertebrates and invertebrates, respectively. Although they may cause serious damage to the livestock industry and to agricultural production in general, their role, except for some groups, such as the Hemisarcoptidae, which are parasites and predators of armoured scales (Gerson *et al.*, 2003), is of secondary importance in plant production. Further information may be found in Krantz (2009b). Some Dermanyssidae, such as *Dermanyssus gallinae* (De Geer) are the most economically deleterious ectoparasites affecting egg-laying hens worldwide. In his extensive review, O'Connor (1994) articulates the treatment of the parasitic species, focusing attention on two fundamental topics, the 'host association' and the 'evolution of the parasitism', respectively. We report here briefly what O'Connor discussed.

Host associations

O'Connor (1994) reported that the quality, duration and specificity of association with other organisms, represent a fundamental aspect of the bio-ecology of many astigmatid mites. Until recently, it was thought that the association between astigmatid mite, deutonymphs and other arthropods was a simple case of commensalism, where only the mite benefitted from phoretic dispersal. Houck and O'Connor (1991) pointed out that in some species, this association may be more complicated because deutonymphs may demand a period of time spent on a host before activating the next development. Fain and Bafort (1967) and Lukoschus *et al.* (1972) found that some deutonymphs collected in mammalian hair follicles, and the subcutaneous tissues of birds and mammals are true parasites; they acquire material from their hosts in an as yet unknown manner.

Specificity of association differs largely among the astigmatid species. Species which live in environments only visited by other organisms, and not created by them, show a tendency to reduce specificity. Thus, Türk and Türk (1957) found in Germany many hosts of species living in decaying organic matter and fungal fruiting bodies. O'Connor (1990) collected *Histiogaster arborsignis* Berlese (Acaridae), a mite inhabiting subcortical spaces, from 22 species of insects included in three orders. The same author observed in Michigan that this relative non-specificity represents a general pattern among astigmatid mites, inhabiting subcortical habitats (O'Connor, 1991). Some species living in other habitats, not due to interactive processes, possess a greater specificity. According to Fashing (1975), this adaptation may depend on a reduced number of potential host species visiting the habitat, as in the case of associations of *Naiadacarus arboricola* Fashing with syrphid flies of the genus *Mallota*, and *Hemisarcoptes* species with the coccinellid beetles of the genus *Chilocorus* (Houck and O'Connor, 1990). Obviously, the phenomenon may reflect ecological changes in the habitat, as in the case of different species of

Histiostomatidae, developing in cattle dung in the Netherlands, and able to use for dispersal, different dung-breeding insects (Bongers *et al.*, 1985). These insects visit the dung piles in various periods during the succession of stages of the decomposition of the dung, and the mites may be exploiting the dung in different time periods as well.

Conversely, mites inhabiting host-created environments show the tendency to be specialized for dispersal, on those hosts and few others. Nidicolous mites associated with insects and vertebrates exhibit a high specificity, and in many species, the life histories and their hosts are strongly correlated. This synchronicity of development has been observed in different species, such as the enslinielline associated with Hymenoptera (Krombein, 1967; Cowan, 1984; Klompen *et al.*, 1987), the glycyphagid *Lophioglyphus liciosus* Volgin associated of mammals (Lukoschus *et al.*, 1972), and the hypoderatid *Hypodectes propus* (Nitzsch) associated of birds (Fain and Bafort, 1967). The correlation of life histories between vertebrate hosts and permanently parasitic Astigmata is known. Dubinin (1951) observed the correlation between the size and age structure of feather mite populations, with the breeding and moulting patterns of the host birds.

Evolution of parasitism

O'Connor (1994) reported that applying the method of Brooks (1985) to hypothesize ecological shifts, it is possible to hypothesize that the shifts in fundamental bio-ecological parameters, characterizing the transition from commensalistic associations to true parasitism, occurred in at least 21 different lineages of astigmatid mites. Parasitism of arthropods has evolved in ten distinct cases, seven of which involve a permanent parasitism. They involve the families Canestriniidae, Heterocoptidae, Rosensteiniidae (genus *Rosensteinia*, an undescribed genus also parasitic on roaches related to the non-parasitic genus *Nycterilichus*, and genera *Micronychites* and *Micronychitoides*, forming a monophyletic group), Hemisarcoptidae (genus *Linobia*), and Acaridae (genera *Askinasia*, *Ewingia*, *Hoogstraalacarus*, *Kanakobia* and two undescribed genera living on decapod crustaceans). Parasitism of arthropods by species characterized by the deutonymph in their life cycle and disperse through this stage has evolved twice. The clade includes the genera *Ensliniella*, *Kennethiella* and *Vespacarus* of the family Winterschmidtiidae, and the genus *Hemisarcoptes* of the family Hemisarcoptidae; the last acts either as a predator or a parasite of diaspidid scale insects, according to the number of mites per host, and the stage of the host attacked. Moreover, arthropod parasites may exhibit a shift in host and life history. The family Epidermoptidae includes species commonly parasitic on the skin of birds. The genus *Myialges* of this family shows a singular transformation in the adult females that leave the bird host and parasitise lice or hippoboscid flies, while the males and nymphs remain on bird hosts; curiously, the females lay their eggs on the arthropod host (O'Connor, 1994).

The analysis of the evolution of parasitism of vertebrates suggests the hypothesis of 11 distinct cases. In this context, there are known two distinct bioecology shifts; the first concerns the permanent parasitism by feeding stages, with the concomitant loss of the deutonymph from the life cycle, and the second the parasitism by deutonymphs only, while the other stages remain free-living nidicoles. The first pathway includes at least four lineages. The first lineage numbers the Psoroptidia, which forms a monophyletic grouping of about 40 families, which appears to have originated as parasites of birds, with a secondary radiation onto mammals. The second lineage includes the genera *Loxanoetus* and *Otanoetus* of the family Histiostomatidae, living in the ears of large mammals and forming a monophyletic group. O'Connor (1994) reported that we do not know whether these mites are true parasites, although at least some of their diet is material of host origin. The third lineage numbers the genus *Chiroptoglyphus* of the family Rosensteiniidae, including species associated constantly in the fur of bats. The fourth lineage includes the genus *Cheiromelichus* and *Nycteriglyphus asiaticus* Fain of the family Rosensteiniidae and probably forms a monophyletic lineage, characterized by species attached to the skin of molossid bats of the genus *Cheiromeles*. A shift from simple phoresy by deutonymphs, to an association in which the deutonymph takes materials from the host, confirms the transition to parasitism in seven lineages species. The lack of a functional mouth in the deutonymph makes the mechanism of food uptake unknown; however, the breakdown of host tissue around the mite has been demonstrated in some Glycyphagidae (Lukoschus *et al.*, 1972), and growth of the mite within the deutonymphal stadium can be pronounced. O'Connor (1994) reported that six of the lineages showing this shift are included in the superfamily Glycyphagoidea, characterized by non-deutonymphal instars mammalian nidicoles. The first lineage numbers in detail the monobasic family Pedetopodidae, with the single species *Pedetes capensis* (Forster), living as deutonymphs in the hair follicles of the springhare. The second lineage numbers the family Chortoglyphidae, parasites on the hair follicles of rodents and sometimes other mammalian groups. The third lineage includes the family Echimyopodidae, living in hair follicles or subcutaneous tissues of marsupials, edentates and rodents, and three lineages within the family Glycyphagidae. The fourth lineage numbers the genus *Baloghella* of the subfamily Melesodectinae, deutonymphs of which are associated with the tissues of the ear conchae of European badgers. The fifth lineage numbers the subfamily Metalabidophorinae, associated with hair follicles of rodents and rarely insectivores. The sixth lineage comprises the subfamilies Lophuromyopodinae and Ctenoglyphinae, living in the hair follicles of rodents. Finally, a seventh numbers the non-glycyphagoid family Hypoderatidae, deutonymphs of which occur subcutaneously in birds and rarely in mammals (O'Connor, 1994).

Habitats

Klimov and Tolstikov (2011) reported that acaroid mites live in a variety of habitats, such as the nests of birds and mammals, and of many invertebrates (termites, beetles, bees, wasps, ants), fungal sporocarps, decomposing organic matter (carrion, various plant materials, animal manure and guano), under bark, in treeholes and in seaweeds. A reduced number of species are aquatic or parasitic on gills of fresh water crabs. Acaroid mites may damage stored food and products, and be a common component of house dust, playing an economical, medical and veterinary importance. Some species of the genera *Acarus*, *Mycetoglyphus*, *Tyrophagus*, *Rhizoglyphus* and *Sancassania* (= *Caloglyphus*) may damage subterranean organs and young leaves of various plants, resulting in serious agricultural pests, subject to quarantine in various countries (Olsen, 1983; Shinkaji *et al.*, 1986; Nakao and Kurosa, 1988; Nakao, 1991; Diaz *et al.*, 2000; Fan and Zhang, 2004, 2007a; Kasuga and Amano, 2006; Rojas and Klimov, 2007).

Some acaroid mites inhabit the soil, disperse through it and exploit different patchy habitats. Others use soil to disperse to patchy habitats as adults or as deutonymphs, as the suidasiid *Sapracarus tuberculatus* which lives in the Holarctic in deep soil and sometimes colonizes bird nests. Species of *Sancassania* inhabit fruiting bodies of mushrooms and only reproduce several weeks each year when mushrooms are available; the remaining time, they remain inactive in soil waiting for the new mushroom season (Klimov and O'Connor, 2003). Some species of the genera *Tyrophagus*, *Rhizoglyphus*, *Schwiebea* and *Sancassania* are habitat generalists, and they are common in both soil and patchy habitats, adopting both strategies. Commonly, these species are serious agricultural and stored-product pests, and they may cause respiratory diseases or allergic reactions in individuals who come into direct contact with these materials. They can easily extend their ranges via commercial trade and establish themselves in new countries and habitats (Klimov and Tolstikov, 2011). Studies carried out by Griffiths (1960) in England, highlighted that within the same agricultural habitat, such as a farm, a different distribution of species may occur, associated to a different exploitation of the resource. He collected 47 species from stacks of hay, maize and straw, and permanent grassland. Of these species, 22 were associated with stored food and food stores. Twenty-one stored-product species were collected from hay and straw stacks and five from grassland. *Acarus siro* Linnaeus, a major pest of stored cereals and cereal products, was the dominant mite in stacks and was also found in grassland samples. *T. longior* Gervais was similarly distributed and occurred almost as frequently. The motile hypopial stage of *A. siro* was found in 43% of the stack samples; it was attached to mites, beetles and other invertebrates. *T. palmarum* Oudemans, an uncommon stored-product mite, was the dominant species collected from grassland samples. Griffiths (1960) concluded that all the mites, commonly associated with foodstuffs stored indoors, are present in the outdoor habitat afforded by stacks. *A. siro* and *T. longior* are dominant in stacks, and both appear to be field dwellers. *T. palmarum* appears to be a successful field dweller and its occasional association with stored food may be accidental. Moreover, the dominant warehouse partnership of *A. siro* and *Glycyphagus destructor* (Schrank) does not occur outdoors; here *T. longior* replaces *G. destructor*. The species of the family Glycyphagidae appear to live mainly in haystacks sited in stackyards. Considering how the mites were acquired, grainstacks standing in the open field have a peculiar combination of dominants, namely *A. siro*, *T. longior*, *T. palmarum* and the mesostigmatic *Alliphis helleri* (R. et G. Canestrini).

O'Connor (1994) discussed this subject extensively and reported that astigmatid mites have successfully colonized and specialized on a number of microhabitats, most of which are spatially restricted, so-called 'patchy', and/or temporally restricted, so-called 'ephemeral'. Patchy or ephemeral habitats employed by astigmatid mites derive through biotic processes, or through the interaction of both physical and biotic processes. Patchy habitats created by physical processes (sand dunes, rock clefts, ponds, lakes and springs) rarely host astigmatid mites. Appropriate habitats are associated with concentrations of organic materials used, either directly as food or as medium for microbial or fungal development. Because mites have a reduced ability to disperse directly, the most appropriate habitats for astigmatid mites must also be attractive to other organisms, employed as hosts to phoretic deutonymphs.

O'Connor (1994) identifies two habitat types, the first derives from biotic processes which do not involve modification of the environment by a host (e.g. growth, death, decay), and the second derives from ongoing interaction between the host and the environment (e.g. food provisioning, nest building). The first type includes non-mobile habitats such as flowers, roots, tubers, fungal fruiting bodies, subcortical spaces, treeholes and other plant cavities, sap flows, algal mats, dung and carrion; mobile habitats, such as the bodies of living animals, belong to this type. The second type numbers the nests, burrows, galleries, refuse piles and food stores of solitary, communal and social insects as well as birds and mammals, including humans.

Life-history patterns of astigmatid mites living in these habitats differ in the nature, quality and degree of specificity of the association with the insect or vertebrate, on which the mite depends for dispersal and/or permanent habitat. According to these premises, we summarize below the fundamental traits of the main habitats.

Habitats created by plants and fungi

Several astigmatid mites live on substrates associated with plants or fungi, but few occur on such relatively non-patchy substrates, such as the bark or leaves of living plants. Some astigmatid mites feed living plant tissue, such as *Rhizoglyphus* species (Acaridae), able to burrow in and feed on the living tissues of bulbs and tubers. A few taxa number species are able to feed on living leaf tissue, although most species living on leaf surfaces are primarily grazers on fungi, growing on the leaves, such as species of genera *Czenspinskia* and *Oulenzia* (Winterschmidtiidae) and *Neotropacarus* and some *Tyrophagus* (Acaridae). Some species of the genus *Carpoglyphus* (Carpoglyphidae) and several undescribed genera of Histiostomatidae live in various inflorescences. Specialized organs of plant able to hold water (phytotelmata) host a diversity of taxa of the families Histiostomatidae, Algophagidae and Acaridae. Water-filled leaves of the pitcher plant families Nepenthaceae and Sarraceniaceae harbour specific genera of Histiostomatidae; the first family known for the Old World numbers the genera *Creutzeria*, *Zwickia* and several unnamed genera, while the family Sarraceniaceae of the New World includes the genus *Sarraceniopus*. Species ascribed to the genera *Bromeliaglyphus*, *Naiacus* and some species of *Rhizoglyphus* and *Schwiebea* of the family Acaridae, as well as undescribed species of *Hormosianoetus* of the family Histiostomatidae, inhabit leafaxils of the Bromeliaceae. It is possible to hypothesize that histiostomatids are microbial feeders, while the feeding habits of the Acaridae living in these habitats are unknown. Some species of the genera *Histiostoma* (Histiostomatidae), *Congovidia* (Hemisarcoptidae), *Rhizoglyphus*, *Sancassania* and other species of undescribed genera of Acaridae consume fruiting bodies of fleshy fungi. *Rhizoglyphus* and *Boletacarus* species (Acaridae) and *Nanacaroides* and some species of *Nanacarus* (Hemisarcoptidae) habit the tougher fruiting bodies of polypore and xylariaceous fungi and tend to specialize either on mycelial tissue or on spores (O'Connor, 1984).

Holdfasts of marine algae are occupied by species of Hyadesiidae, and marine algal mats and halophytic plants or freshwater mosses harbour species of the genera *Algophagopsis*, *Algophagus* and *Neohyadesia* of the family Algophagidae. The examen of the gut contents showed that these species are primarily algal feeders (O'Connor, 1994).

Dead and decaying plants and fungi host a greater diversity of taxa. Subcortical spaces in decaying tree trunks and logs are inhabited by several species of Acaridae belonging to the genera *Calvoliella*, *Histiogaster*, *Schwiebea* and *Thyreophagus*. Some species of the family Histiostomatidae are ascribed to the genus *Bonomoia*, and many *Histiostoma* species may be collected in decaying wood. Species in these genera live in the rotting wood itself and are not confined to insect galleries as are some other taxa. In these habitats, the Acaridae are primarily fungal grazers, while the Histiostomatidae filter microorganisms from wet substrates (O'Connor, 1994).

Treeholes also form habitats for different astigmatid mites. Water-filled treeholes host species of the genera *Naiadacarus* (Acaridae), *Hormosianoetus* (Histiostomatidae) and *Algophagus* and another, undescribed genus of family Algophagidae. These taxa have been the subject of exhaustive studies by Fashing (1994) and will not be discussed here further. Species of the genera *Acotyledon*, *Fagacarus*, *Rhizoglyphus*, *Sancassania* and *Schwiebea* of the family Acaridae and *Histiostoma* of the family Histiostomatidae, commonly inhabit treeholes that are not permanently wet. Species of the genera *Fusohericia* and *Hericia* of the family Algophagidae and species of genera *Histiostoma* and *Sellea* of the family Histiostomatidae use tree wounds that exude fermenting sap as substrates. Species of the genera *Boletoglyphus* (Acaridae) and *Allocalvolia* and *Saproglyphus* (Winterschmidtiidae) prefer as habit decaying fungal fruiting bodies. The primary diets of all of these species seem to be fungi, including yeasts (O'Connor, 1994).

Habitats created by animals through non-interactive processes

Several astigmatid mites live in habitats created by animals through non-interactive processes. Dung produced by vertebrates can host many astigmatid mites, such as species of the families Acaridae (*Sancassania*), Histiostomatidae (*Ameranoetus*, *Aphodanoetus*, *Copronomoia*, *Fibulanoetus*, *Glyphanoetus*, *Histiostoma*, *Myianoetus*, *Rhopalanoetus*, *Syringanoetus* and probably others), and Winterschmidtiidae (unnamed genus). Guano accumulations produced by colonies of bats or birds host species of Rosensteiniidae (many genera), Aeroglyphidae (*Aeroglyphus* and *Glycycometus*), Histiostomatidae (*Glyphanoetus* and *Histiostoma*), Guanolichidae (*Guanolichus* and *Neoguanolichus*), Acaridae (*Acarus* and *Troglocoptes*) and Glycyphagidae (*Glycyphagus*) (O'Connor, 1994).

Frass produced by wood-inhabiting scarabaeoid beetles may be inhabited by *Acotyledon* and *Sancassania* species (Acaridae). Vertebrate carrion is a substrate for species of *Acarus* (Acaridae), *Pelzneria* and *Spinanoetus* species (Histiostomatidae) in early decay stages, while *Lardoglyphus* (Lardoglyphidae) species prefer the older carcasses. The above-reported species are primarily fungivores or microorganismal feeders, but the acarid and lardoglyphid species may also feed directly on the substrate material (O'Connor, 1994).

The bodies of living animals represent constant habitats for many astigmatid mites. Over half of all genera of Astigmata include the monophyletic group Psoroptidia, and a number of species that inhabit constantly on or in the bodies of birds or mammals. Psoroptidid mites are 'microhabitat specialists', and many species often cohabit on a single host. Birds possess an exceptional diversity of microhabitats. The greatest diversity of mite species is associated with the flight and tail feathers, living along the rachis, between the barbs or in the spaces at the bases of the barbs; they are known in almost 18 families, including Alloptidae, Avenzoariidae, Caudiferidae, Cheylabididae, Crypturoptidae, Eustathiidae, Falculiferidae, Freyanidae, Gabuciniidae, Kramerellidae, Ochrolichidae, Proctophyllodidae, Pterolichidae, Ptiloxenidae, Rectijanuidae, Thoracosathidae, Vexillariidae and some Analgidae. Species of the families Analgidae, Psoroptoididae and Xolalgidae inhabit plumaceous barbs of contour feathers, or at the bases of flight feathers. The spaces between coverts and flight feathers are the habitat of species of Trouessartiidae. The space within the feather quills is the habitat of species of Dermoglyphidae, Gaudoglyphidae, Oconnoriidae, Ptyssalgidae, Syringobiidae and some Proctophyllodidae and Pyroglyphidae. Species of Epidermoptidae live in the superficial layers of the skin and in feather follicles, while species of Laminosioptidae inhabit deeper subcutaneous tissues and the follicles of developing feathers (O'Connor, 1994). The case of Syringophilidae is very interesting. These mites are commonly called 'quill mites' and are obligatory ectoparasites of birds; they inhabit the feather quills where they feed on subcutaneous tissue and fluids (Glowska *et al.*, 2020). Species of the subfamily Syringophilinae inhabit all but the body feathers (primaries, secondaries, tertials, rectrices and wing coverts), while the species ascribed to the subfamily Picobinae infect the body feathers internally; mites of this family have been collected on hundreds of bird species, belonging to 95 families and 24 orders (Glowska *et al.*, 2015).

Mites of the family Turbinoptidae inhabit the nostrils, and of family Cytoditidae, deeper in the trachea, lungs and air sacs. These species may feed as actual parasites on host skin, pygidial gland secretions, skin exudates, keratin-rich materials from the quills and feather barbs, or, rarely, on blood. Mites living on feather surfaces may also ingest detritus such as fungal spores and algal cells (Dubinin, 1951).

The mammal body represents a rich source of suitable microhabitats. Species of the families Atopomelidae, Chirodiscidae and Listrophoridae are associated with the hair shafts of small- and medium-sized mammals. The family Lobalgidae includes species specialized for the flattened, grooved hairs of Neotropical sloths and spiny rats. Species of Psoroptidae inhabit the skin surface of many groups of terrestrial mammals, and some may also enter the ear openings and live in the ear canals. Species of the family Chirorhynchobiidae fix to the wing membrane edges of Neotropical bats, and species of the family Myocoptidae anchor themselves to body hairs while feeding from the skin. Species of the Rhyncoptidae (including Audycoptidae) inhabit the hair follicles of primates, carnivores and rodents, and species of Sarcoptidae penetrate the skin and inhabit the epidermal layers of many mammal groups, in particular bats (O'Connor, 1994).

Like birds, the respiratory tract is also exploited by mites. Species of Gastronyssidae live in the nasal passages of rodents and bats, while species of Lemurnyssidae live in similar habitats in primates. The Pneumocoptidae numbers a few species associated with the lungs of rodents. The Entonyssidae, Rhinonyssidae and Halarachnidae are respiratory parasites of vertebrates. The Gastronyssidae also includes species specialized for living in the orbits of the eyes, and the stomach of chiropterans. Feeding on host blood is rare. The bodies of birds and mammals are also exploited by other groups of astigmatid mites not included in the Psoroptidia. In the ears of large mammals inhabit species of the genera *Auricanoetus*, *Loxanoetus*, *Otanoetus* and of an undescribed genus of the family Histiostomatidae. Some species of Rosensteiniidae belonging to the genera *Cheiromelichus*, *Chiroptoglyphus* and *Nycteriglyphus* live constantly on the skin of their bat hosts. Feeding habits of these species are unknown, but the histiostomatids probably filter materials from the ear secretions (O'Connor, 1994).

Astigmatid mites may live permanently on many arthropods. All species in the families Canestriniidae and Heterocoptidae and of the genus *Linobia* of the family Hemisarcoptidae live their lives on the bodies of many species of Coleoptera. These species parasitize adult hosts and tend to be specialized for life, either on the ventral body surface or in the subelytral space. It is possible that most of these mites be exudate feeders, though *Linobia coccinellae* (Scopoli) pierces the cuticle to feed on host haemolymph. Species included in several lineages of Rosensteiniidae exhibit a constant association with insect taxa; thus, species of *Rosensteinia* and one undescribed genus inhabit the bodies of cockroaches (*Blattaria*) and show morphology and feeding habits superficially similar to the above Canestriniidae. Species of *Micronychites* and *Micronychitoides* have been found in earwigs of the genera *Arexenia* and *Xenaria* (Dermaptera: Arexeniina), but their feeding ecology is unknown. A group of genera in the family Acaridae live on decapod crustaceans; their morphology has led previous workers to recognize this group as a separate family, Ewingiidae. Species of the genera *Ewingia*, *Hoogstraalacarus* and *Kanakobia* inhabit the gill chambers of hermit crabs (Coenobitidae) and fresh water crabs (Potamonidae). Species of *Askinasia* and one undescribed genus live attached to leg or abdominal setae of hermit crabs, and also in this case, the feeding ecology of these mites is unknown (O'Connor, 1994).

Habitats created by interaction of hosts with the environment

In contrast to the great diversity of habitat types indicated above as harbouring astigmatid mites, habitats that are created by the activities of other animals often contain an even greater diversity of these mites (O'Connor, 1994). Hughes (1976) reported that perhaps owing to the softness of the cuticle and their small size, many species are domestic in their habits and seek shelter in the nests of insects, birds and mammals, where a choice of food is available. The majority of these nest-dwelling species are scavengers and feed on fungi, remains of food, excrement and on desquamated skin scales or feathers. Many of these scavenging species can live on a wide variety of substances and can also live in human food stores.

Habitats created by insects

Many Coleoptera and Hymenoptera and Diaspididae (Homoptera) create habitats exploitable by astigmatid mites. The tunnels of wood-boring beetles of the family Passalidae host astigmatid mites of the genera *Kanoetus* and *Seolianoetus* of the family

Histiostomatidae and *Passaloglyphus* and some *Schwiebea* species of the Acaridae. Galleries of Scolytidae or Bostrichidae host species of *Ameroseius* as well as of *Afrocalvolia*, *Winterschmidtia* and related undescribed genera of the family Winterschmidtiidae, some species of the genera *Histiogaster* and *Thyreophagus* of Acaridae, and some *Histiostoma* of the Histiostomatidae. Other genera of Histiostomatidae have been described only as deutonymphs collected from various other beetles that excavate cavities in wood. Species of the genus *Kaszabanoetus* have been collected, associated with *Uloma* species (Tenebrionidae), and of *Chiloanoetus* from various genera of Pycnocerini. Species of *Cureulanoetus* are associated with palm weevils of the genus *Rhynchophorus*. Beetles of the family Scarabaeidae (Dynastinae, Rutelinae) that produce excavations in wood or brood cells in soil provisioned with dung or other substrates create habitats for *Sancassania* species (Acaridae) (O'Connor, 1994).

Most of the above-discussed species are reported or hypothized as fungivores (Acaridae) or microorganismal feeders (Histiostomatidae); however, *Afrocalvolia nataliae* (Türk *et* Türk) (= *Calvolia fraxini*) preys eggs and young larvae of its scolytid host (Kielczewski and Seniczak, 1972). The other genera of Winterschmidtiinae also show this habitat type and may also be predacious. Mites of the genera *Hemisarcoptes* (Hemisarcoptidae) and *Thyreophagus* (Acaridae) exploit colonies of scale insects of the family Diaspididae; species of *Hemisarcoptes* prey scale insects and their eggs, while *Thyreophagus* species are fungivores (Houck and O'Connor, 1991; Houck, 1994).

Concerning the mites living in habitats due to the action of other arthropods, we can also cite the case of the nests of solitary and social Hymenoptera that host different astigmatid mites. The nests of solitary wasps host most species of the family Winterschmidtiidae. Hosts number species of the families Sphecidae (genus *Crabrovidia*) and Vespidae (genera *Ensliniella*, *Kennethiella*, *Macroharpa*, *Monobiacarus*, *Vespacarus*, *Zethacarus*, *Zethovidia* and other genera). Species belonging to the genera *Ensliniella*, *Kennethiella* and *Vespacarus* feed on haemolymph of the host larva or on provisioned prey, while species of the genus *Monobiacarus* feed only on provisions (Krombein, 1967; Klompen *et al.*, 1987).

Some Hemisarcoptidae (genus *Congovidia*) and Acaridae (genus *Lackerbaueria*) have also been found in solitary sphecid nests. The latter seems a cleptoparasite (Krombein, 1967), while the former is a fungivore. Solitary bees of the family Megachilidae also host mites of the genus *Vidia*, species of which feed on fungi in the nest lining (O'Connor and Eickwort, 1988). Other families of Astigmata are associated with a much greater diversity of solitary bee (Eickwort, 1979; O'Connor, 1988).

O'Connor (1994) summarized these associations and reported that all species in the family Chaetodactylidae live in the nests of bees. So, species of the genus *Chaetodactylus* are primarily associated with Megachilidae and Xylocopinae, and those of the genus *Sennertia* with Xylocopinae. All genera of the subfamily Horstiinae of the family Acaridae are obligatorily associated with bees, the great part of which are solitary, such as the genera *Halictacarus* and *Schulzea* of the family Halictidae, the genera *Megachilopus*, *Neohorstia* and *Sennertionyx* of the family Megachilidae, the genus *Horstia* of the subfamily Xylocopinae, the genus *Horstiella* of the Apidae, and an undescribed genus of the subfamily Anthophorinae. Another acarid genus, *Ctenocolletacarus*, not belonging to the subfamily Horstiinae has been collected in the nests of stenotritid bees in Australia. Species of Histiostomatidae have been collected in solitary bee nests, such as the genera *Anoetus* and some *Histiostoma* from Halictidae; some species of the genus *Glyphanoetus* from Colletidae and Xylocopinae. Species of the genus *Tortonia* of the family Suidasiidae live in the nests of both solitary wasps and bees. Feeding adaptation of these groups ranges from microorganismal filtration, such as in the Histiostomatidae, to provision thieves occurring in the genus *Sennertia* to true cleptoparasites of the genera *Chaetodactylus*, *Horstia*, and some *Tortonia*.

In the nests of social Hymenoptera and Isoptera live a rich fauna of astigmatid mites (Eickwort, 1990a). Species of the genus *Polistes* of the family Vespidae harbour species of *Sphexicozela* of the family Winterschmidtiidae; species included in the genera *Medeus* of Acaridae and *Tortonia* of Suidasiidae have been collected in *Vespula* nests. Some species of the genus *Anoetus* of Histiostomatidae have been collected in social halictid bee nests, and various taxa of several families are associated with social Apidae. Thus, species of the genera *Kuzinia* of Acaridae and *Cerophagus* of Gaudiellidae are associated with *Bombus*, some species of the genera *Carpoglyphus* of Carpoglyphidae and *Forcellinia* of Acaridae with *Apis*, and all species of Meliponocoptidae (genera *Meliponocoptes*, *Meliponoecius* and one undescribed genus) and most Gaudiellidae (genera *Gaudiella*, *Partamonacoptes*, *Platyglyphus* and unnamed genera) with meliponine stingless bees. Many genera of mites, primarily from the family Acaridae, have been collected in the nests of ants (Formicidae). A number of these taxa are also shared with the nests of termites (Isoptera). The list includes the genera *Cosmoglyphus*, *Forcellinia*, *Froriepia*, *Lasioacarus*, *Ocellacarus*, *Rettacarus*, *Sancassania*, *Tyrophagus* and other genera. Moreover, the faunas of ant and termite nests number the Suidasiidae (genus *Lemmaniella*) and Histiostomatidae (genus *Histiostoma*) (O'Connor, 1994).

Mammal hosts

In the nests of mammals live a greater diversity of astigmatina than in the nests of social insects. These associations include some taxa in all major clades, although few species have been studied with respect to their ecology and life history (O'Connor, 1994). Some species of the genera *Myianoetus* and *Psyllanoetus* of the family Histiostomatidae have been found to be associated with mammal nests or nest-inhabiting insects. Species of the genus *Psylloglyphus* of the family Winterschmidtiidae have also been reported from phoretic associations with nest-inhabiting insects. Most taxa of the superfamily Glycyphagoidea are reported or presumed mammal nest associates. These taxa include the families Chortoglyphidae (genera *Alabidopus*, *Aplodontopus*, *Chortoglyphus* and two undescribed genera) associated with rodents, insectivores and primates; the genera *Echimyopus*, *Marmosopus*, *Marsupiopus* and *Oryzomyopus* of the family Echimyopodidae are associated with marsupials and rodents; and many genera of the family Glycyphagidae are associated with marsupials, edentates, rodents, carnivores and insectivores. Some members of the family Acaridae have been reported from mammal nests or nest-inhabiting arthropods. These genera involved are *Acarus*, *Acotyledon*, *Paraceroglyphus*, *Trichopsyllopus* and *Viedebanttia*. Moreover, some species of Pyroglyphidae, such as *Pyroglyphus* and occasionally other genera, have been collected from mammal nests (O'Connor, 1994).

Bird hosts

The nests of birds are also exploited as habitat, by specialized taxa of astigmatid mites. Thus, species of the genera *Comyianoetus*, *Hexanoetus* and some *Histiostoma* of the family Histiostomatidae have been collected from the nests of raptors. Moreover, species of the families Euglycyphagidae and Glycacaridae have been collected in nests of raptors and a petrel, respectively. Some Glycyphagidae (genera *Glycyphagus* and *Lophuromyopus*), Aeroglyphidae (genera *Aeroglyphus* and *Glycycometus*), Acaridae (genera *Acotyledon* and *Notiopsyllopus*), Lardoglyphidae (genus *Lardoglyphus*) and Suidasiidae (genus *Sapracarus*) are also collected in bird nests. All species of Hypoderatidae and most non-parasitic species of the family Pyroglyphidae (many genera) live in bird nests. The feeding adaptation of vertebrate nidicoles probably varies, including filter feeding on microorganisms as in the Histiostomatidae, fungivory as in many glycyphagoids, consumption of actual nest materials, as well as fungi as in the Acaridae and ingestion of prey, remains as in the Lardoglyphidae. Many Glycyphagidae exhibit strong preferences for a restricted number of fungal species as food resources (Sinha, 1966a), and nidicolous species are notoriously difficult to culture away from natural substrates (Fain *et al.*, 1972; Lukoschus *et al.*, 1971, 1972, 1979).

Woodroffe (1953) studied the nest contents of a number of species of birds and collected six species of stored-product mites, together with a number of their predators. He thought that these mites were scavengers in the nests. One species of these mites was *Glycyphagus domesticus* (De Geer) which was not frequently collected in birds' nests and was later found in a bat roost (Woodroffe, 1956) and is also common in houses and in bee hives. Wasylik (1959) studied the fauna of sparrow, *Passer domesticus* Linnaeus, nests in Poland and collected 11 species of Astigmata, of which ten are common pests of stored food. The examen of *Passer montanus* (Linnaeus) highlighted that sexually mature mites predominate in the summer and nymphs in the autumn. The mites leave the nests in summer and autumn, and it is suggested that at this time, the infestation of food stores is likely to occur (Wasylik, 1964).

Habitat shifts in nest-inhabiting Astigmatid mites

An interesting habitat shift is known among species belonging to many genera of vertebrate nest-inhabiting Astigmata. A number of species live in and around human habitations. These 'stored-product' and 'house-dust' mites number the majority of well-studied species of Astigmata (Hughes, 1976). The majority of these taxa are included in the families Acaridae (genera *Acarus*, *Aleuroglyphus*, *Cosmoglyphus*, *Histiogaster*, *Madaglyphus*, *Rhizoglyphus*, *Sancassania*, *Thyreophagus*, *Tyrolichus* and *Tyrophagus*) and Glycyphagidae (genera *Ctenoglyphus*, *Diamesoglyphus*, *Glycyphagus*, *Gohieria*, *Lophuromyopus* and *Tropilichus*). Other families are Histiostomatidae (genus *Histiostoma*), Winterschmidtiidae (genera *Acalvolia* and *Procalvolia*), Hemisarcoptidae (genus *Nanacarus*), Carpoglyphidae (genus *Carpoglyphus*), Chortoglyphidae (genus *Chortoglyphus*), Echimyopodidae (genus *Blomia*), Aeroglyphidae (genera *Aeroglyphus* and *Glycycometus*), Suidasiidae (genus *Suidasia*), Lardoglyphidae (genus *Lardoglyphus*) and Pyroglyphidae (genera *Euroglyphus*, *Dermatophagoides*, *Malayoglyphus* and *Pyroglyphus*) (O'Connor, 1994).

Chemical communication

As in insects, chemical compounds are used in communications between mites and/or with their environment, including their host plants. This section discusses chemical communication between individuals of the same species (by pheromones) and between different species (by allelochemicals). Different types of pheromones have been reported in the literature, and allelochemicals can be classified into three groups: 'allomones', 'kairomones' and 'synomones', which have distinct actions (Vacante, 2016).

Pheromones

In the 1970s, the strong attraction of males of different tetranychid mites, such as the two-spotted spider mite, *Tetranychus urticae* Koch, towards females was shown to be due to the action of chemical compounds with sex-attracting capabilities, which were named pheromones (Cone *et al.*, 1971a). Methods for the preparation of pheromone extracts and bioassay procedures were developed (Cone, 1979) in association with investigations on the bio-ethology of some pest species. Thus, male behaviour towards quiescent deutonymphs was described by Cone *et al.* (1971b), while the quantification of the male response to the female deutonymph was investigated by Penman and Cone (1972). Later, it was found that the deutonymphal web aided males in finding females, with a female sex pheromone acting as an attractant until the first male contact, and thereafter acting as an arrestant with a range of about 1.8–2.3 mm (Penman and Cone, 1974). Different release sites have been suggested for pheromones (dorsal lyrifissures, basal portion of body), but it seems that the moulting fluid, or the fluid contained between the old and new cuticle, plays an active role in attracting the male (Cone, 1985). Research on the chemical nature of pheromones showed that farnesol was one component of the male sex attractant system in the two-spotted spider mite (Regev and Cone, 1975), and that its content was correlated with egg deposition by the female (Regev and Cone, 1976).

Further studies on pheromones in different mite groups demonstrated that there were different types of pheromones, including sex pheromones, such as rosefuran, a sex pheromone from an acarid mite, *Caloglyphus* sp. (Mori *et al.*, 1998) and alarm pheromones, such as the isopiperitenone of the acarid mites *Tyrophagus similis* Volgin (Kuwahara *et al.*, 1987) and *Tyrophagus neiswanderi* Johnston *et* Bruce (Kuwahara *et al.*, 1989a). Astigmatic mites have been found to use a single chemical compound (or different compounds) to achieve two of the three communication functions of pheromones (alarm, sex and aggregation) (Kuwahara, 2010). Each of these communication functions emerges from the conditions the mites are facing and is based on the functional characteristics of the chemical concerned, such as the dose-response relationship, the active dose range, and the amount of the chemical compound(s) in the opisthonotal gland. In the astigmatic mite *Rhizoglyphus setosus* Manson, all three communication functions can be achieved using products of the opisthonotal glands (Kuwahara, 2010).

Ridgway *et al.* (1999) found that *A. siro* Linnaeus produced alkaneundecane, but the confirmation of gas chromatography–mass spectrometry (GC-MS) is missing. The pheromones are produced mainly by opisthosomal glands (Kuwahara, 2004). Among them, Citral, an alarm pheromone, has been detected in mite-infested stored grain (Tuma *et al.*, 1990). The guanine is an ammonium waste deposited in the faeces (Levinson *et al.*, 1991). All the produced compounds probably change the smell of stored products. Although the compounds play an important role in mite communication, the question is still open – whether their concentrations in stored products lead to damage.

The family Acaridae is known to have alarm pheromones (Sonenshine, 1985). Among the species most important, *Tyrophagus putrescentiae* (Schrank) produces alarm pheromones such as Neryl formate [3,7-dimethyl-(Z)-2,6-octadienyl formate] as a defence mechanism that induces escape behaviours and reduces successful capture by predatory mites (Kuwahara *et al.*, 1975; Howard *et al.*, 1988; Midthassel *et al.*, 2013; Rifà and Griffiths, 2018). The topic has been studied by several authors in various acarid mites. So, Leal *et al.* (1988) identified the 2-hydroxy-6-methylbenzaldehyde as the alarm pheromone of the acarid mite, *T. perniciosus*, and studied its distribution among related mites. The male and female sex pheromones produced by *Acarus immobilis* Griffiths and its sex pheromone activity were investigated by Sato *et al.* (1993) and Shibata *et al.* (1998). Kuwahara *et al.* (1989b) and Leal *et al.* (1989) studied the pheromone of *Caloglyphus polyphyllae* (Zachvatkin). Akiyama *et al.* (1997) studied the Neryl formate as the alarm pheromone of *R. setosus*, and the common pheromone component among four *Rhizoglyphus* species. Noguchi *et al.* (1998) identified the beta-acaridial (2(E)-(4-methyl-3-pentenylidene)-butanedial) as the alarm pheromone of *T. longior* Gervais. Ryono *et al.* (2001) identified the 2-Hydroxy-6-methylbenzaldehyde as the female sex pheromone of *Cosmoglyphus hughesi* Samsinak. Tomita *et al.* (2003) studied the Neryl formate (Z)-3,7-dimethyl-1-2,6-octadienyl formate as the alarm pheromone of *Tyroborus lini* Oudemans. Hiroyuki *et al.* (2003) studied the pheromone of *Histiogaster rotundus* Woodring, Mizoguchi *et al.* (2003) identified the Alpha-acaridial, a female sex pheromone from an alarm pheromone emitting mite, *Rhizoglyphus robini* Claparède, and Mizoguchi *et al.* (2005) identified the S-isorobinal as the female sex pheromone from an alarm pheromone emitting mite, *R. setosus*. *Carpoglyphus lactis* (Linnaeus) may be beneficial in the wine industry, as mite alarm pheromones enhance the aroma of pale and dry wines aged under flor yeasts (Marin *et al.*, 2009). Chemical, instrumental analysis and bioassays showed that the chortolure is an aggregation pheromone for both sexes of this species. Several related compounds are released in smaller amounts by the mites. The alarm pheromones of these mites (neral and geranial) can only be found in total extracts of the mites, in which eight occur only in minute amounts. The method of sampling is therefore crucial for pheromone identification. Feeding experiments with deuterated propionate showed that chortolure is a polyketide, formed by successive addition of four propionate units to an acetate starter (Schulz *et al.*, 2004).

It has been observed that *D. gallinae* (De Geer) is attracted by host-specific pheromones emitted by the hens (El Adouzi *et al.*, 2020). A set of five volatile compounds, among the dozens naturally emitted by hens, has been patented for its attraction in the laboratory (Roy *et al.*, 2018). Besides host-related attractants, *D. gallinae* is also attracted by the aggregation pheromones of congeners, causing clustering together (Entrekin and Oliver, 1982; Koenraadt and Dicke, 2010). Though the pheromone has not been fully characterized chemically, a series of compounds isolated from the odour of freshly fed *D. gallinae* have been patented as attractants (Birkett *et al.*, 2010). Studies on the host-searching behaviour of *D. gallinae* highlighted that aggregation pheromones and kairomones mediate the behaviour of the poultry mite (Koenraadt and Dicke, 2010) and that the foreleg tarsal sense organs intervene in stimuli perceiving (Soler Cruz *et al.*, 2005). After the blood meal, various biological stages of *D. gallinae* form aggregations and thigmokinesis, which refers to the increased locomotion in response to changes in contact with the immediate physical environment, and pheromones play a role in this behaviour (Entrekin and Oliver, 1982; Koenraadt and Dicke, 2010).

Skelton *et al.* (2010) identified the neryl formate as the airborne aggregation pheromone, for the American house-dust mite and the European house-dust mite (Epidermoptidae).

Allomones

Allomones benefit the organism that produces them by a modification of the behaviour of the organism that receives them, although sometimes they do not have much effect on the receiver. Chemicals emitted by the opisthonotal glands of the Astigmatina are anti-predator or antifungal allomones. The relationships between pest mites and their predators may be led by 'infochemical' cues to alert the prey, to the presence of their predators. These chemicals are usually produced by the exuviae, excreta, etc. of predators and recognized by the prey. Such infochemicals can act both as chemical signals emitted by alarmed, damaged or dead conspecifics in the prey population, and as allelochemicals emitted by other potential prey (Kats and Dill, 1998; Grostal and Dicke, 2000; Dicke and Grostal, 2001).

Kairomones

Kairomones produce a benefit to the organism that receives them, and a disadvantage to the one that produces them. Schmidt (1976) observed that females of the phytoseiid *Phytoseiulus persimilis* Athias Henriot spent a lot of time in and around the webbing of their prey, *T. urticae* Koch. Later, it was found that this type of behaviour by predatory mites depended on kairomones (Hislop and Prokopy, 1981; Hoy and Smilanick, 1981; Sabelis and Van de Baan, 1983). These are volatile substances, extractable in solvents, which are emitted by the prey and, when present in an airstream, affect the behaviour of predators (Sabelis *et al.*, 1984).

Studies on the host-searching behaviour of *D. gallinae* highlighted that aggregation pheromones and kairomones mediate the behaviour of the poultry mite (Koenraadt and Dicke, 2010) and that the foreleg tarsal sense organs intervene in stimuli perceiving (Soler Cruz *et al.*, 2005).

Synomones

Synomones are the subject of intensive research because of the implications that they have for pest control, so they are treated quite extensively here. Plant tissues that are injured by phytophages produce specific volatiles or 'synomones', also called 'herbivore-induced plant volatiles' (HIPVs) (Hunter, 2002; Frost *et al.*, 2008), that deter pests or attract predators (Cortesero *et al.*, 2000; Farmer, 2001). The chemical signals are emitted only as a result of pest feeding and provide 'a chemical burglar alarm' (Dicke, 1995).

Dispersal

Mites move within and/or between their habitats. These movements may involve single individuals or entire populations and are accomplished through a 'simple locomotion or crawling', a 'passive dispersal' or a 'phoresy' (Jeppson *et al.*, 1975; Kennedy and Smitley, 1985; Schliesske, 1990; Michalska *et al.*, 2010). In addition to the mode(s) of dispersal adopted, the survival of species also depends on the ability to easily find a suitable habitat for its development and, in the case of stored mites, a suitable food kept in a suitable environment. This need relates dispersal to the phenomena of 'host acceptance' and 'host preference and 'host specificity' better known for phytophagous mites.

In the Astigmatina, the phoresy by hypopodes relies to a large extent on the spread of the species and may represent a serious risk of contamination of stored food. The phenomenon assumes a particular importance in the prevention of attacks and in the control in general. We discuss here only the general features of 'simple locomotion or crawling' and 'passive dispersal' of some important mites injurious to stored food and will treat the topic more extensively in the next chapter on phoresy.

Dispersal of stored mites, implicitly due to 'simple locomotion or crawling' and to a 'passive dispersal', has been recently examined by Solarz *et al.* (2022) in Poland, and we report here the fundamental traits of this review. Up until now, many groups of mites have been reported as occupational biohazards for humans (Arlian, 2002). According to the last specialist, Solarz *et al.* (2022) examined the composition of stored mite communities collected from farm buildings in Poland. They found that the greatest exposure to storage mites commonly occurs in agricultural settings, and therefore agricultural work is considered a major risk factor for occupational diseases (van Hage-Hamsten and Johansson, 1998; Arlian, 2002; Berger *et al.*, 2005; Cichecka *et al.*, 2006). Their previous studies suggest that allergenic mites belonging to Acaridae, Chortoglyphidae and Glycyphagidae should be described as occupational risk factors, contributing to respiratory and dermal diseases among farmers in Poland (Solarz *et al.*, 1997; Cichecka *et al.*, 2006; Solarz and Pająk, 2019a). They reported that the predominant species in all the examined farm buildings were *A. siro* Linnaeus and *A. farris* (Oudemans). Additionally, species as *G. domesticus* (De Geer), *Chortoglyphus arcuatus* (Troupeau), *Lepidoglyphus destructor* (Schrank) and *T. putrescentiae* (Schrank) were also frequently reported. These results confirm the presence of allergenic mites in cowsheds, barns, chaff-cutter buildings, pigsties, pigeon and poultry houses (Terho *et al.*, 1982; Mumcuoglu and Lutsky, 1990; Solarz *et al.*, 1997; Stejskal and Hubert, 2008; Solarz and Pająk, 2019a). Thus, all the mentioned farm buildings should be regarded as potential sources of storage mites in rural environments. However, the stated prevalence of acarids, glycyphagids and cheyletids in the investigated buildings is not in accordance with earlier data published by Hallas (1981), Terho *et al.* (1982) and Solarz *et al.*, (1997).

The acarid mite *T. similis* Volgin is introduced into cucumber greenhouses in England via straw or manure (Hussey *et al.*, 1969). Research in Japan on the horizontal and vertical distribution of acarid populations in agricultural soils in greenhouses indicates that the mite inhabits not only the cultivated soil but also the uncultivated soil in the periphery of the cultivated area inside the greenhouse and the soil immediately outside. The main control sites for *T. similis* are therefore in both the cultivated and peripheral soils of the greenhouse (Kasuga and Amano, 2005).

Phoresy

Houck and O'Connor (1991) defined the phoresy for the Astigmata as 'a phenomenon in which one organism (the phoretic) receives an ecological or evolutionary advantage by migrating from the natal habitat while superficially attached to a selected interspecific host for some portion of the individual phoretic's lifetime. Benefit is not conferred as a nutritional or developmental influence on the phoretic stage'. In a simpler way, the phoresy is the dispersal of an animal by other animals. In mites, it occurs in the deutonymphal (or hypopus) or adult stages, and it has been reported for various mite groups (Binns, 1982). The phenomenon has been defined as a temporary symbiosis between a 'phoretic animal', which is usually smaller, not feeding and inactive, and a 'carrier animal', which is usually larger, that transports it (Walter and Proctor, 1999) from a disadvantaged to a more appropriate habitat. Camerik (2010) revised the concept of phoresy by adapting Athias-Binche's (1994) classification, in which phoresy may be 'facultative' or 'obligatory'. Facultative phoresy is usually related to the density of populations and/or environmental perturbations. Obligatory phoresy occurs when a species living in a relatively stable environment adopts a seasonal migration cycle using a coexisting carrier. Many examples of phoresy involve the hypopodes (heteromorphic deutonymphs) of acarine species from various taxa (Mesostigmata, Astigmatina) that have phoretic relationships with arthropods and sometimes also with vertebrates. The hypopus may attach to its carrier by several mechanisms, which include the 'anal pedicel' (Uropodina), 'posteroventral suckers' (in different Astigmatina) and 'posteroventral claspers' (Glycyphagidae). In different families (Tarsonemidae, Pyemotidae, Scutacaridae, etc.), adult females are phoretic on a variety of arthropod carriers and sometimes on vertebrates (Evans, 1992; Palevsky *et al.*, 2001).

The Astigmatina have a higher diversity in ecological strategies than the other Oribatida. Many species are phoretic on all kinds of carriers (insects, birds, mammals, reptiles), just as some oribatids, but the Astigmatina managed to develop their phoretic behaviour as an art. Not only is the vast majority of the species phoretic, but some of those made their carrier their host. They adopted a parasitic way of life and spent all their life on the host (Siepel, 1994). This behaviour has led to enormous radiation in species diversity, resulting in an ecologically very diverse group. Free-living species are found in most of the superfamilies, but a few of them are exclusively 'free living' and live in various habitats. The Histiostomatoidea live in wet places, where they feed on organic material. The Glycyphagoidea live on stored food and in nests of mammals, the Pyroglyphoidea live in our houses as house-dust mites, and the Acaridoidea contain many species living in stored food, but they are also known as plant pests. Some species in the Hemisarcoptoidea are free living (in stored food, on marine algae and on deciduous trees), but the majority of species of this superfamily are kleptoparasites. Parasitic are the Canestrinoidea feeding on exudates of adult beetles, and the Sarcoptoidea, living in the fur and skin of mammals, while the remaining superfamilies live on birds, either in their nests (Hypoderatoidea) or on their feathers, in the skin or even inside the birds (Pterolichoidea, Freyanoidea and Analgoidea) (Siepel *et al.*, 2016).

Hughes (1976) reported that the ability to form hypopodes is part of genotype of many species of Astigmatina, and this expression is induced by environmental conditions. The degree to which hypopodes are formed in a particular environment varies from species to species. Griffiths (1966) observed that *A. immobilis* Griffiths formed always more hypopodes than either *A. farris* (Oudemans) or a hypopus-forming strain of *A. siro* Linnaeus. The influence of various ecological factors on hypopus formation, such as temperature, RH, pH of food, overcrowding, accumulation of waste products and quantity and quality of food has been studied by various authors (Michael, 1901; Polezhaev, 1938, 1940; Sorokin, 1951; Perron, 1954; Wallace, 1960; Woodring, 1963; Griffiths, 1966, 1969). Perron (1954) studied *Histiostoma laboratorium* Hughes and Woodring (1963) *Rhizoglyphus echinopus* (Fumouze *et* Robin), respectively, and both authors hypothesized that the quality of food was more important than the quantity. It has been observed that in *A. immobilis* the scarcity and imbalance of basic nutrients favour the hypopus formation (Griffiths, 1966, 1969). Moreover, ergosterol is necessary for hypopus formation; it is probably a precursor of a hormone.

Okabe (1999) reviewed the characteristics of heteromorphic deutonymphs of non-psoroptid astigmatid mites. He reported that the deutonymph was called hypopus, since recognized as belonging to the genus 'Hypopus' in Mesostigmata, then as one of the immature stages in Astigmata. Deutonymphal formation is generally induced by adverse environmental conditions, which it should resist or escape from. Although stimuli to moult to the next stage might be different in each species, environmental changes including food supply are necessary for development. The deutonymph, which may represent specific relationships between phoretic hosts, exploits other organisms to migrate into a new environment. In applied fields, the importance of the deutonymph has not been fully understood.

It has been observed that the protonymphs from a starved culture of hypopus-forming strain of *A. siro*, lack the neurosecretory cells present at ecdysis in the anterodorsal region of the central nervous system, and that similar cells are present in the salivary glands (Hughes, 1976). These cells appear at normal ecdysis during the first 30 h of the protonymphal instar; during this time the fate of the protonymph is fixed, and the developmental period of protonymph becoming deutonymph is much extended, probably to allow time for the accumulation of a food reserve (Griffiths, 1966).

The hypopodes of acarid mites are carried by insects. They may appear only at a time when a particular vector is likely to be available. Thus, the acarid mite *Hericia hericia* (Kramer) living in sap exuding from trees, the hypopodes, form in May and early June, coinciding with the egg-laying activities of flies visiting the sap, and the hypopodes use these flies as a means of dispersal (Robinson, 1953, as cited in Hughes, 1976).

Moulting from deutonymph to tritonymph needs water. Baker (1964b) reported that in *Histiostoma feroniarum* (Dufour), living in sewage bacteria beds, the hypopodes are in high numbers where the active zoogloeal film had been allowed to dry out in the atmosphere. In *H. laboratorium*, Perron (1954) found that yeast must be present for moulting, although this need not be in contact with the hypopus. Vitamins, such as thiamin and more particularly, riboflavin, are also said to play an important role.

Cutcher and Woodring (1969) observed in *Caloglyphus boharti* (Cross) (= *Sancassania boharti*) that hypopodes will moult at temperatures over 15°C and a RH above 97% when they are isolated, but not if they are retained in old stock cultures. They hypnotized that the salt concentration of the environment affects the moulting.

Different examples concern about the relation between hypopodes of acarid mites and cultivated plants. Thus, the hypopodes of *M. fungivorus* (Oudemans), which have been reported from mushrooms and various other economic plants (Mahunka, 1961; Manson, 1983; Nakao and Kurosa, 1988; Nakao, 1989), are probably carried by ants (Türk and Türk, 1957), while those of *Sancassania berlesei* (Michael), which have been reported from edible fungi (Hussey *et al.*, 1969; Clancy, 1981), are carried by stored-product insects (*Tenebrio molitor* Linnaeus, etc.) (Hughes, 1976). The hypopodes of different Histiostomatidae are carried by insects. Adults, juvenile stages and hypopodes of *H. feroniarum*, which have been recorded on edible fungi (Hussey *et al.*, 1969) and are responsible for serious crop losses in various mushroom species (Kim and Hwang, 1996; Greenslade and Clift, 2004), have been collected from associations with the weevil *Otiorhynchus cribricollis* Gyll, on strawberry, in a protected crop in Sicily (Vacante, 1989b). Hypopodes of *Histiostoma sapromyzarum* (Dufour), which are commonly found on decaying fungi (*Lactarius*, *Russula*, *Tricholoma*, *Amanita*, *Scleroderma*) (Hughes, 1976) and are also recorded on cultivated mushrooms (Boness, 1968), have been collected on decaying hyacinth bulbs and on damp timber. The hypopus of this species is carried by several beetle species, flies and myriapods (Hughes, 1976).

Phoretic dispersal by deutonymphs

O'Connor (1994) refers that one major modification in the life cycle of astigmatid mites is the shift from a limited dispersal ability of the adult instar to a much greater dispersal potential achieved in a preadult instar, the deutonymph. The deutonymphal instar in the Astigmata has a highly modified morphology and behaviour, which function both in dispersal and typically in resisting adverse conditions. Dispersal in the Astigmata can occur in several modes. The ancestral condition in the group, as observed in the Schizoglyphidae, is dispersal by deutonymphs via phoretic association with an arthropod. The phoresy in the Astigmata has been recently discussed by Houck and O'Connor (1991); the phoresy by deutonymphs represents the primary dispersal mode in the non-psoroptidid Astigmata, with phoretic hosts, including many groups of arthropods and vertebrates.

Non-phoretic dispersal by deutonymphs

O'Connor (1994) reported that a second dispersal mode known in the Astigmata is non-phoretic dispersal by deutonymphs. This type of dispersal adopts two mechanisms. The first is characterized by deutonymphs which retain the ancestral morphology, including the attachment structures used by phoretic individuals, but dispersal is typically under the mites' own power. This type of dispersal includes species of *Sancassania* (= *Caloglyphus*) (Acaridae), associated with scarabaeid beetles and their larvae. Feeding stages of the mites live on dead beetles or larvae; deutonymphs disperse in the soil, encounter beetle larvae and remain with them until their death (Chmielewski and Lipa, 1967). Similarly, the feeding stages of *Histiostoma murchiei* Hughes *et* Jackson (Histiostomatidae) prey on the egg cocoons of earthworms (Oliver, 1962). The phoresy is unknown in this species, and probably the deutonymphs disperse through the litter in search of suitable habitats (O'Connor, 1994).

The phoresy is poorly observed in the histiostomatid taxa associated with pitcher plants. A personal communication of Robert Naczi to O'Connor (1994) reported that species of *Sarraceniopus* only leave the water-filled leaf pitchers as deutonymphs. Deutonymphs move about the plant and congregate at the tips of unopened leaf pitchers, and when the pitchers open, the mites enter. These deutonymphs maintain the modifications associated with phoresy, and probably use the phoresy for long-distance dispersal, but most individuals disperse to new pitchers within a single plant or members of a single clone.

A significant modification of the deutonymph, associated with highly regressive morphological changes, has allowed the passive dispersal via air currents in two taxa of Glycyphagidae. Two species, *Glycyphagus* (*G.*) *domesticus* (De Geer) and *G.* (*Lepidoglyphus*) *destructor* (Schrank), possess deutonymphs with body setation and legs strongly regressive or lacking altogether. These deutonymphs do not emerge from the protonymphal cuticle which dries around the cyst-like deutonymph (Hora, 1934; Knülle, 1959, 1987). Contrarily to other taxa forming regressive deutonymphs (*Acarus*, *Alabidopus*, *Chaetodactylus* and *Hericia*) and not dispersing, the deutonymphs of the two *Glycyphagus* species do disperse via air currents (O'Connor, 1994).

Chmielewski (1970a) observed that a small percentage of mites belonging to the species *C. lactis* (Linnaeus) and *Thyreophagus entomophagus* (Laboulbène) will survive passage through the alimentary canal of mice and the domestic sparrow, the eggs and the immature stages being more resistant than the adults. These animals may therefore contribute to the dispersal of this mite species.

Loss of the dispersal (deutonymphal) morph

Several taxa of astigmatid mites do not form deutonymphs and, in these taxa, all dispersal probably must be accomplished by other instars. They number the family Hyadesiidae, including species restricted to marine, intertidal and tidal habitats; arboreal leaf-inhabiting species of the genera *Czenspinskia* and *Oulenzia* of the family Winterschmidtiidae, and species of the genus *Neotropacarus* of the family Acaridae; most species of the genus *Tyrophagus* (Acaridae) living in soil, litter and synanthropic habitats; the Aeroglyphidae-Rosensteiniidae which occur in bat colonies; and many taxa with non-deutonymphal stages parasitic on arthropods or vertebrates, although a few exceptions to the last occur (O'Connor, 1994).

Deutonymphs are unknown in a number of other lineages of Astigmata, and the fragmentary nature of the data does not permit a statement as yet. O'Connor (1994) suspected that some of these taxa where deutonymphs are never formed are the subfamily Algophagine (Algophagidae), the genera *Fusacarus*, *Gohieria* and *Tropilichus* of the family Glycyphagidae, the genus *Suidasia* of Suidasiidae and *Aleuroglyphus* of Acaridae.

Dispersal of permanent parasites

Few data are available concerning the details of dispersal by non-deutonymphal stages of astigmatid mites (O'Connor, 1994). Dubinin (1951) summarized a high number of observations on the dispersal of astigmatid mites inhabiting bird feathers and concluded that three basic dispersal modes are possible. The first mode concerns the direct transmission from parent to offspring, with a preponderance of dispersers as preadults or teneral adults; the second mode is the direct transfer between adult birds, which accounts for the presence of specific parasites on birds which do not rear their own young. The third transfer occurs via the nest and, although it is documented for some individuals of some species as *Gabucinia delibata* (Robin) (Gabuciniidae) on crows, it is not considered an important mode of dispersal for any species. The direct contact between hosts seems to be the necessary mode of dispersal of astigmatid mite parasites of mammals (Meleney, 1985), although some species such as *Sarcoptes scabiei* (De Geer) (Sarcoptidae) can survive for short periods away from a host (Mellanby, 1972).

The bio-ecology of permanently parasitic mites of arthropods ascribed to the families Canestriniidae, Heterocoptidae, and some Rosensteiniidae and Hemisarcoptidae is unknown. These mites parasitise only adult insects, and the dispersal is realized through the direct contact between individuals, probably during the copulation. A potential exception has been observed in the genus *Megacanestrinia* of the family Canestriniidae, in which, the deutonymph is retained in the life cycle. The feeding stages of the mites live in the subelytral surface of their hosts, the African carabid beetles of the genus *Tejjfus*, while deutonymphs congregate on the venter of the thorax. It seems possible that dispersal to new hosts still involves the deutonymph, resulting in this stage set, for easier access to another individual during mating (O'Connor, 1994).

Dispersal and adverse changes in the environment

In astigmatid mites, the protonymphs possess morphological specializations for dispersal to withstand adverse changes in the environment which would be reductive to other ontogenetic stages. Changes in environmental conditions induce individuals to enter the deutonymphal stage (Wallace, 1960; Woodring, 1963; Griffiths, 1966). In the taxa lacking the ability to form deutonymphs, the tritonymph takes over the function of surviving adverse periods. This adaptation is known for *Czenspinskia transversostriata* (Oudemans) (= *lordi* Nesbitt) of the family Winterschmidtiidae (Dosse and Schneider, 1957), *T. longior* (Gervais) of Acaridae (Sorokin, 1948) and populations of *C. lactis* (Linnaeus) of Carpoglyphidae, which are unable to form deutonymphs (Zyromska-Rudzka, 1964). Populations of *Dermatophagoides* of the family Pyroglyphidae survive adverse conditions in either the protonymphal or tritonymphal instars (Wharton, 1976). Interestingly, *N. arboricola* Fashing of the family Acaridae survives adverse periods as tritonymphs (Fashing, 1976), the deutonymph in this species apparently functioning only in dispersal (O'Connor, 1994).

Factors Influencing Population Dynamics

The most important factors influencing the dynamics of storage mite populations are temperature, RH, food resource and natural enemies. Among these factors, the adaptations to weather variations play a key role. The small size and the delicate structure of the mites' body expose them to severe ecological pressure – to extreme annual cycles in temperature, to extreme weather conditions and to large fluctuations of humidity in warehouse of stored food. In response, mites have established specific adaptive mechanisms to mitigate these pressures. Several mechanisms have been adopted, including diapause, migration to protected niches and the stimulation of plant growth to produce galls or succulent parts (Jeppson *et al.*, 1975). Adaptation to cyclic weather modifications needs structural and behavioural changes. For instance, high temperatures and low humidity are easily overcome by sheltering in moist ground crevices, or under bud scales, or by developing more resistant forms. In some species, females stop their development or produce resistant eggs. As a result, the populations of each species develop optimally within a range of more suitable climatic conditions.

The most important factors that influence mite population development appear to be combinations of T, RH and MC. A rise in T, RH and/or MC increases mite productivity, whereas a decrease in T, RH and/or MC content decreases productivity, with interactions occurring between the variables (Dunn, 2003). Atmospheric humidity levels influence the development of mite populations according to mite species and their ecological adaptations. From a physiological point of view, this phenomenon is related to water balance. Boudreaux (1958) first found a negative influence of high humidity on the fecundity and survival of tetranychid pest mites.

O'Connor (1994) reported that conversely to other groups within the suborder Sarcoptiformes, the Astigmata may be characterized by a series of bio-ecological adaptations related to their exploitation of temporary habitats. The phenomenon calls into question the so-called 'r-selected' characteristics, including a relatively short generation time, high fecundity, relatively short adult lifespan, internal fertilization and dispersal ancestrally as deutonymphs. Despite demographic information available for different taxa of astigmatid mites, the studies on these mites concern primarily species associated with human habitations, and food stores or parasites of domestic animals, and the majority of published data on the bio-ecology deals with these species. However, it is possible to find a diversity of evolutionary lineages among these studies, and the generalizations may have some validity. Many of the ecological categories discussed above, are completely unrepresented or have few species which have been studied. For example, due to the difficulties of maintaining parasitic astigmatid mites away from their hosts, almost no demographic information is available for this great diversity of taxa (O'Connor, 1994).

The available demographic data of astigmatid mites highlighted different general patterns. The first and most common pattern includes taxa with a generation time ranging from 10 to 15 days (egg to egg, or deutonymph to deutonymph, in a temperature regime of 22–27°C), and an average female reproductive period ranging from 15–40 days. This category includes three subgroups, with respect to the total egg production per female. Some taxa in this group show a moderate fecundity in the range of 100–300 eggs per female; they include soil and litter-inhabiting species of the family Acaridae such as *T. putrescentiae* (Schrank) (Rivard, 1959, 1961; Barker, 1967b), *Schwiebea falticis* Woodring and *Schwiebea rocketti* Woodring (Woodring, 1969b), insect nidicoles such as *C. lactis* (Linnaeus) of the family Carpoglyphidae (Chmielewski, 1971d), and *Acotyledon formosani* Phillipsen *et* Coppel of the Acaridae (Phillipsen and Coppel, 1977a), and *G. domesticus* (De Geer) (Glycyphagidae) (Barker, 1968b), a species which exhibits a wide habitat range. Different mites species characterized by a nidicolous ancestry show a similar length of life cycle and female reproductive period but a lower total fecundity (10–100 eggs per female); they are common among vertebrate nidicoles and on stored product, and include *Aleuroglyphus ovatus* (Troupeau) (Acaridae) (Hsin and Chen, 1964), *Gohieria fusca* (Oudemans) (Glycyphagidae) (Boulanova, 1937), *Aeroglyphus robustus* (Banks) (Aeroglyphidae) (Barker 1967a) and *Dermatophagoides pteronyssinus* (Trouessart) (Pyroglyphidae) (Hart and Fain, 1988). Some acarid mites living on organic matter such as dung, compost, treeholes, bulbs, tubers, fungal fruiting bodies and subcortical spaces have a higher fecundity (300–1400 eggs per female). These species number *S. berlesei* (Michael) (Hughes, 1976; Timms *et al.*, 1981), *Sancassania anomalus* (Nesbitt) (Woodring, 1969b; Pillai and Winston, 1969; Barker, 1974), *Sancassania rodriguezi* Samšiňák (Rodriguez and Stepien, 1973), *R. robini* Claparédè (Gerson *et al.*, 1983), *R. echinopus* (Fumouze *et* Robin) (Woodring, 1969b), *H. arborsignis* Woodring and *H. rotundus* Woodring (Woodring, 1969b). Moreover, *Kuzinia laevis* (Dujardin) (Acaridae), living in bumblebee nests, exhibits a high fecundity (Chmielewski, 1969a).

The second demographic category numbers most species of the family Histiostomatidae, living on ephemeral resources, and characterized by shorter generation times, commonly ranging from 4 to 8 days, female oviposition period and lifespan are shorter (5–15 days), with a total egg production between 20 and 120 eggs per female. O'Connor (1994) reported that data are available for *Histiostoma heinemanni* Hill *et* Diahl (Hill and Deahl, 1978), *Histiostoma julorum* (Koch) (Hughes and Jackson, 1958) and *Histiostoma polypore* (Oudemans) (Behura, 1957), inhabiting decaying vegetation or fungal fruiting bodies; *Histiostoma cataglyphi* Yousef *et* Metwally (Yousef *et al.*, 1979), and *Histiostoma formosana* Phillipsen *et* Coppel (Phillipsen and Coppel, 1977b), from the nests of an ant and a termite, respectively; *H. laboratorium* Hughes (Perron, 1954) from *Drosophila* cultures, and *H. murchiei* (Oliver, 1962), a predator in annelid egg cocoons. The last species has a much higher fecundity (up to 500 eggs per female). O'Connor (1994) observed that North American species of the carrion, inhabiting the genera *Spinanoetus* and *Pelzneria* exhibit the same pattern, with a short generation time and female lifespan, and moderate egg production.

A third pattern occurs among species living in some types of insect nests and galleries. Data concerning species belonging to some genera of the family Winterschmidtiidae, such as *Afrocalvolia*, associated with bark beetle galleries, and *Ensliniella*, *Kennethiella* and *Vespacarus*, all living in the nests of solitary vespid wasps, confirm this aspect. In these species, the period between the original infestation by deutonymphs and the production of the next generation of deutonymphs or tritonymphs ranges from 20 to 30 days. The female oviposition period is short (4–7 days) and the egg production ranges from 25 to 125 eggs per female, depending on the number of foundresses per cell (Krombein, 1967; Cowan, 1984; Klompen *et al.*, 1987).

The last category includes species which have longer generation times (30–45 days), longer female reproductive periods (40–80 days) and commonly low to moderate fecundity. Among these species, *N. arboricola* (Fashing) (Acaridae) lives in water-filled treeholes which may persist for years. The life history of this species has been examined by Fashing (1975, 1979, 1994). Other examples of this pattern number species of Pyroglyphidae living in stored products and house dust, such as *D. farinae* Hughes, *Euroglyphus longior* (Trouessart) and *E. maynei* (Cooreman) (Hart and Fain, 1988). The natural habitats of most *Dermatophagoides* species are birds' nests. *Euroglyphus* species have not been collected regularly from natural habitats, but a related species, *Pyroglyphus morlani*, lives in the nests of wood rats (*Neotoma* spp.), many of which are perennial, and may be inhabited by a series of hosts for centuries. Fashing (1979) reported that with reference to *N. arboricola*, mites exhibiting this type of life-history pattern may be described as 'K-selected' (in comparison with other astigmatid mites). These species show shorter generation times and lifespans when compared with more common life-history pattern observed in the Sarcoptiformes, which have generation times of 1–3 years, and female reproductive periods measured in years. The few species of astigmatid mites parasitic on vertebrates studied exhibit different demographic parameters. *S. scabiei* (De Geer) (Sarcoptidae), living in the upper epidermal layers of mammalian skin, and *Psoroptes* spp. (Psoroptidae), living on the skin surface, have a generation time similar to the common pattern among free-living Astigmata (10–12 days) (Mellanby, 1972; Meleney, 1985; Arlian and Vyszenski-Moher, 1988). The normal temperature under which these mites develop corresponds to the host body temperature, and it is much higher than that used in laboratory studies of free-living mites. This led O'Connor (1994) to take into account two serious caveats in order to interpret the studies on astigmatid mite demographics.

The first remark regards the fact that these studies were carried out in laboratory under constant environmental conditions. Many of the studies cited above report variation in life-history parameters under different temperatures and humidity, but most of these studies do not duplicate the natural conditions in which the mites live. Studies carried out on several species of Acaridae show that the lifespans may be extended when they are reared under suboptimal conditions (Boczek, 1973; Boczek and Czajkowska, 1973; Boczek and Sosnowska, 1975). In laboratory, most species exhibit the reaction of poikilothermic animals, characterized by a decreasing developmental time as temperatures increase to the thermal death point. Response to humidity is also predictable in that species whose normal habitat is relatively dry such as *A. robustus* Banks and *G. domesticus* (De Geer) do better under drier conditions than nearer saturation. An opposite situation occurs in the Histiostomatidae and other taxa which commonly live in saturated environments. An appropriate food source is also important, and more success has been registered with relative generalists such as the Acaridae and Histiostomatidae than with specialists, a category which apparently includes most Glycyphagoidea and Hemisarcoptoidea. Sinha (1966a) has found that species of Glycyphagidae inhabiting stored food are much more particular about the fungal species upon which they will feed than are species of Acaridae in the same habitats. The difficulties in culturing nidicolous Glycyphagidae (*Baloghella melis* Mahunka, *Marsupialichus marsupialis* Fain, de Cock *et* Lukoschus) and Echimyopodidae (*Marsupiopus zyzomys* Lukoschus, Janssen Duijghuijsen *et* Fain) have been documented (Lukoschus *et al.*, 1971; Fain *et al.*, 1972; Lukoschus *et al.*, 1972, 1979). Similarly, despite the high level of taxonomic diversity in the family, there are few reports of species of Winterschmidtiidae being successfully cultured away from their natural substrates (O'Connor, 1994).

A second remark regards the interpretation of generation time. The life cycle of most astigmatid mites is similar to the ancestral pattern observed in acariform mites (Fashing, 1994), except that in most species, the formation of the deutonymph is facultative, or it is not formed at all. The demographic data discussed above is derived from laboratory studies, and the generation time has been calculated for individuals which bypassed the deutonymphal instar. This short life cycle due to facultative or permanent lack of an ontogenetic stage, has been called as a characteristic of 'r-selected' species. Some astigmatid mites overwinter as deutonymphs, and this behaviour may represent a general pattern, such as in the case of the Winterschmidtiid *A. nataliae* (Türk *et* Türk) which develops two generations per year in the galleries of its scolytid host, *Leperisinus fraxini* (Panzer) (Kielczewski and Seniczak, 1972). The summer generation develops in July and individuals bypass the deutonymphal stage. However, offspring of the summer generation enter the deutonymphal stage in August and overwinter on the adult bark beetle host until the following May, when they enter the feeding and reproductive phase of the life cycle. The generation time of this overwintering generation lasts 11 months, while the summer generation time lasts 1 month. The same behaviour has been observed among other species of the Winterschmidtiidae (subfamily Ensliniellinae), associated with the nests of solitary vespid wasps (O'Connor, 1994).

Temperature and Relative Humidity

Moshkin and Brygadyrenko (2022) reported that temperature and humidity are important environmental factors for poikilothermic organisms. Those parameters determine the rates of biochemical reactions inside cells and the metabolism of organisms in general (Ebssa *et al.*, 2004; Guschina and Harwood 2006; Waqas *et al.*, 2020; Ramachandran *et al.*, 2021). Temperature and humidity have effects on behaviour reactions, such as the search for food, reproduction and migration or dispersion (Buczek *et al.*, 2017; Requena-García *et al.*, 2017; Martynov *et al.*, 2019). Spread of species in the ecosystem and density of populations of many invertebrates are determined by those parameters (Brygadyrenko, 2015a, b; Aupic-Samain *et al.*, 2021; Rawhat *et al.*, 2021). Moreover, global climate changes make studies of ranges of temperature and humidity preferred by the species especially relevant (Avtaeva *et al.*, 2021; Brygadyrenko *et al.*, 2021; Makaida *et al.*, 2021).

Hughes T. (1943) reported that *A. siro* Linnaeus possesses a high resistance to carbon dioxide accumulation and to lack of oxygen. High MC in home-grown barley is sometimes stored in silos, and it was observed that sufficient carbon dioxide had been evolved to prevent the development of insect pests but not to inhibit the development of the flour mite, whose level of attack made the barley unpalatable to cattle (Hughes, 1976). Temperature is probably the abiotic factor that has been most studied in relation to mite population development and is the most important abiotic factor affecting this. High or low temperatures may have the same influence according to the mite species, its life cycle and season. Temperature acts on different biological stages, eggs, adults, etc. whose tolerances vary with species and life cycle. Different mite species also have different requirements for development, although there is generally no upper limit for humidity (Cunnington, 1969). *A. siro* has a greater rate of development in grains of higher MC and is less frequent in drier warmer climates (Cunnington, 1976). Under laboratory conditions, the lower and upper T limits for *A. siro* are around 2.5°C and 31°C respectively; with a lower RH limit of 62.5% (dependent on T) (Cunnington, 1965), although Aspaly *et al.* (2007) found an upper T limit of 38°C (at 85% RH). *L. destructor* (Schrank) is less tolerant of low humidities than *A. siro* and cannot complete development below 65% RH (dependent on T) (Cunnington, 1976). Its lower and upper T limits are around 3°C and 34°C, respectively (Cunnington 1976). *T. putrescentiae* (Schrank) is more tolerant to high temperatures and can survive and breed readily above 30°C but is less tolerant to low temperatures and cannot develop much below 10°C (Cunnington, 1969). As with the other storage mites, *T. putrescentiae* develops most rapidly at conditions of high RH, with a lower limit of 65% (dependent on T) (Cunnington, 1969). Guimarães (1970) observed on dried figs that when the figs are exposed to high humidities, yeasts grow rapidly and an important loss of sugar may occur. This loss has been detected from a low as 66% RH. Above 80% RH, there is spontaneous exudation of syrup. In Algarve (Portugal) the fig MC ranges from 17 to 25%. Once humidity equilibrium is reached, this corresponds to a range of 60–75% of atmospheric humidity. Complete development of *C. lactis* (Linnaeus) on dried figs is impossible if yeasts are not present. The positive influence on mite reproduction, when the brewer's yeasts were added to fig paste medium in laboratory mite cultures, confirms this aspect. Observations carried out in the laboratory by Guimarães (1970) at a temperature ranging from 12 to 30°C and RH from 70% to 90% showed that the developmental period is 12 days at 12°C and 71% RH and 5–7 days at 30°C and 90% RH. Temperature is important in the mite development, and atmospheric humidity plays a strong importance at lower values. In these temperatures and humidity, the intrinsic rate of increase is r = 1.261. The mite appears to be a species possessing a large thermal range for completing their development, since this is possible at temperatures above 30°C, while the lower threshold of development lies near 8°C. The RH of 67% represents the lowest value that allows the adult development. The development of adults is not possible below 65% of RH.

The optimum conditions for rapid development in the laboratory are 25°C and 90% RH for *A. siro* and *L. destructor*, and 30°C and 90% RH for *T. putrescentiae* (Cunnington, 1976). Due to the wide range of preferred temperatures, *A. siro* and *L. destructor* can be classified as eurythermic, whereas *T. putrescentiae* is stenothermic, and more thermophilous than the other species (Hubert *et al.*, 2010). The species *T. putrescentiae* obviously gravitates in its life strategy to an r-strategy. It is thermophilic, is first to appear in stored products, rapidly increases in numbers, quickly spreads over the entire substrate, tends to consume substrates rich in nutrients, because of peculiarities of its digestive enzymes, and is first to disappear (Akimov, 1985). In addition, this species has no hypopial stage and is oriented towards establishing itself in new substrates. Cunnington considers limits of T and RH within which *A. siro* is able to complete its development are fairly restricted and this explains why, although often loosely described as a cosmopolitan species, its occurrence and distribution are largely confined to countries with a cool, moist climate (Cunnington, 1965).

Mite species can form a hypopial stage, which does not feed, and can be resistant to unfavourable conditions such as desiccation (Griffiths, 1964a) and low temperatures. For instance, there is a report of their survival at temperatures of −18°C (Sinha, 1964c). Interestingly, humidity is the main factor in bringing about termination of the hypopial stage which increases as levels rise above 70% (Stratil and Knulle, 1985). Mite growth is affected by a combination of physical and biological factors that have been established under current climatic conditions. These factors interact in a complex manner (Sinha and Wallace, 1973), and knowledge of the factors important in regulating mite development and infestations can allow for the most appropriate control measures to be undertaken (Collins, 2012). However, global climate patterns have shown to change notably (IPCC, 2018) and grain storage itself can be heavily affected by climate change. The conclusion that seasonal variations would affect the mite population (Sinha, 1968b) indicates that climate change would also have similar effects (Moses *et al.*, 2015). Because of these concerns, there is a need to develop approaches aimed to understand the relative importance of current factors, as a baseline in shaping mite communities. One tool to do this is species distribution modelling (SDM) which employs suitability indices. Suitability indices describe the relationship between habitat suitability score and a given environmental variable of a target species. Habitat suitability is a way to predict the suitability of habitat at a certain location for a given species or group of species, based on their observed affinity for particular environmental conditions (Yi *et al.*, 2016; Ma and Sun, 2018).

Studies carried out by Barker (1967a) on the biology and reproductive rates of *A. robustus* (Banks) at two values of RH, and different temperatures showed that the life cycle is shorter at 70–75% RH than at 85–90% RH. At 28°C, more eggs are laid at 85–90% RH than at 70–75% RH. Moreover, the same author found the low net reproduction rates for the species.

Barker (1967b) studied the biology and reproductive rates of *T. putrescentiae* at different temperatures and high humidity. The most rapid development took place at 32.2°C, whereas 33.9°C was slightly less favourable. A normal life cycle was accomplished at 11.2°C. It was calculated that the major portions of populations of *T. putrescentiae*, with stable age distributions are the non-mobile stages: the eggs comprise over half of the population and one-third of the immature stages are quiescent. The adults account for close to 10% of the population. These theoretical data were confirmed by the examination of a young culture. It was concluded that sampling techniques should account for the immobile stages to give a true picture of a population of this species.

Woertz *et al.* (2002) studied the effects of RH, temperature, pH and vapour-phase toluene concentration on *T. putrescentiae* growth on *Cladophialophora* sp. in controlled environmental chambers. The mites reproduced readily at RH between 90% and 97%, as well as on porous perlite support material pre-soaked in nutrient media of pH 2.5, 4 and 7. The presence of toluene at gas-phase concentrations of 500–2000 mg m^{-3} was found to be non-toxic to the mites. The mites, however, were unable to maintain a large population when the temperature was maintained at 14°C, and overpopulation of the living space led to declines in mite population over time. Overall, it was found to be relatively simple, to rear mites that may be used for fungal biomass control measures in biofilter applications.

Aspaly *et al.* (2007) reported that the pest potential of stored-product mites depends on the reproduction rate, which is affected by the environmental conditions. They investigated the effect of temperature, ranging from 5°C to 35°C, on the population growth of three important mite species, *A. siro*, *T. putrescentiae* and *A. ovatus* (Troupeau) at 85% RH. Starting with ten individuals, the population increase of mites was observed after three weeks of cultivation, or after six weeks for those kept at low temperatures (5°C, 10°C, 12.5°C and 15°C). The rate of increase was calculated for each temperature and species. The data obtained were fitted with polynomial models. The mite population growth rates increased with increasing moderate temperatures until 25°C, when *rm*-values were 0.179, 0.177 and 0.190 for *A. siro*, *A. ovatus* and *T. putrescentiae*, respectively. The lower development threshold was 10.2°C in all three species. The estimated upper-temperature threshold was higher in *T. putrescentiae* (49°C) than in *A. siro* and *A. ovatus* (38°C). Simulation of the rate of population increase under ideal conditions, using real temperature records obtained from Czech grain stores, showed that the pest mite populations increase only during 3.5 months within a typical 9-month storage season in Central Europe. These results indicate that control of mites, be it chemical, physical or biological, is recommended during the months when allergens and pests are produced, that is from September to mid-November and in May (Aspaly *et al.*, 2007).

Sánchez-Ramos *et al.* (2007c) studied the developmental rate of immature stages and the reproduction of adults of *T. putrescentiae*, *T. neiswanderi* Johnston *et* Bruce and *A. farris* (Oudemans) at 70%, 80% and 90% RH and a constant temperature of 25°C. At 70% RH, *T. putrescentiae* and *A. farris* immature stages failed to reach the protonymph stage as 100% of the larvae died, whereas *T. neiswanderi* was able to complete development. The developmental time of all immature stages for the three species was significantly increased as RH was reduced. The mobile stages were particularly susceptible, as the time needed to complete their development at lower RH suffered greater increase than the egg stage. At 70% RH, *T. putrescentiae* and *A. farris* were not able to lay eggs and only 24% of *T. neiswanderi* pairs were fertile. The reproductive parameters of the three species at the RH at which they were able to lay eggs showed significant differences, except for the percentage of fertile mating at 80% and 90% RH. As RH increased, preoviposition period was reduced, and fecundity and daily fecundity was increased, whereas the oviposition period showed different patterns for the three species. The intrinsic rate of increase (r_m) of *T. neiswanderi* at 70% RH was negative, indicating that, under these conditions, the populations of this species will diminish until they disappear. As RH increased from 80% to 90% this parameter was almost twofold for both *Tyrophagus* species. The r_m obtained for *A. farris* at 90% RH was similar to that of *T. neiswanderi* at the same humidity, while at 80% RH, it was very small so that the population doubling time was more than 84 days. The influence of RH on biology of these mites and its practical application as control measure are discussed.

Hubert *et al.* (2013) studied the rate of population increase of *A. siro*, *L. destructor* and *T. putrescentiae* on various types of barley and at various combinations of temperature and humidity. The mites were added into the chambers and incubated for 21 days on seven different kinds of barley, coming from four sites, including six cultivars and a mixture. The population increase of all species was higher on the mixture than on any other cultivar, except for Sebastian and Calgary. The increase of mites was studied at constant temperatures ranging from 5°C to 35°C and RH ranging from 50% to 90%. A positive rate of increase was found above 70% RH for all species. The optimal humidity was at 85% RH for *A. siro* and *L. destructor* and at 90% RH for *T. putrescentiae*. As concerns the temperature, a positive rate of increase was found at temperatures higher than 10°C, 15°C and 20°C for *A. siro*, *L. destructor* and *T. putrescentiae*, respectively. The temperature optima were at 23°C, 25°C and 30°C for *A. siro*, *L. destructor* and *T. putrescentiae*, respectively. Model estimated on laboratory data was then fitted to temperature and humidity records from August to November in the Czech grain store. The estimated population rate of increase was rarely positive; for *A. siro* it was for 24%, for *L. destructor* for only 1%, and for *T. putrescentiae* for only 7% days of the study period. It is concluded that in the climatic conditions of the Czech Republic, the population increase of the three mite pests is negligible during autumn and winter.

Tytar and Oksentiuk (2019) made an attempt to highlight important variables shaping the current bioclimatic niche of a number of mite species associated with the infestation of stored products, by employing a species distribution modelling (SDM) approach. Using the ENVIREM data set of bioclimatic variables, the performance of the most robust models was mostly influenced by the indices based on potential evapotranspiration, which characterize the ambient energy and are mostly correlated with temperature variables, moisture regimes, and the strong fluctuations in temperature reflecting the severity of climate and/or extreme weather events. Although the considered mite species occupy man-made ecosystems, they remain more or less affected by the surrounding bioclimatic environment and, therefore, could be subjected to contemporary climate change. In this respect, investigations are needed to see how this will affect future management targets, concerning the safety of food storage (Tytar and Oksentiuk, 2019).

Food Resource

Exhaustively summarizing the knowledge on the relationship between the various species of mites infesting the stored products and the trophic resources is difficult, at least due to the complexity of the fauna that characterizes the Astigmatina. The literature is characterized by numerous reports on the bio-ecology of the most important species, but to understand the complexity of the subject, it is sufficient to consult Hughes (1976) and the rest of this book. As an example, we report the scientific testimony of Mullen and O'Connor (2019), which shows that the 'acarid mites infest a wide range of stored materials such as grains, milk products, dried fruits, straw, and animal hides, in both households and commercial storage facilities. They are also common contaminants of culture media in which insects and other invertebrates are reared; of bedding materials for mice, guinea pigs, hamsters, and other vertebrates; and in animal feed and animal-holding cages in pet stores and zoos. Their numbers can build rapidly, especially when the infested materials are damp enough to support the growth of fungi on which they typically feed'.

Mullen and O'Connor (2019) made reference to most important acarid mite associated with stored products being *A. siro* Linnaeus, a species characterized by a worldwide geographical distribution, and injurious especially to processed cereal products, such as flour, rather than whole grains or hay and commonly infesting cheese. The amount of damage caused to grains by this mite depends on the MC, and the germ is infested only when the water content is 14% or higher (Mullen and O'Connor, 2019). Another common acarid mite in stored products is the cosmopolitan *T. putrescentiae* (Schrank). Infestations of this mite is particularly a problem in foods with a high protein and fat content such as hams, cheeses, nuts, seeds, dried eggs and fish meal. This mite feeds primarily on fungi (*Aspergillus*, *Eurotium*, *Penicillium*) that tend to thrive on foods stored at warm temperatures (>30°C) and relatively high humidities (>85%). *T. putrescentiae* possesses a wide geographical distribution and has been collected in a wide range of situations, including grasslands, soil, old hay, mushrooms, and the nests of bees and ducks (Mullen and O'Connor, 2019). Another species is *Tyrolichus casei* Oudemans, a cosmopolitan mite reported from stored food, grain, flour, cheese, dog food, old honeycomb and insect collections (Hughes, 1976). *Suidasia nesbitti* Hughes is reported from Europe, Africa, North and South America, and the West Indies and, although it is associated with wheat pollards and bran in England (Hughes, 1976), it also has been collected in rice, whale meat infested with dermestid beetles, dried bird skins, and milking machinery. The above examples clearly show that each species is adapted to a specific exploitation of the resource, and that every technical decision must take this attitude into account.

Some reports highlight the importance of knowledge regarding the ability of some species to adapt to a specific exploitation of the trophic resource. Li *et al.* (2015) reported that *T. putrescentiae* is an alternative prey for the predatory mites used in biological control. In order to maximize the food supply and maintain the population of predators, it is very important to understand the effects of bran moisture content (BMC) and initial population density (IPD) on the mass production of *T. putrescentiae*. They evaluated the population increase of *T. putrescentiae* under six BMCs and five IPDs at optimal temperature and humidity with suitable photoperiod conditions and found that the population growth rates of *T. putrescentiae* were significantly higher with higher BMC. The population increased by 8, 32, 72, and 304 times within five weeks under different BMC which was 8.3%, 10%, 15% and 20%, respectively. Unfortunately, when BMC reached 25%, there was a great amount of mildew growing other than the mites. In the case of IPD, about 10,000 adult mites per 100 g bran resulted in the largest final population of *T. putrescentiae*. However, either lower or higher IPD led to a smaller final population. They concluded that, population growth rate decreased while the initial density was increasing, ranging from 1, 055, 601, 397, 266, and 109 times under initial density of ~5000, 10,000, 15,000, 20,000, and 25,000 adult mites per 100 g bran, respectively. Moreover, they also described a method in detail for the effective isolation and counting of mites from bran. They found that population growth grew to a maximum (over 50,000 mites/bran per gram) under BMC of 20%, IPD of ~50 adults/bran per gram, temperature of 28±1°C, 80 ± 5% RH and photoperiod of all darkness, which could be the best condition for mass production of *T. putrescentiae*. The results of this study provide basic biological information for the mass rearing of the mould mite, which will maintain a stable and controllable food source for the predatory mites.

Hughes (1976) reported that competition occurs between fungi and mites for the same substrate. The phenomenon is well known by the Acarologists that observed in their cultures, the development of moulds rather than mites. So, Solomon *et al.* (1964) showed that the xerophilic moulds *Sporendonema sebi* Fr. and *Eurotium restrictus* Smith on wheat germ flakes caused populations of *A. siro* to decline owing to reduce reproduction, high mortality and retarded development. Griffiths *et al.* (1959) observed that *A. siro* may feed and digest spores of *E. restrictus*, and it is possible that the metabolism of the fungus leads to the accumulation of substances in the substrate having a deleterious effect on the mites. In this sense, Solomon (see Hughes, 1976) found that resterilized food previously infected with *E. restrictus* was consumed by the mites with reluctance. Ždárková (1967) carried out in the years 1962–1965 an entomological investigation in 126 food processing factories and found that acarid mites are still a great problem of stored foodstuffs and food products. They mainly infest flour, grain and poppy seed, all kinds of imported and inland oilseeds, especially groundnuts, soyabeans, sunflower seed, cotton seed and protein poultry feeding mixtures. The most important pests are the species *A. siro*, *G. destructor* (Schrank), *T. putrescentiae* and *G. fusca* (Oudemans). Each of these predominant species was found on a different substrate; so, *A. siro* on protein mixtures, *G. destructor* on cereals, *T. putrescentiae* on oilseeds and *G. fusca* on flour.

The relationship between the development of *C. lactis* (Linnaeus) on dried figs and yeast has been showed by Guimarães (1970). Smrž *et al.* (2016) reported that mycophagy should not be considered a single and homogeneous category of nutritional biology, due to the specific symbiotic chitinolytic bacteria associated with mites and fungi. To test the interaction among mites, fungi and chitinolytic bacteria, they carried out experiments on the model species *T. putrescentiae*. *Mucor* sp, *Alternaria alternata* (Fr.) Keissl., *Penicillium claviforme* Bainier, *Penicillium griseofulvum* Dierckx and *Verticillium* sp. were plated onto malt agar and offered to *T. putrescentiae* in the laboratory. Mites were evaluated utilizing microanatomical examination based on histology, excrement analysis using fluorescence microscopy, bacterial plating, the impact of mite homogenate on fungi in Petri dishes, reproduction of mites feeding upon each fungus and isolation of associated bacteria inside mites. There were clear differences regarding the digested spores of different fungi passing through the gut and subsequently in the faeces. Abundances of bacterial cells in excrement also corresponded to the fungi offered. The extracts from mites had chitinolytic activity, and the plated bacteria are known to produce exochitinases. The various feeding patterns observed were caused by differences in the cell wall structures of the tested fungi. The study illustrates that mycophagy in saprophagous mites does not consist of a single pattern, but rather that it can be classified into several sub-patterns depending upon the digested fungal species and its parts. The results point to a nearly symbiotic relationship between chitinolytic bacteria and digested fungi in mycophagous microarthropods.

Puspitarini *et al.* (2021) investigated the effect of the physicochemical characteristics of stored products on *A. siro*. They studied the host preference and biology of the mite in three varieties of rice, two varieties of sticky rice and three types of cheese. They considered the differences in grain granulometry (whole, broken and flour), as variations in physical characteristics of the stored products, and proportions of basic nutritional content as variations in chemical characteristics. The results indicated that, when offered to grain with different granulometry, *A. siro* selected broken grain as their host and oviposition site. Furthermore, the mite's host preferences were also influenced by the nutritional content variation of the stored products.

The total development of *A. siro* lasted from 16.78 ± 0.13 to 20.60 ± 0.51 days, and its fecundity ranged from 17.65 ± 1.75 to 53.05 ± 0.09 eggs per female. The results suggested that faster development and higher fecundity of the mite were generally found in diet(s) with higher protein, fat and water content. *A. siro* is better adapted to flour products, indicating the influence of particle size on its development and reproduction. Moreover, the physicochemical characteristics of the stored products strongly affected the host selection and biology of *A. siro*, which, in turn, determines the degree of product susceptibility upon mite infestation.

Natural Enemies

The natural enemies of storage mites include pathogens and predators. Pathogens number different species. So, in the case of the grain mite, *A. siro* Linnaeus, the natural enemies include some Eurotiales of the family Trichocomaceae (*Aspergillus restrictus* G.Sm.) and Wallemiales of the family Wallemiaceae [*Wallemia sebi* (Fr.) Arx]. Pathogens of the poultry mite, *D. gallinae* (De Geer), include different entomopathogenic fungi, some nematodes and bacterial endosymbionts. Research works on their employ against *D. gallinae* in poultry houses are not advanced, and they are not yet commercially available for use against *D. gallinae*. Entomopathogenic fungi are frequently used worldwide, and mainly in Latin America to control pests in crops (Harrington *et al.*, 2011; Oliveira D.G.P *et al.*, 2014). Fungi infect insects and mites with their spores that adhere to the hosts' cuticle, germinate, and penetrate and spread into the hosts' body (Oliveira D.G.P *et al.*, 2014). Studies under lab-scale conditions have shown that *D. gallinae* is susceptible to *Beauveria bassiana* (Balsamo) Vuillemin, *Metarhizium anisopliae* (Metschn.) Sorokīn, *Trichoderma album* Preuss and *Isaria fumosorosea* Wize, previously *Paecilomyces fumosoroseus* (Wize) (Steenberg and Kilpinen, 2003; Kaoud, 2010; Tavassoli *et al.*, 2011; Immediato *et al.*, 2015; Tomer *et al.*, 2018; Oliveira D.G.P *et al.*, 2020). Unfortunately, some studies in semi-commercial conditions showed that the action of these fungi is unsatisfactory against *D. gallinae* (Steenberg *et al.*, 2006). The humidity levels in poultry houses are too low to ensure fungal transmission (Harrington *et al.*, 2011), and the effectiveness of these pathogens significantly depends on the fungal strain (Immediato *et al.*, 2015). Moreover, these fungi not being selective for *D. gallinae* only, they may negatively affect the environment, producing an environmental disequilibrium (Pritchard *et al.*, 2015). These means are harmless for poultry, eggs and humans (Oliveira D.G.P *et al.*, 2020), and they could be used within traps placed in the poultry house (Oliveira D.G.P *et al.*, 2014), for example, within an attract-and-kill strategy. Nascimento *et al.* (2020) showed the successful use of *B. bassiana* in autoinoculation devices (traps with corrugated cardboard or loofah sponge) to control *D. gallinae* in both laboratory conditions and poultry houses. This employment allows for the overcoming of environmental problems.

Recently, Park *et al.* (2021a) investigated the control of *D. gallinae* with the fungus *B. bassiana*. A total of five highly virulent *B. bassiana* isolates were selected from the first bioassay. They showed 70–100% mortality in 10 days which was higher than cypermethrin, showing 40% mortality. Considering their fungal virulence, conidial production and thermotolerance, isolate JEF-410 was established to be the best candidate for industrialization. JEF-410 showed a dose-dependent virulence against red mite nymphs. JEF-410 conidia from millet granule-based mass culture were applied to a commercial chicken farm, and it showed significant suppression of population increase under high humidity conditions. *B. bassiana* JEF-410 having higher control efficacy than cypermethrin, could be effectively used to control *D. gallinae* in the fungus-friendly humid chicken farms. This work solved the challenges of mass production of *B. bassiana* and management of the red mite in poultry farming under the right conditions of RH and temperature (Park *et al.*, 2021a).

Predatory species include various mites ascribed to the families Laelapidae, Cheyletidae, Ascidae, Blattisociidae, Melicharidae and Acarophenacidae. Members of the family Laelapidae are free-living predators in soil, as well as nidicolous commensal, facultative or obligate ectoparasites in nests and colonies of vertebrates and insects, while the possible endoparasitic association to marsupial mammals has not been fully investigated (Evans and Till, 1966; Radovsky, 1985; Strong and Halliday, 1994; Gettinger *et al.*, 2005; Lindquist *et al.*, 2009b; Faraji and Halliday 2009; Dowling and O'Connor, 2010). This family currently comprises nine subfamilies, 90 genera and more than 1300 described species of parasites of vertebrates and invertebrates, as well as free-living forms and arthropod symbionts (Lindquist, 2009b; Beaulieu *et al.*, 2011). Trach and Joharchi (2018) reported that the family Laelapidae comprises 146 genera and over 1500 species. Few species of this family are biocontrol agents (Kazemi, 2014). Ten species ascribed to seven genera (*Androlaelaps* Berlese, *Eulaelaps* Berlese, *Haemogamasus* Berlese, *Geolaelaps* Evans *et* Till, *Hypoaspisella* Bernhard, *Stratiolaelaps* Berlese and *Euandrolaelaps* Bregetova) are collected on stored food.

The Cheyletid mites adapted to predation occupy a wide variety of habitats including plants, soil litter, tree bark, stored products and colonies and galleries of insects or ephemeral substrates, requiring dispersal by phoresy on insects or vertebrates; they are mainly free-living predators that feed on various microarthropods, particularly on herbivorous, fungivorous and saprophagous on acarid mites (Žďárková, 1979). A few species, however, can be beneficial as biological control agents controlling acarid mites, which are serious pests damaging agricultural stored food (Gerson and Smiley, 1990; Bochkov and Fain, 2001; Fain and Bochkov, 2001). On stored food are reported about 38 species ascribed to 21 genera [*Acaropsellina* Summers (4 species), *Alliea* Yunker (1 species), *Bothrocheyla* Volgin (1 species), *Caudacheles* Gerson (1 species), *Chelachecaropsis* Attiah (1 species), *Chelacheles* Baker (3 species), *Cheletogenes* Oudemans (1 species), *Cheletomimus* Oudemans (1 species), *Cheletomorpha* Oudemans (1 species), *Cheletonella* Womersley (1 species), *Cheyletia* Haller (1 species), *Cheyletus* Latreille (8 species), *Cunliffella* Volgin (1 species), *Eucheyletia* Baker (2 species), *Hemicheyletia* Volgin (3 species), *Lepidocheyla* Volgin (1 species), *Mexecheles* De Leon (1 species), *Neoeucheyla* Radford (1 species), *Nodele* Muma (3 species), *Oudemansicheyla* Volgin (1 species) and *Zachvatkiniola* Volgin (1 species)]. Natural enemies of the grain mite, *A. siro* are fundamentally included in the family Cheyletidae (Solomon, 1962); known species are *Cheyletus eruditus* (Schrank), *Cheyletus aversor* Rudendorf, *Cheyletus trouessarti* Oudemans and *C. malaccensis* Oudemans (Ždárková, 1967, 1979; Lukás *et al.*, 2007). *Cheyletus* mites feed primarily on mites but also on the eggs of stored-product insects (Ždárková, 1998). Ždárková (1967) collected in food processing factories in Czechoslovakia, associated with acarid mites infesting stored foodstuffs and food products specimens of *C. malaccensis* and *C. trouessarti*. He reported that, among species of the family Cheyletidae commonly found in foodstuffs, the most common was *C. eruditus*. These *Cheyletus* mites prey primarily on mites but also on the eggs of stored-product insects (Ždárková, 1998).

Blattisociid species have been described from a wide range of countries and habitats. Moraes *et al.* (2016) provides information about the world distribution and habitats occupied by species of this family, which should be of interest to research works dealing with the taxonomy, biology and ecology of these groups, and even to applied biological control activities (Santos *et al.*, 2020b). On stored food, nine species are reported; of these, five belong to the genus *Blattisocius* Keegan, namely *Blattisocius dentriticus* (Berlese), *Blattisocius keegani* (Fox), *Blattisocius mali* Oudemans, *Blattisocius quadridentatus* Haines and *Blattisocius tarsalis* (Berlese) respectively, and four are ascribed to the genus *Lasioseius* Berlese, named *Lasioseius allii* Chant, *Lasioseius berlesei* (Oudemans 1938), *Lasioseius penicilliger* Berlese and *Lasioseius sugawarai* Ehara, respectively. The genus *Blattisocius* Keegan includes almost 15 mite species often found in storage facilities (Hughes, 1976; Haines, 1978; Gerson *et al.*, 2003; Britto *et al.*, 2012). A synthesis of the studies evaluating the possible use of *Blattisocius* species for the control of different insect and mite species was presented by Thomas *et al.* (2011).

Several species of Melicharids are predatory, others feed on fungi, pollen and nectar, and one species is known as parasitic on cockroaches (Egan and Hunter, 1975). A complete characterization of Melicharidae was given by Moraes *et al.* (2016). Eight Melicharid mites are reported on stored food. They are *Melichares agilis* Hering, *Proctolaelaps bickleyi* (Bram), *Proctolaelaps hystrix* (Vitzhum), *Proctolaelaps pomorum* (Oudemans), *Proctolaelaps pygmaeus* (Müller), *Proctolaelaps regalis* De Leon, *Proctolaelaps scolyti* Evans, *Proctolaelaps ventrianalis* Karg (Hagstrum *et al.*, 2013; Ebrahimi and Noei, 2022). *Melichares agilis* was also reported as *Typhlodromus domesticus* Oudemans and *Garmania domestica* (Oudemans) (Nesbitt, 1951) and in old literature was confused with *Blattisocius*.

Ascid mites live commonly in soil, litter, stored grains, stored food, on flowers or leaves, and under tree barks (Halliday, 1998; Gerson *et al.*, 2003; Lin, 2003; Moraes *et al.*, 2016), and have received a concrete interest for their possible employment in biological control of pest mites and insects (Moraes *et al.*, 2016). Lindquist *et al.* (2009b) reported that species of the family Ascidae are relatively unspecialized, free-living inhabitants of both soil and arboreally suspended litter and humus layers. Various species are free-living predators inhabiting upper soil layers, plants and stored products, where they feed on nematodes and small arthropods. A reduced number of species parasitise cockroaches and moths (Egan and Moss, 1969; Treat, 1975; Lindquist *et al.*, 2009b). Kinn (1983) reported that Ascid mites are often carried in high numbers by the flies and moths on whose eggs (and young larvae) they feed, and other species are dispersed by beetles. A wide number of pollen-feeding species live in tropical flowers and are dispersed by hummingbirds (Colwell and Naeem, 1994). Other Ascid mites feed on fungi (Ibrahim *et al.*, 1989), or exhibit omnivorous feeding habits. The Ascid mites reported from stored food are *Arctoseius butleri* (Hughes), *Arctoseius insularis* (Willmann), *Gamasellodes bicolor* (Berlese) and *Zerconopsis remiger* (Kramer).

Other predators belong to the Acarophenacidae, whose females are parasitoids of insect eggs and feed only on the eggs of insect hosts, without toxic effects on people or domestic animals. This adaptation suggests the consideration of these mites as potentially useful as biological control agents on stored products (Steinkraus and Cross, 1993).

Obviously, a rich biodiversity can be found in any habitat investigated. So, in Iraq, Al-Kubaisi and Al-Dulaimi (2020) found associated with poultry droppings, various species such as *Androlaelaps casalis*, *Sejus temperaticus*, *Parasitus paraconsanguineus*, *Eugamasus butleri* Hughes, *Macrocheles merdarius* (Berlese), *Macrocheles glaber* (Müller), *Macrocheles muscaedomesticae* (Scopoli), *Macrocheles matrius* (Hull), *Kleemannia plumosus* (Oudemans), *C. eruditus*, *C. malaccensis* and *Pyemotes herfsi* (Oudemans).

6 Collecting, Detecting, Preserving, Preparing and Rearing

Collecting and Detecting

Collection methods vary depending on the behaviour of species, and the needs of the collector (laboratory rearing, systematic study, bio-ecological observations, control, etc.). For identification purposes, it is desirable to collect the largest possible number of specimens, preferably from all biological stages, noting their colour, the damage symptoms caused and the host plant, as well as the date, location of collection, and name of the collector. Armitage and Chambers (2002) reported that stored mites are just visible to the naked eye and are seen as white moving specks or as a moving pink dust on the grain surface or in sieving. They may accumulate in traps at the surface or in bait bags, mite traps and floor traps in empty stores.

The collection and detection of terrestrial mites from substrates can be realized through 'hand collection', 'funnel extraction', 'flotation separation' and 'trapping' (Walter and Krantz, 2009).

The most common methods for mite detection in stored food include sieving with visual inspection, extraction with Tullgren-Berlese funnels, filth flotation and trapping (Krizkova-Kudlikova *et al.*, 2007). In the last years, other methods have been developed, such as the immunochemical methods (ELISA), the near-infrared detection (NIRS), the hyperspectral imaging system (HIS) and the electronic nose gas sensors (E-nose) (Ridgway and Chambers, 1996; Ridgway *et al.*, 1999; Krizkova-Kudlikova *et al.*, 2007; Krizkova-Kudlikova and Hubert, 2008; He *et al.*, 2020).

Rapid detection of contamination enables the interception of an early grain infestation by the pest, which is an important step in stored food protection. At this aim, Krizkova-Kudlikova *et al.* (2007) compared the usability and efficiency of various detection approaches. Under laboratory conditions, grain samples of various sizes were infested with different levels of eggs, adults and faeces of *Acarus siro*. The samples were analysed by enzyme-linked immunosorbent assay (ELISA) by using anti-*A. siro* polyclonal antibody (immunochemical method) and extracted using the conventional methods (Berlese-Tullgren funnels, sieved and processed by filth flotation). The adults or juveniles of *A. siro* could be detected by all three tested conventional methods and ELISA, with detection limits in the range 221–1157 mites/kg grain. Eggs were detected by filth flotation only; the detection limit was 1950 eggs/kg grain. The faeces of *A. siro* were detectable by ELISA test, only. ELISA enabled the detection of the faeces with the minimal threshold level of 1.04 μg faeces /g grain; it means the assay allowed to trace less than one metabolically active mite per gram of grain. The results of the study showed that reliable *A. siro* detection in grain can only be obtained by combining different detection methods. European Union and US administrations dictate zero or near-zero tolerance levels for mite infestation in stored products. This demand is difficult to fulfil because every detection method is limited by its detection limit; thus, it is hard to reliably detect infestation levels lower than obtained detection limits. This methodical limitation is discussed in context with the determined detection limits of the tested methods. The common purpose of the application of these tools is to make the monitoring of the mites present in the foodstuffs easy and reliable, and regardless of these limits, we discuss here briefly the general features of the tools available today.

Sieves

The sieves are tools able to sift food substrates and to separate them from any infesting mites. The dimensions of the meshes are related to the species to be collected. The sifted mites are deposited in containers, either with or without liquid attractants, and collected either manually or with the help of aspirators. There is the possibility of integrating its employ

(V. Vacante and C.H.W. Flechtmann)
DOI: 10.1079/9781786390912.0006

with a series of accessories (pans, rings, lid-sieving aids). Mites observed may or may not be counted for practical application and/or prepared for microscopic examination for identification.

Retsch (2022) reported that the sieves consist of a high stability frame (stainless steel sieve, etc.) and a mesh specific to stored mites; the frame diameter commonly ranges between 200–450 mm, and a mesh between 125–20 μm and up to 1 mm. Commonly, the fabric of each sieve is accurately joined into the frame and tautened. Individual laser engraving of each test sieve provides a clear labelling with full traceability.

Armitage *et al.* (1996) examined the removal of insects, mites and insecticide residues from grain passed through a grain cleaner, comprising a de-awner and aspiration chamber and a sequence of inclined, reciprocating sieves. The cleaner was evaluated for its ability to remove adult insects from 1 kg grain samples by comparing the removal of 30 and 200 insects of each of *Oryzaephilus surinamensis* (Linnaeus) (Coleoptera, Silvanidae), *Sitophilus granarius* (Linnaeus) (Coleoptera, Dryophthoridae) and *Cryptolestes ferrugineus* (Stephens) (Coleoptera, Laemophloeidae) using 1.0 and 2.0 mm sieves. Pest-free samples were reliably produced using a 2.0 mm sieve at the cost of a loss of about 4.5% in weight of dockage or cleanings removed. Cleaning produced a 90% reduction in live mite numbers, but populations partially recovered afterwards, although still reaching only one-fifth of the uncleaned controls. After cleaning, 90% live and dead *Glycyphagus destructor* (Schrank) were recovered from the cleanings but only 30% of *A. siro* Linnaeus were in the cleanings with a further 10% in the prime grain, leaving about 60% unaccounted for. Removal of about 1.6% dockage or cleanings resulted in a loss of approximately 20% of pesticide in half of the tests using etrimfos and pirimiphos-methyl emulsifiable concentrates and dust at 2 and 4 mg/kg (Armitage *et al.*, 1996)

Hand-Operated Vacuum Apparatus

Mites living on the soil surface can be collected using a hand-operated vacuum apparatus to suck up specimens living on plants (weeds), soil, rocks, wood, tree trunks, etc. It is usually preferable to employ mechanical suction, because it avoids any toxicity caused by biological and chemical agents and/or allergic reactions due to mites, fungi, etc. The specimens collected on the filter can either be observed directly with a hand lens, or dissecting microscope, or they can first be washed with water (or not) and poured onto a coloured mat and then observed (Vacante, 2016). The same procedure may be applied for the collection of the mites of stored food.

Tullgren-Berlese Funnels

The use of the Berlese-Tullgren funnel allows the collection of mites (also nematodes and other microarthropods) from the ground vegetation, and the extraction of living mites from soil and litter (Krantz, 1978; McSorely and Walter, 1991; Walter and Krantz, 2009; Upton and Mantle, 2010), as well as from stored food. The funnel extraction technique was developed by Antonio Berlese, and originally used copper coils heated by hot water or steam, to produce a temperature and moisture gradient in the substrate. The design was subsequently improved with the replacement of the hot water, with an overhead incandescent light bulb. Many modifications are reported in the literature (Walter and Krantz, 2009) which aimed to improve the apparatus as a quantitative tool. The modern version consists of a funnel which has at the top (larger side) a screened support, on which the substrate (soil, litter, vegetable matter, etc.) lies. The bottom (smaller) end is usually connected to a vial. The upper surface of the substrate is exposed to a heat source that produces a gradient of humidity that encourages the arthropods that are present to move below towards the vial, into which they fall and are collected either in a preservative mixture (70–80% ethanol with 1–3% glycerine, alcohol, etc.) or on a mixture of moistened plaster of Paris mixed with charcoal, on which they can be kept alive (Walter and Krantz, 2009). The use of alcohol may hamper specimen mounting and some of the media used, such as polyethylene glycol or glycerine/glacial acetic acid (Krantz, 1978), are not safe. The number of specimens collected depends on their activity and the dryness of the substrate, and the preservative mixture containing particles of soil, various detritus and/or arthropods, as well as the mites. The use of a layer of wide mesh cloth or fine screening, located under the original sample, can prevent such items from falling into the funnel.

Mite Traps

Mite monitoring is usually done by placing baited traps in facilities and inspecting the traps for mites.

Different types of traps may be used (Walter and Krantz, 2009). Dunn (2000) reviewed the literature to identify chemicals for testing as candidate attractants for mite pests of stored grain. Several trapping efficiency and improvement studies have been carried out in stored-product mites. Some compounds identified as 13 potential attractants for storage mites are fungal volatile 1-octen-3-ol and food volatile 2-nonanone (Dunn, 2000). These attractants were tested and reported to increase trap capture of *Lepidoglyphus destructor* (Schrank), *A. siro* Linnaeus and *Tyrophagus longior* (Gervais) in wheat or oilseed rape when incorporated in BT® traps and PC™ traps (Wakefield, 2006). However, these attractants have not been investigated for *Tyrophagus putrescentiae* (Schrank). The BT® trap is the first trap designed specifically for the detection and monitoring of storage mite populations, and infestations in food or feed storage and production premises. It has also been used for detecting mites in both laboratory and field studies (Thind and Ford, 2004; Wakefield, 2006). In stored grain and oilseed, the PC™ trap is effective in detecting storage mites and may provide an early indication of their presence when compared to traditional methods (Clarke, 2002; Dunn *et al.*, 2005). Despite these results, Thind (2005) reported that methods for trapping storage mites are either rudimentary or have several limitations. These prevent effective early detection or reliable monitoring of mites in the food, animal feed and associated industries. The BT Trap is made up of a mixed food base lure held in a robust adjustable housing. It is user-friendly, requires an exposure period of only 4 days and can be used in a dusty environment. The performance of the BT Trap was evaluated in the laboratory, and the mean minimal capture rate of mites was 62%. In a comparative field trial, this trap detected significantly more mites than the bait bag or the fishmeal trap. The trap's performance was also assessed in four different types of premises. These were premises engaged in the production of cereal based dry pet food, specialist animal feed, traditional cheese and in the finished cereal section of a large transit storage warehouse. A total of 17 genera or species of storage and predatory mites was detected in these four premises, with at least seven species detected in each of these premises. This study confirms that the BT Trap was able to function under the rigours of use in the field. It has demonstrated its potential as a useful tool for early detection and monitoring of mites in production premises. It also provided information that was vital for improving existing mite control measures, and in implementing effective management strategies (Thind, 2005).

Wakefield and Dunn (2005) studied these traps extensively to provide early warning of hidden pest infestations. However, they highlighted that there is only one type of trap on the market in the UK for storage mites, namely the BT mite trap, or monitor. Laboratory studies have shown that under the test conditions (20°C, 65% RH) the BT trap is effective at detecting mites for at least 10 days for *L. destructor*, *T. longior* and *A. siro*. Further, tests showed that all three species reached a trap at a distance of approximately 80 cm in a 24-hour period. In experiments using 100 mites of each species, and regardless of either temperature (15°C or 20°C) or relative humidity (65% or 80% RH), the most abundant species in the traps was *T. longior*, followed by *A. siro* then *L. destructor*. Trap catches were highest at 20°C and 65% RH. Temperature had a greater effect on mite numbers than humidity. Tests using different densities of each mite species showed that the number of *L. destructor* found in/on the trap was significantly reduced, when either of the other two species was dominant. It would appear that there is an interaction between *L. destructor* and the other two mite species, which affects the relative numbers found within the trap (Wakefield and Dunn, 2005).

Erban *et al.* (2009) (in Callaghan *et al.*, 2012) developed a method for bioassay of Btt for pests of stored grain, *T. putrescentiae*, *Dermatophagoides farinae* Hughes and *L. destructor* using toxins of Btt. They used a diet based feeding assay, where the commercially available Btt product Novodor FC (Valent BioSciences) was incorporated into a fish food-rearing diet. The diet was composed of dog food, wheat germ, dried fish food, pangamin and gelatine in ratios of 10:10:3:2:1 (w/w). Concentrations of Btt product ranged between 0.001–100 mg/g of diet, which was mixed, lyophilized and ground to a powder. The powder was rehydrated for 24 h before use and 50 mg was fed to 50 mites at 75–85% RH. The number of mites was assessed after 25 days, so the bioassay used population suppression as the measure of efficacy.

Amoah *et al.* (2016) reported that it is possible to use the BT® trap, the first trap designed specifically for the detection and monitoring of storage mite populations, and infestations in food or feed storage and production premises. The food-baited traps are fabricated from disposable petri dishes and a dog food based bait, and they were also used for detecting mites in both laboratory and field studies (Thind and Ford, 2004; Wakefield, 2006). In stored grain and oilseed, the PC™ trap is effective in detecting storage mites and may provide an early indication of their presence when compared to traditional methods (Clarke, 2002; Dunn *et al.*, 2005).

Amoah *et al.* (2016) used for capturing *T. putrescentiae* food-baited traps fabricated from disposable petri dishes and a dog food based bait. Mite activity over time and space was monitored in five dry-cured ham aging rooms from three commercial processing facilities that differed in their fumigation frequencies. Weekly sampling of the mite was conducted from June 2012 to September 2013. There were significant differences in the average weekly trap captures in all facilities, especially before and after fumigation, with the majority of mites in traps prior to fumigation. Mite numbers had a pattern of sharp decline after fumigation, followed by a steady increase until the next fumigation. Average trap captures varied due to trap location, over the study period at all study sites, indicating that traps could be used to identify specific locations within an aging room, where mite infestation of hams was more likely to occur. The findings can inform facility managers of mite population changes that can be used as one factor toward making pest management decisions and assessing the impact of fumigation or other pest mitigation actions.

Amoah *et al.* (2017) studied that the spatial and temporal dynamics of pest populations are an important aspect of effective pest management. However, absolute sampling of some pest populations such as the ham mite, *T. putrescentiae*, a serious pest of dry-cured ham, can be difficult. In this study, sampling methods were compared and sampling plans were developed for the ham mite. In one experiment, the number of food-baited traps was maintained at four and the number of mites varied from 500 to 10,000 individuals in environmentally controlled 18 m^2 rooms. The number of mites captured in traps increased linearly as the number of mites released increased. Experiments conducted in simulated ham aging rooms suggested that mites preferred the proximal (closer to the pig's body) to the distal side of the ham. The number of mites captured in traps also varied with trap location relative to hams hung from shelves. There were varying degrees of relationships between the number of mites captured in traps and the sampling methods used, such as counts of mites from the storage rack, vacuum sampling of the floor and ham sampling. Relationships between absolute counts of mites and the number of mites captured in traps or on specific areas of ham were both moderately strong and significant. Two sampling plans were developed for the mite, and it was determined that an economic threshold close to one mite per sample unit on the ham can be estimated, with 21 and 40 samples at precision levels of 0.35 and 0.25, respectively, or with 10 ham samples, when classifying the mite's infestation level as being above or below an action threshold of 0.43. These sampling plans for the ham mite can facilitate the development and evaluation of cost-effective integrated pest management (IPM) interventions (Amoah *et al.*, 2017).

The protocol of Fera, suggested by BPCA (British Pest Control Association) (2019), reports that this trap is specifically designed and developed for monitoring storage mite pests. It contains an attractant made from non-toxic ingredients, allowing its use in a diverse range of premises and environments. The trap detects live mites and other storage pests and provides an early indication of problem areas pinpointing hidden mite infestations and refuges. This gives the customer the opportunity to target remedial treatments and resources for best effect. Subsequent routine monitoring of premises for mites allows the effectiveness of control/hygiene measures to be assessed. Traps are supplied 'ready to use' with instructions from BPCA (British Pest Control Association) (2019).

Flotation Test

The flotation test was developed to identify contaminating matter of animal origin (e.g. whole or fragmented mites and insects, or rodent hairs) in a wide range of food, feed, veterinary and dust samples, and is accredited by AOAC (Association of Official Agricultural Chemists).

It is also possible to use other techniques, such as autosegregators, but flotation techniques seem to be better (Walter and Krantz, 2009). The technique of specific gravity flotation exploits both the specific gravity of mites with respect to soil particles and the marked affinity of their cuticles for petroleum derivatives (McSorley and Walter, 1991). These properties permit the separation of mites from their substrate, by mixing them with a concentrated sugar or salt solution (Hart and Fain, 1987; Pugh, 1996). The mites are collected in suspension and the various soil particles are deposited at the bottom of the container; the mites may then be collected using a Buchner funnel. Another flotation technique, 'hydrocarbon flotation', is based on the tendency of liquid hydrocarbons, such as kerosene, to stick to the arthropod cuticle (Walter *et al.*, 1987; Geurs *et al.*, 1991; Kethley, 1991).

Flotation test represents an accurate method to confirm the presence of mites in commodities and can easily differentiate between dead and live mites. The method can be used to assess mite infestations in a diverse array of products such as fine powders (flour, dried milk, yeast), granular materials (animal feed, wheatgerm), compacted and pelleted materials (animal feed pellets, dairy nuts, pet food), seeds and grain (wheat, barley, oilseed rape, maize, soya, rice), dried fruits, dust (house dust, mattress dust, floor sweepings, etc. for the detection of house-dust mites), plant material (hay, straw, wood or paper-based animal bedding), flower bulbs (tulips, hyacinths) and veterinary samples (e.g. animal dander).

Samples are suspended in a careful balance of aqueous alcohol, acids and salts before percolating kerosene upwards from the base. Any animal matter present is collected on a filter paper, allowing easy examination away from the bulk sample. Thind (2000) studied the determination of low levels of mite and insect contaminants in food and feedstuffs by a modified flotation method. He separated extraneous material from feed and food products by a modified technique, in which kerosene is used in a specially designed flotation flask. This technique, although effective for analysing feed and foods, presented limitations in the analysis of finely powdered materials. Some procedural modifications and an increase in the capacity of the flotation flask from 500 ml to 750 ml allowed a larger sample weight (20 g) to be analysed for mites, insect fragments and rodent hairs, with considerably reduced residue interference. In trials with a variety of products seeded with known numbers of mites, the average recovery was 83%. Recoveries of 89% were obtained from flour samples seeded with insect fragments and rodent hairs. A new process of suspending extracted mites in a mixture of industrial methylated spirit (46%) and glycerol (54%) by volume was used to allow rapid and more precise estimates of mite populations in heavily infested samples.

Immuno-diagnostic Method

Armitage and Chambers (2002) reported that HGCA-funded R&D is developing an immunoassay based pocket diagnostic kit to detect *A. siro* Linnaeus and other common species in processed cereal products. Dunn *et al.* (2002) reported that, to counter this mite, a method is needed by which mites can be detected at an early stage to help prevent the build-up of populations. Immuno-diagnostics have the potential to meet just such requirements as they are reliable, rapid, relatively cheap, user-friendly and can be developed into a kit form for use in the field. In a previous HGCA project, Dunn and collaborators started investigating the potential of immuno-diagnostics with the only available storage mite antibody. They showed that it could be used to detect the cosmopolitan food mite, *L. destructor* (Schrank), but not in the presence of grain. Therefore, the purpose of this study was to build on the experience gained in that project and to raise antibodies that were specific to a model species of mite, namely the flour mite, *A. siro*, but did not cross-react to wheat. This was successfully achieved by developing a new method for antigen extraction, and careful raising and screening of antibodies. The resultant working assay was shown to satisfy all the necessary criteria. It was sensitive to *A. siro* even in the presence of wheat. It was selective to *A. siro* with no cross-reactivity to other storage mites, not even to species such as *A. farris* (Oudemans) or *A. immobilis* Griffits of the same genera. Neither was there any response with any of the five storage insect pest species tested, or to the five species of storage fungi. In addition, the assay did not differentiate between different strains of *A. siro* , such as field and laboratory strains with differing degrees of pesticide resistance and susceptibility. All life stages of *A. siro*, were also detected as were dead adult mites and mite faeces, but there was no response to mite food (yeast and flour). These results suggest that the assay would be useful for the detection of contamination by mites or their products. To test the feasibility of further development of the assay, known numbers of mites were seeded in different quantities of unmilled wheat. The antigen was extracted using a novel rinsing method, which avoided the use of homogenization and its associated problems such as releasing proteins from the wheat itself. These tests showed that the assay could successfully quantify numbers of mites, ranging between 0 to 1000, in 0 g, 10 g, 25 g and 50 g of grain. To validate the immunoassay, a replicate set of samples was analysed, using a standardized flotation test for grain. The results from these two methods were comparable (linear regression of $r^2 = 0.9091$). The results of this work proved the potential of an immuno-diagnostic method as a tool for the rapid and sensitive detection of mites in grain samples and is a significant advance on previous immunoassays which have focused on detecting mites in dust only. According to these results, Dunn *et al.* (2002) proposed to develop this method further by the production of lateral flow devices which are user friendly, suitable as field kits, and have been fully validated by users in different premises. They also plan to extend the applicability of the kit to a wider range of mite species, and also construct a model for the distribution of mites in grain bulks in order to guide the use of the kits and to interpret their results. It is envisaged that the kit could be used throughout the food supply chain to establish the degree of contamination by mites.

Successively, Dunn *et al.* (2008) studied a rapid immuno-diagnostic test for the detection of storage mite infestations in cereals and cereal products. The study was articulated into three phases. The first phase, the proof of concept, concerned the production of a species-specific enzyme-linked immunoassay (ELISA) for the flour mite *A. siro*. The second phase, species- and genera-specific ELISAs, was developed for a range of mites injurious to stored food, such as *L. destructor*, *G. domesticus* (De Geer), *T. longior* (Gervais), mites of the genera *Tyrophagus* and *Glycyphagus* genera and all storage mites. All tests were demonstrably specific to target species or genera, with no cross-reactions observed to other storage pest contaminants or cereals. The validation phase concerned a comparative assessment of the species-specific *A. siro* and the genus-specific *Tyrophagus* ELISAs, with the flotation technique using laboratory and field samples. Both ELISAs were quantitative (0–30 mites per 10 g wheat) and produced good comparative data with the flotation technique (*A. siro* r^2=0.91, *Tyrophagus* spp. r^2=0.99) (Dunn *et al.*, 2008).

Near-Infrared Reflectance Spectroscopy

Ridgway and Chambers (1996) proposed first, followed some years later by Dowell *et al.* (1998) the detection of external and internal insect infestation in wheat, by near-infrared reflectance spectroscopy. Infrared spectroscopy (NIR) is used to detect the amount of water, protein and oil in the food industry (Williams and Norris, 2001). The method can use wavelengths between 700 nm to 2500 nm and detects the vestigial organs of insects in the food stuffs. Haemolymph-type fats and chitin of insects can affect the suitable wavelength. Since chitin is also in the fungal tissue, this method is used to detect fungal infections in stored agriculture crops too. Near-infrared reflectance spectroscopy (NIRS) method has been used also for detection of mites and various insect pests, including *A. siro* Linnaeus (Latifian and Rad, 2022).

Dowell *et al.* (1999) examined the possibility of near-infrared spectroscopy for taxonomic purposes, based on the premise that every species may have a unique chemical composition. They carried out tests with 11 species of beetles commonly associated with stored grain. Spectra from individual insects were collected by using a near-infrared diode array spectrometer. Calibrations were developed by using partial least squares analysis and neural networks. The neural networks calibration correctly identified > 99% of test insects as primary or secondary pests, and correctly identified > 95% of test insects to genus. The absorption characteristics of cuticular lipids may contribute to the classification of these species, highlighting that the technology could be used for rapid, automated identification of many other organisms (Dowell *et al.*, 1999).

Latifian and Rad (2022) confirmed that NIRS can detect different cuticular lipids of insect species. and according to Williams and Norris (2001), the compounds of the insect cuticle that can be identified are pigments, proteins, catalase and oxalates. There are many differences between the spectrophotometer wavelength absorption based on melanin and sclerotin quantity in the insect cuticle (Burks *et al.*, 2000). Therefore, the energy level and wavelength absorbed will be different depending on the cuticle chemical composition, and the population density of the different life cycle stages of each pest species, in the stored agricultural products (Ridgway *et al.*, 2001).

Hyperspectral Imaging System

Hyperspectral imaging system collects and processes information from across the electromagnetic spectrum (Chilton, 2013). The goal of hyperspectral imaging is to obtain the spectrum for each pixel in the image of a scene, with the purpose of finding objects, identifying materials or detecting processes (Chang, 2003; Grahn and Geladi, 2007). He *et al.* (2020) reported an automatic method for detecting mites in flour, established using hyperspectral imaging system (HSI) coupled with chemometrics. Reflectance differences among flour, *T. putrescentiae* (Schrank) and *Cheyletus eruditus* (Schrank), were relatively distinct. The majority of the shape features, including area, perimeter, the major and minor axis in *C. eruditus*, were remarkably larger than flour and *T. putrescentiae*. Textural features reached the level of maximum significance, except for contrast with one-way analysis of variance. Images under the ant colony optimization (ACO) wavebands, in random forests (RF) classification, showed at least 89% recognition rates, better than the successive projections algorithm wavebands. Artificial neural networks (ANN) with ACO wavebands gave higher recognition accuracies in training and validation sets than RF. Further analysis verified hyperspectrum with ACO-PCA-ANN (PCA, principal component analysis), which showed over 98% accuracy. This study revealed the promising potential of HSI coupled with ACO-PCA-ANN as an accurate and rapid method for detecting mites in flour. Mite infestation in flour is a global problem, and an important research area related to stored food safety and quality control in the food industry. An effective classification method for mite contaminants with automated detection has not been previously established in the literature. Therefore, the current study reports an automated hyperspectral imaging (HSI) system, coupled with chemometrics to detect mites *T. putrescentiae* and *C. eruditus* in flour. This study revealed the promising potential of HSI coupled with ACO-PCA-ANN as an accurate and rapid method for detecting mites *T. putrescentiae* and *C. eruditus* in flour. The knowledge gained in this research could be incorporated into the food storage and processing industry, to promote the accuracy of detecting and identifying mites, by using HSI system combined with chemometric algorithms (He *et al.*, 2020).

Electronic Nose

Ridgway *et al.* (1999) studied the detection of mite infestation in wheat by electronic nose (E-nose) with transient flow sampling. They scanned a total of 71 wheat samples of 70 g each, to which were added 0–70 flour mites (*A. siro*), 1–2 weeks after preparation, by a conducting polymer-type electronic nose with transient flow sampling. The device uses an array of non-selective sensors to produce a digital fingerprint (odour profile) relating to the volatile composition of the sample. The results demonstrate the ability of the device to detect the mites in wheat at concentrations of relevance to the cereal trade, of the order of just 1000 *A. siro* kg^{-1} (around 0.001% by weight). One week from preparation, discrimination between samples with no added mites and samples with 70 added mites achieved a classification accuracy of over 83%. Evidence is presented, which suggests the known mite-produced volatile undecane as the origin of the electronic nose response.

Recently, the advances in gas sensors and electronic nose technologies for agricultural cycle applications have been discussed by Seesaard *et al.* (2022).

Packaging Tests

The risk of infestation in packaged commodities is likely to vary according to several factors, such as the nature and constituents of the product. This aspect suggests examining the case, taking into account that some kinds of foods are more prone to infestations than others. Although mites commonly prefer cereal based products and also dried fruits and dried plants, some species such as *Tyrophagus* sp. also infest products with a high fat and protein content such as linseed, dried egg, groundnuts, cheese and ham. The form of packaging is thus important. Packaging is normally tailored to fit the product and designed to last throughout the storage life of the product. The integrity of the packaging and the materials employed are fundamental in preventing infestations. Mites may pass through perforated plastic material with perforations of less than 80 microns in diameter.

Monitoring and Threshold

In the UK, most of Europe and Australia, there is no tolerance for live insects in traded grain; however, thresholds for mites vary, and informal limits for many pet foods are below 10 mites per kilogramme. The Codex standard (CODEX STAN 199-1995) for wheat and durum wheat, intended for processing for human consumption, states that it should be 'free from abnormal flavours, odours, living insects and mites'. European Union and US administrations dictate zero or near-zero tolerance levels for mite infestation in stored products. This demand is difficult to fulfil, because every detection method is limited by its detection limit; thus, it is hard to reliably detect infestation levels lower than the detection limit. These legal constraints shift the concept of fight towards prevention, and every form of monitoring is used to evaluate the degree of hygienic-sanitary suitability of the stored products. In this sense, there is no threshold value. This restriction unequivocally characterizes each IPM strategy.

Monitoring of other mites, such as the poultry red mite, *Dermanyssus gallinae* (De Geer), an important parasite in the poultry industry, it is necessary and may be carried out with a trap. Monitoring and evaluation of interventions against this injurious mite have historically proven difficult. In 2012, AVIVET launched a project to develop a reliable, cheap way to monitor infestation of this mite in poultry houses. Currently, the AVIVET Red Mite Traps are used in most countries with large poultry production industries. Lammers *et al.* (2017) validated the AVIVET trap, a device to quantitatively monitor dynamics of *D. gallinae* infestations. They used the device to study *D. gallinae* in fully equipped cages with two white specific pathogen-free Leghorn laying hens, experimentally exposed to three different infestation levels of *D. gallinae* (low to high). The AVIVET trap was successfully able to detect *D. gallinae* at high (5000 *D. gallinae*), medium (2500 *D. gallinae*) and low (50 *D. gallinae*) levels of *D. gallinae* infestation. The linear equation Y= 10^-10^-(0.47 + 1.21X) with Y = log10 (total number of *D. gallinae* nymphs and adults) in the cage and X = log10 (total number of *D. gallinae* nymphs and adults) in the AVIVET trap explained 93.8% of the variation. The weight of *D. gallinae* in the AVIVET trap also appears to be a reliable parameter for quantifying *D. gallinae* infestation in a poultry house. The weight of *D. gallinae* in the AVIVET trap correlates 99.6% ($P < 0.000$) to the counted number of all stages of *D. gallinae* in the trap (i.e. eggs, larvae, nymphs, and adults), indicating that the trap is highly specific. From this experiment, it can be concluded that the AVIVET trap is promising as a quantitative tool for monitoring *D. gallinae* dynamics in a poultry house as the experimental validation of the AVIVET trap (Lammers *et al.*, 2017).

Preserving, preparing and rearing

The considerable morphological and structural diversity of the various groups of mites involved in the book does not allow us to present a common treatment for their preserving, preparing and rearing, in order to the fact that each systematic group requires a specific protocol. For this purpose, we invite the interested reader to consult specialized texts on the topic.

As concerns the techniques commonly employed for preserving, preparing and rearing mites, it is possible to consult Walter and Krantz (2009) and Vacante (2016). In these works are discussed methods for clearing/maceration, temporary and permanent mounting, and slide preparation including the protocols employed to study specimens with various tools and techniques – Transmission Electron Microscopy (TEM), Scanning Electron Microscopy (SEM) and Low-Temperature Scanning Electron Microscopy (LTSEM). Moreover, Molecular Methods, DNA extraction and rearing are discussed.

Identification of Storage Mites

The identification of storage mites is a task for a specialist using a low-power microscope. The morphological identification of mites collected on stored food is difficult and requires extensive systematic knowledge due to the simultaneous occurrence of several systematic groups. The identification of the most important group, the astigmatid mites, requires specialized knowledge of their morphological characteristics, and is made difficult by the similarity of their organs, especially in immature mites. This constraint requires specialist skills and complicates the monitoring of food mites in ordinary management. To overcome this limitation Yang *et al.* (2011) proposed the use of a biotechnological examination, based on the fact that the second ribosomal internal transcribed spacer (ITS2) and the mitochondrial cytochrome oxidase subunit I (COI) regions are highly conserved in the eukaryotes and are usually used as barcodes. The ITS2 and COI regions of six species of astigmatid mites – *Aleuroglyphus ovatus* (Troupeau), *Blomia tropicalis* Bronswijk, Cock *et* Oshima, *D. farinae* Hughes, *D. pteronyssinus* (Trouessart), *Euroglyphus maynéi* (Cooreman) and *T. putrescentiae* (Schrank) – were obtained by polymerase chain reaction and sequenced. The lengths of the ITS2 sequences varied from 316 bp to 488 bp, while the COI regions were 377 bp or 378 bp long. Considering the ITS2 genes, the intraspecific genetic distance was in the range of 0.00–0.077844, whereas the interspecific genetic distance was 0.202426–0.912959. The values were 0.000–0.029748 and 0.138403–0.279304 for intra- and interspecific genetic distances, when COI genes were used. The phylogenetic trees inferred from the ITS2 and the COI regions, by using maximum parsimony and neighbour-joining methods, were identical to those based on their morphological classification. Thus, the ITS2 and COI regions can be applied as barcodes to identify different species of astigmatid mites (Yang *et al.*, 2011).

Molecular diagnosis is highly valuable for the species identification of microscopic mites. Khaing *et al.* (2014) collected some economically important mites of the superfamily Acaroidea from various stored products in Korea. They determined by PCR, using universal primers, the nucleotide sequences of ribosomal second internal transcribed spacer (ITS2) and the mitochondrial cytochrome oxidase subunit I (COI) regions. In nucleotide sequence comparison at GenBank database, seven species including *Rhizoglyphus robini* Claparède, *R. echinopus* (Fumouze *et* Robin), *Sancassania* sp. *A. siro* Linnaeus, *T. putrescentiae* (Schrank), *T. similis* Volgin in Acaridae and *Suidasia medanensis* Oudemans in Suidasiidae were identified. COI sequences of *R. robini*, *R. echinopus*, *Sancassania* sp. and *T. similis* were first determined. The results suggested that the phylogenetic relationships inferred from the ITS2 region, rather than COI region, were similar to those derived based on their morphological classification. The study provides molecular information for the identification and phylogenetic relationship of acaroid mites in Korea (Khaing *et al.*, 2014).

7 Damage

Damage produced by storage mites to the human economy is direct and indirect. The first type consists of modifications of the standards qualitative and quantitative of stored food, associated with economic loss. The indirect damage concerns the role of these mites as vectors of micromycetes in stored products, their relationship with some bacteria, and their ability to cause allergies to humans and animals.

Direct Damage to Stored Food

Hubert (2011) reported that the synanthropic mites are classified into two artificial groups. The first group, called as house-dust mites, includes species of Pyroglyphidae and Echimyopodidae families, living in the house dust, and the second group, called stored-product mites, includes species of the families Acaridae, Glycyphagidae, Chortoglyphidae and Carpoglyphidae, inhabiting stored food and feed (Bronswijk and Sinha, 1971). However, sometimes, the stored-product mites are collected in house dust and the dust mites are found in the stored food. More than 20 species of mites have been found in households all around the world (Solarz, 2001a), and more than 100 species are associated with the stored product (Hughes, 1976). Clemmons and Douglas (2016) reported that more than 50 species of mites have been found in association with stored products, and among these, *Acarus siro* Linnaeus is one of the most common and well-known species (Campbell *et al.*, 2004; Henszel *et al.*, 2011).

Storage mites are common in stored food products, hay, straw, granaries, barns and other farming and occupational environments, in dust from railway freight wagons, as well as in samples of house dust. The most common species are *A. siro*, *A. farris* (Oudemans), *Tyrophagus longior* (Gervais) and *T. putrescentiae* (Schrank) from Acaridae, *Lepidoglyphus destructor* (Schrank) and *Glycyphagus domesticus* (De Geer) from Glycyphagidae and *Chortoglyphus arcuatus* (Troupeau) from Chortoglyphidae (Boström *et al.*, 1997; Franz *et al.*, 1997; Mehl, 1998; Sánchez-Ramos *et al.*, 2004, 2007c; Pike and Wickens, 2008; Wong *et al.*, 2011; Solarz, 2012). Clemmons and Douglas (2016) reported that among the 'stored-product pests' various mites of the families Acaridae and Glycyphagidae produce serious damage (Olsen *et al.*, 1995; Sallam, 2013), and the most important species are *T. putrescentiae* of the family Acaridae and *Lepidoglyphus destructor* and *G. domesticus* of the family Glycyphagidae (Olsen *et al.*, 1995). Grain mites infest a wide range of food products, including grain, flour, cereal, vegetables, cheese and animal feed (Campbell *et al.*, 2004), and their identification can be difficult; many mites described in the literature as *A. siro* might actually be of the species *A. farris* or *A. immobilis* Griffiths (Colloff, 1998; Webster *et al.*, 2004). However, *A. farris* and *A. immobilis* occur primarily in outdoor environments, and morphologic distinction from the primarily indoor species *A. siro* is possible (Webster *et al.*, 2004). *A. siro* mites are generally not visible to the naked eye, measuring approximately 0.5 mm long with almost colourless, oval bodies and brown legs (Campbell *et al.*, 2004; Solomon, 1962). Interestingly, they are described as having a pungent 'minty' smell, especially when crushed (Solomon, 1962). Estimates of the amounts of stored grain lost to insect damage are 5% to 10% in developed countries and as much as 35% in developing countries (Campbell *et al.*, 2004). In the developed world, contamination of food resulting in diminished aesthetic value is of greater concern than the direct loss of material due to insect consumption (Campbell *et al.*, 2004). Aside from aesthetic issues, stored-product pests have been described as harmless to humans and animals. These organisms can, however, induce allergic reactions, alter the taste of food and potentially serve as mechanical vectors for human pathogens (Campbell

DOI: 10.1079/9781786390912.0007

et al., 2004). Stored-product pests have been identified in laboratory animal facilities but, given that they are regarded as 'pseudoparasites', they are not typically the primary targets of pest management programmes or addressed during animal importation quarantine (Pennanen and Harju, 2003; Charles River Laboratories International, 2011). Even when research integrity and animal health are not likely affected negatively by infestations, the discovery of these pests on animals or food can have undesirable consequences. The untrained eye can easily confuse some species such as grain mites with rodent ectoparasites of concern. If mistaken for ectoparasites, costly additional diagnostics may be undertaken, and animals could be culled or treated unnecessarily (Charles River Laboratories International, 2011). In addition, there is some concern that these pseudoparasites may be fomites or vectors for adventitious pathogens (Hubert *et al.*, 2003; Baker, 2007). Furthermore, the presence of these pests in storage and housing areas can lead to food wastage and negative human health consequences such as allergic hypersensitivity (Campbell *et al.*, 2004; Sublett and Bernstein, 2011).

Hubert (2011) reviewed the pest importance of stored-product mites and suggested three types of pest risks. The first type concerns the direct consumption of human food, animal feed or other products, changing the quality of infested products; the second group concerns the interaction with microorganisms leading to the transfer of mycotoxins, producing fungi or pathogenic bacteria and the third group concerns the production of hazardous compounds, among them, the allergens are of the highest importance.

According to what above referred, stored mites are a common problem in foodstuffs. Their development at high density induces a loss of weight and quality of the stored product; furthermore, stored mites are responsible for spreading various fungal spores in a grain mass and adjacent containers, and when present in high numbers, flour or grain mites promote perspiration and impart an unpleasant odour to the grain.

These statements have not always been shared over time. In this regard Hughes (1976) wrote that heavy infestations of animal and human feeding material with *A. siro* are commonly encountered, and conflicting opinions have been expressed as to their possible harmful effects. So, Szwabowicz and Miedzobrodski (1957) and Szwabowicz *et al.* (1957, 1958a, b) fed chicks, ducks, pigs, sheep and horses with food contaminated with *A. siro* but could find no difference in weight and condition between these animals and the control ones; post-mortem examination also failed to show any pathological changes. Conversely, other specialists, such as Parrish (Brett, 1955, in Hughes, 1976) found that feeding for a long period of time with mite-infested food causes degeneration of liver, kidney, adrenal gland and testicle in cockerels, rabbits and rats. Moreover, it has been observed, one effect of the mites on the food, is the reduction of the available amounts of vitamins A and D. Studies carried out in the laboratory showed that rabbits, pigs and cockerels fed with clean cereals and dosed with many millions of mites were unaffected by them, whereas control animals fed on mite-contaminated food showed vitamin A and D deficiencies.

Mites of stored products infest commonly, products containing large amounts of starch (a compound of carbohydrates in plants) or glycogen (which is the compound of storage of carbohydrates in fungi and bacteria) (Hubert *et al.*, 2014a; Ball and Morell, 2003). In grains they use starch as the main power source, degrading it into glucose by the combined action of α-amylases and α-glucosidase (Erban *et al.*, 2009). Some studies have shown that the enzymatic activity of α-amylase and α-glucosidase is present in distinct species of storage mites, and that they are able to utilize starch, amylopectin and α-glycosides to obtain energy from grains (Morgan and Arlian, 2006; Erban *et al.*, 2009; Hubert *et al.*, 2014a). For having the ability to feed on starch, mites of stored products can destroy the germ of grains, decreasing their quality (Palyvos *et al.*, 2008). Not only the sanitary quality but also the nutritional composition of grains is compromised, especially in relation to B vitamins and iron (Krantz, 1955). Besides the sanitary and nutritional loss, mites may cause loss in quantity (weight) in some cereals. When a high infestation by mites occurs in wheat, they attack the endosperm and germ of the grain, and they feed on starch, reducing the weight of the grain. Another damage in stored products caused by the activity of these mites is the quite characteristic odour observed in storage units, which is caused by the secretion of lipids by mites (Palyvos *et al.*, 2008). In agriculture, they also have importance because they attack the embryo of some grains, causing a significant reduction in the stock of seeds and their germination power (Sinha, 1979).

Moreover, storage mites are strongly allergenic, and there is growing evidence that mites and their by-products cause an increase in allergic responses in human and companion animals and can contribute to poor performance in livestock. There are also reports of anaphylaxis and anaphylactoid reactions due to ingestion of mites from mite-contaminated food. We discuss these aspects briefly here.

Solomon (1946) first studied the methods of assessing the damage of stored mites. It is confirmed that mites cannot penetrate grains if the grain-coat is intact; but usually less than 10% of the grains are so intact. He observed in laboratory experiments that mites consumed up to 3% by weight of the grain before dying out; they were able to destroy the germ completely, but consumed very little of the remainder; the maximum rate of consumption observed over a 5-week period was 0–29% of grain weight per week; when added to grain at 75% RH in the proportion of 1% by weight, mites consumed 1% of the grain in 27 days at 21°C, or in 37 days at 10–15°C. Although mites consumed the grain more rapidly at higher temperatures or humidities, they survived longer at lower temperatures or humidities, and finally did more damage. Mites consumed wheat germ flakes more rapidly than grain, and finally reduced its weight by 74–85%.

Solomon (1946) presented a system of symbols to represent degrees of damage to grain by mites, assessed visually, and gave approximate values for the equivalent weight losses. The relationship between the product, number of mites per 100 ex. grain × time in weeks, and the degree of damage to grain, was calculated from experimental data; the damage done per unit of this density × time product was 2–2.5 times higher at 25°C than at 5–9°C, and about 1–4 times higher than 70°C at 90% RH. The mean density × time value corresponding to 1% loss of grain weight was 526,000 at 25°C, and 1,765,000 at 5–9°C.

Data from granary investigations show the degree of damage associated with various levels of population density and grain moisture. In infested storages, the amount of embryo material destroyed ranged from less than 1% of the total (bulk grain) to about 15% (grain in bags). Mites may cause a rise in the temperature or moisture content of grain, or an increase in the growth of microorganisms; but there is no evidence to suggest that serious damage from these causes is likely, except that dense infestations can cause heating (Solomon, 1946).

Also, Jacobs and Calvin (1988) reported that grain mites primarily attack the germ. However, they will feed on other parts of the kernel, as well as mould growing on the grain. The stored mites are responsible for the spread of various fungal spores throughout a grain mass and into adjoining bins. When present in large numbers, the flour or grain mites promote sweating and impart a disagreeable odour to the grain. Grain mites can cause 'grocer's itch' in humans exposed to the mites. Some persons may be allergic to mites.

Thind and Clarke (2001) examined food retail outlets in the UK for the presence of mites, by analysing 20 g samples, using a flotation method. Mites were found in 21% of 571 samples, which were examined soon after purchase, and in 38% of 421 samples, derived from the 571 samples which were examined after 6 weeks of storage in volunteer' homes. Most of the samples where mites were detected had fewer than five mites. However, a few samples contained more than 20 mites with a maximum of 428 mites detected in a single sample. Sixteen families, genera or species of mites were recovered. The most common species were *A. siro*, *T. putrescentiae*, *L. destructor* and *G. domesticus*. The level of contamination was broadly similar for each of the seven categories of products examined. It is likely that infestation occurs at every stage of food processing and storage. This was the first study of its kind, and there is no reason to believe that results from similar studies carried out in other temperate countries would be markedly different. An increase in the percentage of contaminated samples following storage in a domestic situation suggests that the domestic environment is an important factor in developing infestation.

In Brazil, de Sousa *et al.* (2005) studied the mite fauna assessed in bean, maize and pet food in supermarkets and markets. A 500 g sample of each grain and food was collected in each establishment at two weekly intervals. The sample was divided into two 250 g sub-samples. One of them was submitted to mite evaluation shortly after collection, and the other was incubated at 26°C and 78% RH for 30 days. At the end of this period, the second sub-sample was processed in the same way as the first. 11,956 mites were collected belonging to the Acaridae, Ebertiidae, Glycyphagidae, Cheyletidae, Stigmaeidae, Pyemotidae, Tarsonemidae, Tydeidae, Cunaxidae, Ameroseiidae, Ascidae e Phytoseiidae families. Among the primary mites, *Suidasia medanensis* Oudemans specimens were predominant in the survey (3035 specimens) followed by *Caloglyphus hughesi* (Samsinak) (436 specimens).

Metapronematus sp. was predominant (3417 specimens) among the secondary mites, followed by *Tarsonemus granarius* Lindquist (3003). The number of mites collected in the incubated samples was always greater compared to the non-incubated samples, in both the market and supermarket samples (de Sousa *et al.*, 2005).

Hubert (2011) reported that, the pest importance of mites was suggested as similar to stored-product insects – their feeding directly reduced the weight of stored commodity. However, the mites are of small size and weight. In laboratory experiments, the weight losses of the infested commodities have been observed, but the mites destroyed 3% of the weight of the grain (Solomon, 1946). Žďárková and Reska (1976) studied the weight losses of groundnuts (*Arachis hypogaea* Linnaeus) after infestation by *A. siro* and *T. putrescentiae*. After 3 months, the population increased from 20 mites to 16,500 and 1000 individuals for *A. siro* and *T. putrescentiae*, respectively (Žďárková and Reska, 1976). The estimation of consumption was 12, 3 and 0.2 μg or 6, 0.4 and 6.5 μg per *A. siro* or *T. putrescentiae* after 21, 30 and 90 days, after beginning of experiment, respectively. The fresh mass was 9 μg and 6 μg for adults of *A. siro* and *T. putrescentiae*, respectively. It means that the risk level of a population of mites about 1000 individuals per kg of grain has fresh mass of 6–9 mg. Hypothetically, if the mites feed the same mass of food as their weight, such a population can consume 9 mg per day, and the losses are negligible. In spite of that, the mite feeding on the grain is selective and can reduce the nutrient aspects. The interesting finding is the reduction of fatty acid values in *L. destructor* infested grain (White *et al.*, 1979). Such changes in nutritive quality of the diet are suggested as an explanation of deleterious effect of mite-contaminated feed for pigs. The pigs feeding on *A. siro* contaminated feed showed a lower weight increase than control pigs (Braude *et al.*, 1980). The effect of lower weight increases of pig feeding on *A. siro* contaminated feed was confirmed by Žďárková and Horak (1993). However, the pigs showed more histopathological features, which can be considered a reaction to mite allergens. The mites destroyed the germs, and only a little of the endosperm in wheat (Solomon, 1946). White *et al.* (1979) observed a decrease in wheat seed germination. After 26 weeks, the germination decreases up to 68% and 38% in control and *L. destructor* infested grain, respectively. Žďárková (1996) confirmed in laboratory studies that *A. siro* and *T. putrescentiae* decreased germination. After 6 months, the decrease was from 20% to 70% for cereals (maize, oats, wheat) and from 4% to 100% for vegetables (beet, carrot, kohlrabi, lettuce, raids and spinach). The decrease in germination can be important for malting barley when it is necessary. However, if the malting barley is stored in bins, the mites inhabited the surface of the grain (Armitage *et al.*, 2002), but the mass of barley is low and the effect on germination is not high. The results showed that the mites influence the weight and quality of stored product, and the effect depends on their density and the length of infestation. If it was evaluated separately, the effect can be neglected, because the losses caused by mite contamination are quite higher, and infested commodities are unsuitable for consumption due to contamination by hazardous compounds.

In the Czech Republic, Stejskal *et al.* (2014) reviewed the historical and current pest risks and research concerning seed storage. He reported that stored seed pests, causing injuries to the germ and endosperm, are a high risk of economic damage, due to the high value of seeds coupled with long-term seed storage in small storage units (e.g. boxes, satchels). Among these, rodents represent a significant risk to all types of seeds, especially seeds stored in piles or bags. Mites, psocids and moths are the main pests of stored grass and vegetable seeds; mites can decrease seed germinability by 52% and psocids caused 9.7% seed weight loss in broken wheat kernels after 3 months of infestation under laboratory conditions.

Nayak (2017) reported that both *T. putrescentiae* and *A. siro* are responsible for direct damage, by eating the germ of the grain and spreading fungi in the commodities. Heavily infested commodity becomes tainted with off-flavours, caused by squashed mite bodies when it is moved or processed. High infestations of *T. putrescentiae* are characterized by the dead and live mites settling visibly, as a brownish dust layer on stacks of commodities and floors of stores and emitting a lemon-flavoured odour, leading to the popular name 'Lemon-scented mite'. They can cause discomfort to workers handling infested grain and packaged commodities. The straw itch mite, *Pyemotes tritici* (Lagrèze-Fossat *et* Montagné) can cause severe dermatitis and allergic conditions in people handling infested commodities. Infested feedlots are also reported to have caused feeding problems and allergic conditions in stock. Obviously, the markets refuse infested commodities.

Armitage and Chambers (2002) reported that in the UK at least 12 mite species infest stored grain. The most common species belong to the genera *Acarus*, *Tyrophagus* and *Lepidoglyphus*. Mites can cause direct damage by eating out the germ and can taint produce. The most serious effects potentially result from allergens that mites contain or produce. Mites may also transmit fungal spores and act as vectors of human pathogens. Living and dead mites have been found in foods ranging from biscuits to baby food. Such contamination is causing increasing concern. Mites can survive inside grain, and populations build up quickly. Since mites respire through their outer surface, they are very vulnerable to moisture loss. Free-living mites are usually crushed during conveying. They can be removed physically by grain cleaners, but these do not remove all allergens. The BPCA (British Pest Control Association) (2019) reported that the damage caused by a mite infestation can lead to direct financial losses through deterioration of food quality, downgrading of products, end-user complaints and rejection of stock. In addition, storage mites are strongly allergenic, and there is growing evidence that mites and their by-products cause an increase in allergic responses in human and companion animals and can contribute to poor performance in livestock. There are also reports of anaphylaxis and anaphylactoid reactions due to ingestion of mites from mite-contaminated food.

Franzolin *et al.* (1999) reported that *T. putrescentiae* is prevalent in hot and humid climates. It causes serious economic damage, as well as the reduction of the nutritional content, and of the germination capacity of the seeds (Krantz, 1955; Žďárková, 1991a). It has been collected in a wide variety of foods of high protein and lipid content (Hughes, 1976), such as cheese, bacon, egg powder and groundnut flour (Sánchez-Ramos and Castañera, 2000; Aygun *et al.*, 2007; Jeon *et al.*, 2009). In Spain, it has a serious economic importance, because it feeds on the superficial layer of the typical Spanish cured ham, where the humidity and the heat are essential for the maturation process of the product (Sánchez-Ramos and Castañera, 2000; García, 2004). In Brazil, there was reported the presence of *T. putrescentiae* in soybean crops in Minas Gerais (Oliveira C.M. *et al.*, 2007) and samples of rice and beans sold in bulk in supermarkets in Sao Paulo (Franzolin and Baggio, 2000).

In addition to the damage discussed above, the stored mites change the quality of infested food by the production of secretions and faeces. High levels of infestation change the smell of stored products. Mite-specific volatiles decane, undecane, tridecane and perilen are reported by Curtis *et al.* (1981). Tuma *et al.* (1990) detected tridecane in bin-stored wheat, infested by stored-product mites. Ridgway *et al.* (1999) found that *A. siro* produced alkaneundecane, but the confirmation of gas chromatography-mass spectrometry (GC-MS) is missing. The pheromones are produced mainly by opistosomal glands (Kuwahara, 2004). Among them Citral, an alarm pheromone has been detected in mite-infested stored grain (Tuma *et al.*, 1990). The guanine is an ammonium waste deposited in the faeces (Levinson *et al.*, 1991). All the produced compounds probably change the smell of stored products. Although the compounds play an important role in mite communication, the question is still open whether their concentrations in stored products lead to damage.

Mites as Vectors of Micromycetes in Stored Products

O'Connor (1979) suggested the hypothesis that the ancestors of stored-product mites were free living in the soil, and fungivory was their prevailing feeding strategy. The vectoring potential of stored-product mites is given by their feeding association to fungi and bacteria. Stored-product mites occurred associated with micromycetes, because of their common requirements of moderate temperature and high RH (Sinha 1964a, 1979a). During the feeding, the fungal spores or bacteria are stuck to the bodies and are moved with the mites. The second possible way is that the fungal spores are distributed via faeces, if are not digested.

Grain stores which harbour stored-product mites provide a suitable habitat for many species of fungi, and Machacek *et al.* (in Sinha, 1964a) have isolated over 100 species of fungi from stored grain. Griffiths *et al.* (1959) found that *A. siro* Linnaeus and *T. putrescentiae* (Schrank) can carry spores in the digestive tract, as well as on the outside of the body; these spores are shed with the faeces and distribute the fungi from one stack of grain to another. Moreover, the mites were selective in their feeding, and when given a free choice, they sought out the preferred fungi and digested a greater percentage of spores of the preferred group than they did of the non-preferred fungi.

Sinha (1966a, 1968c) and Sinha and Mills (1968) reared a number of species of stored-product mites on potato-sucrose-agar plates of fungi and Actinomycetes obtained from grain. They found that *T. putrescentiae*, *Tyrolichus casei* Oudemans, *A. siro* and *A. gracilis* Hughes would feed and multiply on a wide variety of fungal cultures, while other species, such as *Suidasia* sp. and *Glycyphagus* sp. would only flourish on a few species. The presence of moulds on mite-infested food may increase the rate of reproduction of the mites (Sinha and Mills, 1968).

Sinha (1968c) reported that closely related species of insects, and closely related species of mites occur sympatrically in granaries and also survive on characteristic groups of fungal species. Taking into account this aspect, he carried out studies with 16 species of arthropods associated with stored products, to evaluate the possible adaptive significance of feeding and survival on 24 species of seed-borne fungi, with which the arthropods are naturally associated in granaries. Three pairs of species of insects and five pairs of species of mites, the members of each pair being related, were compared according to the nutritional adequacy of fungus species of different genera (*Absidia* Tiegh., *Alternaria* Nees, *Aspergillus* Micheli, *Chaetomium* Sacc., *Curvularia* Boedijn, *Fusarium* Link, *Helminthosporium* Link, *Hormodendrum* Bonord., *Mucor* Fresen., *Nigrospora* Zimm., *Penicillium* Link, *Rhizopus* Ehrenb., *Scopulariopsis* Bainier, *Stemphylium* Wallr., *Streptomyces* Waksman et Henrici (Bacteria, Streptomycetaceae), *Trichoderma* Pers. and *Trichothecium* Link). Sinha (1968c) used the number of fungal species on which both species of arthropod survived, as the index of relationship. Assuming that the ancestors of these arthropods were polyphagous-mycophagous, he postulated that the more specialized the species, the smaller is the number of fungal species upon which it can survive. The arthropod species listed first in each pair is the more specialized, and indicated the number of fungal species on which both members of the pair survived: *Tribolium confusum* du Val, *Tribolium castaneum* Herbst (Coleoptera, Tenebrionidae), 5; *Oryzaephilus surinamensis* (Linnaeus), *Oryzaephilus mercator* (Fauvel) (Coleoptera, Silvanidae), 10; *Cryptolestes turcicus* (Grouvelle), *C. ferrugineus* (Stephens) (Coleoptera, Laemophloeidae), 5; *G. domesticus* (De Geer), *G. destructor* (Schrank) (Glycyphagidae), 3; *A. siro*, *A. gracilis* (Acaridae), 14; *Caloglyphus rhizoglyphoides* (Zakhvatkin), *Caloglyphus redikorzevi* (Zakhvatkin) (Acaridae), 3; *Suidasia nesbitti* Hughes, *S. medanensis* Oudemans (Suidasiidae), 5; *T. putrescentiae*, *Tyrophagus casei* (Acaridae), 20. It is suggested that interspecific and intergeneric competition, with the possibility of extinction of the weaker species, are avoided, at least partially through separation of ecological niches by selective mycophagy (Sinha, 1968c).

The fungivory can be characterized by the digestive capability of mites given by the activity of digestive enzymes (Siepel and de RuiterDijkman, 1993). The enzymes are produced by mesothelial cells, when high secretion activity is visible on microanatomical level (Smrz and Catska, 1989). The debris of fungal cells are visible in the food bolus.

Hubert (2011) reported that the activity of digestive tract confirms also fungivory and digestion of bacteria. However, there are differences in the activity of mesenteral cells among fungal and mite species. *T. putrescentiae* is able to reproduce on several fungal species as *A. siro* and *L. destructor* (Schrank), and the growth rate of *T. putrescentiae* is higher on fungus than *A. siro* and *L. destructor*. Moreover, the growth rate was higher on the rearing diets than on fungi (Nesvorna *et al.* 2012). This adaptation confirms that the astigmatid species, commonly considered to be fungivorous, highlights sometimes higher preferences for wheat germ and yeast (Sinha 1964a, 1984; Pankiewicz-Nowicka *et al.* 1984, 1986; Parkinson *et al.*, 1991a, b). The fecundity and population growth of the fungivorous *A. siro* was higher on wheat germs and yeast (Bowman, 1984; Parkinson *et al.*, 1991a). The population development of mites is influenced by the fungal species according to the fact that the fungal species are differently palatable and suitable for their development (Sinha 1964a, Pankiewicz-Nowicka *et al.*, 1984, 1986; Parkinson *et al.*, 1991a). The phenomenon is confirmed by rearing studies that highlighted differences in attractiveness among microfungal species (Daneshvar and Rodriguez, 1979; Smrž and Catska, 1987). Some micromycetes are a food source for mites, while others are less suitable for their development (Griffiths *et al.*, 1959; Solomon, 1964a; Rodriguez *et al.*, 1984; Pankiewicz-Nowicka *et al.*, 1986, Smrž and Čatská, 1987).

Levinson *et al.* (1991) observed that in *A. siro*, the gaseous ammonia, a dietary degradation product of fungi, act as a kairomone, and that the guanine, a non-volatile excrement, acts as an assembly pheromone. The fungi produce gaseous ammonium as a dietary degradation product, which attracts mites. The mites migrate into fungus and start to graze, and during grazing produce the guanine as their waste metabolite. The guanine attracts the other mites to the grazed fungus, but the fungal attractiveness cannot correlate to the suitability. Studies carried out on the feeding and development of *A. siro*, *L. destructor* and *T. putrescentiae* showed that the mite food preference is not correlated with fungal suitability for development (Hubert, 2011). The suitability of fungi for development of mites depends on the mites' ability to digest the fungal nutrients and overcome fungal toxins, as well as the fungal nutrient richness. Hubert *et al.* (2004b, 2004c) defined four fungal groups based on mite preferences, fungal attractiveness and fungal suitability for the mite development. The first group includes the 'species preferred and suitable' for growth. The second group numbers the 'preferred species but unsuitable'. The third group includes the species 'avoided but suitable', and the fourth group numbers the species 'avoided and unsuitable' (Hubert *et al.*, 2004b, c). The fungi 'preferred and suitable' or the fungi 'preferred but unsuitable' attract their mite grazers by volatile compounds. This behaviour is most likely adaptive, and it is unknown whether the attractants produced by these fungi enable them to achieve higher metabolic rates and fitness. The 'preferred and suitable' fungi should profit from the dispersal of their spores (via mite faeces or surface attachment) to new habitats, such as plant litter or uncolonized grain. *T. putrescentiae* consumed a large quantity of fungal mycelium (Hubert *et al.*, 2004b, c), but high population densities of mites may not be advantageous for fungi, since this may result in overgrazing. The 'preferred but unsuitable' fungi may profit from spore dispersal by mites; even though they are grazed, they are not at risk of being overgrazed by mites. Only one fungus was preferred but unsuitable for *T. putrescentiae*, which can be explained by the mite's ability to digest a wide number of fungi. Mites have the least effect on 'avoided and unsuitable' fungal species, whereas preferred fungi seem to be the most influenced by mites in terms of both mite grazing and dispersion of fungal spores (Hubert, 2011).

The preference and suitability of fungi for mite food, in turn, influences fungal dispersion. This last occurs, as fungal spores adhere to the surface of mites or pass through the digestive tract and faecal pellets. The pathogenic fungus *Sclerotium cepivorum* Berk had viable mycelium and sclerotia after passing of the digestive tract of *Rhizoglyphus robustus* Nesbitt (Ramiréz-Suárez *et al.*, 2002). The spore dispersal via the mite faeces could be significant, especially if large quantities of spore-containing faecal pellets are produced. Hubert (2011) hypothesized that this faecal-dispersal mechanism could be more important than the dispersal of spores via mite surface adherence. The transport of fungi is important if they are known as mycotoxin producers. Among them, transport of *Aspergillus flavus* Link was confirmed in laboratory experiments with *T. putrescentiae*. Moreover, *T. putrescentiae* may influence the increase of aflatoxin production from *A. flavus* (Franzolin *et al.*, 1999). Except that, *R. robustus* is able to transmit the fungus *S. cepivorum*, which is a causal agent of garlic onion white rot (Ramiréz-Suaréz *et al.*, 2002). The authors suggest that it participates in fungal epidemiology by creating ports of entry for the pathogen and by accumulating inoculums at infection site. The fungi *A. flavus*, *Aspergillus niger* van Tieghem, *Penicillium crustosum* Thom, *Penicillium aurantiogriseum* Dierckx and *Penicillium chrysogenum* Thom were isolated from the mites in heavy-infested grain stores. These fungi produce mycotoxins such as aflatoxin B1, cyclopiazonic acid, roquefortine C, malformins, penitrem A, verrucosidin and vimellein (Hubert *et al.*, 2004c).

Armitage and George (1986) investigated the effect of three species of mites upon fungal growth in wheat. They compared the growth of naturally occurring grain storage fungi on wheat, infested with *A. siro*, *G. destructor* and *T. longior* (Gervais) with that on uninfected wheat, weekly for a period of 20 weeks. The colonies of the *Aspergillus glaucus* Chevalieri were always smaller on grain infested with mites than on grain without any mite infestation. *Penicillium* spp. were always less numerous on grain which was infested with *A. siro* but it is necessary to say that other mites didn't have any effect on the growth of these fungi species. In contrast, two fungi that are pathogenic to mites, *A. restrictus* Smith and *Wallemia sebi* (Fries), were more abundant in the presence of certain mites. The former was associated with *G. destructor*, the latter with *G. destructor* and *A. siro*. They concluded that the three species of mites either feed on *A. glaucus* and *Penicillium* spp. or inhibit them by an unknown secretion. Parkinson (1990) investigated the population increase and damage by the same three mite species on wheat at 20°C and two humidities (75% and 90% RH). Both *A. siro* and *T. longior* had lower populations at lower humidity; at 75% RH, a maximum number of these two mites colonies were 3000 and 1000, respectively in comparison with 500 and 600 mites, respectively, at 90% RH. But *G. destructor* did make obviously higher growth than at 90% RH at the end of 20 weeks. At both humidities, visible fungus was always less abundant on the infested grain than on the uninfected grain, and our data are in accordance with their findings that, mites significantly limited the fungal growth.

Hubert *et al.* (2003) reported that mites are well documented as vectors of micromycetes in stored products. Since their vectoring capacity is low due to their small size, they can be serious vectors only where there is selective transfer of a high load of specific fungal species. Therefore, the aim of their work was to find out whether the transfer of fungi is selective. Four kinds of stored seeds (wheat, poppy, lettuce, mustard) infested by storage mites were subjected to mycological analysis. They compared the spectrum of micromycete species isolated from different species of mites (*A. siro*, *L. destructor*, *T. putrescentiae*, *C. rhizoglyphoides* and *Cheyletus malaccensis* Oudemans) and various kinds of stored seeds. The fungi were separately isolated from the surface of mites, the mites' digestive tract (= faeces) and the stored seeds, and were then cultivated and identified. The study showed that the fungal transport via mites is selective. This conclusion is supported by different aspects. A lower number of isolated fungal species have been found from mites than from seeds, as well as a lower Shannon-Weaver diversity index in the fungal communities isolated from mites than from seeds. Moreover, a significant effect of mites/seeds as environmental variables on fungal presence in a redundancy analysis (RDA) was observed, and differences in the composition of isolated fungi between mite species are shown by RDA. The results of this study support the hypothesis that mite-fungal interactions are dependent on mite species. The fungi attractive to mites seem to be dispersed more than others. The selectivity of fungal transport via mites enhances their pest importance (Hubert *et al.*, 2003). Storage mites also endanger public health because allergens and transmit mycotoxin-producing fungi (Hubert *et al.*, 2005).

T. putrescentiae can also carry spores of bacteria and fungi such as *Aspergillus* spp. and *Penicillium* spp. (Franzolin *et al.*, 1999; Lee *et al.*, 2006; Jeon *et al.*, 2009). It has been found that the presence of *A. flavus* contributed to the growth of the *T. putrescentiae* mite, which has helped to disperse the spores of fungi in samples from uncontaminated corn (Franzolin *et al.*, 1999).

Hubert *et al.* (2018) reported that insects and mites are common inhabitants and accidental invaders of food, including durable commodities, and their presence can have both direct and indirect effects on human health. The most common direct effect is contamination of food with arthropod fragments and related contaminants, which may be allergenic or even carcinogenic. The most important indirect effect is that their presence can change the storage microenvironment, making durable products suitable for the rapid development of fungi and other microorganisms. Some of these fungi can produce toxins (e.g. aflatoxins) that endanger human health. Insects may actively or passively contribute to the spread of microorganisms, increasing product contamination, and they may host bacteria that have developed antibiotic resistance, contributing to their spread in food. Several species also may host, attract or transmit tapeworms, predators or parasitoids that may affect health. A comprehensive discussion on the topic may be found in Hubert (2011) and Hubert *et al.* (2018).

Storage Mites and Bacteria

Few reports are available about the stored-product mites as transmitters of bacteria. We know that they may ingest and utilize bacteria due to the presence of digestive lysozyme (Erban and Hubert, 2008). Some studies showed that the plating of mite homogenates from *T. putrescentiae* (Schrank) may contain bacteria of the genera *Klebsiella* Trevisan, *Pseudomonas* Migula and *Staphylococcus* Rosenbach. Relatively poor bacterial communities were observed in *Dermatophagoides pteronyssinus* (Trouessart) and *D. farinae* Hughes (Oh *et al.*, 1986; Saleh *et al.*, 1991; Valerio *et al.*, 2005). Bacterial isolation and cultivation methods allowed to determine different bacterial communities in the gut and on surface of *T. putrescentiae* (Smrž, 2003). Smrž and Catska (2010) suggested that the bacteria associated with this mite species are symbiotic and participate in chitin digestion and utilization.

Species belonging to the genera *Serratia* Bizio, *Pseudomonas* Migula, *Brevundimonas* Segers *et al.* and *Stenotrophomonas* Palleroni *et* Bradbury were identified by cultivation. The studies carried out to date seem to indicate that the storage mites can transmit bacteria, but analyses are still missing (Hulbert, 2011). Bacteria of the genus *Bartonella* Strong *et al.* are carried by haematophagous mites, ticks, fleas and flies, and attack the erythrocytes of mammals. Kopecky´ *et al.* (2014) described a *Bartonella*-like clade, a distinct group related to Bartonellaceae, collected in stored-product mites and a predatory mite *Cheyletus eruditus* (Schrank), based on the analysis of cloned 16S rRNA gene sequences. By using the clade-specific primers, closely related *Bartonella*-like 16S rRNA sequences were amplified from both laboratory colonies and field strains of three synanthropic mite species (*A. siro* Linnaeus, *L. destructor* (Schrank) and *T. putrescentiae*) and a predatory mite. Sequences of *Bartonella*-like bacteria were found in 11 strains but were not detected in *D. farinae* Hughes and *D. pteronyssinus* (Trouessart) and two strains of *L. destructor*. All obtained sequences formed a separate cluster branching as a sister group to Bartonellaceae and related to other separate clusters comprising uncultured bacterial clones from human skin and hemipteran Lygaeidae (*Nysius plebeius* Distant and *Nysius* sp.). The classification of sequences into operational taxonomic units (OTUs) showed a difference between *A. siro* and *T. putrescentiae* suggesting that the *Bartonella*-like bacteria are different in these two mite species. However, species-specific sequences in separate OTUs were observed also for *C. eruditus*. Possible symbiotic interactions between *Bartonella*-like bacteria and their mite hosts may occur (Kopecky´ *et al.*, 2014).

Hubert *et al.* (2021) reported that arthropod-associated microorganisms are important because they affect host fitness, protect hosts from pathogens, and influence the host's ability to vector pathogens. Astigmata mites often develop high levels of populations in various types of food items, damaging the food by direct feeding and introducing contaminants, including their own bodies, allergen-containing faeces and associated microorganisms. Hubert *et al.* (2021) access the microbial structure and abundance in rearing diets, eggs, faeces fraction and mite bodies of 16 mite populations belonging to three species [*Carpoglyphus lactis* (Linnaeus), *A. siro* and *T. putrescentiae*] using quantitative PCR and 16S ribosomal RNA (rRNA) gene amplicon sequencing. The mite microbiomes had a complex structure dominated by the following bacterial taxa (OTUs): (i) Intracellular symbionts of the genera *Cardinium* and *Wolbachia* in the mite bodies and eggs. (ii) Putative gut symbionts of the genera *Solitalea*, *Bartonella* and *Sodalis*, abundant in mite bodies and also present in mite faeces. (iii) Faeces-associated or environmental bacteria of the genera *Bacillus*, *Staphylococcus* and *Kocuria* in the diet, mite bodies and faeces.

Hubert *et al.* (2021) observed that the differences between microbial communities in various conspecific mite populations were higher than those between different mite species. To explain some of these differences, these authors hypothesize that the intracellular bacterial symbionts can affect microbiome composition in mite bodies, causing differences between microbial profiles. Microbial profiles differed between various sample types, such as mite eggs, bodies and the environment (spent growth medium-SPGM). Low bacterial abundances in eggs may result in stochastic effects in parent-offspring microbial transmission, except for the intracellular symbionts. Bacteria in the rearing diet had little effect on the microbial community structure in SPGM and mite bodies. Mite fitness was positively correlated with bacterial abundance in SPGM and negatively correlated with bacterial abundance in mite bodies. Hubert *et al.* (2021) demonstrated critical host-microbe interactions, affecting all stages of mite growth and leading to alteration of the environmental microbiome. Correlational evidence based on absolute quantitation of bacterial 16S rRNA gene copies suggests that mite-associated microorganisms are critical for modulating important pest properties of mites by altering population growth.

Storage Mites Allergy

Under favourable conditions of RH, temperature, type of deposit, storage time and type of product stored, mites can multiply and reach a high density of population causing direct or indirect harm to people (Hubert *et al.*, 2014b). Mullen and O'Connor (2019) reported that people who handle mite-infested stored products may become sensitized to the mites on subsequent contact, resulting in an immunological response called 'storage-mite allergy'. Although the precise nature of the allergens is unknown, these substances include components of both live and dead mites and material produced in the mite alimentary tract. Sensitized persons may experience either contact dermatitis or respiratory allergy, depending on the type of exposure.

Allergic contact dermatitis results from exposure to mites in grains, dried fruits, flour and other stored products, causing itching and redness at the contact sites. The families of mites most commonly involved are the Acaridae, Carpoglyphidae and Glycyphagidae. In addition, what was probably *D. pteronyssinus* (Trouessart) has been associated with cases of feather pillow dermatitis. Contact with this mite-infested feather pillows is known to cause red papular lesions and pruritus about the scalp, eyes, ears and nostrils (Traver, 1951; Aylesworth and Baldridge, 1983). A similar allergenic response to *D. farinae* Hughes associated with buckwheat-husk pillows has been reported in China (Hong *et al.*, 1987).

Inhalational allergy results when airborne mites and associated allergens are drawn into the respiratory tract. The mucosal membranes lining the nasal and bronchial passages become irritated and inflamed, causing allergic rhinitis and asthma. The mucous membranes lining the eyelids also may be affected, causing conjunctivitis. These responses involve a T cell-type reaction and both immediate and delayed hypersensitivity. Such reactions to mites present an occupational hazard, especially among farmers and other agricultural workers who handle mite-infested grains and other stored materials. Among the more common storage mites that cause inhalational allergy are *Aleuroglyphus ovatus* (Troupeau) and *T. putrescentiae* (Schrank) (Acaridae), *L. destructor* (Schrank) (Glycyphagidae) and *Blomia tropicalis* Van Bronswijk, de Cock *et* Oshima (Echimyopodidae). Further information on storage-mite allergy can be found in Cuthbert (1990).

The inhalation, ingestion or contact with live or dead mites, by-products of metabolism or faeces can damage the human health, causing respiratory diseases such as asthma, allergic rhinitis, conjunctivitis (Palyvos *et al.*, 2008), contact dermatitis (Sinha, 1979), anaphylaxis and acute enteritis by the ingestion of food contaminated by fungi and bacteria conveyed by the mites (Hughes, 1976; Palyvos *et al.*, 2008). However, 95% of mite allergens are associated with faecal particles related to the digestion process of the mite (Hubert *et al.*, 2004a).

Anaphylactic reactions caused by food infested with mites are increasing, especially in the climatic areas favourable for their development (Sánchez-Borges *et al.*, 2005), and the allergological importance of different storage mites, such as *L. destructor*, *G. domesticus* (De Geer), *T. putrescentiae*, *A. siro* Linnaeus, *A. ovatus*, *S. medanensis* Oudemans and *Thyreophagus entomophagus* (Laboulbéne) increased in the past years. Many allergens from these species have been purified, sequenced and cloned, and main allergens include fatty acid binding proteins, tropomyosin and paramyosin homologues, apoliphorine like proteins, alfa-tubulines and other, such as group 2, 5 and 7 allergens. Besides the purification and characterization of allergens, the allergenicity of other species such as *A. farris* (Oudemans), *Austroglycyphagus malaysiensis* Fain *et* Nadchatram, *Blomia kulagini* (Zakhvatkin) and *Blomia tjibodas* (Oudemans), *C. eruditus* (Schrank), *C. arcuatus*, (Troupeau), *Gohieria fusca* (Oudemans), *Tyrophagus entomophagus* and *T. longior* (Gervais) has been studied. The allergenicity of mites present in agricultural environments has been investigated. Cross-reactivity studies have also been performed to elucidate to what extent all these mites share common, or species-specific epitopes. Vogel *et al.* (2015) published a comprehensive review on the problem of allergies, as well as Fernandez-Caldas *et al.* (2007) discussed extensively the allergenicity of mite species which have been implicated in human respiratory and/or dermatological diseases.

Mites produce thousands of different proteins and other macromolecules. Their extracts for testing are made from aqueous extraction of a variable mixture of whole mites, nymphs, faecal pellets, eggs and spent culture media; they contain over 30 different proteins that can induce IgE antibody production in patients allergic to mites. Out of the 19 denominated groups of allergens, major IgE binding has been reported for the group 1, 2, 3, 9, 11, 14 and 15 allergens. The high-molecular-weight group 11, 14 and 15 allergens have recently been described. The group 1 and 2 allergens represent dominant specificities, which can account for much of the allergenicity of extracts (Thomas *et al.*, 2002). Data on storage mites could be extrapolated from house-dust mites, which have been the main focus of research. Mite extracts contain a variety of biochemically active enzymes, including trypsin, chymotrypsin, carboxypeptidase A and B, glucoamylase and lysozyme. Marked differences in the relative concentrations of some of these enzymes, particularly trypsin and carboxypeptidase A, occur in different mite extracts (Stewart *et al.*, 1992c,

1994). Storage mites and *D. pteronyssinus* have their own unique allergen or allergens, and the cross-reactivity among the storage mites is more common. *L. destructor*, *G. domesticus* and *T. putrescentiae* are allergenically more closely related to each other than to *A. siro* (van Hage-Hamsten *et al.*, 1987). Other studies showed that the major allergens of *T. putrescentiae* have a strong cross-reactivity with *D. pteronyssinus* extracts, and that *D. pteronyssinus* allergens have a partial cross-reactivity with *T. putrescentiae* extracts (Tee *et al.*, 1992; Munhbayarlah *et al.*, 1998). Extensive cross-reactivity occurs among Gly d 2, Lep d 2 and Tyr p 2, and little cross-reactivity between these allergens and Der p 2 (Gafvelin *et al.*, 2001). Spider mite extracts contain species-specific allergens as well as allergens commonly shared with house-dust mite and citrus red mite (Jee *et al.*, 2000; Park *et al.*, 2002).

At clinic level, it is necessary to know the phylogenetic relationship of the allergenic proteins of these mites in order to determine the best combination of extracts for its use. To assess the phylogenetic relationship between the main allergenic proteins of mites, the groups 1, 2 and 5 of Glyciphagidae, Pyroglyphidae, Chortoglyphidae and Acaridae families were compared according to the sequence of mRNA and amino acids with the validated sequences of the National Center for Biotechnology Information, and through bioinformatic alignment analyses for the construction of the trees; the method of neighbour-joining, with bootstrap support with 500 replications, was used as a measure of reliability and robustness. From 15% to 87% of identity was found in the three groups of allergens; the highest was between Der p2 and Der f2 (86.98%) and, the lowest, between Der f5 and Gly d5 (17.87%) Pyroglyphidae showed the highest relationship between the species. The longest branching distance was identified in Glicyphagidae, especially in *B. tropicalis*. Some allergenic proteins have a high identity between the different species of mites, unlike *B. tropicalis*. These results can be taken into consideration when the diagnosis and treatment of allergic diseases are being defined (Sánchez *et al.*, 2019).

According to literature data, there are three kinds of allergenic mites, the 'storage mites', the 'house-dust mites' and the 'plant mites'; the latter are fundamentally the red spider mites and citrus mites. However, these three types are not the only causes of mite allergy, such as *Hemisarcoptes cooremani* (Thomas), a predator of scale insects, can cause allergic reactions (Arlian *et al.*, 1999a).

Steinman and Ruden (2007) reported that some mite species play a significant role among allergens. Like pollens, they are invisible and are from very common environmental sources, forming ambient mediators of allergic disease, but their sources are not even as distinct as certain widespread species of trees. These mites are basically part of the atmosphere of human civilization and prevail in domestic settings, agriculture and certain industries – especially domestic settings, where most people spend most of their time. A great number of factors play a role in exposure to these mites. Mite allergens may spread where mites do not. Public transport vehicles and schools can be sources of mite allergens (Pereira *et al.*, 2004; Taketomi *et al.*, 2006; Abramson *et al.*, 2006).

Humans can be carriers as well as victims of allergenic mites. Li *et al.* (2003) observed that among 1994 individuals with different occupations, 8.07% carried mites in their urine and/or stool. The species of mites in the samples were consistent with those found in the individuals' working environments. A significant association was observed between mite infection and occupation, in particular among workers in medicinal-herb storehouses and rice storehouses and mills; and in miners, railway workers, pupils and teachers (Li *et al.*, 2003).

The primary habitat of storage mites is grains and cereals, and consequently they are most common on farms, in granaries and in factories. At high density of population, they appear as 'the walking dust' consisting of millions of mites moving on barn or stable surfaces. This dust is a strong trigger of occupational allergy. Curiously, storage mite populations are not homogenously widespread in grains and cereals. Athanassiou *et al.* (2005) reported that in Greece, in a flat storage facility for wheat, most mites were found in the upper 0.5 m of the bulk. Exposure to these mites has been identified as a cause of asthma and rhinitis, and according to recent studies, some storage mites have penetrated into the habitats of house-dust mites. Studies from several countries have shown that IgE-mediated allergy to storage mites is of considerable importance in both rural and urban populations; sensitization is not restricted to those with occupational exposure (Franz *et al.*, 1997). Populations in damp urban housing would be at special risk (van Hage-Hamsten and Johansson,

1998). Moreover, these mites have been found even in dry regions, as reported after an examination of house dust in Egypt (Koraiem and Fahmy, 1999). Some of the same allergens are produced by both house dust and storage mites; therefore, the allergic diseases that they cause should be very similar (Olsson and van Hage-Hamsten, 2000; Kanceljak-Macan *et al.*, 2000). The allergens from different species of mites can, however, be quite different, and clinicians should consider non-cross-reactive allergens also as possible causes of reactions. The full range of possible environmental factors should always be kept in mind. Exposure to storage mite allergens can be by ingestion (van Hage-Hamsten and Johansson, 1992), which can cause devastating allergic reactions. Different cases of anaphylaxis were known. A 13-year-old boy experienced a sore throat and tight chest approximately a minute after eating homemade wheat flour hotcakes; this reaction was followed by a stuffy throat and paleness of the face. A 7-year-old girl experienced coughing, wheezing, streaming eyes and skin flushing an hour after eating homemade 'tako-yaki'. Fifteen months later, sneezing, coughing, wheezing, generalized pruritus and streaming eyes occurred after she ate homemade buns. Both children had high levels of IgE antibodies to *D. farinae* Hughes. Mites were found in the boy's hotcake. Three of 176 packages of wheat flour from local retail outlets and seven of the 127 from homes were found to be infested with mites (Matsumoto and Satoh, 2004).

Cui (2014) reported that although allergic disorders due to domestic mite species are well known, little is known about human acariasis, in which mites invade and parasitise the human body in various tissues, from the gastrointestinal apparatus to the lung. However, it is well possible that the clinical symptoms of human acariasis of pulmonary, intestinal, oral (anaphylaxis), urinary, otic and vaginal apparatus often overlap with other disease symptoms leading to frequent misdiagnosis, making the requirement to pay more attention to these infections.

Clemmons and Douglas (2016) stated that *A. siro* is one of several mites that have been identified as a cause of intestinal, pulmonary and urinary acariasis in humans (Cui, 2014; Li *et al.*, 2003). It has been observed that *A. siro* is potentially allergenic and may cause cross-sensitization with house-dust mites. The amino acid sequences of *A. siro* and *D. pteronyssinus* showed 66% of comparability (Pytelkova *et al.*, 2012). Sanchez-Machin *et al.* (2011) found that through the skin test, 14% of the sensitized patients to house-dust mites, showed a concomitant sensitization to the storage mite *T. entomophagus*. Morales *et al.* (2013) observed that both mites of Pyroglyphidae family as of Glycyphagoidea possess a similar protein sequence, though the enzymatic properties of each species differ significantly from the activity of the proteases. *A. ovatus* is also found especially in warm climates, in foods such as wheat grain and its derivatives and preparations with chicken, as well as in dust of storage units, lairs of rodents, chicken coops and barns (Geary *et al.*, 2000), and like *A. siro* it presents great activity on starch digestion (Erban *et al.*, 2009). Results of studies carried out on the evaluation of the immunogenic sites of *A. ovatus*, suggested that both the inhalation as the contact with mites and eggs, as faeces would be immunogenic (Tang *et al.*, 2011); it has been observed that in humans this mite is associated to respiratory and contact dermatitis, especially in the grain industry workers (Geary *et al.*, 2000).

Depending on which organ system is affected, acariasis can lead to a variety of clinical signs, including pain, diarrhoea, coughing and fever (Cui, 2014). In humans, mites have been isolated from both urine and stool samples and have been associated with intestinal ulceration and bladder wall inflammation (Li *et al.*, 2003). The risk of these outcomes is seemingly very low in the laboratory animal setting, given the routine use of personal protective equipment. Mullen and O'Connor (2019) treated extensively the topic and reported that sometimes the mites may penetrate in the natural body orifices, causing temporary internal acariasis. The phenomenon requires commonly the ingestion of food infested by mites and the inhalation of airborne mites or mite-contaminated dust via oral or nasal openings. Mites that are swallowed or inhaled can lead to acariasis involving the alimentary tract, whereas mites that are inhaled also can invade the respiratory apparatus. Cases well documented are the 'pulmonary acariasis' and 'enteric acariasis', while cases of mites infesting the urinary apparatus are rare (Ma and Wang, 1992, in Mullen and O'Connor, 2019).

Samung *et al.* (2006) found in the external auditory canal of a 57-year-old woman in Thailand over 20 specimens of *Suidasia pontifica* Oudemans and concluded that possible sources may be dried chilies, shallots and garlic bulbs, as well as house dust and other materials related to their cultivation, such as straw from the patient's house. *Suidasia pontifica* were recovered from house dust, shallots and garlic bulbs. It was suspected that the mites entered the ear during the handling of these infested materials.

Pulmonary acariasis

It has been observed that mites spread in the lungs, most frequently among individuals exposed to mite-infested stored grains and dried herbs. This aspect can be a serious problem among workers in grain storage facilities and medicinal-herb warehouses in China (Chen *et al.*, 1990; Li and Li, 1990). Clinical signs and symptoms include cough, expectorated phlegm and blood, difficult breathing, chest pain, low-grade fever, restlessness and marked eosinophilia. Pulmonary lesions have been documented on X-ray film, as shadows and nodular opacities in lung tissues. Acarid mites have been collected in the *sputum* of affected individuals (A. siro Linnaeus, T. *putrescentiae* (Schrank), *A. ovatus* (Troupeau), *Sancassania berlesei* (Michael) and what was probably *Sancassania mycophaga* (Mégnin)). A *Carpoglyphus* species, identified as C. *lactis* (Linnaeus), was associated with a case of pulmonary acariasis in Spain (Taboada, 1954), whereas a *Tyrophagus* species and other unidentified mites were collected from *sputum*, bronchial washings and needle-aspirated lung specimens in routine examinations of patients suffering from respiratory ailments (Farley *et al.*, 1989). Different biological instars (eggs, larvae and adults) were found in cytology specimens, with evidence of their being surrounded by acute inflammatory cells in several cases.

Enteric acariasis

Sometimes mites were collected in human excreta, suggesting their presence in the digestive tract. In most cases, they are acarid mites belonging to the genera *Acarus*, *Suidasia* and *Tyrophagus*. Among these mites, *Suidasia pontifica* was collected in the faeces of a woman and two infants in Mexico (Martinez Maranon and Hoffman, 1976), whereas various stages of an *Acarus* or *Tyrophagus* species, together with eggs, were found in bile of a Romanian patient with chronic cholecystitis (Pitariu *et al.*, 1978). The phenomenon may depend on the fact that the woman probably ingested the mites with the infested food, and that the mites aggravated her pre-existing cholecystitis by causing inflammation of the digestive tract, until the mites were eliminated with the bile. Enteric acariasis have been reported also in children with chronic digestive disorders in Russia (Prisich *et al.*, 1986) and in Brazil, under the name of intestinal tyrogliasis (Machado, 1935).

Urinary acariasis

The few cases reported concern acarid mites of the genus *Tyrophagus*. *T. putrescentiae* and *T. longior* have been collected from the human urinary tract (Harwood and James, 1979). Many of these cases may involve a contamination of containers in which the urine was collected or examined. Pitariu *et al.* (1979) reported the recovery of unidentified mites in urinary samples from several patients in Romania during acute attacks, associated with inflammation of the kidneys and urinary bladder. Although many specimens of acarid mites and their eggs have been found in the urinary sediments, with other specimens dead and encrusted with salts, the rule of probable contamination is unclear. Other cases of urinary acariasis have been reported in Japan (Harada and Sadaji, 1925) and China (Chen *et al.*, 1992; Ma and Wang, 1992, in Mullen and O'Connor, 2019). Cases of digestive and urinary acariasis in humans involving *T. longior* have been reported (Harwood and James, 1979).

Dermatitis and other disorders

Stored mites of the genus *Tarsonemus* have been found associated with human dermatitis and other skin disorders (Hewitt *et al.*, 1973; Krantz, 1978; Oehlschlaegel *et al.*, 1983) and to invade various organs and body fluids of humans and other animals (Dahl, 1910). The most common species is *Tarsonemus hominis* (Dahl), probably contaminating the glass slides and other materials used in the preparation of tissues for microscopic studies (Hewitt *et al.*, 1973; Samšiňák *et al.*, 1976). Halliday (2003) reported that several mites occurring in stored grain can cause occupational health problems for grain industry workers, including skin contact dermatitis, inhalant allergy and asthma, and painful irritation caused by direct biting of human skin. Sensitization to mites of stored products is more common in rural areas and among workers in the grain storage units, bakeries and supermarkets, by exposure to mites, and it has been suggested as an occupational disease (Arlian and

Morgan, 2003; Estevez, 2006; Kim and Kim, 2002; Spiewak *et al.*, 2001; Storaas *et al.*, 2005; Koistinen *et al.*, 2006). Asthma and allergic rhinitis in fruit farmers and people living near the orchards of citrus, apple and pear induce an increase in cases (Kim and Kim, 2002; Kim S.H. *et al.*, 2002; Kim Y.K. *et al.*, 2002), contact dermatitis in farmers that grow hop (Spiewak *et al.*, 2001) and gardeners who grow pepper (Groenewoud *et al.*, 2002). However, the occurrence of occupational allergy in urban areas among workers in bakeries and supermarkets has also been reported. Storage mites were responsible for 20% of cases of sensitization of allergic rhinitis in bakery workers (Storaas *et al.*, 2005). Another study reported of a supermarket employee who got hives and dermatitis through pork meat, especially ham, infested by *T. putrescentiae* in her workplace (Estevez, 2006). When *T. putrescentiae* is ingested through contaminated food, inhaled or handled, it can cause diarrhoea, urinary tract problems, acute enteritis and allergic reactions (Aygun *et al.*, 2007; Jeon *et al.*, 2009).

Stejskal *et al.* (2014) reported that *A. siro*, *A. farris*, *T. longior* and *T. putrescentiae* from Acaridae, *L. destructor* and *G. domesticus* from Glycyphagidae and *C. arcuatus* from Chortoglyphidae (Boström *et al.*, 1997; Franz *et al.*, 1997; Mehl, 1998; Sánchez-Ramos *et al.*, 2004, 2007c; Pike and Wickens, 2008; Wong *et al.*, 2011; Solarz, 2012) have been identified as responsible for clinically important allergens, causing occupational allergy among farmers, grain-storage workers and other agricultural workers (Fain *et al.* 1990; Revsbech and Dueholm, 1990; van Hage-Hamsten and Johansson, 1998). These mites cause IgE-mediated sensitization among the above-mentioned occupational populations, who develop asthma, rhinitis and conjunctivitis when exposed to organic dust containing mite allergens (Arlian, 2002; Sánchez-Ramos *et al.*, 2004; Berger *et al.*, 2005; Solarz, 2012).

Solarz and Pająk (2019a) confirmed the occurrence of allergenic mites in cowsheds, barns, chaff-cutter buildings and poultry houses (Hughes, 1976; Terho *et al.*, 1982; Fain *et al.*, 1990; Mumcuoglu and Lutsky, 1990; Solarz *et al.*, 1997; Stejskal and Hubert, 2008), and reveal the presence of these mites in pig sties. Consequentially, all these locations should be regarded as a potential source of storage mite allergens in the farming environment. As the occurrence and density of mites in samples from different farm buildings may vary to a considerable extent, further studies are highly desirable.

Grain mites and beetles are well-known sources of allergens for bakers and farm workers, but the role that these insects play in laboratory animal worker allergies has not been extensively evaluated and established (van Hage-Hamsten *et al.*, 1985; Revsbech and Dueholm, 1990; Sublett and Bernstein, 2011). In a study examining two laboratory animal facilities for storage and house-dust mites, 25% of samples collected contained mites. Most were storage mites of Astigmata (Glycyphagidae, Acaridae) or Prostigmata (*Tarsonemus* spp., *Cheyletus* spp.) species. Mites were found in samples collected from feed and bedding-storage floors and lounge chairs. The house-dust mite species, *Dermatophagoides*, was found in only a single sample (Pennanen and Harju, 2003). An additional study found mite-derived material in 21 of 22 dust samples taken in the laboratory animal facility. Skin prick tests in 40 laboratory animal workers with work-related rhinitis revealed that 14 (35%) were sensitized to storage mite species (*A. siro*, *L. destructor* or *T. putrescentiae*). Nineteen (48%) subjects were positive for laboratory animal sensitization, and only six of these subjects were sensitized to both laboratory animals and mites. Conversely, only 1.7% of 300 persons in the general population of the same county were sensitized to any of the three storage mites (Ruoppi *et al.*, 2005). The authors concluded that in addition to laboratory animals and house-dust mites, storage mite species may pose an occupational health risk in animal facilities (Pennanen and Harju, 2003; Ruoppi *et al.*, 2005). Because of the potential for sensitization to mites, the factors of ventilation, cleaning, and personal protective practices are important considerations, not only when handling animals but also when working around feed and bedding materials. The risk for laboratory animal allergy is well characterized, but the possibility of mite sensitization should also be considered in workers with respiratory complaints, especially when testing for laboratory animal allergy is negative (Ruoppi *et al.*, 2005).

Another category is known as 'plant mites' and occurs on living plants in fields and orchards. It is the case of the 'red spider mite', a pest of tomatoes. The mites persist on the product after harvest and survive until reaching the consumer causing misdiagnosed reactions. Commonly confusing reactions are occupational; a person having allergic reactions from work exposure to fresh tomato suspects only a tomato allergy, and a clinician needs to confirm or refute this suspicion through testing and not simply rely on the report (Steinman and Lee, 2003).

Commonly, storage and citrus mites cause only respiratory allergies. House-dust mites commonly affects negatively, eczema and other skin conditions. The comparison among the sensitization to house-dust mite with the sensitization to storage mite showed that the mean age for the onset of sensitization was 6.7 years for house-dust mites and 18.7 years for storage mites. Patients allergic to storage mites show conjunctivitis frequently, whereas patients sensitized to *T. putrescentiae* show perioral syndrome. The differential sensitization to house-dust mites and storage mites are conditioned by damp climatic and indoor conditions and human activity, and not by urban or rural living environments (Boquete *et al.*, 2000).

Although the stored products may contain allergens of vegetal, animal or microbial (Korunić, 2001), infestation by mites is one of the main problems associated with grain storage (Sánchez *et al.*, 2007c; Ždárková, 1991a), despite the tolerance determined by international organizations be zero or close to zero in mite infestation (Palyvos *et al.*, 2008). Thind and Clarke (2001) found contamination in 21% of samples of cereals and cereal products purchased in retail stores in the United Kingdom, rising to 38% after being stored for six weeks at the home of volunteers. However, due to the limitation of the study, it cannot be said that the products were infested during the storage. A study carried out in Greece highlighted the species of mites that infested grains such as wheat, corn, oats and barley, manufactured agricultural products, dried fruit and powder residue in cooperatives, stores, silos and mills. A total of 55.5% of the samples tested showed contamination by mites of four orders and 15 distinct families. In this study, Astigmata corresponded to 92.7% of the total of mites and *L. destructor* was the dominant species in all of the sites studied (Palyvos *et al.*, 2008). Results of studies carried out in the Czech Republic showed that mites are the most important problem in units of grain storage. The most frequent species were *A. siro*, *A. farris*, *T. putrescentiae* e *L. destructor*. The mites are the most important source of allergens in stored grains and that the control, as a prevention strategy, lacking in many countries (Stejskal and Hubert, 2008). In New Zealand, Cotter *et al.* (2011) examined samples of products derived from grain purchased in shops and gathered in homes. The study did not find the presence of mites in the samples purchased in stores, but four samples of products obtained in houses showed mites representing 7.4% of the total. The four samples that had mites were stored in cabinets for over a month, indicating that storing flour for long periods increases the risk of contamination and that the ideal to prevent contamination would be to store it in the fridge (Cotter *et al.*, 2011).

Systemic anaphylaxis

Ingestion of food infested by mites may cause systemic anaphylaxis (Erben *et al.*, 1993). Some patients developed generalized allergic reactions after ingestion of foods containing large amounts of mites (Castillo *et al.*, 1995). Anaphylaxis by ingestion of cornmeal infested by mites was described in Brazil (Guerra Bernd *et al.*, 2001). Another case of anaphylaxis involves the ingestion of *Blomia freemani* Hughes with contaminated pancake flour (Wen *et al.*, 2005). The incidence of anaphylaxis ranges from 0.5 to 2%, mostly by the ingestion of cereals, especially wheat, infested by mites and their allergens. Cases of anaphylaxis commonly occur in patients with allergic rhinitis or asthma, who are sensitized to dust mites, develop allergic reactions, mediated or not by IgE after eating wheat flour infested by storage mites (Posthumus and Borish, 2012). There has also been reported a case of systemic anaphylaxis by the ingestion of food contaminated with the *T. putrescentiae* mite (Sánchez-Ramos and Castañera, 2000; Matsumoto *et al.*, 1996). A study that assessed the prevalence of sensitization by *T. putrescentiae* showed that it is higher in elderly subjects compared to young adults, especially in people with chronic obstructive pulmonary disease (COPD), when the prevalence reached 45.8% of the studied population (Liao *et al.*, 2010) and that the allergenic of the *T. putrescentiae* mite induces histamine release in more than 50% of sensitized individuals (Liao *et al.*, 2009).

Kew *et al.* (2014) examined a problem connected with the edible bird nests (EBNs) from Hong Kong and the Peoples' Republic of China. This industry was greatly affected, when China banned the import of all the EBNs from Malaysia, except for the canned version, due to detection of high levels of nitrites. Several cases of anaphylaxis following ingestion of EBNs were reported, but the source(s) of the allergens remained unknown. Hence, they started a study to quantify, isolate and identify the mites that are associated with EBNs. The raw EBNs were purchased from swiftlet farms in five locations in Peninsular Malaysia, while the commercial nests were purchased from five different Chinese traditional medicinal shops. The average mite density of all the raw nests was 285 ± 603 mites per gram of EBN while the commercial nests had a much lower mean value of 21 ± 32 mites per gram of EBN ($p = 0.082$). Among the raw EBNs, the nests from Kajang had the highest average mite density (946 ± 1443 mites/g of EBN) whereas the nests from Kuala Sanglang had the lowest (54 ± 34 mites/g of EBN). Among the commercial EBNs, the nests from Company D had the highest average mite density (76 ± 18 mites/g of EBN) whereas the nests from Company A were free of mites. Overall, the average densities of mites in the raw nests obtained from southern regions of Malaysia (Selangor and Johor), were higher than those nests obtained from the northern regions (Kedah and Kelantan). Thirty types of mites were isolated from both the raw and commercial nests. Among these, some are probably feather mites (*Eustathia cultrifer* (Robin *et* Mégnin), *Pteroherpus garrulacis* Mironov *et* Proctor, *Pterodectes amaurochalinus* Akashi *et* Valim, *Laminalloptes* sp., *Berlesella alata* (Trouessart) and *Neochauliacia* sp.), house dust and storage mites (*Suidasia* sp., *Austroglycyphagus* sp. and *A. ovatus* (Troupeau)), mesostigmatid mites (*Dermanyssus* sp.), prostigmatid mites (*Cheyletus* sp., tarsonemid and cunaxid mites), astigmatid mites (*Collocalidectes* sp., *Streetacarus* sp. and *Hemisarcoptes* sp.) and oribatid mites. This study provides baseline information on the density and type of mites that are probably associated with EBNs. This study also heightens the importance of mites as a possible source of EBN-associated anaphylaxis (Kew *et al.*, 2014).

Storage Mites and Animal Health

It is well known that the allergens may accumulate on the skin and hair of dogs (Randall *et al.*, 2005).

Storage mites may be considered important allergens in dogs with atopic dermatitis. High sensitization rates to *Tyrophagus*, *Acarus* and *Lepidoglyphus* species have been reported in atopic dogs, and dry pet food has been suggested as a potential source of storage mite exposure. Brazis *et al.* (2008) evaluated the commercial dry dog food for contamination with storage mites, and how storage time and conditions could influence the risk of contamination. Ten different premium commercial dry dog foods formulated for skin disorders were selected. Food bags were opened and stored for 6 weeks under two different environmental conditions. At different time points, samples from each bag were collected and analysed by microscopy, guanine test, storage mite-specific traps and a modified flotation technique. On opening, two storage mites identified as *A. siro* Linnaeus were isolated from one of the 10 bags by flotation technique, indicating that storage mites can be present in packaged dry dog food bags. After 5 weeks of storage under environmental conditions optimal for mite growth (23.2 +/– 2.1°C and 71 +/– 5.6% of RH), mites were detected by microscopic observation in nine of the ten diets. When mites were identified by the flotation technique, *Tyrophagus* spp. was found to be the most common contaminating species. These results show that dry dog food can be a suitable substrate for storage mite reproduction, and that environmental and storage conditions may influence food contamination and mite development (Brazis *et al.*, 2008).

Clemmons and Douglas (2016) reported that mites are well documented as fungal vectors in stored products, with the potential to change fungal populations and indirectly affect mycotoxin content (Hubert *et al.*, 2003). Some laboratory animal species may be susceptible to acariasis, a syndrome described in humans, which is caused by mite invasion of various tissues (Cui, 2014; Li *et al.*, 2003). Experimental injection of five mite species, including *A. siro*, into the trachea of guinea pigs resulted in lung changes similar to those seen in humans with pulmonary acariasis (Cui, 2014). One study fed *A. siro*-infested food to pregnant mice and saw a statistically significant increase in fetal mortality and feed intake. Litter sizes were reduced as well, but not significantly so (Warner and Bohane, 1978). Grain mites ingested by rodents have been found in faeces and mistaken for other mite species, such as *Myocoptes musculinus* (Koch) (Myocoptidae), *Myobia musculi* (Schrank) and *Radfordia* spp. (Myobiidae) (Charles River Laboratories International, 2011). In a laboratory animal facility, mistaken identification of grain mite adults or eggs detected within rodent faeces could lead to inappropriate treatment and control methods. Grain mite infection of various organs, fungal carrying potential and reproductive effects could negatively affect research and production colonies (Hubert *et al.*, 2003; Li *et al.*, 2003; Warner and Bohane, 1978).

Olivry and Mueller (2019) reported that atopic dermatitis is a common, chronic, recurrent pruritic allergic skin disease of dogs; in most patients, it consists of characteristic skin lesion distribution (Hensel *et al.*, 2015), and it is most often associated with the production of IgE specific for environmental and/or food allergens (Pucheu-Haston *et al.*, 2015; Olivry and Mueller, 2017). Worldwide, house-dust mites of the genus *Dermatophagoides* are the most common allergens, recognized by the circulating IgE of atopic dogs. Storage mites represent another group of acarid mites that commonly invade food sources, especially cereals. The most common species are *A. siro*, *L. destructor* (Schrank), *G. domesticus* (De Geer) and *T. putrescentiae* (Schrank) (Nuttall *et al.*, 2006). An IgE reactivity against storage mites is also very common in dogs with atopic dermatitis (Bensignor and Carlotti, 2002; Arlian *et al.*, 2003). The extensive cross-reactivity that exists between house-dust mites and stored mite allergens implies that house-dust mites-specific IgE from sensitized atopic dogs are likely to also recognize homologous allergens in storage mites, and vice versa (Arlian *et al.*, 1984a; Saridomichelakis *et al.*, 2008). Such allergen cross-reactivity is probably clinically relevant, as beagles experimentally sensitized to the *D. farinae* Hughes house-dust mites exhibited a flare of clinical signs when environmentally or orally challenged with the storage mite *T. putrescentiae* (Marsella and Saridomichelakis, 2010). Consequently, atopic dogs with high levels of *D. farinae* house-dust mites-specific IgE are likely to have a flare of clinical signs, if eating a food item contaminated with stored mites; such recurrence of signs would lead to a false positive diagnosis of food allergy.

Olivry and Mueller (2019) suggested that it is best to keep commercial dry dog foods indoors and in temperate conditions, and to seal bags to decrease the risk of their contamination with *Tyrophagus*. When performing restriction (elimination) and provocation trials for the diagnosis of food allergies in dogs (and likely cats), it seems preferable to choose newly purchased bags of both testing and original diets to reduce the probability of their contamination with *T. putrescentiae*. Indeed, the allergenic cross-reactivity between the *Tyrophagus* and the *Dermatophagoides* is very high (Arlian *et al.*, 2003; Saridomichelakis *et al.*, 2008), and such cross-reactivity could affect the clinical interpretation of dietary trials in atopic dogs, sensitized to house-dust mite (Marsella and Saridomichelakis, 2010). Indeed, atopic dogs with high levels of *D. farinae* house-dust mites specific IgE are likely to have signs of flaring after eating a food contaminated with stored mites, thus leading to a false positive diagnosis of food allergy. In case of doubt about pet food contamination with stored mites, one should perform dietary challenges with homecooked ingredients previously preserved indoors and in temperate conditions. As it appears that the main mite naturally growing on commercial dry dog foods is *Tyrophagus*, and this stored mite is the only one whose cross-reactivity with house-dust mites was shown to be clinically relevant, one should reconsider the validity and clinical relevance of performing IgE serological and intradermal testing, with the other less common stored mite species. *G. domesticus* was also found to be the intermediate host of *Catenotaenia pusilla* (Goeze), an intestinal tapeworm of mice and rats (Hughes, 1976).

Recently, it has been developed and proposed commercially – the 'Artu-vetrin Skin Test' able to highlight the sensitivity of dogs, cats and horses to different antigens, including *A. siro* and *T. putrescentiae*. The use of the test is preparatory to the employ of monoclonal antibodies such as 'Artu-vetrin Therapie'.

Forensic Importance of Storage Mites

Solarz *et al.* (2022) recently summarized this topic in a comprehensive review. They reported that beyond their negative role as allergic agents, stored mites may also be employed as professional markers in forensic analyses (Perotti *et al.*, 2009; Solarz, 2011). Several genera of the family Acaridae can be encountered in forensic situations, particularly as allergenic taxa, in relation to workers, who are occupationally exposed to mites. However, many mites are also involved in the decomposition of animal carcasses and human corpses at every stage. Some of these species will be represented by just a few specimens, whereas others, such as *T. putrescentiae*, *A. farris* (Oudemans), *A. immobilis* Griffiths and *S. berlesei* (Michael), identified in the study of Solarz *et al.* (2022), may increase in number to millions, as confirmed by many works (O'Connor, 2009a; Perotti, 2009; Braig and Perotti, 2009; Perotti and Braig, 2009; Russell *et al.*, 2004; Solarz, 2011; Szelecz *et al.*, 2018). It is widely accepted that the first arthropod scavengers who colonized a dead body were flies with phoretic mite species. The flies complete their life cycle in and around the corpse, while the mites may feed on their immature stages. Since mites reproduce much faster than their carriers, they may offer themselves as valuable timeline markers. The mites collected in the study of Solarz *et al.* (2022) included *A. siro* Linnaeus, *A. farris*, *A. immobilis*, *T. putrescentiae*, *T. longior* (Gervais), *S. berlesei* (Michael), *Rhizoglyphus echinopus* (Fumouze *et* Robin), *L. destructor* (Schrank), *G. domesticus* (De Geer) and other *Glycyphagus* spp., that are often found at every stage of decomposition on animal carcasses and human corpses. Many of them are known from forensic investigation cases (Megnin, 1894; Russel *et al.*, 2004). Members of the Astigmata are more frequently found during the dry decay stage of decomposition. Some of them belong to the phoretic mites of scavengers, while the majority travel on late-arriving carrion insects, such as moths, hide and skin beetles (Perotti and Braig, 2009). Even though insects are larger, and hence more easily identifiable, in particular situations, an informative diversity of phoretic mites may be found on a decaying carcass at any given time, and their composition may provide valuable information in forensic investigations (O'Connor, 2009a). The genus *Tyrophagus* was one of the first mite taxa found on human corpses. It has been used to estimate a post-mortem interval (PMI). The question of how this genus arrives on a carcass is still unresolved (O'Connor, 2009a; Perotti 2009). Mites of the genus *Tyrophagus* are not considered to be phoretic, but their long setae might allow them to be lifted and carried by mammals (Serpa *et al.*, 2004; Marzouk *et al.*, 2006; Perotti and Braig, 2009). It should be stressed that except for *Tyrophagus formicetorum* Volgin, the remaining species of this genus do not go through the hypopial stage (heteromorphic deutonymph) in their life cycle. *Tyrophagus* species can be easily confused with other Acaridae species, such as *T. longior*. This species normally occurs together with another acarid mite, *A. farris*, which presents the deutonymphal phoretic stage (or hypopus). *A. farris* hypopi is even attached to the adults of *T. longior* (Hughes 1976; O'Connor, 2009a). *S. berlesei*, another reported mite species, may produce the hypopial stage, which could attach to chaffers (Scarabaeidae) and stored-product beetles (Tenebrioniidae), such as the flour beetle, *Tenebrio molitor* Linnaeus (Szelecz *et al.*, 2018). Moreover, mites of the genus *Rhizoglyphus* may also develop heteromorphic deutonymphs, which are morphologically similar to those of *Sancassania*. They are phoretically associated with several species of Scarabaeidae (Coleoptera), such as *Osmoderma eremicola* (Knoch), *Bothynus gibbosus* (De Geer) and *Phyllophaga* spp., which belong to the group of opportunistic colonizers of human and animal remains (Rai *et al.*, 2020). Many of the mites described above belong to the most numerous species in the analysed farm buildings. Further studies are needed to clarify the relationship between indoor conditions in farm buildings and the species composition of the storage mite acarofauna. Our study may increase awareness of the occurrence of many allergenic mite species in the locations investigated in the Żywiecki district, as well as being of use in forensic medicine cases (Solarz *et al.*, 2022).

8 Control

Recently, Bingham and Hagstrum (2024) highlighted that the importance of sanitation in food storage areas is sometimes overlooked. Sanitation is fundamental to ensure safe storage of food products. In the absence of a concrete strategy of sanitation, the stored commodities may be damaged by mould, rodents, birds and insect pests. According to the importance of the topic, they examined the history of sanitation legislation, courses and conferences, and research on the costs, benefits and efficacy of sanitation. Although they analysed mainly the role of insect pests, their conclusions may be applied to mites injurious to stored commodities. At this aim, Hoy (2011) reported that the use of traditional insecticides to manage mites has become increasingly difficult, and effective alternative treatments are urgently needed.

Control procedures in use may be divided into measures based on prevention and curative actions, respectively. In this context, physical, chemical or biological tools play a key role. The most important methods provide the use of extreme temperatures, atmosphere modification, inert dusts, natural and synthetic chemical pesticides, fumigants and predators (Bell, 2014). Already in the 1970s in Portugal, Guimarães (1970) reported that different factors are liable for affecting the start or the development of the populations of *Carpoglyphus lactis* (Linnaeus) on dried figs, and among these the most important is the aspect of hygienic factors, such as keeping storage and other contiguous compartments free from dust and other contaminating materials and the previous cleaning of any objects and cloths contacting figs. We discuss here these tools and the strategy of their application.

Food Preservation Techniques

Food preservation techniques play a fundamental role in preserving the quality of foodstuffs. The use of adequate structures able to intervene positively also in the control of pests is fundamental in the production chain. The technical discussion of these aspects is unfortunately beyond the competence of the authors, and the interested reader is invited to consult specific contributions. In general, the use of advanced technologies depends on the socio-economic conditions of the countries concerned.

Various examples may be reported. One among many examples concerns the study of Guimarães (1970) which reported that the type of storage compartment, on its own, represents a means of preventing infestation. In chambers that can be easily sealed, it is possible not only to prevent to a certain extent the entry of mites but also to maintain a long-lasting fumigant atmosphere at low concentration.

An interesting review by Shepherd (2012) discussed extensively the problem of grain storage in Africa. He reported some of the activities that have been promoted by governments and donors in the last few decades to improve grain storage at all stages of the chain, together with problems that have resulted from those actions. Concerning the storage at the farm level, farmers have consistently resisted innovations in storage, usually for sound economic reasons. These last include that traders buy close to their base after harvest and gradually move to distant areas. New storage methods are based on technological aspects, with little attention to farmers' economics. High losses justify storage improvements to researchers but farmers do not share their perception of the problem. Adoption of new storage technologies at the farm level has sometimes been disappointing. Many village stores remain unused, indicating a lack of consideration of their function within supply chains. As the assumptions behind cereal banks were questionable, it is unsurprising that most failed. Smaller traders usually work on the basis of rapidly turning around stock. Many countries continue to believe that intervention in the grain market is necessary. One ambitious trader plans a network of small

 The Handbook of Mites of Stored Products (V. Vacante and C.H.W. Flechtmann)
DOI: 10.1079/9781786390912.0008

rural silos to feed a central silo complex. There are a few examples of food security reserves having a major role in overcoming the consequences of crop failure. Instead of donors supporting new storage construction, governments should seek their assistance for field research. The South African Futures Exchange has revolutionized grain trading in that country. An enabling environment that is conducive to private investment is essential. Improved storage cannot be promoted in isolation from the main users of storage (Shepherd, 2012).

Prevention

Priority, the strategy of control of stored mites provides a rigid prevention. Jacobs and Calvin (1988) reported that prevention is the best strategy to avoid mite problems in stored grains. Proper bin sanitation before the introduction of new grain reduces the need for pesticides. Good sanitation involves the removal of old grain and dust in and around the grain bin. This includes the removal of old grain from corners, floors, and walls and grain that may have spilled on the exterior of the bin. Any grain remaining when a bin is emptied can harbour insect infestations that will move into the new grain.

After the bin is cleaned, and all needed repairs have been made, the floor and wall surfaces both inside and outside the bin should be treated. It is necessary to take special care to treat all cracks, crevices, and areas around doorways and other places where insects hide or enter. Spray the bins about 4–6 weeks prior to storing grain, if the grain is to be stored longer than 6 months (Jacobs and Calvin, 1988). Before grain is placed in a bin, it should be screened to eliminate fine materials and broken kernels. Grain placed in a clean bin should be checked at intervals of 2 weeks during warm months, and at intervals of 1 month during cooler months for the presence of hotspots, mouldy areas and mites. If any of these conditions occur, the grain should be aerated to lower the moisture level and temperature (Jacobs and Calvin, 1988).

Armitage and Chambers (2002) reported that drying cereal grain to less than 14.5% MC is the key to preventing mite problems. This level is in equilibrium with air at RH of 65% at 25°C. Low MC is critical for mite control. At temperatures below 5°C, mites stop breeding.

Nayak (2017) recommended a control starting with hygiene for controlling mites. Strict hygiene practices in and around grain or produce storage and handling areas are essential for control. Remove and destroy unwanted, infested produce by burning or burying. If infested produce is to be kept, fumigate as described below. Clean up and destroy grain dust and grain residues at least weekly. Clean grain storages and handling equipment before handling new uninfested produce.

Physical Methods

Physical control measures are the most effective way to avoid and tackle pest management issues. Main physical techniques include 'aeration', 'cold' and 'heat'. The more the suitable conditions for mite development, the more likely the products are at risk. Mites will develop faster at higher temperatures, and most infestations are due to prolonged storage under less-than-optimal warehouse conditions.

Amoah (2016) reported that physical methods involve the manipulation of the mites' environment to make it less suitable for their proliferation. These methods include manipulating light, temperature and humidity of the environment. García (2004) investigated some physical methods in five Iberian ham factories in the Iberian Peninsula and observed a reduction in the growth rate of mites on ham surfaces when humidity was reduced to levels below 60%. However, mites were forced to seek refuge in cavities inside the hams where they could feed and reproduce. Boczek (1991) reported that ham mite eggs are killed within 24 h when exposed to −15°C but will survive at −10°C. At −5°C, a longer exposure of 24 days is required to obtain similar results. Other researchers stated that freezing at −28°C for 48 h did not kill the eggs (Arnau and Guerrero, 1994). A more recent study suggested that 90% of all stages of *Tyrophagus putrescentiae* can be controlled by freezing to −8°C for 5 h (Eaton and Kells 2011). Rentfrow *et al.* (2006) suggested that hams are frozen to control mites on ham. However, repeated freezing and thawing and high heat may affect the quality of ham. Although manipulation of the environment may be effective, facility managers may require professional assistance to create and/or maintain the new environment. Environmental conditions that are cost-effective and do not significantly affect the quality and/or organoleptic properties of the products should be investigated.

Nayak (2017) highlighted the importance of managing temperature and moisture. Because of their need for warm, moist conditions, mite numbers can be kept low by reducing temperature below 20°C, for example, by aeration, or reducing moisture below 55% RH or 12% moisture for cereal grains, for example, by drying.

Hendrix *et al.* (2018) reported that since methyl bromide is an ozone-depleting substance, there is a need to find effective alternative methods to control mite infestations on dry-cured hams. They studied to determine the most effective RH and/or temperature combination to minimize mite reproduction and mould growth on dry-cured hams in untreated and food-grade ingredient-infused nets. Food-grade coating formulations of xanthan gum and propylene glycol, and carrageenan, propylene glycol alginate and propylene glycol nets were infused into nets. Dry-cured ham cubes and slices were wrapped with untreated and treated nets, inoculated with 20 large mixed-sex mites per cube or 50 large mixed-sex mites per slice, and then stored for 14 days at each temperature (24°C, 28°C and 32°C) and RH (55%, 65%, 75% and 85%) combination in an environmental chamber. *T. putrescentiae* on ham slices in untreated nets were reduced from the initial inoculum level of 50 mites per ham slice when exposed to 85% RH at 24°C, 28°C and 32°C. Nets infused with xanthan gum and propylene glycol or carrageenan, propylene glycol alginate and propylene glycol, completely inhibited mite reproduction at 85% RH. In addition, the nets infused with carrageenan, propylene glycol alginate and propylene glycol, controlled mite reproduction below the initial inoculum level on ham cubes and ham slices at all temperatures and relative humidities that were evaluated. Six to eight trained panelists rated the amount of mould on the ham slice surfaces on a 0–100% scale, and both treated nets evaluated were effective at preventing mould growth on the ham slices (Hendrix *et al.*, 2018).

Aeration

It has long been recommended that aeration may be applied to create somewhat low temperatures, to prevent storage insect infestations (Burges and Burrell, 1964). Aeration is commonly used to chill bulk grain in a temperate environment. A fan attached to a specially constructed ducting system creates the effect. This provides airflow to the base of a silo, or to equally spaced channels beneath the grain store's floor, which are covered by fine mesh metal panels to prevent grain infiltration. In the absence of a ventilated floor, reinforced perforated ducts can be installed on the floor, and manifolded to a fan-driven airflow before the arrival of the grain (Navarro, 2012). Grain drying or conditioning is frequently thought to be a secondary application for aeration. Many mite management programmes include aeration, which plays a critical role in preventive control at a cost that is competitive with other disinfestation methods (Bell, 2014). At humidities less than 55–60% (commodity at 12% MC or less), grain mites cannot survive. Grain that is to be stored for longer than 6 months may need a protective application of an approved insecticide (Jacobs and Calvin, 1988). These data are in agreement with what was found by Guimarães (1970), who observed that at 40%, 50% and 60% RH, the larvae of *C. lactis* (Linnaeus) live 4, 6 and 8 days, respectively. At 40% and 50% RH nymphs and adults managed to withstand a period between 10 and 15 days. Some eggs could hatch at 40% RH.

Low air pressure

Guimarães (1970) observed that a pressure of 3 mm Hg for 24 h had little effect on eggs and nymphs but was lethal for larvae and adults of *C. lactis*.

Cold

Cold treatments are commonly employed as part of IPM systems for stored items such as grains, cereals, oilseeds and seeds, in addition to the use of aeration systems. Guimarães (1970) observed a strong resistance of eggs of *C. lactis* at –20°C for 24 h, while other stages were killed. Fields (1992) examined the effects of cold on mites of durable stored products. Their reproduction stops at 10°C, and populations dwindle and finally die out. Various major pest species, such as bruchids, mites and some Lepidoptera, have tolerant immature stages (Fields, 1992; Bell, 2014). But cooling typically necessitates extremely long holding times to be effective.

Heat

Elevated temperatures are a strategy that has been used successfully for many years against stored-product pests. The thermal limitations of insects and mites are usually between 0°C and 45°C, and the rates of population increase are determined by temperatures within these limits (Mourier and Poulsen, 2000). Already in the 1970s, Guimarães

(1970) reported the positive contribution of high temperatures in the control of storage mites. This author observed that at 45°C for 30 min a total mortality of *C. lactis* occurred, 50°C for 8 min and 55°C for 4 min. At 50°C, the eggs were more resistant and required 10 min for complete control. In general, it is well known that more extreme temperatures have an acute influence on the vitality of Astigmatina. High temperatures, at their extreme, disrupt phospholipid membranes and intracellular proteins (Bale and Hayward, 2010). Heat disinfestation requires just that all particles of an infested grain batch be heated to a suitable fatal temperature/time combination, which is a rather straightforward process. Heat treatment affects different mites and bug species and stages differently; however, most species will not survive more than 12 h at 45°C, 5 min at 50°C and 30 s at 60°C. Death is related to the temperature to which the mites are exposed and the duration of exposure, but a recent study indicates that the rate at which the grain is heated is also important, with a faster rate of heating resulting in more mortality (Hallman and Denlinger, 2019).

Washing

Guimarães (1970) reported that washing of dried figs infested by *C. lactis* in rotating machines equipped with strong showers of water jets is a helpful factor in getting the figs rid of mites. However, removing the dead mites after fumigation is not easy. Usually, the wash water is heated, and the temperature employed is moderate and the duration of the treatment too short, about half a minute, to kill the mites. Moreover, the water immersion does not produce a positive effect; all stages exhibited a resistance greater than 10 days, and one nymph reaching a maximum of 73 days.

Inert dusts

Inert dust alters the structure of lipids through adsorption; the phenomenon is thought to be due to the surface migration of fatty molecules into the hollow crystalline structure of the dust particles (Ebeling, 1971), which also interrupts the lipid layers through physical sheering (Vincent *et al.*, 2003). These inert dust particles act via a chemically neutral mechanism and are not associated with any forms of resistance to mite controls; however, their use can be limited by environmental conditions including very high humidity (>80%) and high levels of environmental dust within farming units (Kilpinen and Steenberg, 2009). Refinement of materials selected for dusting could possibly have the potential to extend the longevity of this type of control, as could the use of dust in liquid form. Schulz *et al.* (2014), however, reported no overall significant difference between liquid and dust form silica-based controls.

Armitage and Chambers (2002) of DEFRA (Department for Environment Food and Rural Affairs) in the 'Preventing and Controlling Mites in Stored Cereals (HGCA Topic Sheet No.62)' reported that only organophosphorus (OP) compounds were approved for admixture to grain. In cereals, mites can be discouraged by accurate application to the bulk of a liquid OP at recommended rates. Treatment of existing infestations was more difficult as they are not very susceptible to Ops, and there is little margin for error. Dust formulations are no longer available and will not be approved for admixture to grain.

An HGCA-funded project therefore assessed several alternatives to OPs, particularly for mite control at the grain surface. Diatomaceous earth (DE), a siliceous dust, resulted to be effective. Unlike a conventional pesticide, DE produces physical control by desiccating insects and mites. However, DE may limit grain flow if the entire bulk is treated, so it can only be incorporated at effective application rates to the grain surface. DE application will only give longer-term control when combined with an effective drying and cooling regime.

Cook *et al.* (2008) carried out experiments to examine the accepted modes of action of the desiccant dust, diatomaceous earth (DE), against the flour mite, *Acarus siro* Linnaeus. Adult mites were exposed to three DE products for periods of 3, 18 and 72 h under conditions designed to allow partial desiccation of the mites without causing mortality. After exposure, the DE dust particles were washed off the mite bodies, and both the DE and the mites were examined for the presence of cuticular hydrocarbons by gas chromatography-mass spectroscopy (GC-MS) analysis. GC-MS identified seven major cuticular lipids of chain length C (13)–C (26) that may have a role in the waterproofing of *A. siro*. After 18 h exposure, n tridecane and several different long-chain fatty acid ethyl esters were detected on one of the DE products. After 72 h, n-tridecane was detected on all three DE products. Mite samples retained after the removal of the

DEs were examined by microscopy. Tentative evidence was observed by conventional low-power microscopy that might indicate the uniform removal of the epicuticle. However, a detailed examination by scanning electron microscopy showed no signs of abrasion (Cook *et al.*, 2008).

To prevent mites occurring at the surface of grain, rake DE into the top 0.3 m at 200 g/m^2 (1g/kg). To deal with existing surface infestations treat with 600 g/m^2 (3 g/kg). The effect of DE is to limit population growth. Control of the least susceptible pests, grain weevils, could take several weeks.

According to the above-mentioned report, diatomaceous earth (DE) has long been utilized as stored-grain protectant (Stejskal *et al.*, 2021). Because it has a physical mode of action and is picked up as pests walk over a treated surface, it has proven useful as a grain protectant. The most widely accepted theory of its mode of action is that damage to the cuticle is caused by the removal or sorption of cuticular waxes, resulting in water loss from the body, which, depending on the RH of the air, leads to death by desiccation. Stored-product mites rely entirely on cutaneous respiration and are hence poorly sclerotized and susceptible to desiccation. The key advantages of inert dusts are that they have low mammalian toxicity, do not contain insecticides or knock-down agents, do not leave harmful residues, are efficient against chemically resistant species, and are persistent and stable at high and low temperatures (Ziaee *et al.*, 2021).

Kılıç (2022) evaluated the acaricidal activity of two Turkish diatomaceous earth (DE) formulations (Detech® WP95 and Detech® Dust) applied on a concrete surface at five different concentrations (1, 2.5, 5, 7.5 and 10 g/m^2) and dead individuals were counted at 11 different time intervals (1, 3, 6, 9, 12, 15, 18, 21, 24, 27 and 30 h) at a temperature of 25 ± 1°C and 75 ± 5% RH. Mite mortalities were observed after 6- and 18-hour exposure periods at all concentrations of dust and wettable powder (WP) formulations, respectively. Specifically, 100% mortality for the WP formulation was achieved at the highest concentration of 10 g/m^2 after 15 h of exposure, and after 27 and 30 h for the lowest concentration. In the case of dust formulation, mortalities were observed after 3 h of exposure at all concentrations except at 1 g/m^2, while a 100% mortality rate was achieved after 21 h of exposure to all concentrations, and after 18 h of exposure for 7.5 and 10 g/m^2. The results of this study showed that both WP and dust formulations of local diatomaceous earth can cause 100% mortality in 24 h on average and can be a promising alternative to conventional chemical acaricides.

Screening, Sorting and Sanitation

Sanitation is currently regarded as the foundation for many IPM programmes for stored products by industry experts. Increased stock rotation, crack and crevice treatments with residual chemical products, regular sweeping, vacuuming, and removing food dust and debris, closing doors and windows, using air curtains near entrances, screening, and space treatments that include fogging with aerosols are common IPM tactics for improving sanitation in food facilities. Other parts of excellent warehouse practice, such as stock rotation and, where appropriate, insect-proof packaging, also help to lessen pest population pressure. Other methods include sieving, screening, projection separation and aspiration (Trematerra and Fleurat-Lessard, 2015).

Chemical Methods

Chemical control was the mainstay of the agricultural and food sectors throughout the majority of the twentieth century, but there has recently been an increased demand to reduce chemical use and reliance on pesticides to minimize long-term health and environmental effects. Chemicals continue to play an important role in the preservation of food commodities and products, although the number of compounds accessible for usage is decreasing, and increasing focus is being directed to non-chemical alternative control approaches (Mohapatra *et al.*, 2017). Various pesticides have been recommended to control storage mites and a number of them have been shown to be effective on the ham mite. They include 'organophosphorus compounds and other chemicals' and 'fumigants'. Regardless of the validity of action of the various active ingredients discussed below, it remains the obligation to comply with the legislation of each country before their possible use. This obligation also applies to trade between one country and another.

Organophosphorus Compounds and Other Chemicals

Studies on fenitrothion and pirimiphos-methyl have shown contradicting results (Hartmannova *et al.*, 1973; Chmielewski, 1987; Nayak, 2006). Other pesticides tested by Nayak (2006) included chlorpyrifos-methyl, deltamethrin, pyrethrin + piperonyl butoxide, s-methoprene and spinosad. Out of these, only pyrethrin + piperonyl butoxide, s-methoprene and spinosad controlled the mite population after at least three weeks of exposure to treated wheat. In a more recent study, seven commercial residual pesticides were tested by exposing mites to treated surfaces. Among the tested pesticides, deltamethrin (Centynal®), chlorfenapyr (Phantom®) and chlorpyrifos-methyl + deltamethrin (Storcide II®) showed promising control on mite populations (Abbar *et al.*, 2013). Although some pesticides may be effective in controlling mites under laboratory conditions, their use is restricted in ham facilities due to concerns about residues (Sánchez-Ramos and Castañera, 2003a). One suggestion would be to treat surfaces in ageing rooms before the cured hams are transferred to these rooms (Abbar *et al.*, 2013).

Nayak (2017) reported that methoprene (e.g. Rizacon S®, IGR®) is registered as a stored grain treatment for insect control and is known to be effective against stored mites. However, the European Pesticides database did not approve this active ingredient. A pyrethrum insecticide product (Pestigas®) and spinosad (Conserve Plus®) can also control these mites. Commonly employed protectant insecticides that are known to be ineffective against these mites are pyrimiphos-methyl (Actellic®), chlorpyrifos-methyl (Reldan®) and fenitrothion (various trade names).

Fumigants

Amoah (2016) reported that chemical control methods are commonly used in stored grain, unfinished food products (Stará *et al.*, 2011c) and ham processing facilities (Rentfrow *et al.*, 2008). Registered postharvest fumigants for stored-product pests include methyl bromide, phosphine and sulfuryl fluoride, while ozone and carbon dioxide are likely allowed in most or all countries. In some countries, carbon sulfide (OCS) has been used, and recently the use of nitric oxide (NO) has been proposed.

Methyl bromide

This chemical has been employed to control pests in commodities and buildings worldwide since the 1930s, because of its large spectrum application and rapid toxicity (Fields and White, 2002). It is non-flammable and non-corrosive (Bond, 1984) and does not leave residues on products. Methyl bromide (MB) kills mites, insects, nematodes and microflora rapidly by damaging the membrane of nerve cells (Fields and White, 2002). Guimarães (1970) reported that in Portugal, this fumigant was used by fig exporters with apparent good results. Methyl bromide is the only known fumigant effective at eradicating existing *T. putrescentiae* (Schrank) infestations. Typically, methyl bromide fumigation occurs one to five times per year, when used by a ham facility (Rentfrow *et al.*, 2006). With limited availability, ham producers will soon need to rely on seven alternative fumigants combined with better management programmes to help effectively manage the pest (Amoah, 2016). Despite these dated reports, it is necessary to mention that methyl bromide fumigation, despite being an effective treatment to ensure that products pose a minimal phytosanitary risk, when released into the atmosphere, depletes the ozone layer and allows increased ultraviolet radiation to reach the earth's surface. To resolve this problem, many countries signed the Montreal Protocol on Substances that Deplete the Ozone Layer in 1987, which establishes legally binding controls on the national production and consumption of ozone-depleting substances. Specifically, MB has been listed as such a substance since 1992, and the European Union has prohibited its employment in accordance with Regulation (EC) No. 1005/2009 since 2010.

Phosphine

Phosphine is reported as the most widely employed fumigant alternative to methyl bromide (Fields and White, 2002) due to its easy application, effectiveness, low cost, lack of residues (Zuryn *et al.*, 2008) and its ability to disperse easily during fumigation (Nath *et al.*, 2011). Total mortality of all life stages has been achieved under laboratory conditions (Zhao *et al.*, 2012). In a simulated ageing house that contained ham mite assays and dry cured ham inoculated with ham mites, 99.8% mortality was recorded two weeks post fumigation (Zhao *et al.*, 2015). Phillips and Schilling (2013) placed mite colony jars in treated and untreated ham ageing rooms and confirmed high mortality in treated ageing rooms. However, strong corrosion to electrical wiring and equipment occurred in the fumigated room about three weeks after fumigation. Corrosion to metals is the major drawback that prevents phosphine from being an ideal fumigant in all structures (Price and Mills, 1988; Brigham, 1998; Adler, 2004). If ham ageing houses could be well sealed and all electrical devices removed or protected, phosphine could be considered a potential alternative to control mite infestations. It is an important control strategy used, when an infestation is discovered in bulk commodities in storage or during shipment, as well as for whole-site treatment of food processing facilities. Adult and nymph mortality was determined by a lack of movement in response to physical stimulation after a two-day recovery period, and unhatched eggs were deemed dead 4 days after fumigation (Abbar *et al.*, 2018). Phosphine fumigation is particularly successful at controlling both the mobile stages and the eggs of *T. putrescentiae* at label rates and normal operating temperatures, with little effect on market quality (Zhao *et al.*, 2016). The temperature has a significant impact on insect control fumigation efficacy. Higher temperatures stimulate insect metabolism and respiration, causing treated insects to consume more fumigant at a faster pace. Chemical approaches such as fumigation and spraying with organophosphorus chemicals are used to control *T. putrescentiae*. However, persistent exposure to organophosphorus chemicals has resulted in resistance in some cases.

Despite these limits, Nayak (2017) highlighted the importance of disinfestation with phosphine fumigation. Fumigation with phosphine killed mites only if it is done in a sealed, gas-tight storage. Dosages lower than recommended on the label often allow mite eggs to survive during the fumigation. Reinfestation due to subsequent hatchings can then occur after the fumigation period is over. Apply 1.5 tablets per cubic metre and leave the enclosure sealed for 7 days if the temperature of the produce is above 25°C, or for 10 days at lower temperatures. Air the treated produce before it is handled.

The Eurofins Food & Feed Testing laboratory network (2022) offers the detection of fumigants like phosphine and sulfuryl fluoride in cereals and oilseeds as well as in pseudo cereals, legumes, spices, nuts, herbs and dried fruits at one location. Levels of phosphine are determined as the total sum of phosphane and phosphide salts, according to the maximum residue limit definition (Regulation EC No 396/2005). For conventional foods, matrix-based limits between 0.01 and 0.7 mg/kg are applied in the EU. In organic cultivation, the use of phosphane in empty containers and silos is allowed, but not for the food itself. The recommended BNN (German Association of Organic Processors, Wholesalers and Retailers) orientation value is <0.01 mg/kg. The acceptance of organic food containing over 0.001 mg/kg is currently being questioned in Switzerland (see Bio-Suisse, federation of Swiss organic farmers).

Unfortunately, pests with high levels of resistance toward phosphine have become common in Asia, Australia and Brazil. High-level resistance is also likely to occur in other regions but has not been as closely monitored. Genetic variants that contribute to high-level resistance to phosphine have been identified in the dihydrolipoamide dehydrogenase gene (Schlipalius *et al.*, 2012). Identification of this gene now allows rapid molecular identification of resistant insects.

Sulfuryl fluoride

In the United States, sulfuryl fluoride has been registered for use in dry cured ham processing facilities since 2005 (EPA, 2005). However, the use of this fumigant under laboratory conditions only resulted in 95% mortality of mite eggs at 20–22°C, even when applied at approximately 100.3 g/m^3, which is three times the label rate (Phillips *et al.*, 2008). More recently, Phillips and Schilling (2013) reported that sulfuryl fluoride may be effective at controlling all life stages of the mite when tested for 48 h at 35°C under laboratory conditions. Since 2009, the approval of sulfuryl fluoride by the BVL (German Federal Office of Consumer Protection and Food Safety) includes its application in dried fruit, as well as for the disinfection of empty mills and rooms, and its application on nuts, as well as for various types of cereals, which are stored in warehouses or under gas-tight tarpaulins. For sulfuryl fluoride, matrix-based maximum limits between 0.01 and 0.02 mg/kg (excluding 0.05 mg/kg for cereals and 10 mg/kg for nuts) are applied. The orientation value of <0.01 mg/kg recommended by the BNN also applies here.

Ozone

Under laboratory conditions, Hassan *et al.* (2010) found that an ozone concentration of 155 ppm for 48 h produced the total mortality of all stages of *T. putrescentiae*. However, in the practical application, ozone may not work well since it does not penetrate surfaces well.

Carbon dioxide

The use of CO_2 at high pressure, in pressure chambers also produces the total control of mites (Adler, 2004; Navarro, 2006; Riudavets *et al.*, 2010). Treatment of *T. putrescentiae* with CO_2 at two high pressures, 15 and 20 bars, produced over 99% mortality for mixed stages of *T. putrescentiae* on artificial standard diets when treated with CO_2 at 15 bar for 15 min, and 100% mortality when the exposure time was increased to 30 min. Although CO_2 may be effective, increasing the pressure may mean an increase in the cost of treatment as more CO_2 is required (Riudavets *et al.*, 2010). Mortality of all stages of mites has been achieved at a concentration of 60% CO_2 with 144 h of exposure (Hassan *et al.*, 2010). However, this may not be practical since the duration of exposure would require that treated facilities remain closed for longer periods and this would interfere with the operation of the facility.

Carbon sulphide

Guimarães (1970) reported that in Portugal, the control of the moths is carried out also with fumigation of carbon sulphide, but the available data are not sufficient to confirm the efficacy of this active ingredient against *C. lactis* (Linnaeus), although in rare cases the surviving of mites have been observed.

Nitric oxide

Yang *et al.* (2022) reported the efficacy of nitric oxide (NO) fumigation, a recently discovered fumigation treatment under an ultra-low oxygen environment, at various concentrations, and time under the laboratory conditions at 25°C against different mite stages on both dietary media and ham meat. Their results showed that NO fumigation was effective against all mite stages and 100% control was achieved. Generally, the egg was the most tolerant stage and required 48-, 24-, 16- and 8-h treatments to achieve 100% mortality at 0.5%, 1%, 1.5% and 2% NO concentration on dietary media, respectively. *T. putrescentiae* (Schrank) mobile immatures and adult stages were less tolerant, and 100% mortality was achieved after 16-, 8-, 8- and 4-h treatment at 0.5%, 1%, 1.5% and 2% NO, respectively. The median lethal concentration (LC50) of NO on eggs was 0.86%, 0.68% and 0.32% for 8-, 16-, and 24-h treatments. In addition, a confirmatory test was conducted on ham meat at 0.5% and 1.0% of NO and similar efficacy was found. Complete control of eggs was achieved after 48- and 24-h treatment at 0.5 and 1.0% of NO, respectively, and larvae and adult mites were 100% controlled after 16 and 8 h at 0.5% and 1.0% of NO, respectively. Their results demonstrated that NO fumigation was effective against *T. putrescentiae* and can be a potential alternative treatment to methyl bromide for cured-ham pest control.

Biological Control

Biological control is described as the action of natural enemies that check the host's or pest's population density. Predators, parasites (or parasitoids) and pathogens are the three types of natural enemies of mites. Biological agents, which range from microbiological diseases to predatory insects, are often host-specific and are best used as preventative control techniques to prevent pest populations from growing. To overcome pests before they may cause harm, mass release or augmentative approaches are required. Bacteria, viruses, protozoa, nematodes and fungi are examples of mite pathogens. This diverse range of species happens in nature. They must be applied to the exact scenario in which they are being used as control agents. Although entomopathogenic fungi have long been thought to have the ability to control insect pests, there have been questions about their safety and specificity of action. Recently, one species *Beauveria bassiana* (Balsamo) has been brought back into focus, as a subject of the current study for usage against grain pests. *Metarhizium* is another genus of fungal disease having the ability to manage pests in stored-product environments. Predatory mites are another group of predators, preying on insect eggs and small larvae. The ascid mite *Blattisocius dentriticus* (Berlese) and the cheyletid mite *Cheyletus eruditus* (Schrank) prey on stored food mite species (Bell, 2014).

Amoah (2016) reported that contradicting results on the biological control of *T. putrescentiae* by *Cheyletus malaccensis* Oudemans have been registered in different countries (Palyvos and Emmanouel, 2004; Cebolla *et al.*, 2009). *C. eruditus* has been employed to control mites in stored grains in the Czech Republic (Ždárková, 1998). However, the release of natural enemies against stored mites may not be very common because the introduction of natural enemies may contaminate the commodity. Contamination of commodities by natural enemies is as bad as contamination by pests. In the United States, the Federal regulations under the Federal Food, Drug and Cosmetic Act do not distinguish pest fragments contributed by stored-product pests from those contributed by natural enemies. None the less, depending on the commodity, most of the natural enemies can be removed (Hagstrum and Subramanyam, 2006).

Natural enemies of the grain mite, *A. siro* Linnaeus are fundamentally included in the family Cheyletidae (Solomon, 1962); known species are *C. eruditus*, *C. aversor* Rudendorf, *C. trouessarti* Oudemans and *C. malaccensis* Oudemans (Ždárková, 1979; Lukás *et al.*, 2007). Cheyletus mites feed primarily not only on mites but also on the eggs of stored-product insects (Ždárková, 1998).

Pritchard *et al.* (2015) reported that on the poultry red mite *Dermanyssus gallinae* (Dugés) may feed different predatory species including the laelapids *Hypoaspis miles* (Berlese), *Hypoaspis aculeifer* (Canestrini), the phytoseiids *Amblyseius degenerans* (Berlese) and *Phytoseiulus persimilis* Athias Henriot, though feeding success as part of experimental *D. gallinae* controls, has proven to be dependent on environmental conditions and absence of alternative prey (Leśna *et al.*, 2009, 2012; Ali *et al.*, 2012). The phytoseiid mite *P. persimilis* preys mainly on the two-spotted spider mite, *Tetranychus urticae* Koch, and has been shown to be attracted to volatile compounds produced by plants fed on by *T. urticae* (Drukker *et al.*, 2000; De Boer and Dicke, 2004). A hypothetical control of the poultry red mite could be, for instance, the addition of such predator attractants to areas typically inhabited by PRM.

Food-Grade Oils and Preservatives

Amoah (2016) reported that food-grade oils and other commonly used preservatives, 'Generally Regarded as Safe' (GRAS), are effective in controlling *T. putrescentiae* (Schrank). A common method employed several decades ago was to rub lard on cured meat to prevent flies and bacteria (Smith, 1923). This method is still being used today, as coating hams with vegetable oils or hot lard is a common practice in Spain to control mites in dry-cured ham (García, 2004), although the ageing time may be extended since lard may make the ham less permeable to moisture. Lard has also been shown to be effective on ham mites under laboratory conditions (Abbar *et al.*, 2016). Propylene glycol, an odourless, colourless solvent registered as a bacteriostat and fungistat (EPA, 2006), has been tested and reported to be effective in controlling mite infestations by preventing mite reproduction under laboratory conditions (Abbar *et al.*, 2016).

Plant Extracts and Oils

Chemical methods, such as fumigation, spraying with organophosphorus compounds, or treatment with benzyl benzoate, dibutyl phthalate and N, N diethyl-m-toluamide, to control stored mites are prohibited because of human health hazards associated with their consumption (Sushma *et al.*, 2022). Hence, there is an urgent need to find more selective, natural compounds, non-toxic to humans and which do not affect the organoleptic character of the treated product. In this regard, plant-derived acaricides are very important. Plants and their derivatives are being employed effectively against mites worldwide, with the aim to build a new pest control strategy for stored-food mites (Gulati and Mathur, 1995; Divekar *et al.*, 2022). Botanical pesticides are environmentally friendly, quickly biodegradable and harmless to non-target creatures. Furthermore, many plant-derived natural compounds that function against insects can be made using locally accessible raw materials. Few research works have been carried out for managing stored food mites, although extracts and essential oils of some plants, such as *Phoeniculum vulgare* Mill. fruits, are insecticidal agents (Kim and Ahn, 2001). Plant alkaloids, secondary metabolites and essential oils are among the substances generated from plants. A recent study has shown that plant products or extracts, such as *Azadirachta indica* A. Juss, *Satureja hortensis* Linnaeus, *Datura stramonium* Linnaeus, *Citrullus colocynthis* (Linnaeus) Schrad., *Eucalyptus* sp. and *Melia azedarach* Linnaeus, have serious importance (Idrees *et al.*, 2016). Pyrethrins have activity on a wide range of insects and mites, including flies, fleas, beetles and spider mites (Ujvary, 1999).

We summarize below the most important aspect of the topic, basically drawn from the review of Nikolaou *et al.* (2021). The latter authors reported that traditionally the control of mites was based on the use of synthetic acaricides such as organophosphates, pyrethroids, pyridazines, juvenile hormone analogues and chitin synthesis inhibitors (Dekeyser and Downer, 1994; Collins, 2006; Hubert *et al.*,

2007); in addition, elevated CO_2 concentrations are applied (Conyers and Bell, 2003). However, benzyl benzoate, a substance that is not only produced and used industrially but which is also naturally produced by a range of plants – cinnamon and cassia (*Cinnamomum* spp.), carnation (*Dianthus caryophyllus* Linnaeus), hyacinths (*Hyacinthus* spp.), tuberose (*Agave amica* Medik.), common jasmine (*Jasminum officinale* Linnaeus) and Santos mahogany (*Myroxylon balsamum* (Linnaeus) Harms) is often used as a commercial acaricide or as a reference substance in acaricidal tests. Unfortunately, this substance is regarded as an allergen (Ong *et al.*, 2012).

Azadirachtin (neem), a commercially available plant-derived pesticide obtained from the plant *A.* indica, is one of the most commonly used natural substances that cause lethal and sublethal insecticidal and miticidal effects. It limits the development of populations and causes mortality of *T. putrescentiae* (Schrank), *A. siro* Linnaeus and *Lepidoglyphus destructor* (Schrank) (Collins, 2006), but it proved to be less effective against five species of stored-product mite pests than synthetic commercial miticides (Stará *et al.*, 2011b).

The components of clove bud (*Eugenia caryophyllata* Thumb.) oil-methyleugenol (median lethal dose (LD_{50}) = 1.18 μg/cm^2), isoeugenol, ß-caryophyllene, eugenol and α-humulene (LD_{50} = 12.90 μg/cm^2) showed lethal toxicity against *T. putrescentiae*, with methyleugenol and isoeugenol being more toxic than benzyl benzoate (Kim *et al.*, 2003).

Rosmarinus officinalis Linnaeus essential oil (EO) and its constituent compounds were active against *T. putrescentiae*, both as a fumigant (LD_{50} = 8.24 μg/cm^3) and in contact toxicity (LD_{50} = 5.49 μg/cm^2). The constituent camphor appeared to be successful in combating the pest (fumigant toxicity LD_{50} = 2.25 μg/cm^3, contact toxicity LD_{50} = 1.34 μg/cm^2) more effectively than benzyl benzoate (LD_{50} = 12.56 μg/cm^3, and 9.03 μg/cm^2) (Jeon *et al.*, 2014). Other constituent substances in rosemary oil are α-pinene, 1,8-cineole and camphene, all exhibiting miticidal activity (Jiang *et al.*, 2009).

Lee (2015) proved the fumigant and contact toxicity of the essential oil of *Ligustrum japonicum* Thunb. leaves against *T. putrescentiae* and calculated the respective LD_{50} values to be 16.48 μg/cm^3 and 8.02 μg/cm^2, respectively. α-pinene is one of the compounds found in *L.* japonicum oil, showing the highest percentage (Lee, 2015). These data indicate that camphor may be effective, as well as other compounds. However, camphor seems to be more toxic to mites than α-pinene (Kim *et al.*, 2003).

Ottoboni (1992) reported that caraway (*Carum carvi* Linnaeus) essential oil was a significant candidate to combat *L. destructor*, *Gohieria fusca* (Oudemans), *A. siro* and *T. putrescentiae* (Afifi and Hafeez, 1988; Ottoboni, 1992). Furthermore, 3,4-methylenedioxybenzene and its derivatives were described as successful miticides against *T. putrescentiae* (Song *et al.*, 2011).

Apiol, contained in the seeds of parsley (*Petroselinum sativum* Hoffm), was not toxic against *T. putrescentiae*, although it has been found to be active against *Dermatophagoides* species.

Essential oils obtained from both the aerial parts and seeds of the forget-me-not plant (*Myosotis arvensis* (Linnaeus)) or ingredient compounds used individually, namely 2,4,5-trimethylbenzaldehyde, 2,4-methylbenzaldehyde, 2,5-dimethylbenzaldehyde, 2-methylbenzaldehyde, 2,3-dimethylbenzaldehyde, 3-methylbenzaldehyde, 4-methylbenzaldehyde, 3-octanone, butyl isothiocyanate and nonanal, showed significantly greater contact and vapour toxicity against *T. putrescentiae* than benzyl benzoate (Park *et al.*, 2017).

T. putrescentiae was proven to be susceptible to the essential oil of garlic (*Allium sativum* Linnaeus), basil (*Ocimum basilicum* Linnaeus) and fenugreek (*Trigonella foenum-graecum* Linnaeus), in descending order, when exposed for 1–3 days (Alkot *et al.*, 2020). However, the sulphide-rich garlic essential oil was toxic to *C. malaccensis* Oudemans, a predatory mite and natural enemy of *A. siro*, *T. putrescentiae* and *L. destructor* (Satyal *et al.*, 2017). None the less, garlic essential oil and its active compounds can be used as possible miticides against a range of mites (Mossa *et al.*, 2018).

Lee *et al.* (2006) compared the acaricidal activity of *Phoeniculum vulgare* (fennel) seed oil components with benzyl benzoate, dibutyl phthalate and N, N-diethyl-m-toluoamide against *T. putrescentiae* and obtained promising results.

De Assis *et al.* (2011) evaluated the fumigant toxicity of eugenol and essential oils from cinnamon (*Cinnamomum zeylanicum* Blume), Surinam cherry (*Eugenia uniflora* Linnaeus), uvalha (*Eugenia* uvalha Cambess.), weeping paperbark (*Melaleuca leucadendra* (Linnaeus)), cake bush (*Piper marginatum* Jacq.) and Brazilian peppertree (*Schinus terebinthifolia* G. Raddi) against *T. putrescentiae*. The lowest median lethal concentration (LC_{50}) values were obtained for eugenol and *C. zeylanicum* essential oil. These data are in agreement, since eugenol is the major component of *C. zeylanicum* essential oil, and its content may exceed three-fourths of all the volatile ingredients of this oil (De Assis *et al.*, 2011). The next most abundant component – linalool is present in about nine times lower amounts (Raina *et al.*, 2001). Although it is not as important, linalool may increase toxic effects since it appears to be toxic against *T. longior*, in contact and fumigant toxicity studies. Similarly, the toxicity of menthol, menthone, fenchone, linalyl acetate and eucalyptol, the most abundant substances in the essential oils of lavender (*Lavandula angustifolia* Mill, *Lavandula* stoechas Linnaeus), peppermint (*Mentha x piperita* Linnaeus) and eucalyptus (*Eucaliptus globulus* (Labill.)), should be assessed for their lethality (Perrucci, 1995).

The observation of the variable susceptibility of various species of mites to natural substances is also seen in the case of natural aldehydes used as miticides. This phenomenon was studied for three natural aldehydes, namely, (2E)-hexenal, (2E, 6Z)-nonadienal and (2E)-nonenal, produced in plants from organic acids, when tested in feeding tests from 36 to 314 mg/g against *A. siro*, *A. ovatus* and *T. putrescentiae*. Specifically, the susceptibility of *A. siro* was similar for all three aldehydes, whereas *T. putrescentiae* was about eight times more susceptible to nonadienal than to hexenal and was not significantly affected by nonenal (Hubert *et al.*, 2008). In another study, benzaldehyde (LD_{50} = 4.23 μg/cm^2) isolated from the peach (*Prunus persica* (Linnaeus) Batsch), as well as salicylaldehyde (LD_{50} = 1.02 μg/cm^2), cinnamaldehyde (LD_{50} = 1.66 μg/cm^2) and phthaldialdehyde (LD_{50} = 5.16 μg/cm^2), were tested against *T. putrescentiae*. All aldehydes exhibited better efficacies than benzyl benzoate (LD_{50} = 9.75 μg/cm^2). However, the values calculated for *P. persica* essential oil (LD_{50} = 11.23 μg/cm^2) were higher than those for benzyl benzoate (Sung *et al.*, 2006). This research shows that aldehydes and essential oils can be applied in grain and food protection. However, their application must be carefully adjusted to the tested species.

There are suggestions that jasmonic acid (JA) can be used as a miticide against mite-pests of grains and stored food. This compound may affect mite reproduction and limit losses. The tomato mutants that were unable to accumulate JA were characterized by a higher rate of egg-hatching of mite pests compared to the wild type. Therefore, JA was suggested as a substance of ovicidal activity (Ament *et al.*, 2004). Most interestingly, JA additionally attracts predatory mites and therefore it may decrease the level of pests (Smart *et al.*, 2013). These data seem to open a field of interesting further research on the species of interest.

Angiosperms also deliver bioactive substances that may be useful in limiting losses caused by mites. For instance, essential oils obtained from gymnosperm plants such as *Pinus pinea* Linnaeus, *Pinus* halepensis Mill., *Pinus pinaster* Aiton and *Pinus nigra* Arnold were described as toxic for *T. putrescentiae* (Macchioni *et al.*, 2002), with *P. pinea* being the most effective. Moreover, 1,8-cineole and limonene showed miticidal activity when tested at 8 or 6 μl on 6 cm of filter paper. Likewise, *Juniperus chinensis* Linnaeus essential oil and its respective components were applied in impregnated disc biotests against *T. putrescentiae*, with the LD_{50} values calculated at 38.1, 15.33 and 42.85 μg/cm^2 for the essential oil, bornyl acetate and α-eudesmol, respectively (Lee *et al.*, 2009). Based on the effect and the content of the substances in the oil, the authors suggested that bornyl acetate is the major substance responsible for this acaricidal activity. The authors also reported that sabinene and α–thujene were not toxic to the mites.

Tests carried out by Heikal (2015) in Egypt against adult stages of *Dermatophagoides farinae* Hughes and *Dermatophagoides pteronyssinus* (Trouessart) showed that lemon grass oil gave the highest toxicity in comparison with geranium or thyme oils. The mortality percentage of both mites was about 100% at 8000 ppm concentration; the LC_{50} of lemon grass was 228.992 and 293.615 ppm against the two mites, respectively. Results of the research suggest that it is preferable to apply the control during the summer season when the mite population density is the least; moreover, the botanical oil extracts effectively controlled both dust mites and can be employed in the biological control programme, as well as play an effective role in the integrated management programmes (Heikal, 2015). Sánchez-Ramos and Castañera (2000) tested 13 monoterpenes, and seven of them including pulegone, menthone, linalool and fenchone were effective against the mobile stages of *T. putrescentiae* by vapour action. However, none of them were effective against eggs.

The above-mentioned results prove that plant-derived extracts and single compounds may become interesting alternatives to commercial miticides. In many cases, they are obtained from plants that are non-toxic to humans, since they are part of the human diet, for example, garlic or parsley. Therefore, it is reported that they are safe for humans, of course depending on the concentration used (Nikolaou *et al.*, 2021).

Unfortunately, some botanical extracts are lethal for beneficial insects, such as citronella, eucalyptus, garlic, pyrethrum and neem. Sometimes, their effects may be non-lethal, such as inhibiting natural enemies from utilizing prey, reducing prey availability, decreasing reproduction, inhibiting the ability of natural enemies to recognize prey, influencing the sex ratio and reducing mobility. None the less, detailed knowledge of the lethal or non-lethal effects of botanical pesticides on beneficial insects is essential for the sustainable control of insect pests and pollination activities, for improved and sustainable agricultural production (Ndakidemi *et al.*, 2016).

Hubert *et al.* (2005) studied the activity of a novel acaricidal compound with inhibitory effects on the digestive enzymes of arthropods and reported that they are a safe alternative to the traditional neurotoxic pesticides used for control of the stored-product pests. They explored the properties of acarbose, the low molecular weight inhibitor of α-amylases (AI), as a novel acaricide candidate for protection of the stored products, from infestation by *A. siro* Linnaeus. *In vitro* analysis revealed that AI blocked efficiently the enzymatic activity of digestive amylases of *A. siro* and decreased the physiological capacity of the mite's gut in utilizing a starch component of grain flour. *In vivo* experiments showed that AI suppressed the population growth of *A. siro*. The mites were kept for 3 weeks on an experimental diet, enriched by AI in the concentration range of 0.005–0.25%. Population growth of *A. siro* was negatively correlated with the content of AI in the treated diet, with a half-population dose of 0.125%.

Genetic Control

Studies on the development of mites in wheat (*Triticum aestivum* Linnaeus), naturally possessing an inhibitor of α-amylase which protects the grain from attack by arthropods, suggested that the quality of the grain is known to affect the growth of mite populations, and that the inhibitory effect of α-amylase on the population density of mites depends on the stored grain, on the species of mites and on the combination of these two factors, thus affecting the growth dynamics of mites (Hubert *et al.*, 2014b). A previous study evaluated the use of bean (*Phaseolus vulgaris* Linnaeus) in the controlling of storage mites and in domestic dust, since its use appears to suppress the growth of mites. However, some people have a food allergy to beans, making its commercial and large-scale use (Hubert *et al.*, 2007) unfeasible.

Management of Storage Mites

The control of stored mites cannot be entrusted to a single means but to different means harmoniously integrated with each other. In this way, an integration of the available means is obtained, and an integrated control system (IPM) is produced. This technical possibility can disregard the natural presence and/or the release of beneficials, mainly cheyletid mites. However, the limits of the technical possibilities available today do not allow other solutions.

The main physical agents that determine the quality of the stored products are air, humidity and temperature (Ždárková, 1991a). Thus, it is essential that there is better control of these variables in the grain storage units because of the sanitary importance that contamination by mites represents to human health and economy (Franzolin and Baggio, 2000). Foods that have a high content of water are more difficult to control and should be monitored more frequently (Sánchez-Ramos *et al.*, 2007c). The Brazilian law states the maximum moisture for flour, cereal starch and bran (15%), potato starch (21%) and manioc starch (18%) (ANVISA, 2005). Guimarães (1970) reported that keeping the atmospheric humidity in storage compartments at 65% RH or less may be a good procedure to control *C. lactis* (Linnaeus) on dried figs. This technical choice could solve by itself the whole problem of mite infestation, provided figs are not stored with a water content exceeding 20%. Such a system would have added advantages, namely a reduced sugar loss through exudation, weak fermentation process and a better conservation. Guimarães (1970) suggested some fundamental rules to ease the control of *C. lactis* on dried figs in Portugal. These rules include the cleaning of storage sites before receipt of the figs, preferably with total fumigation. The delivery of figs must be followed by fumigation, rejecting after the figs are infested or affected by fermentation. It is very important: the washing of rooms, benches, tools, hands and clothes of the staff and elimination of useless or discarded figs. Finally, it is necessary to reduce the storage period to 2 or 3 months following harvest (Guimarães, 1970). The importance of the above discussed aspects has been highlighted by various authors. Poorly maintained facilities and the lack of preventive hygiene measures before storage of products can be considered the main reasons for the high presence of mites in these environments (Palyvos *et al.*, 2008). Therefore, change in the physical factors is a simple and environmentally safe method for controlling infestations of mites on stored food (Ždárková, 1991a). For consumers, it is recommended to observe the conditions of hygiene of places of commercialization, as well as to store the grain for a maximum of 6 months in closed containers in a cool, dry place. And if possible, the grains should be kept in a refrigerator (0°C to 7°C), with humidity not exceeding 12%. The grains should be washed in running water, and the acts of 'choosing and sifting' the grains can spread the mites through the environment (Franzolin and Baggio, 2000). The studies on changes in quality, nutritional composition of food infested by mites, and implications on human health are still limited, especially those concerning nutritional composition. As mites of stored products use starch as an energy source, there is a loss of quality and nutritional composition, but it would be of utmost importance, to estimate the loss in order to develop more specific techniques and mechanisms of control. Mites can cause serious implications to human health, but the mechanisms of how this occurs are not well established yet, mainly by the wide variety of species and by cross-reaction that seems to exist among some species.

Treatments can be applied as the grain is loaded into the bin, through the use of a metering device, calibrated to apply the proper amounts. After the grain is binned and leveled, a surface dressing can be applied to prevent insects from entering the grain on the surface (Jacobs and Calvin, 1988). If infestation occurs in spite of these precautions, fumigation of the grain will be necessary. Because of the higher tolerance of mite eggs to fumigation, the concentration of gas introduced will need to be 50% greater than that for insect control. Fumigants are highly toxic, and technical knowledge is required for their proper use. A qualified, licensed pesticide applicator should be contacted to perform the fumigation (Jacobs and Calvin, 1988).

Armitage and Chambers (2002) reported that damp ambient air above the bulk can moisten the surface layers of dried grain during winter. Increasing surface temperatures can result in both insects and mites proliferating. If the grain is stored below 13% MC, moisture uptake is considerably limited, effectively preventing mites breeding. Dry grain is to be stored below 14.5% MC to prevent mites from developing in bulk and at less than 13% MC for long-term storage to limit surface populations caused by moisture uptake in winter. Cool grain to below 5°C, using aeration, to prevent mite increase and to kill insects.

Monitoring of mite populations in stored grain may be carried out by examining sievings from grain samples, particularly from the surface, or by examining the contents of traps. Monitor mite populations in empty stores using mite traps or bait bags. Combine a top dressing of DE (diatomaceous earth) with a normal cooling and drying regime to prevent surface infestations of mites. Mites proliferate in moist conditions when the grain is in equilibrium with air at over 65% RH (14.5% MC). To prevent mites, grain should be dried to below 14.5% MC before cooling. Surface infestation can be discouraged by lowering the bulk MC below 13% where possible, or by applying a diatomaceous earth (DE) to the surface of dried, cool grain (Armitage and Chambers, 2002).

Collins (2012) reported that although many factors influence mite growth, the change and combinations of the physical conditions (temperature, RH and/or MC) during the storage period are likely to have the greatest impact, with biological factors (e.g. predators and commodity) playing an important role. There is limited information on the effects of climate change, light, species interactions, local density-dependent factors, spread of mycotoxins and action thresholds for mites. A greater understanding of these factors may identify alternative control techniques. The ability to predict mite population dynamics over a range of environmental conditions, both physical and biological, is essential in providing an early warning of mite infestations, advising when appropriate control measures are required, and for evaluating control measures. This information may provide a useful aid in predicting and preventing mite population development, as part of a risk-based decision support system (Collins, 2012).

The BPCA (British Pest Control Association) (2019) reported that the main factors regulating the levels of mite population in stored products and food processing areas are temperature (3–35°C), the MC of the medium >12.8%, the food and the intrinsic rate or fecundity of the species.

Amoah (2016) reported that chemical methods of control of *T. putrescentiae* include the use of fumigants and acaricides. Other methods are biological and physical methods, and the use of food-grade oils, and plant extracts and oils. It should be noted that the future of any alternative management method lies in its ability to be incorporated into a decision-based IPM programme. The combination of these methods in IPM could help avoid infestations, minimize impacts on foods susceptible to the mite, and more importantly, delay the development of resistance to one particular control method (Amoah, 2016). Although several authors have discussed the effectiveness of stand-alone management options in the control of *T. putrescentiae* (Clarke, 2002; García, 2004; Mueller *et al.*, 2006; Rentfrow *et al.*, 2006). However, very few authors have described the possibility of implementing IPM programmes. Rentfrow *et al.* (2006) proposed the implementation of the hurdle approach within an IPM programme, which may be able to help prevent ham mite infestations in ham processing plants. Their approach involved the survey and auditing of 20 ham processing plants based on which the authors proposed that immaculate sanitation practices especially in both the processing rooms and the ageing rooms should be carried out. Grass should be prevented from growing within 9 m of the ageing house and wooden racks and structures must be sealed to prevent mites from hiding in cracks and crevices. The use of breathable paper bags for the hams, as well as freezing hams, at 6 months of ageing could also help to keep mites away. A successful IPM programme relies on an effective monitoring system that provides information on pest infestations and their incidence over time, especially before and after an IPM programme is implemented (Campbell and Hagstrum, 2002; Thind and Ford, 2004). However, less research on monitoring has been carried out on mites compared to insects. In most facilities, early detection is not effective because personnel usually do not have reliable equipment to inspect for and detect mites, particularly during the early stages of an infestation, or when they are present in fine dust or debris (Thind and Ford, 2004).

According to what we discussed above, we summarize below as an example the lines of management of control of stored mites on grain. Usually, the control includes 'preventive control measures' and 'curative control measures'. The preventive control measures provide the preparation of warehouses before receipt and storage of grain. This step needs cleaning and subsequent disinfestation by wet or aerosol treatment. Obviously, it is necessary to realize a comprehensive examination of all items for contamination. Grain preparation requires drying the grain to a state of dry or medium dryness, cleaning from foreign bodies and broken grains, maximum decrease in grain temperature, and spraying grain with contact insecticides (see below). As concerns the control of pest mites (and other organisms), monitoring of mite infestation should be carried out on an ongoing basis.

The curative control measures in cereals and different food products provide the use of physical, mechanical and chemical methods. Physical methods number cooling of grain and products of its processing, heating of the grain and products of its processing, and grain cleaning and products of its processing. Cooling grain and products, as well as heating grain according to established regimes, leads to the death of pests, and cleaning ensures a decrease in infestation.

The chemical methods include the use of phosphine and or contact insecticides. According to environments and warehouse structure the phosphine may be employed as 'fumigation on the go', 'fumigation under a synthetic film' and 'fumigation in a fumigation chamber'. The contact insecticides are sprayed as aqueous solutions or as emulsion concentrates.

In summary, the chemical control is based on the gaseous disinfestation of the premises through fumigation or gasification, wet disinfestation and aerosol disinfestation of the premises. These means can be used in combination with each other. It is obvious that the use of different chemical products is regulated by the legislation of the different counts.

PART II

This second part of the book deals with the fundamental characteristics of mites living associated with the stored food, or sometimes collected in them. The main morphological characters of fundamental taxa and a key for the identification of adults of the different families are presented. In particular, 62 families and 24 genera are discussed, focusing the attention on injurious species, of which their morphological characteristics, geographical distribution, bio-ecology, symptomatology, and damage and control are highlighted.

9 Classification and Key for the Identification of Major Taxa

As discussed in previous chapters, the Acari are members of the Arthropoda, a wide assemblage of terrestrial and marine invertebrates, considered to compose a monophyletic taxon (Weygoldt, 1998). They do not possess a conspicuous primary segmentation or segmentally arranged tergites (Krantz, 2009). At present, there is no definitive version of the higher classification of the Acari, because of a lack of knowledge of the fossil record, and the existence of many unclear relationships between groups. The following studies consider the Acari to be a subclass of the class Arachnida (Krantz, 1978; Johnston, 1982; Lindquist, 1984; Evans, 1992; Walter and Proctor, 1999; Zhang, 2003). Zhang (2003) divides the subclass into three superorders (Opilioacariformes, Parasitiformes and Acariformes) and seven orders; within these three superorders, the Opilioacariformes include the order Opilioacaridida, the Parasitiformes include the orders Holothyrida, Mesostigmata and Ixodida, and the Acariformes include the orders Prostigmata, Astigmata and Oribatida. Lindquist *et al.* (2009a) consider the Acari a subclass and recognize in it only two superorders, the Acariformes and the Parasitiformes. Hoy (2011) follows the classification of Evans *et al.* (1961) and considers the Acari to be a subclass including two orders – the Acariformes and the Parasitiformes. Here, we adopt the classification proposed by Lindquist *et al.* (2009a). Further information on pest mite families, including mite species injurious directly or indirectly to stored food, to consumers of food and/or to workers who have a relationship with these organisms, is given in the key to families that follows.

Superorder Acariformes

The main morphological feature of the Acariformes is the absence of stigmata posteriorly to coxae II. Moreover, the coxae are integrated with the podosoma and sometimes form coxisterna. Tarsus I has paired dorsal setae distally and subdistally, and tarsi II–IV lack a peripodomeric fissure and slit organ. This superorder is divided into two orders, the Trombidiformes and the Sarcoptiformes.

The order Trombidiformes is divided into two suborders, the Sphaerolichida and the Prostigmata, which include about 38 superfamilies. The Sphaerolichida have two monogeneric families and about 16 species; they are closely related to the Prostigmata but do not have any economic importance. No species of Sphaerolichida have been collected in stored food, and this taxon has been inserted in the key to advocate its understanding. The suborder Prostigmata shows various evolutionary adaptations, and the species that are known are terrestrial, aquatic and marine predators, phytophages, saprophages, paraphages and parasites (Walter *et al.*, 2009). The Prostigmata have an internal respiratory system and paired stigmata, set at or near the cheliceral bases or at the humeral angles of the propodosoma. The chelicera may be external or set within the gnathosomatic capsule; they may be chelate or lacking a digitus fixus and with the digitus mobilis are transformed into a stylet. The cheliceral bases may form a stylophore or be fused to the subcapitulum. The palp segments may be blended or reduced. In different groups, the palptarsus may be displaced to the base of the tibia and form the ‘thumb-claw complex’. The prosoma and opisthosoma may bear plates or dorsal shields. The latest estimates indicate that the Prostigmata include about 17,170 described species, 1348 genera, 36 superfamilies and 131 families, which are included in four supercohorts (Labidostomatides, Eupodides, Anystides and Eleutherengonides) (Walter and Proctor, 1999). On stored food, the supercohort Labidostomatides does not include any species. The supercohort Eupodides is divided into five superfamiles, some of which are important pest mites, such as the Eriophyoidea, which numbers three important families of pest mites, the Phytoptidae, Eriophyidae and Diptilomiopidae; this supercohort

 The Handbook of Mites of Stored Products (V. Vacante and C.H.W. Flechtmann)
DOI: 10.1079/9781786390912.0009

numbers on stored food the families Bdellidae, Cunaxidae (superfamily Bdelloidea), Eupodidae (superfamily Eupodoidea) and Tydeidae of the superfamily Tydeoidea. The supercohort Eleutherengonides is divided into two cohorts, the Raphignathina and the Heterostigmatina. The Raphignathina includes five superfamiles, among which occurs the Tetranychoidea, with the families Tetranychidae, Tenuipalpidae and Tuckerellidae, all of which include pest mites. On stored food are reported the superfamily Pterygosomatoidea, with the family Pterygosomatidae, the superfamily Cheyletoidea with the family Cheyletidae, the superfamily Raphignatoidea with the families Stigmaeidae, Raphignathidae, Camerobiidae (=Neophyllobiidae) and Caligonellidae. The cohort Heterostigmatina includes the superfamilies Pyemotoidea, with the families Acarophenacidae, and Pyemotidae and Tarsonemoidea, with the family Tarsonemidae.

The order Sarcoptiformes is divided into two suborders, the Endeostigmata and the Oribatida. The suborder Oribatida numbers five supercohorts, among which the Desmonomatides includes the cohorts Nothrina, Brachypilina and Astigmatina. The cohort Astigmatina numbers ten superfamilies, among which the Hemisarcoptoidea, with the family Hemisarcoptidae, includes important natural enemies of armoured scale insects, the Glycyphagoidea (with the family Glycyphagidae), the Acaroidea (with the family Acaridae) and the Histiostomatoidea (with the family Histiostomatidae). The Acaridae and the Histiostomatidae number several mite pests.

According to Norton and Behan-Pelletier (2009), the following key does not include the supercohort Desmonomatides and the cohort Nothrina. To advocate their identification, we report here the fundamental morphological characters adapted from the above-cited specialists. The species of the first taxa have a holoid body form, and either a macropyline or brachypiline venter. The prodorsum is stegasime, and the bothridial seta and bothridium have a strong proximal bend; the subcapitulum is stenarthric, diarthric or secondary anarthric. The chelicerae have a vestige of trochanter, with the basal portion inserted into the body as an apodeme; the opisthonotal glands are commonly present; the palptibia are articulates fully with the tarsus in Brachypylina and with a dorsal fusion in other groups; the pretarsal claws are commonly supported by two rod-like struts. Species included in the cohort Nothrina have bothridial seta, various but never pectinate, rarely highly regressed or lacking; bothridium with or without respiratory saccules or tubules; ventral region normally macropiline with large plates and little or no distance between genital and anal apertures; chelicerae commonly robust and chelate dentate, commonly with Trägårdh's organ; subcapitulum commonly stenarthric, rarely anarthric or diarthric; rutella well developed, with distal teeth or carina; lyrifissure *iad* present, *ian* commonly present.

Superorder Parasitiformes

The Parasitiformes have 1–4 dorsolateral or ventrolateral stigmata set posteriorly to coxae II. Commonly, all of the coxae are free and movable. Tarsus I bears a dorsal cluster of solenidiform setae subdistally, and tarsi II–IV have a peripodomeric fissure and slit organ. This superorder includes four orders (the Opilioacarida, Holothyrida, Ixodida and Mesostigmata). The order Mesostigmata is divided into three suborders (the Sejida, Trigynaspida and Monogynaspida). The Mesostigmata collected on stored food number in total 37 species, ascribed to the suborders Monogynaspida and Sejida, respectively. The Monogynaspina reported from stored food belong to the two cohorts, the Gamasina and the Uropodina; the former includes the subcohorts Dermanyssiae and Parasitiae, and the latter, the subcohort Uropodiae. The suborder Sejidae numbers only the superfamily Sejoidea and the family Uropodellidae. The cohort Gamasina of the suborder Monogynaspida numbers seven superfamilies (Ascoidea, Dermanyssoidea, Rhodacaroidea, Eviphidoidea, Phytoseioidea, Parasitoidea, Uropodoidea) and ten families (Ameroseiidae, Ascidae, Dermanyssidae, Laelapidae, Digamasellidae, Eviphididae, Macrochelidae, Phytoseiidae, Parasitidae, Uropodidae).

Key for the Identification of the Higher Categories and Families (Adults) of Mites of Stored Food or Sometime Occurring on Stored Food

We present below a key to identify the adults of the pest mite families of the superorders Acariformes and Parasitiformes. The key is fundamentally drawn from Lindquist *et al.* (2009a), integrated with the suggestions of Walter *et al.* (2009) for the Trombidiformes, Lindquist *et al.* (2009b) for the order Mesostigmata, Norton and Behan-Pelletier (2009) for the suborder Oribatida and O'Connor (2009) for the cohort Astigmatina.

Taxa marked with an asterisk were not found in stored food and were included to make the key use easier. Also, the key may appear incomplete at some point, such as the treatment of Oribatida for the lack of indications on superfamilies. This lack depends on the partial information produced by interested specialists. In any case, the interested reader can add the specialized contributions from Krantz and Walter (2009).

1. Without visible stigmata posterior to coxae II; coxae integrated with venter of podosoma and often forming coxisterna; tarsi II-IV without peripodomeric fissure and slit organ; tarsus I with sparse pairings of dorsal setae distally and subdistally .... **order Acariformes...2**

– With 1–4 pairs of dorsolateral or ventrolateral stigmata posterior to coxae II; coxae free and commonly movable; tarsi II–IV with peripodomeric fissure associated with slit organ; tarsus I with a dense cluster of solenidiform setae subdistally .... **superorder Parasitiformes...6**

2. One pair of stigmata opening between bases of chelicerae or on anterior prodorsum commonly present and sometimes associated with peritremes, dorsally on the cheliceral bases or on the anterior margin of prodorsum; chelicerae rarely chelate, digitus fixus often regressed and digitus mobilis commonly a hook, knife, needle or stylet-like structure; cheliceral bases sometimes fused medially; palpi simple or modified into thumb-claw process, sometimes reduced; subcapitulum without rutella; ambulacra of at least legs II and III with two lateral claws and with median empodium (rarely lacking) pad-like or rayed and often with tenent hairs; opisthosoma lacking lateral glands; opisthosomatic setal row *c* commonly with 2 pairs of setae (c_1-c_2), sometimes with 3 pairs or hypertrichous .... **order Trombidiformes...3**

– Stigmata and peritremes not present between cheliceral bases or on prodorsum; chelicerae chelate, commonly dentate, rarely attenuate or stylet-like; cheliceral bases separate; palpi simple, lacking thumb-claw process; subcapitulum commonly with rutella or pseudorutella; ambulacra of legs II–V with one or three claws, empodium claw-like or sucker-like, not pad-like, rarely rayed; opisthosoma commonly with one pair of lateral glands; opisthosomatic setal row *c* commonly with 3–4 pairs of setae or hypertrichous; tracheal system lacking or arising from bases of legs or as brachytracheae on various parts of the legs or idiosoma... .... **order Sarcoptiformes**........4

3. Tracheal system with one pair of stigmata opening between bases of chelicerae or on anterior prodorsum commonly present; prodorsum commonly with four or fewer pairs of setae or hypertrichous, sometimes including 1–2 pairs of bothridial sensilla; chelicerae rarely chelate, commonly with digitus fixus sheath-like or regressed; coxal fields contiguous or II–III distinct .... **suborder Prostigmata...9**

– Tracheal system lacking; prodorsum with three or six pairs of setae, including two pairs of filamentous bothridial sensilla, sometimes in a common pit; chelicerae chelate and with elaborate dentition, coxal fields contiguous........ **suborder Sphaerolichida** (*)

4. Typically small, soft-bodied mites with at most an indistinct prodorsal shield; genital valves present but not sclerotized in adults; leg tarsi with unpaired empodial claw usually present, simple or rayed (paired lateral claws present or absent); opisthosoma without lateral glands and with setae typically relatively short and often highly branched, wedge-shaped, dendritic or stellate. Either oval to globose mites with 5–6 pairs of prodorsal setae (or sometimes hypertrichous), of which 1–2 pairs usually are bothridial. Or elongate to worm-like mites (often with annulations), commonly with reduced prodorsal setation, no bothridia, and sometimes with an unpaired rostral seta .... **suborder Endeostigmata;**

prodorsum with one pair of filiform trichobothria (botridial setae) (*bo*); body shape various but never worm-like; lateral and/or median eyes present; naso well developed and bearing setiform setae *ro*; rutella present and dentate; pretarsi II–IV with paired claws and empodium; each chelicera with two setae; three pairs of genital papillae; body setae simple, dendritic, globose, rod-like, or leaf-like; prodorsum with six pairs of setae, external bothidial seta *exp* present; tarsus I with both claws and empodium .... **family Terpnacaridae**

– Minute to large mites commonly with a distinct prodorsal shield or fully sclerotized prodorsum and sometimes with extensive idiomatic sclerotization; prodorsum with six or fewer pairs of setae, of which 0–1 pair is bothridial; genital valves absent or present and usually well sclerotized; leg tarsi with unpaired empodial claw and never rayed, or reduced and often integrated into a pulvillar pad, or absent (paired lateral claws present or absent); opisthosoma usually with lateral glands, and with setae sometimes elongate and variously ornamented, but rarely highly branched.. 5

5. Prodorsum without specialized sensory organs other than setiform setae; genital aperture exposed or partially covered by paragenital flaps, inversely 'V', 'U' or 'Y'shaped, with usually two pairs of genital papillae in the adult or these variously reduced or modified; anal aperture lacking of conspicuous paired plates; idiosoma weakly sclerotized, with epimeral plates undeveloped or weakly formed between bases of legs; pretarsus with empodial claw and fleshy pulvillus or sucker-like and true paired claws lacking; palpi 2-rarely 3-segmented, males with a sclerotized aedeagus, and often with a pair of copulatory adanal suckers .... **suborder Oribatida, cohort Astigmatina**...44

– Prodorsum usually with a pair of specialized setae arising from sensory pits or bothridia (pseudostigmatic organs); genital aperture usually covered by paired, bomb bay-like sclerotized valves, covering usually three pairs of genital papillae in the adult; anal aperture similarly covered by sclerotized valves; adult idiosoma usually well sclerotized, with well-developed epimeral plates between bases of the legs; leg pretarsi various but often with paired true claws with a median claw-like empodium (tridactyly), or with only the empodial claw present (monodactyly), occasionally bidactylous; pretarsi III–IV usually not conspicuously modified; palpi usually with five segments, rarely 2–4; adult males without a sclerotized aedeagus and without adanal suckers .... **suborder Oribatida, excluding Astigmatina**...30

6. Subcapitulum without hypostomatic modifications but with a pair of horn-like, setiform, or pliable corniculi; tarsus of palp usually with a tined apotelic claw; opisthosomatic stigmata commonly with elongate peritremes extending anteriorly; dorsum of tarsus I without distinct Haller's organ, at most with a single pit or depression.. 7

– Subcapitulum with hypostome modified into a piercing organ with retrorse teeth but without corniculi; tarsus of palp without a tined apotelic claw; opisthosomatic stigmata without elongate peritremes; dorsum of tarsus I with a sensory complex (Haller's organ) consisting of a deep posterior pit and a less well-defined anterior depression, both with sensory setae .... **order Ixodida** (*)

7. Venter of subcapitulum with five or more pairs of setae (excluding the corniculi); tritosternum absent or, if present, consisting of a pair of flagellate laciniae on a small base; anal valves of adults with two or more pairs of setae; bases of chelicerae enclosed by pliable cuticle, epistome (gnathotectum) absent .... **order Holothyrida** (*)

– Venter of subcapitulum with a maximum of four pairs of setae (excluding the corniculi); tritosternum commonly present, with a distinctive base and 1–2 setulose laciniae (reduced or absent in some parasitic taxa); anal valves of adults nude or at most with one pair of setae; bases of chelicerae enclosed by a sclerotized ring, commonly with projecting epistome ....**order Mesostigmata**...8

8. Oviporus covered by a single plate (epyginal shield) with 0–1 pair of setae (rarely with 4 or 5 setae); tarsus of leg IV adults with a maximum of 18 setae, setae av_4 and pv_4 and ventral intercalary sclerite absent ....**suborder Monogynaspida**...55

– Oviporus covered by a single large shield (epigynal shield) bearing six (rarely 2–4) to many setae and often notched or excavated anteriorly; chelicerae without excrescences; tarsus IV with a minimum of 20 setae, setae

av_4 and pv_4 present, commonly on a ventral intercalary sclerite between the basitarsus and telotarsus...**suborder Sejida**; corniculi small, horn shaped; chelicerae moderate in size, denticulate, and with setiform pilus dentilis; idiosomatic shields tuberculate to coarsely reticulate; anal opening small; apex of podonotum produced into a spiky knob; cheliceral digits with few teeth; female epigynal shield rectangular, with numerous setae irregularly inserted over most of surface; opisthonotum of female with a pair of lateral shields and a pygidial shield that extends onto the mesonotal region and lacking horn-like processes posteriorly .... **family Uropodellidae**

9. Adults and deutonymphs often densely hypertrichous, typically free-living predators, larvae commonly strongly heteromorphic, parasitic on arthropods or vertebrates, protonymphs and tritonymphs calyptostatic; terrestrial forms with one or two pairs of bothridial setae on a prodorsal sclerite, aquatic forms highly diverse, without bothridial setae, commonly with multiple modified idiosomatic lyrifissures (glandularia) that open on small platelets or amalgamated shields...**cohort Parasitengonina**...10

– Adults and nymphs with or without dense hypertrichy, bothridial setae and prodorsal sclerite present or absent, without glandularia; larvae more or less similar to other stages, free-living or parasitic; alternating calyptostasis rare.. 11

10. Chelicerae with movable digit commonly blade-like and not retractable .... **subcohort Trombidiae** (*)

– Chelicerae with movable digit styletiform and retractable, either independently or with entire gnathosoma...**subcohort Erythraiae**; prodorsal area bearing two pairs of bothridial organs ....**superfamily Erythraeoidea**...11

11. Gnatosoma entirely retractable into idiosoma; idionotal setae commonly flattened, and serrate .... **family Smarididae**

– Gnathosoma with only chelicerae retractable into idiosoma; idionotal setae usually setiform ....**family Erythraeidae**

12. Females commonly with a pair of stigmata and associated tracheae anterolaterally on prodorsum and commonly with a pair of capitate bothridial setae (both commonly absent in immatures and males); without peritremes; opisthosoma commonly with a longitudinal series of four dorsal plates (tergites), often overlapping; tarsi II–III with stalked, smooth, pad-like empodium commonly flanked by paired claws, or rarely without empodium but with paired claws; sexual dimorphism pronounced, males often with a genital capsule .... **cohort Heterostigmatina**...26

– Both sexes with a pair of stigmata between cheliceral bases or on posterodorsal margin of gnathosoma and often opening into a simple to complex peritreme, or absent; prodorsal trihobothria if present of various forms and on immatures and adults of both sexes; tarsi of legs II–III generally with setulose, rayed, or claw-like empodium, often flanked by paired claws; sexual dimorphism strong or weak, but males without genital capsule 13

13. With 1–2 pairs of variously shaped prodorsal bothridial setae generally inserted in distinctly cup-shaped bothridia 14

– Without prodorsal bothridial setae; ovipositor absent; genital papillae absent in all stages; dorsal shields various ....**cohort Raphignathina**...15

14. Prodorsal peritremes present (sometimes emergent), palptibia usually with at least one claw-like seta distally and often apposed to reduced thumb-like palptarsus; if thumb-claw process absent, then body elongate and with one or more postpedal furrows or oval with a postpedal constriction .....**supercohort Anystides (*pars*)** (*)

– Peritremes absent, palptibia without claw-like setae or thumb-like palptarsus; postpedal furrows absent.. **supercohort Eupodides (*pars*)**...16

15. Pretarsal claws well developed or reduced but always with tenent hairs; empodium, if present, with or without tenent hairs; palpal thumb-claw process usually well developed; chelicerae independently movable,

elongate, with swollen bases and hook-like movable digit with one or more teeth; empodia absent; peritreme emergent .... **superfamily Pterygosomatoidea, family Pterygosomatidae**

– Pretarsal claws lacking tenent hairs; empodia, when present, usually with tenent hairs; palpal thumb-claw process well developed, reduced or absent 20

16. Leg trichobothria absent; chelicerae fused or free but with little potential for lateral motion, inserted on or fused with short subcapitulum... 17

– Trichobothria on tibia IV, and sometimes on tibia I, tarsus III, and tarsus IV; chelicerae free and capable of considerable lateral motion, inserted on elongate subcapitular 'snout'......................**superfamily Bdelloidea**...19

17. Chelicerae freely articulating, digits variously formed; tarsus I with a famulus and at least 1 recumbent solenidion; idiosomatic chaetome sometimes with postlarval hypertrichy; naso usually present...**superfamily Eupodoidea**; soft-bodied, without V- or Y-shaped furrow or epirostrum; anal opening ventral or terminal; intercoxal region with fewer than 10 pairs of setae; femora IV normal or strongly swollen, used in jumping .... **family Eupodidae**

– Chelicerae closely contiguous or fused medially, fixed digit regressed and movable digit stylet-like to whip-like; tarsus I without famulus, solenidia erect; idiosoma never hypertrichous, naso absent .... **superfamily Tydeoidea**...18

18. Tibia I with the 'ereynetal organ', a deeply recessed solenidion in saclike structure connecting to the surface by narrow duct; legs and prodorsum often net-like ornamentation .... **family Ereynetidae**

– Tibia I without 'ereynetal organ', a deeply recessed solenidion in saclike structure connecting to the surface by narrow duct; movable digit of chelicerae stylet-like; palptarsus with no markedly elongated setae; males without aedeagus; legs I with claws; median eyespot absent; epimeral seta *3d* usually present .... **family Tydeidae**

19. Adults with three pairs of genital papillae (deutonymphs with two pairs) and two or more eugenital setae; palpi usually ending in 1–2 elongate setae and often elbowed, without strong spines or knobs; tarsi III–IV and tibia I typically with trichobothria .... **family Bdellidae**

– Adults with two pairs of genital papillae and usually without eugenital setae; palpi usually ending in a stout spine and sometimes with strong spines or knobs; tarsi III–IV and tibia I without trichobothria........... .... **family Cunaxidae**

20. Chelicerae separate, partially fused, or completely coalesced into a stylophore; tibiae and often genua of some legs with more than three setae; palpi with five free segments, thumb-claw process usually well-developed, but palptarsus sometimes reduced and palptibial claw sometimes reduced or absent; legs well-developed, ambulatory and each with a pair of claws and an empodium with tenant hairs .... **superfamily Raphignathoidea (*pars*)**...21

– Chelicerae fused into a stylophore, or fully integrated into a gnathosomatic capsule; all tibiae and genua with three or fewer setae; palpi often with fewer than five segments, palptibial claw various, typically absent in parasitic groups; capsulate forms with well-developed legs, usually with paired claws and with tenant hairs on empodia (sometimes absent on legs I) and well-developed palpi; stylophoric forms with legs flattened, telescoped, or otherwise reduced, often with highly modified empodia; claws may be absent **superfamily Cheyletoidea**; cheliceral stylophore fused to subcapitulum to form a gnathosomatic capsule; peritremes elaborated on dorsal surface of capsule; genu I with solenidion; body oval to rounded, occasionally elongated; palptibial claw present and well developed.. **family Cheyletidae**

21. Soft bodied or with an array of dorsal shields separated by striate cuticle, or with 1–2 dorsal shields; prodorsal collar absent, gnathosoma not deeply retractable.. 22

– Fully armored mites with a holodorsal shield produced anteriorly into a hyaline or ornamented collar into which the narrow, elongate gnathosoma can be retracted..................................... **family Cryptognatidae** (*)

22. Stigmata opening at bases of chelicerae; peritremes simple to highly convoluted, produced on cheliceral bases or on anterior edge of prodorsum 24

– Peritremes and stigmata absent 23

23. Cheliceral distinct and fused medially, at least in basal one-third; mostly soft bodied, or with one or more dorsal plates separated by striate cuticle, rarely with a holodorsal shield but, if present, lacking arched groove and internal structures as above .... **family Stigmaeidae (*pars*)**

– Cheliceral bases insensibly fused into stylophore; palpi robust, subequal to chelicerae in length .... **family Stigmaeidae (*pars*)**

24. Chambered peritremes embedded in dorsal surface of stylophore; coxae II narrowly to widely separated from coxae III 25

– Peritremes linear on prodorsum, usually emergent; coxae II–III contiguous .... **family Raphignathidae**

25. Leg tibiae and tarsi subequal in length; soft-bodied, with at most a small prodorsal shield and/or small sclerites around the setal bases; all 11–12 pairs of dorsal setae considerably shorter than idiosoma; palptibial claw usually well developed .... **family Caligonellidae**

– Leg tibiae about three times longer than respective tarsi; oval to subcircular mites with very long legs .... **family Camerobiidae**

26. Cheliceral bases fused into a subconical stylophore that is not consolidated with the subcapitulum; palpi pronounced, three-segmented, extending well beyond the gnathosomatic extremities, and with well-developed apical claw; female genital opening large, flanked by conspicuous pair of genital valves 27

– Cheliceral stylophore consolidated with subcapitulum to form a capsulate gnathosoma; palpi reduced, commonly with no more than two readily distinguishable non-articulating segments, extending to or slightly beyond other gnathosomatic extremities, and with reduced or vestigial apical claw; female genital opening commonly small and not flanked by genital valves 28

27. Prodorsum of both sexes with a pair of stigmata and associated tracheae, and a pair of capitate bothridial sensilla laterally; tarsi II–IV with paired claws and a stalked empodium, tarsus I with paired claws and no empodium; gnathosoma with two pairs of dorsal setae .... **superfamily Tarsocheyloidea, family Tarsocheylidae** (*)

– Prodorsum of female only with a pair of stigmata and associated tracheae laterally and a pair of ampulliform organs posteromedially; male prodorsum without stigmata and tracheae, and with a pair of normal setae instead of ampulliform organs; tarsi II–IV without claws but large discoid empodium, tarsus I without claws or empodium; gnathsoma without dorsal setae .... **superfamily Heterocheyloidea, family Heterocheylidae** (*)

28. Legs IV of female with separate femur and genu, with a total of 5–12 setae on tibia and tarsus, and often with paired claws and empodium; legs IV of male 5-segmented, with symmetrically or asymmetrically paired claws or single or vestigial claw .... **superfamily Pyemotoidea**...29

– Legs IV of female typically with fused femur and genu, with total of 2–3 setae on tibia and tarsus, and without claws and empodium; leg IV of male, if present, usually 4-segmented, with a single sessile or vestigial claw...**superfamily Tarsonemoidea**; leg IV 3-segmented, usually much more slender than legs II–III; males always with four pairs of legs; legs IV 3- or 4-segmented, always inserted ventrally; male genital capsule caudal .... **family Tarsonemidae**

29. Prodorsum of female lacking bothridia and bothridial setae; coxisternal plates I–II together with maximally two pairs of setae; gnathosoma somewhat hypognathous, enclosed and partly hidden in propodosomatic

camerostome; palpi not differentiated from lateral walls of gnathosoma; male genital capsule ventrocaudal in position .... **family Acarophenacidae**

– Prodorsum of female with pair of bothridia and capitate bothridial setae; coxisternal plates I and II together with four pairs of setae; gnathosoma prognathous, well exposed from propodosoma; palpi delineated from lateral walls of gnathosoma; female coxisternal plates III–IV separated from each other medially by soft cuticle that bears a separate triangular sternal plate; male legs IV differentiated from legs II–III in being somewhat stouter and ending with a single sessile claw .... **family Pyemotidae**

30. Genua of legs shorter than tibiae, differently shaped and functioning as knee; legs articulate with body in deep pockets (acetabula), as a 'ball and socket' joint; trochanters I–II almost totally contained within acetabula; venter comprising a unified rigid plate in which only the subcapitulum and the paired genital valves and anal valves are distinct and movable; subcapitulum commonly diarthric .... **cohort Brachypylina**...38

– Genua of legs similar to tibiae in size and shape, not as knee-like segment; leg articulation not in deep acetabula, at most in shallow depressions; trochanters I–II small but clearly external; venter of various form but rarely as above; paired aggenital and adanal plates often distinguishable; coxisternal region often transversely divided by sejugal articulation and commonly fused to aggenital region; y subcapitulum commonly stenarthric 31

31. Body form ptychoid; legs relatively short, those on each side adjacent, attached to a rather inconspicuous, narrow coxisternal region that connects to a voluminous, soft deformable pleural region; genua of legs similar to tibiae in size and shape, not reduced to knee-like segment; trochanters I–II small but clearly external; venter of various form but rarely as Brachypylina; coxisternal region often transversely divided by sejugal articulation and commonly not fused to aggenital region; subcapitulum commonly stenarthric 32

– Body form not ptychoid; legs longer, or if short, not capable of being withdrawn and covered by prodorsum; coxisternal region conspicuous and wide; pleural region sclerotized, or, if not, noticeable sclerotization generally absent from whole body 33

32. Notogaster clearly of multiple parts; dorsal plate (notaspis) subdivided transversely by a telescoping (type L) articulation; anterior plate (pronotaspis) only with row *c* with four pairs of setae; posterior plate (pygidium) has setae of rows *e* and *f* associated with two transverse carinae; setae of row *d* minute or lacking; setae *p1* and row *h* on ventral margin; p_2 and p_3 set posteriorly on pair of large lateral plates (pleuraspis); epimere II with two pairs of setae; genua I–IV without solenidia .... **superfamily Protoplophoroidea; family Protoplophoroidae**

– Hysterosoma covered with notogaster of typical structure, bearing 14 or 15 pairs of setae (rows *c*, *d*, *e*, *h*, *p* present and row *f* as inconspicuous alveoli); no separate, seta-bearing ventral plate in which anal or genital valves insert; subcapitulum stenarthric; rutella broad, massive; genua I with two solenidia...**supercohort Mixonomatides** (***pars***); combined anogenital region narrow, length commonly greater than three times maximum width; structures of ventral plates otherwise; paired, narrow plicature plates (without setae) present between notogaster and aggenital/adanal plates; pretarsi monodactylous or tridactylous; three pairs of genital papillae well-formed, or only posterior pair reduced...**superfamily Euphthiracaroidea**; paired ventral plate connected across midline in preanal position by a short region of interdigitating ridges; genital plates not touching medially, genital setae spread the length of genital region, set on medial edge, all genital papillae of approximately equal size; palp 3-segmented .... **family Euphthiracaridae**

33. Opisthonotum subdivided transversely by 1–3 complete scissures (if dorsum is otherwise sclerotized) or grooves (if dorsum is rather soft); notogaster with 1–3 transverse scissures or grooves (if only one, the mites are more strongly pigmented, not soft-bodied; scissure runs either anterior to row *d* or posterior to *e* or actually contains row *e* on narrow intercalary sclerite); rutella various, but rarely as above; legs with 5 free segments past coxisternum; femora undivided; opisthonotal glands lacking; pretarsi various, commonly monodactylous .... **supercohort Enarthronotides** (***pars***)...35

– Opisthonotum represented by a single notogaster, not subdivided transversely, commonly well sclerotized; if superficial transverse grooves appear in sclerotized cuticle, then they number more than three and are commonly incomplete medially; with full sejugal articulation; epimera II–III not fused separated by band of unsclerotized cuticle 34

34. Dichoid mites, with full sejugal articulation, with the prodorsum is capable of horizontal and lateral movement with respect to hysterosma, often slightly retractable into it; with two pairs of exobothridial setae; with fewer than ten pair of genital setae, normally set in median row; with 2–3 pairs of aggenital setae, aggenital region variously formed; with 3–4 pairs of anal setae; opisthonotal gland, lyrifissure *iad*, and often *ian* present .... **supercohort Mixonomatides (*pars*)** (*)

– Holoid mites, epimera II–III commonly entirely fused, with sejugal articulation present only in dorsal and sometimes lateral region; proterosoma not retractable into hysterosoma; with 1 or 0 pair of exobothridial setae...**superfamily Crotonioidea**; body often with adherent organic debris; prodorsum separated from notogaster by a band of soft cuticle; anal plate wider or only slightly narrower than adanal plate, not greatly reduced; medial margin of genital plates bearing genital setae, delimited from rest of plane by distinct line or carina; aggenital setae present .... **family Camisiidae**

35. Notogaster with 2–3 transverse scissures or grooves, separating 3–4 dorsal plates or regions 36

– Notogaster with a single transverse scissure of type S, which lies posterior to setal row *c*, *d*, or *e*; with intercalary sclerite containing large, erectile setae of row *e*; setae f_1 and f_2 also erectile, stout with long branches, inserted on intercalary sclerites in pair of circular depressions; setae p_1 and p_2 filiform; prodorsum astegasime; leg pretarsi bidactylous and empodial claw lacking.........................**family Arborichthoniidae** (*)

36. Two or three scissures simple, type E, or only grooves; notogastral setae generally homogeneous, similar in shape and size, none extremely broad (plate-like) or unusually large or erectile; bothridial seta clavate 37

– At least one (of two or three) scissure of type S, with intercalary sclerite and large, erectile setae; or notogastral setae extremely broad, plate-like, covering body-like shields; bothridial setae of various forms; notogaster and prodorsum normally with reticulate or foveolate sculpturing; rostrum distinctly fenestrate; setae of rows *c-f* occupy anterior ½ of notogaster; no large area without setae; setal row *d* not enlarged, similar to row *c*; genital plate not divided; aggenital plate small, not fused to coxisternum; adanal plates fused posteriorly to form a U-shape; pretarsi bi-or tridactylous .... **family Cosmochthoniidae**

37. Body more than three times longer than wide; legs II–III widely separated; pretarsi without distinct claw; empodium highly modified, with hyaline flap on either side; prodorsum astegasime, with a small, narrow anterior lobe bearing setae *ro*; bothridial seta smooth .... **family Pediculochelidae**

– Body length little more than twice maximum width; legs II–III almost adjacent; pretarsi with empodium developed as typical claw; prodorsum stegasime, without obvious, narrow naso; bothridial seta barbed or ciliate; notogaster with three transverse typer E scissures; four pairs of setae (*c* row only) on anteriormost plate; with cupule immediately lateral to each mid-dorsal seta (c_1, d_1, e_1, f_1) in addition to normal complement; genital plates twice the size of anal plates; aggenital seta lacking, anal plates with four pairs of setae; peranal plate and seta lacking .... **family Haplochthoniidae**

38. Notogaster without octotaxic system of porose areas or saccules, pteromorphs absent, dorsophragmata and pleurophragmata normally lacking; epimera II–IV normally rectangular in shape; apodemes two and sejugal apodeme transverse; discidium and circumpedal carina nornally lacking; humerosejugal porose organ Am lacking, Ah normally absent 39

– Notogaster commonly with octotaxic system of porose areas or saccules, pteromorphs present or absent, dorsophragmata and pleurophragmata present; epimera II–IV not rectangular in shape; apodemes 2 and sejugal apodeme angled toward the genital plate; discidium and circumpedal carina normally present; humerosejugal porose organs Am and Ah present 41

39. Prodorsum with true lamellae 40

– Prodorsum without true lamellae; with or without costulae; notogaster commonly convex, without scalps, with 9–15 pairs of setae or setal alveoli, only three pairs of setae set posteromarginally; ornamented cerotegument commonly lacking; tibiae and tarsi without porose areas; epimeral border IV absent; epimera III–IV long, extending well posterior of genital plates; notogaster without anteromedial sclerotized elevation between two oval concavities... family **Oppiidae** (***pars***); epimeral border present; ventral plate without postanal saccule; posterior margin of notogaster undivided; costulae if present shorter than 1/2 length of prodorsum; crista if present not longer than one-third length of notogaster .... **family Oppiidae** (***pars***)

40. Subcapitulum diarthric; chelicerae chelate dentate; notogaster with long, triangular or subrectangular humeral process, projecting anteriorly to bothridia or projecting dorsally over bothridium; genital neotrichy lacking; subcapitular mentum with tectum; palptarsal eupathidium *acm* on tubercle, associated with solenidion, or solenidion set at the base of tubercle...**family Cymbaeremaeidae** (***pars***); notogaster without long, triangular humeral process, if humeral process present, not reaching the level of bothridium; subcapitular mentum with tectum; lamellar cusps touching medially, fused or not, covering the rostrum medially; lamellar setae arising anteroventrally on lamellar cusps; genital neotrichy absent; euphatidium acm of palptarsus on tubercle .... **family Cymbaeremaeidae** (***pars***)

– Subcapitular mentum without tectum; palptarsal eupathidium *acm* not on tubercle; lamellar cusps if touching medially not covering the rostrum; lamellar setae arising anteriorly or dorsally on lamellae or lamellar cusps; tarsus I with two solenidia, pretarsus with normal size claw, and tarsal pulvillus lacking; legs I not modifies for jumping; cerotegument without globules; adanal setation three pairs set adanally and postanally; notogaster with two pairs of setae set centrodorsally, or notogastral setae other than *p* series lacking, with or without alveoli; integument distinctly sculptured with coarse ridges, tubercles, or areoles; lamellae commonly broad, almost parallel, close to lateral marginof prodorsum; genal notch and circumpedal carina lacking; anal plate rectangular; leg tarsi functionally fused to tibiae; interlamellar setae commonly set well anterior of bothridia or on lamellae; rostrum rounded; lamellar cusps present; cerotegument of notogaster with foveae or tubercles; tutorium without free cusp .... **family Carabodidae**

41. Bothridium with internal spiral thickenings; genu IV long thin segment, concave dorsally, longer than genu III, usually longer than tibia IV; pteromorph present, with or without long, knife-like anterior projection; border of epimere IV without tectum; prodorsum shorter than notogaster, and the latter longer than wide; chelicerae dorsally bacilliform tubercle; tarsi II–IV with setiform setae (*p*); bothridial seta well developed; epimera II–IV weakly delineated; cerotegument developed as thin layer, not blocky in structure, never birefringent .... **family Achipteriidae**

– Bothridium without internal spiral thickenings; genu IV straight dorsally, short segment, subequal in length to genu III, commonly shorter than tibia IV; pteromorph if present without knife-like projection; tibia IV with solenidion; tibiae and tarsi without retrotecta; sejugal apodeme and apodemes 3–4 not meeting medially anterior to genital plate.. 42

42. Pedotectum I at most weakly developed as small lamina; prodorsum without genal notch; axillary sacculus of sucapitulum and postanal porose area lacking; pteromorphs or humeral projections lacking; distal eupathidia on paltarsus not in the same plane; notogaster with 4–5 pairs of porose areas; 4–5 pairs of genital setae; prodorsum without prolamellar and sublamella; sternal furrow lacking; marginoventral porose areas present; tarsus I without dorsal porose area .... **family Oribatulidae**

– Pedotectum I well developed as large lamina; prodorsum with genal notch; axillary sacculus of subcapitulum commonly present; postanal porose area commonly present.. 43

43. Notogaster with posterior tectum, and 10 pairs of notogastal setae; pretarsi I and IV with the same number of claws; lamellar setae set on prodorsal surface, and lamellar cusps lacking; postanal porose areas lacking .... **family Chamobatidae**

– Notogaster without posterior tectum, and 10–15 pairs of notogastal setae; pteromorph commonly well developed, projecting ventrally; custodium and postanal porose areas present; subcapitular mentum without tectum; epimere IV with two or three pairs of setae; with 10–15 pairs of minute to long notogastral setae .... **family Ceratozetidae**

44. Body cuticle commonly at least partially striate; genital papillae greatly reduced or absent; tibiae I–II with a dorsal solenidion and 1 (*v'*) or rarely 0 ventral setae; unguinal setae lacking from all tarsi; pretarsi often enlarged, with ambulacral stalk and disc well developed, empodial claws normally reduced or incorporated into ambulacral disc as central sclerite...**Psoroptidia**; chelicerae well developed; proral setae lacking; genua III with solenidion ς present; tibia IV with one ventral seta; tarsus I with solenidion ω_1 set subapically, very near the solenidion ω_3; tarsus II whit solenidion ω set in the basal half of the segment; tarsus IV with more than one seta; pretarsi I–II with ambulacral stalks shorter than the tarsi .... **family Pyroglyphidae**

– Body cuticle smooth or striate; genital papillae normal in form; strongly enlarged and ring-like or reduced; tibiae I–II with dorsal solenidion and 0–2 ventral setae 45

45. Chelicerae laterally compressed, often attenuate, commonly with many finely divided teeth, non-chelated; palpi with a terminal solinidion and commonly at least one elongate, eupathidial seta, normal setae strongly reduced, palp commonly reflexed outward; female genital valves fused to body posteriorly, open anteriorly forming a transverse oviporus; genital papillae of both sexes often in the form of large rings on ventral body surface; genital papillae ring-like, commonly large, not arising from a common furrow and not directly associated with the genital opening; body relatively larger and not dorsoventrally flattened .... **family Histiostomatidae**

– Chelicerae normally chelate, with few teeth, rarely vestigial or fixed digit lacking; palpi reflexed inward, commonly bearing three filiform setae and terminal solenidion; female oviporus longitudinal with genital valves fused in body anteriorly, free posteriorly; genital papillae always associated with genital opening and never in the form of large rings 46

46. Prodorsum with lamellar (*le*) setae absent; dorsal setae may be elongate, but never heavily barbed; males lacking paranal suckers or sucker-like setae on tarsus IV 53

– Prodorsum with lamellar (*le*) setae present, or if absent, then pretarsi with short, strong condylophores, or some dorsal setae long and heavily barbed or empodial claws absent and opisthosoma bilobed posteriorly; males with or without paranal suckers and sucker-like setae on tarsus IV 47

47. Female and sometimes male with empodial claw bifurcate; male with legs III enlarged, ending in a large, straight, empodial claw and an enlarged, straight spine-like seta similar in length and form to empodial claw; male with paranal suckers and sucker-like setae on tarsus IV .... **family Lardoglyphidae**

– Both sexes with empodial claws simple or absent; male with legs III similar to legs IV, or if legs III enlarged, then empodial claw shorter and curved and without an enlarged seta similar in form to empodial claw 48

48. Ventral subcapitulum with a prominent pattern of external transverse and oblique ridges not to be confused with internal pharyngeal sclerotization; empodial claws small or lacking; condylophores very thin or lacking; female commonly with an epigynal apodeme and often with an external copulatory tube; male without paranal suckers or modified setae on tarsus IV .... **family Glycyphagidae**

– Ventral subcapitulum without external ridges 49

49. Body cuticle almost partially striated, often in dorsal propodosomatic region or posteroventrally, or with a pattern of striations broken into scale-like structures; if striations or scale lacking, then with all dorsal setae elongate and heavily barbed except c_1, which is long and nude; solenidion ω_2 of tarsus I more distal than ω_1; tarsus IV of both sexes without tectal setae; opistosoma with a row of small, closely spaced microtrichia extending between the bases of most lateral body setae; most dorsal setae elongate and heavily barbed; pretarsal ambulacra short, relatively simple; empodial claws reduced or lacking .... **family Aeroglyphidae**

– Body cuticle without striations; small, rounded protuberances or small triangular microtrichia may be present; solenidion ω_2 variable in position, commonly more basal than ω_1; tectal setae present on all tarsi or one member of pair may be absent on all tarsi 50

50. Discrete coxal apodeme III and sometimes IV absent; discrete propodosomatic sclerite absent 51

– Discrete coxal apodemes III–IV present, projecting obliquely from bases of trochanters; propodosomatic sclerite commonly present 52

51. Body cuticle covered by small microtrichia; male without paranal suckers or modified sucker-like setae on tarsus IV (a basal sucker may be present in tibio-tarsal articulation) .... **family Echimyopodidae**

– Body without microtrichia; male with paranal suckers and often with sucker-like setae on tarsus IV .... **family Chortoglyphidae**

52. Tarsi with tectal setae asymmetrical, with *tc"* commonly spine-like, rarely filiform, unguinal setae usually larger and stouter than proral setae; prodorsal sclerotization in the form of a rectangular shield that may be incised posteriorly, or prodorsal sclerotization absent; dorsal setae variable but not normally both elongate and heavily barbed; cuticle smooth or rarely with small rounded protuberances; male with legs III similar to legs IV, or legs III enlarged and bearing a large empodial claw and reduced tarsal setation .... **family Acaridae**

– Tarsi with both tectal setae filiform, similar in length; proral setae spine-like or enlarged and claw-like; unguinal setae spine-like, reduced or absent, never distinctly larger than proral setae; propodosoma with lamellar (*le*) setae present, empodial claws present, or if absent, then tibiae I–II with only one ventral seta; body cuticle often with rounded protuberances; opisthosoma not bilobed posteriorly; body distinctly longer than wide; dorsal setae filiform, unbarbed; tarsi with proral setae enlarged and claw-like, unguinal setae reduced or absent .... **family Suidasiidae**

53. Tarsi I–II more elongate, without a greatly enlarged apical seta; pretarsi variable in form but similar on all legs; supracoxal setae short; without lyriform organs on prodorsum; supracoxal gland opening not associated with a large sclerotized region; prodorsal sclerite absent; rostral (*ro*) setae set about hallway between anterior edge of propodosoma and interlamellar (*in*) setae; propodosomatic region with a pair of large ocelli; coxal apodeme I fused medially with coxal apodemes II closing coxal fields I in both sexes; condilophores elongate and separate in female, asymmetrical in male; male with genital (*g*) setae and coxal setae *4b* present .... **family Carpoglyphidae**

– Prodorsal sclerite well developed; rostral (*ro*) setae at the anterior edge of propodosoma or absent; ocelli present or absent; apodemes I not fused medially with apodemes II, coxal fields open; condylophores fused to each other or absent; male with genital setae vestigial or absent, setae *4b* filiform, or absent but with alveoli modified into a sucker 54

54. Empodial claws present, condylophores normally fused in a V-shaped sclerite or a more elongate sclerite in base of ambulacral stalk; tibiae I–II with 1–2 ventral setae; male with all traces of genital setae and coxal setae *4b* absent; male normally with pretarsal ambulacra I–II arising from ventral apex of tarsus, often modified as a sucker .... **family Winterschmidtiidae**

– Empodial claws absent; condylophores apparently absent; tibiae I–II with 0–1 ventral seta; both sexes with genital opening between or posterior to coxal fields IV; female oviporus often confluent with anal opening; male with alveoli of setae *4b* fused medially forming a sucker anterior to aedeagus .... **family Hemisarcoptidae**

55. Epyginal shield of female flask or wedge-shaped or subtriangular, often extending into the opisthogastric region and bearing setae st_5 or flanked by those setae in the postcoxal region; epyginal shield reduced or obliterate in some endoparasitic groups; shield normally not surrounded by fused sternal-endopodal-ventral elements; adults with 1–2 dorsal shields, lacking marginal shields or their platelets; peritreme typically linear, sometimes reflexed anteriorly or posteriorly, rarely vestigial; tritosternal base cylindrical or flattened, never

subrectangular or columnar, femur IV of deutonymph and adult typically with six setae .... **cohort Gamasina** 56

– Epyginal shield of female oval, subtriangular, or tongue-shaped, commonly nude and partially or entirely enclosed by fused sternal-endopodal-ventral shielding bearing sternal setae 1–5 (st_5 rarely on separate shields that may be partially fused to epyginal shield) and commonly confined to podosomatic region, rarely displaced posteriorly or fused with ventral elements; adults with one to several dorsal shields, marginal shields or platelets generally present; peritreme often sinuous, sometimes on horn-like projections, rarely vestigial; tritosternal base often enlarged, subrectangular or columnar; femura IV of deutonymph and adult with 7–8 setae (if only 6 setae present, then hypostomatic setae h_2 and h_3 longitudinally aligned ....**cohort Uropodina...**66

56. Female with sternal setae st_4 inserted on large metasternal plates flanking an anteriorly acuminate epyginal shield anterolaterally; female sperm access system opening via an unpaired, commonly strongly developed endogynum beneath epigynal shield, often with ancillary sclerotized processes; male chelicera with a spermatotreme, a sperm-holding process that is coalesced distally with movable digit; males commonly with high developed, sexually dimorphic, spine-like setae, spurs and apophyses on legs II .... **subcohort Parasitiae; family Parasitidae**

– Female with sternal setae st_4 inserted on small metasternal plates or on soft integument or on posterolateral corners of sternal shield; female sperm access system opening by a pair of small solenostomes in region of coxae III or IV; male chelicera with a spermatodactyl, a sperm-holding process that projects distally, commonly free from movable digit; males with or without sexually dimorphic structures on leg II .... **subcohrt Dermanyssiae...**57

57. Female sternal shield normally fused with metasternal plates so as to include 4th pair of sternal setae; female epigynal shield separate from ventrianal or anal shield; male with holoventral shield or separate sternitigenital and ventrianal shields, when separate, anterolateral margins of ventrianal shield often incised or eroded, and posterolateral margins often fused caudally with posterior margin of dorsal shielding; microtrichia of postanal cribum sometimes extending onto posterior margin of dorsal shield ....**superfamily Rhodacaroidea...** 58

– Female sternal shield normally not fused with metasternal plates, leaving 4th of sternal setae free on soft integument or on metasternal plates that are free or attached to endopodal plates; if sternal setae *st4* on endopodal-metapodal plates that are incorporated with sternal shield, then epigynal shield expanded into a genitiventral shield; male with holoventral shield or separate sternitigenital and ventrianal shields; when separate, anterolateral margins of ventrianal shield intact, and posterolateral margins free caudally from posterior margin of dorsal shielding; microtrichia of cribum confined to postanal region of anal or ventrianal shield.. 60

58. Dorsal shield of adults divided, the anterior shield normally with setal pair j_2 nearly transverse alignment with vertical and paravertical pair j_1 and z_1, and commonly with 2–4 distinctive internal sclerotic nodules in mid-posterior region; anterior portion of the female sternal shield and of male sternitigenital shield weakly defined, commonly carrying sternal setae st_1; spermatodactyl of male often recurved basally 59

– Dorsal shield of adults divided or entire, the anterior shield with setal pair j_2 in usual position well behind vertical and paravertical pairs j_1 and z_1, and without sclerotic nodules; anterior portion of female sternal shield and male sternitigenital shield undifferentiated from more posterior portion; spermatodactyl of male not recurved basally; exopodal elements commonly distinct as sclerotized strips alongside coxae II–IV, endopodal elements usually present and fused with sternal shield in the female and with sternitigenital shield in the male; tibia IV with 10 setae (2 1/1, 3/1 2), tarsi II–IV without elongate dorsal distal setae ....**family Ologamasidae**

59. Paptarsal apotele 2-tined; leg IV commonly with seven setae on genu and seven on tibia; male with sternal setae st_5 on separate platelets .... **family Digamasellidae**

– Palpal apotele 3-tined; legs IV with 10 setae on genu (2 2/1, 3/1 1) and 10 on tibia (2 1/1, 3/1 2); male with sternal setae st 5 on sternitigenital shield .... **family Rhodacaridae**

60. Arthrodial envelope at base of movable digit of chelicerae with one or two plumose or filamentous processes or with a brush and an adjacent fringed coronet; female typically with an epyginal and a ventrianal shield (if anal shield is present, it is widely separated from epyginal shield); male typically with holoventral shield 61

– Arthrodial processes, if present, at base of movable cheliceral digit with at most a fringed coronet but without distinct filamentous or brush-like processes; female with an epyginial and an anal or a ventrianal shield, or rarely with a genitiventral shield narrowly separated from or fused with anal shield; male with separate sternitigenital and ventrianal or anal shields or with a holoventral shield 62

61. Peritremes generally looped proximally, joining the stigmata posteriorly; genu I commonly with two ventral setae; tarsus I commonly without claws; paired paradactyli of pretarsi II–IV commonly broad and extending to apex of claws and divided or deeply serrated distally when observed obliquely; female with a pair of conspicuous accessory sclerites beneath the lateral margins of epyginial shield .... **family Macrochelidae**

– Peritremes normal, joining the stigmata anteriorly; genu I commonly with three ventral setae; tarsus I with or without claws; paired paradactyli of pretarsi II–IV commonly setiform, undivided distally; female with lateral accessory sclerites faint or absent beneath the lateral margins of epigynial shield .... **family Parholaspididae**

62. Female sternal shield with one or typically two pairs of setae; sternal setae 3 often on adjacent platelets or on soft cuticle, rarely on sternal shield; opisthonotum without caudal setae j_5 and marginal caudal setae; posterior row of deutosternal denticles extending laterally beyond insertions of capitular setae; corniculi often divided distally or entire; femur II with 10 setae, including four dorsals.......... **family Ameroseiidae**

– Female with zero to typically three pairs of sternal setae on the sternal shield; opisthonotum commonly with caudal setae j_5 and commonly with one or more marginal *R* setae; posterior row of deutosternal denticles extending laterally beyond insertions of capitular setae; corniculi commonly entire, rarely divided distally; femur II with 10–11 setae, including five dorsals 63

63. Female with epyginial shield truncate or weakly convex posteriorly and either narrowly separated from or abutting a ventrianal shield or widely separated from an anal shield that is round or oval but commonly not inversely subtriangular; male with sternitigenital shield commonly delineated from, though often abutting, a ventrianal shield or separate ventral and anal elements.......... 64

– Female with epyginial shield broadly or narrowly rounded posteriorly, commonly widely separated from inversely subtriangular anal shield, or epyginal shield expanded into a genitiventral or genitiventrianal shield; male with either a consolidated holoventral shield or a sternitigenital shield widely separated from an anal shield.......... 65

64. Deutonymphs and adults with less than 20 pairs of dorsal shield setae; setae j_1 lacking; female with less than four pairs of marginal setae on soft integument; female with phytoseid-type sperm access system, including normally a sclerotized calix and often an associated minor duct; fixed and movable cheliceral digits normally developed, similar in length and apposed; tristosternum present, with well-developed laciniae; anal opening posteroventral, commonly in a ventrianal shield.......... **family Phytoseiidae**

– Deutonymphs and adults with more than 20 pairs of dorsal shield setae; setae j_1 present; female commonly with more than four pairs of marginal setae on soft integument; female with phytoseiid-type or laelapid-type sperm access system, with or without a sclerotized calyx and an associated minor duct; female with third pair of sternal poroids on posterolateral corners of sternal shield and with sternal setae st4 commonly free on soft cuticle; male with endopodal strips beside coxae III–IV commonly free or narrowly connected to sternitigenital shield; movable cheliceral digit commonly bidentate .... **family Ascidae**

65. Tibia and genu I each with 1 anterolateral seta; tectum commonly produced into an elongated lance-like process, rarely multidenticulate; dorsal shield entire; female and male with anal shield .... **family Eviphididae**

– Tibia and genu I each with 2 anterolateral setae; tectum with smooth or denticulate anterior margin, commonly not produced into an elongated process; dorsal shield entire or with lateral incisions or divided; female with anal or, rarely, genitiventrianal shield, male commonly with holoventral shield .... **superfamily Dermanyssoidea (*pars*)**...66

66. Peritremes rarely absent, occasionally reduced; legs I–IV of comparable thickness or coxae I–IV contiguous; sternal shield well developed or reduced but distinctly tanned and easily visible; opisthogaster without spur-like or flattened setae; tristosternum well-developed, with laciniae; chelicera of female whip-like, styliform; cheliceral digits minute, chelate; corniculi membranous, indistinct; elongate second cheliceral segment of female far exceeding basal cheliceral segment 1 in length; male chelicera with second segment of normal length; idiosoma broadly rounded posteriorly .... **family Dermanyssidae**

– Chelicera variously produced but not styliform, corniculi variously developed; tibia I commonly with two ventral setae, including av_2; genu commonly with two posterolateral setae, including pl_2; subcapitulum with internal malae commonly well developed, with fimbriate lateral margins, and equal to or longer than corniculi .... **family Laelapidae**

67. Deutonymphs and adults with more than one shield dorsally, adults usually with one or more pygidial shields or marginal shields in addition to dorsal shield; dorsal shield without differentiated pairs of greatly elongated marginal setae and paranal setae not elongated, although one pair of ventrocaudal setae sometimes so elongated; palptarsal claw present, palpgenu with 5–6 setae; tibiae 1–4 with at least three dorsal setae; femur IV with six setae in various configuration...**subcohort Uropodiae**; coxae of legs I widened and flattened, inserted together in a gnathopodal cavity and often widened so that the tritosternum is partly covered; tritosternum with variously shaped laciniae, commonly not wider than long; pedofossae commonly well developed, sometimes lacking...**superfamily Uropodoidea**; chelicera without internal sclerotized node; hypostomatic setae h_1 typically long and setiform; female genital shield set between coxae II–IV .... **family Uropodidae**

– Chelicera with a distinctive internal sclerotized node associated with the levator tendon 68

68. Internal malae of hypostome simple, without marginal fimbriations or distal moustache-like excrescences, dorsal shield of adults often notched marginally... **family Trematuridae**

– Internal malae with short marginal fimbriations and/or with elaborated distal moustache-like excrescences; dorsal shield of adults without conspicuous ornamentation or T-shaped dorsal setae; movable cheliceral digit shorter than twice its basal width .... **family Oplitidae**

Superorder Acariformes, Order Trombidiformes, Suborder Prostigmata (predators)

10 The Family Acarophenacidae Cross (Fig. 10.1)

The body shape of Acarophenacidae is similar to that of tarsonemid mites, it is elongate to oval, and long from 100 μm to 320 μm. The gnathosoma is partially or completely fused into the propodosoma and the palpi are indistinct. The anterior ventral plate carries three or fewer pairs of setae and the females do not bear prodorsal sensilli. The leg I is thicker than the other legs, and many have a strong sessile claw (Gerson *et al.*, 2003).

The family numbers a few species, some of which have been ascribed to the family Pyemotidae and successively transferred into Acarophenacidae. Magowski (1994) reported a list of six genera and 18 species ascribed to this family, while Walter *et al.* (2009) include in this family seven genera and 35 species.

The females of Acarophenacidae are parasitoids of insect eggs, and feed only on the eggs of insect hosts without toxic effects on people or domestic animals. This adaptation suggested to consider these mites potentially useful, as biological control agents on stored products (Steinkraus and Cross, 1993). The list of epiparasitic species associated with graminivorous arthropods on stored food includes almost six species. *Acarophenax assanovi* Livshits *et* Mitrofanov is reported from Russia, on grain, by Livshits and Mitrofanov (1974). *Acarophenax dermestidarum* Rack is known from Germany, on *Dermestes* sp., common stored-product beetles (Rack, 1959). *Acarophenax lacunatus* Cross *et* Krantz (Cross and Krantz, 1964) is reported on wheat from Iraq and the United States (Cross and Krantz, 1964; Steinkraus and Cross, 1993), and its role of parasitoid has been studied by Faroni *et al.* (2001), on the Bostrichid *Rhyzopertha dominica* (F.) and Oliveira C.R.F. *et al.* (2007), the Tenebrionid *Tribolium castaneum* (Herbst) and the Cucujid *Cryptolestes ferrugineus* (Stephens). *Acarus mahunkai* Steinkraus *et* Cross is reported from the United States, as egg parasite of the lesser mealworm (Coleoptera: Tenebrionidae) (Steinkraus and Cross, 1993). *Acarophenax rackae* Mahunka *et* Zaki, an egg parasitoid of *Tribolium confusum* du Val (Coleoptera, Tenebrionidae), is known from Egypt (Mahunka and Zaki, 1990) and Iran, where it has been collected on storage products (Hajiqanbar *et al.*, 2011; Sayadi *et al.*, 2012). *Acarus tribolii* Newstead *et* Duval (Fig. 10.1) is known from Australia, Brazil, Canada, Egypt, England, Greece, India, Indonesia, Iran and the United States on stored wheat, straw and dust in silos, flour meal grain, maize cob (rotting), groundnut, rice, stored rice, rice dust and debris (Rack, 1959; Cross and Krantz, 1964; Hughes, 1976; Steinkraus and Cross, 1993; Eliopoulos *et al.*, 2002; Ardeshir *et al.*, 2006; Noei, 2007). A comprehensive list of hosts and habitats and distribution of species of the genus *Acarophenax* has been reported by Rahiminejad and Hajiqanbar (2015).

DOI: 10.1079/9781786390912.0010

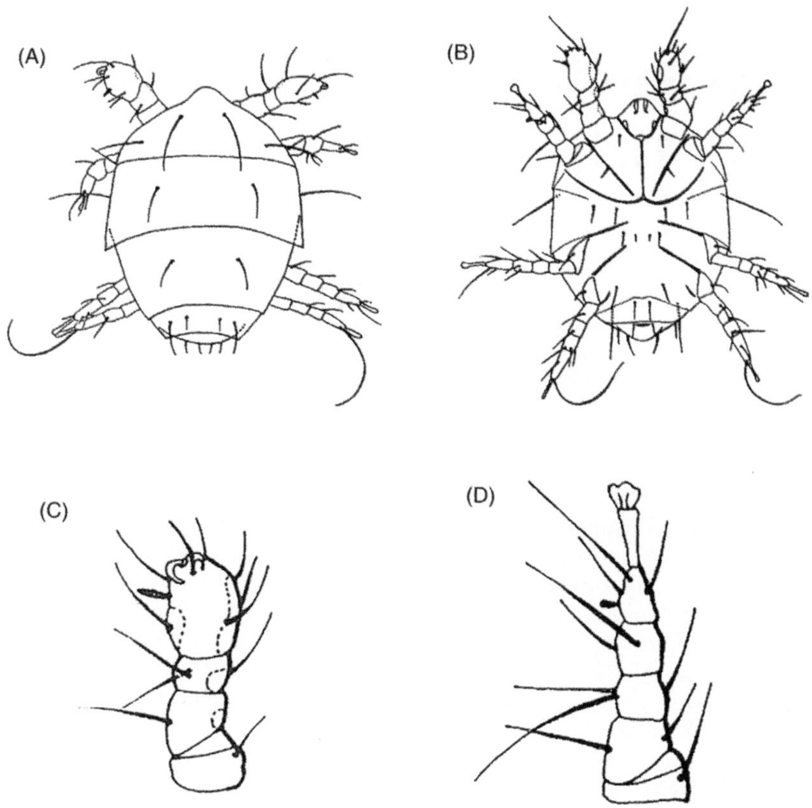

Fig. 10.1. *Acarophenax tribolii* Newstead *et* Duvall. Dorsal view of female (A); ventral view of female (from Hughes, 1976, modified) (B); ventral view of right leg I of female (from Hughes, 1976) (C); ventral view of right leg II of female (from Hughes, 1976, modified) (D).

11 The Family Bdellidae Dugès (Fig. 11.1)

The Bdellidae are moderate to large-sized predatory mites characterized by an elongated, snout-like gnathosoma, and by elbowed pedipalps bearing two (one in *Monotrichobdella* Baker *et* Balock) long terminal setae. The body is mostly weakly sclerotized, commonly red, green, purple or yellow. Some species are long over 3000 μm, but the length is commonly in the range from 500 μm to 2400 μm. Bdellid mites live in soil and litter in various habitats, including dry deserts and cool forests. They are collected also on mosses, lichens and algae on rocks. Some species are collected on leaves, leaf litter and intertidal rocks. Walter *et al.* (2009) reported 16 genera, five subfamilies and about 135 species. Hernandes *et al.* (2016) recognized in their 'Catalogue of snout mites (Acariformes: Bdellidae) of the world' 278 valid species, 11 genera and 5 subfamilies. Some bdellids play a role in regulating populations of economically important arthropods (Gerson and Smiley, 1990). Some species feed on arthropod eggs and on nematodes (Sorenson *et al.*, 1983; Pugh and King, 1985).

On stored food the family includes seven species ascribed to the genera *Spinibdella* Thor (2) (Fig. 11.1), *Bdella* Latreille (1), *Biscirus* Thor (1), *Cyta* von Heyden (1) and *Odontoscirus* Thor (2), respectively. Moreover, Eliopoulos *et al.* (2002) reported from Greece, Bdellidae on grain and dried fruits in Greece. As concerns the species above reported, *Spinibdella cronini* (Baker *et* Balock) was first recorded in China and Japan on different commodities (barley, condiment, dried fruit, dried vegetable, grain, mushroom, processed food, wheat) (Baker and Balock, 1944), and later by Ebrahimi and Noei (2022) in various countries such as Australia, Brazil, Bulgaria, China, Egypt, Hungary, Iran, Japan, Mexico, Slovakia, Syria, Ukraine and the United States (Atyeo 1960, 1963; Sosnina *et al.*, 1965; Wallace and Mahon, 1972; Soliman, 1975; Soliman and Zaher, 1975; Kuznetsov and Livshits, 1979; Lehman, 1982; Swift and Goff, 1987; Li *et al.*, 1992a; Ostovan and Kamali, 1995a; Kamali *et al.*, 2001; Jalaeian *et al.*, 2005; Ripka *et al.*, 2005; Li and Fan, 2007; Ueckermann *et al.*, 2007; Noei, 2007; Kaluz, 2008; Noei *et al.*, 2008a; Abbaszadeh Rad *et al.*, 2010; Daneshnia and Akrami, 2013; Majidi and Akrami, 2013; Masoudian and Khanjani, 2013; Hagstrum *et al.*, 2013). Moreover, the list of stored products includes grain in storage, silos, flour, barley, condiment, dried fruit, dried vegetable, mushroom, processed food, wheat, stored rice, rice dust and debris (Noei, 2007; Noei *et al.*, 2008a; Hagstrum *et al.*, 2013; Khaleghabadian *et al.*, 2015; Abbasi *et al.*, 2016).

Spinibdella lignicola (G. Canestrini) is reported from England (Hull, 1918), Egypt (Trägårdh, 1905; Thor, 1931; Abdel-Shaheed *et al.*, 1971), Italy (Canestrini, 1886; Thor, 1931), Spain (Mihelčič, 1958b), Switzerland (Schweizer and Bader, 1963), Taiwan (Tseng, 1978), China (Sichuan) (Li *et al.*, 1992a) and Czech Republic (Stejskal and Hubert, 2008) on various commodities such as garlic, grain, grass seed residue, sugar beet seed, sweeping and wheat in different facilities (commercial grain storage, farm, flour-mill, flour warehouse) (Hagstrum *et al.*, 2013). *Bdella muscorum* Ewing is characterized by a large geographical distribution (Ebrahimi and Noei, 2022), and reported by Ueckermann *et al.* (2007) and Abbasi *et al.* (2016) in Iran on grain in storage, in silos and on flour. *Biscirus iranensis* Paktinat-Saeej *et* Bagheri, reported from Iran on grain in storage, in silos and on flour by Abbasi *et al.* (2016). *Cyta latirostris* (Hermann) is a cosmopolitan species (Ebrahimi and Noei, 2022) reported from Iran on grain in storage, in silos and on flour (Abbasi *et al.*, 2016). *Odontoscirus kazeruni* (Ostovan *et* Kamali) reported from Iran on stored barley, hay and wheat (Ostovan, 1993; Ostovan and Kamali, 1995a). *Odontoscirus longirostris* (Hermann) is a cosmopolitan species (Ebrahimi and Noei, 2022) reported from Iran on stored products (Ostovan and Kamali, 1995a).

DOI: 10.1079/9781786390912.0011

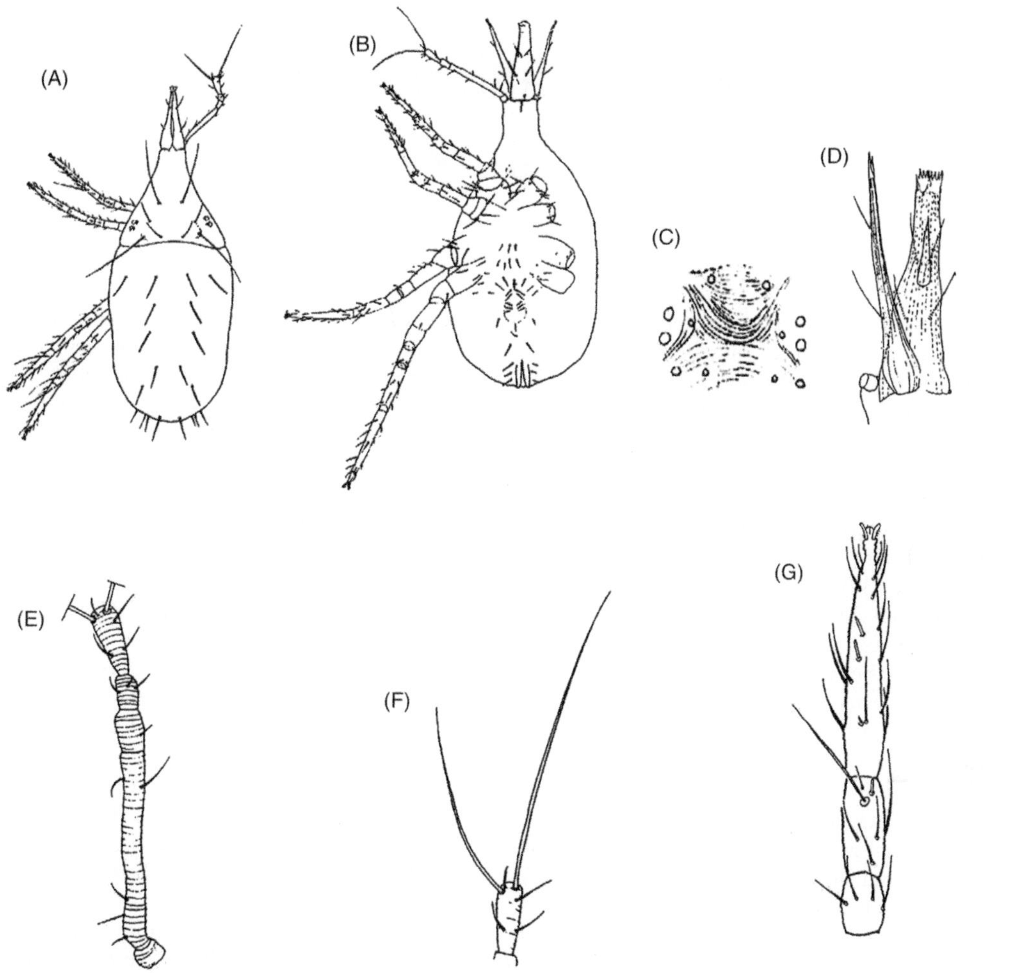

Fig. 11.1. *Spinibdella* sp. Female. Dorsal view (A); ventral view (B); eyes and propodosomal insertion of setae (C); dorsal view of gnathosoma (D); dorsal view of right palp (E); palpal tibiotarsus (F); dorsal view of right leg I (G) (from Hughes, 1976, modified).

12 The Family Caligonellidae Grandjean (Fig. 12.1)

The Caligonellid mites are characterized by peritremes chambered and embedded in the dorsal surface of the stylophore; the coxae II are narrowly to widely separated from coxae III. Tibiae and tarsi of legs are subequal in length. The body is soft, with at most a small prodorsal shield and/or small sclerites around the setal bases; all 11–12 pairs of dorsal setae are considerably shorter than idiosoma, and the palptibial claw is usually well-developed (Walter *et al.*, 2009).

Species of this family are relatively small and free-living predatory mites feeding on small arthropods. They inhabit various habitats such as soil, leaf litter, bark of trees, moss, storehouses and bird nests (Summers and Schlinger, 1955; Meyer and Ueckermann, 1989; Fan, 2000; Doğan, 2003a; Ahaniazad and Bagheri, 2012; Akyol and Koç, 2012). The genera of this family are distinguished by the peritremal arrangement, and the configuration on the dorsal surface of the stylophore (Swift, 1996; Ueckermann and Khanjani, 2003). On stored food are reported four species. The first two species belong to the genus *Paraneognathus* Fan, found in cereal stores and rice hulls in Iran (Ardeshir *et al.*, 2014), and collected by Horn *et al.* (2015) in poultry farms in southern Brazil. The last two species belong to the genus *Molothrognathus* Summers *et* Schlinger, a genus found in stored products in Egypt (Zaher *et al.*, 1986), and very common in poultry nests (Krantz and Walter, 2009). Silva *et al.* (2018) collected species of this genus in nests of *Troglodytes musculus* (Naumamm) (Passeriformes, Troglodytidae) in Brazil.

Paraneognathus wangae (Fan *et* Li) (Fig. 12.1) has been collected in China on various commodities such as condiments, dried fruit, dried vegetables, mushrooms and processed food (Fan and Li, 1995; Fan, 2000), in Iran on cereal store and rice husk (Ardeshir *et al.*, 2015), and in Turkey on whole wheat Mardin and stored wheat (Çobanoğlu *et al.*, 2021). *Paraneognathus oblongus* (Soliman) is known from Egypt, and Iran (Beron, 2020), and collected in Iran on storage products, stored rice, rice dust and debris (Noei, 2007; Noei *et al.*, 2008a; Khaleghabadian *et al.*, 2015). The third species is *Molothrognathus mehrnejadi* Liang *et* Zhang (1997); it has been described from Iran and reported also by Ueckermann and Khanjani (2003) and collected in the same country on storage products by Sayadi *et al.* (2012) and Khaleghabadian *et al.* (2015). The fourth species is *Molothrognathus brasiliensis* Silva, Brentano *et* Ferla, mentioned by Horn *et al.* (2015) in the nests of birds and laying poultry.

DOI: 10.1079/9781786390912.0012

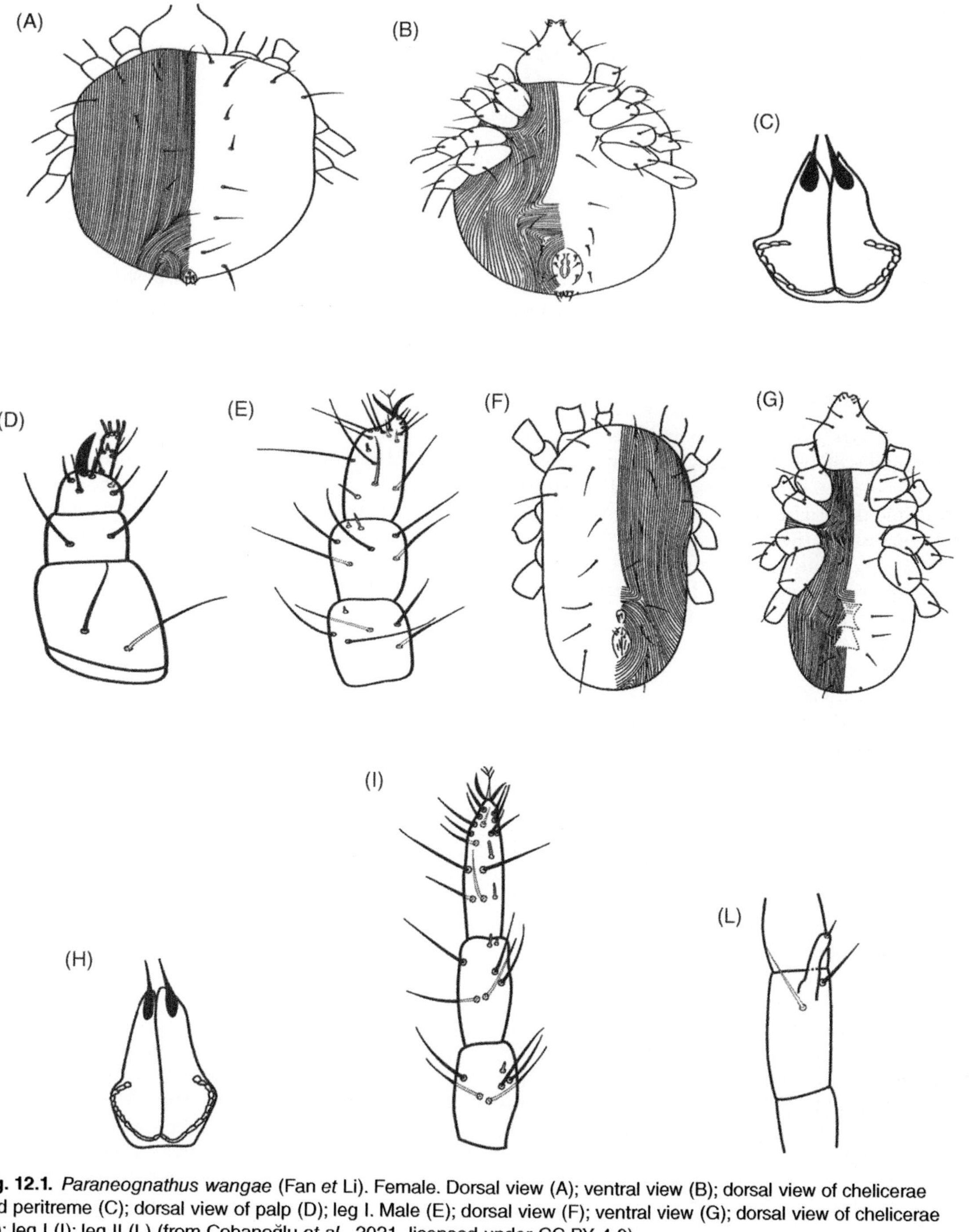

Fig. 12.1. *Paraneognathus wangae* (Fan *et* Li). Female. Dorsal view (A); ventral view (B); dorsal view of chelicerae and peritreme (C); dorsal view of palp (D); leg I. Male (E); dorsal view (F); ventral view (G); dorsal view of chelicerae (H); leg I (I); leg II (L) (from Çobanoğlu *et al.*, 2021, licensed under CC BY 4.0).

13 The Family Cheyletidae Leach (Figs. 13.1–13.6)

Cheyletid mites have a worldwide distribution, and they occur on all continents. The mites of this family occupy a great variety of habitats. Some of them are free-living predators inhabiting plants, soil and plant debris; other groups are nidicolous predators living in nests of birds, mammals and insect colonies. A small group of cheyletids are dwelling in the feather quills of birds. Investigation on cheyletid mites is very interesting because, it could help us to understand the origin and the mechanism of evolution in parasitism (Beklemishev, 1970).

Some representatives of this family are also quite important for the agricultural economy and the health of human and domestic animals. Many predaceous Cheyletidae feed on microarthropods causing damage to agricultural plants and grain supply (Volgin, 1969).

Other mites are parasites of domestic animals (rabbits, cats, dogs, etc.).

The Cheyletidae mites have the oviparous and the genital opening or genital and anal openings of the female approximate or fused, the genua and the tibiae I–IV with three or fewer setae, and the larval seta *rd* lacking. The fused cheliceral bases form a stylophore fused to the subcapitulum, to form a gnathosomatic capsule, with the peritremes located on its dorsal surface (Summers and Witt, 1971; Bochkov, 2002). The family numbers 440 species ascribed into 75 genera, and is divided into different subfamilies (Volgin, 1969; Bochkov, 2004; Gerson *et al.*, 1999; Bochkov and Fain, 2001; Bochkov and Abramov, 2016; Beron, 2021a). In particular, about 78% of cheyletid species are predators, the remaining species are permanent parasites of mammals and birds. Sometimes, they cause allergies and papular dermatitis in humans that have close contact with infested pets (Keh *et al.*, 1987; Gerson *et al.*, 1999; Bochkov and Fain, 2001; Doğan *et al.*, 2011). In Brazil, Ezequiel *et al.* (2003) studied a case of a 6 months old female child showing signs, 4 weeks earlier, of papular macular erythematous lesions associated with edematous pruriginous plaques, with post-inflammatory discolouration. It was observed that removing the child from the environment led to regression of the lesions referred to, with the appearance of new dermatological manifestations when she was again returned home. While at home, the child remained in an old, upholstered kitchen chair for most of the time. The most affected regions were the back and, in lower intensity, the abdominal region. There were no significant lesions in exposed areas, such as the face, and upper or lower limbs. The use of antihistamines and topical corticoids led to temporary improvement. After evaluation of the house dust sample gathered at the site, three species of the mite family Cheyletidae, *Cheyletus malaccensis* Oudemans, *Cheyletus fortis* Oudemans and *Cheletonella caucasica* Volgin, were identified. The advice to sanitize the space, associated with the symptomatic treatment, led to a definitive cure of the patient's clinical case. A further study carried out by Ezequiel *et al.* (2008) evaluated the prevalence of mites of the family Cheyletidae, in house dust in Brazil, and evidenced that 13 (0.85%) of the mites found were of the family Cheyletidae, with 6.9% of positive domiciles. *C. malaccensis*, *Ker bakeri* Zaher *et* Soliman, *C. fortis*, *Cheletonella vespertillionis* Womersley and *Cheletonella caucasica* were the found species. The diversity found in mite species and its presence in close human contact, along with reported related dermatitis due to them, deserve further studies on this topic.

The predatory species occupy a wide variety of habitats including plants, soil litter, tree bark, stored products and colonies and galleries of insects or ephemeral substrates requiring dispersal by phoresy on insects or vertebrates; they are mainly free-living predators that feed on various microarthropods, particularly on herbivorous, fungivorous and saprophagous on acarid mites (Žďárková, 1979). A few species however can be beneficial as biological control agents, controlling acarid mites, which are serious pests damaging

DOI: 10.1079/9781786390912.0013

agricultural stored food (Gerson and Smiley, 1990; Bochkov and Fain, 2001; Fain and Bochkov, 2001).

On stored food are reported about 38 species ascribed to 21 genera [*Acaropsellina* Summers (four species), *Alliea* Yunker (1 species), *Bothrocheyla* Volgin (1 species), *Caudacheles* Gerson (1 species), *Chelachecaropsis* Attiah (1 species), *Chelacheles* Baker (3 species), *Cheletogenes* Oudemans (1 species), *Cheletomimus* Oudemans (1 species), *Cheletomorpha* Oudemans (1 species), *Cheletonella* Womersley (2 species), *Cheyletia* Haller (1 species), *Cheyletus* Latreille (8 species), *Cunlifella* Volgin (1 species), *Eucheyletia* Baker (2 species), *Hemicheyletia* Volgin (3 species), *Lepidocheyla* Volgin (1 species), *Mexecheles* De Leon (1 species), *Neoeucheyla* Radford (1 species), *Nodele* Muma (3 species), *Oudemansicheyla* Volgin (1 species) and *Zachvatkiniola* Volgin (1 species)].

We discuss here briefly, the fundamental traits of the species belonging to these genera and reported from stored food. *Acaropsellina sollers* (Kuzin) is known from Czech Republic, Egypt, England (Hughes, 1976), Germany, Greece (Eliopoulos and Papadoulis, 2001; Eliopoulos *et al.*, 2002), Iraq (Mahmood, 1992), Iran (Ostovan, 1993; Noei, 2007; Hajizadeh *et al.*, 2011), Poland, Russia (Volgin, 1987; Hughes, 1976), Scotland (Hughes, 1976), Turkey, United States (Summers and Price, 1970; Hughes, 1976), Yemen and Wales on different commodities, such as barley, broad bean, wheat, buckwheat, buckwheat product, chickpea, cottonseed, dried fruit, grain, maize, poultry diet, protein mix, residue, rice, wheat, wheat flour, sweeping, stored rice, rice dust and debris (Eliopoulos *et al.*, 2002; Noei, 2007; Noei *et al.*, 2008a; Hagstrum *et al.*, 2013; Maroufpoor and Ostovan, 2017). *Acaropsellina docta* (Berlese) has a cosmopolitan geographical distribution and has been reported as predacious on *Trogoderma* sp. eggs, on stored products, in soil samples, animal feed, barley, clover seed, cottonseed, cocoa bean, ginger, grain, herb, maize, oat, oat grass, protein mix, soybean, sweeping, wheat, wheat flour and wheat flour sweeping (Ostovan, 1993; Fathipour, 1994; Fathipour *et al.*, 1999; Hagstrum *et al.*, 2013). *Acaropsellina tyrophagus* (El Badry) has been described from Egypt on wheat (milled) (El Badry, 1969). Another species of this genus is *Acaropsellina volgini* (Gerson), reported from Egypt (Wafa and Soliman, 1968), Israel (Gerson, 1967) and Turkey (Koç, 1998), where the mite has been collected on barley, cottonseed and wheat. Its biology has been studied by Attiah and Soliman (1970) and Wafa *et al.* (1970).

Alliea prasadi Corpuz-Raros is reported from Philippines on groundnut (Corpuz-Raros, 1998). *Bothrocheyla pavlovskii* (Volgin) was reported from Czech Republic and Russia on sweeping (Volgin, 1963). *Caudacheles lieni* Tseng has been found in Taiwan on bean (Tseng, 1977). *Chelachecaropsis bakeri* Attiah has been collected in Egypt on rice mill (Attiah, 1971). The genus *Chelacheles* numbers the species *Chelacheles hellenicus* Eliopoulos *et* Papadoulis, collected in Greece, on floor litter in flour-mill (Eliopoulos and Papadoulis, 2001), *Chelacheles robustus* Corpus-Raros is reported from Philippines on cassava flour (Corpus-Raros, 1998), and *Chelacheles strabismus* Baker, known from Portugal (Baker, 1958) and Iran where it has been collected on stored rice, rice dust and debris (Noei, 2007; Noei *et al.*, 2008a).

Cheletogenes ornatus (Canestrini *et* Fanzago) (Fig. 13.1) was described by Canestrini and Fanzago (1876) as *Cheyletus ornatus*. The biology of this mite has been investigated by Zaher and Soliman (1971a) and by Zaher *et al.* (1981). Its geographical distribution is cosmopolitan and includes Australia, China, Egypt, Hawaii, Italy, Poland, Turkey, United States and West Indies (Volgin, 1987; Summers and Price, 1970). The list of commodities numbers apricot (dried) and herb (Hagstrum *et al.*, 2013).

Cheletomimus (*Hemicheyletia*) *gracilis* (Fain, Bochkov *et* Corpuz-Raros) has been collected in strawberry plants in Brazil (Silva *et al.*, 2013) and for the first time in stored products in the same country (Siegert *et al.*, 2018).

Cheletomorpha lepidopterorum (Shaw) (Fig. 13.2) was described by Shaw (1794) as *Acarus lepidopterorum* and by Koch (1839a) as *Cheyletus venustissimus*. Its biology has been studied by Taha *et al.* (1988) and Xia *et al.* (2007). The geographic distribution of *C. lepidopterorum* is cosmopolitan and includes Australia, Canada, China, Czech Republic, England, France, Germany, Greece, India, Indonesia, Iran, Ireland, Italy, Japan, Kenya, Mexico, Nigeria, Peru, Pakistan, Philippines, Poland, Portugal, Russia, Taiwan, United States, Wales (Griffiths, 1960; Summers and Price, 1970; Caceres and Fain, 1977; Sepasgozarian, 1978; Tseng, 1979; Genç and Özar, 1986; Özer *et al.*, 1986; Volgin, 1987; Özkan *et al.*, 1988; Aheer *et al.*, 1997; Emekçi and Toros, 1999; Eliopoulos *et al.*, 2002; Akpınar *et al.*, 2017) and other countries. The list of the commodities of the mite numbers bean, candied fruit, Chinese

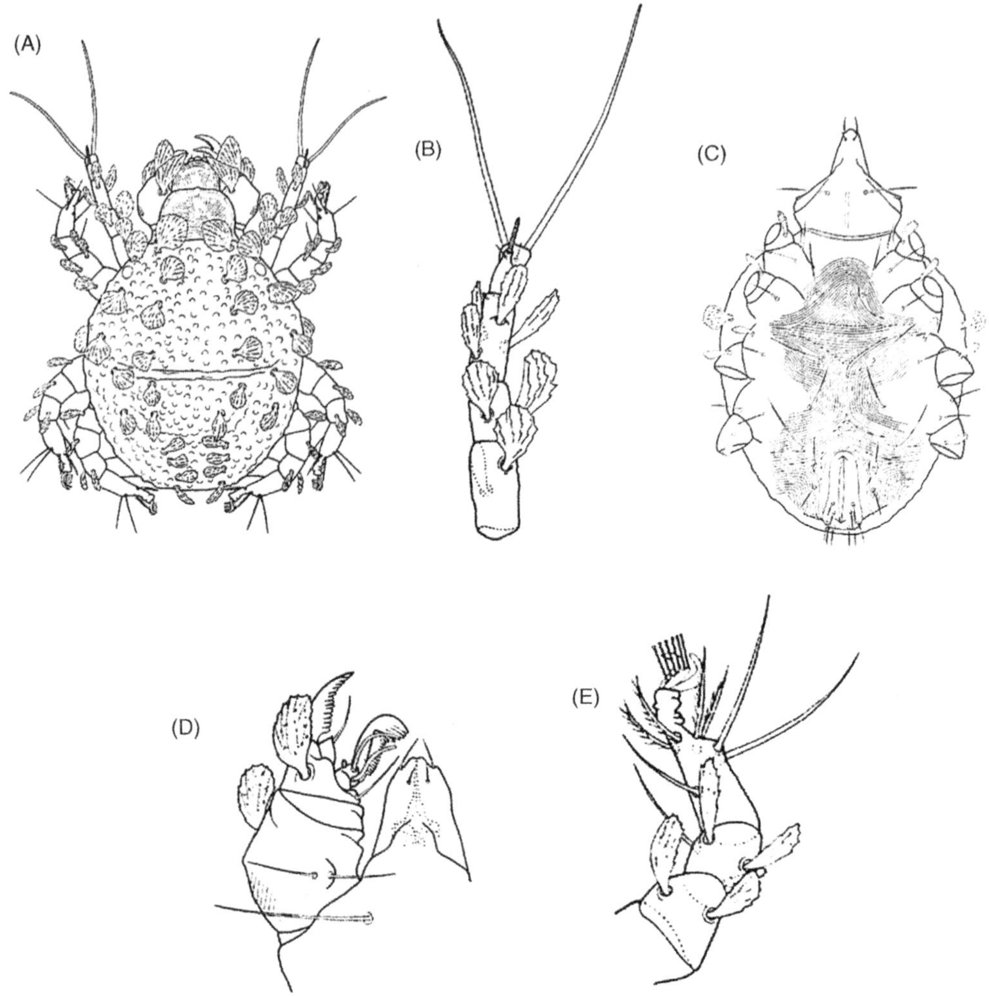

Fig. 13.1. *Cheletogenes ornatus* (Canestrini *et* Fanzago). Female. Dorsal view (A); leg I (B); ventral view (C); ventral view of propodosoma and palp (D); leg IV (E) (from Vacante and Nucifora, 1984-85, licensed under CC BY 4.0).

medicine, condiment, cottonseed, dried fish, dried fruit, dried vegetable, garlic bulb, grain, hay, herb, honey, kaoliang, maize, maize residue, milk (dried), millet, millet debris, manure, poultry waste, mushroom, oat, *groundnut*, processed food, rice, rice (husked), rice (old), decayed rice bran, sugar, trefoil, vegetable (dehydrated), wheat, and the facilities number commercial grain storage, farm, off-farm storage and retail shop (Eliopoulos *et al.*, 2002; Hagstrum *et al.*, 2013; Salarzehi *et al.*, 2018). Moreover, the mite has been found also on plant leaves, tree bark, soil, plant litters, house dust, and bird nests, stored rice, rice dust and debris (Treat, 1967; Van Eyndhoven, 1964; Summers and Price, 1970; Bochkov and Fain, 2001; Noei, 2007; Noei *et al.*, 2008a; Perotti and Braig, 2009; Hajizadeh *et al.*, 2011).

Cheletonella pilosa Tseng is reported from Taiwan on kaoliang, millet and wheat (Tseng, 1977). *Cheletonella iraniensis* Salarzehi, Hajizadeh *et* Ueckermann is reported from Iran on stored rice, decayed rice bran, bran and dust of rice warehouse (Salarzehi *et al.*, 2019).

Cheyletia papillifera Volgin has been collected on starch in Czech Republic and the Netherlands (Volgin, 1987).

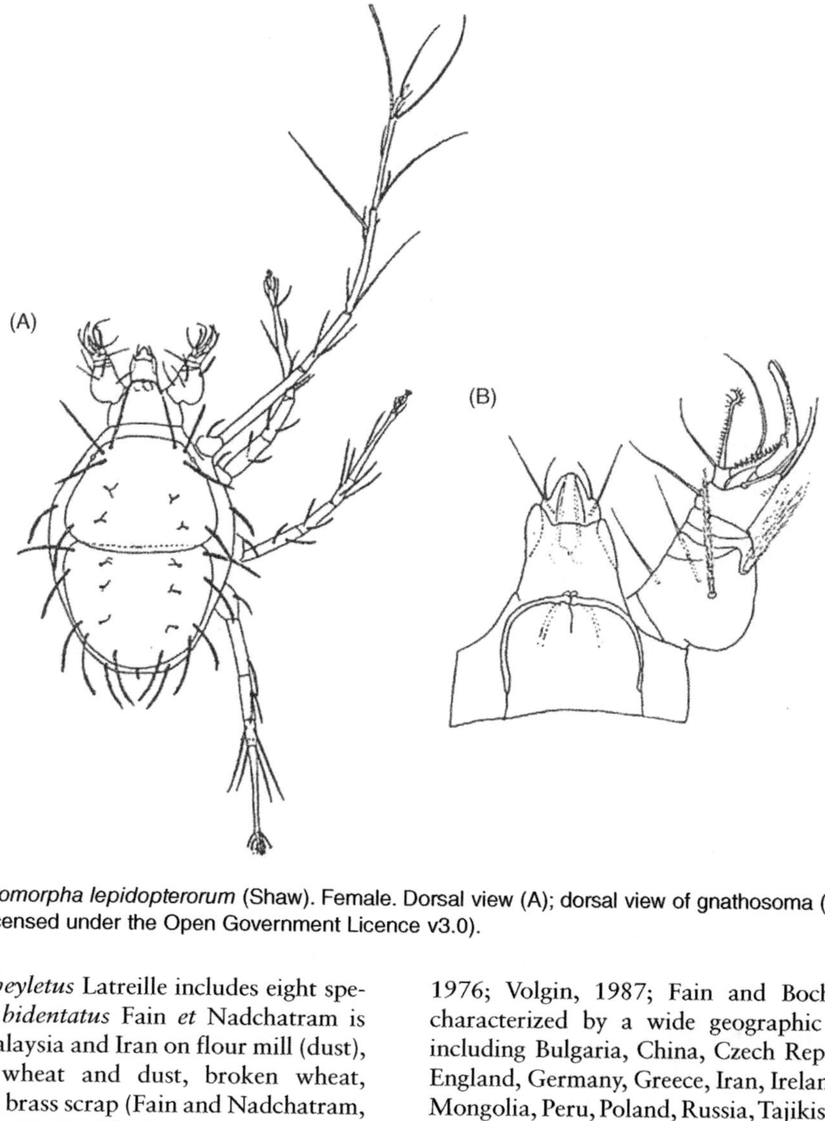

Fig. 13.2. *Cheletomorpha lepidopterorum* (Shaw). Female. Dorsal view (A); dorsal view of gnathosoma (B) (from Hughes, 1976, licensed under the Open Government Licence v3.0).

The genus *Cheyletus* Latreille includes eight species. *Cheyletus bidentatus* Fain *et* Nadchatram is known from Malaysia and Iran on flour mill (dust), grain, broken wheat and dust, broken wheat, paddy, bran and brass scrap (Fain and Nadchatram, 1980b; Ardeshir, 2017). *Cheyletus cacahuamilpensis* Baker is a cosmopolitan species reported from Algeria, Crimea, Greece (Fain and Bochkov, 2001), Iran (Mirfakhraii, 1994; Salarzehi *et al.*, 2018), Mexico, Peru, South Africa and Ukraine (Baker, 1949; Fain and Bochkov, 2001) on flour-mill, rice and decayed rice bran, dust of warehouse, and off-farm storage (Hagstrum *et al.*, 2013; Salarzehi *et al.*, 2018). The biology of this mite has been studied by Mohamed *et al.* (1982). *Cheyletus carnifex* Zachvatkin was described in 1935 (Zachvatkin, 1935) and reported by Rohdendorf (1940) as *C. aversor*. Different specialists have discussed this mite (Summers and Price, 1970; Hughes, 1976; Volgin, 1987; Fain and Bochkov, 2001), characterized by a wide geographic distribution including Bulgaria, China, Czech Republic, Egypt, England, Germany, Greece, Iran, Ireland, Kirghizia, Mongolia, Peru, Poland, Russia, Tajikistan, Uganda, Ukraine, Uzbekistan and the United States (Zachvatkin, 1935; Hughes, 1976; Faraji and Kamali, 1993; Fain and Bochkov, 2001; Elioupolos *et al.*, 2002; Hagstrum *et al.*, 2013; Salarzehi *et al.*, 2018). The mite possesses a wide list of commodities (aquatic product, barley, beet seed, chicken feed, Chinese medicine, clover seed, condiment, cottonseed, flax seed, grain, grain product, grain residue, herb, maize flour, oat, protein mix, residue, rice, decayed rice bran, soya bran, sunflower seed, sweeping, wheat, wheat flour) and facilities (farm, flour-mill, off-farm storage, old distillery silo) (Noei, 2007; Noei et al., 2008a; Hagstrum *et al.*, 2013; Salarzehi *et al.*, 2018).

Cheyletus eruditus (Schrank) (Fig. 13.3) is a well-known species, described as *Acarus eruditus* by Schrank (1781) and by Packard (1870) as *Cheyletus seminivorus* and by Hughes (1948) as *Cheyletus butleri*. The biology of *C. eruditus* has been investigated by Boczek (1959) and Barker (1991). The species has a cosmopolitan geographic distribution including the Afrotropical region (Fain, 1979), Brazil (Porto *et al.*, 2017), Australia, Canada (Liscombe and Watters, 1962), China (Lung-Shut, 1984),

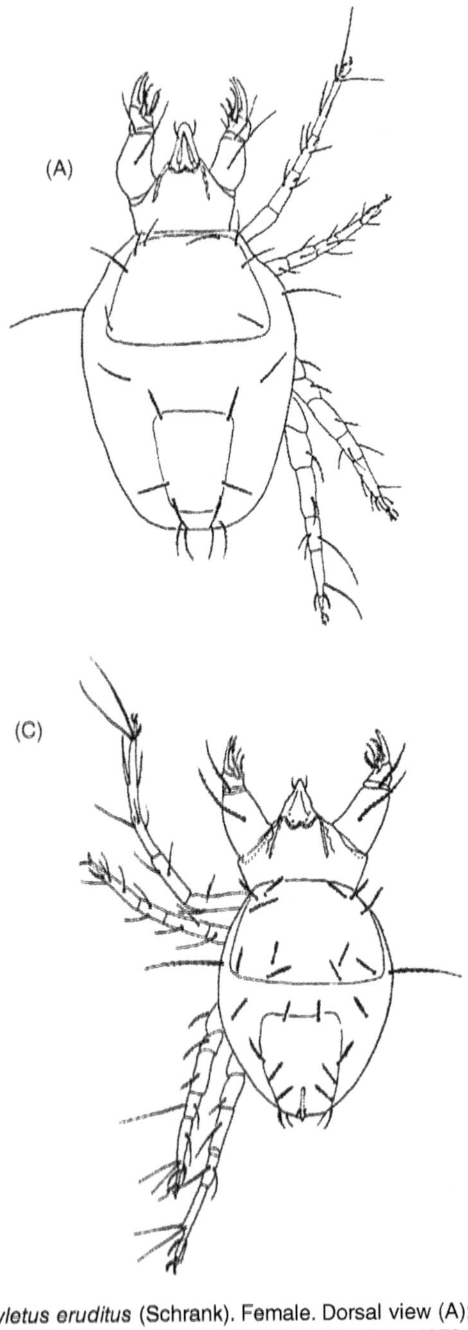

Fig. 13.3. *Cheyletus eruditus* (Schrank). Female. Dorsal view (A); dorsal view of gnathosoma (B). Male. Dorsal view (C); gnathosoma of heteromorphic male (D) (from Hughes, 1976 licensed under the Open Government Licence v3.0.).

Croatia (Pagliarini, 1979), Czech Republic, England (Griffiths, 1960), Germany, Greece (Elioupolos *et al.*, 2002), India, Iran (Sepasgozarian, 1978; Noei, 2007; Mossadegh, 1997), Ireland (Cusack *et al.*, 1975), Japan (Sinha, 1968a), Poland, Scotland, Taiwan (Tseng, 1979), Turkey, United States and Wales (Hagstrum *et al.*, 2013), and is associated with a long list of commodities (animal feed, aquatic product, barley, barley debris, beet seed, buckwheat, buckwheat product, candied fruit, cheese, Chinese medicine, clove seed, condiment, confection, coriander seed, cotton seed, dried fruit, dried vegetable, grain, egg (dried), flax seed, grain, grain residue, grass seed, hemp seed, herb, kale seed, kohlrabi seed, lettuce seed, maize, maize flour, manure, mushroom, oat, onion seed, parsley seed, groundnut, radish seed, rapeseed, residue, stored rice and decayed rice bran, rotten fig fruit, runch seed, rye, seed cotton, soya flour, spinach seed, sugar beet seed, sugar (red), sugar (white), sunflower seed, swede seed, sweeping, wheat, wheat flour, wheat residue, wheat spillage). The facilities include bread van, commercial grain storage, drier, empty grain bin, farm, farm grinder, flour-mill, off-farm storage, provender mill, provender warehouse and railcar (Ardeshir *et al.*, 2006; Elioupolos *et al.*, 2002; Noei, 2007; Hagstrum *et al.*, 2013; Khaleghabadian *et al.*, 2015). This species is known as a potential predator of *Acarus siro* Linnaeus, *Glycyphagus destructor* (Schrank) and *Tyrophagus putrescentiae* (Schrank) (Pulpan and Verner, 1965). *C. eruditus* has been mentioned as usually occurring in stored foods and house dust, feeding on pest mites and reducing pest populations (Horn *et al.*, 2015; Dutra *et al.*, 2018). The quantitative relationship with the stored-product mites has been investigated by Gause *et al.* (1936), followed in the next decades, by commercial efforts to control pest mites in store food by inundative releases (Ždárková and Horak, 1990). Pérez Lozano (1979) studied in Venezuela 40 asthmatic patients with age ranges from 5 to 46 years with three extracts in intradermal skin test: *Dermatophagoides* 1/10, *Cheyletus* (reported as *Cheylatus*) 1/100 and house dust 1/500 w/v, respectively. Eighty-five per cent of patients resulted sensitive to house dust, 57.8% to *Dermatophagoides*, and 55% to *Cheyletus*. Forty per cent resulted positive for the three antigens and 12.5% negative to these antigens. As Cheyletus have been fed with *Dermatophagoides*, the positive reactions observed for the former could be influenced by the type of feeding. Therefore, *Cheyletus* must not be employed in patients' home for eradication of mites. Porto *et al.* (2017) reported a dermatitis by *C. eruditus* in a man in Brazil. Sun *et al.* (2023) evaluated the predation potential of the predatory mites *C. malaccensis* and *C. eruditus* and carried out a study by comparing the life histories of two predatory mites and the functional responses on *Liposcelis bostrychophila* Badonnel (Psocoptera, Liposcelididae), a pest of stored food. The results indicate that compared to *C. eruditus*, *C. malaccensis* had a shorter development time and longer adult survival time at 28°C and 75% RH and showed higher predation ability against *L. bostrychophila* eggs. It has been proven from both its life history of artificial breeding and the predation ability that *C. malaccensis* has much greater biocontrol potential than *C. eruditus* against stored-grain pests.

C. malaccensis (Fig. 13.4) was described by Oudemans in 1903b and reported successively as *Cheyletus egypticus* El Badry, *Cheyletus fortis* Oudemans, *Cheyletus monroi* Hughes and *Cheyletus polymorphus* Volgin. Its biology is widely investigated (Zaher and Soliman, 1971b; Yousef *et al.*, 1982; Saleh *et al.*, 1986; Rout and Misera, 1987; Cebolla *et al.*, 2009; Palyvos and Emmanouel, 2005, 2009, 2011). The geographic distribution is cosmopolitan (Australia, Brazil, Canada, China, Colombia, Czech Republic, Ecuador, Egypt, England, Ethiopia, Germany, Greece, Grenada, India, Indonesia, Iran, Iraq, Ireland, Japan, Malaysia, Mali, Myanmar, New Guinea, Peru, Philippines, Poland, Portugal, Russia, Rwanda, Sierra Leone, Sri Lanka, Tajikistan, Taiwan, Turkey, United States, Uganda, Venezuela, Wales and Yemen) (Fain and Bochkov, 2001; Elioupolos *et al.*, 2002; Hagstrum *et al.*, 2013; Beron, 2021a). In literature, currently 68 commodities are known [apple (dried), aquatic product, barley, biscuit, candied fruit, cereal, cocoa bean, coffee bean, condiment, copra, cotton seed, decayed rice bran, dried fruit, dried vegetable, fereek, fish (dried), fish (minced-dried), fishmeal, fodder, fungus (dried), garden bean, garlic, grain, grain spillage, grass seed, ham (dried), legume, lentil, macaroni, maize, maize grit, maize flour, manure, malt flower, milk, mungo, mushroom, nutmeg, oat, onion bulb, onion seed, palay, illipeanut, pollard, poultry diet, poultry waste, processed food, radish seed, residue, rice, rice (husked), rice (milled), rice (red), rice straw mat, rotten fruit of fig, stored rice, seed cotton, shallot, shea nut, sorghum, soybean, soybean meal,

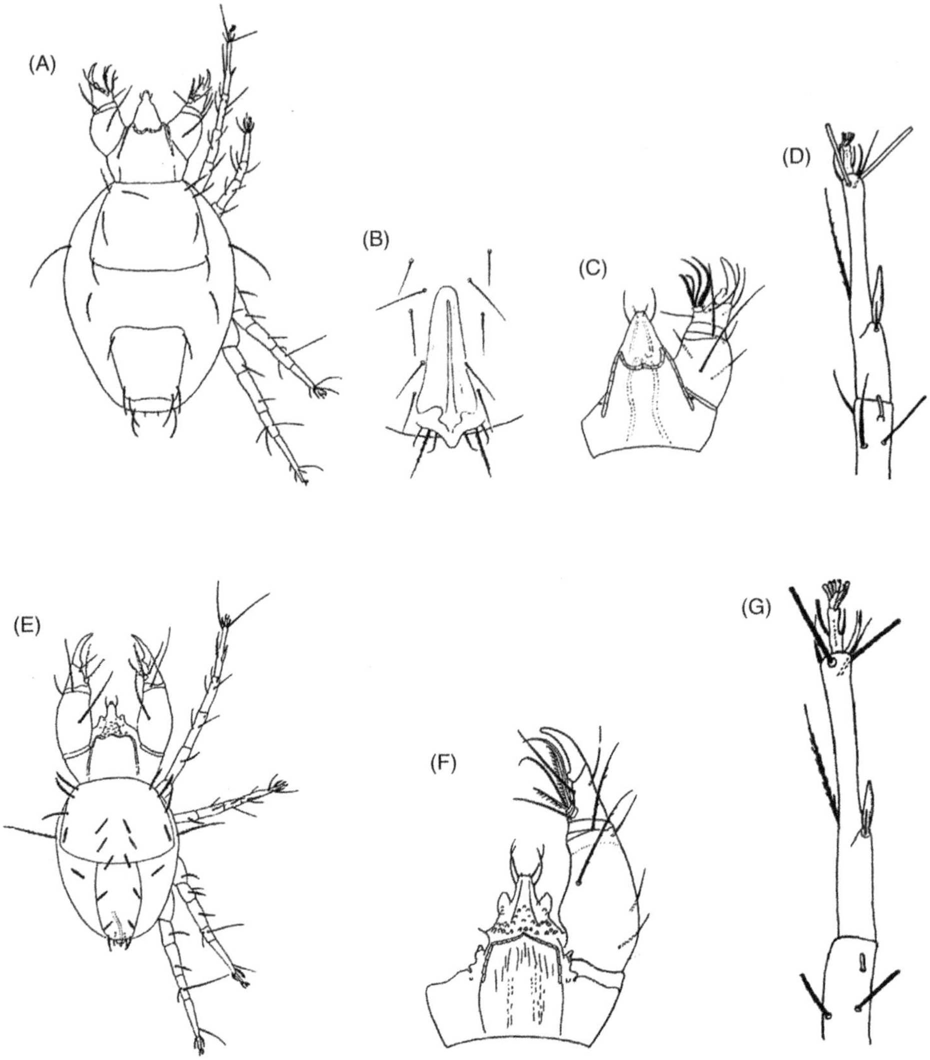

Fig. 13.4. *Cheyletus malaccensis* Oudemans. Female. Dorsal view (A); anal region (B); dorsal view of gnathosoma (C); leg I (D). Male. Dorsal view (E); dorsal view of gnathosoma (F); leg I (G) (from Hughes, 1976, modified, licensed under the Open Government Licence v3.0.).

soybean sauce, soybean spillage, sugar beet seed, sugar (red), sugar (white), sunflower seed, tiki-tiki, tulip bulb, wheat, wheat bran, wheat flour, wheat (milled)] to be collected on ten facilities (commercial grain storage, farm, flour-mill, grain stores, grain warehouse, granary, market, off-farm storage, provender warehouse, ship); moreover, the mite has been collected in honey bee hive, houses and soil (Sepasgozarian, 1978; Elioupolos *et al.*, 2002; Noei, 2007; Hagstrum *et al.*, 2013; Khaleghabadian *et al.*, 2015). A study aimed at describing the mite fauna associated with environmental factors and several production-line environments inside animal feed factories showed that *Cheyletus malaccensis* is a potential predator of *D. farinae* Hughes and *T. putrescentiae* (Schrank) on package depositing, rice bran milling machine, corn conveyor belt, soybean silo, pigeon nests, animal product elevator, pellet cooler, and from the roof of an animal feed factory in Brazil (Siegert *et al.*, 2018).

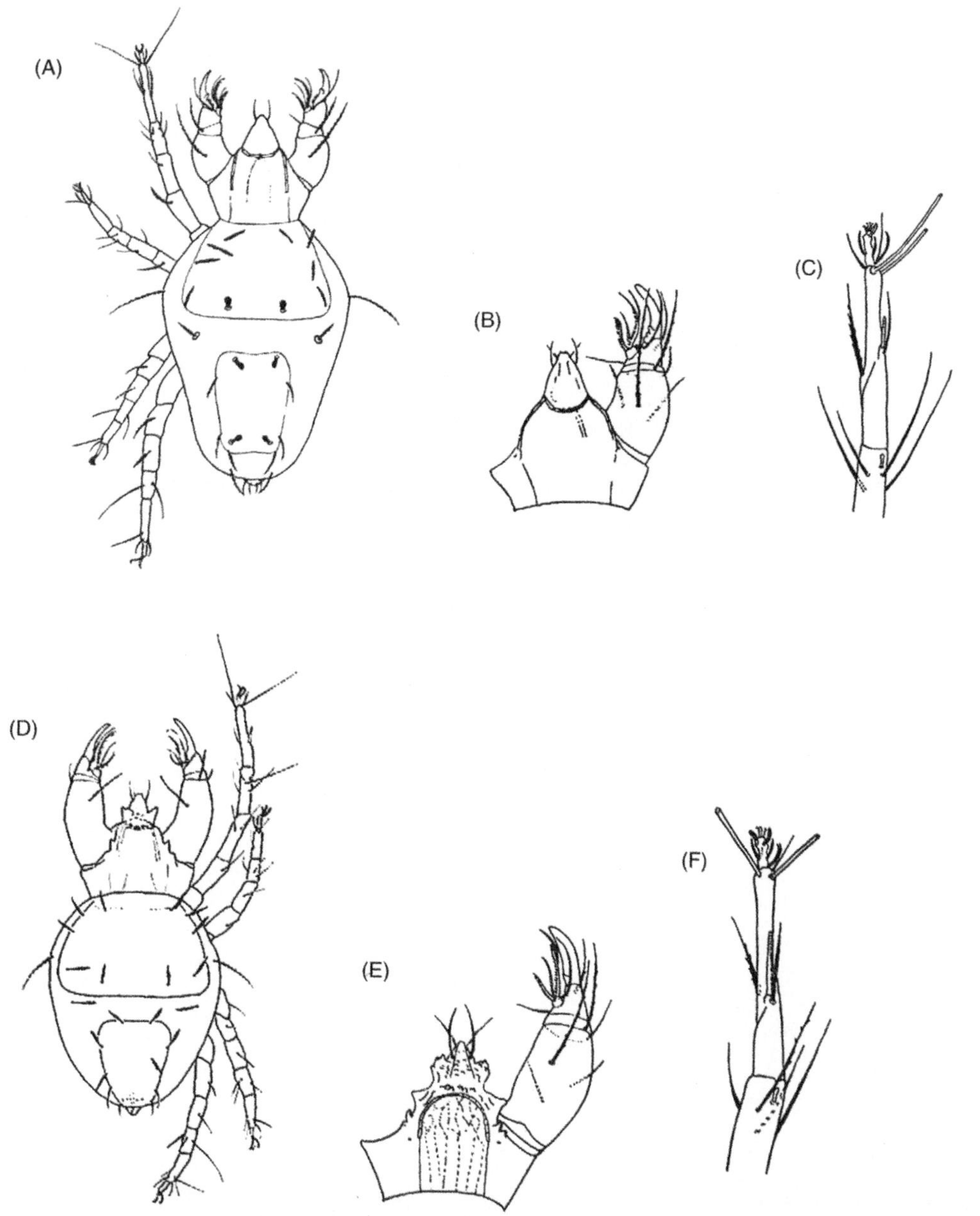

Fig. 13.5. *Cheyletus trouessarti* Oudemans. Female. Dorsal view (A); dorsal view of gnathosoma (B); dorso-lateral view of right leg I (C). Male. Dorsal view (D); gnathosoma (E); dorsal view of right leg I (F) (from Hughes, 1976, licensed under the Open Government Licence v3.0.).

This species has been shown to be an effective predator of stored-product pests *T. putrescentiae*, *A. siro*, *A. ovatus* (Troupeau), *Caloglyphus redickorzevi* Zachvatkin, *G. destructor* (Schrank), of ectoparasites, such as *D. gallinae* (De Geer) and *Megninia ginglymura* (Mégnin) and has become an important tool in the biocontrol of these pests (Cebolla *et al.*, 2009; Al-Shammery, 2014; Granich *et al.*, 2016). In Brazil, this Cheyletid mite has been found on rice, beans, cassava flour, soybean meal and poultry feeds (Baggio *et al.*, 1987; Franzolin and Baggio, 2000; Sousa *et al.*, 2005; Galvão *et al.*, 2011). It has been reported in house dust samples (Ezekiel *et al.*, 2008; Dutra *et al.*, 2018), nests of birds and laying hens (Silva *et al.*, 2013; Horn *et al.*, 2015).

Cheyletus malayensis Cunliffe is known from Russia, Hawaii, Iran, Malaysia, and Philippines (Cunliffe 1962; Summers and Price 1970; Sepasgozarian 1978; Corpuz-Raros 1988), and has been collected in Iran on stored barley, rice, flour, wheat, and in house dust (Sepasgozarian, 1978; Ostovan, 1993; Fathipour *et al.*, 1999; Sayedi *et al.*, 2006). *C. trouessarti* Oudemans (Fig. 13.5) has been investigated by Zhu *et al.* (2000). Its geographic distribution is cosmopolitic and includes Azerbaijan, Canada, China (Lung-Shut, 1984), Czech Republic (Stejskal *et al.*, 2003), England (Griffiths, 1960), France, Germany, Greece (Eliopoulos and Papadoulis, 2001; Eliopoulos *et al.*, 2002), Iran (Sepasgozarian, 1978; Ardeshir, 2017), Ireland, Japan, Malaysia (Fain and Nadchatram, 1980b), The Netherlands (Volgin, 1987), Peru (Caceres and Fain, 1977), Poland, Russia, Scotland, Taiwan (Tseng, 1979), Turkey, United States and Wales (Hagstrum *et al.*, 2013). The list of commodities numbers animal feed, barley, flax seed, grain, grain residue, grass seed, herb, kaoliang, maize flour, protein feeding mixture, rice, rice (husked), rice barn, soya sweeping, sugar beet seed, sweeping, wheat, wheat bran, wheat flour, wheat flour sweeping, and wheat residue), and facilities (barn, commercial grain storage, drier, empty grain bin, farm, farm grinder, flour-mill, flour warehouse, provender mill, provender warehouse and railcar (Sepasgozarian, 1978; Ostovan, 1993; Eliopoulos *et al.*, 2002; Hagstrum *et al.*, 2013; Khaleghabadian *et al.*, 2015). *Cheyletus trux* Rohdendorf has been described as *Cheyletus hendersoni* by Baker (1949). Its geographical distribution includes Belgium, the Czech Republic, Greece, Poland, Russia, Switzerland, Ukraine and the United States. The list of commodities numbers barley, cattle feed residue, flax seed, herb, lettuce seed, maize, millet, mummy, sweeping, wheat and wheat bran collected in flour-mills and off-farm storage (Hagstrum *et al.*, 2013).

Cunlifella variegata (Barilo) is known from Uzbekistan, Russia (Barilo, 1985; Bochkov and Mironov, 1997) and Iran where it has been collected on flour (Ardeshir and Nematollahi, 2008).

The genus *Eucheyletia* Baker numbers two species, named *Eucheyletia flabellifera* (Michael) and *Eucheyletia taurica* Volgin, respectively (Fig. 13.6). The first species was described in 1878 (Michael, 1878), as *Cheyletia flabellifera* and reported from

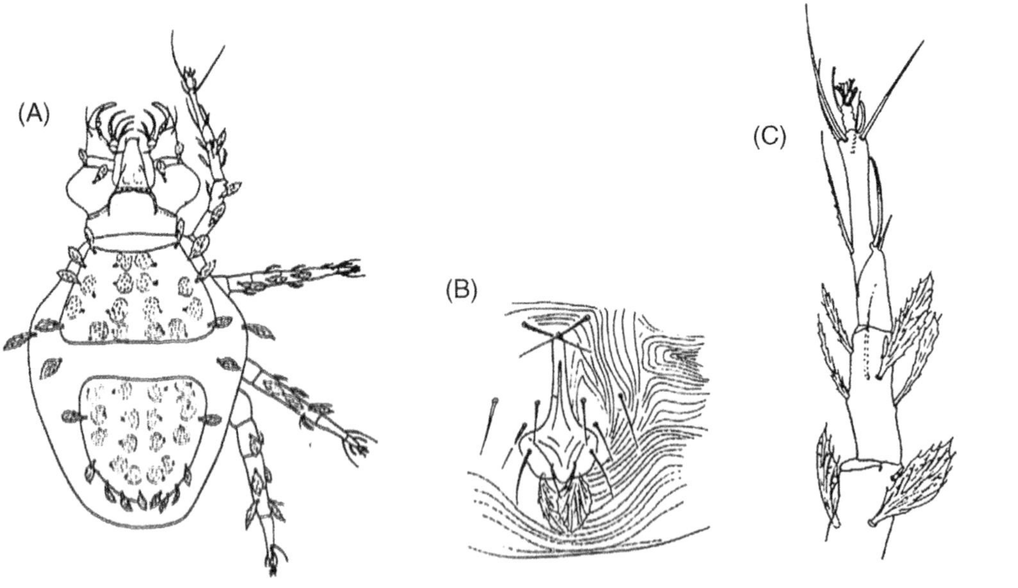

Fig. 13.6. *Eucheyletia taurica* Volgin. Female. Dorsal view (A); anal region (B); leg I (C) (from Hughes, 1976, licensed under the Open Government Licence v3.0.).

Belgium, Canada, England, Iran, Ireland, Japan and Poland (Fain and Bochkov 2001; Salarzehi *et al.*, 2018; Hagstrum *et al.*, 2013) on barley, buckwheat, buckwheat product, grain, oat spillage (old), stored rice and decayed rice bran in cellar, commercial grain storage (Hagstrum *et al.*, 2013; Salarzehi *et al.*, 2018).

E. taurica (Fig. 13.6) is reported from the Czech Republic, England, Russia, and Wales (Hughes, 1976; Summers and Price, 1970; Volgin, 1987). The species was found by Volgin (1987) in cereal stores in the Czech Republic. Zdarková and Horak (1997) observed that this mite feeds on *A. siro* in the laboratory. The species has been collected in stored products in Brazil (Siegert *et al.*, 2018).

The genus *Hemicheyletia* Volgin numbers three species, named *Hemicheyletia reticulata* Jeffrey *et* Campbell, *Hemicheyletia tumidus* Qayyum *et* Chaudhri, and *Hemicheyletia vescus* Qayyum *et* Chaudhri, respectively. *Hemicheyletia reticulata* has been discussed by Thewke and Enns (1979), and reported from Scotland, on cheese stored at creamery, combine harvester and farm (Jeffrey and Campbell, 1975; Thewke and Enns, 1979). *H. tumidus* is known from Pakistan on wheat in grain market and warehouse (Qayyum and Chaudhri, 1979). Also *H. vescus* is reported from Pakistan, where it has been found on rice and wheat in grain market (Qayyum and Chaudhri, 1979).

The genus *Ker* Muma includes two species, *Ker bakeri* Zaher *et* Soliman and *Ker mercedesae* Corpus-Raros, respectively. The first mite is reported from Brazil, Cameroon, the Czech Republic and Egypt, on coffee bean in market (Zaher and Soliman, 1967). *K. mercedesae* is known from Philippines on cassava flour (Corpus-Raros, 1998). Eliopoulos *et al.* (2002) reported from Greece an unidentified species, *Ker* sp., on grain.

Lepidocheyla gracilis Volgin is known from Eastern Europe (Gerson *et al.*, 1999) and Iran (Khaleghabadian *et al.*, 2015), and has been collected on mills, food stores and grain (Bochkov *et al.*, 2005; Khaleghabadian *et al.*, 2015; Abbasi *et al.*, 2016).

Mexecheles votandinii Jeffrey is reported from Scotland in farm, and whisky distillery (Jeffrey, 1976a).

Neoeucheyla iranica Fain *et* Ardeshir is known from Iran on floor debris, wheat grain, dust and straw (Fain and Ardeshir, 2000; Ardeshir, 2004).

Nodele calamondin Muma is reported from the United States, Egypt and Iran, on silo and flour-mills, and grain (Bochkov *et al.*, 2001). The mite has been studied by Summers and Price (1970) and Volgin (1969), and its biology by Barker (1992). *Nodele mu* Haines is reported from Java on cassava (Haines, 1988). *Nodele* sp. has reported from Iran on various stored products (Khalilmanesh, 1973; Sepasgozarian, 1977, 1978; Faraji and Kamali, 1993; Ostovan, 1993; Fathipour, 1994; Modarres Awal, 1994, 1997).

Oudemansicheyla denmarki (Yunker) is reported from Taiwan on betel nut (Yunker, 1961).

Zachvatkiniola reticulata (Cunliffe) is reported from Russia, Iran, China and Taiwan on dried fruit, dried vegetable, grain, kaoliang, maize, mushroom, groundnut, processed food, rice (husked), stored rice and decayed rice bran and wheat (Cunliffe, 1962; Volgin, 1969; Bochkov *et al.*, 2001; Salarzehi *et al.*, 2018).

14 The Family Cunaxidae Thor (Figs. 14.1 and 14.2)

The body of Cunaxidae is small to moderately large from 400 µm to 1000 µm long, whitish or translucent, and frequently red, orange or yellow coloured. The palpi may be raptorial with a claw-like distal palptarsus and armed internally of spines and spur that enable them to grasp their prey. In some species, such as *Parabonzia* sp., they terminate in two long sensory setae and one solenidion (Walter *et al.*, 2009).

They are fundamentally predatory mites, living in most terrestrial habitats including forest systems, grasslands, agricultural fields and anthropogenically disturbed areas. Skvarla *et al.* (2014) reported that the family numbers six subfamilies, 20 genera and 400 species, and published a key to world's species.

Cunaxid mites occur also on stored food products (Hughes, 1976; Tseng, 1980; Fan, 1992), where are known almost 10 species ascribed to the genera *Armascirus* Den Heyer (2 species), *Cunaxa* Von Eyden (6 species), *Coleoscirus* Berlese (1 species) and *Pulaeus* Den Heyer (1 species), respectively.

Armascirus taurus Kramer was described in 1881; the same year the species was described by the same author as *Cunaxa taurus* (Kramer, 1881) and successively identified by Smiley (1992). Its geographical distribution is cosmopolitan (Canada, China, Costa Rica, Cuba, Czech Republic, England, France, Germany, Honduras, Ireland, Malaysia, Mexico, Panama, Philippines, Poland, South Africa, Taiwan, United States, Zaire). Its list of commodities includes barley meal, grain, grass seed, maize and oat, collected in commercial grain storage and farm (Hagstrum *et al.*, 2013). Other species of this genus is *Armascinus hastus* Shiba reported from Iran, on food stores and grain (Abbasi *et al.*, 2016).

The genus *Cunaxa* includes the species *Cunaxa capreolus* (Berlese) (Fig. 14.1), initially described by Berlese (1889) as *Scirus capreola* and correctly identified by Smiley (1992). Soliman *et al.* (1975) and Zaher *et al.* (1975) studied its biology. The mite has a cosmopolitan geographic distribution (China, Egypt, Italy, Hong Kong, Japan, Mexico, Philippines, Spain and the United States) and has been collected on aquatic products, barley, broad bean, condiment, cuttlefish (dried), dried fruit, dried vegetable, mushroom, onion, sorghum, soybean and sunflower seed, stored rice, rice dust and debris (Noei, 2007; Noei *et al.*, 2008a; Hagstrum *et al.*, 2013; Khaleghabadian *et al.*, 2015). *Cunaxa denheyeri* (Fan) is reported from China on aquatic product, condiment, dried fruit, dried vegetable and mushroom (Fan, 1992). *Cunaxa nankanaensis* Ashfaq *et al.* reported from Pakistan on rice husk in rice mill (Ashfaq *et al.*, 2011). *Cunaxa setirostris* (Hermann) (Fig. 14.2) was described by Hermann (1804) as *Scirus setirostris* and correctly identified by Smiley (1992). The mite has a cosmopolitan geographic distribution (Australia, Brazil, Germany, Greece, Guam, India, Indonesia, Israel, Italy, Madeira, Morocco, Philippines and the United States). The list of its commodities includes apple, potato tuber, cotton, soil samples, barley meal, grain, wheat, stored rice, rice dust and debris in farm, food store and off-farm storage (Khalilmanesh, 1973; Sepasgozarian, 1977; Modarres Awal, 1994, 1997; Fathipour, 1994; Eslamizadeh, 1996; Haddad Irani-Nejad, 1998, Haddad Irani-Nejad *et al.*, 1999; Eliopoulos *et al.*, 2002; Noei, 2007; Noei *et al.*, 2008a; Hagstrum *et al.*, 2013). *Cunaxa sinensis* (Fan) is reported from China on aquatic product, condiment, dried fruit, dried vegetable and mushroom (Fan, 1992). *Cunaxa womersleyi* Baker *et* Hoffman is reported from Canada, China, Japan, Taiwan and United States on condiment, dried fruit, dried vegetable, grain, mushroom and red bean (Baker and Hoffman, 1948; Smiley, 1992). Moreover, another unclassified species of the same genus (*Cunaxa* sp.) has been reported on stored products by Noei (2007) and Noei *et al.* (2008a) in Iran on stored products.

 The Handbook of Mites of Stored Products (V. Vacante and C.H.W. Flechtmann)
DOI: 10.1079/9781786390912.0014

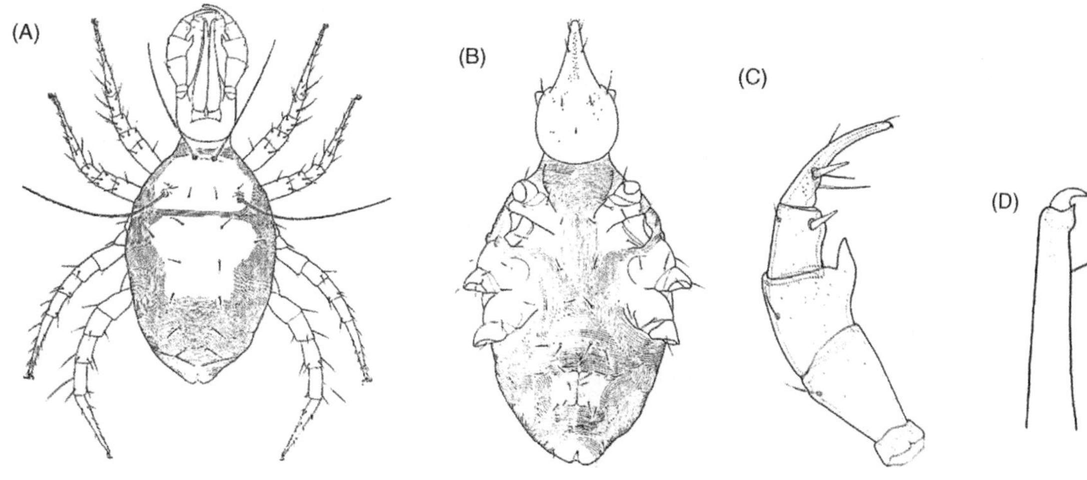

Fig. 14.1. *Cunaxa capreolus* (Berlese). Female. Dorsal view (A); ventral view (B); palp (C); distal part of chelicerae (D) (From Vacante and Nucifora, 1984-85, licensed under CC BY 4.0).

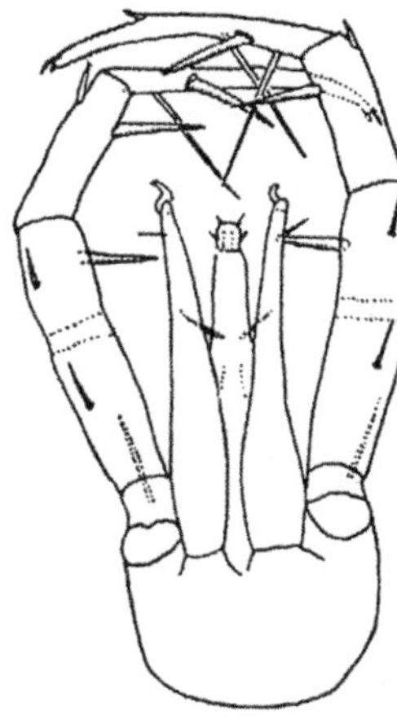

Fig. 14.2. *Cunaxa setirostris* (Hermann). Female. Dorsal view of gnathosoma (from Hughes, 1976, licensed under the Open Government Licence v3.0).

Coleoscirus simplex (Ewing) has a large geographical distribution including the United States, China, Italy, Egypt and Iran (Walter and Kaplan, 1991; Beyzavi *et al.*, 2013; Corpuz-Raros, 1996; Noei, 2007; Noei *et al.*, 2008a; Bashir *et al.*, 2014) on stored rice, rice dust and debris (Noei, 2007; Noei *et al.*, 2008a).

Pulaeus glebulentus Den Heyer is known from China and South Africa on aquatic product (Den Heyer, 1980; Smiley, 1992).

15 The Family Eupodidae Koch (Fig. 15.1)

The eupodid mites inhabit soils, forest litter, mosses, lichens, seaweeds, mushrooms, etc. (Walter *et al.*, 2009). Khaustov (2014) recently reviewed the taxonomy of the superfamily Eupodoidea and provided a key to genera of the family Eupodidae.

The Eupodidae have the anal opening ventral or terminal; the intercoxal region bears fewer than 10 pairs of setae. The femur IV is normal or strongly swollen, employed in jumping. They are a cosmopolitan group of mycophagous, phytophagous and predatory mites living in the soil, forest litter, mosses, lichens, seaweeds, mushrooms, etc. They belong to the superfamily Eupodoidea Koch (Walter *et al.*, 2009). The classification of this superfamily lacks stability (Baker and Lindquist, 2002); Khaustov (2014, 2015) provided a key to the 12 genera of the superfamily, and Walter *et al.* (2009) reported over 80 species.

Only two species have been reported on stored food. They are *Eupodes variegatus* Koch (Fig. 15.1), known from Canada on empty grain bins, and *Eupodes voxencollinus* Sig Thor reported from China on aquatic product (Hagstrum *et al.*, 2013), and from Iran on stored potatoes, and tea (Ostovan, 1993; Taghavi, 1996; Modarres Awal, 1997; Taghavi *et al.*, 1998).

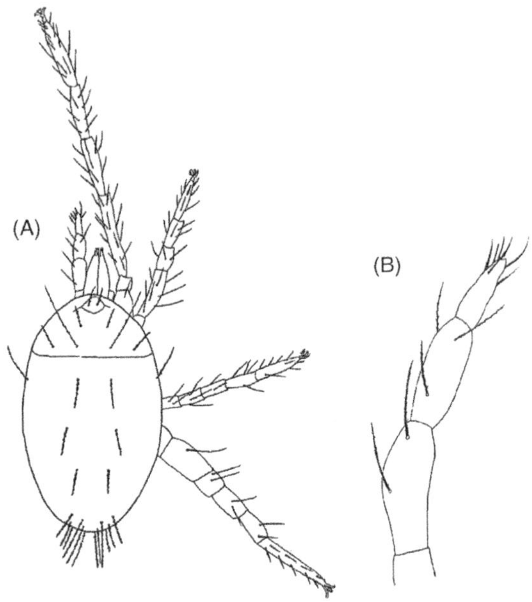

Fig. 15.1. *Eupodes variegatus* Koch. Dorsal view of adult (A); palp (B) (from Meyer and Ryke, 1960, modified).

DOI: 10.1079/9781786390912.0015

16 The Family Camerobiidae Southcott (Fig. 16.1)

Morphological characters of Camerobiids consist of ovoid or nearly circular idiosoma, dorsoventrally compressed. The legs are mostly longer than idiosoma and stilt-like, with long femora and tibiae. The gnathosoma is small, almost covered by the prodorsum and directed ventrally; the palpus is small, five-segmented, with tibial claw vestigial or lacking; the peritreme bears one or four loops, and the chelicerae are fused (Bolland, 1986; Toit *et al.*, 1998).

Camerobiids have a worldwide geographic distribution, and number about 161 species ascribed to seven genera (Fan and Walter, 2011; Zhang *et al.*, 2011; Khanjani *et al.*, 2012a, b, 2013a, 2014; Akyol, 2013; Khanjani and Hoseini, 2013; Zeity and Gowda, 2013; Uluçay and Koç, 2014). They are predatory mites usually collected in various microhabitats, such as humus, tree bark, vegetation, or inside epiphytes in temperate and tropical ecosystems. Some species feed on plant pests such as scale insects (Coccoidea), plant-parasitic mites (Eriophyidae and Tenuipalpidae), and on other small mites that scavenge or feed on fungi on plant surfaces (Tarsonemidae and Tydeidae) (McGregor, 1950; De Leon, 1967; Bolland, 1983; Gerson *et al.*, 2003; Walter *et al.*, 2009). Sepasgozarian (1996) reported that species of this family live on trees, shrubs and grasses but appear to prefer plant litter. Bolland (1986) refers that Camerobiids are rare in soil samples.

This family includes only *Neophyllobius saxatilis* Halbert (Fig.16.1), reported from Ireland in farms (Halbert, 1923).

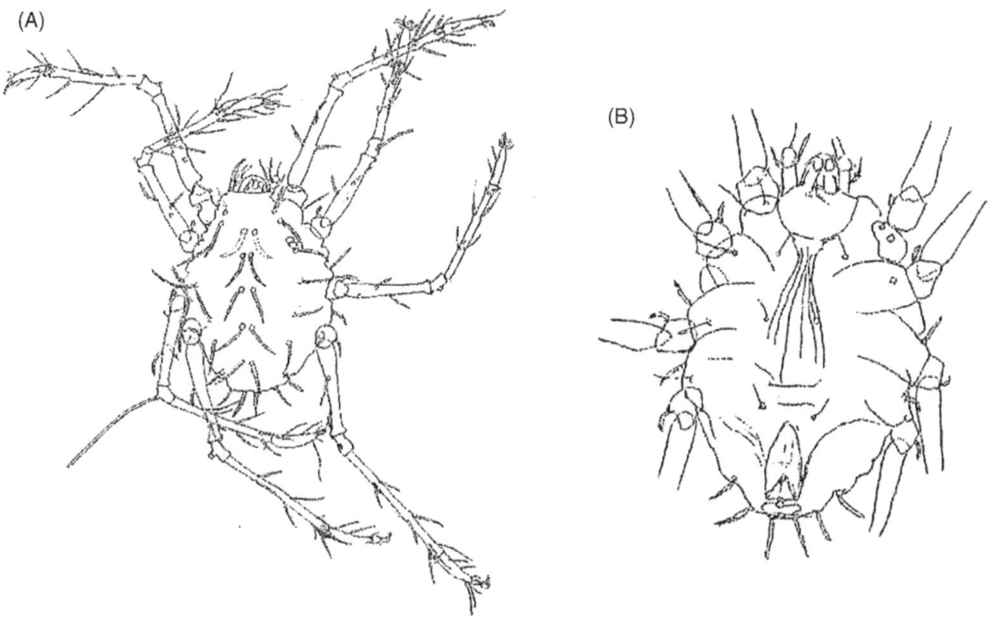

Fig. 16.1. *Neophyllobius saxatilis* Halbert. Dorsal view (A); ventral view (B) (from Van Eyndhoven, 1938).

DOI: 10.1079/9781786390912.0016

17 The Family Pterygosomatidae Oudemans (Fig. 17.1)

Pterygosomatidae mites are red in colour, from small to medium size, characterized by external and emergent peritremes, chelicerae chelate and palpal thumb-claw complex. Sometimes, the body is twice as wide as long.

These mites have been collected on lizards, tortoises, and arthropods in Africa, Asia, Australia, and the Americas. However, although most pterygosomatids parasitised reptiles, some species may be associated with other host taxa (Walter *et al.*, 2009).

The family numbers 10 genera and 177 species of ectoparasitic mites commonly associated with lizards (Bertrand *et al.*, 2013; Fajfer, 2013a, b; Fajfer and González-Acuña, 2013; Paredes-León, 2013; Paredes-León *et al.*, 2013; Fajfer and Melnikov, 2014), with the exception of the genus *Pimeliaphilus* Trägårdh, including 20 species parasitizing arthropods (Paredes-León *et al.*, 2012).

On stored food, this family numbers only *Pimeliaphilus podapolipophagus* Cunliffe (Fig. 17.1), reported from Malawi on dried fish (Cunliffe, 1952).

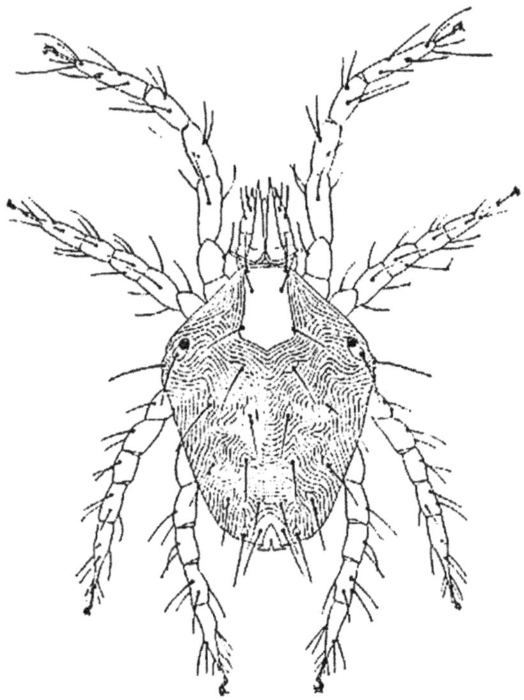

Fig. 17.1. *Pimeliaphilus podapolipophagus* Cunliffe. Adult (from Baker and Wharton, 1952).

DOI: 10.1079/9781786390912.0017

18 The Family Raphignathidae Kramer (Figs. 18.1–18.4)

Raphignathidae mites of the genus *Raphignathus* Dugés possess well-developed dorsal shields, more extensive in males than females; conversely, females of the genus *Neoraphignathus* have no distinct dorsal shields (Smiley and Moser, 1968). They are mostly collected underneath tree bark, in lichens, moss, leaf litter, bird nets, intestines of wedded seals, soil, on many plants and in house dusts or stored grain (Khanjani and Ueckermann, 2003; Krantz and Walter, 2009), where they are predators on small arthropods (Khanjani *et al.*, 2013b).

They are predaceous mites and can be found underneath tree bark, in lichens, moss, leaf litter, pigeon nests, and intestines of wedded seals, in soil, on a wide range of plants and in house dusts (Fan and Yin, 2000; Khanjani and Ueckermann, 2003).

The family is divided into two genera based on the presence (*Raphignathus* Dugès) or absence (*Neoraphignathus* Smiley *et* Moser) of the dorsal shields. The genus *Raphignathus* numbers 72 species and possesses a cosmopolitan geographic distribution, while the genus *Neoraphignathus* Smiley *et* Moser numbers only one species (*Neoraphignathus howei* Smiley *et* Moser) reported only from Nearctic region (Fan and Zhang, 2005). Nasrollahi *et al.* (2018) reported a key of the world species of the genus *Raphignathus* Dugés.

On stored food are reported seven raphignathid mites, all ascribed to the genus *Raphignathus* Dugès. namely as *Raphignathus conspicuus* (Berlese), *Raphignathus giselae* (Fig. 18.1) Meyer *et* Ueckermann, *Raphignathus gracilis* (Rack) (Fig. 18.2), *Raphignathus hecmatanaensis* Khanjani *et* Ueckermann (Fig. 18.3), *Raphignathus orientalis* Fan *et* Li, *Raphignathus zhaoi* Hu, Jing *et* Liang (Fig. 18.4), and an unidentified species, respectively.

Raphignathus conspicuus is reported from England and Ireland on wheat in commercial grain storage and farm (Berlese, 1910). *Raphignathus giselae* is known from Iran, South Africa, Turkey, Yemen and Zimbabwe (Akyol and Koç, 2007; Beron, 2020), and has been collected on stored rice, rice dust and debris in Iran (Noei, 2007; Noei *et al.*, 2008a). *R. gracilis* is known from China on aquatic product and condiment (Rack, 1962). *R. hecmatanaensis* is known from Iran where it has been collected in storage products, stored rice, rice dust and debris (Noei, 2007; Noei *et al.*, 2008a; Khaleghabadian *et al.*, 2015). *R. orientalis* has been found in China on aquatic products and condiment (Fan and Li, 1993a). *Raphignathus zhaoi* is known from China (Fan and Yin, 2000) and Iran, where it has been collected on storage products (Sayadi *et al.*, 2012). *Raphignathus* sp. has been reported in stored products and in bird nests from Turkey (Dönel and Doğan, 2011) and Brazil (Siegert *et al.*, 2018).

DOI: 10.1079/9781786390912.0018

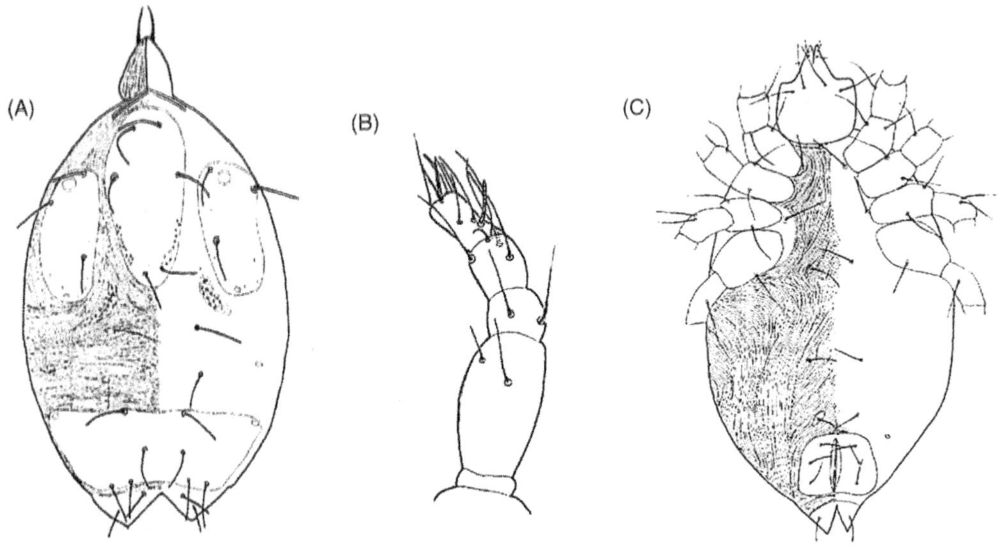

Fig. 18.1. *Raphignathus giselae* Meyer *et* Ueckermann. Female. Dorsal view (A); palp (B); ventral view (C) (from Ghorbani *et al.*, 2013, modified).

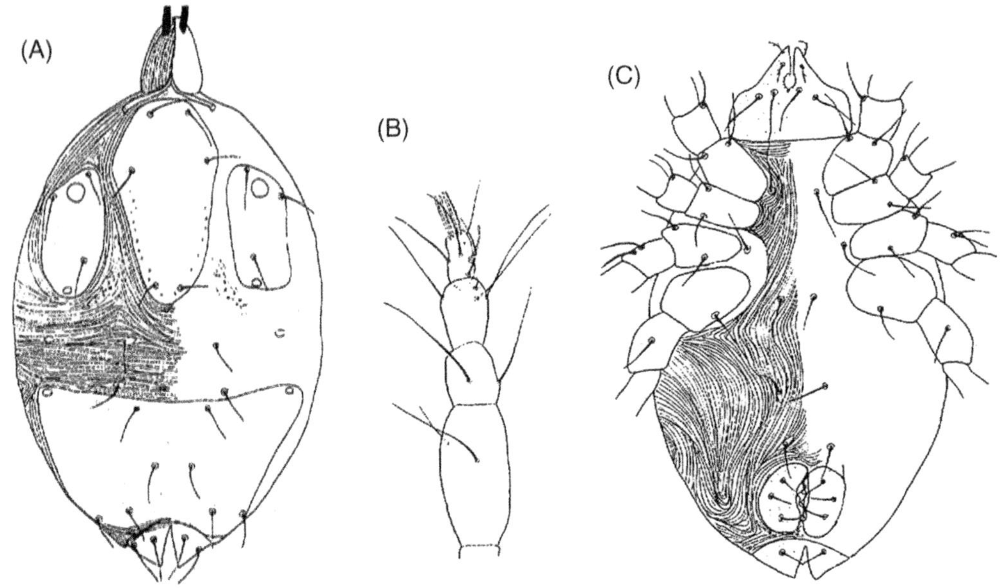

Fig. 18.2. *Raphignathus gracilis* (Rack). Female. Dorsal view (A); palp (B); ventral view (C) (from Ghorbani *et al.*, 2013).

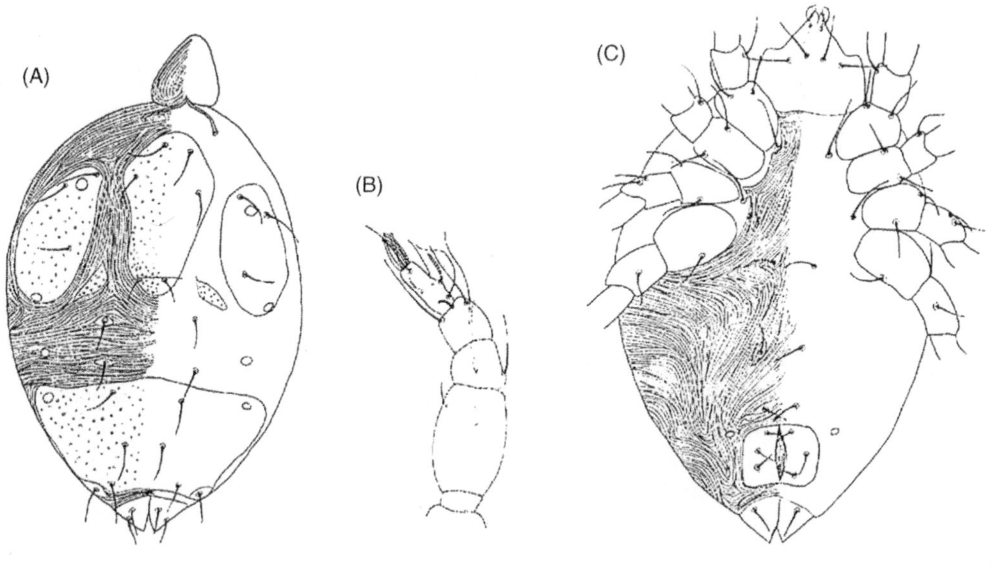

Fig. 18.3. *Raphignathus hecmatanaensis* Khanjani *et* Ueckermann. Female. Dorsal view (A); palp (B); ventral view (C) (from Ghorbani *et al.*, 2013).

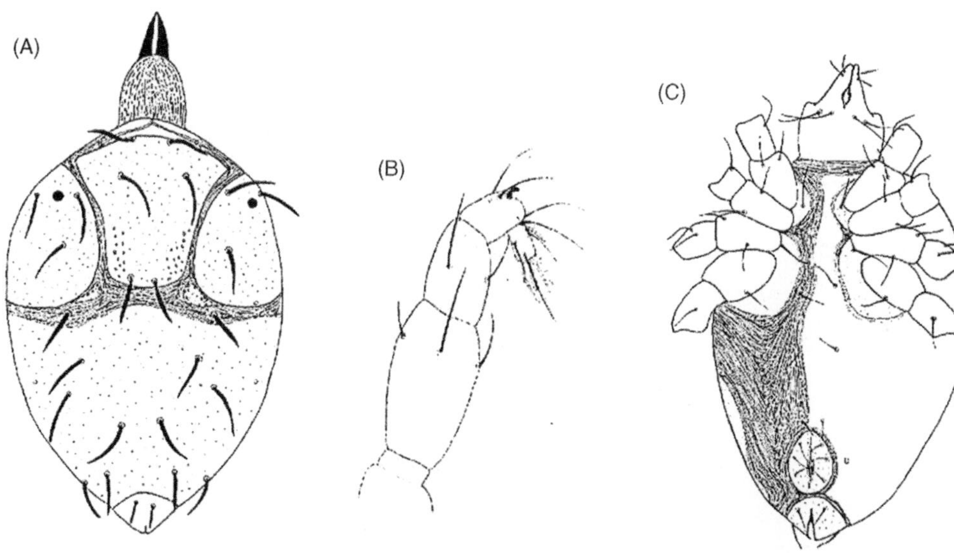

Fig. 18.4. *Raphignathus zhaoi* Hu *et al.* Female. Dorsal view (A); palp (B); ventral view (C) (from Ghorbani *et al.*, 2013).

19 The Family Smarididae Vitzthum (Fig. 19.1)

Smarididae are large predatory mites with long oval bodies, distinctively pointed in front. They are usually red in colour and densely hairy with slender legs, sometimes very long. They have either one or two pairs of eyes. The Smarididae is a relatively small family, usually found in grassland and litter or under the bark of trees (Southcott, 1996; Krantz and Walter, 2009) and numbers 10 genera and 53 species (Zhang *et al*., 2011) (Fig 19.1).

The family includes two subfamilies, Smaridinae and Hirstiosomatinae, the former with five genera, including *Fessonia* von Heyden. This genus is reported from Europe, Australia, North America, Asia and Africa and hitherto encompassed ten known species, nine of which were described from adults. Previously, the larval stage of only one species, *Fessonia papillosa* (Hermann) from France, Italy, Croatia, Germany, Greece and Hungary, was known (Wohltmann, 2010). On stored food, the family numbers two species, *F. papillosa*, reported from China on aquatic products (Li and Fan, 1997) and *Smaris* sp. reported from Iran on dried fruits and nuts (Ardeshir *et al*., 2012). We report here an example of *Smaris mamillatus* (Say).

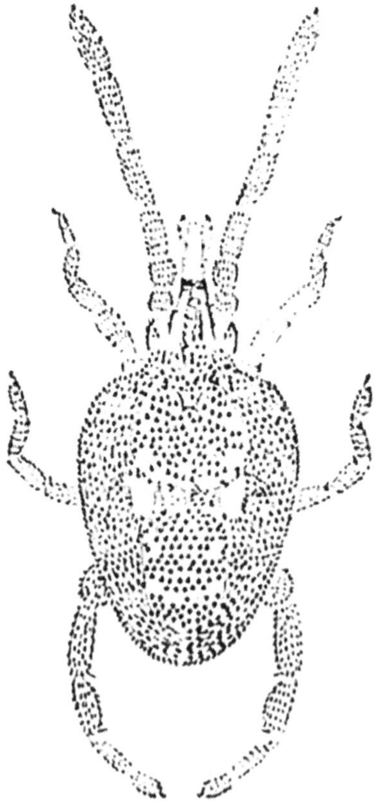

Fig. 19.1. *Smaris mamillatus* (Say). Dorsal view of female (from Baker and Wharton, 1952).

  The Handbook of Mites of Stored Products (V. Vacante and C.H.W. Flechtmann)
DOI: 10.1079/9781786390912.0019

20 The Family Erythraeidae Robineau-Desvoidy

Mites of the family Erythraeidae have the body oval and rather large, commonly reddish coloured and densely hairy. The legs, especially the first and fourth pairs, are long and adapted for running. They have either one or two pairs of eyes, and can be distinguished from related families by the presence of a single claw on the tibia of the palp. The Erythraeid mites are commonly predators at post-larval stages, feeding upon various arthropods. However, larvae of most erythraeids are parasites of different arthropods including insects, such as bugs, grasshoppers, flies and aphids. The larvae bite a hole into the cuticula of the host and use a stylostome, which acts like a drinking straw, to drink body fluids and dissolved tissues (Southcott, 1961, 1991; Goldarazena et al., 2000; Gerson et al., 2003; Saboori and Cobanoglu, 2010). The family numbers five subfamilies, including 55 genera and 771 species (Vercammen-Grandjean, 1973; Zhang *et al.*, 2011). The genus *Erythraeus* Latreille includes two subgenera, *Erythraeus* Latreille and *Zaracarus* Southcott. The subgenus *Erythraeus* numbers 93 species, and 45 of these species are known from larvae (Khanjani et al., 2012c; Mąkol and Wohltmann, 2012, 2013). The subgenus *Zaracarus* includes 27 species, all described from larvae (Mąkol and Wohltmann, 2012, 2013). More than 50% of all larval species of subgenus *Erythraeus* have been recorded as parasites on *Heteroptera*, *Thysanoptera*, *Neuroptera* and other insects, whereas others were captured free-living on herbaceous plants (Haitlinger, 2012; Khanjani et al., 2010, 2012c; Kamran et al., 2013; Stroiński et al., 2013). Species of the genus Leptus Latreille are reported from bees; an unidentified genus is reported from the Philippines on *Apis cerana* Fabricius (Malabanan and Corpuz-Raros, 1998); Leptus ariel Southcott lives on the European honey bee in Guatemala and Peru (Southcott, 1989) and *Leptus* monteithi Southcott is a parasite of a Leioproctus species (Hymenoptera, Colletidae) in Tasmania (Southcott, 1993).

On stored food are reported four species. *Erythraeus* (*Erythraeus*) *garmsaricus* Saboori, Goldarazena *et* Khajeali is reported from Iran on stored wheat, straw and dust in silos, flour-mills and mills (Ardeshir *et al.*, 2006); *E.* (*E.*) *shojaii* Saboori *et* Babolmorad collected on wheat in Iran (Maroufpoor and Ostovan, 2017); *E.* (*E.*) *southcotti* Goldarazena *et* Zhang, known from Spain (Khanjani *et al.*, 2012c) and Iran, where it has been collected on wheat (Maroufpoor and Ostovan, 2017), and *E.* (*Zaracarus*) *coleopterus* Mortazavi, Hajiqanbar *et* Saboori is reported from Iran on wheat, dried fruits and nuts (Ardeshir *et al.*, 2012; Maroufpoor and Ostovan, 2017; Noei *et al.*, 2017).

 The Handbook of Mites of Stored Products (V. Vacante and C.H.W. Flechtmann)
DOI: 10.1079/9781786390912.0020

21 The Family Stigmaeidae Oudemans (Figs. 21.1 and 21.2)

The body of Stigmaeidae mites is red to yellow in colour, ovoid or elongate, and the dorsum is often covered by one or more plates. The stigmata or peritremes are not obvious, and the chelicerae are short and stylet-like; the palpi possess a thumb-claw complex. The tarsus bears often an apical three-tined seta (Gerson *et al.*, 2003).

It is one of the most studied mite family in Acarology, with the number of described species increasing by 44% in the last 20 years. Fan *et al.* (2016) reported 577 described species in 34 genera and presented a pictorial key to genera. Khaustov (2019) reported a list of contributions that increased the number of species to about 598 species of 34 valid genera.

Stigmaeidae mites are free-living predators on plants (leaves, buds, branches and fruits) and in the soil, where feed on eggs and motile forms of other mite species (Tetranychidae, Tenuipalpidae, Eriophyidae, etc.). Some species feed on moss, and other species on sand flies which they wound (Shehata and Baker, 1996).

On stored food, this family numbers six species ascribed to the genera *Mediolata* Canestrini, *Storchia* Oudemans, and *Stigmeus* Koch, respectively.

Mediolata granaria Gonzalez-Rodriguez (Fig. 21.1) is known from Ireland and the United States where it was collected in farm (Gonzalez-Rodriguez, 1965) and poultry litter (Brady, 1970), and from Turkey from moss on bark of *Pinus*, lichen, decayed bark of *Quercus* and from lichen on *Pinus* (Dönel and Doğan, 2012). Recently, the mite has been collected by Khaustov (2022) in Russia on the bark of birch (*Betula pendula* Roth). *Mediolata* sp. has been reported on stored products from Brazil (Siegert *et al.*, 2018).

The genus *Storchia* numbers three species, named *Storchia annae* Fan et Li, *Storchia pacifica* (Summers) (Fig. 21.2) and *Storchia robusta* (Berlese), respectively. *S. annae* is reported from China on dried fruit, dried vegetable and mushroom (Fan and Li, 1993a). *S. pacifica* was first included into the genus *Apostigmaeus* Grandjean (Summers, 1964) and recently included into the genus *Storchia*. Summers (1964) reported the mite from Indonesia (intercepted at Hawaii) on rice (*Oryza sativa* Linnaeus) and from the Philippines on cassava (*Manihot esculenta* Crantz) and *Oryza sativa*; moreover, specimens of this mite were collected on oat (*Avena sativa* Linnaeus) in France, on tuberose (*Polianthes tuberosa* Linnaeus) in Tahiti (Hawaii) and on porcupine [*Coendu rothschildi* (Thomas)]in Corundú (Canal zone, Panama) (Summers, 1964). Moreover, it is reported from China, Indonesia, Panama and Philippines on condiment, dried fruit, dried vegetable, mushroom (Summers, 1964; Fan and Yan, 1997; Miranda *et al.*, 2002), and from Iran on rice, rice bran, stored rice, rice dust and debris (decaying) (Noei *et al.*, 2007). Seeman *et al.* (2023) redescribed this species from specimens, intercepted by the Australian Department of Agriculture, Fisheries and Forestry from Cambodia, Fiji, India, Indonesia, Kiribati, Laos, Pakistan, Papua New Guinea, New Zealand, Nigeria, Samoa, Singapore, Solomon Islands, Tonga, Uganda, Vanuatu and Vietnam. Most of the specimens were intercepted on products made from woven natural fibre such as mats and baskets. *S. robusta* is known from China (Fan and Yan, 1997), France (Corsica) (Grandjean, 1944), Greece (Wood, 1973), Hawaiian Islands (Goff, 1987), Hungary (Komlovsky and Jenser, 1992), Iran (Khanjani and Ueckermann, 2002), Israel (Gerson, 1968), Italy (Berlese, 1885), Japan (Wood, 1973), Namibia (Ueckermann and Meyer, 1987), New Zealand (Wood, 1967), Pakistan (Chaudhri *et al.*, 1979), Solomon Islands (Wood, 1971), South Africa (Meyer, 1969), Turkey (Doğan and Ayyildiz, 2003), USA (Wood, 1973), former USSR (Wainstein and Kuznetzov, 1978) and Iran where it has been collected on rice, rice bran (decaying), food stored, stored rice, rice dust and debris (Noei, 2007; Noei *et al.*, 2007).

 The Handbook of Mites of Stored Products (V. Vacante and C.H.W. Flechtmann)
DOI: 10.1079/9781786390912.0021

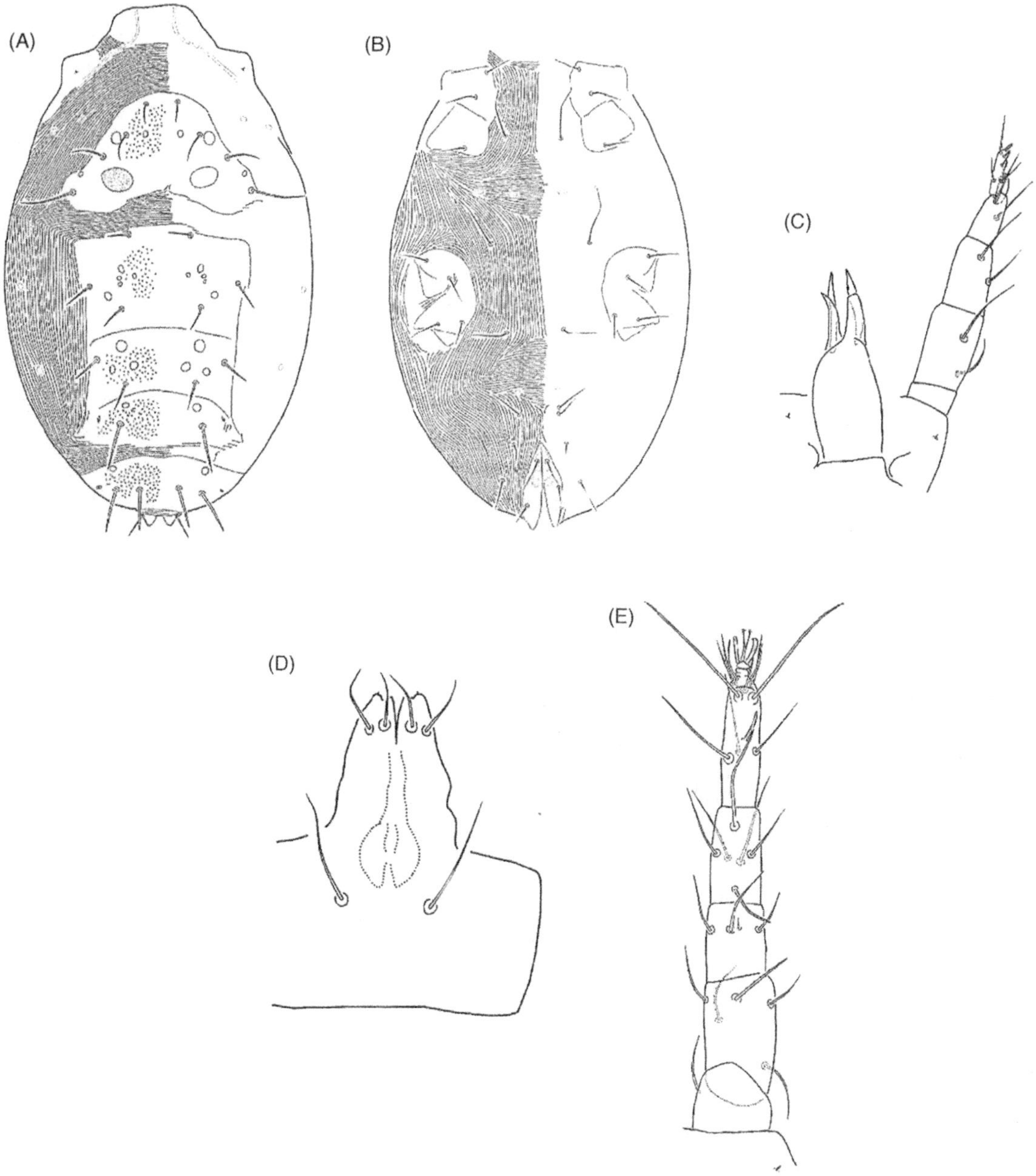

Fig 21.1. *Mediolata granaria* Gonzalez-Rodriguez. Female. Dorsal view (A); ventral view (B); gnathososma (C); subcapitulum (D); leg I (E) (from Khaustov, 2022).

The genus *Stigmaeus* includes the species *Stigmaeus candidus* Fan *et* Li and *Stigmaeus elongatus* Berlese. The geographic distribution of *S. candidus* includes China, where it has been collected on condiment, dried fruit, dried vegetable, mushroom, rice and rice bran (decaying) (Fan and Li, 1993a), and Iran where it has been found on stored rice and decayed rice bran (Noei *et al.*, 2007). *S. elongatus* has a wide geographical distribution including Argentina (Wood, 1973), Iran (Khanjani and Ueckermann, 2002; Noei, 2007), Italy (Berlese, 1886; Canestrini, 1889), Philippines (Wood, 1973; Summers, 1962),

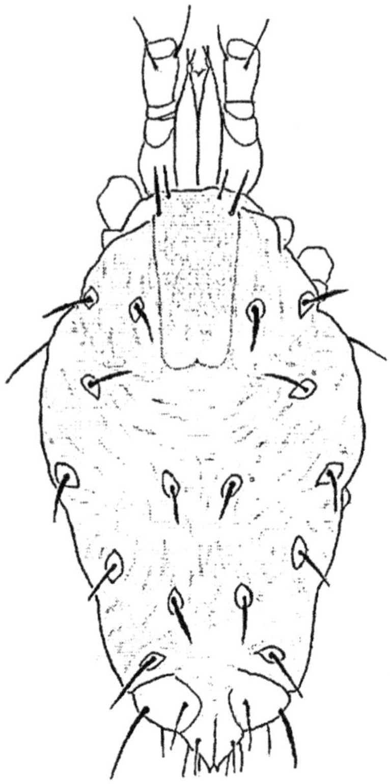

Fig 21.2. *Apostigmaeus pacificus* Summers. Female. Dorsal view (from Summers, 1964, modified).

Turkey (Doğan, 2003b), Ukraine (Crimea) (Kuznetsov, 1978; Wainstein and Kuznetsov, 1978), the United States (Summers, 1962) and other countries; the list of commodities includes food stored, rice, stored rice, rice dust and debris (Noei, 2007; Noei *et al.*, 2007; Hagstrum *et al.*, 2013; Khaleghabadian *et al.*, 2015).

Superorder Acariformes, Order Sarcoptiformes, Suborder Endeostigmata

22 The Family Terpnacaridae Grandjean (Fig. 22.1)

The Terpnacarid mites have various body shapes, but never worm-like. The body setae are simple, dendritic, globose, rod-like or leaf-like; the prodorsum has six pairs of setae, with the external bothidial seta *exp* present. Each chelicera bears two setae. The prodorsum has one pair of filiform trichobothria (botridial setae) (*bo*), lateral and/or median eyes, and the naso well developed and bearing the setiform setae *ro*. The rutella is present and dentate. Tarsus I with both claws and empodium, as well as the pretarsi II–IV. Three pairs of genital papillae (Walter, 2009b).

On stored food, this family includes only the species *Terpnacarus glebulentus* Theron (Fig. 22.1), reported from South Africa (Theron, 1976), the United States (McDaniel and Theron, 1979) and from China, where it has been found on aquatic product, candied fruit, condiment, dried fruit, dried vegetable and mushroom (Hagstrum *et al.*, 2013). Theron (1976) collected different specimens of this mite in South Africa from soil in an *Acacia baileyana* biotope, from soil covered by Kikuyu grass [*Pennisetum clandestinum* = *Cenchrus clandestinus* (Hochst. ex Chiov.) Morrone], from pasture soil, from debris under blue gum trees (*Eucalyptus globulus* Labill.), from cultivated soil, from soil in an *Acacia karroo* [= *Vachellia karroo* (Hayne) Banfi *et* Galasso] biotope, and from soil in a bamboo wood. Moreover, he reared the mite under laboratory conditions, on several fungal species, as *Rhinocladiella anceps* Saccardo *et* Ellis, and unidentified *Aspergillus* and *Penicillium*.

 The Handbook of Mites of Stored Products (V. Vacante and C.H.W. Flechtmann)
DOI: 10.1079/9781786390912.0022

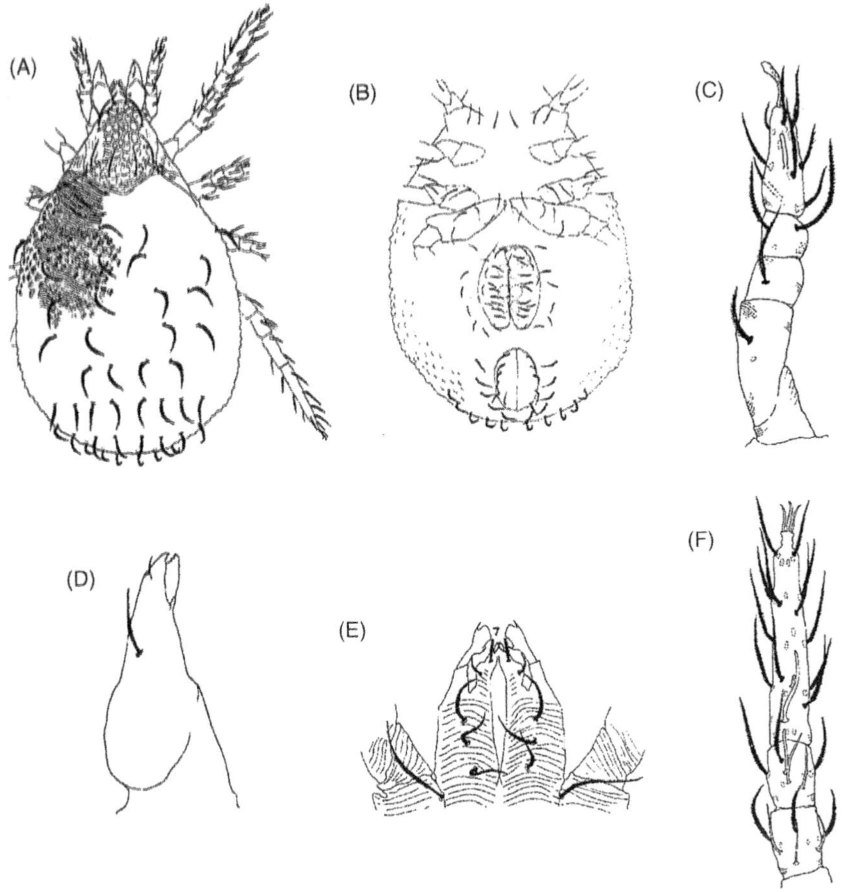

Fig. 22.1. *Terpnacarus glebulentus* Theron. Female. Dorsal view (A); ventral view (B); palp (C); chelicera (D); hypognathum (E); leg I (F) (from Theron, 1976).

Superorder Acariformes, Order Trombidiformes, Suborder Prostigmata (Injurious)

23 The Family Pyemotidae Amerling (Figs. 23.1 and 23.2)

Morphological Characteristics, Systematics and Bio-ecology

The Pyemotidae have a milky white body, spindle-shaped or rounded, segmented, with stylettiform chelicerae; the adults are long from 85 µm to 325 µm and the males are smaller than females. The prodorsum of female bears three pairs of tactile setae (*ve*, *vi*, sc_2), and one pair of capitate trichobothria (sc_1). The gnathosoma is prognatous, and the palpi are distinguishable. The legs I lack empodia, the legs II–IV bear claws and empodia, the trochanter IV is subtriangular, and the leg IV of the male ends with a strong claw. The male has the genital capsule caudal. The physogastry in adult females may be extensive, but nevertheless is limited to the opisthosoma (Walter *et al.*, 2009). Lindquist (1986) reported that the family Pyemotidae numbers three genera, Kethley (1982) includes in this family, two genera with 25 species, and Zhang *et al.* (2011) only one genus and 24 species. A fine review on this family was recently published by Barbosa *et al.* (2021), which discussed the morphology and taxonomy, the biology, the medical and veterinary importance, the list of species, and the control of injurious Pyemotid mites. Recently, Yu and Zhang (2019) described further three new species from New Zealand.

Adult females of Pyemotidae parasitize all stages of holometabolous insects, especially subcortical beetles and stored-product insects. When the female sucks the haemolymph of the host, her opisthosoma becomes greatly swollen because of developing eggs (physogastric female). The eggs develop within the female, and the adults represent the only developmental stage. Adult males emerge first, pierce the swollen opisthosoma with their mouthparts, and remain near the opening. Males copulate with females as the latter emerge from the mother's body. The copulation triggers searching behaviour in females, which immediately leave to attack a suitable host. Species of the genus *Pyemotes* Amerling are recorded parasitizing bees' larvae, pupae, and sometimes adults (O'Connor and Klimov, 2012). Two species groups of females occur according to their toxicity to hosts and their degree of host specificity (Cross and Moser, 1975; Moser, 1975; Moser *et al.*, 1987). So, species of the *Pyemotes scolyti* group exhibit a restricted host relationship and habitat and are commonly associated with bark beetles and attack their eggs, larvae and pupae. Both sexes possess heteromorphy and females are non-venomous. The phoretic females or phoretomorphs are broader, shorter, with thicker and more heavily clawed legs I, than non-phoretic females (Cross and Moser, 1975). *Pyemotes parviscolyti* Cross *et* Moser is phoretic on the pine bark beetle *Pityophthorus bisulcatus* Eichhoff and is not reported on other pine bark insects (Moser *et al.*, 1971). The genus is divided into two groups, *scolyti* and *ventricosus* (Cross *et al.*, 1981). *P. scolyti* (Oudemans) may be retained as an assemblage of cryptic species, probably with ecological barriers between morphologically similar populations associated with *Scolitus multistriatus* (Marsham) on elms and *Scolitus ventralis* on conifers (Moser and Roton, 1972). Conversely, species belonging to the *Pyemotes ventricosus* group possess a wide host range and live in a wide range of habitats; the females are monomorphic, highly venomous, and attack all instars of their hosts. The species belonging to this group are commonly reported as pests in

DOI: 10.1079/9781786390912.0023

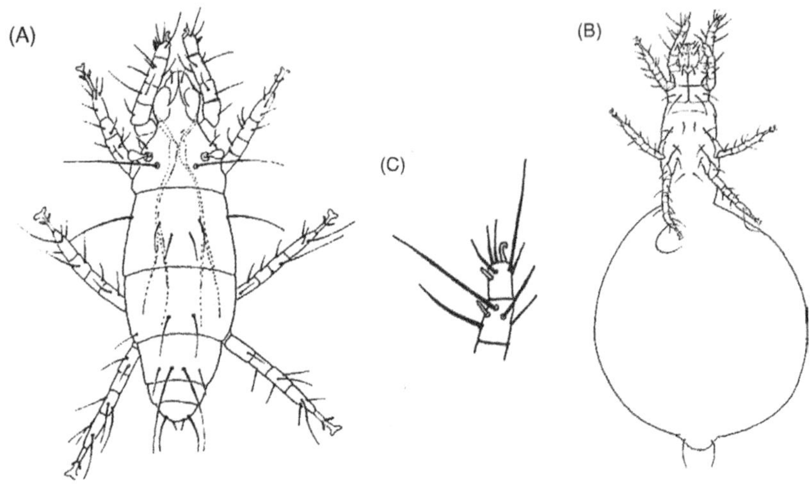

Fig 23.1. *Pyemotes herfsi* (Oudemans). Young female. Dorsal view (A); gravid female (B); leg I (C) (from Hughes, 1976, modified).

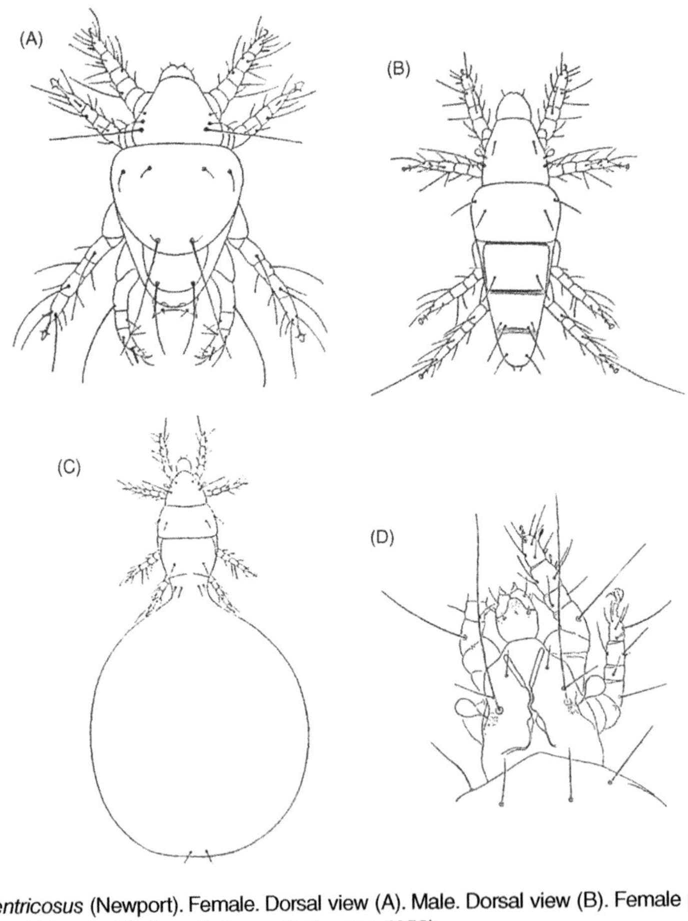

Fig 23.2. *Pyemotes ventricosus* (Newport). Female. Dorsal view (A). Male. Dorsal view (B). Female partially gravid (C); respiratory system of gravid female (D) (from Baker and Wharton, 1952).

laboratory insect cultures and stored foods. The toxin injected by biting females of the ventricosus group, such as *P. herfsi* (Oudemans) (Fig. 23.1), *P. ventricosus* (Newport) (Fig. 23.2) and *Pyemotes tritici* (Lagrèze-Fossat *et* Montagné), induces paralysis and eventual death of the insect hosts. The same toxin causes 'grain itch', consisting in pruriginous dermatitis individuals routinely exposed to these mites, such as food handlers and granary workers. The bites of *P. tritici* can cause severe dermatitis in people handling infested material such as hay. Contact with this mite can also produce asthma or nausea.

P. ventricosus, the type species of the genus, is poorly described and probably was misidentified by subsequent acarologists. Oudemans (1937) speculated that the fungal disease of the bee larvae reported by Frison (1923) for *Anthophora abrupta* Say in the United States was wrongly identified, as they were physogastric females of *P. ventricosus*. However, Cross and Moser (1975) suggested that *P. ventricosus* is only known from the original description and has never been recollected. Following Eickwort (1988), the records of *P. ventricosus* attacking honeybees may refer to *P. tritici*. *Pyemotes anobii* or *P. tritici* may be involved here (Eickwort, 1990b). Eliopoulos *et al.* (2002) reported this species from Greece on grain and legumes. Another species, *Pyemotes islandicus* Sellnick, was collected on grain in Canada (Sellnick, 1940), but we do not dispose of adequate information, and we only discuss *P. herfsi* and *P. tritici*. *Pyemotes* sp. is reported from Iran on dried fruits and nuts by Ardeshir *et al.* (2012).

Selected Websites

http://insects.ummz.lsa.umich.edu/beemites/Species_Accounts/Pyemotes.htm

https://www.google.com/search?q=pyemotidae&tbm=isch&source=univ&sa=X&ved=2ahUKEwjh3cGFt47lAhUlxosKHR1dCb4QsAR6BAgFEAE&biw=1151&bih

The Genus *Pyemotes* Amerling

The general morphological characters of the genus *Pyemotes* are those of the family. This genus includes polixenous or monoxenous insect parasites, exhibiting a wide variety of morphological polymorphisms, and is divided into two groups, *scolyti* and *ventricosus* (Cross *et al.*, 1981). *Pygmephorus islandicus* Sellnick is a secondary species reported from Canada on grain (Sellnick, 1940). Further updates on the topic may be found in a review published by Stingeni *et al.* (2003).

Pyemotes herfsi (Oudemans) (Fig. 23.1)

Common names

Oak leaf gall mite, itch mite.

Diagnostic characteristics

FEMALE. The non-gravid female is about 223 μm long, and the diameter of gravid female is about 2 mm. The gnathosoma is rounded, and the chelicerae are stylet-like, with the palp segments indistinguishable between them. The stigmata are set laterally at the base of the gnathosoma and are related with two tracheae, which converge and open into air reservoirs, which supply the posterior region of the body. The young female has the idiosoma flattened and oval, and grey or yellowish in colour. The propodosoma has straight lateral margins and a pair of dorsal protuberances, with posterior pseudostigmatic organs. The hysterosoma is more than twice as long as the propodosoma and divided into five opistosomal segments decreasing in size towards the posterior end of the body. Apodemes I joined to a long sternum, apodemes II joined to epimerites I, apodemes III and IV, and epimerites IV inserted obliquely in the body. The genital opening lies at the end of the body. The idiosomal setae are smooth and fine, with the verticals (*vi* and *ve*) projected forward over the gnathosoma, and the external scapulars (*sce*) set in front of and behind the pseudistigmatic organs (*sci*), with the external very long. Three pairs of setae (c_1, c_2, c_3) set on the first hysterosomal segment (C), with c_1 and c_2 about the same length and c_3 longer. One pair of setae (*d*) on the second segment (D), the third segment (EF) bears two pairs of setae (*f*, *e*), with f almost three times as long as setae *e*. The opisthosomal segment H bears two pairs of setae of different length (h_1, h_2), with h_2 longer than h_1. Coxae I with one pair of setae, and coxae II with two pairs of setae. The legs are five-segmented. Tarsus I blunt and with a stout hooked claw, with different fine setae set distally, and one short solenidion set on the external margin. The tibia I has a similar solenidion. The tarsi II–IV possess a bifid claw, set on a slender pretarsus ending into a bilobed pulvillus. The femur IV possesses a short basifemur and a long telofemur (Hughes, 1976).

MALE. Similar macroscopically to male of tarsonemid mites and are commonly attached to females. The body bears on the hinder margin a telson-shaped appendage, on this structure is set the sucker for the copulation. Mouthparts and first three pair of legs similar to those of the female, while the legs IV have long femora not subdivided. The tarsus IV has a distal consistent claw (Hughes, 1976).

Geographical distribution

Holarctic and Oriental regions (Oudemans, 1936; Hughes, 1976; Dinabandhoo and Dogra, 1980; De Lillo and Porcelli, 1993; Broce *et al.*, 2006; Marutani *et al.*, 1992).

Commodities, facilities and animal hosts

The mite attacks various insect species belonging to different orders (O'Connor and Klimov, 2012), among which the pink bollworm, *Anobium punctatum* De Geer, the peach moth, *Grapholita molesta* (Busck), the common clothes moth, *Tineola bisselliella* (Hummel), and the Asiatic honey bee, *Apis cerana* Fabricius, periodic cicadas, and various pests of stored grain, as well as humans and their pets (Jones, 2005; Broce *et al.*, 2006; Broce and Kalisch, 2007). Cross and Moser (1975) reported in addition to *A. punctatum*, *Calandra granaria* (L.), *Calandra oryzae* (L.), and probably *G. molesta*. Their usual hosts in the United States are the oak midge larvae (*Contarinia* sp., Cecidomyiidae) (Broce *et al.*, 2006). The mite is reported also on different commodities such as barley, bean, broad bean, cowpea, grain (Alexander *et al.*, 2002), maize, sorghum, watermelon seed and wheat (Al-Safadi, 1987).

Bio-ecology

P. herfsi (Oudemans) is an ectoparasitic mite on a number of insect species, and bites humans, causing red, itchy and painful wheals. The itch mite has a great reproductive potential, and high capacity for dispersal by wind, making them difficult to control or avoid (Broce *et al.*, 2006). De Lillo and Porcelli (1993) found the mite in Italy associated to *Melanaspis inopinata* (Leon.) (Coccoidea, Diaspididae) on *Cotoneaster* sp. and gave some information about its action, retaining the mite as a predator.

Young and mated females inject a neurotoxin-containing saliva into their hosts, which paralyses them and enables the gravid female mites to feed on the host's haemolymph. The opisthosoma of the female enlarges as its progeny develops inside, and, within a few days, up to 250 adult mites emerge from each gravid female (Broce *et al.*, 2006). Bruce and Wrensch (1990) found that progeny of the straw itch mite averaged 254 offspring of which 92% were females. Males emerge before the females, position themselves around the mother's genital opening, and mate with emerging females. Then, mated females disperse to find new hosts. These mites often are dispersed by wind, and when they land on vertebrate hosts, they attempt to feed, resulting in the bites. The mite develops a life cycle within 7 days, and the progeny emergence can be extended to 15 days (Broce *et al.*, 2006).

Cooler, more humid conditions favour population growth of *P. herfsi*, and outbreaks of attacks on humans in the United States have been attributed to fluctuations in the supply of oak midge larvae (Broce *et al.*, 2006).

Symptomatology and damage

This mite parasitises a variety of insect hosts, including bees. Dinabandhoo and Dogra (1980) report that colonies of *Apis cerana* infested by *P. herfsi* in Kulu, India show symptoms of restlessness or lethargy, and do not produce brood. *P. herfsi* causes high incidences of red, itching, and painful welts on people in the midwestern United States, where the mite is normally preying on gall-making midge larvae on oak leaves (Broce *et al.*, 2006). However, rashes resulting from the bite of *P. herfsi* were first documented in Europe in 1936, and the first documented outbreak in the United States is thought to have occurred in 1994 in Kansas City, Kansas. The closely related straw itch mite *P. tritici* (Lagrèze-Fossat et Montagné) was initially suspected, but no specimens of this species were found. The reports showed that the bites occurred on people after being outdoors in or near wooded areas. Based on this information, a search was initiated and resulted in the discovery of *P. herfsi* preying on midge larvae in leaf galls on pin oak trees. In Czechoslovakia, Samsinák *et al.* (1979) discussed *P. herfsi* as a causative agent of another mass dermatitis in Europe.

The United States Centers for Disease Control and Prevention estimated that during an outbreak in August 2004, 54% of the population of Crawford County, Kansas, or about 19,000 people, suffered from its bites (Hansen *et al.*, 2005). During the

2000s, humans have also been affected in other states (Nebraska, Oklahoma, Missouri, Tennessee Shelby County, Texas and Illinois), and in 2008 an outbreak was reported in the northern suburbs of Cincinnati, Ohio, by the Hamilton County Public Health Department.

Humans typically report itching from mite bites within 10–16 h after contact. The victims often do not recall being bitten. The rash that results from the bites is usually described as a red patch, with a small blister in the centre, most often found on the neck, face, arms, or upper torso. A secondary bacterial infection sometimes arises when the bite is scratched (Hansen *et al.*, 2005).

Control

Colonies of *Apis cerana* infested by *P. herfsi* in India were cured by three treatments of fumigation with dichlorvos (Dinabandhoo and Dogra, 1980). Ten ml of dichlorvos solution (50 ppm) was applied to a strip of blotting paper, stapled to an empty frame. To avoid direct contact by bees, the strip was enveloped by perforated paper. A treatment frame was put into the brood chamber in the spring (three treatments each for 12 h, at intervals of 2 days) or in the autumn (five treatments each for 24 h on alternate days). Post-treatment examination revealed that the mite was controlled following three fumigations. No mites were found on bees examined after treatment. The uptake of dichlorvos by the bees was calculated to be 250–500 ng per colony (5000–10,000 bees), which is considered a safe level (Dinabandhoo and Dogra, 1979).

The suggested treatments against the mite bites include the application of calamine lotion, an antihistamine cream such as diphenhydramine, or a corticosteroid cream; a suggested preventive measure is the application of DEET before outdoor activity. Anecdotal reports from the Kansas outbreak, however, suggest that DEET might not provide complete protection against *P. herfsi* (Hansen *et al.*, 2005). The bites are not life threatening, but a few individuals suffering 100 or more bites have undergone brief hospitalizations.

Pyemotes tritici (Lagrèze-Fossat *et* Montagné)

Common names

Grain itch mite, straw itch mite

Diagnostic characteristics

FEMALE. Aspect similar to *P. herfsi*. The body size of a newly laid male or female is 280 μm long and 85 μm wide (Oh *et al.*, 1985), white to yellowish-white in colour. The cheliceral stylets are curved and when protruded unequally, a gap is formed between them near the distal tip. The toxin produced in the modified propodosomal salivary glands and possibly the infracapitular gland may be injected via the tube like structure formed by the medial groove of the cheliceral stylets. The mite feeding occurs only when the cheliceral stylets are inserted into the host tissue fluids and protruded unequally (Bruce *et al.*, 1993). The prodorsal setae *vi* are shorter than *ve*, the pseudostigmatic organ (*sci*) is ovoid and the setae *sce* are set behind and are as long as the setae *ve*. The opisthosomal segment C possesses two pairs of setae (c_1, c_2), with the setae c_2 longer, the segment D has one pair of setae (*d*), the segment EF bears two pairs of setae (*e*, *f*), of which f as long as about two times than setae *e*. The setae h_2 of segment H are longer than setae h_1.

MALE. Similar to male of *P. herfsi*, about 160 μm long.

Geographical distribution

Cosmopolitan (Moser *et al.*, 1987; CAB International, 2019).

Commodities, facilities and animal hosts

P. tritici parasitizes about 150 species of insects belonging to different orders (Coleoptera, Diptera, Hemiptera, Hymenoptera, Lepidoptera, Neuroptera, Strepsiptera) (Gerson *et al.*, 2003). In Brazil, the straw itch mite attacks different managed stingless bee species, such as *Tetragonisca angustula* (Latreille), *Frieseomelitta varia* (Lepeletier), *Melipona subnitida* Ducke and *Melipona asilvai* Moure (Menezes *et al.*, 2009). Moreover, Cross and Moser (1975) reported *Cryptotermes brevis* (Walker), *Araeccrus levipennis* Jordan, *Lixophaga* sp., *Anthonomus grandis* Boheman, 'Ichneumonidae', 'odynerine wasp', *Oryzaephilus surinamensis* (Linnaeus), *Sitotroga cerealella* (Olivier), and *Galleria mellonella* (Linnaeus). Ostovan (1997) and Hagstrum *et al.* (2013) reported the mite on straw, hay, stored grain, and cowpea, and as parasitic of *Ephestia* spp.

Bio-ecology

P. tritici is commonly reported as the grain itch mite (Berenbaum, 1993) or the straw itch mite (Schuster, 2012). It is a cosmopolitan species collected on straw, hay, stored grain and cowpea; it is a parasite of small arthropods and has been investigated as a possible biological control of stored-product pests. It is associated with dermatitis in humans, and causes an itchy rash (Rosen *et al.*, 2002).

The mite is ovoviviparous, and the embryos fully develop inside the female emerging from the genital opening as adults. Males are born after a gestation period that is 2 days less than that of the females; they help the females to emerge and copulation takes place as soon as the females are born. Unmated females are unable to copulate later in life. Fewer than 10% of the offspring are male, although the proportion rises under conditions of gross overcrowding. On average, each female produces about 250 offspring, but after mating with about fifteen females, male fertility rates decline (Schuster, 2012).

Studies carried out in Korea on the life cycle of *P. tritici* on larvae of cigarette beetle, *Lasioderma serricorne* F., a pest of stored tobacco, cigar and cigarette, showed that one generation time takes 20.9 ± 0.7 days, out of which 9.5 ± 0.3 days are spent for feeding, and 10.3 ± 0.8 days for reproduction of progenies. A female of the straw itch mite generates 56.7 ± 6.9 progenies during her reproduction period. As female sucked on its opisthosoma, it grew larger and larger to reach 825 μm in width and 0.346 mm^3 in volume after 9~10 days. By sucking the humour of a host, the abdomen of a female mite became almost a global shape in 2 days. The increased rate of abdominal width is the biggest on the second or the third day of feeding, while the volume of opisthosoma reaches to the largest on 6–8 days after feeding. The largest number of the daily young is produced on the fourth day after a female begins to reproduce (Oh *et al.*, 1985).

Research works carried out in Iran on the biology and population fluctuation of *P. tritici* in laboratory conditions at 85% RH at 27±1°C, rearing mated females on the fifth instar larvae of *S. cerealella*, showed that the mite attacks the hosts and paralyses them after 2–5 min. The female feeds on host haemolymph, and during this feeding phase begins the development of the offspring. The mite reproduces by arrhenotoky parthenogenesis, and the reproduction begins 8–10 days after the attachment to the host; the first individual to emerge is a male. The emerged males wait for emerged virgin females around the genital opening. The mean duration of male life is 54 (18–78) h and in this period each male can fully inseminate 14 (max 45) consecutive virgin females. The time of each insemination is about 30 s. The female is capable of force searching and feeding. The mean duration of female life is 29 (20–35) days and each female continues the reproduction for 19 days. The mean number of the offspring per gravid female is 274.9 (max 346), of which 248.1 (max 322) are female and 268 (max 52) male. The maximum females produced on a day is 30 and for male is 7. Maximum number of reproduction is on the tenth day of the gravid female life. The maximum number of the males is produced near the end of gravid female life (20th, 23rd days). This period is according to 15th–18th days of reproduction (Vafaei *et al.*, 2002).

Survival of adult females varied inversely with temperature and directly with humidity. When a host is parasitised, the percent RH did not appear to be as critical as temperature in the production of mite offspring. Reasonably high numbers (ca. 100) of offspring/female are obtained in a temperature range from ca. 19°C to 28°C and above 30% of RH. Time of emergence or birth of the mite ranges from 4 days at 30°C to 17 days at 18°C. Above or below this temperature range, few offspring emerged. Time of emergence, at the temperatures tested, did not appear to be influenced by percent RH (Bruce, 1984).

The females puncture the host with her stylet, injecting a toxic venom that paralyses it, which eventually dies (Schuster, 2012). Its effectiveness as a parasitoid results from several characteristics, such as the high reproductive potential, short life cycle (4–7 days), no intermediate hosts or food sources; all development occurs within the opisthosomal sac of the adult female, and all offspring are born live as sexually mature adults, only females are parasitic and represent about 90–95% of the population, females mate immediately at birth and begin host-seeking activity, populations are easily reared and synchronized in the laboratory, they are cosmopolitan in distribution (Bruce, 1989).

The above mentioned adaptation suggested to investigate the mite as a potential biological pest control agent for stored-product pests; its high rate of population growth is likely to exceed that of its host arthropod, and it has been found possible to mass rear it in the laboratory (Schuster, 2012).

The Angoumois grain moth (*S. cerealella*) is a pest of stored grain, laying its eggs on the seedheads in the field or on the grain in the silo. Until 1882, it was thought that *P. tritici*, which is sometimes found swarming over stored straw or hay, was feeding on the crop; however, it turned out that the mites are carnivorous and are feeding on the larvae of the moth and other insect pests (Berenbaum, 1993). Today it is believed that the mite has a great potential as a biological control agent for insect pests (Kethley, 1982), and represents the best-studied pyemotid as natural enemy in the biological control and was mass-produced on a commercial basis (Gerson *et al.*, 2003).

The straw itch mite parasitizes the goldspotted oak borer, *Agrilus auroguttatus* Schaeffer, an invasive exotic species to California, United States, and the Mexican goldspotted oak borer, *Agrilus coxalis* Waterhouse, during surveys for natural enemies for a classical biological control programme for A. auroguttatus. *P. tritici*, caused low levels of mortality to each species of flatheaded borer. But it will likely not be a good candidate for a biological control programme because, it is a generalist parasitoid with deleterious human health effects (Coleman *et al.*, 2015).

In previous work, *P. tritici* was induced to probe and pierce a synthetic membrane and feed on a homogenate of cigarette-beetle pupae. Later work with commercially prepared semi-synthetic insect diets, or combinations thereof were judged to be completely unsatisfactory as few offspring were produced. Subsequently, 150 semi-synthetic diets were formulated and tested, of which one repeatedly produced near-normal numbers of offspring or about 150/female. The improved diet is of an appropriate consistency for feeding and contains comparatively high levels of lipoproteins thought to be essential for development (Bruce, 1983).

The straw itch mite has generated considerable interest as a biological control agent of the red imported fire ant, *Solenopsis invicta* Buren. To evaluate the efficacy of the mite against *S. invicta* under the conditions of central Texas, a large-scale field trial was conducted in 1985. A total of 244 *S. invicta* colonies was mapped and colony size and activity ratings recorded. During May, two applications of mites cultured on Angoumois grain moth-infested wheat kernels were introduced into 103 colonies, and 141 control colonies were treated with Angoumois grain moth-infested wheat kernels not infested with mites. The colonies were also twice treated in October. Throughout the test period, colonies were periodically inspected and rated. Analysis of the data using t-tests did not reveal significant differences in ratings between mite-treated and control colonies. Therefore, the straw itch mite was not found to be an effective biological control agent for the red imported fire ant (Harlan *et al.*, 1987). In laboratory and field tests *P. tritici* acted as a biological control agent against the red imported fire ant, when the mite was placed on the nest and taken beneath the soil surface by the ants. All stages of the ant, including the queen, are capable of being parasitized by the mite (Bruce and LeCato, 1980).

Symptomatology and damage

In humans, *P. tritici* may cause severe skin lesions, asthma, or nausea. When the mites come into contact with human skin, they attempt to feed there, causing an intense itchiness, often over the whole-body surface. After about 16 hours, blister-like vesicles erupt on the skin surface and other symptoms can include headache, joint pain, fever and nausea (Berenbaum, 1993).

Control

The straw itch mite caused paralysis and reduced longevity in eucalyptus longhorned borer, *Phoracantha semipunctata* F., under laboratory rearing conditions. The application of dusting sulfur to logs that contained pupating borers, greatly reduced the densities of mites on emerging adult beetles and increased the beetle survivorship. Uniform application to all logs in a glasshouse effectively eradicated the mite infestation. A bioassay showed that sulfur may physically impede the dispersal of immature mites by adhering to the cuticle, but sulfur vapour did not act as a toxin (Hanks *et al.*, 1992). *P. tritici* was reported to kill four colonies of stingless bees *Tetragonisca angustula* (Latreille) and one colony of *Frieseomelitta varia* (Lepeletier) in Brazil, but the infestation was controlled by isolating the dead colonies from the meliponary (Menezes *et al.*, 2009).

24 The Family Tarsonemidae Canestrini *et* Fanzago (Figs. 24.1–24.5)

Morphological Characteristics, Systematics and Bio-ecology

The Tarsonemidae are small mites, from 100 µm to 300 µm long, translucent, pale or whitish, but with their colour influenced by the food they ingest. Species adapted to phytophagy may appear greenish, depending on the green leaves that they have attacked. They are usually dimorphic, with the males different from the females in size and morphological characteristics. The females are usually ovoid, and sometimes elongate, such as in *Steneotarsonemus* spp., while the males are smaller than the females and possess a characteristic subterminal 'genital capsule' and a strong leg IV, particularly conformed and with a single claw. The body consists of a gnathosoma, prodorsum and opisthosoma. The gnathosoma includes parts, such as the stylophore, palpi, infracapitulum and chelicerae. The stylophore is derived from the merged bases of the chelicerae and integrated with the infracapitulum, forms the gnathosomal capsule. The palpi are 2-articulate and usually closely appressed, and the chelicerae are retractable and of different lengths in different species, with the movable digit as a fine stylet. The prodorsum of the female has a sclerotized dorsal shield that is variously ornate, one pair of stigmata, two pairs of setae (v_1, sc_2), one pair of bothridia supporting the pseudostigmatic organs, and one pair of pits (v_2). The prodorsal shield of the male has vertical (v_1, v_2) and scapular setae (sc_1, sc_2) that are variously expressed. The opisthosoma has five dorsal tergites (C, D, EF, H and Ps) that are related to one another in a telescopic manner. Tergite C of the female is usually entire, well separated from the next tergite D, and bears the setae c_1 and c_2. Tergites C and D of the male are fused to form tergite CD. Tergite D of the female is usually entire, with one pair of setae (*d*) and cupules (ia). In the female, tergite EF derives from the fused tergites E and F, and bears two pairs of setae (*e*, *f*) and one pair of cupules (*im*). In the male, tergite EF normally has one pair of setae and one pair of pits, except in some species such as the broad mite, *Polyphagotarsonemus latus* (Banks), which has two setae. Tergite H bears the setae *h* and the cupules ih. The pseudanal tergite Ps is a small structure, caudal or ventrocaudal, little sclerified, and with one pair of caudal setae (*ps*) in the adult female and no setae in the male. Ventrally, the female has an expanded and consistent aggenital shield which sometimes has one pair of aggenital setae; the shield is absent in the male. The ventral podosoma consists of four different coxisternal plates that are generally not ornate. The first two plates are fused to each other in the middle position, and internally they form a prosternal apodema, and apodemes 1 and 2; they stretch around the edge of the rear trochanters I and II in the dorsolateral position above the legs, and they have one pair of setae (*1a*, *2a*) with a pit each (*1b*, *2b*). The coxisternal metapodosomal plates III and IV have three apodemes (3, 4 and poststernal); those of the fourth pair of apodemes are normally associated with the poststernal, which is unequal and median. Two pairs of setae (*3c*, *3b*) are set near apodemes 3 and 4. The legs have six articles. Tibia and tarsus I of the female are fused into a single article, while in the male they are always separated. In most species, leg I of the male and female has a single claw, which in the female may exhibit profound changes and is sessile, large and uncinate. Pretarsus I, unlike pretarsi II and III, lacks a true empodium. Trochanter III of the female is long, subelliptic and fixed; femur III is fused with the genu. Leg IV of the male is usually four-segmented, with the trochanter short and femur and patella fused. Leg IV of the female has three segments, with the trochanter short, the second segment resulting from the fusion of the femur and genu, and the last from the fusion of the tibia with the tarsus; it ends with a long seta and is devoid of

DOI: 10.1079/9781786390912.0024

a distal claw. The chaetotaxy and setal notation are typical; for an extensive discussion of the topic, it is suggested that Lindquist (1986) is consulted. The life cycle of tarsonemids develops through the egg, larva and adult stages, with a quiescent nymph stage inside the larval cuticle. The absence of active nymphs reduces the length of the life cycle, which depends on the host plant, temperature, RH and other ecological factors. Members of this family have various feeding habits, including algivorous, fungivorous, insectophilous, nidicolous and phytophagous, and the symptoms that result from feeding vary according to the species and are sometimes unclear or in doubt (Walter *et al.*, 2009).

Vacante (2016) discussed extensively the tarsonemids injurious to economic plants and to mushrooms. These species may be collected on agricultural products, but their presence originates in the field and must be retained accidentally on stored food.

The Tarsonemidae family numbers about 530 described species ascribed to 40 genera worldwide (Lindquist, 1986; Lin and Zhang, 2002) included in three subfamilies (Pseudotarsonemoidinae Lindquist, Acarapinae Schaarschmidt and Tarsoneminae Canestrini *et* Fanzago) and seven tribes.

Tarsonemids do not cause direct damage to stored food, resulting their presence correlated mainly with their adaptation to feed on fungi. On stored grain are commonly reported different species of the genus *Tarsonemus*, such as *T. granarius* Lindquist (Fig. 24.1) and/or other tarsonemids. Populations of *T. granarius* may be associated with those of *Tarsonemus fusarii* Cooreman, which inhabits various niches hosting fungi, such as hornet nests, humus of oaks and linden, withered grass, etc. (Lindquist, 1972). Hagstrum *et al.* (2013) reported on stored foods seven species of tarsonemid mites. They are *Tarsonemus bilobatus* Suski (Fig. 24.2) collected on Chinese medicine, *Tarsonemus confusus* Ewing (Fig. 24.3) on Chinese medicine and grain, *Tarsonemus floricolus* Canestrini *et* Fanzago (Fig. 24.4) on fish (dried), fish (minced-dried), maize, milk (dried), rice (husked), and soybean sauce, *T. fusarii* on buckwheat, buckwheat product, rice bran, rice (milled), tiki-tiki, and tobacco leaf (cured), and *T. granarius* on aquatic product, barley (Alexander *et al.*, 2002), candied fruit, carrot seed, cereal spillage, Chinese medicine, condiment, coriander seed, dried fruit, dried vegetable, and grain (Al-Safadi, 1987), grass seed, kohlrabi

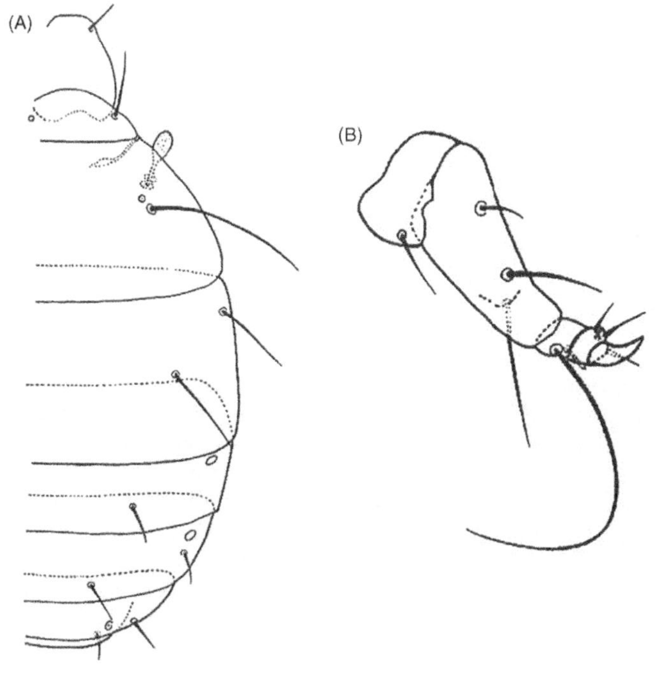

Fig. 24.1. *Tarsonemus granarius* Lindquist. Female. Dorsal view partially drawn (A). Male. Leg IV (B).

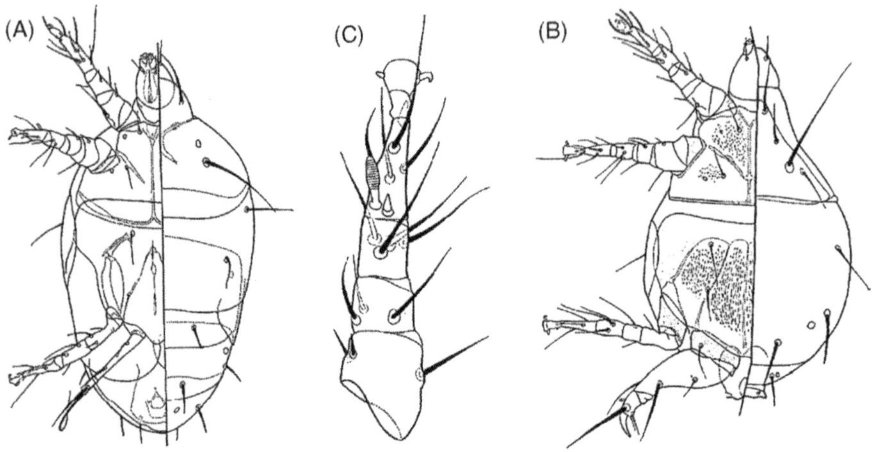

Fig. 24.2. *Tarsonemus bilobatus* Suski. Female. Ventral (at left) and dorsal view (at right) (A). Male. Ventral (at left) and dorsal view (at right) (B); leg II (C) (from Nucifora and Vacante, 2003-2004).

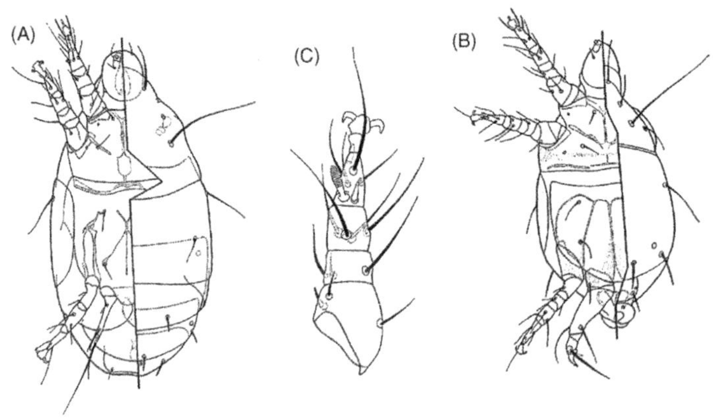

Fig. 24.3. *Tarsonemus confusus* Ewing. Female. Ventral (at left) and dorsal view (at right) (A). Male. Ventral (at left) and dorsal view (at right) (B); leg II (C) (from Nucifora and Vacante, 2003-2004).

seed, lettuce seed, onion seed, processed food, radish seed, rice, spinach seed, sweeping, and wheat (Arlian *et al.*, 1998), *Tarsonemus rakoviensis* Kropczynska on sugar beet seed, *Tarsonemus waitei* Banks (Fig. 24.5) on Chinese medicine, cotton fibre, cottonseed, gin trash, and seed cotton, and wheat. However, with the exception of *T. granarius*, the presence of other mites in stored foods seems occasional and for this reason we discuss here only this last mite. In Iran, it has been collected a *Tarsonemus* sp. on stored rice, rice dust and debris (Noei, 2007; Noei *et al.*, 2008a).

Several tarsonemid species are also considered as allergenic and some authors believe them to be a significant group of allergenic domestic mites (Korsgaard and Hallas, 1979; Hallas and Korsgaard, 1997). On stored products, *Tarsonemus* mites have been reported associated with human dermatitis and other skin disorders (Hewitt *et al.*, 1973; Krantz, 1978; Oehlschlaegel *et al.*, 1983) and to invade various organs and body fluids of humans and other animals (Dahl, 1910). The most commonly implicated species is *Tarsonemus hominis* (Dahl). *However*, it is generally believed that these reports represent contamination of glass slides and other materials used in preparation of tissues for microscopic examinations (Hewitt *et al.*, 1973; Samšiňák *et al.*, 1976).

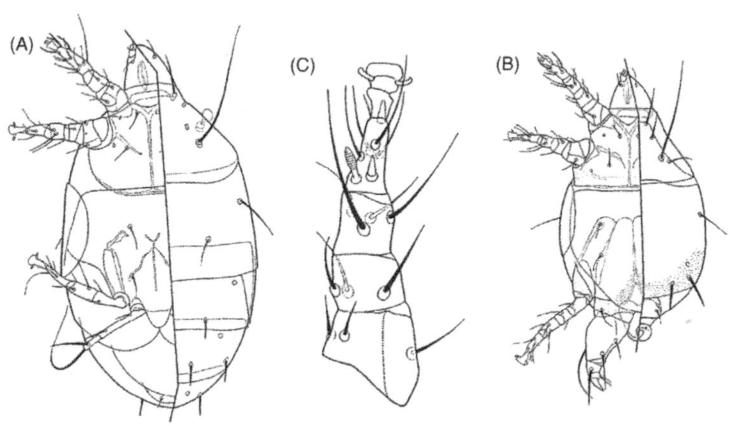

Fig. 24.4. *Tarsonemus floricolus* Canestrini *et* Fanzago. Female. Ventral (at left) and dorsal view (at right) (A). Male. Ventral (at left) and dorsal view (at right) (B); leg II (C) (from Nucifora and Vacante, 2003-2004).

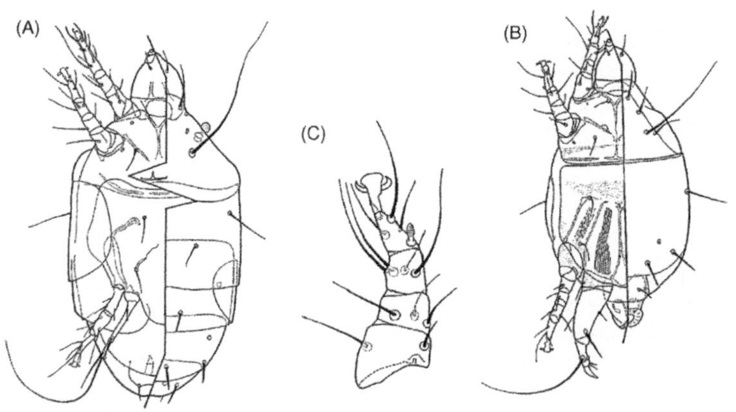

Fig. 24.5. *Tarsonemus waitei* Banks. Female. Ventral (at left) and dorsal view (at right) (A). Male. Ventral (at left) and dorsal view (at right) (B); leg II (C) (from Nucifora and Vacante, 2003-2004).

Selected websites

https://idtools.org/bee_mite/index.cfm?packageID=1&entityID=149

https://www.gbif.org/occurrence/gallery?taxon_key=6059

The Genus *Tarsonemus* Canestrini *et* Fanzago

Adults with unmodified gnathosomal structures, with capsule and palpi not prolonged and beak-like; pharynx and cheliceral bases not greatly enlarged. Femora I and II and tibial sensory cluster with setation complete. Female with stigmata well-spaced posterolaterally from setae *v1*, and with the tegula usually rounded and set between the moderately spaced bases of legs IV. Male with the femorogenu of leg IV often lacks a flange, with tibia and tarsus, may be discrete or coalesced, and the terminal claw well developed, curved, anguiform or sometimes blade like (Lindquist, 1986).

Tarsonemus granarius Lindquist (Fig. 24.1)

Common names

Glossy grain mite, granary mite

Diagnostic characteristics

FEMALE. Both sexes of this species may be confused with *T. fusarii*, from which it may be distinguished by the lack of one pair of ventral apodemes on the gnathosoma. The idiosoma is about 160 µm long, yellow in colour, and with a shining cuticle. The gnathosoma is longer than wide and bears a dorsomedian apodeme, and it is devoid of basal ventral apodemes. The palps are cylindrical, almost parallel, and the cheliceral stylets are short, with the dorsal and ventral setae set near the base of the palps. The dorsal shield is triangular in shape and extends to the bases of the gnathosoma and the pseudostigmatic organs. The idiosoma possesses four overlapping tergites, with the first one larger. The lateral margins of the tergites overlap on the ventral surface. The ventral surface bears coxal plates and overlapping sternites; the posterior of these forms a caudal lobe at the hind margin of the body and possesses one pair of setae. The apodemes I form medially a short sternum, and apodemes II and III are free internally. The anterior median apodeme is set between the legs II and between legs III; the anterior part of this last apodeme is bifurcate, and to it are joined the apodemes of the legs IV. The posterior margins of the coxal plates are marked by a transverse apodeme, separated from the anterior margin of the first ventral shield by a strip of striated cuticle. The propodosoma possesses one pair of vertical setae, one pair of scapular setae and one pair of pseudostigmatic organs, covered distally by a fine process. One pair of pits are set anteriorly to the scapular setae, and the stigmata are opened at the base of legs I. Each hysterosomal tergite bears one pair of setae and the coxal shield I–IV bear coxal setae. As concerns the legs, the chaetotaxy is as other species of the genus (Lindquist, 1972).

MALE. The male of *T. granarius* differs from *T. fusarii* by the anterior connections of the apodemes III and IV with the posterior-median apodeme not interrupted, and for the surfaces of coxae II characterized by a striated area lacking in the *T. fusarii*. Moreover, differences in leg chaetotaxy, setal length, etc. occur. The idiosoma is about 116 µm long. The propodosomal shield is weakly sclerotized, and the hysterosoma bears dorsally a large anterior tergite, and a smaller posterior tergite extending over the genitalia. Four coxal shields on the ventral surface, the anterior separated from the posterior by a surface of striated cuticle. Coxal shields I and II with striated area. The apodemes I form a median sternum, and the apodeme II are free. Apodemes III and IV are long and unite anteriorly and with a median sclerite. No stigmata or pseudostigmatic organs. Two pairs of vertical setae and two pairs of scapulars setae. The anterior scapular setae are long, about twice as long as the posterior. The anterior hysterosomal tergite is large and bears three pairs of setae and one pair of pores, and the posterior tergite has one pair of setae and one pair of pores. Coxal shields I and II with setae, and the third coxal shield bears two pairs of setae. The leg IV has the femora elongated, with margins almost straight internally and slightly arcuate outer externally. The distal segments are short and bent sharply inwards, and the tarsus ends in a well-developed claw (Lindquist, 1972) (Fig. 24.1).

Geographical distribution

Cosmopolitan, reported in Brazil (Sousa *et al.*, 2005), Czech Republic (Žďárková and Horák, 1990), New Zealand (Somerfield, 1980), British Isles, Canada, Japan, Wales (Sinha, 1968b; Lindquist, 1972; Hughes, 1976; Piao, 1990), China (Piao, 1990), Greece (Palyvos *et al.*, 2008), India (Gupta and Chatterjee, 2004), Iran (Faraji, 1993; Modarres Awal, 1997), Italy (Bernini *et al.*, 1995) and West Bengal (Gupta and Chatterjee, 2004).

Commodities, facilities and animal hosts

Spilt buttermilk, debris (Somerfield, 1980), granaries where the mite feeds on fungi that deteriorate grain stored for prolonged periods; this adaptation disseminates the fungi; cereal dust and spillage (Sinha, 1968b; Lindquist, 1972). Aquatic products, barley, candied fruit, carrot seed, cereal spillage, Chinese medicine, condiment, coriander seed, dried fruit, dried vegetable, grain, grass seed, kohlrabi seed, lettuce seed, rice, spinach seed, sweeping, wheat (Faraji, 1993; Modarres Awal, 1997; Hagstrum *et al.*, 2013); stored Chinese herb safflower (Zhao *et al.*, 2015a). Moreover, the mite has been collected in farm, flour-mill, market and off-farm storage. Sousa *et al.* (2005) reported the mite on bean, maize and pet food in supermarkets and markets in Recife (Brazil). Moreover, the mite grain is involved in sanitary problems. Hsu and Li (1985) described adult female and male of *T. granarius* collected from the urine of nephrotic children. Cui (2014) published an interesting review on the human acariasis (pulmonary, intestinal- and urinary) involved *T. granarius* and other mites.

Bio-ecology

T. granarius (and other *Tarsonemus*) is known to be mycophagous and its occurrence on stored food is correlated with the presence of seed-borne fungi (White and Sinha, 1979; Smrz and Catska, 1987), such as *Alternaria alternata* (Fr.) Keissler, *Chaetomium* sp., *Curvularia tetramera* (McKinney) Boldyn, *Fusarium moniliformae* Sheldon, and *Hormodendrum* (*Cladosporium*) *cladosporoides* (Fr.) Saccardo (Sinha, 1964a), and to a lesser extent on fungi of various *Aspergillus* spp. (Sinha *et al.*, 1969a, 1969b).

The mite prefers to develop into granaries, though the fungi on which they develop may live in other habitats. In Canada, Sinha and Wallace (1966) observed *T. granarius* in bulk grain stored indoors, with a preference for wheat and less for oats, and no preference for barley. In Japan, the mite was found often on granary floors, and no forms were collected in stored wheat and rice. The age of the grain on the floors, the decaying organic debris and fungal flora are most important than temperature in the population dynamics. The mite reproduces by parthenogenesis (Sinha, 1968b; Sinha *et al.*, 1969a).

White and Sinha (1981) investigated at 30°C and 90% RH on the fungus *A. alternata* the life history and the population dynamics of the granary mite. They found an intrinsic rate of increase of 0.162 offspring/day, a net reproductive rate of 5.42 times and a mean generation time of 10.43 days. The fertilized females generate females and unfertilized produce males. Observations on agar slands inoculated with *A. alternata* carried out weekly for 8 weeks at 30°C, 25°C, and 20°C and 90% of RH showed a rapid increase of populations at 30°C and 90% RH, and that three species of *Aspergillus glaucus* (L.) link group were an inferior food source for the mite.

In the Czech Republic, *T. granarius* was found associated to both grain mass and grain residues, but although the residue samples had more mites and higher species diversity than the stored grain mass, no correlation in mite abundance and species numbers between samples from grain residues and grain mass was found, thereby indicating low connectivity of these two habitats (Hubert *et al.*, 2006).

Studies on mites associated on specimens of 10 genera of beetles of farm-stored grain or grain residues showed the presence of *Tarsonemus* spp. on external surfaces of beetles. Moreover, the mites can occasionally be found in fungal mycelia attached to beetles (Barker, 1993).

Wu *et al.* (2009b) studied the influence of environmental characteristics and climatic factors on mites in the dust of air-conditioner filters and found that mites present in the dust of air-conditioner filters are potential indoor threat to asthma and allergy sufferers. Moreover, they found that the storage mite *T. granarius* is the predominant species of mites in the dust of air-conditioner filters (MACF). Thus, the possible clinical importance of *T. granarius* should call more of our attentions in the future. It has been found that the cross-reactivity allergens of *T. granarius* mostly occurred in the midgut tissue, gut contents, cuticle and reproductive system in the immunostained sections. The body of *T. granarius* has allergens which can trigger IgE antibodies of the human body and the distribution of allergens in its body is consistent with that in the house-dust mite, *D. pteronyssinus* (Trouessart) (Zhao, 2011).

Simptomatology and damage

Lindquist (1972) reported that, in granaries *T. granarius* feeds on, and perhaps disseminates fungi that deteriorate grain stored for prolonged periods. The adaptation as mould feeder and the absence of its practical importance in the stores was also confirmed successively by Žďárková and Horák (1990) and other specialists. Moreover, *T. granarius* breed in the stored Chinese herb safflower, and this condition requires our attention to prevent and control (Zhao *et al.*, 2015a). The mite was also investigated as probable responsible agent of human allergies (Zhao, 2011).

Control

T. granarius does not produce direct damage to seeds in granaries; its presence is associated with fungi, and for this reason it is feared by farmers and subject to indirect control. The means used in the control of various pests infesting wheat, maize, oats, etc. can also exert some action against the populations of *T. granarius* and has been valued in the past by various specialists. Chemical means have played in this context a key role.

CHEMICAL CONTROL. White and Sinha (1990) studied the effectiveness of chlorpyrifos-methyl on hulled and hulless oats (*Avena sativa* Linnaeus) in a farm granary in southern Manitoba, and found that in both types of oats, the populations of *T. granarius* were unaffected by the insecticide. The insecticide has no impact on microflora, seed germination, and other mite species and the rate of its degradation showed that the ecosystems could largely reestablish themselves in 2 years under western Canadian storage conditions, although the predator diversity was reduced.

White *et al.* (1994) studied the impact of pirimiphos-methyl and cold temperatures on arthropod populations in stored wheat, including various mites among which *T. granarius*. White *et al.* (1997a) monitored the insecticide degradation and the pest survival in treated stored wheat and maize in southern Manitoba. They found that under western Canadian conditions chlorpyrifos-methyl applied on wheat degraded with time, but at 4.6 ppm, offered effective insect control (>90%) for 10 months. Pirimiphos-methyl (5.6 ppm) on maize killed > 90% of insects for 8 months. It did not degrade at either the surface or 1 m depth, but became less effective with time, based on short-term exposure of adults, possibly because of translocation into the seed. Pirimiphos-methyl residues on maize did not decrease; however, after 8 months, pirimiphos-methyl began to appreciably lose effectiveness against the insect *Tribolium castaneum* (Herbst) in 24 h laboratory bioassays, and after a year against *T. castaneum* and the mite *T. granarius* within the granaries.

White and Jayas (1991) evaluated the action of carbon dioxide at concentrations of 15–50% by volume, for its ability to control insect and mite populations in non-airtight bins of wheat stored at 12–15°C for 84 days. The mites *T. granarius*, *Paratriophtydeus coineaui* André, *Lepidoglyphus destructor* (Schrank), and *Aeroglyphus robustus* (Banks) were sharply reduced by day 28 in bins having the two highest carbon dioxide levels. Seed germination and fungal infection by the *A. glaucus* group were not affected by the carbon dioxide. Frequent additions of carbon dioxide can compensate for loss of gas in incompletely sealed bins and provide effective control for arthropod pests.

Despite the possibilities mentioned above, currently the control of seed pests is becoming difficult because the efficient pesticides (e.g. methylbromide, dichlorvos, drinking anticoagulant rodent baits) for seed protection have been lost without the development of adequate substitutes. New research on seed protection in the Czech Republic using biological control (mite predators *Cheyletus* sp.), low pressure, modified atmospheres and hydrogen cyanide is overviewed (Stejskal *et al.*, 2014).

PHYSICAL CONTROL. The above-cited obstacle proposed investigation into the influence of the temperature on the insect and mite populations, and for this purpose Muir *et al.* (1977) carried out a granary trial to evaluate the effects on the insects and mite populations, of turning and transferring a bulk of heavily infested wheat. They found that turning controlled insect populations, but five kinds of mites maintained low population levels throughout most of the experimental period. The turning of grain could reduce the grain temperature at the centre of the bin from above, to below the minimum temperatures required by pests for their development.

25 The Family Tydeidae Kramer (Figs. 25.1 and 25.2)

Morphological Characteristics, Systematic and Bio-ecology

Tydeidae mites are generally small mites with not flashy colours. The body is soft, often with complex striated or reticulated patterns. The idiosoma is striated or reticulated, with two sensilli and a pair of eyes. The bases of chelicerae are fused or contiguous, the movable digit is stylet-like, and the palptarsus lacks elongated setae. The median eyespots are absent, and the epimeral setae *3d* are usually present. The adults possess two pairs of genital papillae, and the male lacks an aedeagus. The legs I possess the claws.

The systematics of the family is not fully defined, and during the last 50 years, the relatively few genera recognized by Baker (1965, 1968a, b, 1970) or by Kuznetsov and Livshits (1973) increased significantly to over 50 genera and seven subfamilies, with the revisions of André (1979, 1980, 1981a, b, 2005). However, some of these last genera are based on permutations and combinations of setal attributes not demonstrably apomorphic, and a further redefinition of the family of André and Fain (2000) reduced the number of subfamilies to three, and at the same time recomposed the complex of genera. Regardless of this, the family numbers a high number of species summarized in the catalogue of da Silva *et al.* (2016a), including 328 species belonging to 30 genera with news on the distribution and type habitat of each species. Walter *et al.* (2009) confirmed these data and reported 30 genera and 340 described species and the paraphyletic condition of the family. The genus *Brachytydeus* Thor comprises 200 species, followed by the genus *Tydeus* Koch with 50 species and the genera *Pretydeus* André and *Pseudolorryia* Kaźmierski with 12 species each.

The Tydeidae is a cosmopolitan family. These mites live in a wide range of habitats and are one of the most commonly encountered arboreal mites, inhabiting leaves, stems, trunks, suspended soils, and specialized habitats such as sporocarps and nests. Algae, fungi and pollen are sources of food for these mites, though some species are facultative predators (Walter and O'Dowd, 1995). The extent of facultative predation among canopy scavenger-microbivores has not been well explored, but many more typical predators often take non-prey foods, especially pollen, honeydew, extrafloral nectar and fungi (McMurtry and Croft, 1997; Walter and Proctor, 1999). A wide list of reports indicates species of tydeids as predators, fungus, pollen and plant feeding, or scavengers (Oatman, 1963; Baker, 1965; Gerson, 1968b). Some species of the old genera *Lorryia* (Oudemans) and *Tydeus* are reported as phoretic on noctuid moths (Treat, 1970). Tydeid mites are often collected in moss, litter, straw, soil, humus, fungi, bird nests, stored food products and on plants (Marshall, 1970; Kazmierski, 1998). With the exception of some species, the interaction of these mites with their habitats is little or poorly known. This difficulty probably depends on the habits of unspecialized feeders of different species, such as *Tydeus californicus* (Banks), retained a predatory mite by some authors (Baker and Wharton, 1952; Rizk *et al.*, 1979) and a phytophagous mite by others (Fleschner and Arakawa, 1953; Zaher and Shehata, 1963 Larvae and protonymphs of *T. californicus* feed primarily on aphid honeydew, but other instars of the same species develop on aphid-free apple leaves. The species *Lorryia formosa* Cooreman reported by many specialists has been named by André (2023) as *Brachytydeus formosa* (Cooreman); this mite has been retained fungivory by Jeppson *et al.* (1975), while Aguilar *et al.* (2001) retained it injurious to branches and citrus fruits. The same species was retained beneficial on citrus in Israel, where it feeds on sooty mould and scale-excreted honeydew on leaves and fruits of citrus (Mendel and Gerson, 1982). What

DOI: 10.1079/9781786390912.0025

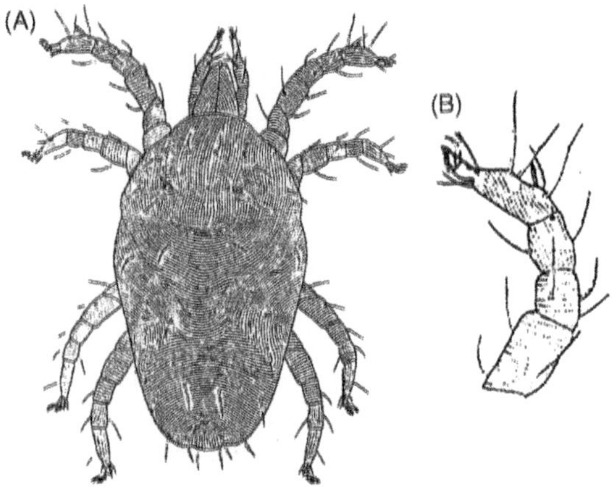

Fig. 25.1. *Brachytydeus australensis* (Baker). Female. Dorsal view (A); leg I (B) (from Vacante and Nucifora, 1984-85, licensed under CC BY 4.0).

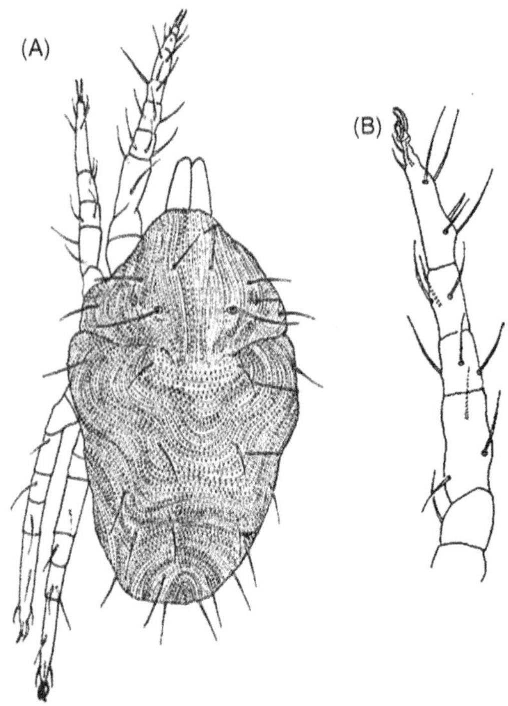

Fig. 25.2. *Tydeus interruptus* Thor. Female. Dorsal view (A); leg I (B) (from Hughes, 1976).

has been briefly reported above, indicates that many systematic and bio-ecological aspects of Tydeids still require further studies today, highlighting that some approaches used are largely approximate.

At present, no tydeid mites are reported as injurious to stored food, despite ten species of this family having been collected in various countries on stored foods (Hagstrum *et al.*, 2013). We do not dispose of information on the risk related to presence of tydeid mites on the stored foods, and to their hypothetical damage. In fact, this interesting group of mites lives probably as fungus feeding or adapted to other feeding habits.

A *Lorryia* sp. has been collected in Iran on stored rice, rice dust and debris (Noei, 2007; Noei *et al.*, 2008a). In the same country, three species of the genus *Proctotydaeus* Berlese have been reported; they are *Proctotydaeus sinhai* Momen, found in Canada and Iran on stored barley and wheat (Ostovan, 1993; Momen and Sinha, 1991; Hagstrum *et al.*, 2013), *P.* sp. near *sinhai* reported from Iran on stored rice, rice dust and debris (Noei, 2007; Noei *et al.*, 2008a), and *Proctotydaeus* sp. from Iran on stored rice, rice dust and debris (Noei, 2007; Noei *et al.*, 2008a).

Seven of these tydeid mites belong to the genus *Brachytydeus*. *Brachytydeus australensis* (Baker) (Fig. 25.1) are known from China and Indonesia on aquatic product, candied fruit, condiment, dried fruit, dried vegetable, mushroom, and processed food collected in food store. *Brachytydeus bedfordiensis* (Evans) is reported from England and Ireland on Chinese medicine, wheat in bread van, commercial grain storage found in farm and flour-mills, and in Canada from stored grain, litter under *Fagus grandifolia* Ehrh., moss, apple tree, grain spill (Silva *et al.*, 2016a). *Brachytydeus brusti* (Momen *et* Sinha) has been reported from Canada on grain spill and wheat (Momen and Sinha, 1991). *Brachytydeus hughesae* (Momen *et* Sinha) has been collected in Canada on hay straw, oat and wheat (Momen and Sinha, 1991), *Brachytydeus manitobensis* (Momen *et* Sinha) is reported from Canada on rapeseed and wheat (Momen and Sinha, 1991), *Brachytydeus oregonensis* (Baker) has been found in China and Ireland on barley (mouldy), candied fruit, condiment, dried fruit, dried vegetable, mushroom, oat, processed food, wheat in bread van, commercial grain storage, farm, flour-mill in provender mill, provender warehouse and railcar; the mite is reported also in stored products (Momen and Sinha, 1991; Siegert *et al.*, 2018) and in poultry nests from Brazil (Horn *et al.*, 2015), and *Brevipalpus reticulatus* (Oudemans) is known from Canada on grain spill and wheat (Momen and Sinha, 1991; Momen and Lundquist, 1995).

Other species are *Perafrotydeus meyerae* (Baker), known from China on Chinese medicine, *Tydaeolus sphaeroclaviger* Kuznetzov, reported from China on Chinese condiment, dried fruit, dried vegetable, mushroom and processed food, *Tydeus kochi* Oudemans has been collected in Canada, Egypt, and Greece on wheat, wheat bran, wheat flour in farm, flour-mill and off-farm storage (Hagstrum *et al.*, 2013), and *Tydeus interruptus* Thor reported from Australia, Canada, Czech Republic, England, Ireland, Norway, Poland, Scotland, United States and Wales on barley (Alexander *et al.*, 2002), grain, grain debris, grain spillage, spinach seed, sugar beet seed, sweeping, wheat (Al-Safadi, 1987), wheat residue in empty grain bin and flour-mill. Hughes (1976) reported that the mite was first found in moss, and it is uncommon in grain debris in association with acarid mites and sometime in large numbers. In Iran, a *Tydeus* sp. on stored product (Noei, 2007; Noei *et al.*, 2008a; Ardeshir *et al.*, 2012) has been collected. Hughes (1976) reported *T. interruptus* (Fig. 25.2) on grain debris in association with acarid mites and sometimes collected in large numbers; the mite is known from Norway and England.

Moreover, Momen and Sinha (1991) reported on wheat and oilseeds, from Canada, *Paratydaeolus canadensis* Momen *et* Sinha, *Paratriophtydeus coineaui* André, *Oakvillea eglalea* Momen *et* Sinha and *Orthotydeus kochi* Oudemans.

Selected website

https://www.google.co.il/search?q=Tydeidae&espv=2&biw=823&bih=527&tbm=isch&tbo=u&source=univ&sa=X&ved=0ahUKEwj9qLTsz5rKAhUCzxoKHWqZA5U

26 Family Ereynetidae Oudemans (Fig. 26.1)

The family Ereynetidae is characterized by a sensory complex of unknown function, called ereynetal organ, set on the tibia I (Fain, 1962, 1985). This organ is composed of three parts: a deeply inserted solenidion enclosed in a sclerotized pouch, a fine canal connecting the pouch to the exterior face of the tibia and a famulus whose alveolus is contiguous with the solenidial pouch. Ereynetid mites live in humid or wet habitats, whether they are cavities of hosts colonized by parasitic species or soils of temperate or tropical regions (André and Fain, 2000); rarely, they are found in exposed foliar habitats or arid regions. The family numbers 180 described species belonging to three subfamilies. Some species of the genus *Ereynetes* Berlese are subcortical and phoretic associates of bark beetles (Hunter *et al.*, 1989), nidicolous associates of native bees (Hunter and Cross, 1968; Flechtmann *et al.*, 1985) or phoretic associates of dung-inhabiting beetles and flies; two species of this genus are associated with pagurid land crabs (Hunter and Poe, 1971; Fain, 1985; Fain and Camerik, 1994). *Ereynetes* spp. were collected on stored rice, rice dust and debris from Iran (Noei 2007; Noei *et al.*, 2008a). A wide list of habitats of various species of this genus was reported by Fain and Camerik (1994). We treat here as an example, *Ereynetes ueckermanni* Fain *et* Camerik (Fig. 26.1), reported by the above specialists from South Africa on cow dung and horse dung.

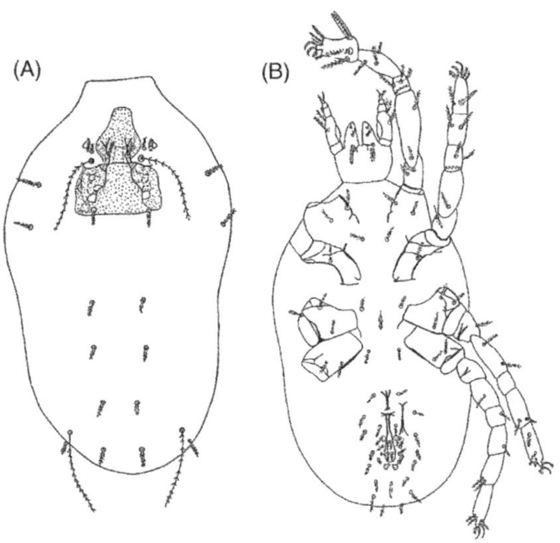

Fig. 26.1. *Ereynetes ueckermanni* Fain et Camerik. Male. Dorsal view (A); ventral view (B) (from Fain and Camerik, 1994).

DOI: 10.1079/9781786390912.0026

Superorder Acariformes, Order Sarcoptiformes, Suborder Oribatida, Cohort Astigmatina

27 The Family Lardoglyphidae Oudemans (Figs. 27.1 and 27.2)

Morphological Characteristics, Systematics and Bio-ecology

Species of this family have the setae *ve* curve and pectinate, arising at the same level as *vi*. The setae *sce* are longer than *sci*. The claws of all the legs are bifid in the female. None of the dorsal setae of the legs is thickened as a spine. Only heteromorphic males are developed (Hughes, 1976). The leg III of male is enlarged, the tarsus ends in a spine-like process, the seta s is spine-like and may be fused to the segment and the pretarsus is lacking (Oconnor, 2009b). It is a small monotypic family including only the genus *Lardoglyphus* Oudemans with seven species. Lardoglyphids are reported as pests of stored food, especially stored meat and seafood products (Hughes, 1976) or bird nests (Philips and Norton, 1978; Asmat and Kader, 1990) and vertebrate carrion, such as human mummies (Early and Goff, 1986; Baker, 1990b; Iverson *et al.*, 1996). Baker (1990) described all post-larval stases of *Lardoglyphus robustisetosus* Baker and the hypopus of *Lardoglyphus radovskyi* Baker from material obtained from the gut contents of human mummies excavated, respectively, in Chile and the United States and provided the keys to the hypopi of all seven species of the genus and to the adults of all five species known as adults. Both keys include all species known as pests of stored food.

The Genus *Lardoglyphus* Oudemans

General morphological characteristics as the family. The genus numbers seven species, of which only three are reported on stored food. They are *Lardoglyphus zacheri* Oudemans, *Lardoglyphus angelinae* Olsen and *Lardoglyphus konoi* (Sasa *et* Asanuma). *L. zacheri* is known from Australia, China, England, Germany, Hong Kong, Mexico and South America on bone, butcher offal, Chinese medicine, fish (dried), hide, shark fin (dried) (Hughes, 1976; Smiley, 1987) and on preserved duck gizzard, bacon, squids, Chinese yam (Li *et al.*, 2015a). *L. angelinae* Olsen is reported from Hong Kong on fish (dried) (Olsen, 1982). The species most important is *L. konoi* (Sasa *et* Asanuma). This last species has the setae f_1 equal in length to e_1 (Fig. 27.1), while *L. zackeri* (Fig. 27.2) has the setae f_1 more than three times as long as the setae e_1 (Hughes, 1976).

Lardoglyphus konoi (Sasa et Asanuma) (Fig. 27.1)

Common names

Dried fish mite, lard mite.

Diagnostic characteristics

FEMALE. The idiosoma is smooth and translucent creamish-white and is 400–550 µm long; the chaetotaxy is similar to that of male, with pa1 and sae relatively short. The *receptaculum seminis* is triangular (Hughes, 1976). The legs possess paired claws.

MALE. The idiosoma is 330–440 µm long. The dorsal setae f_1, f_2, ps_1 and ps_2 are equal in length to e_1 instead of being more than twice as long. The chelicerae have several small teeth on fixed and movable digits. The chitinous sclerite encircling the anal

DOI: 10.1079/9781786390912.0027

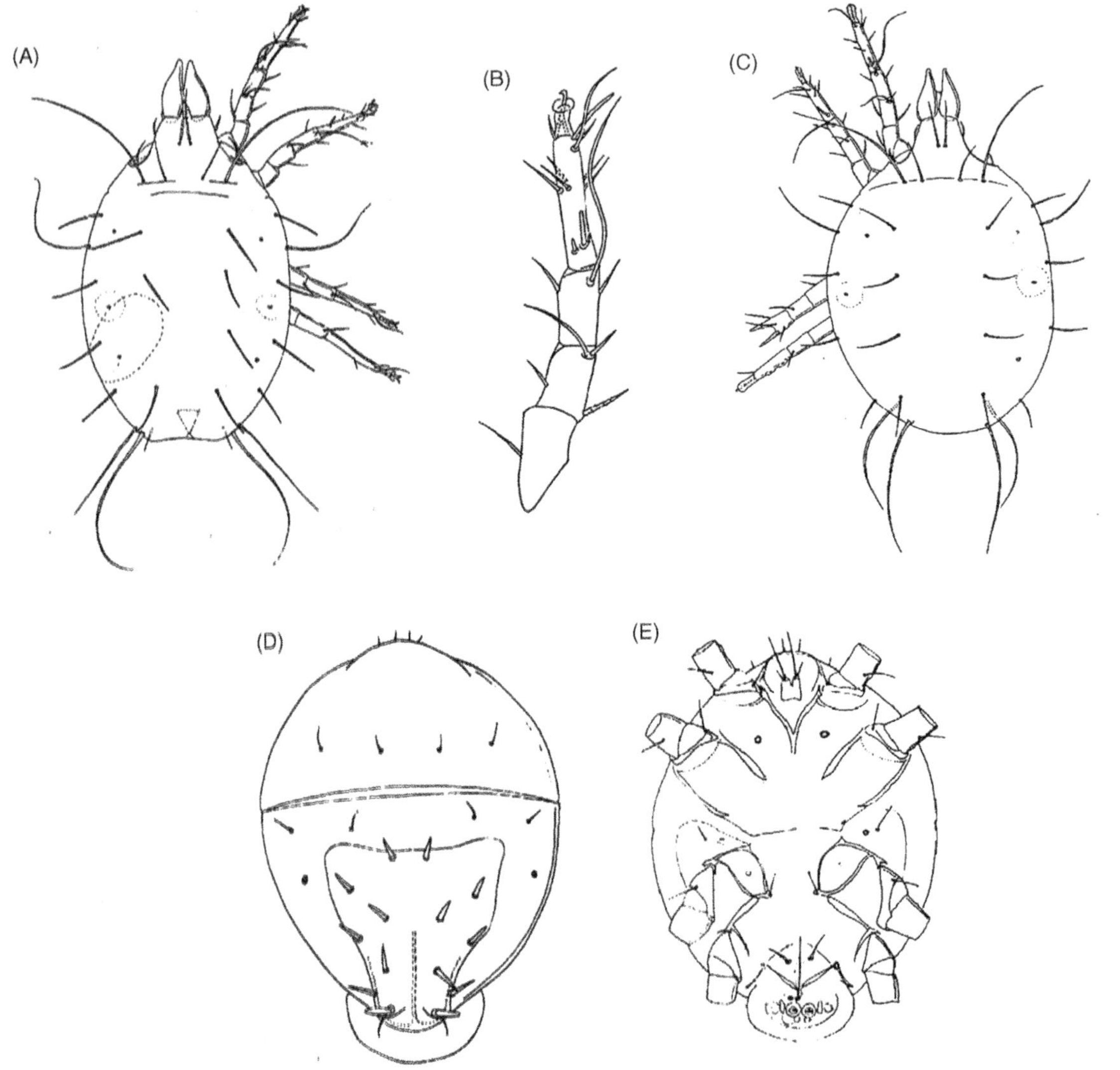

Fig. 27.1. *Lardoglyphus konoi* (Sasa et Asanuma). Female. Dorsal view (A). Male. Leg I (B); dorsal view (C). Hypopus. Dorsal view (D); ventral view (E) (from Hughes, 1976, licensed under the Open Government Licence v3.0).

suckers in the male is bent more sharply towards the posterior margin of the body, and the anal setae lie on either side of the anterior end of the anus, instead of midway along its length. Legs I, II and IV bear undivided claw. The tarsus of leg III is shorter and the setae are terminal. On leg IV, the copulatory suckers are in the middle region of the tarsus (Hughes, 1976).

HYPOPUS. The idiosoma is 215–260 μm long. The setae of the histerosomal shield are thickened to form spines. The posterior projection of apodemes III extends as far back as the setae between apodemes IV. The margin of the sucker plate is wider, and the two central suckers are smaller. Two of the encircling suckers A and D are replaced by horn-like projections, and the auxiliary suckers are transparent. The anterior suckers I and K vary in position and are frequently asymmetrical. The pretarsi of legs I–III are longer and slender. On tarsi I and II, the seta *f* is leaf-shaped instead of being simple, and on tarsi III, all the setae except d are expanded distally into transparent laminae. Tarsus IV bears two similar leaf-like structures (*e* and *p+u*), and *rw* is also present (Hughes, 1976).

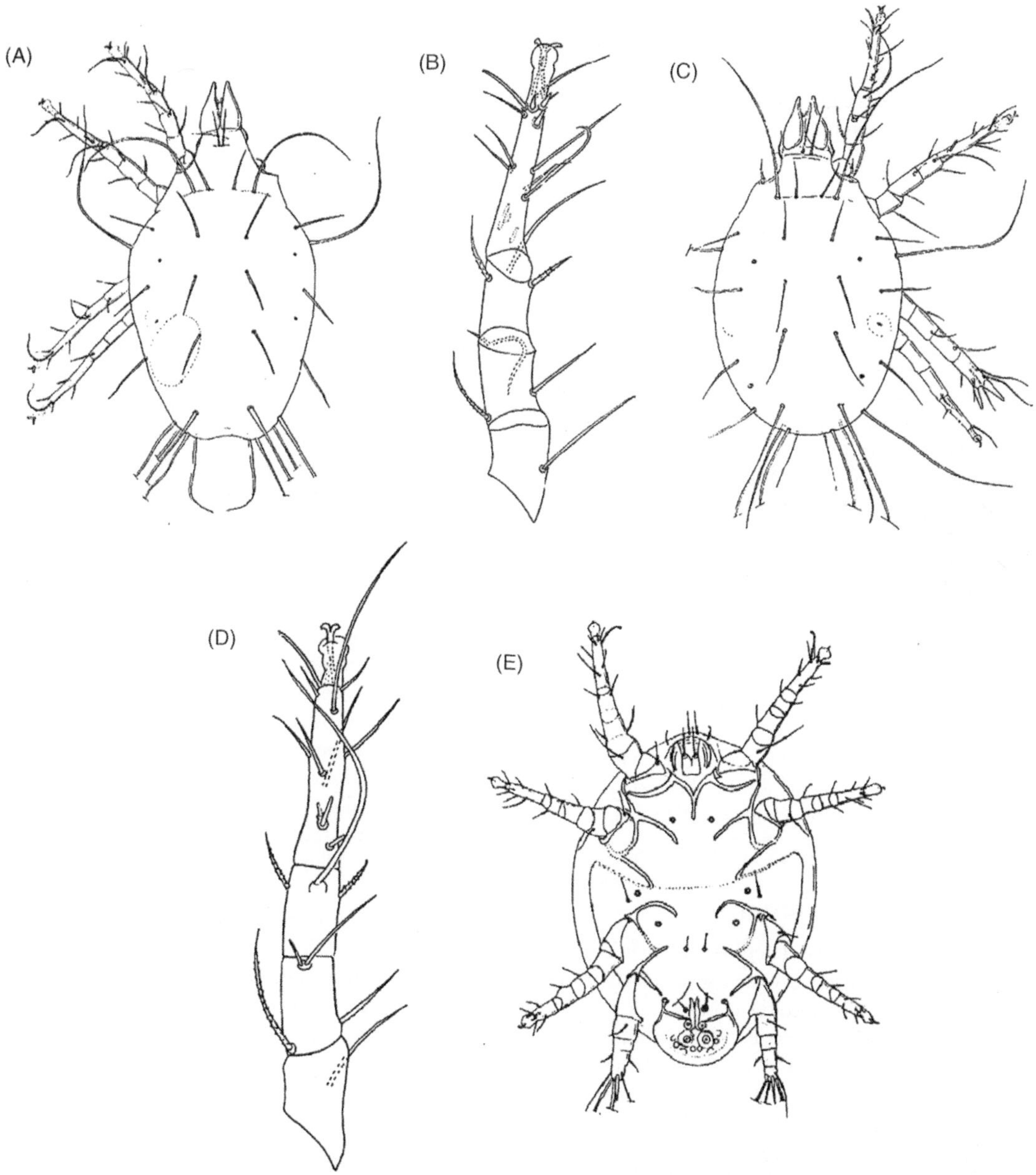

Fig. 27.2. *Lardoglyphus zackeri* Oudemans. Female. Dorsal view (A); leg I (B). Male. Dorsal view (C); leg I (D); ventral view of hypopus (E) (from Hughes, 1976, licensed under the Open Government Licence v3.0).

Geographical distribution

The geographical distribution includes Australia (Halliday, 1998), Bangladesh (Bhuiya and Khawja, 1983), Brazil (Flechtmann and Castelo, 1982), China (Xin, 1965; Fang and Cui, 2009), England, Germany (Rack, 1968a; Hughes, 1976), Egypt (Yassin *et al.*, 2016), Korea (Kim and Song, 1968), Japan (Sasa and Asanuma, 1951), India (Pillai, 1957), Hong Kong (Olsen, 1983, 1988), Indonesia (Esser, 1988; Haines, 1995), Iran (Keshtmand *et al.*, 2020), Japan (Sasa and Asanuma, 1951; Matsumoto, 1970, 1973), Poland (Chmielewski, 1971b), Senegal

(Gueye-Ndiaye and Fain, 1987; Ndiaye and Marchand, 1989), Taiwan (Tseng, 1972) and the United States (Smiley, 1987). Moreover, Cui (2014) included the mite in the group of mites responsible for pulmonary acariasis in humans in East Africa, Korea, China and other African and Asian countries.

Commodities, facilities and animal hosts

The list of commodities includes Chinese medicine (Fang and Cui, 2009), fish product (salted), fish (dried), fish (minced-dried) (Pillai, 1957; Olsen, 1983; Bhuiya and Khawja, 1983; Ndiaye and Marchand, 1989), food grains (Haines, 1995), mushroom growing compost (Keshtmand *et al.*, 2020), prawn cracker, prawn (minced), shark fin (dried), shrimp (dried), squid (dried) collected in food storehouses (Hughes, 1976; Smiley, 1987; Hagstrum *et al.*, 2013) and stored onion bulbs (Yassin *et al.*, 2016). Moreover, it has been collected on butchers' offal (Griffts, see Hughes, 1976), and the hypopi have been collected on *Dermestes carnivorus* Fabricius (Rack, 1968a) and *Dermestes maculatus* De Geer (Coleoptera, Dermestidae) (Hughes, 1976). Another host is *Solenognathus* (Pisces) (Yong *et al.*, 2016a). Olsen (1988) recorded *L. konoi* on dried lizard carcasses from Hong Kong intercepted in California. Moreover, the mite is reported on groundnut, fillet, jerky, dried pork slice and sesame oil slag in China (Li *et al.*, 2015a).

Bio-ecology

The dried fish mite was first described as *Hoshikadania konoi* Sasa *et* Asanuma (1951) and successively transferred by Hughes (1956) into the genus *Lardoglyphus*.

The mite has wide geographical distribution and is well known as pest of dried fish. Adults and nymphs of dried fish mite feed on drying or dried fish; they become very injurious in many tropical and subtropical conditions, especially where habitats are humid, or the fish is not well dried. Infestations of *L. konoi* on cured fish are usually associated with those of *Dermestes* spp. and *Necrobia rufipes* (Fabricius) (Coleoptera, Cleridae). This is partly due to the similarities in the physical and nutritional requirements of all these pests, but it is mainly because the beetle pests are the main agents of dispersal of the *Lardoglyphus* hypopi.

The female lays eggs on the stored fish; the biological cycle develops through the six-legged larva, the eight-legged protonymph, the tritonymph stages and the adult. At 23°C and 87% RH, the life cycles take 9–11 days, but at optimum temperatures, the development may be faster (Hughes, 1956). Optimum conditions for the development of *L. konoi* are 30–35°C and 87% RH. The mite cannot develop at 66% RH and below, but its populations increase at humidities of 75% and above (up to 98% RH).

Some protonymphs do not develop into tritonymphs but instead turn into hypopi first. The hypopi do not feed but are in a dispersal stage. They climb onto the larvae and adults of beetle pests, and use their suckers to attach themselves to the cuticle of insects. On larvae, they are found on the soft membrane between segments, especially near the front of the body; on adults, they are most commonly found on the upper surface of the abdomen under the elytra. They may remain on the insect for several days and are therefore often carried to new infestation sites before they detach themselves and change into tritonymphs. Although Vijayambika and John (1974) have questioned the importance of environmental factors in controlling the production of hypopi, other acarologists (Hughes, 1956; Matsumoto, 1970, 1973) have shown that the proportions of hypopi in *Lardoglyphus* populations are influenced by factors such as unfavourable temperature and humidity levels, or shortage of food or overcrowding.

In Indonesia, Esser (1988) reported that the poor quality of infested fish suffered a 60% reduction in retail value. Heavy mite infestation can cause a drop in retail value of 30–35%.

The aggregation pheromone (1,3,5,7-tetramethyldecyl formate) of *L. konoi* has been isolated from its culture medium. The compound has been assigned the trivial name lardolure. The synthesized diastereomeric mixture of racemates showed one-tenth of the activity of the natural pheromone against the test mite, *Carpoglyphus lactis* (Linnaeus) (Kuawara *et al.*, 1982).

Symptomatology and damage

The dried fish mite is a serious pest of dried fish and salted fish products, reported first in Japan (Sasa, 1964) and Southern India (Pillai, 1957), and successively in various countries of the world. It has also been found on butchers' offal. Feeding by the adults and nymphs of *Lardoglyphus* spp. will cause quantitative loss of dried cured fish. Quality loss may result from the contamination of live and dead bodies, which can be very numerous. The extent and value of losses caused to dried fish by *Lardoglyphus* spp. have not been assessed in either the laboratory or the field.

The hypopi have been collected clinging to *D. carnivorus* Fabricius (Coleoptera, Dermestidae) (Rack, 1968a) and *D. maculatus* De Geer (Hughes, 1976). The mite has been found in Senegal in transformed fishes called Keccax (Ndiaye and Marchand, 1989).

Yong *et al.* (2016a) collected from a Chinese medicine shop in Wuhu City *Solenognathus* (Pisces) and isolated from 500 g of *Solenognathus*, four species of mites. Of 256 mites collected in total, 219 were *L. konoi*, 12 *Tyrophagus longior* (Gervais), 21 *Tyrophagus putrescentiae* (Schrank) and 4 *Euroglyphus maynéi* (Cooreman). The breeding density of *L. konoi* was 0.438/g. Effective measures should be taken to prevent and control it.

Control

Control of dried fish mite infestations provides mainly the choice of preventive measures and comes from the provision of physical barriers (networks, regulation of openings, etc.), and the taking of all initiatives aimed at regulating the thermo-hygrometric conditions.

Initial infestation is often caused by hypopi that have been carried to the fish by beetle pests; consequently, the reduction of initial infestation by beetles also reduces infestation by *Lardoglyphus* spp. The fragmentation of the fish flesh caused by beetle and fly attack, and by handling damage, increases the surface area of the flesh, and may thus encourage heavier mite infestations.

The response of *Lardoglyphus* spp. to salt has not been investigated. *L. konoi* has been found infesting salted fish products, as well as unsalted dried fish, but the salt contents of the infested products were not recorded.

The minimum humidity threshold for *L. konoi* is between 66% and 75% RH If the fish is thoroughly dried, mite problems will be greatly reduced or eliminated. The temperature limits for development are not known, but temperatures above 40–45°C would be expected to kill these mites (Matsumoto, 1970, 1973; Sen, 2005).

28 The Family Acaridae Latreille (Figs. 28.1–28.18)

Morphological Characteristics, Systematics and Bio-ecology

The body of the members of the Acaridae is soft, of medium length and pale whitish to brownish coloured. Some species are weakly pigmented. The integument is smooth, rough or finely furrowed; sometimes completely sclerotized, or ornamented with cuticular tubercles. A sejugal furrow separates the idiosoma into the propodosoma and hysterosoma. The dorsal surface of the gnathosoma is not extended and forms a small lobe between the cheliceral bases. The chelicerae are chelate and dorsally free, laterally compressed and with a large basal region stretching forward, and forming the digitus fixus articulated with the digitus mobilis; at the base of this articulation, there is a small cheliceral seta (*cha*). Both chelicerae move in a vertical plane and are often toothed distally. The chelicerae are located above the labrum or true epistome, and partially involved in the mouth structure. The labrum is grooved and extends posteriorly, and its side walls are related to the ventral area of the gnathosoma; on it are inserted the pharynx dilator muscles. The ventral gnathosoma, called the hypostome or infracapitulum, originates mainly from the fusion of the palpcoxae; it has two whip-like subcapitular ventral setae (*m*) and extends anteriorly into two inner lobes or malae. The palpi have two freely articulating segments (articles); each coxa bears dorsally a palpal supracoxal seta (*elcp*). The first free palpal segment has two simple setae, and the distal segment bears one simple seta, a solenidion and a button-like ventral eupathidium. Several species have a Grandjean's organ – a thin cuticular plate embracing the base of gnathosoma – on the anterolateral propodosoma. From the base of this organ, a lateral sclerite extends around the base of leg I, and at the posterior end of this sclerite is set the supracoxal seta (*scx*). In several species, there is a crack or pore between the lateral sclerite and the supracoxal seta, the opening of which is similar to those of the opisthonotal glands (Grandjean, 1939b; Griffiths, 1970; Hughes, 1976). The prodorsum usually bears a prodorsal sclerite or shield that is variously sized, shaped and structured. The idiosomal setae may vary in number, size and shape, but are distributed according to the same pattern for the group. The prodorsum bears four pairs of prodorsal setae (*vi, ve, sci, sce*); the inner vertical setae (*vi*) are medioanterior, and the external vertical setae (*ve*) are lateral; the scapular setae (*sci, sce*) are posterior. The hysterosoma has 12 pairs of setae (c_1, c_2, cp (third pair of setae in first series or row), c_3 (fourth pair of setae in first series or row), d_1, d_2, e_1, e_2, f_2, h_1, h_2, h_3), four pairs of lyrifissures (*ia, im, ip, ih*) and one pair of opisthonotal glands. The gland opening (*gla*) is commonly posteriorad (towards the rear) of d_2 (Hughes, 1976; Griffiths *et al.*, 1990). Ventrally, the coxae are fused with ventral idiosoma, and each is bounded by the sclerotized apodeme. The medial apodemes of coxae I are fused together along the midline and form a sternum. Coxa I has one seta (*1a*), coxa III has two setae (*3a, 3b*) and coxa IV has one seta (*4a*). The genital opening is located between coxae III and IV, with two pairs of genital papillae covered by the genital valves and 1–3 pairs of genital setae (g_{1-3}). Genital folds are commonly present. The anal opening is posteriorad of the genital opening and has 1–6 pairs of setae (ad_3, ps_3, ps_2, ad_2, ps_1, ad_1). The copulatory opening is posteriorad of the anal opening, connecting a small shield, and leads by a thin or progressively broadened spermathecal duct, into the sac-like inner part of the spermatheca or receptaculum seminis. The base of the inner part of the spermatheca is often sclerotized at its tips with a pair of 'Y'-, 'V'- or 'U'-shaped sclerites of the oviducts. The male has its aedeagus and anal suckers lateral to the anal opening or on tarsi IV (Hughes, 1976; Griffiths *et al.*, 1990; Klimov and O'Connor,

DOI: 10.1079/9781786390912.0028

2003; Fan and Zhang, 2004). The legs are six-segmented, with the last five articles free and the following chaetotaxy: coxae: 1, 0, 2, 1; trochanters: 1, 1, 1, 0; femora: 1, 1, 0, 1; genua: 2 + 2 solenidia (σ_1 and σ_2), 2 + 1 solenidion (σ), 1 + 1 solenidion (σ), 0; tibiae: 2 + 1 solenidion (φ), 2 + 1 solenidion (φ), 1 + 1 solenidion (φ), 1 + 1 solenidion (φ); tarsi: 12 + 3 solenidia (ω_{1-3}) + 1 famulus (ε), 12 + 1 solenidion (ω), various numbers. Solenidia ω_{1-3} of tarsus I are set in distal, median and proximal positions, respectively. On the tibiae, the solenidion j is dorsal, as well as the solenidia s on genua I–III (Grandjean, 1939b; Griffiths, 1970). The sexes commonly differ in the genital regions but have minimal sexual dimorphism, although they are variable between species, and commonly characterized by the males, which in some species are homoeomorphic and heteromorphic with respect to the females (Michael, 1903; Turk and Turk, 1957; Woodring, 1969a). This adaptation is called 'andropolymorphism', in which there are different body shapes, setal lengths and shapes of leg III, and there are four types of males: 'omotype homoeomorphic', 'heteromorphic', 'bimotype bimorphic' and 'pleomorphic' (Woodring, 1969a). In the genus *Rhizoglyphus*, the adult homoeomorphic males are similar to the adult females, from which they differ in a few characters. They have the genital opening set between coxae IV, have an aedeagus, one pair of anal suckers and three pairs of pseudanal setae (ps_{1-3}); moreover, they lack adanal setae (ad_{1-3}), the solenidion φ on tibia IV is spine-like, and tarsus IV bears two suckers. The leg chaetotaxy is as in the adult female, excepting tarsi IV. The adult heteromorphic male is similar to the adult homoeomorphic male, but both sides or one side of leg III is about doubled in size. The body setae are slightly longer. The leg chaetotaxy is as in adult homoeomorphic male, excepting tarsi III (Klimov and O'Connor, 2003; Fan and Zhang, 2004). The life cycle develops through the egg, a six-legged larva, protonymph and/or tritonymph stages. In some species, the protonymph may develop into the deutonymph or hypopus, a heteromorphic stage which does not eat, and allows the survival and dispersal of the species. Several *Tyrophagus* spp. have abandoned deutonymphal dispersal and colonize many habitats. Other genera, for example, *Rhizoglyphus*, have retained the hypopus stage.

Fan and Zhang (2007a) reported more than 90 genera, and about 400 described species; Schatz *et al.* (2011) number over 541 species and 88 genera. These mites are widely spread in various habitats and play various ecological roles – as scavengers, coprophages, micophages or phytophages (Zakhvatkin, 1941; Johnston and Bruce, 1965; Hughes, 1976; O'Connor, 2009b).

Some acarid mites damage ornamentals and vegetables, or are associated with mushrooms. Seven of these species are included in the genus *Tyrophagus* Oudemans: *Tyrophagus communis* Fan *et* Zhang, *Tyrophagus curvipenis* Fain *et* Fauvel, *T. longior* (Gervais), *Tyrophagus neiswanderi* Johnston *et* Bruce, *Tyrophagus perniciosus* Zakhvatkin, *Tyrophagus putrescentiae* (Schrank) and *Tyrophagus similis* Volgin.

Five other species are included in the genera *Mycetoglyphus* Oudemans (*Mycetoglyphus fungivorus* (Oudemans)), *Rhizoglyphus* Claparède (*Rhizoglyphus robini* (Claparède)), *Rhizoglyphus echinopus* (Fumouze *et* Robin), and *Rhizoglyphus setosus* (Manson) and *Sancassania* Oudemans (*Sancassania berlesei* (Michael)). In particular, *M. fungivorus* is injurious to various plants. *R. robini*, *R. echinopus* and *R. setosus* damage the underground organs of various plants in the field and store (Ho and Chen, 1987; Zhang, 2003; Fan and Zhang, 2004). *S. berlesei* is reported from stored food and other substrates (Hughes, 1976), as well as on edible fungi in various countries (Zakhvatkin, 1941; Hussey *et al.*, 1969; Clancy, 1981; Greenslade and Clift, 2004); it has also been collected on melon in greenhouses (Laffi and Ponti, 1997), and on fruiting bodies and hyphae of *Lentinus edodes* (Berk.) Singer [= *Lentinula edodes* (Berk.) Pegler] in Japan, along with the acarid *Histiogaster rotundus* Woodring (Okabe, 2006), but there is no information available on the damage done. Some (and other Astigmatina) cause human allergies (Hage-Hamsten *et al.*, 1985; Terho *et al.*, 1985). In addition to direct damage to crops, unidentified *Tyrophagus* spp. (or other acarid mites) can pose access problems to markets for horticultural crops and stored foodstuffs when intercepted by an importing country. Bass and Hays (1976) observed that *T. putrescentiae* may feed in the laboratory, on eggs of the fire ant and on organic debris.

On stored food are known about 47 species ascribed to 12 genera, five of them belong to the subfamily Acarinae Nesbitt [*Acarus* Linnaeus (six species), *Aleuroglyphus* Zachvatkin (three species), *Tyroborus* Oudemans (three species), *Tyrolichus* Oudemans (one species) and *Tyrophagus* Oudemans (15 species)], and six genera to the subfamily Rhizoglyphinae Zachvatkin [*Histiogaster* Berlese (one species), *Mycetoglyphus* Oudemans (one species), *Rhizoglyphus* Claparédè (three species), *Sancassania* Oudemans (10 species), *Schwiebia* Oudemans (one species) and *Thyreophagus* Rondani (two species)]. Moreover, one genus is placed in *incertae sedis* (*Madaglyphus* Fain (one species)) (O'Connor, 1998). We present below a key to the genera partly following that of Hughes (1976) and discuss only the most important species.

Acarid mites may be involved in allergic diseases. The topic has been discussed extensively by Mullen and O'Connor (2019), who reported that in some conditions, the mites may penetrate the natural body orifices, causing temporary internal acariasis. The phenomenon usually involves the ingestion of food infested by mites, and the inhalation of airborne mites, or mite-contaminated dust via oral or nasal openings. Mites that are swallowed or inhaled can lead to acariasis involving the alimentary tract, whereas mites that are inhaled also can invade the respiratory tract. Cases of mites infesting the urinary tract are rare. For a general discussion of the topic, one may consult Ma and Wang (1992).

Selected websites

https://www.google.it/#q=acaridae+pictures

Spinnen Forum Wiki. Available at: http://wiki.eu-arachnida.de/index.php?title=Tyrophagus

http://www.miteresearch.org/db/main.php?genus=7058

https://www.researchgate.net/publication/221916198_House_Dust_Mites_Other_Domestic_Mites_and_Forensic_Medicine

https://idtools.org/bee_mite/index.cfm?packageID=1&entityID=157

Key for the Identification of Genera

1. External scapular setae *ve* set near the anterior angles of the dorsal propodosomal shield at the same level as *vi* or slightly posterior 2

– Setae *ve* rudimentary or absent or when present are set near the middle of the lateral edge of the propodosomal shield 7

2. Solenidion σ_1 on genu I more than three times longer than σ_2; femur I of male enlarged, with a ventral conical process .... genus ***Acarus*** Linnaeus

– Solenidion σ_1 on genu I less than three times longer than σ_2; femur I of male not enlarged and not bearing a ventral conical process 3

3. Setae *sci* longer than the external scapular setae *sce*; chelicerae and legs poorly tanned 4

– Setae *sci* shorter than *sce*, chelicerae and legs reddish-brown in colour .... genus ***Aleuroglyphus*** Zachvatkin

4. Setae *ve* shorter than the genu, set posterior to setae *vi* .... genus ***Mycetoglyphus*** Oudemans

– Setae *ve* as long as or longer than the genu, set almost on a level with setae *vi* 5

5. Setae d_1 and la about equal in length, shorter than d_3 and d_4 6

– Setae *la* four to six times longer than d_1 .... genus ***Tyrolichus*** Oudemans

6. Five ventral distal tarsal setae, of which the three central ones are thickened; on genu I σ_1 is always less than three times as long as σ_2 .... genus ***Tyrophagus*** Oudemans

– Distal ends of the tarsi with three ventral spines (*u*, *v* and *s*) .... genus ***Tyroborus*** Oudemans

7. Setae *sci*, c_1 and d_1 present 8

– Setae *sci*, c_1 and d_1 absent 9

8. Tarsus I with seta *ba* large, spine-like and adjacent to ω_1 and ω_2 .... genus ***Rhizoglyphus*** Claparède

– Tarsi I–II each with seta *ft'* filiform, not associated with solenidion ω_1 .... genus ***Sancassania*** Oudemans

9. Tarsus without spine-like seta in front of ω_1 and ω_2 ....genus ***Thyreophagus*** Rondani

– Tarsus with spine-like seta in front of ω_1 and ω_2 10

10. Tarsus I short and stubby, about as long as wide, opisthosomal shield absent .... genus ***Schwiebea*** Oudemans

– Tarsus I elongate, more than twice as long as wide; opisthosomal shield present in the male ....genus ***Histiogaster*** Berlese

The Genus *Acarus* Linnaeus

Diagnostic characteristics

The setae *ve* are less than half the length of *vi*. The setae c_1 and d_2 are always short. The solenidion σ_1 on genus I is more than three times longer than σ_2. The male possesses an enlarged first pair of legs, the femur of which bears a stout apophysis or process on its ventral surface. Small thickenings of cuticle may also be present on the ventral side of the genu I. The female reproductive system has a characteristic structure (Hughes, 1976).

Solarz (2011) reports the following differential diagnosis. The setae *ve* arising near the anterior angles of the dorsal propodosomal or slightly posterior; the ventral apex of tarsus I with proral (*p*, *q*) and unguinal (*u*, *v*) setae as thin, not short, stout spines; the setae *sce* about equal in length to *sci*; the seta *scx* generally expanded basally and thickly pectinate. The structure of female reproductive system has chitinized depression in body wall, external opening of expansible tube of bursa copulatrix, chitinized bell-shaped structure leading into receptaculum seminis and two-armed, flame-like chitinized opening to thin-walled oviduct; the claws of female are never bifid.

Economic importance

This genus has a worldwide distribution and includes various species associated with stored food. Most important species are *Acarus farris* (Oudemans), *Acarus immobilis* Griffiths and *Acarus siro* Linnaeus. Chmielewski (2004) reported that the most frequent and numerous mites, on stored grain and other buckwheat products in Poland are *A. farris*, *A. immobilis*, and *A. siro*, followed by *T. longior* (Gervais), *T. putrescentiae* (Schrank), *Glycyphagus domesticus* (De Geer) and *Lepidoglyphus destructor* (Schrank). These *Acarus* species are commonly retained to be known pests of economic and sanitary importance. Other *Acarus* species play a secondary importance, such as *Acarus gracilis* Hughes, known from Argentina, China, England, Germany, Greece, Primorskiy Kray, Taiwan and UK on biscuit, bread, Chinese medicine, grain, grain residue and rice in farm (Hughes, 1957, 1976; Griffiths 1964a; Hagstrum *et al.*, 2013) and in poultry droppings in Iraq (Al-Kubaisi and Al-Dulaimi, 2020), *Acarus mirabilis* Volgin, reported from China and Poland on buckwheat and buckwheat product collected in condiment (Volgin, 1965; Hagstrum *et al.*, 2013), and *Acarus chaetoxysilos* Griffiths, found in cheese store in England and Japan (Griffiths, 1970). We discuss here only the first three species. A systematic key of the species of this genus may be found in Griffiths (1970).

Acarus farris (Oudemans) (Fig. 28.1)

Common name

Unknown

Diagnostic characteristics

FEMALE. The body is larger than that of the male. On leg I, the solenidion ω_1 has the sides expanding gradually from base, then narrowing to an indistinct neck before expanding into terminal head. Width of widest part of head equal to width of widest part of stem. The spine (*v*+*u*) is slender, about half the length of the tarsal claw, ventro-posterior margin convex, the tip directed forwards. The anal setae ad_1, ad_2 and ad_4 almost equal in length, ad_3 are about one-third and ps_3 are about twice as long (Hughes, 1976).

MALE. The body is about 365 µm. On leg I and II, the solenidion ω in lateral vision increases slightly in diameter from the base upwards, then narrows almost imperceptibly before expanding a little into a terminal head; the angle between its anterior margin and the dorsal surface of the tarsus, is nearer to 90° than 45°; the spine (*v*+*u*) on tarsi II, III and IV (and I to IV of the female) is laterally about one-half to two-thirds the length of the tarsal claw and points anteriorly (Hughes, 1976).

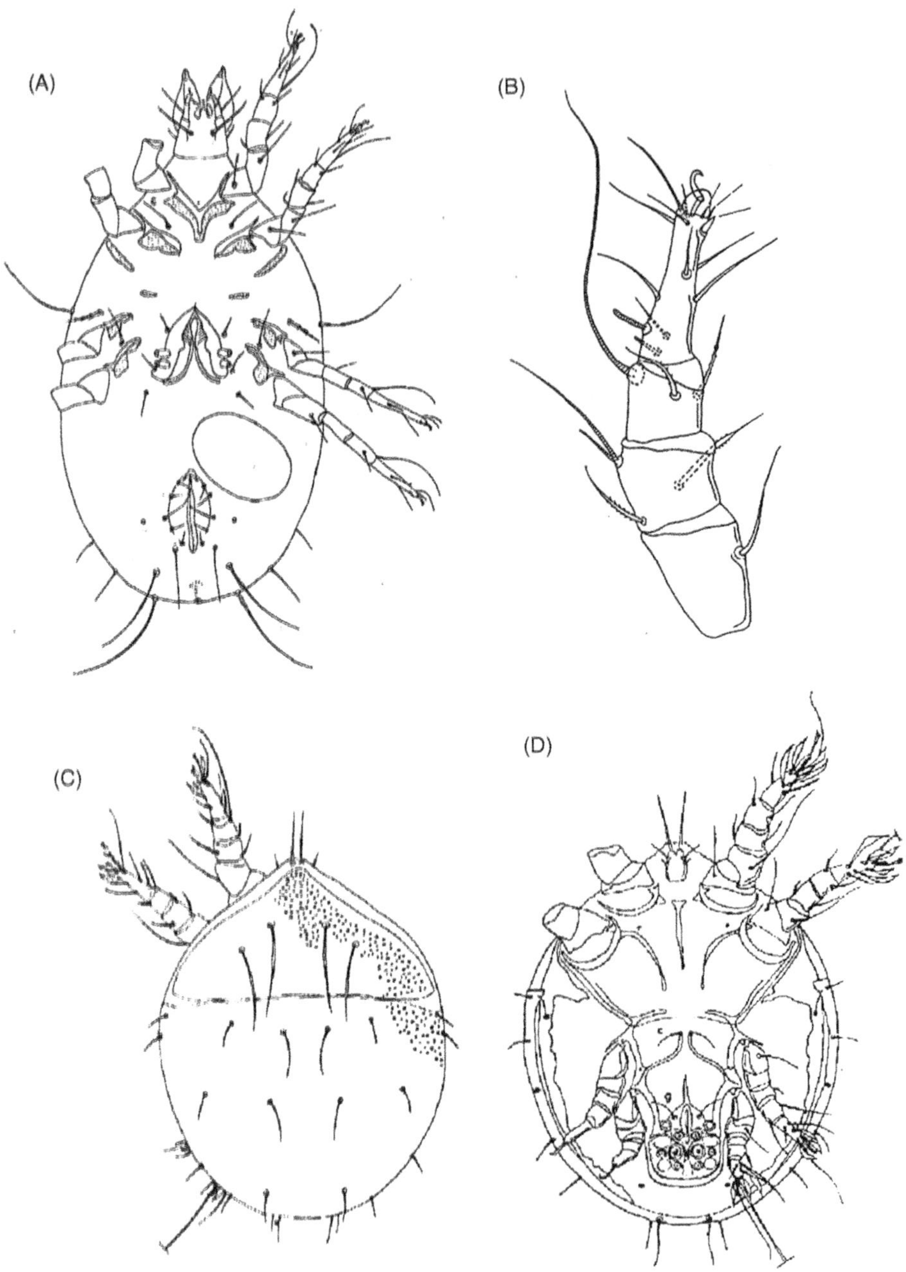

Fig. 28.1. *Acarus farris* (Oudemans). Female. Dorsal view (A); leg I (B). Hypopus. Dorsal view (C); ventral view (D) (from Hughes, 1976, licensed under the Open Government Licence v3.0).

HYPOPUS. The idiosoma is 240 μm long. The setae c_2, d_2, and e_1 are almost equal in length. The distance between the bases of sucker and seta is about the width of the setal base, and the sucker is markedly posterior and external to it. The apodemes IV curve gently forwards, towards the midline, but just before meeting, they ½½ bend forwards and run parallel for a short distance before ending (Hughes, 1976).

Solarz (2011) reports the following differential diagnosis. In both sexes, the setae *sce* about equal in length to *sci*; setae d_1 intermediate in length between c_1 and e_1, never more than three times longer than c_1; most dorsal setae short, setae d_1 and e_1 not extending to base of next posterior seta. Seta *s* of tarsus II of male slender, about ½ – ⅔ the length of empodial claw and points anteriorly; ventro-posterior margin of the seta convex, seta tip directed forward. Tarsi I–II of female with seta *s* slender, about ½ – ⅔ the length of empodial claw; ventro-posterior margin of seta convex, seta tip directed forward; setae ad_3, ad_2 and ad_1 almost equal in length, ps_3 about ⅓ and ps_2 about twice as long; vulva as in *A. siro*. The hypopus has hysterosomal setae c_1, d_1, d_2 and e_1 appreciably shorter than scapulars, and not extending to bases of next most posterior setae; setae *sci* at least two times length of c_1 and d_1; setae c_1 and d_1 about equal in length to h_1; bases of genital setae just forward of coxal sucker bases; distance between base of sucker and base of seta about equal to width of setal base (Solarz, 2011).

Geographic distribution

Azores Islands (Cardoso *et al.*, 2010), England, Scotland, Wales, the Netherlands, Germany (Griffiths, 1964a; Rack, 1968a), Kenya, United States, Poland, Czech Republic (Griffiths, 1964a; Ždárková, 1967; Hubert *et al.*, 2006a), Canada (Sinha and White, 1982; Beaulieu and Knee, 2014), China (Jiang, 1992; Fan *et al.*, 2010a), Egypt (Azza *et al.*, 2019), Finland (Thero *et al.*, 1982), India (Gupta and Chatterjee, 2004), Iran (Sepasgozarian, 1978; Sayedi *et al.*, 2006; Ardeshir, 2010; Sayedy *et al.*, 2012), Malta (Pfliegler *et al.*, 2017), Ireland (Clarke *et al.*, 2015), Japan (Sinha, 1968b), Spain (Sánchez-Ramos *et al.*, 2007a), Italy (Bernini *et al.*, 1995), India (Navpreet and Sakshi, 2018), Poland (Solarz and Pająk, 2019a) and Ukraine (Akimov and Oksentiuk, 2018; Oksentiuk *et al.*, 2022).

Commodities, facilities and animal hosts

This mite is a field species, living also on various commodities, such as barley, barley (germinating), bread, buckwheat, buckwheat product, cereal, cheese, Cabrales cheese, Chinese medicine, cottonseed meal, dry grass, English cheeses, grain, grain residue, grass seed, hay, insect (dried), oat, oilseeds, potato, poultry feed, radish seed, rapeseed, soybean, soya meal, sugarbeet seed, sweeping, straw and wheat in various facilities (cheese store, commercial grain storage, drier, farm, farm granary, farm milling machine, grinder, malthouse, mill and cheddar store) (Griffiths, 1964a; Hughes, 1976; Wilkin, 1979; Smiley, 1987; Gupta and Chatterjee, 2004; Sánchez-Ramos *et al.*, 2007a, Sánchez-Ramos and Castañera, 2007; Sayedy *et al.*, 2012; Hagstrum *et al.*, 2013; Beaulieu and Knee, 2014; Pfliegler *et al.*, 2017; Navpreet and Sakshi, 2018; Akimov and Oksentiuk, 2018). Moreover, the mite has been collected in barns, pigsties, cowsheds and chaff-cutter buildings (Solarz and Pająk, 2019a), and on bean cake, broad bean, dried sweet potato, old silk noodles, cake, rice, feed and starch in China (Li *et al.*, 2015a).

In Poland, the mite has been found in nests of domestic and tree sparrows (Wasylik, 1971, 1973; Sandner and Wasylik, 1973; Chmielewski, 1982), in barns and chaff-cutters (Solarz *et al.*, 1997), in house dust samples (Dzięciołowski, 1994; Chmielewski, 1995), on houseflies, *Musca domestica* (Linnaeus) (Chmielewski, 1983b), in samples of dust and residues from railway trucks (Solarz, 1995), in pollen traps, beehives and honey (Chmielewski, 1971a; 1983a, b, 2002a), spices and grocery articles of consumption (Chmielewski, 1971c), stored herbs (Karnkowski, 1990) and on cereals (grain) (Sandner and Wasylik, 1973).

Solarz *et al.* (2004) found 1734 specimens of *A. farris* in cages, and two specimens in runs of the Silesian Zoo. The mite was also found associated with birds (41.1% of the total) and felids (1.4%). Moreover, 84 specimens have been collected from four samples from barns in Majdan Górny. Moreover, the mite has been collected in poultries houses. In Ukraine, Oksentiuk *et al.* (2022) collected the mite in mills, granaries and warehouses, barns with animal fodder storage and outbuildings.

Bio-ecology

A. farris was first described as *Aleurobius farris* by Oudemans (1905a). The mite infests a number of commodities in various facilities and may represent a problem for the stored food. The species is widely distributed and probably is cosmopolitan; it is often abundant on farms, particularly during August and/or September (in the northern Hemisphere), when it is associated with other acarid mites, such as *T. neiswanderi* Johnston *et* Bruce, *T. longior* (Gervais), etc. (Hughes, 1976; Terho *et al.*, 1982; Hallas and Solberg, 1989; Hallas *et al.*, 1991; Franz *et al.*, 1997). Navpreet and Sakshi (2018) found these mites in 3% of total fauna collected on stored grain. In Poland, Solarz and Pająk, (2019a) found in a study on the exposure of a selected rural population that the most abundant species of mites in farm were *A. siro* (especially in cowsheds and barns), followed by *A. farris* (in barns and pigsties), *L. destructor* (mainly in pigsties) and *G. domesticus* (mainly in barns).

In Spain, Sánchez-Ramos *et al.* (2007a) reported infestations of *A. farris* on the Cabrales cheese in an Asturian maturing cave, and found that the mite is the prevalent species with respect to other acarid mites. Studies carried out in laboratory on the developmental rate and survival of immature stages of *A. farris,* at constant temperatures ranging from 7–29.7°C, and a RH of 90±5%, showed that the larval stage of *A. farris* is particularly susceptible to low and high temperatures, with 81.7% and 95.2% mortality at 7°C and 29.7°C, respectively. The optimal temperature for the development is 27–28°C, and the non-linear logan type-III model, provided the best fit for the relationship between developmental rates and temperature (R(a) 2>0.99), for all immature stages. The lower and upper developmental threshold temperatures, predicted for each stage of *A. farris* are 3–4°C. The differential temperature-development rate for each species might explain the greater abundance of *A. farris*. Furthermore, manipulation of temperature, based on modelling predictions, may well be used to control mite populations during the cheese maturing process.

Sánchez-Ramos and Castañera (2007) investigated the best combination of RH and time of exposure, to be applied on Cabrales cheeses infested with *A. farris*. Laboratory assays reveal that less than 30 h are required to obtain 90% mortality of mites at 50% and 60% RH. Males are more susceptible than females to low RH, since their lethal period values are lower than those obtained for females at the same RH. Moreover, the response within sexes to low moisture treatments changes as exposure time increases, since the LP50 obtained for each sex at 50% and 60% RH are statistically different, whereas the LP90 shows no significant differences within sexes.

Laboratory studies carried out by Sánchez-Ramos *et al.* (2007c), on developmental rate of immature stages and the reproduction of adults of *T. putrescentiae* (Schrank), *T. neiswanderi* and *A. farris,* at 70%, 80% and 90% RH, and a constant temperature of 25°C showed that at 70% RH. *T. putrescentiae* and *A. farris* immature stages failed to reach the protonymph stage, as 100% of the larvae died, whereas *T. neiswanderi* was able to complete the development. The developmental time of all immature stages for the three species significantly increased, as RH was reduced. The mobile stages were particularly susceptible, as the time needed to complete their development at lower RH suffered greater increases than the egg stage. At 70% RH, *T. putrescentiae* and *A. farris* were not able to lay eggs, and only 24% of *T. neiswanderi* were fertile. The reproductive parameters of the three species at the RH, at which they were able to lay eggs, showed significant differences, except for the percentage of fertile mating at 80% and 90% RH. As the RH increased the pre-oviposition period was reduced, and fecundity and daily fecundity were increased, whereas the oviposition period showed different patterns for the three species. The intrinsic rate of increase (*rm*) of *T. neiswanderi* at 70% RH was negative, indicating that at these conditions mite populations of this species will diminish until they disappear. As RH increased from 80% to 90% RH, this parameter was almost twofold for both *Tyrophagus* species. The *rm* obtained for *A. farris* at 90% RH, was similar to that of *T. neiswanderi* at the same humidity, while at 80% RH, it was very small so that the population doubling time was more than 84 days. The threshold temperature for the individual development has been calculated 1.4°C, with estimation based on non-linear model at 90% RH.

Further studies carried out by Sánchez-Ramos and Castañera (2008), on the reproductive parameters and the longevity of *A. farris,* at six constant temperatures, ranging from 7–29°C, and a RH of 90±5% showed that as the temperature increases above 25°C, the pre-oviposition period rapidly decreases to a minimum. The mite fecundity is adversely affected by extreme temperatures, while the oviposition period increases as temperature is reduced. The relationship between reproductive parameters and temperature has been fitted using different polynomial and non-linear models. Male and female longevity increases as temperature decreases, except for the males at 7°C, which shows significantly lower longevity than at 10°C. Moreover, the males show significantly greater longevity than females. Observations on the effect of temperature on the capacity for increase of *A. farris* populations showed that the non-linear Lactin model provides a reliable fit of the relationship between intrinsic rate of increase and temperature (R2>0.99). The optimum temperature for development is 25°C, and at this temperature the population doubling time is 2.8 days. The lower and upper thermal thresholds for *A. farris* populations correspond to 6.4°C and 29.9°C, respectively (Sánchez-Ramos and Castañera, 2008).

Symptomatology and damage

Like other mites of stored product, *A. farris* causes economic loss by feeding on stored grain and imperil the human health by contamination of food and by allergens. Hughes (1976) reported that *A. farris* causes skin irritation to farm workers handling infested bales of hay in England. Moreover, it is the main species of mite infesting Cabrales cheese, a traditional Spanish product, on which this species may reach population density levels of 260 mobile mites/cm^2, resulting in high economic losses (Sánchez-Ramos and Castañera, 2009).

Control

The control involves fundamentally the use of chemical and physical means.

PHYSICAL CONTROL. Sánchez-Ramos and Castañera (2007) tried two modifications of the traditional maturing process to assess the efficacy of low moisture treatments to control the mite on infested cheeses. The first modification consisted of one single exposure of 48 h at 50% RH, and the second one consisted of two exposures of 48 h at 50% RH, separated by a time interval of 15 days. They have not found significant differences in final population density for both low moisture treatments compared to control cheeses. Low-humidity treatments resulted not effective to control *A. farris* in Cabrales cheese, despite the good results obtained in laboratory assays (Sánchez-Ramos and Castañera, 2007). Further trials concerned the physical approach based on lowering the usual cheese maturing temperature (15°C) to 6°C, 4°C or 2°C. As temperature decreased *A. farris* density drastically reduced from 17,433 mites/cm^2 at the control temperature (15°C) to 14, 11 and 1 mites/cm^2 for 6°C, 4°C and 2°C, respectively, though the maturing time was considerably extended (Sánchez-Ramos and Castañera, 2009).

CHEMICAL CONTROL. Sánchez-Ramos *et al.* (2004) tested protease inhibitors of different specificity *in vivo*, to evaluate their action as control agents. The development of *A. farris* was significantly retarded when the immature stages were fed on artificial diet containing inhibitors of serine and cysteine proteases and aminopeptidases, whereas no such effect was found with inhibitors of aspartyl proteases and carboxypeptidases. The most significant effects on *A. farris* occurred when a combination of inhibitors targeting different enzyme classes was supplied mixed in the diet, suggesting a synergistic toxicity. Several plant lectins were also tested, but only wheat germ agglutinin and concanavalin A affected development.

Different chemicals were tested by Sánchez-Ramos and Castañera (2009) to control the mite without affecting the organoleptic quality of Cabrales cheese. The efficacy of four short-chain fatty acids (caproic, caprylic, pelargonic and capric) and two sugar alcohols (ribitol and xylitol), were tested under laboratory conditions against the mobile stages of *A. farris*. Only the pelargonic acid produced significative mortality, in comparison with controls (63.4% vs 26.1%, respectively). Moreover, the contact effect of the monoterpene eucalyptol was assessed in cheese maturing rooms at doses of 2.5 and 1.25 ml/cm^2. The first dose controlled *A. farris*, but its application negatively affected the organoleptic values of the treated cheeses. The application of the waxy food coating READOMCBR to Cabrales cheeses in maturing rooms kept the mite population low but negatively modified the external appearance of the cheese, an important quality parameter (Sánchez-Ramos and Castañera, 2009).

Acarus immobilis Griffiths

Common name

Unknown.

Diagnostic characteristics

ADULTS. Adults and immature stages are indistinguishable from the corresponding stages of *A. farris*, except for a difference in shape of ω_1 on tarsi I and II of adults. In lateral view, the side of ω_1 is virtually parallel, expanding into a distinct egg-shaped terminal head. The two species are genetically different (Griffiths, 1964b).

HYPOPUS. The idiosoma is about 210 μm long, broadly oval, white and translucent in colour. The dorsal surface is arched and covered with shallow punctuations, while the ventral surface is concave. A transverse groove divides the propodosoma from the hysterosoma. The gnathosoma is reduced, only consisting of paired protuberances, each ending in a blunt process. The dorsal chaetotaxy as in *A. siro* and *A. farris*, with exception of the absence of setae *ve* and c_1 and one pair of setae on the posterior margin of the hysterosoma. All setae are short. One pair of circular pores on the hysterosoma, one behind each of the hi setae and one pair of glands on a level with the coxae IV.

On the legs, the number and the size of setae and solenidia are reduced, although ω_1 is more than half the length of tarsi I and II and has an egg-shaped terminal expansion. The solenidia *б* and *φ* of leg I are short and blunt. The ventral spine complex and the setae *e* lacking from the legs I and II, and the seta f from the leg II (Hughes, 1976).

Solarz (2011) published the following differential diagnosis. In both sexes, the setae *sce* about equal in length to *sci*; setae d_1 intermediate in length between c_1 and e_1, never more than three times longer than c_1; most dorsal setae short, setae d_1 and e_1 not extending to base of next posterior seta; tarsus II with solenidion ω_1, with sides almost parallel, expanding into a distinct egg-shaped terminal head, which is wider than widest part of shaft. The angle between the dorsal surface of the tarsus and the anterior face of the solenidion generally about 45–50°. Seta *s* of tarsus II of male slender, about half the length of empodial claw; ventro-posterior margin of the seta convex, seta tip directed forward. Seta *s* of tarsi I–II of female slender; ventro-posterior margin of the seta convex, seta tip directed forward. The hypopus has inert form, legs short (in dorsoventral mounts tarsi of legs I and II are the only segments completely visible), setae very reduced or absent; gnathosoma rudimentary, flagelliform bristles (solenidia) not present; posterior ventral attachment organ rudimentary, with at most one pair of well developed suckers on sucker plate, the central pair of suckers vestigial, anterior peripheral pair moderately well developed; all setae on tarsus III and IV shorter than length of tarsus, spine-like, never leaf-shaped; solenidion ω_1 on tarsus I long, at least twice the length of empodial claw (Solarz, 2011).

Geographic distribution

This mite is reported mainly from temperate countries, such as Canada (Beaulieu and Knee, 2014), China (Zou and Wang, 1989; Fan *et al.*, 2010a), Czech Republic (Hubert *et al.*, 2006a), Germany (Franz *et al.*, 1997), Greece (Palyvos *et al.*, 2008), Italy (Bernini *et al.*, 1995), Malta (Pfliegler *et al.*, 2017), Poland (Solarz and Pająk, 2019a), Turkey (Cilbircioğlu and Çobanoğlu, 2019) and UK (Griffiths, 1964b; Hughes, 1976).

Commodities, facilities and animal hosts

The list of commodities includes baby food, biscuit, Chinese medicine, grain, cheese, grain, grain residue, sugarbeet seed, straw, sweepings, wheat and wheat flour in cheese store, commercial grain storage, farm, farm milling machine, off-farm storage, provender warehouse and railcar (Griffiths, 1964b; Hughes, 1976; Hagstrum *et al.*, 2013; Beaulieu and Knee, 2014; Pfliegler *et al.*, 2017). Moreover, the mite has been collected also in cowsheds and chaff-cutter buildings (Solarz and Pająk, 2019a) and on rice bran (Li *et al.*, 2015a).

Bio-ecology

A. immobilis is an economically important graminivorous pest, feeding directly on various commodities, and characterized by biological traits similar to those of *A. siro*. The mite has been investigated also as source of airborne allergens, especially on farms, and it has been observed that *A. farris* is third last and *A. immobilis* penultimate, among the total number of species collected in farms in Germany (Franz *et al.*, 1997). The male and female sex pheromones produced by *A. immobilis* and its sex pheromone activity were investigated in Japan (Sato *et al.*, 1993; Shibata *et al.*, 1998).

Symptomatology and damage

A. immobilis is a graminivorous pest feeding, causing severe damage (Solomon, 1946). Chmielewski (2004) reported that the most frequent and numerous mites, on stored grain and other buckwheat products in Poland, are in order of importance *A. farris*, *A. immobilis* and *A. siro*, followed by *T. longior*, *T. putrescentiae*, *G. domesticus* and *L. destructor*.

Control

As *A. siro*.

Acarus siro Linnaeus (Fig. 28.2)

Common names

Flour mite, grain mite.

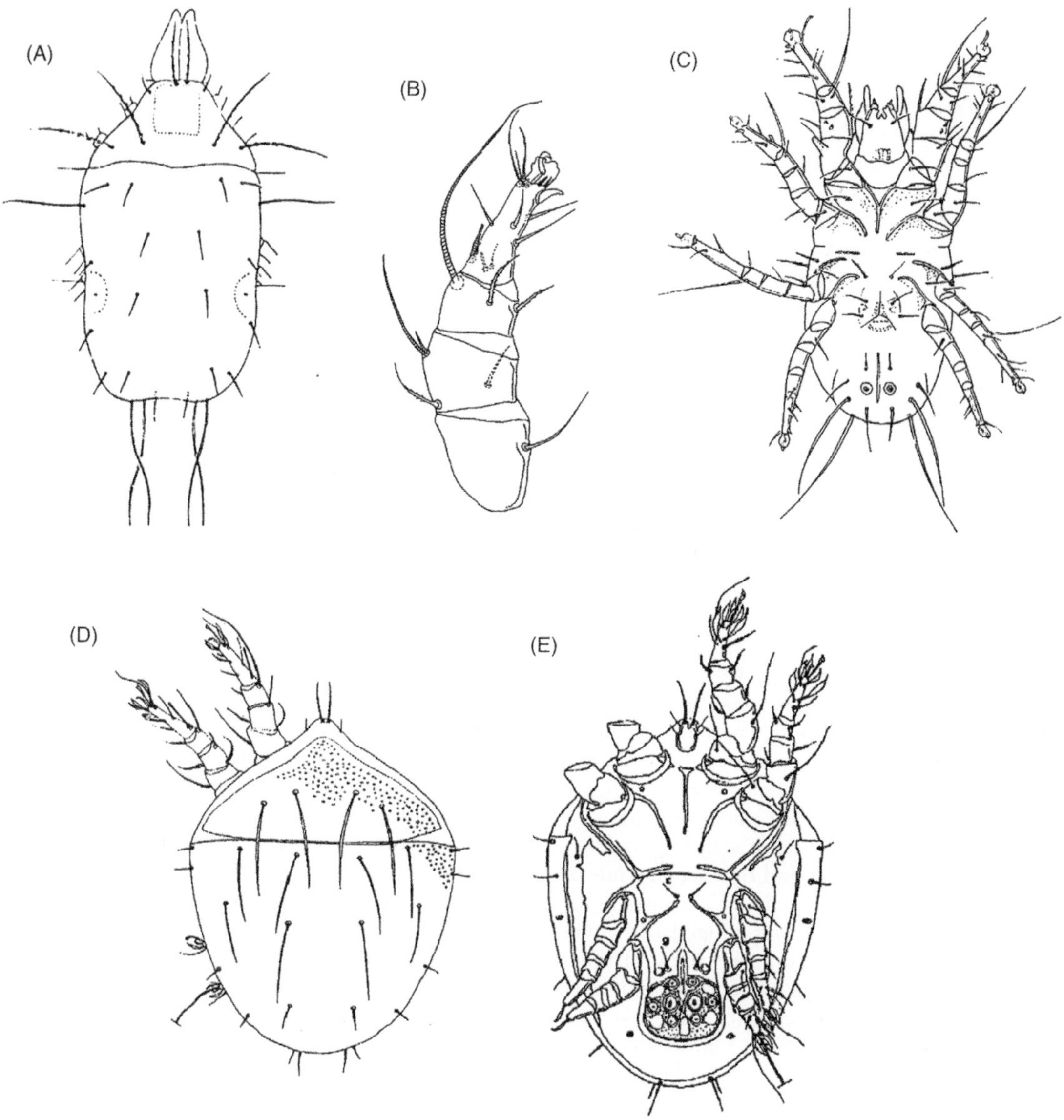

Fig. 28.2. *Acarus siro* Linnaeus. Female. Dorsal view (A); leg I (B). Male. Ventral view (C). Hypopus. Dorsal view (D); ventral view (E) (from Hughes, 1976, licensed under the Open Government Licence v3.0).

Diagnostic characteristics

FEMALE. The idiosoma is 350–650 µm long, oval, and with the posterior margin slightly indented near the bursa copulatrix. The dorsal chaetotaxy is that of male, with setae more sparsely pectinate. The anal opening is surrounded by five pairs of setae, of which ad_3 are twice and ps_3 almost four times the lengths of ad_1, ad_2 and ad_4; ps_1 and ps_2 extend well beyond the posterior margin of the body. The genital opening is located between the coxae III and IV. The bursa copulatrix opens into a narrow expansible tube, which joins a sclerotized bell-shaped structure. This expands into a large thin-walled *receptaculum* that communicates with the oviducts through two openings. The first pair of legs is no wider than remaining ones, and no spur arises from the femur. The chaetotaxy is the same as that of the male, except that on tarsus I, the spines *p* and *q* are distinct and much narrower than unguinal (*u+v*), present in all the legs, large with a backward curve (Hughes, 1976).

MALE. The idiosoma is 320–460 µm long, colourless, with gnathosoma and legs from pale yellow to reddish-brown, according to diet and age; the hind margin is smoothly rounded. The idiosomal setae are fine and some are sparsely pectinate, with pectinations very visible on vi and the distal halves of sc.

The dorsal propodosomal shield is a broad plate extending back, almost as far as the scapular setae. Setae *vi* extend almost to the tips of the chelicerae, and *ve* are considerably shorter (less than one-quarter the length of *vi*) and arise close to the lateral corners of the propodosomal shield. Setae *sc* are about 25% of the idiosoma in length, and arranged in a transverse row; *sce* are slightly shorter than *sci*. The supracoxal seta is expanded basally and thickly pectinate. The Grandjean's organ is a colourless fold of cuticle, whose distal edge is drawn out into a number of filaments of unequal length.

The dorsal hysterosomal setae vary in length and thickness, according to the nutritional quality of the food received in pre-adult stages; poor diet produces adults with short and thin setae, while a rich diet produces longer and thicker ones. The posterior margin of the body is fringed with two pairs of short setae f_2, h_1, and two pairs of long ones h_2, ps_2; the long setae are curved and trail along the ground. The apodemes of leg I are united in the middle line to form a short *sternum*, those of legs II, III and IV are free. The genital opening is located between the coxae IV; the lateral arms of the penis support diverge posteriorly, and the penis is an arc-shaped tube with a blunt end. A pair of anal suckers lies on either side of the posterior end of the anus, and anterior to these are set the p_3 setae.

The chelicerae are distinctly toothed, and at base of the fixed limb is a mandibular spine and posterior to this a conical spur-like outgrowth.

All the legs end in a well developed pretarsus and stalked claw. The leg I has the genu and the femur enlarged, and a spur projecting from the ventral side of femur from which arises the seta *vF*. Two blunter thickenings of the cuticle are also present on the ventral side of the genu but are not easily seen on lightly tanned specimens.

The solenidion ω_1 on tarsi I and II is recumbent, and the angle between the dorsal surface of the tarsus and the anterior face of the solenidion rarely exceeds 45°. It is broadest at the base, then narrows before swelling out into a terminal expansion. The famulus (ε) is shaped like a minute hump and arises from a small pit anterior to ω_1. At the distal end of the tarsus I, *u* and *v* are united to form one large spine. The spine *s* on tarsi II to IV is large, and when the tarsus is viewed laterally, the length of the longest edge (the concave margin) is equal to that of the tarsal claw. On the genu I, σ_1 is more than three times as long as σ_2. The suckers on tarsus IV are separate from one another by a distance equal to their diameter and are nearer to the base than to the apex of the segment (Hughes, 1976).

HYPOPUS. The body is about 230 µm long, pinkish brown in colour, dorsally arched and covered with scattered pits. The ventral surface is concave. The propodosomal shield is distinctly separated from the hysterosoma and projects forward, almost completely concealing the gnathosoma. No eyes are present.

The setae *vi* are strongly pectinate, the *ve* much shorter. The supracoxal setae arise above the coxae I and are almost the same length as *vi*. The bases of the four scapular setae, traverse the propodosoma, *sci* being slightly longer than *sce*. These setae are flattened and slightly expanded in their middle regions and extend posteriorly beyond the junction of propodosoma and hysterosoma. Of the hysterosomal setae c_1 are set between the c_2 setae d_1, and e_1 are in linear series with c_1. Two pairs of humerals project from the side of the body, on the same transverse line as the bases of c_2 and c_1. Three pairs of lateral setae are present. The setae d_2 is lateral to d_1; the setae e_2 and f_2 are set from the body margin between d_2 and e_1. The setae c_1 and d_1 are almost as long as *sci* and approximately equal to c_2 and d_2 in length. The setae c_2 and d_2 are three times longer than e_1.

The lateroopisthosomal glands open on the sides of the body, on a level with coxae IV. Ventrally, the coxae II are well defined laterally by the epimeres II, which are joined to apodemes III; the *sternum* and apodemes II are free, and the coxae III are only partially separated from one another in the midline. The apodemes IV are barely curved and do not meet; just in front of their termination arise the coxal setae. Coxae II, III and IV are distinctly thickened along their edges.

The genital setae arise almost line with a pair of suckers that are anterior to the sucker plate; the setal base is separated from the edge of the sucker by a distance less than the width of the setal base. The sucker plate is small and well removed from the posterior edge of the body, its large central suckers are surrounded by three pairs of peripheral suckers, separated from one another by clear areas.

The legs I and II are almost completely visible from above, but only the terminal setae of III and IV project beyond the edge of the body. All four pairs of legs have well-developed claws and a reduced pretarsus. The chaetotaxy differs from that of other stages, the difference involving change in shape, as well as the expression or suppression of certain setae and solenidia.

Omega 2 and one *б* on leg I and a *б* on leg III are not developed. The seta l on leg III and IV is found in hypopus only. The ventral spine complex of the other stages is replaced by two expanded leaf-like setae *vsc*. Omega 1 on tarsi I and II is longer and slender than in other stages, it is slightly 'goose-necked' with a distinct terminal expansion. Omega 2 is terminal in nymphal, and adult stages arise from the middle dorsal region of the tarsus I. On tarsi I and II, the seta *e* is expanded at its distal end into a sucker-like structure, on tarsus III it is leaf-like and on tarsus IV it is about half the length of the idiosoma. The seta *f* is thin, transparent and leaf-like on all the legs; *ra* or *r* is also leaf-like on all the legs except on IV where it is simple; *la* or *l* is long and leaf-shaped on tarsi I to III; *wa* is broad, flattened and pectinate on tarsus IV. The solenidion φ exceeds the length of the tarsus on leg I, and on tibia II it is about the same length as tarsus II (Hughes, 1976).

Solarz (2011) reports the following differential diagnosis. In both sexes, setae *sce* slightly shorter than *sci*; setae d_1 intermediate in length between c_1 and e_1, never more than three times longer than c_1; seta *scx* expanded basally and thickly pectinate; solenidion ω_1 on tarsi I and II long, recumbent and the angle between the dorsal surface of the tarsus and the anterior face of the solenidion rarely exceeds 45° (generally about 30°). The males have the tarsus II with seta *s* large, about equal in length to length of empodial claw; ventro-posterior margin of this seta concave, claw tip directed backwards; seta *scx* expanded basally and thickly pectinate; lateral arms of the penis support diverge posteriorly; aedeagus as an arc-shaped tube with a blunt end. The females have the tarsi I–II with seta *s* large, about equal in length to the length of empodial claw; ventro-posterior margin of the seta concave, claw tip directed backwards; setae more sparsely pectinate as that of the male; bursa copulatrix opens into a narrow expansible tube, which joins a sclerotized bell-shaped structure; setae ps_3 are twice and ps_2 almost four times the lengths of ad_3, ad_2 and ad_1; vulva characteristic, positioned between coxae III and IV. Hypopus with mobile form; gnathosomal remnant with well developed solenidia; posterior ventral attachment organ (sucker plate) well developed, with eight distinct suckers; legs well developed, with normally formed setae and solenidia; scapular setae relatively long; length of sc is greater than distance between alveoli; hysterosomal setae c_1, d_1, d_2 and e_1 long, extending beyond base of the next most posterior seta; *sci* about 1.5 times length of c_1 and 1.2 times length of d_1; setae c_1 and d_1 about three times longer than h_1; bases of genital setae and flanking pair of coxal suckers almost in line, distance between base of sucker and base of seta is less than width of setal base (Solarz, 2011).

Geographical distribution

This mite is cosmopolitan, and is common in cool and humid regions, rare in tropical and subtropical parts of the world. It is reported from Azores Islands (Cardoso *et al.*, 2010), Australia (Halliday, 1998), Brazil (Barbosa *et al.*, 2022), Canada (Nesbitt, 1945; Beaulieu and Knee, 2014), Chile (Griffiths, 1964a), China (Griffiths, 1964a; Li and Fan, 1997; Fan *et al.*, 2010a), Croatia (Rozman and Kalinović, 2001), Czech Republic (Ždárková, 1967; Hubert *et al.*, 2006a), Denmark (Hallas *et al.*, 1991), Egypt (Gentry, 1965; El-Sayed and Ghallab, 2007), Finland (Terho *et al.*, 1982; Lehtinen and Oksala, 1991), Germany (Rack, 1968a), Greece (Athanassiou *et al.*, 2002, 2005), India (Putatunda, 2002; Gupta and Chatterjee, 2004), Iran (Farahbakhsh, 1961; Gentry, 1965; Saiedy *et al.*, 2009), Iraq (Al-Kubaisi and Al-Dulaimi, 2020), Ireland (Cusack *et al.*, 1975), Italy (Ottoboni *et al.*, 1993), Japan (Sasa, 1965; Sinha, 1968a), Malta (Pfliegler *et al.*, 2017), the Netherlands (Siepel *et al.*, 2016), New Zealand (Robertson, 1946), Poland (Boczek, 1961), Primorskiy Kray (Tareev, 1970), Russia (Zachvatkin, 1941), South Korea (Lee and Cho, 1980; Kim *et al.*, 1988), Spain (Armentia *et al.*, 1997; Moreno, 2015), Switzerland (Büchi, 1991), Taiwan (Tseng, 1972),

Tunisia (Gentry, 1965), Turkey (Emekçi and Toros, 2000), Ukraine (Dudinska, 2003; Akimov and Oksentiuk, 2018; Oksentiuk *et al.*, 2022), UK (Griffiths, 1960, 1964a; Hughes, 1976), United States (Griffiths, 1964a; Hagstrum *et al.*, 2013) and other countries.

Commodities, facilities and animal hosts

The list of commodities includes many products (aquatic product, baby food, barley, barley sweeping, bean curd, beet seed, biscuit, bread, buckwheat, buckwheat product, cake, castor bean, cereal, cereal debris, cereal product, chaff, cheese, chicken feed, chicory, Chinese medicine, cinnamon, clover seed, coltsfoot leaf and root, condiment, cottonseed, dill seed, dried fruit, vegetable (dried), ergot, fenugreek, flax seed, fodder waste, grain, grain residue, grass seed, halva, hay, hemp seed, herb, kale seed, kohlrabi seed, lentil, lettuce seed, linseed, maize, lupine seed, maize groat, meat (dried), milk (dried), mint leaf, mushroom, nettle leaf, oat, oat flake, offal, oilseed cake, onion seed, pea (dried), pea flour, groundnut, poppy seed, potato flour, processed food, protein poultry feed, radish seed, rapeseed, residue, rice, rice (husked), crunch seed, sesame, shrimp (dried), soup (dried), soybean, soybean meal, spinach seed, squid (dried), straw, sugarbeet seed, sunflower seed, sweeping, wheat, wheat bran, wheat flour, wheat flour sweeping, wheat residue). The list of facilities includes commercial grain storage, cheese storage, drier, empty grain bin, farm, farm granary, farm grinder, farm milling machine, flour mill, flour warehouse, grain warehouse, oatmeal mill, mushroom growing cellar, off-farm storage, provender mill, provender warehouse, railcar and wheat cleaning area (Hughes, 1976; Hagstrum *et al.*, 2013). Ždárková (1967) found *A. siro* on protein mixtures in food processing factories in Czechoslovakia. Moreover, the mite has been collected in barns, pigsties, cowsheds, chaff-cutter buildings, and poultry houses (Solarz and Pająk, 2019a). In Malta, it has been collected in flourmill and silo (Pfliegler *et al.*, 2017). In Spain, it is reported on stored foods (Moreno, 2015), and on ham (Sánchez-López, 2002), and in China, has been recently found on rice, wheat, starch, ginseng, groundnut cake, feed, gammon, and cake (Li *et al.*, 2015a). In Egypt, the mite has been collected on fenugreek and lentil (Saleh *et al.*, 1985), and in Iraq, in poultry droppings (Al-Kubaisi and Al-Dulaimi, 2020). In Ukraine, Oksentiuk *et al.* (2022) collected the mite in mills, granaries and warehouses, compound fodder factories, barns with animal fodder storage, outbuildings, honeybee beehives and in vegetable storages.

Bio-ecology

The grain mite was described as *Tyroglyphus farinae* by Linnaeus (1758 and successively identified as *A. siro* by various specialists (Griffiths, 1964a; Attiah, 1969; Hughes, 1976; Smiley, 1987). Prior to 1970 the bibliographical references on *A. siro* may be inaccurate due to the *A. siro* complex including *A. siro*, *A. farris* and *A. immobilis* (Griffiths, 1970). The difficulty to separate the three species morphologically, the possibility of the gene exchange between some of them, although each species displays environmental preferences, and their contemporary presence in some environments, suggested to Webster *et al.* (2004) to investigate their species status, whether genetic data supported the separate species status of these forms. Both nuclear (the second internal transcribed spacer region [ITS-2] of the ribosomal cistron) and mitochondrial (cytochrome oxidase subunit I, mtcoxI hereafter) loci were employed for this purpose. Mtcox1 data does not conflict the differentiation into three separate species, but the ITS-2 data are problematic for this group of mites and suggests that a congener, *A. gracilis* Hughes 1957 is basal to the *A. siro* species complex (Webster *et al.*, 2004).

The grain mite has a cosmopolitan geographic distribution, and it has been collected on a number of commodities; it's the most important pest of storage premises and found more often on processed cereal products such as flour than cereals such as whole grain, or on hay in farm store (Griffts, 1964a). The mite is reported also on cheese, the deep litter of broiler house and in disused beehives. On the flour surface, it occurs to the depth of a few inches. Sinha and Mills (1968) reported *A. siro* as pest of *Penicillium* sp., and Griffiths *et al.* (1959) observed that when the mite and *T. putrescentiae* were maintained on grain contaminated with fungi of the genera *Aspergillus* and *Eurotium,* they preferred to feed on the species of *Eurotium*, such as *Eurotium amstelodami* (L. Mangin) Vuill., *Eurotium repens* (Corda) Sacc. *and Eurotium ruber* Thom *et* Church, which develop in grain of moisture content of 13–15%, instead than on *Aspergillus candidus* Link, *Aspergillus ochraceus* Wilhelm and *Aspergillus flavus* Link, which demand a higher moisture content. The grain mite develops on its preferred fungi. Moreover, *Aspergillus restrictus* G.Sm. and the mould *Sporendonema sebi* (Fr.) v. Arx affect adversely, the cultures of *A. siro*; they reduce its reproduction and induce a high mortality and retard the development (Solomon *et al.*, 1964).

The biological cycle develops through the egg, larval, nymphal, hypopus and adult stages. The hypopus is rare. Cunnington (1985) studied the biology of the grain mite at 5–30°C and 62.5–90% RH. He found that at and above 20°C and 80% RH, the mating and the oviposition occur soon after emergence, but at lower temperatures and RH, the egg-laying is progressively delayed from one to several days. The females need to mate repeatedly in order to achieve the maximum egg production, and the optimum conditions for which are 15°C and 90% RH, where the total output per female average 435 with a maximum of 858. The oviposition rates are highest at higher temperatures, the mean daily rate at 20°C and 25°C and 90% RH, rising to maximum levels of 28/29 eggs per female per day, on day 6. The oviposition follows defined patterns, favourable conditions producing rapid increases in the mean daily oviposition rate to high peak levels reached at an early stage in the oviposition period. Less favourable conditions result in reduced outputs and lower, more uniform rates of egg-laying. The mean oviposition period, varying with RH fell from 72–122 days at 5°C to 9–13 days at 30°C, and the mean incubation period from takes 42–70 days at 5°C to 3–4 days at 30°C. The egg viability increases with increasing RH but is little affected by temperature and unaffected by the age of the female at time of oviposition. The males tend to live longer than females at most conditions; longevity depending on humidity averaging 13–15 days at 30°C and 129–175 days at 5°C. Adult life for females averages 12–19 days at 30°C and 88–169 days at 5°C (Cunnington, 1985). Emekçi and Toros (1989) observed that the temperature range necessary for the mite development lies between 25°C and 31–32°C, with an RH between 62% and saturation. At 25°C and 90% RH, the egg duration is 4.3 days, the immature live 4.6 days, the adult female life is 25.5. days, the pre-oviposition period takes 1.1 days, the oviposition period takes 18.3 days, the mean fecundity is 315 eggs, each female lays 17.2 eggs/day, with an egg mortality of 14% (Emekçi and Toros, 1989). The eggs can tolerate several months at 0°C (Cunnington, 1965). Populations of the grain mite can respond to gradients of RH tending to congregate where the RH is between 80% and 85%; in these conditions associated with suitable temperature occurs the highest rate of reproduction (Solomon, 1962). The temperature of 28°C and RH of 85% are optimal for the population growth of *A. siro*, determining an Index of Suitability (I) of 13.6, according to a finite rate of increase (k) of 1.2 and to a doubling time (T_d) of 5.1 (Aspaly *et al.*, 2007).

Cunnington (1985) reported that the 'index of suitability' (I), calculated from egg number, viability and duration of the egg stage and oviposition period, indicates that the most favourable conditions for oviposition and hatching are 20–25°C and 80–90% RH. This index was developed by R.C. Fisher (1938) and is represented by the following relation $I = NV/L+T$, where N is the number of eggs laid, V is the eggs' viability (the inverse of eggs mortality), L is the duration of the oviposition period and T is the duration of egg stage. This index relates to oviposition and hatching, and the most favourable conditions are those that produce the highest values. It provides a utilitarian of eggs laying performance and survival and was used by Cunnington (1985) for determining the physical limits for complete development of *A. siro*, at 27 combinations of temperature and humidity (Colloff, 2009a).

Studies carried out in China showed that at 25°C and 85% RH, the grain mite develops one generation in about 15 days (egg in 3.5 days, larva in 2.5 days, protonymph in 2.5 days, tritonymph in 2.0 days and prematurity in 4.0 days). The females overwinter on stored products and become active in mid-June and start to lay the eggs 2–3 days after mating, and a female could lay 45–50 or 68–75 eggs when provided with flour or broken rice (Sun *et al.* 2002).

In the Czech Republic, Fejt and Ždárková (2001) studied the bionomics of *A. siro* at 18°C and 20°C and 70% and 80% RH, on wheat germs, linseed, rapeseed and sunflowers to acquire data for the optimization of biological control of stored food mites on oilseeds. The wheat germs were employed as a control substrate as they were considered optimal for development of *A. siro*. The development of eggs and larvae was longer than that of protonymphs and tritonymphs. Wheat germs resulted in a better substrate for the development of mites than oilseeds. Commonly, the lifespan of females as well as the oviposition period was shortest on wheat germs and longest on sunflower; also, at higher temperature, the female lifespan was shorter. The maximum number of eggs laid by a single female (356) was recorded on wheat germs, at 20°C and 80% RH. The development was significantly shorter at 20°C than at 18°C, regardless of humidity. There were no significant differences between the effects of the tested oilseeds (Fejt and Ždárková, 2001).

Szlendak and Kraszpulski (1991) have calculated that immature individuals of *A. siro* ingest a total of 0.1408 J of food energy, 14.1% of which is consumed by the larva, 27.1% by the protonymph and 58.5% by the deutonymph. The total ingestion is much greater (2.8169 J ± 0.0981 SE). Much energy, 0.7031 J ± 0.0231 SE, is used for egg production, which is concentrated in the first 20 days of adult life. Assimilation efficiency during this period varies from 63% to 71%. Most energy ingested between the 20th and 40th days of adult life is used in biomass production, and biomass increases significantly up to the 30th day. Afterwards, old individuals expend a large amount of assimilated energy on respiration due to the increased cost of maintenance (Szlendak and Kraszpulski, 1991).

Parkinson *et al.* (1991) compared the egg production of isolated pairs of *A. siro*, *Glycyphagus destructor* and *T. longior* feeding on a control diet of wheat germ and yeast with that on mycelial pellets from shake cultures of *Cladosporium cladosporiodes* (Fresen.) G.A. de Vries, *Aspergillus repens* (Corda) Saccardo, *Aspergillus ruber* Thom *et* Church and *Penicillium cyclopium* Westling, as well as spores of *P. cyclopium* and *A. repens*. The mites always produced fewer eggs on the fungal diets than on the ideal diet. *T. longior* usually did best of the three mite species on the fungal diets, and *G. destructor* worst. *A. ruber* proved the most suitable fungus for all mite species, while *C. cladosporiodes* was the least suitable. Spores were a less suitable diet than mycelial pellets from shake cultures, which were predominantly mycelium (Parkinson *et al.*,1991).

Puspitarini *et al.* (2021) studied the effect of physicochemical characteristics of stored products on *A. siro*. The host preference and biology of the mite were studied in three varieties of rice, two varieties of sticky rice, and three different types of cheese. They considered the differences in grain granulometry (whole, broken, and flour) as variation in physical characteristics of the stored products, and proportions of basic nutritional content as variation in chemical characteristics. The results indicated that when offered grain with different granulometry, *A. siro* selected broken grain as their host and oviposition site. Furthermore, the mite's host preferences were also influenced by the nutritional content variation of the stored products. The total development of *A. siro* lasted from 16.78 ± 0.13 to 20.60 ± 0.51 days, and its fecundity ranged from 17.65 ± 1.75 to 53.05 ± 0.09 eggs per female. Faster development and higher fecundity of the mite were generally found in a diet(s) with higher protein, fat, and water content. Moreover, *A. siro* is better adapted to flour products, indicating the influence of particle size on the development and reproduction of the mite. Therefore, physicochemical characteristics of the stored products, strongly affected the host selection and biology of *A. siro*, which in turn determines the degree of product susceptibility upon mite infestation (Puspitarini *et al.*, 2021).

The natural enemies include pathogens and predators. The first number *A. restrictus* G.Sm. (Trichocomaceae) and *Wallemia sebi* (Fr.) Arx (Wallemicaceae). The list of predator includes various species, such as *Amblyseius arutanjani*, *Amblyseius barkeri* (Hughes), *Amblyseius herbarius* Wainstein, *Androlaelaps casalis* (Berlese), *Cheyletus eruditus* (Schrank), *Cheyletus malaccensis* (Oudemans), *Hypoaspis aculeifer* (Canestrini), *Hypoaspis miles* (Berlese), *Neoseiulus cucumeris* (Oudemans) and *Pergamasus diversus* (Halbert) (CABI, 2020).

Symptomatology and damage

The grain mite occurs on a number of commodities and is the most important pest of storage premises, and it is found more often on processed cereal products such as flour than cereals such as whole grain or on hay in farm store (Griffiths, 1964a). In addition to above-reported facilities, the mite has been collected also on tea, spices, dried fruit and pulses. On whole cereal grain and on other seeds, the mite infests severely the germ causing losses in germination viability. *A. siro* has been found also on cheese, the deep litter of broiler house and in disused beehives. On the flour surface, it occurs to the depth of a few inches. The moisture content of grain regulates the density of mite population and the grain damage. In particular, damaged grain with a water content of about 14% or over, had the germ eaten away, while grain containing 13% water or less is unaffected. Few mites are found in very damp grain as the germ has already been consumed, and the mites had left the site because of overcrowding and food shortage (Solomon, 1946).

Heavy mite populations are characterized by typical odours produced by mites' lipid secretions. This musty or minty smell and the large numbers of live and dead mite bodies is a major factor in rejecting foodstuffs from sale. These effects on the quality and hygienic condition of stored foodstuffs are discussed by Solomon (1946, 1962), Pulpan and Verner (1965), Smith (1969), Griffiths et al. (1976) and Sinha (1984).

In England, infestation of *A. siro* on barley was retained responsible for scouring in cattle (Hughes, 1976). Handling of infested products can cause a contact dermatitis, respiratory allergy and asthma. The dermatitis experienced by food handlers in contact with *A. siro* is commonly known as **'grocer's itch'**. Grain mites that contaminate grains, flour and animal feedstuffs, produce allergens in the dust produced, and also transfer pathogenic microorganisms. Foodstuffs acquire a sickly sweet smell and an unpalatable taste (Stored Grain Pest, 1955). The mites sometimes bite humans, which can cause an allergic reaction known as 'Baker's itch' (Hill, 2003; Rietschel and Fowler, 2008). Halliday (2003) discussed the health and safety issues related to mites in stored grain and highlighted the importance among the various allergenic species of *A. siro*.

When fed infested feeds, animals show reduced feed intake, diarrhoea, inflammation of the small intestine and impaired growth. Pigs have their live weight gain, feed-to-gain ratio and nitrogen retention markedly reduced by infested feeds (Braude *et al.*, 1980). The presence of mites in cereal-based foods destined to human consumption, and the possible consequences of infestation are discussed by Thind and Clarke (2001). Eriksson *et al.* (1999) identified and characterized two allergens from *A. siro*, homologous with fatty acid-binding proteins. The contamination of the grain by allergens and the transferring of pathogenous microorganisms were discussed by Krizkova-Kudlikova *et al.* (2007).

Farm animals given infested feeds eat less, may sicken and their growth is affected. Grain mites are intentionally inoculated into 'Mimolette cheese' to improve the flavour. When used for this purpose, they may be referred to as 'cheese mites'. Cheese mite species inoculated on Mimolette cheese were identified through cryogenic scanning electron microscopy (Melnyk *et al.*, 2010).

Kopecky *et al.* (2014) reported that bacteria of the genus *Bartonella* are carried by haematophagous mites, ticks, fleas and flies, and attack the erythrocytes of mammals. They described a *Bartonella*-like clade, a distinct group related to Bartonellaceae, in stored-product mites, and a predatory mite *C. eruditus* based on the analysis of cloned 16S rRNA gene sequences. By using the clade-specific primers, closely related *Bartonella*-like 16S rRNA sequences were amplified from both laboratory colonies and field strains of three synanthropic acarid mite – *A. siro*, *L. destructor* and *T. putrescentiae* – and a predatory mite. Altogether, sequences of *Bartonella*-like bacteria were found in 11 strains, they were not detected in *Dermatophagoides farinae* Hughes and *Dermatophagoides pteronyssinus* (Trouessart) and two strains of *L. destructor*. All obtained sequences formed a separate cluster branching as a sister group to Bartonellaceae, and related to other separate clusters comprising uncultured bacterial clones from human skin and hemipteran insects, such as *Nysius plebeius* Distant and *Nysius* sp. The classification of sequences into operational taxonomic units (OTUs) showed a difference between *A. siro* and *T. putrescentiae*, suggesting that the *Bartonella*-like bacteria are different in these two species. However, species-specific sequences in separate OTUs were observed also for *C. eruditus*. These authors discussed the possible symbiotic interactions between *Bartonella*-like bacteria and their mite hosts (Kopecky *et al.*, 2014).

Control

Commonly, *A. siro* and other mites infesting stored food, are controlled by application of preventive measures (aeration of bulk grain with cool dry air, etc.), by the fumigation of insecticides and acaricides as sprays or powders and by biological control. Regardless of the means applied, the monitoring of the mite populations is prior. We report here the main technical possibilities reported in literature, taking into account that every choice must be oriented by the knowledge of the risk of toxic contamination of stored food, and by the European Union and US regulations which demand zero or near-zero tolerance levels for *A. siro* infestations in stored products, being a major pest of stored foods, and responsible of the transmission of diseases and allergy.

PREVENTIVE CONTROL. Preventive measures of control include keeping stocks in storage cool and dry, ensure good ventilation and keep storage areas clean and dry. The use of controlled atmosphere is suggested for infestation in product, while the heat treatment is suggested for infestation in building.

MONITORING. The risk of contamination of the grain by allergens, and transferring of pathogenous microorganisms by *A. siro*, suggests the rapid detection of contamination to intercept an early grain infestation. The topic has been studied by Krizkova-Kudlikova *et al.* (2007) who, under laboratory conditions, infested grain samples of various sizes with different amounts of eggs, adults and faeces of *A. siro* and analysed the samples using three methods, the enzyme-linked immunosorbent assay (ELISA), using the anti-*A. siro* polyclonal antibody (immunochemical method), the extraction in Berlese-Tullgren funnels, sieved and the conventional method processed by filth-flotation. The results of this study highlighted that adults and juveniles of the grain mite could be monitored using the three methods and ELISA with detection limits in the range from 221 to 1157 mites/kg grain. However, the eggs may be detected by filth-flotation only; the detection limit is 1950 eggs/kg grain. The faeces of *A. siro* are detectable only by the ELISA test. This last method allows the detection of the faeces with the minimal threshold level of 1.04 microg faeces/g grain, and it's possible to locate less than one metabolically active mite per gram of grain.

In conclusion, the *A. siro* monitoring in grain can be possible by combining different methods only. However, the situation is complicated by European Union and US administrations dictate zero or near-zero tolerance level for mite presence in stored products. This constraint is difficult to overcome, because every monitoring method is limited by its detection limit; thus, it is hard to reliably detect infestation levels lower than obtained detection limits (Krizkova-Kudlikova *et al.*, 2007).

The rapid immunodiagnostic tests for the detection of storage mite infestations in cereals and cereal products were investigated also by Dunn *et al.* (2008). The study was articulated into three phases. The first phase, the proof of concept, concerned the production of a species-specific enzyme-linked immunoassay (ELISA) for the flour mite *A. siro*. The second phase, species- and genera-specific ELISAs, was developed for a range of mites injurious to stored food, such as *L. destructor* (Schrank), *G. domesticus* (De Geer), *T. longior* (Gervais), mites of the genera *Tyrophagus* and *Glycyphagus* genera and all storage mites. All tests were demonstrably specific to target species or genera, with no cross-reactions observed to other storage pest contaminants or cereals. The validation phase concerned a comparative assessment of the species-specific *A. siro*, and the genus-specific *Tyrophagus* ELISAs with the flotation technique using laboratory and field samples. Both ELISAs were quantitative (0–30 mites per 10 g wheat) and produced good comparative data with the flotation technique (*A. siro* r (2) = 0.91, *Tyrophagus* spp. r (2) = 0.99) (Dunn *et al.*, 2008).

The effectiveness of a mite trap to monitor *A. siro*, *L. destructor* and *T. longior* was discussed by Wakefield and Dunn (2005). There is only one type of trap on the market in the UK for storage mites, namely the BT mite trap, or monitor. Laboratory studies have shown that under laboratory conditions at 20°C and 65% RH, the BT trap is effective at detecting mites for at least 10 days, for all three mites *L. destructor*, *T. longior* and *A. siro*. Further tests highlighted that all three species reach a trap at a distance of approximately 80 cm in a 24 h period. In experiments using 100 mites of each species, and regardless of either temperature (15°C or 20°C) or RH (65% or 80% RH), the most abundant species in the traps is *T. longior*, followed by *A. siro* then *L. destructor*. Trap catches are highest at 20°C and 65% RH. Temperature produces a greater effect on mite numbers than humidity. Tests using different densities of each mite species show that the number of *L. destructor* found in/on the trap is significantly reduced when either of the other two species is dominant. It would appear that there is an interaction between *L. destructor* and the other two mite species that affects relative numbers found within the trap (Wakefield and Dunn, 2005).

Moreover, a practical method of monitoring consists of sampling the foodstuff, and using sieves of appropriate mesh to separate the mites from the food material. A 20X hand lens allows to observe the presence of mites, and low-power stereoscopic microscopy can be useful (Krizkova-Kudlikova *et al.*, 2007). Another method provides monitoring the presence of grain mites on small quantities, which can be proved by wiping a little bit of flour together and smoothing it. If the surface, after a while, gets 'rough', then this indicates the presence of mites.

A simulation model of *A. siro* was developed to predict the maximum rate of its population growth and survival by interpolating daily rates of growth of its population derived from published studies. The model simulates the population dynamics at −10 to 40°C and 20–90% RH, without resource limitation or natural enemies. The model's predictions were in reasonable agreement with naturally occurring mite populations in bulk-stored wheat in granaries, in Manitoba, England and Czechoslovakia (Kawamoto *et al.*, 1991).

PHYSICAL CONTROL. Navarro *et al.* (1985) evaluated the mortality of newly emerged adults of *A. siro* by exposing them to 2%, 4%, 6%, 10% and 21% O_2, in N_2, to 10%, 20%, 30% and 40% CO_2 in 21% O_2, and with the balance of the atmosphere made up of N_2. Tests were conducted at 75% RH and 15°C or 26°C, in a specially designed apparatus. Mean duration of mite survival in air was 11 and 7 days at 15°C and 26°C, respectively. Exposure to 2% O_2 was required to obtain 100% mortality, with 72 h at 15°C; the same result was obtained within 120 h at 26°C by exposure to 10% O_2. At 15°C, 30% CO_2 was required for 96 h to achieve 100% mortality; at 26°C, 20% CO_2 or above produced the same result within 72 h (Navarro *et al.*, 1985).

Boczek and Davis (1985) compared fecundity, longevity and survival to adulthood of *A. siro* at constant and alternating temperatures. They found that both fecundity and longevity are affected by alternating temperatures, but the effect of the frequency of alternations is not significant. Significantly higher fecundity and longer lifespans occur at constant temperatures of 14°C and 21°C than at 28°C or alternating temperature regimes. Alternating temperature regimes, in comparison to constant regimes, did not significantly change the number of males and females surviving to adulthood, or the sex ratio. However, the lowest number of adults surviving is recorded, when temperatures are changed every 12 h, and the highest at a constant temperature of 14°C. Effects of temperature and humidity were investigated by Cunnington also (1965).

Diatomaceous earth can be used as a grain protectant, as part of an integrated strategy based on cooling and drying (Cook and Armitage, 2002; Cook and Armitage 2003; Collins and Cook 2006). Cook and Armitage (2000) evaluated the effectiveness of 'Dryacide', a diatomaceous earth, in association with low temperature and high humidity. Under laboratory conditions they treated 25 kg of wheat, at 16% and 17% moisture content (MC) with 'Dryacide' at 3 and 5 g kg^{-1} and stored in metal bins. The bins were then infested with mixed populations of mites (*A. siro* and *L. destructor*) and insects (*Sitophilus granarius* (Linnaeus) and *Oryzaephilus surinamensis* (Linnaeus)) at approximately 100 and 10 per kg, respectively. The grain was held at a constant temperature of 15°C and RH of 75% and 80%, chosen to simulate bulk surface conditions in UK stores, and monitored for 42 weeks. At 16% MC, the mite and insect populations were suppressed within 2 and 13 weeks, respectively, at both doses. At 17% MC, the mites took 5 weeks to die out, and F1 adults of *S granarius* emerging within 22 weeks were controlled after a further 5 weeks. After 23 weeks, the bins were re-infested with mites, and these were completely suppressed after a further 3 weeks. There was no decline in efficacy throughout the experiment (Cook and Armitage, 2000). Integrated pest management for stored grain in the UK, incorporating diatomaceous earths to prevent surface infestations of insects and mites, has been proposed by Cook and Armitage (2002).

The aeration may be a promising alternative to the use of pesticides, but its effectiveness against mite pests is neither fully understood nor optimized. For this reason, the productivity of three species of storage mites, *A. siro*, *L. destructor* and *T. longior*, was studied in a laboratory-based experiment, at four combinations of temperature and humidity (10°C and 70% RH, 10°C and 80% RH, 20°C and 70% RH, 20°C and 80% RH), with and without an airflow (at 10 m^3/h/tonne, equalling 2.5 1/s/tonne, in tubes containing 15g of grain). This is the first time that a study has examined the three principal components of aeration separately from each other. The effect of these factors was different for each species. For *A. siro*, temperature was the most important factor, while airflow and humidity were of similar, but lesser importance. For *T. longior*, temperature was more important than humidity, while the reverse was true for *L. destructor*. For these two species, airflow was the least important factor. The airflow decreased the productivity of *L. destructor* and *T. longior but* increased the productivity of *A. siro*. This increase in productivity confirms that, in practice, prevention of mite infestations, in particular *A. siro*, will require storage of grain at low temperature, relative humidity and MC (Thind and Dunn, 2002).

Hubert *et al.* (2013) studied in Czech Republic, the rate of population increase of *A. siro*, *L. destructor* and *T. putrescentiae,* on various types of barley, and at various combinations of temperature and humidity. The mites were added into the chambers and incubated for 21 days on seven different kinds of barley, coming from four sites, including six cultivars and one mixture. The population increase of all species was higher on the mixture than on any other cultivar, except for Sebastian and Calgary. The increase of mites was studied at constant temperatures ranging from 5–35°C and RH, ranging from 50% to 90%. Positive rate of increase was found above 70% RH for all species. The optimal humidity was at 85% RH for *A. siro* and *L. destructor* and at 90% RH for *T. putrescentiae*. As concerns the temperature, positive rate of increase was found at temperatures higher than 10°C, 15°C and 20°C for *A. siro*, *L. destructor* and *T. putrescentiae*, respectively. The temperature optima were at 23°C, 25°C and 30°C for *A. siro*, *L. destructor* and *T. putrescentiae*, respectively. Model estimated on laboratory data was then fitted to temperature and humidity records from August to November in the Czech grain store. Estimated population rate of increase was rarely positive: for *A. siro* it was for 24%, for *L. destructor* for only 1% and for *T. putrescentiae* for only 7% days of the study period. In conclusion, in the climatic conditions of the Czech Republic, the population increase of three mite pests, is negligible during autumn and winter (Hubert *et al.*, 2013).

Szlendak *et al.* (1985) examined the effect of gamma-irradiated males on egg production in *A. siro*. They observed mating of unirradiated *A. siro* females with normal and gamma-irradiated males and found that females laid eggs throughout their lifespan if they were mated at least once a week with normal males. Conversely, females paired with treated males contained spermatophores but did not lay any eggs. Previously mated fertile females of A. siro subsequently mated with irradiated, sterilized males ceased egg-laying during the next 1–2 weeks. There was no significant difference when compared to the untreated control in the number of eggs laid, when females were mated for 1 week with irradiated males and then with normal males. Weekly alternation of mating with irradiated and normal males reduced the number of eggs laid by the females and their longevity. Because this mite species mates many times during its lifespan, it is doubtful that it could be eradicated solely by the introduction of sterile males into a natural population. However, the presence of such males would decrease the number of eggs laid by normal females, and it would lead to the decrease in the number of offspring in the next generation (Szlendak *et al.*, 1985).

BIOLOGICAL CONTROL. Different predators attack the grain mite, such as the blattisociid mite *Blattisocius dentriticus* (Berlese) and the cheyletids *C. malaccensis* and *C. eruditus*. The biological control using the predatory mite *C. eruditus* is well known from time (Gause *et al.*, 1936), and it has been discussed by various authors. A biological model of predation was developed by Burnett (1977), using granular food held in closely packed screen trays to propagate the grain mite, *A. siro*, and two of its predators, *B. dentriticus* and *C. eruditus*. Both predators limited prey abundance. Cannibalism among predators, particularly *C. eruditus*, was an important factor in ensuring the survival of the prey and predator populations. *C. eruditus* eliminated *B. dentriticus* when the two species were propagated in the same experimental universe. Cyclicity and dispersion of the interacting species appeared to result more from the initial age structure, and from dispersion of the prey than from predator attack. A population model was used to assess the relative importance of the population components of *A. siro* and *C. eruditus* in the simplified predator-prey interactions (Burnett, 1977).

Despite some practical limitation, a commercial product of *C. eruditus* has been proposed in Europe, to control the pest in grain bins. The use of the predator on stored food has certain limits, provided for the lack of selectivity and persistence of the a.i. used against the harmful insects (Norris, 1958; Pulpán and Verner, 1965). So, White and Sinha (1990) reported that the high toxicity of chlorpyrifos-methyl to predatory mites results in increased pest mite populations. Ždárková and Horák (1990) suggested that the elimination of pest mites in stored food may be obtained by the releases of the predator, before the placement of the grain. This constraint does not allow an easy use of the predator and opens a scenario of uncertainty, which has forced the various specialists to search integrative and/or alternative methods to the use of the beneficial. The release of *C. eruditus* has been examined by Pekar and Ždárková (2004). They examined the development of a stage-structured within-patch model, describing predator-prey interactions of *C. eruditus* and *A. siro* on stored grain, with a parameterization of the model based on data taken from the literature. According to strong variance in these data, they used statistical methods to extrapolate the parameters and compared the results of the simulations with the experimental control of *A. siro* by *C. eruditus* on wheat. They found that all life-history studies of *A. siro* yielded overestimated parameters, leading to an unnaturally rapid population growth of *A. siro* on grain, according to the use of high-quality food (wheat germs), which considerably increased fecundity. A new model with reduced fecundity was found to correspond well with the observed dynamics in a granary. The model of *C. eruditus*, including parameters from different sources, produced dynamics quite similar to the observed one. A simulation of biological control under laboratory conditions, on 45 g of wheat, showed that *C. eruditus* is able to control *A. siro* within the observed period. In contrast, simulation of the control in a granary showed a marked difference, in comparison with the observed dynamics. The discrepancy is attributed to the effects of spatial factors, such as immigration. The predator-prey model developed appeared valid for the prediction of the local (within-patch) population dynamics of *A. siro* and *C. eruditus* on wheat grain, at 20°C and 80% RH (Pekar and Ždárková, 2004).

Lukas *et al.* (2007) reported that among the *Cheyletus* species, occurring in grain stores in Central Europe, only *C. eruditus* is employed for the biocontrol of stored-pest arthropods. Taking into account that field data on the differential natural performance (frequency, abundance and predator-to-prey density dependence) of various *Cheyletus* spp. are missing, they investigated the relations between the population densities of the acarid mites and *Cheyletus* spp., in 147 grain stores in the Czech Republic. They extracted more than 1,000,000 individual pest mites, and 40,000 predatory mites, highlighting that 29% of samples did not contain mites, 41% contained only pest mites, 4% only predatory mites, and in 26% both groups were present simultaneously. The most abundant predatory mite was *C. eruditus* (79%), followed by the 'minor' species *Cheyletus aversor* (Cheyav) (10%), *Cheyletus trouessarti* (Oudemans) (9%) and *C. malaccensis* (2%). It has thus been ascertained a significant positive correlation between the occurrence and population density of the predatory and pest mites, except for *C. malaccensis*. *C. eruditus* resulted in the predatory species with the highest natural performance in the field, indicating that it was the most pre-adapted species for biocontrol in central European grain stores (Lukas *et al.*, 2007).

Huber and Pekar (2009) tried to enhance the predator activity of *C. malaccensis*, with the combination with the antifeedant bean flour. They investigated the suppression of three pest mites, by combination of the predator *C. malaccensis* and bean flour (*Phaseolus vulgaris* Linnaeus), which is an antifeedant. Wheat grain (100g) was contaminated with 100 *A. siro*, *Aleuroglyphus ovatus* (Troupeau) or *T. putrescentiae*. The bean flour was added in concentrations of 0,1, 5 and 10 g kg^{-1} (wt/wt) and/or the predator with the initial predator/prey ratios 0, 1/50, 1/25 and 1/10. The experiment was carried out under the abiotic conditions optimal for pest development. After 21 days, pest and predator numbers were recorded. Application of flour significantly reduced populations of *A. siro* slightly reduced the populations of *T. putrescentiae*, and the efficiency increased with increasing concentration. *A. ovatus* was not affected by bean flour. The populations of all mite species were successfully reduced by the sole addition of *C. malaccensis*, with a higher efficiency at higher *Cheyletus* ratios. The additive effect of the predator and flour was mainly apparent in the reduction of *T. putrescentiae*. The population of *C. malaccensis* was not affected by the presence of bean flour. It is therefore recommended to use bean flour for the reduction of *A. siro*, a combination of bean flour and *Cheyletus* for the reduction of *T. putrescentiae*, and only *Cheyletus* for the reduction of *A. ovatus* (Huber and Pekar, 2009).

Pekar and Hubert (2008) reported that in stored grain *Cheyletus* spp. may be used to control *A. siro*. The efficiency of control depends on many factors, particularly ambient temperature. At this aim, they investigated the effects of temperature and initial prey density on the prey-predator system under laboratory conditions. Ratio-response models were fitted to estimate the efficiency of control for three temperatures. Results showed that at 15°C, a 90% reduction of *A. siro* is achieved by releasing 9 *C. malaccensis* individuals into a population of 100 *A. siro* individuals in 1 kg of grain. At 20°C, 90% reduction requires 7 *C. malaccensis* individuals, and at 25°C requires 3 *C. malaccensis* individuals. Without the predator, the intrinsic rates of increase of *A. siro* populations increases with temperature and is highest for an initial density of 100 individuals, revealing some form of positive interaction among *A. siro* individuals during food processing. The intrinsic rates of increase of *C. malaccensis* populations also increase with temperature and decrease with increasing density of the predator, presumably, as a result of interference competition among predators. At 15°C, the rate of increase for *A. siro* is higher than that for *C. malaccensis*, while at higher temperatures, it is the other way around. Lower developmental thresholds are 10°C for *A. siro* and 13.6°C for *C. malaccensis*. In order to find, when an artificial release of *C. malaccensis* is most efficient, were simulated, a population increase of *A. siro* using temperature records from one grain store. In Central Europe, this type of biological control can be efficient, only when the predator is released at the beginning of storage, that is, in September and October (Pekar and Hubert, 2008).

CHEMICAL CONTROL. Organochlorides and organophosphates are employed from time in the chemical control of *A. siro*. So researches carried out by Wilkin (1975) reported about 50 years ago that farm-stored barley infested with *A. siro*, *A. farris*, *G. destructor* (Schrank) and *T. longior* and admixed with lindane + malathion or pirimiphos-methyl dusts, controlled in both treatments the infestation, and protected the grain for 3 months. Pirimiphos-methyl was the more persistent of the two treatments, and despite the barley having an MC of more than 18%, virtually no breakdown of this compound was detected. In addition, the effects on mites of moving infested barley, through farm conveying systems were examined. Trials at two sites showed that this method gave between 81% and 96% kill, but offered no long-term protection (Wilkin, 1975). Usually, in the control of *A. siro*, three organophosphates, pirimiphos-methyl, chlorpyrifos-methyl and chlorpyrifos, focus the attention. Unfortunately, these a.i. may affect negatively, the populations of the natural enemies of the grain mite. The problem has been investigated by Ždárková (1994) in the Czech Republic, which tested on a laboratory strain of *C. eruditus*, a cheyletid mite, employed for the biological control of stored food mites, and on *A. siro* and *T. putrescentiae*. Biological control is often preceded by a chemical treatment with organophosphates and is fundamental to know the persistence of acaricides on the predators. Results of the researches highlighted that chlorpyrifos-methyl was the most toxic organophosphate on *C. eruditus*. The effectiveness of pirimiphos-methyl and chlorpyrifos was approximately equal and was three times lower than the effectiveness of chlorpyrifos-methyl. The organophosphates were nearly equally effective on both pest mites, and *A. siro* resulted in slightly more susceptible than *T. putrescentiae*. The results suggested the use of pirimiphos-methyl or chlorpyrifos rather than chlorpyrifos-methyl, for protection of empty stores or stored grain against resilient populations of stored food mites (Ždárková, 1994). Szlendak *et al.* (2000) detected pirimiphos-methyl resistance in *A. siro* and *A. farris* by impregnated paper bioassay and esterase activity assays.

Currently, fumigants or contact insecticides are employed. Among the first, aluminium phosphide and magnesium phosphide are largely used after several days that grain is stored or mite infestation is detected. Fumigants in bags are preferred to prevent phosphine residues. Sometimes, in appropriate silos, the gas may circulate inside the mass of grain, by special fan that produces even gas distribution throughout the commodities. On cereals are registered pyrethrum extracts, pyrethroids or organophosphates. In China, it was observed that the grain mite was effectively controlled by using aluminium phosphite (Lin *et al.*, 1992). Chlorpyrifos-methyl was more toxic than pirimiphos-methyl to both mites, *A. siro* and *C. aversor* (Tian and Sun, 1991). An acaricide, Mie Man Ling, was also proved to be an effective acaricide for the mite control (Chen *et al.*, 2002). Weaver and Petroff (2004) reported a list of 'Insecticides labelled for use as empty bin treatments', including cyfluthrin, chlorpyrifos-methyl, diatomaceous earth, malathion, chlorpyrifos-methyl + cyfluthrin, chloropicrin, methyl bromide and phosphine. The 'Liquid insecticides labelled for use as grain protectants' number, chlorpyrifos-methyl, malathion, DDVP, methoprene, chlorpyrifos-methyl +cyfluthrin and pyrethrins. Moreover, it's possible to use 'Dust Insecticides Labelled for Use as Grain Protectants'. Grain protectants are insecticides applied directly onto grain, going into the storage or already in storage. Grain protectants do not kill insects inside the kernels. The insecticides labelled as protectants include malathion and diatomaceous earth. Moreover, Weaver and Petroff (2004) reported that pesticides must be applied legally, complying with all label directions, and precautions on the pesticide container, and any supplemental labelling and rules of state and federal pesticide regulatory agencies. State rules and regulations and special pesticide use allowances may vary from state to state: contact your State Department of Agriculture for the rules, regulations and allowances applicable in your state and locality. Regardless of the action of the individual a.i., their possible choice must be related to the rules of each country.

Conyers and Bell (2003) carried out laboratory investigations on the effects of modified atmospheres (MA) on the eggs of *A. farris*, *A. siro*, *L. destructor* and *T. longior*, variously collected in many facilities, including grain and cheese stores. They employed three low oxygen (O_2) MA mixtures based on carbon dioxide (CO_2), nitrogen (N_2) or simulated burner gas (0.5% or 2% O_2, 10% CO_2, balance N_2), plus 60% CO_2 in air (8% O_2). The mites were exposed at 15°C and 80% RH, a combination of conditions occurring at the surface of stored grain during the autumn and promoting the mite population development. The exposure periods required to prevent egg hatch, for each species in every mixture are given. *T. longior* was the most tolerant species, followed by *A. siro* and *A. farris*, with *L. destructor* the most susceptible. Burner gas was the most effective mixture overall, with 0.5% O_2, but with an increase in the O_2 level to 2% for all the mixtures, CO_2 became the more effective control agent. With 60% CO_2 in air, some loss of efficacy was observed against the three most tolerant species, and even more so for *L. destructor*. Sublethal exposures to MA for at least 4 days in *L. destructor*, 6 days in *A. farris* and *A. siro* and 8 days for *T. longior* caused a delay in egg hatch. *A. siro* is killed by a 3-day exposure to 100% carbon dioxide (Conyers and Bell, 2003).

Collins and Cook (2006) assessed the efficacy of Silico-sec® and Diasecticide™, applied as dry dust and slurry applications to wooden surfaces in the laboratory, against adult *S. granarius*, and *L. destructor* and against larvae of *Ephestia kuehniella* (Zeller).

Legume proteins showed an insecticidal activity against stored-product pests. Grain enriched by bean (*Phaseolus vulgaris* Linnaeus) flour inhibits the growth of the mites. Hubert *et al.* (2006) tested the toxicity of bean flour to storage mites under optimal conditions for their population growth (i.e. rearing diet, temperature 25°C and humidity optimum 85% RH). Bean flour was added to the diet in one of eight concentrations (0%, 0.01%, 0.1%, 0.5%, 1%, 2.5%, 5%, 10%). The population growth of *T. putrescentiae*, *A. siro* and *A. ovatus* (Troupeau), initiating from a density of 50 mites per 0.2 g of diet was recorded for 21 days. The enrichment of grain with bean flour suppressed the population growth of the tested species. These differed in their sensitivity to bean flour. Population growth was decreased to 50%, in comparison to the control (rC50) by the bean flour concentration of 0.02% in *T. putrescentiae*, 0.04% in *A. siro*, and by 4.87% in *A. ovatus*. The concentration of 5% bean flour in diets kept populations of *A. siro* and *T. putrescentiae* at the initial level. The results suggest to apply the bean flour in the integrated control of stored-product mites (Hubert *et al.*, 2006).

Kavallieratos *et al.* (2018) evaluated three zeolite formulations, as stored-wheat protectants regarding the management of larvae, nymphs and adults of *T. putrescentiae* and *A. siro* at different exposure intervals (1, 2, and 5 days) and doses (200, 500 and 1000 ppm). The zeolite formulations tested resulted effective in the control of both mites. The mortality of all stages of both species was considerably affected by the dose and exposure intervals. They found that higher doses and longer exposure times result in higher mortality. Nevertheless, 100% mortality was not recorded on any of the life stages and formulations tested, even in the case of the highest dose and the longest exposure interval. All zeolite formulations provided the same efficacy level, which was similar for immature stages and adults of both mite species. The results suggested that longer exposures of the zeolite formulations are required, to increase the efficacy on the control of *T. putrescentiae* and *A. siro*.

Despite the above-mentioned reports, there is no common protocol for the chemical control of this mite, due to the variable regulations around (de-) registration of pesticides. This constraint suggested for the moment does not include any specific chemical control recommendations. For further information, we suggest you to visit the following resources, such as 'EU pesticides database' (http://ec.europa.eu/food/plant/pesticides/eu-pesticides-database/), the 'PAN pesticide database' (www.pesticideinfo.org) and your national pesticide guide.

INTEGRATED PEST MANAGEMENT. This strategy is commonly based on the integration of different means, free of reductive impact on the beneficial organisms. So, in the UK, the integrated pest management (IPM) strategy for stored grain provides the incorporation of diatomaceous earths to prevent surface infestations of insects and mites (Cook and Armitage, 2002), as part of an integrated strategy based on cooling and drying (Cook and Armitage, 2003; Collins and Cook, 2006). Researches in this sense were carried out by Cook and Armitage (2000), which studied the effectiveness of 'Dryacide', a diatomaceous earth, in association with low temperature and high humidity. Application of bean flour in the integrated control of stored-product mites was suggested by Hubert *et al.* (2006).

Selected websites

https://www.google.co.il/search?q=Acarus+siro&espv=2&biw=821&bih=527&tbm=isch&tbo=u&source=univ&sa=X&ved=0ahUKEwie-sHm4v7JAhWB-2BQKHbrmBsYQsAQIIQ

https://en.wikipedia.org/wiki/Mimolette

https://www.cabi.org/isc/datasheet/2522

The Genus *Tyrolichus* Oudemans

Diagnostic characteristics

This genus includes a single species, *Tyrolichus casei* Oudemans. Following Hughes (1976) the morphological characters are as those of the genus *Tyrophagus* except that the dorsal setae *c1* is the only short dorsal seta of the hysterosoma, and setae d_2 are more than twice as long as c_1. On the tarsi, the dorsal terminal seta *e* is a short thick spine, and all five ventral terminal setae (*p*, *q*, *s*, *u*, *v*) are thickened to form spines of equal size.

According to Klimov *et al.* (2016), the following characters complete the genus identification. The setae *ve* are located near the anterior lateral corners of the prodorsal sclerite, they are barbed, about 40–50% of the length of setae *vi*. The setae *sci* are longer than the setae *sce*. All hysterosomal setae, including c_1, are longer than the distance to the next posterior seta. The setae c_1 are barbed, set distinctly anterior to setae d_1. The genu I has the solenidia σ" and σ' subequal. The tarsi lack of dorsal ridge. All paraproctal setae (ad_{1-3} and ps_{1-3} surrounding the anus) of female are longer than the distance between them.

Economic importance

This genus numbers only the species *T. casei* Oudemans, a cosmopolitan species, collected in stored food on various commodities (Hughes, 1976), but with the exception of cheese, which can be seriously damaged, its role in foodstuffs is secondary compared to that of other acarid mites. The mite is known to cause human dermatitis (Henschel, 1929).

Tyrolichus casei Oudemans (Fig. 28.3)

Common name

Cheese mite

Diagnostic characteristics

FEMALE. The idiosoma is 500–700 µm long. General features as for the male. The anal opening is removed from the end of the body, and the opening to the bursa copulatrix is terminal, and a narrow tube connects this with a saccular receptaculum seminis (Hughes, 1976).

MALE. The idiosoma is 450–550 µm long, with well-tanned legs and gnathosoma. The propodosomal shield is almost square in outline and faintly pitted. The cuticle is smooth, and the supracoxal seta is drawn out into a point at its distal extremity, it is expanded at the base and its lateral edges are barbed, the barbs arising at almost an acute angle.

The hysterosomal setae are arranged as in *T. putrescentiae* (Schrank), the longer ones sparsely pectinate. The setae c_1 are short, and long from 5–7% of the idiosoma; the setae d_1 are two to three times longer than c_1, and d_2 are four to six times longer than c_1. The other setae are also long and arranged in the characteristic fan-shaped 'train'.

The legs are stout and ornamented with a faint reticulated pattern. The solenidion ω_1 is almost cylindrical, slightly expanded in the middle, and arises from the same chitinous depression as the famulus; seta *e* is a stout conspicuous spine at the distal end of all the tarsi, and five spines encircle the base of the claw on the ventral side. On tarsus IV, the suckers arise from the median part of the tarsus. The support of the aedeagus is curved inwards, and the penis itself is almost straight and tapering (Hughes, 1976).

LARVA. The setae d_1 are almost five times longer than *c1*. The coxal rods are present (Hughes, 1976).

Geographical distribution

Cosmopolitan. It's reported from various countries such as Australia, Azerbaijan, Canada, China (Fan *et al.*, 2010a), Czech Republic, England, Germany (Rack, 1968a), the Netherlands (Siepel *et al.*, 2016) India (Nangia, 1986), Iran (Ardeshir *et al.*, 2006), Italy (Bernini *et al.*, 1995), Poland (Wasylik, 1963, 1973), Primorskiy Kray (Tareev, 1970), Spain (Suñer *et al.*, 1985), Taiwan (Tseng, 1979), United States, Ukraine, western and eastern of Russia (Oudemans, 1910; Zakhvatkin, 1959; Hughes, 1976; Dudinska, 2003; Klimov *et al.*, 2016; Akimov and Oksentiuk, 2018; Oksentiuk *et al.*, 2022).

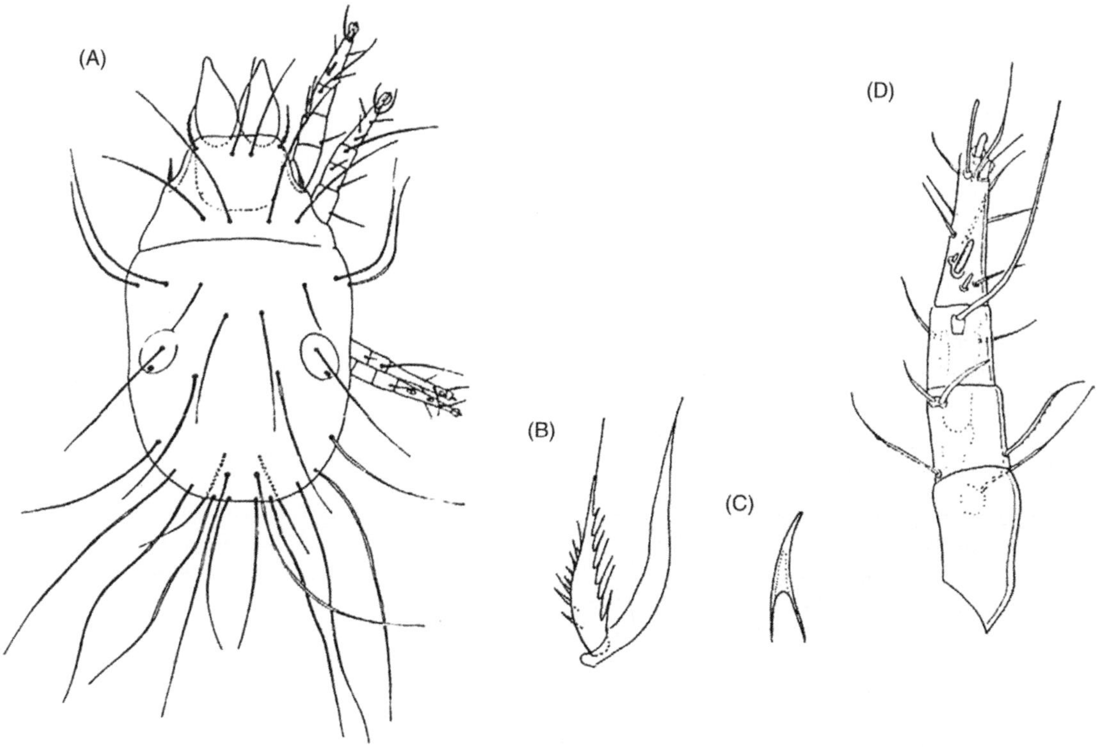

Fig. 28.3. *Tyrolichus casei* Oudemans. Male. Dorsal view (A); supracoxal seta (B); aedeagus (C); leg I (D) (from Hughes, 1976, licensed under the Open Government Licence v3.0).

Commodities, facilities and animal hosts

The mite is known on various commodities (barley, brans, cereals, cheese, Chinese medicine, condiment, cottonseed, dog meal, ergot of rye, flours, grain, ham, herb, honeybee (*Apis*) hives, honeycomb, honeycomb (old), insect (dried), meat (dried), millets, oilseeds, groundnut, pulses, pork (dried), rice (husked), sunflower seed, wheat, wheat flour) found in farm and flour mill (damp) (Hughes, 1976; Suñer *et al.*, 1985; Nangia, 1986; Sánchez-López, 2002; Ardeshir *et al.*, 2006; Hagstrum *et al.*, 2013; Beaulieu and Knee, 2014; Klimov *et al.*, 2016; Akimov and Oksentiuk, 2018). In China, the mite has been collected on rice, semen juglandis, grapefruit and fermented soybean (Li *et al.*, 2015a). Moreover, in Canada the mite has been collected on bee brood (Beaulieu and Knee, 2014). In Ukraine, Oksentiuk *et al.* (2022) reported the mite in mills, granaries and warehouses and outbuildings.

Bio-ecology

In addition to above-reported commodities, *T. casei* has been collected under bark, soil, decaying plant materials, bird and mammal nests. Several records reported the mite from beehives (*Apis mellifera* Linnaeus and *Apis cerana* Fabricius) throughout the world. The mites can reproduce in high numbers in debris from old combs (Delfinado-Baker and Baker, 1987). Under laboratory conditions, it prefers bee bread, pollen, beehive debris and, to a lesser extent, honey, dead brood bees and mould; they did not reproduce on royal jelly or propolis (Chmielewski, 1991a).

The hypopus is unknown (Hughes, 1976), and the species disperses as feeding stages by active movements, air currents or with a host.

The reproduction takes 15–18 days at 23°C and with RH of 87% (Hughes, 1976). Suñer *et al.* (1985) observed that at 16°C and 80% of RH, the life cycle takes 22 days; at 25°C and 80% RH takes

14 days; at 30°C and 95% RH, takes 9 days and at 25°C and 60% RH, takes 17 days. At 15°C and 50% RH., the reproduction does not occur. Nangia and ChannaBasavanna (1989a) studied the influence of temperature and RH on the development of *T. casei*. They found that the life cycle of *T. casei* takes 12–15 days at temperature of 26.5°C and 70% RH; at 30±1°C and with 70% RH are the most suitable for the growth and development of the species, and the rm and X values were high, and so also was the fecundity.

On the Milbenkäse cheese, the mite was identified through cryogenic scanning electron microscopy (Melnyk *et al.*, 2010)

Symptomatology and damage

T. casei infests cheese, grain, damp flour, old honeycombs, dog meal and insect collections (Hughes, 1976). Moreover, it has been reported on ergot of rye (Michael, 1903), under the bark of hold tree stumps (Zachvatkin, 1941), and in mouse nest (Türk and Türk, 1957). However, except for cheese, infestation rate in other stored foods is relatively low. Infestations of this mite are undesirable for many cheeses, with exception of some cheeses on which the culture of the presence of cheese mites (i.e. the Altenburger cheese) produces a 'piquant' taste. High levels of populations of this mite induce the presence, on the cheese, of greyish powder, consisting of very high density of living and dead mites, cast skins and faeces, and producing in the people a sensation of 'ripe' and detectable.

In New Zealand, *T. casei* has been recorded as a major pest of the cheese industry, responsible for severe infestations of cheese in curing rooms (Robertson, 1946). In France and Germany this species has been traditionally used to produce 'mite cheese': Milbenkäse and Mimolette, for gourmet consumption. Research carried out in Punjab showed that *T. casei* has been found in the samples of cheese only. Only 14 out of 3240 total samples examined were occupied by *T. casei* (0.43% of total and 1.14% of mite-infested samples). A total of 202 specimens (1.36%) of this mite were collected of the total mite population (14,812 specimens) (Parminder and Mohd, 2020).

Oudemans (1910) reported a case where the mite was collected in a vessel containing human urine, and Henschel (1929) as responsible of human dermatitis. Moreover, the mite is known to cause respiratory diseases to individuals handling infested cheese (Molina *et al.*, 1975, 1977). Voorhorst *et al.* (1969) reported *T. casei* associated with dermatitis and cause moreover rashers with itching while handling stored food products. In 2013, the US Food and Drug Administration restricted import of Mimolette cheese to the United States due to mite infestation.

Control

Garcia (2004) carried out a study aimed at the development of a method for integrated mite control in the industrial production of Iberian and Alentejan ham. The research was carried out on five factories in the southwest of the Iberian Peninsula. *T. longior* (Gervais) was predominant on hams in the drying rooms, *T. putrescentiae* in cellars, and *T. casei* was collected only in one factory, together with the Ascid *Melichares dentriticus* (Berlese), a predator of tyroglyphid mites. Results showed that RH levels below 60% reduced the development of mite populations on the ham surface, but these conditions forced the mites to shelter; so, *T. putrescentiae* and *T. casei* were collected in cavities inside the ham. At the end of the trial period, 11% of the hams were totally disintegrated by mite damage inside. Mites survived high-temperature treatments which are detrimental to the hams: 71°C, 11 min; 57°C, 30 min and 40°C, 96 h. The mites also survived a long low-humidity period, such as 1 month at 18°C and 32% RH (Garcia, 2004). The cheese may be protected using a thin layer of paraffin wax.

Selected website

https://idtools.org/bee_mite/index.cfm?packageID=1&entityID=156

The Genus *Tyrophagus* Oudemans

Diagnostic characteristics

The idiosoma is oblong, 300–600 µm long and whitish to semi-transparent in colour. The prodorsum has four pairs of setae, and the hysterosoma 12 pairs of setae; all dorsal setae are barbed. The external vertical setae (*ve*) are long and set on anterior lateral sides of the prodorsal shield. The internal scapular setae (*sci*) are longer than the external scapular setae (*sce*). The hysterosomal setae c_1 and d_1, or d_2 are shorter than the distance to the setae in the next row. The supracoxal setae (*scx*) are commonly pectinate, rarely smooth. The distal end of the Grandjean's organ is finger-like. The legs are light brown, with tarsi I–IV slender, more than twice as long as wide. The dorsal distal spine *e* on tarsi I–II is slender. The proral setae on tarsi I–II are thin, longer than unguinal setae. Tarsus I with two long distal setae and one distal solenidion (ω_3); tarsi II–III each with two long distal setae; tarsus IV with two long distal setae in the female (*d*, *f*) and one long distal seta in the male (Fan and Zhang, 2007a).

Solarz (2011) reported the following distinctive characters. In both sexes, the setae *ve* distinctly barbed, relatively long, longer than the length of the genu I, and usually positioned near anterior lateral corners of propodosomal shield; setae *ve* at almost the same level as *vi* and curve downwards; setae *sci* longer than *sce*; setae c_1 and d_2 usually almost equal in length, and shorter than e_1 and h_1; the dorsal terminal seta *e* on tarsi I needle-shaped; presence of five ventral setae on tarsi I, of which the three central ones are thickened; solenidion φ_1 on tibia I always less than three times as long as φ_2; ventral apex of tarsi with proral and unguinal setae usually in the form of short, stout spines, occasionally one or both pairs, strongly reduced or absent; tarsi I–II more than twice as long as basal width; proral setae thinner than unguinal setae but similar in length; Grandjean's organ finger-like. Males without modifications of leg I; leg I not enlarged and the femur does not bear a ventral apophysis (Solarz, 2011).

Economic importance

Tyrophagus spp. are abundant and economically important mites of the family Acaridae inhabiting stored foods and products (Hughes, 1976). The genus *Tyrophagus* was erected by Oudemans, 1924a with the type species *Acarus putrescentiae* Schrank, 1781. Up to now, about 35 species of the genus *Tyrophagus* Oudemans were recorded from around the world (Fan and Zhang 2007a). Some species are injurious to economic plants, ornamentals and vegetables grown both in the open field and in greenhouses (Vacante, 2016). On stored foods are known different injurious species, such as *T. curvipenis* Fain *et* Fauvel, *Tyrophagus fanetzhangorum* Klimov *et* OConnor *T. longior* (Gervais), *T. neiswanderi* Johnston *et* Bruce, *Tyrophagus palmarum* Oudemans, *T. perniciosus* Zachvatkin, *T. putrescentiae* (Schrank) and *T. similis* Volgin. When ingested, the acarid mites, such as *T. putrescentiae*, may play a serious medical and veterinary importance, causing poisoning in humans and animals (Blanco *et al.*, 1997; Matsumoto *et al.*, 2001; Sánchez-Borges *et al.*, 2005; Liao *et al.*, 2009; Canfield and Wrenn, 2010), such as acute enteritis, and may be related to the transmission of bacteria, yeasts and pathogenic fungi through food (Serpa *et al.*, 2004).

Other *Tyrophagus* species are occasionally collected on stored foods in various countries, and according to their secondary role, we report below only the fundamental traits of their bio-ecology to complete the discussion. *Tyrophagus bambusae* (Tseng) is reported from Taiwan on bamboo shoots (rotting) (Tseng, 1972). *Tyrophagus brevicrinatus* Robertson is known from China, England, Germany, Ghana on Chinese medicine, condiment and copra in farm (Robertson, 1959; Johnston and Bruce 1965; Hughes, 1976; Lynch, 1989). *Tyrophagus dimidiatus* Hermann is commonly named mushroom mite and reported from Australia, Poland, Taiwan and United States, where it has been collected on bean (green), fish (minced-dried), garlic, grain, ham, mushroom, sugarbeet seed; its biology has been studied by Miura and Sasa (1961). *Tyrophagus miripes* Athias Henriot has been collected in Czech Republic and Indonesia, on grain and sweeping in food store (Athias Henriot, 1961a; Johnston and Bruce, 1965). *Tyrophagus oudemansi* Robertson has been collected on barley, Chinese medicine, grain, mushroom and rye in China, England, Germany and United States (Robertson, 1959). *Tyrophagus savasi* Lynch is known from England on barley, barley (rolled), biscuit, grain and honey (Lynch, 1989). *Tyrophagus zachvatkini* Volgin has been described from Egypt (Volgin, 1948; Johnston and Bruce, 1965; Attiah, 1969).

We discuss below only the most common *Tyrophagus* species collected on stored food.

Selected website

https://idtools.org/bee_mite/index.cfm?packageID=1&entityID=157

Tyrophagus curvipenis Fain *et* Fauvel (Fig. 28.4)

Common name

Unknown

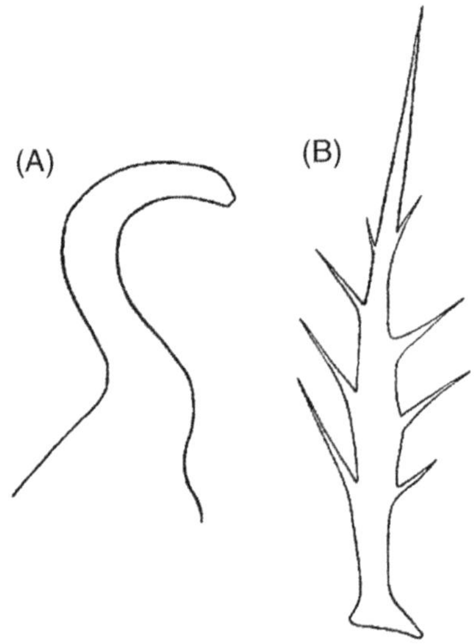

Fig. 28.4. *Tyrophagus curvipenis* Fain *et* Fauvel. Male. Aedeagus (A). Female. Supracoxal seta (B).

Diagnostic characteristics

FEMALE. The idiosoma is 367 μm long and 212 μm wide. The prodorsal shield has prominent eyespots, nearly pentagonal, with the lateral side behind the eyespots concave. Grandjean's organ is finger-like, smooth, with the basal lobe bearing 2–3 spiniform teeth. The setae *scx* are slender, with 6–8 pectinations. The setae d_1 are about 3.0 times longer than the setae c_1, and 2.6 times longer than the setae d_2; the setae d_2 are about 1.2 times longer than the setae c_1. The coxal plates I extend just beyond the apex of the apodeme; the coxal plates II are broadly triangular, extending distally beyond the apex of apodeme 2, with one-third of the posterior margin slightly concave. The spermathecal duct narrows gradually from the copulatory opening, for a distance of about 1.5 times the distance between the sclerites of the oviducts, and then extends to the base of the spermathecal sac over a distance of about 2.8 times the distance between the sclerites of the oviducts; the base of the spermathecal sac is 'r'-shaped. The solenidion ω_1 of tarsus I is slender, cylindrical and slightly widened at the apex; the solenidion ω of tarsus II is slender and almost cylindrical. The setae *w* and *r* of tarsus IV are setiform (Fan and Zhang, 2007a).

MALE. The idiosoma is 321 μm long and 205 μm wide. The prodorsal shield is similar to that of the female. Grandjean's organ is as in the female. The coxal plates I and II, and the solenidia *ω1* and *ω*, of tarsi I and II, are as in the female. The setae d_1 are about three times longer than the setae c_1 and two times longer than the setae d_2; the setae d_2 are about 1.5 times longer than the setae c_1. The aedeagus is 'S'-shaped, the distal two-fifths are reverse curved, tapering from the base to the tip and with the lateral arms turning outwards. The setae *w* and *r* of tarsus IV are setiform (Fan and Zhang, 2007a).

Geographical distribution

This mite is known from Australia, France, New Zealand (Fan and Zhang, 2007a), the Netherlands (Sipel *et al.*, 2016), Portugal (Fain and Fauvel, 1993; Fan and Zhang, 2007a), Greece (Badieritakis *et al.*, 2012, 2014) and New World (United States) (Murrillo *et al.* 2018). Moreover, Murrillo *et al.* (2018) reported that the actual distribution of this mite is much broader. The limited number of records of *T. curvipenis* could be due to difficulties is distinguishing it from other related species, such as the widespread species *T. putrescentiae*. Observation carried out recently on the drawings made over time by Vacante of the specimens of *T. putrescentiae* collected in the 1980s on fruits and bark of citrus showed that the species lives also in Sicily (Italy).

Commodities, facilities and vegetal and animal hosts

Plants and their products on which this mite has been found include apples, the bark of apricot, cherry and nectarine, barley, the bird-of-paradise flower, blackcurrant stems, blackcurrants, capsicum, cherry foliage, clover roots, *Coleophora* tubes, cucumber leaves, *Cymbidium* sp. pollinia, garlic, Granny Smith apple calyxes in sooty mould in store, grapefruit (unsprayed fruit/leaves), hydrangea leaf, kiwi fruit, lemons, mandarin, mummified apricot fruit, mummified loquat fruit, Nashi (Japanese pear) leaves, onions, passion fruit, persimmon, poplar (*Populus nigra* Linnaeus var. *italica*) galls, shallots, strawberries and tree onion bulbs (Fan and Zhang, 2007a). The mite has also been found on algae covering the wooden structures of orchids (Fain and Fauvel, 1993). Badieritakis *et al.* (2012, 2014) recently reported the mite on foliage and litter of *Medicago* species in Greece. Vacante (not published) collected specimens of this mite on fruits and bark of citrus.

Animals and animal products on which the mite has been found include, coccid scale on *Buxus* sp., dead eggs and embryos of *Miomantis caffra* Saussure, dead larva of the citrus borer in wood, dead larva of *Stathmopoda*, dead *Rattus norvegicus* Berkenhout, *Disonycha* and *Agasicles* beetles, dried milk powder, egg cluster of Avondale spider, eggs of *Ceroplastes destructor* (Newstead), honeybees, honeybee hive, nests of the bee *Exoneura concinnula* Cockerell, *Rattus exulans* (Peale), sparrow, starling (*Sturnus vulgaris* Linnaeus) and swallow, scrapings from pallets and logs, wasp [*Vespula germanica* (Fabricius)] (Fan and Zhang, 2007a).

Bio-ecology

T. curvipenis Fain *et* Fauvel is known from a wide range of habitats (Fan and Zhang, 2007a), but there is no information available on its bio-ecology, damage and economic importance. Murrillo *et al.* (2018) sequenced 1.2 kb of the mitochondrial COI gene for 38 individuals belonging to seven species of *Tyrophagus* (*T. curvipenis*, *T. putrescentiae* (Schrank), *T. fanetzhangorum* Klimov *et* O'Connor, *T. longior* (Gervais), *T. perniciosus* Zachvatkin and *T. similis* Volgin) and found that molecular phylogenetic analyses recovered two major clades corresponding to the presence or absence of eyespots and separated all included morphological species. In particular, *T. curvipenis* and *T. putrescentiae* showed the lowest between-species genetic distances, and the highest within-species variation was found in *T. putrescentiae*. Moreover, in this last species they recovered two distinct groups, but no geographical or ecological dissimilarities were found between them. Based on this analysis, they document important morphological differences between *T. curvipenis* and *T. putrescentiae*. For the first time, they recorded the occurrence of *T. curvipenis* in the United States, and it has suggested that it may be an emerging pest as it is currently spreading in agricultural produce (Murrillo *et al.*, 2018).

Ye and Zhang (2014) studied the biology of this mite under laboratory conditions using dry yeast (*Saccharomyces cerevisiae* Meyen), rolled oats (*Avena sativa* Linnaeus) and wheat flour (*Triticum aestivum* Linnaeus), at 25 ± 1°C and 80 ± 5% RH. The egg-hatching success rates were similar in all diet treatments, averaging about 90%. Survival rates of larvae and protonymphs were 100% in all treatments (no deutonymphs were observed in this species), and the survivorship of tritonymphs was 100%, when fed on yeast, but were lower on two other diets. The overall survival rates from egg to adult were high, and not significantly different among the treatments. The development of larvae was fastest on protein-rich yeast (2.2 days for females and 2.0 days for males), intermediate on oats (3.7 days for females and 3.6 days for males), and the slowest on wheat flour (6.1 days for females and 4.7 days for males). Males developed faster only when mites were fed on wheat flour. The duration of protonymphal stage was affected by food, but there was no significant difference between males and females on all three diets, 2.1 days for females and 2.0 days for males on yeast, 4.5 days for females and 4.4 days for males on oats, and 5.4 days for females and 5.2 days for males on wheat flour. The food effect on tritonymphal development was significant, but the gender effect was not, with the exception for mites reared on rolled oats, 2.1 days for females and 2.0 days for males on yeast, 7.8 days for females and 6.0 days for males on oats, and 9.3 days for females and 9.4 days for males on wheat flour. The age at maturity was affected by food type, with males developing slightly faster than females, 10.4 days for females and 10.0 days for males on yeast, 20.0 days for females and 18.1 days for males on oats, and 24.9 days for females and 23.2 days for males on wheat flour (Ye and Zhang, 2014).

T. curvipenis is usually found outdoor in field rather than in stored products, where its presence may be considered sporadic. It prefers habitats or hosts with irregular surfaces, such as the natural openings of chayote fruits or crevices formed on other plants. The preference of *T. curvipenis* for these areas is probably due to the high humidity retained there and the proliferation of fungi and algae that are available as a source of food. Moreover, whereas *T. longior*, *T. putrescentiae*, *T. similis* and *T. neiswanderi* Johnston *et* Bruce are known as injurious pests responsible of direct damage to cultivated plants (Buxton, 1989; Fisher, 1993; Kasuga and Amano, 2003, 2006; Zhang, 2003; Kasuga and Honda, 2006; Oliveira C.R.F. *et al.*, 2007), no reports are available on the responsibility of *T. curvipenis* of direct damage to cultivated plants or their fruits. However, the above-mentioned authors speculated the possibility that *T. curvipenis* can feed on young leaves, flowering buds or fruits. In any case, *T. curvipenis* is regularly intercepted at international ports of entry by quarantine inspectors in natural cavities or crevices of fruits and leaves, causing substantial disruptions in international trade (Barquero, 2015).

Symptomatology and damage

No reports for stored foods.

Control

No reports for stored foods.

Tyrophagus fanetzhangorum Klimov et O'Connor

Syn. *Tyrophagus putrescentiae* (Schrank) *sensu* Fan and Zhang (2007a).

Common name

Unknown

Diagnostic characteristics

FEMALE. The idiosoma is 492 μm long and 314 μm wide. The prodorsal shield is nearly pentagonal, with eyespots and lateral sides slightly concave. Grandjean's organ is finger-like, and the basal lobe bears 1–2 large and two small spiniform teeth. The setae *scx* are widened at the bases of the pectinations, long and number 10–14. The setae d_1 are about 2.1 times longer than the setae c_1 and 2.4 times longer than the setae d_2; the setae d_2 are long, about 0.9 times the length of the setae c_1. Coxal plates I are just beyond the apex of the prosternal apodeme; coxal plates II have a sinuous posterior border, narrowing sharply medially and becoming contiguous with the apex of their apodeme. The spermathecal duct narrows rapidly from the copulatory opening for a distance of about 0.6 times the diameter of the distance between the sclerites of the oviducts and then gradually widens to the base of the spermathecal sac over a distance about 2.2 times that between the sclerites of the oviducts; the base of the spermathecal sac is flat. The distal two-thirds of solenidion $\omega 1$ of tarsus I obviously widens; the solenidion ω of tarsus II slightly widens at the apex. The setae *w* and *r* of tarsus IV are setiform (Fan and Zhang, 2007a).

MALE. The idiosoma is 378 μm long and 233 μm wide. The prodorsal shield is as in the female. Grandjean's organ is finger shaped, and basally has 1–4 spiniform teeth. Eyespots, coxal plates I and II, solenidia ω_1 and ω are as in the female. The setae d_1 are about two times longer than the setae c_1 and 1.9 times longer than the setae d_2; the setae d_2 are about one time the length of the setae c_1. The aedeagus is 'S'-shaped, with the distal one-third reverse curved, tapering from the base to the tip, with the lateral arms turning outwards. The setae *w* and *r* of tarsus IV are setiform (Fan and Zhang, 2007a).

Geographical distribution

T. fanetzhangorum is known from Australia, China, Ecuador, Germany, Japan, the Netherlands and the United States (Fan and Zhang, 2007a).

Commodities, facilities and vegetal and animal hosts

According to Fan and Zhang (2007a), the mite has been collected on banana, sour orange (*Citrus aurantium* L.), garlic, *Gladiolus*, humus, hyacinth, muscari bulbs, mushroom, *Narcissus* bulbs, orange, orchids, palm seed, *Ranunculus* sp. bulbs and tulip bulbs. Moreover, it has been collected also from the burrows and nests of mutton birds (*Puffinus tenuirostris* Temminck) and dead tench (*Tinca* Linnaeus) (Cyprinidae).

Bio-ecology

The description of *T. fanetzhangorum* has been at the centre of an interesting systematic case, started with the revision of Fan and Zhang (2007a), who, analysing the Robertson's (1959) concept of *T. putrescentiae* (Schrank), pointed out a mixture of two species described as *T. putrescentiae* and *T. communis* Fan *et* Zhang, respectively. *T. putrescentiae* differs from *T. communis* in that two-thirds of the posterior margin of coxal plates II is strongly concave, and the solenidion ω_1 is widened in the distal two-thirds. Klimov and O'Connor (2009) did not share this systematic view and stated that the prevailing usage of the name *T. putrescentiae* includes almost exclusively one of these species, and this usage is not in taxonomic accord with the neotype designated by Robertson (1959). They propose that this neotype fixation for the species *T. putrescentiae* be set aside under Article 75.6 (ICZN, 1999), and a neotype consistent with the prevailing usage be designated. Moreover, they verified the synonymy of *T. putrescentiae* (based on the new neotype), after examining the following taxa with extant types: *Tyrophagus americanus* (Banks, 1906), *Tyrophagus breviceps* (Banks, 1906) n. syn, *Tyrophagus cocciphilus* (Banks, 1906) n. syn, *Tyrophagus castellanii* (Hirst, 1912), *Tyrophagus australasiae* (Oudemans, 1916) (tentative synonymy), *Tyrophagus neotropicus* (Oudemans, 1917), *Tyrophagus amboinensis* Oudemans (1925), *Tyrophagus nadinus*) and *T. communis* (Fan and Zhang, 2007), n. syn. For the taxon *T. putrescentiae sensu* Fan and Zhang (based on the neotype designated by Robertson), they proposed the new name *T. fanetzhangorum* Klimov *et* O'Connor, n. sp. Moreover, they designed the lectotypes of the species *T. americanus*, *T. breviceps*, *T. cocciphilus*, *T. castellani*, *T. australasiae*, *T. neotropicus*, *T. amboinensis* and *T. nadinus*.

In conclusion, Klimov and O'Connor (2010) believe that the prevailing usage of the name *T. putrescentiae* comprises almost exclusively the common species, and the name *T. communis* is a junior synonym of eight previously named taxa, with extant types and propose to nominate the rare species *T. fanetzhangorum*. The stability of zoological nomenclature is therefore threatened by three different aspects. The prevailing usage of the name *T. putrescentiae* was not maintained by Fan and Zhang (2007a). The name *T. communis* proposed for the common species is a junior synonym – and therefore not valid – and besides the eight taxa for which synonymy with the common species was verified by them, types of older taxa may also be discovered in the future, thus posing another nomenclatural challenge. They propose to conserve the prevailing usage of the name *T. putrescentiae* by designation of a new neotype, from a culture currently maintained in a research institution, which was cited in many published works, and started from specimens collected close to the type locality of *T. putrescentiae*. Vacante (2016) in *The Handbook of Mites of Economic Plants*, adopted the systematic vision of Fan and Zhang (2007a) in order not to increase the confusion about the case, and waiting to have further data on the topic that would help to find a definitive solution. Today, the developments of the topic suggest to adopt the indications of Klimov and O'Connor (2009, 2010) and Klimov and Tolstikov (2011), and reporting the two species under discussion as *T. fanetzhangorum* and *T. putrescentiae*, respectively. The first species is not common on the stored food and is discussed only for completeness of information. A further discussion on this topic may be found in Fan and Zhang (2014).

Symptomatology and damage

No reports concerning damage on stored food. Filatov *et al.* (2022) reported a case of infestation of a long-term laboratory colony of soft ticks, *Ornithodoros moubata* Murray (Argasidae), with specimens molecularly identified as *T. fanetzhangorum,* using the COI gene and discussed the possible nature of the interaction between these two acarines.

Control

No reports.

Tyrophagus longior (Gervais) (Fig. 28.5)

Common names

Seed mite, mould mite.

Diagnostic characteristics

FEMALE. The idiosoma is 525 μm long and 344 μm wide. The prodorsal shield lacks eyespots is nearly rectangular, with the lateral sides slightly concave, and the posterior broadly round. Grandjean's organ is finger-like and barbed, with the basal lobe bearing one large tridentate, and 2–3 small spiniform teeth. The setae *scx* are slender, tapering from the base to the apex, with 8–12 short pectinations. The setae d_1 are about 1.3 times longer than the setae c_1, and 1.5 times longer than the setae d_2; the setae d_2 are about 1.0 times as long as the setae c_1. The coxal plates I are triangular and do not reach to the posterior apex of the prosternal apodeme; the coxal plates II are triangular, with the posterior margin slightly sinuous, and not reaching the apex of the apodeme 2 distally. The spermathecal duct is broad, widening gradually from midway to the base of the spermathecal sac, about 1.5 times longer than the distance between the sclerites of the oviducts, and the base of the spermathecal sac is nearly flat or bends slightly backwards. The solenida ω_1 and w of tarsi I and II are slender, almost cylindrical, but taper slightly from the base to the apex. The setae *w* and *r* of tarsus IV are spiniform (Fan and Zhang, 2007a).

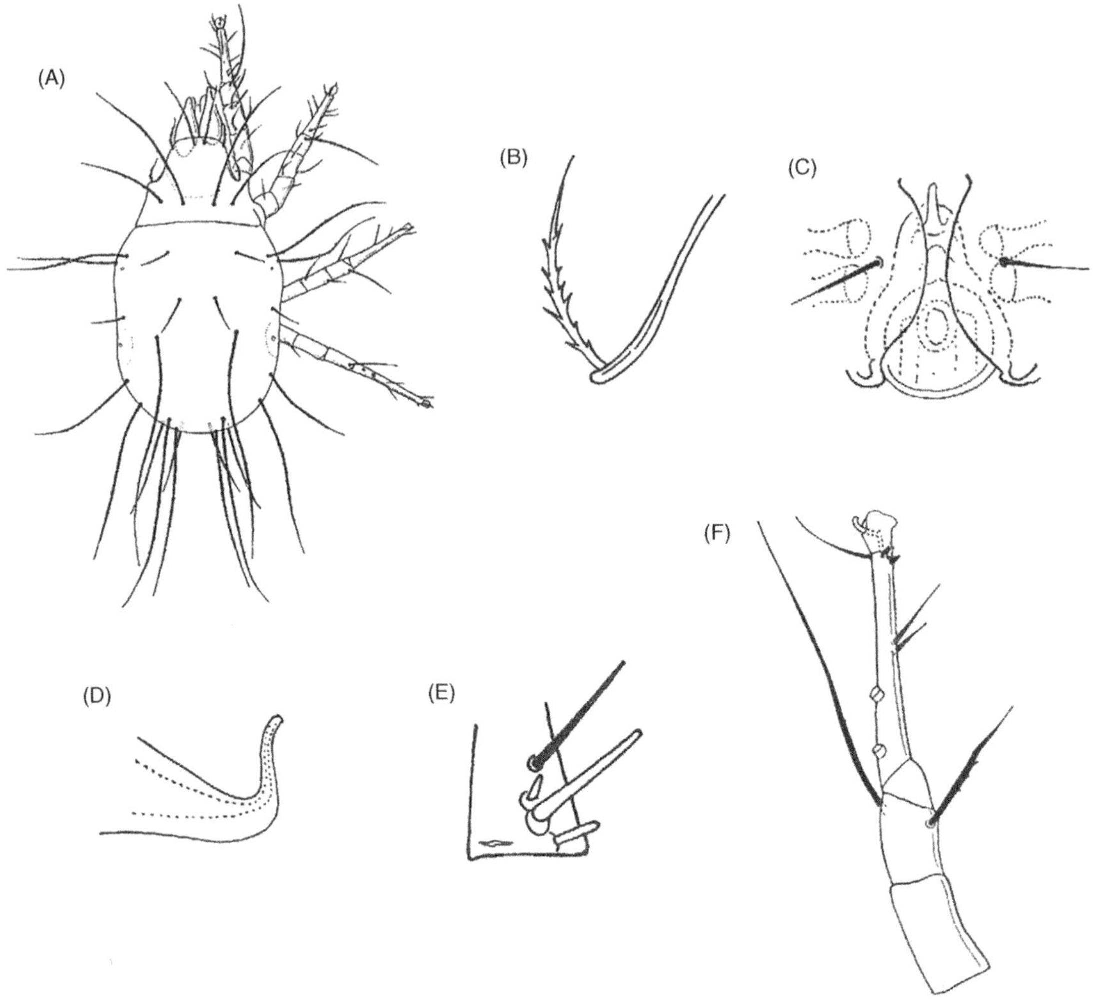

Fig. 28.5. *Tyrophagus longior* (Gervais). Male. Dorsal view (from Hughes, 1976, modified) (A); supracoxal seta (from Hughes, 1976) (B); genital region (from Hughes, 1976, modified) (C); aedeagus (D); basal part of tarsus I (from Hughes, 1976) (E); leg IV (from Hughes, 1976, modified) (F).

MALE. The idiosoma is 563 µm long and 366 µm wide. The prodorsal shield is similar to that of the female. The coxal plates I and II, and the solenidia ω_1 and ω of tarsi I and II, are as in the female.

The setae d_1 are about 1.4 times longer than the setae c_1, and 1.6 times longer than the setae d_2; the setae d_2 are about 0.9 times the length of the setae c_1. The aedeagus has one major curve, narrowing gradually in the basal half, the distal end is straight or slightly reverse curved, and the lateral arms turn inwards. The setae *w* and *r* of tarsus IV are spiniform (Fan and Zhang, 2007a).

Solarz (2011) reported the following distinctive characters. In both sexes the anterior margins of propodosomal sclerite are without pigmented spots (corneae); supracoxal seta (*scx*) thin, not expanded basally, pectination variable; setae d_2 short, at most only slightly longer than c_1; setae d_1 about 1.3–2 times as long as c_1 and d_2; tarsi I–II with solenidion

ω_1 long, slender, tapering towards tip; tip pointed and not expanded apically. The males possess supracoxal seta (*scx*) thin, with short lateral barbs of about equal length; the lateral sclerites supporting the penis (aedeagus) turned inwards; aedeagus straight only slightly curved distally and shaped like the spout of a teapot; aedeagus long, slender and tapering towards its free end; tarsus IV longer than the combined length of the genu and tibia; both tarsal copulatory suckers positioned nearer to the base of tarsus than the apex. The females have the anterior margins of propodosomal sclerite without pigmented spots (corneae); internal spermathecal apparatus large, spermathecal duct moderate or large, base of spermatheca moderately long or very wide, distinctly wider than spermathecal duct; setae ad_2 considerably longer than ad_3, ps_3, ps_2 and ad_1 (Solarz, 2011).

Geographical distribution

T. longior is found worldwide (Robertson, 1959; Zakhvatkin, 1959; Griffiths, 1960; Johnston and Bruce, 1965; Rack, 1968a; Hughes, 1976; Emmanuel and Evans, 1979; Tseng, 1979; Fathipour, 1994; Vacante, 1997a; Noguchi *et al.*, 1998; Putatunda, 2002; Gupta and Chatterjee, 2004; Hubert *et al.*, 2006; Sayedi *et al.*, 2006, 2012; Fan *et al.*, 2010a; Moreno, 2015; Siepel *et al.*, 2016; Pfliegler *et al.*, 2017; Masoudian *et al.*, 2017; Oksentiuk *et al.*, 2022).

Commodities, facilities and vegetal and animal hosts

Hughes (1976) reported *T. longior* on stored home-produced grain, hay and straw, haystacks and grain stacks standing in the open field, barley, grass cleaning plant (tools), cucumber plants, beetroots, tomatoes and *Cyclamen* seed.

Hagstrum *et al.* (2013) reported a long list of commodities including animal feed, barley, beetroot, biscuit, buckwheat, buckwheat product, cheese, Chinese medicine, cow hoof (decaying), cucumber, currant, Cyclamen seed, fish (minced-dried), flax seed, grain, grain residue, grass seed, ham, hay, herb, herring meal, hyacinth bulb, melon, mushroom, oat bran (damp), peach (dried-rotten), soybean flour, stock fish, raisin, rape seed, rye grass seed, sugarbeet seed, sugar (red), sugar (white), sweeping, table jelly (fermenting), tobacco (fermenting), wheat, wheat bran and wheat flour. Moreover, the mite has been collected on cheese (Robertson, 1961; Lozzia and Ottoboni, 1989; Pagani and Zanotti, 1994) and salami (Pagani, 1987; Pagani and Ciampitti, 1991).

Facilities number bread van, cheese factory, cheese store, commercial grain storage, drier, farm, farm granary, farm milling machine, flour mill, grinder, mill, off-farm storage, provender warehouse, railcar and slaughterhouse (Hagstrum *et al.*, 2013). Recently, the mite has been collected on rice, wheat, cake and malted milk in China (Li *et al.*, 2015a). Moreover, this mite is reported in flour mill and silo (Pfliegler *et al.*, 2017). Griffiths (1960) reported that in England the mites commonly associated with foodstuffs stored indoors, are present in the outdoor habitat afforded by stacks. Among these *T. longior* is a dominant species in stacks, and both appear to be field dweller. Host plants of *T. longior* have been discussed by various authors (Lange and Bacon, 1958; Chmielewski, 1979, 1984; Hussey, 1985; Voigt, 1990; Laffi and Ponti, 1997; Czaikowska *et al.*, 1988; Buxton, 1989; Vacante, 1997a, 2016; Ciampolini *et al.*, 1985; Fan and Zhang, 2007a). Oksentiuk *et al.* (2022) collected the mite in Ukraine in barns with animal fodder storage and honeybee beehives.

Animals and their products include cheese (Robertson, 1961), stock fish (Gigja, 1964), bee frames, beehives and surveyed bees and *Bombus* nests, hives of *Apis cerana indica* Fabricius, *A. mellifera* Linnaeus, *Apis dorsata* Fabricius and *Apis florea* Fabricius, honey, bird's nests (including one in a poplar tree), nests of seabirds [*Eudyptula minor albosignata* Finsch and *Pachyptila turtur* (Kuhl)] and the dipper [*Cinclus aquaticus* (Bechst.)] and the starling (*Sturnus vulgaris* Linnaeus) (Fan and Zhang, 2007a).

Moreover, the mould mite has been collected on cheese shelves, dead *Prionoplus reticularis* White (Coleoptera, Cerambycidae), ham, house dust, humans, litter in a densely populated urban area, scrapings from pallets and bags (dried milk powder), ham in a meat works, scrapie-infected sheep, shepherds and wedding cake (Fan and Zhang, 2007a).

Bio-ecology

In several European countries, *T. longior* has been found on cucumber (Chmielewski, 1979, 1984; Hussey, 1985; Voigt, 1990; Laffi and Ponti, 1997) and ornamentals (Czaikowska *et al.*, 1988; Buxton, 1989) in greenhouses. Vacante (1997a) collected the mite on melon, strawberry and tomato in protected crops of the Mediterranean area but did not observe any damage. On orchids (*Cymbidium* spp.) in greenhouses, the mite prefers *Cymbidium clauboda* Goliath and vectors fungi that prevent the flowers from opening, while on other *Cymbidium* spp., the problem is secondary (Ciampolini *et al.*, 1985). In California, the mite damages spinach in spring and autumn crops (Lange and Bacon, 1958), and in the United States, it has been found associated with *Histiostoma* and *Rhizoglyphus* mites. At 23°C and 87% RH, the life cycle of the seed mite takes from 2–3 weeks (Hughes, 1976). At 20°C on wheat grain, the mite populations increase from 5 to 2200 mites before a population crash when the RH is 90%, while at 75% RH, the population peaks at 1000 mites and then decreases (Parkinson, 1990). Dry, warm weather and rapid growing conditions often circumvent more severe damage by the mite (Lange and Bacon, 1958). Chmielewski (1984) reported that biological parameters of the species depend on environmental factors (temperature, RH, kind of food). The mite produces an alarm pheromone (Noguchi *et al.*, 1998).

Parkinson *et al.* (1991) compared the egg production of isolated pairs of *A. siro* Linnaeus, *G. destructor* (Schrank) and *T. longior*, feeding on a control diet of wheat germ and yeast with that on mycelial pellets, from shake cultures of *C. cladosporiodes* (Fresen.) G.A. de Vries, *A. repens* (Corda) Saccardo, *A. ruber* Thom *et* Church and *Penicillium cyclopium* Westling as well as spores of *P. cyclopium* and *A. repens*. The mites always produced fewer eggs on the fungal diets than on the ideal diet. *T. longior* usually did best of the three mite species on the fungal diets, and *G. destructor* worst. *A. ruber* proved the most suitable fungus for all mite species, while *C. cladosporiodes* was the least suitable. Spores were a less suitable diet than mycelial pellets from shake cultures, which were predominantly mycelium (Parkinson *et al.*, 1991). The mite is reported as injurious to ham in Spain (Sánchez-López, 2002).

Symptomatology and damage

Damage of *T. longior* to economic plants has been described by Vacante (2016). On stored food, the mite occurs occasionally and produces a type of damage similar to that of *A. siro*, but the attack densities are normally lower. Chmielewski (1984) reported that *T. longior* is a polyphagous pest of stored food products, seed material and greenhouse plants (cucumber). High infestation by this mite causes damage or even total destruction of food products, seeds and plants.

Cases of digestive and urinary acariasis in humans involving *T. longior* have been discussed (Harwood and James, 1979). Sarathep and Phonkaew (2018) reported that an outbreak investigation of copra itch revealed a cluster of six suspected cases with skin dermatitis, with 11–32 years of age, belonging to a single family, in Phang Nga province, Southern Thailand. Epidemiologic investigation of these suspected cases revealed, five probable cases developing multiple discrete erythematous papules with intense pruritus on the body rather than the extremities, and one confirmed case whose skin was infested with domestic mite, *T. longior*. This mite was also found in unused coir mattresses outside their bedrooms. Household infestation with *T. longior* mites rendered these family members to become more susceptible to expose indoor biting of *T. longior* adult mites that were adapted well to the domestic environments with poor hygienic conditions. Human exposure to mite bites was more likely to be a direct contact than an indirect contact. Findings from this copra itch outbreak investigation provided understanding of natural disease of copra itch and factors that favoured the outbreak, and could guide diagnosis for physicians, surveillance and response for surveillance and rapid response teams, and prevention and control for entomologists and public health personnel.

Control

Control of seed mite requires a meticulous prevention care in the warehouse before storage of goods. In this case a scrupulous cleaning of the rooms must be organized and if necessary, it is possible to use appropriate active ingredients. Monitoring of its populations and both preventive and curative actions.

MONITORING. Early monitoring of mite presence on stored foods is crucial as other stored mites (i.e. *A. siro*).

PHYSICAL CONTROL. On stored foods, aeration is a promising alternative to the use of pesticides for the control of storage insects by cooling bulk grain, but its effectiveness against mite pests is neither fully understood nor optimized. For this reason, the productivity of three species of storage mites, *A. siro*, *L. destructor* and *T. longior*, was studied in a laboratory-based experiment at four combinations of temperature and humidity (10°C and 70% RH, 10°C and 80% RH, 20°C and 70% RH, 20°C and 80% RH), with and without an airflow (at 10 m^3/h/tonne, equalling 2.5 1/s/tonne, in tubes containing 15g of grain). This is the first time that a study has examined, the three principal components of aeration separately from each other. The effect of these factors was different for each species. For *A. siro*, temperature was the most important factor, while airflow and humidity were of similar but lesser importance. For *T. longior*, temperature was more important than humidity, while the reverse was true for *L. destructor*. For these two species, airflow was the least important factor. The airflow decreased the productivity of *L. destructor* and *T. longior* but increased the productivity of *A. siro*. This increase in productivity confirms that, in practice, prevention of mite infestations, in particular *A. siro*, will require storage of grain at low temperature, relative humidity and MC (Thind and Dunn, 2002).

Conyers and Bell (2003) investigated the effects of MA, on the eggs of mite pests of grain and cheese *A. farris*, *A. siro*, *L. destructor* and *T. longior*. All mites were collected in many habitats including grain and cheese stores. They used low oxygen (O_2) MA mixtures, based on carbon dioxide (CO_2), nitrogen (N_2) or simulated burner gas (0.5 or 2% O_2, 10% CO_2, balance N_2) plus 60% CO_2 in air (8% O_2). The mites were exposed at 15°C and 80% RH, a combination of conditions that occurs at the surface of stored grain, during the autumn, which promotes mite population growth. *T. longior* was the most tolerant species, followed by *A. siro* and *A. farris*, with *L. destructor* the most susceptible. Burner gas was the most effective mixture overall, with 0.5% O_2 but with an increase in the O_2 level to 2% for all the mixtures, CO_2 became the more effective control agent. With 60% CO_2 in air, some loss of efficacy was observed against the three most tolerant species and even more so for *L. destructor*. Sublethal exposures to MAs for at least 4 days in *L. destructor*, 6 days in *A. farris* and *A. siro* and 8 days for *T. longior* caused a delay in egg hatch.

Conyers and Bell (2007) examined the use of controlled atmospheres (CA) to disrupt the life cycle of mites. A different approach was required, if the exposure time for CAs with low oxygen (O_2) in nitrogen (N_2) was to be reduced. Rather than attempting to kill all stages, the O_2 content of the CA to be used would not affect egg hatch but would kill the far more susceptible larvae. This would be technically easier to achieve and more economic to run but would require treatment starting from the moment the crop was placed in storage. Different levels of O_2, 4, 5 and 6%, were used with N_2 as the main constituent, as well as 5% O_2 and 7.5% carbon dioxide (CO_2) in N_2, and the addition of 10% or 20% CO_2 to 6% O_2 in N_2. The two mite species tested were *A. siro* L. and *T. longior*, both of which, as eggs, show a high level of tolerance to CAs. The laboratory tests were carried out at 20°C and 25°C on whole wheat, with RH of 75%, 80% and 85%. The exposures were determined by the generation time of each species, at each combination of temperature and RH. Comparison of numbers at the start and end of the exposure, allowed the reduction in population growth to be assessed.

CHEMICAL CONTROL. Old reports on the control of *T. longior* confirm the good response to the parathion dust, however, sprays were effective only on small surfaces (Lange and Bacon, 1958). Moreover, the effectiveness of etrimfos, heptenophos, methacrifos, profenofos and propargite was verified in the laboratory, while cycloprate, fenbutatin oxide and isoprothiolane failed to give complete mortality; several compounds reputedly effective against pest mites or ticks were ineffective against *T. longior* (Stables, 1980). Pirimiphos-methyl sprays were an effective and less persistent alternative to drenches and sprays of parathion (Hussey, 1985). A high-volume spray of dichlorvos gave good control of the mite (Buxton, 1989).

Tyrophagus neiswanderi Johnston *et* Bruce (Fig. 28.6)

Common name

Pollen-cap mite

Diagnostic characteristics

FEMALE. The idiosoma is 493 μm long and 327 μm wide. The prodorsal shield has a pentagonal shape, with prominent eyespots, and lateral sides

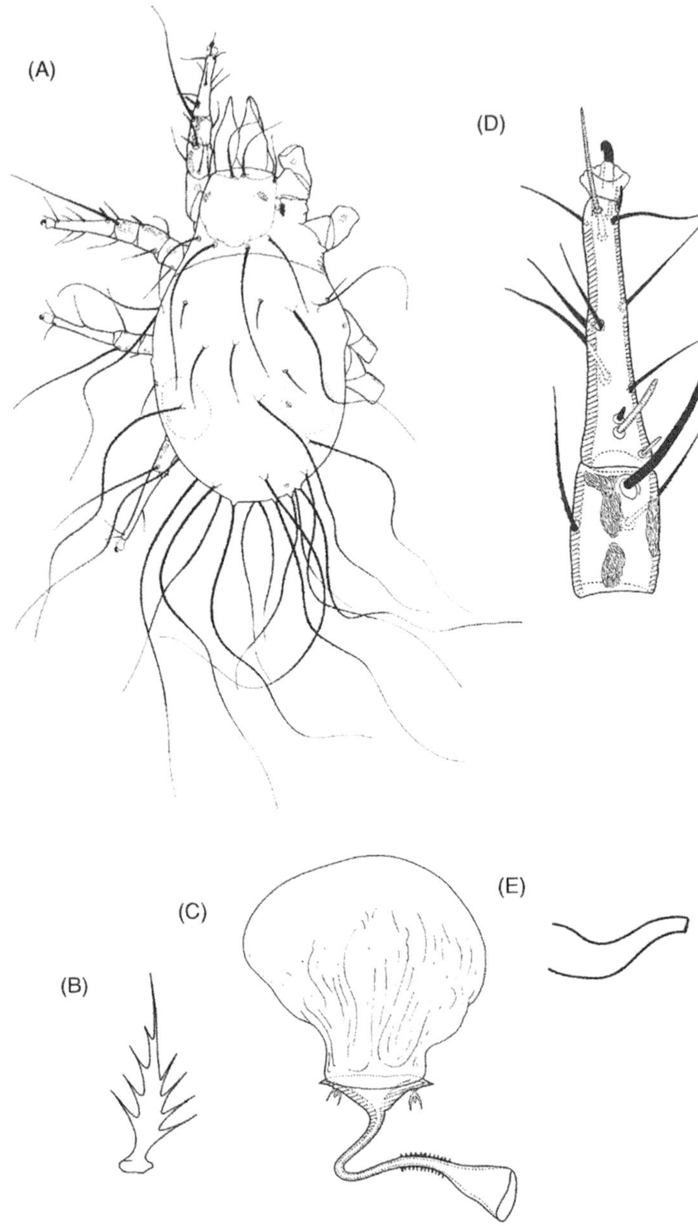

Fig. 28.6. *Tyrophagus neiswanderi* Johnston *et* Bruce. Female. Dorsal view (A); supracoxal seta (B); copulatory opening and spermatheca (C); tibia an tarsus I (D). Male. Aedeagus (E) (from Vacante, 1988).

slightly concave and posterolateral sides strongly concave. Grandjean's organ is finger-like, with a basal lobe with two basally confluent, and one separate spiniform teeth. The setae *scx* have a shaft widened at the point where the pectinations begin, with 8–11 moderate or short pectinations. The setae d_1 are about 1.6 times the setae c_1, and 1.7 times longer than the setae d_2; the setae d_2 are about 0.9 times the length of the setae c_1. Coxal plates I reach to the posterior apex of the prosternal apodeme; coxal plates II narrow sharply along the distal three-fifths, then become contiguous with the apex of apodeme 2, with two-thirds of the posterior margin strongly concave. The spermathecal duct narrows gradually from the copulatory opening to the base of the spermathecal sac and is about 2.6 times longer than the distance between the sclerites of the oviducts; the base of the spermathecal sac is triangular and funnel shaped. The solenida ω_1 and ω of tarsi I and II are slender, almost cylindrical. The setae *w* and *r* of tarsus IV are setiform (Fan and Zhang, 2007a).

MALE. The idiosoma is 397 μm long and 242 μm wide. The prodorsal shield and Grandjean's organ are as in the female. Coxal plates I and II, and solenidia w1 and w of tarsi I and II, are as in the female. The setae d_1 are about 1.5 times longer than the setae c_1 and 1.3 times longer than the setae d_2; the setae d_2 are about 1.1 times longer than the setae c_1. The aedeagus has one major curve, with the distal end almost straight, tapering from base to tip and lateral arms turning outwards. The setae *w* and *r* of tarsus IV are setiform (Fan and Zhang, 2007a).

Geographical distribution

T. neiswanderi is a cosmopolitan species reported from various regions, such as Australia (Fan and Zhang, 2007a), China (Teng *et al.*, 1988; Fan *et al.*, 2010a), Czech Republic (Hubert *et al.*, 2009), Germany (Fan and Zhang, 2007a), Ireland (Hagstrum *et al.*, 2013), Italy (Vacante, 1988), Japan (Nakao and Kurosa, 1988), Kenya (Hagstrum *et al.*, 2013), Mexico (Estebanes-Gonzalez, 1997), the Netherlands (Fan and Zhang, 2007a; Siepel *et al.*, 2026), New Zealand (Martin and Workman, 1985; Fan and Zhang, 2007a), Poland (Czaikowska *et al.*, 1988), Serbia and Montenegro (Petanovic, 2001), Spain (Sánchez-Ramos *et al.*, 2007a, b), South Africa (Meyer Smith and Rodrigues, 1966; Fan and Zhang, 2007a), Switzerland (Fischer, 1993), Turkey (Kirişik *et al.*, 2018), UK (Griffiths, 1979; Fan and Zhang, 2007a) and United States (Johnston and Bruce, 1965; Fan and Zhang, 2007a).

Commodities, facilities and vegetal and animal hosts

The list of commodities of the pollen-cap mite includes barley, Chinese medicine, *Chrysanthemum* cutting, grain collected in bread van and farm (Robertson; 1959; Johnston and Bruce 1965; Hughes 1976). The mite has also been found in Cabrales cheese and nests (Sánchez-Ramos *et al.*, 2007a, b). Moreover, it has been found in the respiratory tract of humans, in the short-tailed bat and in scale culture rooms (Fan and Zhang, 2007a).

Fan and Zhang (2007a) and Vacante (2016) reported the list of host plants and vegetable products infested by *T. neiswanderi*.

Bio-ecology

The pollen-cap mite inhabits various organic substrates, stored food and cultivated plants, such as ornamentals and vegetables grown in greenhouses. At 25°C, the development from egg to adult takes 2–3 weeks, and is faster on corms of *Freesia* and *Crocus*, than on tulip or other plants (Czajkowska and Kropczynska, 1991).

The developmental rate and survival of immature stages of *T. neiswanderi* has been studied on Cabrales cheese, at constant temperatures ranging from 10°C to 31°C, and a RH of 90 ± 5%. At 7°C and 29.7°C, the larvae showed the greatest mortality at extreme temperatures among immature stages, and the optimal temperature for development appeared to be 27–28°C. The mite development is well-described by the Lactin model (R(a)2>0.97). Furthermore, manipulation of temperature, based on modelling predictions, may well be used to control mite populations during the cheese maturing process (Sánchez-Ramos *et al.*, 2007a). At temperatures ranging from 10°C to 31°C, and with an RH of 90 ± 5%, the preoviposition period, the fecundity and the daily fecundity are adversely affected by extreme temperatures, while the oviposition period increases as the temperature decreases. Male and female longevity increase as the temperature decreases, and longevity in males is greater than in females. A non-linear Lactin model describes the effect of temperature on the intrinsic rate of natural increase of mite populations, and the optimum temperature for development is predicted at 26.6°C. At this temperature, the population doubling time is 2.8 days; the lower and upper thresholds are 7.4°C and 31.7°C, respectively (Sánchez-Ramos *et al.*, 2007b). The mite produces a pheromone alarm (Kuwahara *et al.*, 1989a).

Symptomatology and damage

Although the symptomatology and damage of *T. neiswanderi* on stored foods are similar to those of other *Tyrophagus*, usually they are of secondary importance with the exception of cheese that may become not marketable.

Control

Usually, the control is based on prevention, followed by curative actions. Prevention provides the rigid cleaning of the environments and the integrity of foodstuffs. Physical, chemical and/or biological means are curative means.

PHYSICAL CONTROL. Research carried out on the tolerance of *T. neiswanderi* and other acarid mites to high temperatures (40°C for 24 h, 45°C for 3 h and 50°C for 1 h) showed that the tolerance of *T. similis* to high temperature was lower than that of *T. neiswanderi*. *T. similis* may not have developed high-temperature tolerance because, in contrast to *T. neiswanderi*, the species dwells in the soil, where temperature fluctuation is not generally strong (Kasuga and Honda, 2006a). The benefit of this control method for *T. neiswanderi* is only partial.

BIOLOGICAL CONTROL. Releases of *Hypoaspis* spp. (Laelapidae) into the orchid media hindered the development of *T. neiswanderi* populations, thereby preventing injury to the flowers (Workman and Martin, 2002).

CHEMICAL CONTROL. The need to maintain high standards of quality make the use of chemical means on ornamental and flower plants a priority. The choice of acaricides is not easy for reasons of efficiency, cost and phytotoxicity, and many chemicals have been tested. Thus, on orchids in New Zealand, it has been observed that carbaryl, deltametrin, dimethoate and methiocarb initially reduce pollen-cap mite numbers, but after 14 days, only methiocarb sustains this reduction to a similar level to that achieved by the predator mite *Hypoaspis* sp. Dimethoate gave the least effective control of the pollen-cap mite and the greatest reduction in *Hypoaspis* sp. populations. The ability of these four pesticides, plus taufluvalinate, acephate and methamidophos, to reduce movement of the pollen-cap mite was low, and after 48 h, only methiocarb reduced the number of mites moving up the flower stem (Workman and Martin, 2002).

***Tyrophagus palmarum* Oudemans (Fig. 28.7)**

Common name

Unknown

Diagnostic characteristics

FEMALE. The idiosoma is 350–550 µm long. The general aspect similar to that of male (Hughes, 1976). Eyespots absent; the supracoxal setae scx are slender, tapering from base to tip, its shaft bearing ten short pectinations; the setae d_1 are about 4.1× length of c_1, and 5.4× length of d_2; d_2 is about 0.8× length of c_1. The spermathecal duct is cylindrical, widening slightly as it reaches the base of the spermathecal sac, whose base is circular, ring-shaped, and the distance between the sclerites of the oviducts is 12. On the tarsus I ω_1 clavate, on tarsus II ω 'banana'-shaped, medially widened and tapering at the base and apex. The setae w and r of tarsus are filiform (Fan and Zhang, 2007b).

MALE. The idiosoma is 330 µm to 450 µm long (Hughes, 1976). Eyespots, solenidia ω_1 and ω as in the female; d_1 about 2.9× length of c_1 and 3.3× length of d_2; d_2 are about 0.9× length of c_1; aedeagus distinctly S-shaped, distal 1/3 reversely curved, tapering from base to tip, internal diameter linear, lateral arms supporting aedeagus turning inwards; the setae w and r of tarsus IV filiform; ratio $(a+b)$: c=1.5 (Fan and Zhang, 2007b).

Geographical distribution

This mite has a cosmopolitan geographic distribution and is known from various countries, such as Azores Islands (Cardoso *et al.*, 2010), Belgium (Fain and Wauthy 1979), Canada (Delfinado-Baker and Baker, 1987), China (mainland) (Jiang, 1989; Li, 1999), Czech Republic (Ždárková, 1967), Germany (Rack, 1968a; Franz *et al.*, 1998), Ghana, India (Rao and Prakash, 1985, 1986), Indonesia (Haines, 1995), Ireland, Israel (Gerson *et al.*, 1979), Italy (Lozzia and Ottoboni, 1989; Vacante, 1989a), Malta (Pfliegler *et al.*, 2017), the Netherlands (Oudemans, 1924a; Fan and Zhang, 2007b; Siepel *et al.*, 2016), New Zealand (James *et al.*, 1975; Fain and Galloway, 1993), Nigeria, Poland (Studzinski and Malachowska, 1975), Puerto Rico (Delfinado-Baker and Baker, 1987), Russia (Volgin, 1975), Sweden (Bostrom *et al.*, 1997), Spain (Recio-García

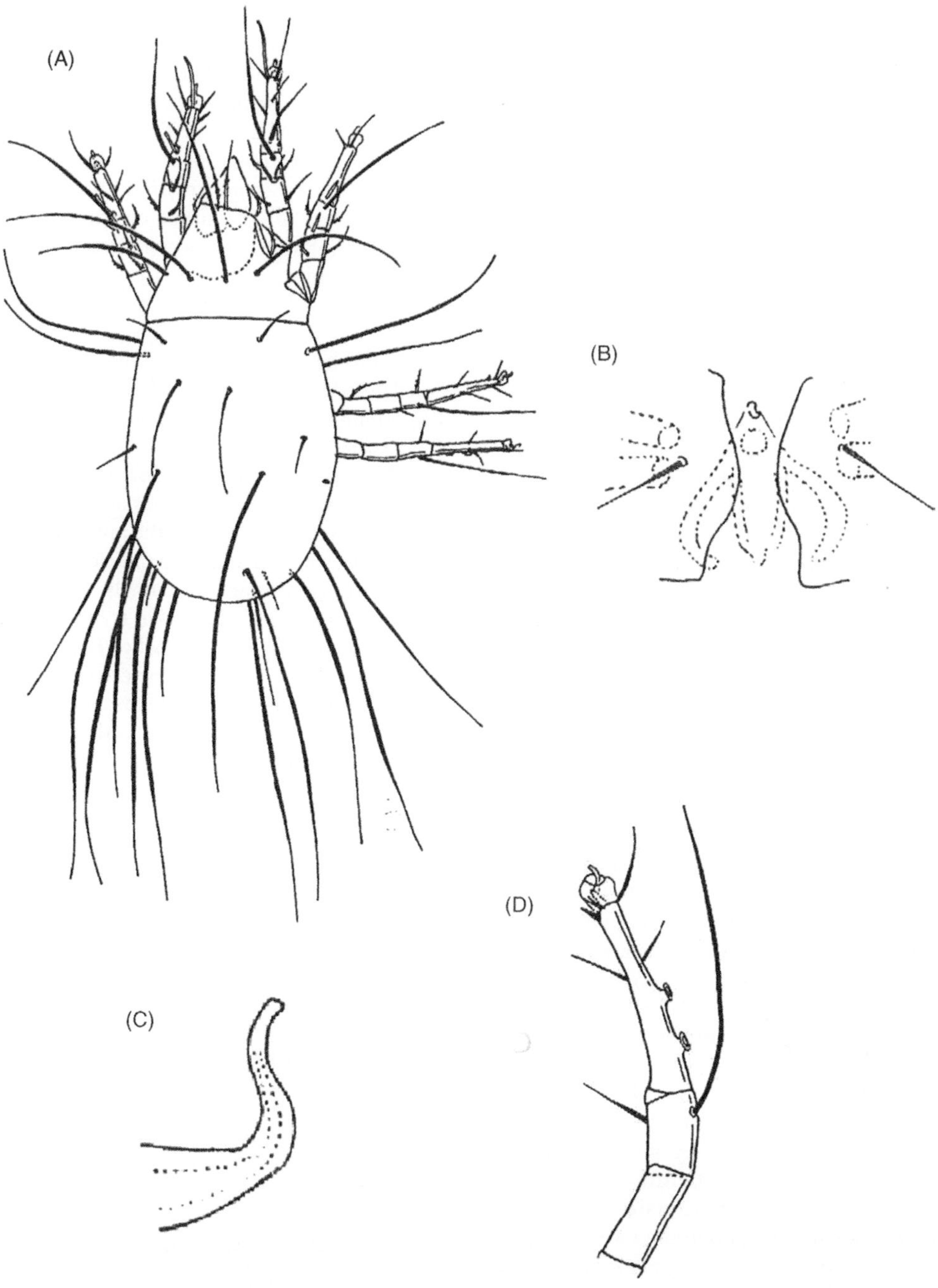

Fig. 28.7. A. *Tyrophagus palmarum* Oudemans. Male. Dorsal view (A); genital region (B) (from Hughes, 1976, modified); aedeagus (from Hughes, 1976) (C); leg IV (from Hughes, 1976, modified) (D).

and García-Cachán, 2002), Taiwan (Tseng, 1972, 1979), UK (Robertson, 1959; Griffiths, 1960; Williams, 1961; Hewitt and Turk, 1974; Wilkin, 1979; Fain *et al*,. 1991; Dunn *et al.*, 2002), United States (Delfinado-Baker and Baker, 1987). Fan and Zhang (2007b) reported that 'presumably, the above distribution is not for *T. palmarum*, but for *Tyrophagus vanheurni* Oudemans'; pending definitive information, we adopt here the above indications.

Commodities, facilities and animal hosts

The list of commodities includes baby food, barley, biscuit, cheese, Chinese medicine (Li, 1999), copra, flour sweeping, grain, ham, hay, honeybee comb, maize, palm kernel dust, pineapple fruit, rice, rice (husked), sausage, straw stack, sultana, tobacco, wheat, collected in various facilities (bread van, cheese store, commercial grain storage, farm, farm granary, farm milling machine, flour mill, food store, grinder, provender mill, provender warehouse and railcar; Oudemans, 1924a; Robertson, 1959, 1961; Griffts, 1960; Johnston and Bruce, 1965; Hughes, 1976; Lynch, 1989; Recio-García and García-Cachán, 2002; Hagstrum *et al.*, 2013). Moreover, it is reported in the surface soil of grasslands, and in stacks standing in open fields (Grifftts, 1960), and in spider webs, where it feeds on the bodies of dead insects (Hughes, 1976). Wasylik (1963) reported the mite from birds' nests. Pfliegler *et al.* (2017) collected the mite on straw.

Bio-ecology

Fan and Zhang (2007) reported that *T. palmarum* Oudemans (1924d) (*sensu* Robertson, 1959) is synonym of *T. vanheurni* Oudemans (1924d). Masoudian *et al.* (2018) in our revision of the genus *Tyrophagus* in Iran accepted this systematic vision, and report the species as *T. vanheurni*. This conclusion contrasts with the systematic identity traditionally assigned to the species, named by Hughes (1976) as *T. palmarum* and cited by many authors in the literature. We prefer to adopt here the Hughes" vision to avoid generating further confusion, while waiting for a broader review of the family, to bring order to the problem.

Fundamental aspects of bio-ecology similar to those of *T. longior* (Gervais). Grifftts (1960) collected mites from stacks of hay, maize and straw and from permanent grassland in the East Midland counties of England. Twenty-two species were associated with stored food and food stores, and of these 21 stored-product species were collected from hay- and straw stacks, and five from grassland. *A. siro* Linnaeus, a major pest of stored cereals and cereal products, was the dominant species in stacks, and was also found in grassland samples. *T. longior* was similarly distributed and occurred almost as frequently. The motile hypopial stage of *A. siro* was found in 43% of the stack samples; it was attached to mites, beetles and other invertebrates. Its distribution pattern resembled that of the fungus beetle *Enicmus minutus* Linnaeus. *T. palmarum* Oudemans, an uncommon stored-product mite, was the dominant species collected from grassland samples. Results suggested that all the mites commonly associated with foodstuffs stored indoors are present in the outdoor habitat afforded by stacks. Moreover, *A. siro* and *T. longior* are dominant in stacks, and both appear to be field dwellers, while *T. palmarum* appears to be a successful field dweller, and its occasional association with stored food may be accidental. The dominant warehouse partnership of *A. siro* and *G. destructor* (Schrank) does not occur outdoors, and *T. longior* replaces *G. destructor*. Glycyphagid mites seem to live mainly in haystacks sited in stackyards, and according to how the mites were acquired, grainstacks standing in the open field show a peculiar combination of dominants, namely *A. siro, T. longior* and *T. palmarum*. *T. palmarum* is reported as injurious pest on ham in Spain (Sánchez-López, 2002). The mite is less important on tropical stored products and is commonly found in association in small number, in mixed populations with *T. putrescentiae* (Schrank) (Kumar, 2017).

Symptomatology and damage

As *T. putrescentiae* or *T. longior*.

Control

As *T. putrescentiae* or *T. longior*.

Tyrophagus perniciosus Zakhvatkin (Fig. 28.8)

Common name

Unknown

Diagnostic characteristics

FEMALE. The body is oval, including the gnathosoma is 549–687 µm long and 263–375 µm wide. The prodorsal shield is punctate, with two pairs of setae (*vi* and *ve*), almost pentagonal in shape with lateral margins, slightly concave. The eyespots are absent. The basal lobe of Grandjean's organ has one large tooth and four small teeth. The supra-

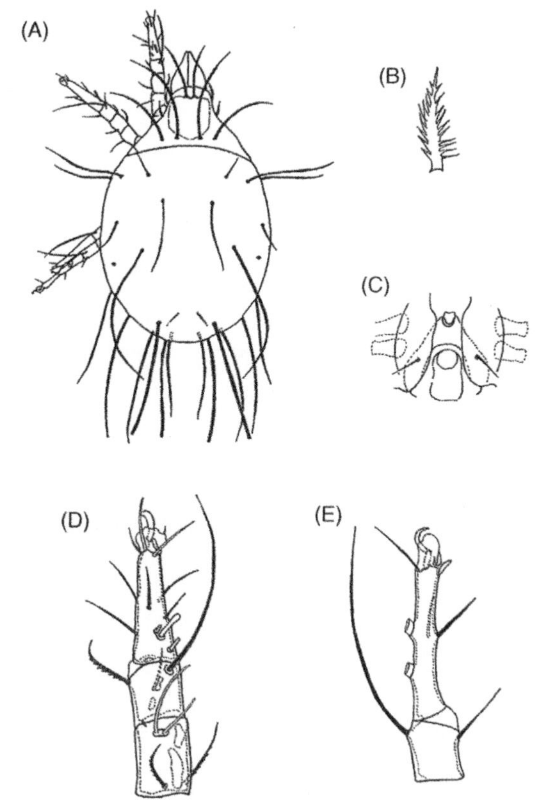

Fig. 28.8. A. *Tyrophagus perniciosus* Zachvatkin. Male. Dorsal view (from Hughes, 1976, modified) (A); supracoxal seta (from Hughes, 1976) (B); genital region and aedeagus (C); leg I (D); leg IV (E) (from Hughes, 1976, modified).

coxal seta *scx* is pectinate with eight rays on each side, and broadly expanding from the base but gradually tapers to a fine point distally. All dorsal setae are finely serrated, and all opisthosomal setae are whip-like except c_1, d_1 and d_2. The opisthosoma possesses three pairs of lyrifissures (*ia*, *im* and *ip*), and one pair of opisthosomal glands (*gla*) set at level of seta e_1. The seta c_1 is the shortest, whereas the seta h_1 is the longest dorsal setae. Ventrally, the coxal plates I are divided with the apodemes I anteriorly, on each side with three nodules; the coxal plates II are broadly triangular, and with a well-developed apodeme near the base of trochanter seta; with a pair of thin sclerotized sejugal apodemes between the coxae II and III; the coxal plates III–IV each with apodemes. The genital region (posterior to sejugal apodemes to coxae IV) with two pairs of genital papillae, a pair of setae (*g*) and genital folds. The anal region with three pairs of adanal setae (ad_{1-3}) and three pairs of pseudoanal setae (ps_{1-3}), the seta ps_1 is the longest anal setae. A pair of lyrifissures (*ih*) set at level of seta ad_2 base. The copulatory opening is set posterior to anal opening, and the spermathecal duct is very wide along its entire length and joins the base of spermathecal sac and with a long neck (Masoudian *et al.*, 2017).

MALE. The body is long, including gnathosoma 433–502 µm and width 210–225 µm. The prodorsal shield is punctate, with two pairs of setae (*vi* and *ve*) nearly pentagonal in shape with lateral margins slightly concave. The eyespots are absent. The basal lobe of Grandjean's organ with one large tooth and four small teeth. The supracoxal seta *scx* is pectinated with 9–10 branches on each side. All dorsal setae are finely serrated. All opisthosomal setae are whip-like, except c_1 and d_2. The opisthosoma bears three pairs of lyrifissures (*ia*, *im* and *ip*) and one pair of opisthosomal glands (*gla*) at level of seta e_1; the setae c_1 are the shortest and the setae h_{1-2}, the longest. Punctate; the palp is two-segmented, and the palp tarsus with a simple seta (*pt*) and one solenidion (ω); the palp tibia bears two simple setae; the infracapitulum has one simple seta (*m*). The rutellum is distinct and developed; the cheliceral seta *cha* is spine-like, cheliceral movable and fixed, and digits with 3–4 teeth. The palp coxa with one *elcp* setae. The coxal plates I are divided with apodemes I anteriorly with four nodes on each side; the coxal plates II with a well-developed apodeme near the base of trochanter seta; between the coxae II and III, with a pair of thin sclerotized sejugal apodemes, 50–55 long; the coxal plates III–IV each one with apodemes; the genital region (between the coxae IV) has two pairs of genital papillae, and one pair of setae (*g*); the aedeagus obviously curves, sickle-shaped, the lateral arms supporting aedeagus turning inwards. The anal region bears three pairs of pseudoanal setae (ps_{1-3}), the seta ps_1 and ps_3 the longest and the shortest anal setae, respectively, a pair of anal suckers 20µ, and a pair of anal discs 4µ in diameter (Masoudian *et al.*, 2017).

Geographical distribution

T. perniciosus is known from Australia (Hughes 1976; Fan and Zhang, 2007a), Bulgaria (Hughes, 1976), Canada (Beaulieu and Knee, 2014), China (Fan *et al.*, 2010a), Czech Republic (Verner *et al.*,

1982; Hubert *et al.*, 2006), Germany (Kazhdaya, 1996; Franz *et al.*, 1997), Greece (Palyvos *et al.*, 2008), Iran (Masoudian *et al.*, 2017), Ireland (Cusak *et al.*, 1975), Japan (Nakao and Kurosa, 1988), Kazakhstan (Sadieva, 1984), the Netherlands (de Jonge, 1988; Siepel *et al.*, 2016), Poland (Szilman *et al.*, 2006), Primorskiy Kray (Tareev, 1970), Russia (Zakhvatkin, 1941), Turkey (Cobanoğlu and Bayram, 1998), UK (Hughes, 1976), Ukraine (Akimov and Oksentiuk, 2018; Oksentiuk *et al.*, 2022) and the United States (USDA, 1981).

Commodities, facilities and vegetal and animal hosts

Commodities of *T. perniciosus* include apricot (dried), barley, beehives, cheese, Chinese medicine, flax seed, grain, grain product, hemp seed, herb, oats, oilseed sweeping, seed, sweeping, and wheat collected in bread van, commercial grain storage, farm, flour mill, dirty seed in a budgerigar cage. Facilities include dust granary, granary, off-farm storage, provender mill and warehouse (Robertson, 1959; Zakhvatkin, 1941, 1959; Johnston and Bruce, 1965; Hughes, 1976; Lynch, 1989; Beaulieu and Knee, 2014; Akimov and Oksentiuk, 2018; Oksentiuk *et al.*, 2022).

Fan and Zhang (2007a) published an extensive list of various hosts of *T. perniciosus*, including cauliflower seedlings, melon, oil-producing seeds and probably many other plants stored in granaries and warehouses, pumpkin, soil and dead plant residues, spinach, tomato (seed and newly germinating plants), various grain products (cereals, etc.). The mite has also been collected on mushrooms (*Morchella conica* Persoon and *Morchella esculenta* (L.) Pers.), and it occurs on protected crops (Zhang, 2003) and is a pest of cucumbers, melons, pumpkins and spinach (Sadieva, 1984; Nakao and Kurosa, 1988; Nakao, 1991). Sadieva (1984) observed the mite feeding on the mycelium of *Sporidesmium mucosum* var. *pluriseptatum* P. Karst. *et* Har. on cucumber. Akimov and Oksentiuk (2018) reported the mite on spoiled root crops. In the same country, Oksentiuk *et al.* (2022) number as commodities and facilities mills, granaries and warehouses, barns with animal fodder storage, honeybee beehives and vegetable storages.

Animals and their products on which *T. perniciosus* has been found include bird cages (e.g. budgerigar cages), humans (causing acariasis), nests of house martins (*Delichon urbica* (L.)) and starlings (*Sturnus vulgaris* Linnaeus).

Bio-ecology

The bio-ecology of *T. perniciosus* is little investigated. According to the information discussed above, the mite may occur in protected crops (Zhang, 2003), where it causes damage to cucumbers, melons, pumpkins and spinach (Sadieva, 1984; Nakao and Kurosa, 1988; Nakao, 1991), and on different stored food. Leal *et al.* (1988) have identified its pheromone alarm. The mite is less important on tropical stored products and is commonly found in association in small number in mixed populations with *T. putrescentiae* (Schrank) (Kumar, 2017).

Symptomatology and damage

The symptomatology and damage caused by *T. perniciosus* is as for other *Tyrophagus* spp.

Control

See most important *Tyrophagus* spp.

Tyrophagus putrescentiae (Schrank) (Fig. 28.9)

Syn. *T. communis* Fan *et* Zhang (2007)

Common names

Mould mite, copra itch mite, copra mite

Diagnostic characteristics

FEMALE. The idiosoma is 432 µm long and 285 µm wide. The prodorsal shield has eyespots, nearly pentagonal with apex towards the rear margin. Grandjean's organ is finger-like, with the basal lobe ornamented with small spiniform teeth. The setae *scx* expand from the base to pectinations, and the shaft bears 8–14 pectinations of various length. The setae d_1 are long – about 2.7 times longer than the setae c_1 and 2.4 times longer than the setae d_2; the setae d_2 are about 1.1 times longer than the setae c_1. Coxal plates I are beyond the apex of the prosternal apodeme; coxal plates II are broadly triangular, with a very slightly concave posterior margin, and the distal point of the triangle stands slightly proud of the apodeme apex. The spermathecal duct narrows rapidly from the copulatory opening, for a distance of about 1.2 times the distance between the sclerites of the oviducts; it forms a neck and

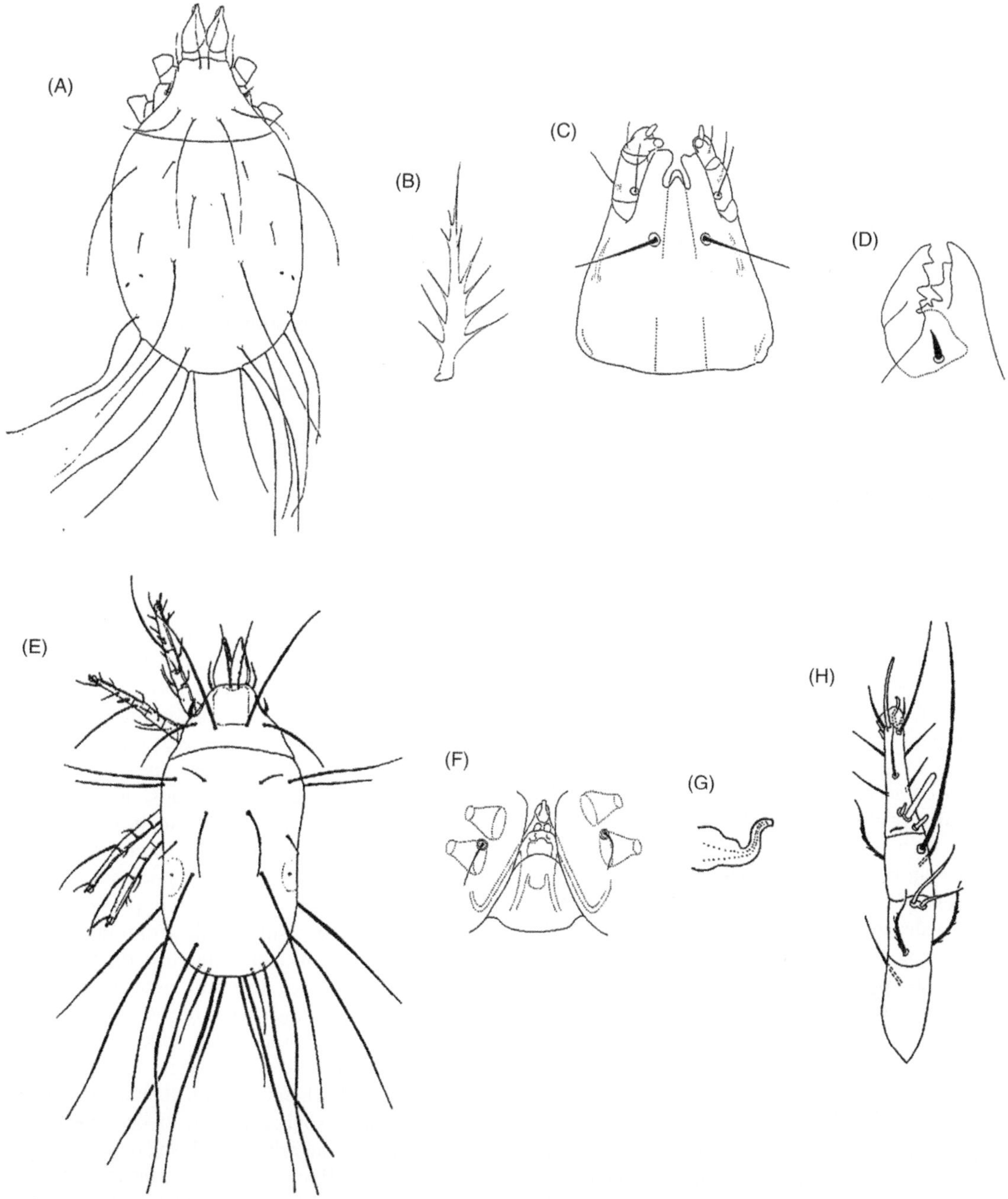

Fig. 28.9. *Tyrophagus putrescentiae* (Schrank). Female. Dorsal view (A); supracoxal seta (B); ventral view of gnathosoma (C); chelicera (D) (from Vacante, 1989). Male. Dorsal view (from Hughes, 1976) (E); genital region (from Vacante, 1989) (F); aedeagus (G); leg I (from Hughes, 1976) (H).

then gradually widens to the base of the spermathecal sac, over a distance of about 1.5 times the distance between the sclerites of the oviducts. The sclerotized base of the spermathecal sac is nearly flat or bends slightly forwards. The solenidion ω_1 of tarsus I is slightly widened at the apex, and the solenidion w of tarsus II is also slightly widened at the apex; the setae w and r of tarsus IV are setiform (Fan and Zhang, 2007a).

MALE. The idiosoma is 429 µm long and 280 µm wide. The prodorsal shield and Grandjean's organ are as in the female. The eyespots, coxal plates I and II, solenidia of legs I (ω_1) and II (ω) are as in the female; the setae d_1 are long, about 2.8 times as long as the setae c_1, and 2.3 times as long as the setae d_2; the setae d_2 are about 1.2 times longer than the setae c_1. The aedeagus is 'S'-shaped, with the distal one-third reversely curved, and tapering from base to apex, and the lateral arms turning outwards. The setae w and r of tarsus IV are setiform (Fan and Zhang, 2007a).

Solarz (2011) reported the following distinctive morphological characters. In both sexes, the anterior margins of propodosomal shield bear pigmented spots (corneae); coxal plate II with a sinuous posterior margin, so that the plate narrows sharply along the distal 1/3; tarsi I and II with solenidion ω_1, terminating in a distinctly expanded tip; distal 2/3 of solenidion ω_1 on tarsi I widened. The female has the genital seta g longer than vulva; paraproctal setae ad_1 much shorter than ad_2; setae ad_2 distinctly longer than anal slit; internal spermathecal apparatus large, base of spermatheca expanded, flat, spermathecal duct with distinct constriction about halfway along length; proximal part of spermathecal duct gradually widened. The male has the aedeagus short (<20µm), S-shaped with two deep curves, one at base, the other in apical third; distal 1/3 of aedeagus curved at an angle of about 75–100° to its median part; tarsal copulatory suckers on tarsus IV equidistant from the base and apex of the segment; setae ps_2 more than 1.8–2 times longer than the anal slit (Solarz, 2011).

Geographical distribution

T. putrescentiae is found worldwide, known from Africa, Argentina, Australia, Brazil, Canada, Chile, China (mainland, Hong Kong), Croatia, Czech Republic, Taiwan, Cook Islands, Crete, Czech Republic, Ecuador, England, Fiji, Greece, Germany, Hong Kong, Hungary, India, Indonesia, Iran, Iraq, Ireland, Italy, Jamaica, Japan, Kenya, Madagascar, Malawi, Malta, Nepal, the Netherlands, New Zealand, Papua New Guinea, Peru, Philippines, Poland, Portugal, Samoa, Singapore, Solomon Islands, South Korea, Spain (Iberian peninsula, Azores and Canarias Islands), Swaziland, Taiwan, Thailand, Tokelau Islands, Tonga, Turkey, UK, Ukraine, United States, Vanuatu, West Africa, Venezuela and Wales (Ždárková, 1967; Guimarães, 1970; Khalilmanesh, 1973; Sepasgozarian, 1978; Ripka *et al.*, 2005; Fan and Zhang, 2007a; Hagstrum *et al.*, 2013; Beaulieu and Knee, 2014; Akimov and Oksentiuk, 2018; Oksentiuk *et al.*, 2022).

Commodities, facilities and vegetal and animal hosts

The mould mite is reported on a long list of commodities, such as apple (rotten), apricot kernel, aquatic product, baby food, bacon, bamboo shoot, banana (dried), barley, barley (germinating), bean, bean curd, beehives, bee pollen, beet seed, biscuit, bread, broad bean, buckwheat, buckwheat product, cabbage (dried), cake, candied fruit, caraway, carob, carpet, carrot seed, cereal, cereal spillage (damp), cheese, chicken feed, chicory root, Chinese medicine, cocoa bean, common bean, condiment, copra, cotton fibre, cottonseed, cottonseed sweeping, cowpea, cumin, date, dried fruit, dried vegetable, egg (dried), feed pellet, freekeh fig (dried), fish (dried), flax seed, fungus (dried), garlic, garlic bulb, ginger, gin trash, gladiolus corm, gluten, grain, grass seed, ham, herb, herring meal, insect (dried), jelly, kaoliang, lettuce seed, lily (rotting), logan (dried), maize, maize flour, maize (old), malt, malt spillage, marmalade, meat product, milk (powdered), millet, mushroom (dried), narcissus, nut, oat, oilseeds, onion, onion seed, palm kernel, parsley (dried), parsnip seed, Paspalum seed, groundnut, pepper, pine seed, pineapple fruit, plum (dried), poppy seed, potato (old), potato (rotten), potato (sprouting), processed food, radish seed, raisin, rape seed, rape seed sweeping, residue, rice, rice (husked), rice (milled), runch seed, rye, sausage, seaweed (dried), seed cotton, shark fin (dried), shrimp (dried), sorghum, soybean, soybean groat, soybean meal, squid (dried), starch, sugarbeet seed, sugar (crude), sugar (red), sugar (white), straw mat, sultana, sunflower seed, tobacco leaf (cured), tobacco seed, tomato relish, tulip, vegetable (pickled), vinegar, wheat, wheat bran, wheat flour, wheat flour sweeping, wheat spillage, yeast (dried) and wool (Ždárková, 1967; Hughes, 1976; Hagstrum *et al.*, 2013). Moreover, the mite has been collected in flour mill and silo (Pfliegler *et al.*, 2017), and in Egypt on cumin, wheat bran, Turkish cheese, fereekeh, caraway, rice, raisin, dry figs, pepper, flour, cowpea, carob and pine seeds (Saleh *et al.*, 1985). In Ukraine Oksentiuk *et al.* (2022) found the mite in mills, granaries and warehouses, barns with animal fodder storage, outbuildings and honeybee beehives.

The facilities include bread van, cheese store, maize processing plant, farm, farm granary, farm milling machine, flour mill, food store, market, mill, off-farm storage (Hagstrum *et al.*, 2013).

The list of plants or vegetable products hosting this mite includes, agar tissue cultures of orchids, almonds, asparagus, aubergines, bananas, bark scrapings, beans in the kitchen cupboard, buffet, *Caladium*, the calyx end of oranges, caper seeds, capsicum, chaff, water vine (*Cissus antarctica* Vent.) seeds, citrus, lemon (*Citrus limon* (Linnaeus) Burm.), cocoa beans, coconuts, Arabian coffee (*Coffea arabica* Linnaeus), elephant ear (*Colocasia esculenta* (Linnaeus) Schott) bulbs, copra, *Cryptes baccarum* [?], cucumber leaf, cultures of *Tribolium*, curry leaves, cycad palm seed, *Dahlia* bulbs, damp grain, debris in containers, decaying coconut, dormant bulbs, dried figs, dried food, dried fruit, dried lychees, dried vegetable leaves, dust in grain silos, feijoas, figs, foodstuffs in household pantries, *Freesia* bulbs, garlic, gladioli bulbs, grain, groundnuts hyacinth, *Iris* bulbs, kangaroo diet, kentia palm, lemon, bay laurel (*Laurus nobilis* Linnaeus) leaves, lychees, matua on nutmeg, mixed grain horse fodder and young stock foods, maize, molactrate molasses blocks, mould in hazel nuts, muscari bulbs, *Narcissus* bulbs, *Nipa* palm leaves, onions, orange, orchid plants, Orchidaceae, palm seeds, persimmons, pimentos, pine tree seeds, pineapple, plant roots, *Plumeria* sp., potato, pollen from hives, *Ranunculus* bulbs, raspberry jam, red dates, rice plant, roses, *Saintpaulia*, seeds of sycamore (*Acer pseudoplatanus* Linnaeus), seeds and spices, soil, soybeans, stored sorghum, strawberries, taro, *Theobroma cacao* Linnaeus (cacao), tomatoes, tulip bulbs, vanilla beans, vanilla pods, *Virgilia divaricata* Adamson, water chestnuts (*Scirpus tuberosus* Desf.), wheat grains ex silo, wheat stores, ya (Chinese white) pear and yams (Fan and Zhang, 2007a; Vacante, 2016). Recently, the mite has been collected also on abalone, apple, barley, bean paste cake, bacon, bedding, biscuits, black fungus, bread, butter, cake, carpet dust, carrot, casing for sausages, cattail mat, chilli sauce, coconut meat, maize, cottonseed hulls, cowpea, cracknel, *Dityophora*, dried bamboo shoots, dried longan pulp, dried persimmon, dried pork flakes, dried pork slice, dried turnip, millet, fermented bean curd, fermented soybean, *Fammulina*, flour, fruit of ginseng, Chinese wolfberry, Chines yam, conch, cotton lint, gammon, grapefruit, fillet, hawthorn, haw jelly, haw flakes, honey lotus root dry, jellyfish, laver, light straw, malted milk, melon seeds meal, mildewed rapeseed, milk powder, mushroom, old silk noodles, onion, groundnut cake, pickled cucumbers, pineapple, potato, preserved duck gizzard, rot leather, saqina, sausage, sea-tent, scallop, shelled fresh shrimps, soy sauce residue, squids, spiced pork, rape seeds cake, rice, sesame, sesame oil slag, soybean, soybean power, stark, straw, sugarcane, sunflower seeds, pine nut, fried pork flakes, wheat koji, jellied meat, jerky in China (Li *et al.*, 2015a).

Animals and their products hosting the mite include beehives, *Calolampra* (Blaberidae), cheese, Cheshire cheese, cockroach colonies, dead larvae in wood infested by the citrus borer, the dead cerambycid *Prionoplus reticularis* White, dried milk powder, dried shellfish, dust beneath banana leaf matting, fish houses, Gouda cheese, human urine, 'Kai-iwi' foil-wrapped cheese, labarotary culture of *Psocoptera*, *Locusta migratoria* (Linnaeus), mealworm cultures, millipedes (*Polyconoceris alaskis*?), *Nauphota* (cockroach), passalid beetle, rabbit, rooster feathers 'Tintinhull', sheep faeces, silverfish cultures, *Tineola biselliella* Hum (Fan and Zhang, 2007a; Vacante, 2016).

The mite has also been collected from fungal cultures (*Sporotrichum*), mushrooms, cracks and the floor, dust beds, glass surfaces in a house, houses, a kitchen in a farmhouse, pallet and bag scrapings and a spoon from the kitchen of a house (Fan and Zhang, 2007a; Vacante, 2016).

Bio-ecology

This mite was originally described as *Acarus putrescentiae* by Schrank in 1781, and successively as *Tyrophagus noxius by* Zachvatkin in 1935. More recently, Fan and Zhang (2007a) analysed the Robertson's (1959) concept of *T. putrescentiae* (Schrank) and pointed out a mixture of two species named *T. putrescentiae* (Schrank) and *T. communis* Fan *et* Zhang, the first rare and the last very common, respectively. *T. putrescentiae* differs from *T. communis* in that two-thirds of the posterior margin of coxal plates II is strongly concave, and the solenidion w1 is widened in the distal two-thirds. Successively, Klimov and OConnor (2010) reported that 'the prevailing usage of the name *T. putrescentiae* comprises almost exclusively, the common species, and the name *T. communis* is a junior synonym of eight previously named taxa with extant types and proposed the new name *T. fanetzhangorum* Klimov *et* OConnor 2009 for the rare species'. According to this interpretation, they 'proposed to conserve the prevailing usage of the name *T. putrescentiae* by designation of a new neotype from a culture currently maintained in a research institution, which was cited in many published works and started from specimens collected close to the type locality of *T. putrescentiae*'. Later, Klimov and Tolstikov (2011) reported again that *T. putrescentiae* Schrank is synonym of *T. communis* Fan *et* Zhang (2007a) and that *T. fanetzhangorum* is a probable synonym. Vacante (2016) retained that the vision of Klimov and OConnor (2010) does not simplify the topic and followed the choice of Fan and Zhang (2007a) that assigned to *T. communis* a worldwide distribution and a wide list of host plants, animals and their substrates, and identified the species with *T. putrescentiae auctorum*. The further development of the topic suggested here to adopt the systematic vision of Klimov and OConnor, and to consider *T. communis* as synonym of *T. putrescentiae* (Schrank).

T. putrescentiae is a cosmopolitan mite of particular importance in tropical regions. It's considered a severe pest in agriculture, and an important species in human and veterinary medicine. It infests a number of stored commodities, with high fat and protein content, throughout world. It can also be collected in cultivated mushrooms, soil, mosses, litter, store houses, barns and farm buildings due to their ability to tolerate low humidity and a wide range of temperatures (Hughes, 1976). The mite has also been seen on pollens.

The life cycle of *T. putrescentiae* develops through the egg, larva, protonymph, tritonymph and adult stages. One male is able to fertilize about 450 females, and the type of food affects the reproductive rates of females. Females reared on wheat germ or yeast mate twice, and in a month can lay about 500 eggs, until 60 eggs/ day, and an average of four eggs/day. About 70% of the eggs are laid during the first 3 weeks of adult life (Boczek, 1991). The oviposition starts above 8°C, with the highest values between 22–26°C (Boczel, 1975), and no development occurs at temperatures below 8.5°C or above 36°C (Cunnington, 1969). Kevan and Sharma (1963a) found that at low temperature, mating was delayed and prolonged and it did not occur at 0°C. According to Cunnington (1985), at 20°C and 80% RH, mating and oviposition occurred soon after emergence, but egg-laying was delayed at low temperatures in *T. putrescentiae* and *A. siro* Linnaeus. *T. putrescentiae* is photonegative and showed that with increase in light exposure, fecundity, egg viability was significantly reduced and hatching period, life cycle was prolonged significantly (Kohli and Mathur, 1993). Ždárková and Voracek (1993) also reported the effect of physical factors on the survival of mites in stored grains. The fecundity depends by the number of males per female; so, females lay fewer eggs if mated by numerous males, and this value can be reduced by up to 50% when the male: female ratio is 4:1. The females predominate, and the sex ratio does not depend on food, parental age or frequency of copulation (Boczek, 1975). The development time from egg to adult, usually varies from 1–3 weeks, and according to the environmental conditions, and the type of food on which the mite is reared, can achieve 118 days (Kheradmand *et al.*, 2007a). Sánchez-Ramos and Castañera (2005) investigated the effect of temperature on reproductive parameters, and longevity of the mould mite at seven constant temperatures, ranging from 10°C to 34°C and a RH of 90 ± 5%. They found that the pre-oviposition period and the fecundity are adversely affected by extreme temperatures, and the oviposition period increases as temperature reduces. Moreover, they observed different patterns for males and females longevity, with greater longevities for males at intermediate temperatures, and more similar values for both sexes at extreme temperatures. Polynomial and non-linear models provided a good fit of the relationship of reproductive and longevity parameters with temperature. The effect of temperature on the intrinsic rate of increase of *T. putrescentiae* populations, was established by the non-linear Lactin model. The optimum temperature for development, was observed at 30°C. At this temperature, the population doubling time is 1.75 days. The lower and upper thresholds for *T. putrescentiae* populations are 10.4°C and 34.8°C, respectively. At 30°C and 90 ± 5% RH, the highest intrinsic rate of natural increase is 0.397, the lowest mean generation time is 12.6, and the lowest population doubling time is 1.75 days (Sánchez-Ramos and Castañera, 2005). Results of other studies showed that optimal conditions of development are 23.2 ± 2.1°C and 71 ± 5.6% RH (Adler, 2004; Riudavets, 2004).

The developmental time of *T. putrescentiae* from egg to adult, in optimal conditions, has been found to be 9.4, 7.2 and 8.5 days at 25°C, 30°C and 32.5°C, respectively. It has been calculated that 100 mites can render about 100 g of dog food to dust, in less than 4 weeks, under ideal conditions (Sánchez-Ramos *et al.*, 2007c). Favourable conditions agevolate the development of the mite populations. which cause a significative depletion of commodity associated with an economic loss (Ždárková, 1991a; Thind and Clarke, 2001; Sánchez-Ramos *et al.*, 2007c). Availability of additional protein or saccharide sources in diet, had an impact on the population increase of *T. putrescentiae*. The mite populations increased by 319, 317 and 180 times within 6 weeks, when yeast powder, glucose or sugar was added to the basic wheat bran diet (diet mass: additive mass 10:3), respectively as compared to basic diet, where it increased by only 70 times (Huang *et al.*, 2013).

Studies carried out in Egypt by Mostafa *et al.* (2013), highlighted the effect of some food stuffs on the biology of *T. putrescentiae*, at 25°C and 75% RH. The study showed that there was a slightly significant difference between the incubation period of mite females and food types. The female life cycle takes 11.1, 12.5, 13.3 and 14.0 days when fed on wheat flour, milk powder, fish powder and granular chicken feed, respectively. Female longevity was the longest on wheat flour (39.0 days), while it was recorded the shortest time on fish powder (25.1 days). The female needed 34.6, 30.4, 22.8 and 21.0 days for egg deposition, respectively. The highest number of deposited eggs was recorded on wheat flour (39.0 eggs), while the lowest was on fish powder (27.8 eggs), as feeding source. The male life cycle duration was significantly affected by the food type, as it averaged 9.8, 10.3, 12.4 and 13.2 days on the employed diets, respectively. Male adulthood was the longest on wheat flour (28.7 days), while decreased to 17.0 days on fish powder. The male lifespan lasted 38.5, 34.2, 31.2 and 30.2 days, when fed on the mentioned diets, respectively (Mostafa *et al.*, 2013).

All immature stages, female longevity and reproductive periods of *T. putrescentiae* are significantly affected by the mushroom species, as total immature developmental time varies from 4.41 ± 0.20 (*Auricularia polytricha* (Mont.) Saccardo) to 9.79 ± 0.92 days (*Ganoderma lucidum* (Curtis) P. Karst.) (Qu *et al.*, 2015). The longest period of the immature stages is the egg stage, whereas shortest stages are the protonymph and tritonymph stages (about 2 days). The intrinsic rate of increase (*rm*) and net reproductive rate (*R*0) are significantly different on nine mushroom hosts (Qu *et al.*, 2015).

Studies carried out in China highlighted that the threshold temperature for the development of egg, larva, protonymph and tritonymph feeding on yeast was 10.47, 8.09, 9.09 and 10.59, respectively; the effective cumulative temperature required for the development of egg, larva, protonymph and tritonymph was 41.79, 58.06, 36.27 and 39.53 degree-days, respectively. It required 221.72 degree-days above 8.38°C for the completion of a generation (Yu *et al.*, 2002). It took 48.04 or 8.41 days to complete a generation at 12.5 or 30°C on yeast, but the generation time lasted longer, 78.79 days at 12.5°C and 10.77 days at 12.5°C, when fed with maize (Liu, T. *et al.* 2006). The finite rate of increase (λ) and intrinsic rate of natural increase (*rm*), for this species were 0.0556 and 1.0572 at 12.5°C, 0.2638 and 1.3019 at 25°C, and 0.0568 and 1.0585, at 30°C, respectively (Liu, T. *et al.*, 2008). *T. putrescentiae* are often closely related to fungi. Xuan *et al.* (1993) observed that it was strongly attracted by and fed on two common parasitic fungi, *Penicillium frequentans* Westling and *Aspergillus fumigates* Fresenius. The mite could not complete its generation when fed on walnut without fungi (Xuan *et al.*, 1993).

Yassin *et al.* (2018a) studied the different biological aspects of *T. putrescentiae* and *R. robini* Claparède, reared at 25+2°C and 75+5% RH, on different food kinds, such as pure white wheat flour, normal wheat flour, wheat germ, rice flour, maize flour and semolina. Obvious differences between incubation period, life cycle, longevity and lifespan of both mite species were observed. The longest lifespan periods of *T. putrescentiae* male and female occurred when the individuals fed on wheat flour (41.61 and 50.9 days, respectively). However, the longest lifespan periods of *R. robini* were observed on maize flour (49.2 and 41.2 days for male and female, respectively). The food type's suitability clearly affected the number of laid eggs by the adult female of the mite species, and revealed a high significance between, the feeding on wheat flour for the first mite and maize flour for the second one (the most suitable feeding sources for mites), than any other feeding types. The highest number of *T. putrescentiae* eggs was, 40.2 eggs on wheat flour, and the lowest number was recorded on wheat germ (27.8 eggs). However, the highest deposited number of *R. robini* eggs was noticed when adult female fed on maize flour (140.2 eggs), but the lowest number of laid eggs was found when the individuals reared on wheat germ (70.4 eggs) (Yassin *et al.*, 2018a). Amoh (2016) referred that in the United States *T. putrescentiae* is a major pest of southern dry-cured ham during ageing. The ham may be cold-smoked with hardwood such as hickory after curing, and ageing occurs at 24–35°C and 55–65% RH in ageing rooms, for up to 2 years. Unfortunately, these conditions may agevolate the growth and development of *T. putrescentiae* populations (Graham *et al.*, 2011). When the ham is aged for less than 3 months little or no infestation occurred. The possibility of infestation is greater if ham aged for more than 5 months. However, short ageing duration of less than 5 months does not assure processors that mite infestation would not occur (Rentfrow *et al.*, 2006), and ageing times longer than 6 months are necessary to obtain the desired product flavour and quality that meets the niche markets (Toldrá *et al.*, 1997). The mite produces a pheromone alarm (Kuawara *et al.*, 1975).

Erban *et al.* (2016a) found symbiotic bacteria in compared mite populations, such as *Wolbachia* (two populations), *Cardinium* (five populations), *Bartonella*-like (five populations), *Blattabacterium*-like symbiont (three populations) and *Solitalea*-like (six populations). Of these, only *Solitalea* was found in all populations. The next most frequent and abundant sequences were *Bacillus, Moraxella, Staphylococcus, Kocuria* and *Microbacterium*. They reported that some bacterial species may occasionally be ingested with food. The bacteriocytes were observed in some individuals, in all mite populations. Bacteria were not visualized in food boli by staining, but bacteria were found by histological means in ovaria of *Wolbachia*-infested populations. They conclude that the presence of *Blattabacterium*-like, *Cardinium, Wolbachia*, and *Solitalea*-like in the eggs of *T. putrescentiae* indicates, mother to offspring (vertical) transmission. Results of this study indicate that diet and habitats influence not only the ingested bacteria, but also the symbiotic bacteria of *T. putrescentiae*.

Carnevari *et al.* (2012) studied in the laboratory, the potential of *T. putrescentiae* to control *Lasioderma serricorne* (F.) (Coleoptera: Anobiidae), a destructive insect pest ever found on stored tobacco. The experimental delineation had 33 arenas, being each experimental unit with 11 arenas of experiments with eggs, larva of *L. serricorne* with the density of *T. putrescentiae* maintained constant, where from 11 of these arenas, eight were for the analysis of predation and three to evaluate the natural mortality of the insect. The highest predation rate was found during the larval stage with, approximately, 54%, 68% and 78% mortality of *L. serricorne* from the fourth until the 60th day of predation. They concluded that it was possible to use the mite *T. putrescentiae* in pest management programmes of *L. serricorne* in the storage units of tobacco.

Symptomatology and damage

T. putrescentiae is a serious pest of different stored foods (Solomon, 1946; Hughes, 1976), decaying vegetable and animal matter (Krezeczkowski, 1961). It is a major stored food pest of commodities of high protein, high fat and/or MC of 15–45%, and it has been collected on over 140 commodities (Hagstrum *et al.*, 2013), including dried fruits, spices, cultured cheeses and other high value foods (Rentfrow *et al.*, 2006). It is also the most predominant mite in processing and packing areas of dry pet food factories (Thind, 2005; Nayak, 2006; Brazis *et al.*, 2008), cereal-based food products and other organic materials that are susceptible to storage mite infestation (Thind and Clarke, 2001; Kheradmand *et al.*, 2007a). It will also infest laboratory cultures of insects and microorganisms (Kheradmand *et al.*, 2007a). It is reported as an injurious pest on ham in Spain (Sánchez-López, 2002) and other countries.

Malik *et al.* (2018) published an exhaustive review on the relationship between *T. putrescentiae* and stored grains, due to its ability to tolerate low humidity and a wide range of temperatures (Hughes, 1976). It can cause problems to many foodstuffs, ranging from weight reduction and degradation of stored foods to accumulation of harmful residues (fungi, dead mites, faeces, eggs and bits of food), through their activities (Ždárková, 1971; Hughes, 1976). According to the indications of Stejskal (2001), for mites and other arthropods infesting stored foods, the outbreaks of the mould mite on grain may be associated with three types of damage, such as significant grain weight losses and decrease of germinability (Solomon, 1946; Ždárková and Reska, 1976), direct damage to human health due to allergenic contamination of food (Arlian, 2002; Matsumoto *et al.*, 1996) and diffusion as vectors of toxicogenic fungi (Hubert *et al.*, 2003) contaminants indirect of food and feed with mycotoxins (Franzolin *et al.*, 1999).

T. putrescentiae infestations induce direct and indirect damage to many commodities, as physical damage (Solomon, 1946; Ždárková and Reska, 1976), change of chemical composition (White *et al.*, 1979), allergic reactions in humans (Arlian, 2002; Vidal *et al.*, 2004; Jeong *et al.*, 2005; Wakefield, 2006) and in dogs fed with infested food (Vollset *et al.*, 1986) and dissemination of toxigenic fungi, such as *Aspergillus* spp. and *Penicillium* spp. (Franzolin *et al.*, 1999).

Moreover, infested commodities may acquire a minty odour due to lipid secretions by mites. Matsumoto *et al.* (1996) retained them responsible for allergic diseases among farmers and workers handling heavily infested stored foods. They can cause acute enteritis, diarrhoea, dermatitis, asthma, pulmonary acariasis and urinary acariasis (Yadav *et al.*, 2006). From the medical and veterinary standpoint, if ingested, the mite can cause poisoning in humans and animals (Blanco *et al.*, 1997; Matsumoto *et al.*, 2001; Sánchez-Borges *et al.*, 2005; Liao *et al.*, 2009; Canfield and Wrenn, 2010), such as acute enteritis, and may be related to the transmission of bacteria, yeasts and pathogenic fungi through food (Serpa *et al.*, 2004). In heavy infestations, *T. putrescentiae* emit a damp pungent smell, earning them the common name of 'lemon-scented mite' (Nayak, 2006). Feeding of mite induces the change of colour of grains from shiny to dull, which slowly turns into blackish colour. High density of populations causes serious damage to the stored foods (Kleih and Pike, 1995) and causes alarm to workers appearing as moving carpets of brown dust on the commodities, silos and sheds.

The mould mite feeds on embryo causing a germination loss in the grains, deterioration in quality of seed that are useless for the flour production and as feed for livestock (Wilkin and Stables, 1985). The direct damage of *T. putrescentiae* to stored grains is through contamination and penetration into seeds/embryo, the consumption of the grain germ and some extent, the endosperm, which unavoidably decreases the vitality and germination of the seeds. The grain becomes useless for seed (Ždárková, 1991b), brewing purpose and unacceptable to the miller (Solomon, 1946). Moreover, the mites can feed on the germ of kernels, thereby reducing the content of iron and vitamins of the B complex and germination ability (Krantz, 1955). The problem of loss in weight and seed germination is treated in several reports (White *et al.*, 1979). It has been observed that the mites cannot penetrate the grains if the coat is intact, and only during the storage the unavoidable scarification of the grains facilitates their attack and penetration. Solomon (1946) has found that the tyroglyphid mites are able to destroy the germ completely consuming very little of the remainder and showing their preference for the germinal part of the grains. Ždárková and Reska (1976) observed that on a limited amount of substrate, one mite consumes on average an amount of grains indirectly proportional to the number of individuals in the population. Moreover, the mites prefer the flour form of the stored grains, and fewer individuals develop in whole and broken grains, and only when the population develops on the whole grains, starts the contamination with its excreta associated with the pungent odour and their conversion into a powdery mass (Yadav, 1989; Singh, 1990).

Krezeczkowski (1961) tested 61 foods and found that *T. putrescentiae* preferred dried mushrooms, wheat germ, yellow cheese, biscuits, dehusked hemp seeds, fruits of pepper and dried leaves of some medicinal plants over the others. Although *T. putrescentiae*, *A. siro* Linnaeus, *Carpoglyphus berlesei* (Michael) and *Leiodinychus krameri* (G. et R. Canestrini) prefer cereals, they consumed preferably the bread (Radinovsky and Krantz, 1961). Moreover, the addition of methyl p-hydroxy benzoate in the diet was the best for *T. putrescentiae* (Bot and Meyer, 1961)., as well as the addition of inorganic salts affect the development of mites in stored grains (Ignatowicz, 1981). A synergistic attractancy of cheese components is reported by Yoshizawa *et al.* (1971).

Ždárková (1971) tested different materials characterized by a distinct odour, such as brick cheese, blue cheese, dates, solutions of lactic acid, cinnamaldehyde and anisaldehyde, and found that *T. putrescentiae* prefers cheese as compared to dates. Yoshizawa *et al.* (1972) and Vanhaelen *et al.* (1980) highlighted the role of some volatile compounds present in the food stuffs, in different ratios in the attractiveness of food products to *T. putrescentiae*. The amount of these compounds is the determining factor for the attractancy of different food stuffs. So, oat flakes are more susceptible to *T. putrescentiae* infestation than whole grains (Anita, 2010).

The use of three varieties of green coffee beans, Cocoa beans and three kinds of legumes (pea, beans and lentils), as food for *T. putrescentiae* and *A. siro*, highlighted a natural resistance towards the food. Wheat grain mixed with bean flour (0.01%), reduced by 50% the population development of *A. siro*, *Caloglyphus redickorzevi* (Zachvatkin), *L. destructor* (Schrank), and *T. putrescentiae*, as compared with the control (Hubert *et al.*, 2007).

At 30°C and 75% RH, the mould mite populations will increase by 500 times a month (Hughes, 1976). These outbreaks cause serious damage to the stored foods (Kleih and Pike, 1995), and are responsible of the emission of a damp, pungent smell (Nayak, 2006). High density of mite populations causes the unhygienic status of stored grain, inducing, world over, an increasing trend among grain buyers towards zero tolerance to these contaminants. The international markets reject commodities infested with mites, and the economic loss varies depending upon the case. If infestation is detected in a shipment of Australian wheat export, it costs the exporter approximately AU$ 1.2–1.8 million, depending on the infestation level (Abare, 2000). The mites are important pests of wheat during its storage, and accountable of both the qualitative and quantitative losses (Mahmood *et al.*, 2011). Weight loss due to *T. putrescentiae* population in oat grains (2.2%) (Anita *et al.*, 2013), and marked decrease in the thousand kernel weight of maize (Farhan *et al.*, 2013) and wheat (Bashir *et al.*, 2013) have been also discussed.

Baker and Wharton (1952) reported that *T. putrescentiae* can transmit plant diseases in the field. The mite is an important vector of dispersing weed fungi throughout mushroom cultivation facilities (Okabe *et al.*, 2001), and also feeds on different fungi including *Eutorium*, *Penicillium*, *Fusarium*, *Alternaria*, *Geotrichum*, *Mucor* and *Trichophyton* (Duek *et al.*, 2001). *T. putrescentiae* can affect fungi by grazing and is involved in the dominance of mycotoxigenic fungal species, due to their role in dispersal of fungal spores and influence the increase of aflatoxin production from the fungus (Franzolin *et al.*, 1999; Hubert *et al.*, 2003).

Zhang (2003) described the symptomatology and the damage for ornamentals and vegetables in greenhouses. They could be similar for other crops, but there is no adequate information available.

Erban *et al.* (2016b) reported that *T. putrescentiae* is an emerging source of allergens in stored products and homes. Faeces proteases are the major allergens of astigmatid mites. In addition, the mites are carriers of microorganisms and microbial adjuvant compounds that stimulate innate signalling pathways. They analysed the mite faeces proteome, proteolytic activities, and mite-bacterial interaction in dry dog food (DDF). Proteomic methods comprising enzymatic and zymographic analysis of proteases and 2D-E-MS/MS were performed. The highest protease activity was assigned to trypsin-like proteases; lower activity was assigned to chymotrypsin-like proteases, and the cysteine protease cathepsin B-like had very low activity. The 2D-E-MS/MS proteomic analysis identified mite trypsin allergen Tyr p3, fatty acid-binding protein Tyr p13 and putative mite allergens ferritin (Grp 30) and (poly)ubiquitins. Tyr p3 was detected at different positions of the 2D-E. It indicates presence of zymogen at basic pI, and mature-enzyme form and enzyme fragment at acidic pI. Bacillolysins (neutral and alkaline proteases) of *Bacillus cereus* Frankland *et* Frankland symbiont can contribute to the protease activity of the mite extract. The bacterial exo-chitinases likely contribute to degradation of mite exuviae, mite bodies or food boluses consisting of chitin, including the peritrophic membrane. Thus, the chitinases disrupt the faeces, and facilitate release of the allergens. *B. cereus* was isolated and identified based on amplification and sequencing of 16S rRNA, and *mot B* genes. *B. cereus* was added into high-fat, high-protein (DDF) and low-fat, low-protein (flour) diets to 1% and 5% (w/w), and the diets palatability was evaluated in 21-day population growth test. The supplementation of diet with *B. cereus*, significantly suppressed population growth, and the suppressive effect was higher in the high-fat, high-protein diet than in the low-fat, low-protein food. Thus, *B. cereus* has to coexist with the mite in balance to be beneficial for the mite. The mite-*B. cereus* symbiosis can be beneficial-suppressive at some level. The results increase the veterinary and medical importance of the allergens detected in faeces. The *B. cereus* enzymes/toxins are important components of mite allergens. The strong symbiotic association of *T. putrescentiae* with *B. cereus* in DDF was indicated (Erban *et al.*, 2016b).

Control

As concerns the stored grains, Malik *et al.* (2018) reported that their managing requires the use of various techniques to ensure that the quality during the storage facility does not deteriorate over time. These measures include use of sanitation, storing sound and dry grain, managing temperature and aeration, use of chemical protectants, MA, fumigants, inert dust and regular sampling (Stará *et al.*, 2011a).

As concerns the control of the mould mite infestations, Amoh (2016) has investigated many mite pest management practices, and/or implemented to help prevent and/or control the mite outbreaks. Chemical methods include the use of fumigants and acaricides. Other methods consist of the use of biological and physical methods, the use of food grade oils, and plant extracts and oils. Obviously, a correct monitoring of the mite populations is an obligatory premise.

MONITORING. The correct implementation of IPM programme relies preliminarily on an effective monitoring system able to inform on pest presence, and on its dynamics of populations before and after the IPM programme implementation. However, not much research on monitoring of mites has been carried out. In most environments, early detection is not effective because personnel usually do not have reliable equipment to inspect for and detect mites particularly during the early stages of an infestation or when they are present in fine dust or debris (Thind and Ford, 2004).

Management of *T. putrescentiae* requires deep understanding of its biology, as different management practices are needed for different commodities. From a practical point of view, the mite monitoring is usually done by placing baited traps in facilities and inspecting the traps for mites. The BT® trap is the first trap designed specifically for the detection and monitoring of mite populations in stored foods, and infestations in food or feed storage and production premises. It has also been used for detecting mites in both laboratory and field studies (Thind and Ford, 2004; Wakefield, 2006). In stored grain and oilseed, the PC™ trap is effective in detecting storage mites, and may provide an early indication of their presence, when compared to traditional methods (Clarke, 2002; Dunn *et al.*, 2005).

However, different scientific aspects still today require adequate answers, such as enhancing of traps and correlation of trap capture with other measures of sampling. About this Amoh (2016) reported that although several trapping efficiency and improvement studies have been carried out in stored-product insects, fewer findings are available for mites infesting stored foods. Some chemical compounds, identified as potential attractants for storage mites are, fungal volatile 1-octen-3-ol and food volatile 2-nonanone (Dunn, 2000). These attractants were tested and reported to increase trap capture of *L. destructor* (Schrank), *A. siro* Linnaeus and *T. longior* (Gervais) in wheat or oilseed rape when incorporated in BT® traps and PC™ traps (as reported in Wakefield, 2006). However, these attractants have not been investigated for *T. putrescentiae*. Research has also shown that pheromone-alone traps only detect a small percentage of pests present. There has also been no reported study to compare and or correlate ham mite sampling techniques or develop sampling plans.

Amoh (2016) employed food-baited traps, realized with Petri dishes and a dog food based bait. It has been observed that the average trap captures vary according to trap location, showing that they could be employed to identify specific sites within an ageing room, where mite infestation occurs. Findings inform on mite population dynamics, and constitute a possibility to making pest management decisions, and assessing the impact of fumigation or other pest mitigation actions. The same author evaluated the effect of trap placement, the trapping duration, and the trap design that would maximize the mould mite detection. A greater number of mites were captured in the standard trap, when compared singly or side by side with other traps. Studies carried out in controlled rooms showed that mites could be trapped up to 6 m from a mite source within 24 hours. A greater number of mites were caught in traps placed along a wall, than those away from a wall. Moreover, trap capture varied under different light conditions. More mites were detected in fully lighted rooms, respect to those in fully dark rooms.

Sampling techniques and sampling plans for the mould mite were compared by Amoh (2016). Amoh (2016) compared food-baited trap number which was maintained at 4 and mite density varied from 500 to 10,000 individuals in environmentally controlled 18 m^2 rooms. The number of mites detected in traps increased as mite density increased, indicating a strong positive correlation ($F1, 262 = 736.32$, $P < 0.01$; $r = 0.86$). Observations carried out in simulated ham ageing rooms showed that mites preferred the proximal to the distal side of the ham, and the trap capture also varied with trap location ($F3, 124 = 6.88$, $P < 0.01$). Varying degrees of correlations between trap capture and sampling techniques (mite counts from the storage rack, vacuum sampling of the floor and ham sampling to count mites) were found. Correlations between absolute mite counts and trap capture or ham sampling resulted both moderately strong and significant ($P < 0.01$). Two sampling plans were developed for the mite, and it was determined that an economic threshold close to one mite/sample unit on the ham can be estimated with 21 or 40 ham samples at precision levels of 0.35 and 0.25, respectively, or with ten ham samples if classifying the mite's infestation level with respect to an action threshold of 0.43.

PHYSICAL CONTROL. Physical methods involve the manipulation of the mites' environment to make it less suitable for their proliferation. These methods include manipulating light, temperature, and humidity of the environment. García (2004) investigated some physical methods in five Iberian ham factories in the Iberian Peninsula and reported a reduction in the growth rate of mites on ham surfaces, when humidity was reduced to levels below 60%. However, mites were forced to seek refuge in cavities inside the hams, where they could feed and reproduce. Boczek (1991) reported that mould mite eggs are killed within 24 h when exposed to –15°C, but will survive at –10°C. At –5°C, a longer exposure of 24 days is required to obtain similar results. Other researchers stated that freezing at –28°C for 48 h did not kill the eggs (Arnau and Guerrero, 1994). A more recent study suggested that 90% of all stages of *T. putrescentiae* can be controlled by freezing to –8°C for 5 h (Eaton and Kells, 2011). Rentfrow *et al.* (2006) also suggested that hams are frozen to control mites on ham. However, repeated freezing and thawing and high heat may affect the quality of ham. Although manipulation of the environment may be effective, facility managers may require professional assistance to create and/or maintain the new environment. Environmental conditions that are cost-effective, and do not significantly affect the quality and/or organoleptic properties of the products should be investigated.

BIOLOGICAL CONTROL. The above discussed problems highlight the need to develop new strategies for the control of mites in stored food. Various studies tried to keep acarine population below the economic threshold values through the use of biocontrol agents, physical and mechanical practices and grain protectants (Kohli and Mathur, 1993).

The release of biocontrol agents, such as cheyletid mites seems to wait for a definitive solution. The cheyletid mite *C. eruditus* (Schrank) is a natural enemy of mites infesting stored grains, and it has been suggested to control the mite infestations (Ždárková, 1998). Unfortunately, the release of natural enemies may contaminate the commodity, and their use in the control of stored food mites may find serious obstacle. Contamination of commodities by natural enemies is as bad as contamination by pests. Federal regulations under the Federal Food, Drug and Cosmetic Act do not distinguish pest fragments contributed by stored-product pests, from those contributed by natural enemies. None the less, depending on the commodity, most of the natural enemies can be removed (Hagstrum and Subramanyam, 2006). Moreover, results on the biological control of *T. putrescentiae* by the predatory mite *C. malaccensis* Oudemans are contradicting (Palyvos and Emmanouel, 2004; Cebolla *et al.*, 2009).

Application of new technologies alternative to chemical control in the stored foods is still limited; new approaches could be the 'food grade oils and preservatives' and the 'plant extracts and oils'. Food grade oils and other usually employed preservatives, 'Generally Regarded as Safe', are effective in the control of *T. putrescentiae*. An old method consisting of rubbing lard on cured meat, to prevent flies and bacteria (Smith, 1923) is still being applied today, as coating hams with vegetable oils or hot lard, represents a common practice in Spain to control mites in dry-cured ham (García, 2004); however, we must take into account that the ageing time may be extended, since lard may make the ham less permeable to moisture. Lard is effective on mould mites under laboratory conditions (Abbar *et al.*, 2016). Moreover, propylene glycol, an odourless, colourless solvent registered as a bacteriostat and fungistat (EPA, 2006), has been reported effective in controlling mite infestations by preventing its reproduction under laboratory conditions (Abbar *et al.*, 2016).

Weng *et al.* (1995) found that the population of *T. putrescentiae* could be suppressed by *H. aculeifer* (Canestrini) at an application rate of 1:5 or 1:10.

CHEMICAL CONTROL. Chemical control includes the traditional use of various chemical compounds. The strategy is commonly employed in stored grain, various stored foods (Stará *et al.*, 2011a) and ham processing facilities (Rentfrow *et al.*, 2008).

Common postharvest fumigants for stored foods have included, until the recent past, methyl bromide, phosphine and sulfuryl fluoride; the use of ozone and carbon dioxide is probably free in most or all countries.

Methyl bromide has been employed to control mites (and other pests) in commodities and buildings worldwide since the 1930s, according to its typical and advantageous properties, such as broad-spectrum application, rapid toxicity (Fields and White, 2002), lack of flammability and corrosive properties (Bond, 1984) and absence of residues on treated products. This a.i. kills mites and other living organisms (Fields and White, 2002), and is the only known fumigant effective at eradicating mould mite infestations. In the ham facility, the methyl bromide fumigations are applied from one to five times per year (Rentfrow *et al.*, 2006), and the restriction to its use poses the need to identify alternative fumigants. In China, Zhang and Jin (1963) found that chloropicrin was an effective fumigant against the active stages of *T. putrescentiae* but had little effect on the eggs. Ethylene dibromide and methyl bromide were highly toxic to all stages of the mite, and ethylene dichloride was effective with higher dosages and longer exposures.

Phosphine is a fumigant alternative to methyl bromide (Fields and White, 2002). It is easily applied, effective, cheap, devoid of persistent toxic residues (Zuryn *et al.*, 2008) and easily dispersant during the fumigation (Nath *et al.*, 2011). Trials carried out in laboratory showed complete mortality of all life stages of mites (Zhao *et al.*, 2012), and in a simulated ageing house containing mould mites and dry-cured ham infested with the mites, 99.8% mortality was observed after 2 weeks of the treatment (Zhao *et al.*, 2015). However, the employ of this a.i. may involve massive corrosion to electrical wiring and equipment of the fumigated room about 3 weeks after fumigation (Phillips and Schilling, 2013). Corrosion of metals is the major drawback that prevents phosphine from being an ideal fumigant in all structures (Price and Mills, 1988; Brigham, 1998; Adler, 2004), and the best preservation of electrical systems in the facilities would indicate phosphine as an unsuitable alternative to control mite infestations. Other authors reported that phosphine has severe restrictions due to safety and environmental considerations and repeated use of chemicals led to resistant populations of *T. putrescentiae* (Szlendak *et al.*, 2000). Moreover, it has undesirable effects on non-target organisms, which fostered environmental and human health concerns. Further trials carried out by Hasan *et al.* (2020) with phosphine or methyl bromide for 48 h exposure at 23°C showed that eggs of *T. putrescentiae* were slightly more tolerant than mobile stages for both chemical compounds. Under laboratory conditions, complete control was achieved for the mould mite with concentrations of 0.85 and 4.0 g/m^3 of phosphine and methyl bromide, respectively, at 48 h exposure. The results give new information for rational use of the existing stocks of methyl bromide, whether for pest mitigation or to help in developing a quarantine treatment schedule with that gas. Phosphine shows good potential as an effective alternative to methyl bromide, but if it was to be adopted as a fumigant in the dry-cured ham industry, methods to prevent metal corrosion would need to be designed, and effectively implemented (Hasan *et al.*, 2020).

Another fumigant, sulfuryl fluoride, has been registered in the United States for employ in dry-cured ham processing facilities since 2005 (EPA, 2005). However, Phillips *et al.* (2008) reported that the use of this fumigant under laboratory produced only 95% of mortality in mite eggs at 20–22°C, even when employed at about 100.3 g/m^3, equal to three times the label rate. Phillips and Schilling (2013) referred that sulfuryl fluoride may be effective at controlling all life stages of the mite, when tested for 48 h at 35°C under laboratory conditions.

Acetylene containing a small amount of PH_3 and H_2S was used to control *T. putrescentiae* by Chen *et al.* (1993), and 97.78% adult mites were killed with the mixture of acetylene at 21.20 mg/l and phosphine at 0.063 mg/l. Other fumigants characterized by some potential under laboratory conditions include ozone (O_3) and carbon dioxide (CO_2). In order to improve the control effect, and reduce the dosage of pesticides, carbon dioxide was often mixed with fumigates. *T. putrescentiae* consumed more oxygen when the concentration of ambient carbon dioxide was added from 0–8% and 16%, and the uptake of phosphine by the mite increased (Jian *et al.*, 1995). High CO_2 was the most important factor that resulted in the mortality of *T. putrescentiae*, temperature the second, and low O_2 the last (Li *et al.*, 1992b).

Hassan *et al.* (2010) found under laboratory conditions that ozone concentration of 155 ppm for 48 h caused the complete mortality of all biological stages of *T. putrescentiae*. Despite these findings, ozone does not penetrate surfaces and its employ may result optimal. Different specialists

(Adler, 2004; Navarro, 2006; Riudavets *et al.*, 2010) confirmed that the use of CO_2 at high pressure in pressure chambers produces a total control of mites. Riudavets *et al.* (2010) treated *T. putrescentiae* with CO_2 at two high pressures, 15 and 20 bars. There was over 99% mortality for mixed stages of *T. putrescentiae* on artificial standard diets when treated with CO_2 at 15 bar for 15 minutes, and 100% mortality, when the exposure time was increased to 30 minutes. Although CO_2 may be effective, increasing the pressure increases the cost of treatment, as more CO_2 is required (Riudavets *et al.*, 2010). The concentration of 60% CO_2 with 144 h of exposure produces the mortality of all stages of mites (Hassan *et al.*, 2010). However, the long duration of exposure would result in practical inconveniences, such as closing facilities, with negative interference with the general operations.

Results of trials carried out with the organophosphate fenitrothion and pirimiphos-methyl have showed unclear results (Hartmannova *et al.*, 1973; Chmielewski, 1987; Nayak, 2006). Other trials have concerned chlorpyrifos-methyl, deltamethrin, pyrethrin + piperonyl butoxide, s-methoprene and spinosad, and of these only pyrethrin + piperonyl butoxide, s-methoprene and spinosad assured a positive control of the mite population, after about 3 weeks of exposure to treated wheat (Nayak, 2006). Yu *et al.* (2006) evaluated the toxicity of 14 insecticides/acaricides to *T. putrescentiae* and found that the most and the least effective were abamectin and dichlorovos, respectively, and none of the tested insecticides/acaricides was effective against eggs.

Abbar *et al.* (2013) suggested treating surfaces in ageing rooms before the cured hams are transferred to these environments. At this aim, they carried out experiments exposing the mites to treated surfaces with deltamethrin, chlorfenapyr and chlorpyrifos-methyl + deltamethrin, and registered some control on mite populations. Sánchez-Ramos and Castañera (2003) refer that despite some pesticides resulting effective under laboratory conditions, the limitations due to their chemical residues restricted their employ in ham facilities. Other reports highlighted that various chemical compounds exercise varying degrees of effectiveness (Hartmannova *et al.*, 1973; Chmielewiski, 1987; Stables, 1980; Pagliarini and Hrlec, 1982). Stables (1980) found that *T. putrescentiae*, *T. longior*, *A. siro* and *G. destructor* were controlled by etrimfos and profenofos. Moreover, phenothrin, fenopropathrin and permethrin resulted effective against *T. putrescentiae*, but at a very high rate of 500 ppm (Chisaka *et al.*, 1985).

Plant-derived extracts, powders (Gulati and Mathur, 1998) and essential oils may be options for mite control (Lee *et al.*, 2006). Plant-derived alkalis, alcohols, aldehydes, terpenoids and some monoterpenoids show fumigant properties (Macchioni *et al.*, 2002). There are several studies which showed the effectiveness of plant essential oils for control of stored products pests (Lee *et al.*, 2006). Among botanicals, *Allium sativum* Linnaeus, *Curcuma longa* Linnaeus, *Azadirachta indica* A. Juss., *Glycyrrhiza glabra* Linnaeus, *Ocimum* sp., reported to have toxic and repellent effects on storage mites (Gulati, 2007a, b; Anita *et al.*, 2014). The essential oils were also used for the control of this mite, and their toxicity was evaluated using 10 ppm of lemon oil, citronella oil or camphor oil. These products could kill all individuals of *T. putrescentiae* in laboratory and lemon grass oil, fir oil, geranium oil, lavender oil were also effective on the mite (Yan *et al.* 1989).

Potts and Rodriguez (1978) showed the inhibitory effect on the development of the populations of *T. putrescentiae,* of spice oil when introduced as an addition to a casein and wheat germ diet. Sánchez-Ramos and Castañera (2003b) tested 13 natural monoterpenes against *T. putrescentiae* and reported that some of these, such as pulegone, menthone, linalool and fenchone showed high acaricidal activity, yielding LC90 values of 14 µl/l or below.

Kim (2002) found the acaricidal activity of *Eugenia* sp., against *T. putrescentiae* effective. Lee *et al.* (2006) studied the acaricidal activity (direct contact application) of 12 fennel seed oil extracts against *T. putrescentiae* and found naphthalene (4.28 $\mu g/cm^2$) and Carvone (4.62 $\mu g/cm^2$) to be the most toxic. Gulati (2007b) reported that garlic products reduced the *T. putrescentiae* and *Sarcocystis nesbitti* Mandour population by 96–100% at 0.2%, respectively. Garlic oil was the most effective, followed by oleoresin and powder. Very high control of both pests occurred at 0.5%, 1% and 2% concentrations of *Curcuma* oil and powder, though some mites were collected from *Curcuma* oleoresin treated grains (Gulati, 2007a). More pronounced inhibitory effects were observed on eggs and larvae, than on nymphs and adults. *G. glabra* and *Ocimun sanctum* Linnaeus extract produced clear effects on the population of the mite. It provided 71.5–94.7% and 66–92% protection against *T. putrescentiae* at different durations (Anita *et al.*, 2014).

Lee *et al.* (2006) found promising results in the comparation of the acaricidal activity of *Foeniculum vulgare* Mill (fennel) seed oil with benzyl benzoate, dibutyl phthalate and N, N-diethyl-m-toluoamide in the control of *T. putrescentiae*. The acaricidal activity of *Pinus pinea* essential oil was also reported to induce complete mortality on adults of *T. putrescentiae* (Macchioni *et al.*, 2002).

INTEGRATED PEST MANAGEMENT. The future of any alternative management method lies in its ability to be incorporated into a decision-based IPM programme. The set of these integrated methods into IPM programme could help to avoid infestations, reduces the impacts on foods susceptible to the mite infestations and the development of resistance to different control methods (Amoh, 2016).

Several authors have discussed the effectiveness of management options in the control of *T. putrescentiae* (Clarke, 2002; García, 2004; Mueller *et al.*, 2006; Rentfrow *et al.*, 2006), but a few of them reported the possibility of implementing IPM programmes. The implementation of the 'hurdle approach' within an IPM programme, could be able to prevent mould mite infestations in ham processing plants. The approach involves the application of rigid sanitation practices in both the processing rooms and the ageing rooms. It is necessary to eliminate grass from growing within 9 m of the ageing house, and wooden racks and structures sealed to prevent mites hiding in cracks and crevices. Breathable paper bags for the hams, as well as freezing hams at 6 months of ageing could also help to prevent mites attacks (Rentfrow *et al.*, 2006).

The efficacy of spinosad and spinetoram, both bacterium-derived pesticides, and their combination with the cheyletid mite *C. malaccensis* was evaluated against *T. putrescentiae* and *A. ovatus* (Troupeau). Trials were carried out under optimal environmental conditions for pest development, and the estimate of pest and predator mites was made after 21 days. Both astigmatid mites were exposed to diet (experiment I) treated with either spinosad or spinetoram (concentrations range of 0.01–2 ppm), and diet (experiment II) treated with either spinosad or spinetoram (0.5 ppm), and/or the predator at initial predator/prey ratio (0.02). Application of spinosyns significantly reduced population of both mite species. Total control of *T. putrescentiae* and *A. ovatus* was achieved by the application of spinosad at 1 and 2 ppm, respectively. As measured by rC_{50} and rC_{90} (concentration for 50% and 90% suppress of population in comparison to control), spinosad was more toxic to *T. putrescentiae* and *A. ovatus* than spinetoram. Furthermore, *T. putrescentiae* was more susceptible to spinosad than *A. ovatus*. Conversely, it was less susceptible to spinetoram than *A. ovatus*. The populations of both mite species were successfully suppressed, by the sole application of *C. malaccensis*. Although the density of predatory mites was not affected by the presence of 0.5 ppm spinosad, it was almost eradicated by spinetoram at 0.5 ppm. A combination of spinosad at 0.5 ppm with two individuals of *C. malaccensis* mites (ratio 0.02) outperformed spinosad used alone at the same former concentration, in reduction efficiency of the pest populations by 12% for *T. putrescentiae,* and 25% for *A. ovatus* within 21 days (Bakr and Selim, 2019).

Tyrophagus similis Volgin (Fig. 28.10)

Common name

The French fly

Diagnostic characteristics

FEMALE. The idiosoma is 522 µm long and 324 µm wide. The prodorsal shield lacks eyespots is obviously punctate, nearly rectangular in shape, with lateral sides slightly concave and the posterior broadly round. Grandjean's organ is finger-like, and the basal lobe bears three obvious spiniform teeth. The setae *scx* are slender, tapering from the base to tip, with 12–16 short pectinations. The setae c_1, d_1 and d_2 are micro-setae, much shorter than others; the setae d_1 are about 1.0 times the length of the setae c_1 and 1.4 times longer than the setae d_2; the setae d_2 are about 0.8 times the length of the setae c_1. Coxal plates I do not reach to the posterior apex of the prosternal apodeme; coxal plates II do not extend posteriorly beyond the apex of apodeme 2, with the posterior margin sinuous and distal, one-third becoming concave and narrow. The spermathecal duct is broad, widening gradually from the copulatory opening to the base of the spermathecal sac, about 1.4 times the distance between the sclerites of the oviducts; the base of the spermathecal sac is broadly round, bending backwards. The solenidion ω_1 of tarsus I is stout and obviously widened at the apex; the solenidion ω of tarsus II is strong and widened at the apex. The setae *w* and *r* of tarsus IV are spiniform (Fan and Zhang, 2007a).

MALE. The idiosoma is 526 µm long and 372 µm wide. The prodorsal shield is laterally concave and

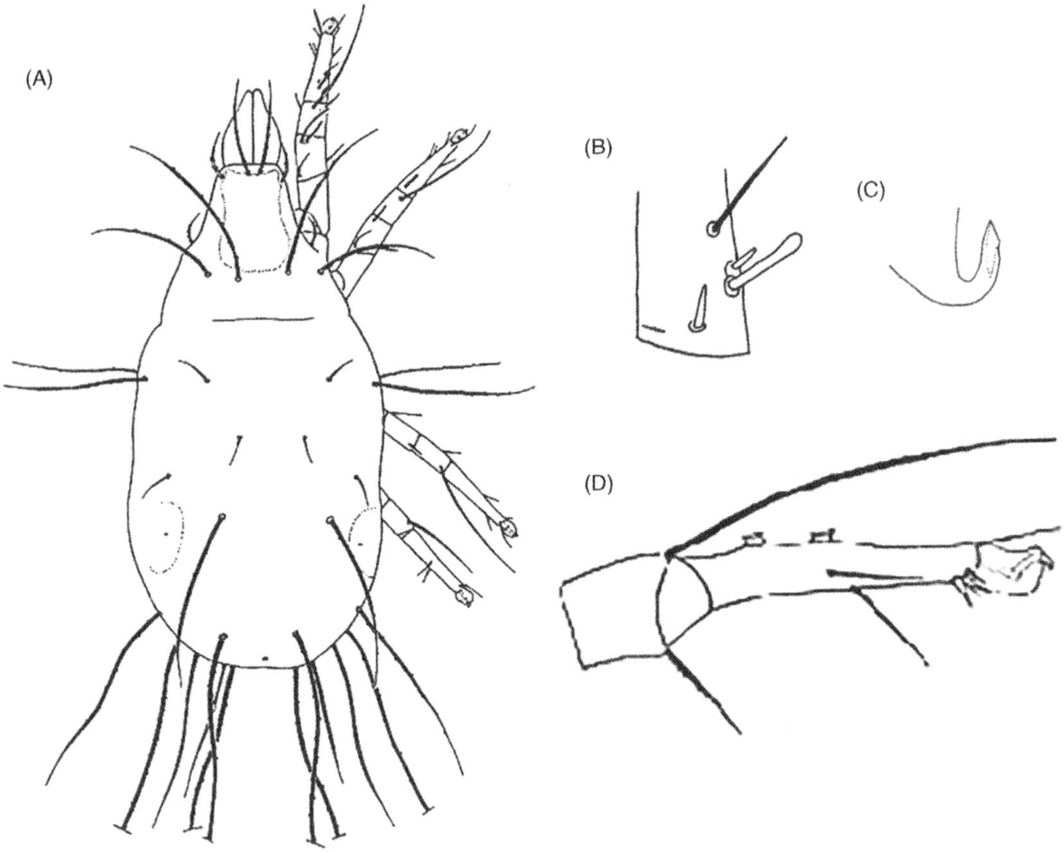

Fig. 28.10. A. *Tyrophagus similis* Volgin. Female. Dorsal view (A); setae and solenidia of basal part of tarsus I (B) (from Hughes, 1976, modified). Male. Aedeagus (from Hughes, 1976) (C); leg IV (from Hughes, 1976, modified) (D).

posteriorly convex. Grandjean's organ is as in the female. Coxal plates I and II, and the solenidia ω_1 and ω, are as in the female. The setae d_1 are about 1.0 times the length of the setae c_1, and 1.1 times longer than the setae d_2; the setae d_2 are about 0.9 times the length of the setae c_1. The aedeagus has one major curve, narrowing gradually in the basal half, and almost cylindrical in the distal half, with the distal end truncated, the internal diameter tapering rapidly from the distal end to mid-length, and the lateral arms turning inwards. The setae w and r of tarsus IV are spiniform, set anteriorly or close to the distal sucker (Fan and Zhang, 2007a).

Geographical distribution

T. similis has a worldwide distribution, and is reported from Australia (Robertson, 1959; Halliday, 1998; Fan and Zhang, 2007a), Belgium (Hughes, 1976), Canada (Beaulieu and Knee, 2014), China (Li, 1999), England (Fan and Zhang, 2007a), Germany (Robertson, 1959), Greece (Athanassoiu *et al.*, 2002; Palyvos *et al.*, 2008), Faroe Islands (Hallas and Solberg, 1989), France (Giustina, 1981), Malta (Pfliegler *et al.*, 2017), Hungary (Ripka *et al.*, 2005), Iceland (Hughes, 1976; Hallas and Gudleifsson, 2008), Iran (Hajiqanbar *et al.*, 2002), Italy (Laffi, 1980), Japan (Nakao and Kurosa, 1988), Mexico (Estebanes-Gonzalez and Rodriguez-Navarro, 1991), Korea (Jung *et al.*, 2010), the Netherlands (Hughes, 1976; Fan and Zhang, 2007a; Siepel *et al.*, 2016), New Zealand (Robertson, 1959; Ramsay and Paterson, 1977; Fan and Zhang, 2007a), Romania (Cindea, 1978), South Africa (Fan and Zhang, 2007a), Sweden (Bostrom *et al.*, 1997), Turkey (Kılıç and Toros,

2000; Çobanoğlu, 2009; Madanlar and Önder, 1996), UK (Robertson, 1959; Fan and Zhang, 2007a), United States (Walter *et al.*, 1986) and Yemen (Al-Safadi, 1987).

Commodities, facilities and vegetal and animal hosts

On stored food, the mite has been collected on apricot (dried), condiment, cottonseed, gin trash, grain, hay, seed cotton collected in farm and off-farm storage (Volgin, 1949; Robertson, 1959; Johnston and Bruce, 1965; Hughes, 1976; Kılıç and Toros, 2000; Athanassiou *et al.*, 2002; Kasuga and Honda, 2006a, b; Çobanoğlu, 2009). Hughes (1976) reported *T. similis* on grassland, soil, *Phlox*, spinach, mushrooms, old hay, chaff, in an eider duck nest and in a *Bombus ionellus* (Kirby) nest. Beaulieu and Knee (2014) reported the mite on stored barley and in pasture.

Fan and Zhang (2007a) and Vacante (2016) listed a number of plants and plant-associated products/substrates, such as acorns (*Quercus* sp.), agricultural soil, barn dust, bean plumules, cucumber, dahlia, garlic, kiwi fruit, *Lilium* bulbs, maize, melon, melon seedbeds, moss, mushrooms, *Narcissus* bulbs, orchid, pasture, pumpkin, rotten straws of rice, sandy soil in pots of *Rama* ryegrass, seeds and debris from a tent, spinach, spinach buds, stored hay, sugarbeet fields and tea tree scrub.

The list of animals and their products on which *T. similis* has been found includes bat debris, black-backed gull colony nests and debris, a dead *Rhopoea* larva (grub), a larva of *Zenarge turneri* Rohwer, nematode culture, the nematode *Heterodera avenae* (Wollenweber), nests of seabirds, starlings (*Sturnus vulgaris* Linnaeus), the Norway rat [*Rattus norvegicus* (Berkenhout)], Polynesian rat [*Rattus exulans* (Peale)], rabbit (*Oryctolagus cuniculus* Linnaeus) and white heron (*Egretta alba modesta* (Gray)) (Fan and Zhang, 2007a; Vacante, 2016).

Bio-ecology

The biology of *T. similis* is known from laboratory cultures (Al-Safadi, 1987, 1990, 1991; Kasuga and Amano, 2000a). In Iceland, the French fly has the deutonymphal stage that is responsible for dispersal. The eggs laid by females from the midsummer generation may overwinter, and hatch as winter generation eggs the next spring. The formation of the deutonymph takes place in the cold winter; the summer generation does not develop this stage. In this environment, the species has a male:female ratio of 1:3 (Hallas and Gudleifsson, 2008). On wheatgerm at 25°C, development from egg to adult takes less than 10 days (Al-Safadi, 1991). At 10°C, 15°C, 20°C and 25°C, and with a 14 h light (14L):10 h dark (10D) photoperiod, the developmental time of immature stages is shorter at 25°C, and the highest fecundity occurs at 10°C (663.4 eggs/female). At 10°C, the intrinsic rate of natural increase (r_m), the net reproductive rate (R_0) and the mean generation time (T) are 0.051, 190.05 and 102.03, respectively; at 15°C, they were 0.085, 190.04 and 61.43, respectively; at 20°C, they were 0.127, 125.37 and 37.98, respectively, and at 25°C, they were 0.160, 36.57 and 22.46, respectively. The species exhibits the highest growth potential at 25°C, but at this temperature the fecundity is the lowest. A limited survival of females occurs at 35°C, and at higher temperatures. The survival of females was investigated at 53%, 66%, 76%, 87% and 100% RH, 20°C and with a 14L:10D photoperiod; limited survival was observed at 53% and 66% RH (Kasuga and Amano, 2000a). On spinach in greenhouses, in central Japan, the mite density in the soil decreases from May to August, and increases rapidly in late autumn, remaining at a high level from December to February; it then increases in April and rapidly decreases in May. The high temperatures registered in greenhouses from spring to early autumn is the main cause of the population decrease. The numbers of mites on spinach buds increase, after the increase of those on and in the soil. The attack on spinach buds occurs mostly in late autumn and early spring (Kasuga and Amano, 2003). In the greenhouse, the mite inhabits either cultivated or uncultivated soil at the periphery of the cultivated area inside the greenhouse, and in the soil immediately outside. Its density in the cultivated and peripheral soils increases in the cooler months and decreases in the warmer months. Only small numbers of *T. similis* are collected from the adjacent outside soil. In the cultivated area, most of the mites inhabit the soil at depths of 0–5 cm throughout the year, and their numbers decrease with depth (Kasuga and Amano, 2005). In England, the mite is introduced into cucumber greenhouses in straw or manure, and in the warm environment they become active and wander on to the newly set-out plants. Attacks decrease naturally within a few weeks, after one or two generations have been developed (Hussey *et al.*,

1969). The species produces an alarm pheromone (Kuwahara *et al.*, 1987). In Yemen, a positive correlation between the density of *Gamasellodes/Protogamasellus* spp. (predator mites), and that of the most abundant mite species present (*T. similis*) was found (Al-Safadi, 1987). Eight genera of bacteria (*Bacillus*, *Nocardia*, *Corynebacterium*, *Erysipelothrix*, *Lactobacillus*, *Arachnia*, *Micrococcus*, *Staphylococcus*) and fungi have been isolated from dead mites. Acaricidal activity has been confirmed in 77.8% of the bacterial isolates, and 100% of the fungi isolates. The microorganisms isolated are deemed effective in biological pest control of the mite on spinach in the greenhouse (Kikuta and Takagi, 2001).

Symptomatology and damage

Symptomatology and damage produced by *T. similis* infestation, on various economic plants have been discussed by Vacante (2016). Few reports are available on stored foods, where symptoms and damages are similar to those of other *Tyrophagus*, though the mite frequency is lower than other *Tyrophagus* species.

Control

Like other acarid mites, the control requires preventive and curative actions. The curative actions are mainly the use of chemical and biological methods. Early monitoring of the mite populations is fundamental to detect the early presence.

BIOLOGICAL CONTROL. Release of the laelapid mite *H. aculeifer* (Canestrini) has been suggested for the biological control of the French fly. The bioecology of this predator has been investigated by Kasuga *et al.* (2006).

CHEMICAL CONTROL. No reports are available on the control of this mite in stored foods, where for its occasional control, the protocol suggested for other stored mites may be adopted. We report below, some report concerning the mite control, in field on vegetal plant host.

Smoking with parathion or spraying both plants and substrates with the same chemical, if high infestations occur, has been reported on cucumber (Hussey *et al.*, 1969; Nakao, 1989, 1991), but usually, mite infestation in protected crops can be controlled by soil sterilization (Misayo *et al.*, 2005). In Italy, the best chemicals for lowering populations of *T. similis* on spinach were benzoximate, endosulfan and fenpropathrin, but none of them gave good results in controlling the eggs, and they all had little selectivity on predacious mites (Ippolito and Triggiani, 1988). In Japan, dichlorvos did not reduce the numbers of mites in the greenhouse, even though it was highly toxic under laboratory conditions, which indicates that the mites present in the soil and on spinach buds had little direct contact with the chemicals. This highlights the fact that once crop damage by mites is detected, it is usually too late to use chemicals, and that mites living in buds are in some way protected from the action of agrochemicals (Kasuga and Amano, 2003). On spinach in Japan, DDVP EC (dichlorvos) gave high acaricidal activity, in all populations of *T. similis*, but had only a low control effect under field conditions (Kasuga and Amano, 2000b), due to the chemical particles not sufficiently reaching mites in the soil and on spinach plants (Kasuga and Amano, 2002).

Tyrophagus tropicus Robertson (Fig. 28.11)

Common name

Unknown

Diagnostic characteristics

FEMALE. Idiosoma about 443 μm long, and 281 μm wide. Eyespots absent, but marginal area of shield faintly punctate; *scx* broadly widened at bases of pectinations, with 16–22 moderate or short pectinations; setae d_1 about 2.2× length of c_1, and 1.2× length of d_2; d_2 about 1.9× length of c_1; coxal plates I and II obscure. Adanal setae ad_1 very long, 4.8× length of ad_3 and ps_3; ad_2 may also be extended. Spermathecal duct narrowing gradually from copulatory opening for a distance nearly 1.3× distance between sclerites of oviducts, and then forming a thin tube leading to base of spermathecal sac over a distance of about 2.0× distance between sclerites of oviducts, base of spermathecal sac large and flat. Tarsus I ω_1 stout, slightly clavate, tarsus II ω stout, obviously clavate; setae w and r of tarsus IV setiform (Fan and Zhang, 2007a).

MALE. Idiosoma 367 μm long, 243 μm wide. Eyespots, coxal plates, solenidia I ω_1 and II ω as in female; d_1 about 2.2× length of c_1, and 1.2× length of d_2; d_2

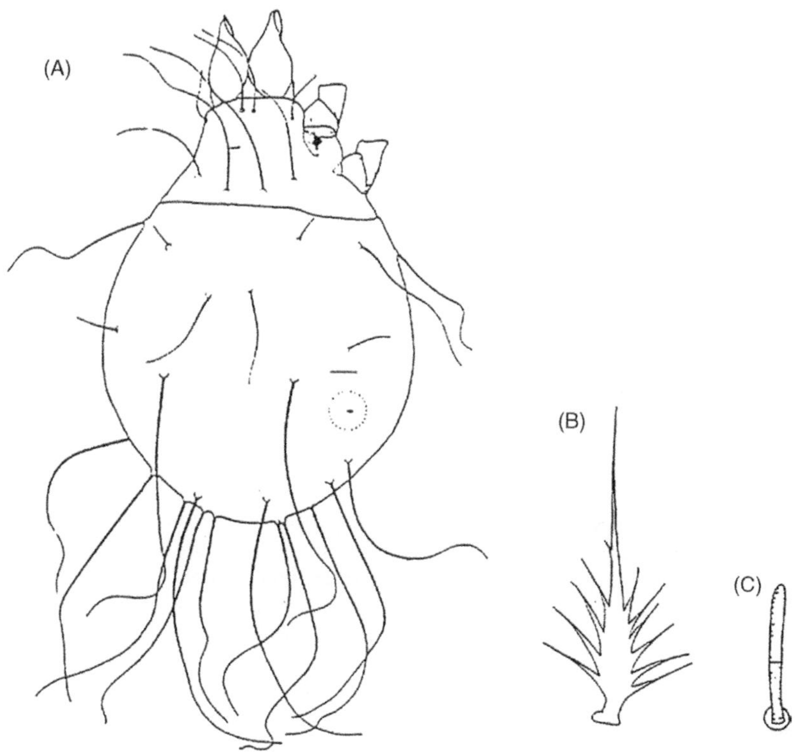

Fig. 28.11. *Tyrophagus tropicus* Robertson. Female. Dorsal view (A); supracoxal seta (B); solenidion of tarsus I (C) (from Vacante, 1989).

about 1.9× length of c_1; aedeagus very short, with one major curve, its internal diameter of ⅓ distal spindle-form and of ⅔ basal linear, lateral arms supporting aedeagus turning outwards; setae *w* and *r* of tarsus IV setiform; ratio (a+b): c = 1.8 (Fan and Zhang, 2007a).

Geographical distribution

The mite is known from Africa (Johnston and Bruce, 1965), China (Wang, 1985; Li *et al.*, 1992a), Czech Republic (Hubert *et al.*, 2009), Egypt (Osman *et al.*, 2016), England (Hughes, 1976), Ghana (Hughes, 1976), Germany (Wildies, 2000), India (Putatunda and Abrol, 2003), Indonesia (Haines, 1995), Italy (Vacante, 1989a), Lagos (Hughes, 1976), the Netherlands (Siepel *et al.*, 2016), New Britian (Hughes, 1976), New Guinea (Johnston and Bruce, 1965), Nigeria (Robertson, 1959) and United States (Hughes, 1976). Moreover, Fan and Zhang (2007a) examined specimens collected in Africa, China (Hong Kong), Malaysia, Nigeria, Samoa and West Africa.

Commodities, facilities and vegetal host

It is reported on various commodities (barley, candied fruit, Chinese medicine, condiment, copra, dried fruit, dried vegetable, grain, mushroom, palm kernel dust, rice, tobacco) collected in food store (Robertson 1959; Johnston and Bruce 1965; Hughes 1976; Lynch, 1989; Hubert *et al.*, 2009). Moreover, Fan and Zhang (2007a) reported the mite on bananas, cocoa (*Theobroma cacao* Linnaeus) beans, dried fruit, dried lychee and palm kernel dust. Recently, it has been collected on adzuki bean, and cucurbitae semina in China (Li *et al.*, 2015a).

Bio-ecology

The life history and morphology of *Tyrophagus tropicus* Roberstom was examined in the laboratory at 30°C, RH 75% and 80%. The mite development provides the biological stages of egg, six-legged larva, protonymph, tritonymph and adult. There is no hypopus stage, but a quiescence existed at the end of each nymphal stage. At 30°C, RH 80% and 75%, the development from egg to adult took about 254 hours and 286 hours, respectively (Lin *et al.*, 1992). The mite has been recorded from four honeybees as *Apis cerana* (Fabricius), *A. dorsata* Fabricius, *A. florea* Fabricius and *A. mellifera* Linnaeus in India (Putatunda and Abrol, 2003). The mite is less important on tropical stored products and is commonly found in association in small number, in mixed populations with *T. putrescentiae* (Schrank) (Kumar, 2017).

Symptomatology and damage

This mite seems of secondary importance among the mite pests of stored foods, where it occurs sometimes, but never at high density of population. The symptomatology of attack is presumably as that of other *Tyrophagus*.

Control

As other *Tyrophagus*.

The Genus *Tyroborus* Oudemans

Diagnostic characteristics

The palpal supracoxal seta (*elcp*) of female is distally bi- or trifurcate; the idiosoma is 300–800 µm long and whitish to semi-transparent in colour. The dorsum bears four pairs of prodorsal setae (*vi*, *ve*, *sci* and *sce*) and 12 pairs of hysterosomal setae (c_1, c_2, c_p, c_3, d_1, d_2, e_1, e_2, f_2, h_1, h_2 and h_3), all dorsal setae barbed. External vertical setae obvious, set on the anterior lateral margins of the prodorsal shield. The internal scapular setae are longer than the external scapular setae. The Grandjean's organ is finger-like, basally with 3–6 teeth. The supracoxal setae *scx* is broadly widened with dense pectinations. The hysterosomal setae c_1 and d_2 are very short, obviously less than distance to setae in next row; d_1 close to c_1 and overlapping bases of e_1. Coxal plates I not reaching to posterior extremity of prosternal apodeme; coxal plate II reduced to an approximately 'L'-shaped sclerite formed by remnant of lateral and anteromedian parts. Genital opening situated between coxae III and IV, with two pairs of genital papillae covered by genital valves and one pair of genital setae (*g*). Genital folds present. Anal opening posteriorad of genital opening, with six pairs of setae (ad_3, ps_3, ad_2, ad_1, ps_2 and ps_1). Spermathecal duct narrowing from copulatory opening for a distance, forming a neck and then gradually widening to base of spermathecal sac. Legs light brown. Tarsi I–II stout, about twice as long as basal width. Dorsoapical seta *e* on tarsi I–II conical. The male is similar to adult female, except the genital opening situated between coxae IV; aedeagus present; with a pair of anal suckers; without adanal setae; tarsus IV having two suckers. Complement of setae on legs as in adult female, except tarsi IV with 3 + 1 dorsoapical spine + 3–5 ventroapical spines (Fan and Zhang, 2006).

Economic importance

The genus *Tyroborus* Oudemans includes only three species, named *Tyroborus lini* Oudemans, *T. miripes* (Athias Henriot) and *Tyroborus ueckermanni* Fan *et* Zhang, respectively (Fan and Zhang, 2006). On stored foods, only *T. lini* is reported. The other species are not injurious. *T. miripes* was collected on eggs of the desert locust *Schistocerca gregaria* (Forskål) in Algeria (Athias Henriot, 1961b) and *T. ueckermanni* was found in feed of pigs in a piggery at Dalton in South Africa (Fan *et* Zhang, 2006).

Tyroborus lini Oudemans (Fig. 28.12)

Common name

Unknown

Diagnostic characteristics

FEMALE. The idiosoma is 400–650 µm long, and the general aspect is as for the male. Ventrally, the anus is removed from the posterior margin of the body (Hughes, 1976).

MALE. The idiosoma is 350–470 µm long. The chelicerae are stout with well-defined teeth on both digits. The propodosomal shield is pentagonal, extending as far back as the setae *sci*, and having a faintly pitted surface, while the surrounding cuticle is smooth. The ventral surface of the body has the coxalsternal skeleton, composed of thick sclerites and distinct epimeral plates. Idiosomal setae arranged as in *T. putrescentiae* (Schrank), the longer ones, especially *ve* and *vi*, being pectinate. The supracoxal seta is large, becoming wider towards the base, with its lateral edges produced into barbs. The dorsal idiosoma

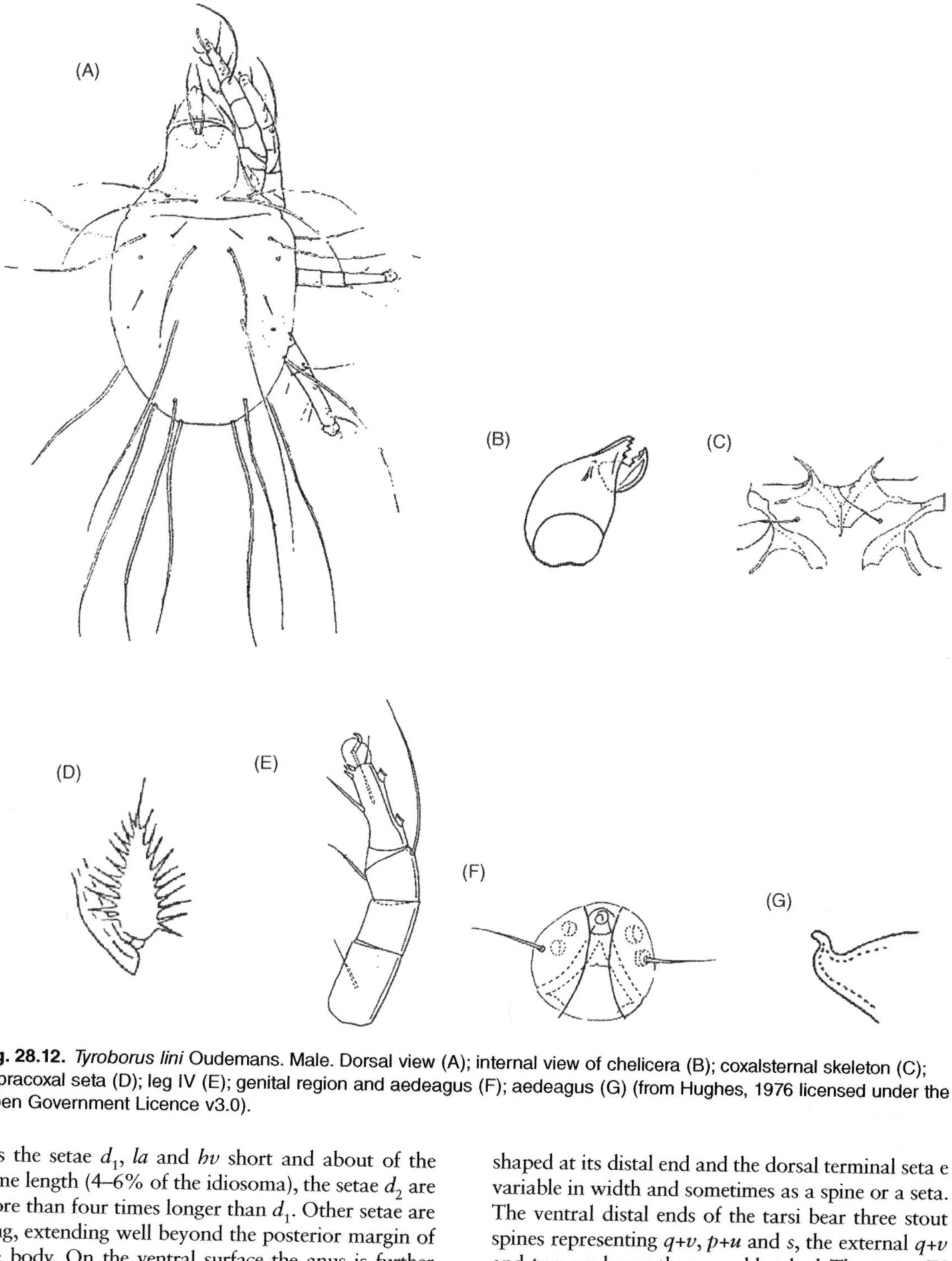

Fig. 28.12. *Tyroborus lini* Oudemans. Male. Dorsal view (A); internal view of chelicera (B); coxalsternal skeleton (C); supracoxal seta (D); leg IV (E); genital region and aedeagus (F); aedeagus (G) (from Hughes, 1976 licensed under the Open Government Licence v3.0).

has the setae d_1, *la* and *hv* short and about of the same length (4–6% of the idiosoma), the setae d_2 are more than four times longer than d_1. Other setae are long, extending well beyond the posterior margin of the body. On the ventral surface the anus is further removed from the posterior edge of the body than in *T. putrescentiae*. Tarsi I and II with ω_1 slightly club-shaped at its distal end and the dorsal terminal seta e variable in width and sometimes as a spine or a seta. The ventral distal ends of the tarsi bear three stout spines representing *q+v*, *p+u* and *s*, the external *q+v* and *p+u* are larger than s and hooked. The tarsus IV is shorter than the combined length of the tibia and genu, the suckers equidistant from the base and apex of the segment. The struts supporting the aedeagus curve outwards, the aedeagus is small and S-shaped, not drawn out into a point (Hughes, 1976).

Geographic distribution

China (Li, 1999), Cuba (Suárez, 2004), Egypt (El-Lakwah and Halawa, 1997), England (Brady, 1970), India (Mathur, 1979), Indonesia (Haines, 1995), Iran (Irani-Nejad *et al.*, 2007; Kordestani Mahani *et al.*, 2014), Mexico (Quintero and Acevedo, 1991), the Netherlands (Oudemans, 1924b; Siepel *et al.*, 2016), New Zealand (Robertson, 1946), Spain (Estrada-Peña *et al.*, 1981), Turkey (Hughes, 1961). Moreover, Hagstrum *et al.* (2013) reported Japan and United States.

Commodities, facilities and vegetal and animal hosts

Chinese medicine (Li, 1999), stored rice, flour, bean cake, dried *Auricularia auricula-judae* (Fr.) Quél., dried *Zanthoxylum bungeanum* Max. (Fan *et al.*, 2001), barley (Suárez, 2004), stored commodities (El-Lakwah and Halawa, 1997), poultry litter (Brady, 1970), Bengal gram (*Cicer arietinum* Linnaeus) (Mathur, 1979), fermented liquid foods (Tuba) (Quintero and Acevedo, 1991), food grains (Haines, 1995), linseed (Oudemans, 1924b), linseed (Robertson, 1946), ham (Estrada-Peña *et al.*, 1981). Moreover, Hagstrum *et al.* (2013) reported condiment, dried fruit, dried vegetable, grain, mushroom, processed food, wheat, wheat flour (old), in food store. Irani-Nejad *et al.* (2007) collected the mite on lucerne fields.

Bio-ecology

Oudemans (1924c) proposed the monotypic genus and designated a type species, *T. lini* Oudemans, that was subsequently described (1924b). Hughes (1976) revived *Tyroborus* based on a single important character: ventroapical setae *p* and *q* on tarsi, are merged with *u* and *v*, respectively (precisely, *p* and *q* are absent or vestigial according to Fain, 1986 and present study). Fain (1986) agreed with Hughes (1976) and distinguished *Tyroborus* from *Tyrophagus* and *Tyrolichus* by a combination of the characters: (i) dorsoapical seta *e* on tarsi spine-like, (ii) proral setae *p* and *q* on tarsi absent or reduced, (iii) supracoxal setae *scx* with numerous projections and (iv) setae *sci* in larva much shorter than *sce*. Fan and Zhang (2006) redescribed the genus and the type species, *T. lini*, and separated the genus *Tyroborus* from *Tyrophagus* by different morphological characters. The species of the genus *Tyroborus* have the setae *elcp* apically bi-or trifurcate, the coxal plate II is reduced to an approximately 'L'-shaped sclerite formed by remnant lateral and anteromedian parts, the setae *e* of tarsi are thick and conical, the tarsi I and II are stout and about twice as long as basal width, the setae *scx* possess more projections and the setae *sci* of the larva are shorter than *sce*. Conversely, the species of the genus *Tyrophagus* have the setae *elcp* not branched, the coxal plate II is broad or concave in the posterior margin but not L-shaped, the setae *e* on tarsi are thin and setiform, the tarsi I and II are slender, more than 3× based width, the setae *scx* have fewer projections, the setae *sci* are longer than *sce*.

T. lini was first described from old linseeds by Oudemans (1924c) and has since been found infesting stored products or indoor materials from countries in Palaearctic, Nearctic, Afrotropical, Oriental and Australian Regions. It takes 14–17 days to complete its life cycle at 23°C, 87% RH on wheat germ (Hughes 1976). The life cycle develops through the egg, larva, protonymph, tritonymph and adult stages, and no deutonymphal stage (hypopus) is known. At 23°C and 87% RH on wheat germ as food. the life cycle takes from 14 to 17 days (Hughes, 1976). Tomita *et al.* (2003) studied the alarm pheromone of this species.

Symptomatology and damage

This mite is known from various stored foods, including ham, on which its presence is sporadic (Estrada-Peña *et al.*, 1981). Symptomatology and damage as other acarid mites, but the levels of populations are commonly less high than other injurious acarid mites.

Control

As other acarid mites. As concerns the chemical control, Stara *et al.* (2011b) tested in laboratory the toxicity of acaricides targeted against house dust mites, on five species of stored food mites, such as *A. siro* Linnaeus, *A. ovatus* (Troupeau), *Carpoglyphus lactis* (Linnaeus), *L. destructor* (Schrank) and *T. lini*. They diluted the formulations of benzyl benzoate, benzyl benzoate/permethrin/pyriproxyfen and neem in water, and applied it to filter paper in an unventilated chamber. The mortality of mites was observed after 24 hours of exposure to acaricide-impregnated filter paper. All of the tested acaricides resulted toxic to all of the mite species. There were significant differences in mortality among the species and the acaricides. Benzyl benzoate/permethrin/pyriproxyfen was the most effective, followed by benzyl benzoate and neem (Stará *et al.*, 2011b).

The study tested the effect of the chitin metabolic effectors, teflubenzuron, diflubenzuron, and calcofluor, and a combination of a chitinase and soybean trypsin inhibitor (STI), on the population growth of eight species of stored-product mites under laboratory conditions. The compounds were incorporated into the diets of the mites in concentrations ranging from 0.01 to 50 mg/g. The final populations of mites were observed after 21 days of growth in controlled conditions. Diflubenzuron and calcofluor suppressed the growth of all the tested species more effectively than the other compounds. The doses required to suppress the mite populations to 50% [rc (50)], in comparison to the control, ranged from 0.29 to 12.68 mg/g for diflubenzuron, and from 1.75–37.7 mg/g for calcofluor, depending on the mite species. When tested at the highest concentration (10 mg/g), teflubenzuron also suppressed all of the tested mite species in comparison to the control. The addition of chitinase/STI into the diet influenced population growth in several ways. When the highest concentration of chitinase in a cocktail of chitinase and STI (12.5 mg/g of diet) was compared to the control populations of *A. siro* Linnaeus, *A. ovatus* and *Aëroglyphus robustus*, Banks decreased significantly, whereas populations of *T. lini* and *Sancassania rodionovi* (Zachvatkin) increased significantly. The sensitivity of species to the tested compounds differed among species. The most tolerant species was *S. rodionovi*, the most sensitive was *A. ovatus*. The results confirmed that calcofluor and diflubenzuron have a toxic effect on stored-product mites (Stará *et al.*, 2010).

The Genus *Aleuroglyphus* Zachvatkin

Diagnostic characteristics

The setae *ve* are long and pectinate, more than half the length of *vi* and arising at about the same level. The setae *sci* are shorter than *sce*. The supracoxal seta is conspicuous and thickly pectinate. The dorsal terminal seta *e* of the tarsus is hair-like. There are three conspicuous ventral spines on the tarsi (*q+v*), (*p+u*) and *s* which lie close together (Hughes, 1976).

Economic importance

This genus includes some species on stored food; of these, the most important is *A. ovatus* (Troupeau). Other species play a secondary importance, such as *Aleuroglyphus siculus* Fumouze *et* Robin, reported from Azerbaijan in farm granary (Zakhvatkin, 1959), and *Aleuroglyphus formosana* Tseng, known from Taiwan on wheat (Tseng, 1972). *Aleuroglyphus beklemishevi* Zachvatkin, and *A. zaheri* Hafez *et* Salem, 1988, collected in house dust at Giza Governorate (Egypt), but both devoid of a phytopathological role.

Aleuroglyphus ovatus (Troupeau) (Fig. 28.13)

Common name

Brown-legged grain mite

Diagnostic characteristics

FEMALE. Body similar to that of male, idiosoma is 580–670 μm long. The anal opening is bordered by four pairs of anal setae; of these a_2 are long and extended beyond the posterior edge of the body. The two pairs of postanals are also long, and almost in line with one another (Hughes, 1976).

MALE. The idiosoma is 480–550 μm long. The general features are as *T. lini* Oudemans, with exception of chelicerae and legs that are intensely tanned, dark reddish-brown in colour and contrast with the glistening white appearance of the rest of the body. The dorsal shield has a pitted surface, oblong in shape, and with slightly concave lateral margins. The supracoxal seta is leaf-like, whose lateral edges are produced into long straight pectinations. The setae *sci* are short, about one-third the length of *sce*. The dorsal hysterosomal setae d_1 (c_1), to d_3 (e_1), *la* (e_2) and *hi* (f_1) are short, about the same length as *sci*, while d_4 (f_1) and *lp* (f_2) are only slightly longer. The only long setae are *sae* (h_1), *sai* (h_2), and two pairs of postanals, which together form a rather sparse 'train'. Al the setae are slightly pectinate and the shorter ones often have bifid ends, the prongs sometimes twisted apart. The legs are short and stout. On tarsi I and II the solenidion ω_1 is long and slender, tapering towards a rounded end and arising from the same depression as the very small famulus. At the distal end of the tarsi, the three ventral spines (*p+u*), (*q+v*), and *s* are stout and massive, the external ones being hooked at

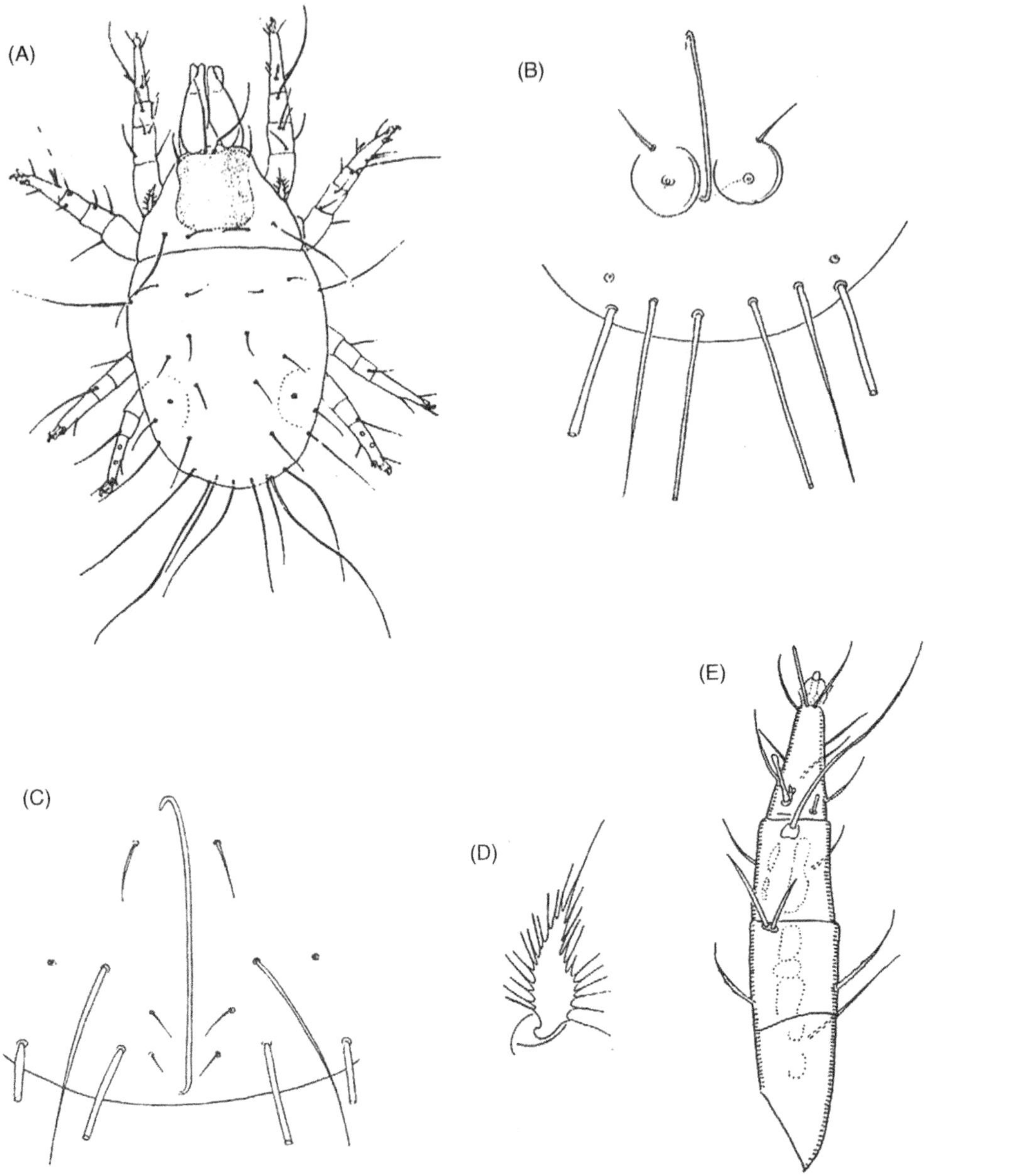

Fig. 28.13. *Aleuroglyphus ovatus* (Troupeau). Male. Dorsal view (A); anal region (B). Female. Anal region (C); supracoxal seta (D); leg I (E) (from Hughes, 1976).

their ends; the dorsal terminal seta *e* is hair-like. On tarsus IV, the two suckers arise in the median part of the tarsus. The genital folds and genital sense organs are slightly tanned and pale yellow in colour; the aedeagus supports are almost straight and diverge posteriorly, the aedeagus itself is a straight tube. The postanal setae are ventral in position and all three pairs are arranged almost in the same straight line (Hughes, 1976).

LARVA. Hexapod, otherwise similar to adult. The setae *sci* are much shorter than *sce,* and the coxal rods are blunt tubes. On tarsus I, ω_1 almost reaches the end of the segment and expands towards its distal end. A pair of long postanal setae is present (Hughes, 1976).

Geographic distribution

The geographic distribution of this mite is worldwide, and is reported in various countries such as Brazil (Flechtmann, 1986), Canada (Hughes, 1976; Beaulieu and Knee, 2014), China (Hsin and Shen, 1963, 1964; Xia *et al.* 2009), Croatia (Pagliarini, 1979; Hughes, 1976), Czech Republic (Ždárková and Voracek, 1993; Aspaly *et al.*, 2007), Egypt (Attiah, 1969), England, France (Hughes, 1976), Greece (Palyvos *et al.*, 2008), India (Putatunda, 2002); Indonesia (Haines, 1995), Iran (Faraji 1993; Noei, 2007; Irani-Nejad *et al.*, 2007), Ireland, Japan (Oshima, 1980), Malta (Pfliegler *et al.*, 2017), Malaysia (Kew *et al.*, 2014), the Netherlands (Hughes, 1976; Sipel *et al.*, 2016), Poland (Solarz, 2012), Primorskiy Kray (Tareev, 1970), Russia (Zakhvatkin, 1959; Smiley, 1987), South Korea (Lee and Choi, 1980), Taiwan (Tseng, 1979), Turkey (Hughes, 1976), United States (Hughes, 1976; Smiley, 1987) and other countries.

Commodities, facilities and vegetal and animal hosts

The list of commodities includes aquatic product, barley, bean, bean curd, candied fruit, carcadic, cheese, chicken meal, Chinese medicine, condiment, coriander, fish (dried product), fodder, grain, herb, lentil, livestock feed, macaroni, maize, maize meal, oat, pepper, pollard, poultry diet, processed food, residue, rice, rye, shrimp (dried), sunflower seed, tiki-tiki, wheat, wheat bran, wheat bran (mouldy), wheat flour, collected in the following facilities, such as bread van, commercial grain storage, farm and off-farm storage (Zakhvatkin, 1959; Hsin and Shen, 1964; Attiah, 1969; Hughes, 1976; Pagliarini, 1979; Smiley, 1987; Faraji, 1993; Xia *et al.*, 2009). Beaulieu and Knee (2014) reported the mite on stored plant products. Moreover, the brown-legged grain mite has been collected in lucerne fields (Irani-Nejad *et al.*, 2007), and in mice burrows, mole nests and broiler houses (Hughes, 1976). The mite has been collected associated with *Apis* sp. (O'Connor and Klimov, 2016), and in lucerne fields of the southwest of east Azerbaijan in Iran (Irani-Nejad *et al.*, 2007). Feeding pigs with feed in feed troughs, in contact with the animal's saliva has been observed (Flechtmann, 1986), being also present in other types of feed and preserved products (Geller *et al.*, 2009). Moreover, *A. ovatus* is reported on house dust (Colloff, 2009b) and also in bran, wheat, chicken meal, dried fish products, flour, pollards, mice burrows, mole nests and broiler houses (Hughes, 1976). Allergens have been isolated from the mite (Colloff, 2009a). Kew *et al.* (2014) isolated from both the raw and commercial nests.

Bio-ecology

This mite has a cosmopolitan geographic distribution and develops on various commodities. It is one of the most prevalent, and serious pests to storage products in various countries. In China, it induces heavy damages to storage grain, and it is also responsible of acariasis in humans (Hsin and Chen, 1963). In Croatia, Pagliarini (1979) reported the mite as very injurious to stored cereals.

Its life cycle develops through the egg, larva, protonymph, tritonymph and adult stages, and no hypopus is known. Hsin and Chen (1964) studied the mite biology under laboratory conditions, in specially designed cells, kept at a room temperature of 25°C and a constant RH of 75%. Before entering into the protonymphal, tritonymphal and adult stages, a quiescent period is experienced in every case of approximately 24 hours. At this temperature and RH, the development of the life cycle takes an average of 16 days and 10 hours; in particular, 80 hours for the egg, 77 for the larval, 115 for the protonymphal, and 122 for the tritonymphal stages, respectively. Mating lasts 2–4 minutes, repeated many times during the whole life. The eggs are laid 1–3 days after mating. Oviposition lasts ca. 4–6 days. The fertilized female will lay 33–78 eggs, with an average of 55.5. No parthenogenesis occurs (Hsin and Chen, 1964). The duration for egg, larva, quiescent larva, protonymph, quiescent protonymph, tritonymph, quiescent tritonymph were 3.3, 1.7, 0.9, 1.2, 0.8, 1.2 and 0.9 days, respectively (Yan *et al.*, 1992).

Xia *et al.* (2009) investigated in the laboratory, the effect of four constant temperatures (20°C, 24°C, 28°C and 32°C) and 85% RH, on the physiological properties (survival, developmental time, sex ratio, fecundity and longevity of females), and population growth of *A. ovatus*. They found that the total developmental time, longevity and oviposition period is shortened, as temperature increased. The shortest developmental time (14.70 ± 0.34

days) was observed at 32°C. At 20°C, 24°C, 28°C and 32°C, mated females laid on average 0.87 ± 0.10, 2.57 ± 0.13, 3.14 ± 0.27 and 1.29 ± 0.16 eggs per day, respectively. Maximum fecundity was recorded at 28°C with 70.83 ± 10.16 eggs per female. The highest intrinsic rate of increase (r_m) was observed to be 0.16, at 28°C (Xia *et al.*, 2009).

Ahmed *et al.* (2012) estimated the influence of three constant temperatures (20°C, 25°C and 30°C) on the damage effect, population growth, life cycle and longevity of *T. putrescentiae* (Schrank), and *A. ovatus* (Troupeau) reared on the dried Shiitake mushroom, *Lentinula edodes* (Berk.) Pegler, at 80% RH and 14:10 (L:D) h photoperiod. The weight loss of the Shiitake mushroom caused by both mites, increased with increasing temperature. The increase in the count of *T. putrescentiae* and *A. ovatus* in the rearing cells, led to a noticeable decrease in the weight of the Shiitake mushroom. The results indicated that *T. putrescentiae* was more harmful than *A. ovatus* on the Shiitake mushroom. The r-values of both mite species increased with increase in temperature up to 25°C, whereas the value of r decreased, when the temperature increased to 30°C. At 20°C, 25°C and 30°C, the values of r were 0.167, 0.182 and 0.159 for *T. putrescentiae* and 0.126, 0.140 and 0.115 for *A. ovatus*, respectively. The developmental time of immature stages, reproductive periods and female longevity of *T. putrescentiae* and *A. ovatus* were adversely affected by the increase in temperature. The life cycle of *T. putrescentiae* varied from 9.26 to 20.64 days and that of *A. ovatus* ranged from 11.65 to 25.21 days, when the temperature increased from 20°C to 30°C. The longest longevity of *T. putrescentiae* and *A. ovatus* (38.37 and 48.18 days) was observed at 20°C, whereas the shortest (20.38 and 28.40 days) was found at 30°C. The results showed that *T. putrescentiae* had lower developmental time and reproductive periods than *A. ovatus*, when reared on the Shiitake mushroom, at all the studied temperatures (Ahmed *et al.*, 2012).

Que *et al.* (2016) determined the complete mitochondrial genome (mitogenome) sequence of *A. ovatus* by long PCR and primer walking methods. The complete mitochondrial genome is 14,305 bp in length and contains 13 protein-coding genes, two ribosomal RNA genes, 22 transfer RNA genes, as well as a displacement loop (D-loop). The overall base composition of the genome is A (23.61%), T (44.60%), C (13.03%) and G (18.76%), respectively. Stop codon was missed for the CYTB gene, and the gene overlaps were suggested between ten pairs of the contiguous genes in *A. ovatus*. The mitogenome would contribute to resolving phylogenetic position and interrelationships of *Aleuroglyphus*.

Aspaly *et al.* (2007) investigated the effect of temperature, ranging from 5–35°C, on the population growth of *A. siro* Linnaeus, *T. putrescentiae* and *A. ovatus* at 85% RH. Starting with ten individuals, the population increase of mites was observed after 3 weeks of cultivation, or after 6 weeks, for those kept at low temperatures (5°C, 10°C, 12.5°C and 15°C). The rate of increase was calculated for each temperature and species. The obtained data were fitted with polynomial models. The mite population growth rates increased with increasing moderate temperatures until 25°C, when r m-values were 0.179, 0.177 and 0.190 for *A. siro*, *A. ovatus* and *T. putrescentiae*, respectively. The lower development threshold was 10.2°C in all three species. The estimated upper temperature threshold was higher in *T. putrescentiae* (49°C), than in *A. siro* and *A. ovatus* (38°C). Simulation of the rate of population increase under ideal conditions, using real temperature records, obtained from Czech grain stores, showed that the pest mite populations increase only during 3.5 months within a typical 9-month storage season in Central Europe. These results indicate that control of mites, be it chemical, physical or biological, is recommended during the months when allergens and pests are produced, that is, from September to mid-November and in May (Aspaly *et al.*, 2007). Xia *et al.* (2009) and Sinha (1968b) reported a value of the threshold temperatures for the population development of *A. ovatus* of 13.3°C and 22.0°C, respectively. Kuwahara *et al.* (1992) studied the sex pheromone of the brown-legged grain mite.

Symptomatology and damage

Although this mite may be injurious, it is rarely analysed for its pathogenicity, although it has been assigned alergogenic properties, in studies carried out in various countries (Ramjan *et al.*, 2005). Damage as other acarid mites.

Siegert (2016) evaluated the nutritional profile of rations infested with *A. ovatus* and *T. putrescentiae*, as well as to identify the preference of these mites for the raw materials, used to prepare the rations. Acarina fauna were evaluated in poultry, cattle and pork rations. In the preference test, the species were offered the possibility of choosing rice bran (FA), soybean (FS) and wheat (FT), meat meal (CF) and milled maize (MM) and fungus (F) Feed colonizers. These products were packed in Petri plates, and in the centre released 20 specimens / species. In order to evaluate the nutritional profile of the rations with *A. ovatus* and *T. putrescentiae,* 1200 grams of ration were separated, placing 100 specimens of each mite that were conditioned in polyethylene terephthalate (PET) bottles, and stored in a greenhouse with a temperature of 25 ± 1°C, at 70 ± 5% RH. Evaluations were performed on the first, 60th and 120th days of exposure. The content of ethereal extract (fat), crude fibre, mineral material, crude protein, riboflavin, thiamine and moisture, besides the presence of fungus and yeasts, were evaluated. The predominant and most abundant species was *T. putrescentiae* (60%), followed by *C. malaccensis* (Oudemans) (20%) and *A. ovatus* (8%). *A. ovatus* preferred meat meal (FC) and MM and rice bran (FA), whereas *T. putrescentiae* preferred meat meal (FC), fungus (F) and rice bran (FA). *A. ovatus* reduced fat content by 72%, protein by 3.93%. The mineral material increased by 28.35% and 40.56%, while *T. putrescentiae* reduced fat by 35.11%, protein by 6.06% and moisture by 13.53%. It increased the concentration of crude fibre by 13.40% and caused vitamin changes at 60 days, increasing thiamine concentration by 38.57% and reducing riboflavin by 50% in diets (Siegert, 2016).

Geller *et al.* (2009) documented in Brazil, a case of severe oral mite anaphylaxis following the ingestion of a pancake made of wheat flour, contaminated with *A. ovatus*. This patient had house dust mite allergic asthma and rhinitis (*Dermatophagoides farinae* Hughes), *D. pteronyssinus* (Trouessart), *Blomia tropicalis* Bronwijk, Cock *et* Oshima and *L. destructor* (Scrank), and developed anaphylaxis immediately after the first bite of this home-made pancake. She also presented intolerance to aspirin, non-steroidal anti-inflammatory drugs and pyrazolones (facial angioedema and laryngeal oedema). Antigenic cross-reactions among several domestic mites are well documented. All grain flours including the ones obtained from wheat should always be stored frozen followed by constant refrigeration, in order to prevent mite proliferation (Geller *et al.*, 2009).

Control

The control of this mite calls for the same protocol of other acarid mites. Reports on physical and chemical means are known.

PHYSICAL CONTROL. Doses of more than or equal to 40 krad completely \r\n inhibited egg development. Eggs of *A. ovatus* were more \r\n resistant to irradiation, than those of *R. echinopus* (Fumouze *et* Robin). The effect\r\n of irradiation on the inert deutonymphs was more detrimental \r\n, than that on young adults. The effect of irradiation on female \r\n lifespan of both species was stimulatory, at doses of >60 krad \r\n and inhibitory at higher doses. Treatment at 5 krad caused at least a 50% \r\n reduction in the fecundity of both mite species (Boczek *et al.*, 1985).

BIOLOGICAL CONTROL. Recently, it has been observed that the biological control of the stored food pests, based on the use of natural compounds and predators, has a high potential for success. At this aim, the suppression of three pest mites by combination of the cheyletid mite *C. malaccensis* (Oudemans) and bean flour (*Phaseolus vulgaris* Linnaeus), a well known antifeedant has been investigated. Wheat grain (100 g) was infested with 100 individuals (*A. siro* (Linnaeus), *A. ovatus* or *T. putrescentiae*). The bean flour was added in concentrations of 0, 1, 5 and 10g kg^{-1}(wt/wt) and/or the predator with initial predator/prey ratios 0, 1/50, 1/25 and 1/10. The experiment was carried out under abiotic conditions that are optimal for pest development. After 21 days, pest and predator numbers were recorded. Addition of flour significantly reduced populations of *A. siro*, slightly reduced populations of *T. putrescentiae*, the efficiency increased with increasing concentration. *A. ovatus* was not affected by bean flour. Populations of all mite species were successfully reduced, by the sole addition of *C. malaccensis*, with a higher efficiency at higher *Cheyletus* ratios. The additive effect of the predator and flour was mainly apparent in the reduction of *T. putrescentiae*. The population of *C. malaccensis* was not affected by the presence of bean flour. It is therefore recommended to use bean flour for the reduction of *A. siro*, a combination of bean flour and *Cheyletus,* for the reduction of *T. putrescentiae*, and only *Cheyletus* for the reduction of *A. ovatus* (Hubert and Pekár, 2009).

CHEMICAL CONTROL. In tests of acaricides for the control of *A.* ovatus infesting cultures of the oriental rat flea, *Xenopsylla cheopis* (Roths.), in jars provided with linen covers and containing Purina dog food in sand, as the rearing medium, Tedion and Kelthane proved ineffective, but light infestations

(100 cc of medium inoculated with 500 mites) were satisfactorily controlled by chlorobenzilate, at 0.2 ppm in the rearing medium or as a 0.05% solution in acetone, painted on to the inner wall of the jar and used to soak the linen cover for 3 s. Heavy infestations (100 cc. medium inoculated with 2000–3000 mites) were controlled by chlorobenzilate, at 20 ppm in the medium or as a 0.5% solution, applied to both the linen cover and the inner rim of the jar; a combination of both treatments (using 2 ppm and a 0.5% solution) gave adequate control for 30 days. The flea larvae were unaffected by 2 ppm chlorobenzilate in the rearing medium; 20 ppm killed larvae, but not adults. Neither stage was affected by treating the linen cover and inner rim of the breeding jar with a 0.5% solution of chlorobenzilate. Measures are recommended for maintaining flea cultures free from mites, based on these experimental results and on a year-long practice (Sternberg and Tahori, 1962).

The efficacy of spinosad and spinetoram, both bacterium-derived pesticides, and their combination with the cheyletid mite *C. malaccensis*, has been evaluated against *T. putrescentiae* and *A. ovatus*. Trials were carried out under optimal environmental conditions for pest development, and the estimate of pest and predator mites was made after 21 days. Both astigmatid mites were exposed to diet (experiment I) treated with either spinosad or spinetoram (concentrations range of 0.01–2 ppm), and diet (experiment II) treated with either spinosad or spinetoram (0.5 ppm) and/or the predator at initial predator/prey ratio (0.02). Application of spinosyns significantly reduced population of both mite species. Total control of *T. putrescentiae* and *A. ovatus* was achieved by the application of spinosad at 1 and 2 ppm, respectively. As measured by rC_{50} and rC_{90} (concentration for 50% and 90%, suppress of population in comparison to control), spinosad was more toxic to *T. putrescentiae* and *A. ovatus* than spinetoram. Furthermore, *T. putrescentiae* was more susceptible to spinosad, than *A. ovatus*. Conversely, it was less susceptible to spinetoram, than *A. ovatus*. The populations of both mite species were successfully suppressed by the sole application of *C. malaccensis*. Although the density of predatory mites was not affected by the presence of 0.5 ppm spinosad, it was almost eradicated by spinetoram at 0.5 ppm. A combination of spinosad at 0.5 ppm with two individuals of *C. malaccensis* mites (ratio 0.02), outperformed spinosad used alone at the same former concentration, in reduction efficiency of the pest populations by 12% for *T. putrescentiae* and 25% for *A. ovatus* within 21 days (Bakr and Selim, 2019).

The Genus *Mycetoglyphus* Oudemans

Diagnostic characteristics

The prodorsal external vertical setae (*ve*) are short and smooth, less than a quarter of the length of the prodorsal vertical internal setae (*vi*) and set posteriorly to setae *vi*. The internal scapular setae (*sci*) are longer than the external scapular setae (*sce*). The dorsal tarsal setae *e* and the ventral tarsal setae *p*, *q*, *s*, *u* and *v*, are spine shaped. On tibia I, the solenidion φ_1 is never more than twice as long as the solenidion φ_2. The leg I of the male is not enlarged, and the femur does not bear a ventral apophysis. The male has a long aedeagus (Hughes, 1976).

Economic importance

The genus includes only two species: *M. fungivorus* (Oudemans) and *Mycetoglyphus sevastianovi* Kadzhaja (O'Connor, 2008a). The first species has been collected on mushrooms growing, and on other substrates (Hughes, 1976), and the second species has been recorded in natural soil near Odessa (Ukraine) (Kadzhaya, 1988). Damage caused by *M. fungivorus* to economic plants is similar to that of *T. similis* Volgin; severe attacks on spinach in greenhouses have been reported (Nakao, 1989; Zhang, 2003).

Mycetoglyphus fungivorus Oudemans (Fig. 28.14)

Common name

Unknown

Diagnostic characteristics

FEMALE. The idiosoma is 500–600 µm long and the general aspect is as for the male (Hughes, 1976).

MALE. The general aspect is similar to that of *T. similis*. The body is 400–600 µm long and with well-tanned appendages. The setae *ve* are very short, less than a quarter as long as the setae *vi*, set

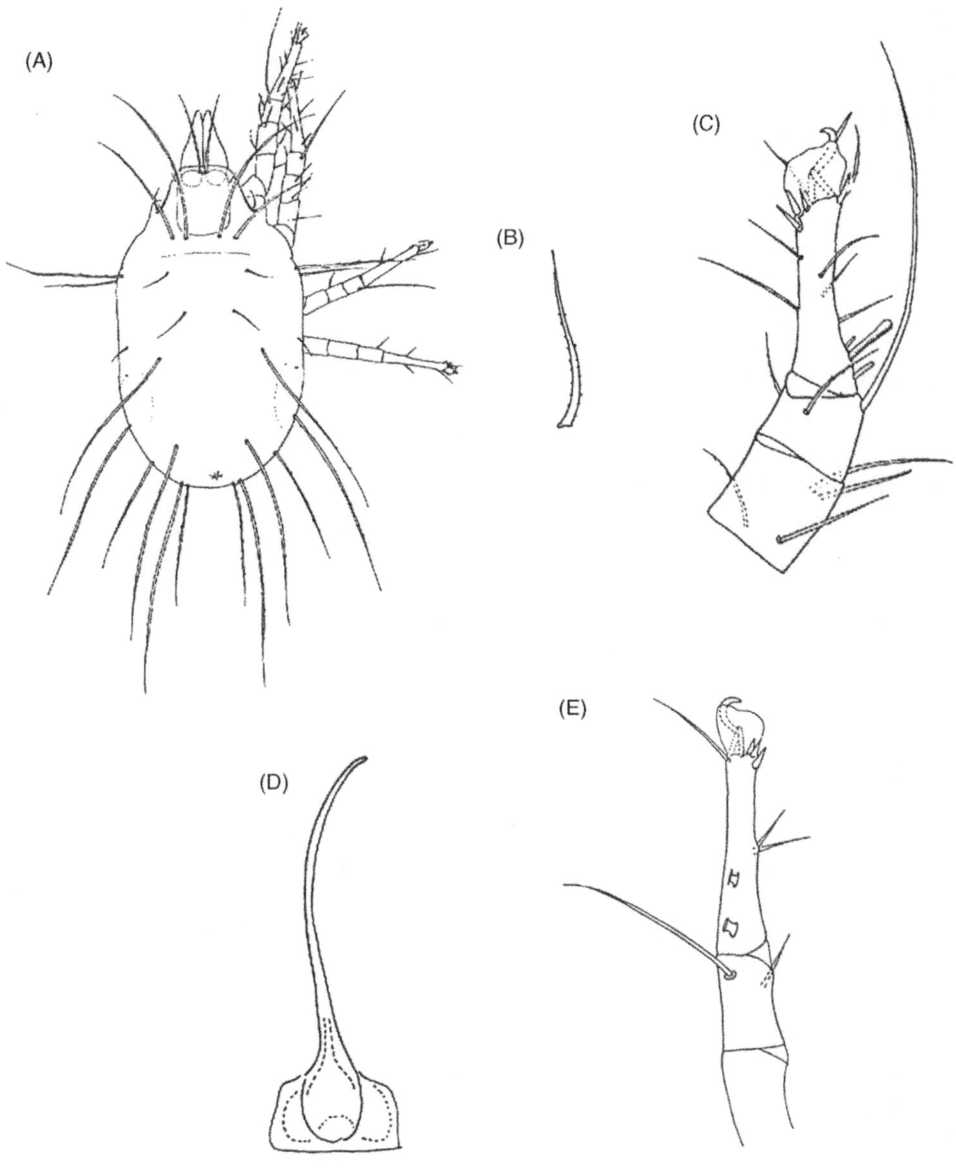

Fig. 28.14. *Mycetoglyphus fungivorus* Oudemans. Female. Dorsal view (A); supracoxal seta (B); leg I (C). Male. Aedeagus (D); leg IV (E) (from Hughes, 1976, licensed under the Open Government Licence v3.0).

posteriorly to the bases of *vi*, never as far back as the midpoint of the lateral edge of the prodorsal shield. The idiosomal dorsal setae d_2 are short, about 6% of the length of the idiosoma; the setae c_1 are 1–1.5 times longer than the setae d_2 and the setae d_1 are 1.5 to twice as long as the setae d_2. The supracoxal seta is curved and slightly pectinate. The solenidia ω_1 and ω of tarsi I and II are club-shaped; on tarsi I–III, there is a distal seta *e* and five ventral spines, slightly unequal in size; on tarsus IV, the suckers are set in the basal half of the article, and the setae *r* and *w* are distal to them. The aedeagus is very long, curved and arising from a basal plate (Hughes, 1976).

HYPOPUS. The body is 250 µm long and yellow-brown in colour, with a widely rounded posterior margin. The prodorsal shield has an almost straight anterior edge, lacking any sign of a rostrum. The setae *vi* are lacking. The gnathosomal base is completely concealed, by the prodorsal shield. Ventrally, the sternum is extended posteriorly, so that coxal plates I and II are completely separate from one another. The sucker plate is well removed from the posterior edge of the body and is almost circular in outline. Tarsus I bears a long expanded and three lanceolate setae, and tarsus IV has two lanceolate setae (Zakhvatkin, 1941).

Geographical distribution

M. fungivorus is reported from South Africa (Hughes, 1976), Canada (Beaulieu and Knee, 2014), China (Chen *et al.*, 1982; Li *et al.*, 2015a), England, Germany, Hungary (Mahunka, 1961; Hughes, 1976), Iran (Sayadi, 2012), Japan (Nakao and Kurosa, 1988), Malta (Pfliegler *et al.*, 2017), the Netherlands (Siepel *et al.*, 2016), Poland (Solarz *et al.*, 2007), Primorskiy Cray (Tareev, 1970), Russia (Zakhvatkin, 1941) Ukraine (Akimov and Oksentiuk, 2018; Oksentiuk *et al.*, 2022) and the United States (Hughes, 1976).

Commodities, facilities and vegetal and animal hosts

M. fungivorus was first collected on mushrooms growing in Sussex (England) (Oudemans, 1932). The mite is reported on grassland, and beneath stacks of hay and straw (Griffiths, 1960), on moles nests and nests of the ant *Camponotus ligniperda* (Latreille), with hypopi probably carried by the ants (Türk and Türk, 1957), on growing mushrooms, celery waste, lettuce, decaying radishes, wet decaying wood debris (Mahunka, 1961), *Phaseolus* sp. (pole beans) (Manson, 1983), spinach (Nakao and Kurosa, 1988; Nakao, 1989), Chinese medicine, rice bran, milk powder, walnut seeds, moon cake, and other stored products (Li *et al.*, 2015a), hay, straw and spoiled root crops (Akimov and Oksentiuk, 2018). Beaulieu and Knee (2014) reported the mite on stored grains. Oksentiuk *et al.* (2022) found the mite outbuildings.

Bio-ecology

The bio-ecology of this mite is not much investigated. Mohamed (2019) observed that the predatory mite *Protogamasellopsis denticus* Nasr successfully developed and reproduced on nymphs of *M. fungivorus* and *Caloglyphus rodriguezi* Samsinak, but significant differences were observed on the developmental time of *P. denticus* Nasr on *M. fungivorus* and *C. rodriguezi* Samsinak nymphs and other biological stages, respectively.

Symptomatology and damage

See *Tyrophagus* spp.

Control

See *Tyrophagus* spp.

The Genus *Rhizoglyphus* Claparède

Diagnostic characteristics

Previously, there had been no revisions of this economically important genus, and about 52 names were in use under the genus *Rhizoglyphus* (Diaz *et al.*, 2000). In 2004, Fan and Zhang examined different *Rhizoglyphus* specimens, mainly from Australasia and Oceania, and described 11 species. In addition, they provided keys to adult females and males of *Rhizoglyphus* from Australasia and Oceania, with lists of hosts and distribution details for each species. The idiosoma is oblong and whitish to semi-transparent. There are four pairs of prodorsal setae, and 12 pairs of hysterosomal setae. The setae *ve* are small or vestigial, set about halfway along the lateral margins of the prodorsal shield. The setae *sci* are shorter than the setae *sce*, and very small, vestigial or lacking. The setae *scx* are smooth, large or small. Grandjean's organ is bifurcate or simply bladed. The legs are reddish-brown. Tarsi I and II have a proximal conical spine (*ba*) set near the bases of the solenidion ω_1 and a famulus ε. Tarsus I bears four long terminal setae, and one apical solenidion ω_3; tarsus II has four long terminal setae; tarsus III has three long terminal setae, and tarsus IV bears two long terminal setae in the female and one long terminal seta in the male. Some tarsal setae are slightly flattened apically (Fan and Zhang, 2004). The homoeomorphic male differs from the female in that the genital opening is set between coxae IV, an aedeagus is present, there are one pair of anal suckers and three pairs of setae (ps_{1-3}); the setae ad_{1-3} are lacking. On tibia IV, the solenidion φ is spine-like, and tarsus IV bears two suckers. The setae on the legs are as in the adult female, except for tarsi IV. The heteromorphic male is similar to the homoeomorphic male but both sides or one side of leg III is about doubled in size. The body setae are slightly longer. The setae on the legs are as in the homoeomorphic male, except for tarsi III (Fan and Zhang, 2004). The hypopus has a highly sclerotized body that is dorsoventrally flat. The gnathosoma is reduced and the chelicerae are lacking. Excepting setae *vi* and h_2, the idiosomal setae are small. The setae *scx* are set ventrally. There are two pairs of genital papillae, and one pair of genital setae. The ventral hysterosomal area forms an anal attachment organ, which bears six pairs of attachment structures (modified setae). Genu I lacks solenidion σ. On tarsus I, solenidion ω_3 is proximal. There is no seta *ba*. Some apical tarsal setae are foliate (Fan and Zhang, 2004).

Economic importance

Diaz *et al.* (2000) highlighted a state of confusion of the systematics of *Rhizoglyphus* and reported that the most serious taxonomic problem relates to the identities of the two most common species, *R. robini* (Claparède) and *R. echinopus.* (Fumouze *et* Robin) Hughes (1948) recognized that there were two common species in agricultural crops, one with short *sci* setae, for which she retained Zachvatkin's concept of *R. echinopus,* and one with longer *sci* setae, for which she used the name *Rhizoglyphus callae* (Oudemans). Interpretations of *echinopus* and *robini* by both Zachvatkin and Hughes persisted until the two species were separated by Eyndhoven (1960, 1968, 1972). After careful study of these two species, including characters from the male and female reproductive systems as well as idiosomal setal lengths, Eyndhoven observed that the species with shorter *sci* setae more closely matched the description of *R. robini,* rather than that of *R. echinopus.* Eyndhoven synonymized *R. callae* with *R. echinopus* and *R. solani* with *R. robini.* These concepts have been followed here, and by most subsequent taxonomic workers (Manson, 1972a; Fain, 1988a), but not by Hughes (1976).

Mites of the genus *Rhizoglyphus* are usually associated with the bulbs and corms of many plants, and are found in different substrates. Some species cause damage to crops and ornamental plants, both in greenhouses and in the field, and their presence may also involve exported/imported horticultural products (carrots, corms of taro, flower bulbs) (Fan and Zhang, 2004). Thirteen *Rhizoglyphus* spp. are reported as pests of crops, or have been described from agricultural settings, usually in close association with cultivated plants (Diaz *et al.*, 2000). However, only *R. robini* and *R. echinopus* are traditionally known as severe pests of economic plants. Recently, the list of injurious *Rhizoglyphus* spp. included *R. setosus* Manson, found damaging, various bulbs and plants of economic importance in different countries. In particular, *R. robini* has been collected on bulbs, leaves and seeds of various cultivated plants, many ornamentals and several vegetables. The mite damages lactis, mainly bulbs of various ornamentals and corms of *Freesia* in storage, greenhouses and in the field, inducing distorted growing tips and leaves, and reducing the bulb to a rotten pulp (Hussey *et al.*, 1969; Manson, 1972a; Hughes, 1976; Meyer Smith, 1981; Ascerno *et al.*, 1983; Okabe and Amano, 1991; Zhang, 2003). *R. echinopus* also infests the bulbs and roots of ornamental plants, and causes similar damage (Manson, 1972a; Fan and Zhang, 2003; Zhang, 2003). Both *R. robini* and *R. echinopus* have quarantine importance, as in exported horticultural products (e.g. carrots) and ornamental bulbs (Fan and Zhang, 2003). Both species were also reported on stored food (Hagstrum *et al.*, 2013). *R. setosus* is associated with the underground organs of various plants (lilies, leeks, garlic, onions, etc.) and different substrates (Ho and Chen, 1987; Chen *et al.*, 2002; Fan and Zhang, 2004), and like *R. robini* and *R. echinopus*, has a quarantine importance. In addition to direct damage, the presence of bulb mites has been associated with pathogenic fungi and bacterial development. *Rhizoglyphus costarricensis* Bonilla, Ochoa *et* Aguilar was collected in Costa Rica, on rice (Bonilla *et al.*, 1990).

Rhizoglyphus robini (Claparède) (Fig. 28.15)

Common names

Robine bulb mite, bulb mite

Diagnostic characteristics

FEMALE. The idiosoma is 676–934 μm long, 482–650 μm wide, whitish and with a smooth

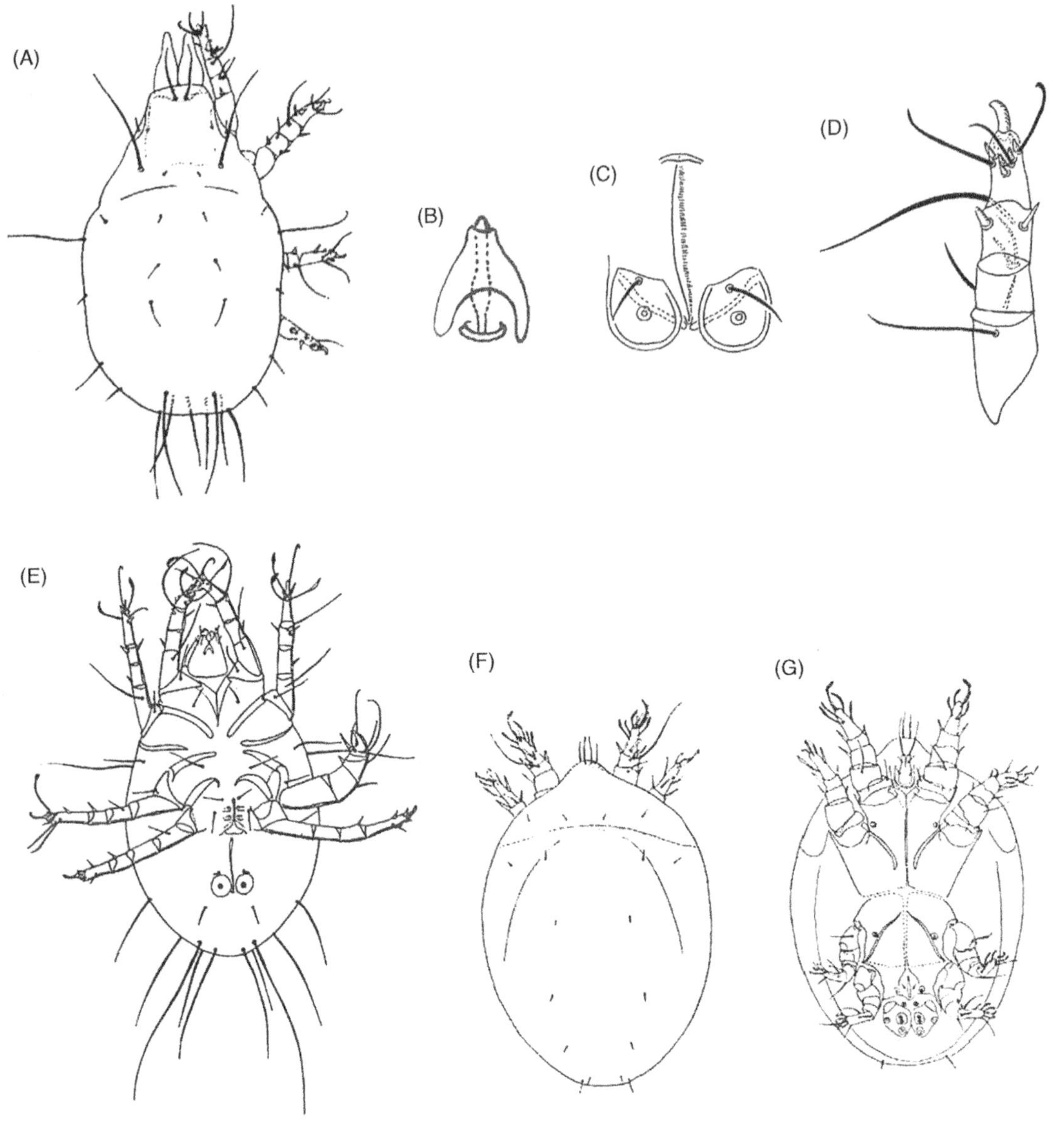

Fig. 28.15. A. *Rhizoglyphus robini* (Claparède). Male. Dorsal view (from Hughes, 1976, modified) (A); base of aedeagus (B); anal region (C) (from Hughes, 1976); leg I (D); ventral view of abnormal heteromorphic male (E) (from Hughes, 1976, modified). Hypopus. Dorsal view (F); ventral view (G) (from Hughes, 1976).

integument. The legs are short and brownish. The prodorsal shield is evenly punctulate. The setae *vi* are blunt, the setae *sci* are about one-third the length of genu I, and the ratio *sci-sci*: *sci-sce* ranges from 0.7–0.8 (i.e. the setae *sci* are not close to *sce*, as the ratio is <2). Grandjean's organ is not bifurcate or bifurcate to small branches. The setae *scx* are slender and pointed. The setae c_1 are from 0.8–1.0 times the length of the setae d_2; the setae c_3 are about as long as the setae c_1; the setae d_1 are from 0.8–1.0 times the length of the setae d_2; the setae d_2 are far from the opisthonotal glands ($_{gla}$); the setae e_1 are from 2.3–3.5 times the length of d_2; the setae e_2 are from 0.8–1.5 times the length of the setae f_2.

The anal opening has six pairs of small setae, subequal in length, less than one-half of the length of genu IV. The copulatory opening has an incomplete circular shield, and the spermathecal duct is long and narrow. The base of the inner part of the spermatheca is minute; the sclerites of the oviducts are 'Y'-shaped, with a distance between them from 6–8 μm, and the basal part is about one-half the length of the branches (Fan and Zhang, 2004).

HOMEOMORPHIC MALE. The body is 603–671 μm long. The setae *sci* are 7–25 μm long. The setae *vi* are 2.4–3.0 times as long as genu I. The setae c_1 are from 0.7–1.0 times the length of the setae d_2; the setae c_3 are about as long as or slightly longer than the setae c_1; the setae d_1 are 0.7–1.0 times the length of the setae d_2; the setae d_2 are far from *gla*; the setae e_1 are 2.1–3.3 times longer than the setae d_2, and the setae e_2 are from 0.9 to 1.3 times the length of the setae f_2. The ratio *sci-sci*: *sci-sce* ranges from 0.6–0.8. The aedeagus is conical, and the canal of the aedeagus lacks lumps. The anal discs lack radiating lines, and their diameter is about one-third of the length of the setae ps_2. The setae ps_1 are from 3.4–4.0 times longer than genu IV, and more than three times as long as the setae ps_2. The setae ps_2 are from 0.8–1.1 times the length of genu IV; the setae ps_3 are thick and pointed. The ratio s¢: s^2 ranges from 0.8–0.9. The solenidion of tibia IV is slightly shorter than the seta *kT* (Fan and Zhang, 2004).

HETEROMORPHIC MALE. The idiosoma is 512–721 μm long and 375–456 μm wide. The prodorsal shield is evenly punctulate, with the posterior margin concave, and the distance between the setae *ve* ranges from 77 μm to 102 μm. The setae *vi* are 2.2–2.6 times as long as genu I. The setae c_1 are 0.7–0.8 times the length of the setae d_2; the setae c_3 are longer than the setae c_1; the setae d_1 are 0.7–0.8 times the length of the setae d_2; the setae d_2 are far from *gla*; the setae e_1 are 2.6–3.3 times longer than the setae d_2, and the setae e_2 are 1.0–1.4 times longer than the setae f_2. The ratio *sci-sci*: *sci-sce* ranges from 0.8–1.1. The aedeagus is conical, and the canal of the aedeagus lacks lumps. The anal discs lack radiating lines, and their diameter is about one-third the length of the setae ps_2. The setae ps_1 are long, 3.0–3.9 times longer than genu IV, and about three times as long as the setae ps_2. The setae ps_2 are 0.5–1.1 times the length of genu IV, and ps_3 is thick and pointed. The ratio s¢: s^2 is 0.8–0.9. The solenidion of tibia IV is shorter than the seta *kT* (Fan and Zhang, 2004).

HYPOPUS. The gnathosoma is 21–22 μm long and 14–15 μm wide. The apical solenidia are 31–32 μm long. The palpal supracoxal setae (*elcp*) are 9–10 μm long; the infracapitular setae (*m*) are 7 μm long. The idiosoma is 302–306 μm long, and 192–195 μm wide. The prodorsal shield is 82–92 μm long. The setae *vi* are 1.9 times the length of genu I. The ratio *c*1-*c*1: *d*1-*d*1: *e*1-*e*1 is 1.3:1.0:1.2–1.3 and h_2: h_3 is 2.2–2.5. The anal disc ad_1 + ad_2 attachment structures are small. The solenidion ω_2 of tarsus I is less than one-half as long as ω_1. The solenidion of genu II is about one-half the length of genu II. The solenidion of tibia II is about two times the length of tibia II. Tarsus II is 2.3–2.8 times the length of tibia II. The solenidion of genu III is less than one-third the length of genu III (Fan and Zhang, 2004).

Geographical distribution

R. robini is found worldwide, in particular from Algeria (Fain, 1988a), Austria (Michael, 1903), Australia (Womersley, 1941; Manson, 1972a), Belgium (Fain, 1988a), Brazil (Barbosa and Moraes, 2020), Canada (Diaz *et al.*, 2000; Beaulieu and Knee, 2014), China (Bu and Li, 1998), Colombia (Fan and Zhang, 2004), Czech Republic (Hrudová and Šafránková, 2017), Egypt (Diaz *et al.*, 2000), Fiji (Fan and Zhang, 2004), France, Germany (Michael, 1903) Greece (Manson, 1972a), India (Fan and Zhang, 2004), Iran (Rahimi, 1991; Fathipour, 1994), Israel (Gerson *et al.*, 1985), Italy (Michael, 1903), Japan (Manson, 1972a), Mexico (Diaz *et al.*, 2000), Nepal (Fan and Zhang, 2004), the Netherlands (Manson, 1972a; Siepel *et al.*, 2016), New Zealand (Womersley, 1941; Manson, 1972a), Poland (Diaz *et al.*, 2000), Primorskiy Kray, Russia (Zakhvatkin, 1941) South Africa (Meyer Smith, 1981), South Korea (Ree *et al.*, 1997), Switzerland (Claparède, 1869), Taiwan (Tseng, 1979), Ukraine (Akimov and Oksentiuk, 2018), UK (Michael, 1903; Hughes, 1948), United States (Manson, 1972a; Diaz *et al.*, 2000) Yugoslavia (Fan and Zhang, 2004).

Commodities, facilities, vegetal and animal hosts

Michael (1903) and Hughes (1948) reported the mite on stored products. Hagstrum *et al.* (2013) numbers as commodities, bamboo shoot, bulb (decaying), Chinese medicine, condiment, garlic, gin trash, grain, rice (husked) and sweet potato tuber. Moreover, the mite has been reported on rice in Taiwan (Tseng, 1979), house dust in Korea (Ree *et al.*, 1997), mushrooms (Fan and Zhang, 2003) and seed head, rotting plant material in New Zealand (Fan and Zhang, 2004), *Secale cereale* in Poland (Diaz *et al.*, 2000) and luggage in Yugoslavia (Fan and Zhang, 2004). Moreover, Ebrahimi and Noei (2022) reported stored potato, onion, shallot, grape, liquorice, darnel, ornamental plants, date palm, saffron maize, cotton field soil samples and honeybee hive, discussed by various Iranian specialists.

A long list of vegetal hosts characterizes this mite (Fan and Zhang, 2004; Vacante, 2016). Animal hosts include a restricted number of species: humans (Fan and Zhang, 2003), Fringillidae (*Fringilla coelebs* Linnaeus), Passeridae (*Passer montanus* Linnaeus), Talpidae (*Scalopus aquaticus* Linnaeus), Turdidae (*Turdus philomelos* C.L. Brehm) and in the United States in the nest of *Scalopus aquaticus* (Linnaeus) (Fain, 1988a).

Bio-ecology

The robine bulb mite is primarily associated with bulbs, corms and tubers or roots of plants, and has been found in seeds, and the lower parts of plants. It is also common in compost, and soil that is rich in organic matter. Sometimes it has been collected also on stored foods. Its life cycle develops through the larva, protonymph (proceeding towards a hypopus), tritonymph and adult stages. The hypopus stage does not feed, resists unfavourable environmental conditions and employs phoresy with insects (narcissus flies, sciarids, etc.) to spread. Two male morphs are known in *R. robini*, the fighter male with a modified third pair of legs, and a nonfighter male with unmodified legs, but the role of fighter males on reproduction in this species, as well as the mechanisms underlying the formation of heteromorphic males, are not fully understood (Radwan, 1995).

The mite severely damages gladiolus, both in the field and in storage (Wang, 1983). In Taiwan, it infests 10% of bulbs within 7 days of planting, and the optimum temperature for development is 24–28°C (Liu and Tzeng, 1994). Different kinds of food influence the reproduction of the robine bulb mite, and temperature affects its development. At 15°C, 20°C, 25°C and 30°C and room temperatures (14–34°C) the life cycle of *R. robini* takes 10.6 ± 2.9, 10.1 ± 2, 9.15 ± 3.1, 8.0 ± 1.48 and 14.6 ± 7.7 days, respectively. The incubation period of eggs, and the developmental period of nymphs are influenced by the different temperatures. The mites also lay gradually a higher number of eggs with the rise of temperature but the length of life, and the number of eggs laid by female are maximum at room temperatures (Raut and Sarkar, 1991). At temperatures from 13°C to 26°C, the life cycle takes from 7–27 days, and at 25°C it takes 9 days (Garman, 1937; Zakhvatkin, 1941; Woodring, 1969b); at 27 ± 1°C, egg to adult development takes 12.1 days (Ostovan and Kamali, 1996). At 15°C, 20°C, 25°C and 30°C, and at room temperature (14–34°C), the life cycle of *R. robini* takes 10.6 ± 2.9, 10.1 ± 2, 9.15 ± 3.1, 8.0 ± 1.48 and 14.6 ± 7.7 days, respectively. The incubation period of the eggs and the developmental period of the nymphs are influenced by temperature. Though the males have a shorter lifespan than the females, the duration of life, irrespective of sex, gradually decreases with increase in temperature from 15°C to 30°C (Raut and Sarkar, 1991). At 27 ± 1°C on potato tubers, the pre-oviposition, oviposition and post-oviposition periods are long 1.0, 22.5 and 8.2 days, respectively, while the reproductive rate (eggs per day), and the fecundity are 58 and 730 eggs (Ostovan and Kamali, 1996). Gerson *et al.* (1983) reared *R. robini* in the laboratory on three diets, under four temperature regimens in individual plexiglass cages. Immature survival was uniformly high on garlic and groundnuts, at 22°C and 27°C, but rather low for mites on garlic at 35°C, or on filter paper at 27°C. The calculated threshold of development was at 11.8°C, and the thermal constant was 184.8 degree-days. The sex ratio was 1:1. Adults kept at 27°C, and offered groundnuts or garlic laid an average of 690 eggs/females and 400 eggs/females, respectively, during about 6 weeks. No eggs were usually laid by females, reared at 35°C or on filter paper. Males lived almost twice as long as females. The intrinsic rates of increase (rm), for mites offered groundnuts or garlic, at 27°C were 0.285 and 0.218, respectively. The compounded effect of various nutrition-affected factors, such as survival, body size and fecundity on changes in mite populations is discussed, as well as the importance of defining diets for polyphagous mites in similar studies (Gerson *et al.*, 1983). Yassin *et al.* (2018a) studied the different biological aspects of *T. putrescentiae* (Schrank) and *R. robini* reared at 25+2°C and 75+5 % RH on different food kinds, such as pure white wheat flour, normal wheat flour, wheat germ, rice flour, maize flour and semolina. Obvious differences between incubation period, life cycle, longevity and lifespan of both mite species were observed. The longest lifespan periods of *T. putrescentiae* male and female occurred, when the individuals fed on wheat flour (41.61 and 50.9 days, respectively). However, the longest lifespan periods of *R. robini* were observed on maize flour (49.2 and 41.2 days for male and female, respectively). The food type's suitability clearly affected the number of laid eggs by the adult female of the mite species and revealed a high significance between the feeding on wheat flour for the first mite, and maize flour for the second one (the most suitable feeding sources for mites) than any other feeding types. The highest number of *T. putrescentiae* eggs was 40.2 on wheat flour, and the lowest number was recorded on wheat germ (27.8 eggs). However, the highest deposited number of *R. robini* eggs was noticed, when adult female fed on maize flour (140.2 eggs), but the lowest number of laid eggs was found when, the individuals reared on wheat germ (70.4 eggs) (Yassin *et al.*, 2018a).

The threshold for development is 11.8°C, and the thermal constant is 184.8 degree-days. On garlic at 27°C, the fecundity is about 400 eggs, male longevity is longer than female longevity, and the intrinsic rate of increase is 0.128 (Liu and Tzeng, 1994). Li *et al.* (2008) reported a threshold temperature and an effective accumulated temperature of 10.01°C and 171.72 degree-days, respectively. To reach the larval stage, the eggs require 43.5 degree-days above 11.7°C and from 9.6–23.9 days to reach the adult stage, depending on the temperature. The fecundity ranges from 99.0 to 203.5 eggs, and the rate of eggs per day, from 4.8–8.9. The longevity of the female and male ranges from 20.8–34.4, and 12.9–42.1 days, respectively (Liu and Tzeng, 1994). In the Kunming region of China, the mite can develop 12 generations a year (Li *et al.*, 2008). The moulting rate of hypopi of *R. robini* is highest, and requires the least number of days, at 24°C and 100% RH, and no moulting occurs below 93% RH. Placing hypopi in distilled water inhibits the process somewhat, and moulting is generally better at lower temperatures than at the higher temperature of 35°C (Capua and Gerson, 1983). *R. robini* can complete its life cycle utilizing filter paper as a sole dietary source. However, the complete cellulase complex that is necessary for crystalline cellulose degradation rarely occurs in arthropods, and in this case, the mite probably derives nutrients from the fungal mycelia, growing on the filter paper rather than from the filter paper itself (Wooddy and Fashing, 1993). Kuwahara *et al.* (1988) showed that neryl formate is the principal component of the alarm pheromone of this mite.

Different predatory mites living in the soil, prey on the robine bulb mite. In Egypt, the ascid *Protogamasellus minutus* Hafez, El-Badry *et* Nasr (Afifi *et al.*, 1986) and the digamasellid mites *Dendrolaelaps sayedi* Hassan, Afifi *et* Nawar (Hassan *et al.*, 1987) were investigated. In the Netherlands, the action of different soil mites collected in association with *R. robini* has been evaluated, such as the laelapid *H. aculeifer* (Canestrini), the ascid *Lasioseius bispinosus* Evans and the parasitic mite *Parasitus fimetorum* (Berlese) (Leśna *et al.*, 1995). The first species gave promising results (Leśna *et al.*, 1996, 2000; Conijn *et al.*, 1997).

Symptomatology and damage

R. robini damages stored bulbs as well those planted in the field or in greenhouses. The mite penetrates inside bulbs and corms through the basal plate and the external layers. Although it feeds on any decaying organic material, it prefers healthy bulbs to rotten ones (Meyer Smith, 1981). Izadi *et al.* (2023) reported that *R. robini*, called also as saffron mite, is one of the most important pests of saffron-producing regions in Iran. It causes yellowing, and decreases saffron growth, and finally it destroys the bulbs. It has been observed that bruised bulbs, and fungus-infected bulbs help its establishment more than healthy bulbs. For example, on *Fusarium*-infected bulbs, the mite populations show exponential growth after 14 days, and on healthy bulbs exponential growth of the population is reached after 60–90 days. The mite is attracted to, and penetrates *Fusarium*-infested bulbs more rapidly than healthy ones (Okabe and Amano, 1990, 1991). The attack produces dark brown streaks on the roots, externally around the neck, and distorted growing tips, leaves and flowers. On damaged bulbs, the population density of *R. robini* increases rapidly, transforming the whole bulb to a rotten pulp. In this phase, it is possible to see thousands of mites and eggs in the damaged tissue. On lilies, the stems can be attacked and broken. Severe attacks induce small and stunted plants, with few or no roots. The use of a combination of fungicides and acaricides, to study the dynamics of the interactions between *R. robini* and root rot (*Pythium ultimum* Trow) on lily bulbs highlighted that a low mite population allowed suppression of the pathogen (Ascerno *et al.*, 1983). On onions, the mite population gathers in the roots, producing plant collapse, and arrest of the growth of the bulbs. The robine bulb mite is closely associated with the spread of the fungus diseases due to *Fusarium*, *Stromatinia* and *Pseudomonas*; the mite feeds on the fungus responsible for *Stromatinia* rot of gladioli, and can be a vector of *Pseudomonas* (Forsberg, 1965; Meyer Smith, 1981). Baker (1983) reported that viable *Fusarium* and *Cylindrocarpon* propagules could be found in the faecal pellets of *R. robini,* feeding on a substrate of these pathogens, and noted the possible role of the mite as a vector of these diseases. Similarly, the acquisition and retention of *Pseudomonas marginata* (McCulloch) Stapp by *R. robini* has been widely established (Poe *et al.*, 1979). In the laboratory, dissemination of the pathogen occurs for a maximum of 3 days, which would reduce the mite's role as a vector to restricted conditions. Disinfestation of surface mites does not influence transmission of pathogen (Noble *et al.*, 1978), so, the propagules are likely to be carried within the digestive system. However, other mechanisms, such as the soil fauna, may be involved (Beute and Benson, 1979) as well, and their role in disease transmission by mites must be evaluated. It is, then, possible that mites play only a minor role in the spread of diseases. Nevertheless, on bulbs, the mite transfers (together with *T. putrescentiae*) the fungal pathogens *Aspergillus niger* van Tieghem, *Haematonectria haematococca* (Berk. *et* Broome) Samuels and Rossman, *Rhizopus stolonifer* (Ehrenb.) Vuill. and *Penicillium chrysogenum* Thom attached to the outside of their bodies, whereas *A. flavus* J.H. Friedrich Link and *A. ochraceus* Wilh. are transferred through the mite's digestive tract, along with their food (Abdel-Sater and Eraky, 2001). The mite has been collected on 85.9% of commercial bulbs, in central and southern Chile (Rojas and Klimov, 2007).

Control

Control involves the use of chemical, physical, cultural and biological methods, applied alone or together. Cleaning of the environment, the handling and storage of bulbs and vegetable material, and the use and marketing of healthy material, free from fungal infections and the presence of mites are all very important. The integration of rigid prophylaxis with biological, physical and chemical means defines an IPM strategy in which the monitoring of mite populations could play a key role.

MONITORING. Traditionally, the assessment provides manual inspection of infested plants or stored products (Latta, 1939; Rawlins, 1955). However, this method may be inadequate because, visible signs of damage may not be apparent until mite outbreaks are well advanced. In Israel, Gerson *et al.* (1985) employed garlic-baited traps to sample and monitor *R. robini* populations, in garlic and onion fields. Traps were an effective sampling tool, and provided information on mite phenology and abundance from fields, under varying conditions. In New York, garlic-baited traps have been useful to detect the presence of *R. robini and* give estimates of their abundance in muck soils (Diaz, 1998). Despite these reports, the results of research on the monitoring of mite populations, do not seem to be conclusive. However, the use of some types of traps might improve this situation. An acarid bait trap constructed from moistened, folded filter paper, containing a small quantity of dry yeast, allows the number of *R. robini* in agricultural soils to be monitored. Under laboratory evaluations, the trap captured 65% of the acarid adults by 3 days after application. Field trials carried out in Welsh onion field soil using both the bait trap and Tullgren funnel traps showed that numbers of *R. robini* obtained from the soil did not differ significantly between the two trap types, suggesting that efficient monitoring of *Rhizoglyphus* is possible using bait traps (Kasuga *et al.*, 2005).

PHYSICAL CONTROL. The high-temperature tolerance of *R. robini* (and other acarids) has recently been reviewed, to explore the possibility of controlling the mite in agricultural soils, by heat treatment at 40°C for 24 h, 45°C for 3 h and 50°C for 1 h. The tolerance of *R. robini* is similar to that of *T. similis* Volgin, and both have lower tolerances than other acarids. *T. similis* and *R. robini* may not have developed high-temperature tolerance because the two species dwell in soil, where temperature fluctuations are not severe (Kasuga and Honda, 2006a). This confirms the effectiveness of control strategies, already used successfully by farmers. In lilies, mite populations are controlled using acaricides, such as pirimiphos-methyl in combination with cultural measures such as hot-water treatment (2 h at 39°C or 41°C), and the storage of bulbs at –2°C. In peat growing media, the combination of hot-water treatment, and biological control by the predatory mite *H. aculeifer* can replace acaricide application and control *R. robini* on lily bulbs, during the propagation phase (Conijn *et al.*, 1997). Flooding the soil with water for 5 days kills 96.1% of the populations in gladiolus, and all mites are controlled after 14 days (Wang and Lin, 1986).

Izadi *et al.* (2023) studied in Iran, the cold tolerance and supercooling point (SCP) of *R. robini*. They measured these parameters in three populations, and two temperature regimes and showed that the mean SCP of *R. robini* was approximately −14.6°C, without significant difference among the populations. On the contrary, acclimation of the mites significantly decreased their SCP to a mean of approximately −16.5°C. Exposure of the mites for 24 h to 0°C and –2.5°C had no significant effect on the survival of the mites, but when the mites were exposed to –5.0°C for 24 h, survival of the three populations reached the lowest level of roughly 60%. By 24 h exposure to –7.5°C, survival of the mites was almost negligible. As a large proportion of mortality was observed above the SCP, and LT50 > SCP, it can be inferred that the *R. robini* is likely a chill-susceptible species. This suggests that this mite lacks the ability to withstand extracellular ice formation. Overall, the results of this study suggest no significant physiological differences between the mite populations (Izadi *et al.*, 2023).

CULTURAL PRACTICES. In garlic cultures in Brazil, a rather large amount of organic matter (compost) added to the soil before 'seeding' the garlic, induced the development of *Rhizoglyphus*. In particular, sugarcane bagasse was readily available and cheap, but as it takes quite some time to compost totally, it has been used 'semi-raw' on garlic fields, which results in the field giving off a slightly sweet odour. This attracts insects, mainly Coleoptera, from a long distance, and many of these carry hypopi of *Rhizoglyphus*, which end up developing into adults, and damaging the garlic plants. A reduction in the amount of organic matter used, and the use of totally composted bagasse solved this problem. Repetitive ploughing of the soil in between subsequent cultures, and exposing the mites to sunlight also helped to eliminate them. So, a preventive cultural procedure to avoid *Rhizoglyphus* problems in the soil could be the use of only a moderate amount of totally composted organic matter, and no excess of moisture (Flechtmann, personal communication).

Palevsky *et al.* (2022) reported that no active ingredients are registered for the control of the garlic bulb mites in the United States during the growing season (Brust, 2022; Kaur, 2023; Smith, 2023), and unfortunately, nor are there known biological means from managed agro-system (California Garlic & Onion Research Advisory Board, 2019). The last phenomenon depends by conventional tillage prior to each crop, lack of organic matter, and reductive impacts of pesticides (Veerman *et al.*, 2020). Cover crops or soil solarization are commonly not applied in traditional management, being few times between crop rotations. Soil fumigants with metam sodium are still suggested as a pre-plant treatment, in fields previously infested by the bulb mite (Dara *et al.*, 2018), but their high cost (California Garlic & Onion Research Advisory Board, 2019) and the environmental impact (Woodrow *et al.*, 2014) limit their use. In the United States, about 90% of all commercially grown garlic are produced in California, and over 92% is grown traditionally on bare soil as a monoculture in large plots, applying a 4-year rotation, commonly associated with a conventional management, based on the use of chemicals for disease, insect, nematode and weed control; in this situation, the resistance is a serious problem, aggravated by a reduced availability of registered chemicals (California Garlic & Onion Research Advisory Board, 2019). Conversely, in the same country, the organic onion and garlic based on cover cropping, associated with crop rotation, is a concrete strategy able to increase soil fertility and pest management, with some employ of organic pesticides (Sullivan, 2003; Adam, 2006; Bachmann and Hinman, 2008). However, soil sterilization methods, solar and chemical, although reduce the pest outbreaks, can induce a negative impact on beneficial microbiota in the soil, such as those organisms able to suppress soil pathogens (Bonanomi *et al.*, 2018). Similarly, weed control by sterilization, exposes the soil to erosion and high temperatures, reduces water retention and intervenes in the continuous depletion of soil organic matter. Reduced tillage, coupled with organic cover represents a good practice able to contain the soil erosion (Eshel *et al.*, 2015; Montanarella *et al.*, 2016). The technique of mix cover crops, named 'service crops', is included into the crop rotation for weed control and species diversification. Service crops adjust the soil moisture and temperature, increase organic matter and enhance soil structure (Eshel *et al.*, 2015; Unc *et al.*, 2021; Schlüter *et al.*, 2022). Free-living nematodes move to bacterial and fungi hotspots (Ackermann *et al.*, 2016), driven by volatiles, released from the microbiota community living on organic matter, and exploit the microbiota as a food source, and may disperse bacteria and fungi (Anderson *et al.*, 2003). They carry a complex of associated microbes to the decomposing substrate, influencing the community dynamics of soil (Jiang *et al.*, 2017). Predatory mites and free-living nematodes cannot dig in the soil, but must be able to move freely in the soil to search food resource (Erktan *et al.*, 2020). Compact soils often hinder this movement, and are devoid of appropriate levels of aeration and moisture to support the activity of these beneficials. Despite their relatively large size, the bulb mites reach easily, their host plants, allowing them to squeeze through tight spaces, and their ability to sense plants infested with the pathogen *Fusarium oxysporum* von Schlechtendal (Ofek *et al.*, 2014).

BIOLOGICAL CONTROL. Biological control is provided mainly by releases of the laelapid mite *H. aculeifer*. The study of this predator began in the 1980s, in propagation lily bulbs and other crops. Experiments carried out in the greenhouse and the field showed that the initial predator-to-prey ratios required to achieve suppression (*c.*1:2) or elimination (*c.*3:1) in the soil were much higher, than those required for bulb mite elimination, when lily bulb scales were embedded in a medium of vermiculite (*c.*1:20) (Leśna *et al.*, 2000). *H. aculeifer* is used in various countries. The biology and potential use in biological control (and bioindication) of free-living mesostigmatic mites, in Australia, has been reviewed by Beaulieu and Weeks (2007).

Palevsky *et al.* (2022) published an interesting review on the control possibilities of these pest mites and number, several groups of natural enemies. Among the natural enemies, they highlight a rich faunistic complex and reported six species of the family Bacillaceae, and six species of the family Morganellaceae (bacteria), five species of entomopathogenic fungi belonging to the family Clacipitaceae, three species ascribed to the family Cordycipitaceae and one species to the family Ophiocordycipitaceae. The entomopathogenic nematodes include seven species of the family Heterorhabditidae. The predatory mites include three species of the family Blattisociidae, four species of the family Laelapidae, three species of Macrochelidae, one species of Parasitidae and two species of the family Rhodacaridae.

As concerns the action of entomopathogenic fungi, Konopická *et al.* (2021) evaluated the efficacy of 17 isolated and three reference strains against *R. robini* females. Results showed high variability in *R. robini* mortality, among the entomopathogenic fungi species and strains tested. The highest efficacy was induced by a strain of *Metarhizium anisopliae* (Metsch.) Sorokin isolated from soil samples collected in the Czech Republic (up to 99.3% mortality), and a strain of *Metarhizium indigoticum* (Kobayasi *et* Shimizu) Kepler, S.A. Rehner and Humber from Israel (98.3% mortality), 4 days from spray application. The concentration-response models showed that the latter strain was more virulent than *M. anisopliae* strains. The median lethal concentration (LC50) in *M. indigoticum* strain was evaluated as 1.01 × 104. *Cordyceps fumosorosea* (Wize) Kepler, B. Shrestha *et* Spatafora strains did not produce mortality higher than 40%. The lowest virulence was then observed in *Beauveria* spp. strains causing mortality of mites between 5% and 25%. Ment *et al.* (2020) showed high efficacy of *Metarhizium brunneum* Petch (isolate Mb7) against *R. robini*, which was susceptible to directly applied Mb7 conidia. Conidia of this fungus applied in vitro at concentration 1 × 107 caused, 43% and 100% mortality of mites, at 3 and 7 days post-inoculation, respectively, and the estimated LT50 value was 4.3 days. Drench application in potted onion experiments also significantly reduced the bulb mite populations. The fungus *Metarhizium* spp. was also effective against *R. robini* in the study by Ko *et al.* (2021). Conidial suspension with a concentration 1 × 107 conidia/mL of fungi *Metarhizium pinghaense* Q.T. Chen *et* H.L. Guo (isolate 3–1–2) and *M. anisopliae* (isolates 3–2–2 and 4–18–3) caused more than 80% mortality after 7 days. An isolate of *Metarhizium pemphigi* (Driver *et* R.J. Milner) Kepler, S.A. Rehner and Humber (isolate 1–1–1) and two isolates of *M. anisopliae* (isolates 4–3–2 and 4–31–2) showed mortality of 90% or more on the fifth day, and 100% mortality after 7 days.

The potential of entomopathogenic nematodes to attack bulb mites, specifically *R. robini*, was examined by Nermut *et al.* (2019). They infected the bulb mites, with infective juveniles of 20 strains of *Steinernema* and *Heterorhabditis* species, at dose of 300 infective juveniles per mite, and the invasion rate and mite mortality were evaluated. Some entomopathogenic nematodes, especially those with small body, may invade and kill adult females of *R. robini*. The most promising species were *Steinernema huense* Phan, Mráček, Půža, Nermut' *et* Jarošová, *Heterorhabditis bacteriophora* Poinar and *Heterorhabditis amazonensis* Andaló, Nguyen *et* Moino, which induced a mortality in *R. robini* up to 30%. The authors concluded that although some entomopathogenic nematodes may invade and kill bulb mites, they are not a viable option for bulb mite biocontrol (Nermut' *et al.*, 2019). However, these beneficials play other important functions in soil, as they can disseminate fungal spores (Nermut' *et al.*, 2020), and represent a prey for invertebrate predators including mites and springtails, or as a host for nematode-trapping fungi, such as *Orbilia oligospora* (Fresen.) Baral and E. Weber, *Monacrosporium eudermatum* (Drechsler) Subram. and *Geniculifera paucispora* (R.C. Cooke) Rifai (Kaya *et al.*, 1998).

The predatory mites proposed for biological control of bulb mites, belong mainly to the family Laelapidae (Moreira and Moraes, 2015). First studies concerned the examen of predator behaviour, and their ability to feed and reproduce on bulb mites. Zedan (1988) observed that protonymphs, deutonymphs, and adults of *Gaeolaelaps aculeifer* (Canestrini) feed and developed on all stages of *R. echinopus*. Reproductive potential of the predator was highest when it fed on adult prey, but less prey was consumed. *G. aculeifer* preferred to feed on immature, rather than adult *R. echinopus*. Leśna *et al.* (1995) found that conversely to *Stratiolaelaps miles* (Berlese), *G. aculeifer* may feed and reproduce on *R. robini*. They also found that local populations of *G. aculeifer* often differ in their feeding preference, and reproductive potential and highlighted the advantage to exploit the 'local' strains, as biological control agents. Moreover, Leśna *et al.* (1996) observed a significant reduction of *R. robini* population by *G. aculeifer* on lilies in laboratory and in a large-scale study carried out in storage rooms. The spatial scale and the structural complexity of the habitat influenced the control of the pest, and the bulb mite populations declined faster in less complex and smaller habitats. An experiment carried out in greenhouse, with intact bulbs grown in peat soil, Leśna *et al.* (2000) observed the elimination of bulb mites with a predator: prey ratio of 3:1. Other Mesostigmata feed on *R. echinopus* (Wu *et al.*, (2009c); Castilho *et al.*, 2009; Moreira and Moraes, 2015) and *R. robini* (Afifi *et al.*, 1986; Leśna *et al.*, 1995; Mowafi, 2005). Wu *et al.* (2009c) reported that a species of *Lasioseius* (Blattisociidae) developed and reproduced on *R. echinopus*. Castilho *et al.* (2009) showed that *Protogamasellopsis zaheri* Abo-Shnaf, Castilho *et* Moraes (Rhodacaridae) fed and reproduced on *R. echinopus*. Moreira and Moraes (2015) observed that *Cosmolaelaps barbatus* Moreira, Klompen *et* Moraes and *Cosmolaelaps jaboticabalensis* Moreira, Klompen *et* Moraes (Laelapidae) oviposited when fed *R. echinopus*. Azevedo *et al.* (2018) observed that *Macrocheles embersoni* Azevedo, Berto *et* Castilho, *Macrocheles muscaedomesticae* (Scopoli) and *Macrocheles robustulus* (Berlese) (Macrochelidae) feed on *R. echinopus*, but with lower predation respect to other prey. In the case of *R. echinopus*, Afifi *et al.* (1986) observed the development and reproduction of *P. minutus* Nasr (Ascidae) when fed with this prey. Also, Leśna *et al.* (1995) reported that *Lasioseius allii* Chant (Blattisociidae) and *P. fimetorum* (Berlese) (Parasitidae) were easily reared on *R. robini*, and the latter suppressed the prey when peat was employed as substrate. Mowafi (2005) observed the development and reproduction of *L. africanus* Nasr (Blattisociidae) feeding on *R. robini*.

All of these studies were carried out in laboratory, and only in the case of *P. fimetorum* were small-scale population experiments conducted in closed flasks. Palevski *et al.* (2022) concluded that further studies are needed to know the potential of these species, as biocontrol agents in potted plant and field trials.

CHEMICAL CONTROL. *R. robini* is commonly sensitive to various active ingredients including azocyclotin, cyhexatin, methidathion, phosmet, pirimiphos-methyl, profenfos, prothiofos and sulfur sprayed on to the soil or directly on to bulbs or used as a soaking solution for bulbs. However, strains that are resistant to one or more chemicals have been recorded in various countries. Using the filter paper method, Kuwahara (1986) evaluated the action of 41 organophosphorus insecticides, and found that strains collected in the field in Japan showed a high level of resistance to a wide range of active ingredients, involving a cross-resistance and major resistance factors. Shinkaji *et al.* (1986) evaluated the susceptibility of *Rhizoglyphus* spp. to dimethoate and disulfoton on rakkyo (*Allium chinense* G. Don) and Chinese chive (*Allium tuberosum* Rottl. ex Sprengel). Wang and Lin (1986) reported that metasystox and neoron failed to control *R. robini* on gladiolus. Chen and Lo (1990) observed high multiple resistance to organophosphorus insecticides, including diazinon, dimethoate, fenitrothion, malathion, methyl parathion, omethoate, parathion and prothiophos. Chen (1990a) improved a method for determining the susceptibility of the robine bulb mite to chemicals; the same author studied the resistance, biology and control of the mite (Chen, 1990b). Kato *et al.* (1993) tested, using a filter paper method, 32 pesticides on females of the mite collected in the field on Welsh onion (*Allium fistulosum* Linnaeus) and found that all collected populations showed a high susceptibility to prothiofos, methidathion, fenthion and pirimiphos-methyl. Among the older chemicals, methyl bromide (which is no longer allowed) is very effective at a rate of 32 g m^{-3} (Meyer Smith, 1981) or 50 g m^{-3}, as is dazomet at 35g m^{-2} (Wang and Lin, 1986). Dip treatment using benzoximate, bromopropylate or demeton-methyl, or fumigation for 48 h with phosphine from aluminium phosphite tablets (1 tablet/m at 25–38°C or 2 tablets/m at 15–20°C) were also effective (Wang, 1983). Similarly, dipping the bulbs of gladiolus before planting and applying a prothiofos solution twice on to the soils 7 and 28 days after planting is effective (Liu, 1995).

Rhizoglyphus echinopus **(Fumouze *et* Robin) (Fig. 28.16)**

Common name

Bulb mite

Diagnostic characteristics

FEMALE. The idiosoma is 791–860 µm long. The prodorsal shield is evenly punctulate. The seta *scx* is thick and 45–50 µm long. Grandjean's organ has a distinct forked tip. The dorsal idiosomal setae are relatively long. The setae *vi* are thick and pointed, and the setae *sci* 1–2 times the length of genu I. The ratio *sci-sci*: *sci-sce* ranges from 0.9 to 1.1. The setae c_1 are 0.8–1.2 times the length of the setae d_2; the setae c_3 are about as long as the setae c_1; the setae d_1 are 0.6–1.1 times the length of the setae d_2; the setae d_2 are far from the *gla*; the setae

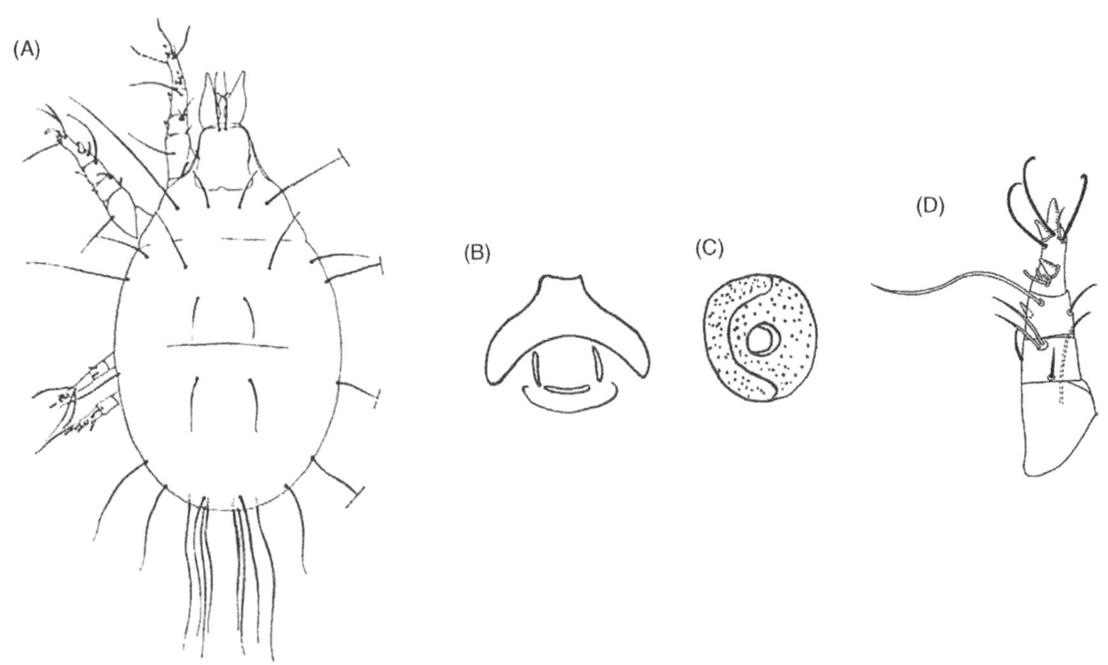

Fig. 28.16. *Rhizoglyphus echinopus* (Fumoze *et* Robin). Male. Dorsal view (from Hughes, 1976, modified) (A); base of aedeagus (from Hughes, 1976) (B). Female. Sclerotized ring surrounding the bursa copulatrix (from Hughes, 1976) (C); leg I (from Hughes, 1976, modified) (D).

e_1 are 1.2–1.5 times the length of the setae d_2, and the setae e_2 0.9–1.3 times the length of the setae f_2. The anal opening has six pairs of small setae, and three pairs are longer than the others; the setae ps_1 are less than one-quarter of the length of genu IV, the setae ps_2 are less than one-half of the length of genu IV, and the setae ps_3 about one-half of the length of genu IV; the setae ad_1 and ad_3 are small. The copulatory opening has a small circular shield; the spermathecal duct is broadly expanded from the copulatory opening to the inner part of the spermatheca. The inner part of the spermatheca is large and Y-shaped; the sclerites of the oviducts are widely expanded, and the distance between them is 97–122 µm, with the basal part subequal to the branches (Fan and Zhang, 2004).

HOMEOMORPHIC MALE. The body is 590–756 µm long. The seta *scx* is thick and 45–50 µm long. Grandjean's organ has a distinctly forked tip. The dorsal idiosomal setae are relatively long, and the setae *sci* are long, being 45–95 µm. The setae *vi* are 2.2–3.2 times as long as genu I. The setae c_1 are 0.8–1.0 times the length of the setae d_2; the setae c_3 are about as long as or slightly shorter than the setae c_1; the setae d_1 are 0.5–1.0 times the length of the setae d_2; the setae d_2 are far from the *gla*; the setae e_1 are 1.0–1.7 times the length of the setae d_2, and the setae e_2 are 0.7–1.3 times the length of the setae f_2. The ratio *sci-sci*: *sci-sce* ratio ranges 1–1.5. The aedeagus is terminally truncated, and its canal lacks lumps. The anal discs lack radiating lines, and their diameter is less than one-third the length of the setae ps_2. The setae ps_1 are 4.5–4.9 times longer than genu IV, and more than three times longer than the setae ps_2. The setae ps_2 are 0.8–1.4 times the length of genu IV, and the setae ps_3 are slightly blunt. The ratio s¢: s^2 is 1.1–1.2. The solenidion φ on tibia IV is pointed and about as long as the ventral seta *kT* (Fan and Zhang, 2004).

Geographical distribution

R. echinopus (Fumouze *et* Robin) is found worldwide (Fan and Zhang, 2003, 2004), and reported from many countries, such as Algeria (Fain, 1988a), Argentina (Diaz *et al.*, 2000), Australia (Manson, 1972a; Fan and Zhang, 2003), Brazil (Ferreira Silva *et al.*, 2016; Barbosa and Moraes, 2020), Canada (Diaz *et al.*, 2000), China (Bu and Li, 1998), Czech Republic (Hrudová and Šafránková, 2017), Egypt (Fan and Zhang, 2004; Mostafa, 2011), England (Hughes, 1976), Fiji (Fan and Zhang, 2003), France (Diaz *et al.*, 2000), Germany (Sturhan and Hampel, 1977), Hong Kong (Manson, 1972a), Hungary (Ripka *et al.*, 2005), India (Sandhu, 1976), Iran (Daneshvar, 1978), Ireland (Hughes, 1961), Italy (Vacante, 1997a), Japan, Korea (Diaz *et al.*, 2000), the Netherlands (Manson, 1972a; Siepel *et al.*, 2016), New Zealand (Manson, 1972a; Fan and Zhang, 2003), Mexico (Diaz *et al.*, 2000), Poland (Labanowski and Soika, 1992), Romania, Russia (Diaz *et al.*, 2000), Scotland, Spain (Diaz *et al.*, 2000), Taiwan (Tseng, 1979), Turkey (Cilbircioğlu and Çobanoğlu, 2019), Ukraine (Oksentiuk *et al.*, 2022), UK (Hughes, 1976), United States (Diaz *et al.*, 2000) and other countries.

Commodities, facilities and vegetal and animal hosts

This mite has been collected on many cultivated plants (Fan and Zhang, 2004; Vacante, 2016). Hughes (1961) reported the mite on stored food, and Hagstrum *et al.* (2013) reported a list of 13 various commodities such as, bamboo shoots, bulb, Chinese medicine, fatty debris, flax seed, garlic, grain, herb, groundnut, shallot, sunflower seed, wheat, wheat spillage (wet, decaying) collected in various facilities such as farm, lard factory, market, off-farm storage and railcar. Moreover, other stored products were reported such as maize (Fain, 1988a), seed in budgerigar cag (Fan and Zhang, 2003), stored wheat (Bu and Li, 1998), rice straw (Tseng, 1979), straw of *Oryza sativa* (Fan and Zhang, 2003), flour mill, damp flour mill, stored products (Hughes, 1976), damp flour mill and flour mill (Fan and Zhang, 2004). Oksentiuk *et al.* (2022) reported the mite from Ukraine in vegetable storages.

The animal hosts include the Ctenophthalmidae (Insecta) *Palaeopsylla minor* (Dale) ex *Talpa europaea* Linnaeus (Fain, 1988a). Ferreira Silva *et al.* (2016) found deutonymphs (hypopus) from *R. echinopus* attached to the head of a subterranean termite belonging to *Coptotermes* genus in Rio de Janeiro, Brazil.

Bio-ecology

This mite also lives on bulbs and corms and has been collected on several ornamentals, vegetables and organic substrates, including animals. Hughes (1961) reported the mite on stored food and Hagstrum *et al.* (2013) reported a list of 13 various commodities.

The life cycle of *R. echinopus* is similar to that of *R. robini* (Claparède). At 27°C on potato tubers, the development from egg to adult takes 13.5 days, the pre-oviposition period is 0.9 days, the oviposition period 31.5 days and the post-oviposition period 13.4 days. The reproductive rate (eggs day–1) is 43, and the fecundity 620 eggs (Ostovan and Kamali, 1996). In the laboratory, at 20°C, 25°C, 27°C and 30°C and 95% RH, the highest survival rate is at 30°C, and this decreases markedly at the lower rearing temperatures. The developmental period becomes shorter, with rise in temperature and each stage becomes longer, as development progresses. The threshold temperature is 9.44 and the thermal constant is 105.52 degree-days. As the temperature increases, adult longevity and the sex ratio increase, and the incubation period, decreases. Temperature does not affect fecundity (Sakurai *et al.*, 1992). At temperatures from 25–34°C, the larva, protonymph and tritonymph live for 4.0, 2.5 and 1.2 days, respectively; adult longevity is 45.8 days (Jiang and Ge, 1994). At 25°C and 95% RH on potato tubers, potato starch and fish powder, the mortality is 15.8%, 33.3% and 29.4%, respectively. On the same foods, fecundity is 202, 161 and 132 eggs and female longevity is 38.6, 20.8 and 32.8 days, respectively. Only on filter paper, does the mite fail to reach the tritonymphal stage (Chen *et al.*, 2006b). Both adults and nymphal stages can overwinter in soil and on damaged host plants. According to the region, in China the mite has from nine generations a year (on *Allium fistulosum*) (Jia *et al.*, 2002) to 10–12 generations a year (Hu *et al.*, 2005, in Fan *et al.*, 2010a). In Jiaxiang, Shandong Province, its populations increase by the end of March to the beginning of April and peak in May and August (Jia *et al.*, 2002), while in Wuhan, Hubei Province, two population peaks have been observed, from late September to late October and from late March to mid-May (Hu *et al.*, 2005, in Fan *et al.*, 2010a). Germinating buckwheat seeds were tested by Chmielewski (2001a) as food of the mite at about +20°C and 95–100% RH. Development cycle, mortality of preimaginal instars, longevity of adults, productivity of females and some other life parameters were studied. Findings obtained allow to calculate (at an average) ontogenesis (12.9 days), outliving of juvenile instars (85.0%), longevity of imagines (33.1 days), fecundity (505.6 eggs per female whole lifespan). Results showed that buckwheat sprouts seem to be an attractive and effective medium, for rearing of bulb mites in the laboratory (Chmielewski, 2001a).

Sturhan and Hampel (1977) reported that the bulb mite was found to feed on the pathogenic nematodes *Aphelenchoides ritzemabosi* (Schwartz), *Ditylenchus dipsaci* (Kuhn), *Ditylenchus destructor* Thorne, *Heterodera avenae* Wollenweber, *Longidorus elongatus* (de Man) and *Xiphinema index* Thorne *et* Allen. Small nematodes were devoured whole and larger ones bitten into pieces and eaten. White Heterodera females with their egg contents appeared preferable to brown cysts.

Symptomatology and damage

The symptomatology of attack and damage are similar to those of *R. robini*. For *R. echinopus*, attack of the roots of *Freesia* and *Gladiolus* also induces the formation of brown streaks, and interior mining is often observed. Corms show distorted growing tips and leaves. While this mite is not a key pest of bulbs, it often causes severe losses, as the slightest injury or bruising of the bulb, will enable the mites to breed. Once *R. echinopus* is established, it extends the damage and reduces the bulb to a rotten pulp. In storage, the mite readily wanders from bulb to bulb, and may act as a vector of different pathogenic fungi. Attack by this mite is similar to that of other common pests of bulbs in that it is associated with mechanical damage or unsatisfactory conditions of storage (Hussey *et al.*, 1969). The bulb mite is often associated with *Fusarium* spp. in gladiolus corms and *Freesia*, and with bacterial or fungal rots of dahlia tubers. When bulbs are intended for forcing, a representative sample should be examined. Attacked freesia and gladiolus corms are usually soft and mushy, and may even be unsuitable for planting. If infestation is widespread, the bulbs should be rejected, and when only a few bulbs are attacked, any that are soft should be rejected (Hussey *et al.*, 1969). The passage of viable propagules of *Verticillium albo-atrum* Reinke *et* Berthold, through the digestive system of *R. echinopus* when the mites fed on colonies, conidia and microsclerotia of the fungus, highlights the possible role of mite in fungal epidemiology (Price, 1976). Gray *et al.* (1975) found that vectoring of smoulder (*Sclerotinia narcissicola* P.H. Gregory) in *Narcissus* by *R. echinopus* had little importance because, the incidence of the disease in commercial stocks was already high. The effects of vectoring are also negligible in soils with a high inoculum of pathogens, as occurs in some muck soils (Abawi and

Lorbeer, 1971). In Chile, the mite has been found in 14.1% of commercial bulbs, and is predominant in central areas (Rojas and Klimov, 2007).

Control

The control of bulb mites follows the same protocol as for *R. robini*, based on the application of physical, biological and chemical methods, alone or together. Also, in this case, control requires a rigorous commercial and agronomic management of the bulbs. The health of the storage environment, and use of healthy material are of primary importance.

PHYSICAL CONTROL. Hot-water treatment gives good results. Immersion in water for 1 h at 38°C is sufficient for all types of bulbs, except probably large lily bulbs. The treatment is unsuitable for forcing tulips, and should be given to narcissi, at the correct developmental stage (Hussey *et al.*, 1969).

BIOLOGICAL CONTROL. See *R. robini.*

CHEMICAL CONTROL. The mite easily develops strains that are resistant to various chemicals, so the active ingredients used, may exhibit a variable action, resulting in the highest known toxicity of old acaricides as cyclodiene GABA (g-aminobutyric acid) antagonists (dieldrin, endrin and aldrin), some organophosphates (chlorpyrifos, diazinon and azinphos-ethyl) and carbamates (carbofuran), anticholinesterases and thiazolidine flubenzimine. Amitraz, fenazaflor, fenazaquin and oxythioquinox are less toxic than the cyclodienes, carbamates and organophosphates, but sodium channel agonists (DDT and pyrethroids) and several acaricides are inactive (Gencsoylu *et al.*, 1998). Fumigation with methyl bromide (3 1b/42.5 m^{-3}) gives good control of *R. echinopus,* with slight phytotoxicity. The application of granular acaricides (e.g. terbufos) in planting trenches (60 kg/ha) gave good control (Liu, 1995), and soil treatments with dazomet are effective (Lee and Leu, 1997). Nawar *et al.* (1990) reported that in Egypt, the bulb mite was the most favourite prey of the ascid mite, *Lasioseius bispinosus* Evans.

The Genus *Sancassania* Oudemans

Diagnostic characteristics

The most important aspects in the history of this genus up to the 1980s were discussed by Timms *et al.* (1981). Further, news may be found in Klimov and OConnor (2003) and Klimov *et al.* (2016), who included in the genus *Sancassania* Oudemans the genera *Caloglyphus* Berlese, *Isoglyphus* Zakhvatkin, *Mycetosancassania* Klimov and *Rhizoglyphoides* Volgin. Moreover, Klimov *et al.* (2016) reported that *Sancassania* is similar to *Cosmoglyphus* Oudemans in their anterior displacement of the anal opening, and morphology of tarsus I. The phoretic deutonymphs of *Sancassania* have the genua III–IV not distinctly longer than the tibiae III–IV (distinctly longer in *Cosmoglyphus*). The adults of *Sancassania* have smooth, filiform or spiniform supracoxal setae (lanceolate, with pectinated margins in *Cosmoglyphus*). The phoretic deutonymphs of *Sancassania* can be distinguished from those of *Ctenocolletacarus* Fain by the non-enlarged rostrum (enlarged in *Ctenocolletacarus*), and the non-striated idiosoma (transversely striated on *Ctenocolletacarus*). The adanal suckers are present in the majority of *Sancassania* species, while *Ctenocolletacarus* males lack of adanal suckers (Klimov *et al.*, 2016).

Knowledge on the systematics of this genus seems not to be conclusive, and in order to avoid creating confusion, the genus is defined here, using mainly the suggestions of Hughes (1976) and Klimov *et al.* (2016). According to this last author the vertical setae *ve* are short and fine, set near the middle of the lateral edge of the prodorsal shield, or are absent. The scapular setae *sce* are longer than the setae *sci*, which are well developed. The supracoxal seta is filiform or short, spiniform; always smooth (never barbed or pectinate). The hysterosoma has a complete set of setae, and the longer setae may be expanded at the base. The anal opening of female is displaced anteriorly, situated in the middle of hysterosoma. The posterior side of the male body does not form a projecting opisthosomal plate. On legs I and II, the seta *ba* is not thickened to form a short conical spine, and is well separated from the solenidion ω_1. The tarsus I bears the seta *aa*. The setae *e* at the distal end of tarsi I–III are shaped like a spine; the setae *ra* and *f* are usually curved, and may have the distal ends expanded into leaf-like plates. Five ventral terminal spines are present, of which *p*, *q*, *u* and *v* are about equal in size, and *s* is a little larger. Heteromorphic males and hypopi are common (Hughes, 1976; Klimov *et al.*, 2016).

Economic importance

Klimov *et al.* (2016) reported that the genus includes more than 80 species and needs a revision. Common, widespread stored-product pests are treated in two older keys to *Sancassania* species from Middle Europe (Türk and Türk, 1957) and the Palaearctic (Zachvatkin, 1941). Both keys use the name *Caloglyphus* (currently a junior synonym of *Sancassania*). Many taxonomic concepts for *Sancassania* species names, including those included in these two keys require revision and careful examination of type material (Klimov *et al.*, 2016).

Some species of *Sancassania* (mainly those previously belonging to the genus *Rhizoglyphoides*) live on fruiting bodies of mushrooms, and only reproduce several weeks each year, when the mushrooms develop; during the remaining time they are inactive in the soil, waiting for the next mushroom season (Klimov and OConnor, 2003). The most common species is *S. berlesei* (Michael), which has been reported on edible fungi in different countries (Hussey *et al.*, 1969; Clancy, 1981). The mite feeds on hyphae and sporocarps of all sizes of the button mushroom, *Agaricus bisporus* Lange, but is not common in mushroom houses (Greenslade and Clift, 2004). In Italy, *S. berlesei* has been found on melon in greenhouses (Laffi and Ponti, 1997), but the damage produced is unknown. The mite has also been found associated with stored products (Hughes, 1976; Hagstrum *et al.*, 2013). Other species play a secondary importance on stored food; for this reason, we treat here only *S. berlesei* and report below, only the main traits of other *Sancassania* species.

Sancassania anomalus (Nesbitt) was collected on navy bean in the United States by Nesbitt (1944), and its bionomy has been successively studied by Woodring (1969b) and Barker (1974). *Sancassania betae* (Attiah) was described as *Caloglyphus betae* by Attiah (1969) in Egypt. *Sancassania hughesi* (Samsinak) was first named *Caloglyphus hughesi* by Samsinak (1966) and reported from Brazil, China, Germany, Indonesia, Japan, Myanmar and Thailand on Chinese medicine, rice, and tamarind in farm and market (Hagstrum *et al.*, 2013). *Sancassania krameri* (Berlese) is reported from Angola, Australia, Brazil, Egypt, England, India, Iran, Ireland, Italy, South Korea, Taiwan, Ukraine and Wales on borecole, fungus (dried), grain, maize, onion, rice (husked), shallot, wheat and wheat flour in farm (Hagstrum *et al.*, 2013). *Sancassania mycophagus* (Megnin) is known from China, Egypt, England, France, Iran, Ireland, Taiwan, Turkey and Ukraine on apricot (dried), Chinese medicine, condiment, garlic bulb (decaying), grain, mushroom and sugarcane in commercial grain storage (Ostaovan, 1993; Hagstrum *et al.*, 2013; Oksentiuk *et al.*, 2022). *Sancassania oudemansi* (Zachvatkin) has a wide geographic distribution including Australia, Brazil, China, Czech Republic, England, Ethiopia, Greece, Indonesia, India, Iran, Italy, Java, Philippines, Russia, Sierra Leone, Sri Lanka, Thailand and Ukraine. This mite has been collected on bean curd, Brazil nut (damp), cereal, Chinese medicine, condiment, copra, dried fruit, dried vegetable, grain, hay, mushroom, maize, groundnut (damp), sweeping, tamarind and wheat bran in farm, flour mill, market, off-farm storage and vegetable storages (Hagstrum *et al.*, 2013; Oksentiuk *et al.*, 2022). *Sancassania redikorzevi* (Zakhvatkin) is known from England, Wales and Iran on grain, and seed (damp, germinating) (Ostovan, 1993; Hagstrum *et al.*, 2013). *Sancassania rhizoglyphoides* (Zakhvatkin) has a wide geographic distribution including China, Egypt, Iran, Ireland, Japan, Russia and Taiwan, and it has been collected on broad bean, Chinese medicine, flax, grain, grape wine, maize, wheat, and wheat spillage (decaying) in commercial grain storage, farm, granary and railcar (Mirfakhraii, 1994; Hagstrum *et al.*, 2013). *S. rodionovi* (Zachvatkin) has a geographic distribution including China, Germany, Iran, Primorskiy Kray, Russia, South Korea, Ukraine and Taiwan. In Iran, *S. rodionovi* (Zakhvatkin) is a serious pest of the mushrooms *A. bisporus* and *Pleurotus ostreatus* Kummer (Kheradmand *et al.*, 2007b); moreover, the mite has been collected on oatmeal, oilseed, onion (decomposing) and wheat flour (Hagstrum *et al.*, 2013). In Ukraine, the mite has been collected in vegetable storages (Oksentiuk *et al.*, 2022).

Moreover, *Sancassania* sp. was collected on individuals of *Microcerotermes diversus* (Silv.) (Termitidae). The presence of large numbers of hypopi on the body of termite workers, especially at the head, chest and limbs, caused weakness and death to workers, due to paralysis and inability to feed. It was also seen that the adults and other stage of mites were feeding on workers cadavers under laboratories condition, leaving corpuses with head capsule and thin layer of cuticle. The effects of different densities of mite (4, 8, 10) mite/worker that may cause death of workers on culture media revealed that (8, 10) caused a similar effect on killing workers, with an advantage on density 4 (mites/worker). Mortality percentage rate was 33.50%, 93.75%, 95.31% after 15 days of infection to densities 4, 8, 10 mite\worker, respectively, incubation condition 30 ± 5°C. Whenever, the number of mites increases, the death of workers is faster, from a biological control point of view, the mites were not good candidates for controlling termites (Al-Jassany and Al-Zubaidy, 2012).

Sancassania berlesei (Michael) (Fig. 28.17)

Common name

Wheat grain mite

Diagnostic characteristics

FEMALE. The idiosoma is rounded and from 800–1000 μm long. The idiosomal dorsal setae are shorter than in the homomorphic male; the setae h_1 are not as long as the setae e_1, slightly pectinate, and with a truncate distal tip. There are six pairs of anal micro-setae, two pairs of which are grouped on either side of the anterior margin of the anus, with four pairs around the posterior margin. The genital sense organs are large and conspicuous. Apart from secondary sexual characters, the leg chaetotaxy is the same as the male. The bursa copulatrix is terminal and surrounded by a small, sclerotized plate; it is connected with the sac-like *receptaculum seminis* by a narrow tube (Hughes, 1976).

HOMOMORPHIC MALE. The idiosoma is 600–900 μm long. The cuticle is smooth, colourless and shining, with pale brown appendages. In damp surroundings, the body is almost spindle-shaped, and widest between the third and fourth pairs of legs. The gnathosoma is narrow and tapering, and the chelicerae with toothed shears and a conspicuous mandibular spine. The Grandjean's organ is a truncate spine with small pectinations. The prodorsal shield is oblong, with a slightly concave or irregular posterior margin and straight sides. All the dorsal idiosomal setae, excepting the vertical setae *vi*, are almost completely smooth, of various lengths and thicker at the base. The vertical setae *ve* are fine and set midway along the lateral sides of the prodorsal shield. The scapular setae *sce* are three to four times longer than setae *sci*. The setae *scx* are slightly pectinate and more than half the length of the setae c_1. The setae c_1 are short, d_1 are two or three times longer than c_1, and d_2 and c_2 are 1.5 to twice as long as c_1; the setae e_1, h_1 and e_2 are long, and the setae h_1 extend well beyond the end of the body. The setae ps_2 are two to five times longer than ps_3, and the setae ps_1 are longer than ps_2 and, like them, stretch beyond the posterior margin. The anal suckers are round and conspicuous. The solenidion ω_1 of tarsus I has a slightly expanded distal end, and arises from the same chitinous depression of the famulus *e*. The solenidion ω_3 is of even diameter, while the seta *d* extends well beyond the end of the claw; the seta *e* is a conspicuous stout spine and *f* and *ra* are falcate, with their distal ends expanded into transparent leaf-shaped blades. The copulatory suckers of tarsus IV arise from the distal half of the segment; the seta *f* is long and slender, slightly expanded at the tip, and *r* and *w* are stout spines (Hughes, 1976).

HETEROMORPHIC MALE. The idiosoma is 700–1000 μm long. The idiosomal setae are longer than those of the homomorphic male and tend to be more obviously thicker towards their bases. The third pair of legs is conspicuously thickened, and each ends in a stout apophysis, and serves as weapons when fighting with other males (Hughes, 1976; Klimov *et al.*, 2016).

HYPOPUS. The idiosoma is 250–350 μm, long and dark brown in colour, with an arched surface, except that the anterior part of the propodosoma is covered with smooth cuticle. The propodosoma is almost triangular in shape and ends anteriorly, in a rostral projection that is broadly angled, with the setae *vi* at its apex, which covers the basal region of the gnathosoma. The vertical setae *ve* are present, short; the vertical setae *vi* arise from the tip of this apex, and two pairs of scapular setae *sc* form a curved transverse row across the prodorsum. The hysterosoma is 4–5 times longer than the propodosoma and bears fine inconspicuous setae. Ventrally, the sternal shield is well defined laterally. The posterior edge of the sternal shield does not touch the anterior edge of the ventrogenital shield. The sternal apodeme does not reach the posterior ends of apodemes 2. The coxal fields II are open, with a short unsclerotized distance, between the anterior and posterior apodemes; the apodemes do not reach the posterior edge of the sternal shield. The coxal setae *3b* and *4a* are conoidal. The claw of tarsi I and II is encircled by five curved leaf-shaped setae, of which two are smaller than the remaining three. On tarsus I the tarsal setae *aa* are

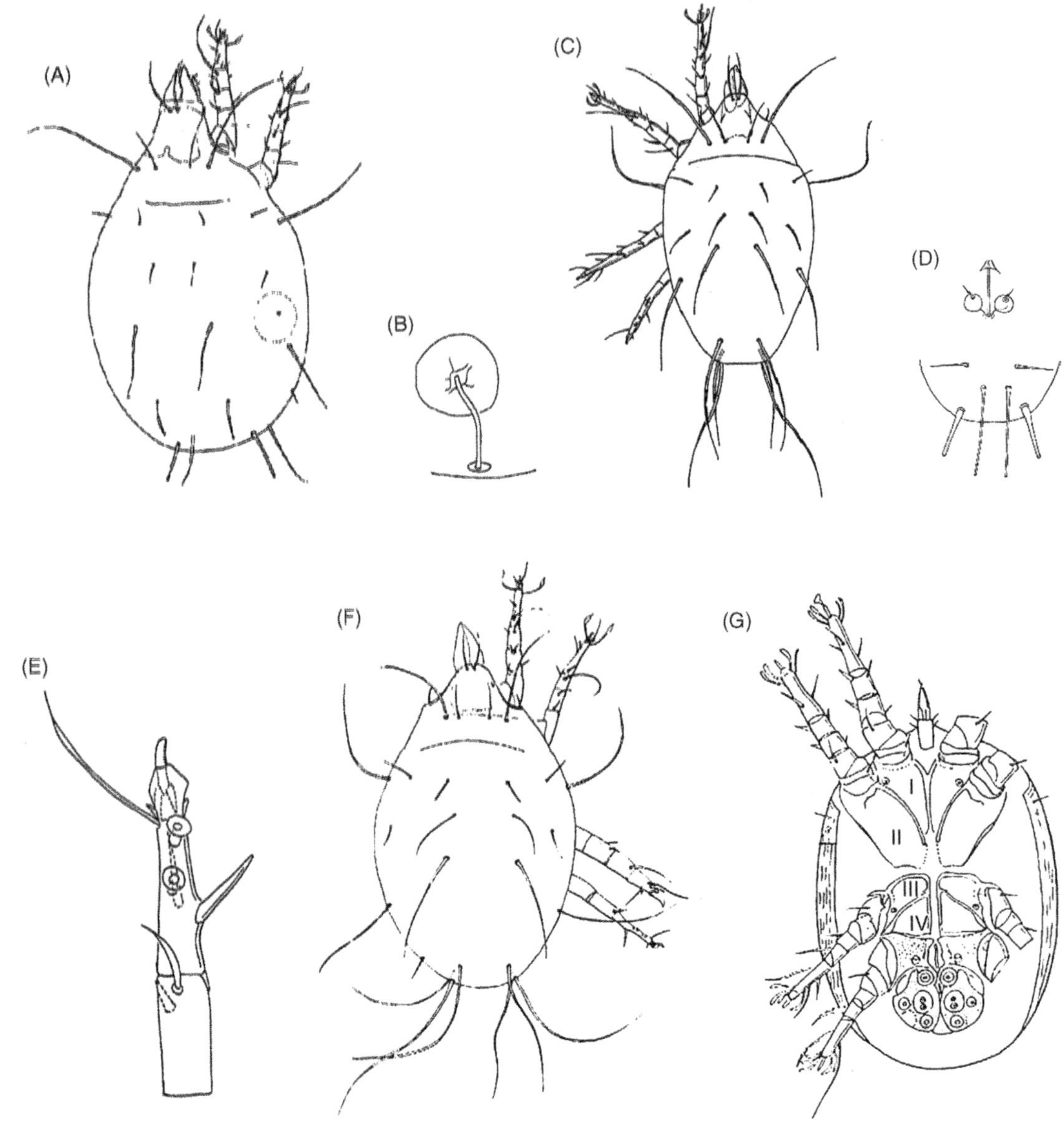

Fig. 28.17. *Sancassania berlesei* (Michael). Female. Dorsal view (A); reproductive system (B). Male. Dorsal view (C); anal region (D); distal segments of leg IV (E). Heteromorphic male. Dorsal view (F). Hypopus. Ventral view (G) (from Hughes, 1976, licensed under the Open Government Licence v3.0).

present, and the tarsal setae *ba* are absent, while on tarsus II the setae *ba* are present. On tarsi I–II the tarsal seta *e* has a saucer-like tip. The genua III–IV are not distinctly longer than the tibiae III–IV. No dorsal solenidion σ on genu III. The conoidal setae ps_2 are lateral to median suckers ad_{1+2} (not anterior). No striate membrane posterior to attachment organ (Hughes, 1976; Klimov *et al.*, 2016).

Li *et al.* (2015b) described morphological variations of *S. berlesei* identified under SEM. The hypopus stage shows the claws and pretarsi well-built, and leaf-like setae, setae of tibia and setae of genu are present. The sucker plate totally contains nine suckers and four shell-like dumplings in which there are symmetric distributions with one pair of central suckers, two pairs of side suckers and one pair of anterior suckers, respectively. One pear-like posterior sucker is located at the back of sucker plate. All suckers are smooth except for anterior sucker with radial stripe.

Geographical distribution

S. berlesei is found worldwide (Hughes, 1976), and reported from Australia (Hughes, 1976; Halliday, 1998), Canada (Beaulieu and Knee, 2014), China (Womersley, 1941; Li *et al.*, 2015b), Czech Republic (Hubert *et al.*, 2004a), England (Brady, 1970; Hughes, 1976), Germany (Hughes, 1976), Korea (Cho *et al.*, 1999), India (Putatunda, 2002), Iran (Daneshvar, 1978; Fathipour, 1994), Iraq (Al-Kubaisi and Al-Dulaimi, 2020), Italy (Zachvatkin, 1941; Laffi and Ponti, 1997), the Netherlands (Hughes, 1976), New Zealand (Timms *et al.*, 1981), Pacific Islands (Womersley, 1941), Philippines (Corpuz-Raros *et al.*, 1988), Russia (Zachvatkin, 1941; Hughes, 1976), Taiwan (Tseng, 1972), United States (Johnston, 1965; Hughes, 1976), Ukraine (Oksentiuk *et al.*, 2022), Wales (Brady, 1970) and other countries.

Commodities, facilities and vegetal and animal hosts

The mite has been collected on various stored products, such as banana, barley, biscuit, cereal spillage (decaying), Chinese medicine, condiment, confection, copra, fish (minced-dried), grain, grain residue (damp, mouldy), linseed, maize, mushroom, orange, groundnut, plum (dried), potato tuber (rotting), rice, rice (husked), wheat and wheat (rotting), all found in farm (Hagstrum *et al.*, 2013).

Host plants and substrates include rotting potato tubers and mushrooms (Zachvatkin, 1941), yams and copra (Womersley, 1941), *Neurospora crassa* Shear *et* Dodge (Ascomycetes) (Pimental *et al.*, 1960), wheatgerm, prematurely baled lucerne hay (Hughes, 1961), damp and mouldy wheat, copra, linseed and groundnuts, insect cultures, deep litter of broiler houses associated with the laelapid mite *A. casalis* (Berlese) (Hughes, 1976), edible fungi (Zachvatkin, 1941; Hussey *et al.*, 1969; Clancy, 1981), hyphae and sporocarps of the button mushroom, *A. bisporus* (J.E. Lange) Imbach (Greenslade and Clift, 2004), melon (Laffi and Ponti, 1997). In Canada, Beaulieu and Knee (2014) reported the mite on stored grains and mushroom bed. The mite has been collected in poultry droppings in Iraq (Al-Kubaisi and Al-Dulaimi, 2020). Oksentiuk *et al.* (2022) reported the mite from Ukraine, in vegetable storages.

Bio-ecology

S. berlesei is a free-living mite inhabiting soil, as well as laboratory colonies of insects and fungi, poultry litter and stored product, poorly maintained. Its life cycle develops through the egg, larva, protonymph, tritonymph and adult stages. Under particular conditions between the protonymphal and tritonymphal stages, this species may form an hypopus stage. The mite cannot reproduce parthenogenetically and mates after the final moult (Timms *et al.*, 1981). With wheat germ as food, under saturated conditions and at 22°C, the life cycle of this mite takes from 8 to 9 days. With high humidity, the female lays 1034 eggs in 39 days, as many as 213 eggs in 24 h. A small number of larvae pass through the hypopal stage, even though they are supplied with abundant food and water (Hughes, 1961). Rodriguez and Stepien (1973) reported a fecundity of 145 eggs, and Timms *et al.* (1981) of 260 eggs, respectively. With *N.crassa* Shear *et* Dodge as food, at 27°C and 73% RH, each mite consumes between 0.01 mg and 0.05 mg a day, the eggs hatch in 24–48 h, and the life cycle completes in 9 days (Pimental *et al.*, 1960).

This mite has been cultured monoxenically on 15 fungal isolates of the genera *Aspergillus*, *Cladosporium*, *Fusarium* and *Penicillium* growing on malt-extract agar. Fungus feeding could serve as more than a simple means of survival for the mite (Daneshvar *et al.*, 1977). At 25°C and 90% RH, and with a yeast diet as food, the development from egg to adult takes 158.8 h (Timms *et al.*, 1981), while with a xenic diet, it takes 193.3 h (Rodriguez and Stepien, 1973). The similar species *S. rodionovi* (Zachvatkin), reared at 25°C on button mushrooms, *A. bisporus*, and on oyster mushrooms, *Pleurotus ostreatus* (Jacq.) P. Kumm., completes its immature development in 17.35 ± 0.58 and 20.17 ± 0.88 days, respectively. On button mushrooms, the rate of development increases gradually from 10–35°C.

The developmental zero is estimated to be 3.5°C with a thermal constant of 357.14 degree-days (Kheradmand *et al.*, 2007b). The hypopi are carried by stored-product insects (*Tenebrio molitor* Linnaeus, etc.) (Hughes, 1976), and by various Scarabaeidae. They are also found on the abdomen and wings of the beetles *Oryctes nasicornis* Linaneus and *Polyphylla fullo* Linnaeus (Zakhvatkin, 1941), and on the termite *Eutermes exitiosis* Hill (Womersley, 1941).

S. berlesei may play a role as a biocontrol agent of nematodes. Abou El-Atta *et al.* (2014) investigated the effects of temperature on the development, reproduction and food consumption of *S. berlesei*, feeding egg masses of root-knot nematodes, *Meloidogyne* spp., an important group of agricultural pests. Mites were reared at 20°C, 25°C or 30°C in the dark. The mites could feed on the nematode egg masses, and their developmental time decreased at higher temperatures. Time from the egg to adult was similar in females and males reared at the same temperature. Adult females lived longer than males at 25°C, but not at 20°C or 30°C. Generally, females showed a higher rate of food consumption than males. Females laid the largest number of eggs at 20°C and 25°C (199.7 and 189.8 eggs/female, respectively), but the intrinsic rate of natural increase, was highest at 30°C (r m = 0.29). Moreover, it has been observed that *S. berlesei* that fed on egg masses of root-knot nematodes, showed a longer developmental time, and a lower reproductive rate than *Sancassania* mites feeding on other diets. None the less, the relatively high value of rm (e.g. at 25°C and 30°C) suggests that this mite may have certain advantages as a biocontrol agent of root-knot nematodes (Abou El-Atta *et al.*, 2014).

Cardinium bacteria, intracellular endosymbiont or parasite frequently occurring in invertebrates including mites, were found associated with *S. berlesei* and other synantropic mites. Amplification of a 776 bp bacterial 16 SrRNA gene fragment, using specific primers, allowed the identification of closely related *Cardinium* sequences, in 13 laboratory-reared populations of mites. In addition, *Cardinium* sequences were identified in three wild mite populations. Large-scale screening of these populations showed 100% prevalence of *Cardinium*, representing the highest incidence, compared to other major Chelicerate groups (Kopecky *et al.*, 2013). Further studies carried out by Zhang *et al.* (2016), provided basic knowledge for understanding the interaction between endosymbionts and mites. Principato *et al.* (2005) reported some cases of human dermatitis caused by *S. berlesei*, as well as Cho *et al.* (1999) examined a case of infestation of external auditory canal of a man.

Symptomatology and damage

The range of food sources informed by various reports, highlights *S. berlesei* as a polyphagous feeder, accepting all kinds of deteriorating plant and animal material. *S. berlesei* occurs in damp situations, such as poorly stored food products with a high MC, barnyard or poultry litter. In food stores, the mite has been occasionally found on damp mouldy wheat, copra, linseed and groundnuts that have been kept under extremely damp conditions (Hughes, 1961).

The mite has been collected on edible fungi (Zakhvatkin, 1941; Hussey *et al.*, 1969; Clancy, 1981). It attacks hyphae and sporocarps of all sizes of *A. bisporus*, but in Australia is not a common species in mushroom houses (Greenslade and Clift, 2004). In Italy, *S. berlesei* has been found on mushrooms (Zachvatkin, 1941) and on melon in greenhouses (Laffi and Ponti, 1997), but on this last crop, the damage caused by the mite is unknown.

Principato *et al.* (2005) reported some cases of human dermatitis, caused by *S. berlesei* infesting poultry farms. The mite appeared unable to determine traumatic lesions on human skin, but it caused itch and inflammation also at the level of mucosae. The mite can be found accidentally, also on reared fowl pecking wounds. Moreover, a case of infestation of *S. berlesei* in the external auditory canal of a 46-year-old male is known. He complained of feeling a foreign body and itching in the left external auditory canal for 1 month, with accompanying otalgia for 3 days. Considering the duration of the patient‘s complaint and the 8–9 day life cycle of the mite, the mites are believed to have lived in the patient's ear for more than three generations (Cho *et al.*, 1999).

Control

In field, the control requires spray application of acaricides. On edible fungi, the mite has high levels of attacks, while on vegetables, it is of secondary importance. On stored products outbreaks, of the populations of *S. berlesei*, depend mainly on poor conditions of storage and require a correct management of the warehouse, eliminating all sources of humidity, and preserving only dry products.

As concerns the chemical control, it is possible to adopt the protocol given for other mites. Also in this case, various experiments carried out in time confirm the validity and the limits of various chemicals. So, Blank and Osborne (1978) tried in the laboratory the toxicity of different chemicals (ethion, parathion-methyl, parathion-ethyl, azinphos-methyl, azinphos-ethyl, demeton-s-methyl, dimethoate, dicrotophos, mevinphos and maldison) against *S. berlesei* and *Tetranychus urticae* Koch and observed that *S. berlesei* was found to be more resistant to organophosphorus acaricides, than a susceptible strain of *Tetranychus urticae*. Resistance varied from twofold for parathion-ethyl to 1840-fold for mevinphos.

The Genus *Thyreophagus* Rondani

Diagnostic characteristics

Both sexes possess a long body and a smooth and non-striated cuticle. Commonly, the female has four pairs of lyrifissures, one out a_1, one between d_4, one between the setae l_1 and 1_2 and one out 1_2. A sejugal furrow is well visible. The gnathosoma and the chelicerae are well developed. The propodosoma bears a large prodorsal shield, punctate or finely striated longitudinally. The posterior region of male is punctate, and forms a ventral median lobe, rounded, extending the body posteriorly. The epimeres I are fused at Y, other epimeres are free, and the epimerites II missing. The anal opening is ventroterminal in the female, and ventral in the male. The genital opening is set between the legs III–IV. The epigynum is little developed, and from it two fine-lines, few sclerified, extend upwards and outwards; they widen into a barely visible structure and extend to the sejugal furrow, then continue forward on each side in a little and dotted shield, longer than wide. This structure is rough also in the male. The bursa copulatrix opens dorsally at the posterior end of the body; a little papilla copulatrix may be present or not. The base of spermatheca is formed by a sclerite of characteristic shape for each species. The male has the sexual apparatus set behind the coxa IV, the anal opening is flanked by two large anal suckers, and the tarsus IV possesses two tarsal suckers. The opening of the opisthonotal gland is located between the setae l_2 and 1_3. The legs are weakly to moderately developed, with the basal part of the claw enveloped in a fleshy pretarsus. The female bears always the setae *vi, scx, sce,* d_4, d_5, 1_{1-3}, l_5, a_1, cx_I, cx_{III}, *ga*, *gm* and *gp*. The setae *scx* is naked and well developed. The setae *ve, sci,* d_{1-3}, *h, sh,* and l_4 are always missing. All males are homeomorphic, and their chaetotaxy as that of female, and with three pairs anal setae (a_{1-3}). Ventrally, the tarsi I and II possess three apical spines (one medial and two paramedial), and two fine setae; dorsally, bear two fine setae and one apical spine. Moreover, 2 very little vestigial spines are present, one apico-anterior, the other apico-posterior, visible only in a certain position of the tarsi. Tarsi III and IV with four spines (three ventral and one dorsal) and four simple setae (two ventral and two dorsal). Tibiae I–II with two setae, of these one is postero-ventral and very short; the tibiae III–IV bear one seta. The genua have 2–2–0–0 setae, the femurs 1–1–0–1 setae, and the trochanters 1–1–1–0 setae. On tarsi I the solenidion ω_1 is basal, ω_2 shorter and narrower and slightly more apical, ω_3 is fine and relatively short. On tarsi II, the solenidion ω_1 is thicker than tarsus I. The famulus of tarsus I is short and thicker, and set on the base of ω_1. The tibiae I–IV have one solenidion. The genu I bear two solenidia, unequal in the female, where the σ_1 (the most anterior) is always longer than σ_2, while in the male is the opposite. The genu II bears a short solenidion (Fain, 1982).

The systematic of this genus has been recently examined by Barbosa *et al.* (2016); they published a key of the species possessing adults, or adequately described and reported that the genus includes 22 species.

Economic importance

The genus numbers about 22 species (Barbosa *et al.*, 2016), among which *Thyreophagus entomophagus* (Laboulbène), a mite with a wide geographical distribution and reported on various stored products (Zakhvatkin, 1959; Sinha, 1964b; Chmielewski, 1975a; Hughes, 1976; Smiley, 1987), as well as in insect collections (Fain, 1982). This species may affect human health when ingested (Blanco *et al.*, 1997; Iglesias-Souto *et al.*, 2009; Sànchez-Machin *et al.*, 2010). Another species, *Thyreophagus gallegoi* Portus *et* Gomez, has been collected on bean curd, condiment, hoisin sauce, wheat flour in food store in China, Hong Kong, Indonesia, Philippines, Spain and Italy (Portus and Gomez, 1979; Vacante, 1989a; Hagstrum *et al.*, 2013). It has been collected in house dust in Spain (Portus and Gomez, 1979; Fain, 1982) and China (Jiang, 1991; Fan *et al.*, 2010a). Moreover, the mite has been intercepted in the United States, from bean curd originating from Hong Kong and Philippines and from hoisin sauce from Hong Kong (Olsen, 1983). *Thyreophagus macfarlanei* (Fain) has been collected on wheat in England (Fain, 1982). *Thyreophagus cracentiseta* Barbosa, Oconnor *et* Moraes was collected on stored human and animal food, and in soy and maize flours in Brazil (Barbosa *et al.*, 2016). Despite these reports we discuss here only the first species, able to play an economic importance.

Thyreophagus entomophagus (Laboulbène et Robin) (Fig. 28.18)

Common name

Collection mite

Diagnostic characteristics

FEMALE. The idiosoma is about 450 μm long and 230 μm wide. The prodorsal shield is 90 μm long and 105 μm wide, almost entirely punctate, and only the postero-median region presents some longitudinal striae. The setae *vi* are 80 μm long, *sce* 160 μm, *scx* 20 μm, d_4 60 μm, d_5 120 μm, 1_1 150 μm, l_2 45μm, l_3 75 μm, l_5 200 μm and *a* 180 μm, respectively. The sternum is 27 μm long. Diameter of genital papillae 8 μm. Spermathecal calyx overturned bell shaped, 21 μm long and 18–20 μm wide. Tarsi I–IV long and wide 29×17μm, 29×17μm, 21×15 μm and 21×15 μm, respectively. Leg chaetotaxy as the genus. Tarsi I and II with apical spines well developed. Tarsi III–IV with the apico-ventral median spine more developed than the two spines apico-ventral paramedian. The solenidion ω_1 of tarsus I is cylindrical-conical and narrow distally; on genu I, the solenidion σ_1 is gradually narrow distally, and long 15 μm; the solenidion σ_2 is thicker and 12 μm long. Solenidion of tibia III 60 μm long, and that of tibia IV from 18 to 25 μm (Fain, 1982).

Barbosa *et al.* (2016) reported that the female has the prodorsal sclerite wider than long, almost entirely punctate, and with a few short longitudinal striations in posteromedian region; the sclerite of spermatheca in form of an inverted bell, with the base 18–20 μm wide; the tarsus III has the seta *pv*" filiform; the solenidion ω_1 of tarsus I is narrowed apically; the posterior ventral seta (*c*") of tibiae I–II is setiform, and about 5 μm long.

MALE. The idiosoma is 375 μm long and 205 μm wide. Dorsal aspect as the female, with exception of the posterior region of the body adorned with a dotted shield. Ventrally, in the posterior region a large, rounded lobe extends the body back. Aedeagus wider (27μm) than long (25 μm). Adanal suckers 21 μm wide. Chaetotaxy of legs I–III as in the female. Tarsus IV with three simple setae, three spines and two suckers. The genu I with σ_1 (anterior) shorter (12μm) than σ_2 (18–21 μm) (Fain, 1982).

Barbosa *et al.* (2016) reported that the male has the body ovoid, 1.5–2 times longer than wide; the tarsus IV has three spine-like setae (*s*, *u*), three filiform setae (*pv*, *tc*') and two suckers (*ft*", *tc*").

HYPOPUS. Hughes (1976) reported that the hypopus of this species was unknown, although Lignières (1893) described it, and Michael (1903) translated this description. Successively, the hypopus has been described by Fain *et al.* (2000).

The body is distinctly widened at the level of the sejugal furrow. The ratio length: width varies between 1.35 and 1.41. The dorsum is finely punctate, except at the level of the sejugal furrow, where the cuticle bears numerous very short and very thin transverse striations. The area situated between the two eye lenses is densely mamillated. Eye lenses have a distinct basal retina, 11 μm wide and separated by a distance of 79–84 μm. The dorsal setae are very thin, from 6–12 μm long, except d_5 (15–18 μm). The setae *sce* are set on or close to the eye lenses. Ventrally, the epimera II are curved outside, longer than the sternum. The epimera III not reaching the midline. The setae *ga* are set on epimera III. The opening of opisthonotal glands is set ventrally, behind the setae *h*. The genital region possesses a pair of conoides (setae *gp*), and a pair of thin setae (setae *gm*). Setae 1_5 are 25 μm long. The suctorial plate as long (38–40 μm) as wide. The anterior suckers are circular, with a diameter of 7.6–8.5 μm, posterior suckers slightly ellipsoidal and longer (10–11 μm) than wide (9.3–9.6 μm). The palposoma is 12–14 μm long, 12–13 μm wide, ending into two very small palps, bearing a solenidion α 40 to 45 μm long. The tarsus I has seven setae of which *la*, *ra*, *f*, *q* and *p* are nanowly foliate; *d* is the longest (60), and *e* is spoon-like. Tarsus II as tarsus I, but the seta *ba* is present near the solenidion ω_1. The

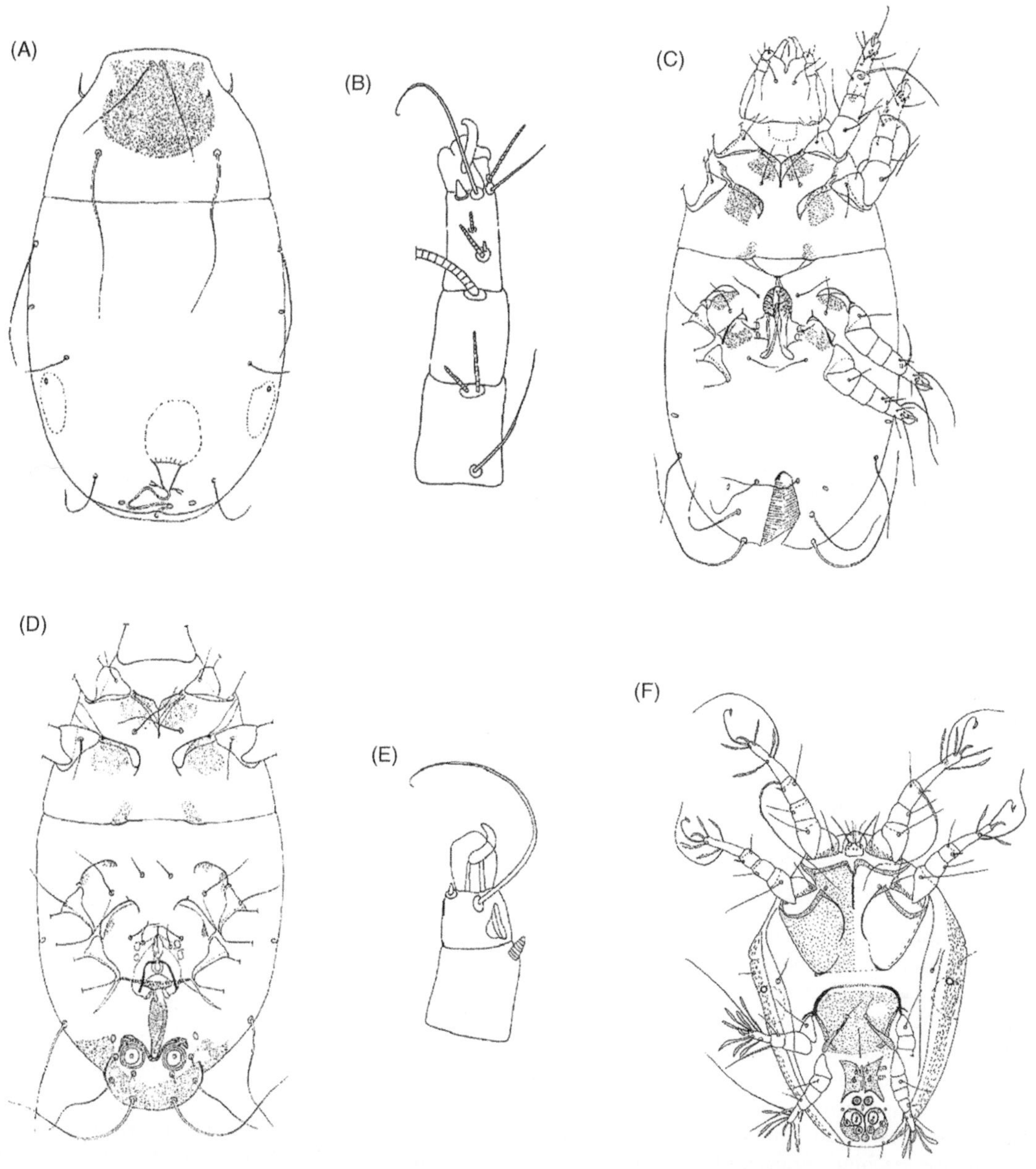

Fig. 28.18. *Thyreophagus entomophagus* (Laboulbène *et* Robin). Female. Dorsal view (A); leg I (B); ventral view (C). Male. Ventral view (D); last two segments of leg IV (E) (from Fain, 1982). Hypopus. Ventral view (F) (from Fain *et al.*, 2000).

tarsus III has eight setae, of which seven are foliate. The tarsus IV has eight setae of which five foliate, two thin simple and thick and forked. Tibiae with 2–2–1–1 setae. Tibia I with a thin and short *hT* (5µm) and a thin *gT* 12–13 µm long. On tibia II *hT* is a narrow spine 6 µm long, and *gT* is thin and 16–20 µm long. The genua I–IV have 2–2–0–0 setae. Genu I with *mG* and *cG* thin, and 18 µm long. Genu II with *mG* and *cG*, 18 and 12 µm long. Tarsus I with ω_1 13, ω_3 17 and ω_2 6 µm. Solenidia ω_1 and ω_3 are both set in the same depressed oval area, the latter also bearing a thin and short famulus *e*.

Tarsus II with ω 13 µm long. Tibia I bearing φ 95 µm. Tibia II with φ 27 µm. Genu I with σ 13 µm. Genu II with σ 4 µm (Fain *et al.*, 2000).

Geographical distribution

The geographical distribution of this mite is cosmopolitan, and includes Australia (Womersley 1941), Belgium (Fain, 1982), Brazil (Mazzuco *et al.*, 2000), Bulgaria (Nachev and Trenchev, 1987), Canada (Sinha, 1964b; Beaulieu and Knee, 2014), China (Xin, 1965; Li, C. *et al.*, 2015a), Côte d'Ivoire (Fain, 1974), Cuba (de la Torre Morin *et al.*, 2006), Czech Republic (Ždárková, 1967), Egypt (Elkawas, 2011), England (Michael, 1903; Newstead and Morris, 1920; Hughes, 1976; Fain, 1982), France (Laboulbène, 1852; Laboulbène and Robin, 1862; Hughes, 1976), Georgia (Reck, 1976), Germany (Fain, 1982; Hughes, 1976), Hong Kong (Olsen, 1983), India (Putatunda, 2002), Iran (Kamali *et al.*, 2001), Italy (Hughes, 1976; Goracci *et al.*, 1985; Vacante, 1989a), New Zealand (Cotter *et al.*, 2011), Philippines (Olsen, 1983), Poland (Chmielewski, 1975a, 1977; Hughes, 1976), Russia (Hughes, 1976; Zakhvatkin, 1941, 1959; Khaustov *et al.*, 2018), Spain (Iglesias-Souto *et al.*, 2009), Taiwan (Tseng, 1972, 1989), UK, United States (Hughes, 1976; Smiley, 1987; Philips, 2009), former Yugoslavia (Chmielewski, 1975a) and other countries.

Commodities, facilities and vegetal and animal hosts

Th. entomophagus is a pest of stored food and insect collections. Hagstrum *et al.* (2013) reported a wide list of commodities infested by this mite, such as baby food, betel nut (decaying), cake, cereal, Chinese medicine, ergot of rye, fish (dried), garlic, ginger (pickled), grain, insect (dried), onion and wheat flour, variously collected in different facilities, such as bread van, farm, farm granary, farm milling machine, flour mill, provender mill and provender warehouse. Many records confirm this adaptation. In Canada, Sinha (1964b) reported *Th. entomophagus* on stored products, and Fain (1982) reported the mite on rye, but not too dry, and on almond in Belgium, and flour in Germany. Various reports reported the mite commonly on the flour (Newstead and Morris, 1920; Zachvatkin, 1941; Ždárková, 1967; Mazzuco *et al.*, 2000; Cotter *et al.*, 2011); moreover, it has been found on other products, such as vanilla pods and saffron for medicinal use (Moniez, 1892), dried medicinal plants, including ergot and cardamom, stimulants, spices, food products and fodders (rye and wheat bran) (Chmielewski, 1975a) and poultry meal (Newstead and Morris, 1920).

Fain (1982) reported that Laboulbène (1852) retained the mite as serious pest of insect collections. In the Mediterranean region, it has been found commonly on death insects, mainly Coleopters Cantharids and also on *Ptinus* in England (Fain, 1982). In Belgium, it has been collected on the collection of the Hymenoptera *Bombus lapidarium* (Linnaeus) (Fain, 1982). *Th. entomophagus* preferentially develops on large insects, resulting very abundant on *Locusta migratoria* (Linnaeus) (Orthoptera) and *Sesia* (Lepidoptera), and is a common pest in southern France (Laboulbène and Robin, 1862), whereas only a few specimens infesting insect collections have been found in Belgium and England (Michael, 1903; Fain, 1982). The mite has also been found in association with scale insects (Nachev and Trenchev, 1987; Vacante, 1989a; Elkawas, 2011). It can feed on larval skins and live coccids (Nachev and Trenchev, 1987), on trogid beetles (Philips, 2009), ergot of rye (Michael, 1903), bird nests (Woodroffe, 1953; Wasylik, 1959), bracket fungi *Fomitopsis betulina* (Bull.) Cui, Han *et* Dai (Pielou and Verma, 1968) and various materials in beehives of *A. mellifera* Linnaeus, such as debris, bee bread, pollen, dead bee brood and mould (Chmielewski, 1977a, 1991a). The mite has been found also on dry Spanish fly *Lytta vesicatoria* Linnaeus harvested for medicinal use (Fumouze and Robin, 1867).

Bio-ecology

Khaustov *et al.* (2018) examined the validity of this taxon and wrote, 'This mite has been reported as a pest infesting dry insect collections in France and named (without a formal description) *Acarus entomophagus*' (Laboulbène, 1852). Later, this taxon was properly described based on adults and nymphs (Laboulbène and Robin, 1862). Although in the latter work, the authorship of '*A. entomophagus*' (*nomen nudum*) is attributed to Laboulbène, the use of the phrase '*Tyroglyphus entomophagus nobis*' (*Tyroglyphus entomophagus* new for ours), without explicitly attributing it to Laboulbène, indicates that this species should have a joint authourship of Laboulbène and Robin. Therefore, the author and year of this species should be reported as *Th. entomophagus* (Laboulbène and Robin, 1862), not '(Laboulbène, 1852)', as cited in the past in literature (Fain, 1982; Fain *et al.* 2000; Barbosa *et al.*, 2016;), conversely to Zakhvatkin (1959), Hughes (1976) and Smiley (1987), which named the mite as *Th. entomophagus* (Laboulbène). According to Khaustov *et al.* (2018), we treat here, the mite as *Th. entomophagus* (Laboulbène *et* Robin).

Th. entomophagus is a probable cosmopolitan species, living as pest of different stored products and insect collections. The life cycle develops through the egg, larva, protonymph, deutonymph and adult (female or male), while hypopus was not observed by (Chmielewski, 1990) and found by (Fain *et al.*, 2000). Brett and Thornback (see Hughes, 1976) reported that the development of the life cycle of the mite takes from 28–38 days at a RH of 75%, 26–32 days at 85%, the temperature in both cases being 18°C and the culture medium, wheat germ and yeast; they consider that *Th. entomophagus* can penetrate into compressed flour more successfully than *A. siro*. Chmielewski (1990) reared under laboratory conditions specimens of *Th. entomophagus* collected from natural beehive debris and from stored rye and wheat bran heavy infested by the mite. Populations of the pest mite were maintained at 20°C, 85% RH, and with wheat germs as food. The influence of different temperature (3°C, 5°C, 15°C, 20°C, 25°C, 30°C, 35°C) and RH (65%, 75%, 85%, 95%) on oviposition, longevity, life history and mortality were studied. Most rapid development was observed at 25–30°C and RH 85–95% and took 14.9–17.8 (11–21) days, respectively. The highest fecundity, on an average 120.4–170.6 (0–340) eggs per female per its whole life examined at 20–30°C and the same RH. The longevity of mites ranged from 2 weeks to 3 months; at lower temperature (3–15°C), the mite lived longer (maximum even higher than 2 years), but at higher temperature (35–40°C) the longevity was very short, especially at lower RH (65%) (from 1 to some days). The reproductive rates calculated for 20°C, RH 85% and wheat germs were as follows: net reproductive rate R = 52,2, innate capacity for increase rm = 0.059 (per day), the finite rate of increase A = 1.060 (female offspring per female per day), the mean length of generation T = 45.18 (days) (Chmielewski, 1990). Moreover, ingesting infested food may produce occupational respiratory disorders and anaphylaxis (Mazzuco *et al.*, 2000; Musken *et al.*, 2000; Iglesias-Souto *et al.*, 2009).

Symptomatology and damage

This species is fairly widely distributed in small numbers in flour, sometimes increasing in numbers sufficient to cause serious damage; Smith (1969) retained that infestation probably originates from the milling machinery. Infestations of this mite on insect collections may cause severe damage. Moreover, it has been observed that ingesting infested food may produce occupational respiratory disorders and anaphylaxis in Spain (Blanco *et al.*, 1997; Iglesias-Souto *et al.*, 2009; Sánchez-Borges *et al.*, 2013), in Germany (Musken *et al.*, 2000) and in Brazil (Mazzuco *et al.*, 2000).

Control

As other acarid mites.

Other Genera

Some genera included into the family Acaridae and reported on stored products play a secondary economic importance. To agevolate their identification, they are treated in the key, but the relative species are not discussed extensively. They are the genera *Schwiebea* Oudemans, *Histiogaster* Berlese and *Madaglyphus* Fain. This last is retained of *incertae sedis* (OConnor, 1998). The genus *Schwiebea* is cosmopolitan and infests cottonseed meal, ginger (rotting), taro and yams Smiley (1987). The genus *Schwiebea* number the species *S. mertzis* Woodring is known from Japan and Taiwan on taro and yam root (Manson, 1972b), and *S.* sp. near *elongatus* (Banks), reported from Iran, on stored products and potato by Ostovan (1993), and from the United States (Hofstetter *et al.*, 2014). The genus *Histiogaster* includes the species *Histiogaster carpio* Kramer, reported from China, and United States on grain (Zachvatkin, 1959; Hagstrum *et al.*, 2013) and the genus *Madaglyphus* numbers the species *Madaglyphus javensis* Haines *et* Lynch collected in Indonesia and Java on rice mill in food store (Haines and Lynch, 1987).

29 The Family Carpoglyphidae Oudemans

Morphological Characteristics, Systematic and Bio-ecology

The Carpoglyphids are free-living mites, usually not associated with a specific habitat, often living on sugary and fermenting materials. In honeybee hives, they feed on various components, including pollen stored in honeycombs (bee bread or bee pollen), plant pollen, honey and nest debris. Phoretic deutonymphs disperse on adult insect hosts to suitable habitats. The phoresy occurs on Lepidoptera and Coleoptera but not on bees.

From the morphological point of view, the Carpoglyphid mites are characterized by the absence of the prodorsal sclerite; the rostral setae (*ro*) are set about halfway between the anterior edge of the propodosoma and the interlamellar (*in*) setae. Moreover, they possess one pair of large ocelli set in the propodosomatic region. The species belonging to the genus *Carpoglyphus* Robin have smooth cuticle; the apodemes of legs I and II are united with the sternum in both sexes. The claw is large, and joined to the end of tarsus by a well-developed pretarsus. The condylophores are elongate and separate in the female and are asymmetrical in the male. The male possesses the genital (*g*) and the coxal setae *4b*. Most of the idiosomal setae are smooth, with exception of the posterior setae (Hughes, 1976; O'Connor, 2009b).

The Carpoglyphidae is a small family, including only two genera and six species (Schatz *et al.*, 2011); five of these species belong to the genus *Carpoglyphus* (Clark *et al.*, 2010). Beron (2021b) listed 4 genera and 12 species.

The life cycle of Carpoglyphidae develops through the egg, larva, deutonymph, tritonymph and adult stages. The phoretic deutonymphs (hypopi) disperse on adult insect hosts to suitable habitats, and the phoresy is known only on Lepidoptera and Coleoptera (but not on bees).

The dried fruit mite, *Carpoglyphus lactis* (Linnaeus), is the most important species, relatively common in beehives, which damages honeycombs and honey production. The mite has been found in old honeycombs, wine barrels in cellars, dried sweet fruits, canned fruits, fruits preserved in sugar, fermenting pulp, dairy products (milk and cheese) and stored honey. It often infests products following the development of yeast. In the field, this species is found in fermenting tree sap flows and burrows of moles and as phoretic deutonymphs on butterflies, moths and scarabaeid beetles (*Gnorimus* sp.) (Klimov *et al.*, 2016). *C. lactis* is retained an alternative food for predatory mites in the Phytoseiidae that are used in biological control (Baxter *et al.*, 2011). At this aim, Zhang (2012) reported that in New Zealand, there has been interest by the biocontrol industry to import this species in the country for mass rearing of predatory mites. Moreover, when large numbers of *C. lactis* mites are ingested with infested food or beverages, they can cause dysentery. These mites also cause dermatitis to handlers of infested materials, such as dried plums (O'Donovan, 1922), and occupational allergies in the biological pesticide industry (Krop *et al.*, 2012). However, they may be beneficial in the wine industry, as mite alarm pheromones enhance the aroma of pale and dry wines aged under flour yeasts (Marin *et al.*, 2009).

Another species associated with bees is *Carpoglyphus munroi* Hughes, few common and first found in a disused bell-tower, and in a mass of cobwebs containing dead insect in England (Hughes, 1952) and successively collected on beehives in the Czech Republic (Haragsim *et al.*, 1978). Another species, *Carpoglyphus wardleorum* Clark, was described later in 2010, but not associated with bees.

Selected websites

https://www.invasive.org/browse/taxthumb.cfm?fam=797

https://idtools.org/bee_mite/index.cfm?packageID=1&entityID=83

DOI: 10.1079/9781786390912.0029

The Genus *Carpoglyphus* Robin

Morphological characteristics

ADULT. Idiosoma slightly flattened, oval and covered with a shining smooth cuticle. Prodorsum with setae *ve* absent. Prodorsal sclerite absent, and no transverse groove separates the propodosoma from the hysterosoma. The body setae are smooth, not heavily barbed. Setae *vi* set about halfway between anterior edge of propodosoma and setae *sci*. Ocelli present. Supracoxal gland opening not associated with a large, sclerotized region. Coxal apodemes I fused medially with coxal apodemes II, closing coxal fields I in both sexes. Tarsi I–II elongate. Pretarsi similar on all legs, and with long, thin condylophores. Empodial claws present. Males without paranal suckers or sucker-like setae on tarsus IV. Condylophores asymmetrical in male. Male with genital setae (*g*) and coxal setae *4b* present (Hughes, 1976; Klimov *et al.*, 2016).

HYPOPUS. Empodial claws arising from membranous ambulacra I–IV (not from tarsal apices). Leg IV with empodial claw present. Leg IV generally similar in form to leg III. Posterior median ventral apodeme absent. Ocelli present, widely separated on propodosoma (Klimov *et al.*, 2016).

Economic importance

The year of the genus *Carpoglyphus* Robin, 1869 is sometimes erroneously reported as 1860, but *Carpoglyphus* Robin, 1860 is not an available name; this is because Robin (1860) quotes some collective characters for '*Carpoglyphus*' and other genera of mites, which does not constitute a valid description, and, thus, the name *Carpoglyphus* Robin, 1860 (*nomen nudum*) is unavailable (ICZN Art. 12.1) (Klimov *et al.*, 2016).

A dichotomous key to adults (males and females) is available in Fain and Rack (1987). This key can be used to identify the two species *C. lactis* (Linnaeus) and *C. munroi* Hughes, both mites associated sometimes with bees. One non bee associated species, *C. wardleorum* Clark, was described successively (Clark, 2010).

The dried fruit mite, *C. lactis*, is the most important species, relatively common in beehives, where damages honeycombs and honey production. The mite has been found in old honeycombs, wine barrels in cellars, dried sweet fruits, canned fruits, fruits preserved in sugar, fermenting pulp, dairy products (milk and cheese) and stored honey. It often infests products following the development of yeast. In the field, this species is found in fermenting tree sap flows, burrows of moles, and as phoretic deutonymphs on butterflies, moths and scarabaeid beetles (*Gnorimus* sp.) (Klimov *et al.*, 2016). When large numbers of *C. lactis* mites are ingested with infested food or beverages, they can cause dysentery. These mites also cause dermatitis to handlers of infested materials, such as dried plums (O'Donovan, 1922), and occupational allergies in the biological pesticide industry (Krop *et al.*, 2012). However, they may be beneficial in the wine industry, as mite alarm pheromones enhance the aroma of pale and dry wines, aged under flor yeasts (Marin *et al.*, 2009).

C. munroi is few common and has been first found in a disused bell-tower and in a mass of cobwebs containing dead insect and from bat roosts in England (Hughes, 1952) and successively reported from the Czech Republic (Haragsim *et al.*, 1978), and Iran on nuts and dried fruits (Kordestani Mahani *et al.*, 2014). We discuss here only the dried fruit mite.

Carpoglyphus lactis (Linnaeus) (Fig. 29.1)

Common name

Dried fruit mite

Diagnostic characteristics

FEMALE. The idiosoma is 380–420 µm long. The general aspect as the male. Ventrally, the sternum and the apodemes II are fused to form an epigynum, covering the anterior end of the genital opening. The genital folds are not well sclerotized, and lie between the coxae II and III. The anal opening extends almost to the posterior margin of the body, and there is only one pair of anal setae. The bursa copulatrix is a round opening on the posterior-dorsal end of the body. The legs are slender than those of the male, with less well-developed pretarsi (Hughes, 1976).

MALE. The idiosoma is 380–400 µm long, oval and slightly flattened, only coloured by the contents of the alimentary canal showing through the translucent cuticle. The legs and the gnathosoma are slightly pinkish. The shoulder region is well definite, and the posterior margin truncate or slightly concave.

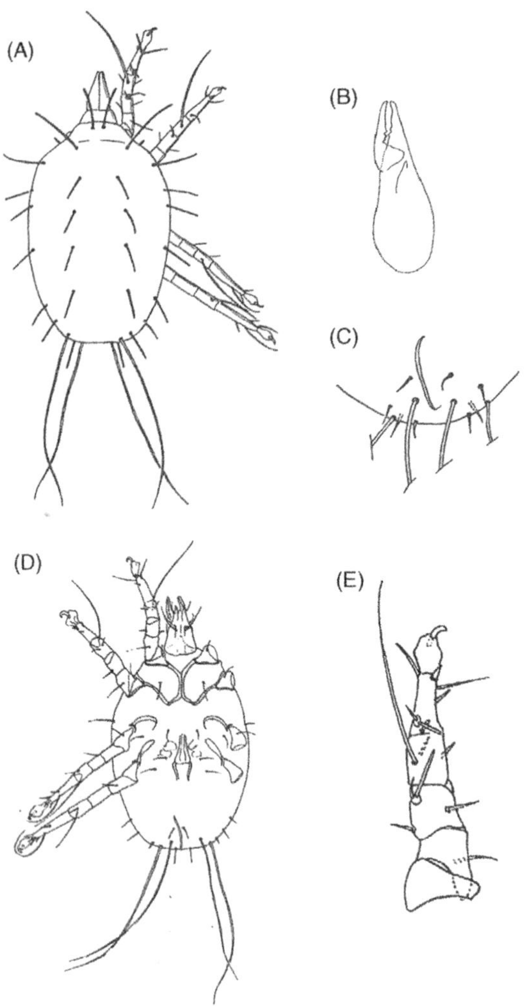

Fig. 29.1. *Carpoglyphus lactis* (Linnaeus). Female. Dorsal view (A); chelicera (B); anal region (C). Male. Dorsal view (D); leg I (E) (from Hughes, 1976, licensed under the Open Government Licence v3.0).

No dorsal propodosomal shield is present. The latero-opisthosomal glands are displaced to the posterior corners of the idiosoma, and contain a coloured substance. A slightly raised cornea is found on either side of the base of the gnathosoma, but there is no pigmented retina. Ventrally, the apodemes are well sclerotized, and apodeme I join in the midline to form a sternum that bifurcates at its posterior end to articulate with the apodemes of leg II. The genital opening is between coxae III and IV. The aedeagus is a blunt tube with its apex direct anteriorly, and the genital sense organ is unusually long. Two pairs of genital setae are present, about equal in length. The anus extends as far as the posterior margin of the body.

With exception of *ve* and two pairs of setae arising posteriorly, all the setae are short (7–12% of the length of the idiosoma), with rounded ends. Setae *vi* do not extend as far as the tips of the chelicerae; *ve* are more posterior and intermediate in position between *vi* and *sci*; d_1 to d_4 and *sai* are arranged in two longitudinal rows down the middle of the idiosoma. The supra-coxal seta is a blunt rod. From the posterior end of the body project two pairs of long setae, pa_1 and *sae*. The gnathosoma is shaped like a cone and very mobile. The chelicerae have thin narrow shears. All the legs terminate in well-developed pretarsi, strengthened by two thin 'tendons', which extend from the end of the tarsus nearly to the sickle-shaped claw. On tarsus I, some of the median and distal setae are spines. Omega 1 is a cylindrical rod frequently bent externally, so as to cover the base of omega 2. On tibia I and II, φ arises from the middle region, and both ventral setae are present (Hughes, 1976).

HYPOPUS. As the genus.

Geographical distribution

Cosmopolitan, reported from various countries, such as Algeria (Fain and Rack, 1987), Argentina (Mullen, 1943; Hughes, 1976), Australia (Womersley, 1941; Halliday, 1998), Belgium (Vitzthum, 1940), Brazil (Endlein and Flechtmann, 1976), Canada (Hughes, 1976), China (Xin, 1965; Shen, 1979, 1989; Zhu and Yang, 2003; Zhan *et al.*, 2017a), Czech Republic (Hubert *et al.*, 2014b), Egypt (Ibrahim, 2006; Hassan *et al.*, 2019), Germany (Oboussier, 1939; Fain and Rack, 1987), Greece (Palyvos *et al.*, 2008), Hong Kong (Olsen, 1983), Kenya (Fain and Rack, 1987), Indonesia (Olsen, 1983), Iran (Sepasgozarian 1978; Sarraf *et al.*, 2015), Korea, Mexico (Olsen, 1983), New Zealand (Zhang, 2012), North America (Treat, 1975; Hughes, 1976; Baker and Delfinado, 1978), Philippines (Olsen, 1983), Poland (Chmielewski, 1967, 1971d), Portugal (Guimarães, 1970), Singapore, Thailand, Taiwan (Olsen, 1983), Turkey (Çobanoğlu, 2008; Dizlek *et al.*, 2019) and other countries.

Commodities, facilities and vegetal and animal hosts

Apricot (dried), bean curd, biscuit, buckwheat, buckwheat product, candied fruit, caramel, cheese, cocoa bean, coconut (fermented), confection, date (red), dried fruit, dried fruit (mouldy), fig (dried), dried bananas, fish (minced-dried), grain, herb, honey, honeycomb, jam (fermenting), groundnut, plum (dried), pollen, potato (rotting), processed food, prune (dried), raisin, shrimp (dried), soybean sauce, soybean paste, sugar (crude), sugar (red), sugar (white), tamarind, wheat flour, wheat flour (old) and wine cork. All commodities have been collected off-farm storage (Hagstrum *et al.*, 2013).

Bio-ecology

The dried fruit mite is a cosmopolitan species, living almost exclusively not only on food products containing sugar, especially dried fruits, honeycombs, pollen in beehives, but also on wine fruit-drinks, cheese, old flour and so on (Guimarães, 1970; Chmielewski, 1970b; Hughes, 1976). Moreover, the mite infests milk products, various sugar-based products (Chmielewski, 1971d) and sweet chow chow (Shen, 1979). In Egypt, *C. lactis* has been collected on stored grains, wheat, maize and rice in granary stores (Hassan *et al.*, 2019). In Japan, the mite is known as the pest of soybeans paste, 'miso' and raw sugar (Okamoto, 1984). Baker and Delfinado (1978) reported infestations of *C. lactis* on stored honeycombs at a bee farm, and its presence in pollen stores in honeycombs, which may be heavily infested after winter storage.

The life cycle of *C. lactis* develops through the egg, larva, protonymph, deutonymph (hypopus), tritonymph and adult stages. Oboussier (1939) was the first to observe the formation of phoretic hypopus in this species, and Vitzthum (1940) redescribed this hypopus and observed the presence of eyes in both adults and hypopus. Chmielewski (1967) obtained the hypopus stage in laboratory cultures. Vitzthum (1940) collected the hypopus of *C. lactis* from the nymphalid *Apatura iris* (Linnaeus) in Belgium; he reported that the mite commonly lives in the flowers, and that its hypopi are veiculate by Lepidoptera. Treat (1975) confirmed the presence of these hypopi on Lepidoptera, and found them on 19 noctuids from Massachusetts and New York, in the United States. He observed that their number varied from one to 28 on each noctuid and that live in the interpalpal area on or behind the base of the proboscis. Treat hypothesized that the mite more readily feeds on decaying material such as fallen fruit or exudate of trees, apparently being more suitable for the mites (Fain and Rack, 1987).

Guimarães (1970) reported that on dried figs the infestation of *C. lactis* occurs only in the storage sites or close premises, both in farm stores of the growers and in the 'fumeiros' (town stores of the industrialists or exporters). The mite does not infest figs during sun drying, but infestation can occur in farm stores if the product is not delivered until October. Infestations of *C. lactis* are frequent from November to the end of the working period (April or May). When stores are empty, a very small number of mites survive in the absence of figs and to the hot and dry summer conditions. These specimens were collected in dust and organic debris as adults and nymphal stages, but never in the hypopial form (Guimarães, 1970). In store compartments, the mite infestation starts in the upper layers. The figs become more compact and mites decrease in number until 0.60 m, and below this level they disappear. When figs are in good conditions, the infestation is restricted to outer surface. But if pulp of figs is altered, mites can get inside. This phenomenon is usual a result of insect attack or deep fermentation.

The adaptation of *C. lactis* to feeding on dried fruits, including the interaction with microorganisms, was investigated by Hubert *et al.* (2014b), which identified potentially pathogenic bacteria carried by the mites to dried fruits for human consumption. They studied the microbial communities associated with *C. lactis* in specimens originating from the packages of dried apricot, plums and figs, and compared to the laboratory strain reared on house-dust mite diet. Genomic sequences of the bacteria genera *Leuconostoc*, *Elizabethkingia*, *Ewingella*, *Erwinia*, *Bacillus* and *Serratia* were found prevailing in specimens sampled from the dried fruits. The treatment and the pretreatment of mites with antibiotics did not reveal the presence of antagonist bacteria, which might slow down the *C. lactis* population growth. The shifts of the microbial community in the gut of the mite were induced by the diet changes. The identified yeasts and bacteria are suggested as the main food source of stored-product mites on dried fruits (Hubert *et al.*, 2014b). Guimarães (1970) reported that the most important factor on the development of mite infestation is the rise of atmospheric humidity; at RH above 70%, the mites can strive. At 25–30°C and 80% RH, the dried fruit mite completes its life cycle within 10 days. The hypopus appears at the deutonymph stage under conditions of high population density, and permits an extension of the developmental period by up to 3 months (Chmielewski,

1971d). Shen (1979, 1989) observed that at 25°C and 75% RH on Cuban sugar, the species needs 13–18 days to complete its life cycle. The female lay 30–70 eggs per week, and the longevity of adults is 40–50 days (Shen, 1979, 1989).

The effects of different ecological factors, as RH, temperature, photoperiod and quality of food resource, on the bio-ecology of *C. lactis* have been variously investigated. The influence of RH has been evaluated at 25°C, various RH (58.5%, 74.0%, 84.0% and 94.0%), and a mixture of sugar and dried yeast (1:1) as food. The hatch rates and the period of egg stage were nearly the same in all humidities. The survival rate from larva to adult was higher as the humidity increased. In 58.5% RH, all hatched larvae were dead at the beginning of larval stage. In 94% RH, the durations of larval, protonymphal and tritonymphal stages were 2.0, 1.3 and 1.7 days, respectively. It took 7.8 days to develop from egg to adult, and was shortest among four conditions. The number of eggs laid per female in the lifespan was 193.4, most in 94% RH. The length and the width of idiosoma at each developmental stage in 74.0% RH was short compared with that of the other conditions. The sex ratio was almost the same in three conditions. The deutonymph (hypopus) was not found during this experiment. The condition of 94% RH was the optimum humidity for individual rearing of *C. lactis* in the experimental conditions (Okamoto, 1984).

The effects of various temperatures (5°C, 10°C, 15°C, 20°C, 25°C, 30°C and 35°C) on the life cycle of *C. lactis* were investigated, using the same protocol above cited, taking into account that the RH varied depending upon temperatures, from 90.5–100%. At 5°C and 35°C, it was impossible for the mite to grow. At low temperatures (within the range between 10°C and 30°C), the completion of the life cycle and the longevity were prolonged, and the number of mites giving rise to adults decreased. Total numbers of eggs laid per female at 15°C and 20°C were larger than those at the other temperatures. At 30°C, the period for the development from an egg to an adult was 6.8 days, shortest and the appearance rate of the adults was highest. These two effects of the high temperatures seem to make amends for the lower fecundity and the shorter longevity, resulting in the fast increase of the mite population (Okamoto, 1986).

The effects of four different photoperiods (darkness, 0, 8, 16 and continuous light, 24 h per day) on the developmental stages and fecundity of *C. lactis* reared on wheat were evaluated under laboratory conditions at 25 ± 2°C and 65 ± 5% RH. The results showed no clear effect of photoperiod on the incubation period. It was 2.9 and 3.4 for male, and 2.7 and 3.2 for female at darkness and continuous light, respectively. The duration of each immature instar, as well as the combined immature stages, increased with increasing hours of light in both male and female. Different patterns were observed for longevity data for male and female. Male longevity was the longest (9.3 days) at continuous light (24 h), whereas the greatest female longevity (15.4 days) was at complete darkness. The oviposition period as well as the total average and mean daily deposited eggs decreased as the photoperiod increased (Ibrahim, 2006).

Güldali and Çobanoğlu (2010) studied the biological parameters, development times, reproductive rates and population growth parameters of *C. lactis* at three different temperatures (18°C ± 1, 23°C ± 1 and 28°C ± 1) and 2 RH (65% ± 1 and 80% ± 1). The thermal constant at 65% and 80% RH were calculated as 83.68 and 92.41 degree days, respectively. The development threshold values were calculated as 7.55°C and 4.90°C, at 65% and 80% RH. According to the life table data, the intrinsic rate of increase (*rm*); ranged from 0.15–0.37 female/female/days, in the different combinations of temperature and RH. The highest *rm* obtained is 0.37 at 28°C and 80% RH. The net reproduction rate (*RO*) ranged from 9.60–21.84 female/female in the different combinations of temperature and RH. Increases in temperatures accelerated the generation period which ranged from 8.06–14.93 days. The shortest generation period was obtained at 28°C and 80% RH (Güldali and Çobanoğlu, 2010).

Sarraf *et al.* (2015) examined life table parameters of dried fruit mite on baker's yeast under laboratory conditions at 20°C and 25°C , 70 ± 5 RH and full darkness, based on age-stage, two-sex life table theory. The intrinsic rate of increase (*r*), the finite rate of population increase (λ) and the mean generation time (*T*) at 20°C were 0.287 day^{-1}, 1.333 day^{-1} and 16.7 days, respectively, while at 25°C were 0.383 day^{-1}, 1.467 day^{-1} and 11.87 days. A significant difference was observed between mentioned parameters in two temperatures ($P<0.01$). Likewise, there was a significant difference between the value of gross reproductive rate (GRR) ($P<0.05$), but no significant difference was observed between net reproductive rate (R_0) in two temperatures ($P>0.05$) (Sarraf *et al.*, 2015).

The effect of dry yeast granules, crushed wheat, crushed rice as a kind of food on biological aspects, fecundity and life table parameters of the grain mite was investigated under laboratory conditions at 25 ± 1°C and 65 ± 5 RH. Results of this study showed the ability of *C. lactis* to feed and develop on the above-mentioned diets as a sole food source. The average of total immature stages lasted (13.4 and 11.6), (16.5 and 15.1) and (20.0 and 18.6) days for female and male reared on the same diets, respectively. Female longevity was 15.1, 18.5 and 19.5 days, while male adulthood averaged 19.3, 24.8 and 25.8 days, when mite fed on the above-mentioned diets. Female oviposition period stayed 11.3, 12.9 and 13.6 days, and the average number of deposited eggs/female was 136.8, 113.9 and 70.5 eggs, with a daily mean of 12.1, 8.8 and 5.2 eggs on the same tested diets. The life table parameters of *C. lactis* were affected by different types of food, whereas the mean generation (*T*) was 20.7, 22.8 and 28.8, while the intrinsic rates of natural increase (*rm*) values were 0.39, 0.32 and 0.24 when mite fed on the same tested diets at the same pattern (Hassan *et al.*, 2019).

The contaminative behaviour of dried fruit mite in stored foods and its fairly high biological potential on sugar compounds, such as honey, have been investigated by different authors. Chmielewski (1999) reported that the honey provides the nourishment necessaries of mites, and ensures their satisfactory development and multiplication. This adaptation is probably one of the reasons of mite infestations of honey that happen from time to time. Honey seems to be one of the appropriate and effective media or valuable addition to diets for laboratory cultures of dried fruit mite and in the production of biological material for experiments as well as for mite demonstration and teaching acarology (Chmielewski, 1999).

More extensive research works have been carried out to estimate the contaminative behaviour of dried fruit mite in stored foods. Zou (2003) separated acaroid mites in stored foods by using lighten, tullgren or water enacopy, and made to sheet and identified. He collected 200 portions in 20 samples of stored foods and found 13 species of acaroid mites belonging to 11 genera in five families from different material, including *C. lactis*. The results highlighted that, in China, stored foods were severely polluted, and much more attention should be paid to control of acaroid mites in stored foods (Zhu and Yang, 2003).

In China, Zhan *et al.* (2017a) collected samples of Fructus Jujubae, and kept them over 6 months in general, from the dried fruit shop and (or) Chinese herbal medicine warehouse, and isolated *C. lactis* from those samples. *C. lactis* was identified in 19 of the 300 samples, with breeding density and breeding rate of 6.52 heads/g and 6.33%, respectively. The constitute ratio at distinct developmental phase was associated with adult (including nymph, 85.71%), larva (12.27%), hypopus (0.56%) and egg (1.45%), respectively. The richness index, diversity index and evenness index were 1.644, 1.644 and 0.923, respectively. *C. lactis* appeared infesting in large quantity in the Fructus Jujubae stored, and the density was higher. The results suggested to take effective measures to prevent *C. lactis* from spreading over other dried products stored in the same room and potential human intestinal acariasis as a result of the biological contamination (Zhan *et al.*, 2017a).

A study carried out on the mite fauna of stored apricots collected mainly from most important cultivated areas and exporting centre of Turkey highlighted 16 mite species, belonging to 11 families and 13 genera. Out of 601 samples, 133 were infested with mites, and the infestation ratio is 22.19%. *C. lactis* was the most abundant species on dried apricots, with infestation rate of 69.14%. The possibility of biological control of stored-product mites by two common predacious species, the Ascid mites *Blattisocius tarsalis* (Berlese) and *Blattisocius mali* (Oudemans) were examined. *B. tarsalis* was the most common and populated beneficial mites, with the 62.30% ratio among the beneficial mites (Çobanoğlu, 2008). Another study carried out in Turkey on dried fig taken from producers' storage, commercial storage, fig processing factories and on dried fig samples kept in storage 8 months, highlighted the presence of eight mites species, belonging to three orders and five families. The pest mites were *Acarus siro* Linnaeus, *C. lactis*, *Glycyphagus destructor* (Schrank), *Glycyphagus domesticus* (De Geer) and *Tyrophagus putrescentiae* (Schrank), and the predatory mites *B. tarsalis*, *B. mali* and *Cheyletus eruditus* (Schrank). Out of 141 samples, 133 were infested by mites, and the infestation ratio was 94.3%. *T. putrescentiae* was the most common species, with an infestation rate of 72.3%, followed by *C. lactis* with 34.8%, *A. siro*, *G. destructor* and *G. domesticus* with 0.7% contamination, respectively. *C. lactis* was found to have the highest density with 10,488 individuals/kg in contaminated fig samples. *B. tarsalis* was the most common predatory mite, with an infestation rate of 77.3%. *B. mali* and *C. eruditus* were found in 9.2% and 4.3% of dried figs, respectively. The mite species composition and their density fluctuated over time, and the dried fig samples kept in storage were always vulnerable to contamination with mites (Dizlek *et al.*, 2019).

Research works carried out in Japan on mites, associated with worker bees and hive debris of *Apis cerana japonica* Fabricius, reared in three regions in Japan showed *Neocypholaelaps favas* Ishikawa, *T. putrescentiae* and *C. lactis* as dominant species and assumed to be scavengers in the hives. Predacious and soil inhabiting mites were also found but in small numbers. There were few mites attached to workers, while hundreds of mites per gram were sometimes collected from hive debris. It was suspected that the mite numbers were influenced by temperature and precipitation. Most mites did not seem to reduce the bee populations or honey production (Okabe *et al.*, 2000).

Insects, rodents, birds and other animal take parts in spreading the mite (Chmielewski, 1971d).

Liu *et al.* (2024) reported that despite the dried fruit mite, *C. lactis* is a serious pest damaging stored products, including dried fruits, fruits preserved in sugar, fermenting pulp, dairy products like milk and cheese, and even jams and honey (Hubert *et al.*, 2021), it has found commercial applications as a prey in the mass rearing of various predatory mite species (Bolckmans *et al.*, 2013), such as *Amblydromalus limonicus* Garman et McGregor, *Amblyseius swirskii* Athias-Henriot (Nguyen *et al.*, 2013) and *Neoseiulus cucumeris* Oudemans (Ji *et al.*, 2015). It has been observed that the different life stages of *C. lactis* can be preyed upon by various phytoseiid species (Nguyen *et al.*, 2013; Ji *et al.*, 2015; Zhang and Zhang, 2022). With this aim, Liu *et al.* (2024) describe an improved method for the collection of *C. lactis* eggs, based on the use of particular mesh and other tools.

Symptomatology and damage

The dried fruit mite is one of the common storage mites with a worldwide distribution. It is considered one of the serious and economic pests of some stored products, and it is also used as an alternative prey for commercially mass rearing of predatory mites (Sarraf *et al.*, 2015).

Vitzthum (1943) reported that *C. lactis* is found almost exclusively on food substances containing sugars from which the bacterial activity produces the lower fatty acid, such as lactic, acetic and succinic acids. The mite causes serious losses and is known as the most important pest species of dried apricots (Güldali and Çobanoğlu, 2010); when it feeds on the fruit, sugar of the dried apricot accelerates the development of mould fungi and causes unwanted flavours and odours, due to its digestive products, secretions and dead bodies. These contaminants cause skin diseases and some disorders of the digestive system in humans (Turgu and Kumral, 2019). Although drying the fruits by reducing the water content prevents the development of microbial factors, the increase of sugar crystals allows the reproduction of the dried fruit mite. Infection and feeding of the mites promote the development of some microbial factors on dried fruits and cause some disorders in the digestive system and allergic reactions in humans (Yildirim Kumral *et al.* 2019). Taboada (1954) reported a pulmonary acariasis in Spain attributed to a species of *Carpoglyphus*. Moreover, Davies (1926) observed *C. lactis* to develop in scabs on a domestic rabbit.

In China, Shen (1979) said that *C. lactis* infests stored sugar, dried fruits and sweet chow chow. It causes severe damage to storage sweetment and acariasis in human. So, it has importance in the sugar storage, dried fruits and in the food hygiene. Its population can easily increase at appropriate conditions. The mite can not only damage and contaminate the food and medicine containing sweet but also make people ill with skin acariasis, lung acariasis, duodenal ulcer, intestines acariasis and so on. The occurrence and damage of *C. lactis* were expatiated and the control measures of this mite were discussed accordingly (Wang *et al.*, 2008).

Hubert *et al.* (2011) reported, in Central Europe, the massive and frequent infestation of *C. lactis* on dried fruit, imported from the Mediterranean region. In 180 samples taken from supermarkets, 13% were contaminated, with the range of contamination level from 0–660 mites per gram of dried fruit. The contamination was found in dried apricots, figs, plums and raisins. To estimate the risks and food preferences of *C. lactis*, its growth rate was examined under laboratory conditions. Starting with a hypothetical population of ten mites per gram of dried fruit, the risk level of 1000 mites per g of dried fruit is reached at 42 days for dried figs, 49 days for dried pineapple and 63 days for dried apricots, dates and plums at 25°C and 85% RH. The mites were able to enter every dried fruit packing material tested, including polypropylene and aluminium foils. This shows that mites can move from package to package in supermarkets. Mites are known as allergen producers and vectors of mycotoxin-producing fungi. These findings indicate that an increased risk of *C. lactis* contamination exists in dried fruit (Hubert *et al.*, 2011).

Dried fruits are among the traditional agricultural products of Turkey, which leads the field both in terms of quality and international trade. Approximately 250,000 tonnes of dried raisins and 50,000 tonnes of dried figs are exported from Turkey annually, which represents some 30–35% and 50–55% of the total world trade, respectively (Emekci *et al.*, 2007). Faunistic research carried out on stored apricots in this country highlighted that *C. lactis* was the most abundant species on dried apricots, with an infestation rate of 69.14% (Çobanoğlu, 2008). Ždárková (1967) reported the mite on fruit drink residue, confection of senna, wine where it clings to pieces of cork, rotting potatoes, cocoa beans and groundnuts.

Marin *et al.* (2009) studied in Spain the contribution of naturally occurring mite populations of *C. lactis* on the aroma of fino wines by sensorial and chemical composition analysis. The diffusion of compounds produced by *C. lactis* during adverse external conditions is possibly achieved due to mites falling inside the oak casks. Fino wines from oak casks colonized by *C. lactis* exhibited a more 'floral' and 'citric' character – such as lemon-like – than wines without mites. These attributes are related to citronellol and neral, respectively. Neral is segregated as an alarm pheromone by this strain of mite. Significant differences were also observed in terms of limpidity, which may be dependent on sample pH. In terms of gustative profile, wines colonized with mites showed higher 'floral and refreshing sensations', and were more bitter (Marin *et al.*, 2009).

Klimov *et al.* (2016) reported that *C. lactis* is a relatively common species in beehives, occurring in the debris on the bottom boards of beehives, honeycombs, dead bees, honey and especially on the bee bread (bee pollen with added honey and bee secretions that is stored in brood cells). The mite can penetrate the brood cells and make burrows in the stored pollen, consuming it and causing the pollen and the debris to spill from the cells. The infested bee bread, mixed with large quantities of dead and live mites, turns to a golden-brown or yellow powdery material covering honeycombs and bottom boards of beehives. The mite damage is especially severe in stored overwintering nests. For example, 250 honeycombs were destroyed by these mites over one winter in a single storage area in Germany (Zander, 1947). A similar phenomenon was observed in Alabama (USA), where stored honeycombs were found heavily infested after winter storage (Baker and Delfinado, 1978). Weak bee colonies are more susceptible to mite attacks (Zander, 1947), while healthy bee colonies usually can clean up the infested pollen (Baker and Delfinado, 1978).

Aside from beehives, *C. lactis* is also found in old honeycombs, wine barrels in cellars, dried sweet fruits, canned fruits, fruits preserved in sugar, fermenting pulp, dairy products (milk and cheese) and stored honey. It often infests products following substantial development of yeast. In the field, this species is found in fermenting tree sap flows, burrows of moles and as phoretic deutonymphs on butterflies, moths and scarabaeid beetles (*Gnorimus* sp.) (Klimov *et al.*, 2016).

The ingestion of large numbers of *C. lactis* with infested food or beverages can cause dysentery. These mites also cause dermatitis to handlers of infested materials, such as dried plums (O'Donovan, 1922), and occupational allergies in the biological pesticide industry (Krop *et al.*, 2012). Li *et al.* (2003) investigated the epidemiology and pathogenic mite species of intestinal and urinary acariasis in individuals with different occupations. Among the 1994 individuals examined 161 carried mites, with 4.61% positive only for stool samples, 1.86% positive only for urine samples and 1.60% for both. The positive rate of mites in stool samples was 6.22%, being 6.84% for males and 5.39% for females. No gender difference was observed (chi (2) = 1.77, $P>0.05$). The mites from stool samples included *A. siro*, *T. putrescentiae*, *Dermatophagoides farinae* Hughes, *Dermatophagoides pteronyssinus* (Trouessart), *G. domesticus*, *Glycyphagus ornatus* (Kramer), *C. lactis* and *Tarsonemus granarius* Lindquist. The positive rate of mites in urine samples was 3.46%. The positive rates for male and female subjects were found to be 3.95% and 2.81%, respectively, with no gender difference observed (chi (2) = 1.89, $P>0.05$). Mite species in urine samples included *A. siro*, *T. putrescentiae*, *Tyrophagus longior* (Gervais), *Aleuroglyphus ovatus* (Troupeau), *Caloglyphus berlesei* (Michael),

Caloglyphus mycophagus (Mégnin), *Suidasia nesbitti* Hughes, *Lardoglyphus konoi* (Sasa *et* Asanuma), *G. domesticus*, *C. lactis*, *Lepidoglyphus destructor* (Schrank), *D. farinae*, *D. pteronyssinus*, *Euroglyphus maynei* (Cooreman), *Caloglyphus hughesi* Samšiňak, *T. granarius* and *Trichomonas hominis* Dahal. The species of mites in stool and urine samples were consistent with those separated from working environment. A significant difference was found among the frequencies of mite infection in individuals with different occupations (chi (2) = 82.55, $P<0.001$), with its frequencies in those working in medicinal herb storehouses, those in rice storehouse or mills, miners, railway workers, pupils and teachers being 15.89%, 12.96%, 3.28%, 2.54%, 5.10% and 2.56%, respectively. The prevalence of human intestinal and urinary acariasis was not associated with gender, and these diseases are more frequently found in individuals working in medicinal herb, rice storehouses or mills and other sites with high density of mites. More attention should be paid to the mite prevention and labour protection for these high-risk groups (Li *et al.*, 2003).

Control

Control of this species is based on the use of physical, chemical and biological means.

PHYSICAL CONTROL. The effects of physical factors (low barometric pressure, high and low temperature, light) on survival of stored-food mites (*A. siro*, *T. putrescentiae*, *A. ovatus*, *C. berlesei*, *G. domesticus*, *C. lactis*, *D. pteronyssinus*, *C. eruditus*) have been studied. Among these, *C. lactis* is the most resistant species, and most specimens (85%) of this species survived the longest exposure (96 h) to the lowest pressure (95 mm Hg) tested. It showed 100% mortality only after 80 h exposure to −15°C, and it was able to withstand 45°C for about 1 h. Mites from the family Acaridae were killed by low pressure of 95 mm Hg after an exposure of only 48 h and after 1 h exposure to −15°C. Constant light has unfavourable effects on development and reproduction of the flour mite, *A. siro*. Our results may have some practical implications. Vacuum of 190 mm Hg will protect the food against the mites. Also, low temperature −15°C could be used to control mites in seed (Ždárková and Voracek, 1993).

BIOLOGICAL CONTROL. Studies carried out on the mite fauna of stored apricots in Turkey allowed to evaluate the possibility of biological control of stored-product mites, by two common predacious species, the Ascid mites *B. tarsalis* (Berlese) and *B. mali* (Oudemans). *B. tarsalis* was the most common and populated beneficial mites with 62.30% ratio among the beneficial mites (Çobanoğlu, 2008). In order to develop a mass production of the phytoseeid mites *A. swirskii* (Athias-Henriot), *N. cucumeris* (Oudemans) and *Amblyseius eharai* Amitai *et* Swirski, the life history parameters, the net reproductive rates (R_0/female) and the intrinsic rates of increase (*rm*/day) of the three species were studied under laboratory conditions at 25 ± 1°C, 90 ± 5% RH and L:D = 14:10 h, with *C. lactis* reared on dry yeast. All predatory species developed normally and reproduced with *C. lactis*. The mean development period of *A. swirskii* females was longest and was different from the other two species. The mean oviposition period, and the fecundity of the three predators did not differ. The mean longevities of *A. swirskii* and *A. eharai* were shorter than that of *N. cucumeris*. The total fecundity among the three predators was not significantly different. Although no differences in R_0 were observed among the three predators, the *rm* of *A. swirskii* was lower than that of *A. eharai* because of its long mean generation time (*T*). The *rm* of *N. cucumeris* was the lowest among the three predators because it had the longest *T*. According to these findings *C. lactis* resulted a suitable extra-guild prey species for the three predators, in comparison with previous reports concerning other food kinds such pollen, insects and other mites. Populations of *C. lactis* increased easily with dry yeast and were very suitable for the predator development (Ji *et al.*, 2015).

CHEMICAL CONTROL. This strategy provides the use of traditional acaricides, and of various fumigants. Fumigants such as methyl bromide and phosphine are used for the control of this harmful species all over the world. However, methyl bromide has been restricted in many countries because the chemical depletes ozone layer and causes acute deaths in humans (Turgu and Kumral, 2019). To remove this limit, Emekci *et al.* (2007) evaluated the effectiveness of high CO_2 applications in flexible PVC envelopes (Grainpro Cocoons), against dried fruit pests under field conditions as an alternative to MeBr. The CO_2 applications were performed in units of 36 m^3 capacity over 3–5 days of exposure. The test insects contained all life stages of different ages of *Cadra cautella* (Walker) (= *Ephestia cautella*), *Plodia interpunctella* (Hübner), (Lepidoptera, Pyralidae), *Oryzaephilus surinamensis* (Linnaeus), (Coleoptera, Silvanidae), *Carpophilus hemipterus* (Linnaeus) (Coleoptera, Nitidulidae) and the mite *C. lactis*. Temperature, RH and CO_2 and O_2 levels were continuously monitored during treatment. The results showed that oxygen concentration in the envelopes decreased below 1%, and CO_2 concentrations reached above 95%. Test insects were completely killed after 5 days of exposure. The cost of CO_2 application is similar to that of MeBr, namely: approximately US$ 1.7–3.5/ton dried fruit. Results of this study showed that dried fruits can effectively be disinfested with high CO_2 applications in flexible units without the use of MeBr (Emekci *et al.*, 2007).

Ferizli *et al.* (2007) evaluated in the Aegean region of Turkey the use of phosphine (FUMI-CELTM, a magnesium phosphide formulation) to control dried fig pests, in stacks under tarpaulin covers. Three stacks built-up with perforated plastic boxes were prepared for about 50 tonnes of dried figs. The stacks were covered with tarps, which were sealed with sandsnakes to the concrete floor. The stacks were dosed at 1, 2 and 3 g PH3/tonne, and the fumigations continued for 7 days at prevailing temperatures during the dried fig harvesting season. Trials were also repeated in the cold season. The fumigations were conducted at the dried fig processing and storage facility of TARIS to test their effectiveness against *C. cautella* (eggs, larvae, and pupae), *C. lactis* (mixed stages) and *O. surinamensis* (eggs, larvae, pupae and adults). Test insects were retrieved from the stacks, on a daily basis after 3 days of fumigations. During the trials, gas concentrations were monitored by an analyser equipped with an electrochemical detector from different locations within the stacks and temperatures were recorded inside the stacks, using T-type thermocouples (Hobo®). All insect bioassays were retained for 2 weeks after removal from the stacks. Complete insect mortality of the test insects and mites was achieved after 3 days of exposure. The trials demonstrated that phosphine holds promise as a replacement for methyl bromide fumigations against the dried fruit pests (Ferizli *et al.*, 2007).

Taking into account that reducing the amounts of both oxygen and humidity negatively affects the population development of *C. lactis*, Turgu and Kumral (2019) studied the effects of different physical agents, namely ferric oxide and ozone gas (oxygen depleters), and calcium chloride and silica gel (humidity reducers), to packaged dried apricot infested with *C. lactis*. The lethal dose concentrations and lethal times of each product were determined by probit analysis. The female mites were successfully controlled with all of the physical agents. A death rate of the females of 99% was most rapidly obtained with ozone gas treatment, followed by ferric oxide. Ozone gas was particularly toxic, causing a significantly high level of mortality after application at 44.4 mg/l (LT_{99}= 39 h). When ferric oxide was applied at a dose of 9000 mg/l volume and higher doses, it killed all the female mites within 3 days. Above the dose of calcium chloride at 3000 mg/l killed all of the females within 81 h. Lastly, the silica gel affected mites over a longer period — 2 weeks and at a higher dosage (56,000 mg/l volume) (Turgu and Kumral, 2019).

The effects of ozone gas as an oxidative agent, and short-term oxygen scavenger, and ferric oxide as an oxygen and moisture absorber against *C. lactis* populations have been shown previously. Unlike previous studies, in this study, the effect on microbial load caused by infestation, and the effect of these applications on the fruit quality parameters and sensory properties were also revealed. According to the results of this study, applications of 44 mg/l of ozone gas, and 35,000 mg/l of ferric oxide, completely suppressed mite populations, but no effect on the number of total mesophilic aerobic bacteria, yeasts and moulds were detected. However, ozone application prevented the discolouration of the dried apricots, caused by the mite infestation. While mite infestation significantly decreased the firmness of the apricots, both ozone and ferric oxide treatments positively affected and improved these parameters. Additionally, no negative effects of ozone and ferric oxide treatments on sensory characteristics of fruits were detected (Yildirim Kumral *et al.*, 2019).

In laboratory tests, the toxicity of acaricides targeted against house-dust mites was tested on five species of stored-product mites, such as *A. siro* (Linnaeus), *A. ovatus* (Troupeau), *C. lactis*, *L. destructor* (Schrank) and *Tyroborus lini* Oudemans. The formulations of benzyl-benzoate, benzyl-benzoate/permethrin/pyriproxyfen and neem were diluted in water and applied to filter paper in an unventilated chamber. The mortality of mites was observed after 24 h of exposure to acaricide-impregnated filter paper. All of the tested acaricides were toxic to all of the mite species. There were significant differences in mortality among the species and the acaricides. Benzyl-benzoate/permethrin/pyriproxyfen was the most effective, followed by benzyl-benzoate and neem. *L. destructor* (LD_{50} 0.01–0.11 μg) was the most sensitive mite species, followed by *A. siro* (LD_{50} 0.04–0.12 μg), *T. lini* (LD_{50} 2–21 μg), *A. ovatus* (LD_{50} 3–18 μg) and *C. lactis* (LD_{50} 4–64 μg). Based on the highly toxic effects of the tested acaricides against the stored-product mites, the acaricides should be considered as a potential tool in the control of stored-product mites, although next screening is necessary (Stará *et al.*, 2011b).

30 The Family Chortoglyphidae Berlese

Morphological Characteristics, Systematic and Bio-ecology

Chortoglyphidae are free-living mites, often found in food stores and granaries, and in the dust of houses (Hughes, 1976). Species of the genera *Chortoglyphus* Berlese and *Aplodontopus* Fain are associated with nests of sciurid, aplodontid and murid rodents in North and Central America, and with sciurid and murid rodents in Australasia, and a mouse lemur in Madagascar (O'Connor, 2009b). The idiosoma has a firm, oval outline and is not divided into propodosoma and hysterosoma. No dorsal propodosomal shield is present. The cuticle is smooth and shining, the body setae are short and nearly all are smooth. The claw is inserted at the end of a fleshy pretarsus. Only one σ is present on genu I. The genital opening of the female is covered by two sclerotized plates, situated between the coxae III and IV, whose posterior margins form a smooth bow-shaped curve. The aedeagus of the male is long and situated between coxae I and II; tarsal and conspicuous anal suckers are present (Hughes, 1976). The adults have highly reduced coxal apodemes III and IV, and the males are the only glycyphagoids to possess the paranal suckers. Some males (not all) have modified sucker-like or spine-like setae on tarsus IV. The deutonymphs of all genera are endofollicular parasites and possess modified, comb-like setae *v*' on tibiae III and IV that engage the lining of the host hair follicle to provide anchorage. The dorsal setae are mostly spine-like, bifurcate apically, anteriorly directed, which also serves to maintain the mite's position in the follicle. Chortoglyphids retain all coxal setae, and setae p_1 and p_2 of the attachment organ take the form of tiny spines at the opisthosomatic terminus. Growth of some deutonymphs is extreme, with body length increasing 10-fold during the engorgement (OConnor, 2009b). The deutonymphs of *Aplodontopus marmotae* Krantz *et al.* cause severe dermatitis in their marmot host (Krantz *et al.*, 2003). *Chortoglyphus arcuatus* (Troupeau) is associated with humans and usually infests stored grain and flour in the world (Hughes, 1976), and may cause severe allergy in humans (Puerta *et al.*, 1993; Muesken *et al.*, 2004; López-Conde *et al.*, 2005; Boquete *et al.*, 2006; Volgel *et al.*, 2015).

The Genus *Chortoglyphus* Berlese

Diagnostic characteristics

The chelicerae are chelatae and conspicuous. The idiosoma is oval and not divided into propodosoma and hysterosoma, with cuticle smooth and shining, and without striations and microtrichae. A discrete propodosomal sclerite is absent. The body setae are short, and nearly all smooth. The genital papillae are normal in form, strongly enlarged and reduced. On tarsus I the solenidion ω_1 is formed as a long curved rod, arising close to a small solenidion ω_2; the seta *wa* of tarsus I is enlarged to form a stout spine, and *ba* as a thin seta; the genu I possesses only one solenidion *б*; the tibiae I–II with dorsal solenidion, and 0–2 ventral setae; the legs are ventrally positioned; the praetarsi are fleshy with inserted claws; the empodial claws is simple; the anus is positioned near the posterior margin of body; both sexes with empodial claws simple or absent; the ventral subcapitulum without external ridges; discrete coxal apodemes III and sometimes IV absent.

The female has the vulva (oviporus) longitudinal with genital valves fused anteriorly to body, free posteriorly, and situated between the coxae III and IV; the genital papillae are always associated with the genital opening; the anal region bears five pairs of setae. The male has a long aedeagus, set between the coxae I and II; the paranal suckers are conspicuous, and often with sucker-like setae on tarsus IV; two pairs of setae in anal region; tarsi IV with tarsal suckers, and leg III similar to leg IV (Solarz, 2011).

DOI: 10.1079/9781786390912.0030

Economic importance

It is a cosmopolitan genus, of which the most important species is *C. arcuatus*, living on various stored food, mainly on wheat, wheat flour (Zachvatkin, 1941; Stejskal and Hubert, 2009; Li *et al.*, 2015a), rice (Attiah, 1969; Saleh *et al.*, 1985), maize (Saleh *et al.*, 1985) and cereals (Ždárková, 1967). It has been collected also in traps, hen nests, feathers, and abandoned bird nests and house dust in Brazil (Moreira, 1978; Flechtmann, 1986; da Silva *et al.*, 2013), in barn dust (Boström *et al.*, 1997), in barns, pigsties, cowsheds, chaff-cutter, buildings, in poultry houses (Solarz and Pająk, 2019a) and in stored poultry mixtures (Ždárková, 1967). Moreover, the mite is reported from sparrow nests (Wasylik, 1959), associated with *Gohieria fusca* (Oudemans) and *Acarus siro* Linnaeus (Bollaerts and Breny, 1951), from the litter of broiler houses (Brady, 1970), in large numbers in the floor dust of a house and in the surface dust of a mattress located in a farmhouse (Hughes, 1976).

Chortoglyphus arcuatus (Troupeau) (Fig. 30.1)

Common name

Unknown

Diagnostic characteristics

FEMALE. The idiosoma is 350–400 µm long. The chelicerae are chelate and conspicuous. The ventral subcapitulum lacks of external ridges. Ventrally, the apodemes I unite to form a short sternum. Apodemes II are long and set transversally across the body, parallel to an elongate sclerite lying between coxae II and III. Apodemes III and IV are not developed. The coxal apodemes III are discrete, and sometimes IV lack or are absent. The idiosoma is oval, and not divided into propodosoma and hysterosoma. A discrete propodosomal sclerite lacks. The body cuticle is smooth and shining, without striations and microtrichae. The body setae are short, and nearly all smooth. Arrangement and length of dorsal setae as the male. The genital papillae are normal in form, strongly enlarged and reduced. Legs I and II are shorter than those of the male, while leg IV are longer, tarsus IV being considerably longer than the two preceding segments taken together. The solenidion ω_1 on the tarsus I has as a long, curved rod arising close to a small solenidion ω_2; the seta *wa* of tarsus I is enlarged to form a stout spine and *ba* as a thin seta; the genu I only with one solenidion *б*; the tibiae I–II with dorsal solenidion, and 0–2 ventral setae; the legs are ventrally positioned; the praetarsi are fleshy and with inserted claws; the empodial claws are simple. The anus is set near the posterior margin of the body. The vulva (oviporus) is longitudinal, and with the genital valves fused to body anteriorly, free posteriorly, and situated between the coxae III and IV; the genital folds are broad plates, with strongly sclerotized and curved posterior borders; the genital papillae are always associated with the genital opening; the anal region possesses five pairs of setae. The bursa copulatrix has a small round opening on the dorsal side of the posterior end of the body (Hughes, 1976; Solarz, 2011).

MALE. The idiosoma is 250–300 µm long, pink in colour and with appendages rather dark in colour. The chelicerae are chelate and conspicuous. The ventral subcapitulum is without external ridges. The idiosoma is oval, and not divided into propodosoma and hysterosoma. The body cuticle is smooth and shining, without striations and microtrichae. The propodosomal sclerite is absent. The body setae are short and nearly all smooth; the setae vi project forward over the gnathosoma, ve are slightly longer and clearly pectinate, set at the same level as vi. The scapular setae are arranged in a transverse row, and are almost equidistant from one another. The supracoxal seta is a thin pectinate rod. The four pairs of dorsal setae are almost in linear series with one another, and only two pairs of laterals are present. The genital opening is between the coxae I and II, so that apodemes I and II are forced apart, and form part of the transparent genital folds; the genital papillae normal in form, strongly enlarged and reduced. No sternum is present, and discrete coxal apodemes III and sometimes IV are absent. The anal slit is some distance from the posterior edge of the body, and is flanked by a pair of conspicuous anal suckers. Each consists of a circular sclerotized base to which is attached, a membranous funnel. A pair of preanal setae lies in front of the suckers, and a pair of postanals is behind. The legs are ventrally positioned, long and slender and ending in a prominent pretarsus, to

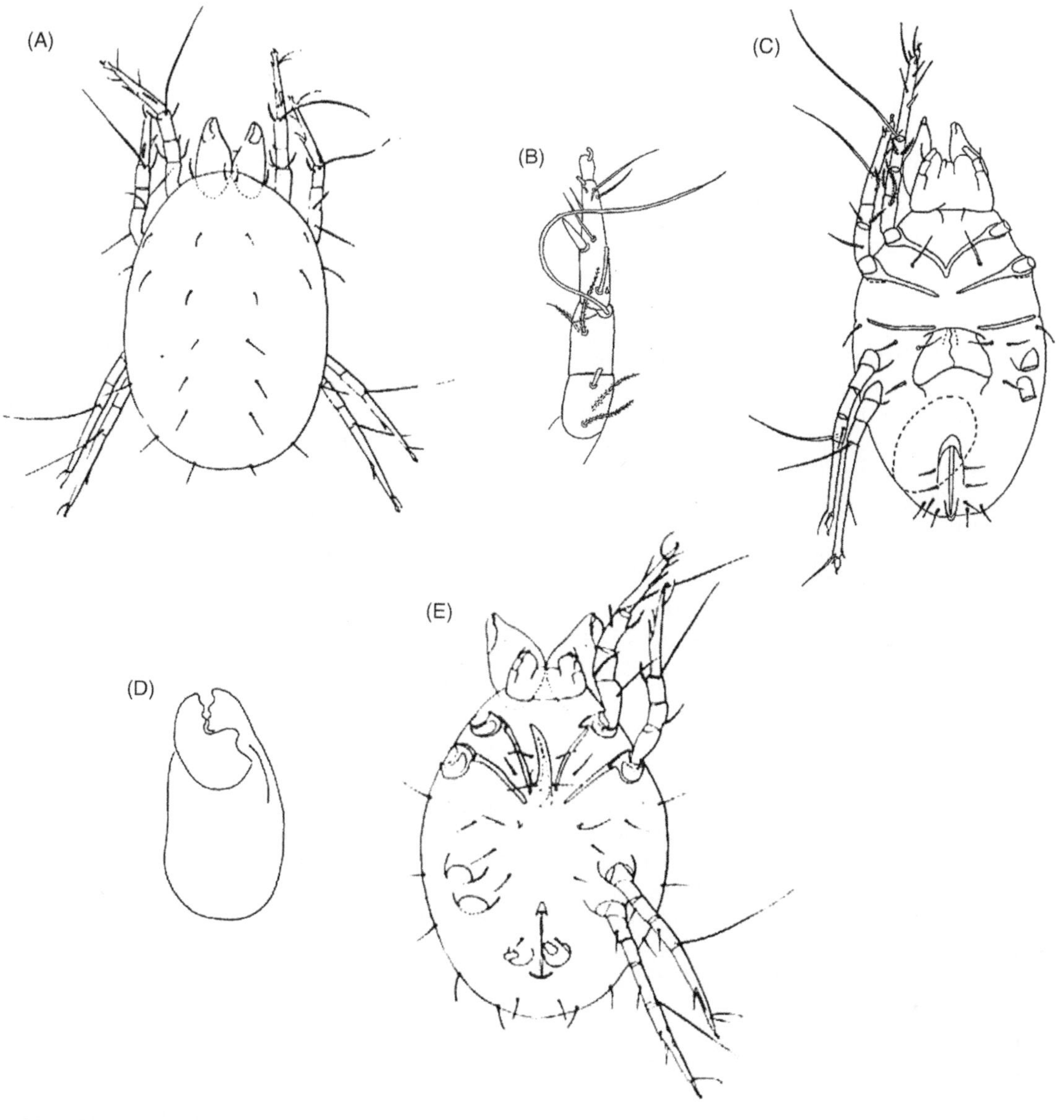

Fig. 30.1. *Chortoglyphus arcuatus* (Troupeau). Female. Dorsal view (A); leg I (B); ventral view (C); chelicera (D). Male. Dorsal view (E) (from Hughes, 1976).

which is attached a small claw. On tarsus I ω_1 is a long, curved rod arising close to a small ω_2, the seta *wa* is enlarged to form a stout spine and *ba* as a thin seta. Only one solenidion σ is on genu I; tibiae I–II with dorsal solenidion, and 0–2 ventral setae. The anus positioned near posterior margin of body; both sexes with empodial claws simple or absent. The aedeagus is long, situated between the coxae I and II; the adanal suckers are conspicuous. The tarsal suckers are present on tarsi IV, and leg III are similar to leg IV (Hughes, 1976; Solarz, 2011).

Geographical distribution

Cosmopolitan, reported from various countries, such as Australia (Halliday, 1998), Barbados (Hughes, 1976), Belgium (Hughes, 1976), Brazil (Moreira, 1978; Flechtmann, 1986; Oliveira and

Daemon, 2003; da Silva *et al.*, 2013; Dutra *et al.*, 2018), Canada (Beaulieu and Knee, 2014), China (Xin, 1965; Li *et al.*, 2015a), Croatia (Pagliarini, 1979), Czech Republic (Ždárková, 1967; Hughes, 1976; Hubert *et al.*, 2006a; Stejskal and Hubert, 2009), Egypt (Attiah, 1969; Saleh *et al.*, 1985; Heikal, 2015), England (Hughes, 1976), France (Hughes, 1976), Germany (Hughes, 1976; Franz *et al.*, 1997; Muesken *et al.*, 2004), Greece (Palyvos *et al.*, 2008), Hong Kong (Olsen, 1983), Iran (Esmaili, 1983; Mirfakhraii, 1994), Italy (Goracci *et al.*, 1985), Malaysia (Mariana *et al.*, 2000), Malta (Pfliegler *et al.*, 2017), the Netherlands (Hughes, 1976; Siepel *et al.*, 2016), New Zealand (Robertson, 1946; Cornere, 1971; Hughes, 1976; Pike and Wickens, 2008), Norway (Mehl, 1998), Poland (Wasylik, 1959; Hughes, 1976; Solarz *et al.*, 1997, 2004; Kosik-Bogacka *et al.*, 2010; Solarz and Pająk, 2019a), Russia (Zachvatkin, 1941, 1959; Hughes, 1976), Spain (López-Conde *et al.*, 2005), Sweden (Boström *et al.*, 1997), Taiwan (Tseng and Chang, 1973; Olsen, 1983), Turkey (Gülkan 2004; Akyazi *et al.*, 2019), U.A.R. (Hughes, 1976), Ukraine (Dudinska, 2003; Oksentiuk *et al.*, 2022) and other countries.

Commodities, facilities and vegetal and animal hosts

The list of commodities of this mite includes barley (Stejskal and Hubert, 2009), cereals (Ždárková, 1967), clover seed (red) (Robertson, 1946), coffee grains and rice straw (Flechtmann, 1986), ginkgo nut (Li *et al.*, 2015a), grass seed (Zachvatkin, 1941), maize (Saleh *et al.*, 1985), malt (Li *et al.*, 2015a), mushroom (Li *et al.*, 2015a), oats (Zachvatkin, 1941), rice, rice (husked) (Attiah, 1969; Saleh *et al.*, 1985), rye (Zachvatkin, 1941), wheat, wheat flour and wheat flour sweeping (Zachvatkin, 1941; Stejskal and Hubert, 2009; Li *et al.*, 2015a). Other commodities reported by Hagstrum *et al.* (2013) are barley, barley sweeping, bean curd, Chinese medicine, dried fruit, dried vegetable, grain, hay, herb, lentil, millet, poppy seed, potato flour, poultry mix, residue, shrimp (dried), soya sweeping, squid (dried), sugarbeet seed, sweeping and sweet potato chip (dried).

Common facilities are barn, commercial grain storage, farm, farm granary, farm milling machine, flour mill, granary, off-farm storage, provender mill, provender warehouse, railcar, silo and stable (Solarz *et al.*, 1997, 2004; Palyvos *et al.*, 2008; Pike and Wickens, 2008; Stejskal and Hubert, 2009; Hagstrum *et al.*, 2013; Pfliegler *et al.*, 2017; Oksentiuk *et al.*, 2022).

Moreover, the mite has been collected in traps, hen nests, feathers and abandoned bird nests (da Silva *et al.*, 2013), in barn dust (Boström *et al.*, 1997), in barns, pigsties, cowsheds, chaff-cutter, buildings, poultry house (Solarz and Pająk, 2019a), stored poultry mixtures (Ždárková, 1967) and house dust (Moreira, 1978). Wasylik (1959) reported the mite from sparrow nests, Bollaerts and Breny (1951) always associated with *G. fusca* (Oudemans) and *A. siro* Linnaeus, Brady (1970) from the litter of broiler houses, and Hughes (1976) in large numbers in the floor dust of a house and in the surface dust of a mattress located in a farmhouse.

Bio-ecology

C. arcuatus is a cosmopolitan mite, living on various stored food, mainly on wheat, wheat flour (Zachvatkin, 1941; Stejskal and Hubert, 2009; Li *et al.*, 2015a), rice (Attiah, 1969; Saleh *et al.*, 1985), maize (Saleh *et al.*, 1985) and cereals (Ždárková, 1967). It has been collected also in trap, Hen's nest, feather and abandoned bird nests in Brazil (da Silva *et al.*, 2013), in barn dust (Boström *et al.*, 1997), in barns, pigsties, cowsheds, chaff-cutter, buildings, in poultry houses (Solarz and Pająk, 2019a) and in stored poultry mixtures (Ždárková, 1967). Moreover, the mite is reported from sparrow nests (Wasylik, 1959), associated with *G. fusca* and *A. siro* (Bollaerts and Breny, 1951), from the litter of broiler houses (Brady, 1970), in large numbers in the floor dust of a house and in the surface dust of a mattress located in a farmhouse (Hughes, 1976).

Its biology has not been studied in detail. The life cycle develops through the egg, larva, protonymph, tritonymph and adult stages. The deutonymph stage is unknown. Some information on the mite biology may be found in the work of Kosik-Bogacka *et al.* (2012), which studied the acarofauna composition in 33 homes in North-western Poland. They measured once a month over 2 years, the temperatures and the RH, and calculated the mean temperature and RH. The highest temperature was recorded from July to September (24.5°C), and the lowest from April to June (19°C). The highest RH was observed from October to December (90%), and the lowest from April to June (68.0%). Number of specimens, the total number of mites collected and coefficients of frequency (F) and dominance (D), for mites isolated from dust in sleeping places, allowed to evaluate the spread of populations of *C. arcuatus*, which occurred in January to March, and in October to December (Kosik-Bogacka *et al.*, 2012).

Studies carried out in Colombia on the prevalence of specific IgE to *Aleuroglyphus ovatus* (Troupeau) and *C. arcuatus* in 77 individuals with allergic asthma and/or chronic allergic rhinitis showed that the sensitization to both mites is common in individuals with respiratory allergies and suggests that they have unique and common allergenic determinants (Puerta *et al.*, 1993).

Skin tests effected in Northern Spain with the mite extracts of different mites, including *C. arcuatus*, showed that the prevalence of positive skin tests to *C. arcuatus was* 58% (two exclusively sensitized to this species). There was no significant correlation between the sensitization to *C. arcuatus* and to the other mites. There was a significant correlation between the number of mites, to which the patients were sensitized, and the time of evolution of the disease ($P < 0.05$). The results of the study highlighted that there is a high prevalence of sensitization to *Glycyphagus domesticus* (De Geer) *and C. arcuatus* in Northern Spain, and in areas where these species are endemic. Sensitization to these species should be considered where these species have been described (López-Conde *et al.*, 2005). Other studies carried out on this topic confirmed the high prevalence of sensitization to *C. arcuatus and that the* sensitization to this species should be considered of clinical significance. There is minimal cross-reactivity between *C. arcuatus* and *Dermatophagoides pteronyssinus* (Trouessart) (Boquete *et al.*, 2006). Muesken *et al.* (2004) demonstrated that *C. arcuatus* is a mite species of allergenic relevance. The allergen pattern of the mite possibly indicates that it possesses characteristic allergenic properties. It is possible that there are allergenic proteins in mite extracts in a molecular weight range of about 52 kDa. Using the patient's sera, allergens of *C. arcuatus* could be found particularly in a high molecular weight range in contrast to the allergen pattern of other domestic mites.

Chemical, instrumental analysis and bioassays showed that the chortolure is an aggregation pheromone for both sexes of this species. Several related compounds are released in smaller amounts by the mites. The alarm pheromones of these mites (neral and geranial) can only be found in total extracts of the mites, in which eight occurs only in minute amounts. The method of sampling is therefore crucial for pheromone identification. Feeding experiments with deuterated propionate showed that chortolure is a polyketide, formed by successive addition of four propionate units to an acetate starter (Schulz *et al.*, 2004).

Control

As for the acarid mites.

31 The Family Aëroglyphidae Zachvatkin

Morphological Characteristics, Systematic and Bio-ecology

The opisthosoma has a row of small, closely spaced microtrichia, extending between the bases of most lateral body setae; most dorsal setae are elongate and heavily barbed; the pretarsal ambulacra are short, relatively simple; the empodial claw is reduced or absent. Live in bat roosts, vertebrate nests, stored products and house dust. The family includes few genera and about 19 species, associated with guano, collected from birds, mammal and bee nests or peridomestic habitats (O'Connor, 2009b). Species of the genus *Aëroglyphus* Zachvatkin have Holartic and Afrotropical distribution, and those included in the genus *Glycycometus* Pillai (= *Austroglycyphagus* Fain and Lowry 1974) are cosmopolitan. The species belonging to this last genus have highly attenuated tarsi, and the ventral setae *pv*' are basally positioned and elongate and wrapping around the length of the tarsus. The species *Aëroglyphus robustus* (Banks) is common in stored food products, where it displays a strong ability to tolerate a relatively low RH that allows to it to survive when other injurious mite cannot (Sinha, 1966b). Species of the genus *Glycycometus* are known inhabiting house dust in tropical countries and may cause allergic reaction in humans (Miranda *et al.*, 2002), and sometimes are collected on stored food, such as the species *Glycycometus kualalumpurensis* (Fain *et* Nadchatram) reported on bean pollard and groundnut, and on house dust in Malaysia (Fain and Nadchatram, 1980a) or *Glycycometus geniculatus* (Vitzthum), reported on Chinese medicine, grain, grain bin residue from Brazil, China, East Africa, England, Germany and Zaire. According to their economic importance, we treat here only *A. robustus* and *G. geniculatus*.

Selected website

https://idtools.org/bee_mite/index.cfm?packageID=1&entityID=75

The Genus *Aëroglyphus* Zachvatkin

Diagnostic characteristics

The prodorsum possesses external vertical setae *ve*. The ventral subcapitulum lacks external ridges. The opisthosoma has a row of small, closely spaced microtrichiae (small hair- or tooth-like processes) extending between the bases of most lateral body setae. The cuticle sometimes possesses larger irregular protuberances. Most dorsal setae, including c_1, are elongate and heavily barbed. The pretarsal ambulacra are short, and relatively simple. The empodial claws are present, claw-like and simple. The anus is located near the posterior margin of the body. The tarsi III–IV lack an elongate, scale-like basal seta. The male without a sclerotized plate posterior to aedeagus. The setae ps_3 are set in a soft cuticle (Klimov *et al.*, 2016).

Economic importance

This genus numbers two species, *Aëroglyphus peregrinans* (Berlese) and *A. robustus*, easily identified with the descriptions of Cooreman (1959). A. robustus is known from Nearctic, where is common, and Palaearctic region. *A. peregrinans* is reported from South East Palaearctic region, Europe and North Africa. Moreover, an undescribed species from an *Apis mellifera* Linnaeus colony in South Africa has been collected by Klimov (Klimov *et al.*, 2016). From the economic point of view, both species are probably neutral; generally, they prefer feeding on dead bees. *A. robustus* exhibits a facultative behaviour, while *A. peregrinans* has a permanent relationship

DOI: 10.1079/9781786390912.0031

with bees. As concerns the host associations, feeding and dispersal all stages of *A. robustus* live in peridomestic habitats (stored food), but it can invade hives of honeybees, where it has a preference towards feeding on dead bees. This mite does not have a phoretic deutonymphs and feeding stages disperse by active movements or on host; the phoresy on a wasp has been observed. *A. robustus* is considered a pest of stored grain, though it is probably a habitat generalist, preferring dry materials. It occurs in stored wheat, oats, barley, stored grain, soybean, hay, nests of birds, beeswax culture of moth *Galleria mellonella* (Linnaeus) (Lepidoptera, Pyralidae), on dead insects and in beehives of *A. mellifera* inside queen cells containing a dead queen (Delfinado-Baker and Baker, 1982). Conversely, *A. peregrinans* is a bee specialist, and all stages presumably live in bee nests. Also, in this species, the phoretic deutonymphs lacks. Feeding stages disperse most frequently on large carpenter bees, bumble bees, or honeybees. In contrast, *A. peregrinans* is a bee specialist and has never been found in other habitats. It disperses as feeding stages on large carpenter bees, bumble bees and honeybees and probably lives in nests of these hosts but documented only for the honeybee *A. mellifera* (Klimov *et al.*, 2016). A few literature records suggest that in beehives, both mite species prefer feeding on dead bees (Grobov, 1978; Delfinado-Baker and Baker, 1982), but the cause of death is unknown for these bees. Blattný (1924) suggested that *A. peregrinans* can switch from being saprophagous on dead bees to parasitic and, thus, irritate bees (Klimov *et al.*, 2016).

Aëroglyphus robustus (Banks) (Fig. 31.1)

Common name

Warty grain mite

Morphological characteristics

FEMALE. The idiosoma is 320 µm long, and oval shaped. No sejugal furrow separates the propodosoma from the hysterosoma. Two lateral shields on the propodosoma, sinuous, extend from the setae *vi* to *sci*; the distal end of these shields is lateral to the setae *vi*. The setae *sci* is shorter than *sce*, relatively close to each other. The opisthosoma bears a row of small, closely spaced microtrichiae (small hair- or tooth-like processes), extending between the bases of most lateral body setae and turning around the setae e_2 and h_1. The cuticle possesses larger irregular protuberances, more or less conical and with dorsal protuberances. Most dorsal setae, including c_1, are elongate and heavily barbed. The epimeres I medially join and form a V. The coxae I and II lack setae. The genital opening is set between the coxae II–IV. The genital opening is set on the posterior part of idiosoma, with anal and adanal setae long. The bursa copulatrix rises in the posterior side of the hysterosoma, with the shape of a triangular appendix. On leg I, the solenidion σ_1 of genu is straight and long, σ_2 short and curved, on tibia the solenidion φ is long more than 80 µm, and on tarsus σ_1 is long 30 µm and parallel to dorsal surface of tarsus, and its distal end passes the seta *aa*. The pretarsal ambulacra are short and relatively simple (Cooreman, 1959).

MALE. The idiosoma is 240 µm long. Two lateral shields on the propodosoma, sinuous, extend from the level of setae *vi* to setae *sci*; the distal end of these shields extend laterally to the setae *vi*. The setae *sci* are shorter than *sce*, relatively close to each other. The opisthosoma bears a row of small, closely spaced microtrichiae (small hair- or tooth-like processes) extending between the bases of most lateral body setae. The cuticle possesses larger irregular protuberances, more or less conical and with dorsal protuberances. All dorsal setae are long and heavily barbed. The epimeres I join medially and form a little sternum, the epimeres II are relatively short, as well as the epimeres III and the IV are shorter and vestigial. The coxal setae of coxae III and IV are long and smooth. The genital opening is located between the coxae II and IV, with the genital setae set in the anterior part and the genital papillae clearly visible. The anal opening is set posteriorly (Cooreman, 1959).

HYPOPUS. Klimov *et al.* (2016) reported that in this species phoretic deutonymphs are absent, while Stingeni *et al.* (2015) identified the so-called hypopial deutonymphs of *A. robustus*; they are typical forms of resistance to untowards environmental conditions. The writers have not seen this stage.

Geographical distribution

Canada (Banks, 1906; Sinha, 1963; Delfinado-Baker and Baker, 1982), Italy (Stingeni *et al.*, 2015), Ohio (Banks, 1906) and United States (Cooper, in Cooreman, 1959).

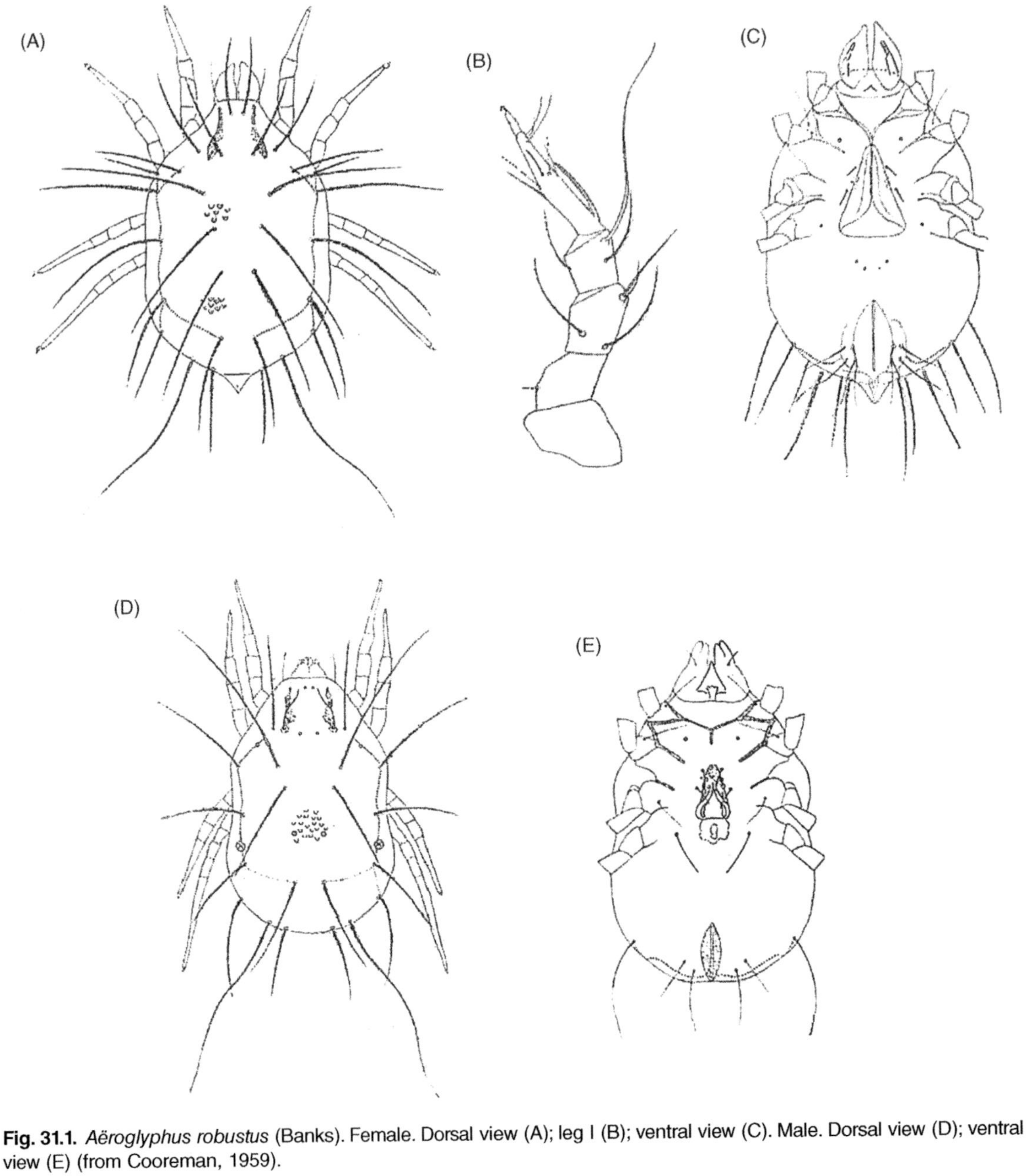

Fig. 31.1. *Aëroglyphus robustus* (Banks). Female. Dorsal view (A); leg I (B); ventral view (C). Male. Dorsal view (D); ventral view (E) (from Cooreman, 1959).

Commodities, facilities, vegetal and animal hosts

Grain, wheat, oats, barley, stored grain, soybean, hay, nests of birds, beeswax culture of moth *G. mellonella* (Linnaeus) (Lepidoptera, Pyralidae), on dead insects and in beehives of *A. mellifera* Linnaeus inside queen cells containing a dead queen (Sinha, 1963, 1966b; Delfinado-Baker and Baker, 1982), on the Hymenopter Euménid *Ancistrocerus antilope* (Panzer) (Cooper, in Cooreman, 1959).

Bio-ecology

A. robustus is reported from North America and Italy, where it has been collected on grain, wheat, oats, barley, stored grain, soybean and hay. Moreover, it has been found in nests of birds, beeswax culture of moth *G. mellonella*, on dead insects and in beehives of *A. mellifera* inside queen cells containing a dead queen (Sinha, 1966b; Delfinado-Baker and Baker, 1982) and *A. antilope* (Cooper, in Cooreman, 1959). The mite is retained, responsible of human allergy (Stingeni *et al.*, 2015).

Sinha (1966b) observed that under laboratory conditions on nine fungus-free diets at 26°C and 75% RH for 2 months, the mite multiplied 88 times on rolled oats and yeast (95:5), 50 times on wheat germ and yeast (95:5), five times on germ-damaged wheat kernels, but did not survive on undamaged wheat kernels. The species is cold-hardy, and can overwinter in an empty granary as hypopus at –39°C. A natural infestation by *A. robustus* developed in a 500-bushel wheat bulk at temperatures from –10°C to 29°C, and a moisture from 12.2% to 15%, during the period February–August. Heaviest infestation occurred during late summer and autumn. It is a potentially serious pest of stored grain in Canada and northern United States (Sinha, 1966b). Studies carried out by Barker (1967a) on the biology and reproductive rates of *A. robustus* at two values of RH and different temperatures showed that the life cycle is shorter at 70–75% RH than at 85–90% RH. At 28°C, more eggs are laid at 85–90% RH than at 70–75% RH. Moreover, the same author found low net reproduction rates for the species.

Moisture content-relative humidity (MC-RH) adsorption and desorption curves were determined for triticale (x – *Triticosecale* Wittmack), cv. welsh, at 22°C and 35–100% RH. The effects of various seed moisture contents on free fatty acid level, microfloral growth and germination loss were determined after 5 weeks of storage, at various RH and 22°C. Infection of seeds by *Penicillium* and *Aspergillus glaucus* (Linnaeus) group occurred above 80–85% RH, and seed germination correspondingly began to decrease. Five common stored-product mites which infested triticale stored at 20°C and 85% RH were *Acarus siro* Linnaeus, *Acarus farris* (Oudemans), *A. robustus*, *Lepidoglyphus destructor* (Schrank) and *Tyrophagus putrescentiae* (Schrank). *A. farris* and *L. destructor* also multiplied on wheat cv. Neepawa which was an inferior substrate to triticale for *A. farris* (Sinha and White, 1982).

House-dust mites and stored-product mites of various species inhabit human homes and stored agricultural products. They are carriers and hosts of microorganisms that enable their survival. In mite species with low bacterial diversity, the bacterial communities are dominated by potential symbiotic or parasitic bacteria—*Cardinium* in *Dermatophagoides farinae* (Hughes) and *A. robustus* and the enteric bacteria *Erwinia* in *Blomia tropicalis* Van Bronswijk, de Cock *et* Oshima and *Xenorhabdus* in *Tyroborus lini* (Oudemans). Among the bacterial species identified, *Staphylococcus*, *Bacillus*, *Kocuria*, *Brevibacterium*, *Corynebacterium* and *Brachybacterium* likely serve as food sources for the mites. The domestic acaridid mites carry high numbers of various bacteria that are potential threats to human health. These results contribute to the general understanding of the ecology of mite adaptation to human-made habitats (Hubert *et al.*, 2016).

As concerns, the natural enemies (Barker, 1968a) studied the life history of *Androlaelaps casalis* (Berlese) at 25.6°C, 26.7°C, 30.0°C, 32.8°C and 36.0°C. Each *A. casalis* female consumed 2.1 ± 0.6 female *Glycyphagus domesticus* (De Geer) per day. Males consumed 0.85 ± 0.2 female of *G. domesticus* per day. An average of 27.4 ± 4.7 *G. domesticus* protonymphs were consumed during development to adulthood of immature *A. casalis* males. Other foods included brewer's yeast, *A. siro*, *T. putrescentiae*, *A. robustus*, the predator *Blattisocius keegani* (Fox) and the young larvae of *Tribolium confusum* (Duval) (Coleoptera, Tenebrionidae).

Symptomatology and damage

White *et al.* (1979) examined stored wheat infested by *L. destructor*, *A. farris* and *A. robustus*, in order to determine the effects of mites on the quality of wheat. *L. destructor* infestation of the grain led to significantly lower fat acidity values ($P<0.01$), and significantly slower loss of seed germination ($P<0.05$), than in the non-infested wheat. They found that seed moisture content, infection of fungi (*Aspergillus candidus* Link and *Alternaria alternatae* (Fr.) Keissler) and bacteria and the electrical conductivity of wheat-steeped water were unaffected by the presence of the mites.

Tridecane, not previously reported in infested bin-stored grain, was associated with *A. siro*, *A. robustus* and *L. destructor*, which were introduced into 15.2% moisture content wheat stored in two unheated experimental bins. The mites and volatiles were sampled at weekly to bimonthly intervals throughout the year. The mites produced tridecane all year, although they overwintered at low numbers. Ventilation of one bin at an airflow rate of 1 l/s/m^3 did not appear to affect mite survival and volatile production. Citral, an alarm pheromone for mites, was detected in mite-infested bin-stored grain. In laboratory studies, all three mite species produced tridecane in wheat incubated at 20°C and 70% RH for 2 weeks. *A. siro* and *A. robustus* reared at 25°C and 90% RH for 4 weeks on wheat and fungus-inoculated wheat also produced tridecane. Production of tridecane by *A. robustus* and *L. destructor* is being reported for the first time (Tuma *et al.*, 1990). Stingeni *et al.* (2015) described an indoor dermatitis in human due to *A. robustus*.

Control

The control strategy includes the employ of physical and chemical means.

PHYSICAL CONTROL. White *et al.* (1997b) evaluated the movement of stored grain by auger, or pneumatic conveyor for control of stored-product insects and mites. In a pilot scale test using nine welded steel bins, each holding 322 kg of wheat (*Triticum aestivum* Linnaeus), movement of grain using an auger in three bins caused 89% mortality of red flour beetle, *Tribolium castaneum* (Herbst) (Coleoptera, Tenebrionidae) adults and 94% mortality of rusty grain beetle, *Cryptolestes ferrugineus* (Stephens) (Coleoptera, Laemophloeidae) adults. Larval mortality of both species, based on grain samples placed in Berlese funnels, was 100%. Pneumatic conveyance of grain in three bins caused 100% mortality of all stages, of both insects, based on grain samples and grain-probe insect traps. In field tests, infested wooden bins of wheat and maize (*Zea mays*) Linnaeus were moved by pneumatic conveyor to a truck, and then hand-poured into welded steel bins. A test carried out by White *et al.* (1997) on wheat and maize, using a screw auger to move the grain showed that, pneumatic wheat movement killed 100% of *A. robustus*. Augering wheat killed 98% of mites, 84% *T. castaneum* adults, 70% *T. castaneum* larvae, 14% *C. ferrugineus* adults and 0% *C. ferrugineus* larvae. Pneumatic movement of maize killed 97% *T. castaneum* adults, 72% *C. ferrugineus* adults and 100% of the fungus beetle *Cartodere constricta* (Gyllenhal) (Coleoptera, Lastridiidae). Augering maize had less effects than pneumatic movement on insect mortality. Pneumatic conveyance of grain, especially wheat, offers an effective physical control for common stored-product insects.

CHEMICAL CONTROL. Chemical control is based on the use of various chemicals, among which the fumigants play a key role (see acarid mites). In the past, a strong interest has been reserved for various traditional synthetic molecules, including the insecticide chlorpyrifos-methyl. White and Sinha (1990) reported that this a.i. was toxic to certain mite species when applied at 3.8–7.7 ppm on hulled oats (*Avena sativa* L.) or 3.1–5.0 ppm on hulless oats in a farm granary. Insecticide residues decreased by about 65% in hulled oats and 50% in hulless oats in 1 year, and by 80% in hulled oats and 60% in hulless oats in 2 years. Bioassay of treated oats with *T. castaneum* adults for 24 h indicated 100% mortality above 4.7 ppm, and negligible mortality below 2 ppm of insecticide. Unfortunately, chlorpyrifos-methyl virtually eliminated the predatory mites *A. casalis* (Berlese) and *B. keegani* Fox in treated, compared with untreated, hulled and hulless oats. Populations of another predatory mite, *Cheyletus eruditus* (Schrank) were depressed, but began to recover when the insecticide decreased below 0.6 ppm on hulled oats. In both types of oats, populations of prey mites *Tarsonemus granarius* (Lindquist) and *A. siro* were unaffected by the insecticide. Other prey mite populations included *A. robustus*, which were moderately reduced, and *L. destructor* and *Paratriophtydeus coineaui* André, which were sharply reduced. The insecticide treatment had no effect on seed germination, or microfloral infection, although these variables changed during 2 years. The lack of impact of the insecticide on microflora, seed germination and several mite species and the rate of insecticide degradation indicated that the original ecosystems could largely reestablish themselves in 2 years under western Canadian storage conditions, although predator diversity was reduced (White and Sinha, 1990).

White and Jayas (1991) evaluated the action of carbon dioxide at concentrations of 15–50% by volume, for its ability to control insect and mite populations in non-airtight bins of wheat stored at 12–15°C for 84 days. Unsealed bins and sealed, untreated bins served as controls. Compressed carbon dioxide was introduced into the bottom of the bins on days 0, 7, 14 and 28, and the bins were sealed to allow 5-, 6- or 14-day half-lives of the gas from loss through diffusion. Levels of carbon dioxide were generally higher at the bottom locations in the bins (range of maximum, 30–50%) than at the top (range of maximum, 15–32%), where the greatest losses occurred, and oxygen levels never fell below 11% at any sample location. Overall, the levels of carbon dioxide in the bins stayed higher in bins that were more airtight. Populations of the insects *C. ferrugineus, T. castaneum* and *Ahasverus advena* (Waltl) (Coleoptera, Silvanidae) and the mites *T. granarius* Lindquist, *P. coineaui* André, *L. destructor* and *A. robustus* were sharply reduced by day 28 in bins, having the two highest carbon dioxide levels. Pests were virtually absent by day 42 in the most airtight bins. There were generally no differences in arthropod numbers in bins with 6- or 14-day half-lives for carbon dioxide loss. Seed germination and fungal infection by the *A. glaucus* group were not affected by the carbon dioxide. It is concluded that frequent additions of carbon dioxide can compensate for the loss of gas in completely sealed bins, and provide effective control of arthropod pests (White and Jayas, 1991). These last authors some years later referred (White and Jayas, 1993) that the carbon dioxide can be used as an effective stored-grain fumigant in relatively airtight bins. They added carbon dioxide to wheat (*T. aestivum*) as a compressed gas, and to barley (*Hordeum vulgare* Linnaeus) as a solid (dry ice) in 322 kg grain bulks. Wheat was stored at temperature decreasing from 18–10°C, over a 12-week period. Bins were left open, sealed without CO_2 added, or with CO_2 supplemented at 25%, 34% and 46% levels. Barley was stored at temperature decreasing from 25 to 20°C, over an 8-week period. Bins were left open, sealed without CO_2 added, or with CO_2 treatments of 23%, 29% and 34%. The wheat and barley moisture content were 14.5–16.3% and 14.5–16.1%, respectively. Oxygen levels in the wheat reflected air displacement with CO_2, but lower O_2 levels in the barley reflected a combination of air displacement by CO_2, and consumption of O_2 by respiring grain and microorganisms at warmer temperatures. The two insects *C. ferrugineus* and *T. castaneum* were controlled in 2 weeks, at 34% CO_2 and 15% O_2 at temperature decreasing from 18–10°C, or 29% CO_2 and 3% O_2, at temperature decreasing from 25–20°C. The mites *T. granarius, L. destructor* and *A. robustus* were killed in less than 2 weeks at these CO_2 levels. Seed germination and microflora were unaffected by all gaseous environments.

The Genus *Glycycometus* Pillai

Diagnostic characteristics

External vertical setae *ve* present on prodorsum, set anteriorly to the setae *vi*, and crista metopica absent. The ventral subcapitulum lacks external ridges. The opisthosoma has rows of small, closely spaced microtrichiae (small hair- or tooth-like processes) extending between the bases of most lateral body setae. Most dorsal setae are elongate and heavily barbed, except the setae c_1, which are elongate and smooth. The anus is set near the posterior margin of the body. The tarsi I–IV bear an elongate scale-like basal seta. The solenidia σ_1 and σ_2 of genu I are about equal in length. The tibiae I and II bear only one ventral seta. The tibiae are short about half the length of the genua. The pretarsal ambulacra are short, and relatively simple; the empodial claws are reduced or absent. The males possess a sclerotized plate posterior to aedeagus, bearing the setae ps_3 (Hughes, 1976; Klimov *et al.*, 2016).

Economic importance

Mites of this genus are reported from Palaearctic, Afrotropical, Neotropical, Oriental and Australian regions, and are more common in the tropics. Records concerning bees are from the Afrotropical regions (Tanzania and Rwanda) and Oriental regions (India, Thailand and The Philippines) (Klimov *et al.*, 2016). *Glycycometus* species have been collected in bird nests, and bat guano (Fain, 1976), stored food (Hughes, 1976; Hagstrum *et al.*, 2013), house dust (Fain, 1976; Fain and Nadchatram, 1980a), bracket fungi (*Polyporus*) and bee nests (*Apis* spp., and *Xylocopa*). Although the genus *Glycycometus* includes species mainly associated with nests of birds and with bat guano (Fain, 1976, 1978), three species have been collected in house dust in tropical countries; they are *G. lukoschusi* Fain from Surinam (Fain, 1976), *Glycycometus malaysiensis* (Fain *et* Nadchatram) and *G. kualalumpurensis* (Fain *et* Nadchatram) from Malaysia (Fain and Nadchatram, 1980a; Ling *et al.*, 2019); the first species has been reported also from Panama (Miranda *et al.*, 2002). The genus has been pretty much neglected from an allergological standpoint, and although there are sporadic records in the literature on dust mite surveys of the tropics, it has probably been confused with the better-known genus *Blomia* Oudemans, and may well be more common and widespread than indicated previously (Colloff, 2009b).

All stages live in various habitats, and only three species have been found associated with bee nests so far and may be bee specialists. The genus does not have phoretic deutonymphs and feeding stages disperse on adult bees and probably by active movements (Klimov *et al.*, 2016). Dispersing mites have been found on Prince Demidoff's bushbaby, *Galagoides demidovii* (Fisher) (Primates, Galagidae), large flying fox, *Pteropus vampyrus* Linnaeus (Chiroptera, Pteropotidae), coffee fruit fly, *Ceratitis* (*Trirhithrum*) *coffeae* Bezzi (Diptera, Tephritidae) and honeybees (*Apis* spp.). Many species found in honeybee nests are generalists and were also reported from other habitats, such as *G. malaysiensis*, *G. molitor* (Volgin *et* Akimov) and *G. thailandicus* (Fain *et* Gerson). However, some species have so far been found only in bee nests where they feed on decomposing organic matter in nest cells, such as *G. geniculatus* (Vitzthum), *G. rwandae* (Fain) and *G. combus* (Chinniah *et* Mohanasundaram). In India, *G. molitor* and *G. thailandicus* have been found in vacated combs, hive debris, brood combs and dead bees in nests of *A. mellifera* and *A. cerana*. *G. molitor* is reported from Ukraine in mills, granaries and warehouses, compound fodder factories, barns with animal fodder storage, outbuildings, honeybee beehives and vegetable storages (Oksentiuk *et al.*, 2022). Phoresy on worker bees has been detected in all three species (Sumangala and Haq, 2001). *G. malaysiensis* and *G. thailandicus* have been identified as saprophagous in beehives (Malabanan and Corpuz-Raros, 1998). *G. combus* was found in stored, damaged comb of *A. florae,* where it lives on the comb material (Chinniah and Mohanasundaram, 1996). On stored food is known *G. kualalumpurensis* reported on bean pollard and groundnut from the Philippines, and *G. geniculatus*, found on Chinese medicine, grain, grain bin residue in farm and market from Brazil, China, East Africa, England, Germany and Zaire. We treat here only *G. geniculatus*.

Glycycometus geniculatus (Vitzthum) (Fig. 31.2)

Common name

Unknown

Morphological characteristics

FEMALE. The idiosoma is 430–500 µm long. General aspect as the male. The genital opening set well in front of the posterior genital setae. The bursa copulatrix is a short wide tube (Hughes, 1976).

MALE. The idiosoma is about 430 µm long. The general aspect as *G. domesticus* (De Geer), but the cuticle is more finely grained, and around the base of the setae *vi*, the cuticle is smooth and forms a dorsal propodosomal shield. The setae *ve* are anterior to *vi*, and enfold the sides of the gnathosoma. All idiosomal setae, with exception of c_1 (that are smooth) are finely pectinate. The setae d_1 and e_1 are about the same length and in linear series with one another. The opisthonotal glands are large and in life coloured blood-red, and red in all developmental stages. All tibiae are less than half the length of the corresponding genua. The tarsi are enveloped in a pectinate subtarsal scale, as in the genus *Aëroglyphus*. On tarsus I, ω_1 is long and curved, and is closely applied to the surface of the segment. The setae *ba*, *la* and *ra* arise from the basal half of the segment, *la* being a long pectinate seta, which reaches almost the base of the pretarsus. On tibia I, φ is exceptionally long and curled into a loose spiral; on tibia II this solenidion is slightly shorter and straight; on tibiae III and IV, φ is less than one-half the length of the corresponding tarsus. Tibiae I and II lack seta *ht*. Genu I has the solenidia σ_1 and σ_2 that are subequal in length (Hughes, 1976).

Geographical distribution

Probably cosmopolitan, reported from Brazil (Sousa *et al.*, 2005), China (Li *et al.*, 2005; Li *et al.*, 2015a), East Africa (Vitzthum, 1919), England (Woodroffe, 1954; Hughes, 1976), India (Chaudhury *et al.*, 2005), Taiwan (Wu, 1999), Zaire (Cooreman, 1942) and other countries.

Commodities, facilities, vegetal and animal hosts

Maize, mildewed fried dough twist, starch, mildewed rapeseed, groundnut cake, wheat koji, haw jelly, *Zizyphus jujuba* Mill. (Li *et al.*, 2015a), Chinese medicine (Li *et al.*, 2005), grain, grain bin residue (Woodroffe, 1954) in farm and market. Moreover, the mite has been collected in bird nests in England (Woodroffe, 1954), in nest of the bee *Xylocopa* (*Koptorthosoma*) *nigrita* (Fabricius) in East Africa

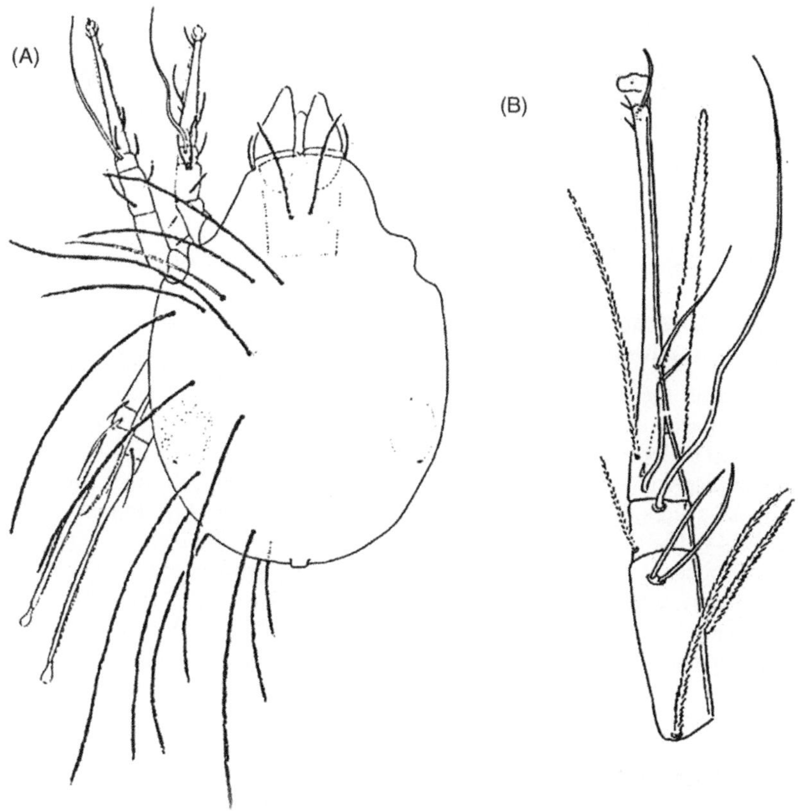

Fig. 31.2. *Glycycometus geniculatus* (Vitzthum). Female. Dorsal view (A); leg I (B) (from Hughes, 1976, licensed under the Open Government Licence v3.0).

(Vitzthum, 1919) and on the fly *Ceratitis* (*Trirhithrum*) *coffeae* (Bezzi) in Zaire (Cooreman, 1942).

Bio-ecology

No data are available on the bio-ecology of this mite.

Symptomatology and damage

No data are available.

Control

No data are available.

32 The Family Echimyopodidae Fain

Morphological Characteristics, Systematic and Bio-ecology

Echimyopodid mites live in the nests of mammals, stored products and house dust in Australian, Nearctic and Neotropical regions. They have a body cuticle covered by small microtrichia. The males do not possess paranal suckers or sucker-like setae modified on the tarsus IV. The tibio-tarsal articulation may bear a basal sucker.

The family numbers five genera and 31 species (O'Connor, 2009b; Schatz *et al.*, 2011). Species injurious to stored food belong to the cosmopolitan genus *Blomia* Oudemans.

The Genus *Blomia* Oudemans

Diagnostic characteristics

The body almost globular in shape, broad across the scapular region and tapering towards its posterior end; no suture between propodosoma and hysterosoma; cuticle rough; the body dorsal setae include two pairs of vertical, two pairs of scapular and 11 pairs of dorsolateral setae (five pairs of dorsal, five pairs of lateral and one pair humeral), all long and pectinated, except the setae c_1 (d_2); six pairs of anal setae in the female; no dorsal shield or crista metopica; *ve* and *vi* inserted near the anterior margin of the propodosoma; pseudostigmatic organ shaped like a pectinate rod; male and female genital openings between coxae III and IV, in the male the opening is largely concealed by two folds, in the female these folds are reduced; epigynum and epiandrum absent; female with copulatory tube; no anal or tarsal suckers in the male; legs long and slender with small pretarsi, and no claws visible; one short solenidion on genu I and II; in the male, the tarsal tibial joint of leg IV has a large arthrodial membrane on the posterior-dorsal surface, so that this joint can bend sharply to the anterior-ventral side (Zachvatkin, 1936; Bronswijk *et al.*, 1973a; Hughes, 1976).

Economic importance

This genus includes few species, some of which have been collected on stored products and/or recognized as sources of allergens that cause respiratory diseases. *Blomia khalilovae* Zachvatkin is reported on barley and millet from Azerbaijan (Zachvatkin, 1949), *Blomia kulagini* Zachvatkin has been found on wheat in granary and in crevices in school gymnasium floor from Russia and Japan (Zachvatkin, 1936, 1941; Bronswijk *et al.*, 1973), *Blomia tjibodas* (Oudemans) in farm granary, farm milling machine, stable and house dust in Germany and Java (Oudemans, 1910, 1914b; Bronswijk *et al.*, 1973), *Blomia thori* Zachvatkin has been collected on stored flax seed and grain product in Russia (Zachvatkin, 1936, 1941; Kadzhaya, 1970) and *Blomia tropicalis* Bronswijk, Cock *et* Oshima has been collected on stored rice, house dusts and in tobacco factory, and reported from pantropic and subtropic regions, including Brazil, India, Indonesia, Japan, Myanmar, Panama, Philippines, Taiwan and Venezuela (Bronswijk *et al.*, 1973; Miranda *et al.*, 2002; Siegert *et al.*, 2018). This mite is commonly found in stored food of animal and plant origin, with high levels of moisture (Flechtmann, 1986). In the tropics, it can be found in house dust with considerable population densities (Ezequiel *et al.*, 2008), and it is one of the main causes of rhinitis and allergy in people, who are in contact with house dust (Colloff, 2009a). The most important species associated with stored products is *Blomia freemani* Hughes, reported from Brazil, China, Egypt, England, Greece, Ireland and Taiwan on Chinese medicine, millet, rice, sweet potato chip (dried) and wheat in commercial grain storage, dock silo, flour mill, market and warehouse (Bronswijk *et al.*, 1973; Hughes, 1948, 1976). Fain *et al.* (1977) *Blomia gracilipes* (Banks), in tobacco infested with the Cigarette Beetle from the United States. Some species have been recognized as sources of allergens that cause respiratory diseases,

DOI: 10.1079/9781786390912.0032

such as *B. tropicalis* (Bronswijk, Cock *et* Oshima) (Arruda and Chapman, 1992; Stanaland *et al.*, 1998; Miranda *et al.*, 2002; Santos da Silva *et al.*, 2017).

Blomia freemani Hughes (Fig. 32.1)

Common name

Unknown

Morphological characteristics

FEMALE. The idiosoma is from 440–520 μm long. Similar to male, with exception for its large size. The genital opening is covered by obliquely placed folds, and on the underside bears two pairs of genital papillae; three pairs of genital setae project from the cuticle on either side of it. The anal opening is set near the posterior margin of the body, and is associated with six pairs of anal setae, and of these two pairs lie on either side of its anterior end, and four pairs at the posterior; the two external posterior pairs are longer than the others and distinctly pectinate. The bursa copulatrix is a long thin tube with a cleft end (Hughes, 1976).

MALE. The idiosoma is 320–350 μm long. The body is globular in shape, widest between the second and the third pairs of legs, and distinctly tapering towards the posterior end. The cuticle is colourless and rough, covered with innumerable fine projections. The dorsal propodosomal shield and the crista metopica are absent. The chelicerae are large and well-sclerotized, with two teeth on the movable digits and, two large and two small on the fixed. The ventral apodemes are slender sclerites, placed obliquely, and those of leg I, only just meeting in the midline. The vertical setae are set close together and extend forwards almost to the tips of the chelicerae, with *vi* posterior to *ve*. The supracoxal seta is a richly branched structure. The setae *sci*, *sce* and *hi* arise almost in the same transverse line; he and d1 also originate in line with one another, and are about the same length. The setae d_2 are close together, arising equidistant from d_1 and d_3, they are much shorter than any of the other dorsal setae and are only spacely pectinate; the setae d_3, d_4, l_1 to l_3, sai and sae are longer setae, the posterior ones exceeding the idiosoma in length. The genital opening is set between coxae IV, and has obliquely genital folds, with internal genital papillae. Three pairs of genital setae, with the third pair close together behind the opening. The aedeagus consists of a curved tube with two basal sclerites. The anal opening extends to the posterior margin of the body; two pairs of smooth anal setae are set on either side of its anterior and posterior end, and a long pair of pectinate postanal setae projects from the hind end of the body. The tarsi are long and slender, exceeding the combined length of the two preceding segments; each terminates in a small lobed pretarsus. The claws lack. Solenidion ω_3 of tarsus I is a slightly curved blunt rod not exceeding the pretarsus in length; the setae *d*, *e* and *f* are short, and three small spines are set on the ventral side of the distal end of the segment. The seta *ba* is nearer to the apex than to the base of the tarsus and arises at the same level as *la* and *ra*, all, together with *wa* are pectinate. The solenidion ω_1 is a rod with a slightly expanded end, whose base is on the same level as a short ω_2. The famulus is not visible. On tarsus II, the solenidion ω_1 is shorter and the base of the setae *ba* is nearer to it. The ventral setae of genua and tibiae I and II are all pectinate. The solenidion φ is very long and slender on all the legs, extending beyond the end of the pretarsus, and on the leg IV, it arises from the middle of the tibiae. Only one solenidion arises from the genu I and none from genua II and III. Tarsus IV is narrow and hinged at an angle to the apex of the tibia by a large arthrodial membrane (Hughes, 1976).

Geographical distribution

China (Xin, 1965; Fan and Cui, 2009; Li *et al.*, 2015a), Egypt (Rezk *et al.*, 1996; Rezk, 2000; Mahgoob *et al.*, 2006), England (Hughes, 1976), India (Gupta and Chatterjee, 2004), Ireland (O'Farrell and Butler, 1948; Hughes, 1976), Taiwan (Wen *et al.*, 2005) and other countries.

Commodities, facilities and vegetal and animal hosts

Flour, broken rice, maize germ, rice contaminated with dust, flour contaminated with dust, fine dried noodles, sweet potato, old silk noodles, cottonseed kernel, cottonseed cake, dried bean curd, beancurd skin, cotton lint, carpet dust (Li *et al.*, 2015a), mixed flour, grain residues (Mahgoob *et al.*, 2006), stored grain (Gupta and Chatierjee, 2004), tradi-

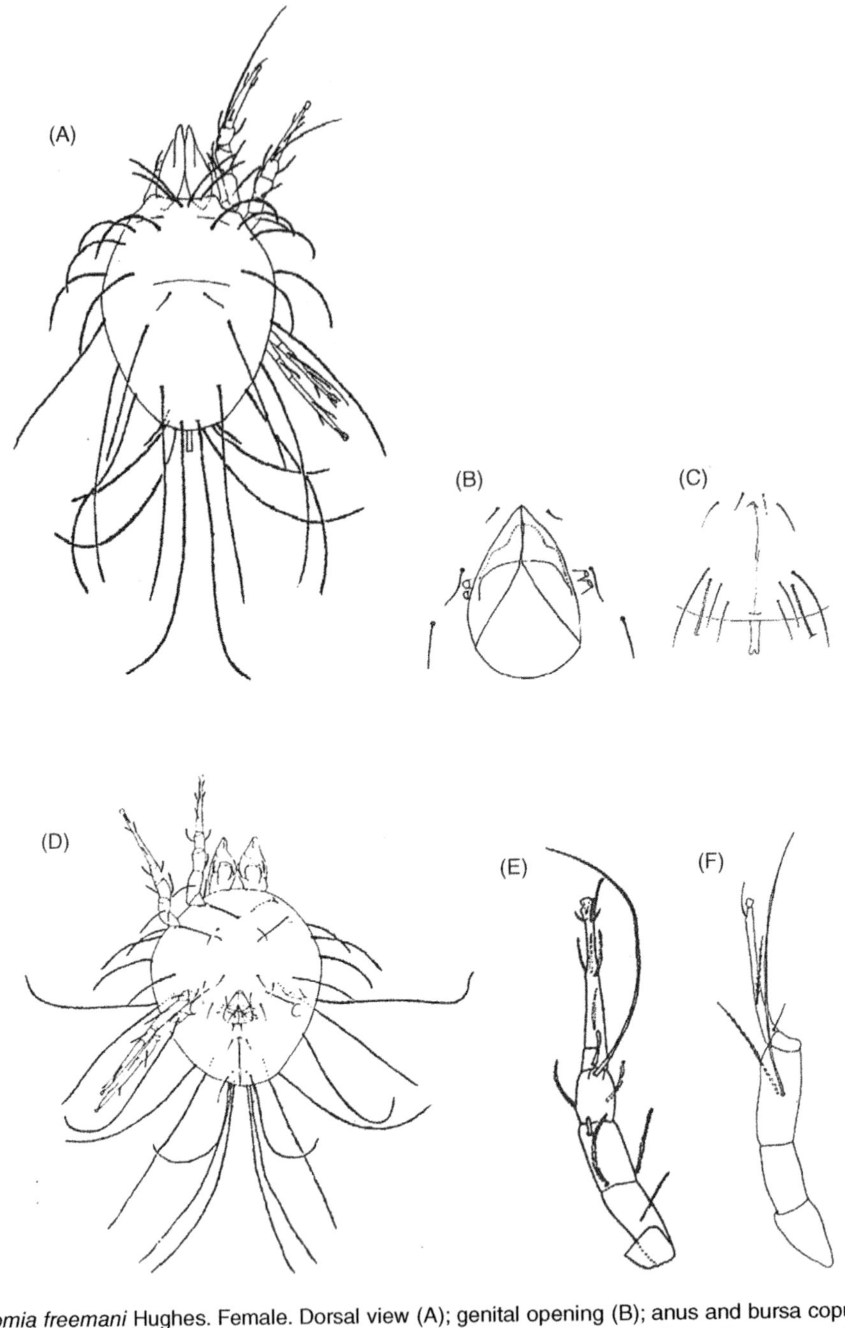

Fig. 32.1. *Blomia freemani* Hughes. Female. Dorsal view (A); genital opening (B); anus and bursa copulatrix (C). Male. Dorsal view (D); leg I (E); leg IV (F) (from Hughes, 1976, licensed under the Open Government Licence v3.0).

tional Chinese medicinal materials (Fan and Cui, 2009), dried dates (Rezk, 2000) and pancake (Wen *et al.*, 2005). Moreover, it has been found also in house dust (Rezk *et al.*, 1996).

Bio-ecology

B. freemani has a large geographical distribution, and has been collected on a number of stored products. The mite is retained also responsible for allergenic diseases. Supposedly, its life cycle is similar to that of *B. tropicalis* that develops through the egg, larva, protonymph, tritonymph and adult stages (Mariana *et al.*, 1996) and the hypopus stage is unknown (Hughes, 1976). The mite has been collected in large numbers in a dock silo on Manitoba wheat, in association with *Acarus siro* Linnaeus and *Lepidoglyphus destructor* (Schrank), while small numbers were collected in bagged Manitoba wheat in a warehouse, and among split wheat in three sections of one large flour mill, where a small permanent population seemed to be established (O'Farrell and Butler, 1948). On stored dried dates in Egypt, the most dominant species were *Tyrophagus putrescentiae* (Schrank), *B. freemani*, *Blattisocius keegani* Fox, *Carpoglyphus lactis* (Linnaeus) and *Cheyletus malaccensis* Oudemans with averages 30.2±2.3, 21.7 ±2.1, 18.5±1.8, 18.3±1.9 and 15.7±2.2 mites/kg dates, respectively (Rezk, 2000).

The effect of the predatory mite, *B. keegani* as biological control agent on two date mites, *T. putrescentiae* and *B. freemani* was examined. The data showed that the daily mean numbers of consumed mites by *B. keegani* were 2.65±0.18 and 1.83±0.06. The total mean numbers of consumed mites were 40.7±1.68 and 35.9±1.1 mites/female when fed on two mite species associated with dried dates, *B. freemani* and *T. putrescentiae*, respectively. The total mean longevity values were 15.3±0.83 and 19.7±0.81 days/female when fed on the two date mites, *B. freemani* and *T. putrescentiae*, respectively. Statistical analysis showed a significant difference between means (Rezk, 2000).

Symptomatology and damage

Symptomatology and direct damage of this mite on stored food are fundamentally unknown. However, it is reasonable to believe that they are similar to those of other stored mites. The most serious aspect associated with the presence of mite on stored foodstuffs is probably that of allergies.

So, systemic anaphylaxis after the ingestion of pancake contaminated with the storage mite *B. freemani* was observed by Wen *et al.* (2005). Unfortunately, Santos da Silva *et al.* (2017) reported that studies to identify marker allergens for *Blomia* and *Dermatophagoides* sensitization are lacking, and sensitization to *Blomia* species has not been fully explored. A few studies recently provided data on sensitization to *B. tjibodas*, *B. kulagini* and *B. freemani* (Garcia-Robaina *et al.*, 1997; Müsken *et al.*, 2000; Wen *et al.*, 2005; Valdivieso *et al.*, 2006; Sade *et al.*, 2010). However, data on cross-reactivity between *Blomia* species are not generally available. Until now, only one study has shown that extracts of *B. tropicalis* and *B. kulagini* are cross-reactive (Cardona *et al.*, 2005).

Control

As other stored mites.

33 The Family Ctenoglyphidae Zachvatkin

Morphological Characteristics, Systematic and Bio-ecology

The body cuticle is usually rough or covered with fine warts. The idiosomal setae are set in a peripheral position, and are widely pectinate or like leaves, forming a fringe encircling the idiosoma. The tarsi are not slender and often bear a dorsal keel. The tibiae I and II possess only one ventral seta (*gT*). The male lack anal and tarsal suckers; the aedeagus is often very long (Hughes, 1976).

O'Connor (1979) reported that Ctenoglyphid mites (genus *Ctenoglyphus* Berlese) number several species inhabiting hay and straw. Many species are described only from hypopi that are endofollicular parasites of rodents. These hypopi were first included into the genus *Rodentopus* Fain and placed in *Ctenoglyphus* by O'Connor, which adopted the method of rearing of hypopi of *Ctenoglyphus plumiger* (Koch) by Chmielewski (1975b). The species infesting the stored products rarely form the hypopi. Some species of this family have been reported in association with stored products. *Ctenoglyphus canestrinii* (Armanelli) was reported on Chinese medicine, grain, hay, and spillage in farm, oat mill, stable from China, England, Hungary, Ireland, Italy, Russia and Wales (Hughes, 1976; Zakhvatkin, 1959); Solaz *et al.* (1997) found the mite in barns from Poland. *C. palmifer* (Fumouze *et* Robin) has been collected on barley (damp), chaff, spillage in farm, oat mill, stable from England, France, Germany, Ireland and Italy (Zakhvatkin, 1959; Hughes, 1976). *C. plumiger* is known on animal feed, barley (damp), buckwheat, buckwheat product, Chinese medicine, condiment, dried fruit, dried vegetable, fishmeal, grain, grass seed (damp), hay, honeycomb, mushroom, oat (damp), processed food, sweeping, wheat (damp), wheatfeed in commercial grain storage, empty grain bin, farm, mill, provender warehouse from Australia, Canada, China, Czech Republic, England, France, Germany, Hungary, Ireland, Italy, the Netherlands, Poland, Russia, Scotland and Wales (Zakhvatkin, 1959; Hughes, 1976). *C. hughesi* Attiah seems a species of secondary importance and, according to Hughes (1976), should be in the genus *Diamesoglyphus* Zachvatkin; it has been found on wheat flour from Egypt (Attiah, 1969). *Diamesoglyphus intermedius* (Canestrini) is reported from Germany on hay in farm.

The Genus *Ctenoglyphus* Berlese

Diagnostic characteristics

Species with pronounced sexual dimorphism; the male is smaller than female, and has a rounded body and less strongly pronounced setae. The female has a flattened idiosoma and overhangs the gnathosoma. In some species, a transverse sejugal furrow separates the propodosoma from the hysterosoma. The cuticle is rough and in the female is covered with irregular projections. The setae arising from the margin of the idiosoma are bipectinate, sometimes leaf-like in shape. Genu I possesses the solenidia σ_1 and σ_2. The male has the aedeagus exceptionally long (Hughes, 1976).

Economic importance

The most important species of this genus listed among the stored-product mites is *C. plumiger*, collected in haystacks, barns and other farm buildings and infesting hay, grain, chaff, straw and various kinds of fodder and dried plant materials (Vitzthum, 1929; Zachvatkin, 1941; Türk and Türk, 1957; Griffiths, 1960; Fain, 1969a; Hughes, 1976). After other reports (O'Farrell and Butler, 1948), it has been found also on fishmeal. Chmielewski (2000b) collected this species in the nest material of birds (*Passer domesticus* Linnaeus), rodents (*Mus musculus* Linnaeus) and bumble bees (*Bombus lapidarius* Linnaeus) (Chmielewski, 1975b). The mite often has been found in sweepings and litter in sheepfolds, cowsheds and stables (mangers). Moreover, it

 The Handbook of Mites of Stored Products (V. Vacante and C.H.W. Flechtmann)
DOI: 10.1079/9781786390912.0033

was collected in hay sweepings from feeding racks for wild forests animals (e.g. bison) (Chmielewski, 1998) and in buckwheat sweepings (leaves, straw) taken from barns (Chmielewski, 2000b).

Ctenoglyphus plumiger (Koch) (Fig. 33.1)

Common name

Unknown

Morphological characteristics

FEMALE. The idiosoma is 280–300 μm long, almost pentagonal in outline, with the cuticle covered with irregular coarse warts on the dorsal side, and finely grained on the ventral. Ventrally, the apodemes I are joined to form a short sternum, and the apodemes II–IV end freely. The epimeres II are short and united with the apodemes III. The genital opening is large and elongated with a well-developed epigynum and extends as far back as the posterior edge of acetabula III. The bursa copulatrix widens towards its base, and is covered with fine warts. The anal slit extends to the posterior edge of the body; two pairs of anal setae arise on either side of its anterior end. The idiosomal setae are longer than in the male, the peripheral ones having more pronounced straight barbs, which arise at an acute angle to the ventral shaft; in life, these setae are smoothly curved, so that the body appears to be outlined by a feathery calyx. The setae arising from the centre of the dorsal surface (i.e. d_1 to d_4 and *sai*) project almost at right angles to the body and are narrow and thickly pectinate. The legs are more slender than those of the male, and the solenidion φ_1 on tibia I is not so well developed (Hughes, 1976).

MALE. The idiosoma is 190–200 μm long, almost pear-shaped in outline, pinkish-brown in colour and without scapular projections. In some places, the cuticle is smooth and in others covered with fine papillae. The ventral surface is well sclerotized, and the apodemes of legs I–IV are united to enclose a triangular area around the long curved aedeagus that is supported on a triangular base lying between the apodemes IV. All the setae arising from the dorsal side of the body are bipectinate, and inserted deeply into the body; in addition, the main axis of each seta is covered with minute projections. The setae arise from the central region of the idiosoma(i.e. d_3 and d_4 are particularly long and pectinate). The setae d_1 are almost equal in length to d_2. In life, these dorsal setae radiate from the body. The legs are long and thick, the tarsi I and II being distinctly keeled on their dorsal sides. Each terminates in a pretarsus and claw, the pretarsus arising from a ventral indentation at the end of the tarsus. The solenidion ω_1 arises from a grove at the base of the keel and is flanked by ω_2 and a very small famulus; the solenidion ω_3 enfolds the pretarsus and the remaining tarsal setae are short and fine. On tibia I, the solenidion φ is long and thick, and arises from a deep cleft in the side of the tibia; it becomes progressively shorter on tibiae II and III, and is a small rod-shaped structure on tibia IV. On genu, σ_2 is much longer than σ_1, and σ_1 is expanded distally. Only one ventral seta arises from tibiae I and II, but both are present on genu I and II (Hughes, 1976).

HYPOPUS. Chmielewski (1975b) described the hypopus of *C. plumiger*, and it is possible to see the drawings of the hypopus published by the same author afterwards (Chmielewski, 2000). The interested reader may consult the above contributions.

Geographical distribution

Probably cosmopolitan, reported from Australia (Womersley, 1941; Halliday, 1998), Canada (Sinha, 1964b; Beaulieu and Knee, 2014), China (Xin, 1965), Czech Republic (Lukás *et al.*, 2007), Egypt (Rahil, 2005), England (Michael, 1901; Hughes, 1976), France (Hughes, 1976), Hungary (Szalay, 1931), Iceland (Denmark) (Hallas, 1981a), the Netherlands, Northern Ireland (Hughes, 1976), Germany (Koch, 1835), Poland (Chmielewski, 1975a; Solarz *et al.*, 1997), Italy (Berlese, 1884; Canestrini, 1888) and Russia (Zachvatkin, 1941, 1959).

Commodities, facilities, vegetal and animal hosts

The mite infests hay, grain, chaff, straw and various kinds of fodder and dried plant materials, and has been collected in haystacks, barns and other farm buildings (Vitzthum, 1929; Zachvatkin, 1941; Türk and Türk, 1957; Griffiths, 1960; Fain, 1969a; Hughes, 1976; Hallas, 1981a; Lukás *et al.*, 2007; Beaulieu and Knee, 2014). After other reports (O'Farrell and Butler, 1948), it has been found also on fishmeal. Solarz *et al.* (1997) found high mite populations in cowsheds and less in barns. Rahil (2005) found the mite on chamomile plants.

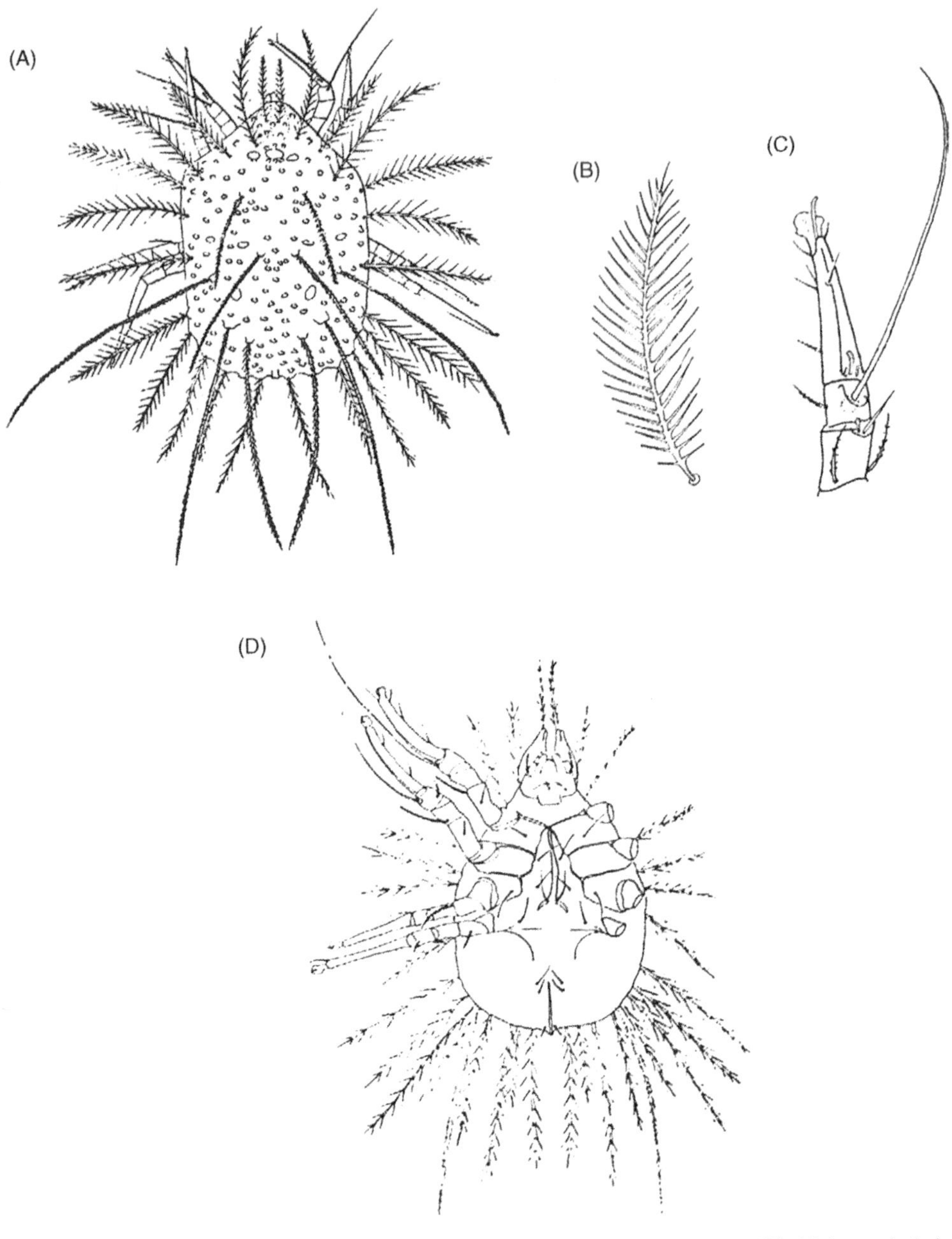

Fig. 33.1. *Ctenoglyphus plumiger* (Koch). Female. Dorsal view (A); dorsal seta (B); leg I (C). Male. ventral view (D) (from Hughes, 1976, licensed under the Open Government Licence v3.0).

C. plumiger has been found also in the nest material of birds (*P. domesticus* Linnaeus) and rodents (*M. musculus* Linnaeus) (Chmielewski, 2000b) and in bumble bees (*B. lapidarius* Linnaeus) (Chmielewski, 1975b). Often, the mite has been found in sweepings and litter in sheepfolds, cowsheds and stables (mangers). Moreover, it was collected in hay sweepings from feeding racks for wild forests animals (e.g. bisons) (Chmielewski, 1998) and in buckwheat sweepings (leaves, straw) taken from barns (Chmielewski, 2000b). *C. plumiger* was also found in apiaries, as an associate of bees, colonizing hive debris and honeycombs (Michael, 1901, 1903; Zachvatkin, 1941; Hughes, 1976; Grobov, 1991). Bee products, especially their mouldy debris, are infested with this mite, which feeds willingly on mycelia of saprophagous fungi.

Bio-ecology

C. plumiger has probably a cosmopolitan geographical distribution. It occurs frequently under hay and straw stacks, and sparingly in damp oats, wheat, barley and grass seed. It has also been collected from spillage of fishmeal, from honeycombs and nests of sparrows (Hughes, 1976; Chmielewski, 1982). In Poland, was found in nests of domestic sparrows (Andrzejewska, 1979; Chmielewski, 1982), in sweepings and litter or debris from cowsheds and barns (Solarz *et al.*, 1994, 1997).

Its life cycle has been investigated in Poland by Chmielewski (2000b). This author maintained cultures of the mite in rearing cages, under controlled laboratory conditions (temperature ca. +20°C, RH – ca. 85%), and provided with bee bread as food. It found that the development cycle (without hypopus instar) takes 26.6 (14–30) days, the mortality of preimaginal instars is 25.3% (17–50%), the percentage of females is 57.2% (40–76%), the longevity of males is 54.5 (35–86) days, the longevity of females is 59.4 (46–120) days, the infecundity before the oviposition is 4.1 (2–8) days, the oviposition period lasts 45.4 (31–59) days, the infecundity after the oviposition period is 11.4 (0–44) days, the fecundity of female is 104.8 (57–175) eggs per female in the lifetime and the productivity per fecundity day is 2.3 (1–10) eggs (Chmielewski, 2000b). Females start to lay eggs about 4 days after the copulation, which is an obligatory condition of the oviposition process. The copulation takes place soon after the eclosion of imagines, and usually is repeated many times during male and female lifespans, or even during the same day. After the oviposition, the females reach an unfecund older age before their natural death. *C. plumiger* form hypopi (heteromorphic deutonymph), which differ in behaviour and morphology from adults and other development stages and were observed in mass cultures of the species from time to time. But in ontogenesis experiments presented herein, when the mites were placed in the containers for individual rearing, hypopi were never found (Chmielewski, 1975b).

Symptomatology and damage

Biological characters of *C. plumiger*, namely fairly high fecundity, longevity and vitality of mites fed bee bread, show that this product is accepted by them as food and explains their occurrence in beehives. Colonizing of hive debris with these mites, their feeding and development on mouldy products (pollen, bees bread, dead bees) are circumstances conducive to spreading of airborne fungi. It suggests the need for initiation of research studies on hygienic and sanitary importance of these mites, especially on their role as vectors of saprophagous microorganisms and infestation of hive products with them. Bee bread might be recommended as an effective natural medium for use in laboratory monocultures of the species and production of biological material for experimental purposes and teaching acarology (Chmielewski, 2000b).

Control

No specific records have been possible to find. If necessary, the control may be similar to that applied for other stored-product mites.

34 The Family Labidophoridae Zachvatkin

Morphological Characteristics, Systematic and Bio-ecology

The word Labidophoridae means or is derived from labia, referring to the two lips bordering the opisthosoma of the hypopus and enabling it to clasp to hairs (Fain, 1969a). The labidophorid mites have the cuticle usually strongly tanned, brown or reddish in colour; it may be smooth, granular or covered with a reticulate pattern. The gnathosoma is usually overhung by an anterior projection of the idiosoma. The coxal sternal skeleton is well developed, often forming a fused ring around the female genital opening. The setae arising from the dorsal and posterior surfaces of the idiosoma are short, smooth or feebly pectinate, and of about the same length. The legs are sometimes modified, and ornamented with ridges and comb-like structures. The leg setae are often pectinate and the claws very small. The sexual dimorphism is usually pronounced. The majority of the species are found in the nests of small mammals (Hughes, 1976). Species of the genus *Gohieria* Oudemans infest stored products, and may be collected in dust house, in mattresses and beds, causing allergenic diseases to humans.

The Genus *Gohieria* Oudemans

Diagnostic characteristics

The propodosoma is extended forwards and overhangs the gnathosoma. The cuticle is evenly sclerotized, brown in colour and ornamented with fine, almost smooth setae. No distinct propodosomal shield or crista metopica is present. The apodemes of all the legs are slender, and unite to surround the genital opening. The genua and tibiae of legs are distinctly ridged, particularly in the females. The distal edges of the femora and genua are expanded. The sexual dimorphism is slight. Female with tracheal tubes (Hughes, 1976).

Economic importance

On stored products, the most important species of this genus is *Gohieria fusca* (Oudemans), probably cosmopolitan and reported on almost 30 commodities, in various facilities (Hagstrum *et al.*, 2013). The mites cause also allergenic diseases in humans (Cornere, 1971; Li *et al.*, 2015c).

Gohieria fusca (Oudemans) (Fig. 34.1)

Common name

Brown flour mite

Morphological characteristics

FEMALE. The idiosoma is 380–420 μm long and pale in colour. Ventrally, all part of the body between the coxae I and IV is occupied by the large genital folds. The apodemes I are united with the epigynum, with a transverse plate at the anterior end of the genital opening. The apodemes II nearly reach the circumgenital ring, and II and IV touch at their internal ends. The small genital papillae project from the posterior end of the genital folds. A pair of well developed tracheae is present which in life are full of air. Each trachea branches into an anterior region which expands into an air sac, and a long winding posterior part which according to van der Hammen (1968) crosses to one from the other side, but does not join it. The folds on either side of the anal opening project beyond the posterior margin of the body. Two pairs of anal setae arise on either side of the anterior end of the opening. The opening of the bursa copulatrix is overhung by a small projection; it is connected with a receptaculum seminis by a tube. The legs are slender than those of the male, the longitudinal ridges better developed. On tarsus I the setae are spaced out along the tarsus, instead of being concentrated at the distal end (Hughes, 1976).

MALE. The idiosoma is 300–320 μm long. Dorsal and ventral surface of the body are tanned and covered

 The Handbook of Mites of Stored Products (V. Vacante and C.H.W. Flechtmann)
DOI: 10.1079/9781786390912.0034

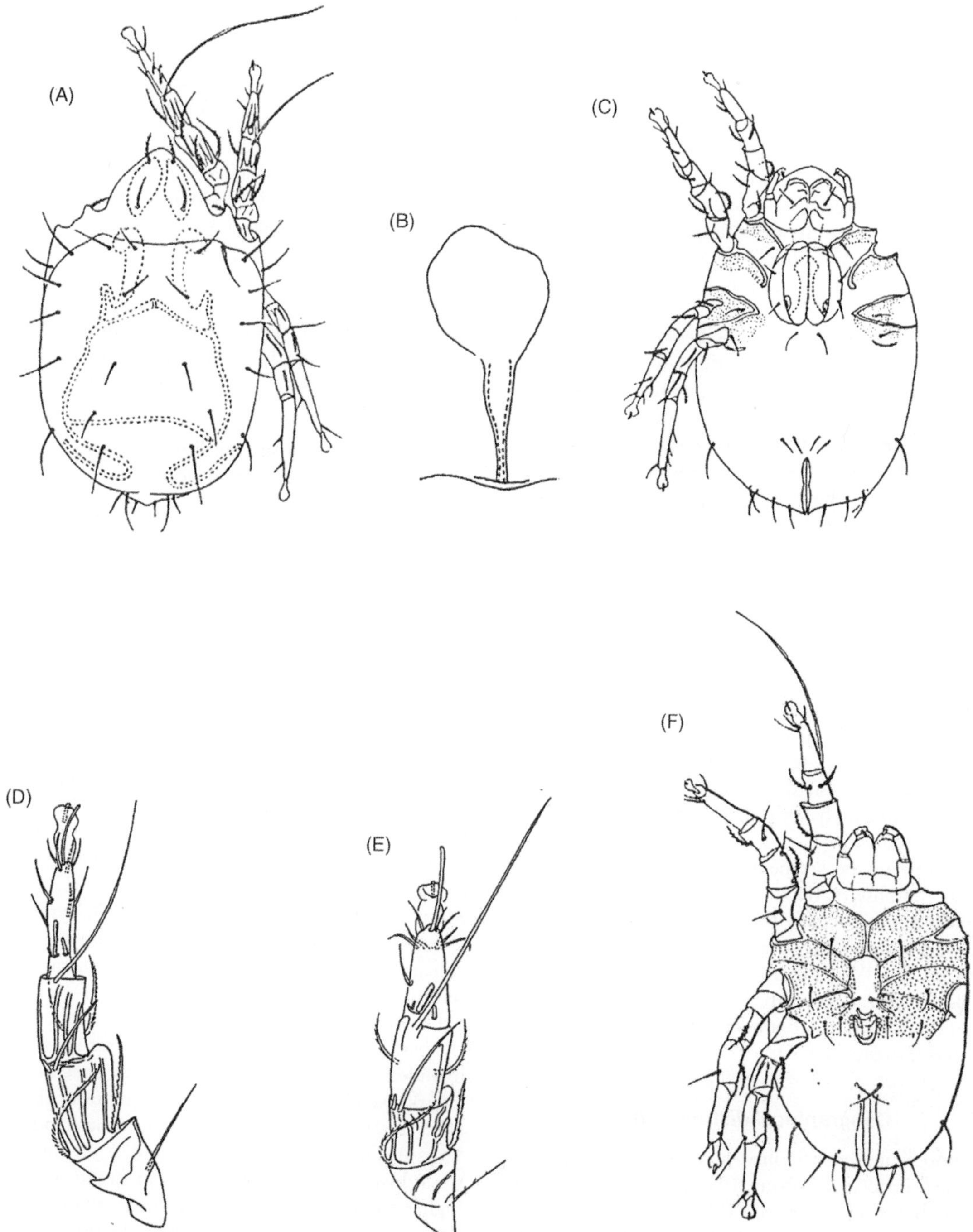

Fig. 34.1. *Gohieria fusca* (Oudemans). Female. Dorsal view (A); receptaculum seminist (B); ventral view (C); leg I (D). Male. Leg I (E); ventral view (F) (from Hughes, 1976, licensed under the Open Government Licence v3.0).

with pinkish-brown pitted cuticle. The idiosoma overlies the gnathosoma with a hood-shaped extension. The anterior edge of the hysterosoma is marked by a transverse pleat, so that, in the living animal, the hysterosoma appears to be covered by a separate shield. All apodemes are slender struts, and those of leg I joining to form a short sternum; this unites the internal ends of apodemes II–IV to enclose a colourless area of cuticle in front of the genital opening. The cuticle of the ventral surface is more coarsely pitted than that of the dorsal, and untanned cuticle joins dorsal and ventral surfaces together. The genital opening is between coxae IV, the aedeagus being a straight tube, whose apex is directed posteriorly. The anal opening extends to the posterior margin of the body, and one pair of anal setae arises at its anterior end. The setae *vi* are pectinate, and the other idiosomal setae are slightly serrated, those arising from the propodosoma being directed forwards, those from the hysterosoma laterally and backwards. The setae *ve* are almost in the same transverse line as the pectinate supracoxal setae, each of which lies in a groove; the setae *sci*, *sce* and *hi* are on a level with one another, and four pairs of dorsals are arranged in linear series. The legs are short and massive, the pretarsus arising from the ventral side of the apex of each tarsus. The distal ends of the genua and femora are expanded, so as to partially envelope the succeeding segment. The dorsal surfaces of genua and tibiae are distinctly ridged. The tarsus I is fore shortened and those setae which normally occupy the middle of the segment (*la*, *ra*, *wa*) arise close to *s* at the base of the pretarsus, while ω_1, ω_2, the famulus and *ba* occupy their normal positions. On tibia I, φ is unusually long, but decreases in length from tibiae I–IV. On genu I, 6_1 is much longer than 6_2. Both ventral setae are present on tibia I. Legs III and IV are noticeably bent, with unusually long pretarsi (Hughes, 1976).

Geographical distribution

Cosmopolitan, reported in various countries, such as Algeria (Louadì and Robaux, 1992), Belgium (Hughes, 1976), Brazil (Flechtmann, 1986; Baqueiro *et al.*, 2006; Barbosa, 2011), Canada (Sinha, 1974; Beaulieu and Knee, 2014), China (Xin, 1965; Shen, 1984; Li *et al.*, 2015c), Colombia (Navarro *et al.*, 2008), Czech Republic (Ždárková, 1967), Egypt (Attiah, 1969; Saleh *et al.*, 1985; Rezk *et al.*, 1996; Metwally *et al.*, 2016), England, France (Hughes, 1976), Germany (Hughes, 1976; Franz *et al.*, 2007), Greece (Palyvos *et al.*, 2008), India (Gupta and Chatterjee, 2004; Pawar *et al.*, 2016; Gill *et al.*, 2018), Iran (Sepasgozarian, 1978; Noei, 2007; Noei and Ostovan, 2012), Ireland (O'Farrell and Butler, 1948), Japan (Hosaya and Kugoh, 1954), Korea (Lee and Choi, 1980), the Netherlands (Siepel *et al.*, 2016), New Zealand (Robertson, 1946; Tenquist and Charleston, 2001), Poland (Kosik-Bogacka *et al.*, 2010), Russia (Zakhvatkin, 1959), Scotland (Sesay and Dobson, 1972), Spain (Cosmes Martín *et al.*, 2003), Switzerland (Mumcuoglu, 1977), Taiwan (Tseng, 1972), Ukraine (Dudinska, 2003; Tytar and Oksentiuk, 2019; Oksentiuk *et al.*, 2022), United States (Smiley, 1987) and other countries.

Commodities, facilities, vegetal and animal hosts

This mite was collected on almost 30 commodities, such as barley, bean (red), Adzuki bean, buckwheat, buckwheat product, Chinese medicine, maize, dried fruit, dried vegetable, fish (dried), flax seed, fungus (dried), grain, grass seed, hemp seed, hay, herb, kaoliang, malt, mushroom, poppy seed, protein poultry feed, rice, rice (husked), soybean bran, soybean sauce, straw, soybean sweeping, stored rice, rice dust and debris, sugar (packaged), sugarbeet seed, wheat, wheat bran, wheat flour (packaged), wheat germ, wheat pollard, wheat flour sweeping (Robertson, 1946; Hosaya and Kugoh, 1954; Sinha, 1974; Hughes, 1976; Shen, 1984; Saleh *et al.*, 1985; Flechtmann, 1986; Noei, 2007; Palyvos *et al.*, 2008; Noei and Ostovan, 2012; Li *et al.*, 2015c; Metwally *et al.*, 2016; Gill *et al.*, 2018; Tytar and Oksentiuk, 2019). In Ukraine, this mite has been collected in barns with animal fodder storage and outbuildings (Oksentiuk *et al.*, 2022).

G. fusca has been recorded also from kiore (Roberts, 1991); Gupta and Chatterjee (2004) included it among the fungivorous species, and Ždárková (1967) found the mite in flour. Pawar *et al.* (2016) collected *G. fusca* in rat house in India. Louadì and Robaux (1992) found it on the mattress dust in Algerie, as well as in Brazil, where a higher frequency of *G. fusca* in beds was observed (Baqueiro *et al.*, 2006). Hughes (1976) collected the mite in the surface dust of a mattress in a very dump house. Rezk *et al.* (1996) found the mite in the house dust in Egypt. Tenquist and Charleston (2001) recorded the species in 14/22 house-dust samples in Auckland (10% of mites recovered). The same mite was collected in house dust in Scotland (Sesay and Dobson, 1972). Cornere (1971) reported that occasionally the mite was associated with allergic reactions in humans.

Main facilities are commercial stores, farm, farm milling machine, flour mill, off-farm storage, food processing factories, provender mill and sweeping (Dudinska, 2003, Palyvos *et al.*, 2008). In China, this mite is prevalent in the screens of air conditioners and its breeding status appears serious (Li *et al.*, 2015b).

Bio-ecology

G. fusca is a cosmopolitan mite, reported on a wide list of stored products, among which barley, highland barley, rice, rice bran, broken rice, wheat flour, maize, feeds, wheat bran and herbal medicines (Ždárková, 1967; Hughes, 1976; Shen, 1984; Li *et al.*, 2015c). Moreover, it has been found on the house dust (Sesay and Dobson, 1972; Rezk *et al.*, 1996; Tenquist and Charleston, 2001; Solarz and Pająk, 2019b), on mattress dust (O'Farrell and Butler, 1948; Louadì and Robaux, 1992) in beds (Baqueiro *et al.*, 2006) and on the dust accumulated in the screens of air-conditioning systems in China, where its breeding status appears serious (Li *et al.*, 2015c). The mite is capable of having mild to moderate cross-reactivity with allergens of house-dust mites (serious source of indoor allergens associated with asthma and other allergic diseases) (Li *et al.*, 2015c).

The life cycle of *G. fusca* develops through the egg, larva, protonymph, tritonymph and adult stages; the hypopus is unknown. The first studies on the mite biology carried out by Boulanova (1937) showed that the female lays from 11 to 24 eggs, and at 24–25°C the life cycle takes from 11 to 23 days. Further studies carried out by Taha *et al.* (2010) under laboratory conditions at 25±2°C and 65% RH with three different food sources (dry yeast granules, crushed wheat and crushed maize) found that developmental stages, fecundity, reproduction and life table parameters of the mite were affected by different food types, whereas total immature duration of stages lasted (31.22 and 29.92), (34.73 and 33.38) and (31.81 and 29.78) days for female and male, when mite fed on the above-mentioned diets, respectively. The female oviposition period lasted 26.8, 30.9 and 29.1 days, which deposited an average number of eggs with a daily rate (132.9 and 4.96), (111.5 and 8.87) and (91.3 and 3.14) eggs, when mite fed on dry yeast granules, crushed wheat and crushed maize, respectively (Taha *et al.*, 2010). Gill *et al.* (2018) found that the incubation period of *G. fusca* fed on crushed maize at temperature of 25°C and 65% RH varies from 5.48 ± 0.9 in females to 4.78 ± 0.49 in males. The duration of total immature duration of stages for females was 20.8 days, while for male this period lasted 18.57 days. The total immature duration of stages for the male is shorter than for females, and the male becoming adult earlier than female. On crushed maize as feeding, the female generation period requires 42.19 days, the longevity period of female is 42.3 days, while that of male is 40.19 days. The oviposition period on crushed maize is about 29.10 days, and the female fecundity is 91.30 ± 8.33 eggs with a daily rate of 3.14 eggs. The female reproduction is 242 individuals (220–294), and the effect of crushed maize food on the life table parameters of *G. fusca* reveals that the net reproduction rate (R_0) is a product of mean total fecundity. The net reproductive rate (R_0), which is a survival rate, showed 58.8. The mean generation time (*T*) is 44.89, while the intrinsic rate of increase (*rm*) is 0.24. The finite rate of increase (*exprm*) is 1.28. The survival rate of sex ratio of progeny (females/total) is 0.62. The stage duration in days of larva, calculated as active (A) and quiescent (Q) stages, is A 7.43 ± 0.58 and Q 2.29 ± 0.45, protonymph A 8.38 ± 0.49 Q 2.33 ± 0.47 and tritonymph A 8.52 ± 0.49 Q 2.86 ± 0.35. The total immature duration is 31.81 ± 1.26, the life cycle is 37.29 ± 1.45, the generation period is 42.19 ± 1.32, the longevity is 42.30 ±1.25 and the lifespan is 79.20±1.93. The incubation period is 4.78±0.49, the larva is A 7.11±0.51 Q 2.00±0.00, the protonymph A 7.89±0.51 Q 2.00±0.00 and the tritonymph A 8.33±0.43 Q 2.44±0.45. The total immature duration is 29.78±1.11, the life cycle is 34.56±1.14, the longevity is 40.20±2.15 and the life cycle is 75.17±2.86 (Gill *et al.*, 2018).

Studies carried out to determine the species composition of allergenic mites in West Pomerania in north-western Poland highlighted that *G. fusca* was found only in rural apartments (Kosik-Bogacka *et al.*, 2010). The mite has been investigated also using modelling the bioclimatic niche of a cohort of selected mite species (Acari, Acariformes) associated with the infestation of stored products (Tytar and Oksentiuk, 2019).

Symptomatology and damage

The mite presence on stored products is inappropriate both for the alterations that it can induce to the foodstuffs and for the pathological implications that can involve the workers who handle the stored products and/or the consumers. Moreover, the mite inhabits dust house.

O'Farrell and Butler (1948) found high densities of *G. fusca* populations in mixed cereal dust and spillage in the mills and from there, it spreads to warehouses storing flour and feeding stuffs. Ždárková (1967) collected the mite in 10.1% of all samples infested by mites in Czech Republic, and it is most abundant in flour where its brown colour makes it easily detected.

Regarding the implications of human disease, Boner *et al.* (1989) evaluated the action of skin test and RAST positivity for different storage mites among which *G. fusca* in children with bronchial asthma and positive prick test to *Dermatophagoides pteronyssinus* (Trouessart). The storage mite with the highest incidence of skin and RAST positivity was *G. fusca*, which caused a positive response in more than 50% of the children evaluated. This antigen appears to possess allergenic determinants, not present in *D. pteronyssinus*. The sensitivity to this antigen might be considered in house-dust mite-allergic children who do not improve after specific immunotherapy with *D. pteronyssinus* extracts (Boner *et al.*, 1989). Solarz and Pajak (2019b) carried out investigation in southern Poland on the occurrence of mites in dust samples from houses in agricultural areas, with particular reference to allergenic and parasitic species as a potential risk factor of diseases among people. Dust taken from beds, floors in bedrooms, upholstery furniture, floors in family rooms and from floors in kitchens host mites in 74.8% of samples collected and *G. fusca* in 25.1%, with a frequency of this last mite of 42.4%.

Control

As for other Astigmata.

35 The Family Glycyphagidae Berlese

Morphological Characteristics, Systematic and Bio-ecology

The glycyphagids are free-living mites, often associated with insects or small mammals, or found in the nests of mammals. The idiosoma is commonly undivided by a transverse furrow into propodosoma and hysterosoma. The dorsal propodosomal shield may be reduced or absent. The cuticle is rarely smooth, commonly rough or covered with small papillae. The ventral subcapitular ridges are prominent. The claw is inserted at the distal end of a fleshy pretarsus, only joined to the end of the tarsus by two thin 'tendons'; it is sometimes absent. The anal and tarsal suckers are usually absent in the male (Hughes, 1976; O'Connor, 2009b).

The systematic structure of the family does not seem definitive. Hughes (1976) recognized five subfamilies called as Ctenoglyphinae Zachvatkin, Glycyphaginae Zachvatkin, Aëroglyphinae Zachvatkin, Labidophorinae Zachvatkin and Nycteriglyphinae Fain, respectively; this view has been outdated and several of these taxa have now become families. Also, some lower categories such as the genera have undergone a rearrangement. For example, Klimov *et al.* (2016) follow the systematic vision of Zachvatkin (1936) and recognize different subgenera within the genus *Glycyphagus* Hering, such as *Glycyphagus* (*Glycyphagus*) and *Glycyphagus* (*Lepidoglyphus*) and so on. At now, the family comprises about 27 genera and 100 species, commonly associated with nests of rodents, insectivores, marsupials and some are associated with other mammalians. Different species are found on stored products, and occur in most domestic premises in Europe feeding on flour, wheat, cheese or ham (Hughes, 1976) and are strongly allergenic in house dust (Maunsell *et al.*, 1968). The life cycle develops through the egg, larva, protonymph, deutonymph, tritonymph and adult stage, and can be completed within 25 days, but hypopal deutonymph resting stages are common, which can last several years. Hora (1934) reported that in *Glycyphagus domesticus* (De Geer) the hypopus appears to be formed independently of external conditions and as many as 50% of the protonymphs may pass through this biological stage. Klimov *et al.* (2016) highlighted that in Astigmata, the deutonymph is phoretic and is very different from any other stage. It is easily identified by the presence of the attachment organ and the lack of functional mouthparts. A few taxa in the families Chaetodactylidae, Glycyphagidae and Acaridae have non-phoretic, immobile deutonymphs. This stage is a featureless sack with greatly reduced legs, attachment organ and mouthparts, and it usually stays under the protonymphal skin. The phoretic deutonymph is an active nymphal stage that serves for dispersal on arthropod or vertebrate hosts, a subtype of heteromorphic deutonymph, does not exist in the genus *Glycyphagus*.

Usually, the glycyphagid mites are common phoretic associates of small mammals, throughout the world. The hypopi occupy four types of habitats: some species are attached to hairs with a clasping apparatus (pilicolous hypopi), others are found in the hair follicles (endofollicular hypopi) or in the wax of the ear, and finally, two species live in the subcutaneous cellular tissue (hypodermic hypopi) (Fain, 1969a).

On stored food, the most important species belong to the genus *Glycyphagus*. Further new information on other secondary species may be found in Hughes (1976) and Hagstrum *et al.* (2013).

Selected websites

https://www.insectimages.org/browse/taxthumb.cfm?fam=733

https://idtools.org/bee_mite/index.cfm?packageID=1&entityID=101

https://www.insectimages.org/browse/taximages.cfm?fam=733

DOI: 10.1079/9781786390912.0035

The Genus *Glycyphagus* Hering

Diagnostics characteristics

The prodorsum lacks external vertical setae *ve* and the internal vertical setae *vi* are long and barbed, and set distinctly posterior to the anterior margin of the propodosoma. The scapular setae (*sci*, *sce*) are arranged in a trapezoid or rectangle. The prodorsal sclerotization or crista metopica is widened or absent. Some dorsal setae are long and heavily barbed. The anal opening is positioned near the posterior margin of the body. The subcapitulum has a distinct pattern of ventral ridges. The tibiae I–II bear two ventral setae. The claws are simple and small. The female usually has a short external copulatory tube. The males lack of paranal suckers and tarsal suckers on tarsus IV. Without a subtarsal scale, *Glycyphagus* (*Glycyphagus*) or with a subtarsal scale, *Glycyphagus* (*Lepidoglyphus*) (Zachvatkin, 1936, 1941; Klimov *et al.*, 2016).

Economic importance

The species of this genus are habitat generalists; all stages live in various habitats, including bee nests, where they feed on various organic materials. Klimov *et al.* (2016) reported the absence of hypopus, and improved the description of Hughes (1976), of this biological stage. The feeding stages disperse on hosts (including bees) by active movements, or by air currents. Some species are reported on rodent and bird nests, other are pests of stored products, or live in house dust, grass and hay and in other substrates. These mites have also been collected in nests of social insects, including bees. Various *Glycyphagus species* cause dermatitis, gastritis and allergies in humans (Klimov *et al.*, 2016). These mites are common in beehives, especially on debris from bottom boards. Infestation rates have been reported ranging from 20% (Grobov, 1978) to 98.8% (Chmielewski, 1991a), to 100% (Vitzthum, 1936). Often, there are hundreds of thousands of mites in a single beehive. They feed on decomposing organic matter, stored pollen, beehive debris and honey, but often prefer dead bees. *G. (G.) domesticus* and *G. (L.) destructor* (Schrank) are the most common and abundant species, while other species are much rarer (*G. (G.) ornatus* (Kramer), *G. (L.) privatus* Oudemans, *G. (L.) michaeli* Oudemans and so on). In Poland, *G. domesticus* has been found to be the most common species of all mites, occurring in 98.8% of samples (Chmielewski, 1991a), while *Glycyphagus destructor* was found only in 4.9% of samples. In German beehives, these two species occurred at nearly the same frequency (Homann, 1933). Under laboratory conditions, *G. domesticus* preferred bee bread, pollen, mould and beehive debris, followed by combs and wax, honey, propolis and dead bees; no reproduction occurred on royal jelly. Conversely, *G. destructor* preferred dead brood bees, mould and beehive debris (Chmielewski, 1991a).

The most important species of this genus is probably *G.* (*G.*) *domesticus*, a widely distributed species, found in large numbers on dried plant, and animal remains in houses and stables, but not in grassland (Griffths, 1960). It is reported in flour, wheat, hay, linseed, tobacco, cheese, ham (Chabasse and Genthorn, 1966), sugarbeet seed (Chmielewski, 1969b) dried calves stomach, and in the nests of both bees and birds, on damp mouldy wallpaper and in the sweepings from barns. It infests commonly the stockfish (Gigia, 1964) and develops high populations on furniture covered with rush or sea grass fibre or upholstered with green Algerian fibre, where it feeds on moulds growing on the fibre (Hughes, 1976). Hagstrum *et al.* (2013) reported a list of 44 commodities, infested by the mite in the world. Moreover, Workers handling cheese infested by *G.* (*G.*) *domesticus* showed dermatitis (Robertson, 1946). Joyeux and Baer (1945) identified this mite as vector of the tapeworm *Catenotaenia pusilla* (Goeze), whose adults live in rodents. Moreover, *G.* (*G.*) *domesticus* has been reported as responsible for ear canker in rabbits (Davies, 1926). The mite has been associated with symptoms of asthma in humans (Deschiens, 1951). *G.* (*G.*) *domesticus* causes the grocer's itch (Baker *et al.*, 1956). So, the mite is feared for its presence on stored food and for its relationship with allergenic diseases. Another injurious species is *G.* (*G.*) *ornatus* (Kramer), characterized by a wide geographical distribution and reported on almost 13 commodities (Zakhvatkin, 1959; Hughes, 1976; Hagstrum *et al.*, 2013). *G.* (*G.*) *privatus* Oudemans has been found in different countries, on various commodities (Hughes, 1976; Hagstrum *et al.*, 2013), *G.* (*L.*) *destructor* (Schrank) is a serious injurious mite, cosmopolitan and reported from a wide list of commodities (Hughes, 1976; Hagstrum *et al.*, 2013). *G.* (*L.*) *fustifer* Oudemans has been collected on various stored products in China (Li *et al.*, 1992a), Czech Republic (Žďárková, 1967), Germany (Hughes, 1976),

Hong Kong (Olsen, 1983), Poland (Solarz and Pająk, 2019), Russia (Zakhvatkin, 1959) and *G.* (*L.*) *michaeli* Oudemans, known from Bulgaria, Canada, China, Czech Republic, England, France, Germany, Hungary, Ireland, Italy, the Netherlands, Poland, Russia and Sweden on various commodities (Hughes, 1976). Secondary species include *G.* (*G.*) *aegyptiacus* Attiah and *G.* (*G.*) *oryzae* Attiah, both reported from Egypt (Attiah, 1969), *G.* (*G.*) *caudatus* found in Germany and Poland on grass seed in farm (Hughes, 1961), *G.* (*L.*) *pilosus* Oudemans reported from Germany (Oudemans, 1906), Poland (Zakhvatkin, 1959), Czech Republic (Ždárková and Pekár, 2007) and Ukraine (Oksentiuk, 2020) on herb (Zakhvatkin, 1959) and cereal crops (Oksentiuk, 2020). The most life-supporting substrate for this species are flour, dry marjorm leaves, different types of feed, garbage, wheat, silage, barley and grain mixture (Ždárková and Pekár, 2007). In Ukraine, *L. pilosus* is known as a pest of cereal products in grain warehouses, mills, silage containers, distilleries and hangars (Oksentiuk, 2020); in the same country, the mite has been collected in barns with animal fodder storage (Oksentiuk *et al.*, 2022). *Lomelacarus faini* Fan *et* Li is known from China on condiment, dried fruit, dried vegetable and mushroom (Fan and Li, 1993a).

Glycyphagus (Glycyphagus) domesticus (De Geer) (Fig. 35.1)

Common names

House mite, furniture mite, grocer's itch mite, hairy grain mite.

Morphological characteristics

FEMALE. The idiosoma is 400–750 µm long. Similar to male, with the genital opening extending to the posterior edge of the acetabula III, shorter than the distance separating it from the anterior end of the anus; a small crescentic epigynum covers the anterior end of the genital folds. The posterior pair of genital setae is inserted just behind the hind end of the genital opening. Two pairs of setae lie on either side of the anterior end of the anus. A tubular bursa copulatrix projects from the hind margin of the body (Hughes, 1976).

MALE. The idiosoma is 320–400 µm long, rounded in shape, with a dull cuticle, owing to the fact that it is covered with minute papillae. The narrow elongated dorsal propodosomal shield (*crista metopica*) extends from the base of the chelicerae to the level of *ve*; it is a narrow sclerite that widens in the middle and surrounds the setae *vi*. Its anterior half is more feebly sclerotized than the posterior. The setae arising from the idiosoma are finely pectinate and in life radiate stiffly from the surface of the body. The supracoxal seta is forked and branched. The four scapular setae are arranged in a transverse row across the idiosoma, *sci* being longer than *sce*. The setae d_1 (d_2) are short, less than half the length of c_1 (d1), their points of origin being almost in the same transverse line as those of e_1 (d_3). As in other *Glycyphagus* and *Lepidoglyphus*, the setae e_1 are probably movable, as the base of each one has an internal projection, which could serve as a muscle attachment. Three pairs of lateral setae are present (l_1 to l_3). From the posterior margin of the body project, three pairs of postanals and *sai* and *sae*. Ventrally, the apodemes of leg I meet to form a short sternum, apodemes II are also well-developed, apodemes III and IV are slender and apodemes III each have an anteriorly directed process. The genital opening lies between the coxae II and III. The legs are long, with tapering segments and each terminates in a pretarsus and small claw. No subtarsal scale is present, but is replaced by a pectinate seta *wa* which is present on all the legs and arises from the middle of the tarsus. The setae *la*, *ba* and *ra* arise between the base of *wa* and the apex of the tarsus. On tarsus I, ω_1 is a slender rod longer than that on tarsus II, and ω_2 is about half its length, much longer than the famulus. On genu I, σ_1 is less than half the length of σ_2, about the same length as ω_1. On tibiae III and IV, *kt* is well removed from the distal edge of the segment (Hughes, 1976).

HYPOPUS. The phoretic deutonymph is absent. Feeding stages disperse on hosts (including bees) by active movements or by air currents (Klimov *et al.*, 2016).

Geographical distribution

Cosmopolitan, reported from various countries among which Australia (Sinha, 1968c; Halliday, 1998, 2003), Canada (Smallman, 1942; Sinha, 1968c; Beaulieu and Knee, 2014), Chile (Klein and Waterhouse, 2000), China (Xin, 1965; Shen, 1984; Li *et al.*, 2015a), Croatia (Tareev and Dubinina, 1985), Czech Republic (Dusbábek, 1995; Lukás *et al.*,

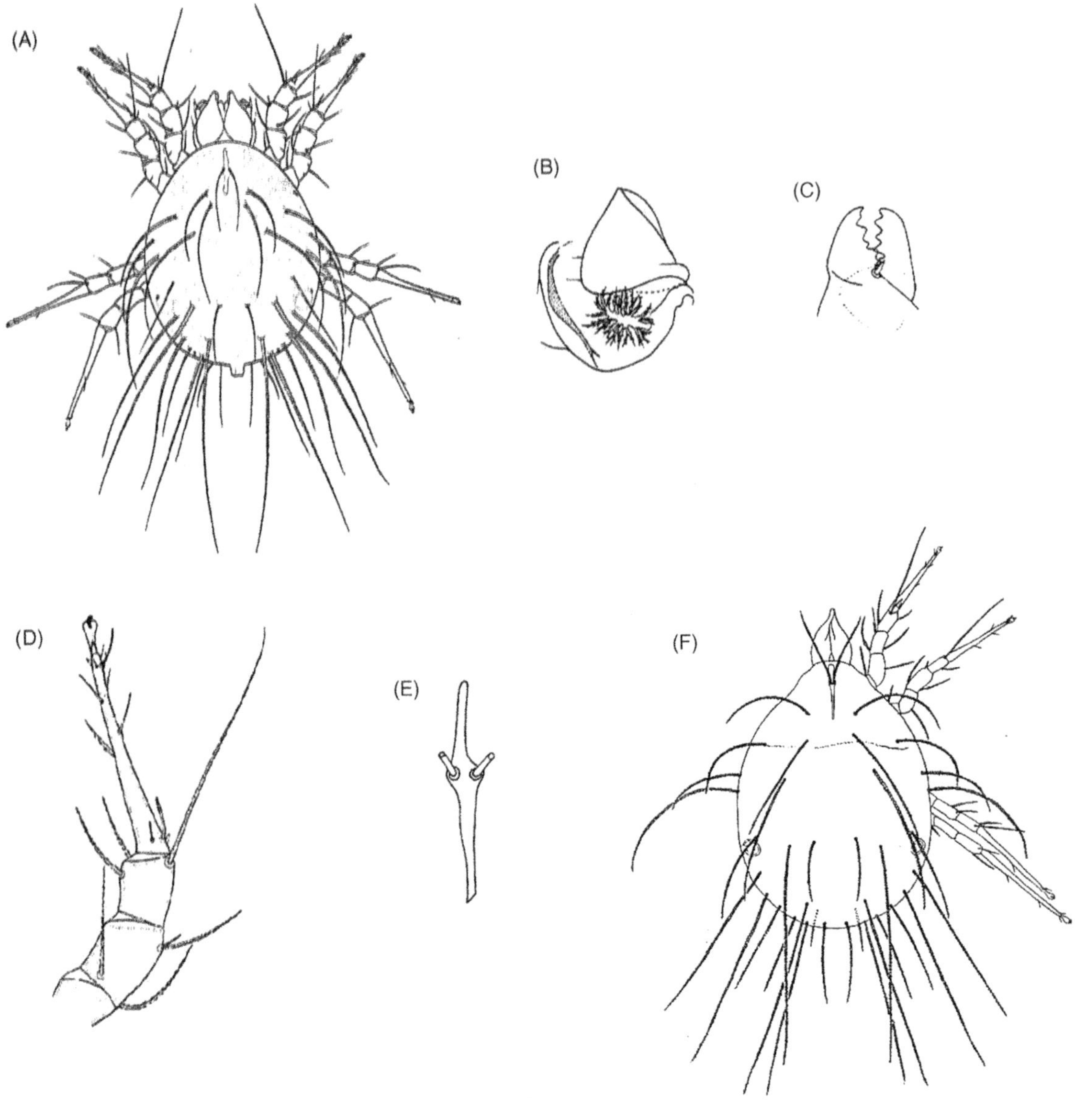

Fig. 35.1. *Glycyphagus* (*Glycyphagus*) *domesticus* (De Geer). Female. Dorsal view (A); supracoxal seta (B); chelicera (C); leg I (D) (from Vacante and Nucifora, 1984-85); crista metopica (E). Male. Dorsal view (F) (from Hughes, 1976, licensed under the Open Government Licence v3.0).

2007), Egypt (Heikal, 2015), England (Hughes, 1976; Blainey *et al.*, 1988), Faroe Islands (Hallas and Solberg, 1989), Finland (Thero *et al.*, 1982), Germany (Rack, 1988; Franz *et al.*, 1998, 2000), Greece (Palyvos *et al.*, 2008), India (Gupta and Chatterjee, 2004), Iran (Sepasgozarian, 1971, 1977, 1978), Italy (Vacante and Calabretta, 1982), Japan (Sinha, 1968a, b, c), Malaysia (Kew *et al.*, 2014), the Netherlands (Siepel *et al.*, 2016), Poland (Chmielewski, 1969b; Solarz *et al.*, 1997, 2004; Pająk and Solarz, 2013; Solarz and Pająk, 2019a), Portugal (Guimarães, 1970), Scotland (Colloff, 1987a, b), Spain (Blasco Sabio *et al.*, 1975), Taiwan (Tseng, 1972), Turkey (Cevizci *et al.*, 2010; Dizlek *et al.*, 2019), United Kingdom (Hughes, 1976; Colloff, 1987b), Ukraine (Dudinska, 2003; Oksentiuk *et al.*, 2022), United States (Cotton and Good, 1937; Smiley, 1991) and other European countries (Sinha, 1968c; Hughes, 1976).

Commodities, facilities, vegetal and animal hosts

The list of commodities includes almond, animal feed, aquatic product, baby food, barley, barley (germinating), bean, bean paste cake, biscuit, bread, buckwheat, buckwheat product, calf stomach (dried), cereal, cheese, Chinese medicine, condiment, crumbs, cucurbitae semina, dateplum persimmon (*Diospyros lotus* Linnaeus), dried figs, dried fruit, dried vegetable, dried sweet potato, dry moss, flax seed, grain, grass seed, ham, hay, haw jelly, haw flakes, hawthorn, hemp seed, herb, herring meal, honey, honey date, fine dried noodles, malt, malt flower (dried), moon cake, mushroom, jam, oat, old silk noodles, orange powder, groundnut, plum powder, poppy seed, potato, processed food, prune, raisin, rapeseed, red dates, rice, saqima, spillage, squid (dried), stark, stock fish, sugarbeet, sugarbeet seed, sugared mandarin, sunflower seed, tangerine, tobacco, wheat, wheat flour, wheat flour sweeping, wheat koji, white granulated sugar, *Zizyphus jujuba* Mill (Sinha, 1968c; Guimarães, 1970; Hughes, 1976; Lukás *et al.*, 2007; Palyvos *et al.*, 2008; Hagstrum *et al.*, 2013; Bealieu and Knee, 2014; Li *et al.*, 2015a). In Ukraine, Oksentiuk *et al.* (2022) collected the mite in mills, granaries and warehouses, compound fodder factories, barns with animal fodder storage, outbuildings, honeybee beehives and vegetable storages.

Moreover, the mite has been found in farming environment, bird nests (Solarz *et al.*, 2004), cowsheds, barns, lofts (Solarz *et al.*, 1997), pigsties, chaff-cutter buildings, poultry houses (Solarz and Pająk, 2019a), farm stores, commercial stores, four mills and v. silos (Palyvos *et al.*, 2008). It has been collected in beehives (Bealieu and Knee, 2014), on edible bird nests (Kew *et al.*, 2014) and on dogs and cats (Eaton *et al.*, 1985). In Iceland, it is a common pest of stockfish (Gigja, 1964).

Colloff (2009c) found *G. domesticus* in the dust, from cloth-covered seats of four passenger trains and bedding in Glasgow (Scotland), and Hagstrum *et al.* (2013) reported a number of facilities as bread van, cheese store, commercial grain storage, drier, farm, farm granary, farm milling machine, flour mill, grinder, house, off-farm storage, provender mill, provender warehouse, railcar, stable and warehouse.

Bio-ecology

This house mite is a cosmopolitan species, common in various European countries, and less in North America, Japan and Australia (Hughes, 1976; Fain *et al.*, 1990). According to the list of commodities above reported, the mite occurs mainly on plant and animal residues in food stores, in pigsties, stables, barns and houses. It has been also collected in different animal and plant food materials (flour, wheat, grain, cheese, ham, linseed, dried meat), and also in tobacco, hay, sugarbeet seed, in beehives, mammal and bird nests, in stacks of hay on an open field, in byres, hay and grain storages, in herbarium (Chabasse and Genthorn, 1966; Barker, 1968b; Chmielewski, 1969b, 1971a; Hughes, 1976; Rosichý *et al.*, 1979; Banaszak, 1980; Mauri and Alzuet, 1980; Terho *et al.*, 1982; Tareev and Dubinina, 1985; Hallas and Solberg, 1989; Hallas and Iversen, 1996; Boczek, 1980, 1999) and in damp houses on mattresses, upholstery furniture with rush, sea grass fibre or green fibre, on wallpapers or in other house-dust samples (Blasco Sabio *et al.*, 1975; Hughes, 1976; Mauri and Alzuet, 1980; Van Bronswijk, 1981; Colloff, 1987b; Dusbábek, 1995; Hallas and Korsgaard, 1997). Solarz *et al.* (2004) reported that in Poland, *G. domesticus* has been collected in warehouses and different home-food products, in debris from cowsheds, barns, rabbit cages and lofts, and on an open field in hay, on bees and in beehives, on beetles (Scarabaeidae), in pollen traps, nests of domestic and tree sparrows, nests of pigeons, blackbirds and barn swallows (Boczek and Gołębiowska, 1959; Janicki *et al.*, 1959; Klimaszewska and Kwiatkowska, 1959; Wasylik, 1959, 1964, 1971, 1973; Boczek *et al.*, 1960, 1961; Chmielewski, 1969b, 1971a, c, 1972, 1982, 1983b, 2000; Boczek and Dutkiewicz, 1972; Sandner and Wasylik, 1973; Andrzejewska, 1979; Banaszak, 1980; Boczek, 1980, 1999; Karnkowski, 1990; Solarz *et al.*, 1994, 1997, 1998, 1999, 2000), in house-dust samples (Romański *et al.*, 1977; Dzięciolowski, 1994; Chmielewski, 1995; Solarz, 2000, 2001a, b), in coal-mine dust (Solarz and Solarz, 1996; Solarz *et al.*, 2002) and in dust from ocean ships (Wegner, 1980).

The life cycle of *G. domesticus* consists of egg, larva, prelarva, protonymph, tritonymph and adult. Various records treated the bio-ecology of the house mites, reporting commonly, the existence of the hypopus stage, and to in order to not change the testimony of these records we discuss here each report and transcribe in full these testimonies. Hora (1934) gave an account of the occurrence of *G. domesticus* in houses, and the conditions under which it is found, describing its life history. He discussed the occurrence of the hypopus, and the marked influence of the humidity and temperature on the life of the mite (on the adult, hypopus and egg). He reported that the life cycle developed in 22 days at 23–25°C and with a RH of 80–90%. Moreover, he discussed the effect of carbon tetrachloride and methyl salicylate on the hypopus. Barker (1968b) studied the life history and the reproductive rates of *G. domesticus* at different temperatures and two RH. He reported that most rapid development of the eggs occurred at 24.4°C, and the optimum temperature for the post-embryonic stages was 22.2°C. A higher percentage of hypopi were formed at 70–75%, than at 95–100% RH. *G. domesticus* has a net reproductive rate of over 20, which was intermediate between those of *Tyrophagus putrescentiae* (Schrank) and *Aëroglyphus robustus* (Banks). Matsumoto *et al.* (1995) examined the effect of temperatures ranging from 0–30°C at intervals of 5°C, and a graded series of RH from 64% up on the life cycle of *G. domesticus*. He employed as food, a mixture of dry yeast powder and fish meal powder in the same ratio by weight. At 30°C and 91.5% RH or 80.6% RH, the eggs did not hatch and at 10°C and 100% RH, egg hatchability was maximal. The moulting rates from egg to adult were high, with combinations of low temperatures and high RH, and low at temperatures above 20°C. The rate of emergence of hypopi increased, as the temperature rose and the RH decreased. At 5°C and 87.1% RH, there was high moulting rate for adults, the same as at 10°C, but the life cycle takes about 130 days. At 0°C and 81.9% RH, the eggs did not hatch until 127 days, but were hatchable. At 25°C and 94.0% RH, the minimum time required to develop from egg to adult was 16.3±1.7 days. The lower temperature and lower RH, prolonged the duration of development from egg to adult. The maximum average number of eggs laid by one female is 118.7±23.3 and occurred at 15°C and 87.3% RH. The eggs were highly resistant to low temperatures. They lived up to 60 days at –10°C but for only 1 day at –20°C. The hypopi moulted to tritonymphs at 15°C, and both 87.3 and 100.0% RH. Comfortable life conditions for *G. domesticus* were temperatures of 10–15°C and RH above 87.3% RH. Chmielewski (2002a) studied the biology of house mite at ca. 20°C and 85% of RH using as food crushed buckwheat seeds. He estimated that the life history takes 18.2 days, the mortality of juvenile instars is about 8.0%, the hatching of adults is 86.0%, the longevity of immature duration of specimens 58.7 days and the fecundity is 224.6 eggs per female lifespan. The results of this research showed that buckwheat is an attractive nutrient for glycyphagids and an effective medium for laboratory cultures of these mites. *G. domesticus* is an abundant mite in the environment of German farms, and a frequent mite species in urban dwellings. Depending on the environmental conditions (temperature, relative humidity, food), the duration of the different developmental stages varies. At constant 25°C and 75% RH, the mite takes 17 days to develop from egg to adult (Franz *et al.*, 2000).

Sinha and Harasymek (1974) exposed *G. domesticus* and *Acarus siro* Linnaeus, normally feeding on grain dust, to other stored foods and fungi, at 17±1°C and 75±2% RH to 19 species of fungi, and six species of bacteria. Both mites thrived on *Syncephalastrum racemosum* Cohn ex J. Schröt. (Syncephalastraceae), *Absidia spinosa* Lendner (*Cunninghamellaceae*) and *Curvularia geniculata* (Tracy *et* Earle) Boedijn (Pleosporaceae), but failed to survive and multiply on *Aspergillus ochraceus* Wilhelm (Trichocomaceae), *Gliocladium roseum Bainier (became Clonostachys rosea* f. *rosea* (Link) Schroers in the Bionectriaceae*)* or *Penicillium* sp. *G. domesticus* multiplied 30 times on *S. racemosum* and *A. siro* multiplied 40 times on *A. spinosa*. Neither of the two mites fed adequately or multiplied on any bacterial species. All mites died in about 35 days, and 50% died in 5 days. *G. domesticus* died sooner (8 days) on *Bacillus subtilis* (Ehrenberg) Cohn (Bacillaceae) than on any other bacteria or fungi. *A. siro* died soonest on *Bacillus megaterium*. They concluded that bacteria, unlike fungi, adversely affect both mite and insect pests of stored products. Hughes (1976) reported that the mite increased to enormous numbers on furniture covered with rush or sea grass fibre, or upholstered with green Algerian fibre, where it feeds on moulds growing on the fibre. About this, Sinha (1968c) showed that the mite is restricted in its fungal diet more than other mites.

Hallas and Korsgaard (1997) examined 3303 dust samples, collected in the dwellings of Danish allergic patients and showed a systematic variation in the weekly average concentration of mites during the year. They explained the variation as a nationwide, annual synchrony of the appearance of resting and active stages in the life cycles of the mites. From this, *Dermatophagoides* spp. develop four annual generations, *G. domesticus* develop three and *Tarsonemus* sp. has six annual generations. As resting stages are synchronous and underrepresented in samples, there is a risk of underestimating mite exposure when it is judged from single samples. The mattress samples had on average 139 mites/0,1 g dust, and in samples from floorings, the figure was 69, but the dust from mattresses in quantitatively a very small part of the dust from a dwelling and patient sampling time correlates with the population dynamics in floor dust. Therefore, it is concluded that beddings are not the primary source of clinically relevant exposure to house-dust mites. The concentration of house-dust mites in the patient dwellings was also associated with the age of the patients and increased with increasing number of persons in the households (Hallas and Korsgaard, 1997). Kuwahara *et al.* (1991) described the Neral as the alarm pheromone of the house mite.

As concerns the natural enemies, Barker (1968c) found the haemogamasid mite *Haemogamasus pontiger* (Berlese) in granaries in Canada infested by *G. domesticus*, *Oryzaephilus mercator* (Fauvel) (Coleopetera, Cucujidae) and *Tribolium confusum* du Val (Coleoptera, Tenebrionidae) and mice. He observed that nymphs and adults fed on *G. domesticus* and on larvae of *T. confusum* had been injured so as to make the haemolymph available. They also fed on heparinised mouse or human blood. The rate of oviposition was lowest when *G. domesticus* was the only prey available, and a maximum of one egg/day was obtained when females were provided with a mixture of injured larvae and *G. domesticus*. Eggs and immature mites that became males developed more rapidly at 27°C than at 24°C, but there was little difference in the development rates of females at the two temperatures.

Symptomatology and damage

Damage of this mite may be direct and indirect. However, although it infests grain dusts and other storage foods and fungi (Sinha and Harasymek, 1974) the direct damage seems of secondary importance in respect to that produced by other storage mites, for example, *A. siro* Linnaeus and other mites. Usually, its populations increase and spread with high level of RH, feeding mainly on fungi, causing a serious indirect damage of sanitary type. A rich literature confirms this aspect. Pillers (1920) found specimens of the mite, taken from the inside of a tame rabbit's ear, and Davies (1926) reports the mite as responsible for ear canker in rabbits. Robertson (1946) reported a number of cases in New Zealand of dermatitis in workers handling infested cheese. Deschiens (1951) has associated *G. domesticus* with other mites with symptoms of asthma in human beings. Blainey *et al.* (1988) discussed the role of storage mites in respiratory symptoms, such as cough and wheeze in farmworkers and farm grain stores. Li *et al.* (2003) found the mite in stool samples and urine samples of the workers The prevalence of human intestinal and urinary acariasis was not associated with gender, and these diseases are more frequently found in individuals working in medicinal herb, rice storehouses or mills and other sites with high density of mites. More attention should be paid to the mite prevention and labour protection for these high-risk groups. A comprehensive review of allergenic involvements of stored and house mites, including *G. domesticus*, has been published by Steinman and Ruden (2007). They reported that various studies highlighted that the storage mite *G. domesticus* may commonly induce symptoms of asthma and rhinoconjunctivitis in sensitized individuals, in both rural and urban settings (Petrunov *et al.*, 1977; Mumcuoglu and Gerber-Hobl, 1977; Wraith *et al.*, 1979; Goracci *et al.*, 1985; Stingeni *et al.*, 1997; Piu and Ballero, 1989; Rasp, 1991; van Hage-Hamsten and Johansson, 1998; Musken *et al.*, 2000). Samples of house dust from beds and carpets made by Colloff (1987a) in homes in Glasgow, Scotland, highlighted as more abundant mites *Dermatophagoides pteronyssinus* (Trouessart) (64.3%) and *G. domesticus* (Colloff, 1987a). In a mite survey conducted in the working environments of 121 farms in five regions of Germany *G. domesticus* resulted among the prevalent mite species (Franz *et al.*, 1997). In Spain, among 50 children with rhinitis and bronchial asthma, 10% were shown by means of skin test to be sensitized to *G. domesticus* (Sanchez Palacios *et al.*, 1991). *G. domesticus* has also been demonstrated to be present in house dust in Norway and other countries of Scandinavia (Mehl, 1998) and in Turkey (Kalpaklioglu *et al.*, 1997). Sensitization to this storage mite in Swedish bakers, was reported to be rare (Brisman *et al.*, 2003). In southwest Spain, a study was done on the prevalence of sensitization to *G. domesticus* in patients, naturally exposed to this mite. Approximately, 50% of the patients with *G. domesticus* at home, were sensitized to this mite. The authors suggest the inclusion of glycyphagid mite extracts in diagnostic tests, in areas where this mite is prevalent (Arias-Irigoyen *et al.*, 2007). In a study of the prevalence of mite species in houses in Bursa, Turkey, *G. domesticus* was found in 16.67% of houses (Gulegen *et al.*, 2005), and a lower prevalence (2.58%) has been reported from Kutahya in Turkey (Akdemir and Gurdal, 2005). In 100 children with a history of mild and moderate asthma who lived in Mexico City, skin tests demonstrated that *Glycyphagus* was positive in 12 (Prieto Ursua *et al.*, 1995). Also, in Chile, *G. domesticus* resulted to be present in homes (Franjola and Malonnek, 1995). The presence of this mite has also been reported from the United States. Among storage mite sensitizations in Wisconsin dairy farmers, sensitization to *Lepidoglyphus destructor* was the most frequently found, followed by *T. putrescentiae* and *G. domesticus* (Marx *et al.*, 1993). In Brunei, Woodcock and Cunnington (1980) reported that the role of storage mites in the causation of asthma in the tropics may have been underestimated. They examined dust from sleeping areas and found that storage mites, especially of Glycyphagid species, were predominant in house dust. On skin tests, on 60 asthmatics with 1% extracts of six mite species, *D. pteronyssinus* caused the greatest number of positive skin reactions (66.7%), with a positive reaction for *G. domesticus* of 40%. Allergy to *G. domesticus* may result in late respiratory reactions. Bronchial provocation tests performed on two separate occasions on a farmer with dust from his own grain led to immediate and late respiratory reactions, followed, without further exposure to grain dust, by severe respiratory reactions during subsequent nights (Davies *et al.*, 1976).

Control

According to the BPCA (British Pest Control Association) (2019), the control of house and storage mites is articulated in different actions, such as monitoring, evaluation of packaging and storage conditions, and direct control based on the application of physical and chemical means.

MONITORING. Early detection and eradication of storage mites are paramount to ensure quality, value and reputation. This action needs the use of mite trap and the flotation test.

MITE TRAPS. Traps are commercially available for catching mites. They contain an attractant made from non-toxic ingredients, allowing its use in a diverse range or premises and environments. Numbers and species of mites collected are assessed by trained personnel, in order to decide the remedial action. The trap detects live mites and other storage pests, and provides an early indication of problem areas, pinpointing hidden mite infestations and refuges. This gives the customer the opportunity to target remedial treatments and resources for best effect.

FLOTATION TEST. A flotation test has been developed to identify contaminating matter of animal origin (e.g. whole or fragmented mites and insects, or rodent hairs) in a wide range of food, feed, veterinary and dust samples and is AOAC accredited. It is the single most accurate and effective method of confirming the presence of mites in commodities, and can easily differentiate between dead and live mites. The flotation test can be used to assess mite infestations in a diverse array of products such as:

1. Fine powders (e.g. flour, dried milk, yeast)
2. Granular materials (e.g. animal feed, wheatgerm)
3. Compacted and pelleted materials (e.g. animal feed pellets, dairy nuts, pet food)
4. Seeds and grain (e.g. wheat, barley, oilseed rape, maize, soya, rice)
5. Dried fruits
6. Dust (e.g. house dust, mattress dust and floor sweepings for the detection of house-dust mites)
7. Plant material (e.g. hay, straw, wood or paper-based animal bedding)
8. Flower bulbs (e.g. tulips, hyacinths)
9. Veterinary samples (e.g. animal dander)

Samples are suspended in a careful balance of aqueous alcohol, acids and salt solutions before percolating kerosene upwards from the base. Any animal matter present, is collected on a filter paper, allowing easy examination away from the bulk sample.

PACKAGING TEST. The risk of infestation in packaged commodities is likely to vary according to the nature and constituents of the product, and to the fact that same foods are most suitable to infestations than others. Mites generally prefer cereal-based products but also dried fruits and dried plants. However, species such as *Tyrophagus* spp. also infest products with a high fat and protein content such as linseed, dried egg, groundnuts, cheese and ham. The form and nature of packaging are very important. Packaging is usually tailored to fit the product, and designed to last throughout the storage life of the product. The integrity of the packaging and the materials used are important considerations in preventing infestations. Mites are capable of passing through perforated plastic material, with perforations of less than 80 microns in diameter.

STORAGE CONDITIONS AND LENGTH. The more suitable the conditions for mite development, the more likely the products are at risk. Mites will develop faster at higher temperatures and most infestations are due to prolonged storage under less than optimal warehouse conditions. As well as the product's processing journey is important, a product can be infested at any stage from initial raw product to finished goods, and during its transit and storage along the way. From this point of view, research structure as FERA (Food & Environmental Research Agency) suggests, packaging penetration tests to help mitigate these risks of infestation. The tests consider the nature of your product, the form of packaging, storage conditions and length of time in storage, with a range of potential pests.

PHYSICAL CONTROL. Ždárková and Voráček (1993) studied the effects of physical factors (low barometric pressure, high and low temperature, light) on survival of main stored-food mites (*A. siro*, *T. putrescentiae*, *Aleuroglyphus ovatus* (Troupeau), *Caloglyphus berlesei* (Michael), *G. domesticus*, *Carpoglyphus lactis* (Linnaeus), *D. pteronyssinus* and *Cheyletus eruditus* (Schrank)). *C. lactis* was the most resistant species and most specimens (85%) of this species survived the longest exposure (96 hours) to the lowest pressure (95 mm Hg) tested. It showed 100% mortality, only after 80 hours exposure to –15°C, and it was able to withstand 45°C for about 1 hour. Acaridae mites were killed by low pressure of 95 mm Hg after an exposure of only 48 hours and after 1 h exposure to –15°C. Constant light has unfavourable effects on development and reproduction of the flour mite, *A. siro*. These results may have some practical implications. Vacuum of 190 mm Hg will protect the food against the mites. Also, low temperature –15°C could be used to control mites in seed (Ždárková and Voráček, 1993).

CHEMICAL CONTROL. As other storage and house-dust mites.

Glycyphagus (Glycyphagus) ornatus Kramer (Fig. 35.2)

Common name

Unknown

Morphological characteristics

FEMALE. The idiosoma is 540–600 μm long. The posterior margin of the genital opening is at the same level as apodemes III, and separated by a space from the posterior pair of genital setae. The bursa copulatrix opens at the tip of a small mound, projecting from the end of the body. The tarsi of all legs are almost straight and *ht* on tibiae I and II are normal setae. On tarsus I, the distal setae *la*, *ra*, *ba* and *wa* are close together. Conversely, in *G. domesticus* these setae are more widely spaced (Hughes, 1976).

MALE. The idiosoma is 430–500 μm long, oval widest between the second and the third pairs of legs, and slightly constricted behind the fourth pair. The *crista metopica*, as that of *G. domesticus*, and the setae *vi* arise from the widened region in the middle of the sclerite. The idiosomal setae are long and densely pectinate, and arise from distinct sclerotized bases. The setae d_1 (d_2) are short, their terminal ends barely reaching the bases of h_1 (d_4), they vary in position in different individuals, and may arise anterior or posterior to e_1 (d_3). The setae e_1 (d_3) exceed the idiosoma in length, and may possibly be movable, as each has a small internal projection at its base. The remaining posterior setae are also extremely long. The supracoxal setae is forked and branked. The tarsi I and II are curved, particularly the tarsus II, and the tibiae and genua are expanded distally. The ventral setae of all the legs are comparatively long and pectinate. On tibia I and II, the seta *ht* is modified into a triangular hairy structure, whose internal edge bears 9–10 teeth on leg I, four to five on leg II. On genu I, σ_1 is shorter than σ_2. On tibiae III and IV, the arthobotrial membrane extends to the base of *kt* (Hughes, 1976).

HYPOPUS. See Knülle (2003).

Geographical distribution

Azores (Portugal) (Borges *et al.*, 2010), China (Xin, 1965; Chen *et al.*, 1982; Tu *et al.*, 2001; Li *et al.*,

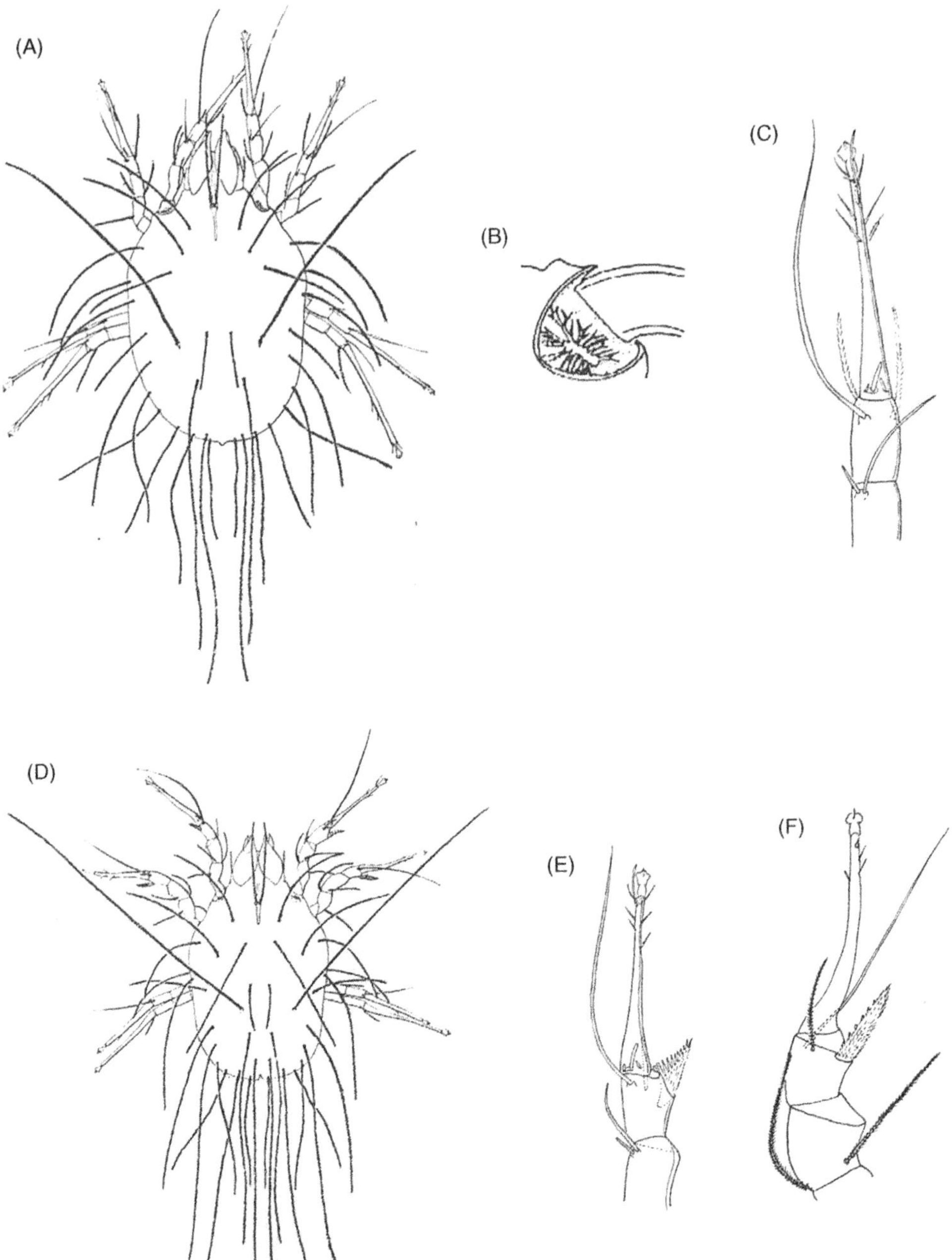

Fig. 35.2. *Glycyphagus* (*Glycyphagus*) *ornatus* Kramer. Female. Dorsal view (A); supracoxal seta (B); leg I (C). Male. Dorsal view (D); leg I (E); leg II (F) (from Hughes, 1976, licensed under the Open Government Licence v3.0).

2015a), Czech Republic (Hughes, 1976), Denmark (Luxton, 1995), Egypt (Bibars *et al.*, 2018), England (Hughes, 1976; Knülle, 2003), France (Hughes, 1976), Germany (Hughes, 1976; Knülle, 2003), India (Podder *et al.*, 2021), Israel, Italy (Hughes, 1976), the Netherlands (Siepel *et al.*, 2016), Poland (Pająk and Solarz, 2013; Solarz and Pająk, 2019a), Turkey (Bayram and Çobanoğlu, 2007), Russia

(Zachvatkin, 1936, 1941; Gilyarov and Krivolutski, 1975; Hughes, 1976) and other countries.

Commodities, facilities, vegetal and animal hosts

Hagstrum *et al.* (2013) reported a list of 13 commodities such as animal feed, Chinese medicine, grain, grass seed, hay, herb, oat, oilseed, processed food, sugarbeet seed, wheat, wheat flour residue and wheat flour sweeping. Bibars *et al.* (2018) include date palm fruits and Li *et al.* (2015a) reported from China, a wide list of commodities, such as crumbs, rape seed cake, beancurd skin, sugarcane, date plum persimmon (*Diospyros lotus* Linnaeus), *Zizyphus jujuba* Miller, honey lotus root dry, preserved peach, jam, dried longan pulp, bean paste cake, biscuit, carrot, winter bamboo shoots and wine lees. *Glycyphagus ornatus* Kramer has also been collected in flour residue from a London wharf, in mud scraped from horses, in animal feeding stuffs, in beech woodland soil and litter in Denmark and in the nests of a number of different species of animals (e.g. in nests of moles, bumble bees, sparrows, Levant voles, shrews and a number of other rodents and insectivores) (Wasylik, 1959; Hughes, 1976; Luxton, 1995). In Poland, this mite was collected in nests of tree and domestic sparrows (Wasylik, 1959, 1963, 1964, 1973), in nests of swallows, in farming environments (cowshed, pigsty) (Andrzejewska, 1979) and in granaries (grain, stored herb mixtures, sweepings) (Janicki *et al.*, 1959; Klimaszewska and Kwiatkowska, 1959; Boczek *et al.*, 1961; Chmielewski, 1972; Sandner and Wasylik, 1973).

The facilities include farm, grain stores, hay loft, stable (Hagstrum *et al.*, 2013) and farming environments (Solarz *et al.*, 2004). Recently, this mite was collected in farm building, barns, pigsties and cowsheds (Solarz and Pająk, 2019a).

Bio-ecology

Mite species known from England, Denmark, Germany, Holland, France, Italy, former USSR, Poland, former Czechoslovakia and Israel (Gilyarov and Krivolutskij, 1975; Hughes, 1976). It is common in the spillage from wheat, hay and oil-containing seeds (Hughes, 1976). *G. ornatus* has also been found in flour residue from a London wharf, in mud scraped from horses (Newstead and Duval, 1920), in animal feeding stuffs (Bollaerts and Breny, 1951), in beech woodland soil and litter in Denmark and in the nests of a number of different species of animals, such as in nests of moles and shrews (Zachvatkin, 1941), bumble bees and sparrows (Wasylik, 1959; Hughes, 1976; Luxton, 1995), a number of other rodents and insectivores (Vysotzkaja, 1961) and Levant voles (*Microtus guentheri*) (Costa, 1961). Occasionally, the mite has been found in the nest of *Bombus terrestris* (Linnaeus) (Zachvatkin, 1936). In Poland, *G. ornatus* was found in nests of tree and domestic sparrows (Wasylik, 1959, 1963, 1964, 1973), in nests of swallows, in farming environments (cowshed, pigsty) (Andrzejewska, 1979) and in granaries (grain, stored herb mixtures, sweepings) (Janicki *et al.*, 1959; Klimaszewska and Kwiatkowska, 1959; Boczek *et al.*, 1961; Chmielewski, 1972; Sandner and Wasylik, 1973). One male was found in a sample from a barn (Solarz *et al.*, 2004).

The life cycle of *G. ornatus* develops as other glycyphagid mites, including egg, larva, protonymph, tritonymph and adult. Knülle (2003) reported also the hypopial stage. This author reported that species of the genus *Glycyphagus* ancestrally inhabited the nest of small mammals, and some of them abandoned the original habitat and invaded plant remains in the field and at peridomestic sites. So, *G. ornatus* occurs in nature in high numbers in the nest of rodents and insectivores, but it can also occupy peridomestic environments. It is probably carried from nests into farm commodities by phoretic hypopodes attached to rodents. The sporadic occurrence of *G. ornatus* in straw, hay, grain spill and other substrates may be extensions of the nest habitat of rodents, according to various reports (Kamenski, 1940; Zachvatkin, 1941, 1953; Bollaerts and Breny 1951; Sorokin, 1953; Zachvatkin and Volgin, 1955; Woodroffe, 1953; Vysotskaja, 1961; Fain, 1969b; Chmielewski 1969b, 1971e, 1972; Kadzhaja, 1970; Jeffry, 1976b; Hughes, 1976; Fain and Mumcuoglu, 1979; Shchur and Golovach, 1981; Luxton, 1981; Lung-Shu, 1984; Daniel, 1988; Nangia and Channabasavanna, 1989b; Piao, 1990; Rosas and Casanueva, 1992).

Knülle (2003) reared *G. ornatus* on a food mixture containing fish meal, dried baker's yeast, blood meal and commercial food for aquarium fishes 'Tetramin', at equal parts, at constant RH of 85% and constant temperature of 20°C. He found that in the progeny of single inseminated females of *G. ornatus* and *Glycyphagus privatus* Oudemans, two distinct morphs of hypopus may occur simultaneously, in the same local population, one morph with attachment structures (claspers) for phoretic dispersal on carrier hosts and another non-phoretic and sedentary morph without such structures. It is a case of genetic dimorphism that poses the need to understand the way genetic heterogeneity (polymorphism) and ecological plasticity (polyphenism) interact and contribute to the expression of each hypopus morph, under beneficial and adverse environmental conditions, and to inquire into the adaptive significance and evolution of the two morphs (Knülle, 2003).

Mostafa *et al.* (2017) determined the effect of yeast and *Fusarium oxysporum* von Schlechtendal on the biological aspects of *G. ornatus*, at different laboratory conditions (15°C, 25°C and 35°C and 75% RH). The developmental period of *G. ornatus* was faster for the male than females, and the temperature of 15°C increased these periods than 35°C. The number of deposited eggs by females of *G. ornatus* differed according to the kind of food source and used temperature. The number of deposited eggs by females of *G. ornatus* was obviously increased, when females were reared on yeast at 25°C than any other diets and temperatures (Mostafa *et al.*, 2017).

Symptomatology and damage

The symptomatology of the attack of this species is poorly documented, probably due to its secondary economic importance. In any case, a correct management of the healthiness of foodstuffs must also take into account this mite, whose attack symptom could be similar to that of other *Glycyphagus* species. In Spain and China (see Cui, 2014), it has been associated with intestinal acariasis.

Control

Like other glycyphagid mites.

Glycyphagus (*Glycyphagus*) *privatus* Oudemans (Fig. 35.3)

Common name

Unknown

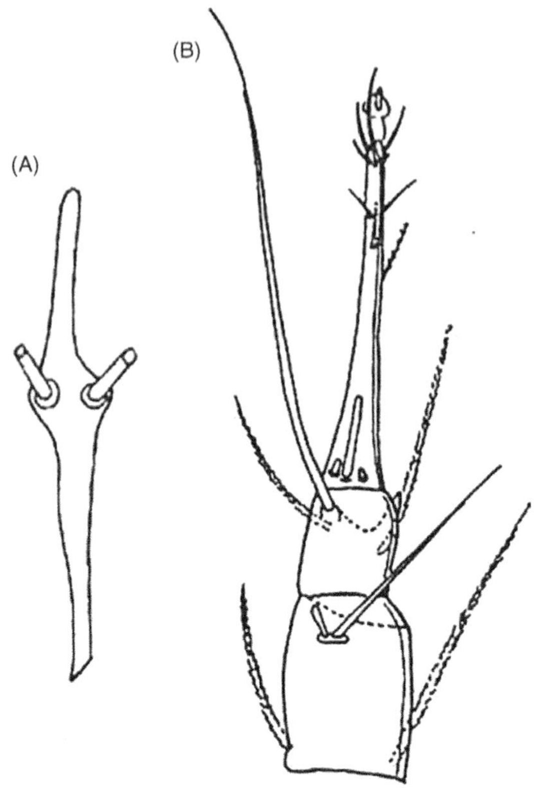

Fig. 35.3. *Glycyphagus* (*Glycyphagus*) *privatus* Oudemans. Female. Crista metopica (A). Male. Leg I (B) (from Hughes, 1976, licensed under the Open Government Licence v3.0).

Morphological characteristics

FEMALE. The idiosoma is 370–450 µm long. The genital opening is longer than in *G. domesticus*, extending as far back as the posterior edge of acetabula IV, its length exceeds the distance between it and the anal slit (Hughes, 1976; Solarz, 2011).

MALE. The idiosoma is 280–360 µm long, with texture of cuticle, length and thickness of setae, length of legs, as other *Glycyphagus* species. The crista metopica extending backwards to the *sc* setae; the setae *vi* arise near its anterior edge, and the region in front of this is indistinctly sclerotized. Of the dorsal setae, d_1 are inserted anterior to e_1, and of the same level as c_1. Apart from this, the arrangement of the dorsal setae is the same as in *G. domesticus*. The distal margin of tibia I and also tibia II is produced as a thin rim, and the setae and solenidia are arranged as in the previous species, any differences being in length of setae and solenidia, rather than in their position. On tarsus I, ω_2 is short, equal in length to the famulus, and on genu I, σ_1 is shorter – much less than half the length of ω_1 (Hughes, 1976; Solarz, 2011).

HYPOPUS. See Knülle (2003).

Geographical distribution

Azores (Portugal) (Borges *et al.*, 2010), Belgium (Knülle, 2003), China (Xin, 1965; Tu *et al.*, 2001), Czech Republic (Hubert *et al.*, 2006a), England (Newstead and Duvall, 1918), Finland (Nordberg, 1936), Germany (Oboussier, 1939), Iran (Noei, 2007; Noei and Ostovan, 2012), Italy (Ottoboni *et al.*, 1984–85; Knülle, 2003), Japan (Oshima, 1970), Poland (Solarz, 2010; Ciechacka *et al.*, 2013), Spain (Boquete *et al.*, 2000), Switzerland (Mumcuoglu, 1977) and other countries.

Commodities, facilities, vegetal and animal hosts

Commodities of this mite include barley, grass seed, Manitoba wheat, dust collected from silos and warehouses (Newstead and Duvall, 1918; O'Farrell and Butler, 1948), grass, clover, flaxseed fields and piles of debris left after winnowing (Zachvatkin, 1941), bran, flaxseed, chaff (Newstead and Duvall, 1918) and aquatic product, Chinese medicine, condiment, dried fruit, dried vegetable, grain, mushroom, sweeping (Fang and Cui, 2009), wheat, stored rice, rice dust and debris (Noei 2007; Noei and Ostovan 2012). Facilities include dock silos warehouses (O'Farrell and Butler, 1948) and barns (Newstead and Duvall, 1918). Moreover, the mite has been collected in nest of birds (Nordberg, 1936), in a sparrow's nest (Wasylik, 1959), in house (Newstead and Duvall, 1918), in house dust (Haarlov and Alani, 1970; Mumcuoglu, 1977) and in bed dust (Ciechacka *et al.*, 2013).

Bio-ecology

This species has a wide geographical distribution, and is reported on various stored products, among which barley, wheat, dried fruit, dried vegetables and other products (Zachvatkin, 1941; O'Farrell and Butler, 1948); moreover, it has been collected in house (Newstead and Duvall, 1918), in house dust (Haarlov and Alani, 1970; Mumcuoglu, 1977), in bed dust (Ciechacka *et al.*, 2013), a sparrow's nest (Wasylik, 1959) and in the nests of birds (Nordberg, 1936). Knülle (2003) reared *G. privatus* at 20°C and 85% RH, on mixed foodstuffs, consisting of three parts of commercial rat food (Altuomin-R, A. Rieper AG, Vinti, Bozen), one part dried baker's yeast and one part wheat embryos. The food was finely ground, and subsequently mixed with the same amount of wheat bran to enlarge the surface area.

The life cycle of *G. privatus* develops through the egg, larva, protonymph, deutonymph, tritonymph and adult stages. Knülle (2003) found that in the progeny of single inseminated females of *G. privatus*, two distinct morphs of hypopus may occur simultaneously in the same local population, one morph with attachment structures (claspers) for phoretic dispersal on carrier hosts, and another non-phoretic and sedentary morph without such structures. He dealt with a case of genetic dimorphism, with the aim to examine in which way genetic heterogeneity (polymorphism) and ecological plasticity (polyphenism) interact and contribute to the expression of each hypopus morph, under beneficial and adverse environmental conditions, and to inquire into the adaptive significance and evolution of the two morphs.

Symptomatology and damage

Direct damage to stored products seems occasional, as well as the ability to cause allergy to humans appears low. Cui (2014) reported in his review of human acariasis, some cases of intestinal acariasis occurred in Spain and China. The mite has been identified in the stools and positively treated with ivermectin.

Control

The need of control is low and eventually how other glycypagids.

Glycyphagus (Lepidoglyphus) destructor (Schrank) (Fig. 35.4)

Common names

Furniture mite, house mite, common hairy mite

Morphological characteristics

FEMALE. The idiosoma is 400–560 µm long. The shape and the arrangement of setae as in the male.

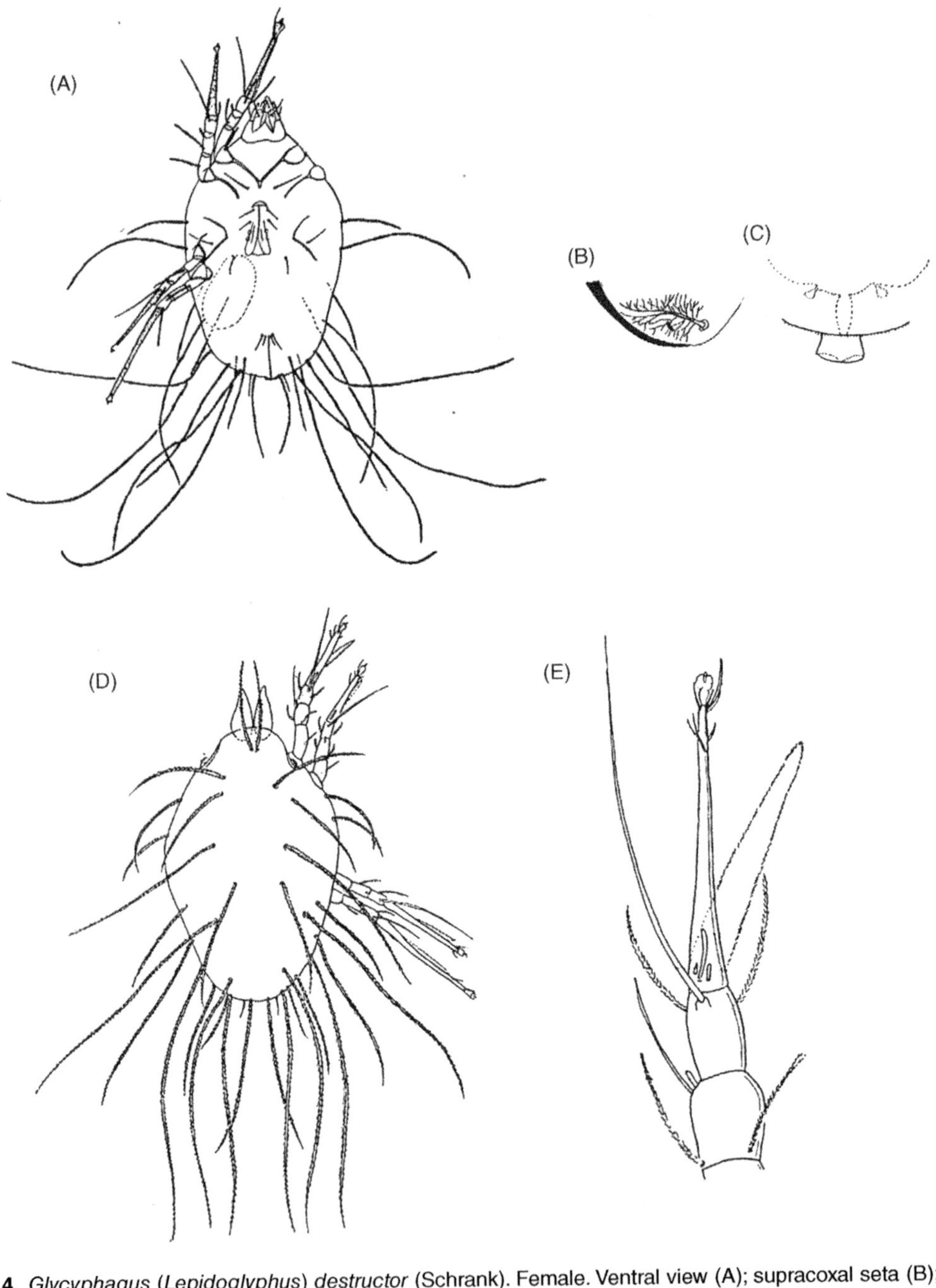

Fig. 35.4. *Glycyphagus* (*Lepidoglyphus*) *destructor* (Schrank). Female. Ventral view (A); supracoxal seta (B); bursa copulatrix and receptaculum seminis (C). Male. Dorsal view (D); leg I (E) (from Hughes, 1976, licensed under the Open Government Licence v3.0).

The genital folds meet for the greater part of their length, and a crescentic epigynum covers their anterior end; the posterior pair of genital setae is on a level with the hind end of the genital opening, between apodemes III and IV. The bursa copulatrix is a short tube with a partially lobed edge. The anus extends to the posterior margin of the body, and two pairs of setae, the outer longer than the inner, are inserted on either side of its anterior end. The setae *4a* arise behind the posterior edge of the genital opening; anus terminal; two pairs of setae (ad_3, ps_3) inserted on either side of its anterior end. The ventral seta *nG* of genua III not widened to form a pectinate scale; solenidion sigma on genu II not thickened (Hughes, 1976; Solarz, 2011).

MALE. The idiosoma is 350–500 μm, long and pear-shaped, constricted behind the fourth pair of legs. The cuticle is dull and whitish, as it is covered with minute papillae. When alive, the animal moves with a jerking gait. The dorsal setae are stiff and densely pectinate, standing away from the surface of the body. The setae vi extend well beyond the tips of the chelicerae, the setae ve are more posterior in position and the same distance apart as *sci*. The scapular setae are arranged in a transverse row across the idiosoma behind leg II, *sci* being almost the same length as *vi*. The supracoxal seta is a much-branched bifid rod. Two pairs of humerals are present; c_1 (d_1) are than d_1 (d_2) which barely reach the posterior margin of the body; e_1 (d_3) are posterior to d_1(d_2) and almost in linear series with c_1(d_1) and f_1(d_4). Three pairs of laterals are present, increasing in length towards the posterior end of the body. From the hind margin of the body, project *sai*, *sae* and three pairs of postanals, one pair of which *id* short and smooth. The setae e_1 (d_3), f_1 (d_4) l_3 and *sai* are the longest setae arising from the idiosoma. Ventrally, the apodemes of leg I meet to form a short sternum, and apodemes II are also well-developed; apodemes III and IV are much reduced, their function of muscle attachment being taken over by epimerites II, each of which has a strong anteriorly directed process. The genital opening lies between coxae III, its anterior end marked by a triangular plate; two pairs of genital setae lie on either side, and a third pair arises behind the posterior edge of the genital opening. The anal opening extends to the posterior margin of the body, it has one pair of setae at its anterior end. The chelicerae are slender, the movable digitus bears four large teeth, and five are present on the fixed digitus. The pedipalp terminates in three small processes.

The legs are long and thin, particularly the third and fourth pairs, and each terminates in a pretarsus and small claw. The tibiae, genua and femora are not expanded and produced into a thin rim at their distal ends. Each tarsus is enveloped in a pectinate subtarsal scale (*wa*), which is attached at the base. When in its natural position, this scale gives the tarsus a ciliated appearance, but in mounted specimens, it tends to get shifted to one side. At the distal end of the tarsus, the pretarsus is encircled by three setae (*d*, *e*, *f*), three small spines and a solenidion ω_3, and close behind these area, *ba* and *ra*. At the base of the tarsus ω_1, ω_2 and the famulus arise close together, ω_1 is a *curved* rod, and ω_2 is less than half the length of ω_1. On genu I, σ_1 is expanded distally, and σ_2 more than four times longer than σ_1; ventral setae *kT* on tibiae III and IV does not arise from the edge of the arthrodial membrane; the ventral setae of the genu and tibia are pectinate. The ventral seta *nG* of genua III not widened to form a pectinate scale; solenidion sigma on genu II not thickened (Hughes, 1976; Solarz, 2011).

HYPOPUS. The body is oval, about 350 μm long, colourless, with reduced legs and a transverse suture across the dorsal surface, dividing the body into propodosoma and hysterosoma. The apodemes of legs I and II are feebly sclerotized, and the rudiment of the genital opening is between leg IV. On legs I to III, the claw is as long as the tarsus, on leg IV is much shorter. A long solenidion arises from the base of the tarsus I. This stage is inactive, oval colourless, with reduced legs, rudiment of the genital slit and apodemes I and II feebly sclerotized; enclosed within the protonymphal cuticle (Hughes, 1976; Solarz, 2011).

Geographical distribution

Cosmopolitan, reported from various countries among which Azores (Portugal) (Borges *et al.*, 2010), Brazil (Flechtmann, 1986), Canada (Smallman, 1942; Sinha, 1963, 1979; White *et al.*, 2011), China (Xin, 1965; Wang, 1989; Li *et al.*, 2015a), Costa Rica (Vargas and Mairena, 1991), Czech Republic (Ždárková, 1967), Denmark (Iversen *et al.*, 1990), Egypt (Mahgoob *et al.*, 2006; El-Naggar *et al.*, 2007; El-Sayed and Ghallad, 2007), England (Griffts, 1960; Brady, 1970; Hughes, 1976), Faroe Islands (Denmark) (Hallas and Solberg, 1989), Finland (Thero *et al.*, 1980), Germany (Homann, 1933; Rack, 1968a), Greece (Athanassiou *et al.*, 2002; Palyvos *et al.*, 2008), Iceland (Denmark) (Hallas, 1981b), Iran (Sepasgozarian 1978; De Saint Georges-Gridelet *et al.*, 2003; Sayedi *et al.*, 2006, 2009; Ardeshir, 2010), the Netherlands (Siepel *et al.*, 2016), Japan (Miyamoto, 1974; Sinha, 1968b), Poland (Boczek and Gołębiowska, 1959; Chmielewski, 1969b), Russia (Zachvatkin,

1941; Dubinina, 1985), Scotland (Colloff, 1987a, 1987b), Spain (Blasco Sabio *et al.*, 1975), Taiwan (Tseng, 1972), Turkey (Cevizci *et al.*, 2010) and Ukraine (Oksentiuk *et al.*, 2022). Moreover, Rack (1968a) found the mite on various hosts from Italy, Bulgaria, Yugoslavia and other countries.

Commodities, facilities, vegetal and animal hosts

Hagstrum *et al.* (2013) reported a wide list of the mite, involving almost 64 commodities. These include animal feed, aquatic product, baby food, barley, barley residue, bean (green), beet seed, biscuit, bread, buckwheat, buckwheat product, calf stomach (dried), candied fruit, cereal, cheese, Chinese medicine, cocoa bean, condiment, cottonseed, dried calves stomachs, dried fruit, dried vegetable, flax seed, gin trash, grain, grass seed, hide (dried), hay, hemp seed, herb, insect (dried), kale seed, kaoliang, lettuce seed, linseed, maize, malt, millet spillage, mushroom, oat, onion seed, poppy seed, poultry diet, processed food, protein mix, radish seed, rapeseed, residue, rice, rice (husked), rye, seed cotton, sorghum residue, soybean, stored rice, rice dust and debris, straw, sugarbeet seed, sunflower seed, sweeping, sweet potato chip (dried), tea, vegetable (dried), wheat, wheat flour, wheat flour sweeping and wheat residue (Zachvatkin, 1941; Boczek *et al.* 1960, 1961; Sinha 1963, 1964b; Žďárková, 1967; Chmielewski, 1969b; Burton in Hughes, 1976; Hughes, 1976; Rosický *et al.*, 1979; Boczek, 1980a, 1999; Solarz, 1986; Hallas and Solberg, 1989; Fain *et al.*, 1990; Boczek, 1999; De Saint Georges-Gridelet *et al.*, 2003; Mahgoob *et al.*, 2006; Noei, 2007; Palyvos *et al.*, 2008; Noei and Ostovan, 2012; Beaulieu and Knee, 2014; Li *et al.*, 2015a).

Moreover, the mite has been collected on unhusked rice, highland barley, mildewed rapeseed, mung bean, pea (Li *et al.*, 2015a) and in raw tobacco and casava meal (Flechtmann, 1986). Siepel *et al.* (2016) reported a list of commodities and facilities from the Netherlands, such as meat meal, companies car seeds, farm stored maize, *Coriandrum* (seeds), potato peels, leg of horse and in crusts round mouth of dwarfgoats. The mite has been collected also on dead insects and dried mammal skins, on rodent and bumble bee nests (Zachvatkin, 1941; Hughes, 1976). Solarz *et al.* (2004) found the mite in barns from a cowshed, in cages and in runs of the Silesian Zoo, where the species was associated with canids, felids, artiodactyls and primates. In Ukraine, the mite has been collected in mills, granaries and warehouses, compound fodder factories, barns with animal fodder storage, outbuildings, honeybee beehives and vegetable storages (Oksentiuk *et al.*, 2022)

Hagstrum *et al.* (2013) reported as facilities commercial grain storage, drier, empty grain bin, farm, farm granary, farm milling machine, flour mill, grinder, mill, off-farm storage, provender mill, provender warehouse and railcar. Palyvos *et al.* (2008) collected the mite in commercial stores, farm stores, mills, silos and grain silos. In Egypt, Mahgoob *et al.* (2006) found this mite in grain residues and mixed flour in warehouses and mills. In the same country, El-Sayed and Ghallad (2007) found it in stored grain. In Iran, Ardeshir (2010) found the mite in nest of the bird white-eared bulbul *Pycnonotus leucotis* (Gould). In the same country, the mite has been collected in wheat grain and refuse sampled in silos and flour mills (De Saint Georges-Gridelet *et al.*, 2003) and was associated with whole wheat, flour and sweepings (residue from the floor, equipment and the outside of sacks) (Sayedy *et al.*, 2009). In Poland, the mite was found in farming environments (sweepings and litter or debris from barns, poultry houses, cowsheds) (Andrzejewska, 1979; Solarz *et al.*, 1994, 1997), in sparrow nests (Wasylik, 1963, 1964, 1971, 1973; Sandner and Wasylik, 1973; Andrzejewska, 1979; Chmielewski, 1982), in beehives (Banaszak, 1980), pollen traps (Chmielewski, 2002), in house dust (Chmielewski, 1995; Solarz 2000, 2001), in dust from warehouses and harbour buildings (Wiěcko, 1986), dust from ocean ships (Wegner, 1980) and in coal-mine dust samples (Solarz and Solarz, 1996; Solarz *et al.*, 2002).

Moreover, the mite was found in stacks of grain, straw and hay (on a field or in farm-houses, storages and barns), in floor debris samples from granaries, byres and stables, in grain dust from grain storages (Hallas, 1981; Terho *et al.*, 1982, 1985; Hallas and Gudmundsson, 1985; Fain *et al.*, 1990), in soil samples (old grassland, agricultural zones), in nests of rodents, birds, bees and bumble bees, on zoological exhibits (insects, mammals), in poultry houses and moreover in dump dwellings – behind wallpapers, in mattress stuffing or in house dust (Rack, 1968a; Blasco Sabio *et al.*, 1975; Hughes, 1976; Rosický *et al.*, 1979; Banaszak, 1980; van Bronswijk, 1981; Chmielewski, 1982; Dubinina, 1985; Colloff, 1987a; Hallas and Solberg, 1989; Hallas *et al.*, 1991; Vargas and Mairena, 1991;

Dusbábek, 1995; Boczek, 1999), and it has been observed that it could increase on poultry dropping in broiler houses (Brady, 1970). In Poland, *L. destructor* (Schrank) is common in stored food products (grain, dried herbs, grinding products, linseeds, sugarbeet seeds, maize-mealy products, spices and grocery articles, fruit-vegetable products, natural honey, home-food products) and in sweepings from warehouses and granaries (Boczek and Gołębiowska, 1959; Janicki *et al.*, 1959; Klimaszewska and Kwiatkowska, 1959; Boczek *et al.*, 1960, 1961; Boczek and Dutkiewicz, 1972; Chmielewski, 1969b, 1971a, c, 1972, 1977b; Solarz, 1986; Karnkowski, 1990; Boczek, 1999).

Bio-ecology

G. destructor is a cosmopolitan species, very common among storage mites and often associated with different stored products, peculiarly in stored grain, linseed, cereals, dried fruits, sugarbeet seed and meal (Zachvatkin, 1941; Chmielewski, 1969b; Hughes, 1976; Rosický *et al.*, 1979; Boczek, 1980a, 1999; Fain *et al.*, 1990). It was found also in stacks of grain, straw and hay (on a field or in farm houses, storages and barns), in floor debris samples from granaries, byres and stables, in grain dust from grain storages (Zachvatkin, 1941; Hallas, 1981; Terho *et al.*, 1982, 1985; Hallas and Gudmundsson, 1985; Fain *et al.*, 1990), in soil samples (old grassland, agricultural zones), in the nests of rodents, birds, bees and bumble bees, in the zoological exhibits of insects and mammals, in poultry houses and, moreover, in damp dwellings behind wallpapers, in mattress stuffing or in house dust (Zachvatkin, 1941; Blasco Sabio *et al.*, 1975; Hughes, 1976; Rosický *et al.*, 1979; Banaszak, 1980; van Bronswijk, 1981; Chmielewski, 1982; Dubinina, 1985; Colloff, 1987a; Hallas and Solberg, 1989; Hallas *et al.*, 1991; Vargas and Mairena, 1991; Dusbábek, 1995; Boczek, 1999).

In the temperate climate, the mite is very common, and an high number of reports on its presence, on various stored food products, occurs in the central European countries, such as the Poland, where the mite has been extensively investigated and collected on grain, dried herbs, grinding products, linseeds, sugarbeet seeds, maize-mealy products, spices and grocery articles, fruit-vegetable products, natural honey, home-food products and in sweepings from warehouses and granaries (Boczek and Gołębiowska, 1959; Janicki *et al.*, 1959; Klimaszewska and Kwiatkowska, 1959; Boczek *et al.* 1960, 1961; Chmielewski 1969b, 1971a, 1971c, 1972; Boczek and Dutkiewicz, 1972; Solarz, 1986; Karnkowski, 1990; Boczek, 1999;). In the same country, it was found in farming environments (sweepings and litter or debris from barns, poultry houses, cowsheds) (Andrzejewska, 1979; Solarz *et al.*, 1994, 1997), in sparrow nests (Wasylik, 1963, 1964, 1971, 1973; Sandner and Wasylik, 1973; Andrzejewska, 1979; Chmielewski, 1982), in beehives (Banaszak, 1980), pollen traps (Chmielewski, 2002b), in house dust (Romański *et al.*, 1977; Chmielewski, 1995; Solarz, 2000, 2001a, b), in dust from warehouses and harbour buildings in Gdynia (Wiĕcko, 1986), in dust from ocean ships (Wegner, 1980) and in coal mine dust samples (Solarz and Solarz, 1996; Solarz *et al.*, 2002). Moreover, Solarz *et al.* (2004) collected the mite from barns from a cowshed, in cages and in runs of the Silesian Zoo, where the species was collected in association with canids, felids, artiodactyls and primates.

In Western Canada, *G. destructor* (and *A. siro* Linnaeus) is a common mite pest species, associated with grain stored in the farm, or in primary elevators (Sinha, 1963). Periodic outbreaks can occur every few years, unaffected by grain ageing (Sinha and Wallace, 1973). Its hypopial stage does not feed and can be resistant to unfavourable conditions, such as desiccation (Griffiths, 1964b). Sinha (1964c) reported its survival at temperatures of –18°C and its inactivity.

Miyamoto (1974) analysed the effects of temperature on the growth and reproduction of *G. destructor* by individual rearing method, starting from hypopi collected from house dust. He exposed the mite to seven different temperatures ranging from 2°C to 37°C and at about 85% RH, using as food a mixture of equal amount of dried yeast and laboratory animal food powder. The mite passes through four stages in a life cycle (egg, larva, protonymph, tritonymph and adult). No hypopal stage (as phoretic deutonymph) was observed. The days required for a life cycle from newly deposited eggs to the first oviposition of the females were 72.8 days, in average at 10°C, 27.8 days at 15°C, 21.5 days at 20°C, 10.0 days at 25°C and 9.6 days at 30°C. The eggs did not hatch at 2°C, as well as at 37°C. The survival rate of mites which completed the development from larva to adult was usually more than 95%, at the temperature levels 15–25°C. The lifespan of the females as well as their oviposition period, was the longest at 10°C, and the period became shorter as the temperature was raised, but the average number of eggs produced per day per female increased as the temperature became higher, with the maximum of 6.8 eggs per day per female in the average at 25°C. The sex ratio of adult which emerged at 25°C was ♀: ♂=1: 0.8, and their 50% survival period of the adult at 25°C was 15 days in females and 11 days in males. The average survival period was 16.1 days in females and 13.3 days in males (Miyamoto, 1974). Stratil *et al.* (1980) investigated the innate capacity for increase of *G. destructor* within a broad range of humidity and temperature (70% RH to close to saturation and 10–30°C), under favourable and standardized nutritional conditions. The innate capacity for increase increased with temperature, and attained its highest value at 25°C; a decrease occurred at 30°C. Concerning the dependence on humidity, the highest values were attained at RH from 85% to 92% RH, at all temperatures. At humidities higher and lower than this optimal range, there is a parallel decrease of the rate of population increase, at all temperatures. At 30°C, however, an increase of the population can occur only at humidities close to 84.5% RH (Stratil *et al.*, 1980).

Barker (1983) studied the life history and reproductive rates of *G. destructor* at 14°C, 20°C, 25°C and 30°C and 75% RH. He found that there was no development at 30°C. A few hypopi were produced at 14°C, but none at 20°C and 25°C. The highest net reproductive rates of 53 and 55 females produced per female per generation were found at 14°C and 25°C (Barker, 1983). Parkinson *et al.* (1991) evaluated the egg production of isolated pairs of *A. siro*, *G. destructor* and *Tyrophagus longior* (Gervais), fed on a control diet of wheat germ and yeast, compared with that on mycelial pellets from shake cultures of *Cladosporium cladosporioides* (Fresen.) G.A. de Vries, *Aspergillus repens* Corda (Sacc.), *Aspergillus ruber* Thom *et* Church and *Penicillium cyclopium* Westling as well as spores of *P. cyclopium* and *A. repens*. The mites always produced fewer eggs on the fungal diets, than on the ideal diet. *T. longior* usually did best of the three mites on the fungal diets, and *G. destructor* worst. *A. ruber* proved the most suitable fungus for all mite species, while *C. cladosporiodes* was the least suitable. Spores were a less suitable diet than mycelial pellets from shake cultures, which were predominantly mycelium (Parkinson *et al.*, 1991). El-Naggar *et al.* (2007) reared *G. destructor* individually on fungal colonies of *Aspergillus niger* van Tieghem, *Alternaria alternata* (Fr.) Keissl. and *F. oxysporum* von Schlechtendal. Tritonymphs were reared on the colony of each fungus, individually at four different temperatures (20°C, 25°C, 30°C and 35°C) and 75% RH. The study showed that *G. destructor* passed through egg, larva, protonymph, tritonymph stages, before reaching adulthood, at 20°C, 25°C and 30°C. The hypopus or deutonymph appeared at 35°C only, which is produced as a response to environmental stress. When the three fungi were compared, the life cycle of female was long when reared on the fungus *F. oxysporum,* 17.21 days, but for male it reached 10.65 days, when fed on *A. alternata* at 20°C. The adult mites lived longer on the colony of *A. niger* (35.8 days for female and 20.25 days for males at 20°C), than on colonies of *A. alternata* and *F. oxysporum*. The fungus *F. oxysporum,* resulted the best host for mite egg laying (351 eggs) at 35°C, while the fungus *A. niger* was the unfavourable one for egg laying (172 eggs) at 35°C (El-Naggar *et al.*, 2007). Chmielewski (2001b) studied life history parameters of *G. destructor* in laboratory, at about 20°C, RH approximately 85% and bruised buckwheat grains used as a mite food. In these conditions, the life cycle develops in 17.7 days, the eclosion of imagines is 71.0%, the percentage of females is 45.1%, the longevity of imagines is 37.7 days, the productivity of females is 78.1 eggs laid per whole lifespan. These parameters provided evidence of the nutritive attractiveness of buckwheat seeds for *L. destructor*, and the effectiveness of this kind of food as a medium for laboratory monocultures of the species (Chmielewski, 2001b). Laboratory experiments with *G. destructor* on a diet of mainly whole wheat were conducted, to study the mite's development and production of a specific allergen (Lep d 2), at four different temperatures (5°C, 10°C, 15°C and 20°C) and three levels of RH (ca. 70–88%). Statistical models were used to analyse the role played by temperature, RH and time, in explaining the observed number of *G. destructor* and the amount of allergen produced. Moreover, the life stage distributions of the mites were determined and related to the population growth. Based on a statistical model, the intrinsic rate of natural increase (*rm*), was computed for a range of different temperatures and RH. High RH in combination with temperatures at about 25°C will lead to the highest *rm* (ca. 0.15 day). The highest concentration of Lep d 2 was 3 mg/g grain, found at 20°C and high RH at a mite density of 254 mites/g grain. The concentration of allergens in the grain was best explained by a model that incorporated both the current and the cumulative numbers of mites (Danielsen *et al.*, 2004).

As concerns the natural enemies, the predators such as the laelapid *Androlaelaps casalis* (Berlese), and the cheyletid *C. eruditus* (Schrank) appear to be dependent on the densities of prey populations within the bin, usually following a similar population growth and crash, as that of the prey, but somewhat delayed (Sinha, 1963; Sinha and Wallace, 1973). Sinha (1979) reported that the two predators are known to feed on them.

Symptomatology and damage

G. destructor (and *A. siro*) is a common mite associated with grain stored on the farm, or in primary elevators in western Canada (Sinha, 1963) and other countries. White *et al.* (2011) consider the mite of primary feeding type directly on grain. Zachvatkin (1941) reported that the nature and degree of damage caused by *G. destructor* had not been investigated, and it was known that the mite is capable of gnawing small indentations in the surface of the seed, but it is incapable of penetrating into the embryo; it has been observed that *G. destructor* lives mainly on debris present among the grain. Its presence in large numbers can make grain and its products unacceptable, and unpalatable for human or animal consumption. In addition, human allergic sensitivities can occur to mite bodies and excreta within grain dust (Revsbech and Andersen, 1987). *A. siro* and *G. destructor* induce allergic reactions like rhinitis and asthma in humans (Fain *et al.*, 1990). They feed on spores and mycelia, and can carry fungi spores to other places, thus distributing the fungi from bird nests to grain stocks (Sinha and Mills, 1968).

G. destructor cannot chew through the hard parts of seeds, preferring instead to feed on grain dust, broken kernels and the field fungus *A. alternata*. Akimov (in Sinha, 1979) has suggested that the digestive system and enzymes of *G. destructor* are adapted to assimilating small particles. Chronic occurrence by this species is linked with seasonal temperatures, and periodic outbreaks tend to be linked with the abundance of *A. alternata* (Sinha and Wallace, 1973). Beaulieu and Knee (2014) reported that in Canada, the most economically important stored-product mites are *A. siro*, commonly called the grain mite or flour mite, and *G. destructor*. They are usually abundant in stored wheat, oats and barley (Sinha 1963, 1964b). Although these species are clearly fungal grazers that can breed on a fungal diet alone, they are able to directly damage cereal grains (Sinha, 1964a, 1979; Griffiths, 1970). *A. siro* can feed directly on the cereal germ, whereas *G. destructor* may only be able to feed on grain dust and broken kernels. In contrast, acarid mites such as *A. siro* have mouthparts that can penetrate grain kernels, causing internal seed changes. *A. siro* prefers to feed on the seed germ, and its digestive and enzyme systems enable it to assimilate hard particles. Economically, it is the most important mite pest of stored grain in Canada (Sinha, 1963). Periodic outbreaks can occur every few years, unaffected by grain ageing (Sinha and Wallace, 1973).

G. destructor is a facultative bee-associated mite, sometimes typically abundant in honeybeehives associated with other mites (*Acarus* spp., *Tyrolichus casei* Oudemans, *Tyrophagus* spp., *G. domesticus* (De Geer) (Eickwort, 1988)).

Control

The protocol of control is similar to other glycyphagid mites. Monitoring, prevention and curative means are adopted. Very interesting is the use of physical means as the aeration and the modified atmospheres.

PHYSICAL CONTROL. Thind and Dunn (2002) reported that aeration is a promising alternative to the use of pesticides for the control of storage insects by cooling bulk grain, and that its effectiveness against mite pests is neither fully understood, nor optimised. At this aim, they investigated the productivity of three species of storage mites (*A. siro*, *G. destructor* and *T. longior*), in a laboratory experiment at four combinations of temperature and humidity (10°C and 70% RH, 10°C and 80% RH, 20°C and 70% RH, 20°C and 80% RH), with and without an airflow (at 10 m^3/h/tonne, equalling 2.5 1/s/tonne, in tubes containing 15g of grain). This is the first time that a study has examined the three principal components of aeration separately from each other. The effect of these factors was different for each species. For *A. siro*, temperature was the most important factor, while airflow and humidity were of similar but lesser importance. For *T. longior*, temperature was more important than humidity, while the reverse was true for *G. destructor*. For these two species, airflow was the least important factor. The airflow decreased the productivity of *G. destructor* and *T. longior* but increased the productivity of *A. siro*.

This increase in productivity confirms that in practice prevention of mite infestations, in particular *A. siro* will require storage of grain at low temperature, relative humidity and moisture content (Thind and Dunn, 2002).

Conyers and Bell (2003) investigated in laboratory, the effects of modified atmospheres (MA) on the eggs of mite pests of grain and cheese. They tested *A. farris* (Oudemans), *A. siro*, *G. destructor* and *T. longior*. Three low oxygen (O_2) MA mixtures were used, based on carbon dioxide (CO_2), nitrogen (N_2) or simulated burner gas (0.5 or 2% O_2, 10% CO_2, balance N_2) plus 60% CO_2 in air (8% O_2). The mites were exposed at 15°C and 80% RH, a combination of conditions that occurs at the surface of stored grain during the autumn, which promotes mite population growth. The exposure periods required to prevent egg hatch, for each species, in every mixture are given. *T. longior* was the most tolerant species, followed by *A. siro* and *A. farris*, with *L. destructor* the most susceptible. Burner gas was the most effective mixture overall, with 0.5% O_2, but with an increase in the O_2 level to 2% for all the mixtures, CO_2 became the more effective control agent. With 60% CO_2 in air, some loss of efficacy was observed against the three most tolerant species and even more so for *G. destructor*. Sublethal exposures to MAs for at least 4 days in *G. destructor*, 6 days in *A. farris* and *A. siro* and 8 days for *T. longior* caused a delay in egg hatch (Conyers and Bell, 2003).

BIOLOGICAL CONTROL. The predators such as the laelapid *A. casalis* and the cheyletid *C. eruditus*, appear to be dependent on the densities of prey populations within the bin, usually following a similar population growth and crash as that of the prey, but somewhat delayed (Sinha, 1963; Sinha and Wallace, 1973). Sinha (1979) reported that the two predators are known to feed on them.

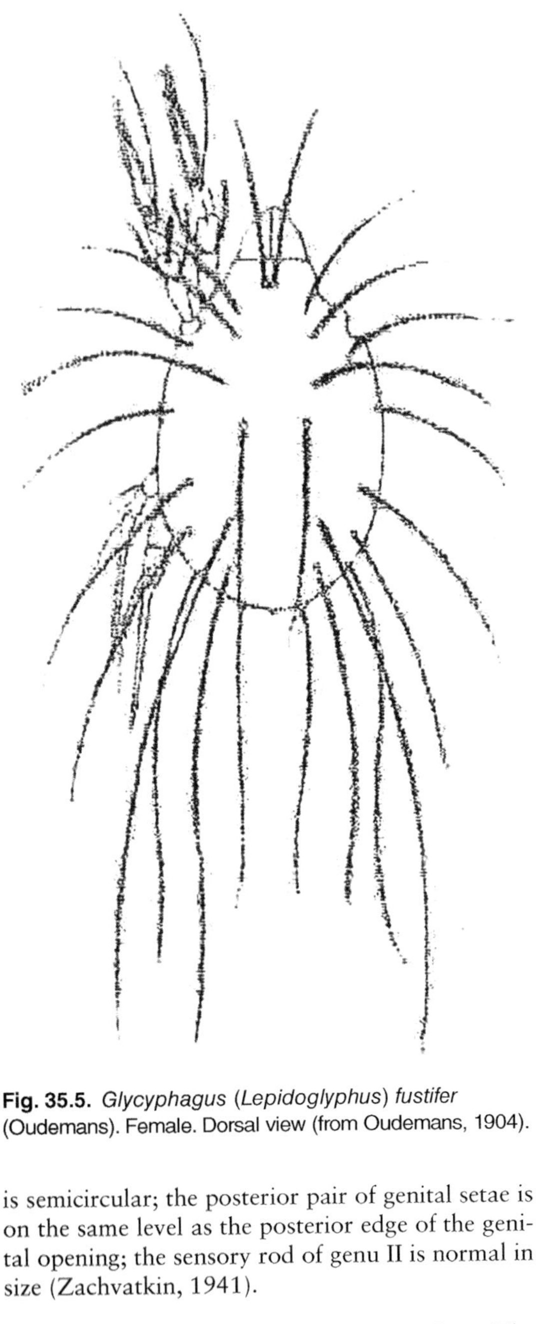

Fig. 35.5. *Glycyphagus* (*Lepidoglyphus*) *fustifer* (Oudemans). Female. Dorsal view (from Oudemans, 1904).

Glycyphagus (Lepidoglyphus) fustifer (Oudemans) (Fig. 35.5)

Common name

Unknown

Morphological characteristics

FEMALE. The idiosoma is 415–480 μm long. The genital opening is very large, much longer than the distance separating it from the anal slit; the epigynum is semicircular; the posterior pair of genital setae is on the same level as the posterior edge of the genital opening; the sensory rod of genu II is normal in size (Zachvatkin, 1941).

MALE. The idiosoma is almost 300 μm long. The epiandrium has a short linear protrusion; the genital and *pra* setae are short; the sensory rod of genu II is enlarged to form a massive spine, equal in length to the genu itself (Zachvatkin, 1941).

The resting hypopus is unknown (Zachvatkin, 1941).

We report here the whole morphological description of this mite by Zachvatkin (1941). The general features of both sexes concern the setae *vp* (*ve*) which are as wide apart as *sci*, so that their bases form a rectangular figure; the distance between *vp* (*ve*) and *sci* is five times smaller than that between the two *sci* setae, and three to four times less than the distance between *va* (*vi*) and *vp* (*ve*); the base of has no projection. The pectinate setae of the idiosoma are about the same length and thickness as in *G. destructor* (Schrank); dg extend slightly beyond the posterior end of the body; all three pairs of *pa* are long and pectinate. The epimerites are the same length as epimeres II. The tibiae and genua are normal in structure, without funnel-shaped widenings; the pectinate setae of the tibiae, and genua I and II are much broader than in *G. destructor*; the basal seta of genu III (*HI*) is represented by a simple, pectinate seta, which is not modified; the ventral setae of tibiae III and IV are noticeably removed from the distal edge of the segment; the apical seta of genu I is extremely short, barely longer than the sensory rod, which is next to it (Zachvatkin, 1941).

Geographical distribution

China (Li *et al.*, 1992a), Czech Republic (Ždárková, 1967), Germany (Oudemans, 1904), Hong Kong (Olsen, 1983), Poland (Boczek *et al.*, 1961; Solarz *et al.*, 2004; Pająk and Solarz, 2013; Solarz and Pająk, 2019), Russia (Zachvatkin, 1941; Gilyarov and Krivolutski, 1975), Ukraine (Akimov and Oksentiuk, 2018; Oksentiuk, 2020; Oksentiuk *et al.*, 2022) and other countries.

Commodities, facilities, vegetal and animal hosts

Hagstrum *et al.* (2013) reported 13 commodities, including bean curd, dried fruit, dried vegetable, dried fungus, grass seed, maize, mushroom and seaweed. Other specialists reported cottonseed (Ždárková, 1967), grain (Akimov and Oksentiuk, 2018), herb (Boczek *et al.*, 1961), poppy seed (Ždárková, 1967), rye (Zachvatkin, 1941), sweeping (Solarz *et al.*, 1994, 1997), turnip seed (Ždárková, 1967). Akimov and Oksentiuk (2018) found this mite as secondary species in Ukraine on oilseeds, grain and beehives. In Poland, Solarz and Pajak (2019) found the mite in various facilities, as farm, especially in pigsties, cowsheds, chaff-cutter building and poultry houses; moreover, it was found in house-dust samples (Solarz, 2001a, b), in stored herbs (Boczek *et al.*, 1961), in sweepings and litter or debris from barns and cowsheds (Solarz *et al.*, 1994, 1997). Solarz *et al.* (2004) found 22 specimens in four samples from barns (2.53% of all mites collected from barns), 26 specimens in three samples from poultry houses (3.12%), and 20 ones in three samples from cowsheds (0.75%). Pajak and Solarz (2013) found that in selected farming environments in Poland, among 960 mites collected from poultry houses, *Lepidoglyphus fustifer* (Oudemans) was predominant (41.2% of the total count) and generally the most abundant species among Astigmatina. The dominance of *L. fustifer* in poultry houses was significantly greater than in cowsheds and chaff-cutters ($\chi 2$ = 15.03 and 19.44; p = 0.0001 and 0.00001, respectively), whereas the difference between poultry houses and barns in relation to the dominance of this species, was statistically nonsignificant ($\chi 2$ = 0.52; p = 0.47). Oksentiuk (2020) reports that in Ukraine, *L. fustifer* belongs to the group of species with a low ecological importance, and it was registered only in the barns with animal forage. Successively, the same author collected this mite in mills, granaries and warehouses, barns with animal fodder storage and honeybee beehives (Oksentiuk *et al.*, 2022).

Bio-ecology

Zachvatkin (1941) reported that this is a synantropic species, first found in an inhabited house, and again in large numbers in the store of a collective farm on rye seed. It is possible that the form found in the former USSR is not identical to that from Bremen, Germany, as Oudemans (1903) describes and figures the hypertrophied solenidion of genu II.

Solarz *et al.* (2004) report that this species is known from Germany, England, Holland, Sweden, former USSR, Poland and former Czechoslovakia. In fact, the species has a wider geographical distribution although it is not so widely distributed as the previous one, but it is more common in stored food. This species was found originally on furniture in a house. It has also been found on rye, cotton, poppy and turnip seeds (Gilyarov and Krivolutski, 1975; Hughes, 1976; Dubinina, 1985; Tareev and Dubinina, 1985). Moreover, in Poland, it was found in stored herbs (Boczek *et al.*, 1961), in sweepings and litter or debris from barns and cowsheds (Solarz *et al.*, 1994, 1997). Solarz *et al.* (2004) evaluated the species as recedent and euconstant in farming environments.

Symptomatology and damage

We do not dispose of comprehensive reports on the damage caused by this mite to stored products. It is ecological role seems of secondary importance, but its presence should be monitored.

Control

When strictly necessary as other Astigmatic mites.

The Genus *Lepidoglyphus* Zachvatkin

This genus includes species without crista metopica on the dorsal propodosoma. The tarsi are ensheathed in a pectinate subtarsal scale. The solenidion σ_2 of the genu I is twice as long as σ_1. Tibiae I and II with two ventral setae. Genital openings of male and female lie between the coxae II and III (Hughes, 1976).

Lepidoglyphus michaeli Oudemans (Fig. 35.6)

Common name

Unknown

Morphological characteristics

FEMALE. The idiosoma is 700–900 µm long. The genital opening is further forward than in *L. destructor*, its posterior edge on a level with the anterior ends of apodemes III, and the posterior pair of genital setae fa behind it; the anterior end of the opening is covered by a crescentic epigynum. The bursa copulatrix is a short inconspicuous tube (Hughes, 1976).

MALE. The idiosoma is 450–550 µm long. The idiosomal setae are densely pectinate. The setae *sci* are much longer than *vi*. The leg segments, particularly the tibia and genu of leg IV are expanded distally into a thin transparent flange, enveloping the base of the succeeding segment. The ventral setae of the tibiae are more 'bushy' than in *L. destructor* and, in both sexes, the ventral seta *nG* of leg III is widened to form a 'furry' scale. On tibiae III and IV, the distal arthrodial membrane extends as far back as the base of *kT* as in *G. ornatus*, and the cuticle on either side is produced into a thin plate so that *kT* arises at the base of a deep cleft (Hughes, 1976).

HYPOPUS. The idiosoma is about 260 µm long. It remains enclosed within the protonymphal cuticle, which becomes dried and covered with a reticulate pattern (Michael, 1901). The hypopus itself has a pear-shaped body with rudimentary limbs; it is without a sucker plate, and is only capable of slight movement (Hughes, 1976).

Geographical distribution

Bulgaria (Zheleva and Angelkova, 1964), Canada (Beaulieu and Knee, 2014), China (Chen *et al.*, 1982; Li *et al.*, 2015; Zhan *et al.*, 2017b), Czech Republic (Ždárková, 1967), England (Michael, 1901), France, Germany (Oudemans, 1903; Hughes, 1976), Hungary (Mahunka, 1961), Ireland, Italy (Hughes, 1976), the Netherlands (Hughes, 1976), Poland (Janicki *et al.*, 1959; Boczek *et al.*, 1960, 1961; Solarz *et al.*, 2004), Russia (Zachvatkin, 1941; Vysotzkaya, 1961), Sweden (Sellnick, 1958) and Ukraine (Oksentiuk *et al.*, 2022).

Commodities, facilities, vegetal and animal hosts

This mite is reported on broken rice, broomcorn (Li *et al.*, 2015), Chinese medicine (Radix *et* rhizoma rhei) (Zhan *et al.*, 2017b), maize, grass seed, millet, oat and soybean (Li *et al.*, 2015). Moreover, it is abundant in dried vegetable matter, in granaries, stables (on hay and fodder) (Michael, 1901). It has been found in house-dust samples, in brewer's yeast and on moist grain (Gilyarov and Krivolutskij, 1975; Hughes, 1976; Van Bronswijk, 1981), in barns (hay, straw, sweepings and litter or debris) (Andrzejewska, 1979), barns with animal fodder storage and outbuildings (Oksentiuk *et al.*, 2022), and in warehouses or granaries (flax-seeds, herbs, grain, sweepings) (Zachvatkin, 1941; Janicki *et al.*, 1959; Klimaszewska and Kwiatkowska, 1959; Boczek *et al.*, 1960, 1961; Sander and Wasylik, 1973; Beaulieu and Knee, 2014). The mite has been collected also in rodent and insectivore nests (Vysotzkaya, 1961), and in beehives but more rarely than other *Glycyphagus* species (Klimov *et al.*, 2016).

Bio-ecology

The mite is widely distributed in Europe, known from France, England, Germany, Holland, Sweden, former USSR, Poland, Hungary, former Czechoslovakia and Bulgaria (Hughes, 1976; Solarz *et al.*, 2004). Although, *L. michaeli* is not so common in stored food as *L. fustifer* under natural conditions, it is widely distributed (Hughes, 1976).

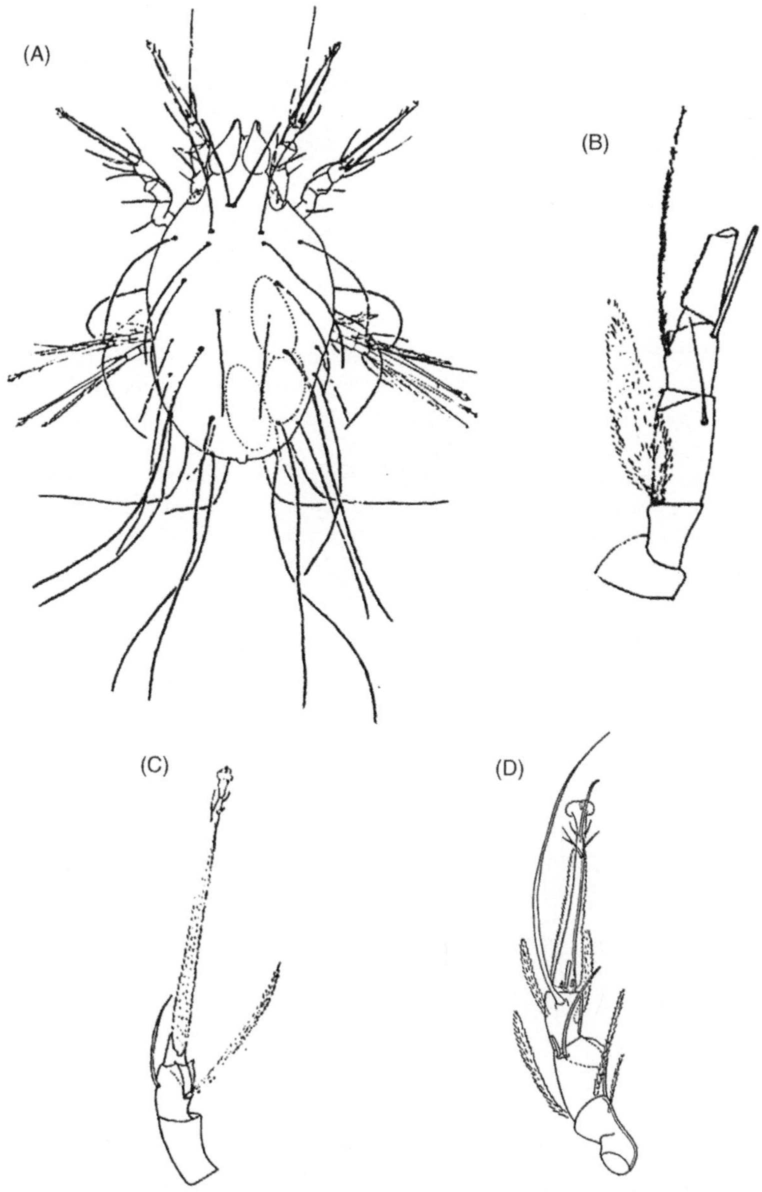

Fig. 35.6. *Lepidoglyphus michaeli* Oudemans. Female. Dorsal view (A); basal segments of leg III (B); leg IV (C). Male. Leg I (D) (from Hughes, 1976, licensed under the Open Government Licence v3.0).

Research, carried out by Solarz *et al.* (2004) in Poland, showed that this species represents 2.30% of all mites collected from cowsheds, 2.53% from barns, and 0.12% in a poultry house. Subdominant and constant in farming environments.

Symptomatology and damage

Like for *G. fustifer*.

Control

As other glycyphagid mites.

36 The Family Hemisarcoptidae Oudemans (Fig. 36.1)

Morphological Characteristics, Systematic and Bio-ecology

The family Hemisarcoptidae has been defined by O'Connor (1982) as follows. These mites possess a small body, whitish in colour and with a prodorsal sclerite. The genital and anal openings are set beyond the coxae IV; they are confluent in the female and separated in the males. The latter also have a median ventral sucker anterior to the genital area. All tarsi bear distally, an elongate sucker-like ambulacrum. The hypopi (facultative deutonymphs), in contrast to the other active stages, are heavily sclerotized and dark brown. They bear a pair of eyes but lack functional mouthparts and have a caudal, ventral attachment plate. Their tarsal appendages lack the sucker-like ambulacra of the other stages, and the fourth pair of legs terminates with two of three long setae.

The family includes 11 genera and 31 species (O'Connor, 2009b), with most deutonymphs being nonspecific associates of cortical and subcortical insects (O'Connor and Houck, 1989). Beron (2021b) listed eight genera and 24 species. Species belonging to the genus *Hemisarcoptes* Lignières, possess a unique, doubly parasitic ecology. Feeding stages are obligate parasites of diaspididae scale insects in most parts of the world but are nonspecific as to host; more than one adult mite feeding on a host typically results in mortality. Some species have been employed in biological control efforts directed against several scale-insect pests with various results (Gerson *et al.*, 1990). Species of the genus *Nanacarus* Oudemans and *Nanacaroides* Volgin and Mironov often inhabit the spore tubes of polypore fungi (Pielou and Varma, 1968; Volgin and Mironov, 1979). *Nanacarus minutus* Oudemans was first reported on *Sorex vulgaris* Linnaeus, *Sorex araneus* Linnaeus (Eulipotyphla, Soricidae), *Eptesicus serotinus* (Schreber) and *Pipistrellus pipistrellus* (Schreber) (Chiroptera, Vespertilionidae) from Germany and *Xylocopa* (*Koptorthosoma*) *tenuiscapa* (Westwood) (Hymenoptera, Apidae) from Java (Oudemans, 1901, 1902a, 1903); successively, the mite was collected on dried fruit, dried vegetable, grain, mushroom and wheat in China (Li *et al.*, 1992a); its role on stored food seems secondary.

DOI: 10.1079/9781786390912.0036

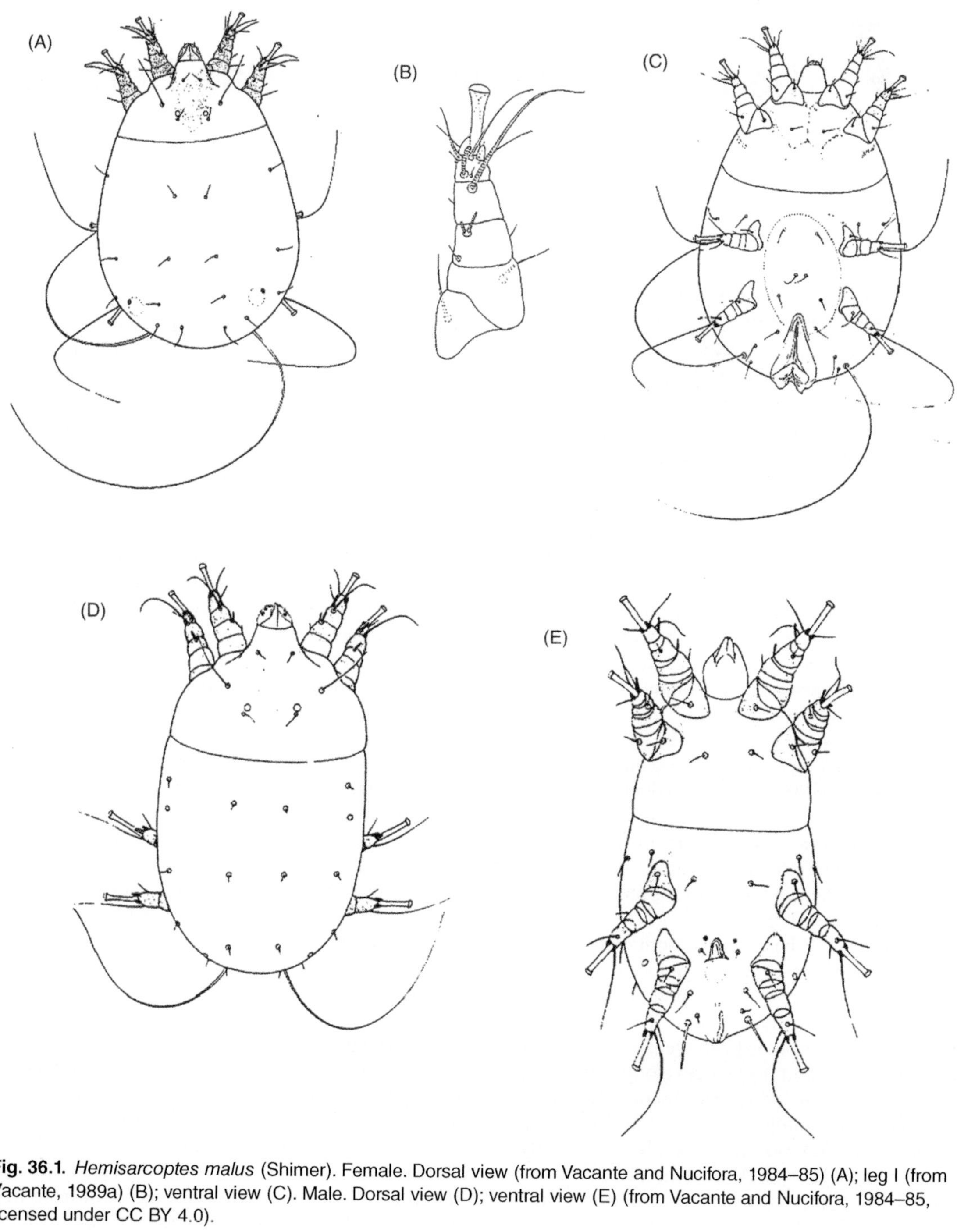

Fig. 36.1. *Hemisarcoptes malus* (Shimer). Female. Dorsal view (from Vacante and Nucifora, 1984–85) (A); leg I (from Vacante, 1989a) (B); ventral view (C). Male. Dorsal view (D); ventral view (E) (from Vacante and Nucifora, 1984–85, licensed under CC BY 4.0).

37 The Family Histiostomatidae Berlese

Morphological Characteristics, Systematic and Bio-ecology

The body of mites of the family Histiostomatidae is large and not dorsoventrally flattened. The gnathosoma is strongly modified and adapted, to feed on minute particles in suspension. The chelicerae are laterally compressed, not chelate, with the digitus fixus comb-like and equipped with many fine teeth. The digitus mobilis is short, fused to the base of the digitus fixus, and with few short distal teeth. The palpi have a freely movable and flattened distal segment, with long tarsal solenidion and eupathididal setae. The prodorsum has four pairs of setae (*ve*, *vi*, *sce*, *sci*) and the hysterosoma ten pairs of dorsal setae (c_1, c_2, c_3, d_1, d_2, e_1, e_2, f_2, h_1, h_2); one pair of setae (h_3) is commonly set ventrally. The opisthonotal glands are commonly located between setae e_1 and e_2. The bursa copulatrix is set between the setae h_1. The genital valves of the female are fused to the body medially and form a transverse oviporus; the genital papillae of both sexes are often seen as large rings on the ventral surface and are not closely associated; the posterior pair is set on the ventral opisthosoma (Hughes, 1976; O'Connor, 2009b).

The deutonymph has a gnathosomatic remnant that is either well developed or reduced. The idiosoma is well-sclerotized, and a sejugal furrow separates the propodosomatic and hysterosomatic sclerites. The ventral coxal surface has a posterior median apodeme, and normal chaetotaxy, with the exception of setae *4b*, which are absent. The attachment organ is commonly developed and bears all structures. The anterior legs are also commonly developed; legs III and IV sometimes have some segments fused and secondary articulation in the midtarsus. The articulations of the trochanter-femur of legs III–IV are anteriorly directed in repose. The pretarsi have a small membranous ambulacrum and empodial claw; empodium IV is sometimes missing (Hughes, 1976; O'Connor, 2009b). The terminology follows mostly that of Houck and O'Connor (1990) and Fashing and O'Connor (1984).

The Histiostomatidae family (formerly the Anoetidae) has been revised by Scheucher (1957), and Hughes R.D. and Jackson (1958) have studied the biology of these mites, and Wirth (2004) investigated the phylogeny, biology and character transformations of the Histiostomatidae. Schatz *et al.* (2011) reported 59 genera and over 567 species. Beron (2021b) reported 58 genera and 522 species.

Histiostomatid mites live in wet substrates and use modified mouthparts to filter organic material and microorganisms from water film (Hughes, 1953; O'Connor, 2009); their genital papillae seem to be an adaptation to regulate their osmotic balance in this aquatic environment (Witaliński *et al.*, 2002), and their size varies with the osmotic pressure of that environment. Various species are also found on wet vegetable matter where decomposition is taking place, and some species are found on decaying mushrooms, on cow dung or on the resins and saps exuding from the damaged bark of trees (*Histiostoma pulchrum* Kramer). Many species live in fully aquatic environments, and they can inhabit water-filled tree holes and leaf axils, often using insects as dispersal agents. Other species have developed an intimate association with animal hosts (OConnor, 2009). *Hormosianoetus laboratorium* (Hughes, 1950) is a common pest of *Drosophila* rearing (Hughes, 1976), and Ashburner (1989) has treated extensively, the Histiostomatidae associated with *Drosophila* cultures. Fain *et al.* (1993) described *Hormosianoetus unidentatum*. Fain *et al.* associated with *Drosophila bipectinata* from Thailand. Klimov *et al.* (2016) reported species found in associations with bees, recorded from the Nearctic, Palaearctic, Oriental and African (including Madagascar) regions. Among bees, this genus has been found in associations with *Heriades*, *Xenoglossa*, *Apis* and *Xylocopa*. Fashing (2010) described two new adaptations for dispersal in this

The Handbook of Mites of Stored Products (V. Vacante and C.H.W. Flechtmann)
DOI: 10.1079/9781786390912.0037

family. Halliday and Collins (2002) described *Histiostoma papillata* from Australia. They found the mite in the fins and gills of juvenile Murray cod, *Maccullochella peelii peelii* (Mitchell), where it appeared to inflict injuries, thought to be responsible for the mortality of fish in a diet trial.

Some species belonging to the genus *Histiostoma* Kramer feed on various bulbs and vegetable substrates (Hughes and Jackson, 1958; Qin and Rohitha, 1996) or developing on mushrooms, where they can cause either direct or indirect damage (Hussey *et al.*, 1969; Hill and Deahl, 1978; Greenslade and Clift, 2004); they are found on decaying mushrooms (*Glyphanoetus fulmeki* Oudemans), on cow dung (*Glyphanoetus phyllotrichus* Berlese). Fungal fruiting bodies are exploited by species of *Histiostoma*, which feed on necrotic tissue and may spread pathogenic microorganisms in commercial mushroom operations (Hill and Deahl, 1978). In all these habitats, the food is of a fluid or semi-fluid nature, and the mouthparts have become modified to gather up particles suspended in the fluid (Hughes, 1953). *Histiostoma feroniarum* (Dufour) is occasionally found on decaying food material such as grain or flour too wet. *Histiostoma sapromyzarum* (Dufour) has also been found on some damp timber. Other species exhibit a specific association with habitats created by insect hosts; so wet subcortical habitats are colonized by *Histiostoma* spp. associated with scolytid beetles, and galleries of Passalidae are home to species of *Histiostoma* species and other species of other genera (Woodring and Moser, 1970; O'Connor, 2009b). Galleries of palm weevils (*Rhynchophorus* sp.) are inhabited by species of Histiostomatidae. Vertebrate nests harbour species of *Histiostoma* and of other genera (Scheucher, 1957; Fain and Philips, 1979).

H. feroniarum may feed in wet, damaged surface of bulbs, tubers and fruits (Hughes and Jackson, 1958; Qin and Rohitha, 1996). *Histiostoma heinemanni* Hill and Dehal can develop large populations in commercial mushroom operations. They are not directly injurious, but may be difficult to remove them from mushrooms, prior to marketing. Six species are reported on stored food. Of these, *H. feroniarum* and *H. sapromyzarum* possess a large geographical distribution and are collected on various facilities. Other species play a secondary role. They are *Histiostoma humidiatus* (Vitzthum), reported from Taiwan (Tseng, 1984) on garlic (rotting), *Histiostoma onioni* Eraky and Shoker (1994) known from Egypt on onions, *H. pulchrum* [= *Seliea pulchrum* (Kramer)], reported from China on dried fruit, dried vegetables and mushrooms (Li *et al.*, 1992a) and *Histiostoma sarrai* Eraky and Shoker (1994) from Egypt on onions.

Further information may be found in Beron (2021b).

Selected websites

https://www.youtube.com/watch?v=G6gmi_UJMa8

https://www.insectimages.org/browse/taximages.cfm?fam=4465

https://www.youtube.com/watch?v=1_Sa7x1AxeU

The Genus *Histiostoma* Kramer

Diagnostic characteristics

The female has the palps well developed and possesses a dorsal sclerite. Ventrally, the anterior genital 'rings' are set laterally between the transverse arms of the coxal apodemes II–III. The posterior genital 'rings' are not associated with the anus. The coxal setae I, III, IV are present. The pretarsi have a membranous ambulacrum, short and not bilobed, and the empodial claws are not elongated, claw-like and hooked.

The male has mouthparts and setation normal. Body form and sclerotization variable; if propodosoma and opisthosoma, each covered by a reticulate sclerite, then setae d_1 far posterior to setae *cp*. The setae h_1 usually are set ventrally. The external vertical setae are short, not extending over the gnathosoma. The dorsal setae are filiform (not heavily barbed or blade-like). The paranal suckers are absent. With two pairs of genital rings. The aedeagus is thin, short and posteriorly directed. Pretarsal ambulacra simple, not bilobed, empodial claws hook like.

The phoretic deutonymph (hypopus) has the gnathosoma elongate or subquadrate, and the anterior prodorsum lacks a pair of brown pigmented areas. The anterior edge of dorsal hysterosoma is entire, smooth and not scalloped, and the anterolateral hysterosoma is devoid of lens-like organs. The hysterosomal setae c_1, d_1, and e_1 are filiform but not lanceolate. The attachment organ is wider than long. The coxal fields III–IV are closed, and the setae of coxae I and III (*1a* and *3a*) are either all filiform or all conoidal. Solenidion ω_1 of leg I set directly on the tibia and associated with the tibial solenidion φ; the claws III and IV are large; the claw III is hooked and the claw IV is hooked or linear. The empodial claws I–IV are simple and not bifurcated. The tarsi III and IV possess a weak, flexible region in middle of segment. The pretarsi III and IV bear empodial claws. The trochanters I–II are devoid of setae (Klimov *et al.*, 2016).

Economic importance

According to Klimov *et al.* (2016) this genus occurs worldwide, including sub-Antarctic islands. It includes about 80 species, but a small number of these cause an economic damage. Some species feed on various bulbs and vegetable substrates (Hughes and Jackson, 1958; Qin and Rohitha, 1996) or develop on fungal fruiting bodies, where they feed on necrotic tissues, and may spread pathogenic organisms in commercial mushroom houses (Hussey *et al.*, 1969; Hill and Deahl, 1978; Greenslade and Clift, 2004). The species of economic importance are *H. feroniarum* (Dufour), *H. sapromyzarum* (Dufour) and *H. heinemanni* Hill and Deahl. Adults, juvenile stages and hypopodes of *H. feroniarum* have been collected on decaying food material such as grain or flour that was too wet (Hughes, 1976) and on edible fungi (Hussey *et al.*, 1969), and they are responsible for serious losses of crops of *Agaricus bisporus* (Lange) Pilat, *Agaricus bitorquis* (Quel.) Sacc. and *Lentinula edodes* (Berk.) Pegler (Greenslade and Clift, 2004). *H. sapromyzarum* has been recorded feeding on edible fungi in Europe (Boness, 1968), but there is no information available on damage in other countries, such as Australia, where the mite has been recorded, but any damage that it causes is unknown (Greenslade and Clift, 2004). *H. heinemanni* develops high populations in commercial mushrooms, and although not directly injurious, its presence is undesirable, and it is difficult to remove before marketing (O'Connor, 2009b). Moreover, six other species are reported on stored food, but they play a secondary role. They are *H. humidiatus* (Vitzthum), reported from Taiwan (Tseng, 1984) on garlic (rotting), *H. onioni* Eraky and Shoker (1994) known from Egypt on onion, *H. pulchrum* [= *Seliea pulchrum* (Kramer)], reported from China on dried fruit, dried vegetable and mushroom (Li *et al.*, 1992a) and *H. sarrai* and Shoker (1994) from Egypt on onion. In Pennsylvania (USA), *H. heinemanni* Hill and Dehal can develop large populations in commercial mushroom operations. They are not directly injurious but may be difficult to remove them from mushrooms, prior to marketing (Hill and Deahl, 1978).

Klimov *et al.* (2016) reported species found in associations with bees, recorded from the Nearctic, Palaearctic, Oriental, and African (including Madagascar) regions. Among bees, this genus has been found in associations with *Heriades*, *Xenoglossa*, *Apis* and *Xylocopa*. It is a large genus with more than 200 described species, but a dichotomous key is not available. Most species have been described based on phoretic deutonymphs. Older, possibly outdated keys to species based on adults and phoretic deutonymphs are available at Hughes and Jackson (1958), Mahunka (1970), Scheucher (1957) and Sevastyanov (1975). We discuss extensively here only *H. feroniarum* and *H. sapromyzarum*.

Histiostoma feroniarum (Dufour) (Fig. 37.1)

Common name

Unknown

Morphological characteristics

FEMALE. The body is pear-shaped, 525 µm long and 338 µm wide, whitish in colour, and commonly containing large quantities of guanine as chalky deposit. The chelicerae have the digitus fixus attenuate, with about 7–9 large spines and about six small teeth near the apex. The cheliceral setae each is shorter than the digitus fixus. Palps each with a long solenidion about 20 µm long, and a short eupathidial seta 8–10 µm long on distal segment. The subcapitulum bears a pair of subcapitular setae. A sejugal furrow divides the dorsum into prodorsum and hysterosoma. The prodorsum possesses a sclerite, slightly ornamented with a pattern of rectangular or irregular areas. The chitinous plate is freely movable, and its movement serves to sweep particles in the fluid medium, towards the anterior end of the gnathosoma. The alternate backward and forward movement of the chelicerae then rakes the solid particles towards the mouth (Hughes, 1953). The idiosoma is covered with fine projections, to which small particles of fungal hyphe and fine debris readily adhere. All prodorsal setae are filiform. The internal vertical setae (*vi*) are 12 µm long, set at anterior margin of the prodorsum; the external vertical setae (*ve*) are 30 µm long, set at lateral area of the prodorsal

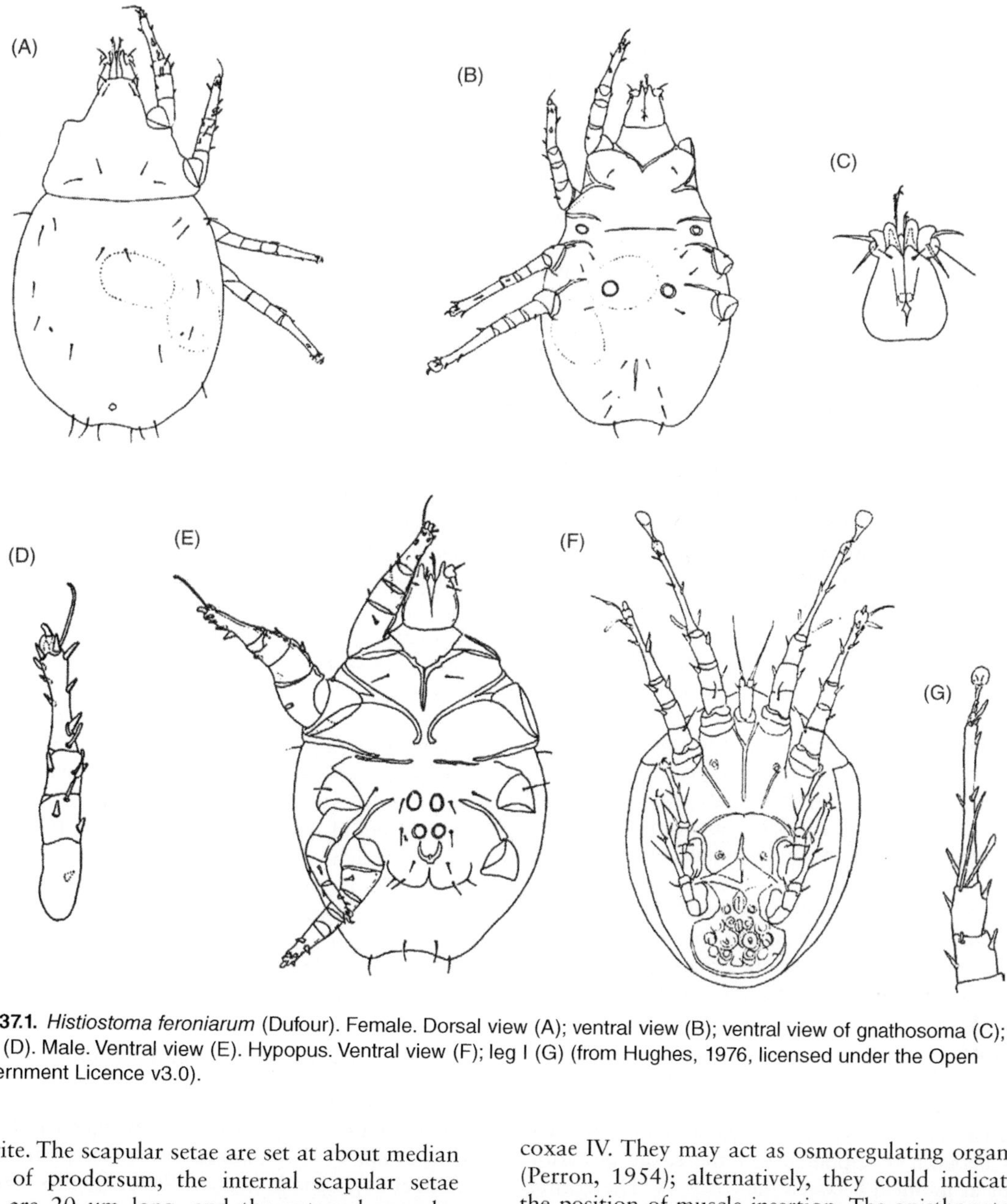

Fig. 37.1. *Histiostoma feroniarum* (Dufour). Female. Dorsal view (A); ventral view (B); ventral view of gnathosoma (C); leg I (D). Male. Ventral view (E). Hypopus. Ventral view (F); leg I (G) (from Hughes, 1976, licensed under the Open Government Licence v3.0).

sclerite. The scapular setae are set at about median area of prodorsum, the internal scapular setae (*sci*) are 20 µm long, and the external scapular setae (*sce*) 28 µm long. Ten pairs of hysterosomal setae (c_1, c_2, c_3, d_1, d_2, e_1, e_2, f_2, h_1, h_2), all filiform, each 24–32 µm long, plus h_3, are set ventrally on the idiosoma. Ventrally, the two pairs of chitinous rings are circular or almost circular, the anterior pair lying between the coxae II and III on either side of the transverse genital opening, the posterior being nearer together and lying at the same level as coxae IV. They may act as osmoregulating organs (Perron, 1954); alternatively, they could indicate the position of muscle insertion. The opisthonotal gland openings are located between the setae e_1 and e_2. The bursa copulatrix is located between the setae h_1. The coxal apodemes are well developed, with the coxal apodemes I fused medially in form of a V. The genital papillae are circular or broadly oval, with the anterior pairs set between the apodemes of the coxae II and III, and the posterior pair between the coxae IV. The genital opening is a transverse slit, located between the posterior apodemes II. The anal opening is a longitudinal slit set at the posterior part of venter. The coxal setae *1a*, *3a* and *4a* hair-like, 12–16 µm long, and the genital setae are laterad of genital opening. Three pairs of pseudoanal setae (ps_{1-3}). The leg setae are spine-like, except for one seta located on apex of legs I and II being slender and long. The setal formula of legs I–IV is: trochanters 1–1–1–0, femora 1–1–0–1, genua 2–2–0–0, tibiae 2–2–1–1, tarsi 12–12–10–10. The solenidion formula is: genua 2–1–0–0, tibiae 1–1–1–1, tarsi 2–1–0–0. Omega 1 arises from the extreme base of tarsus I, and curves backward to overhang the anterior edge of tibia I, the famulus arising from the same rather deep pit. On tarsus II, $\omega 1$ occupies its normal position and is only slightly curved. At the distal ends of all the tarsi, the ventral spines are well developed. On tibia I and II, φ is shorter than tibia II, and on genu I, σ_1 and σ_2 are equal in length. Sigma is missing from genu III. The claws are short and stout (Hughes, 1976; Qin and Rohitha, 1996).

MALE. The body is elongate, oval, 300 µm long and 135 µm wide. Some specimens show variability in size and in thickness of the legs; sometimes the legs and the spines are about the same size as in the female, in others they are considerably thickened, and bear correspondingly well-developed spines, which are particularly noticeable on tarsus II. The gnathosoma as in the female, as well as the shape and the chaetotaxy of the dorsal idiosoma. Ventrally, the four circular chitinous rings are close together, anterior to the genital opening. The coxal apodemes are better developed than in the female; the coxal apodemes I are fused medially in the form of a Y, the coxal apodemes II extending almost to midline, and directed posteriorly in the middle. The coxal apodeme III extending almost to midline but not directed posteriorly. The genital papillae are circular or broadly oval; the anterior pairs are set between the apodemes of the coxae posterior pair at midfield between the coxae IV. The genital opening is located at posterior area between the coxae IV. The anal opening is a longitudinal slit posterior to genital opening. The coxal setae *1a*, *3a*, *3b* and *4a* are hair-like, 12 µm long; the genital setae posterolaterad of genital opening. Three pairs of pseudoanal setae (ps_{1-3}), with ps_2 and ps_3 set on a lobe. The legs are similar to those of female but shorter and stouter (Hughes, 1976; Qin and Rohitha, 1996).

HYPOPUS. The body is 190 µm long and 136 µm wide. The cuticle is well-sclerotized. The gnathosomal base lacks setae. The palps are short, each with a short basal seta, and a long apical solenidion. Dorsally, a sejugal furrow divides the prodorsum from the dorsal hysterosoma. The prodorsum is triangular in outline, devoid of a sclerite. All prodorsal setae are short and hair-like. The internal vertical setae (*vi*) and the external vertical setae (*ve*) lack. The internal scapular setae (*sci*) and the external setae (*sce*) are set on the anterolateral area of prodorsum. The hysterosomal setae are short and hair-like. According to what was observed by Perron (1954) for *Histiostoma laboratorium* Hughes, the contact with the host is first made with the two anterior suckers, the large central suckers then playing a major part in attaching the hypopus to its host. The sucker plate is well developed, and through it project eight suckers arranged in the following order: 2–4–2. The opisthonotal glands and the bursa copulatrix are apparently absent. The coxal apodemes are well developed. The coxal apodemes I are fused medially in the form of a Y. The coxal setae (*1a*, *3a*, and *4a*) in form of small conoids. The legs are slender, with fewer setae than in adults; the legs III and IV are smaller than legs I and II. All legs possess claws, and from the end of leg I projects an expanded seta, at the base of which arises a transparent leaf-shaped seta *d*. A leaf-shaped seta also arises at the end of leg II. On leg I, ω_1 is straight and expanded distally, only slightly shorter than φ of the same leg. Sigma (σ) on leg I is shorter than the spine-like setae arising from genu I. On leg II, ω_1 is only slightly longer than φ and sigma on the tibia and genu, respectively (Hughes, 1976; Qin and Rohitha, 1996).

Geographical distribution

Cosmopolitan, reported from Argentina (Ahmad, 1974; Mauri, 1982), Australia (Womersley, 1941; Clift *et al.*, 1995; Halliday, 1998; Greenslade and Clift, 2004), Canada (Sinha, 1964b; Hughes, 1976), China (Chen *et al.*, 1982; Li *et al.*, 2015a), Columbia (Oudemans in Buitendijk, 1945), England (Hughes, 1976), France (Dufour, 1839; Hughes, 1976), Germany (Hughes, 1976), India (Parveen and Gupta, 2019, 2020a), Iran (Ostovan and Kamali, 1995b; Mossadegh, 1997; Ardeshir, 2002; Haddad Irani-Nejad *et al.*, 2007), Ireland (Hughes, 1976; Curry and Momen, 1988), Israel (Costa, 1967), Italy (Berlese, 1897a; Oudemans in Buitendijk,

1945; Rota and Bolchi Serini, 1974; Hughes, 1976; Vacante, 1989b), the Netherlands (Oudemans in Buitendijk, 1945; Siepel *et al.*, 2016), New Zealand (Womersley, 1941; Qin and Rohitha, 1996), Poland (Lewandowski *et al.*, 1999; Chmielewski, 2009; Karbowiak *et al.*, 2013; Szafranek and Lewandowski, 2017), Portugal (Baker, 1985), South Korea (Kim and Hwang, 1996), Tenerife (Spain) (Sanchez-Covista *et al.*, 1999), United States (Poe, 1971; Hughes, 1976), Venezuela (Hurtado and Parini, 1987) and other countries.

Commodities, facilities, vegetal and animal hosts

H. feroniarum (Dufour) has a wide range of habitats including stored food (Ardeshir, 2002; Hagstrum *et al.*, 2013). Usually, it is widely distributed on all kinds of decaying vegetable matter and is occasionally found on wet grain and decaying mushrooms (Hughes, 1976). Moreover, it is known on Chinese medicine, dried fruit, dried vegetables, grain (wet), mushroom (decaying) on different facilities, such as bread vans, commercial grain storage and farms (Hagstrum *et al.*, 2013; Li *et al.*, 2015a), and on stored rice, rice dust and debris (Ostovan, 1993; Mirfakhraii, 1994; Ostovan and Kamali, 1995b; Modarres Awal, 1997; Mossadegh, 1997; Noei, 2007; Noei and Ostovan, 2012).

In various countries, *H. feroniarum* has been reported from compost and mushrooms (Hussey *et al.*, 1969; Rota and Bolchi Serini, 1974; Clift *et al.*, 1995; Ignatowicz, 2000) as *Agaricus bisporus* (J.E. Lange) Imbach, *A. bitorquis* (Quel.) Sacc. (Greenslade and Clift, 2004), *Lentinula edodes* (Berk.) Pegler (Kim and Hwang, 1996; Greenslade and Clift, 2004), *Pleurotus ostreatus* (Jacq.) P. Kumm. (Kim and Hwang, 1996), *Panaeolus cyanescens* (Berk *et* Broome) Sacc. [= *Copelandia cyanescens*], *Earliella scabrosa* (Pers.) Gilb and Ryvarden (Parveen and Gupta, 2019, 2020a), *Flammulina*, *Pleurotus sajur-caju* (Fr.) Singer and *Hericium erinaceus* (Bull.) Pers. (Li *et al.*, 2015a). Moreover, the mite has been collected on Radix *Codonopsis* and onion (Li *et al.*, 2015a).

Qin and Rohitha (1996) reported the mite from *Sandersonia aurantiaca* (Hook.) (Liliaceae) in New Zealand, as well as Boczek and Blaszak (2005), collected the mite on ornamental plants in Poland. Buitendijk (1945) found specimens of *H. feroniarum* in the Oudemans collection collected in the Netherlands on *Solanum tuberosum* Linnaeus, in Columbia on *Bromelia margarethae* (*Neoregelia margaretae* L.B.) and in Italy on *Dahlia variabilis* (Willd.) Desf. Siepel *et al.* (2016) reported the mite in the Netherlands in champignon culture, in wheat straw, on *Hippeastrum* in glasshouse, on *Gerbera* and on fungus *in vitro* culture (*Dianthus* tissue). In Iran, the mite was collected in lucerne fields (Haddad Irani-Nejad *et al.*, 2007). In Poland, it was reported on seeds and onions (Boczek and Blaszak, 2005).

Moreover, the mite has been found in soil (Curry and Momen, 1988; Boczek and Blaszak, 2005) among the microfauna associated with gladiolus corms, and feeding on diseased and decayed plant tissues (Poe, 1971). Other environments are house dust (Hurtado and Parini, 1987), some insects (Ahmad, 1974) and laboratory cultures (Haddad Irani-Nejad *et al.*, 2007). In Iran, Mossadegh (1997) found *Histiostoma feroniarum* in debris from honeybee hives of *Apis mellifera* Linnaeus. In Poland, it has been collected on the European earwig, *Forficula auricularia* Linnaeus found in apiaries (Chmielewski, 2009) and on the tick *Ixodes hexagonus* (Leach) (Karbowiak *et al.*, 2013). In Australia, the mite is reported as phoretic on the Australian giant earwig, *Titanolabis colossea* (Dohrn) (Anisolabididae) (Tagami and Halliday, 2013). Baker (1985) reported the mite from Portugal on the millepede *Ommatoiulus moreleti* (Lucas). Vacante (1989b) collected many adults and phoretic deutonymph associated with adults of the curculionid *Otiorhynchus cribricollis* (Gyllenhal) on strawberry in protected crops in Italy. Szilman *et al.* (2006) studied occupational exposure to allergenic mites among workers of the Silesian Zoo in Poland and found the mite rare. Moreover, *Histiostoma feroniarum* is the most frequently found mite in stagnant water, residual water and sewage (Mauri, 1982) and in sewage filterbeds (Learner and Chawner, 1998). Animal hosts number *Cormocephalus aurantiipes* (Newport) (Chilopoda, Scolopendridae), *Hister merdarius* (Fabricius), *Hister stercorarius* (Linnaeus), *Hister unicolor* Linnaeus (Coleoptera, Histeridae) and *Spalax ehrenbergi* (Nehring) (Rodentia, Spalacidae) (Beron, 2021b).

Bio-ecology

Histiostoma feroniarum is a generalist cosmopolitan species, known from a variety of habitats with moist, decomposing organic matter such as rotten vegetables, soil, compost, litter and bird nests. Like many other histiostomatids, the mite is a microorganism-feeder that uses its modified chelicerae to filter microorganisms from the substrate. Baker (1964a) observed that the mite is most common to be found in sewage bacteria beds, where it feeds on the zoogloeal film covering the clinker surface, and thus, it helps to prevent the beds from becoming clogged by too much film. The species occasionally enters beehives, probably with their insect carriers, which are attracted to various food sources available at beehives (Klimov *et al.*, 2016).

The life cycle develops through the egg, larva, protonymph, phoretic deutonymph (hypopus), tritonymph and adult stages, and completes it very rapidly, taking from 3 to 3.5 days, with the optimum temperature between 25°C and 30°C; unfertilized eggs give rise to males (Scheucher, 1957). The species is parthenogenetic, and according to Hughes and Jackson (1958), two types of females are generated, one giving rise to males, the other to females. Cooreman (1944) reported that at 20°C to 25°C, the time required for complete development is from 2 to 4 days.

Both thelytokous and arrhenotokous 'varieties' of *Histiostoma feroniarum*, occur in essentially the same habitat. Males of the arrhenotokous strain have 7, and females 14, chromosomes (Heinemann and Hughes, 1969). Meiosis is omitted in males and is replaced by a single mitotic division. Oogenesis is regular, and seven bivalents are present at metaphase I (Oliver, 1971). Liana (2005) investigated the influence of insemination at different life stages, on female fitness. Males guard immature females at the tritonymph stage to inseminate them immediately after the last moulting. Four groups of females were studied. The first group included females inseminated 'naturally' (i.e. mating occurred immediately after guarding), and then the male was removed. The second group includes 'naturally' inseminated females, where after insemination, the male was replaced by two additional males. The third group concerns virgin females reared without males and the fourth group, mature, virgin females to which two virgin males were added 3 days after last moulting. The results showed that females inseminated 'naturally' have higher longevity and fecundity than either virgin or late-inseminated females. Furthermore, longevity and fecundity of the former were also greater than that of females naturally' inseminated and subsequently exposed to males. One may suggest that seminal fluids have a positive effect on female fitness. When delayed insemination occurs, such positive effect may not be observed, due to a change in features of the sperm access system. Harassment may explain decreased longevity and fecundity of females inseminated 'naturally', compared to females that received additional males (Liana, 2005). Males inseminated the virgin females immediately after the last moulting, but there were also other containers for mass culture of mites (Bahrami *et al.*, 2012).

The hypopi will attach themselves to many different kinds of Arthropoda and are particularly plentiful on filter beds where the zoogloeal film has been allowed to dry. Moulting was found to occur in 2–3 days when the hypopi came in contact with a film of water (Baker, 1964b). Phoretic deutonymphs have been found on a great variety of hosts, ranging from Myriapoda to Coleoptera.

Under laboratory condition, and with *Fusarium graminearum* Clade as food, the incubation, larval and nymphal periods and longevity of adults take 1.6, 2.56, 3.09 and 10.48 days, respectively. The sex ratio is 0.65. The preoviposition, oviposition and postoviposition periods last 1.8, 8.27 and 0.66 days, respectively. The gross reproductive rate (*GRR*), net reproductive rate (R_0), intrinsic rate of increase (*rm*), finite rate of increase (λ), mean generation time (*T*) and doubling time (*DT*) are 110.56 females/female, 23.54 females/female, 0.195 day^{-1}, 1.215 day^{-1}, 16.23 days and 3.55 days, respectively. These results show high potential of *Histiostoma feroniarum* as a major pest of laboratory cultures (Bahrami *et al.*, 2012).

Symptomatology and damage

The mite is widely distributed on all kinds of decaying vegetable matter and is occasionally found on wet grain and decaying mushrooms (Hughes, 1976). Moreover, the species can be found in different stored products. Some authors retain that *Histiostoma feroniarum*, like storage mites, can cause significant weight loss of grains and oilseeds (Ardeshir, 2002). This aspect is supported by the fact that some histiostomatid mites have been reported from stored grains and stored products, where they feed on cereal foodstuff and on fungi. Fungi growing on stored products and decaying organic debris assure an adequate food supply for mites in empty granaries, which certainly give stored-grain mites a new variety of food and thus a greater chance of survival in an unbalanced ecosystem (Ashfaq *et al.*, 2000). O'Connor (2009) reported that different mite species of the families Acaridae, Lardoglyphidae and Histiostomatidae were mostly present in stored food products. Pagliarini (1979) found that mostly mites of family Acaridae and Histiostomatidae were more abundant, causing huge losses to all available food products. However, despite the above reports, the authors' views on the real damage caused by *Histiostoma feroniarum* are controversial. Results of some research show that *Histiostoma feroniarum* possesses a high potential as a major pest of laboratory cultures of *Fusarium graminearum* (Bahrami *et al.*, 2012). The writer's opinion is that it is a ubiquitous organism, able to develop in humid environments and on rotting substrates, but unable to cause direct damage, and its presence is fundamentally related to hygienic implications; from this point of view, it is necessary to focus priority on the conservation status of foodstuffs, and the conditions of conservation environments. This does not exclude the need for the control, but it must be directed primarily towards the scrupulous preservation of foodstuffs.

In addition, contamination of stored food by arthropods including this species may seriously damage human health by allergies (Hubert *et al.*, 2003). Skubala *et al.* (2006) studied the accidental acarophagy in humans and animals. They investigated the presence of mites in 90 food samples, including 24 different kinds of fruits, vegetables and mushrooms. *Histiostoma feroniarum* was the most abundant acaridid mite, and it was observed that by washing in running water, only 48.6% and 50% of the total number of mites were removed from fruits and vegetables, respectively.

Adults, juvenile stages and hypopodes of *Histiostoma feroniarum* have been recorded on edible fungi (Hussey *et al.*, 1969), and considered by some authors, causing serious crop losses in various mushroom species, such as *Agaricus bisporus*, *A. bitorquis* and *Lentinula edodes* (Kim and Hwang, 1996; Greenslade and Clift, 2004); other mushrooms infested by the mite are *Pleurotus ostreatus* (Kim and Hwang, 1996), *Panaeolus cyanescens*, *Earliella scabrosa*; the mite was collected from infected fruit body of milk mushroom. The infested fruit body did not develop further and gradually retarded (Parveen and Gupta, 2019, 2020a). *Flammulina*, *Pleurotus sajur-caju* and *Hericium erinaceus* (Li *et al.*, 2015a); in addition, the mite has been collected on Radix *Codonopsis* and on onion (Li *et al.*, 2015a). *Histiostoma feroniarum* has been observed frequently on a mushroom farm in the Northern Italy (Rota and Bolchi Serini, 1974), but its direct damage is unclear. The species was commonly found in the baled straw used to produce mushroom compost in Australia (Clift *et al.*, 1995), and a high number of *Histiostoma feroniarum* was commonly found on mushroom. In South Korea, the mite was observed injurious to the mushroom *Lentinula edodes*, and high levels of populations of the mite caused significant damages to mushroom lamellate (Kim and Hwang, 1996). In Poland, Lewandowski *et al.* (1999) collected the mite in the oyster mushroom houses and recorded the species as common in the substrate used for cultivation of oyster mushroom *Pleurotus ostreatus*. Szafranek and Lewandowski (2017) studied the mites of mushrooms farms in Poland and reported the mite as saprophagous.

The mite is reported as a pest of ornamental plants. Qin and Rohitha (1996) found the species on rotting tubers of *Sandersonia aurantiaca* Hook. (Liliaceae), a commercially important flower in New Zealand, causing damage even under cool storage at 30°C. Poe (1971) collected *Histiostoma feroniarum* during a study on the microfauna of gladiolus corms.

The species was collected in the soil of lucerne fields in Iran (Haddad Irani-Nejad *et al.*, 2007). *Histiostoma feroniarum* resulted in one of the most frequent acarine species, collected in grasslands with 2 years of management (Curry and Momen, 1988).

The species can cause respiratory syndromes like asthma and allergy (Szilman *et al.*, 2006). A study carried out by Sanchez-Covista *et al.* (1999) in Tenerife (Spain) showed the presence of *Histiostoma feroniarum* in the mite fauna of house dust, in homes of patients affected by respiratory symptoms.

This mite is often abundant in sewage filter beds, and it can respond rapidly to changes in the bed environment and is able to persist despite continuous wash-out from the bed (Learner and Chawner, 1998). Ahmad (1974) observed in Argentina that *Histiostoma feroniarum* attacks the larvae of the curculionid *Graphognathus leucoloma* (Boheman). Laboratory cultures are other important food sources for this mite. It attacks different fungal species like *Alternaria* spp., *Phythophthora* spp., *Pythium* spp. and *Rhizoctonia* spp. in laboratory

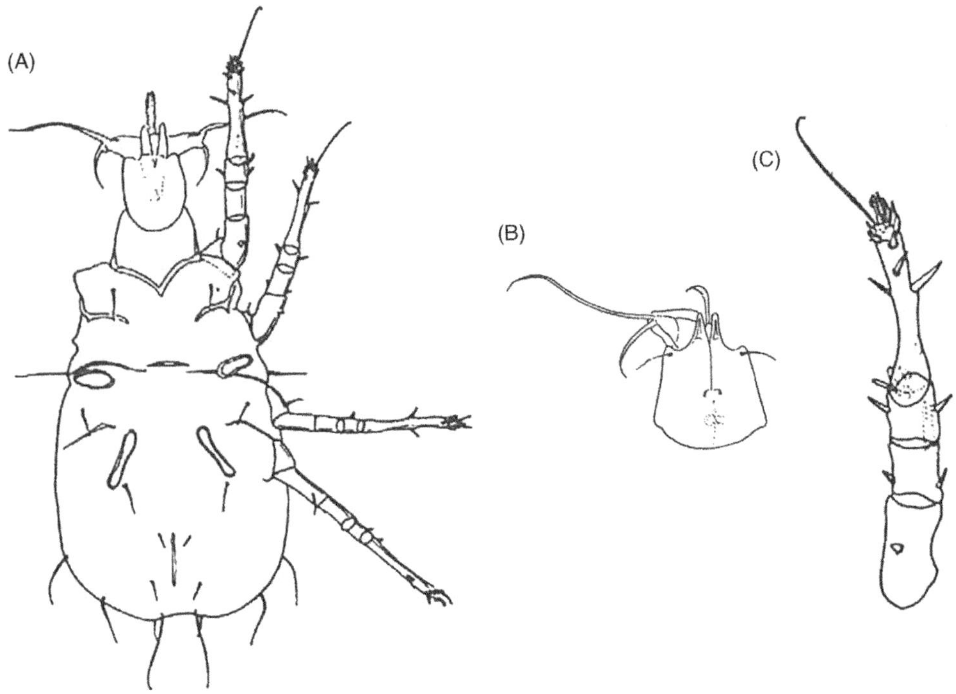

Fig. 37.2. *Histiostoma sapromyzarum* (Dufour). Female. Ventral view (A); ventral view of gnathosoma (B); leg I (C) (from Hughes, 1976, licensed under the Open Government Licence v3.0).

cultures, to such a degree that all cultures had to be destroyed (Bahrami *et al.*, 2012).

Control

The control fundamentally requires a strict hygienic management of the foodstuffs and of facilities where these are kept.

Histiostoma sapromyzarum (Dufour) (Fig. 37.2)

Common name

Unknown

Morphological characteristics

FEMALE. The idiosoma is 300–650 µm long. The palp has its distal end flattened and entire, not bilobed, and in one of the setae arising from this lobe, being more than twice as long as the other. Ventrally, the two pairs of chitinous rings are oval and constricted in the middle, resembling the sole of a shoe. The legs are more slender than in *Histiostoma feroniarum*. The chaetotaxy of leg I is very similar in the two species (Hughes, 1976)

MALE. Similar to female (Scheucher, 1957), with the chitinous rings are kidney-shaped, the convex side directed externally (Hughes, 1976).

HYPOPUS. Closely resembles that of *Histiostoma feroniarum* (Hughes, 1976).

Geographical distribution

Worldwide, known from Argentina (Mauri, 1982), Australia (Hughes, 1976; Halliday, 1998), China (Li, 1987), England (Hughes, 1976), France (Dufour, 1839; Hughes, 1976), Germany (Boness, 1968; Hughes, 1976), India (Parveen and Gupta, 2019, 2020a), Iran (Khaleghabadian *et al.*, 2012), Italy (Oudemans in Buitendijk, 1945; Hughes, 1976), the Netherlands (Oudemans, 1914a; Oudemans, in Buitendijk, 1945; Hughes, 1976; Siepel *et al.*, 2016), Philippines (Hughes, 1976), Poland (Szilman *et al.*, 2006), Turkey (Kiliç *et al.*, 2012) and other countries.

Commodities, facilities, vegetal and animal hosts

The mite is reported on decaying fungi, decaying hyacinth bulb and damp timber (Khaleghabadian *et al.*, 2012; Hagstrum *et al.*, 2013). Hypopodes of *Histiostoma sapromyzarum* were commonly found on decaying fungi (*Lactarius*, *Russula*, *Tricholoma*, *Amanita*, *Scleroderma*) (Hughes, 1976), and also recorded on cultivated mushrooms (Boness, 1968), including *Chlorophyllum hortense* (Murrill) Vellinga, *Laccaria laccata* (Scop.) Cooke, *Ganoderma lucidum* (Curtis) P. Karst. and *Inocybe umbonata* ? (Parveen and Gupta, 2020a).

The mite has been collected on *Dracaena* in glasshouse, in the Netherlands (Siepel *et al.*, 2016), on asparagus in China (Li, 1987) and on fresh onion in Turkey (Kiliç *et al.*, 2012).

Moreover, *Histiostoma sapromyzarum* has been found in considerable numbers, in recirculating industrial waters of a cookie factory in Argentina (Mauri, 1982). Animal hosts include *Melanotus rufipes* (Herbst) (Coleoptera, Elateridae), *Schizotrachelus bakeri* Klein (Coleoptera, Brentidae), *Bubas bubalus* (Olivier) (Coleoptera, Scarabaeidae) and *Myodes* (*Clethrionomys*) *glareolus* (Schreber) (Rodentia, Cricetidae) (Théodoridès, 1955; Beron, 2021b).

Bio-ecology

Adults and hypopodes of *Histiostoma sapromyzarum* are commonly collected on decaying fungi (*Lactarious*, *Russula*, *Tricholoma*, *Amanita*, *Scleroderma*) and are also recorded on cultivated mushrooms (Boness, 1968). The mite has also been found on decaying hyacinth bulbs, and on damp timber. The hypopus is carried by several kinds of beetles, flies and Myriapoda (Hughes, 1976). Szilman *et al.* (2006) studied the occupational exposure to allergenic mites among workers of the Silesian Zoo in Poland and found the mite rare.

Symptomatology and damage

Parveen and Gupta (2019) found this mite on infested bed of oyster mushrooms in India. They observed that the straws of the infested beds turned blackish, and no mycelium growth was noticed. It was earlier recorded from decaying fungi as well as decaying hyacinth branch. The population was moderate along with hypopial stage (Parveen and Gupta, 2019). Behaviour of this mite seems similar to that of *Histiostoma feroniarum*.

Control

As *Histiostoma feroniarum*.

38 The Family Pyroglyphidae Cunliffe

Morphological Characteristics, Systematic and Bio-ecology

The word Pyroglyphidae derives from the old Greek 'pyro' (i.e. fire), according to the fact that some of these mites cause eczema, associated with intense itching and a red rash. The pathology was described as like burns from fire on the skin, with a 'burning' itching sensation. The family is characterized by the anterior propodosoma, extended forward to overhang the gnathosoma. A transverse sejugal furrow separates the propodosoma from the hysterosoma. The body is largely unsclerotized dorsally, except for a propodosomatic sclerite (e.g. *Dermatophagoides*) to being virtually completely sclerotized (e.g. *Pyroglyphus*). The vertical setae are absent, the setae *sci* and *sce* have various sizes, and the coxal seta (*scx*) is set dorsolaterally; the dorsal idiosomal setae numbers the setae c_1, c_2, cp, d_1, d_2, e_1, e_2, f_2, h_2 and h_3. Ventrally, most taxa have well-sclerotized coxal fields and with coxal apodemes; the ventral setae include the setae *1a*, *3a*, *4a* and *4b*, c_3, g, ps_{1-3}. All the legs have a pretarsus. The genital opening of the female ranges from a large, inverted-U to a smaller, almost transverse form, and is completely absent in *Malayoglyphus*. A sclerotized genital shield is present, as well as lateral genital plates. In the male, the anal suckers may be present and encircled by a sclerotized ring. Males of some *Dermatophagoides* and *Sturnophagoides* species possess enlarged legs I, with the degree of enlargement allometrically related to body size. The legs III and IV of the female are usually of equal length and width; the leg IV of the male may be slightly shorter than leg III. The tarsal suckers of the male are short and cylindrical structures. The pretarsal ambulacral disc may be relatively large and sclerotized internally or reduced to a bell-shaped disc with a claw-like empodium centrally. On leg I of both sexes, the solenidion ω_1 as well as ω_3 and the famulus (ε) arise from the distal end of the tarsus. The tarsus II has the solenidion, ω inserted in the basal half of the segment. The genu III possesses the solenidion, σ (Hughes, 1976; O'Connor, 2009b).

The family comprises 18 genera and 35 nominal species of free-living or parasitic mites (Klimov *et al.*, 2015). In his first treatment, Fain (Fain, 1988b, 1990) subdivided the family into five subfamilies (Pyroglyphinae, Dermatophagoidinae, Guatemalichinae, Onychalginae, and Paralgopsinae), while Atyeo and Gaud (1996) recognized only three subfamilies. The systematic organization of the family Pyroglyphidae is still uncertain and is in great need of revision that will unify molecular and morphological evidence. An accurate phylogenetic classification of Pyroglyphidae is important in predicting properties of immunogenic proteins and of ecological aspects (Klimov *et al.*, 2015).

The Pyroglyphidae reproduce sexually. The female stores the sperm in a seminal vesicle and releases it into the oviduct to fertilize the egg during the ovulation process (Walzl, 1992). The structure of the bursa copulatrix and of the seminal vesicle is species specific and can be used for identification of *Dermatophagoides* species (Arlian, 1989; Walzl, 1992). The life cycle consists of five stages (egg, larva, protonymph, tritonymph and adult). The duration of development from egg to adult and population growth is influenced by both R.H. and temperature (Matsumoto *et al.*, 1986; Arlian *et al.*, 1990, 1998a; Hart, 1998; Arlian and Platts-Mills, 2001).

The Pyroglyphidae are free-living mites, mostly living in the nests and burrows of animals and birds, where they feed on detritus from the skin and feathers of their hosts. Other mites in this family feed on stored products (e.g. grain, cereals, nuts, dried fruit, cheeses and pet foods) but only in conditions of high R.H. (Platts-Mills *et al.*, 1997). Some of these mites are cosmopolitan and have been collected in house dust in every part of the world (Fain, 1965, 1966 a, b, 1967a, b; Fain *et al.*, 1969, etc.).

The Handbook of Mites of Stored Products (V. Vacante and C.H.W. Flechtmann)
DOI: 10.1079/9781786390912.0038

Hay *et al.* (1992) reported that *Dermatophagoides* and *Euroglyphus* species are saprophages that feed in the home, primarily on fungi growing on the organic components of floor and mattress dust. This is according to the fact that a number of ingested fungi have been identified in the alimentary tract of house-dust mites, including xerophylic species in the genera *Aspergillus*, *Eurotium*, and *Wallemia* (Hay *et al.*, 1992). Evidence indicates that the mites feed selectively on different fungi and that species differ in their nutritional value. Some fungi (e.g. *Aspergillus penicillioides* Speg.) actually may be detrimental to mite growth and reproduction (Hay *et al.*, 1993). Sloughed skin scales from humans and household pets serve as nutritional substrates for fungi. Mattresses thus provide particularly favourable sites for mites to develop. Human semen associated with sleeping quarters has been shown to be a dietary supplement for house-dust mites and can significantly increase the number of eggs a female produces (Colloff *et al.*, 1989). Van Bronswijk (1981) reported that as well as skin scales, pollen, spores of microorganisms, fungal mycelia and bacteria have been collected in the gut of *Dermatophagoides* spp. This statement is not surprising if it is true that a person commonly produces 0.5–1.0 g of dead skin a day, although a large proportion of this will be removed via the cleaning of clothes and bed sheets and the vacuuming of carpets. De Boer (1998) reported that the food is not normally considered to be a constraint on house-dust mites population growth, although its scarcity after washing or vacuuming could conceivably impede house-dust mites population growth and allergen production. Maunder (1990) wrote that freshly shed skin scales may be dry and indigestible and that mites prefer older skin scales with a higher moisture content; moreover, the fungal growth on skin scales can affect their nutritional value. Van Bronswijk (1981) reported that *Dermatophagoides pteronyssinus* grew better on skin that had previously been partially digested by the fungus *Eurotium amstelodami* Mangin (sin. *Aspergillus amstelodami* Thom *et* Church), than skin with no fungal treatment. However, Douglas and Hart (1989) concluded that no experimental studies provided definite evidence of this, although they did show that small quantities of the fungus *A. penicillioides* Speg. may possess a nutritional value. De Saint Georges-Gridelet (1987) reported that the fungus *Aspergillus repens* (Corda) Saccardo also increases the nutritional value of skin scales to the mites, by reducing the fat content and adding vitamins B and D. Mites kept in the laboratory are normally fed yeast-based food, which is often supplemented with different types of animal protein. Experiments indicate that the type of food used can affect the rate of house-dust mite growth (Hart *et al.*, 2007). This can have a significant impact on the comparability of laboratory trials. Mullen and O'Connor (2019) retained that *Dermatophagoides* and *Euroglyphus* mites are saprophages feeding fundamentally in the home on skin scales shed by humans and indoor pets. Although fungi are commonly collected in the alimentary tract of house-dust mites, they are probably ingested only secondarily and do not represent a significant nutritional component of their diet, according to the fact that some fungi (*A. penicillioides*) may be detrimental to mite growth and reproduction (Hay *et al.*, 1993). Mattresses, and sleeping quarters in general, provide particularly favourable sites for mites to develop. Human semen has been shown to be a dietary supplement for house-dust mites and can significantly increase the number of eggs that a female produces (Colloff *et al.*, 1989).

Three species of house-dust mites live commonly in the human dwellings; they are *Dermatophagoides farinae* Hughes, *D. pteronyssinus* and *Euroglyphus maynéi* (Cooreman) (Fain *et al.*, 1990; Colloff, 1998, 2009a; Arlian, 2001, 2002). They show an adaptation to peridomestic situation in many countries of the world, where they feed on flakes of skin found in dust and typically occur on carpets, around sofas and chairs, and on mattresses; in addition, they infest stored food products. These house-dust mites may be very plentiful in the humid tropics but in temperate climates are more numerous in humid summers than in winters, when the R.H. is usually lower in homes (Platts-Mills *et al.*, 1982). House-dust mites occur in greatest numbers in the more humid living quarters of homes frequented by the occupants, notably wherever dust accumulates in bedrooms and living rooms. Mattresses are especially suitable, apparently due to the accumulation of human squamal cells and other skin debris. Under optimal conditions, as many as 5000 mites have been recovered per gram of mattress dust. The type of floor can influence the species and number of mites. In damp houses, carpeted floors contribute to high populations, whereas in drier homes, there may be little difference in the numbers of mites in carpeted and non-carpeted flooring. When humidity levels are high, even floors covered with linoleum and other plastic materials will support relatively high numbers of house-dust mites. In general, however, drier floors and carpets support higher populations of *D. farinae* than either *D. pteronyssinus* or *E. maynéi*. *E. maynéi* has the highest humidity requirements, occurs primarily in mattresses and bedding, and is the least likely to be found in carpets (Mullen and O'Connor, 2019). *D. pteronyssinus* and *D. farinae* show behavioural differences that enable them to occupy different niches within the same habitat. Van Bronswijk (1981) reported findings from a study by Wharton that noted that although *D. pteronyssinus* and *D. farinae* occupied the same niche, *D. farinae* tended to crawl on top of the substrate, while *D. pteronyssinus* remained beneath it. Usually, both mites live closer to the surface of the mattress, although de Boer (1990a) reported that both species are also present deep inside a mattress, not just near the surface. The R.H. is an important factor regulating the geographic prevalence and the density of *D. farinae*, *D. pteronyssinus* and *E. maynéi*. So, in humid geographic areas, most homes host mite populations, whereas in dry geographic areas characterized by low R.H., few homes host mites. The prevalence and the density of these Pyroglyphid mites, vary both geographically and between homes of the same geographic area. The factors influencing the variation in the mite density between homes are not well understood, but it appears that the mite density is not correlated with the house cleaning practices. It has been observed that a greater mite population live on the carpeted floors than on wood or tile floors. In a home, only a species or more species can coexist. Many homes are coinhabited by more than one species, but one species usually represents the greatest percentage of the total population of Pyroglyphid mites, and the dominant species varies between homes within a geographic area (Arlian and Morgan, 2003).

It has been proved that these mites are important producers of lung allergy in humans and probably the most frequent agents of house-dust asthma (Voorhorst *et al.*, 1964; Spieksma, 1967). Some people suffer a reaction to allergens present in the faeces of house-dust mites. These may cause atopic dermatitis or be inhaled giving rise to asthma or rhinitis (Arlian and Platts-Mills, 2001) in 65 million to 1.2 billion people worldwide (Cunnington and Gregory, 1968; Holt and Thomas, 2005; Colloff, 2009a; Hammad *et al.*, 2009; Lloyd, 2009). Voorhorst *et al.* (1967) found that dust mites are the major source of allergen in house dust. They went on to explain many features of the biology of mites and provided much of the theoretical argument for current strategies used for the control of dust mites. In humid areas of the world, dust mites are ubiquitous, and up to 30% of the population show positive skin test reactions to at least one dust mite species. Sensitization to allergens derived from dust mites is strongly associated with three important diseases, such as asthma, perennial rhinitis, and atopic dermatitis, especially in children and young adults. The most common clinical manifestation is bronchial asthma, characterized by difficulty breathing, inflammation of the nasal passages and conjunctivitis. This may be associated with atopic eczema in some sensitized individuals. Asthmatic attacks tend to occur at night, especially in poorly ventilated bedrooms with old bedclothes and accumulated mattress and floor dust. Occurrence of symptoms is usually seasonal, reflecting the size of the mite populations. The acuteness of the allergic attacks is directly correlated with the number of mites present (Arlian and Platts-Mills, 2001; Mullen and O'Connor, 2019).

The sources of many allergens are mite bodies and mite faeces (Tovey *et al.*, 1981; Arlian *et al.*, 1987). The allergens associated with mite faeces are enzymes that originate from the mite's digestive apparatus. Other allergens include enzymes probably associated with the moulting process that occurs as mites change from one life stage to the next. Some allergens may be components of mite saliva that is left in the environment on food substrates where mites feed. Secretions from the supracoxal glands that are involved in the active uptake of water likely contain proteins, as well as sodium and potassium chloride (Wharton and Furumizo, 1977). After death, soluble protein in body fluids may be released as the body disintegrates. Some proteins from all of these sources could be allergenic (Arlian and Platts-Mills, 2001). *D. pteronyssinus* produces the most potent house-dust allergens, though individuals of *D. farinae and E. maynéi* and of other storage mites can cause allergic responses as well as *D. pteronyssinus*. Each species of mite possesses its own species-specific antigens and allergens, with differences between those associated with the mite body and faeces. The different species exhibit a significant cross-reactivity among the antigens, and makes it difficult to identify the species involved, hindering the clinical diagnosis and the medical treatment. Diagnostic tests are based on the use of skin tests and bronchoprovocation with commercial extracts of specific mite species. Enzyme-linked immunosorbent assays and radioallergosorbent tests resulted less effective than the skin-prick test to identify the allergy in people only mildly sensitized to mite allergens (Mullen and O'Connor, 2019). An extensive treatment of this topic may be found in Colloff (2009a). A comprehensive treatment on the Pyroglyphid mites, injurious to people, may be found in Muhammad Sarwar (2020).

On stored food are reported eight species (Hagstrum *et al.*, 2013). Four of these play a secondary importance; they are *Dermanyssus chelidonis* (Hull) collected on grain in England, and Wales and in nests of swallows, swifts and sparrows (Hull, 1931; Fain *et al.*, 1974), *Dermatophagoides longior* (Berlese), reported from Czech Republic, England, France, Germany and Sweden on barley sweeping and clover (dried) (Berlese, 1897a; Žďárková, 1967), *Dermatophagoides microceras* Griffiths and Cunnington, known from China, England, Spain, United States on Chinese medicine (Griffiths and Cunnington, 1971; Hughes, 1976, Smiley, 1987), *Euroglyphus longior* (Trouessart) reported from Egypt (Saleh *et al.*, 1985), England, France (Hughes, 1976), Germany (Franz *et al.*, 1997), Sweden (Sellnick, 1958; Hughes, 1976), United States (Hughes, 1976) on bean (Saleh *et al.*, 1985), granary debris, hide dust in farm (Sellnick, 1958; Fain, 1965; Hughes, 1976; Franz *et al.*, 1997) and *Pyroglyphus africanus* (Hughes, 1954) reported from Angola on fishmeal, originally named *Dermatophagoides africanus* Hughes (Hughes, 1954, 1976).

We treat below only three species of house dust mites reported worldwide in the human dwellings, named *D. farinae*, *D. pteronyssinus* and *E. maynéi*, and sometimes in stored products or in other situations, such as in household fabrics in Germany, where 18% of the mites were collected in clothing hanging in closets (Elixmann *et al.*, 1991), or also in stored food products. Skoda-Smith *et al.* (1996) reported that in Alabama (USA), an individual experienced sneezing, intense ocular pruritus and facial oedema within minutes of inhaling a puff of dry pizza-dough mix, heavily infested with *D. farinae*.

Selected websites

https://www.fotosearch.ie/photos-images/pyroglyphidae.html

https://www.gettyimages.se/search/2/image?phrase=pyroglyphidae

The Genus *Dermatophagoides* Bogdanov

Diagnostic characteristics

The tegmen is absent. The female does not possess a hysteronotal shield and has the scapular setae *sce* much longer and thicker than the setae *sci*. The legs III and IV are equal or subequal in length. The male has the external scapular *sce* much longer and thicker than the internal scapular setae *sci* and possesses a hysteronotal shield. The perianal ring is simple and not denticulate. The tarsus III lacks spines (Hart and Fain, 1988).

Economic importance

This genus includes few species, reported on bird nests, birds and house dust. It was erected by Bogdanov (1864) for the species *Dermatophagoides scheremetewskyi*, collected on the skin of various people with mange in Moscow. The female of this mite was collected several times on the skin surface of subjects, infested by *Sarcoptes scabiei* (Linnaeus) (Fain, 1967b).

Some species have been collected on stored food, but the importance of this genus is fundamentally associated with the ability of some species to produce allergens in house dust, causing allergic rhinitis and asthmatic symptoms. This possibility was first suggested over 60 years ago by Kern (1921), but the allergenic factor in house-dust remained unresolved for long time until when Voorhorst *et al.* (1964) suggested that a mite may be responsible for the house-dust allergen. In atopic patients, this mite was found to produce similar skin reactions, to those caused by house dust. The mite was identified by Fain (1966a) as *D. pteronyssinus*, a species described by Trouessart (1897) on dried skins in France and showed by subsequent studies as the predominant species in house dust in many parts of the world (Fain, 1966b; Fain *et al.*, 1974). Since then, other mite species have been implicated in house-dust allergies, and in particular other mites of the same family Pyroglyphidae, namely *D. farinae*, found on poultry and pig-rearing meal from England (Hughes, 1961) and *D. microceras* Griffiths and Cunnington (1971) reported in house dust from England (Hart and Fain, 1988; Arlian and Platts-Mills, 2001). Here, we treat only the most important species reported on stored food, although their direct damage to foodstuffs is not very clear and requires appropriate studies.

Dermatophagoides farinae Hughes (Fig. 38.1)

Common name

American house dust mite

Morphological characteristics

FEMALE. The idiosoma is about 360–400 μm long. The hysterosomal shield is not present. The cuticle of the dorsal idiosoma is transversely striated. The sternal region is not so heavily sclerotized. The apodemes I are not bent at right angles. The genital opening is Y-upside down and bounded anteriorly by a crescentic epigynum. The posterior genital plate is only well-sclerotized along its lateral edges. The opening of the bursa copulatrix is on the dorsal surface above the anal region, it communicates with the flask-shaped sclerotized region of the receptaculum seminis by a fine duct. The legs I and II have about the same length and width, and the leg IV are longer than leg III. The chaetotaxy is similar to that of the male, except that two short setae replace the vestigial suckers found on tarsus IV of the male (Hughes, 1976).

MALE. The idiosoma is about 260–360 μm long, rounded and with the sejugal furrow between the propodosoma and the hysterosoma. The propodosomal shield is variable in shape and may have the posterior edge extended laterally, to encircle the scapular setae. The hysterosomal shield does not extend as far forward as the d_1 (d_2) setae. All the idiosomal setae are smooth. The setae *sce* are more than four times longer than *sci*, and a supracoxal seta (*scx*) is present. Two pairs of setae (*cp* and c_3) are present ventro-laterally, *cp* (*hv*) arising lateral to coxae III and being about the same length as *sce*. Four pairs of short dorsal setae are arranged in two longitudinal rows which converge posteriorly, the setae are about the same length as d_2 (*la*), e_2 (*lp*), f_2 (*sae*) and ps_1 (pa_2); h_2 (pa_1) and h_3 (*sai*) are considerably longer, h_3 (*sai*) exceeding half the length of the idiosoma. The coxal region is well-sclerotized and finely punctuate. The apodemes I may remain separate or fuse in the midline to form a short sternum. Apodemes III are long and bent sharply at right angle. The usual coxae setae are present on coxae I and III, and three pairs of setae surround the genital opening, the posterior pair being much shorter than the remaining pairs. The genital opening is between coxae III and IV, and the anus is encircled by an oval perianal ring, which also encloses prominent anal suckers, and a pair of preanal setae. The hind edge of this ring is convex posteriorly. The aedeagus is long and tapering and arises from a triangular base.

All the legs end in prominent pretarsi and small claws. The legs I are considerably thickened and compressed bilaterally. On femur I, a blunt process arises from the ventral surface. On tarsus I, $\omega 1$ arises at the base of the pretarsus on the same level as ω_3. The famulus is subapical. The lateral distal end of this tarsus is produced into a blunt process, and on tarsus III, a similar process has a bifid extremity and a smaller process is present on the opposite side of the segment. Omega$_1$ arises from the base of tarsus II. One ventral seta is present on tibiae I and II. Two solenidia arise from genu I, one much longer than the other. Leg III are noticeably longer and thicker than legs IV. On legs IV, two small, stalked suckers are present at the end of the tarsus (Hughes, 1976).

Geographical distribution

Cosmopolitan and most common in temperate climates, reported from Algeria (Louadi and Robaux, 1992), Argentina (Fain, 1967c; Martínez Canzonieri *et al.* 1996), Australia (Green *et al.*, 1986; Tovey *et al.*, 2000), Barbados (Pearson and Cunnington, 1973), Belgium (Gridelet and Lebrun, 1973), Brazil (Greco *et al.*, 1974; Moreira, 1975; Horn *et al.*, 2018), Brunei (Woodcock and Cunnington, 1980), Bulgaria (Todorov, 1979), Canada (Griffiths and Cunnington, 1971; Murray and Zuk, 1979), Chile (Artigas and Casanueva, 1983; Casanueva and Artigas, 1985), China (Gabriel *et al.*, 1982; Wen *et al.*, 1988; Cai and Wen, 1989; Li *et al.*, 2015a; Zhan *et al.*, 2017b), Colombia (Sanchez-Medina and S.-Gutierrez, 1973; Charlet *et al.*, 1979), Croatia (Blythe, 1976; Macan *et al.*, 2003), Czech Republic (Žďárková, 1967; Samšiňák *et al.*, 1972; Dusbábek, 1975, 1979, 1995), Denmark (Haarlov and Alani 1970; Alani and Haarlov, 1972), Ecuador (Valdivieso *et al.*, 2006), Egypt (Saleh *et al.*, 1985; Rezk *et al.*, 1996; Mahgoob *et al.*, 2006; Ahmed *et al.*, 2020a), England (Hughes, 1961, 1976; Colloff, 1988), Estonia (Eriksson *et al.*, 1998), Finland (Stenius and Cunnington, 1972), France (Vervloet *et al.*, 1982), French Polynesia (Fain, 1988b), Georgia (Dubinina and Plietnev, 1978), Germany (Karg, 1973; Keil, 1983), Greece (Deliargyris *et al.*, 1990; Palyvos *et al.*, 2008), Hungary (Halmai, 1984), India (Dar *et al.*, 1973; Dixit and Mehta, 1973; Dar

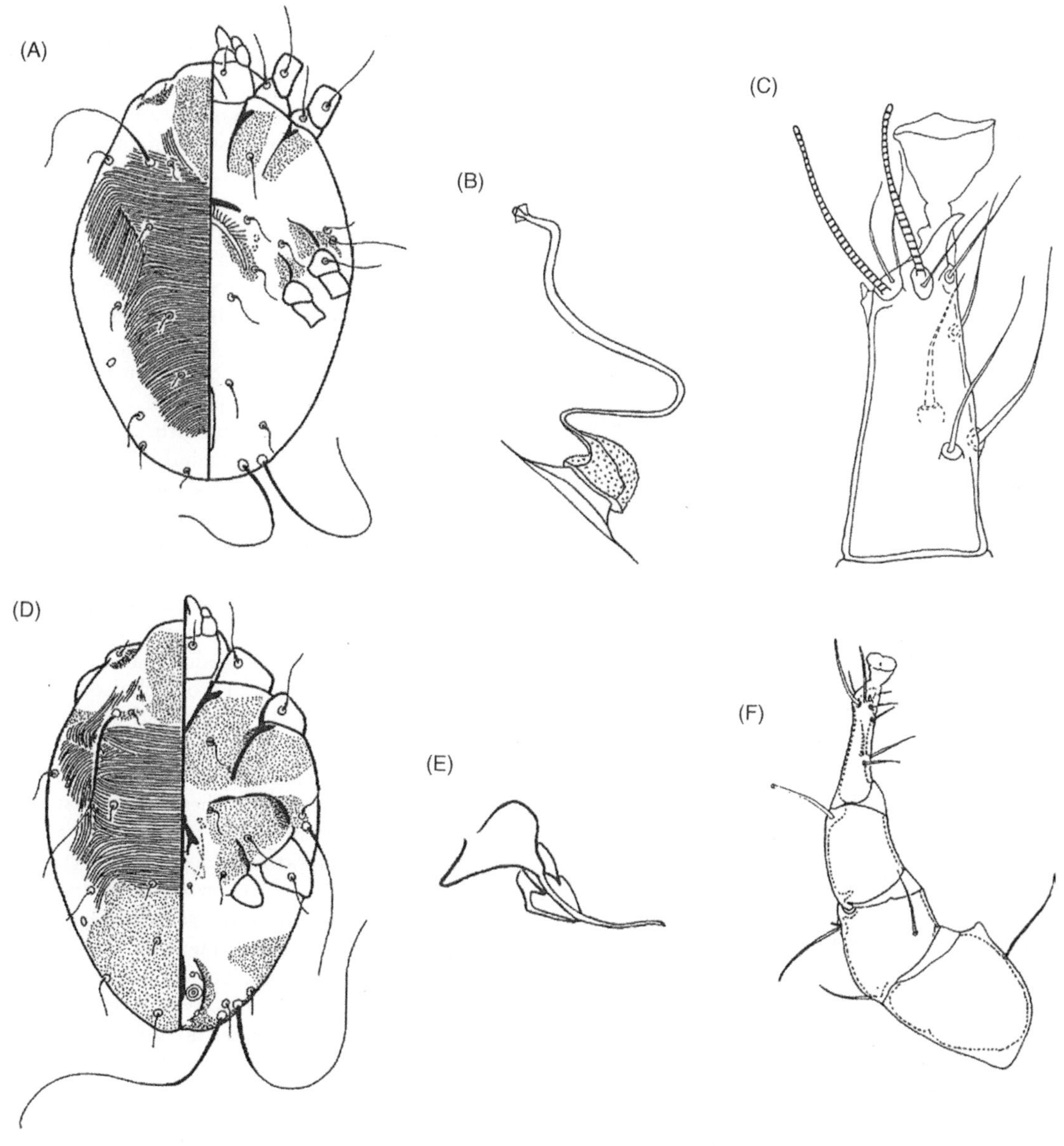

Fig. 38.1. *Dermatophagoides farinae* Hughes. Female. Dorsal (at left) and ventral view (at right) (from Vacante, 1997b) (A); reproductive system (from Hughes, 1976) (B); leg I (from Griffith and Cunnington, 1971) (C). Male. Dorsal (at left) and ventral (at right) (from Vacante, 1997b) (D); aedeagus (E); leg I (F) (from Hughes, 1976).

and Gupta, 1979), Iceland (Eriksson *et al.*, 1998), Indonesia (Baratawidjaja *et al.*, 1998), Iran (Sepasgozarian and Mumcuoglu, 1979; Soltani *et al.*, 2011), Israel (Feldman-Muhsam *et al.*, 1985; Mumcuoglu *et al.*, 1999), Italy (Blythe, 1976; Ottoboni *et al.* 1983; Goracci *et al.*, 1984; Goracci and Goracci, 1990; Vacante, 1997b), Japan (Sasa, 1947; Miyamoto *et al.*, 1970; Oshima, 1970), Kazakhstan (Yagofarov and Galikeev, 1987), Kenya (Blythe, 1976), Lithuania (Dubinina *et al.*, 1984), Malaysia (Ho, 1986), Martinique (Lafosse Marin *et al.*, 2006), Mexico (Gonzalez-Rodriguez and Llorens, 1974), the Netherlands (Spieksma, 1967, 1968; Siepel *et al.*, 2016), New Zealand (Ingham and Ingham, 1976), Poland (Solarz, 1986; Solarz *et al.*, 1995; Solarz, 1997; Jakubas-Zawalska *et al.*, 2013), Portugal (Pinhão and Grácio, 1978), Puerto Rico (Fernandez-Caldas *et al.*, 1995), Russia (Dubinina and Plietniev, 1978), Scotland (Colloff, 1987a), Sierra Leone (Hughes, 1976), Singapore (Griffiths and Cunnington, 1971; Blythe, 1976), South Africa (Ordman, 1971), South Korea (Paik *et al.*, 1992), Spain (Griffiths and Cunnington, 1971), Sweden (Nordvall *et al.*, 1988), Switzerland (Spieksma *et al.*, 1971), Taiwan (Chang and Hsieh, 1989; Wu, 1999), Thailand (Wongsathuaythong and Lakshana, 1972; Malainual *et al.*, 1995), Turkey (Blythe, 1976; *Zeytun et al.*, 2018), United States (De Leon, 1963; Furumizo 1973, 1975; Arlian *et al.*, 1992), Venezuela (Hurtado and Parini, 1987) and Zambia (Buchanan and Jones, 1972, 1974).

Commodities, facilities, vegetal and animal hosts

Commodities of this mite include bean sauce, biscuit meal (De Leon, 1963), bran (Li *et al.*, 2015a), broad bean (El-Sayed and Ghallab, 2007; Li *et al.*, 2015a), carpet dust, cottonseed hull, crumbs, dried bean curd, dried sweet potato, dust in wheat bin, feeds, fine dried noodles, flour contaminated with dust, ginkgo nut (Li *et al.*, 2015a), maize (Saleh *et al.*, 1985), mattes dust, old silk noodles (Li *et al.*, 2015a), poultry and pig-rearing meal (Hughes, 1961, 1976), rice contamined with dust (Li *et al.*, 2015a), rough rice (El-Sayed and Ghallab, 2007), sesame, sesame oil slag, starch, sticky rice, straw (Li *et al.*, 2015a), wheat (Mahgoob *et al.*, 2006) and wheat bran (Saleh *et al.*, 1985; Mahgoob *et al.*, 2006), wheat flour (Yi *et al.*, 2009; Al-Nasser, 2011), wheat pollard (Fain, 1967c), wheat straw (Li *et al.*, 2015a). Moreover, the mite has been collected on Chinese medicine, including *Rhizoma dioscoreae*, *Rhizoma cimicifugae*, *Radix Asteris* (Zi Wan) (Zhan *et al.*, 2017b), in house dust (Spieksma, 1967 and many other reports), in house and mattress dust (Wharton, 1970). In Brazil, the species (*D. farinae*) has been exclusively found on feathers (Horn *et al.*, 2018). Rosa and Flechtmann (1979) have already reported it from house-dust in Brazil. Ždárková (1967) carried out an entomological investigation in 126 food processing factories in Czechoslovakia and found among acarid mites infesting stored foodstuffs and food products, also specimens of *D. farinae*. The facilities include agricultural cooperative unions, commercial stores, farm stores and flour mills (Palyvos *et al.*, 2008).

Bio-ecology

The American house dust mite belongs to the group of 13 species reported from house dust. Among these, *D. farinae*, *D. pteronyssinus* and *E. maynéi* are most common in temperate climates; they are very common in homes worldwide and represent the major source of mite allergen (Arlian and Platts-Mills, 2001). *D. farinae* is reported from a wide list of commodities, but its economic importance is mainly related to its ability to produce allergens. Fain (1967a) hypothesized that the main food source or exclusive of this mite, as well as for *D. pteronyssinus*, consists of human skin scales, which regularly fall from human skin into the floor and accumulate to form dust. Further studies confirmed (also for other Pyroglyphid mites) that the mite mainly feeds on skin scales and other organic detritus that collects in homes (Arlian and Platts-Mills, 2001). Samplings of house dust from 17 flats, three hospitals, two hotels and one students hostel carried out in Poland by Racewicz (2001) showed the presence of *D. pteronyssinus* and *D. farinae*. This last mite was significantly predominant and composed 82.8% of the whole Pyroglyphid collection. Samples from private flats contained significantly more mites than those from hospitals and hotels: mean mite densities per 1 gram of dust were 13.07, 1.03 and 1.00, respectively. These findings suggest that only density of house-dust mites in private flats may be clinically important (Racewicz, 2001).

The reproduction of *D. farinae* is sexual; the female stores the sperm in the seminal vesicle and releases it into the oviduct to fertilize the egg during the ovulation process. The life cycle develops through the egg, larva, protonymph, tritonymph, and adult stages. Furumizo and Wharton (1975) reported a case of postimaginal moulting of a *D. farinae* female.

Arlian *et al.* (1998b) determined the survival, development, and fecundity of *D. farinae* exposed to fluctuating daily regimes of hydrating and dehydrating R.H. The larva emerged from 84, 92 and 94% of eggs incubated at a regime of 2, 4 and 8 hours at 75% R.H., and 22, 20 and 16 hours at 0% R.H., respectively. No emerging offspring completed the life cycle, when exposed to the two and four hours of moist air daily, but 44% and 53% survived for 70 days in the larval or nymphal stages, respectively, and these completed the development to adults, when subsequently held at a constant 75% R.H. Given eight hours of moist air daily, 41% of emerging offspring completes the life cycle, but development is 1.6 times longer compared with development at a constant 75% R.H. For all daily hydrating and dehydrating regimes, a greater percentage of offspring became males than females. Survival of immatures is remarkable at these daily long periods of dehydrating conditions, when a short period of hydrating moisture is provided. When exposed to a daily regime of four hours of 75% R.H. and 20 hours of dry air at 0% R.H., 84% of females survived 28 days, and produced approximately 1/3 of the number of eggs produced in the control at constant 75% R.H. (Arlian *et al.*, 1998b). *D. farinae* is able to survive adverse conditions, particularly low R.H., by forming a prolonged, so-called drought-resistant quiescent protonymph stage. The quiescent protonymph of *D. farinae* consumes 28.5 times less oxygen per hour than an active protonymph. It has a water exchange half-life of 160 days, compared to just 20 hours for an active protonymph, and 28 hours for an adult (Arlian *et al.*, 1983). Hart (1998) reported that the most important limiting factor for house dust mite populations is air humidity. House dust mites osmoregulate through the cuticle and therefore require a high ambient air humidity to prevent excessive water loss. In addition, the supracoxal glands actively take up ambient water vapour, and the protonynph stage of the life cycle is resistant to desiccation. Larger house dust mite populations are found when the absolute indoor air humidity is above 7 g/kg (45% R.H. at 20°C). Studies of the life cycle of cultured *D. farinae* showed that after an initial mating, the females lived for 63.3 ± 64.6 days after their egg production period ended. The long period after the cessation of egg production suggested that *D. farinae* females could mate multiple times and produce eggs continuously for a longer period. Results of this study revealed that the females are capable of more than one successful mating that results in increased egg production compared to that of a single mating. Probably, the females mate multiple times in natural and culture populations; they actively attract males during the reproductive period but not afterward, even though she continues to live a long time (Alexander *et al.*, 2002).

Studies carried out at four constant temperatures (15, 20, 25 and 30°C) under a 14L-10D photoperiod and 80 ± 5 R.H. % showed that the duration period from egg deposition to adult emergence of the American house dust mite averaged 119.2 ± 5.5, 36.9 ± 3.6, 21.1 ± 1.4 and 15.1 ± 1.7 days for female and 129.1 ± 7.1, 35.8 ± 2.9, 19.7 ± 1.8 and 14.6 ± 1.5 days for male, respectively. The average incubation periods decreased with increasing temperature for both females and males. Duration of larval, protonymphal and tritonymphal stages decreased as the temperature increased from 15 up to 30°C for both females and males, while the percentage of time for the quiescent period of each life stage was relatively constant for different life stages and averaging one-third the total duration of each life stage. Under higher temperatures, daily egg productions were higher than lower temperatures, even if the total egg/female was lower. Preoviposition, oviposition and the total adult survival periods generally decreased with increasing temperature (Rezk, 2004). Arlian and Dippold (1996) estimated in laboratory, the duration of the life cycle of the American house dust mite at 16°C, 23°C, 30°C and 35°C and the fecundity at 23°C and 75% R.H. They found that the lengths of the life cycles at 23°C and 30°C were 35.6 +/– 4.4 and 17.5 +/– 1.2 days, respectively. Few eggs completed the development to the adult stage at 16°C and 35°C. At 23°C and 75% R.H., the preoviposition period was 3.7 +/– 1.1 days following emergence of the female from the tritonymph. The average reproductive period was 34.0 +/– 10.7 days with a mean total of 65.5 +/– 17.4 eggs produced per female. Female longevity after cessation of egg production was 63.3 +/– 64.6 days (Arlian and Dippold, 1996).

Various microorganisms modulate the productions of allergens by their mite hosts and/or contaminate immunotherapeutic extracts. Changes in microbiomes relate to the mite population density, microbial microcosm respiration and concentration of guanine, a mite nitrogenous waste metabolite.

The dynamic of populations exhibits an exponential growth, plateau and exponential decline of populations. The intracellular bacterium *Cardinium* and the yeast *Saccharomyces cerevisiae* (Meyen ex Hansen) dominate the internal mite microbiomes, as well as the bacterium *Lactobacillus fermentum* Beijerinck in the mite diet. The reduction in the mite population level during the late phases of culture development depends on the changes in the microbial profiles. The intracellular bacterium *Cardinium* is replaced by *Staphylococcus*, *Oceanobacillus* and *Virgibacillus* and *Saccharomyces cerevisiae* is replaced by the antagonistic fungi *A. penicillioides* (Speg.) and Candida. A positive correlation occurs among the increases in the guanine content and the increases in the *Staphylococcus* and *A. penicillioides* profiles in the culture. The mite microbiome exhibits strong, dynamic alterations in its profiles across different mite culture growth stages (Klimov *et al.*, 2019)

Symptomatology and damage

Ahmed *et al.* (2020a) reported that mites present in house dust can also infest dry-stored food products. Many studies have reported the presence of *D. farinae* in different stored foodstuffs, such as wheat flour (Yi *et al.*, 2009; Al-Nasser, 2011), wheat and grain residue (Mahgoob *et al.*, 2006), and broad bean and rough rice (El-Sayed and Ghallab, 2007). Results of studies carried out by Ahmed *et al.* (2020a) show that *D. farinae* is accessory (occurred in 43.5% of the total number of samples) and influent (constituted 5% of the total number of individuals). However, no reports described the direct damage produced by *farinae* to foodstuffs, in the sense of altering the quantity and quality of the same. Regardless of this, the presence of the mite can be cause for alarm for the manual workers and/or the consumers of foodstuffs, for the presence of allergen, and in this sense, all precautions must be taken to limit its presence in foodstuffs.

Control

The control needs monitoring and involves the use of different means (physical, chemicals, etc.). An accurate review of Arlian and Platts-Mills (2001) gave suggestions for the home environment; however, some of these suggestions can be employed for the mite control, also in the foodstuffs. These technical possibilities include 'reducing indoor R.H., using encasements, washing, drying, and dry cleaning of bedding materials, replacing carpets, draperies, upholstery, vacuuming carpets, freezing soft toys and small items, air cleaning/filtration, duct clearing, using ozone generators, and using chemicals'. In general, limiting the exposure to house-dust mite allergens requires integrated approaches and suitable design (e.g. bio-modified textile materials) may be used as an element of the complex anti-dust mite protection. Crowther and Wilkinson (2008) have published a comprehensive review of the control strategies; they reported that the main actions of control are the prevention, the direct control of dust mites and the control or removal of their allergens. The sampling methods play a fundamental role and expect different options of intervention, such as the mite extraction from dust, the sampling site variations and the monitoring of allergens. Various technical possibilities characterize the control, such as the modifying environmental conditions, the cleaning, the temperature, the R.H. and the chemical control. As concerns this last aspect, it is necessary taking into account that the use of chemicals for the dust mites control on stored food must follow the law of each country.

MONITORING. News on monitoring is taken from Solarz (1997) and mainly referred to the home environment; they are discussed here, only to give a general information to help the solution of the problem of stored foodstuffs. In the case of house-dust mites, the commonly used indicators for assessing exposure and risk are antigens, whole animals (by isolation and microscopy analysis) and guanine levels (Pope *et al.*, 1993). The latter are not fully reliable because of low specificity (Colloff *et al.*, 1991, 1992; Mosbech *et al.*, 1991; Arlian *et al.*, 1992). Analysis can be semiquantitative (e.g. 'absence or presence' and 'low, medium or high level/number') (Pope *et al.*, 1993). Two threshold levels for risk of exposure to house dust mite allergens were proposed: exposure to levels of *Der p* I allergen >2 μg/1 g of dust (= 100 mite specimens per 1 g of dust) found to be associated with increased risk of sensitization, and exposure to levels >10 μg/1 g (= 500 mite specimens per 1 g), found to be associated with a risk of acute asthma symptoms (Platts-Mills and de Weck, 1989; Tovey, 1992).

PHYSICAL CONTROL. Physical control *sensu stricto* includes the management of the temperature, of the R.H. and other means. From a technical point of view, the ventilation by air-conditioning systems is being developed as a means of control. A number of other aspects of the domestic environment are also being manipulated in an integrated approach, to render the habitat less suitable for mites. The potential exists for developing models for house dust mite populations, environmental characteristics and the effects of various approaches to control (Hart, 1998). We treat here, the factors above reported.

TEMPERATURE CONTROL. The temperature control includes different possibilities as the autoclaving (de Boer, 1990b), steam cleaning, electric blankets (de Boer and van der Geest, 1990), bed heaters, freezing (Colloff, 1986; Nagakura *et al.*, 1996) and sunlight (Tovey and Woolcock, 1994).

RELATIVE HUMIDITY CONTROL. Indoor R.H. is the key factor that determines the survival and population development of *D. farinae*. Maintaining R.H. below 50% is one recommendation in a comprehensive plan to reduce house-dust mites and mite allergen levels in homes. Even when mean daily R.H. is reduced below 50%, R.H. may rise above 50% intermittently for brief periods because of activities in the home (e.g. cooking, bathing and ventilation). Chang *et al.* (1998) carried out studies, to evaluate the survival time of randomly selected female *D. farinae* exposed to temperatures from 40°C to 80°C and R.H. between 10°C and 90%. Significant temperature effects were found on the length of time, mites can survive under those conditions. The time needed to induce 100% mortality for female *D. farinae* held at 40°C varied between 39 and 84 h. At 70°C and 80°C, all mites died within five minutes. At 50°C and above, R.H. had little influence and temperature alone was the determining factor for mite survival. The length of survival time decreased precipitously between 40°C and 50°C, suggesting that a loss of physiological integrity of female *D. farinae* probably occurred in this range. The results of this study indicated that the technique of raising temperatures to above 40°C for a short duration, can be used to reduce living house dust mite populations in homes significantly (Chang *et al.*, 1998). Arlian *et al.* (1999b) found that maintaining mean daily R.H. below 50%, even when R.H. rises above 50% for two to eight hours daily, effectively restricts population growth of these mites and thus the production of allergen. To completely prevent population growth of *D. farinae*, the R.H. must be maintained below 35% for at least 22 hours per day when the daily R.H. is 75% or 85% for the remainder of the day.

MODIFYING OF ENVIRONMENTAL CONDITIONS. Crowther and Wilkinson (2008) reported that modifying environmental conditions may be productive, based on the fact that dust mites can be controlled through winter culling in climates that allow it. Controlling mites in this way has two major advantages. First, the hygrothermal conditions that are unfavourable to mite population growth can often be created by relatively minor adjustments to heating and ventilation methods or occupant behaviour. In such cases, the control of mites can be achieved inexpensively and without the use of acaricides (agents that kill ticks and mites). Second, controlling mites by environmental means has potential, both as a curative measure for lleviating symptoms and as a preventive measure – that is, before sensitization occurs. Modelling techniques, thus, make it possible to investigate which modifications to home environments have most impact on mite populations, for any given climatic region, housing type and pattern of occupant behaviour. In this way, it will be possible to establish the most effective, energy efficient and socially acceptable ways of achieving HDM control, by modifying environmental conditions, where this is feasible. In this sense the constructional changes to improve condition help the control; they include different actions, such as the improved maintenance and the avoidance of excess moisture (Eldridge, 1976), improved insulation and ventilation standards (Harving *et al.*, 1991; McIntyre, 1992; Warner *et al.*, 2000), and non-mechanical passive ventilation systems (Palin *et al.*, 1993; McEvoy and Southall, 2004).

CLEANING. This strategy provides different possibilities, such as the use of home disinfectants, washing (McDonald and Tovey, 1992, 1993; Bischoff *et al.*, 1998), dusting, dry cleaning (Kniest *et al.*, 1989), vacuuming (Bischoff *et al.*, 1992; Hay, 1995; Hill, 1998), steam cleaning (Colloff *et al.*, 1995) and carpet cleaning (de Boer, 1990b; Thompson *et al.*, 1991).

CHEMICAL CONTROL. Elixmann *et al.* (1991) investigated five rural houses under identical climatic conditions, and found that the mite population in the house-dust ecosystem consisted of *D. pteronyssinus*, *D. farinae*, *Cheyletus eruditus* (Schrank), *Cheyletus trouessarti* (Oudemans), *Acarus siro* (Linnaeus), *Tyrophagus putrescentiae* (Schrank) and *Tarsonemus* sp. With an average for all investigated textiles of 90 living mites per m^2, the most abundant species was *Dermatophagoides pteronyssinus*. The analysis was done after vacuuming carpets, mattresses and stuffed furniture. They evaluated the possible recolonization after the mite control by an acaricide based on solidified benzyl benzoate (ACAROSAN). During the first 6 months after the mite control, no reactivation of the mite populations arrived. The number of dead mites increased in the first months after the acaricidal treatment. Only 12 months later, a slow recovery of the *Dermatophagoides* species population could be observed. However, even after 2 years, this population reached only about 10% of the original density. A slow recolonization of the mattresses and the upholstery by *Cheyletus* was observed too, whereas storage mites (*Acarus* sp., *Glycyphagus* sp., *Tyrophagus* sp., and *Tarsonemus* sp.) did not reappear after the single acaricidal treatment. Thus, it is to be seen that at least during 2 years, the living conditions of the mites in their biotopes are definitively changed after the application of the acaricide (Elixmann *et al.*, 1991).

Rim and Jee (2006) tested the acaricidal effects of herb essential oils (pennyroyal, ylang ylang, citronella, lemon grass, tea tree and rosemary) at different doses (0.1, 0.05, 0.025, 0.0125 and 0.00625 µl/cm^2) and exposure times (5, 10, 20, 20, 30 and 60 min) on the house-dust mites *D. farinae* and *D. pteronyssinus*. The most effective acaricidal components of pennyroyal (*Mentha pulegium* Linnaeus) were analysed using a gas chromatography-mass spectrometer (GC-MS). Of these essential oils, the most effective was pennyroyal, which is composed essentially of pulegone (> 99%), at a dose of 0.025 µl/cm^2, which at an exposure time of 5 min killed more than 98% of house dust mites. In the pennyroyal fumigation test, the closed method was more effective than the open method, and maximum acaricidal effect was 100% at 0.025 µl/cm^2, 60 min. The results show that herb essential oils, in particular, pennyroyal have potent acaricidal activity (Rim and Jee, 2006).

BIO-MODIFIED TEXTILE MATERIALS. Cieślak *et al.* (2014) studied the effects of non-woven systems modified with triclosan on the development of *D. farinae* in laboratory assay. Two four-layer non-woven polypropylene systems were designed. The first system, designed as A, consisted of two inner melt-blown non-wovens and two outer spun-bonded non-wovens, and the second system, designed as B, consisted of an inner spun-bonded non-woven, an inner needle-felt non-woven, an inner melt-blown non-woven and an outer spun-bonded non-woven. Bioactive microspheres consisting of poly(L-lactide) (95 wt%) and triclosan (5 wt%) were used. Both systems were modified using dispersion with 2.5 wt% and 10 wt% of microspheres, which corresponds to the content of 0.26 wt% and 1.22 wt% of triclosan in non-woven samples. The culture tests were performed for modified and unmodified non-woven samples and the control sample. The anti-dust mite properties were assessed, based on the mortality of mites. All modified non-woven systems showed a strong suppressive effect on the tested *D. farinae*, compared to the control sample. A higher mortality was observed for non-woven system B. It was found that a nearly fivefold increase in the triclosan concentration does not lead to a substantial increase in the decline rate of mite populations (Cieślak *et al.*, 2014).

Dermatophagoides pteronyssinus (Trouessart) (Fig. 38.2)

Common name

European house dust mite

Morphological characteristics

FEMALE. The idiosoma is 334 µm long and 215 µm wide. The chelicerae are very developed and about 75 µm long. The palpi possess two articles, with the basal two times longer than the distal. The propodosoma has a slightly dotted shield, with the anterior side right and the posterior rounded. The sejugal furrow is absent, and sometimes, some specimens have a slight furrow. The middle region of dorsum between the setae d_1-d_1 and e_1-e_1 is longitudinally striated. The setae *ve* and *vi* are absent. The supracoxal seta *scx* is simple, well developed and direct forward. The setae *sce* are strong and long about 180 µm; the setae *sci* are fine and 20–25 µm long. Ventrally, the epimeres I and II are free. The genital apodemes are lateral and the epyginum is well developed. The proximal part of the bursa copulatrix has a sclerite in the form of a daisy, and the distal part is very slightly dilated. The legs are relatively thin, with the tarsi ending with a short pretarsus and a vestigial claw within the pulvillus. The tarsi I and II possess 2 and 1 solenidia, respectively. All tibiae bear 1 solenidion, and the genua possess 2–1–1 solenidia (Hart and Fain, 1988).

MALE. The idiosoma is 285 µm long and 195 µm wide. The propodosomal shield is smaller than that of the female. The hysterosomal shield is long (longer than broad) and extending further forward than hairs *d2*. The seiugal furrow is not very marked or absent. The epimeres I are parallel or diverging. The genital organ is well developed and set at the level of the coxa IV. The anal opening is subterminal, flanked by two large suckers. Chetotaxy as the

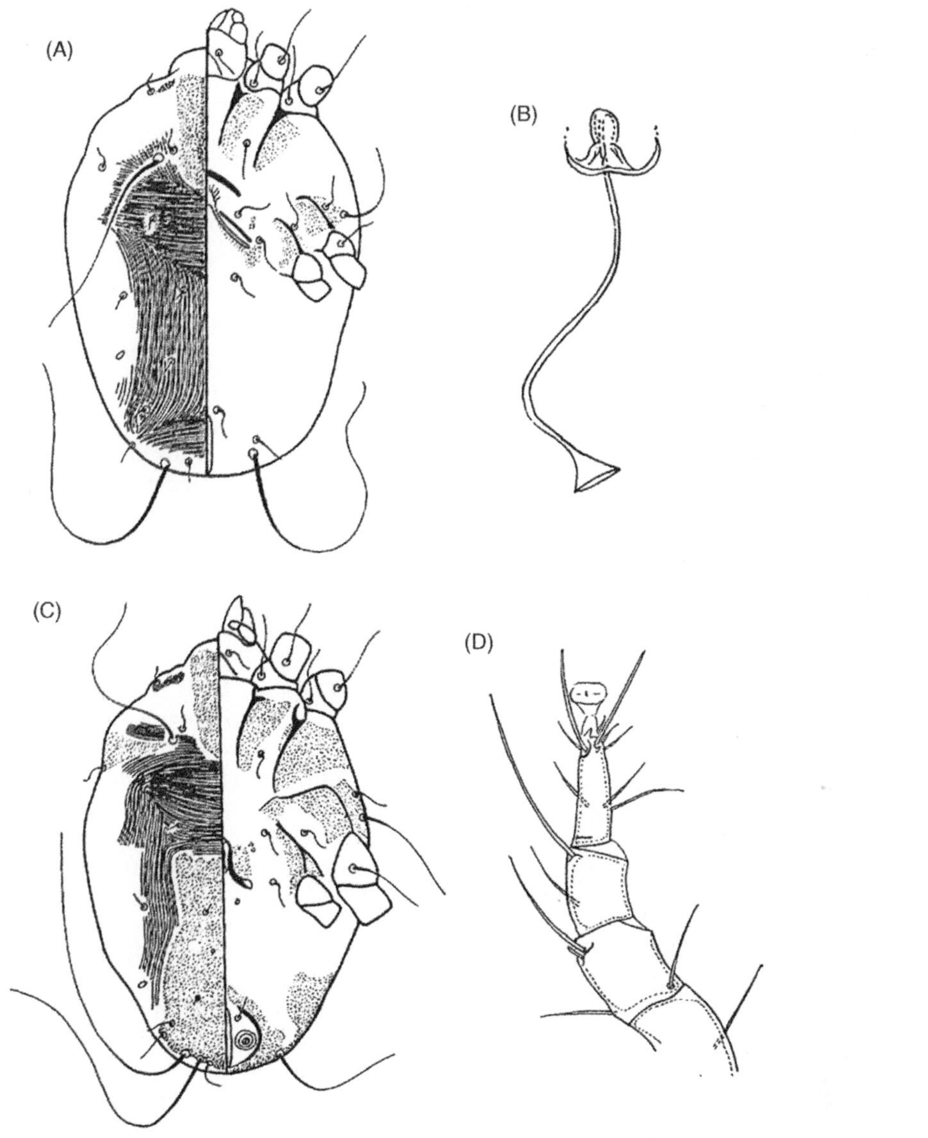

Fig. 38.2. *Dermatophagoides pteronyssinus* (Trouessart). Female. Dorsal (at left) and ventral view (at right) (from Vacante, 1997b) (A); reproductive system (from Hughes, 1976) (B). Male. Dorsal (at left) and ventral view (at right) (from Vacante, 1997b) (C); leg I (from Hughes, 1976) (D).

female. The anal region is encircled by a chitinous ring, piriformis, open front and back. The legs I is normal (Hart and Fain, 1988).

Geographical distribution

Cosmopolitan, reported from Algeria (Louadi and Robaux, 1992), Argentina (Martínez Canzonieri *et al.*, 1996), Australia (Trinca *et al.*, 1969; Green *et al.*, 1986; Colloff *et al.*, 1991; Tovey *et al.*, 2000), Barbados, Belgium (Fain, 1966a; Gridelet and Lebrun, 1973), Brazil (Fain, 1967b; Greco *et al.*, 1974; Moreira, 1975; Rosa and Flechtman, 1979), Brunei (Woodcock and Cunnington, 1980), Bulgaria (Todorov, 1979), Burundi (Fain, 1988b), Canada (Fain, 1966a; Murray and Zuk, 1979), Chile (Artigas and Casanueva, 1983; Casanueva and Artigas,

1985), China (Gabriel *et al.*, 1982; Wen *et al.*, 1988; Cai and Wen, 1989; Li *et al.*, 2015a; Zhan *et al.*, 2017b), Colombia (Sanchez-Medina and S.-Gutierrez, 1973; Charlet *et al.*, 1979), Congo (Fain, 1966a, b), Costa Rica (Vargas and Mariena, 1991), Croatia (Blythe, 1976; Macan *et al.*, 2003), Cuba (Cuervo *et al.*, 1983), Czech Republic (Samšiňák *et al.*, 1972; Dusbábek, 1975, 1979), Denmark (Haarlov and Alani, 1970; Alani and Haarlov, 1972), Ecuador (Valdivieso *et al.*, 2006), Egypt (Rezk *et al.*, 1996), England (Hughes and Maunsell, 1973; Blythe, 1976; Hughes, 1976; Colloff, 1988), Eire (Spieksma, 1967; Fain, 1988c), Faeroes (Korsgaard, in Hallas *et al.*, 2004), Finland (Stenius and Cunnington, 1972), France (Truessart, in Berlese, 1897b; Trouessart, 1901; Araujo-Fontaine *et al.*, 1973; Vervloet *et al.*, 1982), French Polynesia (Fain, 1988b), Georgia (Dubinina and Plietnev, 1978), Germany (Karg, 1973; Keil, 1983), Hungary (Halmai, 1984), India (Alwar and Lalitha, 1961; Dixit and Mehta, 1973; Podder *et al.*, 2009), Georgia (Dubinina and Plietnev, 1978), Germany (Karg, 1973; Keil, 1983), Greece (Deliargyris *et al.*, 1990), India (Fain, 1967b; Dar *et al.*, 1973; Dar and Gupta, 1979), Indonesia (Baratawidjaja *et al.*, 1998), Iran (Sepasgozarian and Mumcuoglu, 1979), Israel (Feldman-Muhsam *et al.*, 1985; Mumcuoglu *et al.*, 1999), Italy (Berlese, 1898; Fain, 1966a; Blythe, 1976; Ottoboni *et al.* 1983; Goracci *et al.*, 1984; Vacante, 1997b), Japan (Sasa, 1947, 1950; Miyamoto *et al.*, 1970; Oshima, 1970), Kazakstan (Yagofarov and Galikeev, 1978), Kenya (Blythe, 1976), Kuwait (Gamal-Eddin *et al.*, 1985), Lithuania (Dubinina *et al.*, 1984), Malaysia (Furumizo and Thomas, 1977; Ho, 1986), Martinique (Lafosse Marin *et al.*, 2006), Mexico (Gonzalez-Rodriguez and Llorens, 1974), Morocco (Fain, 1988c), Mozambique (Ordman, 1971), Namibia (Ordman, 1971), the Netherlands (Fain, 1966a; Spieksma, 1967, 1968; Siepel *et al.*, 2016), New Zealand (Cornere, 1971; Ingham and Ingham, 1976), Norway (Mehl, 1998), Peru (Caceres and Fain, 1977, 1979), Philippines (van Bronswijk *et al.*, 1973b), Poland (Romanski *et al.*, 1977; Solarz, 1986; Solarz *et al.*, 1995; Solarz, 1997), Portugal (Pinhão and Grácio, 1978), Puerto Rico (Fernandez-Caldas *et al.*, 1995), Romania (Popescu and Banescu, 1975), Rwanda (Fain, 1988c), Russia (Dubinina and Plietniev, 1978), Scotland (Colloff, 1987a,b), Singapore (Blythe, 1976), South Africa (Fain, 1966a; Ordman, 1971), South Korea (Paik *et al.*, 1992), Spain (Griffiths and Cunnington, 1971), Surinam (Bronswijk, 1972), Sweden (Nordvall *et al.*, 1988), Switzerland (Spieksma *et al.*, 1971), Taiwan (Chang and Hsieh, 1989; Wu, 1999), Thailand (Wongsathayong and Lakshana, 1972), Turkey (Blythe, 1976), United Kingdom, United States (Fain, 1966a, 1967a; Furumizo, 1973, 1975; Arlian *et al.*, 1992), Uruguay (Schuhl, 1977), Venezuela (Hurtado and Parini, 1987) and Zambia (Buchanan and Jones, 1972, 1974).

Commodities, facilities, vegetal and animal hosts

Hagstrum *et al.* (2013) reported 16 commodities: biscuit, buckwheat, buckwheat product, cake, cereal, Chinese medicine, crabmeat (canned), dried fish, hide dust, jackfruit, jackfruit pickled, dried logan, paper, rice bran, shrimp paste, dried shrimp, frozen shrimp and tamarind in farm. Li *et al.* (2015a) found the mite on paper, rice bran, room dust, bedding, clothing, carpet dust and mattress dust. Zhan *et al.* (2017b) have collected the mite on Chinese medicine including radix of *Angelicae sinensis* (Oliv.) Diels, rhizome of *Atractylodis* DC., radix of *Adenophora* Fisch. and radix of *Asteris* (= *Aster tataricus* Linnaeus).

Fain (1966a) reported the mite on the Scolopacid *Gallinago nigripennis* Bonaparte from South Africa, in the lung of a chimpanzee (*Pan paniscus* Schwarz) from the former Belgian Congo and on *Homo sapiens* Linnaeus on the ear and in human urine. Moreover, he has studied specimens of the mite collected on a buffalo and a mangy dog in India, on the Choloepodid *Choloepus didactylus* (Linnaeus) a slot from Belgium, on the nest of a bird of the genus *Sayornis* Bonaparte from United States and in a bottle containing termites *Macrotermes* collected in the savannah of Foncobele in Congo (Fain, 1966a).

Fain (1967a) reported that the main or exclusive food of the mite appears to be epidermal dandruff, which regularly falls from human skin, and accumulates on the ground to help form dust. In the Netherlands, it has been collected in all homes examined, and in Belgium, in 18 out of 20 homes examined. In some houses in Holland, the number of dust mites exceeded 500 per gram of dust. In Belgium, the maximum number observed was 256 per gram. *D. pteronyssinus* was frequently associated with other mites, but in almost all houses, it was by far the dominant species (Fain, 1966a).

Bio-ecology

Fain (1966a) redescribed the mite and designated a lectotype examining some paratypes deposited in the Berlese collection in Florence. He reported that *D. pteronyssinus* is a cosmopolitan species, whose main habitat seems to be the dust of inhabited places. In some countries, the mite seems to be completely domestic. The high frequency of *D. pteronyssinus* in house dust suggests that it is this environment, which constitutes the true habitat of this mite. The constant presence in this dust, not only of adult forms of this mite but also of many immature ones further reinforces this supposition. It is probable that these mites feed on the numerous epidermal desquamation products (dandruff, scales, hair and other horny waste), which are continuously detached from human skin and contribute to forming dust in inhabited places. Baker *et al.* (1956) reported that Hull bred a related species, *D. scheremetewskyi* (Bogdanow) by feeding them by means of small epidermal threads, detaching themselves around the fingernails and/or fingers. Spieksma achieved the same result with *D. pteronyssinus* using horny dandruff that had fallen from the scalp. In our opinion, it is the horny substance of human skin, which constitutes the main food of this mite, but it is probable that horny waste of animal origin is also used (Fain, 1966a). Successively, Fain (1967a) reported that the main or exclusive food of the mite appears to be epidermal dandruff, which regularly falls from human skin and accumulates on the ground to help form dust. Hay *et al.* (1992) isolated in total nine species of fungi from *D. pteronyssinus* in house dust, including three xerophilic species – *Eurotium amstelodami* Mangin, *Aspergillus penicillioides* Speg. and *Wallemia sebi* (Fr.) Arx, usually collected in laboratory cultures of *D. pteronyssinus*. They concluded that mites do interact with a similar range of fungi in natural dust and in laboratory culture, but that the diversity of fungal species in the laboratory is reduced, and the density of individual fungal species in culture exceeds that of house dust. In a second experiment, dust samples were incubated at room temperature with 75% R.H. The diversity of fungi invariably declined from up to 13 genera to the few species recorded in laboratory culture. This suggests that the dominance of xerophilic fungi in laboratory mite rearings is mediated primarily by low R.H., and the exclusion of airborne spores (Hay *et al.*, 1992). Later, Hay *et al.* (1993) re-examined the possibility that house-dust mites benefit from fungal contamination of the dietary substratum and documented the performance of *D. pteronyssinus* over two successive generations, in the presence or absence of the xerophilic fungus *A. penicillioides*. They found that this fungus reduces survival, development rate, adult length and fecundity of *D. pteronyssinus*. The detrimental effects of *A. penicillioides* were proportional to the fungal density. Despite the antagonistic effects of *A. penicillioides*, a requirement for the fungus, was indicated by the poor performance of fungus-free mites in the second generation; sustained culture of *D. pteronyssinus* in the absence of fungi is probably not possible. They concluded that fungi may alter the particulate nature of the substratum to the detriment of house-dust mites but also provide micronutrients deficient in the diet (Hay *et al.*, 1993).

The life cycle develops through the egg, larva, protonymph, tritonymph, and adult stages. Arlian *et al.* (1990) studied the bio-ecology of the European house dust mite, at various temperatures that normally occur in the mite microhabitat in homes. They found that the egg-to-adult developmental times were 122.8 +/– 14.5, 34.0 +/– 5.9, 19.3 +/– 2.5 and 15.0 +/– 2.0 days at 16°C, 23°C, 30°C and 35°C, respectively. Developmental times of individual life stages were consistent proportions of the total egg-to-adult developmental times, regardless of temperature. Duration of the egg and protonymphal and tritonymphal life stages were 20–26% of the total development times and for the larval stage, 28–33%. At 23°C, the females lived 31.2 +/– 11.1 days and produced 2.5 +/– 0.7 eggs per day, during a reproductive period of 26.2 +/– 10.7 days. At 35°C, the females produced 3.3 +/– 1.3 eggs per day, but female longevity and the reproductive period were reduced to 15.5 +/– 9.6 and 11.6 +/– 6.4 days, respectively. During a lifetime, the females produced 68.4 +/– 30.4 and 48.0 +/– 29.6 eggs at 23°C and 35°C, respectively (Arlian *et al.*, 1990). *D. pteronyssinus* does not form a drought-resistant quiescent protonymph stage similar to *D. farinae*. Arlian *et al.* (1990) found that quiescent protonymphs of *D. pteronyssinus* do not achieve even at low R.H., although they retained that a combination of unfavourable conditions may induce its appearance. De Boer and Kuller (1997) observed that the quiescent phase of *D. pteronyssinus* is relatively brief. De Boer *et al.* (1998) reported that in the carpeted ground floor of Dutch houses, *D. pteronyssinus* adults remain active throughout the winter. Louadi and Robaux (1992) reported that the count and the study of species composition showed that *D. pteronyssinus* represents the 85.57% of total fauna, *E. maynéi* the 3.55% and *D. farinae* the 2.71%, respectively. T Test indicates a significant difference in the density of population between humid and dry zone, and between high and low altitudes. They concluded that the mite population distribution depends on different factors such as climate, altitude and food.

Solarz (1997) studied in Poland, the seasonal dynamics and age structures of Pyroglyphid dust mite populations in three beds from two dwellings in Sosnowiec (Upper Silesia), taking into account the R.H. and the temperature of ambient indoor air over 2 years. He observed an increase in mite density from April to November, related to the increase of indoor R.H. The mites were more abundant in the dwelling with central heating than in that with stove heating. *D. farinae* was the numerically dominant species (62.7% of the total mites), followed by *D. pteronyssinus* (28.8%). The analysis of the age population structure and its dynamics showed the marked differences between particular Pyroglyphid mite species and the beds examined. In total, *D. farinae* populations showed the dominance of immature stages, whereas, in populations of *D. pteronyssinus* and *E. maynéi*, the dominants were adult mites (Solarz, 1997).

Pike *et al.* (2005) reported that house dust mites live in almost all dwellings in New Zealand and are a major risk factor in the development of asthma and perennial allergic rhinitis. They studied the longevity, life stage length, and fecundity of a New Zealand strain of *D. pteronyssinus*, at 23°C and 75% R.H., and the fluctuating conditions typically found in dry (18–25°C, 60–38% R.H.) and damp (18–23°C, 70–55% R.H.), New Zealand dust mite microhabitats in carpets. All the adult mites placed in the 'dry' conditions died within 18 days. Mites in the 'damp' conditions had developmental times, oviposition and death rates that were not significantly different from constant conditions. These mites are tolerant of fluctuating temperatures, but they are more susceptible to environments that strongly fluctuate in humidity (Pike *et al.*, 2005).

Hart *et al.* (2007) compared the life histories of 'wild' *D. pteronyssinus* with laboratory strains, by using a diet consisting of skin and dust, or a laboratory diet consisting of dried liver and yeast. They found that at 25°C and 75% R.H., the fecundity and the rate of reproduction were higher in laboratory strains on both diets, compared with wild mites. They also observed the trends for a shorter pre-reproductive period and more rapid egg development of laboratory mites, compared with wild mites. A little effect of diet, on either strain of mites at 75% R.H. was detected, while at 64% R.H., the fecundity was significantly lower for both strains on both diets, and there were also trends for longer pre-reproductive period, reduced rate of reproduction, reduced adult survival, prolonged egg and juvenile development, or a combination compared with 75% R.H. The egg and the juvenile mortality were significantly higher on the liver and yeast diet, and the skin and dust diet favoured both strains of mites at 64% R.H. On the liver and yeast diet at 64% R.H., the wild mite adults performed significantly better than laboratory mites, and egg mortality was lower. These data highlighted that laboratory mites have stronger reproduction and development than wild mites, except when under environmental stress and that diet is a significant factor, particularly in suboptimal conditions. These results could have serious implications for predictive models of house dust mite populations, in their natural habitat. Ideally, such models should be developed employing data from wild dust mite populations, reared on a natural diet (Hart *et al.*, 2007).

Results of a study carried out by Podder *et al.* (2009) on the mite biology highlighted that the average time from the egg to the adult period is 11.26 days. Fifteen of the 30 eggs observed (50%), became adult. The sex ratio of male and to female is 1: 2.75. The mortality at the larval and protonymph stages is 23% and 12% respectively. They did not observe mortality at the tritonymph stage. The duration of the egg stage is approximately 2.45 days, followed by the larval 1.57, protonymphal and tritonymphal stages each accounting for 2.64 and 2.6 days, respectively. The quiescent protonymphal and resting tritonymphal stages, each one with duration of approximately one day. The preoviposition period is 2.57 days, oviposition period 23.78 days and post-oviposition period 25.73 days. The fecundity ranges from 51–122 eggs. The total oviposition period ranges from 8–22 days, and 1.3 to 6.4 eggs/day/female. Mated males and females live 18–64 days and 20–54 days, respectively, and the difference in longevity between males and females is not statistically significant (t0.05 =2.13) (Podder *et al.*, 2009).

House dust mites are found in almost all dwellings in New Zealand and are a major risk. Arlian *et al.* (1998b) studied the effects of R.H. on the population dynamics of single and mixed species of *D. farinae*, *D. pteronyssinus* and *E. maynéi* at specific R.H. maintained at 20°C, with unlimited food. Population density of single and mixed species (*D. farinae*, *D. pteronyssinus*) increase exponentially when cultured at 65%, 70%, and 75% R.H. The mean population growth rates are 17.3 ± 4.4 SD and 32.5% ± 4.7/wk for *D. farinae* and *D. pteronyssinus*, respectively. The mean population doubling times are 2.2 ± 0.3 and 4.2 ± 1.3 wk for *D. pteronyssinus* and *D. farinae*, respectively. Mixed species cultures, started with equal numbers of *D. farinae* and *D. pteronyssinus*, result in higher percentages of *D. farinae* than *D. pteronyssinus*. In cultures started with 75% of one species and 25% of the other, the more numerous species remained dominant and in similar ratios throughout the experiment. Both *D. farinae* and *D. pteronyssinus* population densities maintained at 85% R.H. decline over a 12-week culture period because of mould growth. *E. maynéi* is unable to survive at 65%, 70%, 75% and 85% R.H., which indicated that their climatic requirements are different from those of *D. farinae* and *D. pteronyssinus*. Population densities of *D. farinae* and *D. pteronyssinus* cultures declines, when held at 21–22°C and R.H. ≤ 50%; however, at 50% R.H., significant proportions of the populations survive for 10 weeks. Half-life for desiccation of *farinae* and *D. pteronyssinus* at 45% R.H., was 11.5 and 1.2 weeks, respectively, but at 50% R.H. was 86.3 and 4 weeks, respectively. The data indicated that a ≤ 50% R.H. would have to be maintained for long periods, to reduce both *D. farinae* and *D. petonyssinus* by desiccation procedures. Both mites have high reproductive potential and population growth rates, which indicated that mite reduction procedures must be thorough, or mite densities will return to high levels, quickly following remediation, if adequate food and suitable microclimate conditions exist.

Symptomatology and damage

See *D. farinae*. Further news on the allergen problems may be found in Colloff *et al.* (1991).

Control

See *D. farinae*.

The Genus *Euroglyphus* Fain

Diagnostic characteristics

The anterior extremity of the body is prolonged by a pointed or forked tegmen (prodorsum), which covers the base of the gnathosoma, in both male and female. The female has the distal part of the bursa copulatrix in the form of a small, oval, strongly sclerified pocket; the male possesses the anal suckers (Hart and Fain, 1988). The trochanter I to III and the tibia IV bear setae, and the tarsus III bears six setae and the tarsus IV, three setae. The female possesses two pairs of anal setae, and the male one pair, and both sexes bear three pair of genital setae. The posterior genital shield of the female covers completely the genital opening (Hughes, 1976).

Economic importance

This genus includes a restricted number of species, such as *E. longior* (Trouessart) reported from Egypt (Saleh *et al.*, 1985), England, France (Hughes, 1976), Germany (Franz *et al.*, 1997), Sweden (Sellnick, 1958; Hughes, 1976), United States (Hughes, 1976) on bean (Saleh *et al.*, 1985), granary debris, hide dust in farm (Sellnick, 1958; Fain, 1965; Hughes, 1976; Franz *et al.*, 1997) and the small house dust mite, *E. maynéi* (Cooreman), cosmopolitan species, reported on various stored products, but funda-

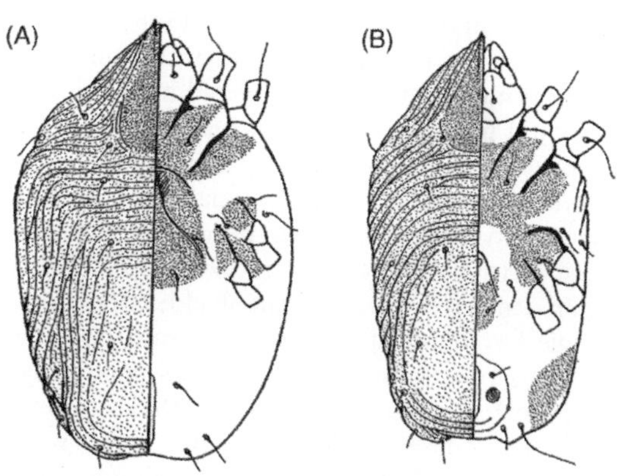

Fig. 38.3. *Euroglyphus maynéi* (Cooreman). Female. Dorsal (at left) and ventral view (at right) (A). Male. Dorsal (at left) and ventral view (at right) (from Vacante, 1997b) (B).

mentally important for its adaptation to living in human dwellings, where it feeds on human skin scales, causing severe allergies. We treat below, only this last mite.

Euroglyphus maynéi (Cooreman) (Fig. 38.3)

Common names

Small house dust mite, Mayne's house dust mite

Morphological characteristics

FEMALE. The idiosoma is about 195 to 280 μm long. The dorsal propodosomal shield is less well-marked than in the male, and its anterior edge is smoothly rounded, not bifid, setae sci and sce about the same length, setae sce short and thin, and tegmen (t) present; tegmen either prominent and triangular, but with apex rounded and not forked, or poorly developed and rounded with a small median notch. No hysterosomal shield is delimited, but this region is devoid of cuticular ridges, and its surface is punctate like that of the propodosomal shield and sternal regions. The posterior edge of idiosoma with 2 minute lobes without hairs; tegmen well developed, triangular with rounded apex (not bifid in the male); cuticle somewhat sclerotized with rather fine-formed markings or creases; hysteronotum within a median shield with margins posterior. The idiosomal chaetotaxy is similar to that of the male, except that the two pairs of postanal setae are of about the same length. The dorsal setae are very short and thin; the setae h_3 are very short (maximum length 50 μm) and thin, and the setae h_2 are very thin and short (not more than 30 μm). The genital opening is only partly covered by the posterior genital plate, whose anterior margin is pointed. The posterior lip of vulva is short and punctate, not covering vulvar slit; or anterior angle of posterior vulvar lip not incised; vulva uncovered; tegmen triangular with rounded, not incised apex; hysteronotum striated with a median shield; copulatory vestibule ovoid, strongly sclerotized and opaque; and tarsi I–IV without apical processes nor spines. The receptaculum seminis is well-sclerotized and oval in shape. From it, paired ducts connect with the ovary, and a long slender tube communicates with the opening of the bursa copulatrix, which is near the posterior end of the anus. Anal external setae and genital anterior setae are absent. The legs are slender, and leg IV is longer than leg III; anterior legs missing chitinous membranes; chaetotaxy condensed; tibials IV, trochanterals I–III; tarsi IV with three setae; tarsi III with five setae; and genu I with one solenidion (Hughes, 1976; Morgan *et al.*, 2015).

MALE. The idiosoma is about 200 μm long. The dorsal propodosomal shield is smaller and pear-shaped, with unforked, rounded apex. The longitudinal ridges extend to the anterior edge, sometimes giving this a bifid appearance. The posterior margin of the body is distinctly incised in the midline. The dorsal setae are variable, and the opisthosoma is slightly but narrowed backwards regularly; the anus is more posterior (anal suckers situated at 25 μm from posterior body margin); posterior body margin wide and straight with two small paramedian lobes; the adanal suckers are well developed; and the tarsi IV lack of suckers.

The apodemes of legs I do not meet to form a sternum. The aedeagus is a short straight tube and small genital sense organs are present. Conspicuous anal suckers are present, bounded by a sclerotized ring. All the body setae are smooth and short, except for the external pair of postanal setae, which is considerably longer than the internal. All legs terminate in a globular pretarsus, but not in a claw. Leg IV are rather shorter and narrower than leg III. Setae are missing from trochanters I to III and tibia IV. Tarsus III bears five setae, tarsus IV three setae, of which one is a peg-like structure at the distal end, which may represent a vestigial sucker. A stout process is found at the end of tarsus III (Hughes, 1976; Morgan *et al.*, 2015).

Geographical distribution

COSMOPOLITAN. Reported from the following countries: Algeria (Louadi and Robaux, 1992), Argentina (Martínez Canzonieri *et al.*, 1996), Australia (Tovey *et al.*, 2000), Belgium (Cooreman, 1950; Gridelet and Lebrun, 1973), Brazil (Greco *et al.*, 1974), Bulgaria (Todorov, 1979), China (Lai and Wen., 1982), Colombia (Charlet *et al.*, 1977), Costa Rica (Vargas and Mairena, 1991), Croatia (Blythe, 1976), Czchech Republic (Samšiňák *et al.*, 1978), Denmark (Korsgaard, 1979), Ecuador (Valdivieso *et al.*, 2006), England (Sesay and Dobson, 1972), Finland (Stenius and Cunnington, 1972), France (Araujo-Fontaine *et al.*, 1973), Germany (Karg, 1973), Greece (Deliargyris *et al.*, 1990), Hong Kong (Gabriel *et al.*, 1982), Hungary (Halmai, 1984), India (Dar and Gupta, 1979),

Iran (Sepasgozarian and Mumcuoglu, 1979), Israel (Feldman-Muhsam *et al.*, 1985), Italy (Noferi *et al.*, 1974; Vacante, 1997b), Japan (Oshima, 1970), Kazakstan (Yagofarov and Galikeev, 1987), Kenya (Blythe, 1976), Lituania (Dubinina *et al.*, 1984), Malaysia (Ho and Nadchatram, 1985), the Netherlands (Spieksma and Spieksma-Boezeman, 1967), New Zealand (Blythe, 1976), Nigeria (Hunponu-Wusu and Somorin, 1978), Norway (Mehl, 1998), Palestine (Mumcuoglu *et al.*, 1994), Papua New Guinea (Anderson and Cunnington, 1974), Peru (Caceres and Fain, 1978), Poland (Solarz, 1986), Portugal (Pinhão and Grácio, 1978), Puerto Rico (Montealegre *et al.*, 1997), Russia (Dubinina and Plietniev, 1978), Slovak Republic (Makovcová *et al.*, 1982), Singapore (Blythe, 1976), South Africa (Fain, 1988a), Spain (Blasco and Portus, 1975), Sweden (Wickman *et al.*, 1993), Switzerland (Spieksma *et al.*, 1971), Taiwan (Oshima, 1970), Thailand (Malainual *et al.*, 1995), Turkey (Blythe, 1976), United States (Yoshikawa and Bennett, 1979) and Venezuela (Hurtado and Parini, 1987).

Commodities, facilities, vegetal and animal hosts

Cottonseed cake decomposed (Cooreman, 1950), Chinese medicine (*Rhizoma homalomenae*, *Radix morindae officinalis*, *Coritex lycii*, *Radix ophiopogonis*, *Radix gentianae macrophyllae*) (Zhan *et al.*, 2017b), bean curd, crabmeat canned, shrimp dried (Hagstrum *et al.*, 2013), stored grain and stored food (Shen, 1984), rice bran broken rice, groundnut, broomcorn, haw sweetmeat, tremella, fermented soybean, room dust, designing fur, cattail mat, light straw, straw, wheat straw, paper, bedding, clothing, carpet dust, mattress dust (Li *et al.*, 2015a) in farm, farm granary and farm milling machine (Hagstrum *et al.*, 2013). Moreover, the mite is well known from time on house dust (Fain, 1965; Spieksma and Spieksma-Boezeman, 1967).

Bio-ecology

The small house dust mite was originally described by Cooreman (1950), as *Mealia maynéi* from Belgium on decomposed cottonseed cake. It is reported from humid geographic areas all over the world and infests various stored products such as cottonseed meal, bean curd, Chinese medicines, crabmeat, shrimps (Hagstrum *et al.*, 2013) and other products. Moreover, *E. maynéi* (Cooreman) is a common species in house dust throughout the world and produces allergens responsible for human diseases (Fain, 1965; Spieksma and Spieksma-Boezeman, 1967; Colloff, 1991a; Arlian *et al.*, 1993). However, the studies of its biology and allergenicity have been relatively scarce because it is difficult to culture, and because species of the genus *Dermatophagoides* have attracted the most attention of research workers. *E. maynéi* has been increasingly recognized as an important source of mite-derived allergens to which, patients with house-dust mite allergy are exposed. In this sense, Colloff (1991a) reviewed published and unpublished data on taxonomy and identification, geographical and habitat distribution, seasonal population fluctuations, life history and the effects of temperature and humidity, culture techniques and allergenicity of *E. maynéi*. The house dust mite is often present in densities greater than 100 mites per gram of dust and usually occurs with *D. farinae* and *D. pteronyssinus*, and it is frequently more abundant than *Dermatophagoides* species. Therefore, its allergenicity and the cross-reactivity between *E. maynéi* and *Dermatophagoides* species are important considerations for diagnosis of mite-induced allergy, and the use of appropriate immunotherapy (Arlian *et al.*, 1993). Walshaw and Evans (1987) stated that there was a strong correlation between the density of *E. maynéi* within a mattress and social class, increasing numbers of mites being found as social class declined. However, they did not examine this relationship with *D. pteronyssinus*. Colloff (1991b) also showed that *E. maynéi* occurred more frequently in samples from homes that he assessed as being damp.

The fundamental aspects of the biology of *E. mainéi*, are similar to those of other Pyroglyphid mites, the sexes are separatee, and the life cycle develops through the egg, larva, protonymph, tritonymph and adult stages. The reproductive biology of *E. maynéi* is not well studied, probably because it is usually less common in homes than *D. farinae* and *D. pteronyssinus*. When it is present, it usually co-inhabits with the *Dermatophagoides* spp. and is more restricted in geographical distribution.

Hart and Fain (1988) found that the pre-reproductive period of *E. maynéi* at 25°C ± 2°C, is 13.80 ± 0.20 days, the reproductive period is 60.20 ± 6.53 days, the fecundity is 84.20 ± 10.32 eggs per female, the rate of reproduction is 1.47 ± 0.18 eggs laid per female per day, and the development from egg to adult lasts 33.00 ± 4.00 days. Arlian and

Morgan (2015) studied the duration of the life cycle from egg to adult, at 23°C and 30°C at 75% R.H., and the fecundity at 23°C and 75% R.H. for *E. maynéi and* compared the results to similar data for *D. farinae* and *D. pteronyssinus*. They found that the adults hatched from eggs after 28 days at 23°C and 20 days at 30°C. The females produced 1.4 eggs/day, during a reproductive period of 24 days at 23°C. *E. maynéi* has a shorter life cycle than *D. farinae* and *D. pteronyssinus* at 23°C, but a longer life cycle at 30°C. *E. maynéi* has a shorter reproductive period and produces fewer eggs than both *D. farinae* and *D. pteronyssinus* (Arlian and Morgan, 2015).

Symptomatology and damage

The direct damage of *E. maynéi* to stored products is unclear, while the relationship between infested foodstuffs and the risk of allergies in the workers involved in their manipulation, is more evident. The problem of the presence of these Pyroglyphid mites affects in the first place, the domestic dwellings, and summarizing the fundamental information in this regard is difficult, due to the considerable number of reports; furthermore, it is beyond the scope of this work. To confirm the hygienic and sanitary importance of the mite, we can consult the work of Colloff *et al.* (1991), which studied the acarofauna and Der p I allergen concentrations in dust samples from mattresses and lounge room carpets, obtained from 20 homes from Perth and Bunbury, Australia. They found that all samples contained mites and the geometric mean numbers of total mites/g of mattress and carpet dust for both cities were 480 and 263, and 585 and 992, respectively. The predominant mite species were *D. pteronyssinus*, *E. maynéi* and *Tarsonemus* spp., while *D. farinae* was absent in all samples. *E. maynéi* was present in the ten Bunbury homes, and in 50% of the Perth homes, ranging from 0 to 81% of mites identified. The arithmetic mean Der p I concentrations in the mattresses and carpets in Perth and Bunbury were 4.2 and 4.1, and 3.8 and 9.2 micrograms/per gram of fine dust, respectively, and Der p I concentration correlated with mite counts ($r = 0.75$; $P < 0.001$). The concentration of Der p I equivalent per 100 mites, was 1.5 micrograms. The data are consistent with the view that asthmatic patients in Western Australia have significant exposure to a variety of house-dust mites, and that *E. maynéi* may be clinically significant (Colloff *et al.*, 1991). Moreover, Cui (2014) produced a detailed review of human acariasis produced by *E. maynéi* in various countries. The first type of acariasis concerns has pulmonary involvements, such as continuous dry cough, wheezing, pain in the chest, increasing dyspnea, bronchiectasis, and it has been reported from East Africa (Carter *et al.*, 1944; Wilson, 1947), Korea (Ryu *et al.*, 2003) and China (Gao *et al.*, 1956). The second type has urinary involvements, such as frequent desire to urinate, pyelonephritis, and pyelocystitis and it has been reported from Canada (McCulloch, 1938), Romania (Pitariu *et al.*, 1979), China (Li *et al.*, 2003), South Africa (Leigh Dini and Frean, 2005).

Control

See *D. farinae*.

39 The Family Suidasiidae Hughes

Morphological Characteristics, Systematic and Bio-ecology

This cosmopolitan family includes species inhabiting invertebrate or insect nests, stored products, house dust and sometimes the soil (O'Connor, 2009b).

The adult has the body distinctly longer than wide; the dorsal setae are filiform and nude. The external vertical setae (*ve*) are present. The propodosoma possesses the lamellar setae (*le*). The body cuticle is often ornated with rounded protuberances; the opisthosoma is not bilobed posteriorly. The tarsi have the proral setae (*p* and *q*) enlarged and claw-like, and the unguinal setae (*u* and *v*) are reduced or absent. The male bears two sucker-like setae on tarsus IV and possesses usually para-anal suckers. The tarsi have both tectal setae (*e* and *f*) filiform, similar in length. The unguinal setae (if present) are shorter than the proral setae (*p* and *q*), the latter spine-like or enlarged and claw-like (O'Connor and Klimov, 2003).

The deutonymph has the propodosoma completely covered by propodosomal shield. The setae *scx* are filiform. The setae *ve* are filiform and short. The leg IV lacks pretarsal empodial claw; the claw III is present. The leg IV is often smaller than the leg III, with some fusion of podomeres and usually with one or more long apical setae. The deutonymphs are known only in the genera *Tortonia* Oudemans and *Sapracarus* Fain and Philips (O'Connor, 2009b). The knowledge on the systematics of this taxon is evolving, but in order not to complicate the discussion in this paper, we prefer to refer to O'Connor (2009). The family Suidasiidae numbers almost 18 known species, inhabiting a variety of habitats, and included in few genera (7?). Species belonging to the genera *Suidasia* Oudemans, *Neosuidasia* Ranganath and Channabasavanna and *Sinosuidasia* Jiang are reported on stored food products and house dust (O'Connor, 2009b). *Neosuidasia faini* Ranganath *et al.* reported on stored food products, such as feed for birds and other products from India (Ranganath *et al.*, 1983; Nangia and ChannaBasavanna, 1989b), and on imported animal feed from Cuba (de la Torre Santana and O'Connor, 2011). *Sinosuidasia orientalis* Jiang (1996) and *Sinosuidasia jinyunensis* Zhang and Li (2002) both reported on stored food products. Two species of the genus *Suidasia* are reported on stored food products; they are *Suidasia nesbitti* Hughes, widespread (Hughes, 1976; Fain and Philips, 1978; Balmes-Pacia and Corpus-Raros, 1998), and the scaly mite, *Suidasia pontifica* Oudemans (sin. *Suidasia medanensis*), restricted to warmer climates (Sierra and Martínez, 1987; de la Torre Santana and O'Connor, 2011). Samung *et al.* (2006) found *S. pontifica* in Thailand.

The Genus *Suidasia* Oudemans

Diagnostic characteristics

The adults have the pretarsal ambulacrum, not greatly expanded. Ventral subcapitulum without external ridges. Body with various patterns: if with small, rounded protuberances, then they are coalescent and at least some unequal. The dorsal setae are smooth. The prodorsum possesses the external vertical setae *ve*, set at sides of prodorsal sclerite, near the transverse level of setae *vi*. The anus located near the posterior margin of the body. The coxal apodemes III–IV are present. The empodial claws is simple. The solenidion, ω_2 is set in basal portion of the tarsus, close to $\omega 1$, or more distal. The tarsi I–IV bear the setae *e* and *f*, both filiform. The supracoxal setae are lanceolate, with fimbriate margins. The hysterosomal setae are short, distinctly not reaching bases of the next setae. The proral setae (*p*, *q*) of tarsi I–IV are enlarged and claw-like. The internal scapular setae (*si*) are microsetae, distinctly anterior to posterior margin of the prodorsal shield. The tarsal setae *aa* are present. This genus does not form the phoretic deutonymphs; they may disperse as feeding stages on insects and other hosts. No records from bees are available (Klimov *et al.*, 2016). A dichotomous key of the

 The Handbook of Mites of Stored Products (V. Vacante and C.H.W. Flechtmann)
DOI: 10.1079/9781786390912.0039

species is available at Fain and Philips (1978). The type species of this genus is *S. pontifica*, and its common synonyms are *Aphelenia* Oudemans, 1923, *Chibidania* Sasa, 1952, and *S. medanensis* Oudemans; this last is an invalid name commonly used, but it is a junior synonym of *S. pontifica*, and should not be employed (Klimov *et al.*, 2016).

Economic importance

The genus has a cosmopolitan geographical distribution, and includes generalist species, with all stages living in various habitats (including bee nests), where they probably feed on pollen and decomposing or fungus-infested organic matter; *S. pontifica* feeds on dead insects and it has been found in bee nests in Sri Lanka (*Apis* spp.) and Indonesia (*Xylocopa*) (Koeniger *et al.*, 1983; Oudemans, 1924b). This species has a host association level facultative, and it has been collected also in nests of honeybees (*Apis*) and large carpenter bees (*Xylocopa*) (Klimov *et al.*, 2016). Two species of the genus *Suidasia* may infest the stored food products; they are *S. nesbitti*, widespread (Hughes, 1976; Fain and Philips, 1978; Balmes-Pacia and Corpus-Raros, 1998), and the scaly mite, *S. pontifica* (sin. *S. medanensis*), restricted to warmer climates (Sierra and Martínez, 1987; de la Torre Santana and O'Connor, 2011); both species inhabit a variety of stored products, peridomestic habitats, including house dust, mostly in the tropics. According to their greater importance and geographical distribution, we discuss here only *S. nesbitti* and *S. pontifica*.

Suidasia nesbitti Hughes (Fig. 39.1)

Common name

Scaly grain mite

Morphological characteristics

FEMALE. The idiosoma is from 300 µm to 340 µm long, and similar to that of male. The anal opening reaches the posterior margin of the body and is surrounded by five pairs of anal setae, the third pair of setae being further from anus than the others. The genital opening extends between the coxae III and IV (Hughes, 1976).

MALE. The idiosoma is from 269 µm to 300 µm long. The body is flattened and broadly oval, its cuticle longitudinally grooved and, at the same time thrown into scale-like wrinkles. This wrinkling extends on to the ventral surface of the opisthosoma and gives the skin an iridescent appearance when the animal is active: it tends to disappear in mounted specimens. A smooth, dorsal propodosomal shield is present, extending as far back as the hysterosoma.

A full complement of setae is present. The setae *vi* extend forward over the gnathosoma; *ve* are short, arising midway along the lateral edge of the propodosomal shield; the supracoxal seta is a flattened structure with well marked lateral pectinations, and Grandjean's organ is a fold of cuticle with a serrate edge. The setae *sce* are more than four times longer than *sci*; the *sci* setae are inserted near to *sce and* separated by the width of the propodosomal shield from each other. On the ventral surface, the apodemes are broad, stout structures.

With the exception of *he* and *sae*, all the setae of hysterosoma are short, about the same length as *sci*, and difficult to distinguish against the wrinkled cuticle. The setae d_1 to d_4 are arranged almost in linear series with one another; *sae* are long limp setae, with exceeded half the length of the idiosoma. Ventrally, the anal opening reaches the posterior margin of the body and is surrounded by three pairs of anal setae. Each chelicera has distinctly toothed shears, and a mandibular spine on its ventral surface.

The legs are rather short and stumpy with the same chaetotaxy in both sexes. On leg I, *d* extends beyond the end of the claw, *e* and *f* are shorter setae, and ω_1 is a curve *d* rod; ventrally *p*, *q*, and *s* are conspicuous bent spines, *s* arising in the middle of the tarsus; *u* and *v* are long and slender and pressed tightly against the pretarsus, so that only the base is clearly visible. The setae and solenidia arising from the base of the tarsus tend to be close together, owing to the shortness of the segment; on tarsus I, ω_1 is a slender curved rod, which extends forward, overlapping the base of *ba*; on tarsus II it is shorter and thicker. The famulus is directed towards the tibia and is often concealed by omega1; *aa*, *ba*, *la*, *ra* and *wa* are all fine setae; ω_2 arises close to *ba*. On the genu, σ_2 is more than three times longer than σ_1, and the setae of this and the remaining segments of leg I are not thickened to form spines. On tarsus IV the copulatory suckers are nearer to the base and apex of the tarsus than to each other. The aedea-

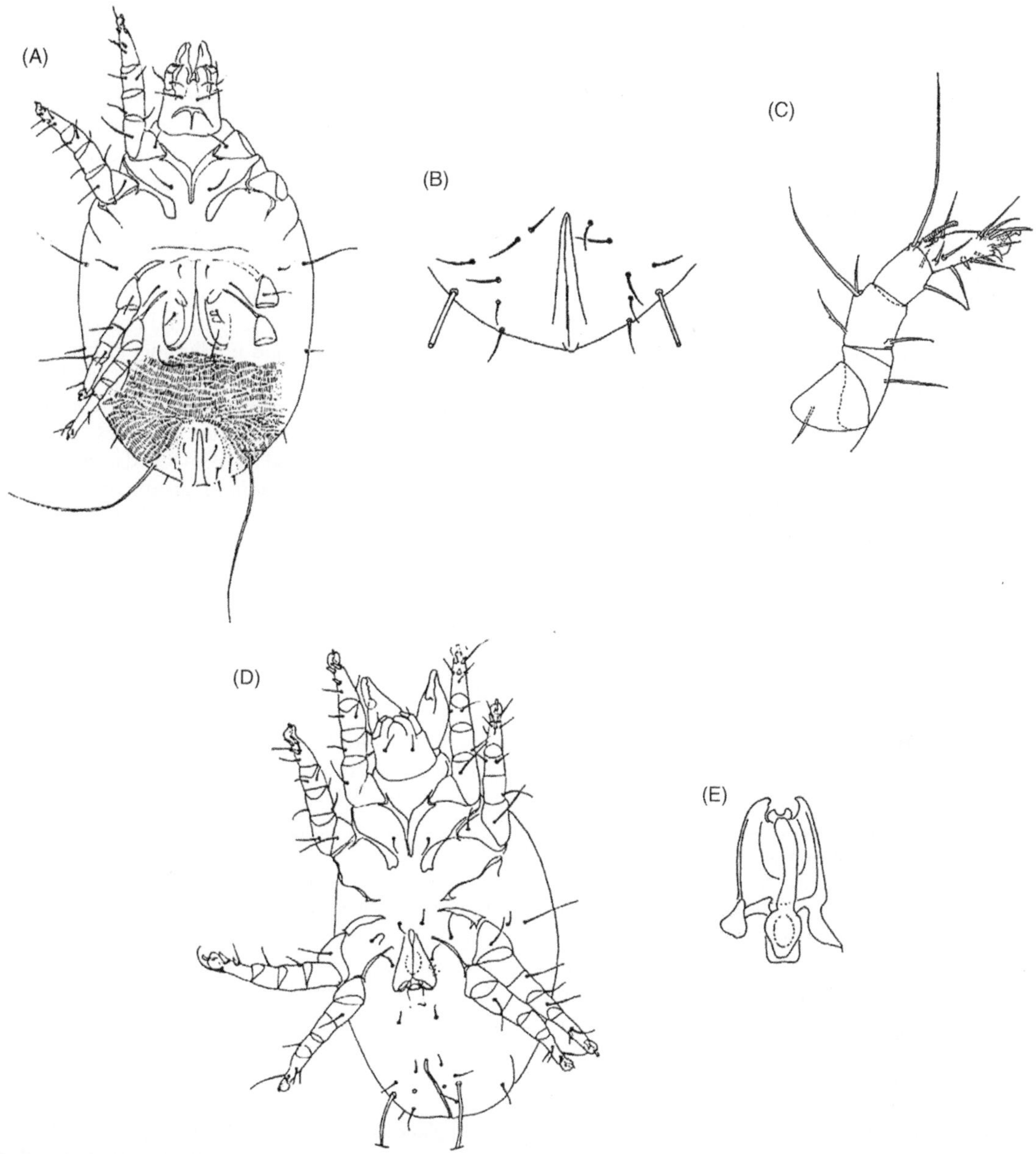

Fig. 39.1. *Suidasia nesbitti* Hughes. Female. Ventral view (A); anal region (B); leg I (C). Male. Ventral view (D); aedeagus (E) (from Hughes, 1976, licensed under the Open Government Licence v3.0).

gus is between the coxae IV, it is a rather large-curved tube (Hughes, 1976).

LARVA. The idiosoma is about 160 μm long. The wrinkling of the cuticle is not so distinct as in the adult. Coxal setae are present, but no coxal rods (Hughes, 1976).

Geographical distribution

Cosmopolitan, reported from Antarctica (Sanyal, 2004; Gantait and Chandra, 2017), Australia (Champ, 1966), Belgium (Hughes, 1976), Benin (Zannou *et al.*, 2013), Brazil (Monteiro and Arruda, 1969; Flechtmann, 1986), China (Wang, 1964; Xin,

1965; Olsen, 1983), Crete (Hughes, 1976), Egypt (Attiah, 1969; Yassin *et al.*, 2018b), England, Finland Hughes, 1976), Germany (Rack, 1968a), Greece (Athanassiou *et al.*, 2002; Palyvos *et al.*, 2018), Hong Kong (Olsen, 1983), Italy (Hughes, 1976), India (Gill and Kaur, 2014), Iran (Sepasgozarian, 1978; Ardeshir *et al.*, 2000; Ardeshir, 2010), Japan (Sasa, 1952), North Africa, North America (Hughes, 1976), Poland (Chmielewski, 1977b), Portugal (Hughes, 1976), Russian Far East (Klimov and Tolstikov, 2011), Senegal (Hughes, 1976), Singapore (Olsen, 1983), Syria (Rack, 1968a), South Africa Hughes, 1976), South Korea (Chu *et al.*, 1967; Lee *et al.*, 1998), Taiwan (Tseng, 1972), Turkey (Ozman and Žďárková, 2000), Ukraine (Oksentiuk *et al.*, 2022), West Indies (Hughes, 1976) and Zaire (Fain, 1971b).

Commodities, facilities, vegetal and animal hosts

The mite is reported on wide list of commodities, including aquatic product, bird skin (Benoit coll., 1962, in Hughes, 1976), bran, broken rice, broomcorn (Li *et al.*, 2015a), broken grains and debris (Ardeshir *et al.*, 2000), caraway, carob, cayenne, cocoa (Saleh *et al.*, 1985), chilli powder (Chakraborty and Gupta, 2016), Chinese medicine, as *Rhizoma chuanxiong*, *Rhizoma acori tatarinowii*, *Radix sanguisorbae*, *Radix codonopsis* (Zhan *et al.*, 2017b), cottonseed (Athanassiou *et al.*, 2002), cowpea (Saleh *et al.*, 1985; Zannou *et al.*, 2013), date dried (Saleh *et al.*, 1985), fereek, fig dried (Saleh *et al.*, 1985; Atwa *et al.*, 2018), fish dried (Olsen, 1983; Chmielewski, 1991b), flour meal (Saleh *et al.*, 1985), ginseng (Li *et al.*, 2015a), ground dust flour (Yong *et al.*, 2016b), macaroni (Saleh *et al.*, 1985), maize debris (Zannou *et al.*, 2013), maize flour (Fain, 1971a), maize grain (Atwa *et al.*, 2018; Gad *et al.*, 2020), milk powder, moong dal (Chakraborty and Gupta, 2016), mushroom (Parveen and Gupta, 2020b), opium seeds (Chakraborty and Gupta, 2016), paddy rice grains (Zannou *et al.*, 2013), persimmon dried (Li *et al.*, 2015a), potato (Gad *et al.*, 2020), rice (Hughes, 1976; Li *et al.*, 2015a), pigeon pea (Chakraborty and Gupta, 2016), pressed rice (Chakraborty and Gupta, 2016), salep (Saleh *et al.*, 1985), sattu (Roasted gram flower) (Chakraborty and Gupta, 2016), seed cotton (Athanassiou *et al.*, 2002), semolina (Chakraborty and Gupta, 2016), shrimp dried (Olsen, 1983), soybean (Yassin *et al.*, 2018b), sunflowerseed (Li *et al.*, 2015a), whale meat infested with Dermestidae (Hughes, 1976), wheat (Atwa *et al.*, 2018; Yassin *et al.*, 2018b), wheat bran (Rack, 1968a; Hughes, 1976; Saleh *et al.*, 1985), wheat flour (Saleh *et al.*, 1985; Chakraborty and Gupta, 2016), wheat pollard (Hughes, 1976), and yeast powder (Saleh *et al.*, 1985) in farms (Palyvos *et al.*, 2008), flour mills (Hughes, 1976; Palyvos *et al.*, 2008), off-farm storage and grain silos (Palyvos *et al.*, 2008). Moreover, it has been collected in containers of rice, cotton and soybean (Flechtmann, 1986). Hagstrum *et al.* (2013) reported the mite also on condiment, dried fruit, dried vegetable, meat product, poultry mash and processed food. The mite has been found in Ukraine, in barns with animal fodder storage (Oksentiuk *et al.*, 2022).

The mite has been collected in old mine underground galleries, as the habitat for mites in Poland (Skubala *et al.*, 2003), and has been found in association with the coleopter *Tribolium castaneum* (Herbst), in Egypt (Metwally *et al.*, 2014). Ardeshir (2010) collected *S. nesbitti* on nest of *Pica pica* (Linnaeus) in Iran. In Belgium, was imported on bird's shins from Central Africa (Benoit coll., 1962, Hughes, 1976), and from milking machinery in a flour mill in Senegal (Freeman coll., in Hughes, 1976). In Turkey, the mite was found in hazelnut processing plants (Ozman and Žďárková, 2000). *S. nesbitti* has been collected in forest and pasture soil in Korea (Lee *et al.*, 1998).

Bio-ecology

S. nesbitti has a cosmopolitan geographical distribution, and is reported in soil, dead insects, stored foods, dust houses, and other environments. Its reproduction is sexual and the deutonymph stage is lacking. Chmielewski (1991b) isolated *S. nesbitti* specimens for laboratory cultures and for biological studies from infested fish meal. He reared the mite at 25°C, 85% R.H., using food-dried baker's yeast. One-day-old females and males were paired and placed into separate rearing cages and observations of longevity, and oviposition of mites were conducted. Fresh laid eggs were observed and mite development cycle was examined. The mean longevity of male (preimaginal + imaginal periods) was about 84.4 days, and that of female on average 67.0 days. Average egg production of female per its whole life was 172.9. Life history took on average 14.4 days, and the natural mortality of various instars was 25%. The frequency of females was slightly higher than frequency of males (Chmielewski, 1991b). Shen (1989, 1993) reported that the duration for egg, larva, quiescent larva, protonymph, quiescent protonymph, tritonymph, quiescent tritonymph of *S. nesbitti* was 4.3, 2.5, 1.0, 2.1, 1.4, 4.1 and 1.1 days, respectively, and the mite completed its life history in about 16.5 days. In Nanchang, Jiangxi province in China, the population reached a peak in mid-May, and it displayed an aggregated spatial pattern (Zou *et al.*, 2003). In Benin, the densities of *S. nesbitti* and *Suidasia* sp. on grains decreased significantly during the dry season (Zannou *et al.*, 2013). Yong *et al.* (2016b) reported that various species of mites are breeding in the ground dust flour in Qiqihaer City in China, and *S. nesbitti* occurs the most. These findings indicate that effective measures should be taken to prevent and control the harm associated with acaroid mite contamination. Mathur and Manju Dalal (1989) studied at 27.5°C and 80 ± 5% R.H., the development and growth of *S. nesbitti* on wheat germ, residue + sucrose + starch, residue + starch, residue + sucrose, and residue. The residue derives from the wheat germ fractionated to exclude the carbohydrates, identified as sucrose and starch. The wheat germ was one most attractive foods. The mite requires the least time (8.3 days) for oviposition when feeds on wheat germ, while it was longest (13.4 days) on residue. The high fecundity was observed on wheat germ (140 eggs). The lower egg mortality was observed on wheat germ (4%), and the incubation period was shortened (3.2 days), on wheat germ. The duration of the development of larva, protonymph and tritonymph stages was shorter with wheat germ (3.5, 3.1 and 2.9 days, respectively), and the complete development from the larva to adult stage took 9.68 days. The lifespan of the adult was the longest (17.5 to 18.6 days) on residue + sucrose + starch (Mathur and Manju Dalal, 1989).

Simptomatology and damage

S. nesbitti is retained as a serious pest of various foodstuffs (Saleh *et al.*, 1985). Moreover, Ardeshir *et al.* (2000) reported that the association between the mite and insects reported by authors (Hughes, 1976) induces to suspect that the high density of *S nesbitti* in broken grain may be positively correlated with the presence of numerous dead insects. In India, the milky white mushroom, *Calocybe indica* Purkay *et* Chandra, was found very heavily infested with *S. nesbitti*, and the population was so high that it completely damaged the mushroom cultivation (Parveen and Gupta, 2020b). Monteiro and Arruda (1969) reported in Brazil, a disease causing the death by starvation within 8–45 days of poultry in April–July, and reduces the amount of food consumed by laying birds by about 70 g per day, has been attributed to the presence in their food during these months of the mites *S. nesbitti* and *S. pontifica*. The mite has been recorded as giving rise to dermatitis in humans (Kilpio and Pirila, 1952), usually named 'pollard itch'.

Control

Similar to other stored mites, the infestations of *S. nesbitti* to the stored foods provide or call for preventive and curative actions. As regards the former, the indications given previously for other Acariforms are valid, and the curative actions mainly based on the use of chemical means. Among these, innovative active ingredients find more space. In this sense, Gulati (2007b) evaluated the efficacy of commercially available garlic products, viz. garlic oil, garlic powder and garlic oleoresin against *Tyrophagus putrescentiae* (Schrank) and *S. nesbitti* in stored wheat grains. All products reduced pest populations by 96 to 100%, at a concentration of 0.2%. Garlic oil was the most potent against two mite species. It significantly reduced the population of *T. putrescentiae* (to average 33.24 mites) and that of *S. nesbitti* (to average 38.52 mites) in treated wheat grains. The population of *T. putrescentiae* in oleoresin and powder-treated wheat grains was 39.37 and 50.31, respectively, while that of *S. nesbitti* was 45.29 and 61.31, respectively, as compared to untreated wheat grains (131 *T. putrescentiae* and 150 *S. nesbitti* mites). Eggs and larval stages were significantly ($p<0.05$) more susceptible to the garlic treatments than nymphs and adults. In garlic oil-treated grains (0.1% concentration), 10.3% and 11.8% of the eggs of *T. putrescentiae* and *S. nesbitti*, respectively, hatched, compared to 29.6% and 34% survival of these adults, at the same concentration. After an exposure of 45 days, garlic oil, oleoresin and powder provided 85.61%, 64.56% and 79.85% protection against *T. putrescentiae*, as compared to 85.6%, 63.62% and 78.96% against *S. nesbitti* in wheat grains (Gulati, 2007b).

Parveen and Gupta (2020b) evaluated in a laboratory study, the bio-efficacy of some green pesticides, Citrus leaf extract and Custard apple leaf extract both at 2% and 5% concentrations, Neem oil + water both as 2 ml in 25 ml water, and 4 ml in 25 ml water and green chilli paste 5mg + 50 ml water and

12 mg + 50 ml water against milky white mushroom *C. indica* revealed that the mean mortality was highest (76.15%) in case of green chilli paste (i.e. 12mg + 50ml water), followed by the same green pesticides with concentration of 5mg in 50 ml water, registering mortality of 68.69%. Citrus leaf extract at 2% concentration, was the poorest among all recording mortality of 35.85%. The mortality achieved in Custard apple, at both the concentrations, is also reasonably good (Parveen and Gupta, 2020b).

Suidasia pontifica Oudemans (Fig. 39.2)

Common name

Unknown

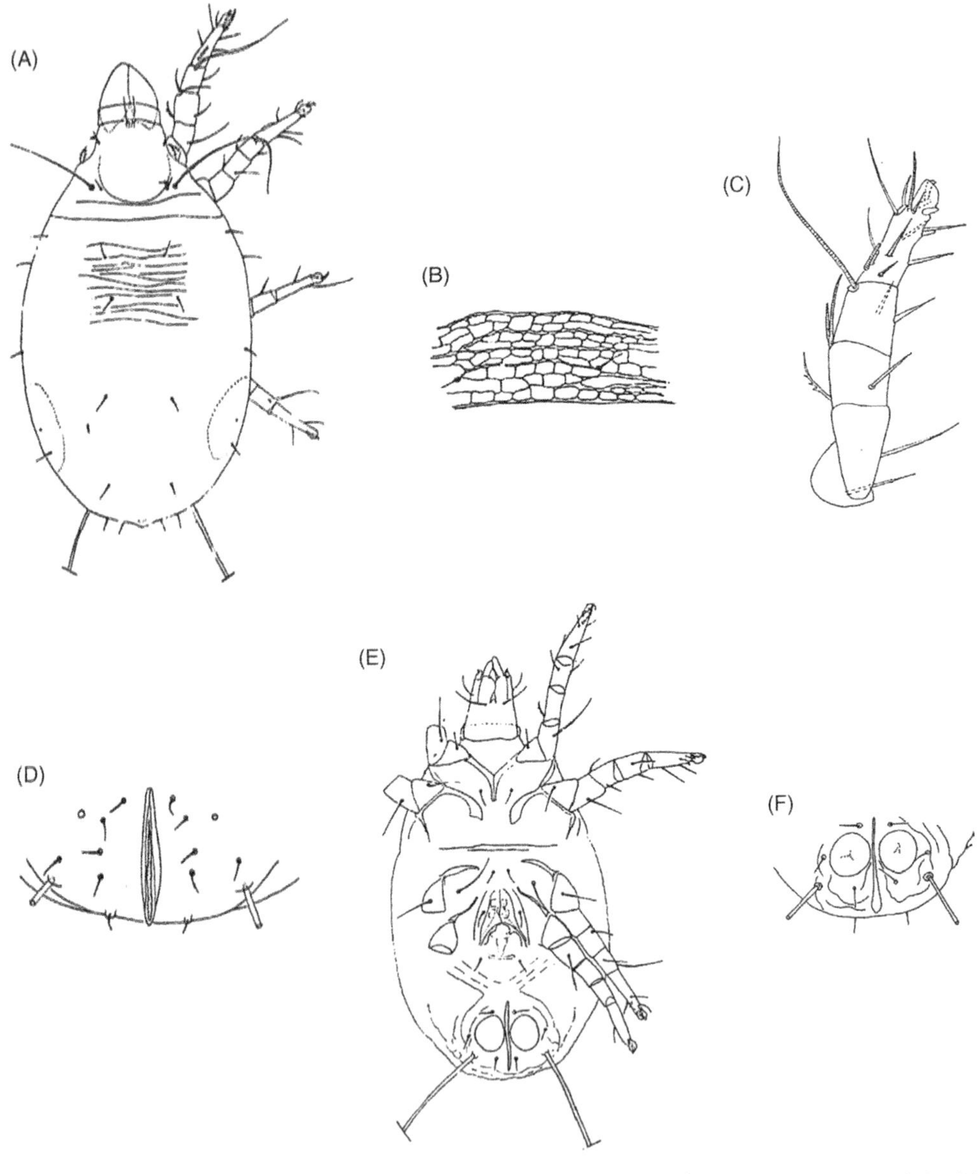

Fig. 39.2. *Suidasia pontifica* Oudemans. Female. Dorsal view (A); surface of cuticle around the setae d_1 (B); leg I (C); anal region (D). Male. Ventral view (E); anal region (F) (from Hughes, 1976, licensed under the Open Government Licence v3.0).

Morphological characteristics

FEMALE. The idiosoma is 290 μm to 360 μm long. The general features as those of the male. The anal setae tend to be arranged in linear series, on either side of the anus (Hughes, 1976).

MALE. The idiosoma is 300 μm to 320 μm long. The shape of the body is the same as in the previous species, but the wrinkling of the cuticle is more scale-like in appearance and lacking in longitudinal folding. The wrinkling can only be seen in freshly mounted specimens. The arrangement and length of the legs resemble those of *Suidasia nesbitti*, the main differences being that *ve* are more anterior in position, intermediate between *vi* and the supracoxal seta, and *he* are equal in length to *hi*. The anal opening is near the end of the body and is flanked by suckers; three pairs of anal setae encircle these suckers. The chaetotaxy of leg I differs from that of *S. nesbitti* in that the ventral spines *u* and *v*, and the famulus are not visible (Hughes, 1976).

LARVA. The idiosoma is about 160 μm long. It differs from that of *S. nesbitti* in having coxal rods as well as coxal setae. Each rod is shaped like a wine bottle, the neck ending in a transparent bulb-like structure (Hughes, 1976).

Geographical distribution

Angola (Hughes, 1976), Brazil (Monteiro and Arruda, 1969; Flechtmann and Castello, 1982; Horn *et al.*, 2018), China (Wang, 1964; Xin, 1965; Olsen, 1983), Colombia (Caraballo *et al.*, 1999), Egypt (Saleh *et al.*, 1985), England (Hughes, 1961), Germany (Hughes, 1976), Greece (Palyvos *et al.*, 2008), Hawai (Sharp and Haramoto, 1970), Hong Kong (Olsen, 1983), Korea (Khaing *et al.*, 2014), India (Manson, 1973; Cicilykutty *et al.*, 1981a,b; Olsen, 1983), Indonesia (Oudemans, 1924b; Koeniger *et al.*, 1983), Iran (Sepasgozarian, 1978), Ivory Coast (Fain, 1974), Japan (Oshima, 1980; Soares *et al.*, 1989), Malaysia (Kurosa, 1976; Mariana *et al.*, 2010), Mexico (Rack, 1968a), New Guinea, New Zealand (Manson, 1973), North Africa (Hughes, 1961), Pakistan (Manson, 1973), Philippines (Olsen, 1983), Puerto Rico (Fox, 1950; Hughes, 1961), Senegal (Gueye-Ndiaye and Marchand, 1989), Singapore (Chew *et al.*, 1999), South Korea (Chu *et al.*, 1967), Sumatra (Oudemans, 1924b), Surinam (Van Bronswijk, 1972), Taiwan (Tseng, 1972; Olsen, 1983), Thailand (Olsen, 1983).

Commodities, facilities, vegetal and animal hosts

The commodities include bean curd (Olsen, 1983), chilli powder (Olsen, 1983), Chinese medicine, as *Rhizoma chuanxiong*, *R. acori tatarinowii* (Zhan *et al.*, 2017b), copra (Horrigan, in Hughes, 1976), cowpea (Hughes, 1976), fish (dried, minced-dried or pickled) (Cicilykutty, 1981a,b; Flechtmann and Castello, 1982; Olsen, 1983; Gueye-Ndiaye and Marchand, 1989), fungus dried (Olsen, 1983), garlic, garlic bulb (Tseng, 1979), honey (Tseng, 1979), insect dried (Hughes, 1961), lily (Tseng, 1979), milk powder (Chakraborty and Gupta, 2016), mushroom, onion (Tseng, 1979), oat meal (Soares *et al.*, 1989), opium seeds (Chakraborty and Gupta, 2016), groundnut, peas (Hughes, 1961), pepper pickled, pickles (Olsen, 1983), rice bran (Hughes, 1961), rice flour (Mariana *et al.*, 2010), polished rice with germ (Soares *et al.*, 1989), pressed rice (Chakraborty and Gupta, 2016), dried shark fin (Olsen, 1983), dried shrimp (Olsen, 1983), red sugar, white sugar (Tjying, 1971), tea, tulip (Tseng, 1979) and wheat bran (Saleh *et al.*, 1985). Moreover, Hagstrum *et al.* (2013) reported a number of other commodities such as barley, bean, bean pollard, bean sprout, biscuit, bread (fermenting), confection, feather, ginger (dried), ham, hog feed, hyacinth, logan (dried) maize, mango (rotting), meat dried, mungo, palay, residue, rice, rice milled, shallot, soybean meal, soybean sauce, tiki tiki, wheat and wheat flour.

The facilities number farm, flour mill, food store, market, off-farm storage and poultry industry (Horn *et al.*, 2018).

The mite has been found by Oudemans (1924b), in nest of a carpenter bee (*Xylocopa* sp.) from Sumatra. The mite has been collected also, on dead mosquitoes in Puerto Rico (Fox, 1950; Hughes, 1961) and on dried water bugs from Mexico (Rack, 1968a), in Hamburg (Germany). Manson (1973) collected the mite on insects. Moreover, the mite has been collected also in house dust in Surinam (Van Bronswijk, 1972), Hawai (Sharp and Haramoto, 1970) and Singapore (Chew *et al.*, 1999) and in soil from Ivory Coast (Fain, 1974) and in soil and litter in West Malaysia (Kurosa, 1976). *S. pontifica* was collected in Korea on mature manure (Khaing *et al.*, 2014). The mite

S. pontifica feeds on dead insects and it has been found in bee nests in Sri Lanka (*Apis* spp.) and Indonesia (*Xylocopa*) (Oudemans, 1924b; Koeniger *et al.*, 1983).

Bio-ecology

Suidasia pontifica Oudemans (1905b) was named *Suidasia medanensis* Oudemans (1924e) until Fain (Fain and Philips, 1978) had the opportunity to examine the typical slides of both mite species. He was not able to find any character that could allow to separate these two species, and they should therefore be considered synonyms.

S. pontifica is a generalist species, it can invade beehives and nests of solitary bees (*Xylocopa*), and it feeds on dead insects, making it a pest in museum collections. The mite has been recorded as part of surveys of bee nest fauna and its biology inside bee nests is unknown. One work reported this species inside honeycombs of *Apis cerana* Fabricius and *Apis florea* Fabricius, in Sri Lanka, and suggested that it is a commensal (Koeniger *et al.*, 1983). In addition, *S. pontifica* inhabits a variety of stored products and peridomestic habitats, including house dust, mostly in the tropics. However, the behaviour of this species appears more complex, and the absence of an appropriate knowledge on its bio-ecology highlights the necessity of appropriate studies. Tseng (1982) summarized otherwise; he referred to mites infesting stored products in Taiwan from 1971 to 1980 and listed 55 species in 18 families (including predacious mites), pointing out that *Tyrophagus putrescentiae* (Schrank) and *S. pontifica* were the most destructive species. Hughes (1976) reported that possibly, its natural habitat is a bee nest.

Also, this species reproduces sexually and lacks the deutonymph stage. Hughes (1954) found that the life history takes from 16 to 18 days at 23°C and 87% R.H., using food wheat germ. The mite will not breed with *S. nesbitti*. Shen (1988) studied the fundamental traits of its biology and reported that under laboratory conditions (25°C and 75% R.H.) the development durations of egg, larva, quiescent larva, protonymph, quiescent protonymph, tritonymph and quiescent tritonymph were 102, 60, 25, 51, 33, 99 and 26 hours, respectively. Adult females repeatedly mated with males during their lifetime. Oviposition began 1–3 days after the first mating and lasted for 4 to 6 days, and the total number of eggs laid was 25 per female on average (Shen, 1988). Mercado *et al.* (2001) studied the life cycle of *S. pontifica* under laboratory conditions at 26°C and 86% R.H. Freshly laid eggs were observed until they developed into adults, and the periods between stages were recorded. Production of eggs by mated females was monitored until they died. The eggs required an average of 12.6± 0.6 days to develop into adults. Mean longevity of mated females and males was similar (48.6 ±13 and 49.1± 20 days, respectively) (Mercado *et al.*, 2001). Soares *et al.* (1989) studied the alarm pheromone of the mite.

Symptomatology and damage

In Cuba, *S. pontifica* affects a large number of products such as wheat bran, semolina, sunflower flour and mill waste, among others. The mite alters the material they feed on, by crushing the particles with their chelicerae, until they turn them into powder. The release of their excreta causes wetting and decomposition of the material, when the mites are in high population (Sierra and Martínez, 1987; de la Torre Santana and O'Connor, 2011).

S. pontifica inhabits a variety of stored products and peridomestic habitats, including house dust, mostly in the tropics. It causes allergic reactions, dermatitis, and rhinitis to humans (Kilpio and Pirila, 1952). It is one of several mite species that is probably responsible for the 'pancake syndrome' (Barrera *et al.*, 2015). This syndrome is known to occur in atopic individuals, when eating starchy foods contaminated by mites, which induces moderate or severe, even fatal, allergic reactions. Ho and Wu (2002) found a case of *S. pontifica* infesting on the tympanum of human in Taiwan. Monteiro and Arruda (1969) reported in Brazil, a disease causing death by starvation within 8–45 days of poultry in April–July, and reducing the amount of food consumed by laying birds by about 70 g per day has been attributed to the presence in their food during these months of the mites *S. nesbitti* and *S. pontifica*.

Control

Control does not always seem necessary. In any case, the indications are similar to those for *S. nesbitti*. Recently, Sansuwan and Chanbang (2020), according to Kunavong *et al.* (2014) that suggested to use heat and formaldehyde, incorporated with propionic acid in chickens feed, suggested the same strategy in swine feed.

40 The Family Winterschmidtiidae Oudemans (= Saproglyphidae Oudemans) (Fig. 40.1)

The prodorsal sclerite is well developed. Apodemes I not fused medially with apodemes II, coxal fields I open. Empodial claw present; condylophores usually fused into a V-shaped sclerite or a more elongate sclerite in base of ambulacral stalk; tibiae I–II with 1–2 ventral setae; male with all traces of genital setae and coxal setae *4b* absent; male usually with pretarsal ambulacra I–II arising from ventral apex of tarsus, often modified as a sucker (O'Connor, 2009b)

The family includes 25 genera and 138 species (O'Connor, 2009b), characterized by a cosmopolitan geographical distribution and a wide range of ecological associations. Beron (2021b) listed 28 genera and 139 species. Some species are fungivorous (*Saproglyphus* spp.) in decaying materials, bark crevices, and woody fungi (Türk and Türk, 1957; Volgin, 1980), and deutonymphs dispersed by a wide variety of insect hosts. Other species are common fungivores on leaf surfaces. Other species prey on the eggs of bark beetles (Kielczewski and Seniczak, 1972; Khaustov, 2000). Other species are found in various types of decaying material such as dung, carrion, and beach wrack and in vertebrate nests. The family Winterschmidtiidae is subdivided into four subfamilies; representatives of two are associated with insects, and those of other two commonly inhabit decaying materials, fungi, plant leaves, vertebrate nests and stored food. In Brazil, species of this family are frequently collected during surveys of arboreal acarofauna (Barbosa and Moraes, 2021). Most of these mites are generalist phoretics. While some associations are more specialized, such as those between some nidicolous species (Fain and Beaucournu, 1993), other species are obligate associates of Hymenoptera (O'Connor, 2009b). *Procalvolia zacheri* (Oudemans) and *Acalvolia squamata* (Oudemans) occasionally infest stored food products (Rack, 1980; Fain and Knülle, 1981). Oudemans (1929a) reported the first species on mouldy cheese in Berlin, and Rack (1980) found the same mite in a newly built, flat-roofed bungalow in north-west Germany, while Fain and Knülle (1981) found all the stages of *A. squamata* in nests of birds in Berlin, Germany. Another species of this last genus, *A. americana* Fan *et al.*, was intercepted on orange from the United States by Fan *et al.* (2010b). Vacante and Nucifora (1984–85) and Vacante (1989a) reported on citrus in Sicily (Italy) *Calvolia hebeclinii* (Sicher) (Fig. 40.1), sometimes associated with the oleander scale, *Aspidiotus nerii* Bouché. There are not many reports on the presence of Winterschmidtiid mites in stored products and the family is reported here for completeness of explanation.

DOI: 10.1079/9781786390912.0040

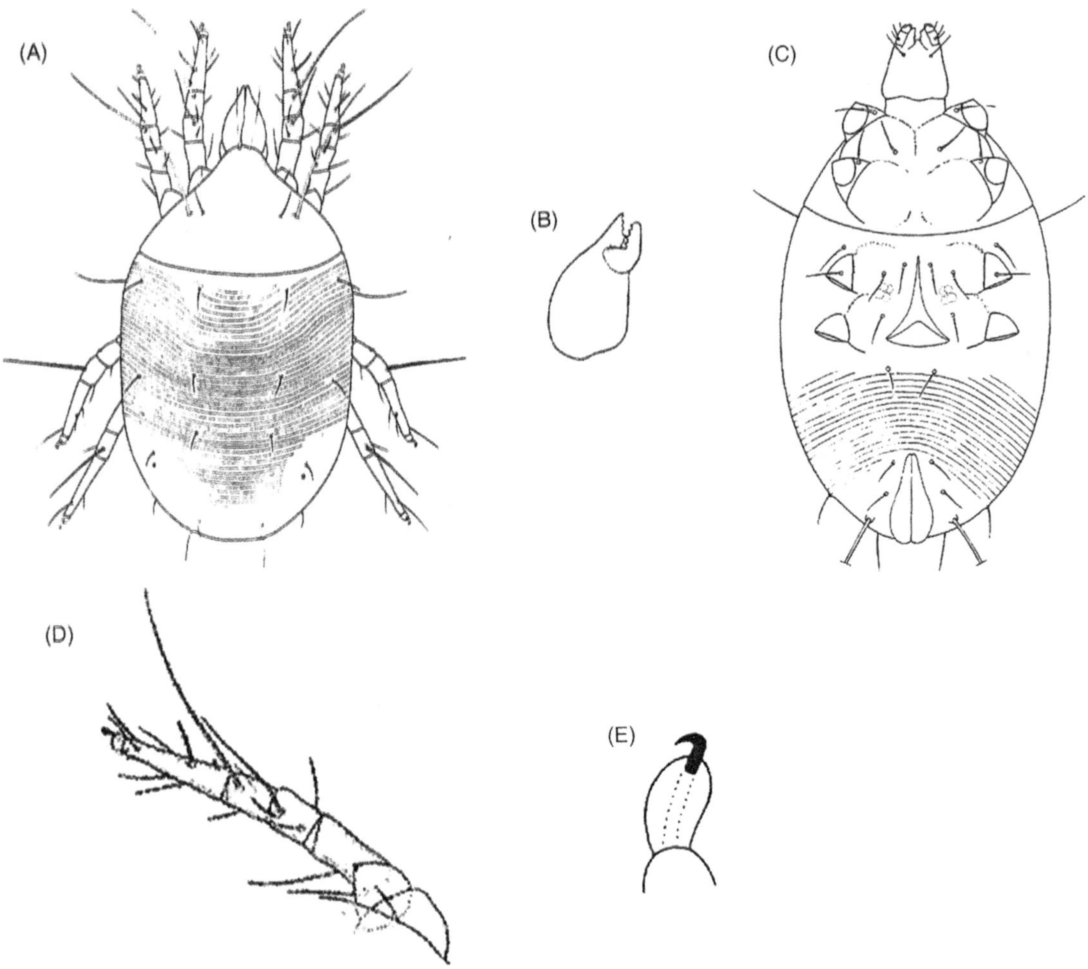

Fig. 40.1. *Calvolia hebeclinii* (Sicher). Female. Dorsal view (A); chelicera (B); ventral view (C); leg I (D); pretarsus (E) (from Vacante and Nucifora, 1984-85, licensed under CC BY 4.0).

41 The Family Achipteriidae Thor

This family is known from the Holarctic, Oriental and Neotropical regions with most species described from North America, Europe, Central America and East Asia. The taxon includes 15 genera, two subgenera and more than 110 species (including 19 subspecies), of which 56 species are known from Palaearctic region (Subías, 2019; Schatz *et al.*, 2011; Bayartogtokh and Ryabinin, 2012). Members of this family are typically collected from forest floor organic soils and rarely in grassland soils.

Adults of Achipteriidae are characteristic among other groups of Oribatid mites, in having the prodorsum with lamellae, adjacent or fused medially. The lamellar cusps, tutorium, and translamella are well developed. The genal notch is present, as well as the dorsophragmata and the pleurophragmata. The notogaster possesses immovable pteromorphs, which bears a knife-like humeral process or not; the notogastral setation is characterized by ten pairs of setae or alveoli. The octotaxic system is developed as porose areas or saccules. The chelicerae are chelate-dentate, and the palpal eupathidium *acm* is attached to solenidion; the discidium and the custodium are present; the pedotectum I is large, extending to base of bothridium, and the pedotectum II is small. The anogenital setation is 6–1–2–3. The genu IV is often concave dorsally, subequal in length or longer than tibia IV; the tibia IV possesses a solenidion. The immatures have plicate cuticle, apheredermous, and unideficient; the humeral organ is absent; the seta *d* is present to tritonymph on tibiae and genua I–IV, or only IV, when respective solenidion exists (Seniczak, 1978; Chistyakov, 1987, 1988, 1994; Seniczak and Seniczak, 2007; Norton and Behan-Pelletier, 2009; Bayartogtokh and Ryabinin, 2012).

Parachipteria punctata (Nicolet) and *P. willmanni* van der Hammen, 1952 are known to be distributed through Holarctic region. Bayartogtokh and Ryabinin (2012) examined 23 specimens (9 males and 14 females) of the first species collected on lichens growing on the bark of larch tree trunk in a cool temperate larch forest (*Larix sibiricus* Ledebour) in the Province of Hövsgöl, Mongolia.

On the stored food, *P. punctata* was collected in Wales, England, on grain (Hagstrum *et al.*, 2013), but its presence is accidental. Another species, *Tectoribates* sp. near *ornatus* (Schuster), has been reported from Iran, on stored rice, rice dust and debris by Noei (2007). The mite is also known from Argentina and Uruguay (Akrami, 2015).

DOI: 10.1079/9781786390912.0041

42 The Family Camisiidae Oudemans

The family Camisiidae has a worldwide geographical distribution, and numbers common, and abundant mites, many of which are camouflated by adherent organic debris (Norton and Behan-Pelletier, 2009). The systematic of this group has been widely discussed by Sellnick and Forsslund (1955), Colloff (1993) and Olzsanowski (1996).

The prodorsum is separated from the notogaster by band of soft cuticle. The bothridial setae, if present, have various forms, but clearly projecting from normal bothridium. The coxisternum has some type of sagittal feature that delineates at least some epimeral halves. The anal plate is wider or only slightly narrower than the adanal plate, and not greatly reduced. The medial margin of the genital plates bearing the genital setae and delimited from the rest of the plate by distinct line or carina; the aggenital setae are present. The body often has adherent organic debris (Norton and Behan-Pelletier, 2009).

The family numbers about 84 species, all parthenogenetic. Travé (1963) reported that species included in the genus *Camisia* von Heyden are primarily arboreal, living on the trunks of the trees or also in the canopy or among mosses and lichens on rock surfaces. Glime (2017) refers that the genus includes saprophagous mites living in semi-aquatic, moss, litter, and canopy habitats. André and Voegtlin (1981) found *Camisia carrolli* André as one of the most abundant mites on Douglas fir, *Pseudotsuga menziesii* (Mirb.) Franco, in the western Cascade Mountains of Oregon. It occurs at maximum density in the upper portion of the canopy; 5–8 year old twigs with needles still attached constitute its optimum microhabitat. This mite is mycophagous and probably feeds on pioneer fungi colonizing young twigs. Fungi and other detritus cover it completely and may help it endure drought. Jeppson *et al.* (1975) referred that *Camisia segnis* (Hermann) according to Grandjean (1936) is arboreal in habit. Seyd and Seaward (1984) reported that *C. segnis* likewise occurs in cooler areas and inhabits both lichens and mosses, including *Sphagnum*, and Glime (2017) wrote that it is known to eat lichens, but it is not known if it eats mosses. Hagstrum *et al.* (2017) reported this mite on wheat in the United States, but we have no news of damage produced by this mite, and we retain that it is an occasional presence.

DOI: 10.1079/9781786390912.0042

43 The Family Carabodidae Koch (Fig. 43.1)

This family includes 48 genera and has been revised by Mahunka (1986, 1987). Norton and Behan-Pelletier (2009) reported that although the carabodids range widely as adults, all species are probably endophages as immature, burrowing in a variety of substrates, including decaying wood. Immature stages of most *Carabodes* Koch feed within fungi, especially bracket fungi, and lichens (Bellido, 1979; Reeves, 1988; Materna, 2000). Major inhabitants of epiphytic lichens include *Carabodes labyrinthicus* (Michael) (Smrž and Kocourkova, 1999), and grazing by adults and immatures on *Evernia punastri* Linnaeus can induce variability in the growth of lichen talli (Prinzing and Wirtz, 1997). *Carabodes femoralis* (Nicolet) feeds on a range of ectomycorrhizal fungi (Schneider *et al.*, 2004). *Carabodes brevis* Banks and *Carabodes radiatus* Berlese prefer arboreal habitats, while *Carabodes granulatus* Banks, *C. labyrinthicus*, *Carabodes niger* Banks and *Carabodes willmanni* Bernini may also be arboreal where lichens, moss, or fungi are present (Norton and Behan-Pelletier, 2009).

The body of carabodid mites is 350 µm to 600 µm long. The rostrum is rounded and the lamellar cusps are present. The cerotegument of the notogaster presents foveae or tubercles. The tutorium lacks free cusps. The genal notch lacks, as well as the circumpedal carina. The anal plate is rectangular. The leg tarsi are functionally fused to tibibiae. The interlamellar setae are usually set well anterior of bothridia or on lamellae (Norton and Behan-Pelletier, 2009).

The mite fauna of stored foods includes the carabodid *Tectocepheus velatus* (Michael) (Fig. 43.1), a cosmopolitan species reported on stored apricot and grain from Turkey and United States (Hagstrum *et al.*, 2013), and from Iran on grain, chickpea, clover, bean, vetch, and soil samples (Fathipour, 1994; Sayedi *et al.*, 2006). We have not found other reports of the mite in this sector, except its discovery in natural environments, and in cultivated soil as has been ascertained in Australia (and in other countries), where it is believed a cosmopolitan Oribatid species that exhibits a preference for cultivated soils. In this country, *T. velatus* appeared adversely affected by ploughing, and to be beneficially affected by increased organic matter at depth through stubble incorporation; this suggested that this species could be a potential indicator of soil organic matter levels in soil (Longstaff *et al.*, 1999). We believe that its presence in stored food is accidental and for this reason, we do not treat this species.

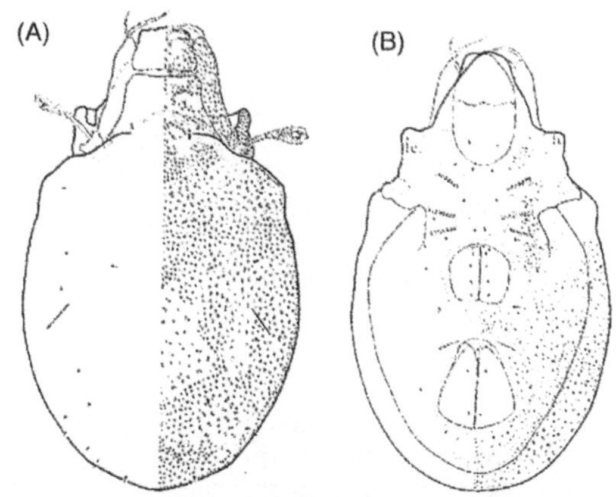

Fig. 43.1. *Tectocepheus velatus* (Michael). Adult. Dorsal view (A); ventral view (B) (Bayartogtokh, 2010).

DOI: 10.1079/9781786390912.0043

44 The Family Ceratozetidae Jacot

Adults of Ceratozetid mites are from 270 µm to 990 µm long. They have the notogastrum without the posterior tectum, and with 10–15 pairs of notogastral setae. The pteromorph is usually developed and projecting ventrally. The custodium is present, as well as the postanal porose area. The subcapitular mentum is without the tectum. The epimere IV possesses two or three pairs of setae (Norton and Behan-Pelletier, 2009).

The Ceratozetidae is a large and cosmopolitan family, including over 310 species distributed in 50 genera. They live commonly in forest, grassland, and tundra habitats from the tropics to warm-temperate or to high arctic and sub-Antarctic regions (Behan-Pelletier, 1999). These mites inhabit mainly litter and soil and may occur in high density in grass (Michael, 1884), canopy habitats (Behan-Pelletier, 2000), mosses and lichens (Materna, 2000), and wet, mesotrophic bogs (Weigmann, 1991). Norton and Behan-Pelletier (2009) reported that the feeding habits of these mites are as diverse as their habitats. The analysis of the gut contents and the feeding experiments showed that species of the genus *Ceratozetes* Berlese and *Fuscozetes* Sellnick are sapophages and mycophages (Luxton, 1972). *Ceratozetes gracilis* Michael feeds primarily on pigmented fungal hyphae (Mitchell and Parkinson, 1976), *Fuscozetes* sp. readily eats dead Collembola and other arthropods (Wallwork, 1958), and *Ceratozetes* sp. feeds on nematodes (Walter, 1988). Meier *et al.* (2002) observed that faecal pellets of *Trichoribates trimaculatus* (Koch) that feed on thalli of the bright yellow lichen, *Xanthoria parietina* (Linnaeus) Th. Fr. contains both viable ascospores and the photobiont (the algal component of a lychen) cells, *Trebouxia arboricola* Puymaly. This suggests that the passage of viable elements in faeces, might provide a common and successful mode of vegetative short and long distance dispersal of lichen-forming ascomycetes and their photobionts.

The stored foods Ceratozetidae number two species, *Trichoribates novus* (Sellnick) and *T. trimaculatus*. The first species is distributed in Holarctic region and is reported from England, France, Germany, Ireland, Italy, the Netherlands, Switzerland and United States. Moreover, it is reported from different Asian countries such as Turkey, Caucasus, Siberia, Far East of Russia, Japan, Kazakhstan and Mongolia (Bayartogtokh and Schatz, 2008). *T. novus* was collected as nymphs and adults in high numbers in meadows, by sweeping the long grass (Michael, 1884), beneath tree bark, in moss under haystacks and in mouldy grain (Hughes, 1976). It does not seem a pest mite. The latter species, *T. trimaculatus* is widely distributed in Holarctic region; in Europe is known everywhere and in Asia from Siberia, Kyrgyzstan and Mongolia; it has been collected in leaf litter of *Quercus robur* Linnaeus and *Aegopodium podagraria* Linnaeus, in leaf litter under dwarf shrubs, in moss and litter (Bayartogtokh and Schatz, 2008). The mite has been collected in Canada, England and Wales on grain. Hughes (1961, 1976) considered this species synonymous to *T. novus*. The presence of both species in foodstuffs is presumably accidental or depending on mouldy grain and for this reason, they are not discussed here.

 The Handbook of Mites of Stored Products (V. Vacante and C.H.W. Flechtmann)
DOI: 10.1079/9781786390912.0044

45 The Family Chamobatidae Thor (Fig. 45.1)

The Chamobatid mites are 290 μm to 700 μm long. The notogaster has a posterior tectum and possesses ten pairs of notogastral setae. The lamellar setae are set on the prodorsal surface, and the lamellar cusps are absent. The postanal porose area is absent. The pretarsi I and IV have the same number of claws (Norton and Behan-Pelletier, 2009).

The family includes few genera and the species *Chamobates cuspidatus* (Michael) (Fig. 45.1) is one of the most common Oribatid mite in central European forests (Sylwestrowicz and Kostecka, 1991). *Chamobates incisus* Van der Hammen was collected in Ireland in a bread van, but it has been determined to be a fungivorous species (Pande and Berthet, 1973). The same mite was collected by Van der Hammen (1952) and other specialists in *Thuja* litter in the Netherlands. Its presence on the stored food appears accidental.

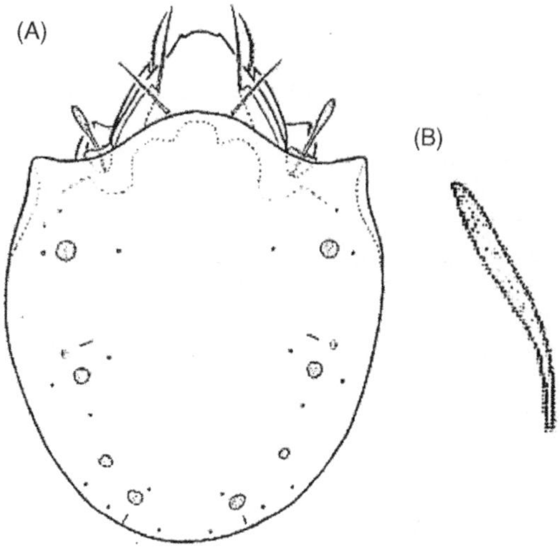

Fig. 45.1. *Chamobates cuspidatus* (Michael). Adult. Dorsal view (A); sensillus (B) (from Bayartogtokh, 2010)

DOI: 10.1079/9781786390912.0045

46 The Family Cosmochthoniidae Grandjean

The Cosmochthoniidae are a family of Oribatid mites living in dry habitats such as dry soils. They are characterized by having four pairs of large erectile setae, arising from transverse scissures on the notogaster. The notogaster is divided by three sutures in *Krivolutskiella pubescens* Gordeeva, *Phyllozetes* Gordeeva and *Cosmochthonius* Berlese, but only by two sutures in *Nipponiella simplex* (Aoki) (Balogh and Balogh, 1992).

The notogaster and the prodorsum have usually a reticulate or foveate sculpturing. The notogaster has three transverse scissures, such that the setal rows *c* and *d* are set on separate plates; the rows *e* and *f* are much longer than others, inserted on intercalary sclerites in two type S scissures. The rostrum is distinctly fenestrate. The setae of the rows *c*–*f* occupy the anterior ½ of notogaster. No large area without setae; the setal row *d* is not enlarged and similar to row *c*. The genital plate is undivided, and the aggenital plates are small and not fused to coxisternum; the adanal plates are fused posteriorly to form a U-shaped. The opisthonotal gland lacking. The pretarsi of legs are bi-or tridactylous (Norton and Behan-Pelletier, 2009).

The family comprises four genera, among which the genus *Cosmochthonius* includes most species (over 32 species and two subspecies) (Subías, 2019). Within this last genus, are included four species reported from stored food, *Cosmochthonius foveolatus* Beck, distributed in littoral, arid, transition and moist vegetation zones, and reported from China on dried fruit, dried vegetable and mushroom (Hagstrum *et al.*, 2013), *Cosmochthonius lanatus* (Michael), cosmopolitan (Ermilov, 2016) reported from China and Ireland on condiment, dried fruit, dried vegetable, and mushroom in commercial grain storage, and farm (Hagstrum *et al.*, 2013), and *Cosmochthonius reticulatus* Grandjean collected in China on dried fruit, dried vegetable and mushroom (Wang *et al.*, 2000). They are not discussed here, as their presence on stored food is accidental. The last species is *Cosmochthonius ponticus* Gordeeva, reported from the Southern Palearctic (western Central Asia and Spain) (Akrami, 2015), and from Iran by Noei (2007) on stored rice, rice dust and debris.

DOI: 10.1079/9781786390912.0046

47 The Family Cymbaeremaeidae Sellnick (Fig. 47.1)

Mites of the family Cymbaeremaeidae are cosmopolitan and inhabit primarily temperate regions. The body is 350 µm to 800 µm long. The notogaster possesses usually 10 or 13 pairs of setae. The subcapitular mentum with or without tectum. The lamellar cusps touching medially, fused or not, covering the rostrum medially; the lamellar setae arising anteroventrally on lamellar cusps. The palpal eupathidium *acm* is free from solenidion, set on tubercle, associated with the solenidion or the solenidion set at the base of the tubercle. The tarsal setation is not reduced and usually (17–20)-15–15–12 (not included solenida) respectively; the femora III–IV have one porose area on paraxial face. The solenidion, *б* is present on the genu III, and tibiae and tarsi possess porosae (Norton and Behan-Pelletier, 2009).

Among the described species, *Cymbaeremaeus cymba* (Nicolet) is reported on dried apricot in Turkey (Grober *et al.*, 2004; Hagstrum *et al.*, 2013), but it has been observed as a dominant corticolous Oribatid mite at mid-height in beech trees in Europe, and immatures were often found in the cast skins of scale insects (Wunderle, 1992); the same specialist cultured successfully, this species on *Protococcus* algae. Another species, *Scapheremaeus palustris* Sellnick (Fig. 47.1), is reported from Ireland, in flour mill (Hagstrum *et al.*, 2013). However, mites of this genus prefer epilithic and epiphytic habitats (Woas, 2002), and are among the dominant corticolous mites in temperate and tropical regions of the world, including the Holarctis, Australia, Mexico, Central and South America, India and the Galapagos Islands (Norton and Behan-Pelletier, 2009). On stored food, the presence of both mites is accidental, and they are not discussed here.

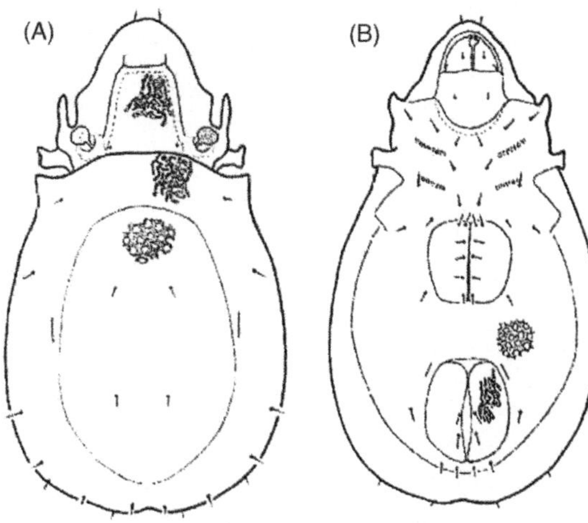

Fig. 47.1. *Scapheremaeus palustris* Sellnick. Adult. Dorsal view (A); ventral view (B) (from Bayartogtokh, 2010).

DOI: 10.1079/9781786390912.0047

48 The Family Euphthiracaridae Jacot (Fig. 48.1)

Species of the family Euphthiracaridae are primarily retained macrophytophagous, feeding on the decaying parts of higher plants. The family numbers six genera, and about 125 species (Norton and Behan-Pelletier, 2009). All immatures studied are endophagic, creating individual burrows in dead wood or woody structures of plants that they leave, only after maturing (Michael, 1888). *Rhysotritia ardua* (Koch) [= *Acrotritia ardua* (Koch)] (Fig. 48.1) is reported on stored food as sweeping, and wheat flour in mill (Seiedy *et al.*, 2000). However, its report is accidental. It is a cosmopolitan mite reported from various countries as Canada, Europe (Koch, 1841; Michael, 1888; Sellnick, 1923; Willmann, 1931; Van der Hammen, 1952), Iran (Sabori *et al.*, 2000), Japan (Aoki, 1958), North America (Jacot, 1930; Walker, 1964), Pacific Islands (Niedbała, 1998), Turkey (Ayyldiz and Özkan, 1988). It is a common species known from Samoa to Easter Island, unknown from the western border of the Pacific, where it lives in natural or partly disturbed lowland habitats (litter of various plants, moss, ferns epiphytic) (Niedbała, 1998). A complete list of geographical distribution is reported by Ebrahimi and Noei (2022).

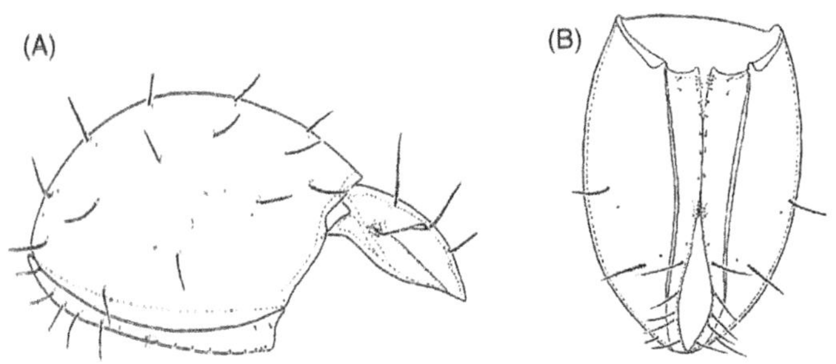

Fig. 48.1. *Rhysotritia ardua* (Koch). Adult partially drawn. lateral view (A); ventral view (B) (from Bayartogtokh, 2010).

DOI: 10.1079/9781786390912.0048

49 The Family Haplochthoniidae Van der Hammen (Fig. 49.1)

This family is characterized by having the notogaster with three transverse type E scissures; four pairs of setae (*c* row only) on anteriormost plate. With cupule immediately lateral to each mid-dorsal seta (c_1, d_1, e_1, f_1), in addition to normal complement. The genital plates are twice the size of anal plates. The aggenital seta lacks, and the anal plates have four pairs of setae; the perianal plate and the seta are absent (Norton and Behan-Pelletier, 2009).

The species *Haplochthonius simplex* Willmann (Fig. 49.1) has been reported on condiment in China (Hagstrum *et al.*, 2013), and on Chinese herb medicine (*Equisetum hiemale* Linnaeus, *Agrimonia pilosa* Ledeb.var. *japonica* (Miq.) Nakai) in storage houses in Japan and China (Piao, 1990). Grandjean (1946b) described accurately, specimens collected in France in house, in soil and plant material. The mite was collected also from moss (*Sphagnum* sp.) on the seepage area below a permanent spring in South Africa (Ermilov *et al.*, 2017), in the humus on the island of Pantelleria (Italy) (Bernini, 1995), and among the house-dust mites in Turkey (Akyazi *et al.*, 2019). Akrami (2015) reported this mite as semi-cosmopolitan, known from Holarctic, Ethiopian (Senegal and Congo), and Oriental regions and Brazil. The species has been collected in Iran on stored rice, rice dust and debris (Noei, 2007). The presence of this Oribatid mite on store food is accidental.

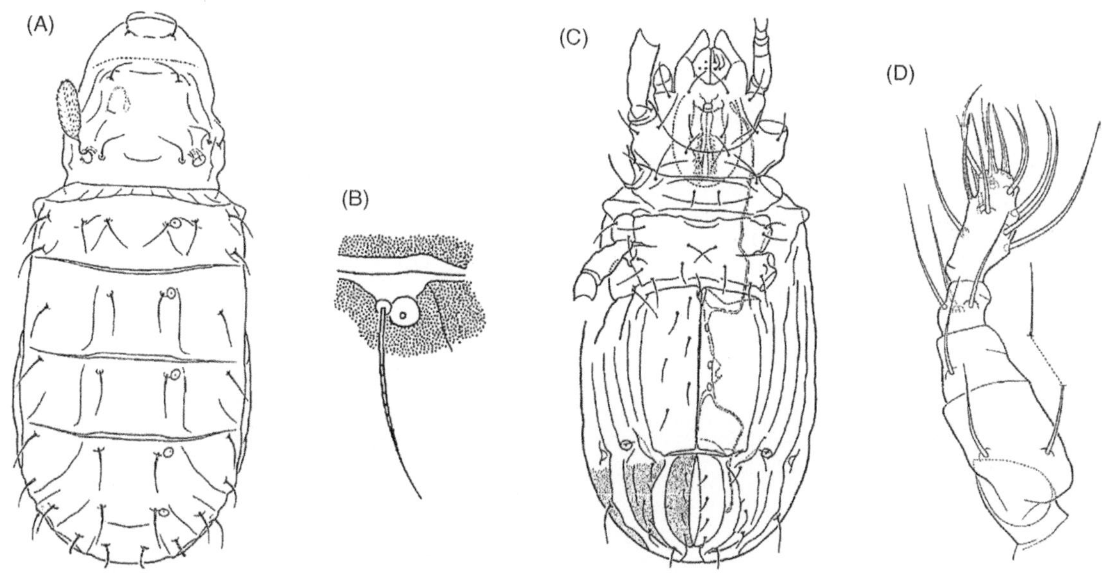

Fig. 49.1. *Haplochthonius simplex* Willmann. Adult. Dorsal view (A); cuticle sculpture (B); ventral view (C); palp (D) (from Grandjean, 1946b).

DOI: 10.1079/9781786390912.0049

50 The Family Oppiidae Sellnick

The family Oppiidae is a large and diverse family of Oribatid mites, including over 1000 known species living in various edaphic habitats across the world. They are common inhabitants of moss, humus, litter and pasture sod, in both moist and dry situations (Northon and Behan-Pelletier, 2009). Masses of fungal spores have been identified from the gut of *Oppia nitens* (Koch) by Luxton (1972); this mite is reported by Hagstrum *et al.* (2013) from farm in Ireland. The mite is known from the United States and the Palearctic and Holarctic regions (see Baran and Ayyildiz, 2004). These authors collected the species in litter and soil under oak tree, and in soil under rock in Turkey. Another species of this family, *O. clavipectinata* (Michael), has been collected on grain in England (Hagstrum *et al.*, 2013).

Different species of this family have been reported from Iran (Ebrahimi and Noei, 2022), such as *Brachioppia* sp. nr. *cuscensis* Hammer, collected on stored rice, barley flour, and wheat (Ostovan, 1993), *Lasiobelba* sp. found on stored rice, rice dust and debris (Noei, 2007), *Oppiella* sp. on grain (Sayedi *et al.*, 2006), and *Ramusella* (*Rectoppia*) sp on stored rice, rice dust and debris (Noei, 2007).

DOI: 10.1079/9781786390912.0050

51 The Family Oribatulidae Thor (Fig. 51.1)

This family is cosmopolitan, and primarily mycophagous, feeding on algae and fungi. The body of the adult is 250 µm to 900 µm long. The notogastral setae are not strongly barbed to plumose. The integument is usually smooth. Adanal setation three pairs. The marginoventral porose areas are absent. The lamellae have different shapes, the translamella is absent or present; the notogaster possesses 13–14 pairs of setae not strongly barbed or plumose, and four pairs of porose areas; the pteromorph is not developed; the legs are tridactylous, and the distal eupathidia on the palptarsus are not in the same plane; the tarsus I without the dorsal porose area (Norton and Behan-Pelletier, 2009; Bayartogtokh, 2010). The family numbers more than 40 genera and over 300 species (Norton and Behan-Pelletier, 2009). The Oribatid are a group of mites living commonly in the soil. The family is relatively rich in the number of species and individuals, and extremely important as decomposers, and common, sometimes abundant in terrestrial ecosystems, especially in forests (Seniczak and Seniczak, 2012). The presence of lots of mites is a sign of healthy soil; so, it is important to investigate soil mites (Yasemin and Ayyildiz, 2017). Moreover, these mites are known to be phoretic on the fir bark beetle *Pityokteines curvidens* (Germar) (Coleoptera, Curculionidae) (Pernek *et al.*, 2012).

The Oribatulidae contains about 16 genera, five subgenera, 209 species and seven subspecies; the family has a worldwide distribution (Subías, 2019). The genus *Phauloppia* Berlese includes about 25 species (Subías, 2019), and of these the window sill mite, *P. lucorum* (Koch) (Fig. 51.1), is the most abundant species in lichens covering the tree trunks (André, 1984, 1985). This species, commonly named window sill mite, is reported on stored food from Bulgaria, England, Germany, the Mediterranean region, Russia, Wales (Willmann, 1931; Hughes, 1976) and Mongolia (Bayartogtokh, 2010), and it has been found on barley, grain and wheat flour (Hughes, 1976). This report contrasts with the ecological adaptation of the species of the genus *Phauloppia* Berlese, known as arboreal or found in lichen patches throughout the temperate Holarctic (Prinzing and Wirtz, 1997). Although the mite is reported on barley, grain and wheat flour (Hughes, 1976), its presence on stored food is accidental. Kim *et al.* (2016) reported that *P. lucorum* has been collected in Korea from the straw mat covering the cherry trees during winter, and that this species inhabiting litter and soil of forest, moss and lichens on tree and stones, is distributed in eastern Asia and areas of Western and Central Europe (Kim *et al.*, 2016).

Different species of the genus *Oribatula* Berlese are reported from stored food; they include *O. tibialis* (Nicolet) and *O. exilis* (Nicolet). *O. tibialis* is cosmopolitan and collected in many environments, even in association with some insectivorous plants (Antor and Garcia, 1995), and known as selective feeder on ectomycorrhyzal fungi (Schneider *et al.*, 2005). In Turkey, it has been found on dried apricot (Çobanoğlu, 2008), but its damage is undescribed. *O. exilis* is reported from England and Wales on grain (Hagstrum *et al.*, 2013). In Iran, has been found *Oribatula* sp. on stored onion, barley, stored rice, rice dust and debris (Ostovan 1993; Sayedy *et al.*, 2006; Noei, 2007), and *Oribatula* (*Zygoribatula*) sp. collected on grain, stored rice, rice dust and debris (Sayedi *et al.*, 2006; Noei, 2007).

Another species, *Scheloribates laevigatus* (Koch), saprophagous and mycophagous and playing a significant impact on decomposition of grass litter through feeding on fungi (Huber *et al.*, 2000; Hubert, 2001), has been collected on stored barley, floor debris, grain, mouldy grain in oat mill, provender mill, rice, and wheat, potato, warehouse; it is reported from England, France, Germany, Iran, Ireland, Japan, the Netherlands, Russia and United States (Ostovan, 1993; Hagstrum *et al.*, 2013). However, its damage is unknown and its presence on stored food is retained accidental.

DOI: 10.1079/9781786390912.0051

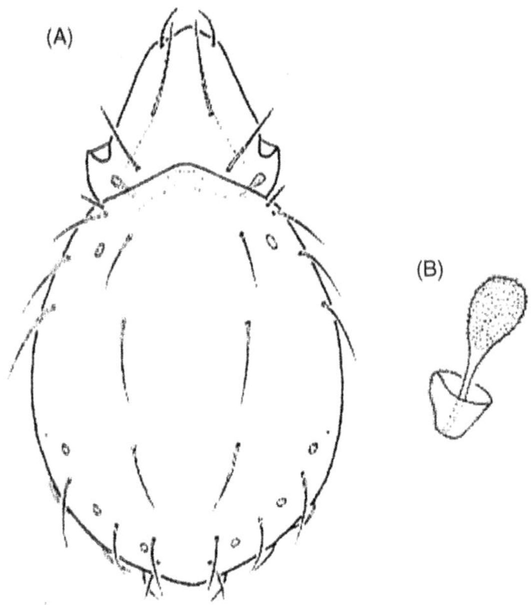

Fig. 51.1. *Phauloppia lucorum* (Koch). Adult. Dorsal view (A); sensillus (B) (from (Bayartogtokh, 2010).

52 The Family Pediculochelidae Lavoipierre

Species of this family are naturally associated with dry soils, probably worldwide, and occur synanthropically in the nests of bees, rats, and chickens. A cladistic analysis showed that it represented a sister group of the Haplochthoniidae and belonged to the superfamily Cosmochthonioidea (Oribatida) (Norton *et al.*, 1983). To date, one genus, *Paralychus* Womersley, and seven species have been described (Schatz *et al.*, 2011), and a key to the species of this genus was produced by Xua *et al.* (2020).

The body is more than three times longer than it is wide. The prodorsum is astegasime, with small, narrow anterior lobe bearing the setae *ro*; the botridial setae are smooth. The legs II–III are widely separated. The pretarsi is without distinct claw; the empodium is highly modified, with a hyaline flap on either side (Norton and Behan-Pelletier, 2009).

On the stored food, two species are reported. The first, *Paralychus chongqingensis* Fan *et al.* was described from China where it was collected on *Auricularia* sp., walnut, star anise, chilli, head of garlic, bean stick, and dried kelp (Fan *et al.*, 1996), and successively reported by Hagstrum *et al.* (2013) on aquatic product, condiment, dried fruit, dried vegetables, mushrooms, processed food from China. The last species, *P. longior* Fan *et al.*, has been described from China on lily fruits, *Auricularia*, tangerine cake, and dried kelp (Fan *et al.*, 1996); the mite was successively reported on aquatic product, candied fruit, dried fruit, dried vegetable, mushrooms from China (Hagstrum *et al.*, 2013). Neither of the two species produces direct damage to the stored food.

 The Handbook of Mites of Stored Products (V. Vacante and C.H.W. Flechtmann)
DOI: 10.1079/9781786390912.0052

53 The Family Protoplophoridae Ewing

Species of the family Protoplophoridae are primitive Oribatid mites, belonging to the major group Enarthronota living in the soil and leaf litter. Their biology and ecology are little investigated. As other species of the superfamily Protoplophoroidea, they have a unique pharyngeal complex that capable of a strong aspiration (Norton and Behan-Pelletier, 2009). Protoplophorid mites are characterized by the ptychoid body form, which is able to fold the apidosoma under the opisthosoma; this is a homoplasic trait found also in another enarthronote superfamily (Hypochthonioidea: Mesoplophoridae), and in the Ptyctima (Euphthiracaroidea and Phthiracaroidea) (Norton, 2001; Niedbała, 2004). Moreover, the notogaster is divided into four parts comprising the anterior plate (or pronotapis), the posterior plate (or pygidium), always with the 2–3 transverse sutures, and two lateral plates (or pleurapis) (Niedbała, 2004; Norton and Behan-Pelletier, 2009). The family numbers 10 genera and 28 species, reported from tropical and subtropical countries (Niedbała, 2004). On stored food, the species *Crytoplophora abscondita* Grandjean has been found on dried fruit, dried vegetables and mushrooms in China (Hagstrum *et al.*, 2013).

Other Oribatid Species

Another secondary species of family Suctobelbidae Jacot belongs to the genus *Flagrosuctobelba* Hammer. It is known from Iran on grain (Sayedi *et al.*, 2006).

The Handbook of Mites of Stored Products (V. Vacante and C.H.W. Flechtmann)
DOI: 10.1079/9781786390912.0053

Order Parasitiformes, Suborder Mesostigmata

54 The Family Ameroseiidae Evans (Figs. 54.1 and 54.2)

The general features and the systematic updatings on this family are based on Mašán (2017). Mites of the family Ameroseiidae are commonly collected in greatly heterogeneous substrates and habitats. A wide number of species is saproxylic and mycophagous, thereby depending upon dead and decaying wood and wood-inhabiting fungi (including ambrosia fungi), or the presence of fresh fruiting bodies (sporocarps) of wood-destroying bracket fungi or other saproxylic animals such as bark and wood-boring beetles. It uses snags, stumps, logs, subcortical area, insect galleries, stems and other forms of woody material for food, foraging, shelter, development, reproduction and other life activities (Westerboer and Bernhard, 1963; De Leon, 1964; Flechtmann, 1985a; Mašán, 1998). In tropical areas, some species are nectar- and pollen-feeders, associated with various flowers and flower-visiting animals, such as bees and wasps, butterflies, and sometimes birds (Evans, 1963c; Allred, 1970; Elsen 1972a,b, 1973; Klimov *et al.*, 2016). Along with saproxylic and plant-inhabiting ameroseiid mites, there is a large group of aerial and potentially cosmopolitan species that scavenge for fungal spores and hyphae, in various synanthropic habitats, such as inside of human houses (especially new buildings and those under construction), office blocks, cellars, farms, granaries, food and product storages, haystacks and arable lands (Rack, 1963, 1972; Hughes, 1976). A small group of edaphic species is associated with humid soils, with high content of raw humus. Phoretic associations with insects and other animals have been established for many different species of the family, but they are not common among the species of the European fauna. It is assumed that many ameroseiids are non-phoretic species, widely distributed by wind and air currents (Mašán, 2017).

The idiosoma is oblong and dorsoventrally flattened. The dorsal shield is entire, holotrichous, having at most 30 pairs of setae; setae J_1 and J_5 are absent. The female has the sternal region with separate sternal and epigynal shields, and the male has the sternitogenital shield, and five pairs of sternal setae are present in both sexes (st_{1-5}). The sternal shield is reduced in size, but entire, with 1–3 pairs of sternal setae (commonly two pairs); the presternal platelets are absent. The metasternal setae (st_4) are set on soft integument or endopodal platelets. The epigynal shield bears one pair of genital setae (st_5); the genital opening of male is located in presternal position. The peritremes are well developed and long. The posterior ventral surface of adults with anal or ventrianal shield, and with at least four pairs of opisthogastric setae. Three circumanal setae are present. The cheliceral digits are short; the arthrodial membrane at base of movable digit is smooth, without brush-like or filamentous processes. The palptarsal apotele has two or three tines. The hypostome with internal malae usually unmodified and lightly fringed. The tibia I has five dorsal setae; the legs II of males without spur-like structures; the tarsus II in the female usually without enlarged spine-like distal setae; the genu IV with normally nine setae, including one ventral (but aberrations exist). Mites with podospermal insemination, spermatodactyl of male present and placed on movable digit of chelicera. The female is monogynaspid (Mašán, 2017).

According to specific microhabitat requirements, it's possible to divide the ameroseiid mites into four groups of species. The first group includes the edaphic species, bound to various kinds of soil substrates, such as highly humic soils, raw humus, decomposing plant substances, leaf and needle litter, moss, etc. The second group includes the saproxilic species,

 The Handbook of Mites of Stored Products (V. Vacante and C.H.W. Flechtmann)
DOI: 10.1079/9781786390912.0054

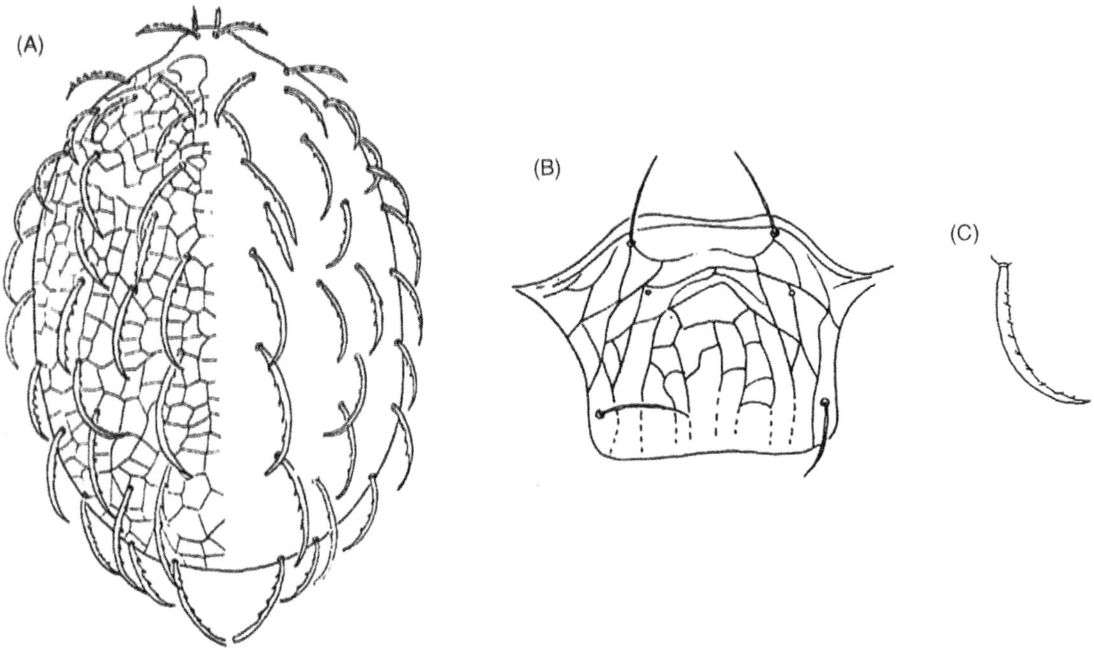

Fig. 54.1. *Klemannia plumigera* Oudemans. Female. Dorsal view (A); sternal shield (B); terminal seta (C) (from Hughes, 1976, licensed under the Open Government Licence v3.0).

adapted to living in rotting wood with growth of wood-inhabiting fungi, including sporocarps of wood-destroying fungi, snags, stumps, stems, subcortical area of dying trees and galleries of xylophagous insects. The third group numbers the saprophilous species, showing affinity to strongly decaying organic matter in dunghills, excrements and compost heaps. They facultatively occur in other substrates consisting of a portion of excrements or decomposing plant and animal substances, such as manured arable soils, bird nests, heterogeneous organic refuses, etc. The fourth group includes the aerial species, especially those spreading by air currents (passive airborne dispersal) and ensures the continued existence of species living in ephemeral and scattered habitats, such as mould growths living on diverse organic substrates. It is a relatively larger group of species that colonize manifold, natural and synanthropic microhabitats, such as interior of human buildings, cellars, farms, stored food, granaries, dumping grounds, haystacks, vertebrate nests and agrocoenoses. The family includes 12 genera and 138 species (Mašán, 2017).

Hagstrum *et al.* (2013) listed on stored food, five species belonging to the genus *Ameroseius* Berlese (*Ameroseius lanceosetis* Livshitz *et* Mitrofanov, *Ameroseius marginalis* Fan *et* Li, *Ameroseius plumigera* (Oudemans), *Ameroseius plumosus* (Oudemans) and *Ameroseius sichanensis* Fan *et* Li). The systematic revision recently proposed by Mašán (2017) reduces the number of the above species to four and includes them in the genus *Kleemannia* Oudemans. Thus, *A. lanceosetis* Livshitz *et* Mitrofanov became synonymous to *Kleemannia pavida* (Koch); the mite has been collected in Russia, Crimea, in grain storage house (Livshitz and Mitrofanov, 1975) and on condiment from China (Hagstrum *et al.*, 2013), in Hungary, German (Szabo *et al.*, 2010; Mašán, 2017), and Iran on straw caches and storage products (Khaleghabadian *et al.*, 2015; Naderi *et al.*, 2016). *A. marginalis* and *A. sichuanensis* are synonymous to *Kleemannia insignis* (Bernhard). This species is reported from Germany on horse dung, rabbit dung and rotting leaves of northern water hemlock (*Cicuta virosa* Linnaeus) (Westerboer and Bernhard, 1963), and from China on bean stick (Fan and Li, 1993b), and as *A. sichuanensis* from China on jelly ear fungus, *Auricularia auricula-judae* (Fr.) Quél. and *A. sichuanensis* from China on condiment, dried fruit, dried

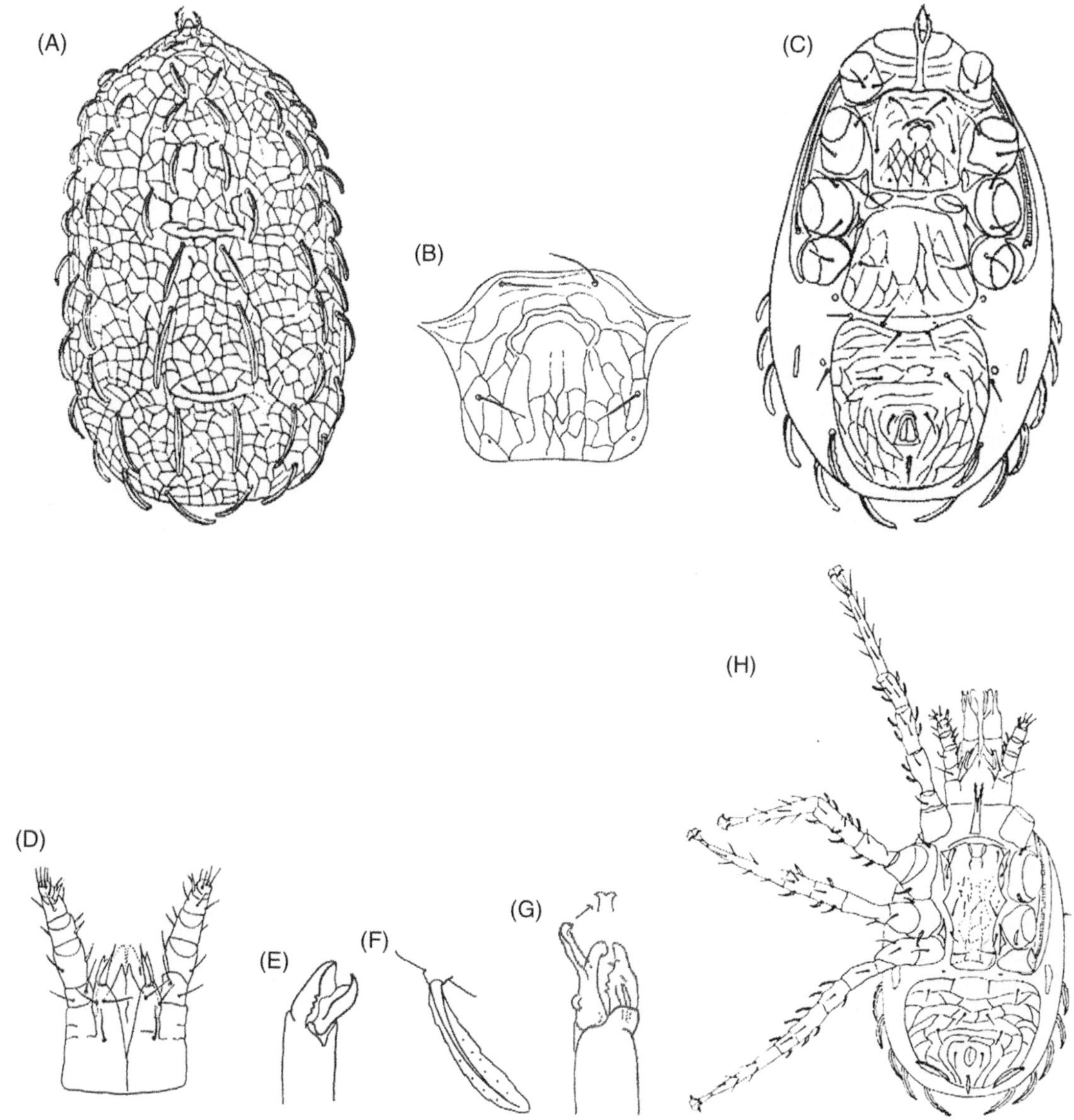

Fig. 54.2. *Klemannia plumosa* (Oudemans). Female. Dorsal view (A); sternal shield (B); ventral view (C); ventral view of gnathosoma (D); chelicera (E); terminal seta (F). Male. Chelicera (G); ventral view (H) (from Hughes, 1976, licensed under the Open Government Licence v3.0).

vegetable, and mushroom (Fan and Li, 1993b). In Slovakia *K. insignis* has been collected in alder forest, nest of *Passer montanus* (Linnaeus), artificial nest box, and straw litter (Mašán, 2017). Moreover, Hagstrum *et al.* (2013) reported the mite (as *A. marginalis*) from China on condiment, dried fruit, dried vegetable, mushroom, processed food. Mašán (2017) stated that *A. plumigera* and *Ameroseius gilarovi* Petrova are synonymous to *K. plumigera* Oudemans. *K. plumigera* (Fig. 54.1) is known from the Netherlands on common eelgrass, *Zostera marina* Linnaeus (Oudemans, 1930a), from the United

Kingdom, and Northern Ireland in damp cracks in a damp stone wall and in siftings from oats and the surface of baled hay (Hughes, 1976), and from Uzbekistan on field with cotton, *Gossypium* sp. in soil substrate (Petrova and Koshchanova, 1986). Rack (1963, 1964) reported large number of this mite in new houses and office building, where it lived as mycophagous on moulds developing on vegetable material, used as insulating. The mite has been reported also from deep litter of broiler houses (Brady, 1970). The mite populations disappeared when the buildings become drier. Moreover, Hagstrum *et al.* (2013) reported the mite from Brazil, Canada, China, Egypt, England, Germany, Greece, Ireland, the Netherlands, Poland, Taiwan, United Kingdom on buckwheat, buckwheat product, condiment, dried fruit, dried vegetables, grain, hay, maize, mushrooms, oat sifting, rice, rice (husked) in commercial grain storage, farm, market and off-farm storage. Moreover, the mite has been collected on *Mus musculus* Linnaeus (Rodentia, Muridae) in Ireland, and on zeegras in Iran. In Slovakia, *K. plumigera* has been collected on shed, soil detritus with sawdust, bottom of cave, organic detritus, alluvial clay substrate, oak forest (*Corno-Quercetum*) with maple (*Acer* sp.), moss on rocky walls in cave entrance, soil detritus with humus, fragments of rock with guano and wood detritus, and soil detritus with humus under stone (Mašán, 2017). *K. delicata* (Berlese) is known from Italy (Mašán, 2017), and Iran on storage products (Khaleghabadian *et al.*, 2015). This mite has been reported as *Ameroseius plumigerus* (Oudemans) from Iran on stored products, buckwheat, buckwheat product, condiment, dried fruit, dried vegetable, grain, hay, maize, mushroom, oat sifting, rice, rice (husked) (Sepasgozarian, 1978; Modarres Awal, 1994, 1997; Kheradmand *et al.*, 2007c). Another species of the same genus, *Ameroseius aegyptiacus* (Nasr *et* Abow-Awad, 1986), is known from Egypt and Iran on rice hulls, and woodchips (Hajizadeh and Karami, 2017). A third *Ameroseius* sp. has been found in Iran, on stored product (Noei, 2007; Noei *et al.*, 2008b).

A. plumosus is synonymous to *Kleemannia plumosa* (Oudemans) (Fig. 54.2). It is a cosmopolitan inhabitant of stored food and the nests of small mammals and social insects, where it feeds on fungi (Hughes, 1976; Halliday, 1997). *K. plumosa* is reported as a nuisance in new houses and office blocks in Germany, a consequence of the vegetable materials used for insulation during construction, absorbing moisture from drying brick or concrete and becoming mouldy, thus permitting large-scale development of mite populations. Obviously, the populations disappeared when the new structures slowly dried out (Rack, 1963, 1972). Moreover, the mite has been collected on pond bat, *Myotis dasycneme* Boie (as *Vespertilio dasycneme*) (Chiroptera, Vespertilionidae) from the Netherlands (Oudemans, 1902b), in rotting leaves from Italy (Berlese, 1918) and in sifted oats and detritus from warehouse floors from England, Scotland and Northern Ireland (Hughes, 1948, 1976). In Iran, the mite has been collected under the elytrae of coleopteran scarabeid *Oryctes nasicornis* Linnaeus. Moreover, in Italy, it has been found in stables, dunghills, chestnut wood, and flowers, and in the Netherlands on the murid *Micromys minutus* Pallas (Mašán, 2017). Hagstrum *et al.* (2013) reported that this mite has a wide geographical distribution, including Australia, Canada, China, Czech Republic, Egypt, England, Germany, Greece, Iceland, Ireland, Israel, Japan, the Netherlands, Poland, Scotland, Taiwan, Thailand, United Kingdom and United States. A long list of commodities includes animal feed, bamboo shoot, biscuit, buckwheat, buckwheat product, confection, dried fruit, dried vegetable, fish (minced-dried), gin trash, grain, hay, maize, mushroom, nuts, groundnut, pork (minced-dried), poultry diet, rice, rice (husked), seed cotton, shrimp (canned), sugar (red), and tuna (canned) and different facilities, such as commercial grain storage, farm, off-farm storage and warehouse (Ardeshir *et al.*, 2012; Hagstrum *et al.*, 2013). Despite this wide list of commodities, the mite does not appear as a strict pest of stored food, where its presence is probably related to poor conditions of conservation environment, and/or to the development of fungi.

Another species of this genus is *Kleemannia pseudoplumosa* (Rack), known from Germany, Russia, South Africa (Mašán, 2017), and Iran (Arjomandi *et al.*, 2013) where it has been collected on straw caches, and stored products (Naderi *et al.*, 2016; Karami *et al.*, 2017).

Another species is *Ameroseius dendrovagans* Flechtmann and Flechtmann (1985), collected from bark beetle galleries in *Pinus* in Brazil.

55 The Family Ascidae Voigts *et* Oudemans (Figs. 55.1 and 55.2)

According to their similarity in external morphology, species of the families Ascidae Voigts and Oudemans, Blattisociidae Garman and Melicharidae Hirschmann were included traditionally into the single family Ascidae (Lindquist and Evans, 1965). Successively, Lindquist *et al.* (2009b) highlighted the paraphyletic structure of previous Ascidae and suggested that it can be divided into three groups: Ascidae *sensu stricto*, Blattisociidae and Melicharidae. Ascidae and Melicharidae are placed in the superfamily Ascoidea, and Blattisociidae in the Phytoseioidea. Sourassou *et al.* (2015) proposed the phylogeny of Ascidae, and the related groups on the basis of molecular data, and the results of their research supported the previous classification of Lindquist *et al.* (2009b). Santos *et al.* (2020a, b, c) recorded three online databases including 381, 392 and 217 valid species in the Ascidae, Blattisociidae and Melicharidae families, respectively, and identified in a wide range of countries and habitats. The analyses based on databases on the distribution and biology of species in those families showed that Ascidae probably originated in the Palaearctic region, whereas Blattisociidae and Melicharidae probably originated somewhere in the Neotropical or Panamanian (Santos *et al.*, 2018). According to above systematic revision, the 18 species reported by Hagstrum *et al.* (2013) on stored food must be variously included into the three families. So, only four species belong to family Ascidae *sensu stricto*, while nine species are assigned to the family Blattisociidae, and five to the family Melicharidae.

Species of this family live commonly in soil, litter, stored grains, stored food, on flowers, or leaves and under tree barks (Halliday, 1998; Gerson *et al.*, 2003; Lin, 2003; Moraes *et al.*, 2016) and have received a concrete interest for their possible employment in biological control of pest mites and insects (Moraes *et al.*, 2016). Lindquist *et al.* (2009b) reported that species of the family Ascidae are relatively unspecialized, free-living inhabitants of both soil and arboreally suspended litter and humus layers. Various species are free-living predators inhabiting upper soil layers, plants and stored products, where they feed on nematodes and small arthropods. A reduced number of species parasitize cockroaches and moths (Egan and Moss, 1969; Treat, 1975; Lindquist *et al.*, 2009b). Kinn (1983) reported that Ascid mites are often carried in high numbers by the flies and moths on whose eggs (and young larvae) they feed, and other species are dispersed by beetles. A wide number of pollen-feeding species live in tropical flowers and are dispersed by hummingbirds (Colwell and Naeem, 1994). Other Ascid mites feed on fungi (Ibrahim *et al.*, 1989) or exhibit omnivorous feeding habits. The female is commonly bisexual and mates to produce eggs; however, 30% of the species listed by Walter and Lindquist (1995) were thelytokous. In the temperate region, the Ascidae have colonized protected artificial habitats, such as food stores, mushroom-growing facilities and greenhouses; they rarely inhabit plants, but on tree crops in tropical north-eastern Australia were the most abundant predatory species found on the leaves of rainforest trees and on the tree crops (Walter and Lindquist, 1997). Zhang and Lin (1991) reported that one Ascid mite was the dominant natural enemy of an acarine pest of rice in southern China.

Ascid mites can be recognized by the podonotal and opisthonotal shields fused or not. The third pair of sternal lyrifissures on posterolateral corners of sternal shield (except *Ectoantennoseius* and some *Anephiasca*, in which it is on unsclerotized cuticle, or *Anystipalpus* and some *Antennoseius*) is absent. The seta st_4 is set usually on unsclerotized cuticle. The peritrematic shield is broadly connected to exopodal shield beside the coxa IV. The fixed cheliceral digit has a setiform *pilus dentilis*, and the movable cheliceral digit is usually bidentate. The spermathecal apparatus is of laelapid-type, but often it is difficult to discern. The genital shield is

 The Handbook of Mites of Stored Products (V. Vacante and C.H.W. Flechtmann)

DOI: 10.1079/9781786390912.0055

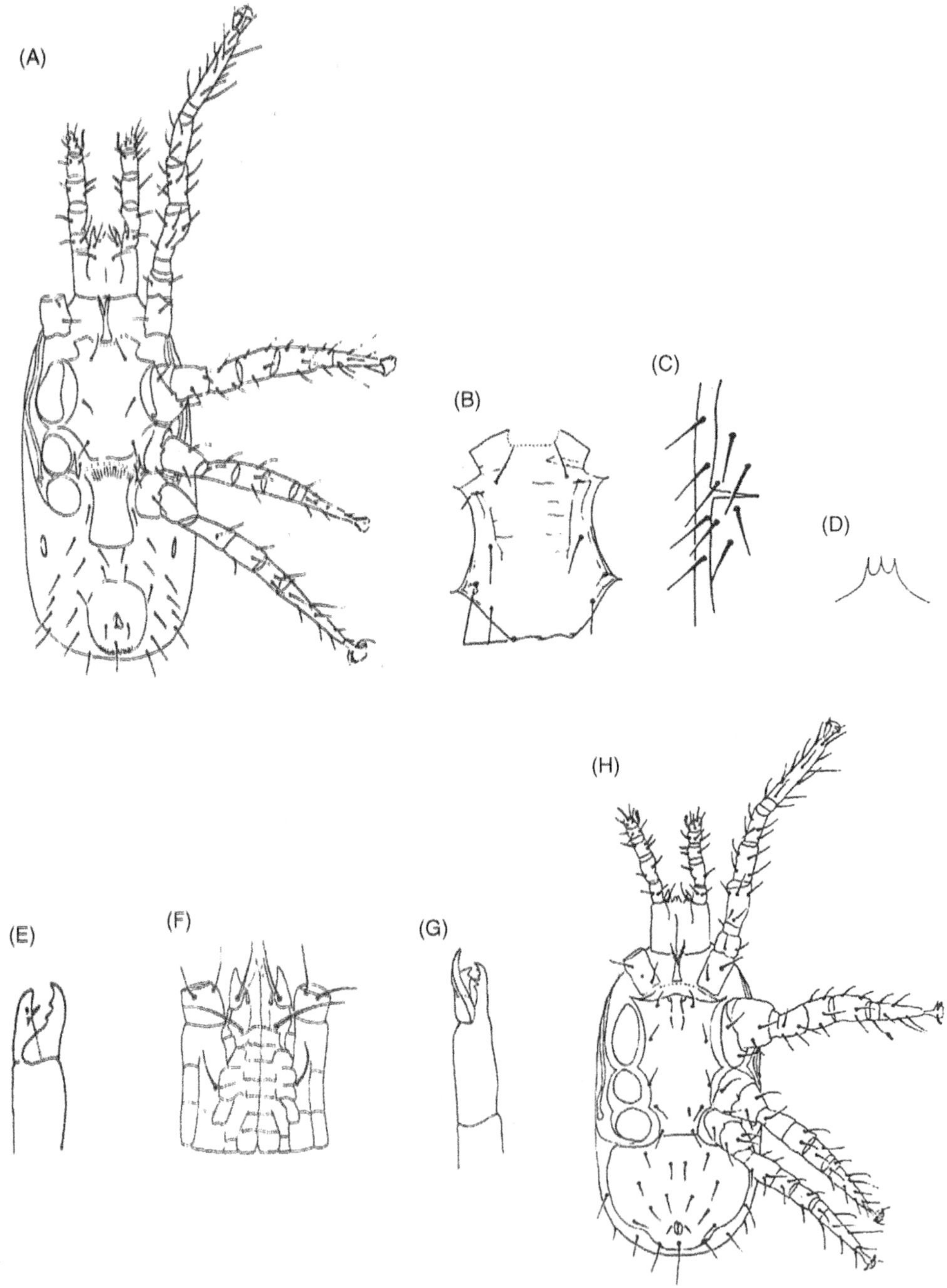

Fig. 55.1. *Arctoseius butleri* (Hughes). Female. Ventral view (A); sternal shield (B); lateral incision of dorsal shield (C); tectum (D); chelicera (E); ventral view of gnathosoma (F). Male. Chelicera (G); ventral view (H) (from Hughes, 1976, licensed under the Open Government Licence v3.0).

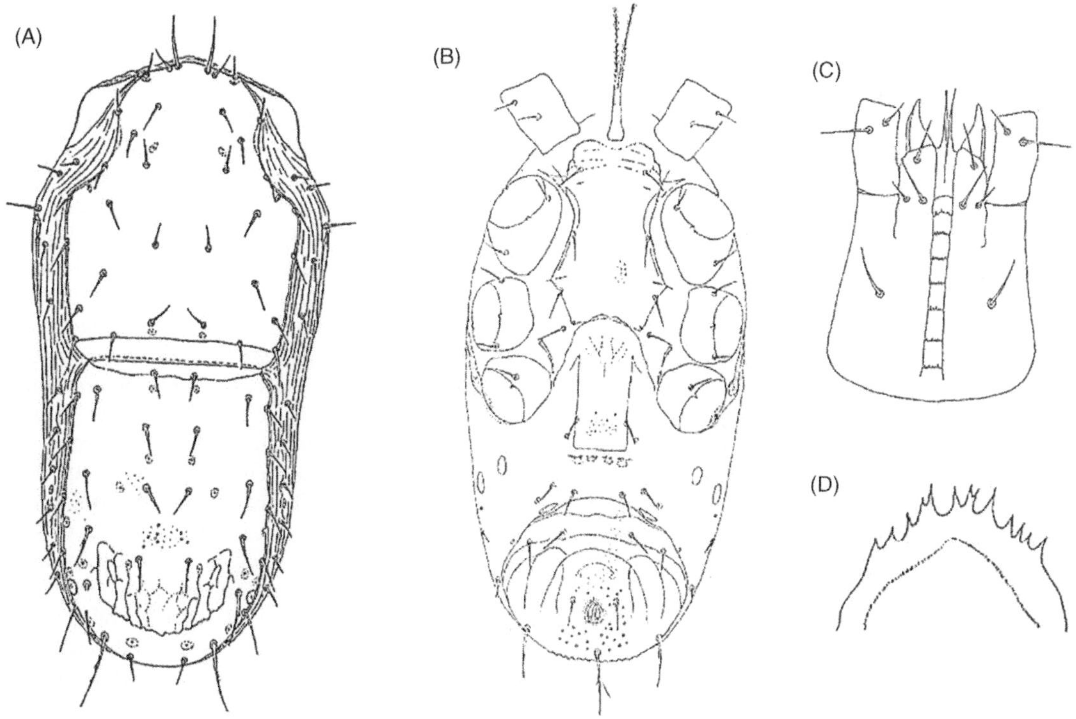

Fig. 55.2. *Protogamasellus hibernicus* Evans. Female. Dorsal view (A); ventral view (B); ventral view of gnathosoma (C); tectum (D) (from Evans, 1982 licensed under CC BY 4.0).

truncate to convex posteriorly (tapering in some *Antennoseius*). With an anal shield bearing only circumanal setae (rare), or with a ventrianal shield bearing additionally up to seven pairs of setae (Moraes *et al.*, 2016).

On stored food are reported the Ascid mites *Arctoseius butleri* (Hughes, 1948) (Fig. 55.1), *Arctoseius cetratus* (Sellnick), *Arctoseius insularis* (Willmann), *Asca* sp., *Gamasellodes bicolor* (Berlese), *Iphidozercon gibbus* (Berlese), *Protogamasellus hibernicus* Evans (Fig. 55.2), *Protogamasellus* sp. nr. *massula* (Athias-Henriot) and *Zerconopsis remiger* (Kramer).

A. butleri has been observed by O'Farrell and Butler (1948) to prey *Drosophila* larvae in rotten dried peaches, and in association with *Acarus siro* Linnaeus in sawdust near a window on the top of a silo and at the base of a wet stack of straw bales (Hughes, 1976). It is known from Ireland (O'Farrell and Butler, 1948) and Scotland (Hughes, 1948, 1976). Damage to stored food is unknown.

A. cetratus has a cosmopolitan distribution and has been collected in a wide list of habitats (see Keum *et al.*, 2015). In Iran, it has been reported from soil (Montazeri *et al.*, 2011; Nemati *et al.*, 2012) and from storage products (Khaleghabadian *et al.*, 2015).

A. insularis is a European species occurring in meadow soil, in decaying plant remains, in sand between roots, under seaweed on shoreline. In Slovakia, the mite was found in arable soils in lowland, in soil of floodplain forest of Salici-Populetum, in littoral reed stand (Phragmition), in nests of *Fulica atra* Linnaeus on water level, in the soil of mesophilous pasture and in rhizosphere of grass; in this country the mite has been found mainly in lowlands (Kalúz and Fend'a, 2005). Hagstrum *et al.* (2013) reported *A. insularis* from China on condiments, dried fruit, dried vegetables and mushrooms. The direct damage produced by this mite on stored food is unknown.

A species of the genus *Asca* von Heyden is reported from Iran on flour, barley, stored rice, pomegranate, and cotton (Ostovan, 1993; Sorush, 1994; Khanjani, 1996; Mossadegh, 1997). *G. bicolor* is a cosmopolitan species inhabiting various habitats with different microclimatic conditions.

Despite its frequent occurrence in the bird nests, this species cannot be considered strictly nidicolous species. Rather preferring mildly wet decaying material. Kalúz and Fend'a (2005) reported that in Slovakia the mite was scarcely collected in arable soil (humid sand, sandy clay with corns, clover) and in meadow soil, regularly occurring in deciduous and pine forests; moreover, it has been found in humus, mouldering straw, moss and tree mould; it has also been collected in horse and rabbit dung, preferring from mildly wet to dry substrates. In this country, the mite has been found in lowland sandy dune with *Pinus silvestris* Linnaeus, lowland reed stands (*Phragmition* sp.), wet alder forest, *Alnion glutinosae* (Ash), willow-poplar floodplain forest (Salici-Populetum), herbaceous vegetation (*Petasites* sp.) along brook, acacia forest, *Robinia pseudoacacia* Linnaeus with moss and sand, oak-hornbeam forest (Querceto-Carpinetum), beech forest, subxerophile meadow, xerothermic forest steppe, spruce forest, moss, rhizosphere of grass, arable soil, leaf litter and soil detritus. Moreover, it commonly inhabits the nests of various birds, such as *Circus aeruginosus* (Linnaeus), *Anas platyrhynchos* Linnaeus, *Fringilla* sp., *Chloris chloris* (Linnaeus), *Passer montanus* (Linnaeus) (combined with *Martes* sp., Mammalia), *Turdus* sp., *Turdus merula* Linnaeus, *Acrocephalus arundinaceus* (Linnaeus), *Sylvia* sp. in spruce forest, combined nest of *Parus major* Linnaeus and *Apodemus* sp. (Aves, Mammalia), *Sturnus* sp., *Lullula arborea* (Linnaeus), *Aquila heliaca* Savigny on beech and *Fulica atra* on water level (Aves).

I. gibbus is reported from Europe, Asia, North America, New Caledonia, Australia, Russia and Ukraine (Berlese, 1903; Chant, 1963; Ishikawa, 1969; Karg, 1996; Gwiazdowicz, 2003, 2007; Sklyar, 2003; Gwiazdowicz and Halliday, 2008; Gwiazdowicz and Marchenko, 2012). In Iran, this mite has been collected on stored decaying onion, starchy materials, animal manure, humus, stored bulbs, soil, bark, rotten wood, fungi and litter (Ostovan, 1993; Ostovan and Kamali, 1994; Modarres Awal, 1997; Kamali *et al.*, 2001; Jalaeian *et al.*, 2006; Khademi *et al.*, 2006; Masnavipour *et al.*, 2011; Naghibinejad *et al.*, 2011; Zare *et al.*, 2012).

Krištofík *et al.* (2002) reported *Leioseius bicolor* (Berlese) (= *Gamasellodes bicolor*) in bird nests of *Lanius collurio* Linnaeus and *Lanius minor* Gmelin. Karami *et al.* (2017) collected the mite from the weed floor of an olive garden in Iran. Moreover, this mite is known on stored food from Iraq, rice in Ireland, and wheat in bread vans (Hagstrum *et al.*, 2013), but also in this case, the damage to stored food is unknown. *Z. remiger* it is a European species, known from Ireland in farms (Hagstrum *et al.*, 2013). Saloña *et al.* (2010) collected it in Northern Spain, among arthropods at different stages of development, from human remains in an advanced stage of decomposition (following autopsy), and from the soil at the scene. In Iran, was collected from soil samples of a citrus garden (Karami *et al.*, 2017). The damage to stored food is unknown.

P. hibernicus is a species reported from Ireland in a grassland/dune area at Carnsore Point (Evans, 1982; Moraes *et al.*, 2016) and Iran, where it has been collected on stored rice and barley (Ostovan, 1993). *P.* sp. nr. *massula* is known from Algeria (Moraes *et al.*, 2016).

56 The Family Melicharidae Hirschmann (Figs. 56.1–56.4)

The family Melicharidae has been classified as a subfamily or a tribe in previous accounts (Hirschmann, 1962; Lindquist and Evans, 1965), and according to a uniquely shared modification of the cheliceral *pilus dentilis* is actually considered a probable sister group of the Ameroseiidae. The family is adapted to a number of terrestrial and aboveground habitats, similar to those of ameroseiids (Lindquist and Evans, 1965; Gilyarov and Bregetova, 1977; Karg, 1993a; Halliday *et al.*, 1998; Gwiazdowicz, 2007). Several species of Melicharids are predatory, others feed on fungi, pollen, and nectar, and one species is known as parasitic on cockroaches (Egan and Hunter, 1975). A complete characterization of Melicharidae was given by Moraes *et al.* (2016). Mites of this family can be recognized by the podonotal and opisthonotal shields usually fused. The third pair of sternal lyrifissures (iv_3) off sternal shield (absent in *Mucroseius* and some *Proctolaelaps*), are commonly inserted on metasternal plates together with the sternal seta st_4, when the metasternal plates are present. The peritrematic shield is free posteriorly from, or narrowly fused with, exopodal shield beside coxa IV (except in *Orthadenella*). The fixed cheliceral digit has a hyaline lobe instead of setiform *pilus dentilis* (except *Orthadenella* and some *Proctolaelaps*). The movable cheliceral digit has usually a pointed process (mucro) on mid-ventral face. The spermathecal apparatus is laelapid-type, and the genital shield is usually gently rounded posteriorly. The anal shield is usually oval or elliptical, bearing only circumanal setae; some genera with small ventrianal shield, bearing 1–3 pairs of setae (broad ventrianal shield bearing 5–6 pairs of setae in *Orthadenella*); with separate ventral and anal shields in *Proctogastrolaelaps*. The family numbers 206 species in 11 genera (Moraes *et al.*, 2016).

On stored food are reported eight Melicharid mites. They are *Melichares agilis* Hering (Fig. 56.1), *Proctolaelaps bickleyi* (Bram), *Proctolaelaps hystrix* (Vitzhum), *Proctolaelaps pomorum* (Oudemans) (Fig. 56.2), *Proctolaelaps pygmaeus* (Müller) (Fig. 56.3), *Proctolaelaps regalis* De Leon, *Proctolaelaps scolyti* Evans (Fig. 56.4) and *Proctolaelaps ventrianalis* Karg (Hughes, 1976; Hagstrum *et al.*, 2013; Ebrahimi and Noei, 2022).

The genus *Melichares* Hering includes two species. Of these, *M. agilis* can be identified following Evans (1958), and another species unidentified is reported from Brazil (Siegert *et al.*, 2018). This genus has been found in wheat products (Chmielewski, 2002a), and Rivard (1962) proved that it can be a future tool for biological control of *Tyrophagus putrescentiae* (Schrank). *M. agilis* was reported also as *Typhlodromus domesticus* Oudemans and *Garmania domestica* (Oudemans) (Nesbitt, 1951) and in older literature was confused also with *Blattisocius*. *M. agilis* occurs in human related habitats and is expected to be a cosmopolitan species. It is reported from various countries, such as Brazil (Pinto da-Rocha, 1994), Czech Republic (Haragsim *et al.*, 1987), England (Hughes, 1961, 1976), Germany (Hering, 1838; Rack, 1968a; Hughes, 1976), Hungary (Kandil, 1983), the Netherlands (Oudemans, 1929 a, b; Hughes, 1976), Poland (Gwiazdowicz, 2007), Taiwan (Tseng, 1982), Turkey (Ozman and Žďárková, 2000; Çobanoğlu, 2008) and other countries. All stages are predatory on acarid mites (and probably other microarthropods) in different habitats, including honeybee hives, where the mite has been found associated with bees, and records from honeybees are from the Western Palaearctic (Klimov *et al.*, 2016). The mite has been found in hives of the European honeybee *Apis mellifera* Linnaeus, and can enter bee nests by walking if a nest is nearby (Haragsim *et al.*, 1978), but the host association level is facultative (Klimov *et al.*, 2016). Moreover, *M. agilis* is known as a widely distributed predator of acarid mites in stored food, particularly dried fruits.

DOI: 10.1079/9781786390912.0056

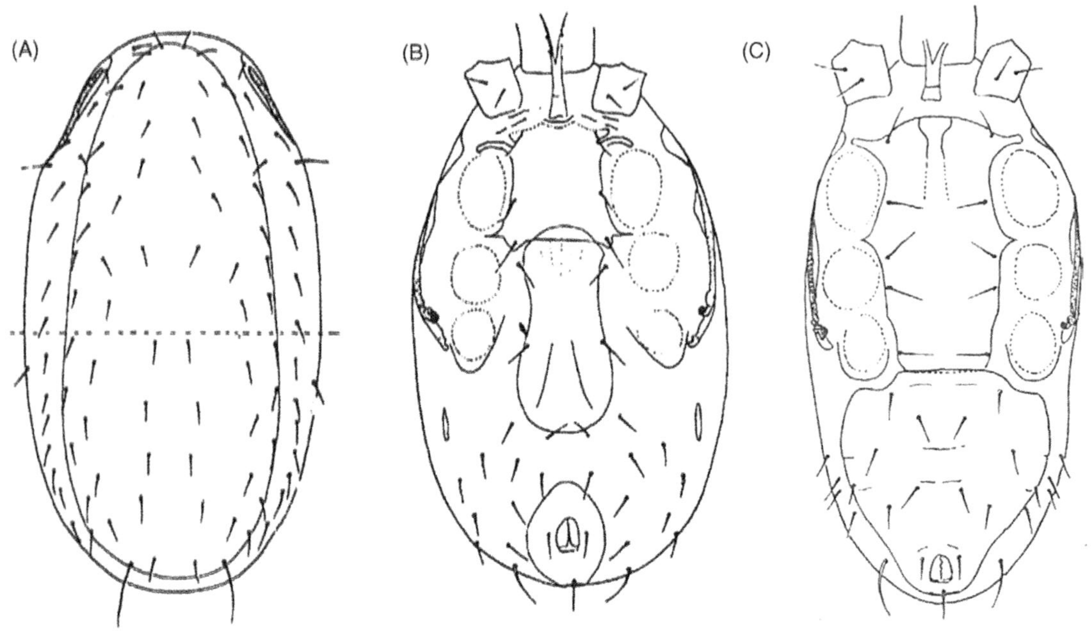

Fig. 56.1. *Melichares agilis* Hering. Female. Dorsal view (A); ventral view (B). Male. Ventral view (C) (from Hughes, 1976, licensed under the Open Government Licence v3.0).

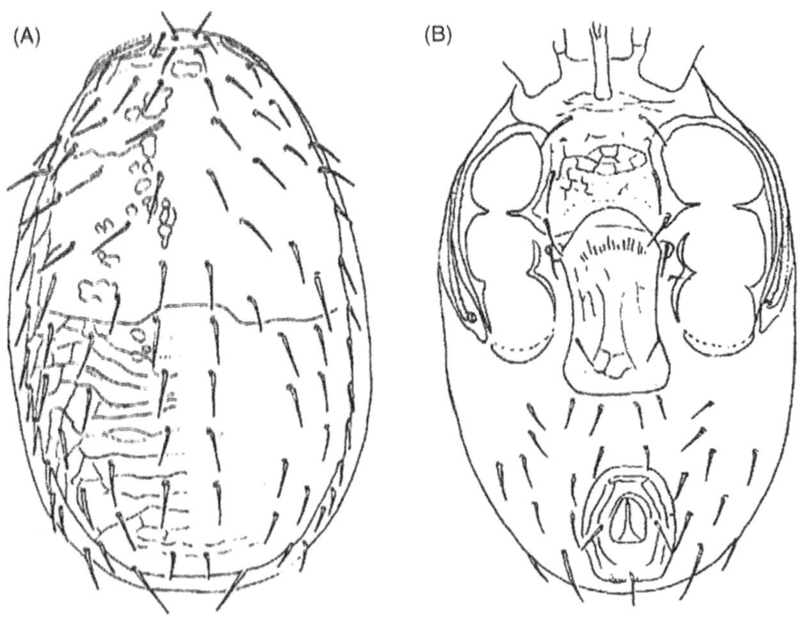

Fig. 56.2. *Proctolaelaps pomorum* (Oudemans). Female. Dorsal view (A); ventral view (B) (from Hughes, 1976, licensed under the Open Government Licence v3.0).

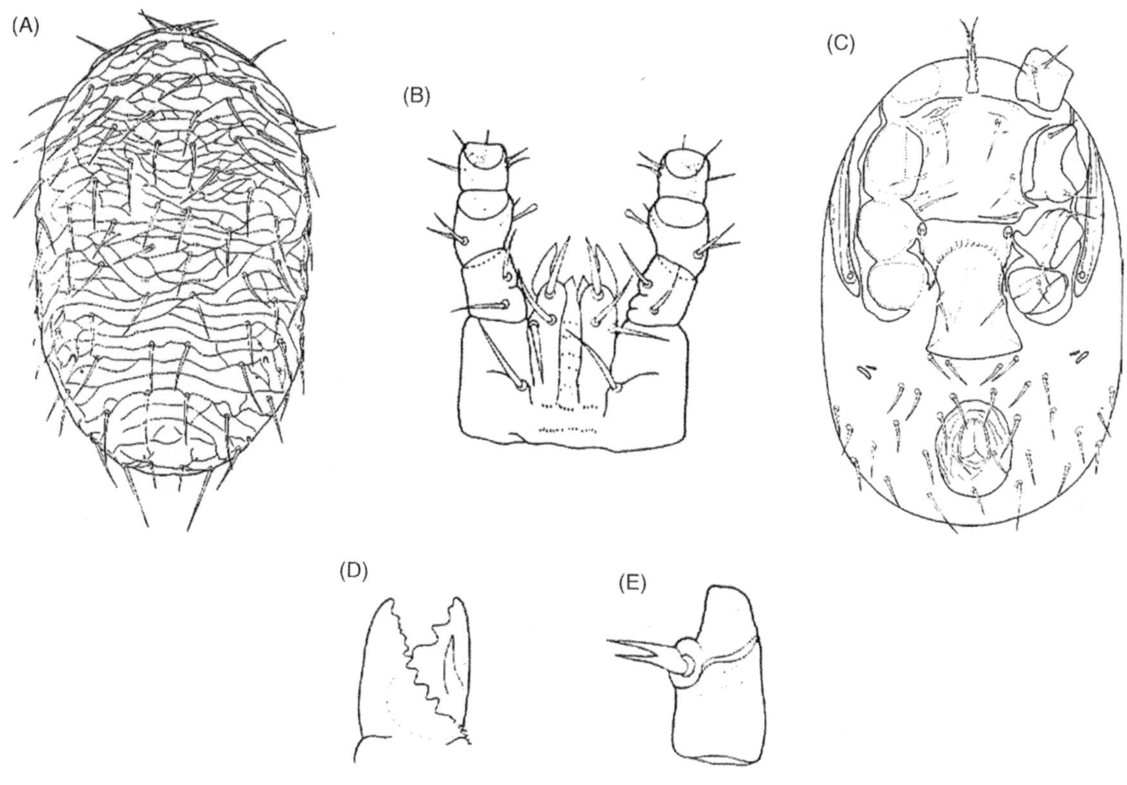

Fig. 56.3. *Proctolaelaps pygmaeus* (Müller). Female. Dorsal view (A); ventral view of gnathosoma (B); ventral view (C); chelicera (D); tibia and tarsus of palp partially drawn (E); E, tibia and tarsus of palp partially drawn (from Vacante and Nucifora, 1984-85).

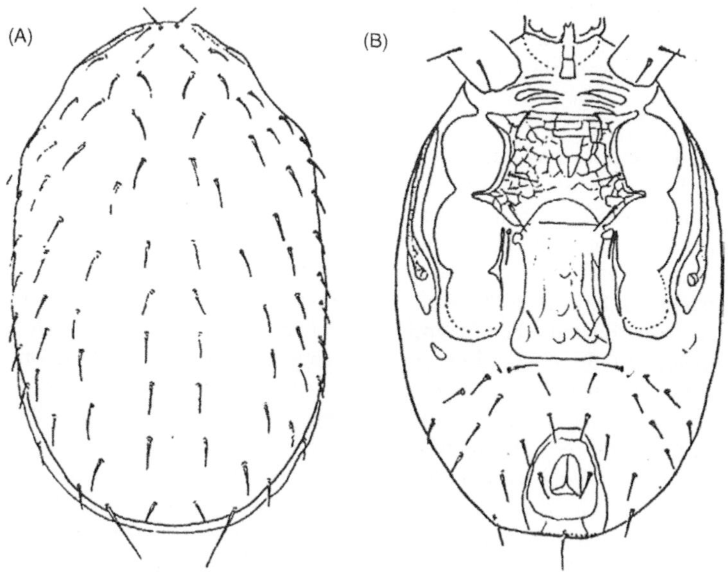

Fig. 56.4. *Proctolaelaps scolyti* Evans. Female. Dorsal view (A); ventral view (B) (from Hughes, 1976 licensed under the Open Government Licence v3.0).

Hughes (1976) reported that the mite takes 7–9 days at 23°C and an R.H. of 87% and *Acarus siro* Linnaeus as food on wheat germ. The eggs are large, and only one develops in the body of the female at the same time. The larvae prey on larvae and eggs of *A. siro* and are sucked dry of their contents. Although nymphs and adults will feed on the larvae and nymphs of *A. siro*, they prefer the eggs. Curiously, these are not sucked dry, but the predator pierces the eggshell, takes a little of its contents and then attacks the next. Hughes (1976) has never seen the adult stages of *A. siro* preyed by *M. agilis*. Infested commodities include dried fruit infested with *Carpoglyphus lactis* (Linnaeus) (Hughes, 1976), dried apricot (Rack, 1968a; Çobanoğlu, 2008), confection, dried fruit (Hering, 1838), grass seed, hog bladder, smoked beef (Oudemans, 1929a, b, quoted in Nesbitt, 1951), honey, red and white sugar (Tseng, 1982). Moreover, the mite has been collected from litter (Kandil, 1983), house dust (Oudemans, 1929a,b), moss (Gwiazdowicz, 2007), in bee hives (Haragsim *et al.*, 1987) and on *Corylus avellana* Linnaeus (Ozman and Žďárková, 2000).

P. bickleyi is known from United States (Moraes *et al.*, 2016; Silva *et al.*, 2020) and Iran (Ostovan and Kamali, 1994); in this last country the mite has been collected on honeybee hive, stored onion and potato contaminated with *Schwiebea* spp. and *Rhizoglyphus* spp. (Sepasgozarian, 1978; Ostovan, 1993; Mirfakhraii, 1994; Modarres Awal, 1994, 1997; Khanjani, 1996; Mossadegh, 1997; Ostovan and Mossadegh, 1999; Khaleghabadian *et al.*, 2015). Flechtmann (1976) reported this species in sugarcane borer laboratory rearings, where it feeds on pollen and fungi, in Brazil.

P. hystrix was originally described as *Lasioseius hystrix* Vitzhum (Vitzthum, 1923) and can be identified following Evans (1958). The mite was described from Austria from the galleries of the Coleoptera Curculionidae *Dendroctonus micans* (Kugelann) and *Hylastes ater* Paykull (Vitzthum, 1923) and is reported from Europe (Fenďa and Mašán, 2003; Salmane, 2005; Gwiazdowicz, 2008), North and Central America (Lindquist and Hunter, 1965; Chaires-Grijalva *et al.*, 2013) and Australia (Halliday *et al.*, 1998); recently the mite has been reported from Asian Russia from galleries of *Dendroctonus micans* under bark of Scots pine (*Pinus silvestris* Linnaeus) (Trach and Khaustov, 2017). On stored food, the mite is reported from Australia on decaying wheat residue (Hagstrum *et al.*, 2013).

P. pomorum was originally described as *Typhlodromus pomorum* Oudemans (Oudemans, 1929a) and *Garmania pomorum* (Oudemans) (Nesbitt, 1951). The systematic history of this species has been widely discussed by Moraes *et al.* (2016). The mite is reported from Australia (Womersley, 1956; Halliday, 1998; Halliday *et al.*, 1998), Brazil (Horn *et al.*, 2018), Canada (Sinha, 1964b), England (Evans, 1958; Hughes, 1961, 1976), Germany (Rack, 1968b), the Netherlands (Oudemans, 1929a), Slovakia (Fend'a, 2010) and Turkey (Çobanoğlu and Bayram, 1998; Gültekin and Özkan, 1999; Bayram and Çobanoğlu, 2005; Çakmak *et al.*, 2011). The commodities include esparto grass, grain, oat (Evans, 1958; Rack, 1968b; Hughes, 1976), vetch and wheat (Gültekin and Özkan, 1999).

Oudemans (1929a) found *P. pomorum* on the fruit and bark of an apple tree. Evans (1958) collected it from a house, where it was associated with *A. siro* and *P. pygmaeus*. According to Hughes (1961, 1976), *P. pomorum* was collected by Thompson on home-grown oats with *A. siro*, and by Smith on bales of mouldy Algerian esparto grass, heavily infested with *A. siro* and *Tyrophagus* sp. In Turkey, the mite has been collected also on mushroom (Çobanoğlu and Bayram, 1998), *Ranunculus damascena* (Linnaeus) (Çakmak *et al.*, 2011) and *Ornithagalum oligophyllum* Clarke (Bayram and Çobanoğlu, 2015).

P. pygmaeus was originally described as *Gamasus pygmaeus* Müller (1859), and successively as *Proctolaclaps hypudaei* (Oudemans, 1902b). *P. pygmaeus* has a cosmopolitan geographical distribution including North America and various countries of Palaearctic, Oriental, Afrotropic, Neotropic and Australian regions (Westerboer, 1963; Karg, 1971; Bregetova, 1977; Haragsim *et al.*, 1978; Zaher, 1986; Vacante and Nucifora, 1987; Casanueva *et al.*, 1994; Halliday *et al.*, 1998; Raut and Bhattacharya, 1999). The species may be identified following Evans (1958), Hughes (1976), Halliday *et al.* (1998) and Abo-Shnaf and Moraes (2016).

P. pygmaeus has been collected in a very wide range of habitats, including soil, stored food, bulbs, laboratory insect cultures, plant roots, decomposing plant material of all kinds and under bark (Chant, 1963; Ehara, 1964; Champ, 1966; McGraw and Farrier, 1969; Treat, 1975; Hughes, 1976; Domrow, 1979). It may be predatory on small arthropods, and also fungivore (Mathys and Tencalla, 1959; Walter and Lindquist, 1989), and unlike most Ascid mites, it is able to ingest solid matter. Many of the specimens examined by Halliday *et al.*

(1998) have solid fungal (conidia, arthrospores, hyphae) or yeast-like inclusions in their gut. There is circumstantial evidence that it can attack humans and that the results can be severe enough to cause extensive papular dermatosis (Andrews and Ramsay, 1982; Halliday *et al.*, 1998). Muma (1955) reared the mite on fungal cultures, but it is also able to feed *Tetranychus urticae* Koch, devouring entirely the prey and does not suck it dry. The female of *P. pygmaeus* lays as many 25 eggs, which hatch in 1–3 days, the larval stage lasts 2–4 days and the nymphal 3–8 days (Mathys and Tencalla, 1959). In laboratory studies, Muñoz Cárdenas (2017) observed that *P. pygmaeus* did not feed on the pre-pupae of the Western flower thrips, *Frankliniella occidentalis* (Pergande), but it feeds on the dried fruit mite, *Carpoglyphus lactis* (Linnaeus) immatures (3.5. ± 0.5 per day).

The mite is reported on stored food from various countries, such as Australia (Champ, 1966; Halliday *et al.*, 1998), Canada (Chant, 1963), China (Li and Fan, 1997), Czech Republic (Müller, 1859; Stejskal *et al.*, 2003), Egypt (Rezk, 2000; Yassin *et al.*, 2016, 2017), England (Hughes, 1961, 1976), Iraq (Mahmood, 1992; Souhaila, 1992), Iran (Ostovan and Kamali, 1994; Shamsi *et al.*, 2008), Japan (Ehara, 1964), Taiwan (Tseng, 1979), Turkey (Özer *et al.*, 1986, 1989) and United States (Chant, 1963) on different commodities such as barley (Hughes, 1976), mouldy barley, rotting bulb (Hughes, 1976), condiment (Li and Fan, 1997), dried dates (Rezk, 2000), dried cyclamen leaf, grain (Mahmmod, 1992; Souhaila, 1992; Ostovan and Kamali, 1994), hay (Yassin *et al.*, 2017), onion bulbs (Yassin *et al.*, 2016), rice, sweeping, mouldy timber (Hughes, 1976), wheat (Hughes, 1976; Özer *et al.*, 1986, 1989; Stejskal *et al.*, 2003) and mouldy wheat in farm. Tseng (1979) found *P. pygmaeus* in Taiwan on stored food, orange and grape. Moreover, the mite has been collected on a wide number of host plants, including *Rosa canina* Linnaeus, *Rosa damascena* Mill. (Çobanoğlu and Bayram, 1999) and *Viburnum opulus* Linnaeus steril (Caprifoliacae) in Turkey (Çobanoğlu *et al.*, 2003) and *Citrus* sp. (Muma, 1955; Ehara, 1964; Vacante and Nucifora, 1987); Ehara (1964) observed *P. pygmaeus* associated with the citrus rust mite, *Aculops pelekassi* Keifer and specimens from potato in laboratory were associated with the white peach scale, *Pseudaulacaspis pentagona* (Targioni). Ryke (1964) collected *P. pygmaeus* on *Protea mellifera* Thumb. flowers. Mathys and Tencalla (1959) found high density of population of the mite on leaves of orchard trees. Garrett and Haramoto (1967) collected the mite on soil surface debris under mixed forest, as well as Hughes (1976), which reported it on soil and leaf mould; moreover, *P. pygmaeus* is well-known from the nests of small mammals (Evans, 1958) and from a wide number of bird species. Hughes (1976) collected large number of *P. pygmaeus* in litter from a squirrel's drey. Rack (1968a, 1970) found the mite in a damp, newly constructed office building, and Brady (1970) from poultry litter in broiler houses. Abo-Shnaf and Moraes (2016) reported the mite from manure, soil, litter underneath mango, apricot, fig, cucumber, poinsettia, *Justicia adhatoda* L. (Acanthaceae), *Gardenia jasminoides* J. Ellis (Rubiaceae), *Acalypha marginata* (Mill.) (Euphorbiaceae), basil, rose, *Clerodendrum splendens* G. Don (Lamiaceae) and *Strelitzia* spp. Shamsi *et al.* (2008) found the mite in Iran on decaying leaves of walnut tree, soil mixed with weeds near an almond tree, soil mixed with weeds at the margin of a sike, decaying apple fruit fallen on the ground in an apple orchard, decaying trunk of a willow tree and on soil mixed with weeds near a walnut.

P. regalis is known from United States (Moraes *et al.*, 2016) and Iran, where it has been collected on stored onion and storage products (Ostovan, 1993; Ostovan and Kamali, 1994; Modarres Awal, 1997; Sayedi *et al.*, 2006).

P. scolyti Evans was collected on *Scolytus* sp. and the chateid *Entodon* sp. by Evans (1958) from England; moreover, Evans (1958) reported that the mite has been recorded from the nests of small mammals, feeding probably, on tyroglyphids in stored food products and in decaying vegetation in the soil. Hughes (1976) reported that the mite was found by Stevens on unsound onions imported from Egypt, on home-grown hay by Smith and in the deep litter of a broiler house in North Yorkshire by Connell, where it was associated with *Caloglyphus berlesei* (Michael). Çakmak *et al.* (2011) reported the mite associated with *Carpophilus hemipterus* (Linnaeus) (Coleoptera, Nitidulidae) on dried fig from Turkey.

P. ventrianalis is known from Central Europe (Moraes *et al.*, 2016) and Iran on storage products (Noei, 2007; Khaleghabadian *et al.*, 2015).

57 The Family Blattisociidae Garman (Figs. 57.1–57.5)

Blattisociid species have been described from a wide range of countries and habitats. A higher number of these mites is reported from China (47), followed by the United States (41), Ecuador (38), India (33) and Poland (32) (Santos *et al.*, 2018). The database produced by Moraes *et al.* (2016) provides information about the world distribution and habitats occupied by species of this family, which should be of interest to research works dealing with the taxonomy, biology and ecology of these groups, and even to applied biological control activities (Santos *et al.*, 2020b). At this aim, Zhang and Xie (2021) examined the lifespans of the potential biological control agents, in the family Blattisociidae. In this review, they summarized the duration of immature stages (egg to adult) and adult lifespans of the mites in the family, based on literature published from 1965 to 2021. We found 57 articles with information on the ecology and biology of the blattisociids. The lifespan and duration of immature stages were only reported in 16 out of 400 species. Most of the experiments were conducted in the laboratory, which may not accurately reflect their lifespans in nature. Temperatures, diets and pesticides are the main factors that affect the lifespan of blattisociids. Low temperature could significantly prolong the lifespan and duration of immature stages of blattisociids. Different prey species or stages also affect the development time and lifespan of blattisociids significantly.

According to Moraes *et al.* (2016), species of this family can be recognized by the podonotal and opisthonotal shields fused or not, the third pair of sternal lyrifissures (iv_3) off sternal shield (rarely on sternal shield in *Lasioseius*, lacking in the genera *Krantzoseius* and *Mucroseius* as well as in some species belonging to the genera *Aceodromus* and *Proctolaelaps*) are usually inserted on metasternal plates together with sternal seta st_4, when the metasternal plates are present. The peritrematic shield is broadly connected to exopodal shield, and curving behind the coxa IV. The fixed cheliceral digit with setiform *pilus dentilis*, and the movable cheliceral digit is often tridentate and without pointed ventral process (mucro). The spermathecal apparatus is phytoseiid-type (not typically so in *Zercoseius*). The genital shield is usually truncated posteriorly. The ventrianal shield usually bears 2–7 pairs of setae in addition to the circumanal setae (Santos *et al.*, 2020b).

On stored food, nine species are reported; of these five belong to the genus *Blattisocius* Keegan–*Blattisocius dentriticus* (Berlese) (Fig. 57.1), *Blattisocius keegani* (Fox) (Fig. 57.2), *Blattisocius mali* Oudemans (Fig. 57.3), *Blattisocius quadridentatus* Haines and *Blattisocius tarsalis* (Berlese) (Fig. 57.4) respectively; four are ascribed to the genus *Lasioseius* Berlese – *Lasioseius allii* Chant, *Lasioseius berlesei* (Oudemans), *Lasioseius penicilliger* Berlese (Fig. 57.5) and *Lasioseius sugawarai* Ehara, respectively. The genus *Blattisocius* includes almost 15 mite species, often found in storage facilities (Hughes, 1976; Haines, 1978; Gerson *et al.*, 2003; Britto *et al.*, 2012). A synthesis of the studies evaluating the possible use of *Blattisocius* species for the control of different insect and mite species was presented by Thomas *et al.* (2011).

Most of the studies refer to *B. dentriticus*, *B. keegani* and *B. tarsalis* (Esteca *et al.*, 2014). In particular, promising results have been reported for *B. tarsalis* in the control of *Ephestia cautella* (Walker) (Pyralidae) (Haines, 1981) and for *B. keegani* in the control of some Coleoptera species (Thomas *et al.*, 2011). Barker (1967b) reported *B. keegani* feeding on eggs of species of the Coleoptera genera *Cryptolestes* (Cucujidae), *Tribolium* (Tenebrionidae), *Trogoderma* (Dermestidae) and *Oryzaephilus* (Silvanidae), all pests of stored grains, while Beavers *et al.* (1972) reported the same predator to reduce egg hatching of *Diaprepes abbreviatus* (Linnaeus) (Curculionidae), a pest of citrus in the United States and on the Caribbean islands. *B. dentriticus*, *B. keegani*,

The Handbook of Mites of Stored Products (V. Vacante and C.H.W. Flechtmann)
DOI: 10.1079/9781786390912.0057

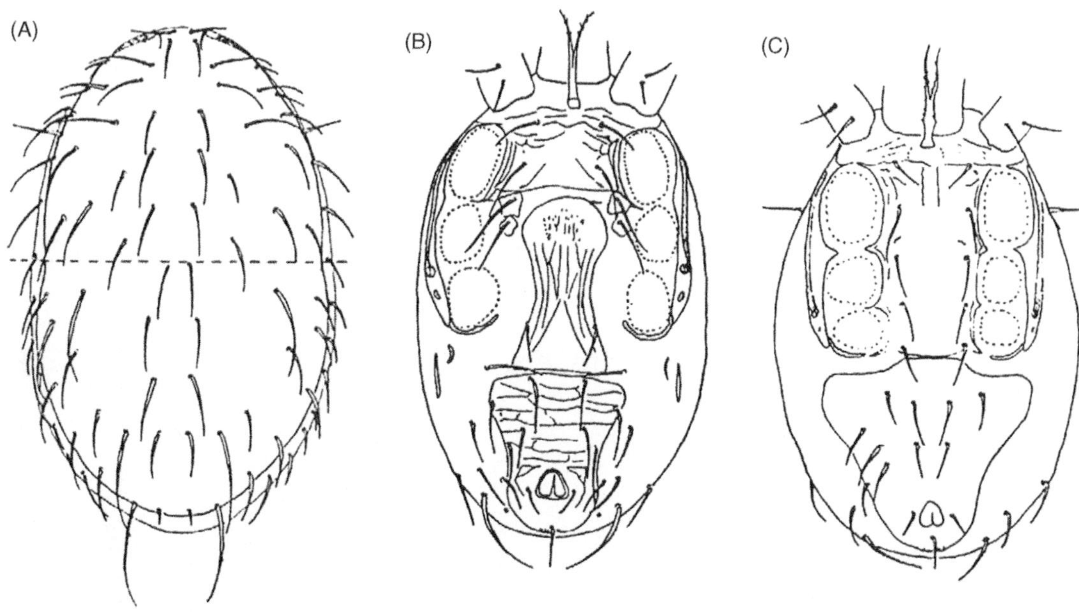

Fig. 57.1. *Blattisocius dentriticus* (Berlese). Female. Dorsal view (A); ventral view (B). Male. Ventral view (C) (from Hughes, 1976, licensed under the Open Government Licence v3.0).

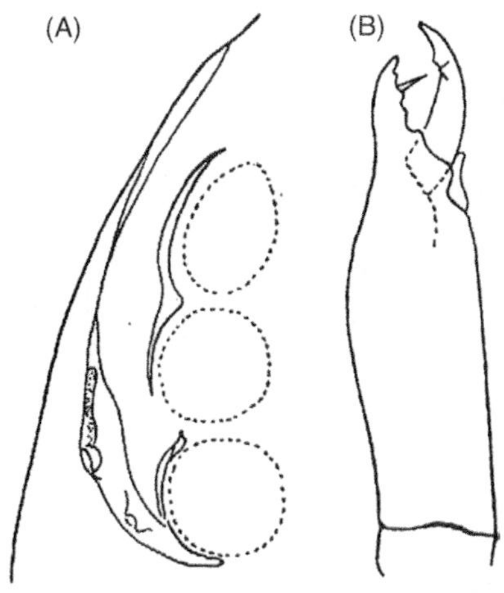

Fig. 57.2. *Blattisocius keegani* (Fox). Female. peritrema (A); chelicera (B) (from Hughes, 1976 licensed under the Open Government Licence v3.0).

B. mali and *B. tarsalis* have been collected in hives of honeybees or on adult bees. Species of *Blattisocius* are also predators in soil, stored food, the nests of small mammals and birds; ripe, dried, and rotting vegetables, on plants (e.g., roses, apple trees and *citrus* trees), and in insect cultures. They may also prey on acarid mites. *Blattisocius patagiorum* Treat is parasitic on adult noctuid moths (Klimov *et al.*, 2016).

B. dentriticus was originally described by Berlese (1918) as *Lasioseius dentriticus* Berlese and retained by Evans (1958) a synonym of *Melichares dentriticus* (Berlese). The species may be identified following Chant (1963), Evans (1958), Hughes (1976), Smiley (1987) and Britto *et al.* (2012). It has a cosmopolitan geographical distribution including Australia (Halliday, 1998; Halliday *et al.*, 1998), Brazil (Treat, 1975; Rosa and Flechtmann, 1979; Flechtmann, 1986), Bulgaria (Deltshet *et al.*, 1998), China (Pan and Teng, 1980; Pan, 1985), England (Hughes, 1961, 1976), Egypt (Zaher, 1986; Atwa *et al.*, 2018; Yassin *et al.*, 2018b), Germany (Rack, 1968a), India (Nanja and Channavasavanna, 1989b), Indonesia (Oudemans, 1925), Iran (Sepasgozarian, 1978; Kamali *et al.*, 2001; Kazemi and Rajaei, 2013), Ireland (Halbert, 1923), Israel (Costa, 1966a), Italy (Berlese, 1918), Japan (Ehara, 1961, 1964), the Netherlands (Oudemans, 1925; Siepel *et al.*, 2018), Peru (Flechtmann, 1980), Poland (Gwiazdowicz, 2007), Slovakia (Fenďa, 2010), South Africa (Grouth and Ueckermann, 1999), South Korea (Kaczmarek and Lee, 2000) and United States (Chant, 1963; McGraw and Farrier, 1969; Muma, 1975; Treat, 1975; Smiley, 1987).

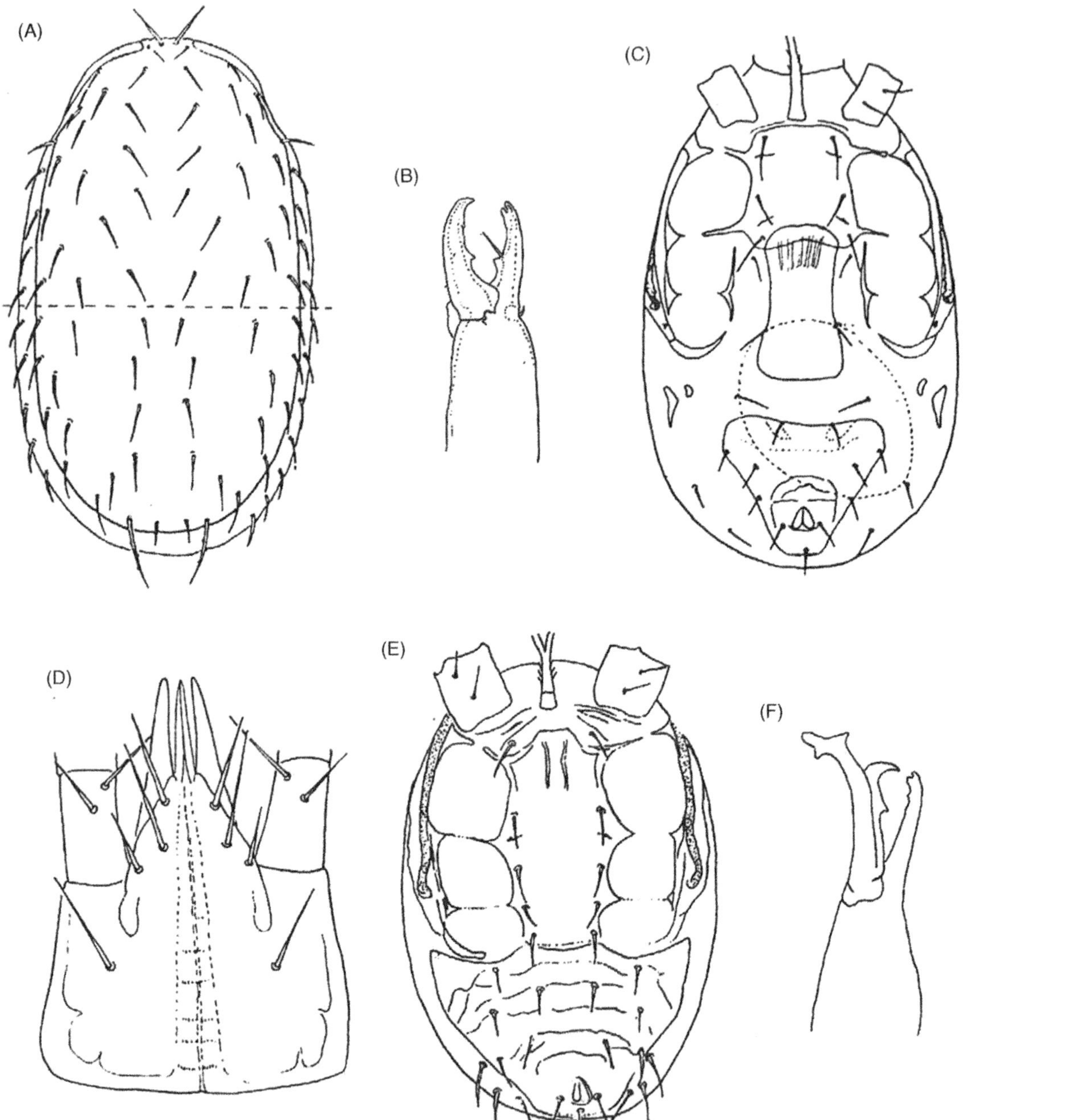

Fig. 57.3. *Blattisocius mali* Oudemans. Female. Dorsal view (A); chelicera (B); ventral view (C); ventral view of gnathosoma (D). Male. Ventral view (E); chelicera (F) (from Hughes, 1976, licensed under the Open Government Licence v3.0).

Various authors reported the mite on stored food, such as Rack (1968a), Hughes (1976), Sepasgozarian (1978), Flechtmann (1986), Zaher (1986), Baggio *et al.* (1987), Nangia and Channavasavanna (1989b), Modarres Awal (1994, 1997) and Britto *et al.* (2012). Commodities include bean (Metwally *et al.*, 2016; Yassin *et al.*, 2018b), dried fruit (Rack, 1968a; Atwa *et al.*, 2018), dried pork (Pan, 1985),

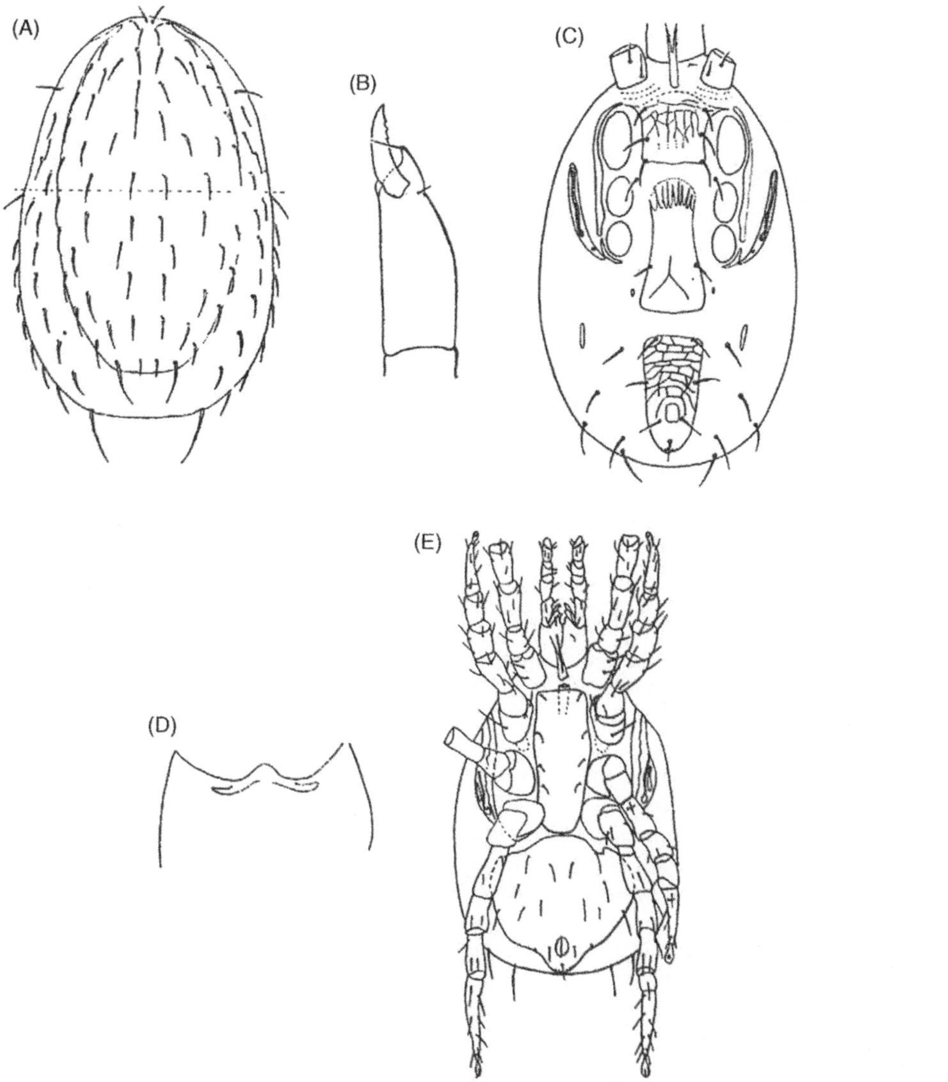

Fig. 57.4. *Blattisocius tarsalis* (Berlese). Female. Dorsal view (A); chelicera (B); ventral view (C); tectum (D). Male. Ventral view (E) (from Hughes, 1976, licensed under the Open Government Licence v3.0).

hay barley (Metwally *et al.*, 2016), grain wheat, wheat flour, stored grain, stored cereals (Chant, 1963; Baggio *et al.*, 1987; Palyvos *et al.*, 2008; Metwally *et al.*, 2016; Atwa *et al.*, 2018), stored garlic, cassava meal (Flechtmann and Zem, 2002), maize (Metwally *et al.*, 2016; Atwa *et al.*, 2018; Yassin *et al.*, 2018b), meat product (Pan, 1985), mouldy coconuts (Powell, in Hughes, 1976), rice (Metwally *et al.*, 2016), sprouting potato (Halbert, 1923), rice bran (Javadpour *et al.*, 2018), stored garlic (Baggio *et al.*, 1987), stored onion (Mohamed, 2013; Atwa *et al.*, 2018), stored soybean (Britto *et al.*, 2012), all found in farm facilities. Moreover, Hagstrum *et al.* (2013) reported on aquatic product, condiments, dried vegetables, maize, mushrooms, pollard, processed food and tea. In Iran, it has been found on potato, zucchini, flour and on the of insectariums (Granpayeh and Ostovan, 2014), and associated with honeybee hives (Ostovan and Mossadegh, 1999; Kamali *et al.*, 2001). Rack (1968a) found the mite in Germany, on dried apple slices from Greece, associated with *Tyrophagus*

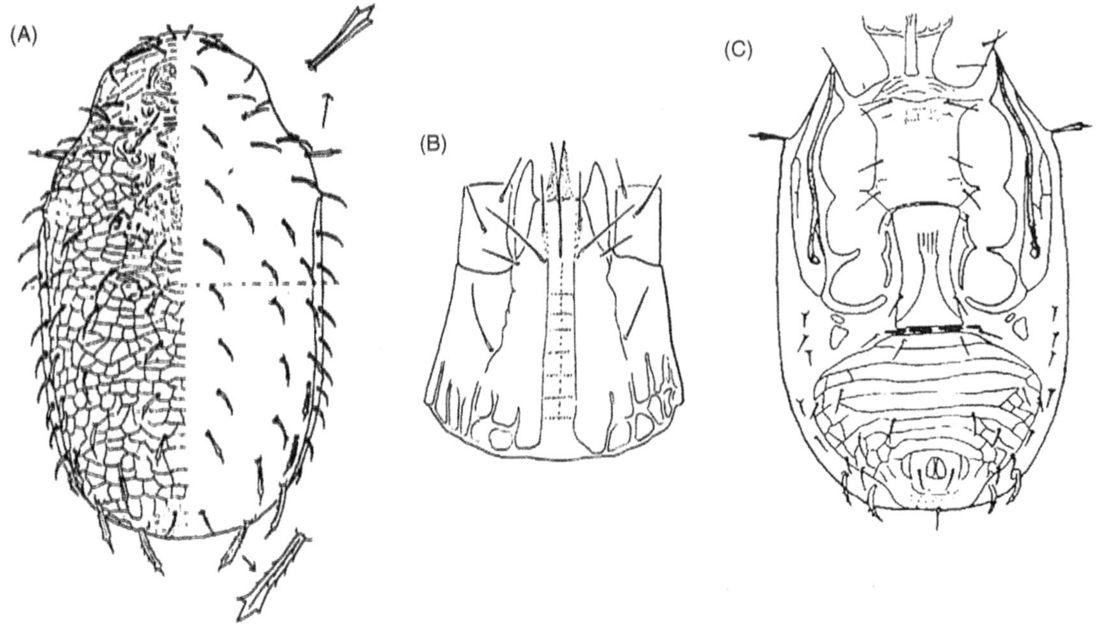

Fig. 57.5. *Lasioseius penicilliger* Berlese. Female. Dorsal view (A); ventral view of gnathosoma (B); ventral view (C) (from Hughes, 1976, licensed under the Open Government Licence v3.0).

longior (Gervais), in a new apartment associated with *Tyrophagus putrescentiae* (Schrank), *Tydeus interruptus* Thor and *Haemogamasus pontiger* (Berlese). Flechtmann (1980) found the mite associated with *T. putrescentiae* in Peru. Moreover, *B. dentriticus* has been found on nests of various bird species; in Italy, Berlese (1918) collected the mite in a pigeon nest. Evans (1958) collected the mite on the thorax of *Caradrina morpheus* (Hufnagel) (Noctuidae). Halliday *et al.* (1998) reported that *B. dentriticus* is a widespread predatory species found in stored food and in bark beetle galleries (Hughes, 1961, 1976; McGraw and Farrier, 1969). In Iran the mite was collected from a colony of *Ephestia* sp. (Pyralidae) (Javadpour *et al.*, 2018). In Australia, the mite has been observed preying on the lychee erinose mite *Aceria litchii* (Keifer) (Eriophyidae) in Queensland (Waite and Gerson, 1994). Siepel *et al.* (2018) reported the mite predating on eggs of *Archips podana* Scopoli (Tortricidae), in soil under carrot treated with DDT, in litterbag with wheat straw, from stem apple tree, from *Calluna vulgaris* (L.) Hull, in mushroom manure, soil under summer wheat after white clover, and from *Sorbus* sp., in col. Rivard (1960) studied the biology of *B. dentriticus* and found that the mite feeds on *T. putrescentiae* and must eat an average of 70 eggs, about 50 young larvae or a small number of protonymphs to complete its development. At 22.2°C and 80% R.H., the male and the female develop in nine and 10 days, respectively, with larvae of *T. putrescentiae* as food. Only the mated females laid eggs. It cannot develop to the adult stage, when mould is the only source of food. Moreover, the mite was found in honeybee hives (*Apis mellifera* Linnaeus) in Poland (Banaszak, 1980).

Mohamed (2013) reared *B. dentriticus* at 30+2°C and 80+5% R.H. on four different foods (the acarid astigmatid mite, *Rhizoglyphus robini* (Claparede), two different fungi (*Fusarium moniliforme* Sheldon, *Botrytis allii* Munn) and larvae of *Musca domestica* Linnaeus, with the aim to evaluate the possibilities of using the gamasid *B. dentriticus* feeding on different foods. In this study, the lowest incubation period of *B. dentriticus* was noticed for females and males fed on immature of acarid bulb mite (1.3 and 0.95 days respectively), but the longest period was recorded when the female and male individuals fed on the fungus *F. moniliforme* (3.7 and 2.6 days respectively). The duration of life cycle for both sexes was affected, by the type of food employed. However, adult longevity of *B. dentriticus* was influenced by the kind of food employed; for female, this period lasted 22.2, 36.8, 32.2 and 24.8 days when the individuals fed on aforementioned foods. These periods changed to (19.2, 30, 25.8 and 21.6 days, respectively) for males that fed on the tested foods, respectively. The study showed that immature of astigmatid mite *R. robini* proved to be the most favourable food, as it gave the highest reproduction rate (47.8 eggs). Conversely, the fungus *F. moniliforme* resulted in the least number of deposited *B. dentriticus* eggs, with an average of 26 eggs. During the lifespan of the predatory mite, the total number of the tested consumed prey was significantly differed. The number was (80.2 and 119 prey) when the predatory mite males and females fed on a mixture of acarid mite *R. robini*, respectively, while the number of consumed preys was (77, 113.4 prey) when predatory mite fed on *M. domestica* larvae, respectively. The mean generation time (T) was 10.35, 26.81, 20.4 and 14.85 days respectively, the doubling time (DT) was 2.34, 0.0753, 5.23 and 3.483 times, respectively, the finite rate of increase (λ) was 1.34, 1.09, 1.14 and 1.22 times/female/day, respectively, and the gross reproductive rate (GRR) was 29.58, 12.45, 18.5 and 22.43 times/female/day, respectively. It could be generally concluded that immatures of bulb mite *R. robini* were the most suitable food for the development and reproduction of predator gamasid mite, *B. dentriticus* (Mohamed, 2013). Silva *et al.* (2016b) observed the development time, reproduction, survival and sex ratio of this predatory mite feeding on *T. putrescentiae* and *Megninia ginglymura* (Mégnin), under laboratory conditions, at a temperature of 25 ± 1°C and 80 ± 5% R.H., in the dark. The length of the preoviposition period was significantly longer when *M. ginglymura* was used as food compared to *T. putrescentiae*. The life-table parameters for *B. dentriticus* differed on the two diets. The results showed that the population of *B. dentriticus* fed on *T. putrescentiae* increased about 7.53 times (Ro = 7.53) every 14.3 days (T = 14.3), corresponding to a daily population growth of about 15% (λ = 1.15) and a production of 0.14 female per female, per day (rm = 0.14). *B. dentriticus* fed on *M. ginglymura* showed lower values (Ro = 2.79; T = 23.76; λ = 1.04; rm = 0.04). The gross reproductive rate (GRR) and the proportion of female offspring were higher on *T. putrescentiae* (GRR = 10.73; ratio of females to males = 0.87). The research highlighted that *B. dentriticus* is a potential predator, developing and reproducing successfully on the two mites *T. putrescentiae* and *M. ginglymura* and *B. dentriticus* performed better against *T. putrescentiae* (Silva *et al.*, 2016b).

B. keegani was described by Fox (1947) as *Melichares keegani* from *Rattus norvegicus* (Berkenhout) in Puerto Rico, and successively found by Day (Hughes, 1976) on baby rats on a ship coming from South America. The mite may be identified following Evans (1958), Chant (1963), Hughes (l.c.), Haines (1978) and Britto *et al.* (2012). Its geographical distribution is cosmopolitan, and includes Australia (Domrow, 1979; Halliday *et al.*, 1998), Brazil (Flechtmann, 1968; Britto *et al.*, 2012; Esteca *et al.*, 2014; Horn *et al.*, 2018), Canada (Chant, 1963; Sinha, 1964b), China (Pan, 1985; Zhang and Fan, 2010), Cuba (Pan and Teng, 1980), Czechia (Stejskal *et al.*, 2003; Hubert *et al.*, 2006a), Egypt (Zaher, 1986; Al-Badry *et al.*, 1980; Saleh *et al.*, 1985; Rezk, 2000; Taha *et al.*, 2007; Metwally *et al.*, 2016; Atwa *et al.*, 2018; Yassin *et al.*, 2018b), England (Hughes, 1976), Greece (Eliopoulos *et al.*, 2002; Athanassiou *et al.*, 2005; Palyvos *et al.*, 2008), Hawaii (Prasad, 1968), India (Nangia and Channa Basavanna, 1989b), Indonesia (Haines, 1995), Iran (Kamali *et al.*, 2001; Shamsi *et al.*, 2008; Kazemi and Rajaei, 2013; Javadpour *et al.*, 2018), Finland (Rack, 1968a), Iran (Modarres Awal, 1994, 1997), Ireland, Israel (Costa, 1966a), Japan (Kamo and Hatsushika, 1962; Treat, 1975), Malaya (Evans, 1958; Shenker, in Hughes, 1976), Mexico (Rack, 1968a), the Netherlands (Siepel *et al.*, 2018), Poland (Chmielewski, 1969b), Puerto Rico (Fox, 1947), Saudi Arabia (Al-Nasser, 2011), Senegal (Freeman, in Hughes, 1976), South America (Day, in Hughes, 1976), Turkey (Özer *et al.*, 1986, 1989; Ozman and Žďárková, 2000; Çobanoğlu, 2008; Çakmak *et al.*, 2011), United States (Chant, 1963; McGraw and Farrier, 1969; Smiley, 1987) and other countries.

B. keegani has been reported on a number of commodities including dried apricot (Özer *et al.*, 1989; Çobanoğlu, 2008), barley (Metwally *et al.*, 2016), beet seed (Chmielewski, 1969b), broad bean (Yassin *et al.*, 2018b), cottonseed (Britto *et al.*, 2012), dog food (Britto *et al.*, 2012), dried fig (Atwa *et al.*, 2018), dried fishes (Flechtmann and Castelo, 1982), dried ham (Pan, 1985), grain (Barker, 1967c; Eliopoulos *et al.*, 2002; Stejskal *et al.*, 2003; Hubert *et al.*, 2006a), faba bean (Metwally *et al.*, 2016), flour (Eliopoulos *et al.*, 2002), decaying hay (Shamsi *et al.*, 2008), onion (Atwa *et al.*, 2018), stored hazelnuts (Ozman and Žďárková, 2000), stored hay (Yassin *et al.*, 2017), stored grain (Özer *et al.*, 1986, 1989; Haines, 1995; Britto *et al.*, 2012), plate maize

(Hughes, 1976), nutmeg (Shenker, in Hughes, 1976), rice bran (Javadpour *et al.*, 2018), salted meat (Pan, 1985), soybean (Yassin *et al.*, 2018b), stored rice (Flechtmann, 1968; Yassin *et al.*, 2018b), dried dates (Rezk, 2000), wheat, wheat bran and wheat flour (Sinha, 1964b; Barker, 1967c; Hughes, 1976; Saleh *et al.*, 1985; Al-Nasser, 2011; Atwa *et al.*, 2018; Yassin *et al.*, 2018b). Moreover, in literature are reported, among the stored food aquatic product, condiments, cotton, dried vegetables, feed mix, processed food, mungo, mushrooms, soybean and sunflower seed (Noei, 2007; Noei *et al.*, 2008b; Hagstrum *et al.*, 2013; Naderi *et al.*, 2016; Maroufpoor and Ostovan, 2017). Facilities include commercial dog food (Esteca *et al.*, 2014), commercial grain storage (Sinha, 1964b; Barker, 1967b), farm, flour mill (Freeman, in Hughes, 1976), provender warehouse (Javadpour *et al.*, 2018) and poultry farms (Horn *et al.*, 2018).

The mite has been found in a sparrow's nest in Scotland (Hoisie, in Hughes, 1976) and in Florida (USA) from *Citrus* trees by Muma (1955). Chant (1963) reported that this species has often been collected in North America. Cunliffe and Baker (1953) recorded it from citrus and *Neotoma* nest in Texas and from rose in Mexico. In Iran, it has been found on horse manure, rice plant remnants, apple leaves and sawdust (Javadpour *et al.*, 2018). Fenilli and Flechtmann (1990) reported the mite on *Araucaria angustifolia* (Bertol.) (Araucariaceae) branches. Britto *et al.* (2012) reported the mite on litter, in Brazil, and Siepel *et al.* (2018) reported it from the Netherlands in moss on roof tiles, soil samples and agricultural grassland on sandy soil.

Pan (1985) observed this species in association with *T. putrescentiae* and the two carabid *Necrobia rufipes* (Fabricius) and *Necrobia ruficollis* (Fabricius) in dried ham in storage in Shanghai (China). In laboratory rearing, he noted that the beetle population did not increase very much and he suspected that *B. keegani* fed on the eggs of the beetles. *B. keegani* has been reported to feed on eggs of several lepidopteran (Stein, 1960; Trivedi *et al.*, 1994; Thomas *et al.*, 2011) and coleopteran (Barker, 1967b; Beavers *et al.*, 1972; Hughes, 1976) species, as well as on astigmatid mites (Barker, 1967b; Hughes, 1976; Rezk, 2000; Taha *et al.*, 2007). In particular, all biological instars feed on the eggs of various species of grain beetles, such as *Cryptolestes turcicus* (Grouvelle), *Cryptolestes ferrugineus*, *Tribolium confusum* (du Val), *Tribolium castaneum* Hbst., *Trogoderma parabile* Beal and *Oryzaephilus surinamensis* Linnaeus, as well as on the eggs of the mites *Glycyphagus domesticus* (De Geer) and *Aëroglyphus robustus* (Banks). At 26.7°C, 70–75% of R.H. and unlimited supply of food, the life cycle takes 4.1 days (Barker, 1967a). Thomas *et al.* (2011) reported that the development from egg to adult at 25°C and 50–60% of R.H., with eggs of *Amyelois transitella* Walker (Lepidoptera: Pyralidae) as prey was 9.2 days. Esteca *et al.* (2014) observed that using as prey different organisms, at 25.0 ± 0.5°C, 80 ± 5% of R.H. and in the dark for a period of 10 days, the predation (number of specimens attacked) of *Thyreophagus* sp. was 145.9 ± 74.6, *Ephestia kuehniella* Zeller was 7.0 ± 4.4 and *Lasioderma serricorne* (Fabricius) was 8.0 ± 6.0, respectively. The oviposition (number of eggs laid) with *Thyreophagus* sp. was 4.0 ± 3.2 eggs, with *E. kuehniella* was 1.0 ± 1.3 eggs, with *L. serricorne* was 0.6 ± 1.3 eggs and with *Zabrotes subfasciatus* (Boheman) was 0.4 ± 0.8 eggs, respectively. The survivorship rates with *Thyreophagus* sp. was 8.7 ± 2.1, with *Z. subfasciatus* was 7.2 ± 3.4, with *E. kuehniella* was 8.6 ± 1.5 and with *L. serricorne* was 7.2 ± 2.6, respectively. In Iran, it has been collected from colony of *Ephestia* sp. (Javadpour *et al.*, 2018). Siegert *et al.* (2018) reported that the genus *Blattisocius* has been reported as predator of stored food pests, including insects and mites (Silva *et al.*, 2016b). *B. keegani* attacks beetles in stored products and shows potential for controlling the navel orangeworm *Amyelois transitella* Walker (Lepidoptera, Pyralidae) (Thomas *et al.*, 2011). It can feed on eggs from insects such as *Cryptolestes*, *Tribolium*, *Trogoderma*, *Oryzaephilus* and mites *G. domesticus* and *Aeroglyphus* sp. (Sinha *et al.*, 1969a). It is considered a natural enemy of storage mites, such as *T. putrescentiae*, and it is also found on rice, soybean meal, poultry feed and corn (Sousa *et al.*, 2005; Galvão *et al.*, 2011).

B. mali was originally described by Oudemans (1929a) as *Typhlodromus mali* and may be identified following various authors (Evans, 1958; Hughes, 1976; Haines, 1978; Smiley, 1987; Britto *et al.*, 2012). Its geographical distribution includes, China (Li and Fan, 1997), Egypt (Saleh *et al.*, 1985; Atwa *et al.*, 2018; Amal *et al.*, 2020), England (Hughes, 1976), Greece (Palyvos *et al.*, 2008), Hungary (Kerezsi *et al.*, 2019), India (Narayanan and Ghai, 1961), Iran (Sepasgosarian, 1978; Kazemi and Rajaei, 2013), Ireland (Türk and Türk, 1952), the Netherlands (Oudemans, 1929a), Taiwan (Tseng, 1982) and Turkey (Özar *et al.*, 1986; Özer *et al.*, 1989; Çobanoğlu, 2008).

The commodities include dried apricot (Çobanoğlu, 2008) associated with *Carpoglyphus lactis* (Linnaeus) (Aitkin, in Hughes, 1976), bean (Saleh *et al.*, 1985), dried currant associated with tyroglyphids (Turk and Turk, 1952), dried fruit (fig) (Özar *et al.*, 1986; Özer *et al.*, 1989), mouldy grain associated with *Acarus siro* Linnaeus (Norris, in Hughes, 1976), granular chicken feed (Atwa *et al.*, 2018), onion (Atwa *et al.*, 2018), processed food, stored raisin associated with *C. lactis* (Crozier, in Hughes, 1976), Burmese rice associated with *T. putrescentiae* (Schrank) (Jocelyn, in Hughes, 1976) and Turkish cheese (Atwa *et al.*, 2018). The facilities include farm, flour mill with *Tyrophagus* sp., off-farm storage (Thompson, in Hughes, 1976; Palyvos *et al.*, 2008).

Oudemans (1929a) took the type specimens on the bark of *Pyrus malus* Linnaeus and in a hole in an apple at Amsterdam and Arnhem respectively. Narayanan and Ghai (1961) collected the mite in a culture of the fruit flies *Dacus cucurbitae* (Coquillett) and *Dacus dorsalis* (Hendel) in India. It's probably a predator of acarid mites and their eggs (Hughes, 1976). Kerezsi *et al.* (2019) found the mite associated with the drosophilid fly *Phortica semivirgo* (Máca) in Hungary.

The biology of *B. mali* has been investigated at 25 ± 1°C and 75 ± 5% R.H. on the immatures of the Astigmatid mite, *R. robini* and the free-living nematodes *Rhabditis sccanica* Allgen. The mean incubation period was 1.85 and 1.55 days when fed on immature stages of *Rhizoglyphus robini* and 1.84 and 1.54 days when fed on *R. sccanica* for females and males, respectively. The duration of the life cycle for both sexes of *B. mali* was significantly affected by the kind of food used, and took 8.68 and 7.73 days, when fed on *R. sccanica* of females and males, However, this duration lasted longest times for *B. mali*, averaged 9.89 and 8.91 days when fed on *Rhizoglyphus robini*, female and male individuals, respectively. Adult longevity of *B. mali* was influenced by the type of food employed, this period lasted 27.87 and 21.8 days on *R. sccanica*. This period was 29.46 and 24.22 days on *R. robini* female and male individuals, respectively. The study showed that the free-living nematode proved to be the most favourable diet, as it gave the highest fecundity rate (86.5 eggs). On the contrary, the least number of eggs were deposited by *B. mali* during feeding on immature of *R. robini* (37.7 eggs). During the development time of lifespan of *B. mali*, the total number of consumed prey significantly differed. It was 138.7 and 90.4 preys when the *B. mali* males and females fed on *R. robini*, respectively (Amal *et al.*, 2020).

Gallego *et al.* (2020a) studied *B. mali* as a potential biological control agent of potato tuber moth *Phthorimaea operculella* (Gelechiidae). A prey acceptance bioassay of Potato tuber moth eggs was carried out. Additionally, two bioassays have been conducted under microcosm conditions, which assessed the densities of mite releases, at two levels of potato tuber moth infestation. The results showed that *B. mali* female adults accept potato tuber moth eggs as prey, and they cause a mortality rate 89.63 ± 2.47%, 48 h later. In addition to this, under microcosm conditions with potato tubers, it has been observed at low infestation level that the effectiveness of the predatory mite varied from 72.50 ± 28.50 to 100%, 28 days later, according to the release rate of mites. Under high infestation level, the effectiveness of biological control of the pest varied from 53.36 ± 25.55 to 92.51 ± 7.88%, also according to the release rate of the mites, and the possible use of biological control with *B. mali* of Potato tuber moth in different types of potato storage was analysed and discussed (Gallego *et al.*, 2020a).

Specimens of *Brachionus quadridentatus Haines* (Rotifera) were collected by Ayasingh from a chocolate factory about 30 km north of Kingston, Jamaica. They were found inside the aluminium foil wrapping, of a sample bar of confectionery chocolate. The chocolate sample had been imported to Jamaica about two years previously (country of origin not recorded) and was infested with the stored-product beetle *Ahasverus advena* Waltl (Silvanidae). It seems probable, by comparison with the feeding habits of other *Blattisocius* species that the mites were preying on the beetle eggs and young larvae (Haines, 1978).

B. tarsalis was originally described as *Lasioseius tarsalis* by Berlese (1918) and may be identified following Evans (1958) Chant (1963), Hughes (1976), Haines (1978), Smiley (1987) and Britto *et al.* (2012). Its geographical distribution is cosmopolitan and includes many countries among which Algeria (Athias Henriot, 1959, 1961c), Australia (Womersley, 1956; Domrow, 1979; Halliday *et al.*, 1998), Brazil (Reis and Paschoal, 1968; Flechtmann, 1981; Britto *et al.*, 2012), Canada (Chant, 1963; Sinha, 1964b; McGraw and Farrier, 1969; Crozier, 1989), China (Pan and Teng, 1980; Li *et al.*, 1992a), Czech Republic (Hubert *et al.*, 2006a; Lukás *et al.*, 2007), Egypt (Metwally *et al.*, 2016; Atwa *et al.*, 2018; Yassin *et al.*, 2018b), England (Hughes, 1976;

Haines, 1978), Germany (Hase, 1933; Westerboer and Bernhard, 1963), Greece (Eliopoulos *et al.*, 2002; Palyvos *et al.*, 2008), Hawaii (Garrett and Haramoto, 1967), India (Gupta and Chetterjee, 1997; Gupta and Chatterjee, 2004), Iran (Kazemi and Rajaei, 2013; Javadpour *et al.*, 2018), Israel (Costa, 1966a), Italy (Berlese, 1918), the Netherlands (Oudemans, 1929a; Siepel *et al.*, 2018), Poland (Banaszak, 1980; Tryjanowski *et al.*, 2001), Saudi Arabia (Al-Nasser, 2011), Slovakia (Fenďa, 2010), Sweden (Sellnick, 1958), Switzerland (Schweizer, 1949, 1961), Russia (Bregetova, 1977), Turkey (Özar *et al.*, 1986; Çobanoğlu, 2008; Çakmak *et al.*, 2011) and United States (Chant, 1963; McGraw and Farrier, 1969). Other countries such as Ecuador, Indonesia, Ireland, Jamaica, Japan, Kenya, Mexico, Peru, Philippines, Singapore and Venezuela are reported by Hagstrum *et al.* (2013).

B. tarsalis is a widespread species reported on many commodities including barley (Metwally *et al.*, 2016), broad bean (Yassin *et al.*, 2018b), crushed rape seed (Domrow, 1979), grain (Chant, 1963; Hubert *et al.*, 2006a; Palyvos *et al.*, 2008; Metwally *et al.*, 2016), granular chicken feed (Atwa *et al.*, 2018), malformed mango (Gupta and Chetterjee, 1997), mushroom, rolled oat (Siepel *et al.*, 2018), pineapple infected by gummosis (Reis and Paschoal, 1968), rice (Yassin *et al.*, 2018b), rice bran (Javadpour *et al.*, 2018), onion (Atwa *et al.*, 2018), stored rice (Britto *et al.*, 2012; Javadpour *et al.*, 2018), stored sorghum (Domrow, 1979), Turkish cheese (Atwa *et al.*, 2018), wheat, wheat bran (Sinha, 1964b; Eliopoulos *et al.*, 2002; Gupta and Chatterjee, 2004; Britto *et al.*, 2012; Yassin *et al.*, 2018b), wheat flour (Eliopoulos *et al.*, 2002; Al-Nasser, 2011) and straw (Javadpour *et al.*, 2018). In Brazil *B. tarsalis* was also reported from commercial bird feed and coffee grains (Flechtmann and Zem, 2002), and in Turkey from apricot, fig, corn, wheat, barley, oat, bran, raisin, hazelnut, wheat and sunflower; in this last country *B. tarsalis* has been collected also on hazelnut (*Corylus avellana* Linnaeus), *Hyacinthus orientalis* Linnaeus, and *Pinus nigra* Arnold (Özar *et al.*, 1986; Özer *et al.*, 1986, 1989; Çobanoğlu, 1996; Özman and Çobanoğlu, 2001; Akyazı and Ecevit, 2003; Bayram and Çobanoğlu, 2005, 2007; Çobanoğlu, 2008). In Greece, the mite has been collected on legumes and dried fruits (Eliopoulos *et al.*, 2002). Moreover, Hagstrum *et al.* (2013) reported as commodities, aquatic product, buckwheat, buckwheat product, chocolate, condiment, corn soya meal, cotton seed, dried vegetable, ginger, grass seed, laying mash, legume, maize, lily bulb, groundnut, processed food, seed cotton, sugar, sugarbeet seed and sweeping. The mite has been found on Spanish muscatel raisins infested with *C.lactis* (Shenker, in Hughes, 1976).

In Poland, the mite has been collected in the red-backed shrike (*Lanius collurio* Linnaeus) (a bird) nests (Tryjanowski *et al.*, 2001), as well as in Slovakia (Fenďa, 2010). In the Netherlands, Siepel *et al.* (2018) reported the mite in soil sample grassland on sandy soil, in culture of predatory mites, in litter in henhouse, in soil under potato, under oats, in mushroom culture and in soil sample. Hagstrum *et al.* (2013) listed as facilities the empty grain bin, farm, flour mill, market, off-farm storage and ship.

B. tarsalis was found also in honeybee hives (*Apis mellifera* Linnaeus), in Canada (Crozier, 1989) and in Poland (Banaszak, 1980). Hughes (1976) reported that the mite is frequently found in stored grain infested with corn moths of the genera *Anagasta*, *Plodia* and *Sitotroga*. It lays its eggs on the silken webbing, and the larvae and nymphs appear to feed on the moth eggs and caterpillars. The females are carried by the moths hanging on to the body, particularly at the base of the wings. Nesbitt (1951) found the mite on cultures of *Tribolium confusum* Jacquelin du Val, *T. castaneum* Herbst (Tenebrionidae), *Attagenus piceus* (Fabricius) (Dermestidae) and *Tineola bisselliella* (Hummel) (Tineidae). Moreover, the mite has been collected with eggs of *Sitotroga cerealella* (Olivier) (Gelechiidae) on stored grains, with eggs of *Anticarsia* (Arctiidae) from a laboratory colony and with eggs of *Anagasta kuehniella* (= *Ephestia kuehniella*) (Pyralidae) on stored grains (Britto *et al.*, 2012); Flanders and Badgley (1963) observed that *B. tarsalis* can only control this pyralid under very special conditions (e.g. the depth of grain in which the moths lived could not exceed 8 mm), and the deviation of these conditions often eliminated the predators. Treat (1969) found that *B. tarsalis* may be associated with the noctuid moth *Epizeuxis aemula* (Hübner), which feeds on the dead leaves of the forest floor. *B. tarsalis* does not commonly infest stored grain, but on this occasion is injurious to corn fodder in Mississippi (Crumb, 1934). The moths are probably used only as a vehicle (Hughes, 1976). In Iran, it has been collected in a colony of *Plodia* sp. (Javadpour *et al.*, 2018). The mite has been found also in soil and litter (Mineiro *et al.*, 2009; Javadpour *et al.*, 2018). This widespread predator sometimes causes annoyance to workers handling insect-infested stored food products (Hughes, 1976).

Haines (1981) investigated the biology of *B. tarsalis* at 27°C and 73% R.H. On eggs of *E. cautella* (Walker), development took 5.9 days, and there was 7% mortality; on eggs of *T. castaneum*, development took 7–4 days, and there was 48% mortality. When offered a choice of eggs of the two pest species, the mites very rarely preyed on those of *T. castaneum*. The number of *E. cautella* eggs consumed was 38 (total) by nymphs, 24/day by ovipositing females, 15/day by non-ovipositing females, and 0.8/day by males. The sex ratio (female: male) was 2.5:1. The adult lifespan average was about 55 days (maximum 153 days), and the oviposition period lasted about 20 days. The females laid up to 5 eggs/day at peak oviposition. The potential rate of increase of *B. tarsalis* on *E. cautella* eggs was very high, far exceeding that of its prey. These data confirm previous postulates concerning the role of *B. tarsalis* in natural control of *E. cautella*, and support the hypothesis that fumigation reduces predator-effectiveness. Modification of control regimes might allow greater natural control by the mite and reduce the increasing pest status of the moth (Haines, 1981).

Studies carried out under the laboratory conditions at 30°C ± 2°C, and 89% ± 5% R.H., with newly deposited eggs of *T. castaneum* and *O. surinamensis* (Linnaeus), the most common insects associated with the predatory mite were used as food showed that predatory larva was a non-feeding stage. *B. tarsalis* females consumed a higher number of prey eggs than males did. Females reached maturity before males. Mites reared on *O. surinamensis* eggs developed faster than those reared on *T. castaneum* eggs. Adult longevity and lifespan of the predator were not affected by type of food. Adult fecundity was high when reared on *O. surinamensis* (Hafeez *et al.*, 1988).

B. tarsalis was studied with positive results in the Danish flour mills, as predatory of eggs of the moth *E. kuehniella*, and resulted suitable for biological control. In this study, the consumption rate of *B. tarsalis* on different age groups (eggs aged 0–1, 1–2, 2–3 and 3–4 days) of *E. kuehniella* eggs was investigated at 21°C and 75% R.H. There were no differences between the numbers destroyed ($p > 0.05$), but a larger amount of the oldest eggs was consumed, compared to eggs 0–1 days old ($p<0.05$) (Nielsen, 1998). The developmental time of the mite was investigated at temperatures of 15°C, 21°C and 25°C and 75% R.H. Eggs 1–3 days old of *E. kuehniella*, killed by freezing, were used as food. The mean developmental times were found to be 22.4, 8.5 and 7.0 days, respectively. Within the investigated thermal limits, the developmental rate showed a linear relationship with temperature, and the corresponding thermal threshold for development was calculated to 10.2°C (Nielsen, 2001).

Taking into account that residual populations of storage mites sheltering in crevices and cracks escape conventional control treatments and are implicated in the infestation of newly harvested grain, Thind and Ford (2006) carried out a series of 24 h laboratory tests to assess the performance of solitary adults of two predatory mite species, *Cheyletus eruditus* (Schrank) and *B. tarsalis*, for controlling small numbers of the flour mite *Acarus siro* (Linnaeus). Tests were carried out in the presence or absence of prey refuges or grain debris to afford shelter to the flour mites. While *C. eruditus* had a significant effect on the motile stages of *A. siro*, in contrast, *B. tarsalis* had a significant effect on the eggs. The maximum percentage of motile stages of *A. siro* eaten by *C. eruditus* was 82%, whereas the minimum percentage of *A. siro* eggs eaten by *B. tarsalis* was 99%. While the performance of *C. eruditus* in predating on motile stages of the flour mite was hindered by the presence of the prey refuge (38% eaten) and grain debris (25% eaten), the performance of *B. tarsalis* in predating on flour mite eggs was unaffected (100% eaten in presence of prey refuge or grain debris). In prolonged exposures (36 days), the performance of two, four or eight adult predators, either a single species or a combination of both, was assessed for their ability to control a population of the flour mite, developing up to F(2) from an initial inoculum of 80 females and 20 males allowed to oviposit for 72 h in the absence of predatory mites. The maximum reduction in prey population of 80% was achieved with eight *B. tarsalis*. Combining the two predatory species did not enhance the reduction of *A. siro* population (Thind and Ford, 2006).

Çobanoğlu *et al.* (2007) investigated the possibilities for the use of *B. tarsalis* as a biological control agent, to control Mediterranean flour moth *E. kuehniella* in Turkey. They investigated the effect of different temperatures (15 ± 1°C and 25 ± 1°C) on the development time of *B. tarsalis* at different food density levels (5, 10, 15 and 20 egg densities). At 15°C, the development times of immature stages were 5.53 (1–6), 4.86 (3–5), 4.53 (0–5) and 4.86 (1–6) days while the adults lifespan were 52.00

(3–67), 42.77 (5–67), 29.85 (1–66) and 27.36 (7–39) days for 5, 10, 15 and 20 moth egg densities. At 25°C, the development times of immature stages were 5.00 (5–5) days at 10 prey eggs and 6.00 (6–6) days at the rest of prey densities; besides this, lifespans of adults were determined 32.00 (20–39), 30.67 (15–40), 42.20 (10–64) and 38.67 (4–69) days, at each food densities consequently. Fecundity and the consumption capacity of predatory mites have been carried out for each temperature at different food densities. The duration of preovipositon, oviposition and postoviposition periods of *B. tarsalis* were also determined. The highest consumption of immature stages was determined (13.27 individuals/for development time) for 20 moth egg densities at 15°C, this value was observed at 10 moth egg densities for adults (172.8 individuals/duration of adult lifespan). The highest consumption of immature stages was determined (34.20 individuals/for development time) for 20 moth egg densities, while this value was observed at 20 moth egg densities for adults (555.1 individuals/duration of adult lifespan) at 25 °C (Çobanoğlu *et al.*, 2007).

Biological control of the Indianmeal moth, *Plodia interpunctella* (Hübner), with *B. tarsalis*, as well as CO_2-comprising modified atmospheres at atmospheric or high pressure, was evaluated in the laboratory in Spain. *B. tarsalis* was able to prey on *P. interpunctella* and to reduce its population. Dosages with modified atmospheres, needed for the control of *P. interpunctella*, were also effective for the control of *B. tarsalis* (Pons *et al.*, 2008).

Al-Nasser (2011) carried out a faunistic survey of mites, in different product stores such as bakeries, warehouses, mills, department stores and houses during a nine-month study period (October 2009 to June 2010), in Egypt. The survey was carried out on wheat flour as different samples 150 g each were collected, monthly from three regions; North, Middle and South. Five species of mites belonging to four families were identified, among which were *B. tarsalis* and *B. keegani*. The mite counts ranged from 0.15 to 20.44 mites/150 g wheat flour. The highest counts for the majority of mites species collected from the three regions were recorded during the months of November, December, January and February, according to the decreasing degrees of temperature and R.H. during these months. *B. tarsalis* and *B. keegani* were detected only in samples from the South. These results indicate the effect of temperature and R.H. on the distribution density of stored product mites (Al-Nasser, 2011).

The genus *Lasioseius* on stored products numbers only four species. The first species, *L. allii* Chant has been described by Chant (1958) on onion bulbs from South Africa, and onions from Portugal. The mite may be identified following Chant (1963). Its geographical distribution concerns Egypt (Zaher, 1986), Finland (Huhta *et al.*, 2010), Portugal, South Africa (Chant, 1958, 1963), Scotland (Evans, 1958), Taiwan (Tseng, 1971b, 1982) and the Netherlands (Siepel *et al.*, 2018). The commodities include garlic bulb, onion bulb, shallot (Chant, 1958, 1963; Tseng, 1971b, 1982), and *Hyacinthus* 'Pink Pearl' (bulb) and *Tulipa* (bulb) (Siepel *et al.*, 2018).

L. berlesei (Oudemans, 1938) may be identified following Chant (1963) and Evans (1958), and its geographical distribution includes Iraq (Mahmood, 1992), Italy (Berlese, 1887; Chant, 1963), the Netherlands (Siepel *et al.*, 2018) and United States (Colorado) (Walter and Lindquist, 1989). Mahmood (1992) collected the mite on rice. Evans (1958) reported that this species has been recorded from a number of localities on the Old Continent. He examined specimens collected in leafmould near a pond (coll. M. E. Bacchus) and under bark of a fallen elm tree (coll. A. H. G. Alston) from England. Chant (1963) studied specimens found on celery stalks from Italy. Siepel *et al.* (2018) reported the mite in soil 0–5 cm., soil samples agricultural grassland, abandoned agricultural land soil sample, harbour mud depot, soil sample grassland sewage farm, soil sample grassland sandy soil, soil sample arable land marine clay soil, soil sample grassland on sandy soil, on stem apple tree and under grass-clover in the Netherlands.

L. berlesei was collected from grassland soils in Colorado, known only from the female, and for the first time in an ascid mite, it was experimentally demonstrated that they reproduce by thelytokous parthenogenesis. The species is omnivorous that will feed on fungi and animal prey, and it shares a multidenticulate ridge opposed to a groove on the paraxial faces of the chelicerae, which appears to function as a mashing and filtering mechanism during mycophagy (Walter and Lindquist, 1989).

L. penicilliger (1916) may be identified following Evans (1958), Chant (1963) and Hughes (1976). Its geographical distribution includes England, Ireland (Evans, 1958; Hughes, 1976), Canada (Crozier, 1989), Czech Republic (Haragsim *et al.*, 1978), Germany (Rack, 1962), Italy (Berlese, 1916), New Zealand (Macfarlane, 2005), Scotland (Sheals, 1956) and United States (Hurlbutt, 1958). Commodities number only the damp barley (Baynes, in Hughes, 1976) reported from commercial grain storage, farm and mill (Burgess, in Hughes, 1976). Berlese (1916) found the mite in Italy in soil under rotten wood. Sheals (1956) found this species in old grassland at Glasgow. Rack (1962) collected the mite from a mole's nest. Walton (in Hughes, 1976) collected the mite from sample bags in a Liverpool office. A number of specimens were found together underside of some machinery in a mill; lacking other mites or insects, it is possible that *L. penicilliger* was feeding on moulds (Hughes, 1976).

Aguilar-Marcelino *et al.* (2014) studied in Mexico the predation by *L. penicilliger* on three nematode species – infective larval stages (L3) of *Haemonchus contortus* (Rudolphi) and adults of *Panagrellus redivivus* (Linnaeus) and *Rhabditis* sp. Experiments were carried out in 5.5-cm-diameter petri dishes containing 2% water-agar over a period of 5 days. Batches of up to 1500 third-stage larvae (L3) of *H. contortus*, and 1000 adult nematodes of *P. redivivus* and *Rhabditis* sp. were exposed to five mites, in separate petri dishes. Upon contact, each mite used its pedipalp and legs to identify and hold its prey and then used its chelicerae to feed upon the prey. Predation by *L. penicilliger* was chance dependent, but mites became aggregated around any injured/damaged prey, thereby suggesting some form of chemoperception. The rate of predation on the three species of nematodes was high, but L3 of *H. contortus* and adult *Rhabditis* sp. were preferred (Aguilar-Marcelino *et al.*, 2014).

García-Ortiz *et al.* (2015) evaluated the predatory behaviour *in vitro* of *L. penicilliger* on three nematode species: *Teladorsagia circumcincta* (Stadelman)(L_3) (a sheep-parasitic nematode), *Meloidogyne* sp. (J_2) (a plant-parasitic nematode), and on various developmental stages of *Caenorhabditis elegans* (Maupas) (a free-living nematode). The confrontation between mites and nematodes was individually assessed in 2% water-agar, placed in plastic petri dishes (2 cm x 1 cm diameter). One thousand nematodes of a species and five mites were placed into each plate (10 replicates) and incubated for 5 days at room temperature (18–25°C). *L. penicilliger* showed voracious feeding activity against the three assessed nematode species. The percentages of predatory activity recorded were 95.1, 80.5 and 79.3 against *Meloidogyne* sp., *C. elegans* and *T. circumcincta*, respectively ($P \leq 0.05$). These results suggest that *L. penicilliger* has important potential as a biological control agent of parasitic nematodes (García-Ortiz *et al.*, 2015).

L. sugawarai was described on pineapple fruit, from Taiwan (Ehara, 1964).

58 The Family Dermanyssidae Kolenati (Fig. 58.1)

Morphological Characteristics, Systematic and Bio-ecology

The family Dermanyssidae is a much more uniform group with all members obligate hematophagous ectoparasites of vertebrates (Lindquist *et al.*, 2009b). The family belongs to the superfamily Dermanyssoidea and includes two genera, *Dermanyssus* (Dugès) and *Liponyssoides* Hirst and 26 species (Beaulieu *et al.*, 2011). The chelicerae of female are whip-like, styliform, with digits minute and chelate. The corniculi are membranous and indistinct. The second cheliceral segment is long, far exceeding in length the basal cheliceral segment 1; conversely, the male has the chelicerae with the second segment normal in length. The idiosoma is broadly rounded posteriorly (Lindquist *et al.*, 2009b).

The most important species of this family is the red mite, *Dermanyssus gallinae* (De Geer), reported on stored food from grain, and from sheepskin rug (Hagstrum *et al.*, 2013). However, it does not cause any direct damage to stored food, and its economic importance is strictly associated to its adaptation as ectoparasite of birds. *D. gallinae* can be misidentified with dermanyssoid parasites such as the northern fowl mite, *Ornithonyssus sylviarum* (Canestrini and Fanzago, 1877), which may, at least in European countries, share the same host species and environment. Another species of this genus is the tropical fowl mite, *Ornithonyssus bursa* (Berlese), ectoparasite of poultry and layers, as well as wild birds and reported in Brazil associated both with rural and urban zones and found in wild birds, such as *Furnarius rufus* Gmelin and *Columbina picui* (Temminck) (Mentz *et al.*, 2015; Silva *et al.*, 2018). In addition, it has economic importance in poultry houses, mainly when there are high infestations, influencing egg development and laying, and might lead to the death of the bird (Flechtmann, 1985b; Mascarenhas *et al.*, 2009). Recently, Silva *et al.* (2023) published a comprehensive review on the control of the poultry hematophagous mites and integrated the discussion with information on their geographical distribution and their economic importance.

Other species of this family of medical and veterinary interest are the American bird mite, *Dermanyssus americanus* Ewing, *Dermanyssus hirundinis* (Hermann) and the house mouse mite, *Liponyssoides sanguineus* (Hirst). This last is a cosmopolitan species living as blood-sucking ectoparasite of synanthropic rodents, and vector of *Rickettsia akari* Huebner, the agent of the zoonotic rickettsialpox (Huebner *et al.*, 1946; Eremeeva and Muniz-Rodriguez, 2020; Szakacs *et al.*, 2020). The transmission of this *Rickettsia* in *L. sanguineus* is transovarial and transstadial (Eremeeva and Muniz-Rodriguez, 2020); Rickettsialpox is commonest in urban areas in crowded living quarters, infested with the house mouse, which is the major reservoir. Moreover, *L. sanguineus* has been reported to transmit the Tularemia once (Strandtmann and Wharton, 1958).

The Genus *Dermanyssus* Dugès

Diagnostic characteristics

According to Moss (1968) the species of the genus *Dermanyssus* have the dorsal shield devoid of the seta *j*3; the sternal shield, a median ventral sclerite set between legs II and III, is narrowed, distinctly wider than long, and bearing –one to two pairs of sternal setae; the tibia I *ad* (anterodorsal) and *pd* (posterodorsal) with two setae; the tibiae II–IV with one seta *al* (anterolateral); the genu IV ad (anterodorsal) and tibia IV *pd* (posterodorsal), each with two setae.

Within the genus *Dermanyssus* is possible to recognize two groups of species. The first group is characterized by a soft body adapted for sporadic and large engorgement, with reduced shielding and slender legs, while the second group possesses a compact and more heavily sclerotized body with shorter and stouter legs. Species belonging to the soft body type are common, and most of them are

 The Handbook of Mites of Stored Products (V. Vacante and C.H.W. Flechtmann)
DOI: 10.1079/9781786390912.0058

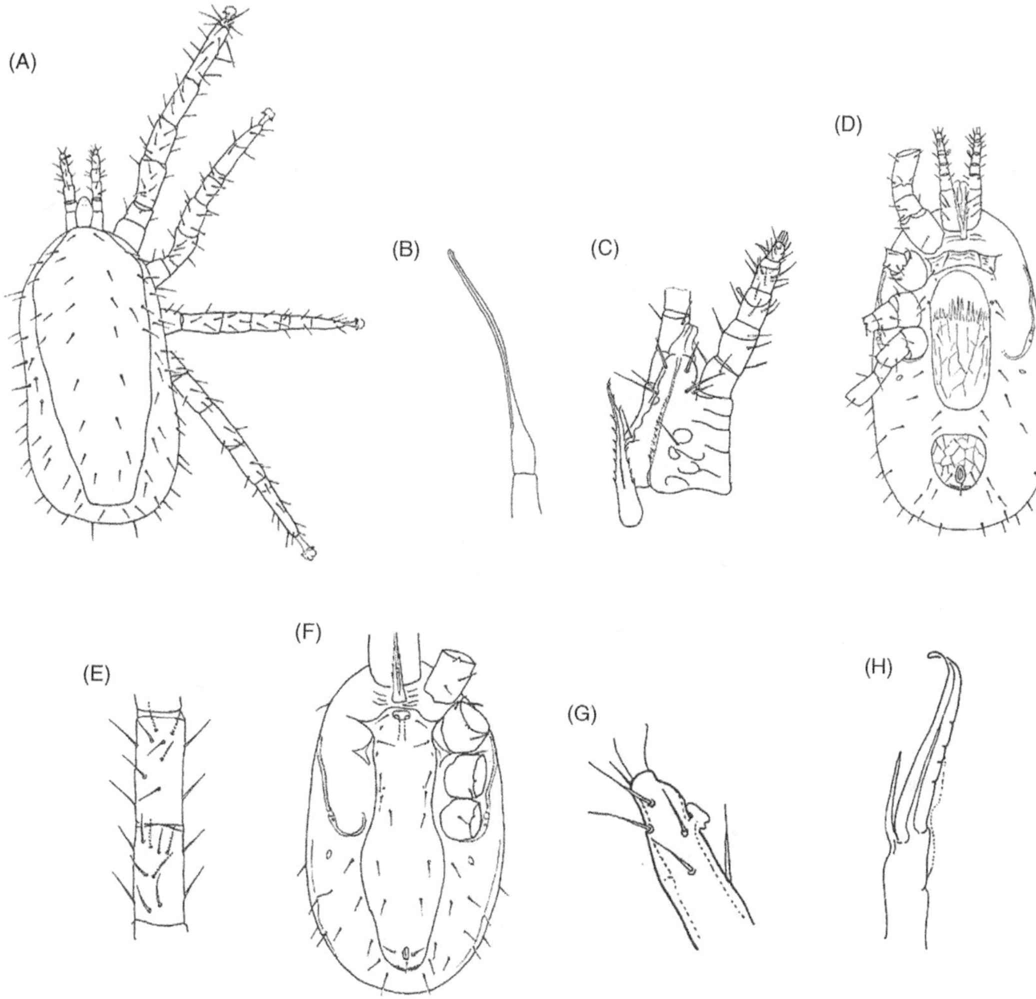

Fig. 58.1. *Dermanyssus gallinae* (De Geer). Female. Dorsal view (A); chelicera (B); ventral view of gnathosoma (C); ventral view (D); genu and tibia I (E). Male. Ventral view (F); distal part of tarsus IV (G); chelicera (H) (from Hughes, 1976, licensed under the Open Government Licence v3.0).

nearly indistinguishable from each other and constitute the gallinae group (Roy *et al.*, 2009a).

Economic importance

Within the family, the genera *Liponyssoides* Hirst and *Dermanyssus* Dugès play a veterinary and medical importance. *Dermanyssus* species are ectoparasites of birds. Most species do not exhibit host specificity and in the absence of birds as hosts, some species have been observed as ectoparasites of mammals, such as man or rodents. Up to 30 bird species, belonging to over 12 families and eight orders (e.g. pigeons, sparrows, rock doves and starlings) are known to be parasitized by *D. gallinae* (Proctor and Owens, 2000; Roy and Chauve, 2007; Sparagano *et al.*, 2014a). Despite this biological adaptation, this last mite is primarily retained a serious pest of chickens. As well as other *Dermanyssus* species, it feeds on blood of their hosts by piercing the skin with their long, slender, extrusible chelicerae, endowed by a reduced chela at their end. They spend most of their time in the nests of their hosts, crawling onto the animals primarily to feed. When they come in contact with human skin, they are prone to bite, typically causing erythematous papules at each puncture site, often accompanied by intense itching (Mullen and O'Connor, 2019).

The recent review of Silva *et al.* (2023) summarized the world economic importance of these mites on the poultry industry.

Dermanyssus gallinae (De Geer) (Fig. 58.1)

Common name

Poultry red mite

Morphological characteristics

The most important identifying characters of *D. gallinae* are the prominent shoulders of the dorsal shield and the jagged edges of the shield reticulations, the position of setae j_1, s_1 and the epigynal pores and the presence on tibia IV of one seta (*pl*). The shape of the dorsal (abruptly narrowed posteriorly) and epigynal (narrowly rounded posteriorly) shields and the chelicera (elongate, with distinct digits) of *O. sylviarum* enable its differentiation from *D. gallinae* (Di Palma *et al.*, 2012). Morphological studies and molecular data (Roy *et al.*, 2009a, 2010; Roy and Buronfosse, 2011) have shown that *D. gallinae* is actually a species complex, including at least two cryptic species. In particular, *D. gallinae s. str.* appears to be the only pest in poultry houses, as well as on other bird species except pigeons, while *D. gallinae* special lineage L1 is restricted to pigeons only (Di Palma and Giangaspero, 2021).

FEMALE. The female possesses a single dorsal shield that tapers posteriorly and has a truncate posterior margin. The chelicerae are long and styliform, and the sternal shield, a median sclerite set between the legs II and III, bears two pairs of setae, and a third pair is located more posteriorly and distinctly separated from the others. The genitoventral or epigynal shield is posteriorly rounded and bears one pair of setae. The anal shield has three setae. Moreover, the dorsal shield has a prominent shoulder; shield reticulations with jagged edges and the setae *j*1 always; *s*1 usually on dorsal shield, epigynal pores on shield and tibia IV with one seta (*pl*) (Di Palma *et al.*, 2012).

The female of *O. sylviarum* has the chelicerae elongate, but with well-developed and distinct fixed and movable digits, while they are whip-like with no evident chela in *D. gallinae*. Moreover, the genitoventral or epigynal shield is attenuated and narrowly rounded posteriorly, while it is broadly rounded posteriorly in *D. gallinae*. Moreover, the dorsal shield of *O. sylviarum* is abruptly narrowed posteriorly and more smoothly narrowed in *D. gallinae* (Di Palma *et al.*, 2012).

MALE. The males have a small presternal genital opening, and the intercoxal region is covered by a sternogenital sclerite. Fusion of the sternogenital and ventroanal elements results in a holoventral shield. The male chelicerae are modified as gonopods for sperm transfer and hence provided with a spermatodactyl (Di Palma *et al.*, 2012).

LARVA AND NYMPHS. The larva is hexapod and with little or no sclerotization and without indication of external genitalia. The nymphs are octopod as the adults but undergo progressive shield differentiation with each moult until adult stage. The epigynal shield in nymphs is reduced compared with the female, and there is no genital opening (Di Palma *et al.*, 2012).

Geographical distribution

Cosmopolitan (Rosen *et al.*, 2002; Haag-Wackernagel, 2005).

Bio-ecology

It is a hematophagous ectoparasitic mite of wild, domestic and synanthropic birds (Rosen *et al.*, 2002; Haag-Wackernagel, 2005), sometimes feeding upon mammalian hosts (DeClercq and Nachtegaele, 1993; Mignon and Losson, 2008). Although *D. gallinae* is fundamentally an avian-specific ectoparasite of farmed birds, such as chickens, turkeys and ducks, as well as of wild and synanthropic birds, such as pigeons, sparrows, starlings, and doves, it sometimes feeds on mammals, such as cats (Guin, 2005; Di Palma *et al.*, 2018), dogs (Ramsay *et al.*, 1975; DeClercq and Nachtegaele, 1993; Friesen *et al.*, 2011), gerbils (Lucky *et al.*, 2001), other rodents (Allymehr *et al.*, 2012; Kowal *et al.*, 2014), goat (Dorny *et al.*, 1994), horses (Mignon and Losson, 2008) and humans (Cafiero *et al.*, 2019). The mite is a serious problem in the Palearctic region and in United States, where it mainly occurs in poultry houses or around buildings, where pigeons, house sparrows or starlings are nesting. It lives in nesting materials, where it spends most of its time moving onto the birds to feed on blood at night. Consequently, workers in poultry operations seldom experience a biting problem, while working during the daytime, even when the houses are heavily infested. However, individuals who enter infested buildings at night may be readily bitten. Sometimes, pet canaries and parakeets are sources of human infestations (Mullen and OConnor, 2019). However, studies carried out in the last decade have highlighted that most of the *Dermanyssus* species actually have more restricted host ranges than first believed (Roy *et al.*, 2009b). In particular, *D. gallinae* might not be the only *Dermanyssus* parasitizing laying hens but that some other closely related species might often have been confused with this species (Roy and Chauve, 2007).

The life cycle of the poultry red mite develops through five stages: the egg, the larva, protonymph, deutonymph and adult (Sparagano *et al.*, 2014a). Pound and Oliver (1976) studied the reproductive morphology and spermatogenesis in *D. gallinae* and reported almost 50 years ago that sperm induction pores, tubuli annulati, rami sacculi, sacculus foemineus, cornu sacculus and acanthomembranous hood in female of *D. gallinae*, support the probability of coxal insemination as reported for some other mesostigmatid species. Development of the circular testis in males (from fusion of two anterolaterally projecting protonymphal testicular arms), and formation of haploid (n = 3) spermatids from haploid (n = 3) spermatogonial cells, is entirely mitotic. Spermatogenesis begins within one hour after protonymphal feeding and continues through adult ecdysis in normally fed individuals. Nutritional deficit causes cessation of spermatogenic divisions in deutonymphs starved for 23 days but is rapidly rejuvenated upon feeding. No rejuvenation of division was observed in adult males given the opportunity to feed, thus adult males might not feed (Pound and Oliver, 1976). More recently, Di Palma and Alberti (2005) reported that neopodospermy, the sperm transfer through accessory insemination pores located close to the legs, occurs in dermanyssoid mites, along with the morphological and functional adaptations related to podospermy. The male genital opening is presternal, and the chelicerae are modified as gonopodes and provided with a sperm transfer process, the spermatodactyl, arising from the movable digit of the chelicerae. The females possess a sperm access system for sperm reception and probably for storage and capacitation. In *D. gallinae*, mating occurs off the host and requires from 14 min to 1 h, during which the male inserts the spermadactyl into the female spermatic orifice and transfers the spermatophor sack containing about 200 spermatozoa. Males can mate up to four times in 4 days, and the longest fertile period of females lasts 3 weeks (Hutcheson and Oliver, 1988). The oviposition starts at 12–24 h after the female has fed, as many as seven eggs being deposited at one time; successive batches of eggs are produced, each batch being laid after a blood meal has been taken (Baker *et al.*, 1956). Females lay eggs in clutches of four to eight, generally laying around 30 eggs in their lifetime (Chauve, 1998). The poultry red mite lays eggs where they hide, in areas such as cracks, crevices and litter. After mating, the female lays eggs (4–8 per day) and takes a blood meal between each batch for 3 days consecutively. A total of 30 eggs are produced in a lifetime, and most are produced after the third, fourth and fifth blood meal. The highest numbers of eggs are laid at temperatures of 20–25°C and 70% RH. At 5°C, eggs do not hatch, but they are still alive, and at 45°C, they dehydrate in a short time (Maurer and Baumgartner, 1992; Nordenfors *et al.*, 1999). Eggs mature into hexapod larvae after 13–51 hours, and the newly hatched larvae moult into octopod protonymphs after about 24 h (Tucci and Guimaraes, 1998) without feeding. After hatching, the six-legged larvae are sluggish, and moult after one day; the larvae do not feed (Axtell, 1999). The eight-legged protonymph feeds and moults to an eight-legged deutonymph, which then feeds and moults to an adult (Chauve, 1998). The protonymphs take a blood meal before moulting into deutonymphs, which then need a blood meal before changing into adults (Maurer and Baumgartner, 1992). The entire cycle can be completed in as few as 7 days (Chauve, 1998). The sex ratio is 1:1. The life cycle can be as short as 5.5–7 days at 25–37°C, and as long as 17 days at 20°C (Maurer and Baumgartner, 1992). Removing the host from an area will not eliminate the mites. The deutonymph and adult are known to resist desiccation and live as long as 8 months without feeding (Chauve, 1998); Baker *et al.* (1956) report that the adults can survive without feeding for 4 or 5 months. The protonymph, deutonymph and adult females of poultry red mite feed on host blood, while males only feed occasionally (Chauve, 1998). The adults and nymphs feed mainly at night and during the day are found in crevices in walls and ceilings of poultry houses (Hughes, 1976).

Red mites are poikilothermic because temperature and humidity can influence the population ecology. Temperatures between 25°C and 30°C are optimal for survival and reproduction; however, the life cycle can take place even at 5°C, although less efficiently. Temperatures of –20°C and 45°C are fatal for red mites. The ideal relative humidity level is 80%, and higher or lower levels stop the development cycle (Nordenfors *et al.*, 1999). *D. gallinae* is widespread also in winter, but more common between May and late October, when the temperatures are ideal for development of the immature forms. In poultry farms, the number of mites rises progressively for 4–6 months, until it reaches a plateau (Nordenfors and Hoglund, 2000). In layer sheds, the density of red mites can be 25–50,000 per chicken; this number can treble in the case of massive infestations. A poultry shed remains infested for 4–5 months after birds are removed. Mites are transmitted by dispersion between farms (on crates, egg trays or even humans) or by direct contact between birds. Tucci *et al.* (2008) evaluated the thermal requirements for development of *D. gallinae* under laboratory conditions at 15°C, 20°C, 25°C, 30°C and 35°C, a 12-h photoperiod and 60–85% RH. The base temperature for the pre-oviposition was 3.4°C, the thermal constant (k) 562.85 degree-hours, the determination coefficient (R^2), 0.59, and the regression equation was Y= −0.006035 + 0.001777x. The base temperature for eggs was 10.60°C, the thermal constant (k) 689.65 degree-hours, the determination coefficient (R^2) 0.94, and the regression equation was Y= −0.015367 + 0.001450 x. For the larva, a base temperature of 9.82°C was found, along with a thermal constant (k) of 464.91 degree-hours, a determination coefficient R^2, 0.87, andaregressionequationofY=−0.021123+0.002151x. The base temperature of protonymph was 10.17°C, the thermal constant (k) 504.49 degree-hours, the determination coefficient (R^2), 0.90, and the regression equation was Y= −0.020152 + 0.001982x. The base temperature of the deutonymph was 11.80°C, the thermal constant (k) 501.11 degree-hours, the determination coefficient (R^2), 0.99, and the regression was equation: Y= −0.023555 + 0.001996 x. The results obtained showed that 15–42 generations of *D. gallinae* may occur during the year in the State of São Paulo (Brazil), as estimated based on isotherm charts. The mite may develop continually in this state, with a population decrease in the winter. There were differences between the developmental stages of *D. gallinae* in relation to thermal requirements (Tucci *et al.*, 2008).

Studies on the host-searching behaviour of *D. gallinae* highlighted that aggregation pheromones and kairomones mediate the behaviour of the poultry mite (Koenraadt and Dicke, 2010) and that the foreleg tarsal sense organs intervene in stimuli perceiving (Soler Cruz *et al.*, 2005). A gradual increase of CO_2, heat and vibrations are activating stimulus (Kilpinen, 2001, 2005), guiding mites toward the host. The recognition of the mites moved close to the potential host may be facilitated by skin lipids of birds, which act as feeding stimulants to poultry red mites (Zeman, 1988). The CO_2 can stimulate a contrast behaviour, inducing mites to 'freeze' or become motionless as response to an increase of CO_2, when in light conditions; this behaviour, represents probably a defensive strategy to avoid being eaten by the host (Kilpinen, 2005). Briefly, like some plant insect pests, *D. gallinae* must find its host at a distance, since it does not live on it. *D. gallinae* is attracted to temperature gradients and to CO_2 puffs, which are features of homeothermal vertebrates that can be detected at a distance but are not host specific. Interestingly, CO_2 is attractive in the dark but induces the mite to freeze in the light, a behaviour that apparently allows *D. gallinae* to avoid being eaten by its host (Kilpinen, 2005). The responses of mites toward CO_2 depend on the dose and context. After the blood meal, various biological stages of *D. gallinae* forms aggregations and thigmokinesis (i.e. increased locomotion in response to changes in contact with the immediate physical environment), and pheromones play a role in this behaviour (Entrekin and Oliver, 1982; Koenraadt and Dicke, 2010). As the nests are discontinuous and temporary microhabitats, an important adaptation of mite parasites involves the use of the host to disperse (Martins-Hatano *et al.*, 2011). Arthropods can contribute to mite distribution in the near vicinity, including houses and farms. Some authors (Flechtmann and Baggio, 1993; Pavlovic *et al.*, 2016) reported examples of phoretic behaviours on insects by *D. gallinae*.

Symptomatology and damage

The three blood-sucking mites belonging to this family, *D. gallinae*, *O. sylviarum* and *O. bursa* stand out, for causing infestations in commercial poultry farms worldwide (Sillos, 2002; Faleiro, 2012; Miguel, 2019; Oliveira, 2021; Bassani-Silva *et al.*, 2022). These mites directly affect the well-being of animals, and consequently cause losses in production that result in significant economic damage for producers (Faleiro, 2012; Nechita *et al.*, 2015; Oliveira, 2021; Soares and Ximenes, 2022).

The official data of the economic importance of this economic sector attest indirectly to the importance of the damage caused by these mites. The United Nation's Food and Agriculture Organization (FAO) reported that the growth in the global poultry sector is driven by a greater demand and purchasing power of the consumer market. Egg production has increased considerably in the past three decades, increasing from 15 million tonnes in 1961 to 87 million tonnes in 2020, and chicken meat production increased from 9 million tonnes in 1961 to 133 million tonnes in 2020 (FAO, 2022). The countries with significant numbers associated with chicken production are United States, China, Brazil and the European Union (FAO, 2023). Brazil exports the most chicken meat to the world (EMBRAPA, 2022). China stands out as the largest egg producer, followed by United States, the European Union, India and Mexico, with Brazil currently in sixth place (Soares and Ximenes, 2022; FAO, 2023).

D. gallinae is considered the most prevalent mite in poultry farms around the world, with emphasis on European breeding models (Kilpinen, 2001; Moro *et al.*, 2011; Murillo *et al.*, 2020; Sparagano and Ho, 2020). It is estimated that approximately 83% of European laying, pullet and breeder farms are infested with *D. gallinae* (Sparagano *et al.*, 2009; George *et al.*, 2015; Pavlićević *et al.*, 2021), with losses ranging between US$130 million and US$231 million per year (Harrington *et al.*, 2008; Sparagano *et al.*, 2014a; Bartley *et al.*, 2015; Makert *et al.*, 2016; Wright *et al.*, 2016; Price *et al.*, 2019; Waap *et al.*, 2020).

D. gallinae is a blood-sucking mite feeding every 2–4 days, mainly at night, although not exclusively (Nakamae *et al.*, 1997); it normally stays on the birds for approximately 30 to 60 min before returning to hide (Kirkwood, 1963; Chauve, 1998; Maurer *et al.*, 1988). Giangaspero *et al.* (2016) reported that the mite attacks areas of the host's body with few feathers, where the skin is thin and the veins are superficial (e.g. the neck, back and under the wing). Dermanyssids are characterized by multiple brief acts of blood-sucking, each associated with salivation. This seems supported by the dynamics of the secretory activity and in the non-simultaneous release of secretion from different cells of the salivary glands (Amosova and Stanyukovich, 2008).

The sanitary and economic importance of *D. gallinae* for the poultry system is also related to its role as a vector of several pathogens. Although various bacteria and viruses are associated with this mite, main scientific knowledge concerns the transmission of some *Salmonella* species, and for different pathogens, the vectorial competence of *D. gallinae* is poorly known (Valiente Moro *et al.*, 2005, 2009; Circella *et al.*, 2011; Chu *et al.*, 2015); its potential should not be underestimated, taking into account that severe infestations suppress the antibody production or reduce antibody titres to some viral vaccines by the host (Sparagano *et al.*, 2014a; Pugliese *et al.*, 2019; Sparagano, 2020).

The role of *D. gallinae* as a vector was shown in *Salmonella* spp. (Flochlay *et al.*, 2017; Cocciolo *et al.*, 2020), *Escherichia coli* Migula (Sparagano *et al.*, 2009; Oliveira *et al.*, 2020) and eastern equine encephalitis virus (Pritchard *et al.*, 2015). The mite also acts as reservoir for some pathogens (*Coxiella burnetii* Philip, *Borrelia afzelii* Manuela Marin Canica *et al.*, *Borrelia burgdorferi* Johnson *et al*; Raele *et al.*, 2018), *Erysipelothrix rhusiopathiae* Migula (Erikson *et al.*, 2010), *Mycoplasma gallisepticum* Edward *et* Kanarek, *Mycoplasma synoviae* Olson *et al.*, *Plasmodium* spp., *Tsukamurella* spp. (Brannstrom *et al.*, 2010) and *Listeria monocytogenes* Murray *et al.* (Sioutas *et al.*, 2023). Infestations of these mites cause in the animals, a state of anaemia, lower feed conversion and weight loss, psychogenic behaviour or somatic stress (irritation, pecking and cannibalism), dermatitis, decreased immunity, and in extreme cases, death by exsanguination (Di Palma *et al.*, 2012; Faleiro, 2012; George *et al.*, 2015; Wright *et al.*, 2016; Bartley *et al.*, 2017; Sparagano and Ho, 2020; Pavlićević *et al.*, 2021). In addition, mite-infested hens suffer from a reduction in egg-laying, and the eggs laid lose their quality (fragile shells and smaller size) (Kaoud and El-Dahshan, 2010; Kasburg, 2016; Thomas *et al.*, 2018; Sparagano and Ho, 2020; Oliveira *et al.*, 2020; Oliveira, 2021; Tabari *et al.*, 2023).

In dogs, cats and horse, the infestations document from severe, mild to absent skin lesions (DeClercq and Nachtegaele, 1993; Guin, 2005; Mignon and Losson, 2008; Friesen *et al.*, 2011; Di Palma *et al.*, 2018), and an immunological reaction in experimentally infested dogs has been reported (Martins *et al.*, 2020). The attacks to the humans can occur within the poultry farm setting, when the mite infestation level is high, and the people (farmers, employees or visitors) are not well protected. Similarly, synanthropic birds (mostly pigeons) when abandoned the nests built close to residential buildings, the hungry mites enter houses, residences, public buildings, hospitals, etc. through crevices and cracks around windows and doors, or via ventilation ducts and air-conditioning systems and attack humans (Cafiero *et al.*, 2019; Sargison *et al.*, 2020). The symptoms of the infestation by

D. gallinae in humans are often recurrent and range from pruritic dermatitis, erythematous rashes, small papules and blisters, urticarial plaques and erythema with consequent histological changes and are mostly described on arms, hands, chest and legs (Cafiero *et al.*, 2019). The importance of this mite species in public health is also associated with their role as reservoirs/vectors of zoonotic pathogens. The adventitious infestations to humans do not reduce the risk of transmitting arthropod-borne disease.

According to what was discussed above, *D. gallinae* causes serious economic loss in the poultry industry and its not negligible interest in human health. It is a significant pest of poultry worldwide (Axtell and Arends, 1990; Axtell, 1999) and a serious economic threat, mainly to the laying hen sector (van Emous *et al.*, 2006; Mul and Koenraadt, 2009) in any farming system (cages, barns, free-range and organic farming), including the recently introduced 'colony' system (Sparagano *et al.*, 2009). High percentages of infested birds are reported in Italy, Serbia, Poland, the UK, Morocco, Japan, Montenegro and the Netherlands (Sparagano *et al.*, 2009). In Europe, *D. gallinae* is one of the major problems in caged laying hens because of their longer production cycle but is less of a problem in broiler industries (Chauve, 1998). In the European Union, egg industry losses associated with production and the control of *D. gallinae* is estimated at €130 million (US$177 million) per year (George *et al.*, 2009a). This agrees with what was reported by other authors (Sparagano *et al.*, 2014b; Van Emous, 2017; Sleeckx *et al.*, 2019), which evaluate the overall cost of attacks of *D. gallinae* to the poultry industry roughly estimated at €231 million per year for the EU egg industry, with similarly significant losses elsewhere imposes strict control strategies. However, when established in a poultry farm, the control of *D. gallinae* is very challenging due to its high reproduction rate, long survival period (even in the absence of the host), feeding and resting behaviour and increasing resistance to acaricides (Sparagano *et al.*, 2014; Sparagano, 2020).

Moreover, the poultry red mite is a known vector (transmitter) for the St. Louis encephalitis virus and has been linked to other illnesses as well (Chauve, 1998). The mites spread other illnesses such as fowl pox virus, Newcastle virus and fowl cholera (Hoy, 2011). Less intensive farming systems, such as barns, free-range and organic farming often have higher infestation rates because *D. gallinae* can find more hiding places and can more easily avoid chemical control methods in these environments. *D. gallinae* causes stress behaviour in its poultry hosts, reduced egg production and egg grade (van Emous *et al.*, 2006), anaemia (Kirkwood, 1967) and diminished disease resistance (Kilpinen *et al.*, 2005). *D. gallinae* is also a vector of several infectious disease agents (Valiente Moro *et al.*, 2005). Though relatively rare, the scientific literature records episodes of hen mortality associated with *D. gallinae*, mainly involving chicks (Fossum *et al.*, 2009).

The poultry red mite is also widely reported as being responsible for dermatological problems of varying severity in humans, both in poultry workers (technicians, farmers, veterinarians) (Rosen *et al.*, 2002; Cafiero *et al.*, 2011) and in urban residents. Military personnel living in barracks, hospital patients and office employees may be exposed to mite-infested synanthropic birds (Auger *et al.*, 1979; Fuentes *et al.*, 2009). Williams (1958) reported that although a marked clinical dermatitis is common in some individuals that become closely associated with *D. gallinae*, the literature presents little evidence that this mite will ingest human blood, and many investigators feel that this species will never ingest it. The invasion of a New York City apartment by these mites resulting in attacks on the occupants was described by Williams (1958), who observed that the finding of mammalian erythrocytes in the digestive tract of mites collected from this apartment, some of which were taken from the bed of the occupants, as well as the appearance of fresh blood splotches on the bed sheets resulting from crushed mites, which had recently fed, offers some factual evidence that *D. gallinae* may, on occasion at least, partake of human blood (Williams, 1958).

We summarize here what has been extensively discussed by Mullen and O'Connor (2019). These authors report that a number of cases have been reported in hospitals and other institutional settings, where employees and patients have been bitten by *D. gallinae*. The sources of the problem generally can be traced to nesting birds, notably pigeons, on windowsills and ledges, under eaves and in air-intake ducts or air conditioners mounted on the outside walls. The mites enter rooms around windows and doors, through crevices and cracks, or via ventilation ducts and air-conditioning systems. In other situations, they may drop onto individuals from roosting or nesting birds in ceilings, or from overhead sites on porches and walkways near buildings. In such cases, close inspection may reveal mites crawling on clothing, furniture or bed linens, particularly at night, when the mites are active (Mullen and O'Connor, 2019).

Human infestations with *D. gallinae* have been variously called chicken tick rash, bird mite disease, psora dermanyssica, pseudogale and gamasoidosis. The term 'fowl mite dermatitis' is likewise used, but it also can be applied to skin reactions caused by other avian mites that attack people (Mullen and O'Connor, 2019).

Most bites tend to occur on the arms and chest protected by clothing rather than on exposed skin such as the hands and face. Only in exceptional cases do bites occur in the axillary and pubic areas. The bites are usually painful and typically result in red maculopapular skin lesions on the upper portions of the body and extremities. During the bite, close examination will reveal the mite to be a tiny red speck at the centre of the papule. Occasionally, the bites produce vesicles, urticarial plaques and diffuse erythema, with dermatographia frequently seen. In multiple bite cases, a pruritic rash may develop and persist until the source of the infestation has been eliminated. Itching tends to be most intense at night. The problem is usually resolved by treatment with antihistamines or topically applied steroids, combined with moving individuals from affected areas (Mullen and O'Connor, 2019).

St. Louis encephalitis, eastern equine encephalitis and western equine encephalitis (WEE) viruses have been isolated from *D. gallinae* infesting wild birds. However, conflicting evidence has been reported regarding the ability of *D. gallinae* to transmit any of these viruses among birds or to humans (Mullen and O'Connor, 2019).

Recently, Schiavone *et al.* (2022) critically described available strategies and methods currently available for conducting trials, as well as outcomes, analysing their possible strengths and weaknesses, with the aim to provide researchers with useful tools to correctly approach the study of the vectorial role of *D. gallinae*. These pathogens mediated by *D. gallinae* as vector. The latter include the avian pathogenic *Escherichia coli* (Migula) (APEC), *Salmonella enterica* Kauffmann *et* Edwards, serovars Enteritidis and Gallinarum and Influenza virus.

Control

We discuss here the means and strategies of control known worldwide, fundamentally taken from the comprehensive review of Decru *et al.* (2020), Di Palma and Giangaspero (2021) and Silva *et al.* (2023), focusing the attention on the limit and the perspectives of main means.

Silva *et al.* (2023) reported that the elimination or control of hematophagous Dermanyssid mites has become increasingly difficult due to their resistance to the chemical products available on the market (Morrone and Mayworm, 2001; Billingsley *et al.*, 2008; Harrington *et al.*, 2008; Faleiro, 2012; Wright *et al.*, 2016; Thomas *et al.*, 2018; Miguel, 2019; Xu *et al.*, 2019; Guerrini *et al.*, 2022), associated with a change in the system of raising animals in battery cages, to that of enriched cages or 'cage-free' systems that hinders the control and observation of the animals, as they start to live free in sheds, and roam freely in pastures (Hartcher and Jones, 2017; Taborda, 2018; Leite *et al.*, 2021). Moreover, Spagarano (2009) reported that the cost of losses due to preventive and control measures against *D. gallinae* are difficult to calculate. The cost in France has been estimated at €4.33/100 poultry and €3.83/100 poultry for cages; in the Netherlands, it ranges from €0.14–€0.29 per poultry (Sparagano *et al.*, 2009; Flochlay *et al.*, 2017) with productivity losses reaching €0.57–€2.50 per poultry, in one year (Hwang, 2023).

This requires the application of new technologies to control the hematophagous mites, without the use of chemical acaricides, such as plant-derived products, entomopathogenic fungi, diatomaceous and silica-based products, semiochemicals and vaccines.

Different chemical, physical and biological strategies to control the development of populations of the poultry red mite appear in this scenario. However, effective solutions to this problem are yet to be found. The control of the mite infestations in field has relied on synthetic acaricides, but the development of resistances and the restricted list of authorized active ingredients make fundamental the development of novel control measure. The combination of alternative control measures, such as monitoring of the mite infestation, botanical acaricides, inert dusts, biological control and vaccines poses the best ways for achieving satisfactory results (Lima-Barbero *et al.*, 2020).

The application of any control methods must be based on the correct identification of the mite pest. After its identification, the control of the red mite infestations could be entrusted to strict preventive and sanitary measures. Regular monitoring of the poultry red mite on a farm is a fundamental step of control and allows to intervene as soon as the red mite population starts to increase (Oh *et al.*, 2020). Application of hygiene regulations based on cleaning and hygiene facilities, training of farmers and technician, interruption between the production cycles and disinfestation before the arrival of new birds and checking of new animals on poultry farms, prevent or limit the infestation (Oh *et al.*, 2020). The choice of products against *D. gallinae* safe for animals, humans and the environment, the use of a correct dose and the low application frequency, the use in empty houses between flocks helps to realize an anti-resistance strategy. The most promising approach for controlling *D. gallinae* would be via vaccination, which offers the advantages of prolonged effectiveness, freedom from chemical residues/environmental pollution, and a reduced risk of resistance. Although improvements in mite *in vitro* rearing and in vivo testing (Bartley *et al.*, 2012; Nunn *et al.*, 2020), application of omics strategies (Price *et al.*, 2019; Lima-Barbero *et al.*, 2019a, b) and delivery methods (Price *et al.*, 2019) are helping to facilitate progress in this field, much still needs to be done to develop a commercially viable vaccine for *D. gallinae* (Di Palma and Giangaspero, 2021).

MONITORING. Monitoring techniques provide information on population dynamics and spatial distribution of the poultry red mite. These techniques agree on the detection of the low numbers of *D. gallinae*, the evaluation of the effectiveness of treatments, the knowledge of the mite population and the critical threshold for treatment (Mul *et al.*, 2016; Lammers *et al.*, 2017). Different tools have been developed to monitor the populations of *D. gallinae* in layer houses. We discuss here the methods most common.

The technique of Mite Monitoring Score (MMS) (Cox *et al.*, 2009) does not include the use of mite traps. A visual observation of the mite infestation is carried out on different points, at different levels in the layer houses, and at each point is observed, an area of 1 m^2 and a score of 0–4 is used to evaluate the infestation rate.

Other techniques provide the visual assessment of mites captured in traps and score them in/on these traps. The Rickstick method (Van Emous and Napel, 2007) consists of a 12 cm long wooden stick placed into a 10 cm long PVC tube attached under the perch. For scoring the infestation, the wooden stick is pulled out of the tube, and the number of mites is visually assessed with reference images, also in this case with a score of 0–4.

The Simplified Passive Tape (SPT) trap (Roy *et al.*, 2014; Chiron *et al.*, 2015) provides the trapping of mites with a piece of adhesive tape and their successive visual scoring. A 4-state rating (Score 0–3), and two 2-state ratings (mites absent/present or <10 mites/>10 mites) have been proposed.

Other monitoring techniques evaluate the mite numbers quantitatively, mostly with the employ of traps. Nordenfors and Chirico (2001) developed the corrugated cardboard traps, one of the most popular techniques. Corrugated cardboard is put inside the plastic/PVC tubes that are attached in the housing system (Van Emous and Napel, 2007; Lammers *et al.*, 2017). Before counting, the traps are usually collected in sealed plastic bags and frozen at −18°C to −20°C for at least 2 days to kill the mites. To count the mites, they are all collected from the plastic bag, trap and dismantled cardboard in a petri dish (Nordenfors and Chirico, 2001; Lammers *et al.*, 2017). After collection, the mites can be counted either manually or with computer-based estimations, such as using the programme ImageJ (Schneider *et al.*, 2012) after taking a photograph (Lammers *et al.*, 2017). The cardboard traps can also be impregnated with acaricides or non-chemical control means (e.g. plant-based products or entomopathogenic fungi) to use them as a control mechanism instead of solely as a monitoring tool (Chirico and Tauson, 2002; Sleeckx *et al.*, 2017; Nascimento *et al.*, 2020).

Schulz (2014) developed a monitoring trap consisting of a blood test tube, filled with corrugated cardboard and compared this trap with the visual Mite Monitoring Score (MMS) method in 16 houses. The monitoring trap of Schulz (2014) detects the presence of *D. gallinae* in the layer house earlier than with the MMS method. Moreover, only one of the three mite-free houses according to the MMS method, was also mite-free according to the method of Schulz (2014). No correlation was observed between the MMS score and the number of mites per trap.

Koenraadt and Dicke (2010) reported that mites can communicate with each other, signalling the discovery among them of a suitable hiding space (i.e. the cardboard trap). However, the number of mites in these traps is the same number of mites present at the place where the trap is exposed. Chiron *et al.* (2015) duped this obstacle using the Semi Attractive Traps (SAT), where mites are attracted to traps with water. They placed in the layer houses, small plastic jars with perforated lids, and filled with water containing 0.01% surfactant. When the mites enter the jar, they are drowned by the water and surfactant, and they cannot communicate with each other. The number of mites captured in these traps is more representative of the mites occurring at that place.

Recently, Lammers *et al.* (2017) proposed the AviVet trap, based on the principle of corrugated cardboard within a barcode-labelled Tylene tube (for identification) and *D. gallinae* specific. The mechanism of collecting mites is similar to that of the regular cardboard traps, but the mites are weighed and not counted. The validation process showed that the weight correlates for 99.6% with the counted number of all stages, and the regression line is Y = 58.8 + 9.56x, where Y = total number of mites in the AviVet trap and x = the weight of all mites in the AviVet trap.

The intensive application required from the above-discussed monitoring techniques suggested to Mul *et al.* (2015, 2016) to develop and validate an 'automated mite counter', placed under the perch, counting *D. gallinae* and no other mite species. The mites enter the counter through a hole in the lid and are detected and counted by a sensor device. The mites detected are removed by air suction into a filter and stay in this filter until removal every week. Validation confirmed the efficacy of the automated mite counter, providing a good estimate of the number of mites, even at low infestation levels. Currently, this equipment is commercialized as MiteAlert®; it does not collect the mites in the filter but blown out by the pump back into the hen house. Regardless of the validity of the discussed methods, some studies showed that farmers agree to methods that are not too labour intensive, like the Rickstick method (Van Emous and Napel, 2007), the SPT trap (Roy *et al.*, 2014), the AviVet trap (Lammers *et al.*, 2017) and the Velcro trap (Tuovinen *et al.*, 2010), in which a section of Velcro wrapped around the perch provides a hiding place for mites (COREMI, 2000; Mul *et al.*, 2020). The automated mite counter (Mul *et al.*, 2015) is also one of the preferred methods of farmers as it is a practical tool for monitoring with minimal workload.

THRESHOLDS. Mul and Koenraadt (2009) reported that preventing the establishment of ectoparasitic poultry red mite populations is a key factor in ensuring welfare and egg production of laying hens and absence of allergic reactions in workers in poultry facilities. Using the Hazard Analysis and Critical Control Point method, a panel of experts identified hazards and associated risks concerning the introduction and spread of this mite in poultry facilities. Together, they provide an overview of possible corrective actions that can be taken to prevent population establishment. Additionally, a checklist of the most critical control points has been devised as management tool for poultry farmers. This list was evaluated by Dutch and British poultry farmers. They found the checklist feasible and useful (Mul and Koenraadt, 2009).

Preventive actions alone often do not produce sufficient control of the mite, and it is necessary to apply curative means. However, also, this choice does not allow the complete eradication of *D. gallinae* population, and the control measures must aim to keep the level of infestation under an economic threshold, in order to reduce negative effects on the hens, humans and production. Towards this aim, in an IPM strategy, it is crucial to apply timing actions able to prevent the increase of mite population from causing damage. The application of this 'action threshold', allows to perform the treatment/action not too soon or too much, avoiding ecological negative effects, high costs, and onset of resistance. The treatments are not made too late, and efficient control is possible. Unfortunately, general thresholds are not yet available for the control of *D. gallinae*. The lack of knowledge of reliable thresholds hampers the application of IPM strategies for layer houses and makes the study of the topic necessary. Different monitoring techniques or treatments are characterized by their own thresholds and state the need of the treatment, though these are not scientifically shown (Decru *et al.*, 2020).

The difficulty in individuating thresholds depends on the determining of the population size and the knowledge of influencing variables. In addition, although production losses due to *D. gallinae* infestations are well known (Sleeckx *et al.*, 2019), the exact relationships between infestations and bird health/economic impact are poorly known. A first promising study toward a general critical action threshold was based on concomitant SAT and SPT trapping (Chiron *et al.*, 2015), and working with the temporal evolution of the trapped *D. gallinae* population size. This method is based on Verhulst's mathematical model, which describes the temporal dynamics of populations of living organisms in three successive phases (latency phase, exponential growth phase, equilibrium phase). For an exhaustive information on the topic, the reader is advised to consult the review of Decru *et al.* (2020).

Chiron *et al.* (2015) reported that the key moment approaches when in three successive monitoring moments, >20% of the placed SPT capture mites. It follows that an SPT-based threshold can easily be applied by the farmer. This could help to define a critical threshold associated with monitoring, by different means (SPT, Ricksticks, Velcro traps, automated counter, AviVet). However, the study on this threshold needs further refinement, before it can be used with confidence. Although the SPT-based threshold prevents the mite population from increasing exponentially reaching the point of economic damage, it does not take this economic impact into account.

Another possibility concerns the Dynamic Adaptive model (DAP) (Mul *et al.*, 2017), able to predict the population dynamics of *D. gallinae* in a layer house, based on monitoring data of the current flock population, temperature data and treatment dates. The model deals with variation in population dynamics by including flock-specific parameters and could be improved including more parameters like flock age or husbandry measures. Further studies on the DAP model were carried out in three different laying hen houses (organic aviary, aviary with winter garden and aviary without winter garden) and resulted in a practical applicable model, forecasting the mite population growth. At the same time, an economic model and an advice algorithm were tested in these farms, to calculate the optimum interval between two treatments, to keep the population under a certain growth level and to determine if an extra treatment is cost-effective by getting lower numbers of mites and therefore better feed conversion. The threshold and the moment when exceeding the threshold were different for all layer houses, as the cost price of the eggs, treatment costs and treatment efficacy were different for all layer houses. This illustrates the difficulty of determining general applicable action and economic thresholds (Mul *et al.*, 2017, 2020).

BOTANICAL ACARICIDES. The use of botanical acaricides (plant-derived products) may be alternative to use of chemicals in the control of *D. gallinae*. They can exercise an acaricidal and toxic activity, or own repellent or attractive effects on *D. gallinae* (George *et al.*, 2009b; Birkett *et al.*, 2011; Na *et al.*, 2011; Miguel, 2019; Nascimento, 2019; Gay *et al.*, 2020). The repellent and attractive effects exhibit much potential in combination with other means. The botanical acaricides have low toxic effects on mammals and a short environmental persistence; consequently, they have a low impact on the environment (Kim *et al.*, 2007; George *et al.*, 2008a, 2009b). The use as repellent requires lower dosages than as toxicant (Ellse and Wall, 2014), and for this reason they are very suitable for use in IPM strategies.

Some products can be sprayed or administered through contact impregnation in traps. These compounds induce mortality or repellence (Tabari *et al.*, 2023). However, few field trials have been carried out. Plant based products are a promising method because they do not contaminate the environment and can be controlled in doses that do not cause health problems in animals and humans, and do not leave residues in food products (Faleiro, 2012; Kasburg, 2016; Miguel, 2019; Nascimento, 2019). In a laboratory in which almost all contact between mites and plant-derived oils and extracts is guaranteed, it was possible to obtain mortality rates above 80%, showing the effectiveness of the products (Morrone and Mayworm, 2001; George *et al.*, 2008c; Nechita *et al.*, 2015; Lee *et al.*, 2019; Alves *et al.*, 2020; Baran *et al.*, 2020; Tabari *et al.*, 2023). Lundh *et al.* (2005) found a good efficacy in tests conducted in the field on laying farms, using oil and vegetable extracts impregnated in traps. However, for the oil or extract to have an effect on the mite, it needs to come into contact with the product, and its effectiveness when used in a trap is limited to captured mites (Lundh *et al.*, 2005). When used as a spray, it is necessary to develop new methods that ensure the dispersion and maintenance of the product in the environment and in animals for a longer period of time (Tabari *et al.*, 2023), in addition to avoiding the formation of oil films (Lima-Barbero *et al.*, 2020).

It has been found that the efficacy of essential oils in the control of *D. gallinae* is mainly due to effects of their volatile components (Kim *et al.*, 2004; George *et al.*, 2009b; Na *et al.*, 2011; Lee *et al.*, 2019); this suggests the hypothesis of a neurotoxic effect rather than a mechanical one (López and Pascual-Villalobos, 2010).

The ability of essential oils to act mainly in their vapour phase may be an advantage, reaching the mites in their hiding places. However, this volatile nature could be the cause of the short effect of many oils (Ellse and Wall, 2014). Moreover, the plant essential oils may lack of standardization in formulation, with differences in chemical composition of different batches, resulting in a poor acaricidal efficacy (George *et al.*, 2009b; Ellse and Wall,

2014). The isolation of the active ingredient of the essential oils, among which the eugenol seems to be promising, showing the highest toxic effect against *D. gallinae* in various studies (Sparagano *et al.*, 2013; Lee *et al.*, 2019), characterized by a repellent effect changing to an attractive effect over time.

The efficacy of these botanical products has been evaluated only at laboratory scale, while in the field their efficacy might be affected by environmental factors (RH, dust and other pesticides employed) (George *et al.*, 2010a; Lee *et al.*, 2019). Moreover, some essential oils (e.g. thyme) do not produce any negative effect on hens, while others (e.g. pennyroyal) have an impact on chicken health and egg production (George *et al.*, 2010b). The short residual toxic effect of botanical products does not allow a stand-alone treatment against *D. gallinae* and suggests their use with other means with longer-term effects (George *et al.*, 2010a).

Botanical acaricides are proposed for a curative employ after passing an action threshold. Some botanical acaricides are commercially available but not allowed in all countries for use in layer houses. Among these is well known the MiteStop® (Alpha-Biocare GmbH), a product based on neem seed extract (*Azadirachta indica* Juss.), which is active against *D. gallinae in vitro* as well as in field (Abdel-Ghaffar *et al.*, 2008, 2009; Locher *et al.*, 2010).

Torres-Cabra and Lagos-Lopez (2019) evaluated the acaricidal activity of essential oil of coriander, *Coriandrum sativum* Linnaeus, at concentrations of 1, 2 and 3 mg/cm^2 through contact bioassays on adults of *D. gallinae*, compared to a chemical acaricide (Neguvon). The mites were collected from backyard hens (*Gallus gallus* Linnaeus). Statistically, mortality percentages were established, and an anova was carried out. The lethal activity percentage was established. Additionally, the median lethal concentration (LC50) and the median lethal time (TL50) were estimated. Coriander caused 100% mortality on *D. gallinae* population, 25 min after contact with a concentration of 2 and 3 mg/cm^2, and after 30 min, with a lower concentration of 1 mg/cm^2. The LC50 of the coriander essential oil was 1.3 mg/cm^2, with a 13 min TL50 on the *D. gallinae* population. The essential oil of coriander is suggested as an alternative to be used instead of the chemical acaricides for the control of the red poultry mite *D. gallinae* (Torres-Cabra and Lagos-Lopez, 2019).

Odaka *et al.* (2019) examined a mixture of substances derived from food and food additives (SFF), including a polysaccharide extract from laver seaweed *Pyropia yezoensis* (Ueda) (Bangiaceae) for acaricidal effects. Treatment with SFF was more effective in killing mites by blocking their respiratory organs than SFF without the polysaccharide extract. In a field trial, treatment with SFF reduced the density of mites in a poultry farm, as well as the proportion of blood-stained eggs. This SFF appears useful for controlling the population of *D. gallinae* in the poultry farms (Odaka *et al.*, 2019).

INERT DUSTS. Inert dusts, or silica-based products, contain silicon dioxide (SiO2) as active ingredient, which is one of the few permitted products, allowed for control of *D. gallinae*. Silicas are both synthetic and natural. Synthetic silicas contain only amorphous silicon dioxide; natural products are mainly based on diatomaceous earth (DE) and contain a small amount (<1%) of crystalline silicon dioxide. In several countries, only the natural silicas are allowed for use at organic farms. However, the amorphous form is considered relatively healthy, while the crystalline form in natural silicas is more harmful to environmental, animal and human health (Merget *et al.*, 2002; Schulz *et al.*, 2014). Synthetic silicas are not non-chemical treatments, but their mode of action against *D. gallinae* is retained to be completely mechanical (not poisonous), drying out the epicuticle of the exoskeleton of the poultry red mites through the absorbent character of the silicon dioxide particles, leading to desiccation of the mites (Maurer and Perler, 2006; Kilpinen and Steenberg, 2009). Moreover, this physical mode of action is less likely to develop than to single-target molecules (Schulz *et al.*, 2014).

Both diatomaceous earth and silica-based products are wettable, or in the form of a dry powder that acts on mites with acaricidal efficacy. The silicon dioxide adheres to the mite's body and immobilizes it, leading to desiccation and death (Harrington *et al.*, 2009; Schulz *et al.*, 2014). The mite remains inside crevices, in organic materials, protected from contact with the products sprayed in the environment (Schulz *et al.*, 2014; Miguel, 2019; Nascimento, 2019). Ulrichs *et al.* (2020) obtained good control of *D. gallinae* with products based on diatomaceous earth in the laboratory but did not obtain concrete results when using the same products in field tests; this phenomenon may have been influenced by the physical properties of the products (Ulrichs *et al.*, 2020), ambient humidity (Harrington *et al.*, 2011) or changes in climate that did not allow the mites to absorb the products

(Ulrichs *et al.*, 2020). Although the method does not pose a risk of intoxication (Schulz *et al.*, 2014), silica varies in purity and size, threatening user and animal safety because it irritates the respiratory tract when inhaled (Takken and Knols, 2010; Flochlay *et al.*, 2017; Hwang, 2023).

These silica-based products are commercially available and widely employed in Europe. However, consisting of dusts of fine particles, they can negatively affect the respiratory tracts of humans. This induces to employ liquid silica-based products, in order to reduce the deleterious effects of airborne silica (Maurer *et al.*, 2009). The liquid distribution of silica does not eliminate these effects, so long as some particles become airborne when the fluid dries and is further dispersed by hens. In laboratory, the diatomaceous earth seems more effective than synthetic silica (Maurer *et al.*, 2009), though the latter produces a longer effect, employed in the field as liquid form with respect to diatomaceous earth (Schulz *et al.*, 2014; Alves *et al.*, 2020). The action against *D. gallinae* varies between different silica-based products (especially amorphous silicas) due to different factors (variation in the absorption capacity of particles, chemical composition, particle size and specific surface) (Schulz *et al.*, 2014). Moreover, the effectiveness of inert dusts greatly diminishes at humidity levels of >85% (which can easily be reached in layer houses) (Kilpinen and Steenberg, 2009; Maurer *et al.*, 2009). The inert dusts are often used preventively, but repeated treatments during the production are often necessary (Kilpinen and Steenberg, 2009). Unfortunately, repeated treatments with silica are often not sufficient to control the populations of *D. gallinae*. Repeated treatments reduce the action of silica and hamper its efficacy, depending probably from the flock age or from the accumulation of dust and debris in laying hen houses over time (Mul *et al.*, 2017). The mechanical cleaning of the surfaces prior to silica application could avoid this aspect (Alves *et al.*, 2020).

Schulz *et al.* (2014) studied the acaricidal efficacy of 12 products containing natural or synthetic silica, nine in powder form, and three for liquid application. Mite mortality was measured at several intervals, and the mean lethal time (LT_{50}) was determined by Probit analysis after Abbott's correction. The LT_{50} values of the products significantly differed (Tukey's HSD $p < 0.05$). LT_{50} values of powdery formulations ranged from 5.1 to 18.7 h and overlapped with those of the fluid ones, which ranged from 5.5 to 12.7 h. In order to explain the differences in efficacy of the tested silica products, further characterizations were carried out. X-ray fluorescence, specific surface, cation exchange capacity (CEC) and water absorption capacity (WAC) were measured. Furthermore, electron microscopy was conducted, and different products were compared. Silicon dioxide content (65–89% for powders and 57–80% for fluids) had no significant impact on efficacy, while specific surface and CEC (2.4–23.2 mEq 100^{-1} g^{-1} for powders and 18–30.8 mEq 100^{-1} g^{-1}) were positively, and WAC (1.3–4.4 wt% for powders and 3.3–4.8 wt% for fluids) negatively related to the acaricidal efficacy. Influence of these parameters on acaricidal efficacy was significant, according to the results of a stepwise regression analysis ($p < 0.01$) (Schulz *et al.*, 2014).

SEMIOCHEMICALS. Hematophagous arthropods use chemical (semiochemical) signals to find their hosts and mates. Chemical signals produced by other mites of the same species (pheromones) attract them (Aufray *et al.*, 2023), and chemicals produced by the hosts (allelochemicals) attract (kairomones) or repel them (allomones) (Takken and Knols, 2010; Hwang, 2023). Volatile organic compounds (VOCs) were isolated from the uropygial gland of a duck (non-host mite) and can be added directly to chicken feed. After ingestion, chickens release these compounds through the uropygial gland and repel hematophagous mites (*D. gallinae* and *Ornithonyssus* sp.), interrupting their feeding and social behaviour (Pageat, 2017; Hwang, 2023). Aufray *et al.* (2023) showed that hungry females of poultry red mite on a French farm were attracted to a complex mixture of five synthetic VOCs (R-1-octen-3-ol, octanal, nonanal, (E)-2-nonenal and nonanoic acid), combined in equivolumetric proportions or with (E)-2-nonenal as the main constituent (Aufray *et al.*, 2023). Cunha (2008) conducted a study on the response of protonymphs to scents from the extracts of conspecific mites and observed that the protonymphs of *D. gallinae* produce a pheromone capable of attracting and stimulating encounters with other protonymphs of *D. gallinae* (Cunha, 2008). Silva *et al.* (2023) conclude that chemical ecology studies involving repellency or attraction in hematophagous mites are still poorly understood, but the results so far indicate a new field for new research and different approaches aimed at developing new products for use against poultry hematophagous mites.

VACCINES. Silva *et al.* (2023) summarized results of the research works carried out on this topic by various specialists, highlighting the limits and the perspectives of this interesting method. Vaccination of hens has a low risk of environmental contamination, and the risk for resistance emergence in mites is very unlikely (Wright *et al.*, 2009). Vaccines are a method of prophylaxis, and may be incorporated in IPM for *D. gallinae*, being the mass administration by injection not possible, once the hens have recovered in the layer house. The first step consists of the identification of suitable vaccine antigen, followed by boost immunity or to administer therapeutic vaccines through, for example, drinking water during the laying should be investigated.

During an infestation, some antigens of the parasite are continuously exposed to the host. These are known as 'exposed antigens'. In the case of *D. gallinae*, the natural immune reaction of the hens against these 'exposed antigens' is not effective for controlling the mites (Wright *et al.*, 2009; Bartley *et al.*, 2015). The development of vaccines against *D. gallinae* is based on the use of 'concealed antigens'; the antigens not exposed to the host during the blood-sucking, like some proteins associated with the midgut of *D. gallinae* (Pritchard *et al.*, 2015). The possibility of using a concealed midgut-related antigen has allowed the development of an effective vaccine, marketed as TickGARD against the cattle tick *Rhipicephalus microplus* (Canestrini) based on the BM86 protein (Wright *et al.*, 2009).

The development of vaccines of *D. gallinae*, including the search for antigens, is still an ongoing process, which has taken advantage of the studies of genomics, and of the transcriptomes and the genome of *D. gallinae* (Schicht *et al.*, 2014; Burgess *et al.*, 2018). Different research works highlighted the potential of both, native (autogenous) and recombinant antigens for vaccination against the mite (Harrington *et al.*, 2009; Wright *et al.*, 2009; Bartley *et al.*, 2015, 2017). In an autogenous vaccine, mites for antigen extraction are used from the same premises for which the vaccine will be used. In autogenous vaccines, the effective antigen(s) are not well-defined and, as such, poorly quantifiable. The efficacy of autogenous vaccines cannot be quantified and may vary between batches. In a recent field evaluation of both an autogenous vaccine and a prototype recombinant vaccine, the autogenous vaccine led to a reduction of 78% of the mite population, while the recombinant vaccine did not show a significant difference in mite numbers compared to the control. The autogenous vaccine is a possibility for use against *D. gallinae* on a small scale, and developing an optimal, standardized recombinant sub-unit vaccine with prolonged efficacy, to avoid the need for boosting, is a long term goal (Bartley *et al.*, 2017).

Market *et al.* (2016) found a method for the development of an immunological control strategy, based on the identification of mite protein antigens, which elicit antibodies with anti-mite activity in the immunized chicken. Hens were immunized with different PRM protein extracts, formulated with two different adjuvants, and IgY-antibodies were isolated from the eggs. A PRM *in vitro* feeding assay that used chicken blood spiked with these IgY preparations was used to detect antibodies, which caused PRM mortality. *In vitro* feeding of mites with IgY isolated from hens immunized with PRM extract formulated with one of the adjuvants showed a statistically significant increase in the mortality as compared to control mites. After the separation of total PRM extracts in two-dimensional gels, several protein spots were recognized by such IgY preparations. Ten protein spots were subjected to mass spectrometry (MS/MS) for the identification of the corresponding proteins. Complete protein sequences were deduced from genomic and transcriptomic assemblies derived from high throughput sequencing of total PRM, DNA and RNA. The results may contribute to the development of an immunological control strategy for *D. gallinae* (Market *et al.*, 2016).

Vaccine research aimed at identifying new forms of poultry mite control has highlighted new perspectives (Bartley *et al.*, 2015; Kasburg, 2016; Miguel, 2019; Xu *et al.*, 2020), driven by an increase in egg and meat production and consumer demand, and it is performed in an environmentally responsible and organic way (George, 2010). Vaccines are retained as solutions that offer protection and can also be used in IPM to cut back on the use of acaricides. Moreover, vaccines are safe for poultry and do not pollute the environment, leave residues in meat and eggs, or pose a risk of triggering resistance in mites (Bartley *et al.*, 2015; Makert *et al.*, 2016; Miguel, 2019; Hwang, 2023). Results of studies have shown the efficacy of vaccines against ectoparasites, as in the case of transmission-blocking vaccines (TBVs) (Bartley *et al.*, 2015; Kasburg, 2016; Miguel, 2019; Xu *et al.*, 2020). The vaccination process induces a memory immune response, characterized by a specific antibody production against the antigenic target, providing in this way, protection against pathogens that can be controlled by the humoral immune response (Brannstrom *et al.*, 2010). TBVs aim is to produce a humoral immune response in the vaccinated host, triggering the specific antibody production that is transferred to ectoparasites during blood feeding (Billingsley *et al.*, 2008; Dinglasan and Jacobs-Lorena, 2008; Acquah *et al.*, 2019; Coelho *et al.*, 2019; Ribeiro *et al.*, 2021). The transferred antibodies bind to proteins essential for ectoparasite's survival, hindering its reproduction and transmission of pathogens (Brannstrom *et al.*, 2010; Shimp *et al.*, 2013; Ribeiro *et al.*, 2021; Mariano *et al.*, 2023). The development of TBVs applied to vector-borne diseases was previously described against the sporozoite forms of *Plasmodium* causing malaria. The specific antibody production, triggered by the recombinant Pfs25 protein, acts within the insect vector, interfering in pathogen transmission (Acquah *et al.*, 2019; Mariano *et al.*, 2023). Antibodies transferred during the vector blood feeding, recognize the gamete surface antigen of *Plasmodium*; they affect the parasite's life cycle inside the mosquito by hindering its sexual development and stopping its biological cycle and transmission (Billingsley *et al.*, 2008; Shimp *et al.*, 2013; Acquah *et al.*, 2019; Coelho *et al.*, 2019; Mariano *et al.*, 2023). Shimp *et al.* (2013) reported that in the first clinical trial, the sporozoite vaccine reduced the incidence of disease over a 14-month period in about 50% of vaccinated infants and children, preventing subsequent blood-stage infection. Another case includes the vaccines (TickGARD® in Australia and Gavac® in Cuba) against the cattle tick, *R. microplus*, developed from the recombinant Bm86 glycoprotein extracted from the tick's gut (Harrington *et al.*, 2009; Wright *et al.*, 2016; Bhowmick and Han, 2020; Ribeiro *et al.*, 2021). The mechanism of action is related to the induction of specific antibodies that target cells in the tick's gut and either hinder their development or induce their death (Xu *et al.*, 2020; Ribeiro *et al.*, 2021).

In a laboratory study, Harrington *et al.* (2008) used an extract of *D. gallinae* effective around 50%; they hypothesized that the feeding chamber interfered with the mite's feeding, reducing the transfer of antibodies induced by vaccination. Xu *et al.* (2020) suggested that this low protection rate could be explained according to Harrington *et al.* (2008) by interference from the feeding chamber. Harrington *et al.* (2009) used against *D. gallinae*, the recombinant antigen Bm86 and Subolesin from *R. microplus*; these were able to generate an immune response but characterized by low efficacy (23.03% and 35.1%, respectively); the antigens Bm86 and Subolesin are not found in *D. gallinae*, and their lack probably hampers the development of an effective immune response (Harrington *et al.*, 2009). Wright *et al.* (2016) and Price *et al.* (2019) obtained results below 50% using proteins extracted from *D. gallinae* macerate to act as a possible vaccine. Price *et al.* (2019) believed that the low efficacy may depend on the lack of a specific humoral immune response to the antigen employed as a consequence of low levels of antigen expression or incorrect folding of the expressed protein. Bartley *et al.* (2015) obtained a potential vaccine, using extracted protein from *D. gallinae* macerate, achieving around 50% protection in laboratory tests, but it was ineffective when using recombinant proteins and submitting it to a field test. The cause may have been the lack of proper selection of antigen, inducing an inadequate protective immune response (Bartley *et al.*, 2017). Other specialists (Xu *et al.*, 2020; Fujisawa *et al.*, 2021; Murata *et al.*, 2021) developed recombinant proteins (rDg-CatD-1, Dg-APMAP-N and Deg-CPR-1), obtaining an efficacy rate above 50% in laboratory tests.

Silva *et al.* (2023) reported that the efficacy of transmission-blocking vaccines against poultry mites must be improved. From a practical point of view, the large numbers of animals make administering vaccines time-consuming and costly for the producer. Consequently, the expense and time would be justified, only if poultry farmers dispose of effective vaccines over a long period of time, representing an advantageous, competitive option, with respect to the low cost acaricides available in the market today (Mariano *et al.*, 2023). Replacing chemicals with vaccines is strongly suggested by consumers who demand sustainable food products. However, control methods currently in use, are still characterized by wide variations in effectiveness, which make it difficult to determine the efficacy of the treatment, especially in the field.

BIOLOGICAL CONTROL. The use of natural enemies against *D. gallinae* is expected within an IPM strategy and commonly applied in various crops. The use of biological control is devoid of reductive impact on the environment and allows to reduce the risk of resistance. *D. gallinae* is an ectoparasite, and the use of enemies that share its living environment is promising (Roy *et al.*, 2017; Zriki *et al.*, 2020). The natural communities of enemies of *D. gallinae* living in the layer house should be investigated, as these can partly produce a control against poultry red mites. Some natural enemies are also artificially reared and commercialized, to mass release them for controlling pest species. This methodology is certainly effective in enclosed systems, where the natural enemies are confined to the release site (Harrington *et al.*, 2011). The strategy is already widely applied in greenhouses (Pilkington *et al.*, 2010), in citrus groves (Vacante and Gerson, 2012) and in other agroecosystems (Benuzzi and Vacante (2021). Other natural enemies are some entomopathogenic fungi, nematodes and bacterial endosymbionts.

PREDATORS. Research works on naturally occurring predator mites in poultry houses and wild bird nests allowed to identify candidate predators for commercialization and mass releases, to control *D. gallinae*. Different mite species have been identified, such as *Androlaelaps casalis* (Berlese), *Cheyletus eruditus* (Schrank), *Hypoaspis aculeifer* (Canestrini) and *Stratiolaelaps scimitus* Berlese (previously *Hypoaspis miles*) as predators of *D. gallinae* (Leśna *et al.*, 2009). In Iraq, Al-Kubaisi and Al-Dulaimi (2020) found associated with poultry droppings, various species such as *A. casalis*, *Sejus temperaticus*, *Parasitus paraconsanguineus* Oudemans *et* Voigts, *Eugamasus butleri* Hughes, *Macrocheles merdarius* (Berlese), *Macrocheles glaber* (Müller), *Macrocheles muscaedomesticae* (Scopoli), *Macrocheles matrius* (Hull), *Kleemannia plumosus* (Oudemans), *C. eruditus*, *Cheyletus malaccensis* (Oudemans) and *Pyemotes herfsi* (Oudemans).

These species may be commonly collected in layer houses, and the risk of an impact on non-target biodiversity is limited, although it must be investigated. However, these predators are not *D. gallinae* specific, also predating on other arthropods and even among predators (Roy *et al.*, 2017). *Hypoapsis* species exhibit a high predation ability against *D. gallinae*, higher than that of *A. casalis*, but they are less mobile (Tirry and Van Leeuwen, 2014). Unfortunately, populations of *Hypoapsis* spp. do not establish in poultry houses, as specimens were never retrieved a week after release (Koppert Biological Systems). In Europe, *A. casalis* and *C. eruditus* are released in commercial layer houses, reared and commercialized by Koppert Biological Systems, the Netherlands, as ANDROLIS R and TAURRUS R, respectively. Both predators can be employed complementarily. *A. casalis* mainly feeds on juvenile stages of *D. gallinae* and is very mobile and active, while *C. eruditus* preys on all stages of *D. gallinae* and is a wait-and-sit predator with a slow dispersion. *A. casalis* prefers more humid microhabitats, while *C. eruditus* prefers drier places such as nests and perches. Both predators are usually released preventively, with the protocol of numbers, frequency and locations of releases, depending on the number of hens and housing system. When the level of populations of *D. gallinae* becomes too high nevertheless, extra releases of both species can also be done curatively. The use of predatory mites is promising as a part of a combination of treatments in an IPM strategy, although it should be taken into account that other treatments for *D. gallinae*, such as synthetic chemical, botanical acaricides or inert substances, can also have deleterious effects on the natural predators of *D. gallinae* (Roy *et al.*, 2017). Faleiro *et al.* (2015) revealed that *C. eruditus* may be a possible new tool for the biological control of the ectoparatites *D. gallinae* and *Megninia ginglymura* (Mégnin) (Analgidae). This predator could be considered the main natural enemy, with the best potential for controlling these ectoparasites associated with laying hens and should be taken into consideration for future biological control studies, becoming an additional strategy for controlling *D. gallinae* principally.

ENTOMOPATHOGENIC FUNGI, NEMATODES AND BACTERIAL ENDOSYMBIONTS. The non-chemical control measures against *D. gallinae* include entomopathogenic fungi, nematodes and bacterial endosymbionts. Researches on their employ against *D. gallinae* in poultry houses are not advanced, and they are not yet commercially available for use against *D. gallinae*.

Entomopathogenic fungi inhabit the environment and act by germinating and penetrating the body of arthropods through the cuticle, producing paralysis of essential organs, resulting in the host's death (Miguel, 2019). These organisms are being studied as a tool, in the control of mites on commercial farms for egg production, through spraying in or impregnating traps (Kasburg, 2016; Miguel, 2019; Nascimento, 2019; Lima-Barbero *et al.*, 2020). Good results were found when using entomopathogenic fungi, mainly under laboratory conditions, to control the mite population (Kasburg, 2016; Nascimento, 2019; Tabari *et al.*, 2023). The efficiency of employing these fungi to combat bird mites in the field may face some challenges, such as the capacity for fungus proliferation in uncontrolled risks to animal and human health, and does not leave any residue in the end products (Kasburg, 2016; Miguel, 2019; Nascimento, 2019).

Entomopathogenic fungi are frequently used worldwide, and mainly in Latin America, to control pests in crops (Harrington *et al.*, 2011; Oliveira *et al.*, 2014). Fungi infect insects and mites with their spores that adhere to the host's cuticle, germinate, penetrate and spread into the host's body (Oliveira *et al.*, 2014). Studies under lab-scale conditions have shown that *D. gallinae* is susceptible to *Beauveria bassiana* (Balsamo) Vuillemin, *Metarhizium anisopliae* (Metschn.) Sorokīn, *Trichoderma album* Preuss and *Isaria fumosorosea* Wize, previously *Paecilomyces fumosoroseus* (Wize) (Steenberg and Kilpinen, 2003; Kaoud, 2010; Tavassoli *et al.*, 2011; Immediato *et al.*, 2015; Tomer *et al.*, 2018; Oliveira *et al.*, 2020). Unfortunately, some studies in semi-commercial conditions showed that the action of these fungi is unsatisfactory against *D. gallinae* (Steenberg *et al.*, 2006). The humidity levels in poultry houses are too low to ensure fungal transmission (Harrington *et al.*, 2011), and the effectiveness of these pathogens significantly depends on the fungal strain (Immediato *et al.*, 2015). Moreover, being these fungi not selective for *D. gallinae* only, they may negatively affect the environment, producing an environmental disequilibrium (Pritchard *et al.*, 2015). Regardless of the above, these means are harmless for poultry, eggs and humans (Oliveira *et al.*, 2020). These fungi could be used within traps placed in the poultry house (Oliveira *et al.*, 2014), for example within an attract-and-kill strategy. Nascimento *et al.* (2020) showed the successful use of *B. bassiana* in autoinoculation devices (traps with corrugated cardboard or loofah sponge) to control *D. gallinae* in both laboratory conditions and poultry houses. This employ allows to overcome the environmental problems.

Recently, according to the warm and humid conditions of chicken farms, Park *et al.* (2021a) investigated the control of *D. gallinae* with an alternative application to the chemical control with the entomopathogenic fungus *B. bassiana*. Based on the characterization and virulence assays of *B. bassiana* isolates against red mite, a red mite-pathogenic fungal library was constructed. A total of five highly virulent *B. bassiana* isolates were selected from the first bioassay. They showed 70–100% mortality in 10 days, which was higher than cypermethrin, showing 40% mortality. Considering their fungal virulence, conidial production and thermotolerance, isolate JEF-410 was established to be the best candidate for industrialization. JEF-410 showed a dose-dependent virulence against red mite nymphs. JEF-410 conidia from millet granule based mass culture were applied to a commercial chicken farm, and it showed significant suppression of population increase under high humidity conditions. *B. bassiana* JEF-410 having higher control efficacy than cypermethrin, could be effectively used to control *D. gallinae* in the fungus friendly humid chicken farms. This work solved the challenges of mass production of *B. bassiana* and management of the red mite in poultry farming under right conditions of RH and temperature (Park *et al.*, 2021a).

As concerns the nematodes, no field experiments on their efficacy against *D. gallinae* have been carried out yet, but using them for controlling flies and beetles in poultry houses has been ineffective, despite positive results in laboratory conditions (Axtell, 1999). Both fungi and nematodes require particular environmental conditions, such as high humidity levels and free water, making it very hard to use them in practice in poultry houses (Harrington *et al.*, 2011; Sparagano *et al.*, 2014a).

Targeting endosymbionts that are vital for mite reproduction and growth is another pathway that is being investigated for control of *D. gallinae*. Some endosymbionts have been identified in *D. gallinae* with DNA sequencing (De Luna *et al.*, 2009), but further studies must be carried out, before it can be employed in the control.

PHYSICAL CONTROL. Physical control includes hygiene measures, fundamentally based on cleaning and disinfection, the modification of light regime, the temperature and the use of electrical current. All these means may be regarded as means able to realize the 'prevention'.

CLEANING. Practices such as cleaning and hygiene facilities between flocks, can greatly reduce the level of populations of *D. gallinae* but are often underestimated as control practices, or too costly, as steam and heat treatments, and in the meantime, these results are unable to, alone, control the mite infestations.

LIGHT REGIME. Disrupting light-dark cycles can affect the mite populations. The host-searching activity of *D. gallinae* starts during the dark period, and the highest activity happens 5–11 h after darkness, so when this dark period is stopped, their host-seeking activity slows down, and their natural feeding cycle is disrupted (Sokół *et al.*, 2008; Wang *et al.*, 2019). Zoons (2004) and Stafford *et al.* (2006) showed that a short cycle of intermittent light/dark periods reduces the numbers of *D. gallinae*; prolonged darkness compared to standard light regimes increases the number of mites (Wang *et al.*, 2019). Disturbing the dark-light rhythm also largely affects hens, but this technique cannot be applied in Europe, as the EU legislation requires a statutory dark period of 8 h (1999/74/EC).

TEMPERATURE. Effects of temperature on adult and nymph survival of *D. gallinae* were first studied by Wang *et al.* (2020) to optimize storage conditions. They studied the fecundity of *D. gallinae* after mites were stored at optimal storage conditions. Results showed the survival rates of protonymphs (42.11%), deutonymphs (8.19%) and females (19.78%) at 5°C after 84 days were higher than those at 0°C, 25°C and 30°C. Thereafter, the fecundity and the capability of re-establishing colonies of *D. gallinae* were evaluated, after they were stored for 40 and 80 days at 5°C. After storage, the mean number of eggs showed no statistical difference between the treated (5°C for 40 or 80 days) and control groups (25°C for 7 days), while the hatching rates of eggs were in all cases above 97%. The dynamic changes of mite populations and egg numbers showed similar trends to the control group, after the stored adult or nymph mites were fed on chicks. *D. gallinae* can be successfully stored at 5°C for 80 days with no interference with the fecundity of mites, and the stored mites could re-establish colonies successfully. Adults and nymphs were two main stages with capability for low temperature storage. These results suggest that low temperature storage is a viable option for colony maintenance of *D. gallinae* under laboratory conditions (Wang *et al.*, 2020).

In conclusion, temperatures over 45°C are lethal for *D. gallinae*. The Thermokill method developed by 'Van Eck Industrial Hygiene BV' takes advantage of this. The method involves gradually heating hen houses, during the empty period, to above 45°C for at least 2 days. By heating up the poultry house gradually, mites are lured out of their hiding places, which improves the killing efficiency of the heat. It has been reported that this method can be very efficient in eliminating PRM; however, it is expensive, and some housing systems can be damaged by the heat.

Suspected or confirmed cases of dermatitis suggest that the patients should shower extensively and wash their clothes at 60°C; infested premises should be intensively vacuum cleaned and disinfested using pyrethroids, while textiles (curtains, carpets, cushions) should be steamed clean or washed at 60°C (Cafiero *et al.*, 2019). The situation related to occupational dermanyssosis by *D. gallinae* in humans, related to the poultry farm infestation, requires improvements in control strategies. To avoid infestation, poultry workers must be covered by suitable protective clothing, before entering the facilities (Cafiero *et al.*, 2019).

Q-PERCH. The Q-perch® (Vencomatic, the Netherlands) is a perch where two electric wires and insulators are located under the cap of the mushroom shaped perch. According to the design, the hens cannot be damaged. A small electrical current run through the wires, killing the mites on their way from their hiding place to the hen's nightly resting place (van Dick, 2016). The efficacy of the Q-perch is not demonstrated scientifically; moreover, installing a Q-perch is rather expensive and needs a pervasive change of the infrastructure. It is used as a preventive measure to control mite population growth, and not as a curative measure.

CHEMICAL CONTROL. Currently, the control of the poultry red mite is dominated by the use of synthetic acaricides. Organophosphates, pyrethrin and pyrethroids are commonly employed; only a few products are licensed in the EU for use against *D. gallinae*, and not many new molecules are marketed by pharmaceutical companies. The recent planetary scandal related to the presence of residues of fipronil (an unlicensed molecule) in eggs, represents the tip of the iceberg (i.e. the continuous need by farmers of new molecules to combat *D. gallinae* in laying hen farms) (Di Palma and Giangaspero, 2021).

The chemical control provides the use of chemical products as a last possibility, when non-chemical means are not sufficient to control the mite (Barzman *et al.*, 2015). The choice of the acaricide is important, taking into account, minimizing the total quantity of active ingredient applied, though respecting the dosages helps avoid resistance emergence in the mite and optimize the results of the treatment. Unfortunately, none of the chemicals available are totally selective for *D. gallinae*, as they also have toxic effects on other insects and arachnids.

The main acaricides used in the control of poultry red mite belong to the class of organophosphates, pyrethroid, formadin, isoxazolines, carbamate, macrocyclic lactones and dichloro-diphenyltrichloroethane (DDT) (Kasburg, 2016; Pavličević *et al.*, 2021; Santos *et al.*, 2022; Kang *et al.*, 2023). Although these active ingredients can be sprayed into the environment, they do not reach some areas, such as crevices, or pass through feathers of the poultry, thus preventing mites from coming into contact with the chemical compounds (Wright *et al.*, 2016; Nascimento, 2019; Lima-Barbero *et al.*, 2020). However, the most recent molecule – isoxazoline – can be administered through drinking water (Soares and Ximenes, 2022).

In general, application of chemicals is reductive for the poultry production industry, since most of them are highly toxic, constituting a risk to both the animals and humans; highly polluting to the environment, leaving residue in water and soil, and can expose consumers to contaminated eggs and meat (Faleiro, 2012; Kasburg, 2016; Miguel, 2019; Nascimento, 2019). Furthermore, one of the main concerns in using acaricides is the selection of resistant mite populations (Kasburg, 2016; Miguel, 2019; Nascimento, 2019). These chemicals are employed indiscriminately, and many producers ignore the restrictions in their countries (Sillos, 2002; Miguel, 2019; Guerrini *et al.*, 2022; Kang *et al.*, 2023), and several active ingredients are banned in Europe and/or in EU member states (Harrington *et al.*, 2009; Guerrini *et al.*, 2022). So, fipronil is forbidden to be used in animals used as food (Guerrini *et al.*, 2022). Products belonging to the class of carbamates, organophosphates and pyrethroids have been banned in the EU for use against *D. gallinae*. There is a growing tendency to restrict the use of chemicals, especially in food production systems, to avoid food contamination, exposing workers to potentially harmful substances (Guerrini *et al.*, 2022).

Products classified as biocides or veterinary medicines allowed in Europe are reported on the database of the European Medicine Agency database (www.ema.europa.eu) for veterinary medicines, and on the European Chemical Agency database (www.echa.europa.eu) for biocides. These lists can change daily, making it necessary to regularly check for allowed products. Every EU country has its own national admissions, which makes it even more confusing, which product is licensed in a specific country. At now, only the organophosphate phoxim (ByeMite®), spinosad (Elector®), fluralaner (Exzolt®) and some products of silica/diatomaceous earth are allowed in most European countries. These products are allowed during production, although with some restrictions.

The organophosphate phoxim (ByeMite®) is licensed as a veterinary medicine against ectoparasites for livestock (including layer hens) but is not allowed for organic farming. This active ingredient is the only veterinary drug registered for the treatment of poultry red mite infestations. However, it is not available in all EU countries (Flochlay *et al.*, 2017; Hwang, 2023). Although some studies report high efficacy of ByeMite® in multiple systems after repeated application (Keïta *et al.*, 2006; Meyer-Kuhling *et al.*, 2007), others report variable results among different regions, and different contact duration (Abdel-Ghaffar *et al.*, 2009). In addition, resistance emergence against phoxim is already reported in several countries.

Spinosad (Elector®) is a biological acaricide and acts upon the nervous system of the mite; it is licensed as a biocide and is also allowed for organic farming. Tests carried out in layer houses showed that the effect of Elector® does not last long and that it is not sufficient as stand-alone treatment (Abdel-Ghaffar *et al.*, 2009). Other trials carried out in laboratory showed increasing efficacy 5 days (adults) and 6 days (nymphs) after exposure. Further trials carried out in field under commercial conditions confirmed the laboratory results. These trials were conducted in three separate houses on farms with high natural mite populations. The first and second houses were sprayed with concentrations of 2000 and 4000 ppm of spinosad (2 and 4 g/L), whereas the third house was used as an untreated control. Spraying was conducted using a sprayer (Stadikopumpe VAAR 252–200 LE, Dinklage, Germany) with a 200 l reservoir with permanent stirring, a 50 m long flexible tube with a double jet system, and jet size of 0.08 mm. The pressure used was 6–10 bar. Sampling was conducted immediately before the application and once a week for 56 days in the first house and 93 days in the second. In the first house, almost 100% reduction in mite numbers was observed until 28 days post treatment; 74.3% reduction was observed 49 days after treatment. In the second house, rates of reduction were higher than 90% until day 77. The untreated control house maintained high red mite infestations throughout the experimental period. The results showed effective control of all mobile stages of *D. gallinae* with spinosad (Liebisch *et al.*, 2011).

Moreover, fluralaner has been widely used to combat ectoparasites, and its use against *D. gallinae* was approved in the EU, but in a restricted way, leading to many producers to resort to and rely on unlicensed chemicals (Guerrini *et al.*, 2022). A fluralaner based product called Exzolt® has been marketed; it is the first systemic acaricide against *D. gallinae*, administered orally through drinking water, and classified and licensed as a veterinary medicine. It is also allowed in organic farming, though with an extended withdrawal period. Fluralaner is an isooxazoline, and inhibits the ligand-gated chloride channels, with a high selectivity for the nervous system of mites, ticks and insects (Gassel *et al.*, 2014). In laboratory conditions, Exzolt® kills the mites within 4 h (Brauneis *et al.*, 2017); field studies showed that mite reduction almost occurred immediately, and reductions of up to 100% were observed in all tested units (Thomas *et al.*, 2017; Sleeckx *et al.*, 2019). The duration of efficacy (of >90%), varied greatly among the tested layer farms, ranging from 56 to 238 days (Thomas *et al.*, 2017).

Studies carried out in laboratory showed the effectiveness of the spraying of mineral oils, diesel oil, petroleum and plant oils, mainly rapeseed and concentrated orange oil, for the poultry red mite control (Maurer *et al.*, 2009). These oils have mainly a physical action by occluding the stigmata and preventing the normal breathing of the mite (Tirry and Van Leeuwen, 2014). The use of diesel oil is associated with the risk of egg contamination (Maurer *et al.*, 2009). The greasy component of the oils can stain the eggs, affect the functioning of parts of the system like the egg belt, and they are also rather difficult to remove by cleaning. Further, these oils must be applied everywhere, as the mites are able to avoid the oil spots. From a practical point of view, different oils have been used by farmers for years, often successfully controlling mite infestations. The use of these oils is not currently allowed against *D. gallinae*, as there is no registration for any of such products.

INTEGRATED PEST MANAGEMENT. Integrated pest management (IPM), or integrated control (Rezende *et al.*, 2014; Pavlićević *et al.*, 2021) provides identification, certification and monitoring techniques to control mites on poultry farms (Harrington *et al.*, 2011; Decru *et al.*, 2020; Pavlićević *et al.*, 2021). The strategy is based on the study of the environment to understand the difficulties and choose techniques, methods or products capable of increasing mite mortality in environments (Sparagano *et al.*, 2014a; Pavlićević *et al.*, 2021). The IPM may include chemical control measures, using chemicals usually sprayed within environments and on animals (Wright *et al.*, 2016; Nascimento, 2019; Lima-Barbero *et al.*, 2020; Pavlićević *et al.*, 2021). Moreover, may be applied mechanical or physical measures, including cleaning the environment to remove organic matter and sanitizing the materials and equipment afterward (Sparagano *et al.*, 2014a). Biological controls can be employed through traps impregnated with oils or plant extracts (Harrington *et al.*, 2008; Rezende *et al.*, 2014; Pavlićević *et al.*, 2021; Mariano *et al.*, 2023).

According to what was discussed above, the problem of *D. gallinae* is still a worrying issue worldwide. It is important to underline that an acceptable control of their populations in poultry systems would be possible only through Integrated Pest Management (IPM) strategies; however, IPM strategies, largely limited to basic biosecurity, synthetic acaricides, sanitation and physically acting substances, remain under utilized in the poultry industry, compared to the agricultural sector, such as horticulture. There is a long way still to achieve a highly effective and sustainable control of *D. gallinae*. In the meantime, the following simple considerations should be highlighted (Di Palma and Giangaspero, 2021). In conclusion, the current control strategies are not sufficient alone to control *D. gallinae* in layer houses; this suggests studies to evaluate the added value of combining multiple means, as some can work additively or even synergistically. The proposal is at the base of Integrated Pest Management (IPM).

Sparagano *et al.* (2014b) proposed a predicted compatibility matrix of existing and emerging control strategies. However, few combinations have actually been scientifically studied, and this matrix should be mainly considered as a prediction of compatibilities, based on the mode of actions of each treatment separately (Decru *et al.*, 2020). Steenberg and Kilpinen (2014) have shown that the combination of the fungus *B. bassiana* and the inert substances (desiccant dusts) has beneficial effects. Although this study was carried out in a laboratory and the combination still needs to be tested on the field, it highlighted the potential added value of combining means of control.

Two projects highlighted the potential of combining the predatory mites *A. casalis* and *C. eruditis* with the acaricides milbemectin and amitraz, using impregnated cardboard traps. The test carried out in layer houses, cage and aviary systems, showed that the combinations had a better efficacy than all the treatments separately. However, the combinations were still not sufficient to control severe infestations of *D. gallinae*. 'Triple' treatments, with the addition of botanical acaricides had higher efficacy (Tirry and Van Leeuwen, 2014; Sleeckx *et al.*, 2017; Zoons and Deschuytere, 2018).

D. gallinae is attracted by host specific pheromones, emitted by the hens (El Adouzi *et al.*, 2020). A set of five volatile compounds, among the dozens naturally emitted by hens, has been patented for its attraction in the laboratory (Roy *et al.*, 2018). Besides host related attractants, *D. gallinae* is also attracted by the aggregation pheromones of congeners, causing clustering together (Entrekin and Oliver, 1982; Koenraadt and Dicke, 2010). Though the pheromone has not been fully characterized chemically, a series of compounds isolated from the odour of freshly fed *D. gallinae* has been patented as attractants (Birkett *et al.*, 2010).

Interfering in the sensory interactions between *D. gallinae* and its host or between mites among each other are promising possibilities to increase IPM, and such approaches possess a minimal environmental impact. The automated mite counter (Mul *et al.*, 2015) was designed to attract mites, as it provides a heat source. Some botanical acaricides also act on the sensors of *D. gallinae*, as they have a repellent or attractive effect on *D. gallinae*. Attractants and repellents have much potential for use in certain combinations (Decru *et al.*, 2020).

The 'attract-and-kill' and/or 'push-pull' techniques possess an interesting potential for a mite that does not live on the host. The principle of the attract-and-kill or lure-and-kill technique, widely employed in crops (Charmillot *et al.*, 2000), involves diverting the pest from its target by competing odours to other sites, equipped to kill it by various means, like acaricides or entomopathogenic fungi. In the case of *D. gallinae*, the mite traps employed are impregnated with eradicating treatments (Chirico and Tauson, 2002). An IPM strategy gives your preference to the use of non-chemical killing means (Barzman *et al.*, 2015); the use of pesticides only on traps reduces the total amount of pesticide and results in a reduced impact on the environment. Moreover, the strategy slowed down the resistance emergence, as the parasites are exposed to effective doses in an environment (Koenraadt and Dicke, 2010). However, the effective implementation of an attractive substance in such a method has many impediments. The push-pull method combines a repellent activity to push the pest away from its target and attract it to another element. In this way the attract-and-kill strategy is enhanced, by adding a repellent into the process. However, so far, it has never been used to control *D. gallinae*. Repellents alone can also be used (e.g. by preventing *D. gallinae* to hide into the cracks and crevices), resulting in *D. gallinae* becoming more reachable with preferably non-chemical contact means (Decru *et al.*, 2020).

Some plant based food or drinking water supplements increase or sustain the health and natural resistance in hens, and also have an effect on the odour of the hen, rendering it less attractive for the mites and as such working as a repellent (El Adouzi *et al.*, 2020). This can be beneficial as mites will feed less, and starved mites seem to be more susceptible to acaricides and possibly also to desiccants (George *et al.*, 2008b). The potential of combining these supplements with other treatments (i.e. predatory mites and acaricide-impregnated traps) has been demonstrated in field conditions (Sleeckx *et al.*, 2017; Zoons and Deschuytere, 2018).

Not all treatments can be used in combination, and some probably have antagonistic effects. Broad-spectrum approaches like silica or heat treatment (e.g. might have adverse effect on the release of natural enemies). The treatments can easily be combined with simple management actions, like cleaning places where hotspots are found with water and soap, to keep the infestation under control, which has proven to be effective (Sparagano *et al.*, 2014a; Mul *et al.*, 2009, 2020). The influence of different housing systems and of parameters like temperature and humidity on the efficacy of any treatment has also been illustrated (Zoons and Deschuytere, 2018). Using systems that are less beneficial for poultry red mites, and limiting the number of potential mite refugia could also help in controlling infestations (Sparagano *et al.*, 2014b). A holistic approach based on integrating biosecurity and prevention measures, monitoring, attention to different conditions in different housing systems and interactions with environmental conditions is indispensable for controlling poultry red mites in layer houses (Decru *et al.*, 2020).

59 The Family Digamasellidae Evans (Fig. 59.1)

Members of this family are predators, insect associates and in few cases fungivores. Digamasellids are widespread in surface and subsurface soils and in accumulations of decaying organic material such as compost, manure and tidal debris, where they feed on collembolas, nematodes, arthropod eggs and probably fungi (Evans and Till, 1979; Shcherback, 1980). The dorsal shield of the adults is divided; the palpi are five-segmented, and the palptarsal apotele is two-tined; the leg IV usually possesses seven setae on genu (1 2/1, 2/0 1) and seven on tibia (1 1/1, 2/1 1); the anterior portion of female sternal shield and male sternitigenital shield are weakly defined; the male has the sternal setae *st5* on separate platelets, and the female has separate epigynal shield (Lindquist *et al.*, 2009b). The sistematic of this family is not characterized by a definitive and uniform organization. According to Lindquist *et al.* (2009b), the family Digamasellidae is equivalent to subfamily Dendrolaelapinae *sensu* Shcherback (1980) and Karg (1993a). Members of this family are classified into relatively few genera by some authors (Lindquist, 1975b; Evans and Till, 1979), and more taxa by others (Shcherback, 1980; Karg (1993a) to accommodate some of many subgenera of narrow scope, recognized by Hirschmann and Wisniewski (1982a, b). The family includes 13 genera and 261 species (Beaulieu *et al.*, 2011). Here, we treat only the species *Digamasellus presepum* Berlese (1918) reported by Hughes (1976) (Fig. 59.1), on the edges of ceiling boards in the lower hold of a ship. The sistematic of the species has been revised in the last years, and recently, it has been reported as *Digamasellus* (*Presepodendrolaelaps*) *presepum* (Berlese, 1918) [=*Gamasellus* (*Digamasellus*) *presepum*] (Taylor, 2018). However, to not complicate the discussion, we prefer to maintain the old nomenclature here.

The geographical distribution of the mite is fundamentally European, and it has been found in England (Brady, 1970; Hughes, 1976), Germany (Hirschmann, 1960; Karg, 1971), Iran (Kazemi and Rajaei, 2013; Farahi *et al.*, 2018a; Nemati *et al.*, 2018), Italy (Berlese, 1918), the Netherlands (Siepel *et al.*, 2018) and Turkey (Çobanoğlu *et al.*, 2003).

Within this family, the phoresy by deutonymphs is common, and the deutonymphs of *D. presepum* are phoretic on geotrupine dung beetles (Hirschmann, 1960).

Hughes (1976) reported the mite from a ship, and Hirschmann (1960) found it from manure and Brady (1970) reported that it is one of the ten species achieving the greatest density in litter and dropping of poultry in broiler houses. In Iran, it has been found in soil (Kazemi and Rajaei, 2013; Nemati *et al.*, 2018), in buffalo manure and in sheep manure (Farahi *et al.*, 2018a). In the Netherlands, it has been found in litter of henhouse, and in henhouse (Siepel *et al.*, 2018). In Turkey, *D. presepum* was previously reported associated with Tetranychidae and Tenuipalpidae species from ornamental plants in Ankara, Turkey (Çobanoğlu *et al.*, 2003).

In Iran, *Dendrolaelaps* sp. has been found on stored onion and potato, and in soil samples (Ostovan, 1993; Fathipour, 1994; Mirfakhraii, 1994).

 The Handbook of Mites of Stored Products (V. Vacante and C.H.W. Flechtmann)
DOI: 10.1079/9781786390912.0059

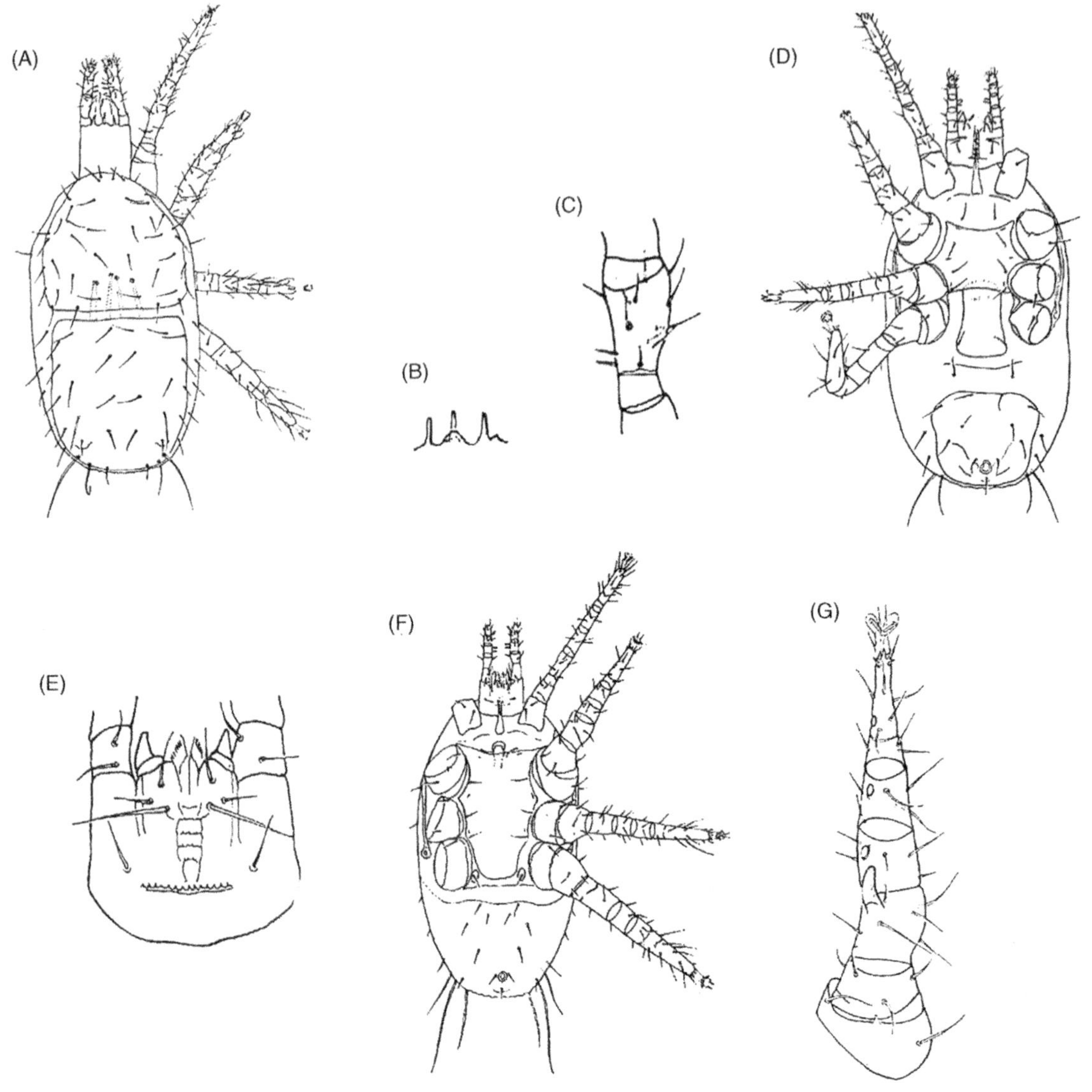

Fig. 59.1. *Digamasellus presepum* (Berlese). Female. Dorsal view (A); tectum (B); femur IV (C); ventral view (D); ventral view of gnathosoma (E). Male. Ventral view (F); leg I (G) (from Hughes, 1976, licensed under the Open Government Licence v3.0).

60 The Family Ologamasidae Ryke (Fig. 60.1)

Mites of the family Ologamasidae are a very conspicuous and abundant group of predators, living in soil and litter, especially in the Southern Hemisphere. Little is known of their biology and behaviour, but they appear to be generalist predators. The family numbers more than 44 genera and 467 described species (Castilho *et al.*, 2016). The family includes only one species collected on stored food, *Gamasiphis lanceolatus* Karg, collected in a glasshouse near Rostock in Germany (Karg, 1987), and reported from Iran on stored rice, rice dust and debris (Noei, 2007; Noei *et al.*, 2008b) (Fig. 60.1).

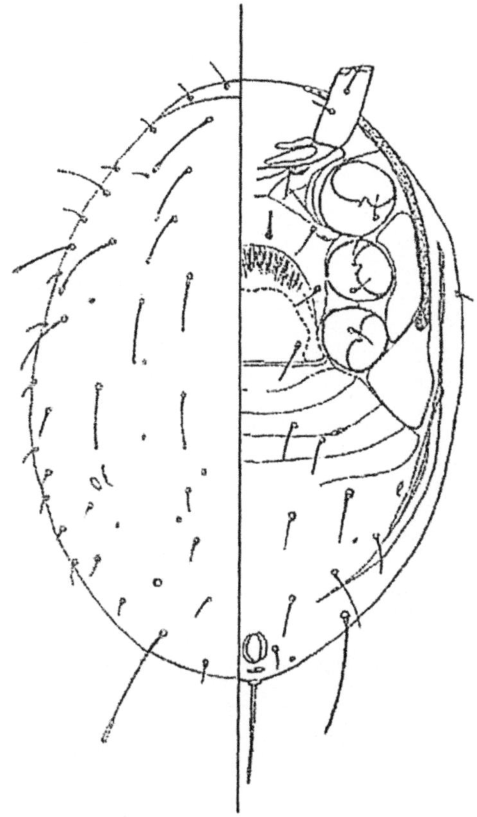

Fig. 60.1. *Gamasiphis lanceolatus* Karg. Female. Dorsal view (at left) and ventral view (at right) (from Karg, 1987, modified).

DOI: 10.1079/9781786390912.0060

61 The Family Rhodacaridae Oudemans

The Rhodacaridae includes mostly small, delicate forms, frequently collected in deeper soil layers, even into underlying groundwater habitats (Krantz, 1986). These mites prey on small collembolans, immature oribatid mites and nematodes (Sardar and Murphy, 1987; Karg, 2000). On stored food, *Protogamasellopsis posnaniensis* Wisniewski and Hirschmann is reported, known from Poland (Wiśniewski and Hirschmann, 1991) and Iran, where it has been collected on stored rice, rice dust and debris (Noei, 2007; Noei *et al.*, 2008b).

DOI: 10.1079/9781786390912.0061

62 The Family Oplitidae Johnston (Fig. 62.1)

Adults of Oplitidae are small, with a length of 750 μm and weakly sculptured. Several species of this family have been described from litter, tree hole and subcortical habitats, and most species are mymecophilous, at least in the adult stage. Many species are tropical. One unidentified species of the genus *Oplitis* sp. is reported on stored rice, rice dust and debris from Iran (Noei, 2007; Noei *et al.*, 2008b). Systematic and morphological aspects of this genus are discussed by Hunter and Farrier (1975). We report here as example the species *Oplitis arboricavi* Hunter and Farrier (Fig. 62.1).

DOI: 10.1079/9781786390912.0062

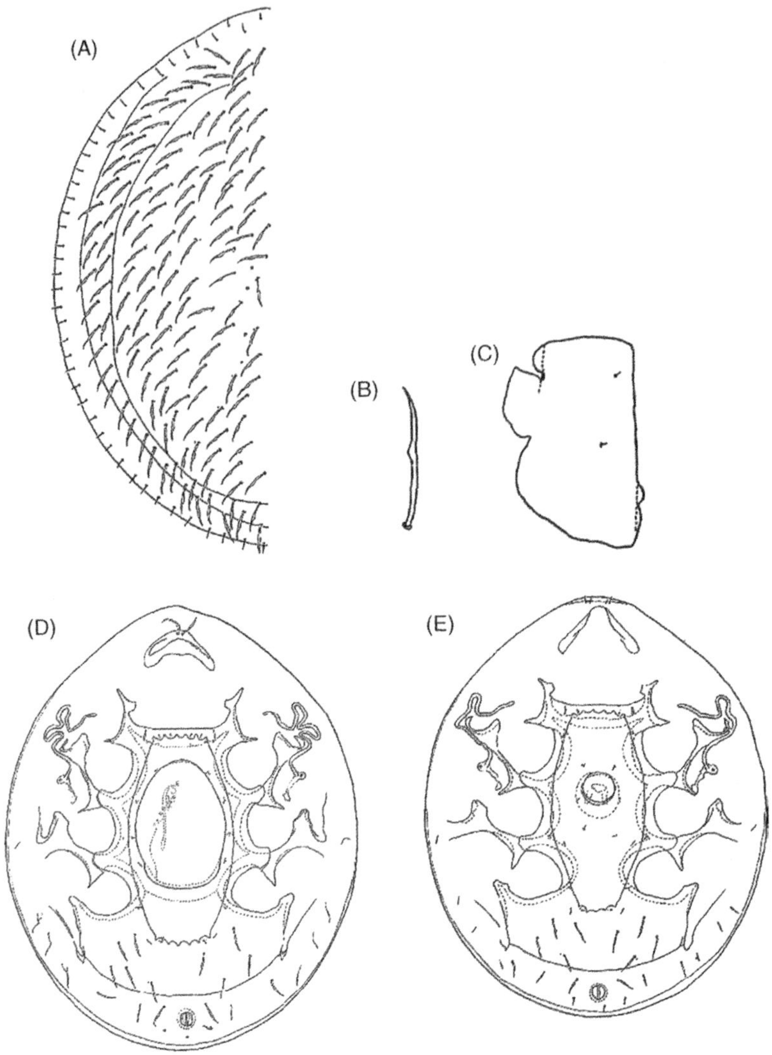

Fig. 62.1. *Oplitis arboricavi* Hunter *et* Farrier. Female. Left side of dorsal view (A); dorsal shield seta (B); coxa I (C). Male. Dorsal view (D); ventral view (E) (from Hunter and Farrier, 1975, licensed under CC BY 4.0).

63 The Family Trematuridae Berlese (Figs 63.1 and 63.2)

The family Trematuridae is cosmopolitan and includes species that occur in bark beetle galleries, soil and litter, and sometimes in the nests of ants and small mammals. The deutonymphs are phoretic on a variety of insects, especially beetles in the families Cerambycidae and Scolytidae. The family includes about 400 species, and a list may be found in Hallan (2010). Hirschmann and Wisniewski (1989, 1993) ascribed all the species in the single genus *Trichouropoda* Berlese and arranged them in a series of groups, which were previously treated as genera or subgenera. The status of these genus-level taxa is unresolved, and a revised future classification will probably divide the species among ten or more genera. The same need arises at a specific level. So, as concerns the species collected on stored food, Mašán (2001) states that *Trichouropoda jacksonia* and *Leiodinychus krameri* (G. *et* R. Canestrini) are junior synonyms of *Trichouropoda patavina* (G. Canestrini), while Hirschmann and Zirngiebl-Nicol (1961), Hirschmann and Wiśniewski (1988) and Bal and Özkan (2007) identified *L. krameri* (G. *et* R. Canestrini) as *Trichouropoda orbicularis* (Koch). Similar uncertainty and need for in-depth studies is evident in the placement of species in higher taxa. In fact, while some authors include some species in the Trematuridae family (Lindquist *et al.*, 2009b), others place them within the Uropodidae family.

In order to avoid confusion, pending a complete review of the group, we prefer to discuss the above-mentioned species, not to face the systematic details, according to the technical nature of the text.

On stored food are known three species [*Trichouropoda elegans* (Kramer), *T. patavina* (G. Canestrini) (Fig. 63.1) and *T. orbicularis* (Koch) (Fig. 63.2)], all ascribed to the genus *Trichouropoda* Berlese.

T. elegans (Kramer) is reported from various countries, such as Germany, France, Great Britain, Sweden, Finland, Poland, Czech Republic, Slovakia, Ukraine, Hungary, Greece and Iran (Błoszyk *et al.*, 2018). In Iran, it has been collected on straw caches, stored rice, rice dust and debris (Noei, 2007; Noei *et al.*, 2008b; Naderi *et al.*, 2016).

T. patavina is synonym of *Trematura jacksonia* Hughes and *L. krameri* (Mašán, 2001). The mite is known from Egypt (El-Sharabasy, 2010; Metwally *et al.*, 2014), England (Hughes, 1976), Ireland (O'Farrell and Butler, 1948; Hughes, 1976) and Iran (Kamali *et al.*, 2001). The mite has been found on stored rice, rice dust and debris in Iran (Noei, 2007; Noei *et al.*, 2008b), on stored foodstuffs (O'Farrell and Butler, 1948), floor dust obtained from oatmeal mills (Butler, in Hughes, 1976), in fine sievings from beer's yeast (Cranham, in Hughes, 1976), in maltings (Ethelston, in Hughes, 1976), in grain residue (Hughes, 1976) and in farm grain stores (Cusack, 1974; Cusack *et al.*, 1975).

T. orbicularis is retained synonym of *L. krameri* (Hirschmann and Zirngiebl-Nicol, 1961; Hirschmann and Wiśniewski, 1988), described by Canestrini, as *Uropoda krameri* (1884). More recently, Bal and Özkan (2007) confirmed the identification of the species as *T. orbicularis* and consider *U. krameri* and other uropodoid mites as synonyms of *T. orbicularis*. The mite may be identified following Hughes (1976), Smiley (1987) and Bal and Özkan (2007).

The geographical distribution of this mite is cosmopolitan and on stored food is reported from different countries, such as Australia (Champ, 1966; Halliday, 1998), Canada (Sinha, 1964b), China (Shen, 1984), Czech Republic (Lukás *et al.*, 2007), India (Gupta and Chatterjee, 2004), Indonesia (White, 1995), Iran (Modarres Awal, 1994, 1997), Ireland (O'Farrell and Butler, 1948), Taiwan (Tjying, 1970), Turkey (Bal and Özkan, 2007), United States (Radinovsky, 1962; Smiley, 1987) and other countries. The infested commodities include damp, mouldy and sprouting barley, wheat, corn (Hughes, 1961), condiment, copra, dried fruit, dried vegetables (Hagstrum *et al.*, 2013), grain (Michael, 1889; Lukás

DOI: 10.1079/9781786390912.0063

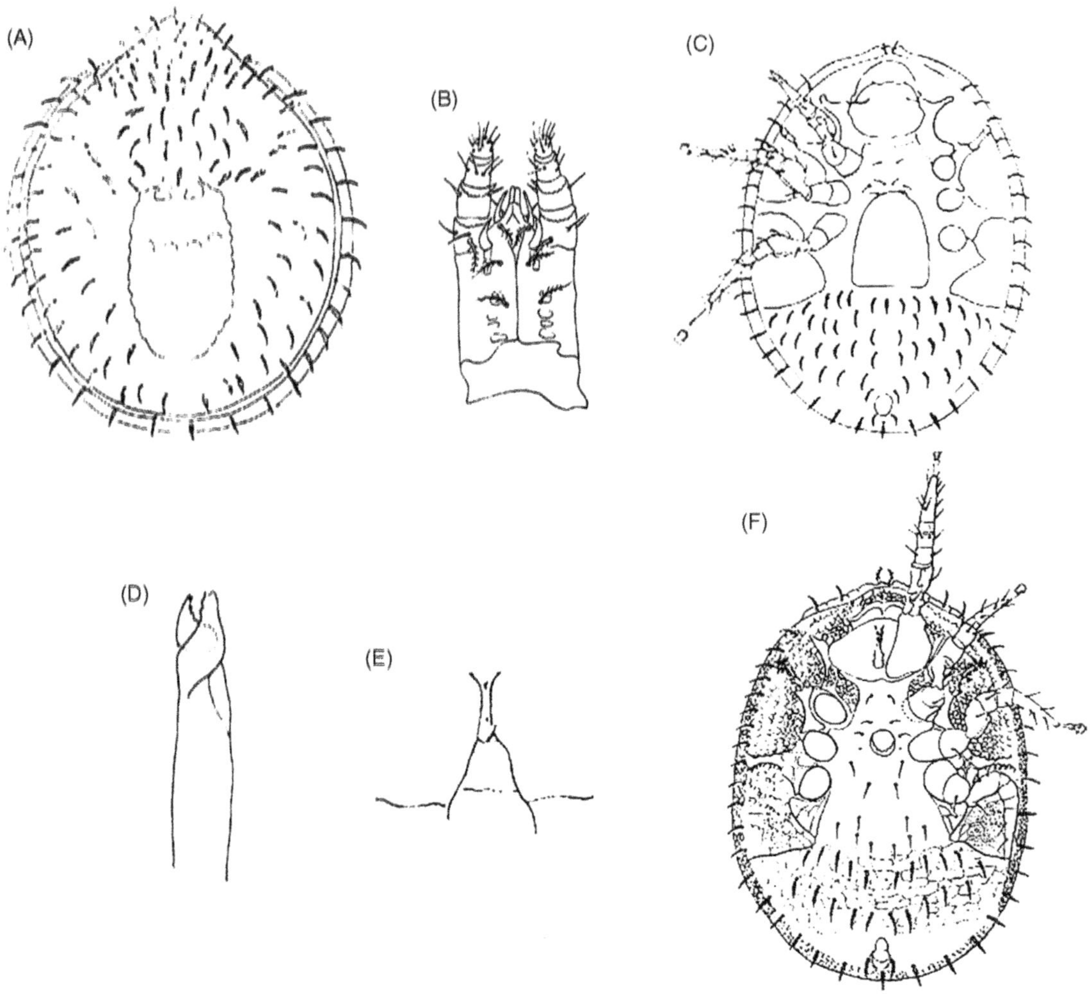

Fig. 63.1. *Trichouropoda patavina* (G. Canestrini). Female. Dorsal view (A); ventral view of gnathosoma (B); ventral view (C); chelicera (D). Male. Tectum (E); ventral view (F) (from Hughes, 1976, licensed under the Open Government Licence v3.0).

et al., 2007), hay (André, 1934), damp, mouldy and sprouting maize, maize residue (Bal and Özkan, 2007), mushrooms, groundnut debris, pulse, rice (Hagstrum *et al.*, 2013), damp, mouldy, sprouting wheat (Tjying, 1970) and wheat bran (Hagstrum *et al.*, 2013). Common facilities are commercial grain storage (Tjying, 1970), farm (Michael, 1889), food store, mill, railcar, ship and warehouse (Hughes, 1961). In Ireland, the mite has been collected on stored foodstuffs (O'Farrell and Butler, 1948); farm and commercial grain stores, flour mills and railway wagons, which contained grain (Cusack, 1974; Cusack *et al.*, 1975; Cusack and Brennan, 1976). Hughes (1976) reported that this mite may be found in large numbers in damp, mouldy sprouting barley, wheat or corn in ships and warehouses. The mite lives commonly in moss. André (1934) reported that *T. orbicularis* occurs in large numbers in private house. Ethelson (in Hughes, 1976) collected *T. orbicularis* from litter of a broiler house. In Iran, the mite has been collected on stored products, stored rice, rice dust and debris (Sepasgozarian, 1978; Modarres Awal, 1994, 1997; Noei, 2007; Noei *et al.*, 2008b), and in Ireland on stored foodstuffs (O'Farrell and Butler, 1948); the species is known from farm and commercial grain

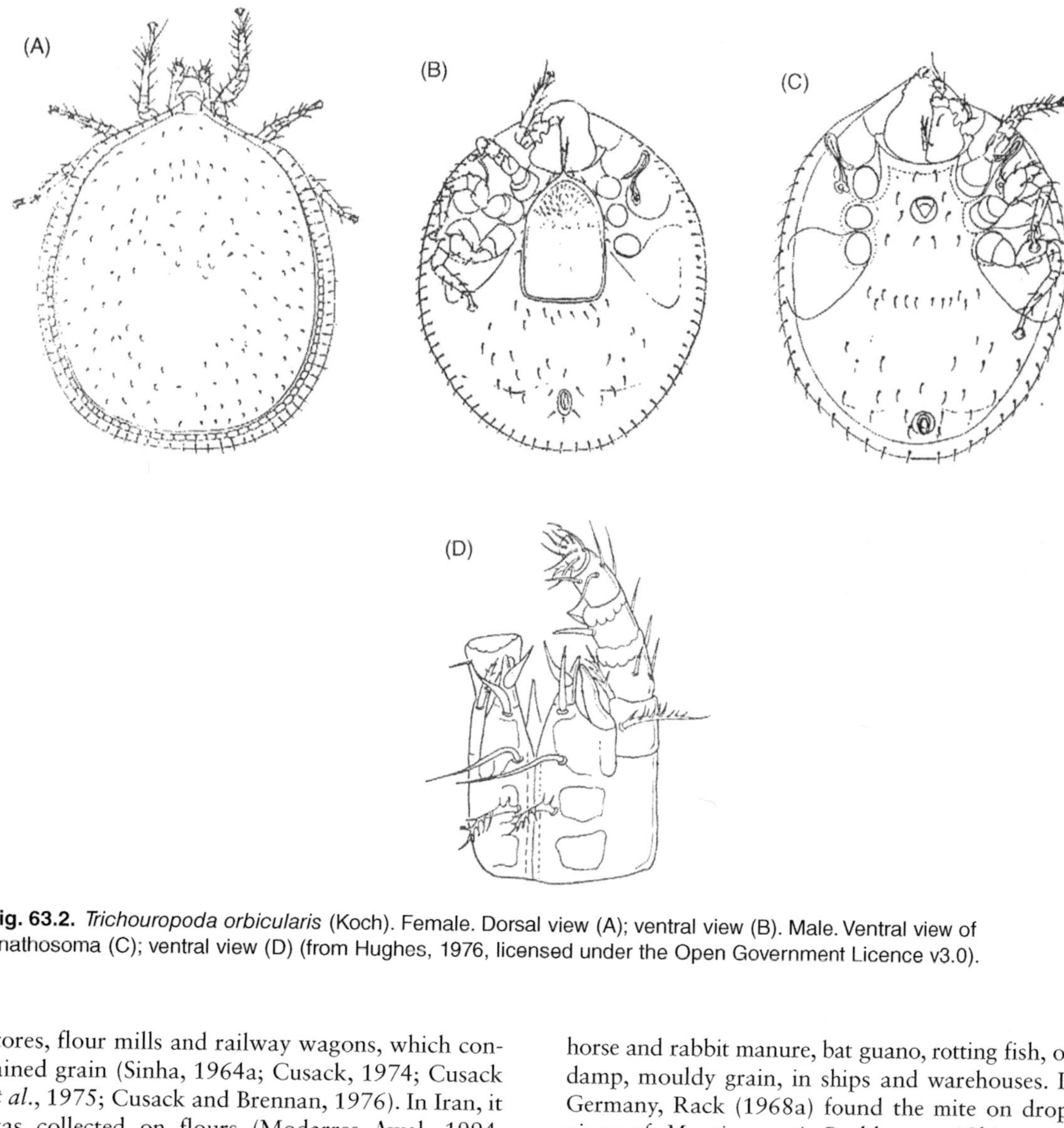

Fig. 63.2. *Trichouropoda orbicularis* (Koch). Female. Dorsal view (A); ventral view (B). Male. Ventral view of gnathosoma (C); ventral view (D) (from Hughes, 1976, licensed under the Open Government Licence v3.0).

stores, flour mills and railway wagons, which contained grain (Sinha, 1964a; Cusack, 1974; Cusack *et al.*, 1975; Cusack and Brennan, 1976). In Iran, it was collected on flours (Modarres Awal, 1994, 1997). André (1934) reported that the mite also was collected in moss, decaying or rotten hay or grass, and in the dirt or mud in stables, and that presumably, it was also herbivorous and lichenivorous. Moreover, he found *L. krameri* in the dust of hay or grasses in a granary. In many countries, including the Pacific Northwest, the same behaviour was observed (Krantz, 1961), and the mite has been described as Tertiary Acariforms or feeders on two fungi and decaying vegetable debris (Krantz, 1961). Hirschmann and Zirngiebl-Nicol (1961) reported the mite in decaying hay, wood, pile, cattle, horse and rabbit manure, bat guano, rotting fish, on damp, mouldy grain, in ships and warehouses. In Germany, Rack (1968a) found the mite on droppings of *Myotis myotis* Borkhausen (Chiroptera, Vespertilionidae) heavily soiled and mouldy animal feed, barn with old bedding, chicken coop. In Israel, the mite has been collected in poultry house (Mumcuoglu and Lutsky, 1990). Błoszyk (1999) noted that *T. orbicularis* prefers unstable microhabitats (especially nests of birds), and Slovak specimens were especially collected from nests of birds (personal communication with Dr. Mašán). Błoszyk (1999) says, it would probably be phoretic species. It is possible that the species eat waste material of birds there. Błoszyk *et al.* (2020) studied the occurrence of mites in bat boxes and focused on

Uropodina, in Western Poland. Guano was collected from 58 bat boxes occupied by ten species of bats. Excrements from particular bat boxes were placed separately in string bags and transported to the laboratory. The study revealed only one species of Uropodina mite on bat guano in the studied bat boxes, namely *Leiodinychus orbicularis*. Mégnin (1894) found the mite associated with human corpses in dry decomposition and skeletal stage.

Radinovsky (1965a) reported that at 20°C and 95% RH with white bread as food, the life cycle takes from 54 to 80 days. At 30°C and the same RH, it is completed in from 31 to 41 days. Mites reared at 20°C were found to be larger than those reared at 30°C. Radinovsky (1965b) observed that male mates with the female by turning her over and, with the aid of chelicerae and legs, pushing a spermatophore under the apical region of the genital shield. The female then rights herself, but the male then climbs on to her back. The spermatophore remains in position and gradually becomes smaller, as the spermatozoa are assimilated into the oviduct. Complete assimilation takes from 40 minutes to 7¼ hours, and during this time, the male is carried around by the female. The spermatozoa are rod-shaped, 18 µm long, 1.5 µm wide. Not more than one egg is usually laid and sometimes the young larvae escape from the egg at the moment of oviposition (Michael, 1889; Radinovsky, 1965b). Sometimes the deutonymphal stage is prolonged (Hughes, 1961; Radinosky, 1965a). This may be due to unfavourable conditions, or it may be because the deutonymph is the stage at which dispersal normally occurs. Very large numbers of this mite are often found in damp, mouldy sprouting barley, wheat or corn in ships and warehouses. It is also commonly found in moss. André (1934) records it as occurring in large numbers in a private house. Ethelston collected it from the litter of a broiler house. Sinha (1964c) states that it is very resistant to cold, and will survive for more than 168 hours at –18°C ±1°C.

64 The Family Eviphididae Berlese (Figs 64.1 and 64.2)

The family Eviphididae is a group of predatory soil mites, characterized by a wide range of ecology and behaviour, cosmopolitan and common in various types of habitats, including stable and continuous microhabitats in soil and leaf litter, as well as temporally and spatially isolated patches of resources, such as dung and carrion. The family numbers a range of insecticolous species that have strong associations with scarab beetles and other arthropods, including some that inhabit the subelytral cavity of their hosts and breed in their subterranean nesting chambers. Phoresy on flying insects ensures the existence of species in ephemeral and scattered habitats (Mašán and Halliday, 2010).

The Eviphididae are widely quoted as being predatory, especially as predators of nematodes, and other possible preys have been mentioned, including insect eggs and larvae. However, direct evidence of feeding behaviour is available for only a few species. Sardar and Murphy (1987) and Murphy and Sardar (1991) found that *Alliphis halleri* (G. *et* R. Canestrini) (Fig. 64.1) fed and reproduced on a diet of three species of nematodes but did not feed on mites, Collembola, enchytraeid worms, or insect eggs and larvae when they were offered to it. These results support the statements by Karg (1971, 1983) that *A. halleri* is a specific predator of nematodes. Karg and Grosse (1983) extended this observation to show that *A. halleri*, referred to as *Alliphis siculus* (Oudemans) (Fig. 64.2), can have a beneficial effect by feeding on pest nematodes in agricultural crops (Mašán and Halliday, 2010). Evans (1969) divided the Eviphididae into two main ecological groups, consisting of the first of non-phoretic species that occur in stable continuous habitats such as soil and forest litter, and the second of species that occur in temporary patches of organic matter, such as dung, and which disperse by phoresy on various insects. The second of these categories actually covers a wide range of different life history strategies (Mašán and Halliday, 2010).

The family Eviphididae was originally proposed by Berlese (1913), as tribe Eviphidini, to accommodate the genera *Eviphis* Berlese and *Copriphis* Berlese. Berlese placed the Eviphidini in the family Laelapidae, together with the Laelapini, Iphiopsini and Phytoseiini, but produced no detailed diagnosis for any of the included genera. These genera were included in the Laelapidae by many later authors, among which Vitzthum (1941), Baker and Wharton (1952) and Turk (1953). Evans (1957) first raised this taxon to full family rank, with the genera *Eviphis*, *Alliphis* Halbert and *Melittiphis* Berlese, and transferred the subfamily Thinoseiinae Evans into the Eviphididae, including the genus *Thinoseius* Halbert. Mašán and Halliday (2010) reported that the family currently includes 19 genera and about 120 species found over all climatic zones of the world. There are 16 genera and 29 well defined and recognizable species in Europe, 14 genera and 19 species in Slovakia. Beaulieu *et al.* (2011) reported 19 genera and 108 species. The family is currently divided into two subfamilies, namely Thinoseiinae, which includes only the genus *Thinoseius*, and Eviphidinae, which includes all the other genera (Mašán and Halliday, 2010).

Females of Eviphididae are podospermal with a holodorsal shield. The sternal shield bears a maximum of three pairs of setae; the metasternal setae are inserted on soft integument, or on distinct metasternal platelets; the epigynal shield bears one pair of setae or none. The anal shield has one pair of para-anal setae and one postanal seta, sometimes surrounded by sclerotized integument that mimics a ventri-anal shield. Tibia I possesses four or five dorsal setae (1–5/3–2, 1–5/2–2, 1–4/2–2 or 1–4/2- 1); tibia I and genu I each have one anterolateral seta (1–5/3–2, 1–5/2–2 or 1–4/2–1) and tibia III has seven setae (1–3/2–1) (Mašán and Halliday, 2010).

On stored food are known two species, *A. siculus* and *A. halleri*, whose correct identification in the past has been controverse. The type species of this

(V. Vacante and C.H.W. Flechtmann)
DOI: 10.1079/9781786390912.0064

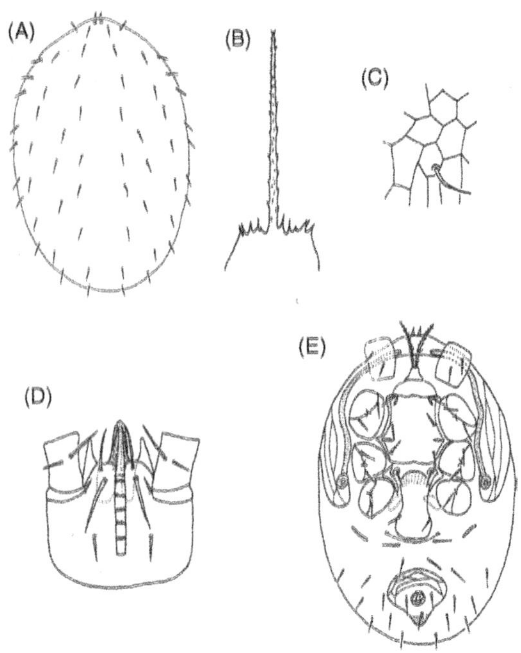

Fig. 64.1. *Alliphis halleri* (G. et R. Canestrini). Dorsal view (A); tectum (B); sculpturing on the dorsal shield (C); ventral view of gnathosoma (D); ventral view (E) (from Russel Shoemake, 1970).

genus is *A. halleri*, collected from 'the decayed roots of henbane and under the bark of cut pine logs', in Ireland (Halbert, 1923). Successively, Karg (1963) in his review of the genus reported *Alliphis* spp. from rotting organic substrates in top soil and also on coprophilous beetles, particularly the Scarabaeidae. Oudemans (1905c) described *A. siculus*, and Karg (1971) stated that all stages of this species employ *Geotrupes* spp. as phoretic hosts. Costa (1963) redescribed the male, female and deuteronymph of *A. halleri* from *Copris hispanus* (Linnaeus) in Israel, and commented that on the basis of literature alone, he was unable to separate *A. halleri* and *A. siculus*. Samšiňák and Daniel (1978) observed that *A. halleri* is the most numerous member of the genus, having been recorded on *Geotrupes stercorarius* (Linnaeus) in Central Europe, on *C. hispanus* in Israel and on *Geotrupes laevistriatus* Motschulsky in Japan. They compared the specimens named by Karg as *A. siculus*, with material from *G. stercorarius* and compost highlighting no differences, and concluded that *A. siculus* to be synonymous with *A. halleri*. In Britain, Hyatt (1959) recorded *A. halleri* from *G. stercorarius*. Evans *et al.* (1961) recorded it from this beetle, as well as from a wide range of soils (Christie, 1983).

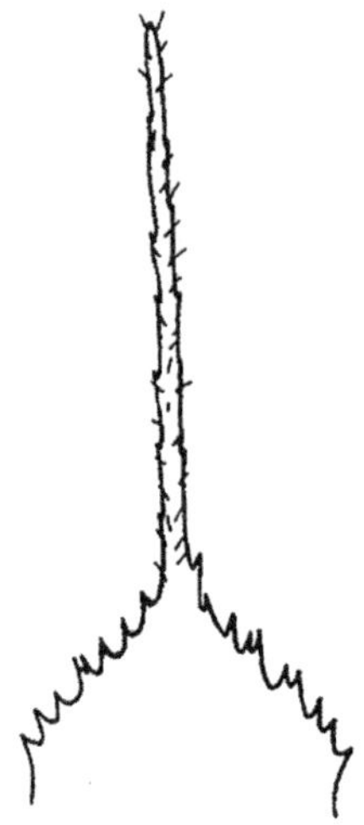

Fig. 64.2. *Alliphis siculus* (Oudemans). Tectum (from Russel Shoemake, 1970).

Successively, Halliday (2008) studied the problem and concluded that the names *A. siculus* and *A. halleri* have been widely used for mites that are common in agricultural soils, and phoretic on dung beetles, especially in Europe. These names have sometimes been regarded as synonyms. However, examination of the type specimens of *A. siculus* showed that this species has often been misidentified. These two species differ in many ways, especially in characters of the dorsal idiosomal chaetotaxy. He redescribed both species and provided details of how they may be distinguished, most easily by the lengths of the setae on the dorsal shield. *A. halleri* is widespread in Europe, North Africa and the United States, while *A. siculus* is known only from Italy (Halliday, 2008).

Most important morphological characters of the females of *A. siculus* are, the posterior dorsal shield with five pairs of *J*-series setae, with an extra pair present between the setae J_2 and J_5; longest posterior dorsal setae 45–65 μm in length, and the length of dorsal shield is 452–507 μm. The mite was described by Oudemans (1905e) and up to 2007 known only from Italy (Mašán and Halliday, 2010). It has been collected on grain in Czech Republic (Lukás *et al.*, 2007). Females of *A. halleri* have the posterior dorsal shield with four pairs of *J*-series setae, without an extra pair between J_2 and J_5, and longest posterior dorsal setae less than 30 μm in length. The peritrematal shields with a sculptural pattern of two longitudinal lines in posterior part; the reticulation of dorsal shield well developed over whole surface; larger species, width of epigynal shield greater than 45 µm, width of anal shield greater than 80 µm, the length of dorsal shield 420–500 µm. Its geographical distribution is Holarctic. It has been collected on bread van in Ireland (Cusack *et al.*, 1975). Sardar (1980, in Christie, 1983) carried out an extensive study of *A. halleri*, including laboratory studies on its biology and food consumption.

65 The Family Laelapidae Berlese (Figs 65.1–65.7)

Members of the family Laelapidae are free-living predators in soil, as well as nidicolous commensal, facultative or obligate ectoparasites in nests and colonies of vertebrates and insects, while the possible endoparasitic association to marsupial mammals has not been fully investigated (Evans and Till, 1966; Radovsky, 1985; Strong and Halliday, 1994; Gettinger *et al.*, 2005; Lindquist *et al.*, 2009b; Faraji and Halliday 2009; Dowling and OConnor, 2010). Mites of the subfamily Laelapinae are the most common ectoparasites of small rodents and frequently found on rodent's bodies or in their nests (Strandtmann and Wharton, 1958; Gettinger *et al.*, 2005; Lindquist *et al.*, 2009b; Mašán and Fenďa, 2010; Kaminskiene *et al.*, 2017, 2020). Species of the genera *Haemolaelaps* Berlese, *Echinolaelaps* Ewing and *Laelaps* Koch) are blood-feeding parasites of rodents and are found worldwide; only a few species can occasionally bite humans (Wall and Shearer, 2001).

Members of the subfamily Hypoaspidinae (Hypoaspididae of Karg, 1993a) are often collected in litter or soil substrates, such as species of the genera *Pseudoparasitus* Oudemans, *Ololaelaps* Berlese and those often included under *Hypoaspis* Canestrini, such as *Gaeolaelaps* Evans and Till, *Cosmolaelaps* Berlese and *Stratiolaelaps* Berlese, while species of these and other genera are routinely or occasionally found in the nests of mammals or arthropods or associated with insects (Evans and Till, 1966, 1979; Karg, 1978, 1993a). Some species are ectoparasites of adult bees or their brood, like *Tropilaelaps* spp., and some of them are also known to feed facultatively on nesting host animals by chewing or punching a hole in the skin (Radovsky, 1967; Lindquist *et al.*, 2009b). Some species have established close phoretic relationships with other arthropods and less commonly with vertebrates (Krantz, 1998a; Lindquist *et al.*, 2009b). Those are common predators in different kinds of habitats, such as freshwater and marine habitats, the littoral-intertidal estuarine milieu and organic matter (Costa, 1962, 1968, 1974; Lindquist *et al.*, 2009b).

Laelapidae possess a dorsal plate with more than 23 pairs of setae. The female has a flask-shaped genital shield, small metasternal shields and the stigmata, with straight peritremes open between the legs III and IV. There are no ventral leg grooves, and the leg I usually carry ambulacra (Gerson *et al.*, 2003). The tibia I usually with two ventral setae, including av_2; the genu IV usually with two posterolateral setae, including pl_2; subcapitulum with internal malae usually well developed, with fimbriate lateral margins, and equal to or longer than corniculi (Lindquist *et al.*, 2009b).

The family Laelapidae currently comprises nine subfamilies, 90 genera and more than 1300 described species of parasites of vertebrates and invertebrates, as well as free-living forms and arthropod symbionts (Lindquist, 2009b; Beaulieu *et al.*, 2011). Trach and Joharchi (2018) reported that the family Laelapidae comprises 146 genera and over 1500 species. Few species of this family are biocontrol agents (Kazemi, 2014). An updating catalogue of free living species has been recently published by Moraes *et al.* (2022).

Arrhenotoky appears to be the main sex-determining mechanism (De Jong *et al.*, 1981; Ruf, 1991), but species in some taxa generate only uniparental progeny. The resultant populations may be exclusively female, whereas others consist of males and females (Mihm and Chiang, 1976).

Species of the genus *Hypoaspis (Stratiolaelaps)* are common predators in soil, mostly generalists, and have been employed in the biocontrol of soil-inhabiting mites and insects. Only two species – *Geolaelaps aculeifer* (Canestrini) (= *H. aculeifer*) and *Stratiolaelaps miles* (Berlese) – have been used in greenhouses to control a number of soil-inhabiting pest insects and mites. Presently, these are commercially available for the control of mushroom flies and western flower thrips (Premachandra *et al.*, 2003).

The Handbook of Mites of Stored Products (V. Vacante and C.H.W. Flechtmann)
DOI: 10.1079/9781786390912.0065

Greenhouse experiments provide convincing evidence that *G. aculeifer* can suppress populations of bulb mites, on intact bulbs in the soil. It is commercially used to control the greenhouse sciarids. It also feeds on thrips pupae, shore fly larvae and acarid mites (*Rhizoglyphus sp.* and *Tyrophagus sp.*), and can be released to reduce densities of these pests in the soil or growing media (Jess and Schweizer, 2009). This evidence showed that there is great potential for soil-dwelling predatory mites in the control of arthropod pests in the soil and in compost (Fathipour and Maleknia, 2016). The medical and veterinary interest in the family Laelapidae is mainly due to, the strong association of some species with commensal rodents. Although some laelapid species are reported as causing occasional discomfort to humans (Mullen and OConnor, 2019), Maaz *et al.* (2018) refer that such events are very rare and mostly doubtful.

Ten species ascribed to seven genera (*Androlaelaps* Berlese, *Eulaelaps* Berlese, *Haemogamasus* Berlese, *Gaeolaelaps* Evans *et* Till, *Hypoaspisella* Bernhard, *Stratiolaelaps* Berlese and *Euandrolaelaps* Bregetova) are collected on stored food.

Many species of the genus *Androlaelaps* develop tight associations with their bird or mammal hosts. Four species have been recorded either from honeybees or bumble bees; they are probably all generalists that opportunistically invade bee nests (Klimov *et al.*, 2016). Within this genus, the species better known is *A. casalis* (Berlese) (Fig. 65.1). It has been described in 1887, in Italy, from house as *Iphis casalis* Berlese and successively as *Haemolaelaps casalis* (Berlese) and *Hypoaspis freemani* Hughes (1948) and ascribed in the past to the family Dermanyssidae. The description of Hughes (1976) and Smiley (1987) allow its identification. The geographical distribution of this mite is cosmopolitan and includes Australia (Halliday, 1998), Brazil (Horn *et al.*, 2018), Canada (Sinha and Wallace, 1973; Sinha, 1979a), China (Tian, 1984; Ma *et al.*, 2015), Czech Republic (Lukás *et al.*, 2007), Egypt (Saleh *et al.*, 1985; Zaher, 1986), England (Hughes, 1976), Greece (Palyvos *et al.*, 2008), India (Gupta and Chatterjee, 2004), Indonesia (White, 1995), Iran (Sepasgozarian, 1978; Kazemi and Rajaei, 2013), Ireland (Cusack, 1974; Cusack *et al.*, 1975), Saudi Arabia (Fouly and Al-Rehiayani, 2011), Scotland (Jeffrey, 1976b), United States (Smiley, 1987) and other countries. The mite is known on various commodities, such as apple (Kamali *et al.*, 2001), barley (Kamali *et al.*, 2001; Palyvos *et al.*, 2008), bran (Palyvos *et al.*, 2008), chaff and dust material (Ardeshir and Nematollahi, 2008), darnel (Kamali *et al.*, 2001), dried fruit and nuts (Palyvos *et al.*, 2008; Ardeshir *et al.*, 2012), flour (Palyvos *et al.*, 2008; Granpayeh *et al.*, 2012), grain (Gupta and Chatterjee, 2004; Lukáš *et al.*, 2007; Ardeshir and Nematollahi, 2008), grain residue (Palyvos *et al.*, 2008), maize (Palyvos *et al.*, 2008), oat (Palyvos *et al.*, 2008), potato (Granpayeh *et al.*, 2012), rice (Saleh *et al.*, 1985; Kamali *et al.*, 2001), stored rice, rice dust and debris

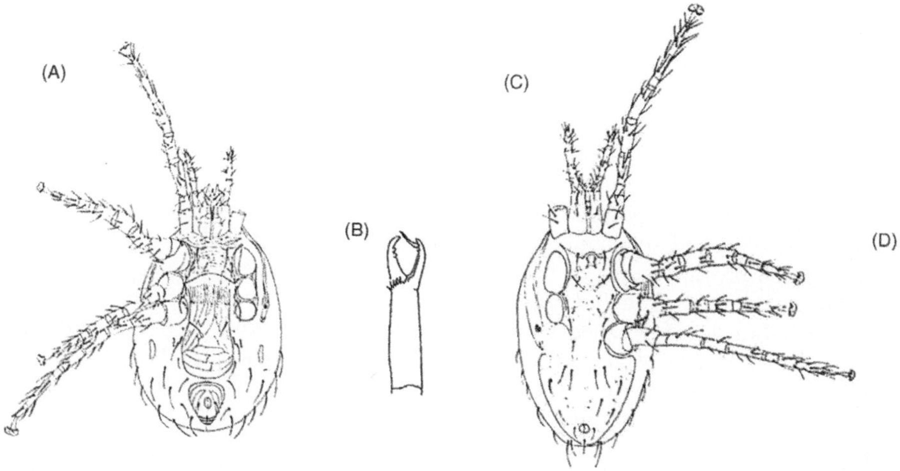

Fig. 65.1. *Androlaelaps casalis* (Berlese). Female. Ventral view (A); chelicera (B). Male. Ventral view (C); chelicera (D) (from Hughes, 1976, licensed under the Open Government Licence v3.0).

(Noei *et al.*, 2008b), stored onion (Kamali *et al.*, 2001), wheat (Sinha and Wallace, 1973; Cusack, 1974; Cusack *et al.*, 1975; White, 1995; Kamali *et al.*, 2001; Palyvos *et al.*, 2008), stored wheat (Ardeshir and Nematollahi, 2008), wheat residue (Palyvos *et al.*, 2008), wild sainfoin (Kamali *et al.*, 2001) and zucchini (Granpayeh *et al.*, 2012). Hagstrum *et al.* (2013) reported the mite also on aquatic product, condiment, dried vegetable, grass seed, hay, mushroom, processed food and sweeping. Moreover, the mite is reported by various authors from Iran on stored products (Sepasgozarian, 1978; Sayedi *et al.*, 2006; Sayedi *et al.*, 2009; Hajizadeh *et al.*, 2010). Kazemi and Rajaei (2013) reported a long list of other habitats, such as soil, litter, humus, plants, nests of insects, where the mite has been collected from other specialists in Iran. Palyvos *et al.* (2008) found the mite in Greece also in dust. In Saudi Arabia, it has been collected on debris of date palms (Fouly and Al-Rehiayani, 2011).

The facilities include commercial grain storage (Sinha and Wallace, 1973; Cusack, 1974; Cusack *et al.*, 1975; Palyvos *et al.*, 2008), drier, empty grain bin, farm (Palyvos *et al.*, 2008), mill (Sayedy *et al.*, 2009), off-farm storage and provender mill (Hagstrum *et al.*, 2013).

The mite is a generalist predatory mite of *Tyrophagus putrescentiae* (Schrank), and *Glycyphagus domesticus* (De Geer), *Blattisocius keegani* (Fox), and of immature stages of *D. gallinae* (Hughes 1976; Leśna *et al.*, 2009, 2012). It has been collected from bodies and also from the nests of mammals and bird species. Baker and Craven (2003) found the mite on roost (pipistrelle) in the West Sussex as well as Woodroffe (1956) in *Plecotus auritus* Linnaeus (Chiroptera, Vespertilionidae) in North Hampshire. Baker and Craven (2003) reported that this species preys on small mites and insects but will also take other food (e.g. egg yolk, fresh meat and blood). It is found in a wide variety of habitats (e.g. the bodies and nests of mammals, particularly rodents, and birds, horticultural peat, broiler house litter, and grain and hay sievings). In addition to the British records of bat associations, *A. casalis* has been found on *Tadarida brasiliensis* (Geoffroy), the Brazilian free-tailed bat, in Alabama, United States (Durden *et al.* 1992). Horn *et al.* (2018) collected *A. casalis* in the nest of the bird, *Zenaida auriculata* (Des Murs) in Brazil. Silva *et al.* (2013) observed this species in laying hen nests and abandoned bird nests in poultry houses. Refaei and El-Naggar (2008) found the mite associated with honeybee in Egypt.

Rosen *et al.* (2002) observed multiple erythematous papules, associated by severe pruritus in humans bitten by the mites *Pyemotes tritici* (Lagrèze-Fossat et Montagné), *D. gallinae* (De Geer), *Ornithonyssus bacoti* Hirst and *A. casalis* (Berlese). They discussed the impact of these species on human health.

Barker (1968a) studied the life history of *A. casalis* at 25.6°C, 26.7°C, 30.0°C, 32.8°C and 36.0°C. He observed that each *A. casalis* female consumed 2.1 ± 0.6 adult female *G. domesticus* per day, and the males consumed 0.85 ± 0.2 female of *G. domesticus* per day. An average of 27.4 ± 4.7 *G. domesticus* protonymphs were consumed during development to adulthood of immature *A. casalis* males. Other foods included brewer's yeast, *Acarus siro* (Linnaeus), *T. putrescentiae*, *Aëroglyphus robustus* (Banks), the predator *B. keegani* and the young larvae of *Tribolium confusum* (Duval) (Barker, 1968a).

Mustafa *et al.* (2016) reared *A. casalis* collected from untreated onion soil, in Egypt, on free-living nematodes, *Rhabditella muscicola* Andrassy and the acarid mite, *T. putrescentiae* at 20°C, 25°C and 30°C and 70 % RH. They found that the egg incubation period reaches 2.0, 1.8; 1.8, 1.72 and 1.69, 1.57 days at 20°C, 25°C and 30°C, respectively, when females were fed on both diets at different temperatures. The female life cycle period was significantly longer when mites fed on nematodes than that recorded by feeding on *T. putrescentiae*. It was 7.98, 6.88 and 6.0 days when fed on free nematodes at 20°C, 25°C and 30°C for female, respectively. The mites reached maturity after 6.98, 6.5 and 5.5 days, respectively when females fed on acarid mite. Statistical analysis of obtained data pointed out that the longevity of the resulted females was significantly differed when the mite fed on different diets, at different temperatures. It was 39.8, 37.8; 37.0, 34.0 and 34.0, 31.0 when the adult female fed on the free-living nematodes, and *T. putrescentiae* at 20°C, 25°C and 30°C, respectively. The pre-oviposition and post-oviposition periods of adult female of *A. casalis* not changed significantly with the free nematodes and the acarid mite. There were obviously significant differences, when the mites during their oviposition time, fed on both of the same foods; these periods took 35.92, 34.08; 33.0, 30.02 and 30.0, 26.98 ± 0.04 days at 20, 25 and 30°C, respectively. Feeding of *A. casalis* female on free-living nematodes increased significantly, the number of deposited eggs, than when fed on the acarid mite. The numbers of laid eggs were 39.8, 38.0; 44.0, 34.8 and 40.0, 31.8

when fed on nematodes and *T. putrescentiae* at 20°C, 25°C and 30°C, respectively. The female lifespan of *A. casalis* lasted 47.8, 44.8 and 43.9, 40.5 and 40.04, 36.6 days, when the mites were fed on nematodes and *T. putrescentiae* at 20°C, 25°C and 30°C, respectively. The incubation period of male lasted 2.0, 1.8; 1.8, 1.72 and 1.69, 1.57 days, when fed on the two diets at 20°C, 25°C and 30°C, respectively. The male completed its life cycle in 7.0 and 6.01 days at 20°C when fed on both free nematodes and the acarid *T. putrescentiae*, respectively. However, these periods changed to 6.0 and 5.8 days at 25°C and 5.5 and 5.0 days at 30°C, respectively. The research highlights that when the male of *A. casalis* fed on the same previously mentioned diets, at the same laboratory conditions, the duration of the longevity took 33.6 and 32.5; 32.5 and 30.5 and 30.1 and 28.2 days at 20°C, 25°C and 30°C, respectively. The lifespan of *A. casalis* male when fed on two diets at 20°C, 25°C and 30°C and 70% RH, differed significantly according to the type of introduced rearing diets as in case of females. The lifespan took 40.5, 38.61 and 38.6, 36.3 and 35.8, 33.2 days, respectively (Mustafa *et al.*, 2016).

Currently, *A. casalis* and *C. eruditus* are introduced as predators in commercial layer houses, under their respective commercial names ANDROLIS® and TAURRUS®(Koppert Biological Systems, the Netherlands) (Roy *et al.*, 2017).

Androlaelaps fahrenholzi (Berlese) was described as *Laelaps fahrenholzi*, by Berlese (1911) and successively as *Laelaps glasgowi* Ewing on wild rat in Illinois (United States) (Ewing 1925). Different names have been used for this species, such as *Haemolaelaps glasgowi* (Ewing) and *Androlaelaps glasgowi* (Ewing), but *A. fahrenholzi* has the priority over other names (Timm, 1972). Vinarski and Korallo-Vinarskaya (2016) reported that in the Russian literature, including the most recent publications, the species has been known under the taxonomic name *Androlaelaps* (or *Haemolaelaps*) *glasgowi*. They followed conventionally this usage for the sake of convenience, though in the vast majority of recent acarological publications, the name *A. fahrenholzi* (Berlese) is used.

A. fahrenholzi is a cosmopolitan species, which has been recorded infesting a great number of mammal species worldwide (Baker *et al.*, 1956; Strandtmann and Wharton, 1958). In central and northern Argentina, it was recorded on mammals of the orders Rodentia, Didelphimorphia, Microbiotheria, Xenarthra and Chiroptera (Lareschi and Mauri, 1998). However, mites identified as *A. fahrenholzi* are often highly variable morphologically, and it is unclear whether this mite species is truly polyxenous or is a composite of species with narrower host ranges. Lareschi *et al.* (2006) collected *A. fahrenholzi* in Uruguay on five rodent species, including new host records from *Scapteromys tumidus* (Waterhouse) and *Oxymycterus nasutus* (Waterhouse). These results are in accordance with previous findings in the Argentinean Buenos Aires province, where this mite has been reported in association with a variety of wild rodents, including *Akodon azarae*, *Oligoryzomys* species, *Scapteromys aquaticus* Thomas and *Oxymycterus rufus* (Fischer) (Lareschi, 1996). Hagstrum *et al.* (2013) reported the species from the United States, on grain.

Androlaelaps fenilis (Mégnin) has been collected on stored food only from Iran (Khalili-Moghadam *et al.*, 2012).

Androlaelaps shealsi Costa is known from Israel (Costa, 1968) and Iran (Haddad Irani-Nejad *et al.*, 2003), where it has been collected on stored food (Sayedi *et al.*, 2006).

Eulaelaps stabularis (Koch) (Fig. 65.2) was originally described by Koch (1839a) as *Gamasus stabularis*. It has a cosmopolitan geographical distribution and is reported regarding stored food from various countries such as Canada, Czech Republic, England, Ireland, Taiwan and the United States. Commodities include grain, sweeping, warehouse debris and wheat on commercial grain storage and farm (Hagstrum *et al.*, 2013). Moreover, the list of stored food includes stored rice, rice dust and debris (Shahhosseini and Kamali, 1989; Modarres Awal, 1994, 1997; Ostovan and Mossadegh, 1999; Noei, 2007; Noei *et al.*, 2008b).

Uchikawa and Rack (1978) referred that *Eulaelaps* mites are blood-suckers and predators and live in various habitats. They are often associated with birds and mammals and their nests. Some mites found on particular animals show specific morphology. *E. stabularis* is a very common mite in the nests and on the bodies of rodents, and insectivores throughout the world. This mite has been incriminated as the cause of dermatitis in man on several occasions (Strandtmann and Wharton, 1958) and has been implicated in the transmission cycle, either found naturally or experimentally infected with *Francisella tularensis* (McCoy *et* Chapin) (Davis *et al.*, 1981; Nutting, 1984).

Haemogamasus pontiger (Berlese) (Fig. 65.3) was originally described as *Laelaps pontiger* from stables in

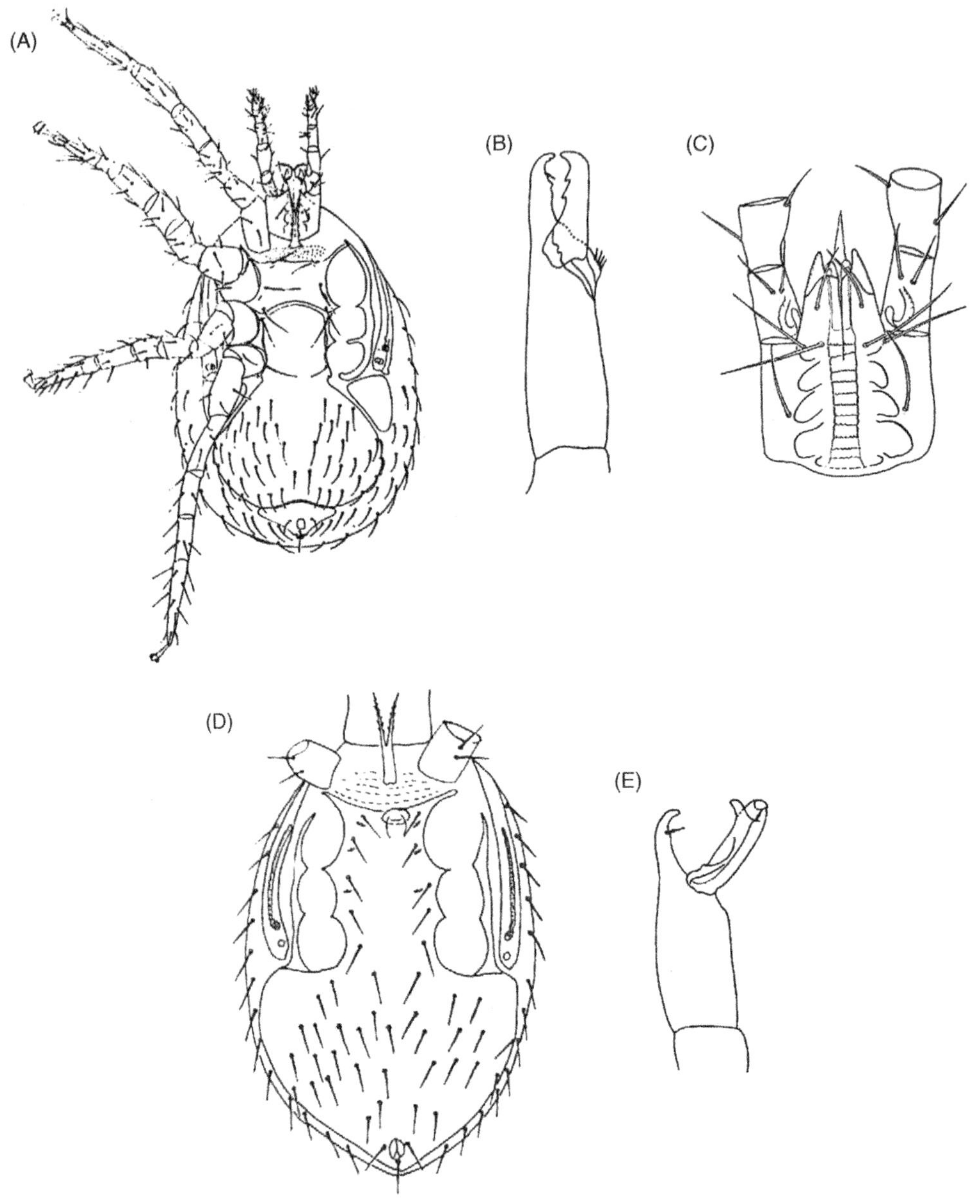

Fig. 65.2. *Eulaelaps stabularis* (Koch). Female. Ventral view (A); chelicera (B); ventral view of gnathosoma (C). Male. Ventral view (D); chelicera (E) (from Hughes, 1976, licensed under the Open Government Licence v3.0).

Italy (Berlese, 1904), and it may be identified following Hughes (1976), and Halliday (2011). The mite has a cosmopolitan geographical distribution and is reported from various countries, such as Argentina (Mosquera, 1988), Australia (Womersley, 1956; Domrow, 1992; Halliday, 2011), Canada (Barker, 1968c), China (Luo *et al.*, 2007), Czech Republic (Kučerová and Horák, 2004; Stejskal and Hubert, 2008), England (Hughes, 1961, 1976), Iran (Modarres Awal, 1994, 1997), Israel (McIntosh, 1990, in Halliday, 2011), Italy (Berlese, 1904), Japan (Asanuma, 1963), Russia (Bregetova, 1949, 1956), Scotland (Jeffrey, 1976b; Colloff, 1987b)

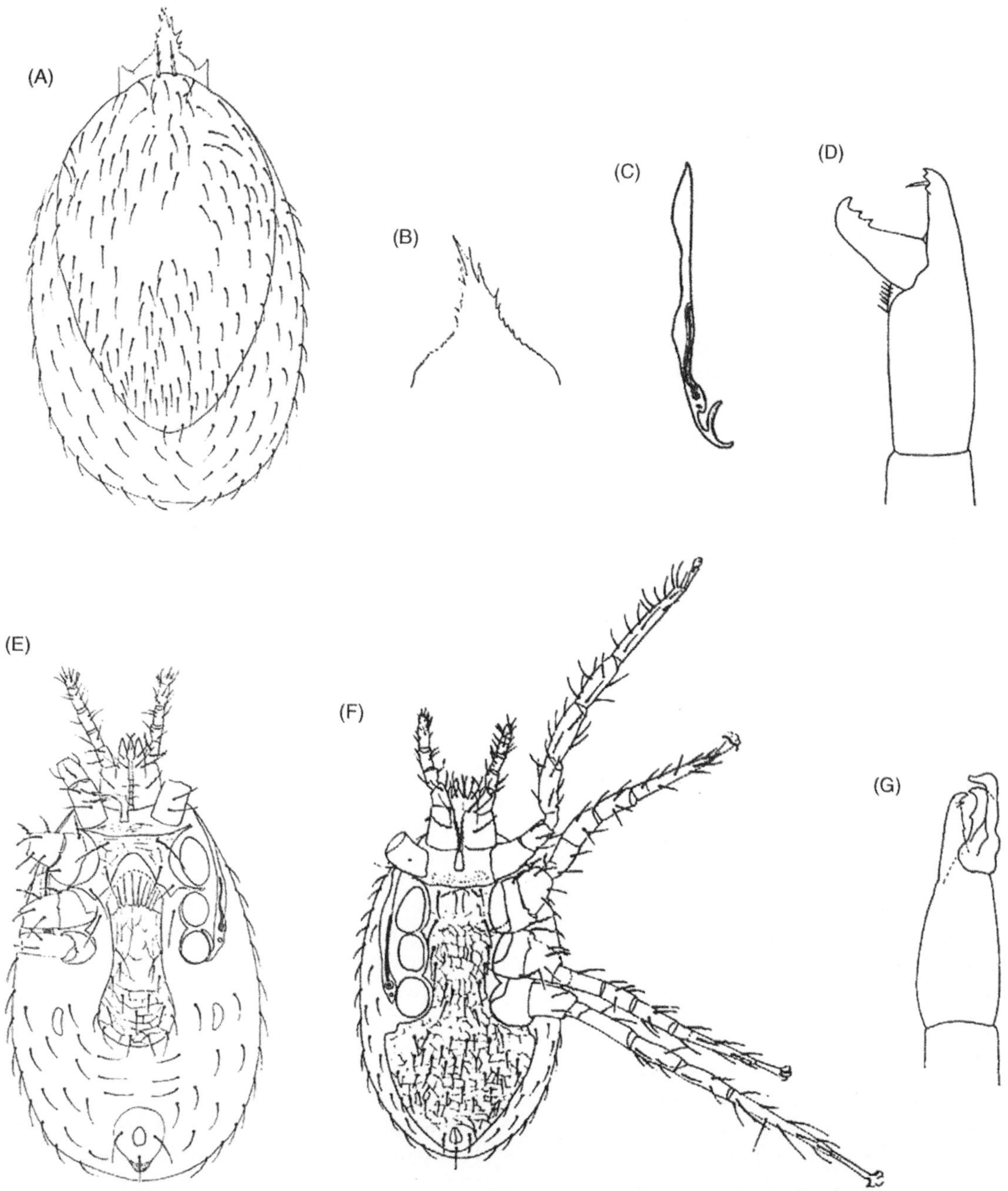

Fig. 65.3. *Haemogamasus pontiger* (Berlese). Female. Dorsal view (A); tectum (B); peritreme shield (C); chelicera (D); ventral view (E). Male. Ventral view (F); chelicera (G) (from Hughes, 1976, licensed under the Open Government Licence v3.0).

and United States (Whitaker and Wilson, 1974; Williams *et al.*, 1978). On stored food, it has been collected on barley (Hughes, 1976), beet seed (Kučerová and Horák, 2004), grain (Sinha, 1964c; Stejskal and Hubert, 2008; Halliday, 2011), oat (Hughes, 1976), rapeseed, sweeping (Hagstrum *et al.*, 2013), walnut (Halliday, 2011), wheat (Barker, 1968c; Hughes, 1976; Halliday, 2011), stored rice, rice dust and debris

(Sayedi *et al.*, 2006; Noei, 2007; Noei *et al.*, 2008b) and wheat flour in commercial grain storage, empty grain bin, farm, flour mill, provender mill, provender warehouse, railcar and wheat elevator (Hughes, 1976; Hagstrum *et al.*, 2013). Halliday (2011) has studied specimens collected in Australia in different habitats, such as beneath stored food boxes, fowl dung and feathers, bran in hen house, controlled temperature room, untreated stored wheat, on man, on asparagus, invading house and on straw bale construction. Records listed by Halliday (2011) show that the species is widespread in stored grain and straw in Australia, at least in the south-eastern part of the continent, and that the habitats in which *H. pontiger* has been found in Australia are typical of records from other parts of the world, such as stored grain, straw, poultry litter and similar habitats; this author referred that the occurrence of the species in stored food and packing materials means that it is transported in these commodities by humans, and it has often been intercepted in quarantine (Keegan, 1951; Williams *et al.*, 1978). The mite is common in the detritus of warehouse floors, where it lives as predator of astigmatid mites, and of weakly sclerotized insect larvae, and it will also feed on heparinized mouse and human blood (Furman, 1959; Hughes, 1976).

Moreover, *H. pontiger* has also been found in the fur or nests of small mammals in many areas of the world, such as China (Luo *et al.*, 2007), Finland (Mrciak and Brander, 1965), Norway (Mehl, 1977), Poland (Haitlinger, 1988), former USSR (Bregetova, 1949), United States (Whitaker and Wilson, 1974) and New Zealand (Tenquist and Charleston, 2001). Moreover, the mite has been collected in beehives (Banaszak, 1980), bird nests (Mehl, 1977), kitchen cupboards (Turner and Bishop, 1998) and upholstered seats on trains (Colloff, 1987b). Sinha (1964a) found that *H. pontiger* is able to live but not reproduce on a number of microorganisms commonly found in the granaries of Canada. The mite has been collected in association with rats, moles and a bat (Evans and Till, 1966) and also in the deep litter of broiler houses (Brady, 1970).

At 24°C and 27°C, the RH from 95% to 100% and *G. domesticus* (De Geer) supplied as food the eggs hatch in an average of 3.2 and 2.6 days, respectively. Under these conditions the life cycle of the female is completed in 7.9 and 7.6 days and that of the male in 7.0 and 6.3 days; the males develop faster at 27°C than at 24°C. Fewer eggs are laid when *G. domesticus* is the only prey available, but when the diet is supplemented with wounded *T. confusum* (du Val) larvae, the rate of oviposition usually increases. *H. pontiger* can complete its development when wheat germ only is supplied as food (Barker, 1968c). A study carried out by Sinha (1964c) on the ability of the major acarine grain pests, such as *A. siro* Linnaeus and *G. destructor* (Schrank), and eight other mites occurring in farm granaries in Canada, to recover from exposures to −18 ± 1°C, showed that 0% of *Tydeus interruptus* Thor, 4% of *G. destructor*, 1% *A. siro* and 2% of *L. krameri* (Canestrini) survived −18 ± 1°C for 7 days, 17% of *A. casalis* for 3 days, 10% of *Haemogamasus glasgowi* (Ewing) for 1 day, 63% of *Cheyletus eruditus* (Schrank), 6% of *Kleemania plumosus* Oudemans, 28% of *H. pontiger* for 1 h, and *Caloglyphus berlesei* (Michael) for less than 1 h.

Halliday (2011) reported that *H. pontiger* is widespread in straw and stored grain in southeastern Australia. It sometimes invades houses and other structures and has been suspected of biting humans and causing dermatitis. However, behavioural and morphological evidence has shown that it is a predator. There is no convincing evidence that this species is a parasite of mammals or that it has any medical significance. According to Halliday (2011), the mite has sometimes been reported as a parasite or facultative parasite, or included by various authors (Hirst, 1926; Willmann, 1952; Allred, 1957; Verts and Carraway, 2002; Watson, 2008) in lists of mammal parasites, but does not exist a clear evidence that the mite actually does feed from the body fluids of vertebrates. Conversely, there is good evidence to show that *H. pontiger* is a predator. Furman (1959) reported that *H. pontiger* fed on heparinized mouse blood but would not feed through the intact or abraded skin of humans or several species of rodents and would not feed on dried blood. Hughes (1961) referred that the mite frequently may occur in the nests of small mammals and in their fur, and in the detritus of warehouse floors, but had never seen it containing blood. Also, the specimens examined by Halliday did not show signs of the blood that is usually visible in the gut of parasitic species of Dermanyssidae and Macronyssidae. Hughes (1961) concluded that *H. pontiger* probably feeds on detritus and on other mites, and that has never observed specimens containing a large amount of blood; indeed, the mouthparts are not adapted for piercing skin, but it would probably feed on blood or tissue fluid if the host's skin was abraded.

H. pontiger has also been suspected of causing dermatitis in humans (Baker *et al.*, 1956; Allred, 1957; Flynn, 1973; Mrciak, 1979; McGarry *et al.*, 2001;

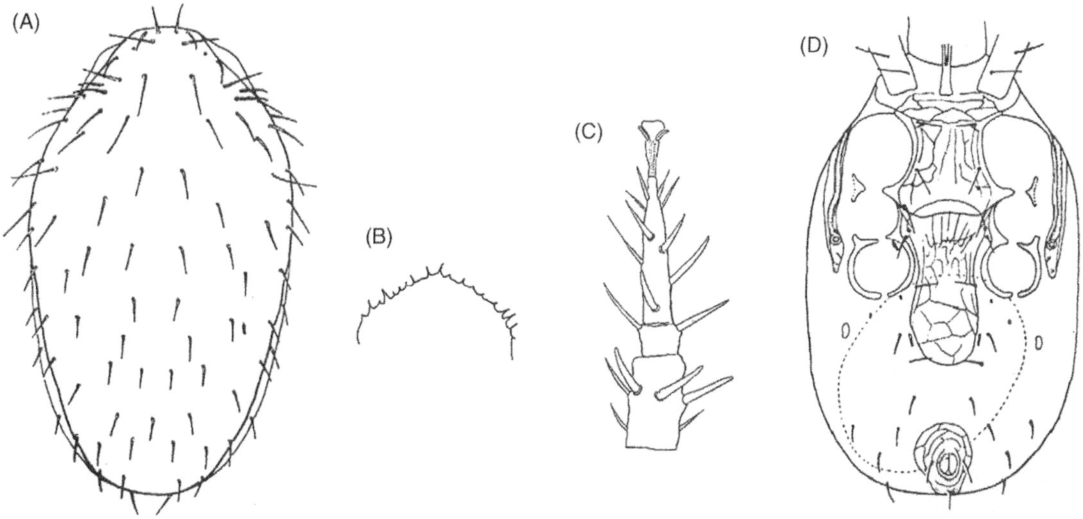

Fig. 65.4. *Gaeolaelaps aculeifer* (Canestrini). Female. Dorsal view (A); tectum (B); leg IV (C); ventral view (D) (from Hughes, 1976, licensed under the Open Government Licence v3.0).

Mullen and O'Connor, 2009). All of these records appear to be based on the work of Hill and Gordon (1945), but the evidence incriminating *H. pontiger* there is not convincing. The topic is well discussed by Halliday (2011) and we suggest consulting it.

G. aculeifer (Canestrini) (= *H. aculeifer*) (Fig. 65.4) was originally described as *Laelaps aculeifer* by Canestrini (1884), from botanic garden of Padua (Italy), and more recently included in the genus *Gaeolaelaps* Evans *et* Till (Beaulieu, 2009). Nomenclatural notes on the names *Gaeolaelaps* and *Geolaelaps* have been examined by Halliday and Lindquist (2007). The mite may be identified following Hughes (1976) and Beaulieu (2009). It is a predatory, soil-dwelling mite, typically generalist predator employed as biological control agents of various soil pests, such as thrips pupae, fly maggots and bulb mites (Walter and Oliver, 1990; Chambers *et al.*, 1993; Bennison *et al.*, 2002; Walter and Campbell, 2003; Freire *et al.*, 2007; Navarro-Campos *et al.*, 2016); moreover, it preys on mushroom flies in mushroom cultivation (Chambers *et al.*, 1993; Ali *et al.*, 1999; Cabrera *et al.*, 2005; Jeon *et al.*, 2007) and attacks the poultry red mite, *D. gallinae* (De Geer) (Ali *et al.*, 2012; Leśna *et al.*, 2012). Recent research has shown that *G. aculeifer* is a natural enemy of *Varroa destructo*r (Anderson *et* Trueman), the honeybee ectoparasitic mite (Rangel and Ward, 2018). It preys on other small soil organisms, such as Collembola, mites, nematodes and Enchytraeidae (Kevan and Sharma, 1964; Sardar and Murphy, 1987; Karg, 1995; Axelsen *et al.*, 1997). Karg (1961) reported that *G. aculeifer* is a polyhagous predator that will feed upon Collembola, such as *Folsomia fimetaria* (Linnaeus) and *Tullbergia kraussbaueri* (Borner), mites (*Tyrophagus* sp.), insect larvae and to some extent on nematodes. Moreover, the abundance of *G. aculeifer* depends upon that of the Collembola and mites and not upon that of the nematodes. Kevan and Sharma (1963a, b) reported that *G. aculeifer* feeds upon *T. putrescentiae*, particularly the larvae and nymphs. Ryke (1959) states that none of the species of hypoaspids collected from soils, stored food products, the nests of insects or from the bodies of mammals or invertebrates appear to be parasitic, but species that occur on myriapods apparently show certain modifications, suggesting a symbiotic mode of existence. Hughes (1976) reported that Lund found *G. aculeifer* on caked flour and suggested that the species was in that case feeding on other mites present, rather than upon the caked flour. Brett (in Hughes, 1976) found the mite on daffodil bulbs where it was feeding on *Rhizoglyphus* sp.

G. aculeifer has a large geographical distribution, including various countries, such as Canada (Kevan and Sharma, 1963a, b), England (Hughes, 1961, 1976; Evans and Till, 1966, 1979), Germany (Karg, 1961), Iran (Kamali *et al.*, 2001), Ireland (Evans and

Till, 1966, 1979), Israel (Evans and Till, 1966), Italy (Canestrini, 1884), India (Sandhu *et al.*, 2005), the Netherlands (Siepel *et al.*, 2018), North America (Kevan and Sharma, 1963a, b, 1964), Russia (Bregetova, 1964) and Sweden (Sellnick, 1958). On stored food, it is reported from daffodil bulb (Brett, in Hughes, 1976), caked flour (Lund, in Hughes, 1976), grain in commercial grain storage (Siepel *et al.*, 2018) and pearl millet (Sandhu *et al.*, 2005). In Iran, the mite has been found on stored rice and barley, grape, darnel, liquorice, lucerne, *Trifolium* sp., nest of small mammals, soil samples, rice dust and debris (Ostovan, 1993; Fathipour, 1994; Khanjani, 1996; Darvishzadeh, 1997; Noei, 2007; Noei *et al.*, 2008b).

G. aculeifer is known from a variety of habitats, such as gardens (Canestrini, 1884), small mammals (Bregetova, 1956), pastures (Sellnick, 1958), potato fields (Karg, 1961) and stored food products (Hughes, 1961, 1976). In eastern Canada, it is known from soil and litter, collected from under sugar maple trees (*Acer saccharum* Marsh) (Kevan and Sharma, 1963a, b), and from pot soil in the greenhouses and from cultivated garden soil. Evans and Till (1966) reported the mite as common in soil and litter and referred that it is recorded from the nest of *Riparia riparia* Linnaeus in Gloucestershire, Britain, from the nest of *Spalax ehrenbergi* Nehring and the nests of a variety of rodents in the former USSR. Siepel *et al.* (2018) reported a long list of habitats, among which are frequent, various kinds of soil. The list of habitats includes arable land, wheat, *Lolium perenne* Linnaeus in greenhouse soil, soil on former compost business, soil samples of agricultural grasslands on sandy soil, agricultural grassland, grassland clay soil, harbour mud depot, grassland sandy soil, arable land on sandy soil, grassland on river clay soil, arable land of marine clay soil, grassland on loess and grassland on sandy soil, as well as bulb of *Eucomis montana* Compton, Lilium 'Triumphator' (bulb), *Freesia*, *Lilium*, under potato, under rye treated with dieldrin, under peas in pot white clover treated with aldrin, beech litter, litter oak forest, under *Hieracium pilosella* (Vaill.) south slope and litter *Ligustrum*.

The biology of *G. aculeifer* has been investigated by various authors, and we report here for the reason of space, only the most significative reports. Usher and Davis (1983) observed that in the absence of males, the females of *G. aculeifer* lay eggs, which develop only into males. However, in the presence of males, the eggs develop into both sexes. The species is therefore presumed to be arrhenotokous (haplodiploid). Kevan and Sharma (1964) studied the biology of *G. aculeifer* under laboratory conditions at 26°C, 17°C and 11°C and 98% RH, and with *T. putrescentiae* as food. In this condition, the oviposition period at 26°C, 17°C and 11°C were 26–32, 35–41 and 17–23 days, respectively, and the number of eggs laid after the first mating were 15–20, 8–17 and 2–7, respectively. The incubation periods and the percentages of eggs hatching were 1–4, 2–6 and 10–14 days, with a percentage of eggs hatching of 92.3, 70.6 and 45.6, respectively. The duration in days of the postembryonic instars, at the three temperatures, was for the larva 1–2, 1–3 and 3–6 days, for the protonymph 5–7,5–8 and 10–15 days, and for the deutonymph 2–5, 6–7 and 15–22 days, respectively. The duration of the egg-laying to adult was 10–13, 17–23 and 40–48 days, respectively. At 26°C, *G. aculeifer* took 10–13 days to reach maturity from the time the eggs were deposited, about a week longer at 17°C, and over a month longer at 11°C. Although no development occurred at temperatures below 11°C, adult *G. aculeifer* survived temperatures of 8°C, 4°C and 0°C for considerable periods; even at 0°C, they remained alive after more than a month of observation, although they were not active.

Barker (1969) studied the effect of two prey species, *T. putrescentiae* (Schrank) and *G. domesticus* (De Geer), on the oviposition of *G. aculeifer*. The females of *G. aculeifer* fed on *T. putrescentiae*, laid more eggs and lived longer than those that fed on *G. domesticus*. Predators fed on *G. domesticus* laid more eggs per day and had a shorter mean generation than predators fed on *T. putrescentiae*. The greatest finite rate of increase, of *G. aculeifer* was obtained when *G. domesticus* was the prey.

Usher and Davis (1983) reported that laboratory experiments showed that *G. aculeifer* is not cannibalistic, except under extreme conditions of hunger, and there is no evidence of oophagy. The predator does not 'overkill' its prey; with large density of *Sinella coeca* (Schott) (Collembola, Entomobryidae) as prey the efficiency of capture increases with the predator density, while with easily caught prey, the efficiency decreases and large density of *Hypogastrura denticulata* (Bagnall) (Collembola, Hypogastruridae) reduce the predator's longevity and fecundity. Usher and Davis (1983) reviewed the criteria for a species to reach the eusocial threshold of Wilson (1971). Although several generations of mites can overlap, there is neither cooperative broad care nor the existence of a reproductive caste. There is, however, apparently no harassment of the young. It is suggested that mites in this family, especially the species associated with vertebral nests, are intermediate subsocial.

Park *et al.* (2021b) evaluated the development time for the different life stages, total preadult duration, adult longevity, adult pre-oviposition period (APOP), total pre-oviposition period (TPOP), reproductive days and fecundity of *G. aculeifer* fed on *T. putrescentiae* at 25 ± 1°C, 90 ± 5% RH, and a photoperiod of 16:8 (L: D) hours. The mean developmental time of the egg stage was 3.71 ± 0.11 days, those of larva, protonymph and deutonymph were 1.05 ± 0.034, 3.32 ± 0.101, 3.44 ± 0.093, respectively, and the preadult duration was 11.51 days. The adult longevity was 66.24 ± 2.20, with female longevity 80.31 days, and male longevity 73.33 ± 3.55. The total longevity was 77.76 ± 2.21. The APOP was 3.23 ± 0.19 and the TPOP 14.88 ± 0.34, respectively. The oviposition days were 24.50 ± 0.73. The fecundity (offspring/individual) was 74.88 ± 2.25, and the fecundity (eggs/day) was 3.06 ± 0.07. The intrinsic rate of increase (*r*) was 0.1733 ± 0.0071 day^{-1}, finite rate of increase (λ) was 1.1893 ± 0.0084 day^{-1}, the net reproductive rate (R_0) was 47.49 ± 5.80 offspring/female, the mean generation time (offspring/individual) (*T*) was 22.27 ± 0.44 days, and the gross reproduction rate (GRR) was 47.53 ± 5.81 offspring/female.

This mite has been assayed as a natural enemy of nematodes, Collembola, thrips and acarid mites. The topic has been extensively discussed by Gerson *et al.* (2003). To this end, Conijn *et al.* (1997) (and other specialists) carried out experiments of biological control of *Rhizoglyphus robini* (Fumouze *et* Robin) based on release of *G. aculeifer*. They compared the release of *G. aculeifer* with different strategies of control, such as hot-water treatment, hot-water treatment and release of *G. aculeifer*, acaricide treatment with pirimiphos-methyl, and hot-water treatment and pirimiphos-methyl treatment. When bulb scales were mixed with vermiculite and packed in plastic, complete control of *R. robini* was achieved by acaricide treatment, whereas biological control led to decimation of the pest population. However, when peat was used as a packing medium, biological control gave more satisfactory results than acaricide treatment. It is concluded that a combination of hot-water treatment and biological control can replace acaricide application to control the bulb mite in lilies during the propagation phase.

Pérez-Rodríguez *et al.* (2018) tested the potential of *G. aculeifer* as a predator of the invasive citrus mealybug *Delottococcus aberiae* (De Lotto) (Hemiptera: Pseudococcidae). The predation and fecundity of mites over a 14-day period were analysed when fed on three different diets under laboratory conditions: *D. aberiae* eggs, first instar nymphs, and without food. *G. aculeifer* preyed on *D. aberiae* and both predation rates and the proportion of females that laid eggs were significantly higher when females preyed on mealybug nymphs than on eggs. These data prompted the evaluation of *G. aculeifer* as a biological control agent of *D. aberiae* under semi-field conditions. Infestation levels of *D. aberiae* on citrus potted plants were reduced in response to *G. aculeifer* releases. The implications of these results on conservation biological control of mealybugs are discussed (Pérez-Rodríguez *et al.*, 2018).

G. aculeifer is reared and commercially proposed by different companies, such as in the form of ENTOMITE-A produced by Koppert B. V. Ltd., Berkel en Rodenrijs, the Netherlands.

G. sp. nr. *glabrosimilis* (Hirschmann) has been collected on stored wheat in Iran (Mirfakhraii, 1994).

Hypoaspisella lubrica (Voigts *et* Oudemans) (Fig. 65.5) was described in 1904 from Bremen (Germany) in a chicken cage and may be identified following Hughes (1976). *H. lubrica* has been found on wheat, condiment, grain, grain debris, oat (rotting), rice (rotten), wheat, stored rice, rice dust and debris (Noei, 2007; Noei *et al.*, 2008b; Hagstrum *et al.*, 2013; Maroufpoor and Ostovan, 2017). Its geographical distribution includes Brazil (Horn *et al.*, 2018), Canada (Evans and Till, 1966; Karg, 1993a), China (Li and Meng, 1992), Czech Republic (Zeman and Jurík, 1981; Fenďa, 2010), England (Hughes, 1976), India (Bandyopadhyay *et al.*, 2023), Ireland (O'Farrell and Butler, 1948; Cusack, 1974; Cusack *et al.*, 1975; Bolger *et al.*, 2018), Poland (Gwiazdowicz *et al.*, 2006), Russia (Bregetova, 1956; Porshakov *et al.*, 2017), Turkey (Bayram and Çobanoğlu, 2005), United States (Evans and Till, 1966) and other countries. *H. lubrica* has been collected on different commodities such as grain debris, rotting oat (Hughes, 1976); moreover, Hagstrum *et al.* (2013) reported condiment, grain, rotten rice and wheat in commercial grain storage, farm, flour mill, oatmeal mill, railcar and warehouse (O'Farrell and Butler, 1948; Cusack, 1974; Cusack *et al.*, 1975). Evans and Till (1966) reported the mite in decaying oats and hay, in the nest of *Riparia*

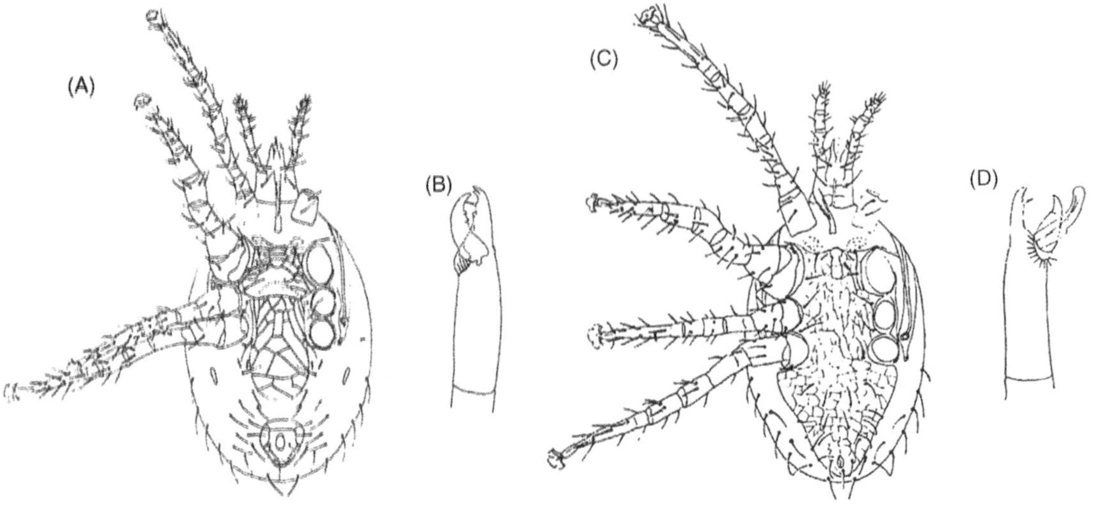

Fig. 65.5. *Hypoaspisella lubrica* (Voigts *et* Oudemans. Female. Ventral view (A); chelicera (B). Male. Ventral view (C); chelicera (D) (from Hughes, 1976 licensed under the Open Government Licence v3.0).

riparia (Linnaeus) (Hirundinidae), on a variety of rodents and in their nests. The mite was found on various bees in Europe and North America (Karg, 1994; Klimov *et al.*, 2016), and according to what was discussed above is associated with acarid mites on grain debris, rotting oats and nests of smal mammals from England, Czezch Republic, Northern Ireland, Russia and the United States (Hughes, 1976). In India, *H. lubrica* has been found in mushroom culture (Parveen and Gupta, 2020c) and in soil in tea plantation (Ahmed *et al.*, 2020b).

Bregetova (1977) and Karg (1993a) reported *H. lubrica* on rotting substrates and on nests of rodents in Europe and North America. More recently, Porshakov *et al.* (2017) collected the mite on small mammals in the semi-desert teritories of Saratof Trans-Volga region in Russia. Moreover, *H. lubrica* has been collected on nests of various birds. Various reports confirm this habitus. Zeman and Jurík (1981) found the mite in the nests of *Parus major* Linnaeus in Czechslovakia. Other studies carried out in Poland highlighted the presence of *H. lubrica* on nests of the white stork (*Ciconia ciconia* Linnaeus) (Bajerlein *et al.*, 2006) and of the white-tailed sea eagle *Haliaeetus albicilla* (Linnaeus) (Gwiazdowicz *et al.*, 2006). Horn *et al.* (2018) captured the mite on traps in poultry industry in Brazil.

In Poland, Madej *et al.* (2011) collected *H. lubrica* in a moderately humid pine-oak forest, markedly transformed by human activity, in the Rybnik Forest District (southern Poland). In Turkey *H. lubrica* has been reported from hazelnut orchards (Çobanoğlu and Özman, 2002) and from *Iris* sp. (Bayram and Çobanoğlu, 2015). In Egypt, the mite has been found. Kalmosh and Yassin (2018) found in the soil cultivated by wheat and soybean crops. Siepel *et al.* (2018) reported the mite on cultures of *Chrysanthemum* and *Frankliniella occidentalis* (Pergande), on litter and on litter henhouse in the Netherlands.

Hypoaspisella linteyini Samšiňák has been collected in Iran on storage products (Sayedi *et al.*, 2006), and horse, sheep, buffalo and cow manure (Farahi *et al.*, 2020). *H. sclerotarsus* Costa was described by Costa (1968) from litter in Israel. Its geographical distribution includes Israel (Costa, 1968), Iran (Ostovan, 1993; Khalili-Moghadam and Saaboori, 2015) and Egypt where it has been collected on rice (Saleh *et al.*, 1985). In Iran, it has been collected in soil of ant (unknown) nests. The list of commodities includes rice, straw caches, stored rice, flour and barley, lucerne, bean, astragalus, rice dust and debris (Ostovan, 1993; Fathipour, 1994; Khanjani, 1996; Noei, 2007; Noei *et al.*, 2008b; Hagstrum *et al.*, 2013; Naderi *et al.*, 2016).

Stratiolaelaps gurabensis (Fox) (Fig. 65.6) has been collected on stored wheat, barley and rice in Iran (Ostovan, 1993). *S. miles* (Berlese) was described as *Laelaps miles* by Berlese (1892) and also named *S. scimitus* (Berlese), or *H. miles* (Berlese). The mite

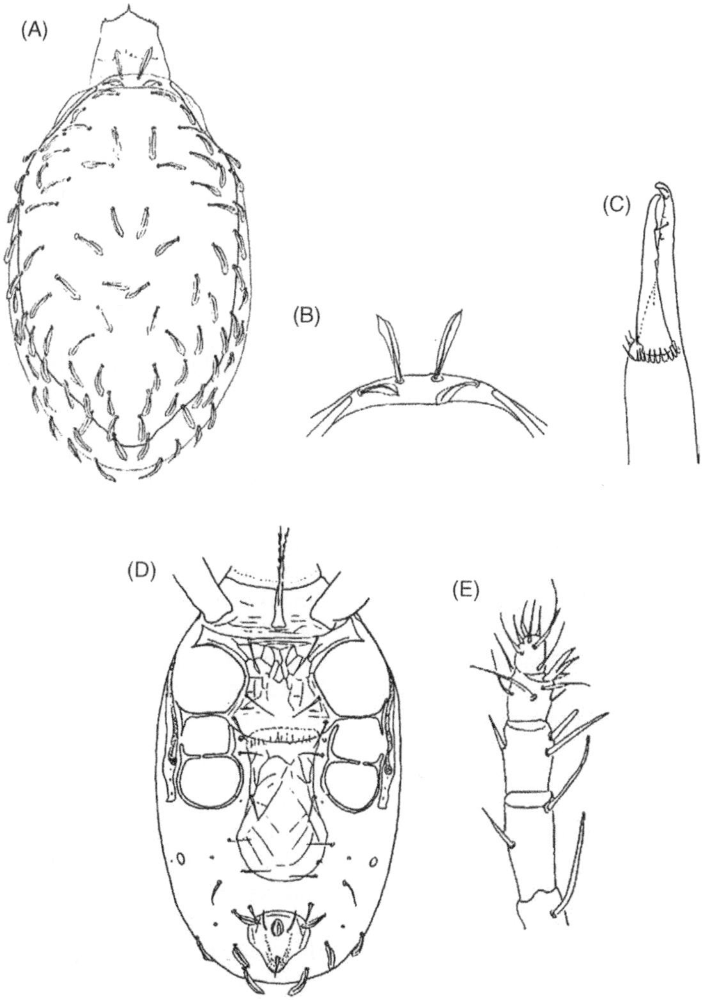

Fig. 65.6. *Stratiolaelaps gurabensis* (Fox). Female. Dorsal view (A); vertex of idiosoma (B); ventral view (C); palp (D) (from Hughes, 1976, licensed under the Open Government Licence v3.0).

may be identified following Evans and Till (1966) and Hughes (1976). Its geographical distribution includes Iran (Nemati *et al.*, 2000), Italy (Berlese, 1892), Russia (Bregetova, 1956), Scotland (Evans and Till, 1966; Hughes, 1976), United States and West Indies (Hughes, 1976). Cuthbert (in Hughes, 1976) found the mite on decaying oatmeal spillage in Leith, and Hosie (in Hughes, 1976) reported it on wheat residues in Glasgow (Scotland). In Iran, *S. miles* has been collected on straw caches, stored rice, rice dust and debris (Noei, 2007; Noei *et al.*, 2008; Naderi *et al.*, 2016). Moreover, the mite is reported from a variety of rodents and their nests in Europe, the USSR and the United States (Evans and Till, 1966; Hughes, 1976). *S. miles* is a generalist soil predatory mite, feeding on various pests, such as the fungus gnat larvae, thrips pupae, pathogenic nematodes, springtails, strawberry root weevil, actus root weevil. It is primarily a predator of fungus gnat larvae in the soil, but it also attacks thrips pupae on the floor and soil surface of the greenhouse. In the absence of thrips, pupae and fungus gnat larvae, it is a scavenger and can feed on soil debris.

Gerson *et al.* (2003) discussed the bio-ecology of *S. miles* and their ability to control glasshouse pests. We mention here the most important reports on its bio-ecology. Wright and Chambers (1994) studied the development rate of egg to adult at different temperatures, fecundity, longevity and rate of predation on sciarids. They report that with an increase in temperature, the time taken for development decreased from 33.7 days at 15°C to 9.2 days at 28°C. The threshold temperature to complete its life cycle was between 10°C and 12°C. When fed on *A. siro* Linnaeus, *S. miles* laid 2–3 eggs a day and unfertilized eggs only developed into males. Adults of the mite (0–1 day old) survived for 24.0 ± 1.4 days without food, and if fed for 6 days and then starved, females lived for 65.4 ± 2.6 days, which was significantly longer than males (45.2 ± 3.0 days). With food, 60% of males and females survived for 142 days. All larval instars of sciarids were attacked by *S. miles*, although the numbers consumed were dependent on the size of the larvae. Egg predation was negligible and pupae were not attacked. In productivity studies of *S. miles* in culture (fed on *A. siro*), the time taken for the mites to reach a density of 45,000–65,000/litre, ranged from 76 to 23 days at 15°C and 28°C, respectively (Wright and Chambers, 1994).

Enkegaard *et al.* (1997) studied the biology of *S. miles* fed on the mushroom sciarid larvae, *Lycoriella solani* Winnertz (Diptera: Lycoriidae) and the mould mite, *T. putrescentiae* (Schrank), under laboratory conditions at 20°C, 75% RH and LD16:D8 hours. *S. miles* had a significantly shorter development time, and a significantly lower juvenile mortality when fed on sciarid larvae than on mould mites, the development time being 14.5 days and the mortality 3.5% on the former prey. The pre-oviposition and post-oviposition periods of *S. miles* were not uninfluenced by the prey species and were 5–9 and 32–37 days, respectively. Oviposition periods of 53.2 and 68.5 days and female longevities of 82 and 109.6 days were observed on diets of sciarid larvae and mould mites, respectively. The male longevity (168–219 days) was uninfluenced by the prey species. The egg production of *S. miles* on sciarid larvae was evaluated to be 44.4 ± 4.33 eggs per female, as compared to 22.43 ± 1.79 eggs per female on mould mites. The sex ratio of the offspring was significantly influenced by the prey species, the ratios (♀/(♀+♂)) being 0.66 on sciarid larvae and 0.54 on mould mites. The net reproductive rate (R_0) for *S. miles* fed on sciarid larvae was approximately 27, which was three times higher than for mites feeding on mould mites. The innate capacity of increase (r_m) was highest (0.0747 day^{-1}) when sciarid larvae served as food, giving a doubling time of 9.3 days, as compared to 12.8 days on mould mites. The generation times were 44.28 on sciarid larvae and 40.67 days on mould mites. The daily food consumption rate of juvenile and adult *S. miles* was 0.24 and 0.86 sciarid larvae, and 10.8 and 21.7 mould mites, respectively. However, as concerns the weight consumed, the consumption of sciarid larvae was 2–3.5 times the weight of mould mites. The ratio of females to males influenced the oviposition period and the egg production of *S. miles*, with virgin females laying fewer eggs over a longer period of time, as compared with females with access to males. The egg production in relation to the sex ratio was described by models predicting a maximum number of eggs per female of 22.3 to be attained at a sex ratio of 0.69 (♀/(♀+♂)), and a maximum daily number of eggs per female of 0.33 to be attained at a sex ratio of 0.37 (♀/(♀+♂)) (Enkegaard *et al.*, 1997).

S. miles is widely marketed for release in greenhouse crops to manage populations of dark-winged fungus gnats, *Bradysia* spp. (Diptera: Sciaridae) and for supplemental control of thrips. Biobest commercializes *Hypoaspis* system 10.000 in 500 ml cardboard tube, containing 10.000 predatory mites using as carrier, a mix of vermiculite, peat and factitious prey. Bioline proposed Hypoline. Other firms market the predatory mite, using more or less the same methods. These companies suggested in labels or through their technicians, the strategies of release.

Cabrera *et al.* (2005) studied the development and reproduction at 21–23°C of *S.* (= *Hypoaspis*) *scimitus* (Berlese) provided *ad libidum* with one of three different preys, the fungus gnat *Bradysia* aff. *coprophila* (Lintner), potworms (Enchytraeidae), or *Sancassania* aff. *sphaerogaster* (Zachvatkin). Developmental duration of the egg and non-feeding larval stages were 2.47 and 1.11 days, respectively; mortalities were 8.3% and 5.5%. The predatory mite failed to develop beyond the protonymphal stage, when provided with *Sancassania* aff. *sphaerogaster* alone, although some feeding was observed. Its development and reproduction were successful on both fungus gnat larvae and enchytraeids, with no influence of prey on protonymphal duration (4.70 days) and mortality (8.3%), or on deutonymphal duration (4.61 days) and mortality (6.1%). Adult female *S. scimitus* feeding on potworms, compared to those feeding on fungus gnat larvae, had a significantly shorter pre-oviposition period (2.69 vs 4.59 days). However, diet did not influence other adult female developmental or reproductive characteristics; the oviposition period was 18.6

days, the post-oviposition period was 6.2 days, the total adult longevity was 27.3 days, and the total number of eggs was 26.5 days. *S. scimitus* reared on potworms tended (p = 0.06) to have a higher intrinsic rate of increase, a higher finite rate of increase and a shorter doubling time (r_m = 0.142 day^{-1}, λ = 1.153, *Dt* = 4.85 days) than those reared on fungus gnat larvae (r_m = 0.105 day^{-1}, λ = 1.110, *Dt* = 6.58 days), but differences in net reproductive rate (R_0) and generation time (*G*) were not significant (Cabrera *et al.*, 2005).

Successively, Xie *et al.* (2018) studied the development, survival and reproduction of *S. scimitus* on four diets (eggs of *Ephestia kuehniella* Zeller (Pyralidae), cysts of *Artemia* sp. (Artemiidae), pollen of *Typha orientalis* Presl (Typhaceae) and *Tyrophagus curvipenis* (Fain *et* Fauvel) in the laboratory at 25 ± 1°C, 80 ± 5% RH and 24 h D: 0 h L. *E. kuehniella* eggs and *T. curvipenis* were very suitable for the overall development of *S. scimitus* to adulthood, with over 90% survival rates. Shrimp cysts were not suitable, with 11.1% survival to adults, and *T. orientalis* pollen were intermediate, with 50% survival to adults. *S. scimitus* immature development was the fastest on *T. curvipenis* (9.67 ± 0.18 days), and it was also very fast on *E. kuehniella* eggs, but relatively slow on *T. orientalis* pollen (18.26 ± 0.68 days) and the slowest on *Artemia* cysts (21.36 ± 1.23 days). The female pre-ovipositional period (APOP) averaged less than 3 days on *E. kuehniella* eggs but over 3 days on *T. curvipenis*, but mated and unmated females had a similar APOP, regardless of diet. Neither diet nor mating had significant effects on the total pre-ovipositional period from birth to oviposition (around 13 days). The oviposition period averaged 1–2 days longer for mated females than unmated ones, regardless of diet, which had no significant effects. Mated females produced 50% more eggs than unmated females when feeding on *E. kuehniella* eggs, but 69% more than unmated females when feeding on *T. curvipenis*; the effects of mating were significant, and the effects of diet were different depending on mating status of the predators: fecundity was higher on *E. kuehniella* eggs (20.21 ± 1.16), than on *T. curvipenis* (17.07 ± 0.49) only for unmated females. Most life table parameters (intrinsic rate of population increase *r*, finite rate of population increase λ, and net reproductive rate R_0) were not significantly affected by diet and mating. However, the mean generation time *T* was shorter in unmated females than in mated ones, and more so when feeding on *T. curvipenis*. *T. curvipenis* is easy to rear, so it is considered a very promising diet for *S. scimitus* as it does not damage plants nor cause allergenic reactions in workers or users. Since *T. curvipenis* is common on plants, future studies should explore the combined use of foliar predators (e.g. Phytoseiidae) and soil predators for the biocontrol of greenhouse pests (e.g. thrips), using a shared non-pest prey such as *T. curvipenis* (Xie *et al.*, 2018).

Euandrolaelaps sardous (Berlese) (Fig. 65.7) (not *sardoa*, see Hughes, 1976) was described from Sardinia (Italy) (Berlese, 1911), and can be identified following Evans and Till (1966), Hughes (1976) and Keum *et al.* (2016). Its geographical distribution includes England (Evans and Till, 1966; Hughes, 1976), Ireland (O'Farrell and Butler, 1948), Israel (Costa and Nevo, 1969), Italy (Berlese, 1911), Korea (Keum *et al.*, 2016), the Netherlands (Hughes, 1976) and Russia (Bregetova, 1956). O'Farrell and Butler (1948) reported the mite from Ireland on foodstuffs, and known commodities are damp oat spillage, and sweeping in warehouse floors (Hughes, 1976).

Evans and Till (1966) reported the mite in litter and in the nests, more rarely on the bodies of moles and field mice, *Apodemus flavicollis* (Melchior) and *Apodemus sylvaticus* (Linnaeus), in Britain, Europe and the former USSR. Bregetova (1956) recorded it from the nest of *Phoenicurus phoenicurus* (Linnaeus) in the former USSR. Keum *et al.* (2016) reported the mite from Korea on from apple orchard, riparian and grassland habitats.

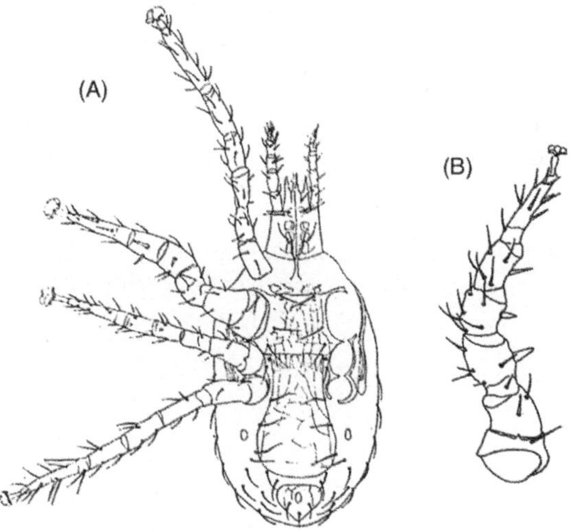

Fig. 65.7. *Euandrolaelaps sardous* (Berlese). Female. Ventral view (A); leg II (B) (from Hughes, 1976).

Cabrera *et al.* (2005) studied the development and reproduction at 21–23°C of *Stratiolaelaps* (= *Hypoaspis*) *scimitus* (Berlese), provided *ad libidum* with one of three different preys, the fungus gnat *Bradysia* aff. *coprophila* (Lintner), potworms (Enchytraeidae) or *Sancassania* aff. *sphaerogaster* (Zachvatkin). Developmental duration of the egg and non-feeding larval stages were 2.47 and 1.11 days, respectively; mortalities were 8.3% and 5.5%. The predatory mite failed to develop beyond the protonymphal stage when provided with *Sancassania* aff. *sphaerogaster* alone, although some feeding was observed. Its development and reproduction were successful on both fungus gnat larvae and enchytraeids, with no influence of prey on protonymphal duration (4.70 days) and mortality (8.3%), or on deutonymphal duration (4.61 days) and mortality (6.1%). Adult female *S. scimitus* feeding on potworms, compared to those feeding on fungus gnat larvae, had a significantly shorter pre-oviposition period (2.69 vs 4.59 days). However, diet did not influence other adult female developmental or reproductive characteristics; the oviposition period was 18.6 days, the post-oviposition period was 6.2 days, the total adult longevity was 27.3 days and the total number of eggs was 26.5 days. *S. scimitus* reared on potworms tended ($p = 0.06$) to have a higher intrinsic rate of increase, a higher finite rate of increase and a shorter doubling time ($r_m = 0.142$ day^{-1}, $\lambda = 1.153$, $Dt = 4.85$ days) than those reared on fungus gnat larvae ($r_m = 0.105$ day^{-1}, $\lambda = 1.110$, $Dt = 6.58$ days), but differences in net reproductive rate (R_0) and generation time (G) were not significant (Cabrera *et al.*, 2005).

Successively, Xie *et al.* (2018) studied the development, survival and reproduction of *S. scimitus* on four diets (eggs of *E. kuehniella*), cysts of *Artemia* sp. (Artemiidae), pollen of *T. orientalis* Presl (Typhaceae) and *T. curvipenis* (Fain *et* Fauvel) in the laboratory at 25 ± 1°C, 80 ± 5% RH and 24 h D: 0 h L. *E. kuehniella* eggs and *T. curvipenis* were very suitable for the overall development of *S. scimitus* to adulthood, with over 90% survival rates. Shrimp cysts were not suitable, with 11.1% survival to adults, and *T. orientalis* pollen were intermediate, with 50% survival to adults. *S. scimitus* immature development was the fastest on *T. curvipenis* (9.67 ± 0.18 days), and it was also very fast on *E. kuehniella* eggs, but relatively slow on *T. orientalis* pollen (18.26 ± 0.68 days) and the slowest on *Artemia* cysts (21.36 ± 1.23 days). The female pre-ovipositional period (APOP) averaged less than 3 days on *E. kuehniella* eggs but over 3 days on *T. curvipenis*, but mated and unmated females had a similar APOP, regardless of diet. Neither diet nor mating had significant effects on the total pre-ovipositional period from birth to oviposition (around 13 days). The oviposition period averaged 1–2 days longer for mated females than unmated ones, regardless of diet, which had no significant effects. Mated females produced 50% more eggs than unmated females when feeding on *E. kuehniella* eggs, but 69% more than unmated females when feeding on *T. curvipenis*; the effects of mating were significant, and the effects of diet were different, depending on mating status of the predators: fecundity was higher on *E. kuehniella* eggs (20.21 ± 1.16) than on *T. curvipenis* (17.07 ± 0.49) only for unmated females. Most life table parameters (intrinsic rate of population increase r, finite rate of population increase λ, and net reproductive rate R_0) were not significantly affected by diet and mating. However, the mean generation time T was shorter in unmated females than in mated ones, and more so when feeding on *T. curvipenis*. *T. curvipenis* is easy to rear, so it is considered a very promising diet for *S. scimitus*, as it does not damage plants nor cause allergenic reactions in workers or users. Since *T. curvipenis* is common on plants, future studies should explore the combined use of foliar predators (e.g. Phytoseiidae) and soil predators for the biocontrol of greenhouse pests (e.g. thrips), using a shared non-pest prey such as *T. curvipenis* (Xie *et al.*, 2018).

66 The Family Macrochelidae Vitzthum (Figs 66.1 and 66.2)

According to Halliday (2000), 'the family Macrochelidae, contains a very large number of species of mites that are found wherever organic matter is decomposing, especially in dung and carrion, compost and leaf litter. In these habitats, they prey on small arthropods, especially the eggs and larvae of flies, but they are also able to feed on other prey, such as nematodes and oligochaetes. The flies that breed in these habitats are sometimes pests of humans and their livestock, and the observation that macrochelids prey on these species has been the motivation for much of the biological and taxonomic work that has been done on the family (see Krantz, 1983). A further factor adding to interest in the family is phoresy. The species that occur in ephemeral patches of substrate, such as dung pads or cadavers, disperse by attachment to other organisms, especially dung- and carrion-feeding beetles and flies, and are often found on these insects in collections (Krantz, 1998b)'.

A rich scientific literature confirms these aspects (Evans and Browning, 1956, Bregetova and Koroleva, 1960; Krantz, 1962; Hyatt and Emberson, 1988; Karg, 1993b; Halliday, 2000; Gerson *et al.*, 2003; Mašán, 2003).

Macrochelid mites possess peritremes looped around the stigmata and set between the legs III and IV. The ambulacra lack on the thin leg I, as well as leg grooves. The females have small metasternal plates, and the spermadactyl digits of the male are distally free (Gerson *et al.*, 2003). Krantz (1998b) included in the family of about 400 species ascribed to 15 genera. The genera were revised by Krantz (1962), and Karg (1993a) keyed the European species, while Halliday (2000) revised the Australian taxa. Further, taxonomic studies reported that the family numbers today over 470 species in 20 genera, and the genus *Macrocheles* Latreille about 325 described species, being the most diverse genus of the family (Krantz, 1962, 1998b, 2018; Lindquist *et al.*, 2009b; Emberson, 2010; Beaulieu *et al.*, 2011; Krantz and Moser, 2012; Makarova, 2012).

The biology and ecology of the Macrochelid mites have been investigated by many workers (Filipponi, 1955, 1960; Axtell, 1961, 1963a, 1966; Rodriguez and Wade, 1961; O'Donell and Axtell, 1965; Mahunka, 1971; Keiback, 1978; Axtell and Edwards, 1983). Most phoretic macrochelids have an haplodiploid sex determination system, the arrhenotoky, where males are parthenogenetically produced from unfertilized eggs (Oliver, 1977; Norton *et al.*, 1993; Manning and Halliday, 1994). A few species can be thelytokous and phoretic, like *Macrocheles similis* Krantz and Filipponi, a species similar to *M. muscaedomesticae* (Scopoli) (Manning and Halliday, 1994), and one species, *Macrocheles mycotrupetes* Krantz and Mellott, phoretic on dung beetles, behaves like a diplodiploid (Krantz and Royce, 1994). Experiments on arrhenotokous and phoretic *M. glaber* indicated that virgin females can easily be fertilized by their sons, allowing the start of a population (Manning and Halliday, 1994). Fertilization by sons, oedipal reproduction, has also been studied experimentally for *M. muscaedomesticae* (Farahi *et al.*, 2018b).

Phoretic associations are reported for macrochelid mites, and knowledge is greatly increasing with the description of new species and new phoretic associations (Haloti *et al.*, 2005; Ács *et al.*, 2017; Azevedo *et al.*, 2017; Hartini and Dwibadra, 2017; Ozbek, 2017; Alatawi *et al.*, 2018; Knee, 2018; Kontschán, 2018). Foretic important *Macrocheles* species arrive on carcasses through phoresy on flies and beetles. Those associated with corpses and carcasses can inform about circumstances of death, environment and habitat, making a link to a site or a location. Carriers for macrochelids linked to decomposing mammals are necrophagous and necrophilous insects and micro-mammal hosts (Leclercq and Watrin, 1973; Korn, 1983; Andreev, 1988; Krantz and Whitaker, 1988; Mašán, 1999a, 2003; Halliday, 2000). In Europe, macrochelids on carrion or burying beetles (Silphinae and Nicrophorinae) are outnumbered by species of

DOI: 10.1079/9781786390912.0066

Parasitidae (Hyatt, 1980, 1990; Mašán, 1999a, 2003). In Hungary, it was found in association with *Cetonia aurata* (Linnaeus) beetle (Ács and Kontschán, 2014), and in Slovakia, it was collected in the nests of *Aquila pomarina* (Brem) (Krištofík *et al.*, 2009).

Klimov *et al.* (2016) reported that within the family Macrochelidae, species of the genus *Macrocheles* are common in a wide range of organic substrates (animal dung, hay, rotting stumps, galleries of bark beetles and subcortical situations, and nests or galleries of mammals, birds and social insects, including bees, such as *Bombus*, *Apis* and *Melipona*). These species commonly form phoretic associations with beetles and flies, to colonize fresh substrates. Commonly, only inseminated females disperse, with the exception of the protonymphs and deutonymphs of *Macrocheles dimidiatus*. *Berlese* species group has also been found to be phoretic. Among mites collected on bees and/or in their nests, there are two very similar species from bumble bees (*Bombus* spp.), *Macrocheles rotundiscutis* Bregetova and Koroleva (= *M. bombophilus* Götz) from Europe and *Macrocheles praedafimetorum* Richards and Richards from North America. These species show a clear preference toward bumble bee nests, although they can be found in other habitats. In nests of bumble bees, the mites often reach large population sizes feeding on a wide range of saprophagous organisms common to dung and decaying matter without any negative effect to the hosts. Cannibalism has also been recorded. It is interesting that dispersal does not occur on bumble bees but on a dung-inhabiting beetle that is apparently attracted to bumble bee nests, by the decaying material on the bottom of the domicile (Richards and Richards, 1977). In contrast, species reported from hives of honeybees (*Apis* spp.) and other social bees are habitat generalists that can probably also use bees as transport (Grobov, 1978; Chmielewski, 1991c). Because *Macrocheles* species may consume harmful microarthropods in bee nests, they can be beneficial to their hosts. Predation on the bee brood is unlikely. *Macrocheles* may be somewhat harmful, if large numbers of mites disperse on a single adult bee. For example, a single adult bee may carry up to 30 females of *M. glaber* (Müller) (Vitzthum, 1926). In this relatively uncommon case, the bees' movements can be hampered (Homann, 1933). Beresford and Sutcliffe (2009) studied the effect of mite phoresy, on stable fly populations in south-central Ontario. They reported that the stable flies disperse from their overwintering farms and re-establish populations at neighbouring dairy and beef farms each spring. They tested the hypothesis that mite phoresy affects the dispersal of their stable fly phorionts *M. muscaedomesticae* and *Macrocheles subbadius* (Berlese). At a beef farm, they observed a higher proportion of parous stable flies with mite scars than expected, based on the proportion of nulliparous females carrying mites. These results were consistent with their hypothesis, indicating that stable flies without mites may be emigrating more, than flies carrying mites. Moreover, they tested the hypothesis by comparing the mite load on dispersing stable flies, sampled with a vehicle-mounted truck trap to the mite load on resident stable flies, sampled from three dairy farms and one beef farm (May to October), and significantly no stable flies caught in the truck trap were carrying mites, compared to the seasonal means of 10% and 5% of female and male stable flies, with mites at the four farms (Beresford and Sutcliffe, 2009).

Pereira and Castro (1945) observed that several species of the genus *Macrocheles* attack juveniles of housefly, *Musca domestica* Linnaeus; face fly, *Macrocheles autumnalis* De Geer; and stable fly, *Stomoxys calcitrans* (Linnaeus), feeding on eggs and first stage larvae; Axtell (1961) and Kinn (1966) studied the rate of predation on housefly eggs and larvae by many species of Macrochelidae. In the absence of housefly immature stages, they feed as alternative food, mainly on small arthropods and nematodes (Evans *et al.*, 1961; Rodriguez and Wade, 1961), with the exception of only a handful of non-phoretic, detritivorous species (Manning and Halliday, 1994). As predators, they influence population growth of other micro-invertebrates (Geden *et al.*, 1988; Perotti, 1999, 2001) and may have effects on the advancement and composition of ephemeral micro-ecosystems (Kamaruzaman *et al.*, 2018). Krantz (1998b) reported many of Macrochelidae occupy specialized and often unstable habitats. Most known species have adapted to life in dung deposits, where prey is plentiful, and the potential exists for rapid population growth. Phoresy on co-occurring flying insects plays a vital role in assuring niche continuity for macrochelids in these ephemeral substrates. He discussed the emergence of phoresy, as a major survival strategy in the Macrochelidae associated with dung beetles and placed a special emphasis on the behavioural and chemical mechanisms that mediate phoretic specificity of macrochelid species, in the unique n-dimensional universes of their scarab hosts. Moreover, Krantz (1998b) reported that phylogenetic analysis of selected phoretic and non-phoretic macrochelid taxa has shown a strong correlation between phoretic state and evolutionary position, indicating that an increasing commitment to phoresy in the Macrochelidae is correlated with an advance from early derivative to terminal taxa. In this context, laboratory and field studies have confirmed the importance of chemical, behavioural and ecological factors in maintaining the integrity of the relationship between phoretically specific macrochelids and their dung beetle hosts (Krantz, 1998b).

This common association with the pests has suggested that they could be used as biocontrol

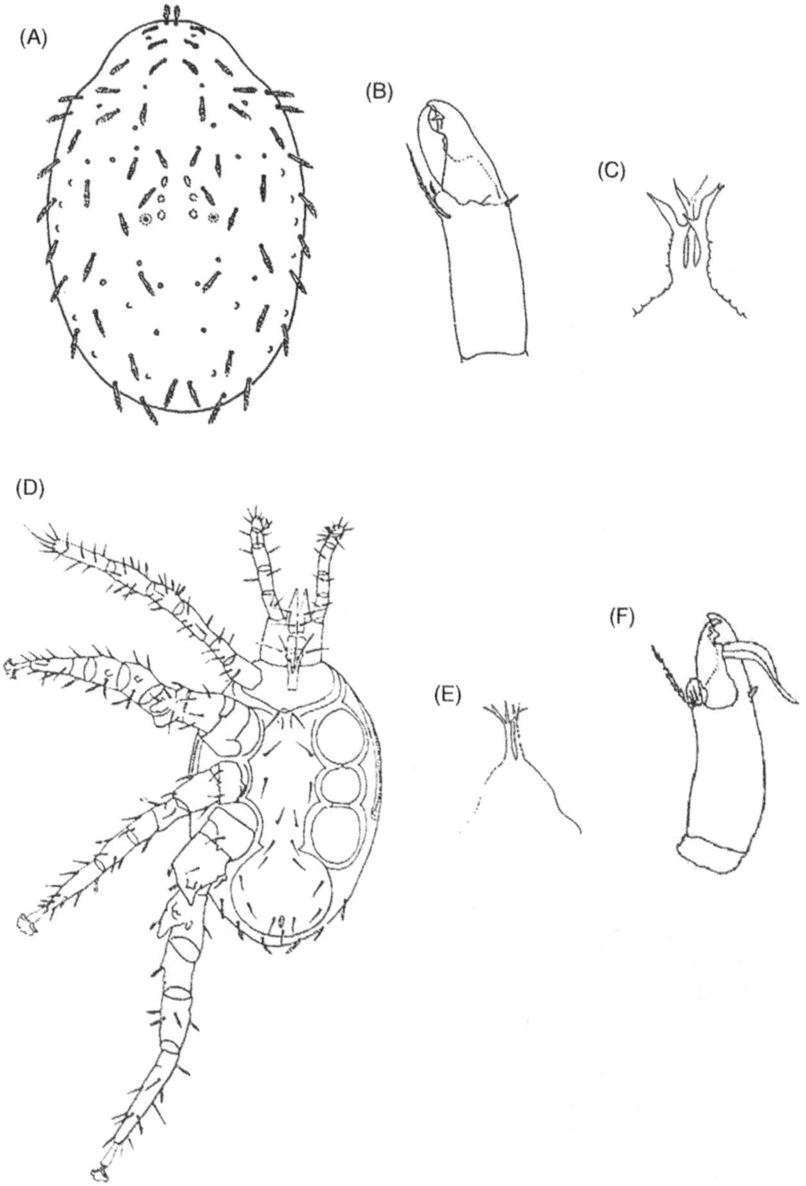

Fig. 66.1 *Macrocheles matrius* (Hull). Female. Dorsal view (A); chelicera (B); tectum (C). Male. Ventral view (D); tectum (E); chelicera (F) (from Hughes, 1976, licensed under the Open Government Licence v3.0).

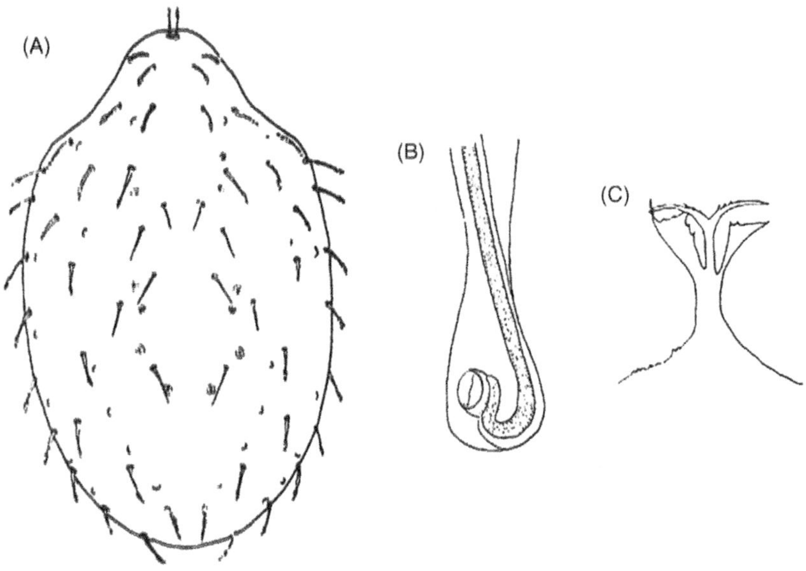

Fig. 66.2. *Macrocheles muscaedomesticae* (Scopoli). Female. Dorsal view (A); stigma and peritreme (B); tectum (C) (from Hughes, 1976, licensed under the Open Government Licence v3.0).

agents (e.g. *Macrocheles robustulus* Berlese), which is presently, commercially available for the control of pre-pupae and pupae of thrips, as well as larvae of *Lyprauta* sp. (Keroplatidae). More intensive studies of the macrochelids may turn possible, the detection of other species with potential for practical use (Azevedo *et al.*, 2015). The potential of *Macrocheles* species as control agents of harmful flies (Diptera) and the biology of some species (e.g. *Macrocheles embersoni* Azevedo, Castilho and Berto) on *S. calcitrans* (Linnaeus) and *M. domestica* (Diptera, Muscidae) were discussed by Azevedo *et al.* (2018).

On stored food are reported *M. glaber*, *M. matrius* (Hull) (Fig. 66.1), *M. merdarius* (Berlese), *M. muscaedomesticae* (Fig. 66.2) and *M. robustulus*.

M. glaber was described by Müller (1860) in association with insects in Czech Republic. The mite possesses a wide geographic distribution (Mašán, 2003), and its morphology has been described by various authors (Evans and Browning, 1956; Bregetova and Koroleva, 1960; Halliday, 1986; Özbek *et al.*, 2015). The mite has been collected in many habitats (Mašán, 2003; Kazemi and Rajaei, 2013; Özbek *et al.*, 2015), and different works show the possibility to rear *M. glaber* on various dipteran pests, such as *Lucilia* sp., *M. domestica*, *M. vetustissima* Walker and *Haematobia irritans exigua* (Linnaeus) (Halliday and Holm, 1987; Azevedo *et al.*, 2015). *M. glaber* may also feed on nematodes, such as *Panagrellus* sp., Rhabditidae (Filipponi and Delupis, 1963; Manning and Halliday, 1994), springtails and phorid flies. Al-Dulaimi (2002) studied the predation rate, and development time of *M. glaber* feeding on frozen housefly eggs. However, there are no experiments on the life table, especially the population parameters, of this species. The topic has become of actuality and has been extensively investigated; Azevedo *et al.* (2019) studied the free-living nematodes as alternative prey for soil predatory mites in the focus of conservation biological control.

Axtell (1963a) found the mite on dairy cattle manure from inside the barns in New York (United States). Its association with dung beetles (Scarabaeidae, Geotrupidae) is well known (Filipponi and Pegazzano, 1962; Cicolani *et al.*, 1981; Halliday and Holm, 1985; Shereef *et al.*, 1990; Halliday, 2000; Mašán, 2003; Niogret *et al.*, 2006). In Central Europe, it has been found in nests of hoopoe (*Upupa epops* Linnaeus) (Krištofík *et al.*, 2013). Howes (1979) recorded a single deutonymph of *M. glaber* on the common noctule, *Nyctalus noctula* (Schreber) (Chiroptera, Vespertilionidae), from England, but is likely to have originated from the bird nests or rot holes in the tree in which the host bat was found. Moreover, species of *Macrocheles* have previously been found in bat guano in caves in Italy (Cicolani and Sabatino, 1991). On stored food, it is reported only on rice from Iraq (Hagstrum *et al.*, 2013).

M. glaber feeds on eggs, newly hatched and small larvae of housefly, *M. domestica*. Al-Dulaimi (2002) reared the mite in the laboratory on housefly frozen eggs, at constant conditions of 28 ± 1°C and 90% RH, using sterilized horse dung substrate. The predation rates of adult female and male on frozen eggs were 18 and 3 eggs/mite/day, respectively, and the number of frozen eggs destroyed by adult female through its life was 185.6 eggs. The mean duration of adult female from egg to adult stage was 2.67 days, the longevity of female was 27.8 days, the mean daily egg production was 2.7 egg with total egg productivity of 72.1 egg (Al-Dulaimi, 2002). Marquardt *et al.* (2015) reported that the oviposition of *M. glaber* can involve either oviparity or egg production, followed by immediate hatching of the larva, a process called ovoviviparity. They observed three phases in the process of ovoviviparity, the pre-oviposition behaviour, the oviposition followed by holding the egg under the gnathosoma, and the hatching of the larva assisted by the female, respectively. Two factors appear to favour the occurrence of ovoviviparity in this species, such as poor feeding and poor substrate quality. They observed egg cannibalism four times, in two of the four females studied and discussed, the consequences of variable oviposition strategy, parental care and egg cannibalism in *M. glaber* in the context of maximization of reproductive success (Marquardt *et al.*, 2015). Wen *et al.* (2017) studied the development time of the different life stages, adult longevity, adult pre-oviposition period (APOP), total pre-oviposition period (TPOP), oviposition period, fecundity, and sex ratio of *M. glaber* fed on *Coboldia fuscipes* Meigen (Diptera, Scatopsidae) in climatic chambers maintained at 25 ± 1°C, 90% ± 10% RH, and no lighting. The eggs development took 1.09 ± 0.04 that of larva-protonymph 1.85 ± 0.12, of deutonymph 1.48 ± 0.09, of preadult 4.40 ± 0.17, of adult 55.88 ± 4.78, of female adult 75.53 ± 5.23, and of male adult 24.45 ± 2.24 days, respectively. The APOP takes 7.84 ± 1.27, the TPOP 12.03 ± 1.26, the longevity is 52.21 ± 4.74, the female longevity is 79.69 ± 5.22, the male longevity is 29.25 ± 2.35 and the oviposition days is 18.65 ± 1.84 days, respectively. The fecundity (eggs/female) is 18.94 ± 1.97, the reproductive female (%) 96.88 ± 3.10 and the sex ratio (% female) is 61.54 ± 6.75. Regarding the population parameters, the predation rate of protonymph + deutonymph is 4.55 ± 0.76, of adult female (first 4 days) is 4.12 ± 0.39 and of adult male (first 4 days) is 1.49 ± 0.15. The intrinsic rate of increase r (d-1) is 0.0699 ± 0.0074, the finite rate of increase λ (d-1) is 1.0724 ± 0.0080, the net reproductive rate R_0 (offspring) is 9.93 ± 1.58, the mean generation time T (d) is 32.84 ± 1.97, the net predation rate C_0 (prey) is 17.51 ± 0.79 and the finite predation rate ω (d-1) is 61 1.148 ± 0.013 (Wen *et al.*, 2017).

M. matrius was described by Hull (1925) from poultry manure in England and can be identified following the morphological descriptions of different specialists (Evans and Browning, 1956; Bregetova and Koroleva, 1960; Hughes, 1976; Qayyum *et al.*, 2016). Its geographical distribution is cosmopolitan and includes different countries, among which Austria (Johnston, 1970), Canada (Sinha, 1964b), Egypt (Zaher, 1986), England (Hughes, 1976; Hyatt and Emberson, 1988), Finland (Mrciak and Brander, 1965), Germany (Krantz, 1972), Iceland (Ólafsson and Ingimarsdottir, 2009), Israel (Costa, 1966a), Italy (Filipponi and Cicolani, 1974; De Lillo and La Notte, 1995), New Zealand (Emberson, 1973), Turkey (Qayyum *et al.*, 2016), USSR (Bregetova and Koroleva, 1960), United States (Axtell, 1963a) and other countries. It is a bisexual, arrhenotokous species, with the males developing from unfertilized eggs and the females from fertilized eggs. Samples collected during various times of the year have always shown a notable excess of females (spanandry) (Cicolani, 1977). Evans and Browning (1956) reported that *M. matrius* is widely distributed in poultry manure, where the dung is mixed with large amount of litter, and dung beetles such as *Geotrupes silvaticus* Panzer (Geotrupidae), aid in its distribution (Götz, 1952; Hirschmann, 1970). Axtell (1963a) found the mite on dairy cattle manure, from inside the barns in New York (United States). In Hungary, it was collected mostly from soil samples but also from dungs of horse (Erőss and Mahunka, 1971), and in Slovakia was collected in the nests of *Aquila pomarina* Brem (Krištofík *et al.*, 2009) and of hoopoe (*Upupa epops* Linnaeus) in Central Europe (Krištofík *et al.*, 2013). Evans and Browning (1956) collected the mite in a nest of the bird *Riparia riparia* Linnaeus. Hyatt and Emberson (1988) reported the mite associated with poultry, also with the mink *Mustela vison* (von Schreber) (Carnivora, Mustelidae), and the water shrews *Neomys fodiens*

(Pennant) (Soricomorpha, Soricidae), in a canary cage and from grass clippings and compost. In Brazil, the mite has been collected on the birds *Gallinago paraguaiae* (Vieillot) and *Turdus leucomelas* Vieillot (Silva, 2014). Bregetova and Koroleva (1960) report *M. matrius*, as an inhabitant of nests of a variety of rodents throughout Russia, and in nests of *Mus spicilegus* Petény (Rodentia, Muridae) in Slovakia (Mašán and Stanko, 2005; Várfalvyová *et al.*, 2011). The mite is also known to occur in association with coprophilous scarabs (Götz, 1952). Specimens from the Hamburg collections were recovered from beetle cultures and from damp straw (Krantz, 1972). Pfliegler *et al.* (2017) collected the mite in silo from Malta.

On stored food, the mite has been collected on floor debris, sieving of grain (Hughes, 1961, 1976) and grass seed in farm (Hagstrum *et al.*, 2013).

M. matrius preys on eggs and larvae of housefly and nematodes (Rhabditidae) (Filipponi and Cicolani, 1974; Cicolani, 1977). Cicolani (1977) studied the influence of temperature on the speed of development, from egg to adult of the mite. The lower temperature limit, for complete preimaginal development was found to be 10–12°C, and the upper limit 36–38°C. The maximum speed of development in both sexes was observed at 34°C, while the optimum was between 32°C and 36°C. The most suitable mathematical expression for the speed of development–temperature relationship was found to be the logistic equation calculated for females ($100/y = 2.01/1 + e^{5.19466-0.23924x}$) and males ($100/y = 2.36/1 + e^{4.86240-0.211609x}$). A comparison was also made between some values obtained at constant temperatures and at fluctuating temperatures. The temperature range for this species is between 12°C and 36°C. Preliminary experiments have indicated a probable optimum range of 24–34°C. Within the 24–34°C range, the temperature relationships for fecundity, longevity and rate of oviposition were determined in females reared at 28°C and 75 % RH and fed on *M. domestica* eggs and nematodes. The optimum range for fecundity is 24–28°C, and for fertility 26–28°C. As concerns the r_m, a population of *M. matrius* can multiply 1.71 times per day at 32°C, 1.92 at 26°C, 2.10 at 28°C, 2.15 at 30°C, 2.20 at 32°C and 2.25 at 34°C. These values compare with 1.17 at 24°C, 1.45 at 28°C, 1.41 at 32°C and 1.37 at 36°C for *M. domestica* (Filipponi and Cicolani, 1974). Soliman *et al.* (1978) found *M. matrius* in great numbers and fed variously on housefly eggs and larvae during its different stages, except larvae, which proved to be unfeeding stage. They observed that the female consumed an average of 12.9 and 118.4 eggs, during immature stages and adult stages, respectively. Feeding on housefly eggs increased the female predator fecundity and prolonged oviposition period. The number of deposited eggs per female averaged 63.8 and 40.7 eggs, when fed on housefly eggs and larvae respectively. The generation period (from egg to egg) averaged 4.9 and 6.0 days, when fed on housefly eggs and larvae, respectively (Soliman *et al.*, 1978).

M. merdarius has a cosmopolitan geographical distribution (Bregetova and Koroleva, 1960; Mašán, 2003; Kılıç *et al.*, 2012), and it is reported on stored onion, storage products, stored rice, rice dust and debris from Iran (Ostovan, 1993; Fathipour, 1994; Noei, 2007; Noei *et al.*, 2008b; Khaleghabadian *et al.*, 2015).

M. muscaedomesticae is a cosmopolitan species (Azevedo *et al.*, 2017), reported from different countries among which Australia (Krantz and Withaker, 1988; Halliday, 1989), Brazil (Marchiori, 2001; Silva *et al.*, 2013), Canada (Banks, 1912; Hennessey and Farrier, 1988), England (Evans and Browning, 1956; Hughes, 1976; Hyatt and Emberson, 1988), Iran (Mossadegh, 1997), Philippines (De Jesus and Rueda, 1990; Takaku *et al.*, 2012), Scotland (Evans and Browning, 1956) and the United States (Ewing, 1913; Hennessey and Farrier, 1988; Krantz and Whitaker, 1988). The mite may be identified following Hughes (1976), and Özbek *et al.* (2015). *M. muscaedomesticae* is the most important and extensively investigated predacious mite of all biological stages, except the puparium, of house flies and allied species in poultry and other confined animal manure (Wade and Rodriguez, 1961; Axtell, 1963b, 1969; Filipponi and Petrelli, 1967; Cicolani, 1979b). So, it lives on substrates such as dung, which provides a suitable habitat for the flies, and the females use the adult fly and also dung beetles to transport from one breeding site to another. Rodriguez *et al.* (1962) found that the nature of substrate influenced the egg production of the mite, and this was found to be highest, when a used standard housefly rearing medium was utilized, and nematodes and fly eggs were added to serve as food. The mite has also been recorded from deep pit poultry manure in Essex (Bills, in Hughes, 1976). The eggs are laid by virgin and mated females, some attempt being mate to conceal them in cracks or under clumps of substrates (Wade and

Rodriguez, 1961), although this does not prevent the occurrence of cannibalism. The larvae and adults feed readily on fly eggs, but when they are available, proto- and deutonymphs prefer nematodes such as *Rhabditella leptura* (Cobb) (Rhabditida, Rhabditidae) and, to a lesser extent, species belonging to the genera *Diplogaster* Schultze (Diplogasteridae) and *Panagrolaimus* Fuchs (Panagrolaimidae) (Rodriguez *et al.*, 1962). Two distinct deutonymphs are formed, one giving rise to males and the other to females. Virgin females always give rise to males, mated females to females and an occasional male (Pereira and Castro, 1945, 1947). According to the above-discussed point, this mite is retained a predator of nematodes, fly eggs and larvae, which can be found in substrates of equine, bovine dung and laying hen systems (Marchiori, 2001; Silva *et al.*, 2013).

The aspects discussed previously show that the mite is mostly associated with synanthropic flies, also with fly breeding places, compost heaps, rotting seaweed and artificial nests of rats, and it is highly prevalent on muscoid flies of the families Muscidae and Fanniidae (Filipponi, 1960; Rodriguez and Wade, 1961; Axtell, 1964; Perotti and Brasesco, 1996, 1997; Sacchi Carmona Rodrigueiro and Pires do Prado, 2004). These flies colonize and reproduce on a particular variety of organic material (e.g. poultry manure), as well as on sources of food decay abundant in urban areas. They are considered highly synanthropic insects (Legner and Bowen, 1973; Perotti, 1998). *M. muscaedomesticae* is less common on other arthropods and mammals, to a point where it has rarely been reported on Calliphoridae, the dominating fly family of animal decomposition. It can occur on adult blowflies under special circumstances like in indoor decomposition, due to loss of phoretic specificity (Perotti and Braig, 2009). The macrochelid is commonly found in dung and as a phoretic of *M. domestica* Linnaeus and related higher Diptera (Evans and Browning, 1956; Axtell, 1963a); it has also been found on coprophagous beetles (Cicolani, 1979a) and on two species of small mammals, the bank vole, *Clethrionomys glareolus* (Schreber) (Rodentia, Cricetidae) and the house mouse, *Mus musculus* Linnaeus (Rodentia, Muridae) in Central Europe (Haitlinger, 1976). Krantz and Whitaker (1988) reported that unpublished records showed that the range of habitats of *M. muscaedomesticae* is broader than that suggested in the literature, and that the mammal associations cited above may be more than chance occurrences. Several specimens of *M. muscaedomesticae* were found on a northern hopping mouse, *Notomys alexis* Thomas (Rodentia, Muridae) at the Melbourne Zoo, in Australia. The mite has been found on or with the white-footed mouse, *Peromyscus leucopus* (Rafinesque) and the hispid cotton rat, *Sigmodon hispidus* Say and Ord (Rodentia, Cricetidae) in North Carolina and *Sciurus carolinensis* Gmelin (Rodentia, Sciuridae) in Indiana (United States). Specimens have also been collected from the downy woodpecker, *Dryobates pubescens* (Linnaeus) in North Carolina and *Sayornis* sp. in Maryland and bird nests (three swallow nest in New York). They cited one collection made from a turtle, *Terrapene carolina* (Linnaeus), in North Carolina, and a unique specimen in the OSU collection was retrieved from the mouth of a crocodile, *Crocodylus johnsoni* (Krefft) (Crocodylia, Crocodyllidae), in Darwin (United States). Clearly, this ubiquitous species has succeeded in utilizing a broad spectrum of invertebrates and vertebrates, including mammals (Krantz and Whitaker, 1988). Hyatt and Emberson (1988) reported that a sample from a deep litter, in which partridges *Perdix perdix* (Linaneus) were being reared, contained thousands of females of this species.

In stored food, the female of *M. muscaedomesticae* are likely to be found anywhere that is open to the visit of flies; the mite is known on dried fish, grain, maize trash, onion bulb, mouldy groundnut, wheat rotting residue in flour mill and wheat elevator (Hagstrum *et al.*, 2013; Khaleghabadian *et al.*, 2015).

Females are diploid and males haploid (arrhenotoky), with unfertilized eggs producing male offspring. Offspring are commonly about 40% female. Males are shorter lived and are not important as predators on housefly. The mite life cycle consists of egg, six-legged larva, protonymph, deutonymph and adult. Most predation is by the adult female, and deutonymph feeding on housefly eggs and first instar larvae, with a preference for the eggs. The mites will feed on immatures of other muscoid flies, if present, as well as on eggs of the poultry house pill beetle, *Carcinops pumilio* (Erichson) (Coleoptera, Histeridae). Free-living rhabditid nematodes are fed on by the nymphal, and adult stages of the mite, and are important in maintaining a high rate of survival and reproduction by the mite. The mites also feed on the eggs and larvae of species of small flies, such as Drosophilidae and Sphaeroceridae, which may be abundant in the manure under some conditions. Acarid mites are another food, although less important than the nematodes and small Diptera. The adult female mite is phoretic on the housefly, and other muscoid filth flies (Axtell, 1964; Farish and Axtell, 1971; Borden, 1989; Ho, 1990). This phoresy is regulated by olfactory responses in the flies and the ageing manure, resulting in dispersal to the most favourable fly breeding areas. The first pair of legs of the mite is not used for walking but rather is waved about in the air like antennae and possess olfactory sensilla on the tarsi (Farish and Axtell, 1966; Coons and Axtell, 1973). The mite can detect odours of manure and adult flies; the balance between the two determines whether or not the mite stays in the manure. If the manure is aged and less odoriferous, the mite will leave by attaching to a visiting fly. Attachment of the mite to the fly is with the chelicerae; normally, no feeding occurs. Macrochelids as well as other predacious mites are able to reproduce, and prey effectively on housefly eggs and first instar larvae in manure that is of reasonable moisture level, so that the mites can freely move and locate prey. Wet, fluid manure is physically unsuitable for mite survival, and limits the populations of suitable prey. It is likely that other species, especially beetles, sometimes prey on macrochelid mites in their early life stages, but this is not well documented (Flies Breeding in Accumulated Organic Wastes, 2022).

Wade and Rodriguez (1961) studied the standardization of general rearing conditions and a substrate (ground steer manure plus 10% soybean oil meal, moistened), as well as the definition of biological performance of *M. muscaedomesticae* under these conditions. They observed that eggs from virgin females are smaller (0.370 by 0.249 mm) than eggs from mated females (0.374 by 0.256 mm) and required a longer incubation period. These individuals spent less time as larvae; neither difference in the first two stages was significant. The protonymphal stage lasted significantly longer in the male progeny from virgin females, while the deutonymphal stage tended to be shorter. However, the total elapsed time from egg to adult for male or female progeny was not significantly different. The pre-oviposition period in fertilized females (53.5 h) exceeded that in virgin females (48 hours), but not significantly. Significantly more eggs were laid by virgin females (91.7) than fertilized females (61.4). The maximum recorded 24-hour egg production from a virgin female was 28, from a fertilized female 25; the average daily production was 7.70 and 5.29, respectively. The average lifespan of fertilized females (23.9 days) was close to that of unfertilized females (24.3 days); the male lifespan was shorter, averaging 15 days (Wade and Rofriguez, 1961).

Farahi *et al.* (2018b) reported that when a virgin female of *M. muscaedomesticae* colonizes a new manure substrate, it produces male offspring through parthenogenesis (arrhenotoky); when her sons reach maturity, oedipal mating takes place and the female begins to produce bisexual offspring. In order to examine the consequence of oedipal reproduction on population development, they designed two separate experiments to compare life history traits, and life table parameters of oedipal versus bisexual cohorts of *M. muscaedomesticae*, using the age-stage, two-sex life table method. Experiments were conducted at 28 ± 1°C, using a photoperiod of 14:10 (L: D) h, and 65 ± 5% RH, with housefly eggs used to feed mites. Mean adult female longevity was 38.63 days, and fecundity 128.51 offspring under bisexual reproduction, and 37.48 days and 68.23 offspring under oedipal reproduction. In the bisexual cohort, the intrinsic rate of increase (*rm*), the finite rate of increase (λ), the net reproduction rate (R_0), the GRR and the mean generation time (T) of *M. muscaedomesticae* were 0.2938 day^{-1}, 1.3415 day^{-1}, 54.216 offspring/individual, 77.7 offspring/individual and 13.5885 days, respectively. Because only male eggs were produced during the first 5.62 days (on average) of the oviposition period in the oedipal cohort, it was theoretically incorrect to compute the population parameters using the survival and fecundity values for this group, even though bisexual reproduction did occur after this period. Results of this study determined that the effect of oedipal reproduction could be correctly defined and analysed by using the age-stage, two-sex life table method and demonstrated that virgin females are able to produce and copulate with their sons (oedipal mating), which then allows those females to produce both sexes. This reproductive system can enable this valuable natural enemy to considerably extend its distribution potential (Farahi *et al.*, 2018b). Durkin *et al.* (2019) studied several aspects of the biology of *M. muscaedomesticae* associated with drosophilid flies collected in Alberta, Canada. They assessed the degree of divergence of our populations from others, compared their life history to other published populations and experimentally tested whether *M. muscaedomesticae* feeds on *Drosophila hydei* (Sturtevant) hosts by comparing the body mass of mites that attached to hosts to those that did not. There was no strong phylogenetic differentiation among any of the *M. muscaedomesticae* specimens, suggesting multiple recent introductions of this species to Canada. Compared to other populations, their mites showed a lower fecundity, which may have been a result of the temperature or nematode only diet, in which they were maintained. The mites that attached to hosts for 4 h weighed significantly more than those that did not. Without direct evidence for host tissue transfer to the mites, it is difficult to determine whether the mites are indeed feeding on their hosts while attached. However, the existing evidence for the costs fly hosts endure, at the expense of these mites, makes this relationship antagonistic (Durkin *et al.*, 2019). Rodriguez *et al.* (1962) reported that during a 2-year ecological study of macrochelid mites in their natural habitat, the manure pile, free-living nematodes were found in close association with them. Field populations of *M. muscaedomesticae* were highest in winter and spring, when free-living nematodes evidently are most prevalent; emerging house flies transport these mites to other oviposition sites, during the fly season. The exact food habits of the macrochelid *Glyptholaspis confusa* (Foa), another regular inhabitant of the manure pile, remain relatively unknown. Of the three nematode species cultured in the laboratory, namely *Diplogaster* sp., *Panagrolaimus* sp. and *R. leptura* (Cobb), proved the most easily reared and the best food for *M. muscaedomesticae*. The adult mites, given equal choice, preferred housefly eggs over nematodes, while proto- and deutonymphs under the same conditions preferred the nematodes. Standard fly larva rearing media (new, fermented and used) were tested as substrates, to replace steer manure, in an effort to sustain a food chain of bacteria, nematodes and mites within one culture. The highest mite reproduction rate (23.11 ± 5.58 progeny per female per day) was achieved on used media (recovered after rearing fly larvae), with both housefly eggs and nematodes added as food. Nematodes alone returned 12.11 ± 1.42 progeny per female per day. On fermented fly larval media, progeny return was as high on nematodes alone as with both nematodes and housefly eggs added. Cultures pre-seeded with nematodes 2 days before the introduction of mites were self-sustaining (Rodriguez *et al.*, 1962).

Luong *et al.* (2015) examined the physical and physiological costs of ectoparasitic mites on host flight endurance, and reported that dispersal is essential for locating mates, new resources and to escape unfavourable conditions. Parasitism can impact a host's ability to perform energetically demanding activities, such as long-distance flight, with important consequences for gene flow and meta-population dynamics. Moreover, the ectoparasites can adversely affect host flight performance, by diminishing flight aerodynamics and/or by inflicting physiological damage while feeding on host tissue. They carried out experimental flight assays using two fruit fly-mite systems based on the first on *Drosophila nigrospiracula* (Patterson *et* Wheeler) and *M. subbadius* (Berlese), and the last on *D. hydei* (Sturtevan) and *M. muscaedomesticae*, respectively. Flies that are burdened by mites are expected to exhibit lower flight endurance compared to uninfected flies, and the presence of mites attached, significantly decreased flight endurance by 57% and 78% compared to uninfected *D. nigrospiracula* and *D. hydei*, respectively. The physiological damage caused by *M. subbadius* was revealed through a 53% decline in flight time, among previously infected flies (mites removed just prior to flight assay). The presumably phoretic *M. muscaedomesticae* also induced a 62% reduction in flight endurance among previously infected *D. hydei*. The results show a strong deleterious effect of ectoparasitic mites on host flight performance, mediated by a reduction in flight aerodynamics and damage to host physiology. Adverse effects on host flight and/or dispersal may have broad implications for gene flow, population genetic structure and local adaptation in both host and parasite meta-populations (Luong *et al.*, 2015).

Azevedo *et al.* (2018) compared the predation and oviposition rates of *M. embersoni* Azevedo, Castilho and Berto, *M. muscaedomesticae* and *M. robustulus* (Berlese) on *S. calcitrans* (Linnaeus), *M. domestica*, *Haematobia irritans* (Linnaeus), *Bradysia matogrossensis* (Lane), *Rhizoglyphus echinopus* (Fumouze *et* Robin) and the free-living nematode *Rhabditella axei* (Cobbold). They evaluated the life cycle of the best performing predator on the most suitable prey species at 30 ± 2 °C, 95 ± 5% RH, and in the dark. The three macrochelid species consumed all evaluated prey species, but *M. embersoni* showed higher predation and daily oviposition rates on larvae of *S. calcitrans* (23.8 larvae consumed and laid about 4.0 eggs, respectively) than other species. Total immature development of *M. embersoni* was completed in at most about

1.3 days on eggs of *S. calcitrans* and *M. domestica* and 1.5 days on *R. axei*. *M. embersoni* laid most eggs (0.28–0.34 female/female/day; the r_m ranged from 0.28 to 0.34) on those same preys. The results of this study suggest that *M. embersoni* is the best performing predator (Azevedo *et al.*, 2018).

Heikal (2020) evaluated the potential use of two predacious mites, *Parasitus fimetorum* (Berlese) and *M. muscaedomesticae*, two for controlling the root-knot nematode, *Meloidogyne javanica* Treub under laboratory and semi-field conditions. Results of this study revealed that the two predators significantly reduced the root-knot nematode numbers. In addition, the highest reduction percentage (57.24%) in nematode juveniles was recorded at the treatment of (1000 nematode + 10 mites). For *M. muscaedomesticae*, the highest mortality percentage (50.83%) in nematode juveniles was recorded at the treatment of (1000 nematode + 50 mites), followed by (1000 nematodes + 20 mites) 48.88%, while the treatment of (1000 nematode + 10 mites) gave (47.13%). The combination of the two mite species (1000 nematodes + 50 mites/species) caused the highest mortality percentages in nematode juveniles (69.29%), followed by (1000 nematodes + 20 mite/species) 50.51% and the treatment of (1000 nematode + 10 mite/species) (37.66%). At the pot experiments, the highest overall mortality percentage in *Meloidogyne javanica* juveniles was recorded at the treatment of *P. fimetorum* + *M. muscaedomesticae* giving (57.07%), followed by the treatment of *P. fimetorum* (39.17%), and then, by *M. muscaedomesticae* alone that recorded only (17.47%) (Heikal, 2020).

What is discussed above relates to danger direct or indirect of some dipteran species. Unfortunately, control strategies for dipterous pest species continue to rely heavily on insecticides, even though these species have developed resistance to most insecticides currently available, due to their widespread use. Increased public and farmer awareness have prompted scientists to investigate alternative methods for managing these filth flies (Shiralizadeh *et al.*, 2021). This need has been formulated from time, such as by Mahmood and Al-Dulaimi (1986), which suggested that *M. muscaedomesticae* may be used as a biocontrol agent against *M. domestica* breeding in manure, due to its efficiency in destroying eggs and first instar larvae. More recently, Heikal (2020) reported that predacious mites can be partially considered a control tool of the parasitic nematodes.

Brazilian researchers play a key role in the research of biological control of the above pests. Azevedo *et al.* (2022) reported that the control of the housefly, *M. domestica*, and the stable fly, *S. calcitrans*, in Brazil relies heavily on the use of chemical products, which very often do not provide adequate population reduction. Predatory mites of the family Macrochelidae have been evaluated under laboratory conditions for the biological control of these organisms, especially of the housefly. They examined the predation capacity of the macrochelids *M. embersoni* and *M. muscaedomesticae* on the housefly and the stable fly, under semi-field conditions (screen-houses) (Azevedo *et al.*, 2022). Shiralizadeh *et al.* (2021) studied the effect of two temperatures on *M. muscaedomesticae*'s functional response to varying egg densities of *M. domestica*. The functional response of *M. muscaedomesticae* female was examined on the housefly eggs in an incubator at two constant temperatures of, 27 ± 1°C and 33 ± 1°C, 65 ± 5% RH, 14:10 h (L: D). Densities of 1, 2, 4, 8, 12, 16, 25, 30, 40 and 80 housefly eggs were offered to mated adult female mites. Before the experiments, the female predators were fed and mated for 48 hours and then starved for 24 hours. Each treatment was replicated ten times. After 24 hours of exposure, the number of preys killed by each predator was recorded. By logistic regression, the functional response of *M. muscaedomesticae*, to varying densities of the housefly eggs was classified as type III in 27°C and type II in 33°C. Rogers' equation was used to calculate the functional response parameters. At 27°C, the attack coefficient and handling time were 0.0204 h^{-1} and 0.6497 h, respectively, and 0.6493 h^{-1} and 0.4769 h at 33°C. At 33°C, the maximum attack rate (T/Th) was 50.32 eggs/day and 36.94 eggs/day at 27°C. R 2 values for *M. muscaedomesticae* at two constant temperatures were estimated to be 0.935 and 0.996 at 27°C and 33°C, respectively. Only at 40 and 80 prey densities did the analysis of variance reveal a significant difference in adult females' daily prey consumption between two constant temperatures. Additionally, there were significant differences between all densities at 33°C, and the majority of densities at 27°C (Shiralizadeh *et al.*, 2021).

In Brazil, Moya Borja (1981) maintained a colony of *Dermatobia hominis* (L. Jr.) in the laboratory for 6 years. Due to the fact that some *Dermatobia* adults from the laboratory colony were infested with the phoretic mite, *M. muscaedomesticae*, studies were carried out to determine the effect of the phoretic mites on sexual behaviour and longevity of *Dermatobia* adults. Results obtained showed that *Dermatobia* adults carrying *M. muscaedomesticae* are unable to copulate; *Dermatobia* females carrying phoretic mites failed in capturing and ovipositing their eggs on the vector *M. domestica* Linnaeus, and the longevity and fertility of *Dermatobia* adults carrying phoretic mites were not affected.

M. robustulus was described by Berlese (1904). Costa (1966b) reported that this species has been redescribed as *Macrocheles punctillatus* by Willmann (1939) who has collected it from cultures of white worms, *Enchytraeus albidus* Henle (a soil annelid that is cultured in moist soil). This habitat is also mentioned by Bregetova and Koroleva (1960), who added *Onthophagus* sp. as host. Evans and Browning (1956) have collected both, males and females from bullock manure. Axtell (1961) has collected it from cow and horse manure in the United States. Costa collected *M. robustulus* from *C. hispanus* (Linnaeus) (Costa, 1963), and many specimens were collected from this beetle in laboratory cultures. He was keeping single beetles in culture jars with fresh manure and moist soil. These were checked monthly, when they were cleaned from mites, and after that fresh manure was added. In one case, as many as 104 females of *M. robustulus* were collected off a single beetle after 1 month. It seems that moist soil (which may be rich in nematodes and enchytraeids) is one of the factors needed for massive development of this mite. In the laboratory, the mites were reared in plaster cells as described by Costa (1966a), they feed readily on nematodes of the genus *Panagrellus*. The females of *M. robustulus* have a longevity of at least 2 months, probably longer (during April the average room temperature is still rather low). A more interesting point is that although *M. robustulus* is arrhenotokous (Filipponi, 1964), females taken in nature produced males as well as females. The males became more abundant, towards the end of the reproductive period of each female. The incubation time for the eggs was 5–6 days during February–March and 1–2 days during May–June. The whole cycle from larva to female took 14 days in April and only 9–11 days for the male. During July, the whole cycle was as short as 5–6 days. The larva seems to be non-feeding all the other stages fed readily on nematodes (Costa, 1966b).

The mite may be identified following Costa (1966b). On stored food, it is reported from Ethiopia and on cereal residue (Hagstrum *et al.*, 2013), and Iran on stored potato (Ostovan, 1993). The predator is frequently observed in greenhouse soils. Messelink and van Holstein-Saj (2008) studied the effects of this predator on western flower thrips, in comparison with *H. aculeifer* (Canestrini). *M. robustulus* controlled thrips significantly better (up to 70% reduction) than *H. aculeifer* (up to 50% reduction). Population build-up of soil-dwelling predatory mites was assessed in a freesia glasshouse. *M. robustulus* and *H. miles* (Berlese) were introduced after soil sterilization. *M. robustulus* reached densities up to 2800/m^2, whereas the maximum observed density of *H. miles* was 340/m^2, only (Messelink and van Holstein-Saj, 2008).

Grosman *et al.* (2011) reported that the predatory mite *M. robustulus* has been commercially available since 2010. The effectiveness of this predator in the control of sciarid flies has, however, not yet been tested. They compare the effectiveness of *M. robustulus*, and the frequently used *H. aculeifer* in controlling sciarids, in potted chrysanthemum under greenhouse conditions. They also evaluated the potential of using a mulch layer to improve establishment, population increase and performance of the predators. Both predators had a significant impact on sciarid densities, with a reduction of 97.1% by *M. robustulus* and 87.1% by *H. aculeifer*. When the predators were introduced in combination with a mulch layer of Biotop®, predator densities increased by a factor 3.1 for *M. robustulus*, and 11 for *H. aculeifer*. The increase of predatory mite density was associated with an increase in the density of astigmatid mites, on which the predators were reared and that were introduced simultaneously with the predators. Sciarid density was reduced by 99.5%, when *M. robustulus* was introduced together with the Biotop® mulch layer, significantly lower compared to treatments with *H. aculeifer* with or without Biotop®. These results demonstrate that *M. robustulus* is an effective predator of *Bradysia* spp., and that in combination with Biotop®, it provides better control than the frequently used *H. aculeifer* (Grosman *et al.*, 2011).

Gallego *et al.* (2020b) examined the potential control of the potato tuber moth, *Phthorimaea operculella* (Zeller), one of the most severe pests affecting the potato, *Solanum tuberosum* Linnaeus, by the predatory mites, *M. robustulus* and *Blattisocius tarsalis* (Berlese). In the two first assays under laboratory conditions, the acceptance rate of the pest eggs was assessed for each predatory mite. Then, in a third assay, the functional response of *B. tarsalis* was studied. The results showed that *M. robustulus* did not prey on the pest eggs (number of eggs surviving = 4.33 ± 0.38), whereas *B. tarsalis* did (number of eggs surviving = 0.5 ± 0.5). Likewise, *B. tarsalis* showed a type II functional response, when it killed the eggs. The results showed the potential use of *B. tarsalis* as a biological control agent of *P. operculella* in potato under storage conditions (Gallego *et al.*, 2020b).

Different companies, such as Agrimag Bio, Fitogest, Koppert, etc. market *M. robustulus*, and propose it for the control of Thrips, Sciaridae and Lepidoptera. The predator feeds on various species of Thrips and is active against the eggs, larvae and pupae of sciarids. In addition, the mite feeds on the eggs of Diptera such as *Lyprauta* sp., of Lepidoptera, such as *Duponchelia fovealis* Zeller and other insects that live in the superficial layer of the soil. The predator can be found on the surface of the soil around the plant, but not on the epigeal part. It can be released at any time of the year, as long as the temperature and the RH allow it; in fact, it is not influenced by the photoperiod and is not characterized by any diapause. It is released in spring, with a temperature not lower than 15°C by distributing it directly on the ground. It is important to ensure a certain humidity of the first soil layers. Koppert markets the mite, as Macro-Mite Packing, as cardboard cylinders, containing 50,000 predatory mites (all stages) in vermiculite and peat. The dosage consists of preventive releases at a rate of 250 individuals/m^2.

67 The Family Parholaspididae Evans

The Parholaspidids are free-living predatory mites collected in various habitats such as soil, leaf litter, moss and the nests of rodents (Petrova, 1977; Chen *et al.*, 1994; Gu and Guo, 1996; Marchenko, 2016). They are characterized by sharp, spear-like gnathosomal parts, a feature assumed to help overpower prey (Walter and Campbell, 2003). The family is not adapted to phoretic dispersal associated with insects, but individuals have strong and long legs to facilitate ambulatory movements. The family is known from the Austral-Asian region (Halliday, 1995) and North America (Krantz, 1963; Marshall, 1964; Hirschmann, 1966), but much less from natural habitats in Europe (Moraza and Peña, 2006; Lindquist *et al.*, 2009b).

The family numbers about 163 species ascribed to 14 genera. The genera *Gamasholaspis* Berlese and *Holaspina* Berlese are the most diverse and include 34 species and 31 species in the world, respectively (Marchenko, 2016; Quintero-Gutiérrez and Halliday, 2021). On stored food is reported *Holaspina alstoni* (Evans), characterized by a cosmopolitan geographical distribution including Iran, Japan, Korea, India, East- and South-Asia and Europe (Ishikawa, 1980; Datta and Bhattacharjee, 1989; Latifi *et al.*, 2005; Kontschán *et al.*, 2014) and found in Iran on storage products (Sayedi *et al.*, 2006). Lee and Lee (1996) reported *H. alstoni* and other species of this family from soil and humus.

DOI: 10.1079/9781786390912.0067

68 The Family Parasitidae Oudemans (Fig. 68.1)

Parasitid mites are predators, free-living or associated with insects, living on a wide variety of primarily ground-level substrates that support microarthropod and nematode populations (Karg, 1993a; Karg and Freier, 1995). Gerson *et al.* (2003) report that mites of this family are common predators in soil, and that some species can be found on stored food products.

In this family, apotele of the palp tarsus has three prongs. Male of this group has spermatodactyl on the movable digit of the chelicera, with its distal end fused with the digit. Parasitid species are common predators in soil. The species of this family can be found on stored food products (Gerson and Smiley, 1990). They are often dispersed by various insects especially beetles and flies Wise, G.U., Henessey, M.K. and Axtell, R.C. (1988) A new species of manureinhabiting mite in the genus *Poecilochirus* (Acari: Mesostigmata: Parasitidae) predacious on house fly eggs and larvae. Ann. Entomol. Soc. Am. 81, 209–224. According to Lindquist *et al.* (2009b), the morphological characters of this family are the same as the subcohort Parasitiae. The female has the sternal setae st_4 set on large metasternal plates, flanking an anteriorly acuminate epigynal shield antero-laterally. The female sperm access system opening via an unpaired, usually strong developed endogynum, beneath epigynal shield, often with ancillary sclerotized process. The male has chelicera with spermatotreme, a sperm-holding process that is coalesced distally with movable digit; the males usually with highly developed, sexual dimorphic, spine-like setae, spurs and apophyses on leg II.

On stored food are reported *E. butleri* Hughes (Fig. 68.1) and *Paragamasus truncus* Schweizer.

E. butleri was described from England in floor detritus, where it was probably feeding on other soft-bodied mites (Hughes, 1948, 1976). The mite may be identified following Hughes (1976) and Hyatt (1980). The geographical distribution includes England (Hughes, 1948, 1976), Iceland (Sellnick, 1940), Ireland (Cusack *et al.*, 1975; Hyatt, 1980), Sea Island of Borkum, Germany (Oudemans, 1903b, 1905d) and Scotland (Brady, 1970; Hyatt, 1980).

The mite has been collected on brewer grain (Hyatt, 1980), cereal, grass seed, sweeping and the facilities include farm, flour mill, railway wagons (Cusack *et al.*, 1975), provender mill, warehouse and flour sievings from brewer's grain (Hyatt, 1980). Sellnick (1940) reported the mite in a barn/granary ('Scheune') and in dry grass and old hay. Moreover, the species has been recorded from the German North Sea Island of Borkum (Oudemans, 1903b, 1905d) (no habitat is reported).

The species has been collected also from deep litter in a poultry house in Lanarkshire, Scotland (Brady, 1970; Hyatt, 1980). Al-Kubaisi and Al-Dulaimi (2020) studied in Iraq the dominance and frequency of mite species associated with poultry droppings including *E. butleri*.

P. truncus was described by Schweizer (1961), and it has been reported from farm grain store in Ireland (Cusack, 1974; Cusack *et al.*, 1975).

P. truncus is not phoretic and reproduces sexually (Karg, 1971). Athias Henriot (1972) redescribed this species as *Paragamasus subtruncus* (Schweizer), but in this work we prefer to maintain the hold nomenclature to avoid confusion. In Ireland, the mite has been collected on pine shelter belt on bog (Longworth, 1973), in soil under Norway spruce (Heneghan, 1994) and in pasture (Piotrowska, 2009). In Norway, Seniczak *et al.* (2021) reported the mite as muscicolous, praticolous and silvicolus. In Latvia, it has been collected in rotting substrates such as in humus and inland meadows (Salmane, 1999). In Romania, Manu and Onete (2016) collected the mite in urban parks.

 The Handbook of Mites of Stored Products (V. Vacante and C.H.W. Flechtmann)
DOI: 10.1079/9781786390912.0068

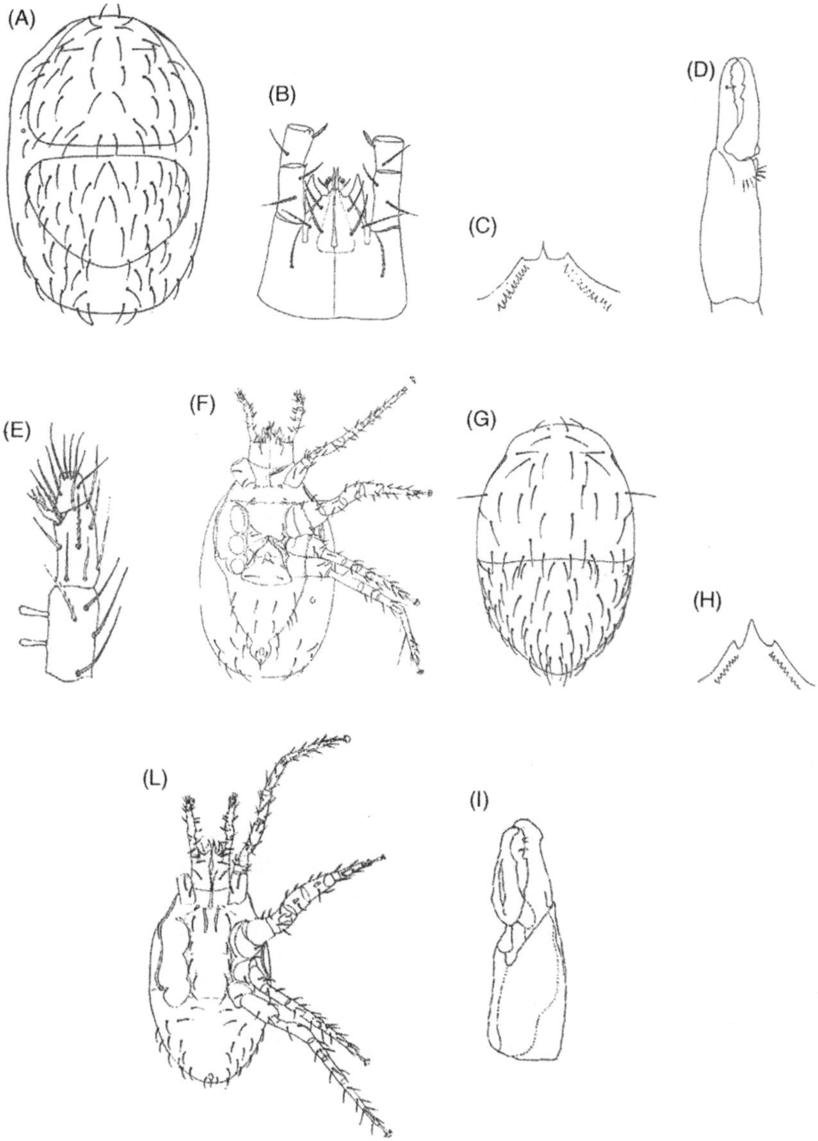

Fig. 68.1. *Eugamasus butleri* Hughes. Female. Dorsal view (A); ventral view of gnathosoma (B); tectum (C); chelicera (D); palp (E); ventral view (F). Male. Dorsal view (G); tectum (H); chelicera (I); ventral view (L) (from Hughes, 1976, licensed under the Open Government Licence v3.0).

69 The Family Phytoseidae Berlese (Figs 69.1–69.3)

The phytoseiids inhabit typically, the vegetation and feed on small arthropods, pollen, honeydew, plant exudates or leaf cell contents, and to a certain extent, fungi (Zemek and Prenerová, 1997). Some species are known as common inhabitants of leaf domatia (Walter, 1996), and few are collected in soil, such as species of the genera *Neoseiulus* Hughes, *Proprioseiopsis* Muma and *Amblyseius* Berlese, or in animal nests, such as species of the genus *Neoseiulus*. Although phytoseiids specialized on spider mites, such as species of the genera *Phytoseiulus* Evans, *Galendromus* Muma, and some *Neoseiulus* are the best known, most are polyphagous and feed on a variety of plant products and small arthropods, including injurious thrips, whiteflies, scale crawlers and Eriophyoid mites (McMurtry and Croft, 1997). Over half of the literature on acarine biocontrol agents pertains to the family Phytoseiidae, and about 20 species are being mass reared, and sold by about 50 companies worldwide (Gerson *et al.*, 2003).

The phytoseiid mites possess an entire dorsal plate (except *Macroseius biscutatus* Chant, Denmark *et* Baker), with fewer than 23 pairs of setae, their stigmata open between legs III and IV, and the metasternal plates are small. They lack leg grooves, leg I carry ambulacra, and the digits of the male spermadactyls are distally free (Gerson *et al.*, 2003). Moreover, the phytoseiid mites possess fixed and movable cheliceral digits normally developed, similar in length and opposed; the tritosternum is present and with well-developed laciniae. The anal opening is posteroventral, usually set in a ventrianal shield (Lindquist *et al.*, 2009b).

Kostianinen and Hoy (1996) reported that most of the 1800-plus described species, about 15% of the known diversity of the Mesostigmata, have been described in the last 50 years, as a direct result of their role as biological control agents. Updatings carried out by Demite *et al.* (2014, 2020) reported that the family consists presently of 2521 valid species, dispatched in three subfamilies and 94 genera.

Denmark *et al.* (1999) reported that the systematics of the Phytoseiidae had a convoluted history that has not yet been entirely resolved, but many people use the system proposed by Chant and McMurtry (1994) that divides the family into three subfamilies, namely, according to descending number of described species, as Amblyseiinae, Typhlodrominae and Phytoseiinae. These subfamilies differ between them in morphology and in biological traits, such as patterns of oviposition (Zhang, 1995).

Subfamilial categories and some genera have remained relatively stable over time, but generic concepts have been varied widely over the years, with as few as eight genera recognized by some authors, and as many as four dozen recognized by others. This has resulted in considerable confusion in the literature (Lindquist *et al.*, 2009b). The catalogue of Moraes *et al.* (1986, 2004) and the revisions by Chant and McMurtry (1994, 2003, 2004, 2005a, b) help sort out the confusion in generic identities.

The life cycle of phytoseiid mites comprises egg, larva, protonymph, deutonymph and adult stages, and feeding by larvae in some species is facultative. Except for a few known thelytokous species, the phytoseiids are parahaploid (pseudoarrhenotokous) (Hoy, 1985). Most phytoseiids are multivoltine, and the adult female is the overwintering or aestivating stage (Veerman, 1992). The life cycle from egg to egg takes 8–10 days, at 22–24°C (Helle and Sabelis, 1985), but varies with subfamily and with genera within subfamilies (Zhang, 1995).

Phytoseiid mites are well known natural enemies of phytophagous arthropods, on cultivated and non-cultivated plants. They have mostly been used to control pest spider mites, but some can control thrips in greenhouses and fields. Certain phytoseiids consume large numbers of prey and maintain plant-feeding mites at low densities. They have a high reproductive rate, a rapid developmental rate comparable to their prey, a female-biased sex ratio equivalent to their prey, allowing them to respond

DOI: 10.1079/9781786390912.0069

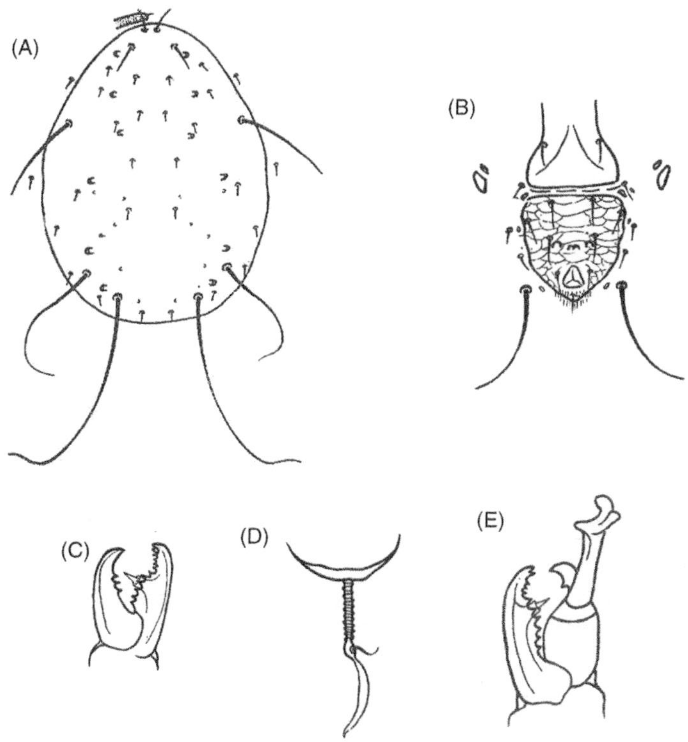

Fig. 69.1. *Amblyseius obtusus* (Koch). Female. Dorsal view (A); posterior ventral surface (B); chelicera (C); *receptaculum seminis* (D). Male. Chelicera (E) (from Livshitz and Mitrofanov, 1972).

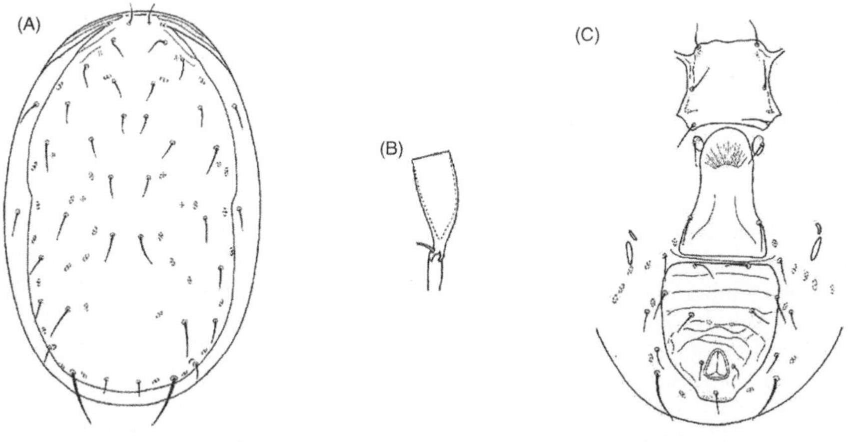

Fig. 69.2. *Neoseiulus barkeri* Hughes. Female. Dorsal view (A); *receptaculum seminis* (B); ventral view (C) (from Vacante and Nucifora, 1987).

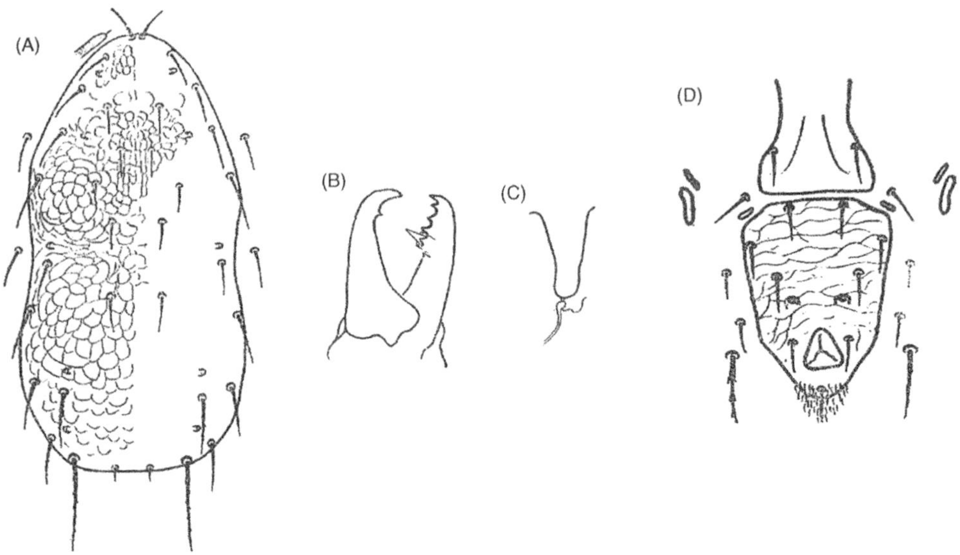

Fig. 69.3. *Neoseiulus cucumeris* (Oudemans). Female. Dorsal view (A); chelicera (B); *receptaculum seminist* (C); posterior ventral surface (D) (from Livshitz and Mitrofanov, 1972).

numerically to increased prey density, and can easily be mass reared (Hoy, 2011).

Phytoseiid mites feed on a variety of food and have developed different feeding habits. Some may feed on pollen, fungal spores and plant exudates, and some can feed, develop and reproduce on these plant materials. Others are obligatory predators that can survive on these alternative foods but fail to deposit eggs or fail to develop on such food. Phytoseiid mites are the most important mites employed in biological control and are usually reared using a tritrophic system that although efficient, is expensive and laborious (Barbosa and Moraes, 2015). The wide number of reports makes it very difficult to summarize this topic, and for further insights may consult Hoy (2009), Sabelis (1996) and Vacante (2016). The best known and investigated phytoseiid mite is probably *Phytoseiulus persimilis* Athias Henriot, typical natural enemy of the two-spotted spider mite, *Tetranychus urticae* Koch severe pest of ornamental and vegetable crops. This predator is reared using as food, all stages of its prey, *T. urticae*. Other phytoseiid mites are reared on different preys. Research works developed by Barbosa and Moraes (2015) on the use of mites of the cohort Astigmatina (Sarcoptiformes), used as factitious prey in the mass rearing of phytoseiids, are demonstrative and represent a simpler production system. These authors evaluated the potential of ten Astigmatina species to serve as factitious food sources for the mites *Euseius concordis* (Chant), *Iphiseiodes zuluagai* Denmark *et* Muma, *Neoseiulus barkeri* Hughes and *N. californicus* McGregor, all phytoseiid species commonly found in different countries. The high fecundity and survival rates obtained suggest that *Thyreophagus* n. sp. is a suitable prey for rearing *N. barkeri* and that *Austroglycyphagus lukoschusi* (Fain) and *Blomia tropicalis* Bronswijk, Cock *et* Oshima is suitable for rearing *N. californicus*. Oviposition by *E. concordis* was negligible, but survivorship was high on most prey species, suggesting that these species may be useful for the maintenance of the predator. *I. zuluagai* had low fecundity and survival on all the astigmatid species evaluated, and none were suitable for its rearing (Barbosa and Moraes, 2015). Similar studies were conducted in the biological control of insect pests. So, phytoseiid mites have been successfully used for the control of biotype B of the whitefly, *Bemisia tabaci* (Gennadius) (Hemiptera: Aleyrodidae). Cavalcante Cerqueira *et al.* (2015) evaluated five Brazilian populations of phytoseiids as candidates for the control of the whitefly, a serious pest of different crops worldwide. They studied the phytoseiid mites *Amblydromalus limonicus* (Garman *et* McGregor), *Amblyseius herbicolus* (Chant), *Amblyseius largoensis* (Muma), *Amblyseius tamatavensis* (Blommers) and *Neoseiulus*.

tunus (De Leon), which are found naturally in Brazil and elsewhere. The work was conducted at 28 ± 1°C, 75 ± 10% RH, and a photoperiod of 12:12 (L:D) hours. All evaluated phytoseiids preyed on eggs of *B. tabaci*, with the highest levels of predation recorded for *A. herbicolus* and *N. tunus* and the highest level of oviposition recorded for *A. tamatavensis*. The results show the Brazilian populations of those three species to be promising as control agents of *B. tabaci*. *Aleuroglyphus ovatus* (Troupeau) (Acari: Acaridae) was found to be a suitable prey for the mass production of those predators. Complementary studies are considered justified, given the positive results of this study (Cavalcante Cerqueira *et al.*, 2015).

From an ecological point of view, McMurtry *et al.* (2013) proposed a classification of phytoseiid mites into four lifestyles:

Type I lifestyle includes specialized mite predators and includes three subtypes (Ia, Ib, Ic); the first numbers of specialized predators of *Tetranychus* (Tetranychidae), the second specialized predators of web nest-producing mites (Tetranychidae), and the third specialized predators of tydeoids (Tydeoidea), respectively. *P. persimilis* is a type I lifestyle (subtype Ia) phytoseiid mite, feeding exclusively on *Tetranychus* species specifically *T. urticae Koch* and whose survival depends on the presence and quality of its prey. *P. persimilis* has been employed as a biocontrol agent in greenhouses since 1968 (Vacante, 2016). It is highly voracious, with adults able to eat 20–30 eggs or 3–5 females and 10–20 juveniles of *T. urticae* per day (Escudero and Ferragut, 2005).

Type II lifestyle includes selective predators of tetranychid mites. *N. californicus* (McGregor) possesses characteristics of both type II specialist predatory mites and type III generalist predatory mites. This phytoseiid prefers tetranychid mites as food, but consumes also other mites, thrips and even pollen in the absence of primary prey. It is often used to control *T. urticae* and other phytophagous mites on various crops in temperate and subtropical regions around the world (Vacante, 2016).

The species belonging to the type III lifestyle are generalist predators and includes five subtypes (IIIa, IIIb, IIIc, IIId and IIIe). The first are generalist predators living on pubescent leaves; the species included in the subtype IIIb, are generalist predators living on glabrous leaves; those belonging to the subtype IIIc are generalist predators living in confined space on dicotyledonous plants. The subtype IIId numbers generalist predators living in confined spaces on monocotyledonous plants, and the subtype IIIe includes generalist predators from soil/litter habitats. *Neoseiulus cucumeris* (Oudemans) is a type III lifestyle (subtype IIIe) phytoseiid mite, which is an efficient natural enemy of small arthropod pests worldwide. Its rearing is easy, as well as its establishment on crops, where its populations may develop at high densities. The use of this phytoseiid is proposed commercially and is successfully employed for the biological control of phytophagous mites and various thrips species. Its primary targets are thrips including western flower thrips, onion thrips and plague thrips. *N. cucumeris* is a generalist that also feeds on pollen and mites, particularly broad mite. *Amblyseius swirskii* (Athias Henriot) is a type III lifestyle (subtype IIIb) phytoseiid mite. It has been proposed commercially since 2005 and is largely used in greenhouse crops in augmentative release against thrips and whiteflies. It can develop and reproduce on a variety of other food sources including spider mites, eriophyid mites, broad mites, pollen and artificial diet (Nguyen *et al.*, 2014; Vacante, 2016). *N. barkeri* Hughes is a type III lifestyle (subtype IIIe) phytoseiid mite. It feeds on mites living on stored food, spider mites, thrips, broad mites, whitefly eggs and various pollens (Jafari *et al.*, 2012). It has been reported in many parts of the world including Asia, Africa, Europe, America and Australia. Furthermore, it has been used in augmentative biocontrol of onion thrips, *Thrips tabaci* Lindeman and *T. urticae* (Jafari *et al.*, 2010). *Iphiseius degenerans* (Berlese) is a type III lifestyle (subtype IIIc) phytoseiid mite. It is considered to be a generalist feeding on both live prey and pollen (Vantornhout *et al.*, 2005). In 1994, it was first commercialized in Belgium to control thrips in greenhouse crops.

The type IV lifestyle includes pollen-feeding generalist predators.

On stored food are collected six species, named *Amblyseius* sp. (Fig. 69.1), *N. alpinus* (Schweizer), *N. barkeri* (Hughes) (Fig. 69.2) and *N. cucumeris* (Oudemans) (Fig. 69.3), *Proprioseiopsis bregetovae* (Abbasova), *Typhlodromus* (*Anthoseius*) *transvaalensis* (Nesbitt), respectively (Hughes, 1948, 1976; Noei, 2007; Noei *et al.*, 2008b; Hagstrum *et al.*, 2013).

Hughes (1976) described *Amblyseius* sp. a species near *Amblyseius obtusus* (Koch) (Fig. 69.1) (*sensu* Chant, 1959; Livshitz and Mitrofanov, 1972) possessing four narrow plates set between the genital and the ventro-anal shield and having two pairs of metapodal plates. The figures of Chant (1959) and Livshitz and Mitrofanov (1972) would lead us to believe that it may be identified as *A. obtusus* (Koch). Obviously, to have a certain confirmation, it would be necessary to study the specimens of Hughes. In England, the mite was collected in hay bales (Hunter, in Hughes, 1976) and on the surface of bin-stored barley (Smith, in Hughes, 1976). Its geographic distribution concerns many countries (Algeria, Argentina, Armenia, Azerbaijan, Azores, Canada, Chile, Costa Rica, Croatia, Cuba, Czech Republic, England, France, Germany, Georgia, Greece, Haiti, Hawaii, Hungary, Iran, Ireland, Italy, Kazakhstan, Latvia, Moldova, Morocco, New Zealand, Norway, Pakistan, Poland, Portugal, Romania, Russia, Slovakia, Spain, Sweden, Tunisia, Turkey, Ukraine, United States and Venezuela) (Demite *et al.* 2020).

N. alpinus was originally described as *Amblyseius alpinus* by Schweizer (1922), which found it on moss in Switzerland. It may be identified following Ferragut *et al.* (2010). The species reproduces by thelytoky and has a cosmopolitan geographical distribution (Moraes *et al.*, 2004). It occurs rarely on crops, sometimes in apple orchards and on strawberries, where it seems able to control the two-spotted spider mite, *T. urticae* and the strawberry tarsonemid mite, *Phytonemus pallidus* (Banks) (Fitzgerald *et al.*, 2008). On stored food, it is reported on rice from Egypt (Hagstrum *et al.*, 2013).

N. barkeri was originally described as *Typhlodromus bakeri* by Hughes (1948), and may be identified following Hughes (1976), Ferragut *et al.* (2010) and Abo-Shnaf and Moraes (2014). The phytoseiid has a worldwide geographical distribution (Moraes *et al.*, 2004; Demite *et al.*, 2014, 2020), and is reported as a predator of *T. urticae* and thrips on strawberry, cucumber and aubergine, in greenhouses. It preys on the western flower thrips, *Frankliniella occidentalis* (Pergande) (Rodriguez-Reina *et al.*, 1992); the onion thrips, *T. tabaci* (Bonde, 1989; Broodsgaard and Hansen 1992); and *T. urticae* in cucumber (Fan and Petitt, 1994a). Moreover, augmentative releases of *N. barkeri* help in the control of the broad mite, *Polyphagotarsonemus latus* (Banks) on peppers (Fan and Petitt, 1994b).

On stored food, Hughes (1976) collected large numbers of this mite on germinating barley, in association with *Caloglyphus* sp. Nesbitt (1951), collected the mite from the debris in the holds of wheat-transporting vessels, associated with *A. siro* Linnaeus and *Tyrophagus* sp. Moreover, *N. barkeri* has been found on grain, seed cotton and wheat in commercial grain storage, farm, off-farm storage, provender warehouse, railcar and ship (Hagstrum *et al.*, 2013). Rack (1970, 1971) observed large numbers of *N. barkeri* associated with six species of mould-eating mites, such as *T. putrescentiae* (Schrank), *G. domesticus* (De Geer), *Kleemannia pavida* (Koch), *K. plumosus* (Oudemans), *Kleemannia plumigera* Oudemans and *Proctolaelaps pygmaeus* (Müller). Hughes (1976) suggested the hypothesis that like other phytoseiid mites, it is a predator, which supplements its diet with moulds.

Bonde (1989) studied among the first the biology of *Amblyseius barkeri* feeding on *T. tabaci* at 25°C. He described some morphological features, egg-laying, mating and predatory behaviour. An average duration of 2.2, 0.8 and 3.2 days for the egg, larval and nymphal stages were observed, with mortalities at 1.0%, 1.0% and 3.1%, respectively. Females represented 63% of the population and required multiple matings for optimal fertility. The oviposition period was 20.3 days and the average oviposition rate 2.3 eggs per day. The intrinsic rate of increase was 0.22 per day. The expected lifespan was 29.6 days for female and 27.4 days for male. Males and females of *A. barkeri*, both consumed 3.3 nymphs of thrips per day (mean value for the feeding stages). The larva does not take up food. In the absence of thrips, *A. barkeri* was able to consume the two-spotted spider mites, *T. urticae* and their eggs, the adult of the broad mites, *P. latus* and the pollen of various plants. Moreover, cannibalism was observed when food was lacking. Baier (1992) reported that the efficacy of the oligophagous predatory mites *A. barkeri* and *Neoseiulus (Amblyseius) cucumeris* (Oudemans) was very variable, especially in the control of thrips in vegetable and ornamental crops. Laboratory investigations of the biology and ecology of these two species showed that *A. barkeri* needed a higher RH for its development than *N. cucumeris*. Adult longevity and egg-laying capacity required a higher RH for female of *A. barkeri*, than for female of *N. cucumeris* (Baier, 1992). Fan and Petitt (1994c) determined the functional responses of adult females of *N. barkeri* Hughes on eggs, larvae and adults of the two-spotted spider mite, *T. urticae* in the laboratory. The study was conducted on pepper leaf discs over a 24-hour period, at 25 ± 1°C, 70–80% RH and 16L:8D photoperiod. Prey densities ranged 5–80 eggs, or 5–40 larvae or 1–8 female adults of *T. urticae* per disc. The predation rate of *N. barkeri* adult females on *T. urticae* eggs was the same as on its larvae, but the predation rate on adult females was much lower. Moreover, they discussed the role of this generalist predatory mite, in the integrated and biological control of greenhouse pests. Jafari *et al.* (2012) studied the effect of seven constant temperatures of 15°C, 20°C, 25°C, 27°C, 30°C, 35°C and 37°C, on developmental time of *N. barkeri* in laboratory conditions under 65% ± 5% RH and a photoperiod of 12:12 (L:D) hours on nymphal stages of *T. urticae* Koch. Total developmental time of females (from egg to adult emergence) at the above-mentioned temperatures was 26.59, 14.43, 6.32, 5.64, 4.59, 3.98 and 4.67 days, respectively. Developmental rate of *N. barkeri* increased as temperature increased from 15°C to 35°C but declined at 37°C. A linear and two nonlinear models were fitted to developmental rate of immature stages of *N. barkeri* to predict the developmental rate as a function of temperature, as well as to estimate the thermal constant (K) and critical temperatures (i.e. T min, T opt and T max). The estimated values of the T min and K for total developmental time, using the linear model were 12.07°C and 86.20 degree-days (DD), respectively. The T min and T max estimated by the Sharpe Schoolfield Ikemoto (SSI) model were 11.90°C and 37.41°C, respectively. The estimated T opt for overall immature stage development of *N. barkeri* by the Lactin and SSI models were 33.89°C and 24.51°C, respectively. Based on the biological criteria of model evaluation, the linear and SSI models were found to be the best models for describing the developmental rate of overall immature stages of *N. barkeri* and estimating the temperature thresholds (Jafari *et al.*, 2012). The same authors (Jafari *et al.*, 2013) reported that in west of Iran *N. barkeri* is an indigenous biological control agent on cucumber and maize infested by spider mites and *T. tabaci,* and it can prevent the outbreak of them. They determined the demographic parameters of the Iranian population of *N. barkeri,* using first instar larvae of *T. tabaci* as prey under laboratory conditions at 30 ± 1°C, 65 ± 5% RH, and photoperiod 8:16 hours (D: L). The results showed that *N. barkeri* successfully developed on first instars larvae of *T. tabaci*. The developmental times for egg, larva, protonymph, deutonymph and all immature stages were 1.37, 1.03, 1.54, 1.71 and 5.68 days, respectively. Also, 96.66% of eggs hatched, and the survival rate for the movable immature stages was 100%. The total prey consumptions by protonymphs, deutonymphs and adult female were 2.70, 3.57 and 75.85 individuals, respectively. The adult longevity of the male was 14.36 days, while the females lived 20.17 days. The female ratio was 0.62%. The intrinsic rate of increase (r_m) and the net reproductive rate (R_0) were 0.252 day^{-1} and 18.70 female offspring, respectively. The results of this study showed that *N. barkeri* could be considered as a biological agent for the control of *T. tabaci* on cucumber.

Elhalawany *et al.* (2021) evaluated *N. barkeri* as a biocontrol agent for *Aceria oleae* (Nalepa) and *Tegolophus hassani* (Keifer), on olive seedlings in Egypt. Three release rates (50, 100 and 150 predators/olive seedlings) were tested, and the numbers of *N. barkeri* and eriophyids were assessed over the 2017 and 2018 seasons. For both years, the densities of *N. barkeri* increased significantly over 4 months, and the densities of the two species of eriophyids were reduced. Significant differences were observed between the three release rates. Early releases of 100 *N. barkeri* per seedling at the first appearance of eriophyids, substantially suppressed populations of the two pests considered (Elhalawany *et al.*, 2021).

N. cucumeris was originally described as *Typhlodromus cucumeris* by Oudemans (1930b), which collected it on *Cucumis melo* Linnaeus, infested with spider mites in France. The mite has a wide geographical distribution (Moraes *et al.*, 2004), and may be identified following Abo-Shnaf and Moraes (2014). On stored food, *N. cucumeris* is reported from Australia and India on mouldy wheat (Hagstrum *et al.*, 2013). Its life history has been investigated by various authors, and the wide numbers of reports makes it difficult to summarize them. The broad host range and the ability of the mite to survive on plant pollen suggested to consider it as a type III predator (McMurtry and Croft, 1997).

N. cucumeris develops through one larval stage, and two nymphal stages (protonymph and deutonymph). The life cycle from egg to adult takes from 9 to 12 days, at 25°C and adults may live up to 28–35 days. The adult female lays an average of 35 eggs during its lifetime. The non-feeding larvae emerge from eggs in about 3 days and moult into protonymphs 2 days later. The two nymphal stages last for 7–10 days before developing into adults. However, these biological parameters change, depending on available food sources. Van Rijn and Tanigoshi (1999) and Sarwar (2016) reported that although this predator could survive and reproduce solely on pollen diets (variable results on different pollen), the survival, longevity and fecundity were higher when prey was included in the diet. Ranabhat *et al.* (2014) studied the juvenile development and survival and demographic parameters of *N. cucumeris* fed on pollen of castor bean, tulip, apple, Christmas cactus, horse-chestnut, maize and birch were assessed under laboratory conditions. Deprivation of food and pollen of castor bean plants resulted in 100% juvenile mite mortality. Feeding mites with tulip and horse chestnut pollen resulted in the shortest development, and the highest total fecundity. Adult mites fed on birch, tulip, maize and apple pollen lived significantly longer, compared with those that fed on pollen of horse-chestnut and Christmas cactus. The intrinsic rate of natural increase ranged between 0.1013 females/female/day for maize and 0.1806 females/female/day for horse-chestnut pollen as food. Net reproductive rate was the lowest when fed with maize pollen, and highest when fed with horse-chestnut pollen. Population doubling time was highest on maize pollen and shortest on horse-chestnut pollen. Their study revealed that birch, tulip, horse-chestnut, apple and maize pollen can be used by *N. cucumeris* from early spring to late summer as a suitable alternative food in periods when prey in the field is scarce or absent (Ranabhat *et al.*, 2014).

Zhang *et al.* (2003) examined the life history of *N. cucumeris* feeding on the bamboo spider mite, *Aponychus corpuzae* Rimando under four constant temperatures. The females of *N. cucumeris* required 20.0 ± 0.5 days to complete its development from the egg to adult at 15 ± 1°C, and only 7.7 ± 0.3 days at 30 ± 1°C. The developmental duration of each stage of *N. cucumeris* decreased, with increasing temperature from 15°C to 30°C. The average oviposition period was 22.2 ± 1.6 days, and the fecundity was 28.9 ± 1.9 eggs. The female longevity (egg to death) was 40.6 ± 0.7 days. Laboratory studies of the predator: prey numerical responses revealed that at the ratios between 3:30 and 9:30, predators quickly brought *A. corpuzae* populations under control, and the prey densities began to decrease from the fourth day after predator introduction. The study showed that *N. cucumeris* can develop, survive and increase its population when feeding on *A. corpuzae* and has good potential as a biological control agent of *A. corpuzae* in moso bamboo forests in Fujian, China (Zhang *et al.*, 2003).

Ji *et al.* (2007) examined in laboratory two aspects of mating effects on the fecundity, sex ratio and longevity of *N. cucumeris*. The first aspect concerned the females mated by one, two or three different males (unmated and 3 days old) at 5-day intervals, while the second aspect concerned the females mated by males with different age/mating statuses (number of females mated previously by the male). Females allowed to mate with a second or third male at 5-day intervals produced 39 eggs on average, but those mated with a single male produced 28 eggs on average. Matings with additional males 5 or 10 days after the first male increased the duration of the oviposition period of these females by 5–7 days, and at the same time, reduced the post-oviposition period by about 10 days. Females with additional matings by one or two different males at 5-day intervals survived a few days shorter than females without additional males. Mating with a different female each day, a male of *N. cucumeris* could mate with 5–8 females, which produced a total of 85–116 eggs: females mated with a male during days 1 and 2 in its adulthood and with a male of the last 2 days of life (days 7 and 8), produced about half as many eggs as females mated with a male during 3–6 days of its adulthood. Females mated with males that are too young or too old had a shorter oviposition period and a longer post-oviposition period and longevity than females mated with middle aged males. In both experiments, rates of oviposition remained similar in females with high or low fecundity. This indicates that in both cases, the increased fecundity is due to the extension of the oviposition period, through additional sperm supplied by the second male and or third male (in experiment 1) or more sperm by males not too young nor too old (experiment 2) (Ji *et al.*, 2007).

Li *et al.* (2021) evaluated the feasibility of using *T. curvipenis* (Fain *et* Fauvel) (Acari: Acaridae) as a supplementary food source for *N. cucumeris* by investigating the survival and development of *N. cucumeris* at different prey densities with the influence of conspecifics. Furthermore, they investigated predation rates and their body size at adult emergence. The results showed that *N. cucumeris* developed from egg to adult in approximately 6 days. Survival rates of immature predators increased significantly with the given prey density. No significant difference in body size was found between the survived adults fed at different prey density, but the females were always larger than males. *T. curvipenis* resulted in an excellent alternative food source for the biological control agent *N. cucumeris* (Li *et al.*, 2021).

Ramakers (1980) studied among the first, the use of this mite to control thrips on ornamentals and vegetables in greenhouses. Today, the mite is reared in insectaries and commercialized worldwide as a biological control agent for thrips and tetranychid mites in greenhouses in Europe, Canada, United States, Australia and New Zealand. So, in the United States, *N. cucumeris* is commercialized by the member companies of the Association of Natural Biocontrol Producers (ANBP) including Tip Top Bio-Control, Bio Control, S.A., Koppert, Rincon-Vitova, Distribuciones IMEX S.A. de C.V., Anatis Bioprotection, Entomology Solutions, Natural Insect Control, Applied Bio-nomics, Hydro-Gardens, Beneficial Insectary, Sound Horticulture, Crop Defenders, Buglogical Control Systems, M&R Durango, IPM Laboratories, Biobest, EcoSolutions, Natural Enemies Bio. Pest Mgmt., Everwood Farm, ARBICO Organics, Evergreen Growers Supply, Green Methods, Syngenta Bioline and Orcon (LeBeck and Leppla 2010).

P. bregetovae is known from Azerbaijan and Ukraine (Moraes *et al.*, 2004; Noei *et al.*, 2009), and Iran (Hajizadeh, 2007), where it has been collected on stored rice, rice dust and debris (Noei, 2007; Noei *et al.*, 2008b).

T. (*Anthoseius*) *transvaalensis* (Nesbitt) has a wide geographical distribution including Argentina, Algeria, Australia, Azerbaijan, Brazil, Cameroon, China, Colombia, Costa Rica, Egypt, Guinea, Hawaii, Iran, Israel, Panama, Philippines, Singapore, South Africa, Taiwan, United States-Florida (Moraes *et al.*, 2004; Noei *et al.*, 2009; Cedola and Castresana, 2014). In Iran, it has been collected on stored rice, rice dust and debris (Noei, 2007; Noei *et al.*, 2008b). In Brazil, this mite was found in native plants and plants associated with strawberry crops (Ferla and Moraes, 2002; Ferla *et al.*, 2007). It has also been found in laying hen systems (Horn *et al.*, 2015) and bird nests (Silva *et al.*, 2013).

70 The Family Uropodellidae Camin (Fig. 70.1)

The Uropodellidae is a monogeneric family of heavily armoured mites, whose members are easily recognized by the anterior margin of the podonotal shield, which is produced into a spiky knob. The cheliceral digits possess few teeth. The female epigynal shield is subrectangular, with numerous setae irregularly set over most of surface. The opisthonotum of female has a pair of lateral shields and a pygidial shield that extends onto the mesonotal region and lacking horn-like processes posteriorly (Lindquist *et al.*, 2009b). The genus *Uropodella* Berlese includes seven little-known species from the Americas, Africa, Asia and Australia, where they are associated with rotting wood, tree holes and bark beetle galleries. Nothing is known of their biology. One undescribed species has been reported from Australia, and it may represent an undescribed genus (Walter and Proctor, 2001; Beaulieu *et al.*, 2010).

Uropodella laciniata Berlese (1888) (Fig.70.1) was described from a series of specimens collected in Argentina, Brazil and Paraguay. In the past, the mite was assigned to the family Uropodidae, and successively Camin (1955) described and included it in the new family Uropodellidae. The mite is reported on stored food from England, Germany (Hagstrum *et al.*, 2013), Indonesia (Haines, 1995) and Taiwan (Tseng, 1979) on decaying plant, fungus-growing rice crop (Tseng, 1979) and food grains (Haines, 1995) in beehive and store (Hagstrum *et al.*, 2013).

DOI: 10.1079/9781786390912.0070

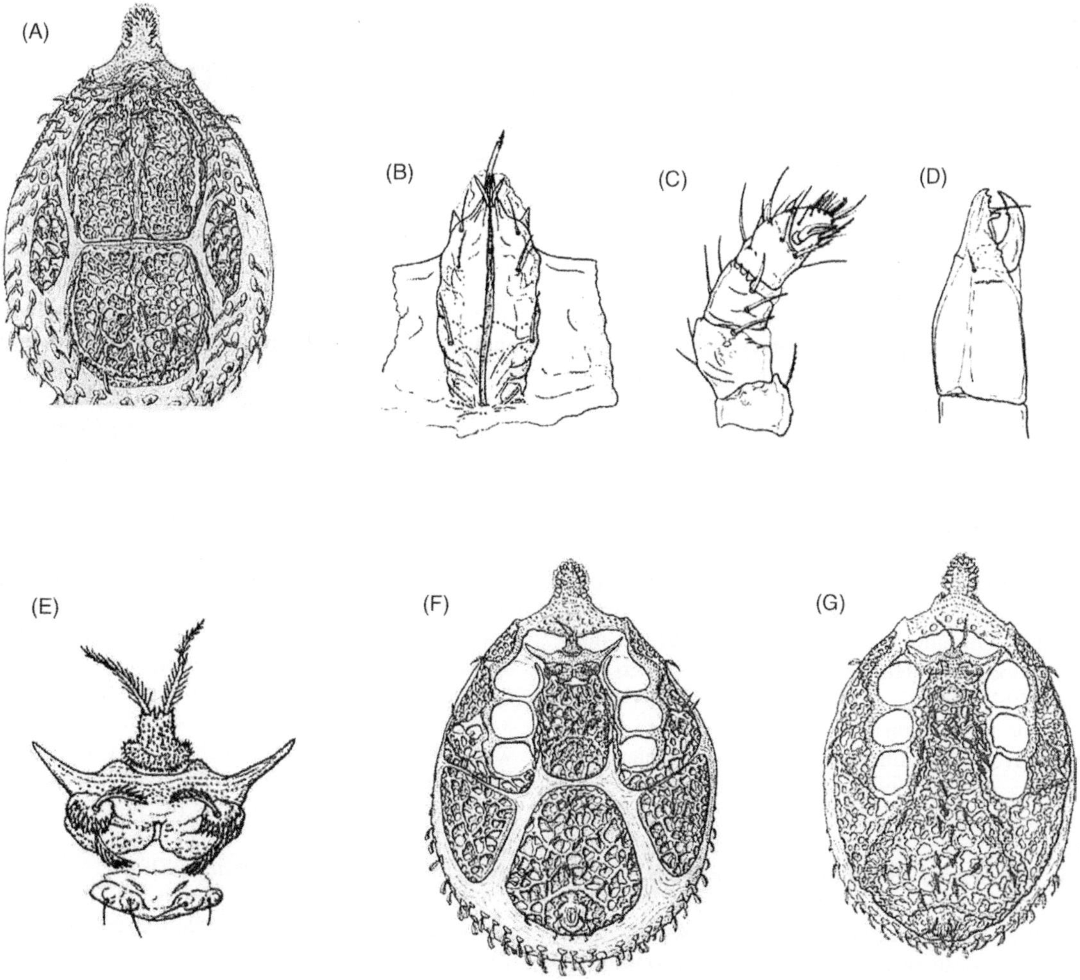

Fig. 70.1. *Uropodella laciniata* Berlese. Female. Dorsal view (A); ventral view of gnathosoma (B); palp (C); chelicera (D); tritosternum and sternal shiel (E); ventral view (F). Male. Ventral view (G) (from Camin, 1955).

71 The Family Uropodidae Kramer (Figs 71.1 and 71.2)

The family Uropodidae belongs to the suborder Uropodina. According to Halliday (2015), species ascribed to this taxon are very abundant in forest litter but can also be found in large numbers in moss, under stones, in ant nests, in the nests and burrows made by vertebrates and in dung and carrion. Most appear to be predators that feed on nematodes or other small invertebrates, but others may feed on living and dead fungi and plant tissue (Lindquist *et al.*, 2009b). None are known to be of direct economic importance, except for a few species that contaminate earthworm cultures and stored food. The higher classification of the group is not stable, and the number of family names used by recent authors varies from 13 (Lindquist *et al.*, 2009b) to 35 (Beaulieu *et al.*, 2011). Wiśniewski and Hirschmann (1993) catalogued over 2000 species, and the number of known species has increased significantly since then (Halliday, 2015). Recently, some species of the family Uropodidae were included in the family Urodinychidae Berlese (Bánki *et al.*, 2022 and others). However, the 'uncertain' systematic of the suborder Uropodina invites to avoid some easy innovation and use the old name of Uropodidae.

The Uropodidae mites are brown-black and turtle-like mites, characterized by leg grooves, stigmata set between the legs II and III and short convoluted peritremes. The family includes about 100 genera and 1500 species (Hirschmann and Wisniewski, 1992).

Most species of the family Uropodidae live in forest soil, humus and decaying organic material in mixed forests, under or among leaves of deciduous trees in forests, grass and decaying material and under bark, leaves and litter of coniferous trees, in animal dung and manure heaps, under stones in tidal coasts, in the nest of ants and moles, and among moss and decaying moss material, wood, soil and agricultural areas (Karg, 1989; Wiśniewski and Hirschmann, 1993; Błoszyk, 1999; Mašán, 1999b).

According to Karg (1986), these mites reproduce usually sexually, and some species are also parthenogenetic. Dispersal is often during the prolonged deuthonymphal stage, when the mites attach to various insects, mostly beetles, by an anal pedicel (Faasch, 1967). A survey of soil fauna in Poland revealed 30 cases of centipedes carrying mites of the suborder Uropodina. The 155 phoretic deutonymphs collected belonged to two species of Uropodina, *Oodinychus ovalis* (Koch) and *Uroobovella pulchella* (Berlese). These mites displayed a high degree of selectivity in their choice of carrier. The only species of centipede transporting mites was *Lithobius forficatus* (Linnaeus), despite the presence of 30 other species in the same habitats. It is possible that the large size and relatively fast speed of movement of this centipede make it a very good mite carrier. The majority of the mites were located on the sides of the centipedes, on segments near the anterior end. The high selectivity in the choice of carrier, as well as the point of attachment, suggests adaptation by the mites for phoresy by *L. forficatus* (Błoszyk *et al.*, 2006).

According to main reports, the Uropodids reported on stored food are *Uroobovella marginata* (Koch) (Fig. 71.1) (syn. *Fuscuropoda marginata*), *Uroobovella obovata* (G. Canestrini *et* Berlese), *Lithobius* (G. *et* R. Canestrini), *Trematura jacksonia* Hughes and *Uroseius acuminatus* (Koch) (Fig. 71.2). Awaiting a definitive revision of the group, to avoid increasing confusion, we here treat the above mentioned species individually and ascribe them to the traditionally employed taxa. In accordance with this reflection, we continue to ascribe *U. marginata*, *U. obovata* and *U. acuminatus* to the Uropodidae family, in contrast to what was reported by Lindquist *et al.* (2009b), which include the first two species in the family Dinychidae and the last in the family Dithinozercoidae. *L. krameri* and *T. jacksonia* are discussed in the family Trematuridae.

U. marginata was originally described by Koch (1839b) as *Notaspis marginatus*. The mite has a worldwide distribution (Karg, 1989) and may be identified following Hughes (1976) and Afifi (1983). The species is common in rotting material

DOI: 10.1079/9781786390912.0071

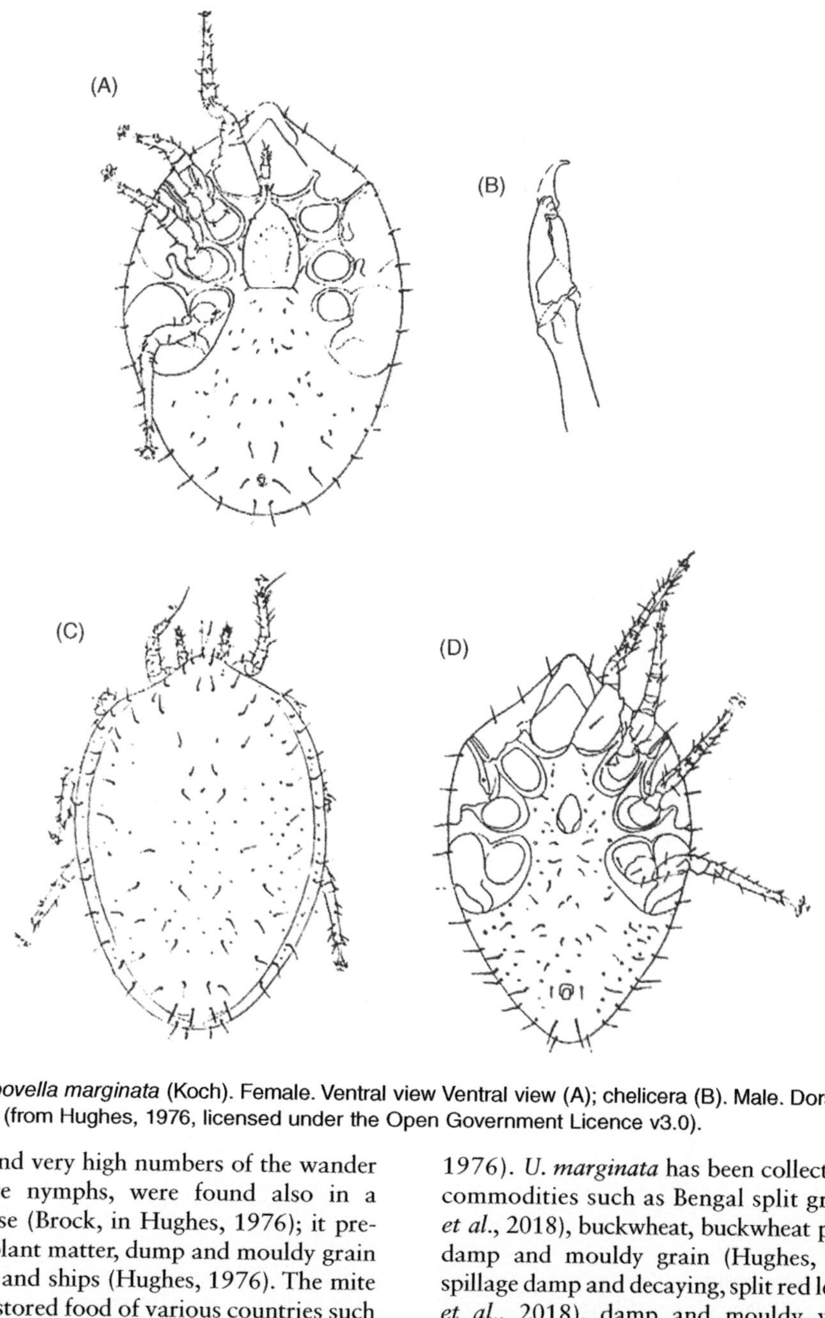

Fig. 71.1. *Uroobovella marginata* (Koch). Female. Ventral view Ventral view (A); chelicera (B). Male. Dorsal view (C); ventral view (D) (from Hughes, 1976, licensed under the Open Government Licence v3.0).

and manure, and very high numbers of the wander or pedunculate nymphs, were found also in a cucumber house (Brock, in Hughes, 1976); it prefers decaying plant matter, dump and mouldy grain in warehouses and ships (Hughes, 1976). The mite is reported on stored food of various countries such as Australia (Champ, 1966; Halliday, 1998), England (Hughes, 1976), Germany (Rack, 1968a; Hughes, 1976), India (Gupta and Chatterjee, 2004; Ganguly *et al.*, 2018), Iran (Noei, 2007), Ireland (O'Farrell and Butler, 1948; Cusack, 1974; Cusack *et al.*, 1975), Italy, the Netherlands (Hughes, 1976), United States (Smiley, 1987) and Yugoslavia (Hughes, 1976). *U. marginata* has been collected on various commodities such as Bengal split gram (Ganguly *et al.*, 2018), buckwheat, buckwheat product, grain, damp and mouldy grain (Hughes, 1961, 1976), spillage damp and decaying, split red lentil (Ganguly *et al.*, 2018), damp and mouldy wheat residue (Gupta and Chatterjee, 2004) in farm, mill, ship and warehouse (Hughes, 1961, 1976). In Ireland, the mite was found in stored foodstuffs (O'Farrell and Butler, 1948), and farm grain stores (Cusack, 1974; Cusack *et al.*, 1975). Rack (1968a) found *U. marginata* in Hamburg in garbage dump, compost soil, heavily mouldy Brazil nuts from Brazil,

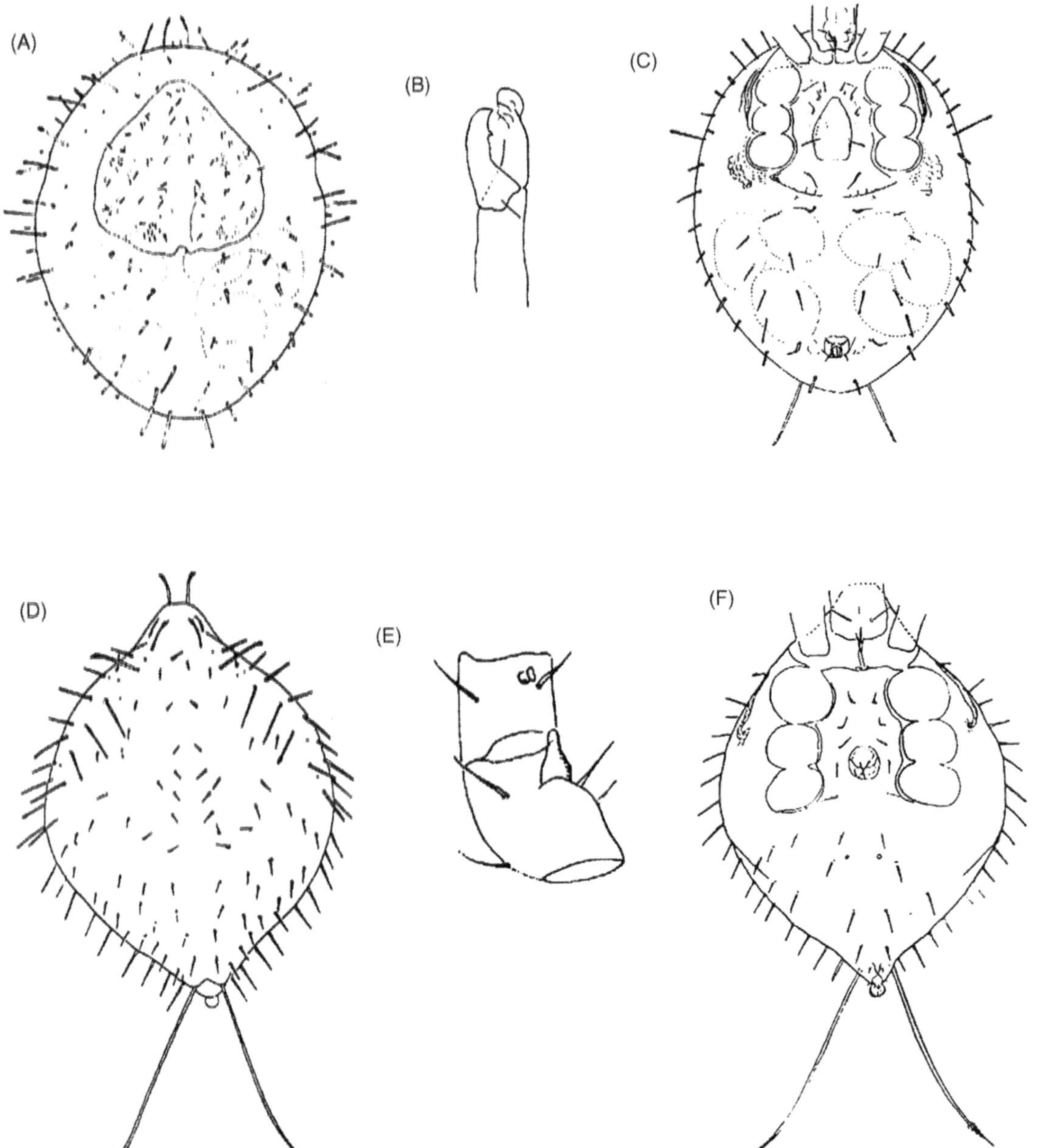

Fig. 71.2. *Uroseius acuminatus* (Koch). Female. Dorsal view (A); chelicera (B); ventral view (C). Male. Dorsal view (D); femur and genu II (E); ventral view (F) (from Hughes, 1976, licensed under the Open Government Licence v3.0).

sugarcane straw (packaging material for exotic cacti), feed mill, oil cake and rice residues. In Iran, the mite has been collected on stored rice, rice dust and debris (Noei, 2007; Noei *et al.*, 2008b).

The life history of *U. marginata* has been investigated by Rohde (1959), Faasch (1967) and Jalil and Rodriguez (1970). The mite is a scavenger, and it feeds also on fungal hyphae and spores, and will attack soft bodied animals such as nematodes, fly larvae and the immature stages of various mites as *Caloglyphus mycophagous* (Mégnin). Rohde (1959) reported that the life cycle of *U. marginata* takes from 30 to 40 days. Jalil and Rodriguez (1970) observed that at 27°C the mite develops a generation in 24 days and lays seven eggs/days on a diet of nematodes, and only four eggs on housefly juveniles. The female has 7 months post-oviposition period. Marquardt and Kaczmarek (2019) observed the egg deposition, incubation and hatching behaviour of *U. marginata*, using continuous video monitoring. The oviposition consists of three distinct phases. In phase one, the species engages in pre-ovipositional activity, prior to the egg being extruded from the genital opening. The phase one lasted approximately about 6 minutes. The phase two involved egg deposition and lasted approximately 35 seconds. Egg covering occurs in phase 3, and lasts about 25 minutes. The incubation time lasts 137 hours.

The spermatozoa are enclosed in a spermatophore, which is manipulated by male into the genital opening of the female, with the help of his palps. The eggs are laid singly in a shallow hole excavated by the second pair of legs and the palps of the female. In the life cycle, one of two deutonymphs develops: a Dauernymphe, non-phoretic, or a Wandernymphe, which is phoretic. The wandernymph is also more resistant to drought and lack of food. The ambulacrum of the wandernymph differs from that of the dauernymph, in having a more rounded pulvillus and proximally directed claws, possibly better adapted for clinging to a smooth surface. The cuticle of posterior idiosoma is thickened, and a gland lying above the hindgut is able to manufacture a secretion that is discharged through the anus and hardens in air to form a stalk. This remains attached to the anus and serves to fix the nymph to passing objects. Moreover, *U. marginata* has been reported in association with Coleoptera, Diptera, Myriapoda (Perotti and Braig, 2009), ground skinks (Mertins and Hartdegen, 2003) and also the nests of different birds in Scandinavia (Nordberg, 1936). This last author found the mite on various birds, such as eagle owl, *Bubo bubo* (Linnaeus), Carrion crow, *Corvus cornix* Linnaeus, Swallow, *Hirundo rustica* Linnaeus, Jackdaw, *Corvus monedula* (Linnaeus), Starling, *Sturnus vulgaris* Linnaeus, Great spotted woodpecker, *Dendrocopos major* Linnaeus, Stock dove and *Columba oenas* Linnaeus. Błoszyk *et al.* (2005) collected the mite in nests of the White stork, *Ciconia ciconia* (Linnaeus). Bahrami *et al.* (2011) found deutonymphs of this mite on *Euonthophagus gibbosus* (Scriba) (Coleoptera, Scarabaeidae) in horse dung in Iran. In the same country, the species was also collected from *Euoniticellus pallens* Olivier (Scarabaeidae) (Kazemi, 2009). Stenhouse (2015) reported the phoresy of *U. marginata* on *Pterostichus madidus* (Fabricius) (Coleoptera, Carabidae).

U. marginata inhabits the below surface, damper zones of manure heaps, strata preferred by newly hatched fly larvae (maggots). Deutonymphs and adults (males and females) prey only first instar fly juveniles; the mite can neither puncture the chorion of the fly egg nor attack its older, larger and more vigourous larvae. The contents of frozen prey eggs, whose chorion had been ruptured, or flesh fly eggs with punctured chorion, were readily consumed (Willis and Axtell, 1968). In accumulating poultry manure heaps, *U. marginata* arrives later than *M. muscaedomesticae*, being the first species carried by dung beetles, and the last disseminated by housefly. The uropodid mite develops a longer biological cycle, becoming numerically dominant in the heaps, only 5–6 weeks after manure accumulation (Willis and Axtell, 1968). Rodriguez *et al.* (1970) obtained 87% mortality of juvenile house flies after adding 200 mites to a container with 1000 fly eggs. The use of 20 predators against 250 fly eggs produced 27% of pest mortality (Willis and Axtell, 1968). Studies carried out in the laboratory against the lesser housefly, *Fannia canicularis* Linnaeus (Diptera, Fannidae), showed that males and females of *U. marginata* consumed an average of seven eggs during 11 days at 21°C, and produced 14–18% mortality (O'Donnell and Nelson, 1967). The mite also attacks the slug *Laevicaulis alte* (Férussac), a serious pest of many crops in tropical East Africa, India and Indonesia. It has been observed that effect depended on the slug/mite ratio and concrete death of *L. alte* occurred when this ratio reached 1/5 or more (Raut, 1996). Updated information on this topic may be found in Gerson *et al.* (2003).

Creuwels and Pieterse (2022) reported *U. obovata* from Poland, Netherland, Romania and Germany. Moreover, the mite has been collected on stored rice, rice dust and debris in Iran (Noei, 2007; Noei *et al.*, 2008b).

U. acuminatus was described as *Uropoda acuminata* by Koch (1847) from Germany. Its geographical distribution of stored food numbers England and Germany (Hughes, 1976). The commodities include mouldy wheat, in a ship infested with flies (Wynter Cross, in Hughes, 1976). Vitzthum (1928) collected the deutonymph of this mite from beehives and attached it to *Aphiochaeta rufipes* (Meigen) (Diptera, Phoridae). The adults are sometimes found in decaying plant material (Hughes, 1976).

References

Abare (2000) Australian Commodities, 7, 4, ACT, Canberra, Australia.

Abbar, S., Zhao, Y., Schilling, M.W. and Phillips, T.W. (2013) Chemical alternatives for suppressing the ham mite, *Tyrophagus putrescentiae*. In: *Proceedings, Annual International Research Conference on Methyl Bromide Alternatives and Emission Reduction, 4–6 November 2013, San Diego, U.S.A.*, p. 36.

Abbar, S., Amoah, B., Schilling, M.W. and Phillips, T.W. (2016) Efficacy of selected food-safe compounds to prevent infestation of the ham mite, *Tyrophagus putrescentiae* (Schrank) (Acarina: Acaridae) on southern dry cured hams. *Pest Management Science* 72(8), 1604–1612.

Abbar, S., Sağlam, Ö., Schilling, M.W. and Phillips, T.W. (2018) Efficacy of combining sulfuryl fluoride fumigation with heat to control the ham mite, *Tyrophagus putrescentiae* (Schrank) (Sarcoptiformes: Acaridae). *Journal of Stored Products Research* 76, 7–13.

Abbasi, H., Ardeshir, F. and Bagheri, M. (2016) Fauna of superfamily Bdelloidea and family Cheyletidae mites in grain stocks of East Azerbaijan province, Iran. *Proceedings of the 22nd Iranian Plant Protection Congress, Karaj, Iran*, 516.

Abdel-Ghaffar, F., Sobhy, H.M., Al-Quraishy, S. and Semmler, M. (2008) Field study on the efficacy of an extract of neem seed (Mite-Stop R) against the red mite *Dermanyssus gallinae* naturally infecting poultry in Egypt. *Parasitology Research* 103, 481–485.

Abdel-Ghaffar, F., Semmler, M., Al-Rasheid, K. and Mehlhorn, H. (2009) In vitro efficacy of ByeMite R and Mite-Stop R on developmental stages of the red chicken mite *Dermanyssus gallinae*. *Parasitology Research* 105, 1469.

Abdel-Sater, M.A. and Eraky, S.A. (2001) Bulbs micoflora and their relation with three stored product mites. *Mycopathologia* 153, 33–39.

Abdel-Shaheed, G.A., Hammad, S.M. and El-Sawaf, S.K. (1971) Survey of mites found on some field crops in Alexandria District, U.A.R. *Journal of Applied Entomology* 69, 106–110.

Abo-Shnaf, R.I.A. and Moraes, G.J. (2014) Phytoseiid mites (Acari: Phytoseiidae) from Egypt, with new records, descriptions of new species, and a key to species. *Zootaxa* 3865(1), 1–71.

Abo-Shnaf, R.I.A. and Moraes, G.J. (2016) *Proctolaelaps* species (Acari: Mesostigmata: Melicharidae) from Egypt, with description of a new species and complementary descriptions of other five species. *Zootaxa* 4162(3), 479–503.

Abou El-Atta, D.A.E., Ghazy, N.A. and Osman, M.A. (2014) Effects of temperature on the life-history traits of *Sancassania* (*Caloglyphus*) Berlesei (Acari: Astigmatina: Acaridae) feeding on root-knot nematodes, *Meloidogyne* Spp. (Nematoda: Meloidogynidae). *Experimental and Applied Acarology* 64(3), 299–307.

Aboul-Nasr, A.E. and Bassal, T.T.M. (1971) Biochemical and physiological study of certain ticks (Ixodoidea). The sugar content and concentration in *Argas* and *Hyalomma* biological fluids. *Journal of Medical Entomology* 8, 521–524.

Abramson, S.L., Turner-Henson, A., Anderson, L., Hemstreet, M.P., Bartholomew, L.K. *et al.* (2006) Allergens in school settings: Results of environmental assessments in 3 city school systems. *Journal of School Health* 76(6), 246–249.

Ackermann, M., Prill, P. and Ruess, L. (2016) Disentangling nematode-bacteria interactions using a modular soil model system and biochemical markers. *Nematology* 18, 403–415.

Ács, A. and Kontschán, J. (2014) Contribution to the macrochelidae vitzthum, 1930 fauna of the Carpathian Basin and the Balkan Peninsula (Acari: Mesostigmata). *Opuscula Zoologica* 45(2), 109–118.

Ács, A., Sutak, A. and Kontschan, J. (2017) New records of macrochelid mites and description of a new phoretic species (Acari: Mesostigmata: Macrochelidae) from Greece. *Acarologia* 56, 63–71.

Adam, K. (2006) Organic Allium Production. Available at: https://attra.ncat.org/wp-content/uploads/2019/05/allium.pdf? (accessed 17 April 2023).

Adler, C. (2004) Pests of dried fruits, cocoa, and spices in Germany. In: *Proceedings, Biocontrol of arthropod pests in stored products, 24–25 May 2004, Athens, Greece*, pp. 19–21.

Afifi, A.M. (1983) Description of developmental stages of *Uroobovella (Fuscuropoda) marginata* (Koch), (Acari: Gamasida: Uropodidae). *Bulletin – Zoological Society of Egypt* 33, 81–85.

Afifi, F. and Hafeez, S. (1988) Effect of different plant extracts on the toxicity and behaviour of *Tyrophagus putrescentiae* Schrank (Acari: Acaridae). *Annals of Agricultural Sciences* 33, 1375–1385.

Afifi, A.M., Hassan, M.F. and Nawar, M.S. (1986) Notes on the biology feeding habits of *Protogamasellus minutus* Hafez, El-Badry & Nasr (Acari: Gamasida: Ascidae). *Bulletin of the Entomological Society of Egypt* 66, 251–259.

Aguilar, H., Childers, C.C. and Welbourn, W.C. (2001) Relative abundance and seasonal occurrence of mites in the family Tydeidae on citrus in Florida. In: Halliday, R.B., Walter, D.E., Proctor, H.C., Norton, R.A. and Colloff, M.J. (eds) *Acarology: Proceedings of the 10th International Congress*. CSIRO Publishing, Melbourne, Australia, pp. 376–380.

Aguilar-Marcelino, L., Quintero-Martínez, M.T., Mendoza de Gives, P., López-Arellano, M.E., Liébano-Hernández, E. *et al.* (2014) Evaluation of predation of the mite *Lasioseius penicilliger* (Aracnida: Mesostigmata) on *Haemonchus contortus* and bacteria-feeding nematodes. *Journal of Helminthology* 88(1), 20–23.

Ahaniazad, M. and Bagheri, M. (2012) A new species of the genus *Molothrognathus* Summers and Schilinger (Acari: Trombidiformes: Caligonellidae) from Iran. *Acarologia* 52, 373–376.

Aheer, G.M., Akbar, S. and Chaudhri, W.M. (1997) New species of the genera *Cheletomorpha* and *Ker* (Acarina: Cheyletidae) from Pakistan. *Acarologia* 48(2), 117–121.

Ahmad, R. (1974) Studies on *Graphognathus leucoloma* (Boh.) (Col.: Curculionidae) and its natural enemies in the central provinces of Argentina. *Technical Bulletin of the Commonwealth Institute of Biological Control* 17, 19–28.

Ahmed, N., Rady, G.H., Abdelnabby, H., Ma, W.H., Liu, S.S. and Wang, M. (2012) Damage effect of two storage mites (Acari: Acaridida) and their biology on the Shiitake mushroom (*Lentinula edodes*) at three constant temperatures. *International Journal of Acarology* 38(3), 230–235.

Ahmed, A.K., Kamal, A.M., Mowafy, N.M.E. and Hassan, E.E. (2020a) Storage mite infestation of dry-stored food products and its relation to human intestinal acariasis in the city of Minia, Egypt. *Journal of Medical Entomology* 57(2), 329–335.

Ahmed, R., Gupta, S.K., Roy, S. and Bora, D. (2020b) Diversity and seasonal variation of mesostigmatid mites in three tea gardens of Assam (India) with different agro-practices. *Proceedings of the Entomological Society of Washington* 122, 750–756.

Ahmedani, M.S., Shagufta, N., Aslam, M. and Hussnain, S.A. (2010) Psocid: A new risk for global food security and safety. *Applied Entomoloy and Zoology* 45, 89–100.

Akdemir, C. and Gurdal, H. (2005) House dust mite in Kutahya, Turkey. *Türkiye Parazitolojii Dergisi* 29(2), 110–115 (in Turkish).

Akimov, I.A. (1973) On the morphological and physiological characteristics of the alimentary canal of the bulb mite *Rhizoglyphus echinopus* (Fumouze *et* Raobin). Proceedings of the 3rd International Congress of Acarology, Prague 1971, Czech Republic, pp. 703–706.

Akimov, I.A. (1985) The biological aspects of harmfulness of acaroid mites. *Naukova Dumka*, Kiev, 1–157 (in Russian).

Akimov, I.A. and Oksentiuk Ya.R. (2018) Functional and ecological adaptations of several acaridid mite species (Acariformes, Astigmata) for feeding on stored produce. *Vestnik Zoologii* 52(4), 323–330.

Akimov, I.A., Badanin, I.V. and Wita, I. (2002) Some aspects of trophic specialization in both free-living and parasitic prostigmatid mites (Acari: Actinotrichida) from a morpho-functional view-point. *Acarina* 10, 65–73.

Akiyama, M., Sakata, T., Mori, N., Kato, T., Amano, H. and Kuwahara, Y. (1997) Chemical ecology of astigmatid mites.46. Neryl formate, the alarm pheromone of *Rhizoglyphus setosus* Manson (Acarina: Acaridae) and the common pheromone component among four *Rhizoglyphus* mites. *Applied Entomoloy and Zoology* 32, 75–79.

Akpınar, D., Çobanoğlu, S. and Öğreten, A. (2017) Traits of Cheyletidae family and their usage possibilities in biological control. *Türk Tarım ve Doğa Bilimleri Dergisi* 4, 9–13 (in Turkish).

Akrami, M.A. (2015) An annotated checklist of oribatid mites (Acari: Oribatida) of Iran. *Zootaxa* 3963(4), 451–501.

Akyazı, F. and Ecevit, O. (2003) Determination of mite species in hazelnut orchards in Samsun, Ordu and Giresun provinces. *OMÜ Ziraat Fakültesi Dergisi* 18, 39–45 (in Turkish).

Akyazi, R., Soysal, M., Klimov, P.B. and Altunç, Y.E. (2019) House Dust Mite Species in Ordu Province, Turkey. *Journal of Agricultural Science* 25, 417–426.

Akyol, M. (2013) Two new species of the genus *Neophyllobius* Berlese (Acari: Camerobiidae) from Turkey. *International Journal of Acarology* 39(7), 542–546.

Akyol, M. and Koç, K. (2007) Two new records of the genus *Raphignathus* (Acari: Actinedida) for the Turkish Fauna. *Journal of Applied Biological Sciences* 1(3), 15–18.

Akyol, M. and Koç, K. (2012) A new species and two new records of the family *Caligonellidae* (Acari: Raphignathoidea) from Turkey. *International Journal of Acarology* 38, 40–45.

Al Safadi, M.M. (1990) The effect of light on the life cycle of the Acari *Tyrophagus similis* (Astigmata: Acaridae). *Revue d'ecologie et de biologie du sol* 27(4), 459–465.

Al Safadi, M.M. (1991) The effects of temperature on the developmental rate of the *Tyrophagus similis* Volgin (Acari: Acaridae). *Revue d'ecologie et de biologie du sol* 28, 341–347.

Alani, M.D. and Haarlov, N. (1972) The house-dust mite: a possible source of allergen in the environment of patients with atopic dermatitis. *Journal of the National Medical Association* 64, 302–305.

Alatawi, F.J., Kamran, M. and Mirza, J.H. (2018) Mesostigmatic mites (Acari: Mesostigmata) of Saudi Arabia (excluding Phytoseioidea), new records and a key to the known species. *Zootaxa* 4388, 373–394.

Al-Badry, E.A., Rizk, G.N. and Hafez, S.M. (1980) Frequency of occurrence of predaceous and parasitic mites inhabiting stored products. *Mesopotamia Journal of Agriculture* 15(1), 223–234.

Alberti, G. (1973) Ernährungsbiologie und Spinnvermögen der Schnabelmilben (Bdellidae, Trombidiformes). *Zeitschrift für morphologie der Tiere* 76, 285–338.

Alberti, G. (1979) Fine structure and probable function of genital papillae and Claparède organs of Actinotrichida. In: Rodriguez, G.R. (ed.) *Recent Advances in Acarology, II*. Academic Press, New York, pp. 501–507.

Alberti, G. (1984) The contribution of comparative spermatology to problems of acarine systematics. In: Griffiths, D.A. and Bowman, C.E. (eds) *Acarology VI*, Volume 1. Ellis Horwood, Chichester, UK, pp. 479–490.

Alberti, G. (1998) Fine structure of receptor organs in oribatid mites (Acari). In: Ebermann, E. and Schuster, R. (eds) *Arthropod Biology: Contributions to Morphology, Ecology and Systematics*. Biosystematics and Ecology Series 14, Österreichische Akademie der Wissenschaften, Vienna, pp. 27–77.

Alberti, G. (2000) Chelicerata. In: Jamieson, B.G.M. (ed.) *Reproductive Biology of Invertebrates, Volume IX, Part B, Progress in Male Gamete Ultrastructure and Phylogeny*. Oxford & IBH Publishing, New Delhi/Wiley, New York, pp. 311–388.

Alberti, G. and Bader, C. (1990) Fine structure of external "genital" papillae in the freshwater mite *Hydrovolzia placophora* (Hydrovolziidae, Actinedida, Actinotichida, Acari). *Experimental and Applied Acarology* 8, 115–124.

Alberti, G. and Coons, L.B. (1999) Mites. In: Harrison, F.W. and Foelix, R.F. (eds) *Microscopic Anatomy of Invertebrates, Volume 8C, Chelicerate Arthropoda*. Wiley, New York, pp. 515–1265.

Alberti, G. and Crooker, A.R. (1985) Internal anatomy. In: Helle, W. and Sabelis, M.W. (eds) *Spider Mites: Their Biology, Natural Enemies and Control. World Crop Pests, Volume 1A*. Elsevier Science, Amsterdam, The Netherlands, pp. 29–62.

Alberti, G. and Di Palma, A. (2002) Fine structure of the phytoseiid-type sperm access system (Acari, Gamasida, Phytoseiidae). In: Bernini, F., Nannelli, R., Nuzzaci, G. and de Lillo, E. (eds) *Acarid Phylogeny and Evolution: Adaptations in Mites and Ticks. Proceedings of IV Symposium of the European Association of Acarologists*. Kluwer Academic Publishers, Dordrecht, The Netherlands, pp. 241–252.

Alberti, G. and Seeman, O. (2005) Ultrastructural observations on Holothyrida (Acari: Anactinotrichida). In: Weigmann, G., Alberti, G., Wohltmann, A. and Ragusa, S. (eds) *Acarine Biodiversity in the Natural and Human Sphere*, 5th Symposium of the European Association of Acarologists, Berlin. Phytophaga 14(2004), pp. 103–111.

Alberti, G. and Storch, V. (1974) Über Bau und Funktion der Prosoma-Drüsen von Spinnmilben (Tetranychidae, Trombidiformes). *Zeitschrift für morphologie der Tiere* 79, 133–153.

Alberti, G., Storch, V. and Renner, H. (1981) Über den feinstrukturellen Aufbau der Milbencuticula (Acari, Arachnida). *Zoologische Jahrbücher. Abteilung für Anatomie und Ontogenie der Tiere* 105, 183–236.

Alberti, G., Moreno, A.I. and Kratzmann, M. (1994) The fine structure of trichobothria in moss mite with special emphasis on *Acrogalumna longipluma* (Berlese, 1904) (Oribatida, Acari, Arachnida). *Acta Zoologica* 75, 57–74.

Alberti, G., Norton, R.A., Adis, J., Fernandez, N.A., Franklin, E. *et al.* (1996) Porose areas and related organs in oribatid mites (Oribatida). In: Mitchell, R., Horn, D.J., Needham, G.R. and Welbourn, W.C. (eds) *Acarology IX: Proceedings, Volume 1.* Ohio Biological Survey, Columbus, OH, pp. 277–283.

Alberti, G., Seniczak, A. and Seniczak, S. (2003) The digestive system and fat body of an early-derivated oribatid mite *Archegozetes longisetosus* Aoki (Acari: Oribatida, Trhypochthoniidae). *Acarologia* 83(1), 149–219.

Al-Dulaimi, S.I. (2002) Predation by the mite *Macrocheles glaber* (Müller) (Acarina: Macrochelidae) on the house fly *Musca domestica* L. with some notes on its biology. *Bulletin of the Iraq Natural History Museum* 9(4), 7–11.

Alexander, A., Fall, N. and Arlian, L. (2002) Mating and fecundity of *Dermatophagoides farinae. Experimental and Applied Acarology* 26, 79–86.

Ali, O., Dunne, R. and Brennan, P. (1999) Effectiveness of the Predatory Mite *Hypoaspis miles* (Acari: Mesostigmata: Hypoaspidae) in conjunction with pesticides for control of the mushroom fly *Lycoriella solani* (Diptera: Sciaridae). *Experimental and Applied Acarology* 23, 65–77.

Ali, W., George, D.R., Shiel, R.S., Sparagano, O.A.E. and Guy, J.H. (2012) Laboratory screening of potential predators of the poultry red mite (*Dermanyssus gallinae*) and assessment of *Hypoaspis miles* performance under varying biotic and abiotic conditions. *Veterinary Parasitology* 187, 341–344.

Al-Jassany, R.F. and Al-Zubaidy, R.A.R. (2012) New record of *Sancassania* mite on workers of termite in Iraq and its effects on mortality. *Iraqi Journal of Agricultural Science* 43(4), 34–56.

Alkot, W.M., Naiem, E.S., El-Ebiary, M.E., Tayesh, M.A.A. and Seif, A.I. (2020) Lethal and sublethal effects of some essential oils against *Callosobruchus maculatus*, *Tyrophagus putrescentiae* and their natural enemies, *Anisopteromalus calandrae* and *Cheyletus malaccensis. Egyptian Journal of Experimental Biology* 16, 33–45.

Al-Kubaisi, F.M.A. and Al-Dulaimi, S.I.H. (2020) Dominance and frequency of mite species associated with poultry droppings. *Iraqi Journal of Science* 61(7), 1603–1611.

Allred, D.M. (1957) Mites found on mice of the genus *Peromyscus* in Utah. II. Family Haemogamasidae. *Proceedings of the Entomological Society of Washington* 59, 31–39.

Allred, D.M. (1970) New ameroseiid mites from birds of New Guinea. *Journal of Medical Entomology* 7, 99–102.

Allymehr, M., Tavassoli, M., Manoochehri, M.H. and Ardavan, D. (2012) Ectoparasites and gastrointestinal helminths of house mice (*Mus musculus*) from poultry houses in Northwest Iran. *Comparative Parasitology* 79, 283–287.

Al-Nasser, A.S. (2011) Quantitative survey of stored products mites infesting wheat flour in Jeddah Governorate. *International Journal of Nematology and Entomology* 3(6), 78–84.

Al-Safadi, M.M. (1987) The life cycle of the Acari *Tyrophagus similis. Journal of Zoology* 213, 141–146.

Al-Shammery, K.A. (2014) Influence of feeding on three stored product pests on rearing of the predatory mite *Cheyletus malaccensis* (Acari: Cheyletidae) in Hail, Saudi Arabia. *Journal of Life Science* 11, 260–266.

Alves, L.F.A., Oliveira, D.G.P., Pares, R.B., Sparagano, O.A. and Godinho, R.P. (2020) Association of mechanical cleaning and a liquid preparation of diatomaceous earth in the management of poultry red mite, *Dermanyssus gallinae* (Mesostigmata: Dermanyssidae). *Experimental and Applied Acarology* 81, 215–222.

Alwar, V.S. and Lalitha (1961) A note on the occurrence of the mite *Dermatophagoides* sp. in domestic animals. *Indian Veterinary Journal* 38(6), 292–294.

Amal, A.A., Yassin, E.M.A., El-Bahrawy, A.F., El-Sharabasy, H.M. and Marwa, S.K. (2020) Biology of *Blattisocius mali* (Oudemans) (Acari: Gamasida: Ascidae) feeding on different diets under laboratory conditions. *Egyptian Veterinary Medical Society of Parasitology Journal* 16, 92–101.

Ament, K., Kant, M.R., Sabelis, M.W., Haring, M.A. and Schuurink, R.C. (2004) Jasmonic acid is a key regulator of spider mite-induced volatile terpenoid and methyl salicylate emission in tomato. *Plant Physiology* 135, 2025–2037.

Amoah, B. (2016) Monitoring populations of the ham mite, *Tyrophagus putrescentiae* (Schrank) (Acari: Acaridae): Research on traps, orientation behavior, and sampling techniques. Thesis, Kansas State University, Manhattan, Kansas, USA, pp. 1–145.

Amoah, B., Schilling, M.W. and Phillips, T.W. (2016) Monitoring *Tyrophagus putrescentiae* (Acari: Acaridae) with traps in dry-cured ham aging rooms. *Environmental Entomology* 45(4), 1029–1039.

Amoah, B., Hagstrum, D., Subramanyam, B., Campbell, J.F., Schilling, M.W. and Phillips, T.W. (2017) Sampling methods to detect and estimate populations of *Tyrophagus putrescentiae* (Schrank) (Sarcoptiformes: Acaridae) infesting dry-cured hams. *Journal of Stored Products Research* 73, 98–108.

Amosova, L. and Stanyukovich, M. (2008) Ultrastructure of salivary glands of the chicken red mite *Dermanyssus gallinae* (Acarina, Gamasina, Dermanyssidae). *Journal of Evolutionary Biochemistry and Physiology* 44, 627–633.

Amrine, J.W. Jr, Stasny, T.A. and Flechtmann, C.H.W. (2003) *Revised Keys to World Genera of Eriophyoidea (Acari: Prostigmata)*. Indira Publishing House, West Bloomfield, MI, pp. 1–244.

Anderson, H.R. and Cunnington, A.M. (1974) House dust mites in the highlands of Papua New Guinea. *Papua New Guinea Medical Journal* 17, 304–308.

Anderson, G.L., Caldwell, K.N., Beuchat, L.R. and Williams, P.L. (2003) Interaction of a free-living soil nematode, *Caenorhabditis elegans*, with surrogates of foodborne pathogenic bacteria. *Journal of Food Protection* 66, 1543–1549.

André, M. (1934) Note sur un Acarien (*Leiodinychus krameri* G. and R. Can.) parasite des habitations humaines. *Bulletin de la Société zoologique de France* 59(2), 112–122.

André, H.M. (1979) A generic revision of the family Tydeidae (Acari: Actinedida). 1. Introduction, paradigms and general classification. *Annales de la Société royale zoologique de Belgique* 108(3–4), 189–208.

André, H.M. (1980) A generic revision of the family Tydeidae (Acari: Actinedida). IV. Generic descriptions, keys and conclusions. *Bulletin et annales de la Société royale belge d'entomologie* 116, 103–168.

André, H.M. (1981a) A generic revision of the family Tydeidae (Acari: Prostigmata). II. Organotaxy of the idiosoma and gnathosoma. *Acarologia* 22, 31–40.

André, H.M. (1981b) A generic revision of the family Tydeidae (Acari: Prostigmata). III. Organotaxy of the legs. *Acarologia* 22, 165–178.

André, H.M. (1984) Notes on the ecology of corticolous epiphyte dwellers. 3. Oribatida. *Acarologia* 25(4), 385–395.

André, H.M. (1985) Association between corticolous microarthropod communities and epiphytic cover on bark. *Holarctic Ecology* 8, 113–119.

André, H.M. (1988) Age-dependent evolution: from theory to practice. In: Humphries C.J. (ed.) *Ontogeny and Systematics*. Columbia University Press, New York, pp. 137–187.

André, H.M. (1989) The concept of stase. In: André, H.M. and Lions J.-C. (eds), L'Ontogenèse et le concept de Stase chez les Arthropodes. *AGAR Publishers*, Wavre, Belgium, pp. 3–16.

André, H.M. (2005) In search of the true *Tydeus* (Acari, Tydeidae). *Journal of Natural History* 39, 975–1001.

André, H.M. (2023) *Lorryia* (Acariformes, Tydeidae): The evolutionary plasticity of an enigmatic genus. *Acarologia* 63(3), 844–855.

André, H.M. and Fain, A. (2000) Phylogeny, ontogeny and adaptive radiation in the superfamily Tydeoidea (Acari: Actinedida), with a reappraisal of morphological characters. *Zoological Journal of the Linnean Society* 130, 405–448.

André, H.M. and Van Impe, G. (2012) The missing stase in spider mites (Acari: Tetranychidae): When the adult is not the imago. *Acarologia* 52, 3–16.

André, H.M. and Voegtlin, D.J. (1981) Some observations on the biology of *Camisia carrolli* (Acari: Oribatida). *Acarologia* 23(1), 81–89.

André, H.M., Noti, M.-I. and Lebrun, P. (1994) The soil fauna: The other last biotic frontier. *Biodiversity and Conservation* 3, 45–56.

Andreev, E.A. (1988) Ecology of mites phoretic on beetles of the genus *Necrophorus* (Silphidae). *Nauchnie dokladi vysshei shkoli. Biologicheskie naucki* 5, 44–47.

Andrews, J.R.H. and Ramsay, G.W. (1982) A case of papular dermatosis in man attributed to an ascid mite (Acari). *Journal of Medical Entomology* 19, 111–112.

Andrzejewska, Z. (1979) Roztocze i zaleszczotki synantropijne okolic Turka. M.B. Thesis, Adam Mickiewicz University, Poznań (Poland) (in Polish).

Animal Welfare Regulations (2013) 9 CFR Parts 1–4, USDA. Available at: https://www.nal.usda.gov/animal-health-and-welfare/animal-welfare-act (accessed 20 June 2023).

Anita (2010) Feeding potential of *Tyrophagus putrescentiae* Schrank (Acari: Acaridae) and its management. M.Sc. Thesis, CCSHAU, Hisar, pp. 1–82.

Anita, A., Gulati, R., Kaushik, H.D. and Arvind, A. (2013) Effect of *Tyrophagus putrescentiae* Schrank on weight loss in stored oats and green gram. *Annals of Plant Protection Sciences* 21(1), 90–93.

Anita, A., Gulati, R., Kaushik, H.D. and Arvind, A. (2014) Efficacy of *Ocimum sanctum* and *Glycyrrhiza glabra* against stored Mite, *Tyrophagus putrescentiae* Schrank in oat flakes. *Biopesticides International* 10(1), 41–49.

Antor, R.J. and Garcia, M.L. (1995) A new mite-plant association: mite living amidst the adesive traps of a carnivorous plant. *Oecologia* 101, 51–54.

ANVISA (Agencia Nacional de Vigilancia Sanitaria) (2005) Resolucao RDC no 263 de 22 de setembro de 2005. Aprova o "Regulamento Tecnico para produtos de cereais, amidos, farinhas e farelos". Available at: http://portal.anvisa.gov.br/wps/wcm/connect/1ae52c0047457a-718702d73fbc4c6735/RDC_263_2005.pdf?MOD=AJPERES (accessed 28 May 2020).

Aoki, J. (1958) Einige Phthiracariden aus Utsuku-shigahara, Mitteljapan (Acarina: Oribatei). *Annotationes zoologicae Japonenses* 31, 171–175.

Aoki, J. (1973) Soil mites (oribatids) climbing trees. In: Daniel, M. and Rosický, B. (eds) *Proceedings of the 3rd International Congress of Acarology Held in Prague, 31 August–6 September 1971*. Dr W. Junk, The Hague, The Netherlands (reprinted by Springer), pp. 59–65.

Acquah, F.K., Adjah, J., Williamson, K.C. and Amoah, L.E. (2019) Trasmission-blocking vaccines: Old friends and new prospects. *Infection and Immunity* 87, e00775–18.

Araujo-Fontaine, A., Wagner, M. and Kremer, M. (1973) Contribution a l'etude des acariens de la poussiere domestique en Alsace. Relations avec les conditions d'habitat. *Comptes Rendus des Seances de la Societe de Biologie et de Ses Filiales* 167, 371–378.

Ardeshir, F. (2002) Étude des acariens des grains de froment stockés au nord de l'Iran. Thése du grade de Docteur (Ph.D.), Gent University, Belgium, pp. 1–154.

Ardeshir, F. (2004) Study of the mite population in stored grains in different seasons in Iran. *Proceeding of 16th Plant Protection Congress, Tabriz, Iran*, 282.

Ardeshir, F. (2010) A preliminary study on mite fauna of bird nests in Iran. *Podoces* 5(2), 112–115.

Ardeshir, F. (2017) Cheyletid mites (Acari: Trombidiformes) in stored grains in Iran. *Persian Journal of Acarology* 6(1), 11–24.

Ardeshir, F. and Nematollahi, M.R. (2008) Mites fauna of wheat stored and new record of *Cunliffella variegata* (Barilo, 1985) in Esfahan Province. In: Manzari, S. (ed.) *Abstract Book of 18th Iranian Plant Protection Congress*. Hamedan, Iran, p. 274.

Ardeshir, F., Yousefi, A. and Saboori, A. (2006) A survey on mite population of stored wheat in Tehran province. *Proceeding of 17th Iranian Plant Protection Congress*. Karaj, Iran, p. 188.

Ardeshir, F., De Saint Georges-Gridelet, D., Grootaert, P., Tirry, L. and Wauthy, G. (2000) Preliminary observations on mites associated with stored grain in Iran. *Belgian Journal of Entomology* 2, 287–293.

Ardeshir, F., Ranji, H., Obaidy, O. and Khany, M. (2012) Mites fauna in dry fruits and nuts in West Azerbaijan Province, Abstract Book of the 20th Iranian Plant Protection Congress, Shiraz, Iran, p. 512.

Ardeshir, F., Bagheri, M. and Paktinat, S. (2014) Redescription of male and female of *Paraneognathus wangae* (Fan & Li) (Acari: Caligonellidae) with a key to all known species of the genus *Paraneognathus*. *Persian Journal of Acarology* 3, 145–151.

Ardeshir, F., Bagheri, M. and Saeej, S.P. (2015) Redescription of male and female of *Paraneognathus wangae* (Fan & Li) (Acari: Caligonellidae) with a key to all known species of the genus *Paraneognathus*. *Persian Journal of Acarology* 3(2), 145–151.

Arias-Irigoyen, J., Lombardero, M., Arteaga, C., Carpizo, J.A. and Barber, D. (2007) Limited IgE cross reactivity between *Dermatophagoides pteronyssinus* and *Glycyphagus domesticus* in patients naturally exposed to both mite species. *Journal of Allergy and Clinical Immunology* 120(1), 98–104.

Arillo, A., Subías, L.S. and Shtanchaeva, U. (2012) A new species of fossil oribatid mite (Acariformes, Oribatida, Trhypochthoniidae) from the Lower Cretaceous amber of San Just (Teruel Province, Spain). *Systematic and Applied Acarology* 17, 106–112.

Arlian, L.G. (1989) Biology and ecology of house dust mites, *Dermatophagoides* spp. and *Euroglyphus* spp. *Immunology and Allergy Clinics of North America* 9, 339–356.

Arlian, L.G. (1992) Water balance and humidity requirements of house dust mites. *Experimental and Applied Acarology* 16, 15–35.

Arlian, L.G. (2001) Dust mites: Update on their allergens and control. *Current Allergy and Asthma Reports* 1, 581–586.

Arlian, L.G. (2002) Arthropod allergens and human health. *Annual Review of Entomology* 47, 395–433.

Arlian, L.G. and Dippold, J.S. (1996) Development and fecundity of *Dermatophagoides farinae* (Acari: Pyroglyphidae). *Journal of Medical Entomology* 33, 257–260.

Arlian, L.G. and Morgan, M.S. (2003) Biology, ecology, and prevalence of dust mites. *Immunology and Allergy Clinics of North America* 23(3), 443–468.

Arlian, L.G. and Morgan, M.S. (2015) Reproductive biology of *Euroglyphus maynéi* with comparisons to *Dermatophagoides farinae* and *D. pteronyssinus*. *Experimental and Applied Acarology* 66, 1–9.

Arlian, L.G. and Platts-Mills, T.A.E. (2001) The biology of dust mites and the remediation of mite allergens in allergic disease. *Journal of Allergy and Clinical Immunology* 107(3), 406–413.

Arlian, L.G. and Veselica, M.M. (1979) Water balance in insects and mites. *Comparative Biochemistry and Physiology* 64A, 191–200.

Arlian, L.G. and Vyszenski-Moher, D.L. (1988) Life cycle of *Sarcoptes scabiei* var. *canis*. *Journal of Parasitology* 74(3), 427–430.

Arlian, L.G., Woodford, P.J., Bernstein, I.L. and Gallagher, J. (1983) Seasonal population structure of house dust mites, *Dermatophagoides* spp. (Acari: Pyroglyphidae). *Journal of Medical Entomology* 20(1), 99–102.

Arlian, L.G., Geis, D.P., Vyszenski-Moher, D.L., Bernstein, I.L. and Gallagher, J.S. (1984a) Cross antigenic and allergenic properties of the house dust mite *Dermatophagoides farinae* and the storage mite *Tyrophagus putrescentiae*. *Journal of Allergy and Clinical Immunology* 74, 172–179.

Arlian, L.G., Geis, D.P., Vyszenski-Moher, D.L., Bernstein, I.L. and Gallagher, J.S. (1984b) Antigenic and allergenic properties of the storage mite *Tyrophagus putrescentiae*. *Journal of Allergy and Clinical Immunology* 74(2), 166–171.

Arlian, L.G., Bernstein, I.L., Vyszenski-Moher, D.L. and Gallagher, J.S. (1987) Investigations of culture medium-free house dust mites. III: Antigens and allergens of body and fecal extract of *Dermatophagoides farinae*. *Journal of Allergy and Clinical Immunology* 79, 457–466.

Arlian, L.G., Rapp, C.M. and Ahmed, S.G. (1990) Development of *Dermatophagoides pteronyssinus* (Acari: Pyroglyphidae). *Journal of Medical Entomology* 27(6), 1035–1040.

Arlian, L.G., Bernstein, D., Bernstein, I.L., Friedman, S., Grant, A. *et al.* (1992) Prevalence of dust mites in the homes of people with asthma living in eight different geographic areas of the United States. *Journal of Allergy and Clinical Immunology* 90, 291–300.

Arlian, L.G., Rapp, C.M. and Fernandez-Caldas, E. (1993) Allergenicity of *Euroglyphus maynéi* and its cross-reactivity with *Dermatophagoides* species. *Journal of Allergy and Clinical Immunology* 91(5), 1051–1058.

Arlian, L.G., Confer, P.D., Rapp, C.M., Vyszenski-Moher, D.L. and Chang, J.C.S. (1998a) Population dynamics of the house dust mites *Dermatophagoides farinae*, *D. pteronyssinus*, and *Euroglyphus maynéi* (Acari: Pyroglyphidae) at specific relative humidities. *Journal of Medical Entomology* 35(1), 46–53.

Arlian, L.G., Neal, J.S. and Bacon, S.W. (1998b) Survival, fecundity and development of *Dermatophagoides farinae* (Acari: Pyroglyphidae) at fluctuating relative humidity. *Journal of Medical Entomology* 35, 962–966.

Arlian, L.G., Morgan, M.S. and Houck, M.A. (1999a) Allergenicity of the mite *Hemisarcoptes cooremani*. *Annals of Allergy, Asthma & Immunology* 83(6 Pt 1), 529–532.

Arlian, L.G., Jacqueline S. Neal, B.S., DiAnn L. Vyszenski-Moher, M.S. (1999b) Reducing relative humidity to control the house dust mite *Dermatophagoides farinae*. *Journal of Allergy and Clinical Immunology* 104(4), 852–856.

Arlian, L.G., Schumann, R.J., Morgan, M.S. and Glass, R.L. (2003) Serum immunoglobulin E against storage mite allergens in dogs with atopic dermatitis. *American Journal of Veterinary Research* 64, 32–36.

Arjomandi, E., Kazemi, S. and Afshari, A. (2013) Fauna and diversity of the manure inhabiting Mesostigmata (Acari) in Kerman County, South Eastern Iran. *Persian Journal of Acarology* 2, 253–263.

Armentia, A., Martinez, A., Castrodeza, R., Martinez, J., Jimeno, A., Mendez, J. and Stolle, R. (1997) Occupational allergic disease in cereal workers by stored grain pests. *Journal of Asthma* 34, 369–378.

Armitage, D. and Chambers, J. (2002) Preventing and Controlling Mites in Stored Cereals (HGCA Topic Sheet No. 62). Available at: http://adlib.everysite.co.uk/adlib/defra/content.aspx?id=000HK277ZX.0AK0W8F4IFIGSR (accessed 12 December 2021).

Armitage, D.M. and George, C.L. (1986) The effect of three species of mites upon fungal growth on wheat. *Experimental and Applied Acarology* 2, 111–124.

Armitage, D.M., Cook, D.A. and Duckett, C. (1996) The use of an aspirated sieve to remove insects, mites and pesticides from grain. *Crop Protection* 15(8), 675–680.

Armitage, D.M., Cook, D.A. and Baxter, D.E. (2002) Farm-scale experiments to compare infestation and quality changes in malting barley stored at three moisture contents. *Journal of the Institute of Brewing* 108(2), 178–186.

Arnau, J. and Guerrero, L. (1994) Physical methods of controlling mites in dry cure ham. *Fleischwirtschaft* 74(12), 1311–1313.

Arruda, L.K. and Chapman, M.D. (1992) A review of recent immunochemical studies of *Blomia tropicalis* and *Euroglyphus maynéi* allergens. *Experimental and Applied Acarology* 16(1–2), 129–140.

Artigas, J.N. and Casanueva, M.E. (1983) Acaros del polvo de las habitaciones en Chile (Acari). *Gayana. Zoologia* 47, 3–106.

Asanuma, K. (1963) Occurrence of a mesostigmatic mite, *Haemogamasus pontiger* (Berlese) in Tokyo, Japan. *Japanese Journal of Sanitary Zoology* 14, 252.

Ascerno, M.E., Pfleger, F.L., Morgan, F. and Wilkins, H.F. (1983) Relationship of *Rhizoglyphus robini* (Acari: Acaridae) to root rot control in greenhouses-forced Easter lilies. *Environmental Entomology* 12, 422–425.

Ashburner, M. (1989) Mites. In: Ashburner, M., Golic, K.G. and Hawley, R.S. (eds) *Drosophila, A Laboratory Handbook*. Cold Spring Harbor Laboratory Press, New York, pp. 1191–1225.

Ashfaq, M., Sarwar, M. and Amjad, A.W. (2000) Two new mite species (hypopi) of the genus *Histiostoma* (Acari: Histiostomatidae) from Pakistan. Pak. *Journal of Agricultural Science* 37(1–2), 33–41.

Ashfaq, M., Bashir, M.H., Kamran, M., Afzal, M. and Raza, A.B.M. (2011) Record of one new species of the genus *Cunaxa* (Acari: Cunaxidae) from rice husk. *Pakistan Journal of Zoology* 43, 1.

Asmat, G.S.M. and Kader, B. (1990) A preliminary survey of mites from the nest of certain birds in the Chittagong University campus. *Bangladesh Journal of Zoology* 18, 177–183.

Aspaly, G., Stejskal, V., Pekar, S. and Hubert, J. (2007) Temperature-dependant population growth of three species of stored product mites (Acari: Acaridida). *Experimental and Applied Acarology* 42, 37–46.

Assis, C.P., Gondim, M.G., Jr., Siqueira, H.A. and Câmara, C.A. (2011) Toxicity of essential oils from plants towards *Tyrophagus putrescentiae* (Schrank) and *Suidasia pontifica* Oudemans (Acari: Astigmata). *Journal of Stored Products Research* 47, 311–315.

Athanassiou, C.G., Palyvos, N.E., Eliopoulos, P.A. and Papadoulis, G.T. (2002) Mites associated with stored seed cotton and related products in Greece. *Phytoparasitica* 30(4), 387–394.

Athanassiou, C.G., Kavallieratos, N.G., Palyvos, N.E., Sciarretta, A. and Trematerra, P. (2005) Spatiotemporal distribution of insects and mites in horizontally stored wheat. *Journal of Economic Entomology* 98 (3), 1058–1069.

Athias Henriot, C. (1959) Phytoseiidae et Aceosejidae (Acarina, Gamasina) d'Algerie. III. Contribution aux Aceosejinae. *Bulletin de la Société d'histoire naturelle de l'Afrique du Nord* 50, 158–195.

Athias Henriot, C. (1961a) Mésostigmates (Urop. exel.) édaphiques méditerranéens (Acaromorpha, Anactinotrichida) (Coll. Prof. H. Franz et C. Athias Henriot. Premièr série. *Acarologia* 2, 159–174.

Athias Henriot, C. (1961b) *Tyrophagus miripes* n. sp. (acariens actinotrichidés, Acaridae). *Annales Ecole Nationale Agriculture Alger* 3(2), 1–4.

Athias Henriot, C. (1961c) Mesostigmates (Urop. excl.) edaphiques mediterraneens (Acaromorpha, Anactinotrichida). *Acarologia* 3, 381–509.

Athias Henriot, C. (1968) L'appareil d'insémination laelapoïde (Acariens Anactinotrichides: Laelapoidea, ♀♀). Premières observations. Possibilié d'emploi à des fins taxonomiques. *Bulletin scientifique de Bourgogne* 25, 229–274.

Athias Henriot, C. (1969a) Contribution aux peuplements acarologiques des biotopes terrestres humides médioeuropéens: Observations sur le genre *Panteniphis* Willm. (Mésostigmates, Rhodacaride). *Revue d'ecologie et de biologie du sol* 6, 69–78.

Athias Henriot, C. (1969b) Les organes cuticulaires sensoriels et glandulaires des Gamasides. Poroidotaxie et adenotaxie. *Bulletin de la Société entomologique de France* 94, 485–492.

Athias Henriot, C. (1969c) Notes sur la morphologie externe des gamasides (Acariens Anactinotriches). *Acarologia* 11, 609–629.

Athias Henriot, C. (1972) Sur la systématique et la géographie de la lignée *Truncus* du sous-genre *Anidogamasus* Athias, 1970 (genre *Paragamasus* Hull, 1918: Arachnides, Gamasides tocospermiques, Parasitina). *Bulletin du Muséum National d'Histoire NaturelleBulletin du Muséum National d'Histoire Naturelle* 3(33), *Zoologie* 27, 317–325.

Athias-Binche, F. (1994) La Phoresie chez les Acariens. Aspects Adaptatifs et Évolutifs. Editions du Castillet, Perpignan, France, pp. 1–178.

Attiah, H.H. (1969) Tyrogliphoid mites associated with stored food in U.A.R. (Sarcoptiformes – Acarina). *Tech. Bull., United Arab Republic, Min. Agric., Plant Prot. Depart* 10, 1–51.

Attiah, H.H. (1971) *Chelachecaropsis bakeri*, a new genus and species associated with stored food mites in Egypt (Acarina, Cheyletidae). In: Daniel, M. and Rosicky, B. (eds) *Proceedings of the 3rd International Congress of Acarology, Prague, August 31–September 6, 1971, Junk, Den Haag*, pp. 349–352.

Attiah, H.H. and Soliman, Z.R. (1970) Life-history of the predator mite, *Acaropsis aegyptiaca* Wafa and Soliman. *Bulletin de la Société entomologique d'Égypte* 54, 155–158.

Atyeo, W.T. (1960) A revision of the family Bdellidae in North and Central America (Acarina: Prostigmata). *University of Kansas Science Bulletin* 40, 345–499.

Atyeo, W.T. (1963) The Bdellidae (Acarina) of the Australian Realm. Part I: New Zealand, Campbell Island, and the Auckland Islands, 113–166; Part II: Australia and Tasmania, 167–210. *Bulletin of the University of Nebraska State Museum* 4(8), 113–210.

Atwa, W.A., El-Naggar, M.E., Khalil, A.M., El-Shaer, M.E. and Mostafa, Z.M.M. (2018) Ecological studies on predaceous and parasitic mites associated with some stored products. *Menoufia Journal of Plant Production* 3, 163–167.

Aufray, T., Arriaga-Jiménez, A., Taudière, A., Roy Lauren, J.M., Lapeyre, B. and Roy, L. (2023) Attractant activity of host-related chemical blends on the poultry red mite at diferent spatial scales. *Journal of Chemical Ecology* 49, 18–35.

Auger, A., Nantel, J., Meunier, N. and Harrison, J.R. (1979) Skin acariasis caused by *Dermanyssus gallinae* (De Geer): An in-hospital outbreak. *Canadian Medical Association Journal* 17, 700–703.

Aupic-Samain, A., Baldy, V., Delcourt, N., Henning Krogh, P., Gauquelin, T., Fernandez, C. and Santonja, M. (2021) Water availability rather than temperature control soil fauna community structure and prey-predator interactions. *Functional Ecology* 35(7), 1550–1559.

Avtaeva, T., Petrovičová, K., Langraf, V. and Brygadyrenko, V. (2021) Potential bioclimatic ranges of crop pests *zabrus tenebrioides* and *harpalus rufipes* during climate change conditions. *Diversity (Basel)* 13, 559.

Axelsen, J.A., Holst, N., Hamers, T. and Krogh, P.H. (1997) Simulations of the predator-prey interactions in a two species ecotoxicological test system. *Ecological Modelling* 101, 15–25.

Axtell, R.C. (1961) New records of North American Macrochelidae (Acarina: Mesostigmata) and their predation rates on the house fly. *Annals of the Entomological Society of America* 54(5), 748.

Axtell, R.C. (1963a) Effects of macrochelidae (Acarina: Mesostigmata) on house fly production from dairy cattle manure. *Journal of Economic Entomology* 56(3), 317–321.

Axtell, R.C. (1963b) Acarina occurring in domestic animal manure. *Annals of the Entomological Society of America* 56(5), 628–633.

Axtell, R.C. (1964) Phoretic relationship of some common manure-inhabiting Macrochelidae (Acarina: Mesostigmata) to the house fly. *Annals of the Entomological Society of America* 57(5), 584–587.

Axtell, R.C. (1966) Comparative toxicities of insecticides to house fly larvae and *Macrocheles muscaedomesticae* a mite predator of the house fly. *Journal of Economic Entomology* 59(5), 1128–1130.

Axtell, R.C. (1969) Macrochelidae (Acarina: Mesostigmata) as biological control agents for synantropic flies. Proceedings Second International Congress of Acarology, Sutton Bonington, England, July 19–25, 1967, Akademai Kiado, Budapest, pp. 401–416.

Axtell, R.C. (1999) Poultry integrated pest management: Status and future. *Integrated Pest Management Reviews* 4, 53–73.

Axtell, R.C. and Arends, J.J. (1990) Ecology and management of arthropods pests of poultry. *Annual Review of Entomology* 35, 101–126.

Axtell, R.C. and Edwards, T.D. (1983) Efficiency and nontarget effects of larvadex as a feed additive for controlling house flies in caged-layer poultry manure. *Poultry Science* 62(12), 2371–2377.

Aygun, O., Yaman, M. and Durmaz, H. (2007) A survey on occurrence of *Tyrophagus putrescentiae* (Acari: Acaridae) in Surk, a traditional Turkish dairy product. *Journal of Food Process Engineering* 78(3), 878–881.

Aylesworth, R. and Baldridge, D. (1983) *Dermatophagoides scheremetewskyi* and feather pillow dermatitis. *Minnesota Medicine* 66(1), 43.

Ayyldiz, N. and Özkan, M. (1988) Erzurum ovasi oribatid akarlari (Acari: Oribatida) üzerine sistematik araştirmalar. I. Likel oribatidler. *Doğa. Türk zooloji dergisi* 12(2), 115–130.

Ayyldiz, N. and Luxton, M. (1990) The genus *Cosmochthonius* Berlese, 1910, (Oribatida, Cosmochthoniidae). *Acarologia* 31(3), 279–284.

Azevedo, L.H. de, Emberson, R.M., Esteca, F.d.C.N. and Moraes, G.J. (2015). Macrochelid mites (Mesostigmata: Macrochelidae) as biological control agents. In: Carrillo, D., Moraes, G. and Peña, J. (eds) *Prospects for Biological Control of Plant Feeding Mites and Other Harmful Organisms. Progress in Biological Control.* Springer 19, pp. 103–132.

Azevedo, L.H. de, Castilho, R.C., Berto, M.M. and Moraes, G.J. (2017) Macrochelid mites (Mesostigmata: Macrochelidae) from Sao Paulo state, Brazil, with description of a new species of *Macrocheles*. *Zootaxa* 4269, 413–426.

Azevedo, L.H. de, Ferreira, M.P., de Campos Castilho, R., Duarte Cançado, P.H. and Moraes, G.J. (2018) Potential of *Macrocheles* species (Acari: Mesostigmata: Macrochelidae) as control agents of harmful flies (Diptera) and biology of *Macrocheles embersoni* Azevedo, Castilho and Berto on *Stomoxys calcitrans* (L.) and *Musca domestica* L. (Diptera: Muscidae). *Biological Control* 123(24), 1–8.

Azevedo, L.H., Leite, L.G., Chacon-Orozco, J.G., Moreira, M.F.P., Ferreira, M.P. *et al.* (2019) Free living nematodes as alternative prey for soil predatory mites: An interdisciplinary case study of conservation biological control. *Biological Control* 132, 128–134.

Azevedo, L.H. de, Borges, V., Mesquita Filho, W., de Campos Castilho, R. and Moraes, G.J. (2022) Semi-field evaluation of the predation of *Macrocheles embersoni* and *Macrocheles muscaedomesticae* (Acari: Mesostigmata: Macrochelidae) on the house fly and the stable fly (Diptera: Muscidae). *Pest Management Science* 78(3), 1029–1034.

Azza, A.M., Seham, A.E.E.-D. and Tawfik, W.A. (2019) Survey of main mites and insect species associated with stored maize in giza governorate. *Journal of Plant Protection and Pathology, Mansoura University* 10(5), 257–260.

Bachmann, J. and Hinman, T. (2008) Garlic: Organic Production. Available at: https://attra.ncat.org/garlic-organic-production-page/ (accessed 17 April 2023).

Bader, C. (1989) Milben Acari: Eine wenig bekannte, vielgestaltige, farbenprächtige, medizinisch, ökonomisch, biologisch wichtige Tiergruppe. *Veröffentlichungen aus dem Naturhistorischen Museum, Basel, Switzerland* 22, 1–47.

Badieritakis, E.G., Thanopoulos, R.C. and Emmanouel, N.G. (2012) Mite fauna in foliage and litter of Medicago species in Greece. *International Journal of Acarology* 38(8), 681–691.

Badieritakis, E.G., Fantinou, A.A. and Emmanouel, N.G. (2014) A qualitative and quantitative study of mites in similar alfalfa fields in Greece. *Experimental and Applied Acarology* 62(2), 195–214.

Baggio, D., Figueiredo, S.M., Flechtmann, C.H.W., Zambon, G.Q. and Miranda, S.H.G. (1987) Avaliação da presença de ácaros em cereais armazenados na grande São Paulo. *Anais da Escola Superior de Agricultura Luiz de Queiroz* 44, 617–626.

Bahrami, F., Arbabi, M., Vafaei Shoushtari, R. and Kazemi, S. H. (2011) Mesostigmatic mites associated with coleoptera and biodiversity calculation of these mites phoretic on dung beetles in Golestan Province (North of Iran). *Middle-East Journal of Scientific Research* 9(3), 345–366.

Bahrami, F., Fathipour, Y. and Kamali, K. (2012) Life history and population growth parameters of the astigmatidmite *Histiostoma feroniarum* Acari: Histiostomatidae) feeding on *Fusarium graminearum*. *Persian Journal of Acarology* 1(2), 147–156.

Baier, B. (1992) A comparative study of the biology and ecology of *Amblyseius barkeri* and *Neoseiulus cucumeris*. *Bulletin OEPP* 22(3), 429–436.

Bajerlein, D., Błoszyk, J., Gwiazdowicz, D.J., Ptaszyk, J. and Halliday, B. (2006) Community structure and dispersal of mites (Acari, Mesostigmata) in nests of the white stork (*Ciconia ciconia*). *Biologia* 61(5), 525–530.

Baker, E.W. (1949) A review of the mites of the family Cheyletidae in the United States National, Museum. *Proceedings of the United States National Museum* 99(3238), 267–320.

Baker, E.W. (1958) *Chelacheles strabismus*, a new genus and species of mite from Portugal. *Proceedings of the Entomological Society of Washington* 60, 234–235.

Baker, R.A. (1964a) Some preliminary observations on the mite faunaof sewage bacteria beds. *Annals and Magazine of Natural History* 13(7), 707–709.

Baker, R.A. (1964b) The further development of the hypopus of *Histiostoma feroniarum* (Dufour, 1839) (Acari). *Annals and Magazine of Natural History* 13(7), 693–695.

Baker, E.W. (1965) A review of the genera of the family Tydeidae (Acarina). *Advances in Acarology* 2, 95–133.

Baker, E.W. (1968a) The genus Lorryia. *Annals of the Entomological Society of America* 61(4):987–1008.

Baker, E.W. (1968b) Two New Genera of Tydeidae (Acarina). *Annals of the Entomological Society of America* 61(4), 968–970.

Baker, E.W. (1970) The genus *Tydeus*: Subgenera and species groups with descriptions of new species (Acarina: Tydeidae). *Annals of the Entomological Society of America* 63, 163–177.

Baker, G.T. (1983) Observations on the biology of *Rhizoglyphus robini* Claparède (Acaridae). *Zeitschrift für Angewandte Entomologie* 95, 162–166.

Baker, G.H. (1985) Parasites of the millipede *Ommatoiulus moreleti* (Lucas) (Diplopoda: Iulidae) in Portugal, and their potential as biological control agents in Australia. *Australian Journal of Zoology* 33, 23–32.

Baker, A.S. (1990) Two new species of *Lardoglyphus* Oudemans (Acari: Lardoglyphidae) found in the gut contents of human mummies. *Journal of Stored Products Research* 26(3), 139–147.

Baker, D.G. (ed.) (2008) *Flynn's Parasites of Laboratory Animals*. 2nd edn. Wiley-Blackwell, USA, pp. 1–840.

Baker, E.W. and Balock, J.W. (1944) Mites of the family Bdellidae. *Proceedings of the Entomological Society of Washington* 46(7), 176–184.

Baker, A.S. and Craven, J.C. (2003) Checklist of the mites (Arachnida: Acari) associated with bats (Mammalia: Chiroptera) in the British Isles. *Systematic and Applied Acarology, Special Publications* 14, 1–20.

Baker, E.W. and Delfinado, M.D. (1978) Notes on driedfruit mite *Carpoglyphus lactis* (Acarina: Carpoglyphidae) infesting honeybee combs. *Journal of Apicultural Research* 17(1), 52–54.

Baker, E.W. and Hoffmann, A. (1948) Acaros de la familia Cunaxidae. *Anales de la Escuela Nacional de Ciencias Biologicas* 5(3–4), 229–273.

Baker, G.T. and Krantz, G.W. (1985) Structure of the male and female reproductive and digestive systems of *Rhizoglyphus robini* Claparède (Acari, Acaridae). *Acarologia* 26, 55–65.

Baker, A.S. and Lindquist, E.E. (2002) *Aethosolenia laselvensis* gen.nov., sp.nov., a new eupodid mite from Costa Rica (Acari: Prostigmata). *Systematic and Applied Acarology, Special Publication* 11, 1–11.

Baker, E.W. and Wharton, G.W. (1952) *An Introduction to Acarology*. McMillan, New York, pp. 1–465.

Baker, E.W., Evans, T.M., Gould, D.J., Hull, W.B. and Keegan, H.L. (1956) *A Manual of Parasitic Mites of Medical and Economic Importance*. National Pest Control Association, New York, pp. 1–170.

Bakr, A.A. and Selim, S. (2019) Selective biorational treatments for managing the storage mites, *Tyrophagus putrescentiae* (Schrank) and *Aleuroglyphus ovatus* (Troupeau) under laboratory conditions. *Systematic and Applied Acarology* 24(3), 337–347.

Bal, D.A. and Özkan, M. (2007) Some biological and ecological remarks on *Trichouropoda orbicularis*, a pest of harvested maize grains and a larviparous uropodid mite (Acari: Mesostigmata: Uropodina). *Munis Entomology & Zoology* 2(1), 119–128.

Bale, J.S. and Hayward, S.A.L. (2010) Insect overwintering in a changing climate. *Journal of Experimental Biology* 213(6), 980–994.

Ball, S.G. and Morell, M.K. (2003) From bacterial glycogen to starch: understanding the biogenesis of the plant starch granule. *Annual Review of Plant Biology* 54, 207–233.

Balmes-Pacia, J.S. and Corpus-Raros, L.A. (1998) Biology of *Suidasia pontifica* Oudemans (Acari: Acaridida: Suidasiidae). *Philippine Entomologist* 12, 137–153.

Balogh, J. (1972) *The Oribatid Genera of the World*. Akademiai Kiado, Budapest, Hungary, pp. 1–188.

Balogh, J. and Balogh, P. (1992) *The Oribatid Mites Genera of the World, 1*. Hungarian Natural History Museum, Budapest, Hungary, pp. 1–263.

Banaszak, J. (1980) Badania nad fauna towarzyszaca w zasiedlonych ulach pszczelich (Investigations of the fauna associated in bee-hives). *Fragmenta faunistica* 25, 127–177 (in Polish).

Bandyopadhyay, P., Karmakar, K. and Halliday, B. (2023) Checklist of Indian mites in the family Laelapidae (Acari: Mesostigmata). *Zootaxa* 5249 (4), 401–424.

Bánki, O., Roskov, Y., Döring, M., Ower, G., Vandepitte, L. *et al.* (2022) Catalogue of Life Checklist (Version 2022-07-12). *Catalogue of Life*. DOI: 10.48580/dfpz.

Banks, N. (1906) A revision of the tyroglyphidae of the United States. *U.S. Department of Agriculture Technical Series* 13, 1–33.

Banks, N. (1912) New American mites. *Proceedings of the Entomological Society of Washington* 14, 96–99, pls. I–II.

Baqueiro, T., Martins Carvalho, F., Freitas Rios, C., dos Santos, N.M., and Alcântara-Neves, N.M. (2006) Dust mite species and allergen concentrations in beds of individuals belonging to different urban socioeconomic groups in Brazil. *Journal of Asthma* 43(2), 101–105.

Baratawidjaja, I.R., Baratawidjaja, P.P., Darwis, A., Yi, F.C., Chew, F.T., Lee, B.W. and Baratawidjaja, K.G. (1998) Mites in Jakarta homes. *Allergy* 53, 1226–1227.

Baran, S. and Ayyildiz, N. (2004) *Oppia nitens* C.L. Koch, 1836, a New Species for the Turkish Fauna (Acari, Oribatida, Oppiidae). *Turkish Journal of Zoology* 28, 111–113.

Baran, A.I., Jahanghiri, F., Hajipour, N., Sparagano, O.A.E., Norouzi, R.N. and Moharramnejad, S. (2020) In vitro acaricidal activity of essential oil and alcoholic extract of *Trachyspermum ammi* against *Dermanyssus gallinae*. *Veterinary Parasitology* 278, 109030.

Barbosa, M.F.C. (2011) Seleção de ácaros da ordem Astigmata (Acari) para uso como alimento na criação de ácaros predatores fitoseídeos (Acari: Phytoseiidae). Dissertação apresentada para obtenção do título de Mestre em Ciências. Área de concentração: Entomologia, Universitade de São Paulo, Escola superior de Agricultura "Luis de Queiroz", Piracicaba, Brazil, pp. 1–57.

Barbosa, M.F.C. and Moraes, G.J. (2015) Evaluation of astigmatid mites as factitious food for rearing four predaceous phytoseiid mites (Acari: Astigmatina; Phytoseiidae). *Biological Control* 91, 22–26.

Barbosa, M.F.C. and Moraes, G.J. (2020) *Rhizoglyphus* mites (Acari: Astigmata: Acaridae) from Brazil, with complementary description of *Rhizoglyphus vicantus* Manson. *Systematic and Applied Acarology* 25(2), 360–378.

Barbosa, M.F.C. and Moraes, G.J. (2021) Mites of the family Winterschmidtiidae (Acari: Sarcoptiformes: Astigmatina) from agricultural habitats in Brazil, with description of a new species and a key to species reported. *Systematic and Applied Acarology* 26(6), 1040–1054.

Barbosa, M.F. C., O'Connor, B.M. and Moraes, G.J. (2016) A new species of *Thyreophagus* (Acari: Acaridae) from Brazil, with notes on species associated with stored food and human habitats and a key to species of this genus. *Zootaxa* 4088, 279–291.

Barbosa, M.F.C., Flechtmann, C.H.W. and Moraes, G.J. (2021) Familia pyemotidae amerling. In: Barros-Battesti, D.M., Machado, R.Z. and André, M.R. (eds) *Ectoparasitofauna brasileira de importância veterinária, 1, Acarofauna de importância veterinária: Acariformes, ed. Jaboticabal, SP, CBPV*, pp. 84–96.

Barbosa, M.F.C., Corrêa, A. and Moraes, G.J. (2022) *Acarus siro* L. (Astigmatina: Acaridae): First confirmed record for Brazil based on morphological and molecular characterization. *Journal of Stored Products Research* 97, 1–6.

Barker, P.S. (1967a) Effect of humidity and temperature on the biology of *Aëroglyphus robustus* (Banka) (Acarina: Glycyphagidae). *Canadian Journal of Zoology* 45(4), 479–483.

Barker, P.S. (1967b) The effects of high humidity and different temperatures on the biology of *Tyrophagus putrescentiae* (Schrank) (Acarina: Tyrogliphidae). *Canadian Journal of Zoology* 45, 91–96.

Barker, P.S. (1967c) Bionomics of *Blattisocius keegani* (Fox) (Acarina: Ascidae), a predator on eggs of pests of stored grain. *Canadian Journal of Zoology* 45, 1093–1099.

Barker, P.S. (1968a) Bionomics of *Androlaelaps casalis* (Berlese) (Acarina: Laelapidae) a predator of mite pests of stored cereals. *Canadian Journal of Zoology* 46(6), 1099–1102.

Barker, P.S. (1968b) Bionomics of *Glycyphagus domesticus* (De Geer) (Acarina: Glycyphagidae), a pest of stored grain. *Canadian Journal of Zoology* 46(1), 89–92.

Barker, P.S. (1968c) Note on the bionomics of *Haemogamasus pontiger* (Berlese) (Acarina: Mesostigmata), a predator on *Glycyphagus domesticus* (De Geer). *Manitoba Entomology* 2, 85–87.

Barker, P.S. (1969) The response of a predator, *Hypoaspis aculeifer* (Canestrini) (Acarina: Laelapidae), to two species of prey. *Canadian Journal of Zoology* 47(3), 343–345.

Barker, P.S. (1974) Bionomics of *Caloglyphus anomalus* Nesbitt (Acarina: Acaridae). *Manitoba Entomology* 8, 41–47.

Barker, P.S. (1983) Bionomics of *Lepidoglyphus destructor* (Schrank) (Acarina: Glycyphagidae), a pest of stored cereals. *Canadian Journal of Zoology* 61(2), 355–358.

Barker, P.S. (1991) Bionomics of *Cheyletus eruditus* (Schrank) (Acarina: Cheyletidae), a predator of *Lepidoglyphus destructor* (Schrank) (Acarina: Glycyphagidae), at three constant temperatures. *Canadian Journal of Zoology* 69, 2321–2325.

Barker, P.S. (1992) Bionomics of *Nodele calamondin* Muma (Acarina: Cheyletidae) fed on *Lepidoglyphus destructor* (Schrank) (Acarina: Glycyphagidae), at two constant temperatures. *Canadian Journal of Zoology* 70, 2333–2337.

Barker, P.S. (1993) Phoretic mites found on beetles associated with stored grain in Manitoba. *Canadian Entomologist* 125 (4), 715–719.

Barilo, A.B. (1985) A new species of the genus *Neoeucheyla* Radford (Cheyletidae: Prostigmata), from Southern region of Uzbekistan. *Uzbekskij biologicheskij zhurnal*, 45–47.

Barquero, M. (2015) Estados Unidos amenaza con limitar ingreso de piña y chayote. *La Nacion* 17 June 2015.

Barrera, O. M., Murgas, I.L., Bermudez, S. and Miranda, R.J. (2015) Oral anaphylaxis by ingestion of mite contaminated food in Panama City, 2011–2014. *Revista Alergia México* 62, 112–117.

Bartley, K., Huntley, J.F., Wright, H.W., Nath, M. and Nisbet, A.J. (2012) Assessment of cathepsin D and L-like proteinases of poultry red mite, *Dermanyssus gallinae* (De Geer), as potential vaccine antigens. *Parasitology* 139, 755–765.

Bartley, K., Wright, H.W., Huntley, J.F., Manson, E.D., Inglis, N.F. *et al.* (2015) Identification and evaluation of vaccine candidate antigens from the poultry red mite (*Dermanyssus gallinae*). *International Journal of Parasitology* 45(13), 819–830.

Bartley, K., Turnbull, F., Wright, H.W., Huntley, J.F., Palarea-Albaladejo, J., Nath, M. and Nisbet, A.J. (2017) Field evaluation of poultry red mite (*Dermanyssus gallinae*) native and recombinant prototype vaccines. *Veterinary Parasitology* 244, 25–34.

Barzman, M., Bàrberi, P., Birch, A.N.E., Boonekamp, P.M., Dachbrodt-Saaydeh, S. *et al.* (2015) Eight principles of integrated pest management. *Agronomy for Sustainable Development* 35(4), 1199–1215.

Bashir, M.H., Mahmood, S.U., Khan, M.A., Afzal, M. and Zia, K. (2013) Estimation of nutritional losses caused by *Rhizoglyphus tritici* (Acari: Acaridae) in stored wheat. *Pakistan Journal of Agricultural Science* 50(4), 631–635.

Bashir, M.H., Afzal, M., Ashfaq, M., Ali, A., Kamran, M. and Honey, S.F. (2014) Subfamily Coleoscirinae (Acari: Trombidiformes: Cunaxidae), with description of one new species from Pakistan. *Journal of Insect Science* 14, 1–14.

Bassani-Silva, R., Castro-Santiago, A.C., Calchi, A.C., Perles, L., Takatsu, J.C. *et al.* (2022) Sleeping with the enemy: Case reports of *Ornithonyssus bursa* (Berlese, 1888) (Mesostigmata: Macronyssidae) causing human dermatitis in Brazil. *Parasitology Research* 121(9), 2641–2649.

Baxter, I., Midthassel, A., Stepman, W., Fryer, R., Garcia, F.P. *et al.* (2011) Field results of a sachet release system using the predator *Amblyseius swirskii* (Athias Henriot) (Acari: Phytoseiidae) and the factitious prey, *Suidasia medanensis* Oudemans (Acari: Astigmata). *IOBC/WPRS Bulletin* 68, 1–4.

Bayartogtokh, B. (2010) *Oribatid mites of Mongolia (Acari: Oribatida)*. KMK Scientific Press, Moscow, Russia, pp. 1–372 (in Russian).

Bayartogtokh, B. and Schatz, H. (2008) *Trichoribates* and *Jugatala* (Acari: Oribatida: Ceratozetidae) from the Central and Southern Alps, with notes on their distribution. *Zootaxa* 1948, 1–35.

Bayartogtokh, B. and Ryabinin, N.A. (2012) The soil mite family Achipteriidae (Acari: Oribatida) in Mongolia and the Russian far east. *Acarologia* 52(2), 135–156.

Bayram, S. and Çobanoğlu, S. (2005) Mesostigmata (Acari) of bulbaceous ornamental plants in Turkey. *Acarologia* 45(4), 257–265.

Bayram, S. and Çobanoğlu, S. (2007) Mite fauna (Acari: Prostigmata, Mesostigmata, Astigmata) of coniferous plants in Turkey. *Türkiye Entomoloji Dergisi* 31, 279–290.

Beaulieu, F. (2009) Review of the mite genus *Gaeolaelaps* Evans & Till (Acari: Laelapidae), and description of a new species from North America, *G. gillespiei* n. sp. *Zootaxa* 2158, 33–49.

Beaulieu, F. and Knee, W. (2014) Plant-feeding mites of the Canadian prairies. In: Cárcamo, H.A. and Giberson, D.J. (eds) *Arthropods of Canadian Grasslands (Volume 3): Biodiversity and Systematics. Part 1.* Biological Survey of Canada, pp. 29–72.

Beaulieu, F. and Weeks, A.R. (2007) Free-living mesostigmatic mites in Australia: Their roles in biological control and bioindication. *Australian Journal of Experimental Agriculture* 47, 460–478.

Beaulieu, F., Walter, D.E., Proctor, H.C. and Kitching, R.L. (2010) The canopy starts at 0.5 m: Predatory mites (Acari: Mesostigmata) differ between rainforest floor soil and suspended soil at any height. *Biotropica* 42, 704–709 + supplementary pages 1–13.

Beaulieu, F., Dowling, A.P.G., Klompen, H., de Moraes, G.J. and Walter, D.E. (2011) Superorder Parasitiformes Reuter, 1909. In: Zhang, Z.-Q. (ed.) *Animal biodiversity: An outline of higher-level classification and survey of taxonomic richness. Zootaxa* 3148, 123–128.

Beavers, J.B., Denmark, H.A, and Sehime, A.G. (1972) Predation by *Blattisocius keegani* on egg masses of *Diaprepes abbreviatus* in the laboratory. *Journal of Economic Entomology* 65, 1483–1484.

Begon, M., Harper, J.L. and Townsend, C.R. (1986) Ecology. In: *Individuals, Populations and Communities.* Blackwell, Oxford, UK, pp. 1–876.

Behan-Pelletier, V.M. (1999) Oribatid mite fauna of northern ecosystems: A product of evolutionary adaptations or physiological restraints? In: Needham, G.R., Mitchell, R., Horn, D.J. and W.C. (eds) Acarology IX. Symposia, Vol. 2. Ohio Biological Survey, Columbus, USA, pp. 87–106.

Behan-Pelletier, V.M. (2000) Ceratozetidae (Acari: Oribatida) of arboreal habitats. *Canadian Entomologist* 132, 153–182.

Behura, B.K. (1957) The life-history of *Histiostoma polypori* (Oud.) (Acari: Tyroglyphoidea). *Journal of the New York Entomological Society* 65, 51–78.

Beklemishev, V.N. (1970) Principles of Comparative Anatomy of Invertebrates, vol. 1, 490 pp., vol. 2, 529 pp. (Russian editions 1951, 1964). Translated in English by Oliver and Boyd, Edinburgh, UK.

Bell, C.H. (2014) Pest control of stored food products: insects and mites, In: Lelieveld, H.L.M., Holah, J. and Napper, D. (eds) *Hygiene in Food Processing: Principles and Practice*, 2nd edn, Woodhead Publishing Series in Food Science, Technology and Nutrition, pp. 494–538.

Bellido, A. (1979) Postembryonic development of *Carabodes willmani* Bernini, 1975 (Acari, Oribatei). *Acarologia* 20(3), 419–432.

Bennison, J.A., Maulden, K. and Maher, H. (2002) Choice of Predatory Mites for Biological Control of Ground-Dwelling Stages of Western Flower Thrips Within a "Push-Pull" Strategy on Pot Chrysanthemum. *IOBC/WPRS Bulletin* 25, 9–12.

Bensignor, E. and Carlotti, D.N. (2002) Sensitivity patterns to house dust mites and forage mites in atopic dogs: 150 cases. *Veterinary Dermatology* 13, 37–42.

Benuzzi, M. and Vacante, V. (2021) Difesa fitosanitaria in ortofrutticoltura biologica – In ambiente mediterraneo, Edagricole, Bologna, 1–440.

Berenbaum, M. (1993) *Ninety-Nine More Maggots, Mites, and Munchers.* University of Illinois Press, USA, pp. 31–32.

Beresford, D.V. and Sutcliffe, J.F. (2009) The effect of *Macrocheles muscaedomesticae* and *M. subbadius* (Acarina: Macrochelidae) phoresy on the dispersal of *Stomoxys calcitrans* (Diptera: Muscidae). *Systematic and Applied Acarology, Special Publications* 23, 1–30.

Berger, I., Schierl, R., Ochmann, U., Egger, U., Scharrer, E. and Nowak, D. (2005) Concentrations of dust, allergens and endotoxin in stables, living rooms and mattresses from cattle farmers in Southern Bavaria. *Annals of Agricultural and Environmental Medicine* 12, 101–107.

Bergstrom, J. (1979) Morphology of fossil arthropods as a guide to phylogenetic relationships. In: Gupta, A.P. (ed.) *Arthropod Phylogeny.* Van Nostrand Reinhold, New York, pp. 1–56.

Berlese, A. (1881) Indagini sulle metamorfosi di alcuni Acari insetticoli. *Atti del Reale Istituto Veneto di Scienze, Lettere ed Arti, Series 5*, 8, 37–81.

Berlese, A. (1882–1893) Acari, Myriapoda et Scorpiones hucusque in Italia Reperta. Ordo Prostigmata (Trombidiidae). Patavii, Portici.

Berlese, A. (1884) Acari, Myriopoda et Scorpiones hucusque in Italia reperta. Patavii, Portici, fasc. 14, 1.

Berlese, A. (1887) Acari, Myriapoda et Scorpiones hucusque in Italia Reperta. Patavii, Portici, fasc. 38, 8, plates 1–10.

Berlese, A. (1888) Acari austro-Americani quos collegit Aloysius Balzan. Manipulus primus. Species novas circiter quinquaginta complectens. *Bollettino della Società Entomologica Italiana* 20, 171–222.

Berlese, A. (1889) Acari Italiani Myriapoda et Scorpiones hucusque in Italia reperta. *Redia* 14, 78–105.

Berlese, A. (1892) *Laelaps (Iphis) miles*. Acari Myriopoda et Scorpiones Hucusque in Italia Reperta. Patavii, fasc. 63, 9.

Berlese, A. (1897a) Acari. Miriapodi Scorpioni Italiani, Order Cryptostigmata, Firenze, 107, 156–157.

Berlese, A. (1897b) Acari, Miriapodi Scorpioni Italiani, Order Cryptostigmata, Firenze, 1, 104.

Berlese, A. (1898) Acari, Myriopoda et Scorpiones Hucusque in Italia Reperta. Patavii. Fasc. 92, no 3.

Berlese, A. (1903) Acari Nuovi I. *Redia* 1, 235–252.

Berlese, A. (1904) Acari nuovi. Manipulus Ilus. *Redia* 1, 258–280.

Berlese, A. (1910) Acari Nuovi. Manipulus V. *Redia* 6, 199–214.

Berlese, A. (1911) Acarorum species novae quindecim. *Redia. Giornale di entomologia, Firenze* 7, 429–435.

Berlese, A. (1913) Acarotheca Italica, Ricci, Firenze, 1–221.

Berlese, A. (1916) Centuria prima di Acari nuovi. *Redia*12, 19–67.

Berlese, A. (1918) Centuria quarta di Acari nuovi. *Redia* 13, 115–192.

Bernini, F. (1995) Acariformes Oribatida. In Massa, B. (ed.) *Arthropoda di Lampedusa, Linosa e Pantelleria (Canale di Sicilia, Mar Mediterraneo)*. Naturalista Siciliano, Tipografia Luxograph, Palermo 19, 63–85.

Bernini, F., Castagnoli, M. and Nannelli, R. (1995) Arachnida Acari. In: Minelli A., Ruffo, S. and La Posta, S. (eds) *Checklist Delle Specie Della Fauna Italiana*, 24. Calderini, Bologna, pp. 1–131.

Beron, P. (2020) Acarorum Catalogus VII. Trombidiformes, Prostigmata, Raphignathoidea (Fam. Barbutiidae, Caligonellidae, Camerobiidae, Cryptognathidae, Dasythyreidae, Dytiscacaridae, Eupalopsellidae, Homocaligidae, Mecognathidae, Raphignathidae, Stigmaeidae, Xenocaligone llidae). Pensoft & National Museum of Natural History, Bulgarian Academy of Sciences, Sofia, pp. 1–306.

Beron, P. (2021a) Acarorum Catalogous IX. Trombidiformes Prostigmata Superfamilia Cheyletoidea (Cheyletidae, Psorergatidae, Demodecidae, Harpyrhynchidae, Syringophilidae) Superfamilia Cloacaroidea (Cloacaridae, Epimyodicidae). Pensoft & National Museum of Natural History, Bulgarian Academy of Sciences Sofia, pp. 1–468.

Beron, P. (2021b) Acarorum Catalogous IX. Acariformes, Acaridida, Schizoglyphoidea (Schizoglyphidae), Histiostomatoidea (Histiostomatidae, Guanolichidae), Canestrinioidea (Canestriniidae, Chetochelacaridae, Lophonotacaridae, Heterocoptidae), Hemisarcoptoidea (Chaetodactylidae, Hyadesiidae, Algophagidae, Hemisarcoptidae, Carpoglyphidae, Winterschmidtiidae). Advanced Books. Available at: https://doi.org/10.3897/ab.e68613.

Bertrand, M., Kukushkin, O. and Pogrebnyak, S. (2013) A new species of mites of the genus *Geckobia* (Prostigmata, Pterygosomatidae), parasitic on *Mediodactylus kotschyi* (Reptilia, Gekkota) from Crimea. *Vestnik Zoologii* 47(2), 1–13.

Beute, M.K. and Benson, D.M. (1979) Relation of small soil fauna to plant disease. *Annual Review of Phytopathology* 17, 485–502.

Beyzavi, G., Ueckermann, E.A., Faraj, F. and Ostovan, H. (2013) A catalog of Iranian prostigmatic mites of superfamilies Raphignathoidea and Tetranychoidea (Acari). *Persian Journal of Acarology* 2(3), 389–474.

Bhowmick, B. and Han, Q. (2020) Understanding tick biology and its implications in anti-tick and transmission blocking vaccines against tick-borne pathogens. *Frontiers in Veterinary Science* 7, 319.

Bhuiya, B.A. and Khawja N.H. (1983) A preliminary study on some pests of dry fishes, prawns and shrimps from Chittagong (Bangladesh). *Chittagong University studies Part. II Sei* 7(1), 85–90.

Bibars, E.A.M., Yassin, E.M.A. and Abdel Khalik A.R. (2018) Survey of different mites and insect pests associated with date palm fruits in different locations of Egypt. *Egyptian Journal of Agricultural Research* 96(3), 909–919.

Billingsley, P.F., Foy, B. and Rasgon, J.L. (2008) Mosquitocidal vaccines: A neglected addition to malaria and dengue control strategies. *Trends in Parasitology* 24, 396–400.

Bingham, G.V. and Hagstrum, D.W. (2024) Importance of sanitation for stored-product pest management. *Insects* 15, 3. DOI: 10.3390/insects15010003.

Binnington, K.C. and Obenchain, F.D. (1982) Structure and function of the circulatory, nervous, and neuroendocrine system of ticks. In: Obenchain, F.D. and Galun, R. (eds) *Physiology of Ticks*. Pergamon Press, Oxford, UK, pp. 351–398.

Binns, E.S. (1982) Phoresy as migration – some functional aspects of phoresy in mites. *Biological Reviews* 57, 571–620.

Birkett, M., Pickett, J., Dewhirst, S., Jespersen, J.B., Steenberg, T. and Kilpinen, O. (2010) Composition Comprising a Volatile Carbocyclic Acid or an Aldehyde and Its Use as Attractant for Mites. Slagelse: Patent no WO2010130990A8. Rothamsted Research Limited, Aarhus Universitet.

Birkett, M.A., Hassanali, A., Hoglund, S., Pettersson, J. and Pickett, J.A. (2011) Repellent activity of catmint, nepeta cataria, and iridoid nepetalactone isomers against afro-tropical mosquitoes, ixodid ticks and red poultry mites. *Phytochemistry* 72, 109–114.

Bischoff, E.R.C., Fischer, A. and Liebenberg, B. (1992) Assessment of mite numbers: New methods and results. *Experimental and Applied Acarology* 16, 1–14.

Bischoff, E.R., Fischer, A., Liebenberg, B. and Kniest, F.M. (1998) Mite control with low temperature washing – II. Elimination of living mites on clothing. *Clinical & Experimental Allergy* 28(1), 60–65.

Blainey, A.D., Topping, M.D., Ollier, S. and Davies, R.J. (1988) Respiratory symptoms in arable farmworkers: Role of storage mites. *Thorax* 43, 697–702.

Blanco, C., Quiralte, J., Castilho, R., Delgado, J., Artega, C., Baber, D. and Carrilho, T. (1997) Anaphylaxis after ingestion of wheat flour contaminated with mites. *Journal of Allergy and Clinical Immunology* 99, 308–313.

Blank, R.H. and Osborne, G.O. (1978) Comparison of the toxicity of organophosphorus acaricides to the mites *Sancassania berlesei* (Tyroglyphidae) and *Tetranychus urticae* (Tetranychidae). *New Zealand Journal of Agricultural Research* 21(3), 527–531.

Blasco Sabio, C., Berenguer, G. and Portus Vinyeta, M. (1975) Estudio de la acarofauna del polvo domestico de Barcelona y poblaciones circundantes. *Allergologia et Immunopathologia* 3, 403–418.

Blasco, C. and Portus, M. (1975) Acarofauna. *Revista de la Real Academia de Farmacia Barcelona* 11, 37–59.

Blattný, C. (1924) Roztoči v úlech a jejich poměr κ Wightské nemoci. *Včelařské rozhledy: měsíčník pro československé včelaře* 2, 90–91 (in Czechoslovak).

Błoszyk, B. (1999) Geograficzne i ekologiczne zr nicowanie zgrupowa roztoczy z kohorty Uropodina (Acari: Mesostigmata) w Polsce. Kontekst Publishing, Poznan, Poland, pp. 1–245 (in Polsk).

Błoszyk, J., Gwiazdowicz, D.J., Bajerlein, D. and Halliday, R.B. (2005) Nests of the white stork *Ciconia ciconia* (L.) as a habitat for mesostigmatic mites (Acari, Mesostigmata). *Acta Parasitologica* 50(2), 171–175.

Błoszyk, J., Climczak, J., Leśniewska, M. (2006) Phoretic relationships between Uropodina (Acari: Mesostigmata) and centipedes (Chilopoda) as an example of evolutionary adaptation of mites to temporary microhabitats. *European Journal of Entomology* 103(3), 699–707.

Błoszyk, J., Adamskic, Z. and Napierałaa, A. (2018) Survey of European mites from the suborder Uropodina: II. Morphology, geographical distribution, biology, and ecology of *Trematurella elegans* (Kramer, 1882). *Acarologia* 58(3), 683–709.

Błoszyk, J., Rutkowski, T., Wojtaszyn, G., Książkiewicz-Parulska, Z., Zacharyasiewicz, M. and Napierała, A. (2020) *Leiodinychus orbicularis* (C.L. Koch, 1839) in bat boxes in Poland. *European Journal of Biological Research* 10(2), 150–155.

Blythe, M.E. (1976) Some aspects of the ecological study of house dust mites. *British Journal of Diseases of the Chest* 70, 3–31.

Bochkov, A.V. (2002) The classification and phylogeny of the mite superfamily Cheyletoidea (Acari, Prostigmata). *Entomol. Obzr. (Entomological Review)* 81, 488–513 (in Russian).

Bochkov, A.V. (2004) Mites of the family Cheyletidae (Acari: Prostigmata): phylogeny, distribution, evolution and analysis of host-parasite relationships. *Parazitologiya* 38, 122–138 (in Russian).

Bochkov, A.V. and Abramov, V.V. (2016) To the fauna of the free-living Cheyletidae (Acariformes: Cheyletoidea) of the European part of Russia. *Systematic and Applied Acarology* 21(3), 335–346.

Bochkov, A.V. and Fain, A. (2001) Phylogeny and system of Cheyletidae (Acari: Prostigmata) with special reference to their host-parasite associations. *Bulletin de l'Institut royal des Sciences naturelles de Belgique* 71, 5–36.

Bochkov, A. and Miranov, S.V. (1997) On a taxonomy of predatory mites of the genus *Neoeucheyla* Radford, 1950 and related genera (Acari: Cheyletidae). *Acarina* 5(1–2), 73–78.

Bochkov, A.V., Fain, A. and Ardeshir, F. (2001) Redescription of *Nodele calamondin* Muma, 1964 (Acari Cheyletidae). *Bulletin de l'Institut royal des Sciences naturelles de Belgique* 37, 123–126.

Boctor, F.N. (1972) Biochemical and physiological studies of certain ticks (Ixodoidea). Free amino acid in female *Argas (Persicargas) arboreus* Kaiser, Hoogstraal and Kohls (Argasisae) analyzed by gas-liquid chromatography. *Journal of Medical Entomology* 9, 201–204.

Boczek, J. (1959) Biologia i ekologia sierposza rozkruszkowca (*Cheyletus eruditus* Schrank 1781 – Acarina, Cheyletidae). *Prace Naukowe Instytutu Ochrony Roślin* 1, 175–230.

Boczek, J. (1961) Investigations on the distribution of mites of importance to stored products under field conditions. *Prace Naukowe Instytutu Ochrony Roślin* 3 (pt. 1), 81–99.

Boczek, J. (1973) Procesy starzenia sie u owadow i roztoczy. *Kosmos* 6, 611–621.

Boczek, J. (1975) Reproduction biology of *Tyrophagus putrescentiae* (Schr.) (Acarina: Acaridae). In: Brady, E.U., Brower, J.H., Hunter, P.E., Jay, E.G., Lum, P.T.M., Lund, H.O., Mullen, M.A. and Davis, R. (eds) *Proceedings of the First International Working Conference on Stored-Product Entomology*, 7–11 October 1974, Savannah, Georgia, pp. 154–159.

Boczek, J. (1980) Zarys akarologii rolniczej. PWN (Polish Scientific Publisher), Warszawa, pp. 1–356.

Boczek, J. (1991) Mite pests in stored food. In: Gorham, J.R. (ed.) *Ecology and Management of Food-Industry Pest.* Association of Official Analytical Chemists. Arlington, VA, pp. 55–79.

Boczek, J. (1999) Zarys akarologii rolniczej. Wydawnictwo Naukowe PWN, Warszawa, pp. 1–388.

Boczek, J. and Blaszak, C. (2005) Mites (Acari). Significance in life and economy of humans. SGGW (Szkoła Główna Gospodarstwa Wiejskiego w Warszawie), Warszawa, pp. 1–267.

Boczek, J. and Czajkowska, B. (1973) Some aspects of aging in *Acarus siro* (L.) (Acarina: Acaridae). *Ekol. Polska* 21, 173–184.

Boczek, J. and Dutkiewicz, J. (1972) Roztocze i owady w pyłach przyczyną alergicznych schorzeń układu oddechowego. *Medyc. Wiejska* 7, 157–165.

Boczek, J. and Gołębiowska, Z. (1959) Investigations on the distribution of mites connected with the storage of food products in Poland. *Roczniki Nauk Rolniczych, A* 79, 969–988 (in Polish).

Boczek, J. and Sosnowska, B.M. (1975) Studies on aging in acarid mites, II. *Zeszyty Problemowe Postepow Nauk Rolniczych* 171, 169–177.

Boczek, J., Gołebiowska, Z. and Krzeczkowski, K. (1960) Roztocze szkodliwe w przechowalniach siemienia lnianego i konopi w Polsce. *Prace Naukowe Instytutu Ochrony Roślin* 2, 57–86.

Boczek, J., Filipek, P., Gołebiowska, Z. and Krzeczkowski, K. (1961) Characteristic of the fauna of stored herbs in Poland. *Prace Naukowe Instytutu Ochrony Roślin* 3, 5–64 (in Polish).

Boczek, J., Jura, C. and Krzysztofowiez, A. (1969) The comparison of the structure of the internal organs of post-embryonic stages of *Acarus farris* (Oud.) with special reference to the hypopus. In: Evans, G.O. (ed.) *Proceedings of the 2nd International Congress of Acarology, Sutton Bonington, England, 19th–25th July, 1967*, Akad. Kaidó, Budapest, Hungary, pp. 265–271.

Boczek, J., Jaminska, D., Bruce, W.A. and Davis, R (1985) Effect of gamma radiation on *Rhizoglyphus echinopus* (Fumouze and Robin) and *Aleuroglyphus ovatus* (Troupeau) (Acari: Acaridae). *Journal of Entomological Science* 20(1), 115–120.

Boczek, J.H. and Davis, R. (1985) Effects of alternating temperatures on *Acarus siro* L. (Acari: Acaridae). *Experimental and Applied Acarology* 1, 213–217.

Bogdanov, A. (1864) Deux acariens trouves par Mr. Scheremeteswky sur l'homme. *Bulletin de la Société impériale des naturalistes de Moscou* 37, 341–345.

Bolger, T., Arroyo, J. and Piotrowska, K. (2018) A catalogue of the species of Mesostigmata (Arachnida, Acari, Parasitiformes) recorded from Ireland including information on their geographical distribution and habitats. *Zootaxa* 4519 (1), 1–220.

Bolckmans, K.J.F., van Houten, Y.M., van Baal, A.E. and Stam, A.T. (2013) Phytoseiid predatory mite releasing system and method for production. World Patent WO/2013/043,050. Koppert B.V.

Bollaerts, D. and Breny, R. (1951) Les acariens nuisibles aux matières entreposées. *Agricultural Reviews (Brussels)* 4, 738–764.

Bolland, H.R. (1983) A description of *Neophyllobius aesculi* n. sp. and its developmental stages (Acari: Camerobiidae). *Entomologische Berichten* 43(3), 42–47.

Bolland, H.R. (1986) Review of the systematics of the family Camerobiidae (Acari, Raphignathoidea). I. The genera *Camerobia*, *Decaphyllobius*, *Tillandsobius*, and *Tycherobius*. *Tijdschrift voor entomologie* 129(7), 191–215.

Bolland, H.R., Gutierrez, J. and Flechtmann, C.H.W. (1998) *World Catalogue of the Spider Mite Family (Acari: Tetranychidae)*. Brill, Leiden, The Netherlands, pp. 1–392.

Bond, E.J. (1984) *Manual of Fumigation for Insect Control*. Food and Agriculture Organization, Rome.

Bonde, J. (1989) Biological studies including population growth parameters of the predatory mite *Amblyseius barkeri* (Acarina: Phytoseiidae) at 25°C in the laboratory. *Entomophaga* 34, 275–287.

Bonanomi, G., Cesarano, G., Antignani, V., Di Maio, C., De Filippis, F. and Scala, F. (2018) Conventional farming impairs *Rhizoctonia solani* disease suppression by disrupting soil food web. *Journal of Phytopathology* 166, 663–673.

Boner, A., Richelli, C. and Piacentini, G. (1989) Skin and serum reactivity to some storage mites in children sensitive to *Dermatophagoides pteronyssinus*. *Annals of Allergy* 63, 82–84.

Boness, M. (1968) Uber Milbenauftreten in Pilzkulturen. *Anz. Schädl* 51, 41–42.

Bongers, M.G.H., OConnor, B.M. and Lukoschus, F.S. (1985) Morphology and ontogeny of histiostomatid mites (Acari: Astigmata) associated with cattle dung in The Netherlands. *Zoologische Mededeelingen* 223, 1–56.

Bonilla, G., Ochoa, R. and Aguilar, H. (1990) *Rhizoglyphus costarricensis* spec. nov. (Acari: Acaridae) asociado con el dano en la semilla de *Oryza sativa* L. en Costa Rica. *Turrialba* 40, 198–204.

Boquete, M., Carballada, F., Armisen, M., Nieto, A., Martin, S., Polo, F. and Carreira, J. (2000) Factors influencing the clinical picture and the differential sensitization to house dust mites and storage mites. *Journal of Investigational Allergology and Clinical Immunology* 10(4), 229–234.

Boquete, M., Carballás, C., Carballada, F., Iraolo, V., Carnés, J. and Fernández-Caldas, E. (2006) In vivo and in vitro allergenicity of the domestic mite *Chortoglyphus arcuatus*. *Annals of Allergy* 97(2), 203–208.

Borden, E.E.R. (1989) The phoretic behavior and olfactory preference of *Macrocheles muscaedomesticae* (Scopoli) (Acarina: Macrochelidae) in its relationship wirh *Fannia canicularis* (L.) (Diptera: Muscidae). *Pan-Pacific Entomologist* 65, 89–96.

Borges, P.A.V., Costa, A., Cunha, R., Gabriel, R., Gonçalves, V. *et al.* (eds) (2010) A list of the terrestrial and marine biota from the Azores. *Princípia, Cascais*, 1–432.

Boström, S., Johansson, E., Härfast, B., Lundqvist, L., Bäckman, I., von Rosen, E. and van Hage-Hamsten, M. (1997) Characterization of the mite fauna (Acari) in Swedish barn dust. *International Journal of Acarology* 23, 127–132.

Bot, J. and Meyer, M. (1961) An artificial rearing medium for Acarid mites. *Journal of the Entomological Society of Southern Africa* 29, 199.

Boudreaux, H.B. (1958) The effect of relative humidity on egg-laying, hatching, and survival in various spider mites. *Journal of Insect Physiology* 2, 65–72.

Boulanova, E. (1937) Notes morfologique et biologique sur *Ferminia fusca* Ouds. *Byulleten' Moskovskogo Obshchestva Ispytatelei Prirody, Otd. Biol. N.S.* 46(6) 304–316.

Bowman, C.E. (1984) Comparative enzymology of economically important astigmatid mites. In: Griffiths, D.A. and Bowman, C.E. (eds) *Acarology 6*. Ellis Horwood Publishing, Chichester, UK, pp. 993–1001.

Brady, J. (1970) The mites of poultry litter. Observations on the bionomics of common species with a species list for England and Wales. *Journal of Applied Ecology* 7, 331–348.

Braig, H.R. and Perotti, M.A. (2009) Carcasses and mites. *Experimental and Applied Acarology* 49, 45–84.

Brannstrom, S., Hansson, I. and Chirico, J. (2010) Experimental study on possible transmission of the bacterium *Erysipelothrix rhusiopathiae* to chickens by the poultry red mite, *Dermanyssus gallinae*. *Experimental and Applied Acarology* 50, 299–307.

Braude, R., Low, A.G., Mitchell, K.G., Pittman, R.J. and Wilkin, D.R. (1980) Effect of flour mite infestation (*Acarus siro* L.) on nutritive value of pig diets. *Veterinary Record* 106(2), 35–36.

Brauneis, M.D., Zoller, H., Williams, H., Zschiesche E. and Heckeroth, A.R. (2017) The acaricidal speed of kill of orally administered fluralaner against poultry red mites (*Dermanyssus gallinae*) on laying hens and its impact on mite reproduction. *Parasites & Vectors* 10, 594.

Brazis, P., Serra, M., Selles, A., Dethioux, F., Biourge, V. and Puigdemont, A. (2008) Evaluation of storage mite contamination of commercial dry dog food. *Veterinary Dermatology* 19(4), 209–214.

Bregetova, N.G. (1949) Contribution to the fauna of mites of the family Haemogamasidae in U.S.S.R. *Parazitologicheskii Sbornik* 11, 161–185 (in Russian).

Bregetova, N.G. (1956) Gamasoid mites. Opred. Faun. SSSR. 61, 1–247 (in Russian).

Bregetova, N.G. (1977) Family Aceosejidae. In: Gilarov, M.S. and Bregetova, N.G. (eds) Opredelitel' obytayshchikh v pochve kleshchey Mesostigmata (Identification key to soil-inhabiting mites Mesostigmata), Nauka, Leningrad, pp. 169–226 (in Russian).

Bregetova, N.G. and Koroleva, E.V. (1960) The macrochelid mites (Gamasoidea, Macrochelidae) in the USSR. *Parazitologicheskii Sbornik* 19, 32–154 (in Russian).

Brigham, R.J. (1998) Corrosive effect of phosphine, carbon dioxide, heat, and humidity on electronic equipment. Phase 1. Ottawa, Ontario: AAFC, Environ, Can. USDA.

Brisman, J., Lillienberg, L., Belin, L., Ahman, M. and Jarvholm, B. (2003) Sensitisation to occupational allergens in bakers' asthma and rhinitis: a case-referent study. *International Archives of Occupational and Environmental Health* 76(2), 167–170.

British Pest Control Association (2019) Mite monitoring and biology for pest management professionals. Technical PPC95 May 2019. Available at: https://bpca.org.uk/Test-News/mite-monitoring-and-biology-for-pest-management-professionals-/227505 (accessed 21 August 2020).

Britto, E.P.J., Lopes, P.C. and Moraes, G.J. (2012) *Blattisocius* (Acari, Blattisociidae) species from Brazil, with description of a new species, redescription of *Blattisocius keegani* and a key for the separation of the world species of the genus. *Zootaxa* 3479, 33–51.

Broce, A. B., Zurek, L., Kalisch, J.A., Brown, R., Keith, D.L. *et al.* (2006) *Pyemotes herfsi* (Acari: Pyemotidae), a mite new to North America as the cause of bite outbreaks. *Journal of Medical Entomology* 43, 610–613.

Broce, A.B. and Kalisch, J. (2007) Oak Leaf Itch Mite, Kansas State University, October 2007.

Brody, A.R. (1970) Observations on the fine structure of the developing cuticle of a soil mite *Oppia coloradensis* (Acarina: Cryptostigmata). *Acarologia* 12(2), 421–431.

Brody, A.R. and Wharton, G.W. (1970) *Dermatophagoides farinae*: ultrastructure of lateral opisthosomal dermal glands. *Transactions of the American Microscopical Society* 89(4), 499–513.

Brody, A.R., McGrath, J.C. and Wharton, G. (1972) *Dermatophagoides farinae*: The digestive system. *Journal of the New York Entomological Society* 80, 152–177.

Brody, A.R., McGrath, J.C. and Wharton, G.W. (1976) *Dermatophagoides farinae*: the supracoxal glands. *Journal of the New York Entomological Society* 84, 34–47.

Bronswijk, J.E.M.H. van (1972) Parasitic mites of Surinam. X. Mites and fungi associated with house floor dust. *Entomologische Berichten* 32, 162–164.

Bronswijk, J.E.M.H. van, Cock, de, A.W.A.M. and Oshima, S. (1973a) The genus *Blomia* Oudemans (Acari: Glycyphagidae). 2. Comparison of its species. *Acarologia* 15(3), 490–505.

Bronswijk, J.E.M.H. van, Cock, de, A.W.A.M. and Oshima, S. (1973b) The genus *Blomia* Oudemans (Acari: Glycyphagidae). I. Description of *Blomia tropicalis* sp.n. from house dust in tropical and sub-tropical regions. *Acarologia* 15(3), 477–489.

Bronswijk, van J.E.M.H. (1981) *House Dust Biology (For Allergists, Acarologists and Mycologists)*. NIB Publishers, Zoelmond, The Netherlands, pp. 1–316.

Bronswijk, van J.E.M.H. Van and Sinha, R.N. (1971) Pyroglyphid mites (Acari) and house dust allergy. *Journal of Allergy* 47, 31–51.

Brooks, D.R. (1985) Historical ecology: A new approach to studying the evolution of ecological associations. *Annals of the Missouri Botanical Garden* 72, 660–680.

Bruce, W.A. (1983) Mites as biological control agents of stored product pests. In: Hoy, M.A., Cunningham, J.L. and Knutson, L.V. (eds) Biological Control of Pests by Mites. Div. of Agriculture and Natural Resources, Univ. of California, Berkeley, Spec. Publ. 3304, pp. 74–78.

Bruce, W.A. (1984) Temperature and humidity: Effects on survival and fecundity of *Pyemotes tritici* (Acari: Pyemotidae). *International Journal of Acarology* 10(3), 135–138.

Bruce, W.A. (1989) Artificial diet for the parasitic mite *Pyemotes tritici* (Acari, Pyemotidae). *Experimental and Applied Acarology* 6, 11–18.

Bruce, W.A. and LeCato, G.L. (1980) *Pyemotes tritici:* A potential new agent for biological control of the red imported fire ant, *Solenopsis invicta* (Acari: Pyemotidae). *International Journal of Acarology* 6(4), 271–274.

Bruce, W.A., and Wrensch, D.L. (1990) Reproductive potential, sex ratio, and mating efficiency of the straw itch mite (Acari: Pyemotidae). *Journal of Economic Entomology* 83, 384–391.

Bruce, W.A., Kethley, J.B. and Kaliszewski, M.J. (1993) Morphology of the gnathosoma of

Pyemotes tritici: Cheliceral stylets and an associated cheliceral structure (Acari: Pyemotidae). *International Journal of Acarology* 19(2), 127–136.

Brust, G. (2022) Garlic Bulb Mites. University of Maryland Extension. Available at: https://extension.umd.edu/resource/garlic-bulb-mites (accessed 17 April 2023).

Brygadyrenko, V.V. (2015a) Evaluation of the ecological niche of some abundant species of the subfamily Platyninae (Coleoptera, Carabidae) against the background of eight ecological factors. *Folia Oecologica* 42(2), 75–88.

Brygadyrenko, V.V. (2015b) Influence of moisture conditions and mineralization of soil solution on structure of litter macrofauna of the deciduous forests of Ukraine steppe zone. *Vìsnyk of Dnìpropetrovsk Unìversity Ser. Bìology, Ekology* 23(1), 50–65 (in Russian).

Brygadyrenko, V., Avtaeva, T. and Matsyura, A. (2021) Effect of global climate change on the distribution of *Anchomenus dorsalis* (Coleoptera, Carabidae) in Europe. *Acta Biologica Slovenica* 7, 237–260.

Bu, G.-S. and Li, L.-S. (1998) Taxonomic notes on and key to known species of the genus *Rhizoglyphus* (Acari: Acaridae) from China. *Systematic and Applied Acarology* 3, 179–182.

Buchanan, D.J. and Jones, I.G. (1972) Allergy to house dust mites in the tropics. *BMJ* 3, 764.

Buchanan, D.J. and Jones, I.G. (1974) Mites and house dust mite allergy in bronchial asthma in northern Zambia. *Postgraduate Medical Journal* 50, 680–682.

Büchi, R. (1991) Folgen zu hoher Feuchtigkeit: Befall durch Staubläuse und Milben. *Mühle + Mischfuttertech* 128(32), 406–408.

Buczek, A., Zając, Z., Woźniak, A., Kulina, D. and Bartosik, K. (2017) Locomotor activity of adult *Dermacentor reticulatus* ticks (Ixodida: Ixodidae) in natural conditions. *Annals of Agricultural and Environmental Medicine* 24(2), 271–275.

Buitendijk, A.M. (1945) Voorloopige Catalogus van de Acari in de Collectie-Oudemans. *Zoologische mededeelingen* 24(12), 281–391.

Burges, H.D. and Burrell, N.J. (1964). Cooling bulk grain in the British climate to control storage insects and to improve keeping quality. *Journal of the Science of Food and Agriculture* 15(1), 32–50.

Burgess, S.T.G., Bartley, K., Nunn, F., Wright, H.W., Hughes, M. *et al.* (2018) Draft genome assembly of the poultry red mite, *Dermanyssus gallinae*. *Microbiology Resource Announcements* 7(18), e01221-18.

Burks, C.S., Dowell, F.E. and Xie, F. (2000) Measuring fig quality using near-infrared spectroscopy. *Journal of Stored Products Research* 36, 289–296.

Burnett, T. (1977) Biological models of two acarine predators of the grain mite, *Acarus siro* L. *Canadian Journal of Zoology* 55(8), 1312–1323.

Buxton, J.H. (1989) *Tyrophagus longior* (Gervais) (Acarina; Acaridae) as a pest of ornamentals grown under protection. *Plant Pathology* 38, 447–448.

CABI (2019) Invasive Species Compendium. Available at: https://www.cabi.org/isc/datasheet/46004#totaxonomicTree (accessed 7 October 2019).

CABI (2020) Invasive Species Compendium. Available at: https://www.cabi.org/isc/datasheet/2522 (accessed 5 May 2020).

CABI (2023) CABI Compendium: Status as Determined by CABI editor. Wallingford, UK.

Cabrera, A.R., Cloyd, R.A. and Zaborski, E.R. (2005) Development and Reproduction of *Stratiolaelaps scimitus* (Acari: Laelapidae) with Fungus Gnat Larvae (Diptera: Sciaridae), potworms (Oligochaeta: Enchytraeidae) or *Sancassania* aff. *sphaerogaster* (Acari: Acaridae) as the sole food source. *Experimental and Applied Acarology* 36, 71–81.

Caceres, I. and Fain, A. (1978) Notes sur la faune acarologique des poussières de maisons du Perou. *Bulletin et annales de la Société royale belge d'entomologie* 114, 301–303.

Caceres, I. and Fain, A. (1979) Notes sur la faune acarologique des poussieres de maisons du Perou. In: Piffl, E. (ed.) *Proceedings of the 4th International Congress of Acarology, Saalfelden, Austria, 12–19 August, 1974*. Akademiai Kiado, Budapest, Hungary, pp. 259–261.

Cafiero, M.A., Galante, D., Camarda, A., Giangaspero, A. and Sparagano, O. (2011) Why dermanyssosis should be listed as an occupational hazard. *Occupational and Environmental Medicine* 68, 628–610.

Cafiero, M.A., Barlaam, A., Camarda, A., Radeski, M., Mul, M.F., Sparagano, O. and Giangaspero, A. (2019) *Dermanysuss gallinae* attacks humans. Mind the gap! *Avian Pathology* 48(1), 22–34.

Cai, L. and Wen, T. (1989) Faunal survey and seasonal prevalence of house dust mites in the urban area of Shanghai. *Acta Ecologica Sinica* 9, 225–229 (in Chinese).

California Garlic & Onion Research Advisory Board (2019) A Pest Management Strategic Plan for Garlic Production in California. Available at: https://ipmdata.ipmcenters.org/documents/pmsps/2019%20Garlic%20PMSP.pdf (accessed 17 April 2023).

Çakmak, I., Faraji, F. and Çobanoğlu, S. (2011) A checklist and key to the Ascoidea and Phytoseioidea (except Phytoseiidae) species of Turkey with three new species records (Acari: Mesostigmata). *Turkiye Entomoloji Dergisi* 35(4), 575–586.

Camerik, A.M. (2010) Phoresy revisited. In: Sabelis, M.W. and Bruin, J. (eds) *Trends in Acarology: Proceedings of the 12th International Congress 21–26 August 2006, Amsterdam*. Springer, Dordrecht, The Netherlands, pp. 333–336.

Camin, J.H. (1955) Uropodellidae, a new family of Mesostigmatid mites based on *Uropodella laciniata* Berlese, 1888 (Acarina: Liroaspina). *Bulletin of the Chicago Academy of Sciences* 10(5), 65–81.

Campbell, J.F., Arthur, F.H. and Mullen, M.A. (2004) Insect management in food processing facilities. *Advances in Food and Nutrition Research* 48, 239–295.

Canestrini, G. (1884) Acari nuovi o poco noti. *Atti del Reale Istituto veneto di scienze, lettere ed arti, Series* 6, 2(5), 693–724, plates VI–IX.

Canestrini, G. (1886) Prospetto dell'acarofauna italiana. 2: Erythraeini, Cheyletini, Bdellini, Eupodini e Analgesini, Padova, 159–311, Tav. X–XXII.

Canestrini, G. (1888) Prospetto dell'acarofauna italiana Famiglie: Tarsonemini Tiroglifini appendice agli Analgesini. *Padova* 3, 309–425.

Canestrini, G. (1889) Prospetto dell'Acarofauna Italiana. *Atti del Reale Istituto veneto di scienze, lettere ed arti, Series* 6, 7, 491–537.

Canestrini, G. and Canestrini, R. (1882) Acari Italiani nuovi o poco noti. *Atti del Reale Istituto veneto di scienze, lettere ed arti, Series* 5, 8, 913–930 + plates VIII–X.

Canestrini, G. and Fanzago, F. (1876) Nuovi Acari Italiani. *Atti del Reale Istituto veneto di scienze, lettere ed arti* 5, 99–111.

Canestrini, G. and Fanzago, F. (1877) Intorno agli Acari Italiani. *Atti del Reale Istituto veneto di scienze, lettere ed arti* 5, 4, 69–208 + plates.

Canfield, M.S. and Wrenn, W.J. (2010) *Tyrophagus putrescentiae* mites grown in dog food cultures and the effect mould growth has on mite survival and reproduction. *Veterinary Dermatology* 21, 58–63.

Capua, S. and Gerson, U (1983) The effects of humidity and temperature on hypopodial molting of *Rhizoglyphus robini*. *Entomologia Experimentalis et Applicata* 34, 96–98.

Caraballo, L., Lagares, A., Mercado, D., Puerta, L. and Femández-Caldas, E. (1999) Sensitization to the mite *Suidasia medanensis* in asthmatic patients (Abstract). *Journal of Allergy and Clinical Immunology* 103, S189.

Cardona, G., Guisantes, J., Postigo, I., Eraso, E., Serna, L.A. and Martinez, J. (2005) Allergenic cross-reactivity between *Blomia tropicalis* and *Blomia kulagini* (Acari: Echymiopodidae) extracts from optimized mite cultures. *Journal of Investigational Allergology and Clinical Immunology* 15, 259–265.

Canevari, G.C., Rezende, F., Silva, R.B., Faroni, L.R.A., Zanuncio, J.C., Papadopoulou, S. and Serrão, J.E. (2012) Potential of *Tyrophagus putrescentiae* (Schrank) (Astigmata: Acaridae) for the Biological Control of *Lasioderma serricorne* (F.) (Coleoptera: Anobiidae). *Brazilian Archives of Biology and Technology* 55(2), 299–303.

Carter, H.F., Wedd, G. and D'Abrera, V.S.E. (1944) The occurrence of mites (Acarina) in human sputum and their possible significance. *Indian Medical Gazette* 79, 163–168.

Casanueva, M.E. and Artigas, J.N. (1985) Distribucion geografica y estacional de los acaros del polvo de habitacion en Chile (Arthropoda: Acari). *Gayana Zoologia* 49, 3–75.

Casanueva, M.E., Berrios, A.M., Peredo, A.A. and Martinez, R.I. (1994) Acaros asociados a equinos estabulados 1. *Androlaelaps casalis* (Berlese), *Haemogamasus horridus* Michael y *Proctolaelaps pygmaeus* (Muller), primer registro para Chile y redescripcion de los machos (Acari: Mesostigmata). *Gayana Zoologia* 58.

Castilho, R.C., Moraes, G.J., Silva, E.S. and Silva, L.O. (2009) Predation potential and biology of *Protogamasellopsis posnaniensis* Wisniewski & Hirschmann (Acari: Rhodacaridae). *Biological Control* 48, 164–167.

Castilho, R.C., Silva, E.S., Moraes, G.J. and Halliday, B. (2016) Catalogue of the family Ologamasidae Ryke (Acari: Mesostigmata). *Zootaxa* 4197(1), 1–147.

Castillo, S., Sanchez-Borges, M., Capriles, A., Suarez-Chacon, R., Caballero, F. and Fernandez-Caldas, E. (1995) Systemic anaphylaxis after ingestion of mite-contaminated flour. *Journal of Allergy and Clinical Immunology* 95, 304.

Cavalcante Cerqueira, A.C., dos Santos, V.L.V., Rossi, L.C.C. and Moraes. G.J. (2015) Potential of five Brazilian populations of phytoseiidae (Acari) for the biological control of *Bemisia tabaci* (Insecta: Hemiptera). *Journal of Economic Entomology* 8, 1–5.

Cebolla, R., Pekar, S. and Hubert, J. (2009) Prey range of the predatory mite *Cheyletus malaccensis* (Acari: Cheyletidae) and its efficacy in the control of seven stored-product pests. *Biological Control* 50, 1–6.

Cedola, C. and Castresana, J. (2014) First record of *Typhlodromus* (*Anthoseius*) *transvaalensis* (Acari: Phytoseiidae) from Argentina. *Revista de la Sociedad Entomológica Argentina* 73(1–2), 61–63.

Ceraul, S.M., Sonenshine, D.E., Ratzlaff, R.E. and Hynes, W.L. (2003) An arthropod defensin expressed by the hemocytes of the American dog tick, *Dermacentor variabilis* (Acari: Ixodidae). *Insect Biochemistry and Molecular Biology* 33, 1099–1103.

Cevizci, S, Gökçe, S., Bostan, K. and Kaypmaz, A. (2010) Depo Gıdalarını ve Peynirleri Enfeste Eden Akarlara Halk Sağlığı Açısından Bakış. *Turkiye Parazitolojii Dergisi* 34(3), 191–199 (in Turkish).

Chabasse, Y. and Genthorn, H. (1966) Les acariens detricoles. Parasitisme d'un jambon par *Glycyphagus domesticus*. *Recueil de médecine vétérinaire. Ec. Alfort* 142, 1049–1054.

Chaires-Grijalva, M.P., Estrada-Venegas, E.G., Equihua-Martínez, A., Moser, J.C., Sánchez-Martínez, G. *et al.* (2013) Mesostigmatid mites (Acari: Mesostigmata) associated to *Dendroctonus rhizophagus* from Chihuahua, Mexic. *Revista Mexicana de Biodiversidad* 84, 1235–1242.

Chakraborty, S. and Gupta, S.K. (2016) A report on mites occurring in household stored products in South Bengal. *International Journal of Science and Research* 5(4), 556–560.

Chambers, R.J., Wright, E.M. and Lind, R.J. (1993) Biological control of glasshouse sciarid flies (*Bradysia* spp.) with the Predatory Mite, *Hypoaspis* Miles on Cyclamen and Poinsettia. *Biocontrol Science and Technology* 3, 285–293.

Champ, B.R. (1966) Insects and mites associated with stored products in Queensland. 4. Acarina and Pseudoscorpiones. *Queensland Journal of Agricultural and Animal Sciences* 23, 197–210.

Chang, C.-I. (2003) *Hyperspectral Imaging: Techniques for Spectral Detection and Classification*. Springer Science & Business Media. ISBN 978-0-306-47483-5.

Chang, Y.-C. and Hsieh, K.-H. (1989) The study of house dust mites in Taiwan. *Annals of Allergy* 62, 101–106.

Chang, J.C.S., Arlian, L.G., Dippold, J.S., Rapp, C.M. and Vyszenski-Moher, D. (1998) Survival of the House Dust Mite, *Dermatophagoides Farinae*, at high temperatures (40–80°C). *Indoor Air* 8(1), 34–38.

Chant, D.A. (1958) Descriptions of six new species of *Garmania* Nesbitt and *Lasioseius* Berlese (Acarina: Aceosejidae). *Canadian Journal of Zoology* 36, 383–390.

Chant, D.A. (1959) Phytoseiid mites (Acarina: Phytoseiidae). Part I. Bionomics of seven species in South-Eastern England. Part 2. A taxonomic review of the family Phytoseiidae, with descriptions of 38 new species. *Canadian Entomologist* 91(12), 1–164.

Chant, D.A. (1963) The subfamily Blattisocinae Garman (=Aceosejinae Evans) (Acarina: Blattisocidae Garman) (=Aceosejidae Baker and Wharton) in North America, with descriptions of a new species. *Canadian Journal of Zoology* 41, 243–305.

Chant, D.A. and McMurtry, J.A. (1994) A review of the subfamilies phytoseiinae and typhlodrominae (Acari: Phytoseiidae). *International Journal of Acarology* 20, 223–310.

Chant, D.A. and McMurtry, J.A. (2003) A review of the subfamily Amblyseiinae Muma (Acari: Phytoseiidae). Part I, Neoseiulini new tribe. *International Journal of Acarology* 29, 3–46.

Chant, D.A. and McMurtry, J.A. (2004) A review of the subfamily Amblyseiinae Muma (Acari. Phytoseiidae). Part III, The tribe Amblyseiini Wainstein, subtribe Amblyseiina n. subtribe. *International Journal of Acarology* 30, 171–228.

Chant, D.A. and McMurtry, J.A. (2005a) A review of the subfamily Amblyseiinae Muma (Acari: Phytoseiidae). Part VII, Typhlodromipsini n. tribe. *International Journal of Acarology* 31, 315–340.

Chant, D.A. and McMurtry, J.A. (2005b) A review of the subfamily Amblyseiinae Muma (Acari: Phytoseiidae). Part VIII, The tribes Macroseiini Chant, Denmark and Baker, Phytoseiulini n. tribe, Afroseiulini n. tribe and Indoseiulini Ehara and Amano. *International Journal of Acarology* 32, 13–25.

Chapman, R.F. (1982) *The Insects: Structure and Function*. 3rd edn. Harvard University Press, Cambridge, MA, pp. 1–919.

Charles River Laboratories International (2011) Pseudoparasites, Technical Sheet. Available at: http://www.criver.com/files/pdfs/infectious-agents/rm_ld_r_pseudoparasites.aspx (accessed 4 December 2021).

Charlet, L.D., Mulla, M.S. and Sánchez-Medina, M. (1977) Domestic Acari of Colombia: abundance of the European house dust mite, *Dermatophagoides pteronyssinus* (Acari, Pyroglyphidae) in homes in Bogota. *Journal of Medical Entomology* 13, 709–712.

Charlet, L.D., Mulla, M.S., Sanchez-Medina, M. and Reyes, M.A. (1979) Species composition and population trends of mites in various climatic zones in Colombia. *Journal of Asthma Research* 16, 131–148.

Charmillot, P.J., Hofer, D. and Pasquier, D. (2000) Attract and kill: A new method for control of the codling moth *Cydia pomonella*. *Entomologia Experimentalis et Applicata* 94, 211–216.

Chaudhri, W.M., Akbar, S. and Rasool, A. (1979) Studies on the predatory leaf inhabiting mites of Pakistan. Univ. Agric., Faisalabad, Pakistan, Project no. PK-ARS 30, 1–234.

Chaudhury, S., Roy, I., Podder, S., Gupta, S.K. and Saba, G.K. (2005) Diversity of synanthropic mites in Kolkata Metropolis, India. *Records of the Zoological Survey of India* 104(3–4), 151–159.

Chauve, C. (1998) The poultry red mite *Dermanyssus gallinae* (De Geer, 1778): Current situation and future prospects for control. *Veterinary Parasitology* 79(3), 239–245.

Chen, J.-S. (1990a) An improved method for determining the susceptibility of *Rhizoglyphus robini* and *R. setosus* (Acarina: Acaridae) to pesticides. *Experimental and Applied Acarology* 8, 175–178.

Chen, J.-S. (1990b) The biology, insecticide resistence, and control countermeasures of bulb mites. *Chinese Journal of Entomology, Special Publication* 3, 93–106.

Chen, J.-S. and Lo, K.-C. (1990) Toxicity of organophosphorus insecticides to diazinon-reversely-selected, pesticide-pressure-relaxed and field-resistant strains of bulb mite *Rhizoglyphus robini* (Acari: Acaridae). *Experimental and Applied Acarology* 8, 243–252.

Chen, S.-B., Jiang, Z.J., Lu, P.-Y. and Ma, E.-P. (1982) China. In: Prasad, V. (ed.) *History of Acarology*. Indira Publishing House, Michigan, pp. 65–113 (in English).

Chen, X.B., Sun, X., Hu, S.F. (1990) Clinical manifestation and treatment of pulmonary acariasis. *Zhongguo Ji Sheng Chong Xue Yu Ji Sheng Chong Bing Za Zhi* 8(3), 217–219 (in Chinese).

Chen, X.B., Fu, C.B., Chen, X.B. and Fu, C.B. (1992). Mites causing pulmonary, intestinal and urinary acariasis. In: Chen, X.B. and Ma, E.P. (eds) *Researches of acarology in China*. Chongqing Publishing House, Chongqing, China, pp. 109–113.

Chen, Q.-Z., Jian, F.-J., Yu, L.-Q. and Sun, M.-H. (1993) The toxicity of acetylene to the adults of copra mite (*Tyrophagus putrescentiae* (Schrank)). *Journal of Zhengzhou Grain College* (3), 1–4.

Chen, W., Guo, X. and Gui, Y. (1994) A new species of *Gamasholaspis* from China (Acari: Parholaspidae). *Acta Arachnologica Sinica* 3, 61–64.

Chen, W.H., Liu, Y.C. and Ho, C.C. (2002a) The life cycle, distribution and host plants of bulb mite, *Rhizoglyphus setosus* Manson (Acari: Acaridae) in Taiwan. *Plant Protection Bulletin (Taipei)* 44, 341–352.

Chen, D.-P., Qian, W.-P. and Zhou, Y.-Z. (2002b) The efficacy and toxicity of Mie Man Ling against *Acarus siro*. *Chinese Journal Hygienic Insecticides Equipments* 8(4), 24–26.

Chew, F.T., Zhang, L., Ho, T.M. and Lee, B.W. (1999) House dust mite fauna of tropical Singapore. *Clinical & Experimental Allergy* 29(2), 201–216.

Childs, M. and Bowman, C.E. (1981) Lysozyme activity in six species of economically important astigmatid mites. *Comparative Biochemistry and Physiology B* 70, 615–617.

Chilton, A. (2013) The Working Principle and Key Applications of Infrared Sensors. AZO Sensors. Available at: https://www.azosensors.com/article.aspx?ArticleID=339 (accessed 31 December 2022).

Chinniah, C. and Mohanasundaram, M. (1996) Three new species of mites associated with insects from Tamil Nadu, India. *Entomon* 21, 157–163.

Chirico, J. and Tauson, R. (2002) Traps containing acaricides for the control of *Dermanyssus gallinae*. *Veterinary Parasitology* 110, 109–116.

Chiron, G., Varescon, A., Lubac, S., Bicout, D.J. and Roy, L. (2015) A decision-making method to anticipate outbreaks of *Dermanyssus gallinae* populations in layer farms. Actes des 11èmes Journées de la Recherche Avicole et Palmipèdes

à Foie Gras, Tours, France, 25–26 March 2015, pp. 179–183.
Chisaka, K., Minamite, Y., Chgami, H. and Katsuda, Y. (1985) Efficacy of various types of pyrethroid compounds against *Tyrophagus putrescentiae* and *Dermatophagoides farinae*. *Japanese Journal of Sanitary Zoology* 36, 7–13.
Chistyakov, M.P. (1987) Ontogenetic variability of *Parachipteria bellus*. Gorkii Pedagogical Institute, *Viniti Press, Gorkii*, 1–17 (in Russian).
Chistyakov, M.P. (1988) Ontogenetic variability of *Achipteria coleoptrata* (Oribatei). Gorkii Pedagogical Institute, Viniti Press, Gorkii, 1–18 (in Russian).
Chistyakov, M.P. (1994) Morphological variability of juvenile instars of oribatid mite (*Achipteria italicus* Oudemans, 1913). In: Animals in Natural Ecosystems. Nizhny Novgorod Pedagogical University, Nizhny Novgorod, 41–47 (in Russian).
Chmielewski, W. (1967) Stadium hypopus *Carpoglyphus lactis* (L., 1758) (Glycyphagidae, Acarina). *Prace Naukowe Instytutu Ochrony Roślin* 9(2), 161–173.
Chmielewski, W. (1969a) Obserwacje nad biologia now ego dla akarofauny polskiej gatunku *Kuzinia laevis* (Dujardin, 1849) (Acarina, Acaridae). *Polskie Pismo Entomologiczne* 39, 603–617 (in Polish).
Chmielewski, W. (1969b) Fauna of mites in stored seeds of sugar beets. *Polskie Pismo Entomologiczne* 39, 619–628 (in Polish).
Chmielewski, W. (1970a) The passage of mites through the alimentary canal of vertebrates. *Ekologia Polska* 35, 741–756.
Chmielewski, W. (1970b) Morphological and biologic-ekological charakteristic of *Carpoglyphus lactis* (L.) – Of kind appearing in natural honey, in warehouses and beehives. *Pszczelnicze Zeszyty Naukowe* 14(1–3), 109–127 (in Polish).
Chmielewski, W. (1971a) Investigations on the species composition of the mites in the beehives and storehouses of honey. *Pszczelnicze Zeszyty Naukowe* 15, 69–80 (in Polish).
Chmielewski, W. (1971b) A study on the influence of some ecological factors on the hypopus formation of stored product mites. In: Daniel, M. and Rosicky, B. (eds) *Proceedings of the third International Congress of Acarology, Prague, Prague (Czechoslovakia), August 31–September 6, 1971, Junk, Den Haag*, pp. 356–363.
Chmielewski, W. (1971c) Mites occuring on food products. *Prace Naukowe Instytutu Ochrony Roślin* 13, 167–186 (in Polish).
Chmielewski, W. (1971d) Morphology, biology and ecology of *Carpoglyphus lactis* (L. 1758) (Glycyphagidae, Acarina). *Prace Naukowe Instytutu Ochrony Roślin* 13, 63–166 (in Polish).
Chmielewski, W. (1971e) Species composition and intensification of occurrence of acarofauna in stored grass seeds. *Prace Naukowe Instytutu Ochrony Roślin* 13, 201–215 (in Polish).
Chmielewski, W. (1972) Nasilenie występowania i skład gatunkowy akarofauny mieszanek ziołowych używanych w lecznictwie. *Prace Naukowe Instytutu Ochrony Roślin* 14, 65–80 (in Polish).
Chmielewski, W. (1975a) Z badań nad *Thyreophagus entomophagus* (Laboulbène). *Zeszyty Problemowe Postępów Nauk Rolniczych* 171, 253–259 (in Polish).
Chmielewski, W. (1975b) Nowo stwierdzone stadium rozwojowe roztoczy z rodzaju *Ctenoglyphus* -*C. plumiger* (Koch, 1835) (*Acarina, Glycyphagidae*). *Zeszyty Problemowe Postępów Nauk Rolniczych* 171, 261–268 (in Polish).
Chmielewski, W. (1977a) Wyniki obserwacji powiazan roztoczy z owadami (Acari-Insecta). *Polskie Pismo Entomologiczne* 47, 59–78 (in Polish).
Chmielewski, W. (1977b) Dane o występowaniu niektórych gatunkóu roztoczy alergogennych (Acaroidea) w Polsce. *Wiadomości Parazytologiczne* 23, 109–113 (in Polish).
Chmielewski, W. (1979) Mites occurring on greenhouse cucumbers. *Ochrona Roślin* 23(10), 5–7.
Chmielewski, W. (1982) Mites (Acarina) inhabiting sparrow (*Passer domesticus* L.) nests. *Wiadomości Parazytologiczne* 28, 105–107 (in Polish).
Chmielewski, W. (1983a) Bee bread as a food of tyroglyphids of *Acarus* genus (*Acarina: Acaridae*). *Zeszyty Problemowe Postępów Nauk Rolniczych* 252, 171–178.
Chmielewski, W. (1983b) The cases of infestation of insects by mites. *Zeszyty Problemowe Postępów Nauk Rolniczych* 252, 179–188 (in Polish).
Chmielewski, W. (1984) *Tyrophagus longior* (Gerv., 1844) (Acarina, Acaridae) – Bioecology, occurrence and harmfulness. *Prace Naukowe Instytutu Ochrony Roślin* 26, 69–87.
Chmielewiski, W. (1987) An attempt of use of insecticide Actellic 50 EC in control of copra mite, *Tyrophagus putrescentiae* (Schr.) (Acarida, Acaridae) – A pest of flax and hemp sow seeds. *Prace Naukowe Instytutu Ochrony Roślin* 28, 139–146.
Chmielewski, W. (1990) Bio-ecology and population development *Thyreophagus entomophagus* (Acarida, Acaridae) – Acarid mite found in beehives.

Pszczelnicze Zeszyty Naukowe 34, 31–42 (in Polish).
Chmielewski, W. (1991a) Stored products mites (Acaroidea) in Polish bee hives. In: Dusbabek, F. and Bulva, B. (eds) *Modern acarology. Proceedings of the 8 International Congress of Acarology, Ceske Budejovice, Czechoslovakia, 6–11 August 1990, vol. I.* SPB Academic Publishing, The Hague, The Netherlands, pp. 615–619.
Chmielewski W. (1991b) Biological observations of allergenic mites *Suidasia nesbitti* (Acarida, Saproglyphidae). *Wiadomosci Parazytologiczne* 37(1), 133–136 (in Polish).
Chmielewski, W. (1991c) Roztocze (Acarida) pszczoly miodnej (*Apis mellifera* L.) w Polsce. *Wiadomosci Parazytologiczne* 37, 91–94 (in Polish).
Chmielewski, W. (1995) Mites (Acarina) in house dust of a basemat flat. Proc. Symp. "Advances of Acarology in Poland", Siedlce, September 26–27, 1995, Drukarnia ISK, Siedlce, 205–209.
Chmielewski, W. (1998) Wyniki wstępnych badań *Acaroidea* i *Psoroptoidea (Acari: Acaridida)* kilku wybranych siedlisk polskiej części Puszczy Białowieskiej. *Parki Narodowe i Rezerwaty Przyrody* 17.3(supl), 137–143.
Chmielewski, W. (1999) Bionomics of *Carpoglyphus lactis* (Acari: Carpoglyphidae) on honey. In: Bruin, J., van der Geest, L.P.S. and Sabelis, M.W. (eds) *Ecology and Evolution of the Acari. Series Entomologica*, Vol. 55. Springer, Dordrecht, The Netherlands, pp. 423–424.
Chmielewski, W. (2000a) Associations of mites with beetles (Acarina - Coleoptera (Insecta)). Proceedings of the 26th Symposium of Acarology, Kazimierz Dolny, Poland, TN, September 24–26, 1999. In: Ignatowicz, S. (ed.) Akarologia polska u progu XXI wieku, Wydawnictwa SGGW (Szkoły Głównej Gospodarstwa Wiejskiego w Warszawie), Warszawa, pp. 79–90.
Chmielewski, W. (2000b) Biological effects of feeding of *Ctenoglyphus plumiger* (Koch) (Acari: Ctenoglyphidae) on bee bread. *Pszczelnicze Zeszyty Naukowe* 64, 223–229.
Chmielewski, W. (2001a) Buckwheat Sprouts as a Food of *Rhizoglyphus echinopus* (f. et r.) (acari: acaridae) Reared under Laboratory Conditions. In: *Proceeding of the 8th ISB*, pp. 681–686.
Chmielewski, W. (2001b) Buckwheat as a nourishment of *Lepidoglyphus destructor* (Schr.) (*Acari: Glycyphagidae*). *Fagopyrurn* 18, 61–64.
Chmielewski, W. (2002a) Bionomics of *Glycyphagus domesticus* (De Geer) (*Acari: Glycyphagidae*) feeding on buckwheat seeds. *Fagopyrum* 19, 105–108.
Chmielewski, W. (2002b) Akarofauna zebranawpo awiaczach py^{3}ku. In: Ignatowicz, S. (ed.) In: *Postêpy polskiej akarologii. Wydawnictwa SGGW (Szkoły Głównej Gospodarstwa Wiejskiego w Warszawie)*, Warszawa, pp. 87–90 (in Polish).
Chmielewski, W. (2004) Preliminary studies on mites (Acarina) infesting stored grain and other buckwheat products. *Fagopyrum* 21, 117–121.
Chmielewski, W. (2009) Phoretic mites (Acarina) on earwigs, *Forficula auricularia* L. (Insecta, Dermaptera), found in apiaries. *Journal of Apicultural Science* 53, 75–80.
Chmielewski, W. and Lipa, J.J. (1967) Biological and ecological studies on *Caloglyphus* mites (Acarina: Acaridae) associated with Scarabaeidae. *Acta Parasitologica* 14, 179–186.
Cho, J.H., Kim, J.B., Cho, C.S., Huh, S. and Ree, H.I. (1999) An Infestation of the Mite *Sancassania berlesei* (Acari: Acaridae) in the External Auditory Canal of a Korean Man. *Journal of Parasitology* 85(1), 133–134.
Christie, Y.J.E. (1983) A new species of *Alliphis* (Mesostigmata: Eviphididae) from Britain. *Acarologia* 24(3), 231–242.
Chu, J.K., Song, S.B., Kim, D.K. and Kim, Y.K. (1967) Epidemiologyical studies on the acaroid mite. *Korean Journal of Parasitology* 5, 71–77.
Chu, T., Murano, T., Uno, Y., Usui, T. and Yamaguchi, T. (2015) Molecular epidemiological characterization of poultry red mite, *Dermanyssus gallinae. The Journal of Veterinary Medical Science* 77, 1397–1403.
Chua, K.Y., Stewart, G.A., Thomas, W.R., Simpson, R.J., Dilworth, R.J., Plozza, T.M. and Turner, K.J. (1988) Sequence analysis of cDNA coding for a major house dust mite allergen, Der p 1. Homology with cysteine proteases. *Journal of Experimental Medicine* 167(1), 175–182.
Chwedorzewska, K.J., Korczak-Abshire, M., Olech, M., Lityńska-Zając, M. and Augustyniuk-Kram, A. (2013) Alien invertebrates transported accidentally to the Polish antarctic station in cargo and on fresh foods. *Polish Polar Research* 34, 55–66.
Ciampolini, M., Lugaresi, C., Rota, P.A. and Cappella, A. (1985) Un acaro nocivo ad orchidee in serra. *Informatore Agrario* 41(33), 29–33.
Cichecka, E., Maniurka, H., Szilman, P., Szilman, E., Solarz, K. and Sieroń A.L. (2006) Sensitization to storage mites in urban and subagricultural population of Upper Silesia. In: Buczek, A. and

Błaszak, C. (eds) *Arthropods. Epidemiological Importance*. Koliber, Lublin, pp. 287–293.

Cicilykutty, A.K., Balasubramanian, N.K. and John, P.A. (1981a) The influence of physical conditions (Moisture content) of the flesh of fish on the intensity of infestation by adult mite *Suidasia medanensis*. *Acarologia* 22(3), 291–293.

Cicilykutty, A.K., Balasubramanian, N.K. and John, P.A. (1981b) The influence of physical conditions (Salt content) of the flesh of fish on the intensity of infestation by adult mite *Suidasia medanensis*. *Acarologia* 22(3), 295–297.

Cicolani, B. (1977) The influence of temperature on the speed of development, from egg to adult, of *Macrocheles matrius* (Acarina: Mesostigmata). *Acarologia* 19(3), 384–394.

Cicolani, B. (1979a) Osservazione faunistiche ed ecologiche sugli acari Coprobionti (Nota I): il genere *Macrocheles* Latr. (Mesostigmata) in Abruzzo. *Biologia Contemporanea, A.N.L.S.B. (Roma)* 6(3), 116–126.

Cicolani, B. (1979b) The intrinsic rate of natural increase in dung macrochelid mites, predators of *Musca domestica* eggs. *Italian Journal of Zoology* 46, 171–178.

Cicolani, B. and Sabatino, A. di (1991) Genetic differentiation between *Macrocheles penicilliger* and *Macrocheles minervae* (Acarina: Mesostigmata). In: Dusbabek, F. and Bukva, V. (eds) *Modern Acarology, 2. Prague*. Academia and The Hague, SPB Academic Publishing, pp. 437–442.

Cicolani, B., Bianchi Bullini, A.L. and Bullini, L. (1981) Morphological and genetic differentiation between *Macrocheles glaber* and *Macrocheles perglaber* (Acarina: Mesostigmata). *International Journal of Acarology* 7, 221–224.

Ciechacka, M., Pająk, C., Solarz, K., Szilman, P., Szilman, E. and Włodarz-Ulman, I. (2013) Abundance of allergenic mites in bed dust samples from one-family houses in the territory of żywiec district. In: Buczek, A. and Błaszak, C. (eds) *Arthropods, The Medical and Veterinary Aspects*. PAN, Koliber, Lublin, Poland, pp. 49–59.

Cieślak, M., Solarz, K. and Kamińska, I. (2014) Effect of bioactive modified nonwoven systems on the development of house dust mites *Dermatophagoides farinae* (Acari: Pyroglyphidae) in laboratory assay. *Textile Research Journal* 85(5), 469–478.

Cilbircioğlu, C. and Çobanoğlu, S. (2019) Phytophagous mite (Acari) species on garlic (*Allium sativum* L.) cultivation areas and storages of Kastamonu, Turkey. *Persian Journal of Acarology* 8(3), 211–224.

Circella, E., Pugliese, N., Todisco, G., Cafiero, M., Sparagano, O. and Camarda, A (2011) *Chlamydia psittaci* infection in canaries heavily infested by *Dermanyssus gallinae*. *Experimental and Applied Acarology* 55, 329–338.

Clancy, G. (1981) Observations of mites associated with the low yielding crops of cultivated *Agaricus bisporus* in Australia. *Mushroom Science* 11, 233–244.

Claparède, E. (1869) Studien an Acariden. *Zeitschrift für wissenschaftliche Zoologie* 18, 445–546.

Clark, J.M. (2010) A new species of *Carpoglyphus* (Astigmatina: Carpoglyphidae) from the bark of black beech (*Nothofagus*) honeydew in New Zealand. *International Journal of Acarology* 36, 453–459.

Clarke, P.G. (2002) The ability of buried PCTM traps to detect stored-product mites in wheat. In: Credland, P.F., Armitage, D.M., Bell, C.H., Cogan, P.M. and Highley, E. (eds) *Proceedings of the 8th International Working Conference on Stored Product Protection, 22–26 July 2002, York, U.S.A.* CAB International, Wallingford, UK, pp. 390–395.

Clarke, D., Dix, E., Liddy, S., Gormally, M. and Byrne, M. (2015) The influence of household pets on the composition and quantity of allergenic mite fauna within Irish homes: a preliminary investigation. *The Royal Entomological Society, Medical and Veterinary Entomology*. DOI: 10.1111/mve.12147. Available at: file:///C:/Users/Utente/Downloads/Theinfluenceofpetsonthecompositionandquantityofallergenicmitefaunain Irishhomes.pdf (accessed 4 May 2020).

Clemmons, E.A. and Douglas, K.T. (2016) Booklice (*Liposcelis* spp.), Grain Mites (*Acarus siro*), and Flour Beetles (*Tribolium* spp.): 'Other Pests' Occasionally Found in Laboratory Animal Facilities. *Journal of the American Association for Laboratory Animal Science* 55(6), 737–743.

Clift, A.D., Terras, M.A. and Elliot, T.J. (1995) Mites as indicators of compost conditioning. In: Elliott, T.J. (ed.) *Proceedings of the 14th International Congress on the Science and Cultivation of Edible Fungi, Oxford, 17–22 September 1995*. Balkema, Rotterdam, The Netherlands, pp. 507–513.

Çobanoğlu, S. (1996) Determination, distribution and hosts of Acarina group harmful and useful on stored products in the province of Edirne (Turkey). *Türkiye Entomoloji Dergisi* 20, 199–210 (in Turkish).

Çobanoğlu, S. (2008) Mites (Acari) associated with stored apricots in Malatya, Elaz and Izmir provinces of Turkey. *Türkiye Entomoloji Dergisi* 32(1), 3–20.

Çobanoğlu, S. (2009) Mite population density analysis of stored dried apricots in Turkey. *International Journal of Acarology* 35(1), 67–75.

Çobanoğlu, S. and Bayram, Ş. (1998) Mites (Acari) and flies (Insecta: Diptera) from natural edible mushrooms (*Mochella*: Ascomycetes) in Ankara, Turkey. *Bulletin et annales de la Société royale belge d'entomologie* 134, 187–198.

Çobanoğlu, S. and Bayram, Ş. (1999) Mites (Acari) species associated with cultivated and wild rose plants in Çamlıdere, Turkey. *Entomologist's Monthly Magazine* 135, 245–248.

Çobanoğlu, S. and Özman, S. (2002) Beneficial mite species of hazelnut orchard ecosystem from black sea region of Turkey. In: *Proceedings of the 2nd Meeting of WG.4, Prague, 30–31 May 2002.* Biocontrol of Arthropod Pests in Stored Products, pp. 91–99.

Çobanoğlu, S., Uysal, C. and Ökten, E. (2003) The complex of the beneficial mite fauna of ornamental trees and shrubs in Ankara, Turkey. *Entomologist's Monthly Magazine* 139, 7–12.

Çobanoğlu, S., Bayram, Ş., Denizhan, E. and Sağlam H.D. (2007) The effect of different temperatures and food densities on development of *Blattisocius tarsalis* (Berlese) (Acari: Ascidae) reared on Mediterranean flour moth *Ephestia kuehniella* Zeller (Lepidoptera: Pyralidae). *Türkiye Entomoloji Dergisi* 31(1), 7–20.

Cocciolo, G., Circella, E., Pugliese, N., Lupini, C., Mescolini, G. *et al.* (2020) Evidence of vector borne transmission of *Salmonella enterica enterica* serovar Gallinarum and fowl typhoid mediated by the poultry red mite, *Dermanyssus gallinae* (De Geer, 1778). *Parasites Vectors* 13, 513.

Çobanoğlu, S., Öğreten, A., Sade, E. (2021) The occurrence of *Paraneognathus wangae* (Fan & Li) (Acari: Caligonellidae) and *Raphignathus gracilis* Rack (Acari: Raphignathidae) of stored products in Turkey. *Journal of Agricultural Sciences* 27(2), 164–169.

Codex Alimentarious, International Food Standards of FAO and World Health Organization (1995) Standard for Wheat and Durum Wheat CXS 199–1995. Available at: https://www.fao.org/fao-who-codexalimentarius/sh-proxy/fr/?lnk=1&url=https%253A%252F%252Fworkspace.fao.org%252Fsites%252Fcodex%252FStandards%252FCXS%2B199-1995%252FCXS_199e.pdf (accessed 22 November 2022).

Coelho, C.H., Rappuoli, R., Hotez, P.J. and Duffy, P.E. (2019) Transmission-blocking vaccines for malaria: Time to talk about vaccine introduction. *Trends in Parasitology* 35, 483–486.

Coineau, Y. (1974a) Introduction a l'Étude des Microarthropodes du Sol et des ses Annexes. Documents pour l'Enseignements Pratique de l'Écologie. Doin, Paris, 1–117.

Coineau, Y. (1974b) Eléments pour une Monographie Morphologique, Écologique et Biologique des Caeculidae (Acariens). *Mémoires du Muséum National d'Histoire Naturelle (N.S.) A* 81(1), 1–299.

Coineau, Y., Theron, P.D. and Valette, C. (1997) Une association de phanères, constituant probablement un système de communication sonore original chez les acariens. *Acarologia* 38, 111–116.

Coleman, T.W., Jones, M.I., Hoddle, M.S., Haavik, L.J., Moser, J.C., Flint, M.L. and Seybold, S.J. (2015) *Pyemotes tritici* (Acari: Pyemotidae): a parasitoid of *Agrilus auroguttatus* and *Agrilus coxalis* (Coleoptera: Buprestidae) in the southwestern United States of America and southern Mexico. *Canadian Entomologist* 147(02), 244–248.

Collins, D. (2006) A review of alternatives to organophosphorus compounds for the control of storage mites. *Journal of Stored Products Research* 42, 395–426.

Collins, D.A. (2012) A review on the factors affecting mite growth in stored grain commodities. *Experimental and Applied Acarology* 56, 191–208.

Collins D. and Cook, D.A. (2006) Laboratory evaluation of diatomaceous earths, when applied as dry dust and slurries to wooden surfaces, against stored-product insect and mite pests. *Journal of Stored Products Research* 42(2), 197–206.

Colloff, M.J. (1986) Use of liquid nitrogen in the control of house dust mite populations. *Clinical & Experimental Allergy* 16, 41–47.

Colloff, M.J. (1987a) Mites from house dust in Glasgow. *Medical and Veterinary Entomology* 1(2), 163–168.

Colloff, M.J. (1987b) Mite fauna of dust from passenger trains in Glasgow. *Epidemiology and Infection* 98(1), 127–130.

Colloff, M.J. (1988) House dust mite ecology in my bed. In: De Weck, A.L. and Todt, A.L. (eds) *Mite Allergy, a Worldwide Problem*. UCB Institute of Allergy, Brussels, Belgium, pp. 51–54.

Colloff, M.J. (1991a) A review of the biology and allergenicity of the house-dust mite *Euroglyphus maynéi* (Acari: Pyroglyphidae). *Experimental and Applied Acarology* 11(2–3), 177–198.

Colloff, M.J. (1991b) Population studies on the house dust mite, *Euroglyphus maynéi* (Cooreman, 1950) (Pyroglyphidae). In: Schuster, R. and Murphy, P.W. (eds) *The Acari: Reproduction, Development and Life History Strategies*. Chapman & Hall, London, pp. 497–505.

Colloff, M.J. (1993) A taxonomic revision of the oribatid mite genus *Camisia* (Acari: Oribatida). *Journal of Natural History* 27(6), 1325–1408.

Colloff, M.J. (1998) Taxonomy and identification of dust mites. *Allergy* 53, 7–12.

Colloff, M.J. (2009a) *Dust Mites*. Csiro Publishing, Oxford Street, Collingwood, Australia, pp. 273–328.

Colloff, M.J. (2009b) Identification and taxonomy, classification and phylogeny. In: Colloff, M.J. (ed.) *Dust Mites, Springer Science & Business Media*, CSIRO Publishing, Collingwood, Australia & Springer: Dordrecht, The Netherlands, pp. 1–44.

Colloff, M.J., Lever, R.S. and McShany, C. (1989) A controlled trial of house dust mite eradication using natamycin in homes of patients with atopic dermatitis: effect on clinical status and mite populations. *British Journal of Dermatology* 121, 199–208.

Colloff, M.J., Stewart, G.A. and Thompson, P.J. (1991) House dust acaro-fauna and Der p I equivalent in Australia: The relative importance of *Dermatophagoides pteronyssinus* and *Euroglyphus maynéi*. *Clinical & Experimental Allergy* 21(2), 225–230.

Colloff, M.J., Ayres, J., Carswell, F., Howarth, P.H., Merrett, T.G. *et al.* (1992) The control of allergens of dust mites and domestic pets: a position paper. *Clinical & Experimental Allergy* 22(Suppl 2), 1–28.

Colloff, M.J., Taylor, C. and Merrett, T.G. (1995) The use of domestic steam cleaning for the control of house dust mites. *Clinical & Experimental Allergy* 25, 1061–1066.

Colwell, R.K. and Naeem, S. (1994) Life-history patterns of hummingbird flower mites in relation to host phenology and morphology. In: Houck, M.A. (ed.) *Mites, Ecological and Evolutionary Analyses of Life-History Patterns*. Chapman and Hall, New York, pp. 23–44.

Cone, W.W. (1979) Pheromones of Tetranychidae. In: Rodriguez, J.G. (ed.) *Recent Advances in Acarology, Volume II*. Academic Press, New York, pp. 309–317.

Cone, W.W. (1985) Mating and chemical communication. In: Helle, W. and Sabelis, M.W. (eds) *Spider Mites: Their Biology, Natural Enemies and Control. World Crop Pests, Volume 1A*. Elsevier Science, Amsterdam, The Netherlands, pp. 243–251.

Cone, W.W., McDonough, L.M., Maitlen, J.C. and Burdajewicz, S. (1971a) Pheromone studies of the two-spotted spider mite, *Tetranychus urticae* Koch. I. Evidence of a sex pheromone. *Journal of Economic Entomology* 64, 355–358.

Cone, W.W., Predki, S. and Klostermeyer, E.G. (1971b) Pheromone studies of the two-spotted spider mite, *Tetranychus urticae* Koch. II. Behavioral response of male *T. urticae* to quiescent deutonymphs. *Journal of Economic Entomology* 64, 379–382.

Conijn, C.G.M., Leśna, I. and Altena, K. (1997) Biological control of the bulb mite *Rhizoglyphus robini* by the predatory mite *Hypoaspis aculeifer* on lillies: Implementation in practice. *Acta Horticulturae* 430, 619–624.

Conyers, S.T. and Bell, C.H. (2003) The effect of modified atmospheres on the survival of the eggs of four storage mite species. *Experimental and Applied Acarology* 31(1–2), 115–130.

Conyers, S.T. and Bell, C.H. (2007) Low oxygen requirements for population control of two mite species of stored grain. In: Donahaye, E., Navarro, S., Bell, C., Jayas, D., Noyes, R. and Phillips, T.W. (eds) *Proceedings of International Conference on Controlled Atmosphere and Fumigation in Stored Products, 8–13 August 2004*. Gold Coast, Australia, pp. 8–13.

Cook, D.A. and Armitage, D.M. (2000) Efficacy of a diatomaceous earth against mite and insect populations in small bins of wheat under conditions of low temperature and high humidity. *Pest Management Science* 56(7), 591–596.

Cook, D.A. and Armitage, D.M. (2002) Integrated pest management for stored grain in the U.K. incorporating diatomaceous earths to prevent surface infestations of insects and mites. *IOBC/WPRS Bulletins* 25, 221–229.

Cook, D.A., Wakefield, M.E. and Bryning, G.P. (2008) The physical action of three diatomaceous earths against the cuticle of the flour mite *Acarus siro* L. (Acari: Acaridae). *Pest Management Science* 64(2), 141–146.

Coons, L.B. and Alberti, G. (1999) Acari: Ticks. In: Harrison, F.W. and Foelix, R.F. (eds) *Microscopic Anatomy of Invetebrates Chelicerate Arthropoda, Volume 8B*. Wiley, New York, pp. 267–514.

Coons, L.B. and Axtell, R.C. (1971) Cellular organization in the synganglion of the mite *Macrocheles muscaedomesticae* (Acarina: Macrochelidae). *Zeitschrift für Zellforschung und Mikroskopische Anatomie* 119, 309–320.

Coons, L.B. and Axtell, R.C. (1973) Sensory setae of the first tarsi and palps of the mite, *Macrocheles muscaedomesticae. Annals of the Entomological Society of America* 66, 539–544.

Coons, L.B., Roshdy, M.A. and Axtell, R.C. (1974) Fine structure of the central nervous system of *Dermacentor variabilis* (Say), *Amblyomma americanum* (L.), and *Argas arboreus* Kaiser, Hoogstraal, and Kohls (Ixodoidea). *Journal of Parasitology* 60, 687–698.

Cooreman, J. (1942) Notes et observations sur les Acariens, 1. *Bulletin du Musée royal d'histoire naturelle de Belgique* 18(33), 1–7.

Cooreman, J. (1944) Notes and observations sur les Acariens. *Bulletin du Musée royal d'histoire naturelle de Belgique* 20(8), 1–16.

Cooreman, J. (1950) Sur un Acarien nouveau, préjudiciable sur matières alimentaires entreposées: *Mealia maynéi* n.sp. *Bulletin et annales de la Société royale belge d'entomologie* 86, 164–168.

Cooreman, J. (1959) Note sur le genre *Aeroglyphus* Zachvatkin, 1941 (Acaridiae, Glycyphagidae). *Bulletin de l'Institut royal des sciences naturelles de Belgique* 35, 1–19.

COREMI (2000) Cost Action Network (FA1404). Improving current understanding and research for sustainable control of the poultry red mite *Dermanyssus gallinae*. Available at: www.coremi.eu (accessed 20 June 2021).

Cornere, B.M. (1971) The incidence of House Dust Mites in Auckland. *New Zealand Journal of Medical Laboratory Technology* 25, 7–9.

Corpuz-Raros, L.A. (1996) Philippine predatory mites of the family Cunaxidae (Acari). Genus *Pulaeus* Den Heyer with records of two species from Central Kalimantan, Borneo and Java, Indonesia. *Philippine Entomologist* 10(2), 119–138.

Corpuz-Raros, L.A. (1998) Twelve new species and one new record of Cheyletidae (Acari) from the Philippines. *International Journal of Acarology* 24, 259–290.

Corpuz-Raros, L.A., Sabio, G.C. and Velasco-Soriano, M. (1988) Mites associated with stored products, poultry houses and house dust in the Philippines. *Philippine Entomologist* 7(3), 311–321.

Cortesero, A.M., Stapel, J.O. and Lewis, W.J. (2000) Understanding and manipulating plant attributes to enhance biological control. *Biological Control* 17, 35–49.

Cosmes Martín, P.M., Domínguez Noche, C. and Moreno Ancillo, A. (2003) Sensibilización a ácaros en el norte de Extremadura. *Alergología e Inmunología Clínica* 18, 75–80.

Costa, M. (1961) Mites recovered from the nests of Levant vole (*Microtus guentheri*) in Israel. *Annals and Magazine of Natural History* 13(4), 257–282.

Costa, M. (1962) Mites from the nests of the mole-rat (*Spalax ehrenbergi*) in Israel. *Annals and Magazine of Natural History* 13(4), 481–503.

Costa, M. (1963) The mesostigmatic mites associated with *Copris hispanus* (L.) (Coleoptera, Scarabaeidae) in Israel. *Zoological Journal of the Linnean Society* 45, 25–45.

Costa, M. (1966a) The present state of knowledge of mesostigmatic mites in Israel (Acari: Mesostigmata). *Israel Journal of Zoology* 15, 69–82.

Costa, M. (1966b) Notes on Macrochelids associated with manure and coprid beetles in Israel. I. *Macrocheles robustulus* (Berlese, 1904). Development and biology. *Acarologia* 8(4), 532–548.

Costa, M. (1967) Parasitic and nest-inhabiting mites. In: Teodor, O. and Costa, M. (eds) *A Survey of the Parasites of Wild Mammals and Birds in Israel. Part One. Ectoparasites*. The Israel Academy of Sciences and Humanities, Jerusalem, Humphrey, H.A. Ltd, London, pp. 104–111.

Costa, M. (1968) Little known and new litter inhabiting Laelapine mites (Acari: Mesostigmata) from Israel. *Israel Journal of Zoology* 17, 1–30.

Costa, M. (1974) Mesostigmatic mites (Acari: Mesostigmata) from the Mediterranean shores of Israel. I, The genus *Hypoaspis* Canestrini, 1884. *Israel Journal of Entomology* 9, 219–228.

Costa, M. and Nevo, E. (1969) Nidicolous arthropods associated with different chromosomal types of *Spalax ehrenbergi* Nehring. *Zoological Journal of the Linnean Society* 48(2), 199–215.

Cotter, M., Siebers, R., Pike, A., Fitzharris, P. and Crane, J. (2011) Storage mites in flour samples in Wellington, New Zealand. *Journal of Investigational Allergology and Clinical Immunology* 21(5), 410–411.

Cotton, R.T. and Good, N.E. (1937) Annotated list of the insects and mites associated with stored grain and cereal products, and their arthropod parasites and predators. *United States Department of Agriculture, Miscellaneous Publications* 258, 1–81.

Cowan, D.P. (1984) Life history and male dimorphism in the mite *Kennethiella trisetosa* (Acari: Winterschmidtiidae), and its symbiotic relationship with the wasp *Ancistrocerus antilope* (Hymenoptera: Eumenidae). *Annals of the Entomological Society of America* 77, 725–732.

Cox, M., De Baere, K., Vervaet, E., Zoons, J. and Fiks-Van Niekerk, T. (2009) Red mites: Monitoring method and treatment. Book of Abstracts 8th European Symposium on Poultry Welfare, 18–22 May, Cervia, Italy, pp. 18–22.

Creuwels, J. and Pieterse, S. (2022) Checklist Dutch Species Register – Nederlands Soortenregister. Naturalis Biodiversity Center. DOI: 10.15468/rjdpzy.

Cross, E.A. and Krantz, G.W. (1964) Two new species of the genus *Acarophenax* Newstead and Duvall 1918 (Acarina: Pyemotidae). *Acarologia* 6, 287–295.

Cross, E.A. and Moser, J.C. (1975) A new, dimorphic species of *Pyemotes* and a key to previously-described forms (Acarina: Tarsonemoidea). *Annals of the Entomological Society of America* 68, 723–732.

Cross, E.A., Moser, J.C. and Rack, G. (1981) Some new forms of *Pyemotes* (Acarina: Pyemotidae) from forest insects, with remarks on polymorphism. *International Journal of Acarology* 7, 179–196.

Crowther, D. and Wilkinson, T. (2008) House Dust Mites. In: Bonnefoy, X., Kanpen, H. and Sweeney, K. (eds) *Public Health Significance of Hurban Pests*. World Health Organization, Regional Office for Europe, Copenhagen, pp. 85–129.

Crozier, L. (1989) *Melittiphis alvearius* (Berlese) and other mites found in honeybee colonies in Nova Scotia. *Journal of Apicultural Research* 28, 166–168.

Crumb, S.E. (1934) A classification of some noctuid larvae of the subfamily Hypeninae. *Entomologica Americana* 14, 159–160.

Cuervo, N., Dusbabek, F., de la Cruz, J. and Abreu, R. (1983) Los acaros (Acarina, Pyroglyphidae, Cheyletidae, Saproglyphidae y Glycyphagidae) de los polvos domesticos en Cuba. *Revista Cubana de Medicina Tropical* 35, 83–103.

Cui, Y. (2014) When mites attack: domestic mites are not just allergens. *Parasites & Vectors* 7, 411–417.

Cunha, L.M. (2008) *Dermanyssus gallinae* (Acari: Dermanyssidae) (De Geer, 1778): Colonização E Resposta de Protoninfas Alimentadas a Correntes de area Odores de Extratos de Ácaros Co-específicos Em Olfatômetro Discriminante, Federal University of Minas Gerais, Belo Horizonte, Brazil.

Cunliffe, F. (1952) Biology of the cockroach parasite, *Pimeliaphilus podapolipophagus* Tragardh; with a discussion of the genera *Pimeliaphilus* and *Hirstiella*. *Proceedings of the Entomological Society of Washington* 54, 153–169.

Cunliffe, F. (1962) New species of Cheyletidae (Acarina). *Proceedings of the Entomological Society of Washington* 64, 209–218.

Cunliffe, F. and Baker, E.W. (1953) A guide to the predatory phytoseiid mites of the United States. *Pinellas Biology Laboratory Publication, Inc., U.S.A.* 1, 1–28.

Cunnington, A.M. (1965) Physical limits for complete development of the Grain mite, *Acarus siro* (Acarina, Acaridae), in relation to its world distribution. *Journal of Applied Ecology* 2, 295–306.

Cunnington, A.M. (1969) Physical limits for complete development of the copra mite, *Tyrophagus putrescentiae* (Schrank) (Acarina: Acaridae). In: Evans, G.O. (ed.) *Proceedings of the second International Congress of Acarology, 19–25 July, 1967*. Sutton Bonington, UK, pp. 241–248.

Cunnington, A.M. (1976) The effect of physical conditions on the development and increase of some important storage mites. *Annals of Applied Biology* 82, 175–201.

Cunnington, A.M. (1984) Resistance of the grain mite *Acarus siro* (*Acarina, Acaridae*) to unfavorable physical conditions beyond the limits of its development. *Agriculture, Ecosystems & Environment* 11, 319–339.

Cunnington, A.M. (1985) Factors affecting oviposition and fecundity in the grain mite *Acarus siro*

L. (Acarina: Acaridae), especially temperature and relative humidity. *Experimental and Applied Acarology* 1, 327–344.

Cunnington, A.M. and Gregory, P.H. (1968) Mites in bedroom air. *Nature* 217, 1271–1272.

Curry, J.P. and Momen, F.M. (1988) The arthropod fauna of grassland on reclaimed cutaway peat in central Ireland. *Pedobiologia* 32, 99–109.

Curtis, R.F., Hobson-Frohock, A., Fenwick, G.R. and Berreen, J.M. (1981) Volatile compounds from the mite *Acarus siro* L. in food. *Journal of Stored Products Research* 17(4), 197–203.

Cusack, D. (1974) Investigations on the biology and economic importance of acarine pests of stored food in Ireland. Ph. D. thesis, National University of Ireland, Dublin, pp. 1–143.

Cusack, P.D., Evans, G.O. and Brennan, P.A. (1975) A survey of the mites of stored grain and grain products in the Republic of Ireland. *Scientific Proceedings of the Royal Dublin Society, Series B*. 3, 273–329.

Cusack, D. and Brennan, A. (1976) Rate of colonisation of clean wheat and oatmeal by mites in infested areas of a flour mill. *Scientific Proceedings of the Royal Dublin Society, Series A*, 3, 341–353.

Cutcher, J.J. and Woodring, J.P. (1969) Environmental regulation of hypopial apolysis of the mite, *Caloglyphus boharti*. *Journal of Insect Physiology* 15, 2045–2057.

Cuthbert, O.D. (1990) Storage mite allergy. *Clinical Reviews in Allergy & Immunology* 8, 69–86.

Czajkowska, B. and Kropczynska, D. (1991) The influence of different host plants on the reproductive potential of *Tyrophagus putrescentiae* (Schrank) and *Tyrophagus neiswanderi* Johnston and Bruce (Acaridae). In: Schuster, R. and Murphy, P.W. (eds) The Acari. Reproduction, Development and Life History Strategies, Chapman and Hall, London, pp. 313–317.

Czaikowska, B., van de Vrie, M. and Kropczynska, D. (1988) Mites of the genus *Tyrophagus* as pests of ornamentals in greenhouses. *Mededelingen van de Faculteit Landbouwwetenschappen, Rijksuniversiteit Gent* 53(2b), 799–809.

Dara, S.K. Natwick, E.T. and Orloff, S.B. (2018) UC IPM Pest Management Guidelines: Onion and Garlic. UC ANR Publication 3453. Available at: https://onvegetables.com/2021/12/09/stored-garlic-might-have-mites/ (accessed 17 April 2023).

Dahl, F.T. (1910) Milben als Erzeuger von Zellwucherungen. *Centralblatt für Bakteriologie und Parasitenkunde* 53(5), 524–533.

Daneshnia, N. and Akrami, M.A. (2013) Mites (Acari) associated with the fig trees (*Ficus carica* L.) in Estahban (Fars Province), Iran. *Persian Journal of Acarology* 2(3), 539–541.

Daneshvar, H. (1978) A study on the fauna of plant mites in Azerbayjan. *Entomologie et Phytopathologie Appliquees* 46(1–2), 1–11.

Daneshvar, H. and Rodriguez, J.G. (1979) Fungi dietary of *Caloglyphus berlesei* (Mich.) (Acarina: Acaridae). In: Piffl, E. (ed.) *Proceeding of 4th International Congress of Acarology, Saalfelden, Austria, 12–19 August 1974*. Akademiai Kiado, Budapest, Hungary, pp. 593–598.

Daniel, M. (1988) Mesostigmatid mites in nests of small terrestrial mammals and features of their environment. Ceskoslowenská Academie VED, Studie CSAV 10, Academia, Prague, Czech Republic, pp. 1–178.

Danielsen, C., Stengård Hansen, L., Nachman, G. and Herling C. (2004) The influence of temperature and relative humidity on the development of *Lepidoglyphus destructor* (Acari: Glycyphagidae) and its production of allergens: a laboratory experiment. *Experimental and Applied Acarology* 32, 151–170.

Dar, N. and Gupta, V. (1979) Studies on the house dust mites of India and their role in causation of bronchial asthma and allergic rhinitis. Part I. The mites. *Oriental Insects* 13(3–4), 261–298.

Dar, N., Menon, M.P.S., Shivpuri, D.N. and Gupta, V.K. (1973) The mite fauna of Indian house dusts. *Aspects of Allergy and Applied Immunology* 6, 51–60.

Darvishzadeh, I. (1997) Biodiversity of grape mites and their predators in Safiabad, Khuzestan. M. Sc. thesis, Islamic Azad University Science and Reserch Branch, Tehran, Iran, pp. 1–102.

Datta, A.K. and Bhattacharjee, P.C. (1989) *Krantzolaspina rebatii*, a new genus and a new species (Acari: Mesostigmata: Parholaspididae) from Dibrugarh, Assam, India. In: Channabasavanna, G.P. and Viraktamath, C.A. (eds) *Progress in Acarology, 1*. Brill, Leiden, The Netherlands, pp. 411–414.

Davies, C.J. (1926) Vent disease in rabbits. *Veterinary Journal* 83, 320–322.

Davies, R.J., Green, M. and Schofield, N.M. (1976) Recurrent nocturnal asthma after exposure to grain dust. *American Review of Respiratory Disease* 114(5), 1011–1019.

Davis, J.W., Karstad, L.H. and Trainer, D.O. (eds) (1981) *Infectious Diseases of Wild Mammals.*

2nd edn, Iowa State University Press, Ames, IA, pp. 1–446.

De Boer, J.G. and Dicke, M. (2004) The role of methyl salicylate in preysearching behavior of the predatory mite *Phytoseiulus persimilis*. *Journal of Chemical Ecology* 30, 255–271.

de Boer, R. (1990a) Effect of heat treatments on the house dust mite *Dermatophagoides pteronyssinus* and *D. farinae* (Acari: Pyroglyphidae) in a mattress-like polyurethane foam block. *Experimental and Applied Acarology* 9, 131–136.

de Boer, R. (1990b) The control of house dust mite allergens in rugs. *Journal of Allergy and Clinical Immunology* 86, 808–814.

de Boer, R. (1998) Reflections on the control of mites and mite allergens. *Allergy* 53(Suppl. 48), 41–46.

de Boer, R. and Kuller, K. (1997) Mattresses as a winter refuge for house-dust mite populations. *Allergy* 52, 299–305.

de Boer, R. and van der Geest, L.P. (1990) House dust mite (Pyroglyphidae) populations in mattresses, and their control by electric blankets. *Experimental and Applied Acarology* 9, 113–122.

de Boer, R., Kuller, K. and Kahl, O. (1998) Water balance of *Dermatophagoides pteronyssinus* (Acari: Pyroglyphidae) maintained at brief daily spells of elevated air humidity. *Journal of Medical Entomology* 35, 905–910.

De Jesus, L.A. and Rueda, L.M. (1990) Some Philippine macrochelid mites (Acarina: Macrochelidae). *Philippine Entomologist* 8, 709–725.

De Jong, J.H., Lobbes, P.V. and Bolland, H.R. (1981) Karyotipes and sex determination in two species of laelapid mites (Acari: Gamasida). *Genetica* 55, 187–190.

de la Torre Morin, F., de la Coba, R.A., Gallego, M., Iraola, V. and Fernandez-Caldas, E. (2006) Sensitization to *Thyreophagus entomophagus* and other mite species in Cuban patients with allergic asthma and/or rhinoconjunctivitis. *Journal of Allergy and Clinical Immunology* 117, 119.

de la Torre Santana, P.E. and OConnor, B.M. (2011) Primer registro de *Neosuidasia faini* Ranganath & ChannaBasavanna, 1983 (Acari: Suidasiidae) en América. *Revista Ibérica de Aracnología* 19, 167–168.

De Leon, D. (1963) A new *Dermatophagoides*: it prevents the rising of self-rising flour (Acarina: Epidermoptidae). *Florida Entomologist* 46, 247–250.

De Leon, D (1964) Four new *Ameroseius* Berlese 1903 from the United States (Acarina: Ameroseiidae). *Florida Entomologist* 47, 213–217.

De Leon, D. (1967) Some mites of the Caribbean area. Part I. Acarina on plants in Trinidad, West Indies. Allen Press Inc., Lawrence, U.S.A., pp. 1–66.

De Lillo, E. and Porcelli, F. (1993) *Pyemotes herfsi* (Oud.) (Acari, Pyemotidae) antagonista di *Melanaspis inopinata* (Leon.) (Coccoidea, Diaspididae) in Puglia. *Entomologica, Bari* 27, 117–124.

De Lillo, E. and La Notte, P. (1995) Indagine sull'Acarofauna in alcune aree cerealicole di Puglia e Basilicata. *Entomologica, Bari* 29, 107–134.

De Luna, C.J., Moro, C.V., Guy, J.H., Zenner, L. and Sparagano, O.A. (2009) Endosymbiotic bacteria living inside the poultry red mite (*Dermanyssus gallinae*). *Experimental and Applied Acarology* 48, 105–113.

de Saint Georges-Gridelet, D. (1987) Vitamin requirements of the European house dust mite, *Dermatophagoides pteronyssinus* (Acari: Pyroglyphidae), in relation to its fungal association. *Journal of Medical Entomology* 24, 408–411.

de Saint Georges-Gridelet, D., Authy, G.W., Ardeshir, F., Grootaert, P. and Tirry L. (2003) Mite community of storage facilities in Iran. *Bulletin de l'Institut Royal des Sciences naturelles de Belgique. Entomologie & Biologie* 73, 67–76.

DeClercq, J. and Nachtegaele, L. (1993) *Dermanyssus gallinae* in a dog. *Journal of Canine Practice* 18, 34–36.

Decru, E., Mul, M., Nisbet, A.J., Vargas Navarro, A.H., Chiron, G. *et al.* (2020) Possibilities for IPM strategies in european laying hen farms for improved control of the poultry red mite (*Dermanyssus gallinae*): Details and state of affairs. *Frontiers in Veterinary Science* 7, 565866. DOI: 10.3389/fvets.2020.565866.

Dekeyser, M.A. and Downer, R.G. (1994) Biochemical and physiological targets for miticides. *Pesticide Science* 40, 85–101.

Delfinado-Baker, M. and Baker, E.W. (1982) A new record for *Aeroglyphus robustus* in beehive. *The American Bee Journal* 122, 110–110.

Delfinado-Baker, M. and Baker, E.W. (1987) Notes on mites new to beehives in Puerto Rico and in North American. *The American Bee Journal* 127(5), 365–366.

Deliargyris, N., Galatas, J. and Kontoy-Fili, K. (1990) House dust mites in Athens, Greece. *Clinical & Experimental Allergy* 20(Supplement 1), 120 (abstract).

Deltshet, C., Andreev, S., Blagoev, G., Golemansky, V., Milojkova, G. *et al.* (1998) Invertebrates (non-Insecta) in Bulgaria. In: Meine, C. (ed.) *Bugaria's Biological Diversity: Conservation Status and Needs Assessment, vol. I and II*. Pensoft, Sofia and Moscow, pp. 109–161.

Demite, P.R., McMurtry, J.A. and Moraes, G.J. (2014) Phytoseiidae database: A website for taxonomic and distributional information on phytoseiid mites (Acari). *Zootaxa* 3795(5), 571–577.

Demite, P.R., Moraes, G.J., McMurtry, J.A., Denmark, H.A. and Castilho, R.C. (2020) Phytoseiidae Database. Available at: www.lea.esalq.usp.br/phytoseiidae (accessed 8 August 2022).

Den Heyer, J. (1980) *Pulaeus*, a new cunaxid genus (Prostigmata: Acari). *Acarologia* 21, 18–33.

Denmark, H.A., Evans, G.A., Aguilar, H., Vargas, C. and Ochoa, R. (1999) *Phytoseiidae of Central America (Acari: Mesostigmata)*. Indira Publishing House, West Bloomfield, MI, pp. 1–125.

Department of Agriculture and Food, Western Australia (2009) Archived 2009-06-09 at the Library of Congress Web Archives.

Deschiens, R. (1951) L'acariase de l'appareil respitarice chez le primates et chez l'homme. *Annales de l'Institut Pasteur, Paris* 80, 107–147.

Di Palma, A. and Alberti, G. (2005) Functional morphology of accessory structures involved in reproduction among gamasid mites (Acari, Anactinotrichida): A review. *Atti Dell'accademia Nazionale Italiana Di Entomologia Ra Liii*, 361–378.

Di Palma, A. and Giangaspero, A. (2021) Laelapid and dermanyssid mites of medical and veterinary interest. In: Caplan, M., Mitchell, R., Bradshaw, R. and McManus, L. (eds) *Reference Module in Biomedical Sciences*. Elsevier, Amsterdam, The Netherlands, pp. 1–14.

Di Palma, A., Giangaspero, A., Cafiero, M.A. and Germinara, G.S. (2012) A gallery of the key characters to ease identification of *Dermanyssus gallinae* (Acari: Gamasida: Dermanyssidae) and allow differentiation from *Ornithonyssus sylviarum* (Acari: Gamasida: Macronyssidae). *Parasites and Vectors* 5, 104.

Di Palma, A., Leone, F., Albanese, F. and Beccati, M. (2018) A case report of *Dermanyssus gallinae* infestation in three cats. *Veterinary Dermatology* 29(348), e124.

Diaz, A. (1998) Aspects of the biology, ecology and behavior of the bulb mite, *Rhizoglyphus robini* (Claparede) (Acari: Acaridae): A pest of onions in New York. Ph.D. Thesis. Cornell University, Ithaca, New York.

Diaz, A., Okabe K., Eckenrode, C.J., Villani, M.G. and OConnor, B.M. (2000) Biology, ecology, and management of the bulb mites of the genus *Rhizoglyphus* (Claparede) (Acari: Acaridae). *Experimental and Applied Acarology* 24, 85–113.

Dick van, de Ven (2016) Q-Perch, electronic control of red mite. Vencomatic Group. In: Abstract Book, 2nd COST ACTION FA1404. Improving Current Understanding and Research for Sustainable Control of the Poultry Red Mite *Dermanyssus gallinae* (COREMI), 1–3 June, Zagreb, Croatia, 24.

Dicke, M. (1995) Plant characteristics influence biological control agents: implications for breeding for host plant resistance. *IOBC/WPRS Bulletins* 19(5), 72–80.

Dicke, M. and Grostal, P. (2001) Chemical detection of natural enemies by arthropods: An ecological perspective. *Annual Review of Ecology, Evolution, and Systematics* 32, 1–23.

Dinabandhoo, C.L. and Dogra, G.S. (1979) Dichlorvos for the control of acarine disease of Indian honey bee, *Apis cerana* Fab. *Indian Bee Journal* 41, 1–6.

Dinabandhoo, C.L. and Dogra, G.S. (1980) Incidence and control of ectoparasitic mite, *Pyemotes herfsi* Oudemans of the Indian honey bee, *Apis cerana* Fab. *American Bee Journal* 120, 44–47.

Dinglasan, R.R. and Jacobs-Lorena, M. (2008) Flipping the paradigm on malaria transmission-blocking vaccines. *Trends in Parasitology* 24, 364–370.

Divekar, P.A., Narayana, S., Divekar, B.A., Kumar, R., Gadratagi, B.G., Ray, A. and Behera, T.K. (2022) Plant secondary metabolites as defense tools against herbivores for sustainable crop protection. *International Journal of Molecular Sciences* 23(5), 2690.

Dixit, I.P. and Mehta, R.S. (1973) Prevalence of *Dermatophagoides* sp. Bogdanov, 1864 in India and its role in causation of bronchial asthma. *Journal of the Association of Physicians of India* 21, 31–37.

Dizlek, H., Karagoz, M., Faraji, F. and Cakmak, I. (2019) Mites in dried figs of Turkey: Diversity, species composition and density. *Systematic and Applied Acarology* 24(6), 992–997.

Doğan, S. (2003a) On caligonellid mites from Turkey (Acari: Caligonellidae). *Archives des Sciences, Genève* 56, 63–67.

Doğan, S. (2003b) A new species and two new records of the genus *Stigmaeus* (Acari: Stigmaeidae) from Turkey. *International Journal of Acarology* 29, 381–387.

Doğan, S. and Ayyildiz, N. (2003) *Stigmaeus zootaxai*, a new species of the genus *Stigmaeus* (Acari: Stigmaeidae) from Turkey with data of other stigmaeid mites. *Archives des Sciences, Genève* 56(1), 1–10.

Doğan, S., Jalaeian, M. and Kamali, H. (2011) New records of two cheyletid mite species (Acari: Cheyletidae) from Iran. *Turkish Journal of Zoology* 35(3), 1–2.

Domrow, R. (1979) Ascid and ameroseiid mites phoretic on Australian mammals and birds. *Records of the Western Australian Museum* 8, 97–116.

Domrow, R. (1992) Acari Astigmata (excluding feather mites) parasitic on Australian vertebrates: An annotated checklist, keys and bibliography. *Taxonomy of Invertebrates* 6, 1459–1606.

Dönel, G. and Doğan, S. (2011) A new species of *Raphignathus* Dugés and a newly discovered male of *R. fani* Doğan & Ayyıldız from Turkey. *International Journal of Acarology* 37, 27–33.

Dönel, G. and Doğan, S. (2012) Three new mite species of the genus *Mediolata* Canestrini (Acari, Stigmaeidae) from Turkey and re-description of *Mediolata granaria* Gonzalez-Rodriguez. *Journal of Natural History* 46(11–12), 683–699.

Dorny, P., Van Wyngaarden, T., Vercruysse, J., Symoens, P.C. and Jalila, A. (1994) Survey on the importance of mange in the aetiology of skin lesions in goats in peninsular Malaysia. Tropical Animal Health and Production. 26, 81–86.

Dosse, G. and Schneider, I. (1957) Biologie und Lebensweise von *Czenspinskia lordi* Nesbitt (Acar., Sarcoptiformes). *Zeitschrift für angewandte Zoologie* 44, 403–418.

Douglas, A.E. and Hart, B.J. (1989) The significance of the fungus *Aspergillus penicilloides* to the house dust mite *Dermatophagoides pteronyssinus*. *Symbiosis* 7, 105–116.

Dowell, F. E., Throne, J.E. and Baker, J.E. (1998) Automated nondestructive detection of internal insect infestation of wheat kernels using near-infrared reflectance spectroscopy. *Journal of Economic Entomology* 91, 899–904.

Dowell, F.E., Throne, J.E., Wang, D. and Baker, J.E. (1999) Identifying stored-grain insects using near-infrared spectroscopy. *Journal of Economic Entomology* 92(1), 165–169.

Dowling, A.P.G. and Oconnor, B.M. (2010) Phylogeny of Dermanyssoidea (Acari: Parasitiformes) suggests multiple origins of parasitism. *Acarologia* 50(1), 113–129.

Drukker, B., Bruin, J., Jacobs, G., Kroon, A. and Sabelis, M.W. (2000) How predatory mites learn to cope with variability in volatile plant signals inthe environment of their herbivorous prey. *Experimental and Applied Acarology* 24, 881–895.

du Toit, B.J., Theron, B.D. and Ueckermann, E.A (1998) A new genus and four new species of the family Camerobiidae (Acari: Raphignathoidea) from South Africa. *International Journal of Acarology* 24(1), 3–19.

Dubinin, V.B. (1951) Feather mites (Analgesoidea). Part I. Introduction to their study. Fauna of the USSR, Arachnida, Vol. 6 (fasc. 5), Akademia Nauk SSSR, Moscow-Leningrad, 1–363 (in Russian).

Dubinina, H.V. (1985) Ecologo-faunistic study of house-dust mites with regard to the problem of house-dust allergy. *Parazitologicheskij Sbornik* 33, 209–229 (in Russian).

Dubinina, H.V. and Plietniev, B.D. (1978) Acarofauna of dust of human dwellings. *Parazitologicheskij Sbornik* 28, 37–46 (in Russian).

Dubinina, H.V., Vaitsekuaskaite, R.L., Baiorinaite, A. and Razgauskas, A.F. (1984) The fauna of house dust mites in the homes of bronchial asthma patients in the Lithuanian SSR. *Meditsinskaia Parazitologiia i Parazitarnye Bolezni* 5, 63–67 (in Russian).

Dudinska, A.T. (2003) The ways of long-term acarocomplex synanthropic storehouse-mites formation in mills under transcarpathian conditions. *Vestnik Zoologii* 37(5), 85–89.

Duek, L., Kaufman, G., Palevsky, E. and Berdicevski, I. (2001) Mites in fungal cultures. *Mycoses* 44, 390–394.

Dufour, M.L. (1839) Description et figure de quelque parasites de 1'Ordre des Acariens. *Annales des Sciences Naturelles 2ª serie, XI, Zoologie*, 274–281.

Dumez, M.E., Herman, J., Campizi, V., Galleni, M., Jacquet, A. and Chevigne, A. (2014) Orchestration of an uncommon maturation cascade of the housedust mite protease allergen quartet. *Frontiers in Immunology* 5, 138.

Dunlop, J.A. and Bertrand, M. (2011) Fossil labidostomatid mites (Prostigmata: Labidostomatidae) from Baltic amber. *Acarologia* 51, 191–198.

Dunlop, J.A. and Selden, P.A. (2009) Calibrating the chelicerate clock: A paleontological reply to Jeyaprakash and Hoy. *Experimental and Applied Acarology* 48, 183–197.

Dunn, J.A. (2000) Literature review identifying chemicals for testing as candidate attractants for mite pests of stored grain. Report no. 109, Central Science Laboratory, Department for Environment, Food and Rural Affairs, Sand Hutton, York, UK, pp. 1–16.

Dunn, J.A. (2003) The effect of fluctuating temperature and humidity, and aeration on population growth of *Acarus siro* (L.) near the surface of a grain bulk. In: Credland, P.F., Armitage, D.M., Bell, C.H., Cogan, P.M. and Highley, E. (eds) *Proceedings of the 8th International Working Conference on Stored Product Protection, 22–26 July 2002, York, UK*. CAB International, Wallingford, UK, pp. 173–178.

Dunn, J.A., Danks, C., Thind, B.B., Banks, J.N. and Chambers, J. (2002) A rapid method for detecting the predominant storage mite pest species, *Acarus siro*, in the presence of grain. HGCA Project Report, No. 284, 1–21. Available at: https://projectblue.blob.core.windows.net/media/Default/Research%20Papers/Cereals%20and%20Oilseed/284_complete_final_report.pdf (accessed 20 June 2020).

Dunn, J.A., Prickett, A.J. and Thind, B.B. (2005) Assessment of monitoring methods for early detection of three species of storage mites in bulk oilseed rape. *Experimental and Applied Acarology* 37, 131–140.

Dunn, J.A., Thind, B.B., Danks, C. and Chambers, J. (2008) Rapid method for the detection of storage mites in cereals: Feasibility of an ELISA based approach. *Bulletin of Entomological Research* 98(2), 207–213.

Durden, L.A., Best, T.L., Wilson, N. and Hilton, C.D. (1992) Ectoparasitic mites (Acari) of sympatric Brazilian Free-Tailed Bats and Big Brown Bats in Alabama. *Journal of Medical Entomology* 29, 507–511.

Durkin, E.S., Proctor, H. and Luong, L.T. (2019) Life history of *Macrocheles muscaedomesticae* (Parasitiformes: Macrochelidae): New insights on life history and evidence of facultative parasitism on Drosophila. *Experimental and Applied Acarology* 79(3/4), 309–321.

Dusbábek, F. (1975) Population structure and dynamics of the house dust mite *Dermatophagoides farinae* (Acarina, Pyroglyphidae) in Czechoslovakia. *Folia Parasitologica* 22, 219–231.

Dusbábek, F. (1979) Dynamics and structure of mixed populations of *Dermatophagoides farinae* and *D. pteronyssinus*. *Recent Advances in Acarology* 2, 173–177.

Dusbábek, F. (1995) Present state of research on house dust mites (Pyroglyphidae) in the Czech Republic. *Wiadomości Parazytologiczne* 41, 337–342.

Dutra, M.S, Roncada, C., da Silva G.L., Ferla, N.J and Pitrez, P.M. (2018) Mite fauna assessment in houses of two distinct socioeconomic groups from Southern Brazil. *Journal of Medical Entomology* 55(3), 620–625.

Dwibadra, D. (2015) Taxonomy and biogeography of the family macrochelidae (Acari: Mesostigmata) associated with dung beetles (Coleoptera: Scarabaeidae) in Kalimantan, Indonesia. Hokkaido University Graduate School of Agriculture. Division of Environmental Resources Doctor Course, 1–167.

Dziêciołowski, R. (1994) Roztocze kurzu domowego Poznania i okolic. M. B. Thesis, Adam Mickiewicz University, Poznań, Poland (in Polish).

Early, M. and Goff, M.L. (1986) Arthropod succession patterns in exposed carrion on the island of O'ahu, Hawaiian Islands, USA. *Journal of Medical Entomology* 23, 520–531.

Eaton, K.K., Downing, F.S., Griffiths, D.A., Hockland, S., Lynch, S. (1985) House dust mites (*D. pteronyssinus*) in pets' beds and their relation to dust allergy. *Clinical & Experimental Allergy* 15(2), 151–154.

Eaton, M., and Kells, S.A. (2011) Freeze mortality characteristics of the mold mite *Tyrophagus putrescentiae*, a significant pest of stored products. *Journal of Economic Entomology* 104(4), 1423–1429.

Ebeling, W. (1971) Sorptive dusts for pest control. *Annual Review of Entomology* 16, 123–158.

Ebssa, L., Borgemeister, C. and Poehling, H.-M. (2004) Effectiveness of different species/strains of entomopathogenic nematodes for control of western flower thrips (*Frankliniella occidentalis*) at various concentrations, host densities, and temperatures. *Biological Control* 29(1), 145–154.

Ebrahimi, N. and Noei, J. (2022) Checklist of mites associated with stored products (Arachnida: Acari) of Iran. *Persian Journal of Acarology* 11(4), 559–631.

Egan, M.E. and Hunter, P.E. (1975) Redescription of a cockroach mite, *Proctolaelaps nauphoetae*, with notes on its biology. *Annals of the Entomological Society of America* 68, 361–364.

Egan, M.E. and Moss, W.W. (1969) The life cycle and behavior of cockroach mite, *Proctolaelaps nauphoetae* (Acari: Mesostigmata: Ascidae). *Not. Nat. Ac. Nat. Sci. Philadelphia* 420, 1–9.

Ehara, S. (1961) On some Japanese mesostigmatic mites (Phytoseiidae and Aceosejidae). *Annotationes Zoologicae Japonenses* 34(2), 95–98.

Ehara, S. (1964) Some mites of the families Phytoseiidae and Blattisocidae from Japan (Acarina: Mesostigmata). *Journal of the Faculty of Science, Hokkaido University, Zoology* 15, 378–394.

Eickwort, G.C. (1979) Mites associated with sweat bees (Halictidae). In: Rodriguez, J.G. (ed.) *Recent Advances in Acarology, Vol. I*. Academic Press, New York, pp. 575–581.

Eickwort, G.C. (1988) The origins of mites associated with honeybees. In: Needam, G.R., Page, E.E., Delfinado-Baker, M. and Bowman, C.E. (eds) *Africanised Honeybees and Bee Mites*. Ellis Horwood Chichester, UK, pp. 327–338.

Eickwort, G.C. (1990a) Associations of mites with social insects. *Annual Review of Entomology* 35, 469–488.

Eickwort, G.C. (1990b) Mites: An overview. In: Morse, R. and Nowogrodzki, R. (eds) *Honey Bee Pests, Predators and Diseases, 2nd edition*. Cornell University Press, Ithaca, pp. 188–199.

El Adouzi, M., Arriaga-Jiménez, A., Dormont, L., Barthes, N., Labalette, A. *et al.* (2020) Modulation of feed composition is able to make hens less attractive to the poultry red mite *Dermanyssus gallinae*. *Parasitology Research* 147(2), 171–181.

El Badry, E. (1969) Two new species of cheyletid mites from milled wheat (Acarina: Cheyletidae). *Journal of Stored Products Research* 5, 157–167.

Eldridge, H. (1976) Common defects in buildings, Her Majesty's Stationery Office, London, pp. 1–486.

Elhalawany, A., Abo-Shnaf, R. and Sanad, A.S. (2021) Release of predatory mite, *Neoseiulus barkeri* (Acari: Phytoseiidae) for its suppression two species of eriophyid mites (Acari: Eriophyidae) on olive seedlings in Egypt. *International Journal of Acarology* 47(1), 1–6.

Eliopoulos, P.A. and Papadoulis, G.T. (2001) New records of mites (Acari: Cheyletidae) from stored products with description of a new species in Greece. *International Journal of Acarology* 27, 29–33.

Eliopoulos, P.A., Athanassiou, C.G., Palyvos, N.E., Stathas, G.J. and Buchelos, C.T. (2002) Dominance and frequency of predatory mites in stored products in Greece. Advances in stored product protection. In: Credland, P.F., Armitage, D.M., Bell, C.H., Cogan, P.M. and Highley, E. (eds) *Proceedings of the 8th International Working Conference on Stored Product Protection, 22–26 July 2002, York, UK*. CAB International, Wallingford, UK, pp. 258–262.

Elixmann, J. H., Bischoff, E., Jorde, W. and Linskens, H.F. (1991) Changements during 2 years in populations of different mite species in house dust before and after a single acaricidal treatment. *Acarologia* 32(4), 385–398.

Elkawas, H. (2011) *Acarines as Biological Control Agents: An Overview of Bio-relationships between Mites and Insects in Egypt*. Lambert Academic Publishing, London, pp. 1–128.

El-Lakwah, F.A. and Halawa, Z.A. (1997) Efficacy of modified atmospheres containing various carbon dioxide concentrations against some mite species infesting stored commodities. *Annals of Agricultural Sciences* 35(1), 619–628.

Ellse, L. and Wall, R. (2014) The use of essential oils in veterinary ectoparasite control: A review. *Medical and Veterinary Entomology* 28, 233–243.

El-Naggar, M.E., Abd Al-Aal, Z.E., El-Basheir, Z.M. and Mohamed, A.E.M. (2007) Developmental periods and fecundity of *Lepidoglyphus destructor* (Schrank) (Acari: Glycyphagidae) when fed on different fungi at different temperatures. In: *Proceedings of the second Scientific Environmental Conference 2007*. Zagazig University, pp. 115–123.

El-Sayed, F.M.A. and Ghallab, M.M.A. (2007) Survey on mites associated with major insect pests infesting stored grains in Middle Delta. *Journal of the Egyptian Society of Acarology* 1(1), 29–38.

Elsen, P. (1972a) Sept nouvelles espèces de *Neocypholaelaps* Vitzthum, 1941 (Mesostigmata: Ameroseiidae) phorétiques d'Hyménoptères africains. *Revue de Zoologie et de Botanique Africaines* 86, 17–30.

Elsen, P. (1972b) *Afrocypholaelaps* gen. n., un nouveau genre pour *Neocypholaelaps africana* Evans, 1963, et redescription de cette espèce (Acari: Mesostigmata). *Revue de Zoologie et de Botanique Africaines* 86, 158–162.

Elsen, P. (1973) Nouvelles espèces d'*Ameroseius* Berlese, 1903 (Acari, Mesostigmata: Ameroseiidae) phorétiques d'Hyménoptères et de Coléoptères. *Revue de Zoologie et de Botanique Africaines* 87, 729–757.

El-Sharabasy, H.M. (2010) A survey of mite species associated with the red palm weevil, *Rhynchophorus ferrugineus* (Olivier) in Egypt. *Egyptian Journal of Biological Pest Control* 20(1), 67–70.

Emberson, R.M. (1973) Macrochelid mites in New Zealand (Acarina: Mesostigmata: Macrochelidae). *New Zealand Entomologist* 5, 118–127.

Emberson, R.M. (2010) A reappraisal of some basal lineages of the family Macrochelidae, with the description of a new genus (Acarina: Mesostigmata). *Zootaxa* 2501, 37–53.

EMBRAPA (2022) Embrapa Suínos e Aves. Brasília-DF. Available at: https://www.embrapa.br/suinos-e-aves (accessed 14 December 2022).

Emekçi, M. and Toros, S. (1989) An investigation on the development of *Acarus siro* L. (Acarina, Acaridae) under the different temperatures and relative humidities. *Turkiye Entomoloji Dergisi* 13(4), 217–228 (in Turkish).

Emekçi, M. and Toros, S. (1999) Depolanmış hububat akarları üzerinde araştırmalar. *Orta Anadolu'da Hububat Tarımının Sorunları ve Çözüm Yolları Sempozyumu, Konya, Turkey*, 483–490 (in Turkish).

Emekçi, M. and Toros, S. (2000) Investigations on the stored grain mites. Orta Anadolu'da hububat tarımının sorunları ve çözüm yolları Sempozyumu, 8–11 Haziran 1999, Konya, Turkey, 483–490 (in Turkish).

Emekci, M., Ferizli, A.G., Tütüncü, S. and Navarro, S. (2007) The applicability of controlled atmospheres as an alternative to methyl bromide fumigation of dried fruits in Turkey. In: Donahaye, E.J., Navarro, S., Bell, C., Jayas, D., Noyes, R. and Phillips, T.W. (eds) *Proceedings of International Conference of Controlled Atmosphere and Fumigation in Stored Products, Gold-Coast Australia, 8–13 August 2004*. FTIC Ltd. Publishing, Israel, pp. 165–172.

Emmanuel, N. and Evans, G.O. (1979) Survival of mites associated with growing barley through harvest and into storage. In: Rodriguez, J.G. (ed.) *Recent Advances in Acarology*. Academic Press, New York, San Francisco, London, I, pp. 299–304.

Emous, R. van (2017) Verwachtte schade bloedluis 21 miljoen euro. Pluimveeweb.nl. 2017. Available at: https://www.pluimveeweb.nl/artikelen/2017/01/schade-bloedluis-21-miljoen-euro/ (accessed 28 October 2020).

Emous, R.A. van, and Napel, J. (2007) Monitoring van bloedluispopulatie op praktijkbedrijven; buis met stokje zeer geschikt voor bewustwording. *De Pluimveehouderij* 37, 8–9.

Emous, R.A. van, Fiks-van Niekerk, T.G.C.M. and Mul, M.F. (2006) Red mites in theory and practice. Praktijkrapport Pluimvee 17. The Netherland, Animal Science Groupe, Lelystad.

Endlein, U.M. and Flechtmann, C.H.W. (1976) Ocorrência de *Carpoglyphus lactis* L., 1758 Acari, Astigmata em bananas-passa em Sao Paulo, Brasil. *Ciência e Cultura* 288, 932–933.

Enkegaard, A., Sardar, M.A. and Brødsgaard, H.F. (1997) The predatory mite *Hypoaspis miles*: biological and demographic characteristics on two prey species, the mushroom sciarid fly, *Lycoriella solani*, and the mould mite, *Tyrophagus putrescentiae*. *Entomologia Experimentalis et Applicata* 82(2), 135–146.

Entrekin, D.L. and Oliver, J.H. (1982) Aggregation of the chicken mite, *Dermanyssus gallinae* (Acari: Dermanyssidae). *Journal of Medical Entomology* 19(6), 671–681.

EPA (Environmental Protection Agency) (2005) Evaluation of the ability of sulfuryl fluoride (ProFume™) to replace methyl bromide in postharvest uses. EPA, Washington, DC. Available at: https://fluoridealert.org/wp-content/pesticides/sf.docket.EPA-HQ-OAR-2005-0122.pdf (accessed 20 July 2020).

EPA (Environmental Protection Agency) (2006) Final Rulemaking: The 2006 Critical Use Exemption from the Phase-Out of Methyl Bromide. Available at: http://www.epa.gov/spd-public/mbr/ EPA, Washington, DC (accessed 20 July 2020).

Eraky, S.A. and Shoker, N.I. (1994) Two new deutonymphs of the genus *Histiostoma* Kramer, 1876 (Acari: Histiostomatidae) existing in stored onions. *Journal of Agricultural Science* 25(2), 163–168.

Erban, T. and Hubert, J. (2008) Digestive function of lysozyme in synanthropic acaridid mites enables utilization of bacteria as a food source. *Experimental and Applied Acarology* 44(3), 199–212.

Erban, T. and Hubert, J. (2010) Determination of pH in regions of the midguts of acaridid mites. *Journal of Insect Science* 10, 1–12.

Erban, T. and Hubert, J. (2011) Visualization of protein digestion in the midgut of the acarid mite *Lepidoglyphus destructor*. *Archives of Insect Biochemistry and Physiology* 78, 74–86.

Erban, T., Erbanova, M., Nesvorna, M. and Hubert, J. (2009) The importance of starch and sucrose digestion in nutritive biology of synanthropic acaridid mites: α-amylases and α-glucosidades are suitable targets for inhibitor based strategies of mite control. *Archives of Insect Biochemistry and Physiology* 71(3), 139–158.

Erban, T., Klimov, P.B., Smrz, J., Phillips, T.W., Nesvorna, M., Kopecky, J. and Hubert, J. (2016a) Populations of stored product mite *Tyrophagus putrescentiae* differ in their bacterial communities. *Frontiers in Microbiology*. Available at: https://www.frontiersin.org/articles/10.3389/fmicb.2016.01046/full (accessed 26 May 2020).

Erban, T., Rybanska, D., Harant, K., Hortova, B. and Hubert, J. (2016b) Feces Derived Allergens of *Tyrophagus putrescentiae* Reared on Dried Dog Food and Evidence of the Strong Nutritional Int eraction between the Mite and *Bacillus cereus* Producing Protease Bacillolysins and Exo-chitinases. *Frontiers in Physiology*. Available at: https://www.frontiersin.org/articles/10.3389/fphys.2016.00053/full (accessed 26 May 2020).

Erben, A.M., Rodriguez, J.L., McCullough, J. and Ownby, D.R. (1993) Anaphylaxis after ingestion of beignets contaminated with *Dermatophagoides farinae*. *Journal of Allergy and Clinical Immunology* 92(6), 846–849.

Eremeeva, M. and Muniz-Rodriguez, K. (2020) Rickettsialpox-A rare but not extinct disease: Review of the literature and new directions. *Russian Journal of Infection and Immunity* 10(3), 477–485.

Erikson, H., Brannstrom, S., Skarin, H. and Chirico, J. (2010) Characterization of *Erysipelothrix rhusiopathiae* isolates from laying hens and poultry red mites (*Dermanyssus gallinae*) from an outbreak of erysipelas. *Avian Pathology* 39, 505–509.

Eriksson, N.E., Holmen, A., Möller, C. and Wihl, Å.J. (1998) Sensitization according to skin prick testings in atopic patients with asthma or rhinitis at 24 allergy clinics in Northern Europe and Asia. *Allergology International* 47, 187–196.

Eriksson, T.L., Whitley, P., Johansson, E., van Hage-Hamsten, M. and Gafvelin, G. (1999) Identification and characterisation of two allergens from the dust mite *Acarus siro*, homologous with fatty acid-binding proteins. *International Archives of Allergy and Immunology* 119, 275–281.

Erktan, A., Or, D. and Scheu, S. (2020) The physical structure of soil: Determinant and consequence of trophic interactions. *Soil Biology and Biochemistry* 148, 107876.

Ermilov, S.G. (2016) Additions to the oribatid mite fauna of Central Ethiopia, with description of a new species of *Scheloribates* (*Bischeloribates*) (Acari, Oribatida). *Spixiana* 39(1), 75–82.

Ermilov, S.G., Hugo-Coetzee, E.A. and Theron, P.D. (2017) New and interesting moss mites (Acari, Oribatida) from South Africa, with description of two new species. *Systematic and Applied Acarology* 22(10), 1560–1573.

Erőss, J. and Mahunka, S. (1971) Investigations on coprophilous and stercoricolous Macrochelids (Acari, Gamasina) in Hungary, as possible agents in the control of synanthropous flies. *Parasitologia Hungarica* 4, 215–226.

Eshel, G., Unc, A., Egozi, R., Shakartchy, E., Doniger, T. and Steinberger, Y. (2021) Orchard floor management effect on soil free-living nematode communities. *Soil Research* 60, 310–319.

Eshel, G., Egozi, R., Goldwasser, Y., Kashti, Y., Fine, P. *et al.* (2015) Benefits of growing potatoes under cover crops in a mediterranean climate. *Agriculture, Ecosystems & Environment* 211, 1–9.

Escudero, L.A. and Ferragut, F. (2005) Life-history of predatory mites *Neoseiulus californicus* and *Phytoseiulus persimilis* (Acari: Phytoseiidae) on four spider mite species as prey, with special reference to *Tetranychus evansi* (Acari: Tetranychidae). *Biological Control* 32, 378–384.

Eslamizadeh, R. (1996) Predators of European red mite in Nazloo, Urmieh region and biology and efficiency of *Orius minutus* (L.) and *Stethorus punctillum* Weise in the laboratory. M.Sc. thesis, Urmieh University, 1–142.

Esmaili, M. (1983) *The Important Pests of Fruit Trees*. Sepehr Publication, Tehran, Iran, pp. 1–578.

Esser, J. R. (1988) Assessment and reduction of insect infestation of cured sh in South East Asia, with laboratory studies on *Chrysomya maegacephala* (fab.), a principal causative agent, Durham theses, Durham University. Available at: http://etheses.dur.ac.uk/6592/ (accessed 6 May 2020).

Esteca, F.C.N., Pérez Madruga, Y., Britto, E.P.J. and Moraes, G.J. (2014) Does the ability of

Blattisocius species to prey on mites and insects vary according to the relative length of the cheliceral digits? *Acarologia* 54(3), 359–365.

Estevez, M.D. (2006) Occupational contact urticaria-dermatitis by *Tyrophagus putrescentiae*. *Contact Dermatitis* 55(5), 308–309.

Estrada Peña, A., Sánchez Acedo, C., Castillo Hernández, J.A. and Gutiérrez Galindo, J. F. (1981) *Tyroborus lini* (Oudemans, 1924) como ácaro contaminante de alimentos. *Rev. Ibér. Parasitology* 41(3), 435–445.

Eurofins (2022) COCERAL Report on the Use of Substances to Protect Stored Grains and Oil Seeds. Available at: https://www.eurofins.de/food-analysis/food-news/food-testing-news/fumigants-phosphane-und-sulfuryl-fluoride/ (accessed 18 January 2023).

Evans, G.O. (1952) Terrestrial Acari new to Britain. II. *Annals and Magazine of Natural History (Series 12)* 5, 660–675.

Evans, G.O. (1957) An introduction to the British Mesostigmata (Acarina) with keys to families and genera. *Zoological Journal of the Linnean Society* 43, 203–259.

Evans, G.O. (1958) A revision of the British Aceosejinae (Acarina: Mesostigmata). *Proceedings of the Zoological Society of London* 131, 177–229.

Evans, G.O. (1963a) Some observations on the chaetotaxy of the pedipals in the Mesostigmata (Acari). *Annals and Magazine of Natural History* 6, 513–527.

Evans, G.O. (1963b) Observations on the chaetotaxy of the legs in the free-living Gamasina (Acari: Mesostigmata). *Bulletin of the British Museum (Natural History) Zoology* 10, 277–303.

Evans, G.O. (1963c) The genus *Neocypholaelaps* Vitzthum (Acari: Mesostigmata). *Annals and Magazine of Natural History* 13(6), 209–230.

Evans, G.O. (1969) A new mite of the genus *Thinoseius* Halbt. (Gamasina: Eviphididae) from the Chatham Islands, New Zealand. *Acarologia* 11, 505–514.

Evans, G.O. (1982) Observations of the genus *Protogamasellus* Karg (Acari: Mesostigmata) with a description of a new species. *Acarologia* 23(4), 303–313

Evans, G.O. (1992) *Principles of Acarology*. CAB International, Wallingford, UK, pp. 1–563.

Evans, G.O and Browning, E. (1956) British mites of the subfamily Macrochelinae Tragardh (Gamasinia-Macrochelidae). *Bulletin of the British Museum (Natural History) Zoology* 4(1), 1–55.

Evans, G.O. and Till, W.M. (1966) Studies on the British dermanyssidae (Acari: Mesostigmata). Part II. Classification. *Bulletin of the British Museum (Natural History) Zoology* 14(5), 109–370.

Evans, G.O. and Till, W.M. (1979) Mesostigmatic mites of Britain and Ireland (Chelicerata: Acari-Parasitiformes). An introduction to their external morphology and classification. *Transactions of the Zoological Society of London* 35(2), 139–270.

Evans, G.O., Sheals, J.G. and Macfarlane, D. (1961) The terrestrial Acari of the British Isles. An introduction to their morphology, biology and classification. 1. Introduction and biology. British Museum, Natural History, London, pp. 1–219.

Ewing, H.E. (1913) A new parasite of the house fly (Acarina, Gamasoidea). *Entomological News* 24, 452–456, pl. XVII.

Ewing, H.E. (1925) New parasitic mites of the genus *Laelaps*. *Proceedings of the Entomological Society of Washington* 27, 1–7.

Eyndhoven, G.L. van (1960) Artuntescheide beim genus *Rhizoglyphus* (Acar.). In: *Proceedings XI International Congress of Entomology, 17–25 August 1960*. Vienna, pp. 274–276.

Eyndhoven, G.L. van (1964) *Cheletomorpha lepidopterorum* (Shaw, 1794) (= *Ch. venustissima*) (Acari, Cheyletidae) on Lepidoptera. Beaufortia, Series of miscellaneous publications, Zoological Museum, Amsterdam, The Netherlands 11, 53–60.

Eyndhoven, G.L. van (1968) *Rhizoglyphus engeli* nov. spec., with notes on the genus *Rhizoglyphus* (Acari: Acaridae). *Beaufortia* 15, 95–103.

Eyndhoven, G.L. van (1972) Some notes on the genitalia of the genus *Rhizoglyphus* (Acari: Acaridae). *Zeszyty Problemowe Postępów Nauk Rolniczych* 129, 23–26.

Eyndhoven, G.L. van and Vacante, V. (1985) The Berlesei-group of the genus *Bryobia* Koch (Acari, Tetranychidae). *Redia* 68, 377–437.

Ezequiel, S., da Silva O., Gazêta, G.S. and Freire, da S.N.M. (2003) Dermatological manifestations caused by mites of the cheyletidae family: A case report. *Anais Brasileiros de Dermatologia* 78(1), 87–90.

Ezequiel, S., Gazeta, G.S., Amorim, M. and Serra-Freire, N.M. (2008) Ácaros da família Cheyletidae

(Acari: Actinedida) em ecossistema domiciliar no município de Juiz de Fora, MG, Brasil. *Revista de Patologia Tropical* 37(1), 70–74.

Faasch, H. (1967) Beitrage sur Biologie der einheimischen Uropodiden *Uroobovella marginata* (C.L. Koch 1839) und *Uropoda orbicularis* (O.F. Müller 1776) und experimentelle Analyse iheres Phoresieverhaltens. *Zoologische Jahrbücher Abt. System* 94, 521–608.

Fain, A. (1962) Les acariens parasites nasicoles des Batraciens. Revision des Lawrencarinae Fain, 1957 (Ereynetidae: Trombidiformes). *Bulletin de l'Institut royal des sciences naturelles de Belgique* 38(25), 1–69.

Fain, A. (1965) Les Acariens nidicoles et détriticoles de la famille pyroglyphidae cunliffe. *Revue de Zoologie et de Botanique Africaines* 72(3–4), 257–288.

Fain, A. (1966a) Nouvelle description de *Dermatophagoides pteronyssinus* (Trouessart, 1897). Importance de cet Acarien en pathologie humaine (Psoroptidae). *Acarologia* 8(2), 302–327.

Fain, A. (1966b) Allergies respiratoires produites par un Acarien *(Dermatophagoides pteronyssinus)* vivant dans les poussières des habitations. *Bulletins de l'Académie royale des sciences, des lettres et des beaux-arts de Belgique* 6(6–7), 479–499.

Fain, A. (1967a) Les hypopes parasites des tissues cellulaires des oiseaux (Hypodectidae. Sarcoptiformes). *Bulletins de l'Académie royale des sciences, des lettres et des beaux-arts de Belgique* 43, 1–139.

Fain, A. (1967b) Le genre *Dermatophagoides* Bogdanov, 1864 – Son importance dans les allergies respiratoires et cutanées chez l'homme (Psoroptidae: Sarcoptiformes). *Acarologia* 9(1), 179–225.

Fain, A. (1967c) Deux nouvelles espèces de Dermatophagoidinae. Rattachement de cette sous-famille aux Pyroglyphidae (Sarcoptiformes). *Acarologia* 9(4), 870–881.

Fain, A. (1969a) Les deutonymphs hypopiales vivant en association phoretique sur les mammiferes (Acarina: Sarcoptiformes). *Bulletins de l'Académie royale des sciences, des lettres et des beaux-arts de Belgique* 45(33), 1–262.

Fain, A. (1969b) Morphologie et cycle evolutif des Glycyphagidae commensaux de la taupe *Talpa europaea* (Sarcoptiformes). *Acarologia* 11, 750–795.

Fain, A. (1971a) Évolution de certains groupes d'hypopes en fonction du parasitisme (Acarina, Sarcoptiformes). *Acarologia* 13, 98–112.

Fain, A. (1971b) Notes sur les Acariens des denres alimentaires à Kinshasa. (Republique Democratique du Congo). *Revue de Zoologie et de Botanique Africaines* LXXXIV(1–2), 175–183.

Fain, A. (1974) Observations sur la faune acarologique du sol dans une savane de Côte d'Ivoire (Acaridae: Sarcoptiformes). *Bulletin de l'Institut français d'Afrique noire. Série A* 36, 69–91.

Fain, A. (1976) Le genre *Austroglycyphagus* Fain et Lowry, 1974 (Acarina, Astigmata, Glycyphagidae) description d'espèces nouvelles. *Acarologia* 17(4), 709–729.

Fain, A. (1979) Acariens du genre *Cheyletus* (Prostigmata: Cheyletidae) recoltes dans la region Afrotropical. *International Journal of Acarology* 6(4), 275–284.

Fain, A. (1982) Revision des genres *Thyreophagus* Rondani, 1874 et *Michaelopus* Fain & Johnston, 1974 (Acari: Acaridae) avec description de neuf especes nouvelles. *Bulletin de l'Institut royal des sciences naturelles de Belgique* 54(7), 1–47.

Fain, A. (1985) Nouvelles observations sur l'organe ereynetal et les solenidions chez les Ereynetidae (Acari, Prostigmata). *Bulletin et annales de la Société royale belge d'entomologie* 121, 247–260.

Fain, A. (1986) A new mite (Acari, Acaridae) from a nest of a wasp *Paragia tricolor* Smith in Australia. *Records of the Western Australian Museum* 12(4), 407–413.

Fain, A. (1988a) Notes sur le genre *Rhizoglyphus* Claparede, 1869 (Acari, Acaridae) avec description de deux especes nouvelles. *Acarologia* 29(1), 53–62.

Fain, A. (1988b) Morphologie, systématique et distribution géographique des acariens responsables d'allergies respiratoires chez l'homme. In: Guérin, B. (ed.) *Acariens et Allergies, Allerbio*, Varennes en Argonne, France, pp. 13–55.

Fain, A. (1988c) Nouvelles localités pour des acariens de la famille Pyroglyphidae. *Bulletin et annales de la Société royale belge d'entomologie* 124, 267–268.

Fain, A. (1990) Morphology, systematics and geographical distribution of mites responsible for allergic disease in man. In: Guerin, B. (ed.), Mites and Allergic Disease, Allerbio, Varennes-en-Argonne, France, pp. 11–134.

Fain, A. and Ardeshir, F. (2000) Note on the genus *Neocheyla* Radford, 1950 (Acari: Cheyletidae) with description of a new species from Iran. *International Journal of Acarology* 26, 329–334.

Fain, A. and Bafort, J. (1967) Cycle evolutif et morphologie de *Hypodectes (Hypodectoides) propus* (Nitzsch), acarien nidicole a deutonymphe parasite tissulaire des pigeons. *Bulletins de l'Académie royale des sciences, des lettres et des beaux-arts de Belgique, Classe des sciences* 53, 501–533.

Fain, A. and Beaucournu, J.C. (1993) Les hypopes des astigmates (Acari) phorétiques sur les puces (Siphonaptera) de mammiferès d'oiseaux. *Bulletin de l'Institut royal des sciences naturelles de Belgique, Entomology* 63, 77–93.

Fain, A. and Bochkov, A.V. (2001) A review of the genus *Cheyletus* Latreille, 1776 (Acari: Cheyletidae). *Bulletin de l'Institut royal des sciences naturelles de Belgique, Entomology* 71, 83–114.

Fain, A. and Camerik, A.M. (1994) Notes on the mites of the genus *Ereynetes* Berlese (Acari: Ereynetidae), with description of five new species from South Africa. *Bulletin de l'Institut royal des sciences naturelles de Belgique, Entomology* 64, 145–164.

Fain, A., and Fauvel, G. (1993) Tyrophagus curvipenis n.sp. from an orchid cultivation in a greenhouse in Portugal (Acari: Acaridae). *International Journal of Acarology* 19, 95–100.

Fain, A. and Galloway, T.D. (1993) Mites (Acari) from nests of sea birds in New Zealand. II. Mesostigmata and Astigmata. *Bulletin de l'Institut royal des sciences naturelles de Belgique, Entomology* 63, 95–111.

Fain, A. and Knülle, W. (1981) The life-cycle of *Acalvolia squamata* (Oudemans, 1909) (Acari, Astigmata, Saproglyphidae). *International Journal of Acarology* 7(1–4), 139–142.

Fain, A. and Lowry, W. (1974) A new genus and two new species of Glycyphaginae from Australia (Acarina: Glycyphagidae). *Bulletin et annales de la Société royale belge d'entomologie* 110, 215–222.

Fain, A. and Mumcuoglu, Y. (1979) *Glycyphagus (Myacarus) helveticus* sp.n. (Acari, Astigmata) vivant dans le nid d'*Arvicola terrestris* en Suisse. *Revue suisse de Zoologie* 86, 445–461.

Fain, A. and Nadchatram, M. (1980a) New house dust mites from Malaysia. 1. Two new species of *Austroglycyphagus* Fain & Lowry, 1974 (Astigmata: Glycyphagidae). *International Journal of Acarology* 6(1), 1–8.

Fain, A. and Nadchatram, M. (1980b) Cheyletid parasites or commensals in Malaysia (Acari: Cheyletidae). *International Journal of Acarology* 6(3), 191–200.

Fain, A. and Philips, J.R. (1978) Notes on the genus *Suidasia* Oudemans 1905 with a description of a new species from Australia (Acari, Astigmata, Saproglyphidae). *International Journal of Acarology* 4, 115–123.

Fain, A. and Philips, J.R. (1979) Astigmatic mites from nests of birds of prey in the U.S.A. V, Four new species of Anoetidae. *International Journal of Acarology* 5, 147–153.

Fain, A. and Rack, G. (1987) Notes on the mites living in the flowers of *Espeletia* spp. (Asteraceae) in Colombia. 1. *Carpoglyphus sturmi* sp.n. (Acari, Carpoglyphidae). *Entomologische Mitteilungen aus dem Zoologischen Museum Hamburg* 9(130), 9–19.

Fain, A. and Wauthy, G. (1979) Les Acaridae (Acari: Astigmata) d'horizons hemiedaphiques dans trente forêts décidues de Belgique. *Bulletin et annales de la Société royale belge d'entomologie* 115, 169–182.

Fain, A., Cunnington, A.M. and Spieksma, F.T.M. (1969) *Malayoglyphus intermediits* n. g., n. sp. récolté dans les poussières de maisons à Singapore et à Djakarta (Pyroglyphidae: Sarcoptiformes). *Acarologia* 11(1), 121–126.

Fain, A., de Cock, A.W.A.M. and Lukoschus, F.S. (1972) Parasitic mites of Surinam XVII. Description and life cycle of *Marsupialichus marsupialis* sp. n. from *Didelphis marsupialis* (Glycyphagidae: Sarcoptiformes). *Acarologia* 14, 81–93.

Fain, A., Lukoschus, F.S., Louppen, J.M.W. and Mendez, E. (1973) *Echimyopus dasypus*, n. sp., a hypopus from *Dasypus novemcinctus* in Panama (Glycyphagidae, Echimyopidae: Sarcoptiformes). *Journal of Medical Entomology* 10, 552–555.

Fain, A., Oshima, S. and Bronswijk, J.E.M.H (1974) *Hirstia domicola* sp.n. from house dust in Japan and Surinam (Acarina: Sarcoptiformes, Pyroglyphidae). *Japanese Journal of Sanitary Zoology* 25(3), 197–203.

Fain, A., Hyland, K. and Tadkouski, T. (1977) *Blomia gracilipes* (= *Chortoglyphus gracilipes*) redescription & status (Acarina: Glycyphagidae). *Entomological News* 88(9–10), 267–269.

Fain, A., Guerin, B. and Hart, B.J. (1990) Mites and allergic disease. In: Guérin, B. (ed.) *Allerbio*. Varennes en Argonne, France, pp. 1–190.

Fain, A., Greenwood, M.T. and Macfarlane, D. (1991) Mites (Acari) found in the nests of the dipper *Cinclus cinclus aquaticus* Bechstein, in Wales (British Isles). *Acarologia* 32(3), 193–204.

Fain, A., Athias-Binche, F. and Carton, Y. (1993) *Histiostoma unidentatum* n.sp. (Acari, Astigmata) associated with a fly *Drosophila bipectinata* originating from Thailand. *Bulletin de l'Institut royal des sciences naturelles de Belgique, Entomologie* 63, 71–76.

Fain, A., Knülle, W. and Wurst, E. (2000) First Description of the Hypopial Stage of *Thyreophagus entomophagus* (Laboulbene, 1852) (Acari Acaridae). *Bulletin de la Société royale de botanique de Belgique* 136, 153–156.

Fajfer, M. (2013a) Two new species of scale mites (Acari: Pterygosomatidae) parasitizing *Agama agama* (Sauria: Agamidae) from Kenya. *Zootaxa* 3745(2), 257–262.

Fajfer, M. (2013b) Three new species of scale mites (Acari: Pterygosomatidae) parasitizing *Agama sankaranica* (Sauria: Agamidae). *Zootaxa* 3700(2), 271–282.

Fajfer, M. and González-Acuña, D. (2013) Pterygosomatid mites of a new species group ligare (Acariformes: Pterygosomatidae: Pterygosoma) parasitizing tree iguanas (Squamata: Liolaemidae: *Liolaemus*). *Zootaxa* 3693(3), 301–319.

Fajfer, M. and Melnikov, D. (2014) New species and records of scale mites (Acari: Pterygosomatidae) from Arabian agamid lizards (Squamata: Agamidae). *Zootaxa* 3764(4), 401–417.

Faleiro, D.C.C. (2012) Ácaros Associados a Ninhos Abandonados Por Pássaros E a Aves de Postura de Ovos Comerciais, No Vale Do Taquari, Rio Grande Do Sul, Univates University Center: Lajeado, Brazil.

Faleiro, D.C.C., Toldi, M., Silva, G.L. and Ferla, N.J. (2015) The ectoparasites *Dermanyssus gallinae* and *Megninia ginglymura*: bioecology and natural enemies in commercial egg-laying hens. *Systematic and Applied Acarology* 20(8), 861–874.

Fan, Q.-H. (1992) Two new species of the genus *Rubroscirus* (Acari: Cunaxidae). In: Fujian Association for Science and Technology (eds) *First Academic Annual Meeting of Youth Proceedings*. Fujian Sciences and Technology Press, Fuzhou, China, pp. 723–726.

Fan, Q.-H. (2000) A phylogenetic analysis of the family Caligonellidae (Acari: Prostigmata) with description of two new species. *Acta Entomologica Sinica* 43, 421–428.

Fan, Q.-H. (2006) Bibliographic analysis of acarological papers published in Zootaxa from 2001 to 2005, with a catalogue of described new taxa. *Zootaxa* 1385, 53–66.

Fan, Q.-H. and Li, L. (1993a) Descriptions of three new species of superfamily Raphignathoidea (Acari: Actinedida). *Fujian Agriculture and Forestry University (Natural Sciences Edition)* 22(3), 321–324 (in Chinese).

Fan, Q.-H. and Li, L.-S. (1993b) Two new species of the genus *Ameroseius* from China (Acari: Ameroseiidae). *Entomological Journal of East China* 2(2), 17–20 (in Chinese).

Fan, Q.-H. and Li, L.-S. (1993c) A new species of *Lomelacarus* from China and proposal of a new subfamily for Glycyphagidae (Acari: Acaridida). *Acta Arachnologica Sinica* 2(1), 1–3.

Fan, Q.-H. and Li, L.-S. (1995) A new genus and a new species of Caligonellidae associated with food from China (Acari: Actinedida). *Acta Arachnologica Sinica* 20, 323–327.

Fan, Y.Q. and Petitt, F.L. (1994a) Parameter estimation of the functional response. *Environmental Entomology* 23, 785–794.

Fan, Y.Q. and Petitt, F.L. (1994b) Biological Control of Broad Mite, *Polyphagotarsonemus latus* (Banks), by *Neoseiulus barkeri* Hughes on Pepper. *Biological Control* 4(4), 390–395.

Fan, Y. and Petitt, F.L. (1994c) Functional response of *Neoseiulus barkeri* Hughes on two-spotted spider mite (Acari: Tetranychidae). *Experimental and Applied Acarology* 18, 613–621.

Fan, Q.-H. and Walter, D.E. (2011) *Acamerobia inflatus* gen. n. & sp. n. from Australia (Acari: Prostigmata: Raphignathoidea: Camerobiidae) with notes on the idiosomal chaetotaxy. *Zootaxa* 3045, 45–56.

Fan, Q.H. and Yan, C. (1997) The genus *Storchia*, with the description of a new species (Acari: Prostigmata: Stigmaeidae). *Systematic and Applied Acarology* 2, 161–166.

Fan, Q.H. and Yin, X.M. (2000) The genus *Raphignathus* (Acari: Raphignathidae) from China. *Systematic and Applied Acarology* 5, 83–98.

Fan, Q.-H. and Zhang, Z. Q. (2006) Revision of *Tyroborus* Oudemans, 1924 (Acari: Astigmata: Acaridae). *Zootaxa* 1152, 1–43.

Fan, Q.-H. and Zhang, Z.-Q. (2003) *Rhizoglyphus echinopus* and *Rhizoglyphus robini* (Acari: Acaridae) from Australia and New Zealand: identification, host plants and geographical distribution. *Systematic and Applied Acarology, Special Publications* 16, 1–16.

Fan, Q.-H. and Zhang, Z.-Q. (2004) Revision of *Rhizoglyphus* Claparede (Acari: Acaridae) of

Australia and Oceania. *Systematic and Applied Acarology Soc., London*, 1–374.

Fan, Q.H. and Zhang, Z.Q. (2005) Raphignathoidea (Acari: Prostigmata). Fauna of New Zealand 52. Mannaki Whenua Press, Lincoln, Canterbury, New Zealand, 1–400.

Fan, Q.-H. and Zhang, Z.-Q. (2007a) *Tyrophagus* (Acari: Astigmata: Acaridae). Fauna of New Zealand 56, Manaaki Whenua Press, Lincoln, New Zealand, 1–291.

Fan, Q.-H. and Zhang, Z.-Q. (2007b) Revision of some species of *Tyrophagus* (Acari: Acaridae) in the Oudemans Collection. *Systematic and Applied Acarology* 12, 253–280.

Fan, Q.-H. and Zhang, Z.-Q. (2014) Comment on the proposed conservation of usage by designation of a replacement neotype for *Acarus putrescentiae* Schrank, 1781 (currently *Tyrophagus putrescentiae*; Acariformes, Acaridae). *Bulletin of Zoological Nomenclature* 71(2), 99–102.

Fan, Q.-H., Chen, Y. and Wu, Z-Y. (2001) A revision of mites (Acari) living on stored food from China. In: Hu, C. and Cheng, J. (eds) *Advances in Urban Entomology, Proceedings of the Sixth National Symposium on Urban Entomology*, pp. 205–220 (in Chinese).

Fan, Q.-H., Li, L.-S. and Xuan J.-Y. (1996) Two new species of the genus *Paralycus* from Sichuan, China (Acari: Pediculochelidae). *Acta Zoologica Sinica* 21(2), 174–178 (in Chinese).

Fan, Q.-H., Chen, Y. and Wang, Z.-Y. (2010a) Acaridia (Acari: Astigmatina) of China: A review of research progress. *Zoosymposia* 4, 225–249.

Fan, Q.-H., Sherly G. and Kumarasinghe, L. (2010b) Genus *Acalvolia* (Acari: Winterschmidtiidae), with the description of a new species from the US. *Zootaxa* 2719, 41–61.

Fan, Q.-H., Flechmann, C.H.W. and Moraes, G.J. (2016) Annotated catalogue of Stigmaeidae (Acari: Prostigmata), with a pictorial key to genera. *Zootaxa* 4176, 1.

Fan, Q.-H., Dayal, S.S., Fong, H.M., Rakuita, P. and Ram, J.A. (2020) A contribution to the fauna of mites (Acari) in Fiji. *Systematic and Applied Acarology* 25(8), 1444–1460.

Fang, W. and Cui, Y.-B. (2009) A survey of stored product mites of traditional Chinese medicinal materials. *The Pan-Pacific Entomologist* 85(4), 174–178.

FAO (Food and Agriculture Organization of the United Nations) (2022) Agricultural Production Statistics 2000–2020, Rome, Italy. Available at: https://www.fao.org/3/cb9180en/cb9180en.pdf (accessed 3 January 2023).

FAO (Food and Agriculture Organization of the United Nations) (2023) Gateway to Poultry Production and Products: Production, Rome, Italy. Available at: https://www.fao.org/poultry-production-products/production/en/ (accessed 3 January 2023).

Farahbakhsh, G. (1961) Checklist of important insects and other enemies of plants and agricultural products in Iran. Ministry of Agriculture, Plant Protection Organization, Tehran, Iran, pp. 1–153.

Faraji, F. (1993) Injurious mites (Acari) associated with stored rice in eastern Mazandaran. In: *Proceedings of the 11th Iranian Plant Protection Congress, 21–24 September 1993*. University of Guilan, Rasht, Iran, p. 257.

Faraji, F. and Kamali, K. (1993) Mites associated with *Citrus* spp. in eastern Mazandaran. In: *Proceedings of the 11th Iranian Plant Protection Congress, 21–24 September 1993*. University of Guilan, Rasht, Iran, p. 186.

Faraji, F. and Halliday, B. (2009) Five new species of mites (Acari: Laelapidae) associated with large Australian cockroaches (Blattodea: Blaberidae). *International Journal of Acarology* 35(3), 245–264.

Farahi, S. Shishehbor, P. and Nemati, A. (2018a) Some mesostigmatic mites (Acari: Parasitiformes) of Khuzestan province, southwestern Iran. *Persian Journal of Acarology* 7(4), 323–344.

Farahi, S., Shishehbor, P. and Nemati, A. (2018b) Bisexual and oedipal reproduction of *Macrocheles muscaedomesticae* (Acari, Macrochelidae) feeding on *Musca domestica* (Diptera, Muscidae) eggs. *Acarologia* 58(2), 430–441.

Farahi, S. Shishehbor, P. and Nemati, A. (2020) Records of Parasitidae and Laelapidae (Acari: Mesostigmata) from domestic animal manure in Khuzestan province, southwestern Iran with a new record for the Asian fauna. *Journal of Insect Biodiversity and Systematics* 6(3), 247–260.

Farhan, M., Afzal, M., Bashir, M.H., Ahmad, T. and Asi, M.R. (2013) Changes in nutritional value of stored maize grains infested with mites (*Rhizoglyphus tritici*) under different storage time. *International Journal of Applied Agricultural Sciences* 5(1), 47–52.

Farish, D.J. and Axtell, R.C. (1966) Sensory functions of the palps and first tarsi of *Macrocheles muscaedomesticae* (Acarina: Macrochelidae), a predator of the house fly. *Annals of the Entomological Society of America* 59, 165–170.

Farish, D.J. and Axtell, R.C. (1971) Phoresy redefined and examined in *Macrocheles muscaedomesticae* (Acarina: Macrochelidae). *Acarologia* 13, 16–29.

Farley, M.L., Mabry, L.C. and Hieger, L.R. (1989) Mites in pulmonary cytology specimens. *Diagnostic Cytopathology* 5(4), 416–426.

Farmer, E.E. (2001) Surface-to-air signals. *Nature* 411, 854–586.

Faroni, L.R.D., Guedes, R.N.C. and Matioli, A.L. (2001) Effect of temperature on development and population growth of *Acarophenax lacuatus* (Cross & Krantz) (Prostigmata: Acarophenacidae) on *Rhyzoptertha dominica* (F.) (Coleoptera: Bostrichidae). *Biocontrol Science and Technology* 11, 5–12.

Fashing, N.J. (1975) Life history and general biology of *Naiadacarus arboricola* Fashing, a mite inhabiting water-filled treeholes (Acarina: Acaridae). 1. *Naturalis Historia* 9, 413–424.

Fashing, N.J. (1976) The resistant tritonymphal instar and its implications in the population dynamics of *Naiadacarus arboricola* Fashing (Acarina: Acaridae). *Acarologia* 18, 704–714.

Fashing, N.J. (1994) Life-history patterns of astigmatid inhabitants of water-filled treeholes. In: Houck, M.A. (ed.) *Mites, Ecological and Evolutionary Analyses of Life-History Patterns.* Springer Science+Business Media, Dordrecht, The Netherlands, pp. 160–185.

Fashing, N.J. (2010) Two novel adaptations for dispersal in the mite family Histiostomatidae (Astigmata). In: Sabelis, M.W. and Bruin, J. (eds) *Trends in Acarology: Proceedings of the 12th International Congress, 21–26 August 2006, Amsterdam.* Springer, Dordrecht, The Netherlands, pp. 81–84.

Fashing, N.J. (1979) *Naiadacarus arboricola*, a "K-selected" acarid mite. In: Rodriguez, J.G. (ed.) *Recent Advances in Acarology, Vol. I.* Academic Press, New York, pp. 525–529.

Fashing, N.J. and O'Connor, B.M. (1984) *Sarracenzopus* – A new genus for histiostomatid mites inhabiting the pitchers of the Sarraceniaceae (Astigmata: Histiostomatidae). *International Journal of Acarology* 10, 217–227.

Fathipour, Y. (1994) Soil mite fauna in orchards of Tabriz and population fluctuation and abundance of important species. M. Sc. thesis, Tarbiat Modarres University, Tehran, pp. 1–172.

Fathipour, Y. and Maleknia, B. (2016) Mite predators. In: Omkar (ed.) *Ecofriendly Pest Management for Food Security*. Elsevier, Academic Press, San Diego, USA, pp. 329–366.

Fathipour, Y., Kamali, K. and Ostovan, H. (1999) Mites of the family Cheyletidae (Acari: Prostigmata) collected from several parts of Iran and a key for their identification. *Journal of Agricultural Science* 1(1), 65–77.

Feiertag-Koppen, C.C.M. and Pijnacker, L.P. (1982) Development of the female germ cells and process of internal fertilization in the two-spotted spider mite *Tetranychus urticae* Koch (Acariformes: Tetranychidae). *International Journal of Insect Morphology and Embryology* 11, 271–284.

Fejt, R. and Ždárková, E. (2001) Bionomics of *Acarus siro* L. (Acarina: Acaridae) on Oilseeds. *Plant Protection Science* 37(3), 111–114.

Feldman-Muhsam, B., Mumcuoglu, K.Y. and Osterovitch, T. (1985) A survey of house dust mites (Acari, Pyroglyphidae and Cheyletidae) in Israel. *Journal of Medical Entomology* 22, 663–669.

Fenďa, P. (2010) Mites inhabiting bird nests in Slovakia (Western Carpathians). In: Sabelis, M. and Bruin, J. (eds) *Trends in Acarology, Proceedings of the 12h International Congress, 1st Edition.* Springer Science + Business Media B.V., Dordrech, The Netherlands, pp. 199–205.

Fenďa, P. and Mašán, P. (2003) Roztoče-Acari (Parasitiformes, ex. Uropodina). In: Mašán, P. and Svatoň, J. (eds) *Pavúkovce Národného parku Poloniny-Arachnids of the Poloniny National Park (Arachnida: Araneae, Pseudoscorpiones, Opiliones, Acari-Parasitiformes).* Štátna ochrana prírody SR Banská Bystrica a Správa Národného parku, Poloniny, Snina, pp. 143–205 (in Slovak).

Fenilli, R. and Flechtmann, C.H.W. (1990) Ácaros do pinheiro-do-Paraná em Lages, Santa Catarina. *Anais da Escola Superior de Agricultura Luiz de Queiroz* 47, 243–250.

Ferizli, A.G., Emekci, M., Tutuncu, S. and Navarro, S. (2007) Studies of Phosphine as a fumigant for dried fruit under tarpaulin covers. In: Donahaye, E.J., Navarro, S., Bell, C., Jayas, D., Noyes, R. and Phillips, T.W. (eds) *Proceedings of International Conference Controlled Atmosphere and Fumigation in Stored Products, 8–13th August 2004.* Gold-Coast, Australia, FTIC Ltd. Publishing, Israel, pp. 477–484.

Ferla, N.J. and Moraes, G.J. (2002) Ácaros predadores (Acari) em plantas nativas e cultivadas no estado do Rio Grande do Sul, Brasil. *Revista Brasileira de Zoologia* 19, 1011–1031.

Ferla, N.J., Marchetti, M.M. and Gonçalves, D. (2007) Ácaros predadores (Acari) associados à cultura de morango (*Fragaria* sp., Rosaceae) e plantas próximas no estado do Rio Grande do S. *Biota Neotropica* 2, 1–8.

Fernandez, N.A., Alberti, G. and Kümmel G. (1991) Spermatophores and spermatozoa of oribatid mites (Acari: Oribatida). Part I. Fine structure and histochemistry. *Acarologia* 32, 261–286.

Fernandez-Caldas, E., Sepulveda, A. and Montealegre, F. (1995) Identification of the domestic mite fauna of Puerto Rico. *Allergy* 50(Supplement 26), 72.

Fernandez-Caldas, E., Iraola, V. and Carnes, J. (2007) Molecular and Biochemical Properties of Storage Mites (Except Blomia species). *Protein & Peptide Letters* 14(10), 954–959.

Ferragut, F., Pérez Moreno, I., Iraola, V. and Escudero, A. (2010) *Ácaros depredadores en las plantas cultivadas*. Família Phytoseiidae. Ediciones Agrotécnicas, Madrid, Spain, pp. 1–202.

Ferreira Silva, A., Teixeira Pinto, Z., Leal Caetano, R., Carriço, C., Pereira Sato, T., Amorim. M. and Salles Gazeta, G. (2016) Bulb Mites *Rhizoglyphus echinopus* (Fumouze and Robin) Associated with Subterranean Termite (Isoptera) in Brazil. *EntomoBrasilis* 9(1), 65–68.

Fields, P.G. (1992) The control of stored-product insects and mites with extreme temperatures. *Journal of Stored Products Research* 28(2), 89–118.

Fields, P.G. and White, N.D.G. (2002) Alternatives to methyl bromide treatments for stored-product and quarantine insects. *Annual Review of Entomology* 47, 331–359.

Filatov, S., Erhart, J., Rego, R.O.M. (2022) Infestation of laboratory colonies of the soft tick, *Ornithodoros moubata* Murray 1877 (Acari: Ixodida: Argasidae) by *Tyrophagus fanetzhangorum* Klimov and OConnor 2009 (Acari: Astigmatina: Acaroidea: Acaridae). *Acarologia* 62(1), 223–227.

Filipponi, A. (1955) Sulla natura dell'associazione tra *Macrocheles muscaedomesticae* e *Musca domestica*. *Rivista di parassitologia* 16(2), 83–102.

Filipponi, A. (1960) Macrochelidi (Acarina: Mesostigmata) Foreticidi di mosche. Risultati parziali di una indagine ecologica in corso nell'agro pontino. *Parassitologia* 2(1–2), 167–172.

Filipponi, A. (1964) Experimental taxonomy applied to the Macrochelidae (Acari: Mesostigmata). *Acarologia* 1964, 92–100.

Filipponi, A. and Cicolani, B. (1974) Influenza della temperatura sulla fecondità, longevità e capacità moltiplicativa nell'intervallo ottimale di *Macrocheles matrius* (Acarina, Mesostigmata). *Rivista di parassitologia* 35, 291–306.

Filipponi, A. and Delupis, G.D. (1963) Sul regime dietetico di alcuni Macrochelidae (Acari: Mesostigmata), associati in natura a muscidi di interesse sanitario. *Rivista di parassitologia* 24, 277–287.

Filipponi, A. and Pegazzano, F. (1962) Specie italiane del *gruppo-glaber* (Acarina, Mesostigmata, Macrochelidae, *Macrocheles*). *Redia* 47, 211–238.

Filipponi, A. and Petrelli, M.G. (1967) Autoecologia e capacità moltiplicativa di *Macrocheles muscaedomesticae* (Scopoli) (Acari: Mesostigmata). *Rivista di Parassitologia* 28, 129–156.

Fisher, R.C. (1938) Studies of the biology by the Death Watch beetle *Xestobium rufovillosum* De Geer. II. The habits of the adult with special reference to the factors affecting oviposition. *Annals of Applied Biology* 25, 155–180.

Fisher, S. (1993) Observation of a new pest of cucumber in western Switzerland, *Tyrophagus neiswanderi* Johnston and Bruce (Acari: Acaridae). *Revue Suisse de Viticulture Arboriculture Horticulture* 25, 103–104.

Fitzgerald, J., Xu, X.M., Pepper, N., Easterbrook, M. and Solomon, M. (2008) The spatial and temporal distribution of predatory and phytophagous mites in field-grown strawberry in the UK. *Experimental and Applied Acarology* 44, 293–306.

Flanders, S.E. and Badgley, M.E. (1963) Prey-predator interactions in self-balanced laboratory populations. *Hilgardia* 35(8), 145–183.

Flechtmann, C.H.W. (1968) Nota sobre ácaros de produtos armazenados. *Solo, Piracicaba* 60, 63–65.

Flechtmann, C.H.W. (1976) Sobre dois ácaros (Arthropoda, Acari) freqüentes em criações da broca da cana-de-açúcar (Insecta, Lepidoptera). Ecossistema, *Pinhal* 1(1), 62–63.

Flechtmann, C.H.W. (1980) Dois àcaros associados à abelha (*Apis mellifera* L.) no Perù. *Anais da Escola Superior de Agricultura Luiz de Queiroz* 37(2), 737–741.

Flechtmann, C.H.W. (1981) New records of mites from Brazil with description of two new species in the genus *Oligonychus* Berlese (Acari: Tetranychidae). *Revista Brasileira de Biologia* 41, 861–866.

Flechtmann, C.A.H. (1985a) On the biology of *Ameroseius dendrovagans* (Acari, Mesostigmata, Ameroseiidae). *Revista Brasileira de Zoologia* 2, 397–399.

Flechtmann, C.H.W. (1985b) *Ácaros de importância médico veterinária.* 3rd edn. Nobel, São Paulo, Brazil, pp. 1–192.

Flechtmann, C.H.W. (1986) *Ácaros de produtos armazenados e na poeira domiciliar*. Piracicaba, FEALQ, Brazil, pp. 1–97.

Flechtmann, C.H.W. and Baggio, D. (1993) On phoresy of hematophagous ectoparasitic Acari (Parasitiformes: Ixodidae and Dermanyssidae) on coleoptera observed in Brazil. *International Journal of Acarology* 19(2), 195–196.

Flechtmann, C.H.W. and Castelo, F.P. (1982) On some insects and mites associated with dried and salted fish in Brazil. *Acta Amazonica* 12(2), 489–490.

Flechtmann, C.H.W. and Flechtmann, C.A.H. (1984) Reproduction and chromosomes in the broad mite, *Polyphagotarsonemus latus* (Banks, 1904) (Acari, Prostigmata, Tarsonemidae). In: Griffiths, D.A. and Bowman, C.E. (eds) *Acarology VI*, Vol. 1. Ellis Horwood, Chichester, UK, pp. 445–456.

Flechtmann, C.H.W. and Flechtmann, C.A.H. (1985) A new species of *Ameroseius* (Acari: Mesostigmata, Ameroseiidae) from Brasil. *Revista Brasileira de Zoologia* 2(6), 393–396.

Flechtmann, C.H.W. and Zem, A.C. (2002) Ácaros de produtos armazenados. In: Lorini, I., Miike, L.H. and Scussel, V.M. (eds) *Armazenagem de grãos*. Instituto Bio Geneziz, Campinas, Brazil, pp. 807–856.

Flechtmann, C.H.W., Fain, A. and Leal, M.C.A. (1985) Acari domum meliponinarum brasiliensium habitantes. VII, *Ereynets* (*Gymnereynetes*) *meliponae* nov. spec. (Acari, Ereynetidae). *Revista Brasileira de Biologia* 45, 95–99.

Flechtmann, C.H.W., Guerrero, B.J.M., Arroyave, J.A. and Costantino, Ch.L.M. (1990) A little known mode of dispersal of *Polyphagotarsonemus latus* (Banks). *International Journal of Acarology* 16, 181–182.

Fleschner, C.A. and Arakawa, K.Y. (1953) The mite *Tydeus californicus* on citrus and avocado leaves. *Journal of Economic Entomology* 45, 1092.

Flies Breeding in Accumulated Organic Wastes (2022) Available at: http://www.faculty.ucr.edu/~legneref/biotact/ch-50.htm (accessed 29 June 2022).

Flochlay, A.S., Thomas, E. and Sparagano, O. (2017) Poultry red mite (*Dermanyssus gallinae*) infestation: A broad impact parasitological disease that still remains a significant challenge for the egg-laying industry in Europe. *Parasites Vectors* 10, 357.

Flynn, R.J. (1973) Parasites of Laboratory Animals. Iowa State University Press, Ames, IA.

Forsberg, J.L. (1965) The relationship of *Pseudomonas marginata*, *Stromatina gladiola*, bulb mites, and chemical control treatments to the occurrence of scab and *Stromatinia* rot of gladiolus. *Phytopathology* 55, 1058.

Fossum, O., Jansson, D.S., Etterlin, P.E. and Vågsholm, I. (2009) Causes of mortality in laying hens in different housing systems in 2001 to 2004. *Acta Veterinaria Scandinavica* 15, 51–53.

Fouly, H. and Al-Rehiayani, S.M. (2011) Predaceous mites in Al-Qassim, Saudi Arabia, with description of two new laelapid species. *Journal of Entomology* 8, 139–151.

Fox, I. (1947) Seven new mites from rats in Puerto Rico. *Annals of the Entomological Society of America* 40, 598–603.

Fox, I. (1950) A new tyroglyphid mite from Puerto Rico. *Proceedings of the Biological Society of Washington* 63, 205–208.

Franjola, R. and Malonnek, M. (1995) Housedust mites in the city of Valdivia, Chile. *Boletín Chileno de parasitología* 50(1–2), 16–20.

Franz, J.T., Masuch, G., Bergmann, K.C. and Müsken, H. (2000) *Glycyphagus domesticus* (De Geer 1778): Scanning electron microscopic studies. *Allergologie* 23(5), 219–225.

Franz, J.T., Masuch, G., Bergmann, K.C.and Musken, H. (1998) Study of the domestic mite fauna of farms in Germany: A cross-section analysis. *Allergologie* 21(8), 371–380.

Franz, J.T., Masuch, G., Bergmann, K.C.and Musken, H. (2000) Development and developmental stages of the mite *Glycyphagus domesticus* (De Geer 1778): Scanning electron microscopic studies. *Allergologie* 23(5), 219–225.

Franz, J.T., Masuch, G., Musken, H. and Bergmann, K.C. (1997) Mite fauna in German farms. *Allergy* 52(12), 1233–1237.

Franzolin, M.R. and Baggio, D. (2000) Contaminacao por acaros em arroz polido e feijao comercializados a granel. *Revista de Saúde Pública* 34(1), 77–83.

Franzolin, M.R., Gambale, W., Cuero, R.O. and Correa, B. (1999) Interaction between toxigenic *Aspergillus flavus* Link and mites *Tyrophagus putrescentiae* (Schrank) on maize grains: Effects on fungal growth and aflatoxin production. *Journal of Stored Products Research* 35(3), 215–224.

Freire, R.A.P., Moraes, G.J., Silva, E.S., Vaz, A.C. and Castilho, R.C. (2007) Biological control of *Bradysia matogrossensis* (Diptera: Sciaridae) in mushroom cultivation with predatory mites. *Experimental and Applied Acarology* 42, 87–93.

Friesen, R., Farias, M.R. and Schenato, L.A. (2011) Acaríase cutânea por *Dermanyssus gallinae* em um cão. *Revista de Educação Continuada em Medicina Veterinária e Zootecnia* 11, 60–61.

Frison, T.H. (1923) Notes on the life history, parasites, and inquiline associates of *Anthophora abrupt* Say, with some comparisons with the habits of certain other *Anthophora*. *Transactions of the American Entomological Society* 48, 137–156.

Frost, C.J., Mescher, M.C., Carlson, J.E. and Moraes, C.M. de (2008) Plant defense priming against herbivores: Getting ready for a different battle. *Plant Physiology* 146, 818–824.

Fuentes, M.V., Saintz-Elipe, S., Saez-Duran, S. and Galn-Puchades, M.T. (2009) Human ectoparasitism due to the poultry red mite, *Dermanyssus gallinae*, in the city of Valencia (Spain) and its surroundings. *Revista Ibero-latinoamericana de parasitología* 68, 188–191.

Fujisawa, S., Murata, S., Takehara, M., Aoyama, J., Morita, A. *et al.* (2021) In vitro characterization of adipocyte plasma membrane-associated protein from poultry red mites, *Dermanyssus gallinae*, as a vaccine antigen for chickens. *Vaccine* 39(41), 6057–6066.

Fumouze, A. and Robin, C. (1867) Mémoire anatomique et zoologique sur les Acariens des genres *Cheyletus*, *Glycyphagus* et *Tyroglyphus*. *Journal de l'anatomie et de la physiologie normales et pathologiques de l'homme et des animaux* 4, 561–601+Pl. XXII–XXV.

Furman, D.P. (1959) Feeding habits of symbiotic mesostigmatid mites of mammals in relation to pathogen-vector potentials. *American Journal of Tropical Medicine and Hygiene* 8, 5–12.

Furman, D.P. and Sousa, O.E. (1969) Morphology and biology of a nest-producing mite, *Bakericheyla chanayi* (Acarina: *Cheyletidae*). *Annals of the Entomological Society of America* 62, 858–863.

Furumizo, R.T. (1973) The biology and ecology of the house-dust mite *Dermatophagoides farinae* hughes, 1961 (Acarina, Pyroglyphidae). PhD thesis, University of California, Riverside.

Furumizo, R.T. (1975) Geographical distribution of house dust mites (Acarina, Pyroglyphidae) in California. *California Vector Views* 22, 89–95.

Furumizo, R.T. and Thomas, V. (1977) Mites of house dust. *Southeast Asian Journal of Tropical Medicine and Public Health* 8, 411–412.

Furumizo, R.T. and Wharton, G.W. (1975) A case of postimaginal molt in the American house dust mite *Dermatophagoides farinae* Hughes, 1961 (Acari: Pyroglyphidae). *Acarologia* 17(4), 730–733.

Gabriel, M., Cunnington, A.M., Allan, W.G.L., Pickering, C.A.C. and Wraith, D.G. (1982) Mite allergy in Hong Kong. *Clinical & Experimental Allergy*12, 157–171.

Gad, M.E.G., Metwally, A.M. and Bream, A.S.M. (2020) Some astigmated, prostigmated and cryptostigmated mites inhabiting some stored products at el-sharqia governorate, Egypt. *Journal of Plant Protection and Pathology, Mansoura University* 11(4), 215–219.

Gaede, K. (1991) On the water balance of *Phytoseiulus persimilis* A.-H. and its ecological significance. *Experimental and Applied Acarology* 15, 181–189.

Gafvelin, G., Johansson, E., Lundin, A., Smith, A.M., Chapman, M.D. *et al.* (2001) Cross-reactivity studies of a new group 2 allergen from the dust mite *Glycyphagus domesticus*, Gly d 2, and group 2 allergens from *Dermatophagoides pteronyssinus*, *Lepidoglyphus destructor*, and *Tyrophagus putrescentiae* with recombinant allergens. *Journal of Allergy and Clinical Immunology* 107(3), 511–518.

Gallego, J.R., Gamez, M. and Cabello, T. (2020a) Potential of the *Blattisocius mali* Mite (Acari: Blattisociidae) as a biological control agent of potato tubermoth (Lepidoptera: Gelechiidae) in stored potatoes. *Potato Research* 63(2), 241–251.

Gallego, J.R., Caicedo, O., Gamez, M., Hernandez, J. and Cabello, T. (2020b) Selection of predatory mites for the biological control of potato tuber moth in stored potatoes. *Insects* 11(3), 196.

Galvão, A.S., Silva, E.A., Sousa, J.M., Sila, R.R., Silva, M.J.S. and Ribeiro, A.E.L. (2011) Acarofauna em produtos alimentícios comercializados em mercados de São Luis-MA. *Magistra* 23, 252–256.

Gamal-Eddin, F.M., Aboul-Atta, A.M., Hamad, M.G.M., Eraki, A.-S.M. and Tubeileh, F.M.S. (1985) House dust mites in Kuwait, preliminary pilot survey to shed some light on diagnosis and treatment of mite allergy. *Journal of the Egyptian Society of Parasitology* 15, 313–322.

Ganguly, A., Gupta, S.K., Debnath, M. and Haqe, R. (2018) On some mites from stored products of grocery shops in Howrah District, West Bengal. *Bionotes* 20(3), 96–98.

Gantait, V.V. and Chandra, K. (2017) Faunal diversity in Antarctica: Contributions of zoological survey of India faunal diversity in Antarctica: Contributions of zoological survey of India. *Proceedings of the National Academy of Sciences, India* 83(2), 353–361.

Gao, J.M., Liu, M.H. and Wei, B.X. (1956) A case of mites in patients with respiratory disease and observation on its life cycle and resistance. *Chinese Medical Journal* 30, 1048–1052.

García, N. (2004) Efforts to control mites on Iberian ham by physical methods. *Experimental and Applied Acarology* 32(1–2), 41–50.

García-Ortiz, N., Aguilar-Marcelino, L., Mendoza-de-Gives, P., López-Arellano, M.E., Bautista-Garfias, C.R. and González-Garduño, R. (2015) *In vitro* activity of *Lasioseius penicilliger* (*Arachnida*: *Mesostigmata*) against three nematode species: *Teladorsagia circumcincta*, *Meloidogyne* sp. and *Caenorhabditis elegans*. *Veterinaria México OA* 2(1), DOI: 10.21753/vmoa.2.1.340.

Garcia-Robaina, J.C., Eraso, E., Martinez, J., Martinez, A., de la Torre Morin, F. *et al.* (1997) Sensitization to *Blomia kulagini* in a general population of a subtropical region of Spain (Canary Islands). *Allergy* 52, 727–731.

Garman, P. (1917) *Tarsonemus pallidus* Banks, a pest of geraniums. *Bulletin of the Maryland Agricultural Experiment Station* 208, 327–342.

Garman, P. (1937) A study of the bulb mite. *Bulletin of the Maryland Agricultural Experiment Station* 402, 889–907.

Garrett, L.E. and Haramoto, F.H. (1967) A Catalog of Hawaiian Acarina. *Proceedings of the Hawaiian Entomological Society* 19(3), 381–414.

Gassel, M., Wolf, C., Noack, S., Williams, H. and Ilg, T. (2014) The novel isoxazoline ectoparasiticide fluralaner: selective inhibition of arthropod gammaaminobutyric acid- and Lglutamate-gated chloride channels and insecticidal/acaricidal activity. *Insect Biochemistry and Molecular Biology* 45, 111–124.

Gaud, J. and Atyeo, W.T. (1996) Feather mites of the World (Acarina, Astigmata): the supraspecific taxa Part 1. *Annales du Musée Royal du Congo Belge Tervuren (Belgique): Serie in-8° - Sciences Zoologiques, Tervuren, Belgique* 277, 1–193.

Gause, G.F., Smaragdova, N.P. and Witt, A.A. (1936) Further studies of interaction between predator and prey. *Journal of Animal Ecology* 5, 1–18.

Gay, M., Lempereur, L., Francis, F. and Megido, R.C. (2020) Control of *Dermanyssus gallinae* (De Geer 1778) and other mites with volatile organic compounds, A review. *Parasitology* 147, 731–739.

Geary, M.J., Knihinicki, D.K., Halliday, R.B. and Russell, R.C. (2000) Contact dermatitis associated with the brown-legged mite, *Aleuroglyphus ovatus* (Troupeau) (Acari: Acaridae), in Australia. *Australian Journal of Entomology* 39, 351–352.

Geden, C.J., Stinner, R.E. and Axtell, R.C. (1988) Predation by predators of the house-fly in poultry manure-effects of predator density, feeding history, interspecific interference, and field conditions. *Environmental Entomology* 17, 320–329.

Geller, M., Hahnstadt, L.R., Rego, R.M. and Caldas, E.F. (2009) Anafilaxia induzida por farinha de trigo contaminada por ácaros. *Rev. Bras. Alergia Imunop* 32(5), 199–201.

Genç, H. and Özar, A.İ. (1986) Preliminary investigations on the mites found on stored products in İzmir. *Türkiye Bitki Koruma Dergisi (Plant Protection Bulletin)* 10, 175–183.

Gentry, J.W. (1965) Crop Insects of Northeast Africa-Southwest Asia. Agriculture Handbook 273, USDA, Washington, USA, pp. 1–210.

George, D.R. (2010) Effect of plant essential oils as acaricides against the poultry red mite, *Dermanyssus gallinae*, with special focus on exposure time. *Veterinary Parasitology* 169, 222–225.

George, D.R., Guy, J.H., Arkle, S., Harrington, D., De Luna, C. *et al.* (2008a) Use of plant-derived products to control arthropods of veterinary importance: A review. *Annals of the New York Academy of Sciences* 1149, 23–26.

George, D.R., Smith, T.J., Sparagano, O.A.E. and Guy, J.H. (2008b) The influence of 'time since last blood meal'on the toxicity of essential oils to the poultry red mite (*Dermanyssus gallinae*). *Veterinary Parasitology* 155, 333–335.

George, D.R., Masic, D., Sparagano, O.A.E. and Guy, J.H. (2008c) Variation in chemical composition and acaricidal activity against *Dermanyssus gallinae* of four eucalyptus essential oils. *Experimental and Applied Acarology* 48, 43–50.

George, D., Smith, T., Shiel, R., Sparagano, O. and Guy, J. (2009a) Mode of action and variability in efficacy of plant essential oils showing toxicity against the poultry red mite, *Dermanyssus gallinae*. *Veterinary Parasitology* 161, 276–282.

George, D.R., Sparagano, O.A.E., Port, G., Okello, E., Shiel, R.S. and Guy, J.H. (2009b) Repellence of plant essential oils to *Dermanyssus gallinae* and toxicity to the non-target invertebrate *Tenebrio molitor*. *Veterinary Parasitology* 62, 129–134.

George, D.R., Sparagano, O.A.E., Port, G., Okello, E., Shiel, R.S. and Guy, J.H. (2010a) Environmental interactions with the toxicity of plant essential oils to the poultry red mite *Dermanyssus gallinae*. *Medical and Veterinary Entomology* 24, 1–8.

George, D.R., Sparagano, O.A.E., Port, G., Okello, E., Shiel, R.S. and Guy, J.H. (2010b) The effect of essential oils showing acaricidal activity against the poultry red mite (*Dermanyssus gallinae*) on aspects of welfare and production of laying hens. *Animal Welfare* 19, 265–273.

George, D.R., Finn, R.D., Graham, K.M., Mul, M.F., Maurer, V. and Moro, C.V. (2015) Should the poultry red mite *Dermanyssus gallinae* be of wider concern for veterinary and medical science? *Parasites Vectors* 8, 178.

Gerson, U. (1967) Some cheyletid and pseudocheylid mites from Israel. *Acarologia* 9, 359–369.

Gerson, U. (1968a) Some raphignathoid mites from Israel. *Journal of Natural History* 2, 429–437.

Gerson, U. (1968b) Five tydeid mites from Israel (Acarina: Prostigmata). *Israel Journal of Zoology* 17, 191–198.

Gerson, U. (1985) Other predaceous mites and spiders. In: Helle, W. and Sabelis, M.W. (eds) *Spider Mites. Their Biology, Natural Enemies and Control. World Crop Pests, Volume 1A*. Elsevier Science, Amsterdam, The Netherlands, pp. 205–210.

Gerson, U. and Smiley, R. (1990) *Acarine Biocontrol Agents*. Chapman and Hall, New York, pp. 1–174.

Gerson, U., Capua, S. and Thorens, D. (1983) Life history and life tables of *Rhizoglyphus robini* Claparede (Acari: Astigmata: Acaridae). *Acarologia* 24, 439–448.

Gerson, U., Fain, A, and Smiley, R.L. (1999) Further observations on the cheyletidae (Acari) with a key to the genera of the cheyletinae and a list of the known species in the family. *Bulletin de l'Institut royal des sciences naturelles de Belgique, Entomology* 69, 35–68.

Gerson, U., Kenneth, R. and Muttath, T.I. (1979) *Hirsutella thompsonii*, a fungal pathogen of mites. II. Hostpathogen interactions. *Annals of Applied Biology* 91(1), 29–40.

Gerson, U., OConnor, B.A. and Houck, M.A. (1990) Acari. In: Rosen, D. (ed.) *Armored Scale Insects: Their Biology, Natural Enemies and Control, vol 4B*. Elsevier, Amsterdam, The Netherlands, pp. 77–97.

Gerson, U., Smiley, R.L. and Ochoa, R. (2003) *Mites (Acari) for Pest Control*. Blackwell Publishing, Oxford, UK, pp. 1–539.

Gerson, U., Yathom, S., Capua, S. and Thorens, D. (1985) *Rhizoglyphus robini* Claparede (Acari: Astigmata: Acaridae) as a soil mite. *Acarologia* 26, 371–380.

Gettinger, D., Martins-Hatano, F., Lareschi, M. and Malcolm, J.R. (2005) Laelapine mites (Acari: Laelapidae) associated with small mammals from Amazonas, Brazil, including a new species from marsupials. *Journal of Parasitology* 91(1), 45–48.

Geurs, M., Bongers, J. and Brussard, L. (1991) Variations of the heptane flotation method for improved efficiency of collecting microarthropods from a sandy loam soil. *Agriculture, Ecosystems & Environment* 34, 213–221.

Giangaspero, A., Camarda, A., Cafiero, M.A., Marangi, M. and Sparagano, O. (2016) *Dermanyssus gallinae*: A never-ending story for the poultry industry and public health. *Nova Acta Leopoldina* 411, 1–17.

Gigja, G. (1964) Pests in stockfish. University Research Institute, Reykjavik, Iceland. Department of Agriculture Reports, Series B. N.18, 5–41.

Gill, N.K. and Kaur, H. (2014) A study on the occurrence, prevalence and species composition of mite fauna in human dwelling of Patiala city, Punjab (India). *Indian Journal of Scientific Research* 8(1), 91–97.

Gill, N.K., Paray, M.Y. Dehar, P.S. and Dhingra, S. (2018) Evaluations of studies on biology, fecundity and life table parameters of stored grain mite *Gohieria fusca* (Oud.) feed on crushed maize. *Research Guru* 12(2), 670–674. Available at: http://www.researchguru.net/volume/Volume%2012/Issue%202/RG86.pdf (accessed 10 October 2020).

Gilyarov, M.S. and Bregetova, N.G. (1977) Key to the soil inhabiting mites. *Mesostigmata. Nauka Press, Leningrad*, 1–718 (in Russian).

Gilyarov, M.S. and Krivolutski, A.D. (1975) Opredelitel obitayushchikh v pochve kleshchej Sarcoptiformes (Key for the identification of soil mites Sarcoptiformes). Izd. Nauka, Moskva, USSR, 1–492 (in Russian).

Glime, J.M. (2017) Arthropods: Mites (Acari). Chapter 9-1. In: Glime, J.M. (ed.) *Bryophyte Ecology. Volume 2. Bryological Interaction. Ebook sponsored by Michigan Technological University and the International Association of Bryologists.*

Available at: http://digitalcommons.mtu.edu/bryophyte-ecology2/ (accessed 1 April 2021).

Glowska, E., Chrzanowski, M. and Kaszewska, K. (2015) Checklist of the Quill mites (Acariformes: Syringophilidae) of the World. Zootaxa 3968(1), 1. DOI:10.11646/zootaxa.3968.1.1.

Glowska, E., Filutowska, Z. K., Dabert, M. and Gerth, M. (2020) Microbial composition of enigmatic bird parasites: *Wolbachia* and *Spiroplasma* are the most important bacterial associates of quill mites (Acariformes: Syringophilidae). *MicrobiologyOpen* 9(5), e964. Available at: https://pubmed.ncbi.nlm.nih.gov/32141700/ (accessed 8 June 2023).

Goff, M.L. (1987) A catalog of Acari of the Hawaiian Islands. *University of Hawaii, Research and Extension Service* 75, 1–75.

Goldarazena, A., Zhang, Z.-Q. and Jordana, R. (2000) A new species and a new record of ectoparasitic mites from thrips in Turkey (Acari, Trombidiidae and Smarididae). *Systematic Parasitology* 45, 75–80.

Gonzalez Rodriguez, J. and Llorens Delgado, M. (1974) Identificacion de los acaros causantes de la sensibilizacion al polvo casero. Primer informe en Mexico. *Revista médica del Instituto Mexicano del Seguro Social* 13, 49–57.

Gonzalez-Rodriguez, R.H. (1965) A taxonomic study of genera *Mediolata, Agistemus* and *Zetzellia* (Acarina: Stigmaeidae). *University of California Publications in Entomology* 41, 1–64.

Goraccci, E., Lazzeri, S., Magnisi, M.T., Sondini, M.L. and Duranti, A. (1984) Gli allergoacari della polvere dei materassi della citta di Livorno. *Folia Allergology and Immunology Clinic* 31, 415–421.

Goracci, E. and Goracci, G. (1990) Gli allergoacari. *Caleidoscop. (Roma)* 8(54), 1–71.

Goracci, E., Lazzeri, S., Zuccherelli, D., Rossetti, M. and Poggianti, A.M. (1985) Acari of food storage facilities. An ecologic and immuno-allergic study. *Quaderni Sclavo di diagnostica clinica e di laboratorio* 21, 436–446.

Götz, H. (1952) Düngerbewohnende parasitiforme Milben und die Gattung *Macrocheles* Latreille 1829. Diss. Friedrich-Alexander-Universität, Erlangen, Germany, 134 pp + Plates 1–148.

Gould, S.J. (1981) Palaeontology plus ecology as palaebiology. In: May, R.M. (ed.) *Theoretical Ecology: Principles and Applications.* 2nd edn, Blackwell Scientific Publications, Oxford, UK, pp. 295–317.

Gould, S.J. (1989) *Wonderful Life. The Burgess Shale and the Nature of History*. W.W. Norton, New York, pp. 1–347.

Graham, P.P., Marriott, N.G. and Kelly, R.F. (2011) Dry curing Virginia-style ham. Virginia Polytechnic Institute and State University, Virginia Cooperative Extension Publication, Publication 458–223. Available at: https://virginia-beach.ext.vt.edu/content/dam/virginia-beach_ext_vt_edu/files/4-h/20120223-AGR-EXT-DryCuringVirginiaStyleHam.pdf (accessed 22 March 2021).

Grahn, H. and Geladi, P. (2007) *Techniques and Applications of Hyperspectral Image Analysis.* John Wiley & Sons Ltd, The Atrium, Southern Gate, Chichester, UK, pp. 1–368.

Grandjean, F. (1935a) Observations sur les oribates (8e série). *Bulletin du Muséum National d'Histoire Naturelle (Paris)* 8(2), 237–244.

Grandjean, F. (1935b) Observations sur les acariens (1e séries). *Bulletin du Muséum National d'Histoire Naturelle (Paris)* 7(2), 119–126.

Grandjean, F. (1936) Les Oribates de Jean Frédéric Hermann et de son père. *Annales de la Société entomologique de France* 105, 27–110.

Grandjean, F. (1938) Sur l'ontogénie des Acariens. *Comptes rendus de l'Académie des Sciences (Paris)* 206D, 146–150.

Grandjean, F. (1939a) Quelques genres d'acariens appartenent au groupe des Endeostigmata. *Annales des sciences naturelles. Zoologie* 11, 1–122.

Grandjean, F. (1939b) La chaetotaxie des pattes chez les Acaridiae. *Bulletin de la Société zoologique de France* 64, 50–60.

Grandjean, F. (1940) Les Poils et les organes sensitifs portés par les pattes et le palpe chez les Oribates. Deuxième partie. *Bulletin de la Société zoologique de France* 65, 32–44.

Grandjean, F. (1941) La chaetotaxie comparée des pattes chez les Oribates (1re série). *Bulletin de la Société zoologique de France* 66, 33–50.

Grandjean, F. (1942) La chaetotaxie comparée des pattes chez les Oribates (2re série). *Bulletin de la Société zoologique de France* 67, 40–53.

Grandjean, F. (1944) Observations sur les acariens de la famille des Stigmaeidae. *Archives des sciences physiques et naturelles* 26, 103–131.

Grandjean, F. (1946a) Les poils et les organes sensitifs portés par les pattes et le palpe chez les Oribates. Troisième partie. *Bulletin de la Société zoologique de France* 71, 10–29.

Grandjean, F. (1946b) Les Enarthronota (Acariens). Première Série. *Annales des sciences naturelles. Zoologie* 11(8), 215–248.

Grandjean, F. (1947) L'origine pileuse des mors et la chaetotaxie de la mandibule chez les Acariens actinochitineux. *Comptes rendus de l'Académie des Sciences (Paris)* 224, 1251–1254.

Grandjean, F. (1954) Etude sur les Palaeacaroides (Acariens, Oribates). *Mémoires du Muséum national d'histoire naturelle (n.s.), Série A., Zoologie* 7, 179–274.

Grandjean, F. (1958) Sur le comportement et la notation des poils accessoires postérieurs aux tarses des Nothroïdes et d'autres acariens. *Archives de zoologie expérimentale et générale* 96, 277–308.

Grandjean, F. (1970) Stase. Actinopiline. Rappel de ma classification des Acariens en trois groupes majeurs. *Acarologia* 11, 796–827.

Granich, J., Horn, T.B, Körbes, J.H., Toldi, M., Silva, G.L. and Ferla, N.J. (2016) Development of *Cheyletus malaccensis* (Acari: Cheyletidae) feeding on mite species found in commercial poultry systems: *Megninia ginglymura* (Acari: Analgidae) and *Tyrophagus putrescentiae* (Acari: Acaridae). *Systematic and Applied Acarology* 21, 1604–1613.

Granpayeh, S. and Ostovan, H. (2014) Mites recorded from insectary cultures in the Shiraz region of Iran. *International Research Journal of Applied and Basic Sciences* 8(5), 612–616.

Granpayeh, S., Ostovan, H. and Haghani, M. (2012) Abundance of mites associated with insectariums in Shiraz (Fars Province), Abstract Book of the 20th Iranian Plant Protection Congress, Shiraz, Iran, p. 457.

Greco, D.B., Moreira, N.S., Filogonio, C.J.B. and Greco, J.B. (1974) Demonstracao da presenca de acaros em po domicilar de Belo Horizonte e outras Cidades de Minas Gerais. 14th Congresso Brasileiro de Alergia e Imunopatologia, Recife, Brazil, p. 1.

Green, W.F., Woolcock, A.J., Stuckey, M., Segewick, C. and Leeder, S.R. (1986) House dust mites and skin tests at different Australian localities. *Australian and New Zealand Journal of Medicine* 16, 639–643.

Greenslade, P. and Clift, A. (2004) Review of pest arthropods recorded from commercial mushroom farms in Australia. *Australas. Mycologist* 23(3), 77–93.

Gridelet, D. and Lebrun, P. (1973) Contribution a l'etude ecologique des acariens des poussieres de maisons. *Acarologia* 15, 461–467.

Griffiths, D.A. (1960) Some field habitats of mites of stored food products. *Annals of Applied Biology* 48(1), 134–144.

Griffiths, D.A. (1964a) A revision of the genus *Acarus* (Acaridae, Acarina). *Bulletin of the British Museum (Natural History) Zoology* 11, 413–464.

Griffiths, D.A. (1964b) Experimental studies on the systematic of the genus *Acarus* Linnaeus 1758 (Sarcoptiformes, Acarina). Proc. 1st Congr. Acarology 1963. *Acarologia* 6, 101–116.

Griffiths, D.A. (1966) Nutrition as a factor influencing hypopus formation in *Acarus siro* species complex (Acarina: Acaridae). *Journal of Stored Products Research* 1, 325–340.

Griffiths, D.A. (1969) The influence of dietary factors on hypopus formation in *Acarus immobilis* Griffiths (Acari, Acaridae). In: Evans, G.O. (ed.) Proc. 2nd International Congress of Acarology, Sutton Bonington, England, 19–25 July, 1967, Akadémiai *Kiadó*, Budapest, Hungary, pp. 419–432.

Griffiths, D.A. (1970) A further systematic study of the genus *Acarus* L. 1758 (Acaridae, Acarina), with a key to species. *Bulletin of the British Museum (Natural History) Zoology* 19, 83–118.

Griffiths, D.A. (1979) The morpho-species and its relationshipto the biological species in the genus *Tyrophagus* (Acaridae, Acarina). In: Rodriguez, J.G. (ed.), *Recent Advances in Acarology*, Volume I, Academic Press, New York, 199–212.

Griffiths, D.A. and Boczek, J. (1977) Spermatophores of some acaroid mites (Astigmata: Acarina). *International Journal of Insect Morphology and Embryology* 6, 231–238.

Griffiths, D.A. and Cunnington, A.M. (1971) *Dermatophagoides microceras* sp. n., a description and comparison with its sibling species, *D. farinae* Hughes, 1961. *Journal of Stored Products Research* 7, 1–14.

Griffiths, D.A., Wilkin, D.R., Southgate, B.J. and Lynch, S.M. (1976) A survey of mites in bulk grain stored on farms in England and Wales. *Annals of Applied Biology*, 180–185.

Grifftts, D.A., Hodson, A.C. and Christensen, C.M. (1959) Grain storage fungi associated with mites. *Journal of Economic Entomology* 52(3), 514–518.

Griffiths, D.A., Atyeo, W.T., Norton, R.A. and Lynch, C. A. (1990) The idiosomal chaetotaxy of astigmatid mites. *Journal of Zoology* 220, 1–32.

Grobler, L., Bayram, S. and Çobanoglu, S. (2004) Two new species and new records of oribatid mites from Turkey. *International Journal of Acarology* 30(4), 351–358.

Grobov, O.F. (1978) Kleshchi medonosnoy pchely (*Apis mellifera* L.): ikh znachenie i osnovnue printsypy bor'by s kleschchevymi porazheniyami. Doctor of Sciences (Habilitation), Thesis. 536. Moscow: All-Union Institute of Experimental Veterinary, All-Union Academy of Agricultural Sciences (in Russian).

Grobov, O.F. (1991) *Kleśći: parazity pćel i vrediteli ich produkcii.* Rosagropromizdat, Moskva, 1–96 (in Russian).

Groenewoud, G.C.M., de Graaf in't Veld, C., Van Oorschot-van Nes, A.J., de Jong, N.W., Vermeulen, A.M. *et al.* (2002) Prevalence of sensitization to the predatory mite *Amblyseius cucumeris* as a new occupational allergen in horticulture. *Allergy* 57(7), 614–619.

Grosman, A., Messelink, G. and de Groot, E. (2011) Combined use of a mulch layer and the soil-dwelling predatory mite *Macrocheles robustulus* (Berlese) enhance the biological control of sciarids in potted plants. Integrated control in protected crops, temperate climate. *Bulletin IOBC/WPRS* 68, 51–54.

Grostal, P. and Dicke, M. (2000) Recognising one's enemies: a functional approach to risk assessment. *Behavioral Ecology and Sociobiology* 47, 258–264.

Grouth, T.G. and Ueckermann, E.A. (1999) Predatory mites (Acari) found under citrus trees in the southern African lowveld. *International Journal of Acarology* 25, 235–238.

Guerra Bernd, L.A., Arruda, L.K. and Barros Antunes, H.B. (2001) Oral anaphylaxis to mites. *Allergy* 56(1), 83–84.

Guerrini, A., Morandi, B., Roncada, P., Brambilla, G., Dini, F.M. and Galuppi, R. (2022) Evaluation of the acaricidal effectiveness of fipronil and phoxim in field populations of *Dermanyssus gallinae* (De Geer, 1778) from ornamental poultry farms in Italy. *Veterinary Sciences* 9, 486.

Gueye-Ndiaye, A. and B. Marchand (1989) *Lardoglyphus konoi* et *Suidasia pontifica* déprédateurs des sardinelles braisées-séchées au Sénégal étude en microscopie électronique a balayage. *Acarologia* 30(2), 131–137.

Gueye-Ndiaye, A. and Fain, A. (1987) Note sur les acariens des denrées stockées au Sénégal. *Revue zoologique africaine* 101, 365–370.

Guimarães, J.M. (1970) *Bases fitosanitárias para a solução do problema dos ácaros do fico seco no Algarve.* Publicacao Subsidiada pela Junta Nacional das Frutas, Lisboa, pp. 1–174.

Guin, J.D. (2005) Contact arthropod reactions: beware the family cat. *Contact Dermatitis* 52, 44–45.

Gu, Y. and Guo, X. (1996) A new species of *Gamasholaspis* from China (Acari: Parholaspidae). *Acta Zootaxonomica Sinica* 21, 309–311.

Gulati, R. (2007a) Turmeric (*Curcuma longa* L.) as acaricidal against *Tyrophagus putrescentiae* Schrank and *Suidasia nesbitti* Hughes in stored wheat. *Journal of Food Science and Technology* 44(4), 367–370.

Gulati, R. (2007b) Potential of Garlic as grain protectant against *Tyrophagus putrescentiae* Schrank and *Suidasia nesbitti* Hughes in wheat. *Systematic and Applied Acarology* 12, 19–25.

Gulati, R. and Mathur, S. (1995) Effect of Eucalyptus and Mentha leaves and Curcuma rhizomes on *Tyrophagus putrescentiae* (Schrank) (Acarina: Acaridae) in wheat. *Experimental and Applied Acarology* 19(9), 511–518.

Gulati, R. and Mathur, S. (1998) Olfactory responses of *Tyrophagus putrescentiae* (Schrank) towards *Cajanus cajan. Bull & Bear Life Sciences* 8, 53–61.

Güldali, B. and Çobanoğlu, S. (2010) Investigation on the life table parameters and development threshold of *Carpoglyphus lactis* (L.) (Acari: Carpoglyphidae) at different temperatures and relative humidities. *Türkiye Entomoloji Dergisi* 34(1), 53–65 (in Turkish).

Gulegen, E., Girisgin, O., Kutukoglu, F., Girisgin, A.O. and Coskun, S.Z. (2005) Mite species found in house dust in houses in Bursa. *Turkiye Parazitolojii Dergisi* 29(3), 185–187 (in Turkish).

Gülkan, B. (2004) Hatay ilindeki ev tozu akar faunasının belirlenmesi ve alerji ile ilişkisinin araştırılması. Sivas Cumhuriyet University, Institute of Health Sciences, Department of Parasitology, Turkey (in Turkish).

Gültekin, N. and Özkan, M. (1999) Investigations on the Acari of stored foods in Erzurum province. *Turkish Journal of Entomology* 23, 289–303 (in Turkish).

Gupta, A. and Chatterjee, M. (2004) Some new records of mites infesting stored grains in Kolkata and its neighbourhood. *Records of the Zoological Survey of India* 102(1–2), 77–82.

Gupta, S.K. and Chatterjee, K. (1997) Acari: Plant mites. State Fauna Se 6, Fauna of Delhi, 485–522.

Guschina, I.A. and Harwood, J.L. (2006) Mechanisms of temperature adaptation in poikilotherms. *FEBS Letters* 580, 5477–5483.

Gutierrez, J. and Helle, W. (1985) Evolutionary changes in the Tetranychidae. In: Helle, W. and Sabelis, M.W. (eds) *Spider Mites: Their Biology,*

Natural Enemies and Control. World Crop Pests, Volume 1A. Elsevier Science, Amsterdam, The Netherlands, pp. 91–107.

Gwiazdowicz, D.J. (2003) Description of *Iphidozercon poststigmatus* sp. n. (Acari, Ascidae) with a key to Palearctic species of the genus *Iphidozercon*. *Biologia* 58, 151–154.

Gwiazdowicz, D.J. (2007) Ascid mites (Acari, Mesostigmata) from selected forest ecosystems and microhabitats in Poland. Wydawnictwo Akademii Rolniczej Im. Augusta Cieszkowskiego, Poznan, Poland, 1–248.

Gwiazdowicz, D.J. (2008) Mesostigmatid mites (Acari) associated with Scolytidae in Poland. In: Gwiazdowicz, D.J. (ed.) *Selected Problems of Acarological Research in Forests*. Wydawnictwo Uniwersytetu Przyrodniczego, Poznan, Poland, pp. 59–95.

Gwiazdowicz, D.J. and Halliday, R.B. (2008) The Australian species of *Iphidozercon* (Acari: Ascidae). *Zootaxa* 1921, 47–68.

Gwiazdowicz, D.J. and Marchenko, I.I. (2012) Two new species of *Iphidozercon* (Acari: Ascidae) with a key to females. *Acta Zoologica Academiae Scientiarum Hungaricae* 58, 41–52.

Gwiazdowicz, D.J., Bloszyk, J., Bajerlein, D., Halliday, R.B. and Mizera, T. (2006) Mites (Acari: Mesostigmata) inhabiting nests of the white-tailed sea eagle *Haliaeetus albicilla* (L.) in Poland. *Entomologica Fennica* 17, 366–372.

Haag-Wackernagel, D. (2005) Parasites from feral pigeons as a health hazard for humans. *Annals of Applied Biology* 147, 203–210.

Haarlov, N. and Alani, M. (1970) House-dust Mites *Dermatophagoides pteronyssinus* (Trt.), *D. farinae* Hughes, *Euroglyphus maynéi* (Cooreman) (Fain) in Denmark (Acarina). *Scandinavian Entomology* 1, 301–306.

Haddad Irani-Nejad, K. (1998) Cotton mites fauna in Moghan plain and evaluating the effects of morphological characteristics of some cotton varieties on biological reactions of the two spotted spider mite *Tetranychus urticae* (Acari: Tetranychidae). Ph.D. dissertation, Tarbiat Modarres University, Tehran, Iran, pp. 1–290.

Haddad Irani-Nejad, K., Kamali, K. and Maleki Milani, H. (1999) Some species of Astigmata in cotton fields of Moghan plain. *Journal of Agricultural Science* 9(1), 21–31.

Haddad Irani-Nejad, K., Hajiqanbar, H. and Talebi Chaichi, P. (2003) Introduction of some mesostigmatic mites of sugarbeet fields in Miandoab Plain. *Journal of Agriculture and Natural Resources Sciences* 10(2), 147–157.

Haddad Irani-Nejad, K., Rahgozar, M. and Valizadeh, M. (2007) Astigmatic mite fauna of alfalfa fields and their distribution in south west of Azarbaijan province. *Journal of Agricultural Science* 17, 127–137 (in Persian).

Hafeez, S.M., Taher, S.H. and Mallawani, M.A. (1988) Biological studies on *Blattisocius tarsalis* Keegan, a predatory mite inhabiting stored food in Egypt. *Annals of Agricultural Sciences* 33, 1387–1393.

Hage-Hamsten van, M. van and Johansson, S.G. (1992) Storage mites. *Experimental and Applied Acarology* 16(1–2), 117–128.

Hage-Hamsten van, M. and Johansson, E. (1998) Clinical and immunologic aspects of storage mite allergy. *Allergy* 53(48 Suppl), 49–53.

Hage-Hamsten van, M. and Johansson, S.G.O. (1992) Storage mites. *Experimental and Applied Acarology* 16, 117–128.

Hage-Hamsten van, M., Johansson, S.G., Johansson, E. and Wiren, A. (1987) Lack of allergenic cross-reactivity between storage mites and *Dermatophagoides pteronyssinus*. *Clinical & Experimental Allergy* 17(1), 23–31.

Hage-Hamsten van, M., Johansson, S.G.O., Hoglund, S., Tull, P., Wiren, A. and Zetterstrom, O. (1985) Storage mite allergy is common in a farming population. *Clinical & Experimental Allergy* 15, 555–564.

Hagstrum, D.W. and Subramanyam, B. (2006) Fundamentals of Stored-Product Entomology. AACC International. St Paul, Minnesota, U.S.A., pp. 1–323.

Hagstrum, D.W., Klejdysz, T., Subramanyam, B. and Nawrot, J. (2013) Atlas of Stored-Product, Insects and Mites, AACC International, Inc., Minnesota, pp. 1–589.

Haines, C.P. and Lynch, S.M.T. (1987) A new species of *Madaglyphus* Fain (Acarina: Acaridae) from rice mill in Java, Indonesia. *Acarologia* 28, 255–263.

Haines, C.P. (1978) A revision of the genus *Blattisocius* Keegan (Mesostigmata: Ascidae) with especial reference to *B. tarsalis* (Berlese) and the description of a new species. *Acarologia* 20(1), 19–38.

Haines, C.P. (1981) Laboratory studies on the role of an egg predator, *Blattisocius tarsalis* (Berlese) (Acari, Ascidae), in relation to the natural control of *Ephestia cautella* (Walker) (Lepidoptera:

Pyralidae) in warehouses. *Bulletin of Entomological Research* 71, 555–574.

Haines, C.P. (1988) New species of predatory mite (Acarina: Cheyletidae) associated with bostrichid beetles on dried cassava. *Acarologia* 29, 361–375.

Haines, C.P. (1995) Grain Storage in the Tropics. In: Jayas, D.S., White, N.D.G. and Muir, W.E. (eds), Stored Grain Ecosystems, Marcel Dekker, Inc, New York, Basel, Hong Kong, pp. 55–99.

Haitlinger, R. (1976) Parasitological investigation of small mammals of Gory Sowie (Middle Sudates). IV. Acarina (Mesostigmata). *Polskie pismo entomologiczne* 46, 771–821.

Haitlinger, R. (1988) Haemogamasidae Oudemans, 1926 (Acari, Mesostigmata). *Polskie pismo entomologiczne* 58, 635–661.

Haitlinger, R. (2012) New records of mites (Acari, Prostigmata, Erythraeidae, Trombidiidae) from Albania, Macedonia, Montenegro and Serbia, with a description of *Erythraeus* (*Erythraeus*) *albanicus* sp. nov. *Systematic and Applied Acarology* 17(3), 339–345.

Hajiqanbar, H., Baradaran, P. and Taheri, M.S. (2011) First record of *Acarophenax rackae* (Acari: Heterostigmata: Acarophenacidae), an egg parasitoid of *Tribolium confusum* (Col.: Tenebrionidae) from Iran. *Journal of Entomological Society of Iran* 31(1), 87–88.

Hajizadeh, J. (2007) Phytoseiid mites fauna of Guilan province, part II: subfamilies Amblyseiinae Muma and Phytoseiinae Berlese (Acari: Phytoseiidae). *Agricultural Research* 7(1), 7–25 (in Persian).

Hajizadeh, J. and Karami, F. (2017) Additional descriptions of *Ameroseius aegyptiacus* (Nasr and Abou Awad) and *Ameroseius lanceosetis* Livshitz and Mitrofanov (Acari: Ameroseiidae), with a revised key to the ameroseiid mites of Iran. *Linzer Biologische Beitraege* 49(2), 1323–1334.

Hajizadeh, J., Faraji, F. and Rafati Fard, M. (2010) Four species of family Laelapidae (Acari: Mesostigmata), new records for the Iran. In: Manzari, S. (ed.) *Abstract Book of 19th Iranian Plant Protection Congress*, 31 July–3 August, Tehran, Iran, p. 355.

Hajizadeh, J., Noei, J., Salehi, L. and Ostovan, H. (2011) Cheyletid mites associated with stored rice in Iran, the first record of *Chelacheles strabismus* from Iran and a key for their identification. *Journal of Entomological Society of Iran* 30(2), 85–88.

Halbert, J.N. (1923) Notes on Acari, with description of new species. *Zoological Journal of the Linnean Society* 35, 363–395.

Hall, C.C. (1959) A dispersal mechanism in mites. *Journal of the Kansas Entomological Society* 32, 45–46.

Hallan, J. (ed.) (2010) Trematuridae Species Listing. *Biology Catalog*. Texas A&M University. Available at: http://insects.tamu.edu/research/collection/hallan/Acari/Family/Trematuridae.txt.

Hallas, T.E. (1981a) Mites of stored hay in Iceland. *Journal of Agricultural Research in Iceland* 13(1–2), 61–67.

Hallas, T.E. (1981b) Mites of stored hay in Iceland. *Journal of Agricultural Research in Iceland* 17, 31–37.

Hallas, T.E. and Gudleifsson, B.E. (2008) Summer occurrence of *Tyrophagus similis* Volgin (Acari, Acaridae) in hayfields in northern Iceland may include a facultative deutonymph. *Journal of the Acarological Society of Japan* 17, 101–106.

Hallas, T.E. and Gudmundsson, B. (1985) Mites of stored hay in Iceland. Related to quality of hay and the storage duration. *Journal of Agricultural Research in Iceland* 17, 31–37.

Hallas, T.E. and Iversen, M. (1996) Sources of exposure to storage mites in the farming environment. *Annals of Agricultural and Environmental Medicine* 3, 9–12.

Hallas, T.E. and Korsgaard, J. (1997) Systematic variations in the appearance of house-dust mites (*Dermatophagoides* spp.), house mites (*Glycyphagus domesticus*) and of *Tarsonemus* sp. in dust samples from dwellings. *Rev. Esp. Allergy Immunology and Clinical* 12, 173–177.

Hallas, T.E. and Solberg, H. (1989) Mites of stored hay on the Faroe Islands (Acari). *Entomologiske Meddelelser* 57, 151–156.

Hallas, T.E., Gislason, D., Bjornsdottir, U.S., Jorundsdottir, K.B., Janson, C., Luczynska, C.M., Gislason, T. (2004) Sensitisation to house dust mites in Reykjavik, Iceland, in the absence of domestic exposure to mites. *Allergy* 59, 515–519.

Hallas, T.E., Iversen, M., Korsgaard, J. and Dahl, R. (1991) Number of mites in stored grain, straw and hay related to the age of the substrate (Acari). *Entomologiske Meddelelser* 59, 57–60.

Halliday, B. (2008) *Alliphis siculus* (Oudemans 1905) is not a synonym of *Alliphis halleri* (G. & R. Canestrini 1881) (Acari: Eviphididae). *Systematic and Applied Acarology* 13(1), 51–64.

Halliday, B. (2011) Occurrence of the predatory mite *Haemogamasus pontiger* (Berlese) (Acari: Laelapidae) in Australia, with a review of its biology. *Australian Journal of Entomology* 50(1), 61–64.

Halliday, B. (2015) Catalogue of genera and their type species in the mite Suborder Uropodina (Acari: Mesostigmata). *Zootaxa* 3972(2), 101–147.

Halliday, B. and Holm, E. (1985) Experimental taxonomy of Australian mites in the *Macrocheles glaber* group (Acarina: Macrochelidae). *Experimental and Applied Acarology* 1, 277–286.

Halliday, R.B. (1986) Mites of the *Macrocheles glaber* group in Australia (Acarina, Macrochelidae). *Australian Journal of Zoology* 34, 733–752.

Halliday, R.B. (1989) Mites of the Macrocheles muscaedomesticae group in Australia (Acarina: Macrochelidae). *Invertebrate Taxonomy* 3(4) 407– 430.

Halliday, R.B. (1995) A new genus and two new species of Parholaspididae from Australia and Papua New Guinea (Acarina: Mesostigmata). *Records of the Western Australian Museum Supplement* 52, 209–217.

Halliday, R.B. (1997) Revision of the Australian Ameroseiidae (Acarina: Mesostigmata). *Invertebrate Taxonomy* 10, 179–201.

Halliday, R.B. (1998) *Mites of Australia. A Checklist and Bibliography*. CSIRO Publishing, Collingwood, Australia, pp. 1–317.

Halliday, R.B. (2000) The Australian species of *Macrocheles* (Acarina: Macrochelidae). *Invertebrate Taxonomy* 14, 273–326.

Halliday, R.B. (2003) Health and safety issues related to mites in stored grain. In: Wright, E.J., Webb, M.C. and Highley, E. (eds) *Stored grain in Australia 2003, Proceedings of the Australian Postharvest Technical Conference, Canberra, 25–27 June 2003*. CSIRO Stored Grain Research Laboratory, Canberra, Australia, pp. 116–118.

Halliday, R.B. and Collins, R.O. (2002) *Histiostoma papillata* sp. n. (Acari: Histiostomatidae), a mite attacking fish in Australia. *Australian Journal of Entomology* 41(2), 155–158.

Halliday, R.B. and Holm, E. (1987) Mites of the family Macrochelidae as predators of two species of dung-breeding pest flies. *Entomophaga* 32, 333–338.

Halliday, R.B. and Lindquist, E.E. (2007) Correspondence Nomenclatural notes on the names *Gaeolaelaps* and *Geolaelaps* (Acari: Laelapidae). *Zootaxa* 1621, 65–67.

Halliday, R.B., Walter, D.E. and Lindquist, E.E. (1998) Revision of the Australian Ascidae (Acari: Mesostigmata). *Invertebrate Taxonomy* 12, 1–54.

Hallman, G.J. and Denlinger, D.L. (eds) (2019) *Temperature Sensitivity in Insects and Application in Integrated Pest Management*. CRC Press, pp. 1–320.

Halmai, Z. (1984) Changes in the composition of house-dust mite fauna in Hungary. *Parasitologia Hungarica* 17, 59–70.

Haloti, S., Glida, H., Niogret, J., Janati-Idrissi, A., Bertrand, M. and Lumaret, J-P. (2005) Acariens-Macrochelidae (Acari: Mesostigmata) phoretiques d'Afrique 1: macrochelides coprophiles du Maroc. *Acarologia* 45,155–159.

Hammad, H., Chieppa, M., Perros, F., Willart, M.A., Germain, R.N. and Lambrecht, B.N. (2009) House dust mite allergen induces asthma via Toll-like receptor 4 triggering of airway structural cells. *Nature Medicine* 15, 410–416.

Hammen, L. van der (1952) The oribatei (Acari) of The Netherlands. *Zoologische verhandelingen* 17, 1–139.

Hammen, L. van der (1961) Description of *Holothyrus grandjeani* nov. spec., and notes on the classification of the mites. *Nova Guinea, Zoology* 9, 173–194.

Hammen, L. van der (1970) La segmentation primitive des acariens. *Acarologia* 12, 3–10.

Hammen, L. van der (1980) *Glossary of Acarological Terminology. Volume 1, General Terminology*. W. Junk, The Hague, The Netherlands, pp. 1–244.

Hammen, L. van der (1989) *An Introduction to Comparative Arachnology*. SPB Academic Publishing, The Hague, The Netherlands, pp. 1–576.

Hänel, H. (1986) Effect of juvenile hormone (III) from the host *Apis mellifera* (Insecta: Hymenoptera) on neurosecretion of the parasitic mite *Varroa jacobsoni* (Acari: Mesostigmata). *Experimental and Applied Acarology* 2, 257–271.

Hanks, L.M., McCelfresh, J.S., Millar, J.G. and Paine, T.D. (1992) Control of the straw itch mite (Acari: Pyemotidae) with sulfur in an insect rearing facility. *Journal of Economic Entomology* 85(3), 683–686.

Hansen, G., Taylor, C., Goedeke, J., Broce, A., Lewis, L. *et al.* (2005) Outbreak of pruritic rashes associated with mites-Kansas. *The Morbidity and Mortality Weekly Report* 54, 952–955.

Harada, S. H. and Sadaji, T. (1925) On a case of mites found in human urine. *Chugai Iji Shimpo* 44(14), 859–866 (in Japanese).

Haragsim, O., Samšiňák, K. and Vobrázková, E. (1978) The mites inhabiting the bee hives in ČSR. *Zeitschrift für Angewandte Entomologie* 87(1), 52–67.

Hardison, J.R. (1959) Evidence against *Fusarium poae* and *Siteroptes graminum* as causal agents of silver top of grasses. *Mycologia* 51, 712–728.

Harlan, G.T., Sherman, A.P., Jr., Sorensen, A.A. and Trostle, M.R. (1987) The Straw Itch Mite, *Pyemotes tritici* (Acari: Pyemotidae), as a Biological Control Agent of Red Imported Fire Ants, *Solenopsis Invicta* (Hymenoptera: Formicidae). *Florida Entomologist* 70(4), 439–444.

Harrington, D., El Din, H.M., Guy, J., Robinson, K. and Sparagano, O. (2008) Characterization of the immune response of domestic fowl following immunization with proteins extracted from *Dermanyssus gallinae*. *Veterinary Parasitology* 160, 285–294.

Harrington, D., Canales, M., de la Fuente, J., de Luna, C., Robinson, K., Guy, J. and Sparagano, O. (2009) Immunisation with recombinant proteins subolesin and Bm86 for the control of *Dermanyssus gallinae* in poultry. *Vaccine* 27, 4056–4063.

Harrington, D.W.J., George, D.R., Guy, J.H. and Sparagano, O.A.E. (2011) Opportunities for integrated pest management to control the poultry red mite, *Dermanyssus gallinae*. *World's Poultry Science Journal* 67, 83–94.

Hart, B.J. (1998) Life cycle and reproduction of house-dust mites: environmental factors influencing mite populations. *Allergy* 53(48), 13–17.

Hart, B.J. and Fain, A. (1987) A new technique for isolation of mites exploiting the difference in density between ethanol and saturated NaCl: Qualitative and quantitative studies. *Acarologia* 28, 251–254.

Hart, B.J. and Fain, A. (1988) Morphological and biological studies of medically important house-dust mites. *Acarologia* 29(3), 285–295.

Hart, B.J., Crowther, D., Wilkinson, T., Biddulph, P., Ucci, M., Pretlove, S., Ridley, I. and Oreszczyn, T. (2007) Reproduction and development of laboratory and wild house dust mites (acari: pyroglyphidae) and their relationship to the natural dust ecosystem. *Journal of Medical Entomology* 44(4), 568–574.

Hartcher, K.M.H. and Jones, B. (2017) The welfare of layer hens in cage and cage-free housing systems. *World's Poultry Science Journal* 73, 767–782.

Hartini, S. and Dwibadra, D. (2017) Macrochelid mites (Acari: Mesostigmata) associated with dung beetles in Mount Gede-pangrango National Park, West Java, Indonesia. *Journal of Species Research* 6, 181–189.

Hartmannova, V., Rupes, V. and Vranova, J. (1973) The effect of some insecticides on *Tyrophagus putrescentiae* (Schrank) and selected species of fungi. *Czech Mycology* 27, 48–54.

Harving, H., Hansen, L.G., Korsgaard, J., Nielsen, P.A., Olsen, O.F. *et al.* (1991) House dust mite allergy and anti-mite measures in the indoor environment. *Allergy* 46(Suppl. 11), 33–38.

Harwood, R.F. and James, M.T. (1979) *Entomology in Human and Animal Health*. Macmillan Pub. Co. Inc, New York, pp. 1–548.

Hasan, M.M., Aikins, M.J., Schilling, M.W. and Phillips, T.W. (2020) Comparison of Methyl Bromide and Phosphine for Fumigation of *Necrobia rufipes* (Coleoptera: Cleridae) and *Tyrophagus putrescentiae* (Sarcoptiformes: Acaridae), pests of high-value stored products. *Journal of Economic Entomology* 113(2), 1008–1014.

Hase, A. (1933) Uber die Mehlmottenmilbe *Typhlodromus tineivorus* Oudms. 1929. *Zoologischer Anzeiger* 104, 237–239.

Hassan, M.F., Afifi, A.M. and Nawar, M.S. (1987) *Dendrolaelaps sayedi*, a new digamasellid mite (Acari: Gamasida) from Egypt, with notes on its biology. *Bulletin of the Entomological Society of Egypt* 66, 227–235.

Hassan, Md. M., Phillips, T. W., Aikins, M.J. and Schilling, M.W. (2010) Controlled atmosphere treatments to control arthropod pests of American dry cured hams. In: Proceedings, Annual International Research Conference on Methyl Bromide Alternatives and Emissions Reductions, November 2–5, Orlando, Florida, USA, p. 50.

Hassan, A.T., Mahmoud, H.I., Hassan, M.I., Omar, N.R. and Nasr, H.M. (2019) Biological studies on the stored grain mite, *Carpoglyphus lactis* (Linne) (Acari: Astigmata: Carpoglyphidae). *Egyptian Journal of Agricultural Research* 97(1), 167–177.

Hay, D. (1995) An in situ coring technique for estimating the population size of house dust mites in their natural habitat. *Acarologia* 36, 341–345.

Hay, D.B., Hart, B.J., Pearce, R.B., Kozakiewicz, Z. and Douglas, A.E. (1992) How relevant are house dust mite-fungal interactions in laboratory culture to the natural dust system? *Experimental and Applied Acarology* 16, 37–47.

Hay, D.B., Hart, B.J. and Douglas, A.E. (1993) Effects of the fungus *Aspergillus penicillioides* on the house dust mite *Dermatophagoides pteronyssinus*: An experimental re-evaluation. *Medical and Veterinary Entomology* 7(3), 271–274.

Hazan, A., Gerson, U. and Tahori, A.S. (1973) Life history and life tables of the carmine spider mite. *Acarologia* 15, 414–440.

Hurlbutt, H. (1958) A study of soil-inhabiting mites from Connecticut apple orchards. *Journal of Economic Entomology* 51, 767–771.

He, P., Wu, J., Wang, J., Ren, Y., Ahmad, W., Liu, R., Ouyang, Q., Jiang, H. and Chen. Q. (2020) Detection of mites *Tyrophagus putrescentiae* and *Cheyletus eruditus* in flour using hyperspectral imaging system coupled with chemometrics. *Journal of Food Process Engineering* 43(6), e13386.

Heikal, H.M. (2015) Studies on the Occurrence, Identification and Control of House Dust Mites at Rural Houses of Shebin El-Kom Locality, Egypt. *Pakistan Journal of Biological Sciences* 18(4), 179–184.

Heikal, H.M. (2020) *Parasitus fimetorum and Macrocheles muscaedomesticae* (Acarina: Parasitidae, Macrochelidae) as Natural Predators of the Root Knot Nematode, *Meloidogyne javanica* Treub. *Egyptian Journal of Biological Pest Control* 30, 33.

Heinemann, R.L. and Hughes, R.D. (1969) The cytological basis for reproductive variability in the Anoetidae (Sarcoptiformes: Acari). *Chromosoma* 28, 346–356.

Helle, W. (1967) Fertilization in the two spotted spider mite (*Tetranychus urticae*: Acari). *Entomologia Experimentalis et Applicata* 10, 103–110.

Helle, W. (1985) Genetics. In: Helle, W. and Sabelis, M.W. (eds) *Spider Mites: Their Biology, Natural Enemies and Control. World Crop Pests, Volume 1B*. Elsevier Science, Amsterdam, The Netherlands, pp. 185–192.

Helle, W. and Pijnacker, L.P. (1985) Parthenogenesis, chromosomes and sex. In: Helle, W. and Sabelis, M.W. (eds) *Spider Mites: Their Biology, Natural Enemies and Control. World Crop Pests, Volume 1A*. Elsevier Science, Amsterdam, The Netherlands, pp. 129–139.

Helle, W. and Sabelis, M.W. (eds) (1985) Spider Mites. *Their Biology, Natural Enemies and Control. World Crop Pests, Volume 1B*. Elsevier Science, Amsterdam, The Netherlands.

Helle, W., Bolland, H.R., van Arrendok, R., DeBoer, R., Schulten, G.G.M. and Russel, V. (1978) Genetic evidence for biparental males in haplodipolid predator mites (Acarina: Phytoseiidae). *Genetica* 49, 165–171.

Helle, W., Bolland, H.R., Jeurissen, S.H.M. and Van Seventer, G.A. (1984) Chromosome data on the Actinedida, Tarsonemida and Oribatida. In: Griffiths, D.A. and Bowman, C.E. (eds) *Acarology VI, Volume 1*. Ellis Horwood, Chichester, UK, pp. 449–454.

Hendrix, J.D., Zhang, X., Campbell, Y.L., Zhang, L., Siberio, L. *et al.* (2018) Effects of temperature, relative humidity, and protective netting on *Tyrophagus putrescentiae* (Schrank) (Sarcoptiformes: Acaridae) infestation, fungal growth, and product quality of dry cured hams. *Journal of Stored Products Research* 77, 211–218.

Heneghan, L. (1994) Studies of soil microarthropod communities, experimentally manipulated by chronic low - level nutrient input, and their impact on some ecological processes. Ph. D. Thesis, University College Dublin, Dublin, pp. 1–312.

Hennessey, M.K. and Farrier, M.H. (1988) Systematic revision of thirty species of free-living, soil inhabiting gamasine mites (Acari: Mesostigmata) of North America. *Technical bulletin (North Carolina Agricultural Research Service)* 258, 1–123.

Henschel, J. (1929) Reizphysiologische Untersuchung der Käsemilbe *Tyrolichus casei* (Oudemans). *Zeitschrift f r vergleichende Physiologie* 9, 802–837.

Hensel, P., Santoro, D., Favrot, C., Hill, P. and Griffin, C. (2015) Canine atopic dermatitis: detailed guidelines for diagnosis and allergen identification. *BMC Veternary Research* 11, 196.

Henszel, L., Kalisinska, E., Kosik-Bogacka, D.I. and Kuzna-Grygiel, W. (2011) Preliminary studies on storage mites in litter and animal fodder collected from farm buildings in Northwestern Poland. *Polish Journal of Environmental Studies* 20, 795–800.

Herfs, A. (1926) Ökologische Untersuchungen an *Pediculoides ventricosus* (Newp.) Berlese. *Zoologica Original-Abhandlungen aus dem Gesamtgebiete der Zoologie* 74, 3–68.

Hering, E. (1838) Die Kratzmilben der Thiere und einige verwandte Arten nach eigenen Untersuchungen beschrieben. *Nova Acta Physico-Medica Academiae Caesareae Leopoldino-Carolinae Naturae Curiosum* 18, 573–624.

Hermann, J.F. (1804) III. Ciron (*Scirus*). *Mémoire aptérologique* 60–62; pl. 3; fig. 12; pl. 6; fig. 12.

Hernandes, F.A., Skvarla, M.J., Fisher, R., Dowling, A.P. G., Ochoa, R., Ueckermann, E. A. and

Bauchan, G. (2016) Catalogue of snout mites (Acariformes: Bdellidae) of the world. *Zootaxa* 4152(1), 1–83.

Hewitt, M. and Turk, S.M. (1974) *Cheyletiella* sp. in the personal environment. With notes on the difference between *C. parasitivorax* Megnin and *C. yasguri* Smiley. *British Journal of Dermatology* 90(6), 679–683.

Hewitt, R., Barrow, G. I., Miller, D.C., Türk, P. and Türk, S. (1973) Mites in the personal environment and their role in skin disorders. *British Journal of Dermatology* 89, 401–409.

Hill, D.S. (2003) *Pests of Stored Foodstuffs and Their Control*. Kluwer Academic Publishers, Dordrecht, The Netherlands, pp. 1–496.

Hill, M. (1998) Quantification of house-dust-mite populations. *Allergy* 53(Suppl. 48), 18–23.

Hill, A. and Deahl, K.L. (1978) Description and life cycle of a new species of *Histiostoma* (Acari: Histiostomidae) associated with commercial mushroom production. *Proceedings of the Entomological Society of Washington* 80, 317–329.

Hill, M.A. and Gordon, R.M. (1945) An outbreak of dermatitis amongst troops in North Wales caused by rodent mites. *Annals of Tropical Medicine & Parasitology* 39, 46–52.

Hiroyuki, H., Naoki, M., Kimiko, O., Ritsuo, N. and Yasumasa, K. (2003) Chemical ecology of astigmatid mites LXIX. Neryl formate [3,7-dimethyl-(Z)- 2,6-octadienyl formate] as the alarm pheromone of an acarid mite, *Histiogaster rotundus* Woodring (Acari: Acaridae). *Applied Entomoloy and Zoology* 38, 379–385.

Hirschmann, W. (1960) Gangsystematik der Parasitiformes. 3. Die Gattung *Dendrolaelaps* Halbert 1915. *Acarologie* 3, 1–27, (28 Tafeln).

Hirschmann, W. (1962) Gangsystematik der Parasitiformes. Teil 5, Gamasiden. Acarol. Schrift. Vergleich. Milbenk. Furth, Hirschmann, Verlag, Germany, pp. 1–80.

Hirschmann, W. (1966) Gangsystematik der Parasitiformes. *Teil 15, Gänge von Litoralmilben und neue Litoralmilbenarten Acarologie* 9, 25–44.

Hirschmann, W. (1970) Acarologie- Schriftenreihe für Vergleichende Milbenkunde. Folge 14. Fürth (Bayern), 1–60.

Hirschmann, W. and Wiśniewski, J. (1982a) Weltweite Revision der Gattungen *Dendrolaelaps* Halbert 1915 und *Longoseius* Chant 1961 (Parasitiformes). Band 1. Beschreibung der Untergattungen und Arten, Bestimmungstabellen, Chätotaxie, Porotaxie. Acarologie. *Schriftenreihe für Vergleichende Milbenkunde* 29, 1–190.

Hirschmann, W. and Wisniewski, J. (1982b) Weltweite Revision der Gattungen *Dendrolaelaps* Halbert 1915 und *Longoseius* Chant 1961 (Parasitiformes). Band II. Artenverzeichnisse, Krankheiten, Missbildungen. Inseminationsapparate, Abbildungen. *Acarologie. Schriftenreihe für Vergleichende Milbenkunde* 29, 1–48 + 94 plates.

Hirschmann, W. and Wiśniewski, J. (1988) Weltweite Revision der Gattung *Trichouropoda* Berlese 1916, Weltbestimmungstabellen, Ergänzungsbeschreibungen *elegans*-Gruppe, *urospinoidea*-Gruppe, *orbicularis*-Gruppe. *Acarologie* 35, 1–201.

Hirschmann, W. and Wisniewski, J. (1989) Gangsystematik der Parasitiformes. Teil 507. Weltweite Revision der Ganggattung *Trichouropoda* Berlese 1916. Geschichte, Bestimmungstabellen, Artenverzeichnisse (Trichouropodini, Uropodinae). *Acarologie. Schriftenreihe für Vergleichende Milbenkunde* 36, 1–65.

Hirschmann, W. and Wisniewski, J. (1992) Gangsstematik der Parasitiformes. Teil 543. Die Teilgange (Larve, Protonymphe, Deutonymphe) von *Polyaspinus schweizeri* und *Polyaspinus nicolae* (Uropidini, Uropodinae). *Acarologie* 39, 161–168.

Hirschmann, W. and Wisniewski, J. (1993) Acari Parasitiformes. Supercohors Atrichopygidiina Hirschmann 1975. Die Uropodiden der Erde. *Acarologie. Schriftenreihe für Vergleichende Milbenkunde* 40, 1–466.

Hirschmann, W. and Zirngiebl-Nicol, I. (1961) Die Gattung *Trichouropoda* Berlese 1916, Cheliceren und System der Uropodiden. Acarologie – Gangsystematik der Parasitiformes, Teil 4. Schriftenreihe für vergleichende Milbenkunde, Folge 4, Fürth/Bayern, pp. 1–41.

Hirst, S. (1912) Report on the mite causing Copra Itch. *J. Trop. Med. Hyg.* 15, 374–375.

Hirst, S. (1923) On some arachnid remains from the Old Red Sandstone (Rhynie Chert Bed, Aberdeenshire). *Annals and Magazine of Natural History (Series 9)* 12(70), 455–474.

Hirst, S. (1926) The principal species of Acari parasitic on rats, with brief notes on hosts and distribution. Ceylon J. Sci., Section D - Med. Sci. 1, 273–276 + Plates XXVII-XXXIII.

Hislop, R.G. and Prokopy, R.J. (1981) Mite predator responses to prey and predator-emitted stimuli. *Journal of Chemical Ecology* 7, 895–904.

Ho, C.-C. and Chen, J.S. (1987) A new record of bulb mite, *Rhizoglyphus setosus* Manson, (Acarina:

Acaridae) from Taiwan. *Journal of Agricultural Research of China* 36, 237–238 (in Chinese).

Ho, C.-C. and Wu, C.-S. (2002) *Suidasia* mite found from the human ear. *Forms – Entomology* 22, 291–296 (in Chinese).

Ho, T.M. (1986) Pyroglyphid mites found in house dust in Peninsular Malaysia. *Tropical Biomedicine* 3, 89–93.

Ho, T.M. (1990) Phoretic association between *Macrocheles muscaedomesticae* (Acari, Macrochelidae) and flies inhabiting poultry manure in peninsular Malaysia. *Experimental and Applied Acarology* 10, 61–68.

Ho, T.M. and Nadchatram, M. (1985) Distribution of *Dermatophagoides pteronyssinus* (Astigmata: Pyroglyphidae) in Cameron Highlands, Malaysia. *Tropical Biomedicine* 2, 54–58.

Hofstetter, R.W., Moser, J.C. and Blomquist, S.R. (2014) Mites associated with bark beetles and their hyperphoretic ophiostomatoid fungi. *CBS Biodiversity Series* 12, 156–176.

Holt, P.G. and Thomas, W.R. (2005) Sensitization to airborne environmental allergens: unresolved issues. *Nature Immunology* 6, 957–960.

Homann, H. (1933) Die Milben in gesunden Bienenstöcken. *Zeitschrift für Parasitenkunde* 6, 350–415.

Hong, C.S., Park, H.S. and Oh, S.H. (1987) *Dermatophagoides farinae*, an Important Allergenic Substance in Buckwheat-Hush Pillows. *Yonsei Medical Journal* 28(4), 274–281.

Hora, A.M. (1934) On the biology of the mite, *Glycyphagus domesticus* De Geer (Tyroglyphidae, Acarina). *Annals of Applied Biology* 21(3), 483–494.

Horn, T.B., Körbes, J.H., Granich, J., Senter, M. and Ferla, N.J. (2015) Influence of laying hen systems on the mite fauna (Acari) community of commercial poultry farms in southern Brazil. *Parasitology Research* 115, 355–366.

Horn, T.B., Granich, J., Horn Körbes, J., Liberato Da Silva, G. and Juarez Ferla, N. (2018) Mite fauna (Acari) associated with the poultry industry in different laying hen management systems in Southern Brazil: A species key. *Acarologia* 58(1), 140–158.

Hosaya, H. and Kugoh, T. (1954) On the new records of *Gohieria fusca* (Oudemans, 1903) from packed bean flour and sugar in Japan. *Japanese Journal of Sanitary Zoology* 5, 154–156.

Houck, M.A. (1994) Adaptation and transition into parasitism from commensalism: A phoretic model. In: Houck, M.A. (ed.) *Mites: Ecological and Evolutionary Analyses of Life-history Patterns*, 1st edn, Chapman and Hall, New York, pp. 252–281.

Houck, M.A. and O'Connor, B.M. (1990) Ontogeny and life history of *Hemisarcoptes cooremani* (Acari: Hemisarcoptidae). *Annals of the Entomological Society of America* 83, 869–886.

Houck, M.A. and OConnor, B.M. (1991) Ecological and evolutionary significance of phoresy in the Astigmata. *Annual Review of Entomology* 36, 611–636.

Howard, R.W., Kuwahara, Y., Suzuki, H. and Suzuki, T. (1988) Pheromone study on acarid mites. XII. Characterization of the hydrocarbons and external gland morphology of the opishonotal glands of six species of mites (Acari: Astigmata). *Applied Entomoloy and Zoology* 23(1), 58–66.

Howes, C.A. (1979) The noctule bat, *Nyctalus noctula* (Schr.) in Yorkshire. *Naturalist* 104, 31–38.

Hoy, M.A. (1985) Recent advances in genetics and genetic improvement of the Phytoseiidae. *Annual Review of Entomology* 30, 345–370.

Hoy, M.A. (2011) *Agricultural Acarology: Introduction to Integrated Mite Management*. CRC Press, Boca Raton, FL, pp. 325–327.

Hoy, M.A. and Smilanick, J. (1979) A sex pheromone produced by immature and adult females of the predatory mite, *Metaseiulus occidentalis* (Acarina: Phytoseiidae). *Entomologia Experimentalis et Applicata* 26, 291–300.

Hoy, M.A. and Smilanick, J. (1981) Non-random prey location by the phytoseiid predator *Metaseiulus occidentalis*: Differential responses to several spider mite species. *Entomologia Experimentalis et Applicata* 29, 241–253.

Hrudová, E. and Šafránková, I. (2017) *Plodia interpunctella* (Hübner, 1813) and *Nemapogon granellus* (Linnaeus, 1758) - Uncommon Pests on Stored Garlic (*Allium sativum* L.) in the Czech Republic. *Acta Univ. Agric. et Silvic. Mendelianae Brun* 65, 1167–1173.

Hsin, K.-L. and Chen, C.-P. (1963) Morphological study on *Aleuroglyphus ovatus* Troupeau (Acarina: Acaridae). *Acta Entomologica Sinica*. Available at: http://en.cnki.com.cn/Article_en/CJFDTOTAL-KCXB196303006.htm (accessed 21 May 2020).

Hsin, K.-L. (= Xin, J.-L.) and Chen, C.P. (=Shen, Z.-P.) (1964) Studies on the life-history of *Aleuroglyphus ovatus* Troupeau (Acarina, Acaridae). *Acta Entomologica Sinica* 13(3), 428–435 (in Chinese).

Hsu, P.-K. and Li, J.-C. (1985) Mites collected from the urine of nephrotic children. *Annual Bulletin of the Society of Parasitology, Guangdong Province* 7, 203–206.

Lin, J.-Z. and Zhang, Z.-Q. (2010) Tarsonemidae of China: a review of progress on the systematics and biology, with an updated checklist of species. In: Zhang, Z.Q., Hong, X.-Y. and Fan, Q.H. (eds) *Xin Jie-Liu Centenary: Progress in Chinese Acarology, Zoosymposia* 4, 175–185.

Huang, H., Xuenong, X., Jiale, L., Guiting, L., Endong, W. and Yulin, G. (2013) Impact of proteins and saccharides on mass production of *Tyrophagus putrescentiae* (Acari: Acaridae) and its predator *Neoseiulus barkeri* (Acari: Phytoseiidae). *Biocontrol Science and Technology* 23(11), 1231–1244.

Huber, J. and Pekar, S. (2009) Combination of the antifeedant bean flour and the predator *Cheyletus malaccensis* suppresses storage mites under laboratory conditions. *Biocontrol* 54(3), 403–410.

Huber, J., Kubátová, A. and Šárová, J. (2000) Feeding of *Scheloribates laevigatus* (Acari: Oribatida) on different stadia of decomposing grass litter (*Holcus lanatus*). *Pedobiologia* 44(5), 627–639.

Hubert, J. (2001) The influence of *Scheloribates laevigatus* (Acari: Oribatida) on decomposition of *Holcus lanatus* litter. *Acta Societatis Zoologicae Bohemicae* 65, 77–80.

Hubert, J. (2011) *The Pest Importance of Stored Product Mites (Acari, Acaridida)*. Nova, Science Publishers, Inc, New York, pp. 1–46.

Hubert, J., Stejskal, V., Munzbergova, Z., Kubatova, A., Vanova, M. and Žďárková, E. (2003) Mites as selective fungal carriers in stored grain habitats. *Experimental and Applied Acarology* 29(1–2), 69–87.

Hubert, J., Kudlíková, I. and Stejskal, V. (2004a) Review of digestive enzymes of stored product and house dust mites. *Phytophaga* XIV, 695–710.

Hubert, J., Jarosik, V., Mourek, J., Kubatova, A. and Zdarkova, E. (2004b) Astigmatid mite growth and fungi preference (Acari: Acaridida): Comparisons in laboratory experiments. *Pedobiologia* 48, 205–214.

Hubert, J, Stejskal, V., Munzbergová, Z., Kubátová, A., Vanova, M. and Zdarkova, E. (2004c) Mites and fungi in heavily infested stores in the Czech Republic. *Journal of Economic Entomology* 97(6), 2144–2153.

Hubert, J., Dolečková-Marešová, L., Hýblová, J. and Kudlíková, I., Stejskal, V. and Mareš, M. (2005) *In vitro* and *in vivo* inhibition of α-amylases of stored-product mite *Acarus siro*. *Experimental and Applied Acarology* 35(4), 281–291.

Hubert, J., Munzbergová, Z., Kucerová, Z. and Stejskal, V. (2006a) Comparison of communities of stored product mites in grain mass and grain residues in the Czech Republic. *Experimental and Applied Acarology* 39(2), 149–158.

Hubert, J., Němcová, M., Aspaly, G. and Stejskal, V. (2006b) The Toxicity of Bean Flour (*Phaseolus vulgaris*) to Stored-Product Mites (Acari: Acaridida). *Plant Protection Science* 42, 125–129.

Hubert, J., Stejskal, V., Munzbergova, Z., Hajslova, J. and Arthur, F.H. (2007) Toxicity and efficacy of selected pesticides and new acaricides to stored product mites (Acari: Acaridida). *Experimental and Applied Acarology* 42, 283–290.

Hubert, J., Stejskal, C.V.A., Aspaly, G. and Nzbergova, Z.M. (2007) Suppressive potential of bean (*Phaseolus vulgaris*) flour against five species of stored-product mites (Acari: Acaridae). *Journal of Economic Entomology* 100(2), 586–590.

Hubert, J., Münzbergová, Z., Nesvorná, M., Poltronieri, P. and Santino, A. (2008) Acaricidal effects of natural six-carbon and nine-carbon aldehydes on stored-product mites. *Experimental and Applied Acarology* 44, 315–321.

Hubert, J., Kučerová, Z., Aulicky, R., Nesvorna, M. and Stejskal, V. (2009) Differential levels of mite infestation of wheat and barley in Czech grain stores. *Journal of Insect Science* 16, 255–262.

Hubert, J., Pekar, S., Nesvorna, M. and Sustr, V. (2010) Temperature preference and respiration of Acarid mites. *Journal of Economic Entomology* 103, 2249–2257.

Hubert, J., Erban, T., Nesvorna, M. and Stejskal, V. (2011) Emerging risk of infestation and contamination of dried fruits by mites in the Czech Republic. *Food Additives & Contaminants: Part A* 28(9), 1129–1135.

Hubert, J., Pekár, S., Aulický, R., Nesvorná, M. and Stejskal, V. (2013) The effect of stored barley cultivars, temperature and humidity on population increase of *Acarus siro*, *Lepidoglyphus destructor* and *Tyrophagus putrescentiae*. *Experimental and Applied Acarology* 60, 241–252.

Hubert, J., Nesvorna, M. and Erban, T. (2014a) Growth-suppressive effect of the a-amylase inhibitor of *Triticum aestivum* on stored-product mites varies by the species and type of diet. *Experimental and Applied Acarology* 62, 57–65.

Hubert, J., Nesvorná, M., Kopecký, J., Ságová-Marečková, M. and Poltronieri, P. (2014b) *Carpoglyphus lactis* (Acari: Astigmata) from various dried fruit differed in associated micro-organisms. *Journal of Applied Microbiology* 118, 470–484.

Hubert, J., Kopecky, J., Sagova-Mareckova, M., Nesvorna, M., Zurek, L. and Erban, T. (2016) Assessment of Bacterial Communities in Thirteen Species of Laboratory-Cultured Domestic Mites (Acari: Acaridida). *Journal of Economic Entomology* 109(4), 1887–1896.

Hubert, J., Stejskal, V., Athanassiou, C.G. and Throne, J.E. (2018) Health Hazards Associated with Arthropod Infestation of Stored Products. *Annual Review of Entomology* 63(1), 553–573.

Hubert, J., Nesvorna, M., Green, S.J. and Klimov, P.B. (2021) Microbial communities of stored product mites: Variation by species and population. *Microbial Ecology* 81(2), 506–522.

Huebner, R.J., Jellison, W.L. and, and Pomerantz, C. (1946) Rickettsialpox, a newly recognized rickettsial disease; isolation of a rickettsia apparently identical with the causative agent of rickettsialpox from *Allodermanyssus sanguineus*, a rodent mite. *Public Health Reports* 61(47), 1677–1682.

Huffaker, C.B., van de Vrie, M. and McMurtry, J.S. (1970) Ecology of tetranychid mites and their natural enemies: a review. II. Tetranychid populations and their possible control by predators: an evaluation. *Hilgardia* 40(11), 391–458.

Hughes T.E. (1953) The functional morphology of the mouth parts of the mite *Anoetus sapromyzarum* Dufour 1839, compared with those of the more typical Sarcoptiformes. *Proceedings of the National Academy of Sciences Amsterdam* 56C(2), 278–287.

Hughes, A.M. (1948) *The Mites Associated with Stored Food Products*. Ministry of Agriculture, Fisheries and Food Technology Bulletin, London, pp. 1–166.

Hughes, A.M. (1952) A new Glycyphagid mite-*Carpoglyphus munroi*. *Zoological Journal of the Linnean Society* 42(284), 168–175.

Hughes, A.M. (1956) The mite genus *Lardoglyphus* Oudemans, 1927 (= *Hoshikadania* Sasa and Asanuma, 1951). *Zoologische mededeelingen* 34(20), 271–285.

Hughes, A.M. (1957) Silk production in the Cheyletidae (Acarina: Arachnida). Nature 180, 815–816.

Hughes, A.M. (1961) The mites of stored food. Ministry of Agriculture, Fisheries and Food, Technical Bulletin no. 9. Her Majesty's Stationery Office, London, pp. 1–287.

Hughes, A.M. (1976) The Mites of Stored Food and Houses. MAFF (Ministry of Agriculture, Fisheries and Food) Technical Bulletin No. 9 (rev. edn). Her Majestry's Stationery Office, London, pp. 1–401.

Hughes, A.M. and Hughes, T.E. (1939) The internal anatomy and postembryonic development of *Glycyphagus domesticus* De Geer. *Proceedings of the Zoological Society of London* 108, 715–733.

Hughes, A.M. and Maunsell, K. (1973) A study of a population of house dust mite in its natural environment. *Clinical & Experimental Allergy* 3, 245–249.

Hughes, R.D. and Jackson, C.G. (1958) A review of the Anoetidae (Acari). Va. J. Sci. 9(1), 5–198.

Hughes, T.E. (1943) The respiration of *Tyrophagus farinae*. *Journal of Experimental Biology* 20(1), 1–5.

Hughes, T.E. (1950) The physiology of the alimentary canal of *Tyrophagus farinae*. *Quarterly Journal of Microscopical Science* 91(1), 54–60.

Hughes, T.E. (1954) Some histological changes which occur in the gut epithelium of *Ixodes ricinus* females during gorging and up to oviposition. *Annals of Tropical Medicine & Parasitology* 48, 397–404.

Hughes, T.E. (1959) *Mites or the Acari*. Athlone Press, London, pp. 1–225.

Huhta, V. (2016) Catalogue of the Mesostigmata mites in Finland. *Memoranda Societatis pro Fauna et Flora Fennica* 92, 129–148.

Huhta, V., Siira-Pietikäinen, A., Penttinen, R. and Räty, M. (2010) Soil fauna of Finland: Acarina, Collembola and Enchytraeidae. *Memoranda Societatis pro Fauna et Flora Fennica* 86, 59–82.

Hull, J.E. (1918) Terrestrial acari of the tyne province. *Transactions of the Natural History Society of Northumberland, Durham, and Newcastle-upon-Tyne* 75, 13–88.

Hull, J.E. (1925) Acari of the family Gamasidae; new and rare British species. *Annals and Magazine of Natural History* (Series 9) 15, 201–219.

Hull, J.E. (1931) A new genus and species of analgesidae (feather-mites). *Vusculum* 17, 145–147.

Hunponu-Wusu, O.O. and Somorin, A.O. (1978) Epidemiological aspects of house dust mite allergy in Nigeria. *Journal of Dermatology* 5, 27–32.

Hunter, J.E. and Farrier, M.H. (1975) Mites of the genus *Oplitis* Berlese (Acarina: Uropodidae) associated with ants (Hymenoptera: Formicidae) in the southeastern United States. Part I. *Acarologia* 17(4), 595–623.

Hunter, M.D. (2002) A breath of fresh air: beyond laboratory studies of plant volatile-natural enemy interactions. *Agricultural and Forest Entomology* 4, 81–86.

Hunter, P.E. and Cross, E.A. (1968) Key to North American Ereynetinae and description of a new species (Acarina: Ereynetidae). *Journal of the Entomological Research Society* 3, 154–158.

Hunter, P.E. and Poe, S.L. (1971) Description of *Ereynetes* (*Anereynetes*) *coenobitus* n.sp. (Acarina: Ereynetidae) associated with hermit crab. *Florida Entomologist* 54, 135–138.

Hunter, P.E., Rosario, R.M.T. and Moser, J.C. (1989) Two new species of *Ereynetes* (Acari: Prostigmata: Ereynetidae) associated with bark beetles. *Journal of the Economic Science* 24, 16–20.

Hurtado, I. and Parini, M. (1987) House dust mites in Caracas, Venezuela. *Annals of Allergy* 59, 128–130.

Hussey, N.W. (1985) Cucumbers. In: Hussey, N.W. and Scope, N.E.A. (eds) *Biological Pest Control. The Glasshouse Experience*. Blandford Press, Poole, UK, pp. 175–179.

Hussey, N.W., Read, W.H. and Hesling, J.J. (1969) *The Pest of Protected Cultivations: The Biology and Control of Glasshouse and Mushroom Pests*. Edward Arnold, London, pp. 1–404.

Hutcheson, H.J. and Oliver, J.H. (1988) Spermiogenesis and Reproductive Biology of *Dermanyssus gallinae* (De Geer) (Parasitiformes: Dermanyssidae). *Journal of Medical Entomology* 9(25), 321–330.

Hyatt, K.H. (1959) Mesostigmatid mites associated with *Geotrupes stercorarius* (L.) (Col., Scarabaeidae). *Entomologist's Monthly Magazine* 95, 22–23.

Hyatt, K.H. (1980) Mites of the subfamily Parasitinae (Mesostigmata: Parasitidae) in the British Isles. *Bulletin of the British Museum (Natural History) Zoology* 38(5), 237–378.

Hyatt, K.H. (1990) Mites associated with terrestrial beetles in the British Isles. *Entomologist's Monthly Magazine* 126, 1512–1515.

Hyatt, K.H. and Emberson, R.M. (1988) A review of the Macrochelidae (Acari: Mesostigmata) of the British Isles. *Bulletin of the British Museum (Natural History) Zoology* 54, 63–125.

Hwang, E.T. (2023) Management of the poultry red mite *Dermanyssus gallinae* with physical control methods by inorganic material and future perspectives. *Poultry Science* 102, 102772.

Ibrahim, G.A., Afifi, A.M. and Abdel-Halim, S.M. (1989) Laboratory observations on the biology of *Proctolaelaps pygmeus* (Muller) and *Protogamasellus mica* (Athias Henriot) (Acari: Gamasia: Ascidae). *Bulletin de la Société entomologique d'Égypte* 68, 43–47.

Ibrahim, W.L.F. (2006) Effect of Photoperiod on the development and fecundity of *Carpoglyphus lactis* L. (Acari: Carpoglyphidae). *Egyptian Journal of Hospital Medicine* 23, 212–218.

ICZN (1999) International Code of Zoological Nomenclature. Fourth edition. London, U.K.

Iglesias-Souto, J., Sanchez-Machin, I., Iraola, V., Poza, P., Gonzalez, R. and Matheu, V. (2009) Oral mite anaphylaxis by *Thyreophagus entomophagus* in a child: A case report. *Clinical and Molecular Allergy* 7, 10–13.

Ignatowicz, S. (1981) Effect of inorganic salts upon biology and development of acarid mites. VII. Rapid desiccation of the copra mite *Tyrophagus putrescentiae* (Schrank), and other mites with tricalcium phosphate and ferric phosphate. *Polskie Pismo Entomologiczne* 51, 471–482.

Ignatowicz, S. (2000) Evaluation of the efficacy of predatory mites in controlling pests of cultivated mushrooms in organic mushroom houses. *Organic Farming Research Foundation Information Bulletin*, 1–11.

Immediato, D., Camarda, A., Iatta, R., Puttilli, M.R., Ramos, R.A., Di Paola, G., Giangaspero, A., Otranto, D. and Cafarchia, C. (2015) Laboratory evaluation of a native strain of *Beauveria bassiana* for controlling *Dermanyssus gallinae* (De Geer, 1778) (Acari: Dermanyssidae). *Veterinary Parasitology* 212(3–4), 478–482.

Ingham, P.E. and Ingham, D.M. (1976) House dust mites and infant-use sheepskins. *Medical Journal of Australia* 1, 302–304.

Institute for Laboratory Animal Research (2011) *Guide for the Care and Use of Laboratory Animals*. 8th edn, The National Academies Press, Washington, DC, pp. 1–220.

Ioffe, I.D. (1965) Distribution of neurosecretory cells in the central nervous apparatus of *Dermacentor pictus* Herm. (Acarina, Chelicerata). *Doklady Akademii Nauk SSSR* 154, 25–29.

IPCC (2018) Summary for Policymakers. In: Masson-Delmotte, V., Zhai, P., Pörtner, H.-O., Roberts, D., Skea, J. *et al.* (eds) *Global Warming of 1.5°C. An IPCC Special Report on the Impacts of Global Warming of 1.5°C above Pre-Industrial Levels and Related Global Greenhouse Gas Emission Pathways, in the Context of Strengthening the Global Response to the Threat of Climate Change, Sustainable Development, and Eff orts to Eradicate Poverty*. World Meteorological Organization, Geneva, Switzerland, pp. 1–32.

Ippolito, R. and Triggiani, O. (1988) Prove di efficacia di alcuni acaricidi nei riguardi di *Tyrophagus similis* Volgin (Acaridida-Acaridae) su spinacio. *Informatore Fitopatologico* 38(2), 43–48.

Ishikawa, K. (1969) Studies on the mesostigmatid mites in Japan. IV. Family Blattisocidae Garman. *Rep. Res., Matsuyama Shinonome Junior College* 4, 111–139.

Ishikawa, K. (1980) Taxonomic and ecological studies in the family Parholaspidae (Acari, Mesostigmata) from Japan. *Bulletin of the National Museum of Nature and Science Series A (Zoology)* 6, 1–25.

Ivanov, V.P. and Leonovich, S.A. (1983) Sensory organs. In: Balashov, Y.S. (ed.) *An Atlas of Ixodid Tick Ultrastructure (English Translation from Russian)*. Special Publication of the Entomological Society of America, Lanham, MD, pp. 191–220.

Iversen, M., Korsgaard, J., Hallas, T.E. and Dahl, R. (1990) Mite allergy and exposure to storage mites and house dust mites in farmers. *Clinical & Experimental Allergy* 20(2), 211–219,

Iverson, K., OConnor, B.M., Ochoa, R., Heckmann, R. (1996) *Lardoglyphus zacheri* (Acari: Lardoglyphidae), a pest of museum dermestid colonies, with observations on its natural ecology and distribution. *Annals of the Entomological Society of America* 89, 544–549.

Izadi, H., Tamanadar, E., Khajehali, J. and Samadieh, H. (2023) *Rhizoglyphus robini*, a pest mite of saffron, is unable to resist extracellular ice formation. *Experimental and Applied Acarology* 91, 29–41.

Jacobs, S. and Calvin, D. (1988) Flour and Grain Mites. Available at: https://extension.psu.edu/flour-and-grain-mites (accessed 9 March 2017).

Jacot, A.P. (1930) Oribatid mites of the subfamily Phthiracarinae of the northeastern Uinted States. *Proceedings of the Boston Society of Natural History* 39, 209–261.

Jafari, S., Fathipour, Y., Faraji, F. and Bagheri, M. (2010) Demographic response to constant temperatures in *Neoseiulus barkeri* (Phytoseiidae) fed on *Tetranychus urticae* (Tetranychidae). *Systematic and Applied Acarology* 15, 83–99.

Jafari, S., Fathipour, Y. and Faraji, F. (2012) Temperature-dependent development of *Neoseiulus barkeri* (Acari: Phytoseiidae) on *Tetranychus urticae* (Acari: Tetranychidae) at seven constant temperatures. *Journal of Insect Science* 19 (2), 220–228.

Jafari, S., Abassi, N. and Bahirae, F. (2013) Demographic parameters of *Neoseiulus barkeri* (Acari: Phytoseiidae) fed on *Thrips tabaci* (Thysanoptera: Thripidae). *Persian Journal of Acarology* 2(2), 287–296.

Jakubas-Zawalska, J., Siuda, K. and Solarz, K. (2013) A comparative study of the house-dust acarofauna of dwellings and single-family houses from urban and rural areas of southern Poland. In: Buczek, A. and Błaszak, C. (eds) *Arthropods-The Medical and Veterinary Aspects*. Koliber, Lublin, Poland, pp. 31–47 (in Polish).

Jalaeian, M., Saboori, A. and Seyedoleslami, H. (2005) Prostigmatid mites (Acari: Prostigmata) associated with fruit trees in the Western area of Isfahan. *Journal of Entomological Society of Iran* 25(1), 67–68.

Jalaeian, M., Saboori, A. and Seyedoleslami, H. (2006) Introduction of some families, genera and species of mesostigmatic mites (Acari: Mesostigmata) from Isfahan Province. In: Manzari, S. (ed.) *Abstract Book of the 17th Iranian Plant Protection Congress, Campus of Agriculture and Natural Resources*. University of Tehran, Karaj, Iran, pp. 1–183.

Jalil, M. and Rodriguez, J.G. (1970) Biology and odor perception by *Fuscuropoda vegetans* (Acarina: Uropodidae), a predator of the house fly. *Annals of the Entomological Society of America* 63, 935–938.

James, M.P., Lake, D.E. and Sinclair, C.G. (1975) Bovine nasal granuloma. *New Zealand Veterinary Journal* 23(4), 63–64.

Janicki, J., Kadłubowski, W., Kiełczewski, B. and Matuszewski, A. (1959) Przyczynek do badań nad składem gatunkowym owadów i roztoczy w niektórych magazynach zbożowych województwa poznańskiego. *Roczniki Wyższej Szkoły Rolniczej (Poznań)* 6, 41–50.

Javadpour, M., HaJizadeh, J. and Hosseini, R. (2018) Blattisociid mites of Guilan province of Iran with a checklist for Iranian Blattisociid mites (Mesostigmata: Blattisociidae). *Entomofauna* 39(2), 697–710.

Jee, Y.K., Park, H.S., Kim, H.Y., Park, J.S., Lee, K.Y. *et al.* (2000) Two-spotted spider mite (*Tetranychus urticae*): an important allergen in asthmatic non-farmers symtomatic in summer and fall months. *Annals of Allergy, Asthma & Immunology* 84(5), 543–548.

Jeffrey, I.G. (1976a) An undescribed species of *Mexecheles* (Acarina: Cheyletidae) from a Scottish farm. *Acarologia* 17, 668–670.

Jeffrey, I.G. (1976b) A survey of the mite fauna of Scottish farms. *Journal of Stored Products Research* 12, 149–156

Jeffrey, I.G. and Campbell, J.B. (1975) New species of *Hemicheyletia* (Acarina: Cheyletidae). *Journal of Stored Products Research* 11, 103–105.

Jeon, H.Y., Kim, H.H., Jung, J.A., Kang, T.J. and Yang, C.Y. (2007) Damage Status of Poinsettia by the Fungus Gnat (*Bradysia difformis*) and Its Control with Predatory Mite (*Hypoaspis aculeifer*). *Korean Journal of Horticultural Science & Technology* 25, 468–473.

Jeon, J.H., Lee, C.H. and Lee, H.S. (2009) Food protective effect of geraniol and its congeners against stored food mites. *Journal of Food Protection* 72(7), 1468–1471.

Jeon, J.-H., Park, J.-H., Chung, N. and Lee, H.-S. (2014) Active monoterpene ketones isolated from *Rosmarinus officinalis* with fumigant and contact action against *Tyrophagus putrescentiae* (Schrank). *Journal of Food Protection* 77, 1355–1360.

Jeong, K.Y., Kim, W.K., Lee, J.S., Lee, J. Lee, I.Y. *et al.* (2005) Immunoglobulin E reactivity of recombinant allergen Tyr p 13 from *Tyrophagus putrescentiae* homologous to fatty acid binding protein. *Clinical and Diagnostic Laboratory Immunology* 12(5), 581–585.

Jeppson, L.R., Keifer, H.H. and Baker, E.W. (1975) Mites Injurious to Economic Plants, University of California Press, Berkeley and Los Angeles, CA, pp. 1–614 + 74 pl.

Jesionowska, K. (1991) Morphology of the eupodidae (actinotrichida, actinedida, eupodoidea). In: Dusbabek, F. and Bukva, V. (eds) *Modern Acarology, Volume 2*. Academia, Prague and SPB Academic Publishing, The Hague, The Netherlands, pp. 265–278.

Jess, S. and Schweizer, H. (2009) Biological control of *Lycoriella ingenua* (Diptera: Sciaridae) in commercial mushroom (*Agaricus bisporus*) cultivation: A comparison between *Hypoaspis miles* and *Steinernema feltiae*. *Pest Management Science* 65, 1195–1200.

Jeyaprakash, A. and Hoy, M.A. (2008) First divergence time estimate of spiders, scorpions, mites and ticks (subphylum: Chelicerata) inferred from mitochondrial phylogeny. *Experimental and Applied Acarology* 47, 1–18.

Ji, J., Zhang, Z.-Q., Zhang, Y. X., Chen, X. and Lin, J.Z. (2007) Effects of mating rates on oviposition, sex ratio and longevity in a predatory mite *Neoseiulus cucumeris* (Acari: Phytoseiidae). *Experimental and Applied Acarology* 43, 171–180.

Ji, J., Zhang, Y.-X., Lin, J.-Z., Chen, X., Sun, L. and Saito, Y. (2015) Life histories of three predatory mites feeding upon *Carpoglyphus lactis* (Acari: Phytoseiidae; Carpoglyphidae). *Systematic and Applied Acarology* 20(5), 491–496.

Jian, F.-J., Chen, Q.-Z. and Lu, A.-B. (1995) Some biochemical aspects of phosphine in combination with carbon dioxide against the adults of *Tyrophagus putrescentiae* (Schrank) (Astigmata: Acaridae). *Acta Entomologica Sinica* 38(1), 13–19.

Jiang, Z.-T. (1989) Three new records of genus *Tyrophagus* from Jiangxi, China. *Journal of Jiangxi Normal University (Natural Sciences)* 13(2), 22–27.

Jiang, Z.-T. (1991) A new species and a genus and three newly recorded species of Acaridae, Glycyphagidae from China (Acariformes: Acaroidea). *Jiangxi Science* 9, 240–246.

Jiang, Z. (1992) A new species of Acaridae from China (Acari: Acariformes). *Journal of Nanchang University(Natural Science)* 16, 239–242, 253.

Jiang, Z.-T. (1996) A new genus and species of Acaridae from China (Acari: Acariformes). *Acta Zootaxonomica Sinica* 21(1), 76–82.

Jiang, Z., Akhtar, Y., Bradbury, R., Zhang, X. and Isman, M.B. (2009) Comparative toxicity of essential oils of *Litsea pungens* and *Litsea cubeba* and blends of their major constituents against the cabbage looper, *Trichoplusia ni*. *Journal of Agricultural and Food Chemistry* 57, 4833–4837.

Jiang, Y., Liu, M., Zhang, J., Chen, Y., Chen, X. *et al.* (2017) Nematode grazing promotes bacterial community dynamics in soil at the aggregate level. *ISME Journal* 11, 2705–2717.

Johnston, D.E. (1965) Comparative studies on the mouth-parts of the mites of the suborder Acaridei (Acari). Dissertation presented in Partial Fulfillment of the Requirements for the Degree Doctor of Philosophy in the Graduate School of The Ohio State University, pp. 1–192.

Johnston, D.E. (1970) Notes on a collection of Austrian Macrochelidae with a description of *Macrocheles beieri* n.sp. *Annalen des Naturhistorischen Museums in Wien* 74, 145–150.

Johnston, D.E. (1982) Acari. In: Parker, S.P. (ed.) *Synopsis and Classification of Living Organisms.* Volume 2, McGraw-Hill, New York, pp. 111–169.

Johnston, D.E. and Bruce, W.A. (1965) *Tyrophagus neiswanderi*, a new acarid mite of agricultural importance. *Research Bulletin of Ohio Agricultural Experiment Station*, 977.

Jones, R. (2005) *Bibliography of Commonwealth Apiculture.* Commonwealth Secretariat, pp. 1–360.

Jonge, J.T. de (1988) Remarkable mites and insects in and around buildings in 1986. *Entomologische Berichten (Amst.)* 48, 18–19.

Joyeux, C. and Baer, G. (1945) Morphologie, evolution et position systématique de *Catenotaenia. Revue suisse de Zoologie* 52, 13–51.

Jung, J.-A., Cho, M.R., Kim, H.-H., Kang, T.-J., Lee, J.-H. and Do, K.-R. (2010) Damages by *Tyrophagus similis* (Acari: Acaridae) in Greenhouse Spinach in Korea. Korean *Journal of Applied Entomology* 49(4), 429–432.

Kaczmarek, S. and Lee, J.H. (2000) Gamasia (Acari) in the soil of some forest habitats in South Korea. *Acta Zoologica Cracoviensia* 43, 293–302.

Kadzhaja, G. Sch. (1970) *Schädliche Tyroglyphen Transkaukasiens. Akademie der Wissenschaften der Grusinischen SSR*. Meznireba Verlag, Tibilisi, Georgia, USSR, pp. 1–90 (in Russian).

Kadzhaya, G.S. (1970) Fauna vrednykh akaroideï zakavkaz'ya. Izdatel'stvo "Metsniereba", Tbilisi, pp. 1–88 (in Russian).

Kadzhaya, G.S. (1988) On the taxonomy of the genus *Mycetoglyphus* (Acariformes, Acaroidea). *Zoologicheskii Zhurnal* 67, 141–142 (in Russian).

Kaestner, A. (1980) Invertebrate Zoology: Anthropod relatives, chelicerata, myriapoda. II, Volume 2, Translated from Russian by Levi, H.W. and Levi, L.R., Interscience, France, pp. 1–472.

Kaya, H.K., Koppenhöfer, A.M. and Johnson, M. (1998) Natural Enemies of Entomopathogenic Nematodes. *Japanese Journal of Nematology* 28, 13–21.

Khaleghabadian, Z., Sadeghi Namaghi, H., Ardeshir, F., Nemati, A. and Hatefi, S. (2015) Fauna of predatory mesostigmatic and prostigmatid mites associated with stored food mites in the Mashhad. *Journal of Iranian Plant Protection Research* 28(4), 555–564.

Kalmosh, F.SH. and Yassin, E.M.A. (2018) Biodiversity of soil mites associated with wheat and soybean crops in Sharkeia and Beheira governatorates. *Egyptian Journal of Agricultural Research* 96 (3), 955–964.

Kalpaklioglu, A.F., Emekci, M., Ferizli, A.G. and Misirligil, Z. (1997) House dust mite fauna in Turkey. *Journal of Investigational Allergology and Clinical Immunology* 7(6), 578–582.

Kaluz, S. (2008) Soil mites (Acari) of the forests in floodplain areas of the rivers Danube and Morava. *Peckiana* 5, 89–103.

Kalúz, S. and Fend'a, P. (2005) *Mites (Acari: Mesostigmata) of the family Ascidae of Slovakia.* Institute of Zoology, Slovak Academy of Sciences, Bratislava, Slovakia, 1–166.

Kamali, K., Ostovan, H. and Atamehr, A. (2001) *A Catalog of Mites and Ticks (Acari) of Iran.* Islamic Azad University Scientific Publication Center, pp. 1–192.

Kamaruzaman, N.A., Mašán, P., Velásquez, Y., González-Medina, A., Lindström, A., Braig, H.R. and Perotti, M.A. (2018) *Macrocheles* species (Acari: Macrochelidae) associated with human corpses in Europe. *Experimental and Applied Acarology* 76, 453–471.

Kamensky, A.F. (1940) Die Getreidemilben in den Ursteppen Kasachstans. *Journal of Zoology (Moskau)* 19, 603–617 (in Russian).

Kaminskiene, E., Radzijevskaja, J., Balčiauskas, L., Gedminas, V. and Paulauskas, A. (2017) Laelapidae mites (Acari: Mesostigmata) infesting small rodents in the Curonian spit, Lithuania. *Biological Theory* 63(2), 169–176.

Kaminskiene, E., Radzijevskaja, J., Stanko, M., Balčiauskas, L. and Paulauskas, A. (2020) Associations between different Laelapidae (Mesostigmata: Dermanyssoidea) mites and small rodents from Lithuania. *Experimental and Applied Acarology* 81, 149–162.

Kamo, H. and Hatsushika, R. (1962) Studies on parasitic mites from western Japan. *Ann. Rep. Stud. Sci. Res. Fund Minis. Educ., Med. Pharmacol. Ser.*, 1961, 183 (in Japanese).

Kang, J.W.; Chae, H.; Hossain, M.A. (2023) Poultry red mite eradication potential of ivermectin and allicin combination treatment. *Veterinary Medicine & Science* 9, 1292–1296.

Kamran, M., Afzal, M. and Bashir, M.H. (2013) A New Species of Genus *Erythraeus* (Acari: Erythraeidae) From Punjab, Pakistan. *Pakistan Journal of Zoology* 45(1), 35–39.

Kanceljak-Macan, B., Macan, J., Buneta, L. and Milkovic-Kraus, S. (2000) Sensitization to non-pyroglyphid mites in urban population of Croatia. *Croatian Medical Journal* 41(1), 54–57.

Kandil, M. (1983) The Mesostigmata (Acari) fauna of the Hortobágy National Park. In: Mahunka, S. (ed.) *The fauna of the Hortobágy National Park II*. Akadémiai Kiadó, Budapest, Hungary, pp. 365–373.

Kaoud, H.A. (2010) Susceptibility of poultry red mites to entomopathogens. *International Journal of Poultry Science* 9, 259–263.

Kaoud, H.A. and El-Dahshan, R.A. (2010) Effect of Red Mite (*Dermanyssus gallinae*) Infestation on the Performance and Immune Profile in Vaccinated broiler breeder flocks. *Journal of American Science* 6, 72–79.

Karami, F., Hajizadeh, J. and Ostovan, H. (2017) Fauna of Ascoidea (except Ameroseiidae) in Guilan province, Iran with two new species record for Iran mite fauna. *Linzer Biologische Beitraege* 49(2), 1309–1321.

Karbowiak, G., Solarz, K., Asman, M., Wróblewski, Z., Slivinska, K. and Werszko, J. (2013) Phoresy of astigmatic mites on ticks and fleas in Poland. *Biology Letter* 50(2), 87–94.

Karg, V.W. (1961) Okologische Untersuchungen von edaphischen Gamasiden (Acarina, Parasitiformes). *Pedobiologia* 1, 77–98.

Karg, W. (1963) Systematische Untersuchung der Eviphididae Berlese, 1913 (Acarina, Parasitiformes) mit einer neuen Art aus Ackerböden. *Zoologischer Anzeiger* 170, 269–281.

Karg, W. (1971) Acari (Acarina), Milben. Unterordnung Anactinochaeta (Parasitiformes). Die freilebenden Gamasina (Gamasides), Raubmilben. *Die Tierwelt Deutschlands* 59, 1–475.

Karg, W. (1973) Hausstaubmilben in der Deutschen Demokratische Republik. *Allergie und Immunologie* 19, 81–85.

Karg, W. (1978) Die Gattung *Pseudoparastius* Oudemans, 1902. *Mitteilungen aus dem Zoologischen Museum in Berlin* 54, 203–212.

Karg, W. (1983) Verbreitung und Bedeutung von Raubmilben der Cohors Gamasina als Antagonisten von Nematoden. *Pedobiologia* 25, 419–432.

Karg, W. (1986) Vorkommen und Ernährung der Milbencohorts Uropodina (Schildkrötenmilben) sowie ihre Eignung als Indikatoren in Agroökosystemen. *Pedobiologia* 29, 285–295.

Karg, W. (1987) Zur Kenntnis der Raubmilbenarten der Gattung *Gamasiphis* Berlese, 1904 (Acarina, Parasitiformes). Mitteilungen aus dem Museum für Naturkunde in Berlin. *Zoologisches Museum und Institut für Spezielle Zoologie (Berlin)* 63(2), 301–308.

Karg, W. (1989) Acari (Acarina) Milben, Unterordnung Parasitiformes (Anactinochaeta), Uropodina Kramer, Schildkrötenmilben. Die Tierwelt Deutschlands, Gustav Fischer Verlag, Jena, Germany 67, 1–203.

Karg, W. (1993a) Raubmilben: Acari (Acarina), Milben Parasitiformes (Anactinochaeta) Cohors Gamasina Leach. Tierwelt Deutsch. 59. Teil, Gustav Fischer Verlag, Jena, Germany, 1–523.

Karg, W. (1993b) Predatory mites of Hypoaspidae, Laelapidae and Phytoseiidae on the Galapagos Archipelago (Acarina, Parasitiformes). *Mitteilungen aus dem Zoologischen Museum in Berlin* 69, 261–284.

Karg, W. (1994) Raubmilben Nutzliche Regulatoren im Naturhaushalt. Lebensweise, Artenbestimmung und Nutzung. Die Neue Brehm-Bucherei bd. 624, Westarp Wisenchaften, Magdeburg, Germany, pp. 1–206.

Karg, W. (1995) Raubmilben als Indikatoren bei der Entwicklung eines Ökologisch Orientierten Pflanzenschutzes. Nachrichtenbl. Dtsch. *Pflanzenschutzdienstes (Braunsch)* 47, 149–156.

Karg W. (1996) Neue Arten aus Raubmilbengattungen der Gamasina Leach (Acarina, Parasitiformes) mit Indikationen zum Entwicklungsalter. *Mitteilungen aus dem Museum für Naturkunde in Berlin. Zoologisches Museum und Institut für Spezielle Zoologie (Berlin)* 72, 149–195.

Karg, W. (2000) Die euedaphische Raubmilbengattung *Multidentorhodacarus* Shcherbak, 1980 mit einer neuen Art aus Mittelamerika. *Abhandlungen der Naturforschenden Gesellschaft zu Görlitz* 72, 143–149.

Karg, W. and Freier, B. (1995) *Parasitiforme Raubmilben als Indikatoren für den ökologisken Zustand von Ökosystemen*. Blackwell Wissenschafts-Verlag, Berlin-Dahlem, Germany, pp. 1–96.

Karg, W. and Grosse, E. (1983) Raubmilben als Antagonisten von Nematoden. *Nachrichtenblatt für den Pflanzenschutz in der DDR* 37, 208–212.

Karl, E. (1965a) Untersuchungen zur Morphologie und Okologie von Tarsonemiden gartnerischer Kulturpflanzen. I. *Tarsonemus pallidus* Banks. *Biologisches Zentralblatt* 84, 47–80.

Karl, E. (1965b) Untersuchungen zur Morphologie und Okologie von Tarsonemiden gartnerischer Kulturpflanzen. II. *Hemitarsonemus latus* (Banks), *Tarsonemus confusus* Ewing, *T. talpae* Schaarschmidt, *T. setifer* Ewing, *T. smithi* Ewing und *Tarsonemoides belemnitoides* Weis-Fogh. *Biologisches Zentralblatt* 84, 331–357.

Karnkowski, W. (1990) Numerousness and specific mite fauna composition in herb material from three stores. *Roczniki Nauk Rolniczych* 20, 73–78 (in Polish).

Kasburg, C.R. (2016) Seleção E Caracterização de Isolados de Fungos Entomopatogênicos Visando Ao Controle Do Ácaro Vermelho *Dermanyssus gallinae* (Acari: Dermanyssidae), State University of Western Paraná, Cascavel, Brazil.

Kasuga, S. and Amano, H. (2000a) Influence of temperature on the life history parameters of *Tyrophagus similis* Volgin (Acari: Acaridae). *Applied Entomoloy and Zoology* 35, 237–244.

Kasuga, S. and Amano, H. (2000b) Survey for the genus *Tyrophagus* and its damage in spinach crop with view to develop its pest management strategy. *Journal of the Acarological Society of Japan* 9, 31–42 (in Japanese).

Kasuga, S. and Amano, H. (2002) Susceptibility of 18 Japanese populations of *Tyrophagus similis* Volgin (Acari: Acaridae) to DDVP EC. *Japanese Journal of Applied Entomology and Zoology* 46, 99–101.

Kasuga, S. and Amano, H. (2003) Seasonal prevalence and susceptibility to agrochemicals of *Tyrophagus similis* (Acari: Acaridae) in spinach buds and agricultural soil under greenhouse conditions. *Experimental and Applied Acarology* 30(4), 279–288.

Kasuga, S. and Amano, H. (2005) Spatial distribution of *Tyrophagus similis* (Acari: Acaridae) in agricultural soils under greenhouse conditions. *Applied Entomoloy and Zoology* 40, 507–511.

Kasuga, S. and Amano, H. (2006) Infestation of *Tyrophagus similis* Volgin (Acari: Acaridae) on spinach during the seed germination period. *Journal of the Acarological Society of Japan* 15(1), 69–73.

Kasuga, S. and Honda, K. (2006a) High-temperature tolerance of *Tyrophagus similis*, and its comparison with *T. neiswanderi*, *T. putrescentiae* and *Rhizoglyphus robini*. *Japanese Journal of Applied Entomology and Zoology* 50, 19–23.

Kasuga, S. and Honda, K. (2006b) Suitability of organic matter, fungi and vegetables as food for *Tyrophagus similis* (Acari: Acaridae). *Applied Entomoloy and Zoology* 41(2), 227–231.

Kasuga, S., Honda, K., Kawai, A. and Amano, H. (2005) A bait trap for monitoring acarid mites (Acari: Acaridae) in agricultural soils. *Applied Entomoloy and Zoology* 40, 303–308.

Kasuga, S., Kanno, H. and Amano, H. (2006) Development, oviposition, and predation of *Hypoaspis aculeifer* (Acari: Laelapidae) feeding on *Tyrophagus similis* (Acari: Acaridae). *Journal of the Acarological Society of Japan* 15, 139–143.

Kato, T., Kadono, F. and Amano, H. (1993) Insecticide susceptibility of the robin bulb mite, *Rhizoglyphus robini* Claparede (Acari: Acaridae), from the Welsh onion fields in Chiba Prefecture. *Proceedings of the Kanto-Tosan Plant Protection Society* 40, 29–31.

Kats, L.B. and Dill, L.M. (1998) The scent of death: chemosensory assessment of predation risk by prey animals. *Ecoscience* 5, 361–394.

Kaur, N. (ed.) (2023) Pacific Northwest Insect Management Handbook. Oregon State University. Onion Bulb Mites. Available at: https://pnwhandbooks.org/ node/7803 (accessed 17 April 2023).

Kavallieratos, N., Athanassiou, C., Boukouvala, M. and Rumbos, C.I. (2018) Acaricidal effect of three zeolite formulations on different life stages of *Tyrophagus putrescentiae* (Schrank) and *Acarus siro* L. (Sarcoptiformes: Acaridae). *Journal of Stored Products Research* 78, 39–44.

Kawamoto, H., Sinha, R.N., Muir, W.E. and Woods, S.M. (1991) Simulation Model of *Acarus siro* (Acari: Acaridae) in Stored Wheat. *Environmental Entomology* 20(5), 1381–1386.

Kazemi, S. (2014) A new mite species of *Pseudoparasitus* Oudemans (Acari: Mesostigmata: Laelapidae), and a key to known Iranian species of the genus. *Persian Journal of Acarology* 3, 41–50.

Kazemi, S.H. (2009) Mesostigmatic mites (Acari) associated with the family Scarabaeidae (Coleoptera) in some regions of north and northeastern Iran. Ph.D. thesis of Tarbiat Modares University, pp. 1–218.

Kazemi, S.H. and Rajaei, A. (2013) An annotated checklist of Iranian Mesostigmata (Acari), excluding the family Phytoseiidae. *Persian Journal of Acarology* 2(1), 63–158.

Kazhdaya, G.S. (1996) Some data on acaroid mitesin Saarland (with description of variability of *Forcellinia diamesa*, Acaroidea, Acariformes). *Zoologicheskii Zhurnal* 75, 620–624.

Kazmierski, A. (1998) Tydeinae of the world: generic relationship, new and redescribed taxa and keys to all species. A revision of the subfamilies Pretydeinae and Tydeinae (Acari: Actinedida: Tydeidae), Part IV. *Acta Zoologica Cracoviensia* 41, 283–285.

Keegan, H.L. (1951) The mites of the subfamily Haemogamasinae (Acari: Laelaptidae). *Proceedings of the United States National Museum* 101, 203–268.

Keh, B., Lane, R.S. and Shachter, S.P. (1987) *Cheyletiella blakei,* an ectoparasite of cats, as cause of cryptic arthropod infestations affecting humans. *Western Journal of Medicine* 146, 192–194.

Keil, H. (1983) Okofaunistische Untersuchungen der Hausstaubmilben in Hamburg. *Mitteilungen aus dem Zoologischen Museum Hamburg* 7, 343–386.

Keïta, A., Pagot, E., Pommier, P., Baduel, L. and Heine, J. (2006) Efficacy of Phoxim 50% E.C. (ByeMite) for treatment of *Dermanyssus gallinae* in laying hens under field conditions. *Revue de Medecine Veterinaire* 157, 590–594.

Kennedy, G.G. and Smitley, D.R. (1985) Dispersal. In: Helle, W. and Sabelis, M.W. (eds) *Spider Mites: Their Biology, Natural Enemies and Control.* World Crop Pests, Volume 1A. Elsevier Science, Amsterdam, The Netherlands, pp. 233–242.

Kerezsi, V., Kiss, B., Deutsch, F. and Kontschán, J. (2019) First record of *Blattisocius mali* (Oudemans, 1929) in Hungary associated with the drosophilid fly *Phortica semivirgo* (Máca, 1977). *Redia (Firenze)* 102, 69–72.

Kerfoot, W.C. (1982) A question of taste: crypsis and warning coloration in freshwater zooplankton communities. *Ecology* 63, 538–554.

Kern, A. (1921) Dust sensitization in bronchial asthma. *Medical Clinics of North America* 5, 751–758.

Keshtmand, K., Ostovan, H. and Asadi, R. (2020) Report of a species of the family Lardoglyphidae (Acari: Sarcoptiformes) from Iran. *Journal of Plant Protection Research* 2(1), 65–68.

Kethley, J.B. (1982) Prostigmata. In: Parker, S.P. (ed.) *Synopsis and Classification of Living Organisms.* McGraw-Hill, New York, pp. 117–145.

Kethley, J.B. (1991) A procedure for extraction of microarthropods from bulk samples of sandy soils. *Agriculture, Ecosystems & Environment* 34, 193–200.

Keum, E., Kang, M. and Jung, CH. (2015) New record of *Arctoseius cetratus* (Sellnick, 1940) (Mesostigmata: Ascidae) phoretic to sciarid fly from mushroom culture in Korea. *Korean Journal of Environmental Biology* 33(2), 209–214.

Keum, E., Kaczmarek, S. and Jung, C. (2016) A new record of *Hypoaspis sardous* (Canestrini, 1884) (Acari: Mesostigmata: Laelapidae) from Korea. *Journal of Species Research* 5(3), 477–482.

Kevan, D.K. and Sharma, G.D. (1963a) The effects of low temperatures on *Tyrophagus putrescentiae.* In Naegele, J.A. (ed.) *Advances in Acarology.* Vol. I, Cornell University, Ithaca, New York, pp. 112–130.

Kevan, D.K. and Sharma, G.D. (1963b) Observations on the Biology of *Hypoaspis aculeifer* Canestrini 1885 (Acarina: Laelaptidae), a mite apparently new to North America. In: *Proceedings XVI International Congress of Zoology, 20–27 Augusta 1963*, Washington, DC, I, 14.

Kevan, D.K.M. and Sharma, G.D. (1964) Observations on the Biology of *Hypoaspis aculeifer* (Canestrini 1884), Apparently New to North America (Acarina: Mesostigmata: Laelaptidae). *Acarologia* 6, 647–658.

Kew, P.E, Wong, S.F., Ling, S.J., Lim, P.K.C. and Mak, J.W. (2014) Isolation and identification of mites associated with raw and commercial farm edible bird nests. *Tropical Biomedicine* 32(4), 761–775.

Khaing, T.M., Shim, J.-K. and Lee, K.-Y. (2014) Molecular identification and phylogenetic analysis of economically important acaroid mites (Acari: Astigmata: Acaroidea) in Korea. *Entomological Research* 44(6), 331–337.

Khademi, N., Saboori, A. and Ueckermann, E. (2006) Fauna of mesostigmata in citrus orchards in Jahrom region, Iran. In: Bruin, J. (ed.) *Abstract Book of 12th International Congress of Acarology.* Amsterdam, The Netherlands, p. 91.

Khaleghabadian, Z., Sadeghi Namaghi, H., Ardeshir, F., Hatefi, S. and Paktinat Saeej, S. (2012) Part of fauna of mites associated with stored food products in Mashhad region, Iran. In: Sarafrazi A., Asef, M.R., Mozhdehi, M., Mozhdehi, M., Solhjouy Fard, S. and Abdollahi T. (eds) *Proceeding of 20th Iranian Plant Protection Congress.* University of Shiraz, Iran, p. 408.

Khaleghabadian, Z., Sadeghi Namaghi, H., Ardeshir, F., Nemati, A. and Hatefi, S. (2015) Fauna of predatory mesostigmatic and prostigmatid mites associated with stored food mites in the Mashhad. *Iranian Journal of Plant Protection Science* 28(4), 555–564.

Khalili-Moghadam, R. and Saaboori, A. (2015) Some mesostigmatic mites (Acari: Mesostigmata) associated with ants in Shahrekord region, Iran. *Ecologica Montenegrina* 2(4), 315–326.

Khalili-Moghadam, A., Riahi, E., Nemati, A., Abotalebian, M. and Mardani, A. (2012) The fauna of some edaphic Mesostigmata (Acari) in Isfahan. In: *Abstract Book of the 20th Iranian Plant Protection Congress*. University of Shiraz, Iran, p. 46.

Khalilmanesh, B. (1973) Phytophagous mite fauna of Iran (I). *Applied Entomology and Phytopathology Tehran* 35, 30–38.

Khanjani, M. (1996) Mites (Acari) associated with Fabaceae plants in Hamedan province and functional responses of *Anystis baccarum* (L.) and *Erythraeus* sp. to developmental stages of *Tetranychus turkestani* (U. and N.). Ph.D. dissertation, Tarbiat Modarres University, Tehran, Iran, pp. 1–437.

Khanjani, M.and Hoseini, M.A. (2013) Three new species of the genus *Neophyllobius* Berlese (Acari: Camerobiidae) from southern and southwestern Iran. *Zootaxa* 3666, 510–522.

Khanjani, M. and Ueckermann, E.A. (2002) The stigmaeid mites of Iran (Acari: Stigmaeidae). *International Journal of Acarology* 28, 317–339.

Khanjani, M. and Ueckermann, E.A. (2003) Two new species of the genus *Raphignathus* Dugés (Acari: Raphignathidae) from Iran. *Acarologia* 43, 299–306.

Khanjani, M., Mirmoayedi, A., Rezai-Nahad, A. and Asali-Fayaz, B. (2010) Two new larval species of *Erythraeus* (*Zaracarus*) (Acari, Erythraeidae) from western Iran. *Zootaxa* 2537, 19–32.

Khanjani, M., Asadabadi, A., Izadi, H. and Doğan, S. (2012a) A new species of *Neophyllobius* (Acari: Raphignathoidea, Camerobiidae) from southeast Iran. *Systematic and Applied Acarology* 17 (1), 67–73.

Khanjani, M., Yazyanpanah, S., Ostovan, H. and Fayaz, B.A. (2012b) Three new species of the genus *Tycherobius* Bolland (Acari: Camerobiidae) from Iran. *Zootaxa* 3266, 23–40.

Khanjani, M., Mirmoayedi, A., Asali-Fayaz, B. and Sharifian, T. (2012c) Two new larval species of the genus *Erythraeus* (*Erythraeus*) (Acari, Erythraeidae) from Iran. *Zootaxa* 3479, 52–68.

Khanjani, M., Hajizadeh, J., Hoseini, M.A. and Jalili, M. (2013a) Two new species of the genus *Tycherobius* Bolland (Acari: Camerobiidae) from north of Iran. *International Journal of Acarology* 39(2), 130–139.

Khanjani, M., Pishavar, S. and Mirmoayedi, A. (2013b) A new species of *Raphignathus* Dugés (Acari: Raphignathidae) from Iran. *Acarina* 21(1), 62–68.

Khanjani, M., Hoseini, M.A., Yazdanpanah, S. and Masoudian, F. (2014) *Neophyllobius lorestanicus* sp. nov. and *N. ostovani* sp. nov. (Acari: Camerobiidae) from Iran. *Zootaxa* 3764, 441–454.

Khaustov, A.A. (2000) Mites of the family Winterschmidtiidae (Acari, Astigmata), associated with bark beetles (Coleoptera, Scolytidae) in Crimea. *Vestnik Zoologii. Supplement* 14(1), 50–59.

Khaustov, A.A. (2014) A new genus and species in the mite family Eupodidae (Acari, Eupodoidea) from Crimea. *ZooKeys* 422, 11–22.

Khaustov, A.A. (2015) A new genus of the family Eupodidae (Acari: Eupodoidea) with the description of a new species from Crimea. *Systematic and Applied Acarology* 20, 927–936.

Khaustov, A.A. (2019) Contribution to systematics of the genus *Eustigmaeus* (Acari: Stigmaeidae) of Russia. *Acarologia* 59(1), 152–173.

Khaustov, A.A. (2022) Review of the Mediolata (Acari: Stigmaeidae) of Russia. *Acarina* 30 (1), 29–56.

Khaustov, A.A., Klimov, P.B., Trach, V.A., Bobylev, A.N., Salavatulin, V.M., Khaustov, V.A. and Tolstikov, A.V. (2018) Review of mites (Acari) associated with the european spruce bark beetle, *Ips typographus* (Coleoptera Curculionidae: Scolytinae) in Asian Russia. *Acarina* 26(1), 3–79.

Kheradmand, K., Kamali, K., Fathipour, Y. and Goltapeh, E.M. (2007a) Development, life table and thermal requirement of *Tyrophagus putrescentiae* (Astigmata: Acaridae) on mushrooms. *Journal of Stored Products Research* 43, 276–281.

Kheradmand, K., Kamali, K., Fathipour, Y., Goltapeh, E.M. and Ueckermann, E.A. (2007b) Thermal requirement for development of *Sancassania rodionovi* (Acari: Acaridae) on mushrooms. *Journal of Economic Entomology* 100(4), 1098–1103.

Kheradmand, K., Kamali, K., Fathipour, Y., Ueckermann, E. and Mohammadi Goltapeh, E.

(2007c) Mites fauna associated with button mushroom (*Agaricus bisporus*) in Karaj region, Iran. *Acta Entomologica Sinica* 50(40), 416–422.

Kielczewski, B. and Seniczak, S. (1972) Cykl rozwojowy drapeznego roztocza *Calvolia fraxini* E. Türk *et* Türk (Tyropglyphidae, Acarina). *Prace z Zakresu Lesnictwa* 34, 83–88 (in Polish).

Kikuta, H. and Takagi, R. (2001) Isolation and acaricidal activity of microorganisms from dead *Tyrophagus similis* Volgin. *Journal of Rakunō Gakuen University. Natural Science* 25, 271–276 (in Japanese).

Kılıç, N. (2022) Efficacy of Dust and Wettable Powder Formulation of Diatomaceous Earth (Detech®) in the Control of *Tyrophagus putrescentiae* (Schrank) (Acari: Acaridae). *Insects* 13, 857.

Kılıç, N. and Toros, S. (2000) Tekirdağ İli ve çevresinde depolanan ürünlerde akarlar, yoğunlukları ve konukçuları ile önemli görülen türün biyolojisi üzerinde araştırmalar. PhD Thesis, Ankara University, Turkey (in Turkish).

Kiliç, T., Çobanoğlu, S., Yoldaş, Z. and Mandanlar, N. (2012) Mite (Acari) species determined in fresh onion fields in Izmir province. *Türkiye Entomoloji Dergisi* 36(3), 401–411 (in Turkish).

Kilpinen, O. (2001) Activation of the poultry red mite, *Dermanyssus gallinae* (Acari: Dermanyssidae), by increasing temperatures. *Experimental and Applied Acarology* 25, 859–867.

Kilpinen, O. (2005) How to obtain a bloodmeal without being eaten by a host: the case of poultry red mite, *Dermanyssus gallinae*. *Physiological Entomology* 30, 232–240.

Kilpinen, O. and Steenberg, T. (2009) Inert dusts and their effects on the poultry red mite (*Dermanyssus gallinae*). *Experimental and Applied Acarology* 48, 51–62.

Kilpinen, O., Roepstorff, A., Permin, A., Nørgaard-Nielsen, G., Lawson, L.G. and Simonsen, H.B. (2005) Influence of *Dermanyssus gallinae* and *Ascaridia galli* infections on behaviour and health of lying hens (*Gallus gallus domesticus*). *British Poultry Science* 45(1), 26–34.

Kilpio, O. and Pirila, V. (1952) A new tyroglyphid mite causing dermatitis. *Acta Dermato-Venereologica Supplement* 32, 197–200.

Kim, E.H. (2002) Acaricidal activity of phenylpropenes identified from essential oil of *Eugenia caryophyllata* against *Dermatophagoides farinae, Dermatophagoides pteronyssillus* (Acari: Pyroglyphidae) and *Tyrophagus putrescentiae* (Acari: Acaridae). MS Thesis, Seoul National University, Suwon, Republic of Korea, pp. 1–105.

Kim, C.M. (2004) Trigynaspida (Acari: Mesostigmata): new diagnosis, classification, and phylogeny. *Acarologia* 44, 157–194.

Kim, D.H. and Ahn, Y.J. (2001) Contact and fumigant activities of constituents of *Foeniculum vulgare* fruit against three coleopteran stored-product insects. *Pest Management Science* 57(3), 301–306.

Kim, K.C. and Hwang, C.Y. (1996) An investigation of insect pest on the mushroom (*Lentinus edode, Pleurotus ostreatus*) in south region of Korea. Korean *Journal of Applied Entomology* 35, 45–51.

Kim, Y.K. and Kim, Y.Y. (2002) Spider-mite allergy and asthma in fruit growers. *Current Opinion in Allergy and Clinical Immunology* 2(2), 103–107.

Kim, D.K. and Song, S.B. (1968) Epidemiological studies on the mites in food. *Korean Journal of Preventive Medicine* 1(1), 37–41 (in Korean).

Kim, K. C., Kim, S.G. and Choi, H.S. (1988) An investigation of insect pests and the period of maximum occurrence of key insect pests in stored rice grain. Korean *Journal of Applied Entomology* 27 (2), 117–124.

Kim, Y.K., Chang, Y.S., Lee, M.H., Hong, S.C., Bae, J.M. *et al.* (2002a) Role of environmental exposure to spider mites in the sensitization and the clinical manifestation of asthma and rhinitis in children and adolescents living in rural and urban areas. *Clinical & Experimental Allergy* 32(9), 1305–1309.

Kim, S.-H., Kim, Y.-K., Lee, M.-H., Hong, S.-C., Bae, J.-M. *et al.* (2002b) Relationship between sensitization to citrus red mite (*Panonychus citri*) and the prevalence of atopic diseases in adolescents living near citrus orchards. *Clinical & Experimental Allergy* 32(7), 1054–1058.

Kim, E.-H., Kim, H.-K., Choi, D.-H. and Ahn, Y.-J. (2003) Acaricidal activity of clove bud oil compounds against *Tyrophagus putrescentiae* (Acari: Acaridae). *Applied Entomoloy and Zoology* 38, 261–266.

Kim, S.I., Yi, J.H., Tak, J.H. and Ahn, Y.J. (2004) Acaricidal activity of plant essential oils against *Dermanyssus gallinae* (Acari: Dermanyssidae). *Veterinary Parasitology* 120, 297–304.

Kim, S.I., Na, Y.E., Yi, J.H., Kim, B.S. and Ahn, Y.J. (2007) Contact and fumigant toxicity of oriental medicinal plant extracts against *Dermanyssus gallinae* (Acari: Dermanyssidae). *Veterinary Parasitology* 145(3–4), 37–382.

Kim, J., Bayartogtokh, B. and Jung, C. (2016) First record of the genus *Phauloppia* Berlese, 1908 (Acari: Oribatida: Oribatulidae) with description of *Phauloppia lucorum* (C.L. Koch, 1841) from Korea. *Journal of Species Research* 5(3), 368–371.

King, C., Simpson, R.J., Moritz, R.L., Reed, G.E., Thompson, P.J. and Stewart, G.A. (1996) The isolation and characterization of a novel collagenolytic serine protease allergen (Der p 9) from the dust mite *Dermatophagoides pteronyssinus*. *Journal of Allergy and Clinical Immunology* 98(4), 739–747.

Kinn, A.N. (1966) Predation by the mite *Macrocheles muscaedomesticae* (Acarina: Macrochelidae) on three species of flies. *Journal of Medical Entomology* 3(2), 221–223.

Kinn, D.N. (1983) The life cycle of *Proctolaelaps dendroctoni* Lindquist and Hunter (Acari: Ascidae): a mite associated with pine bark beetles. *International Journal of Acarology* 9, 205–210.

Kirişik, M. Topuz, E. and Çobanoğlu, S. (2018) *Tyrophagus neiswanderi* (Acari: Acaridae) as a Pest of Greenhouse Spinach in Antalya, Turkey. *Journal of Agricultural Science* 24, 517–522.

Kirkwood, A. (1963) Longevity of the mites *Dermanyssus gallinae* and *Liponyssus sylviarum*. *Experimental Parasitology* 14(3), 358–366.

Kirkwood, A. (1967) Anemia in poultry infested with red mite *Dermanyssus gallinae*. *Veterinary Record* 80, 514–516.

Kleih, U. and Pike, V. (1995) Economic assessment of psocid infestations in rice storage. *Tropical Science* 35, 280–289.

Klein Koch, C. and Waterhouse, D.F. (2000) The distribution and importance of arthropods associated with agriculture and forestry in Chile. ACIAR (Australian Centre for International Agricultural Research) Monograph No. 68, 1–234.

Klimaszewska, H. and Kwiatkowska, B. (1959) Roztocze (Acarina) magazynów zbożowych Łodzi i okolic. *Fragmenta Faunistica* 8, 227–233 (in Polish).

Klimov, P.B. and O'Connor, B.M. (2003) Phylogeny, historical ecology and systematics of some mushroom-associated mites of the genus *Sancassania* (Acari: Acaridae), with new generic synonymies. *Invertebrate Systematics* 17, 469–514.

Klimov, P.B. and O'Connor, B.M. (2009) Conservation of the name *Tyrophagus putrescentiae*, a medically and economically important mite species (Acari: Acaridae). *International Journal of Acarology* 35(2), 95–114.

Klimov, P.B. and O'Connor, B.M. (2010) *Acarus putrescentiae* Schrank, 1781 (currently *Tyrophagus putrescentiae*; Acariformes, ACARIDAE): proposed conservation of usage by designation of a replacement neotype. *Bulletin of Zoological Nomenclature* 67(1), 24–27.

Klimov, P.B. and Tolstikov, A.V. (2011) Acaroid mites of northern and eastern Asia (Acari: Acaroidea). *Acarina* 19(2), 252–264.

Klimov, P.B., Bochkov, A.V. and OConnor, B.M. (2015) Philogenetic position of the house dust mite subfamily Guatemalichinae (Acariformes: Pyroglyphidae) based on integrated molecular and morphological analysesand different measures of support. *Cladistics* 32, 261–275.

Klimov, P.B., OConnor, B., Ochoa, R., Bauchan, G.R., Redford, A. and Scher, J. (2016) Bee Mite ID: Bee-Associated Mite Genera of the World. USDA APHIS Identification Technology Program (ITP), Fort Collins, CO. Available at: http://www.idtools.org/id/mites/beemites (accessed 20 April 2020).

Klimov, P., Molva, V., Nesvorna, M., Pekar, S., Shcherbachenko, E., Erban, T. and Hubert, J. (2019) Dynamics of the microbial community during growth of the house dust mite *Dermatophagoides farinae* in culture. *FEMS Microbiology Ecology* 95(11), fiz 153. DOI: 10.1093/femsec/fiz153.

Klompen, L.S.H., Lukoschus, F.S: and OConnor, B.M. (1987) Ontogeny, life history and sex ratio evolution in *Ensliniella kostylevi* (Acari: Winterschmidtiidae). *Journal of Zoology* 213, 591–607.

Knee, W. (2018) New *Macrocheles* species (Acari, Mesostigmata, Macrochelidae) associated with burying beetles (Silphidae, *Nicrophorus*) in North America. *Zookeys* 721, 1–32.

Kniest, F., Liebenberg, B. and Bischoff, E. (1989) Presence and transport of dust mites in clothing. In: Science Writers' Press Conference Notes. American Academy of Allergy and Immunology 45th Meeting, San Antonio, 24 February–1 March, 1989, Milwaukee, Wisconsin, U.S.A.

Knülle, W. (1959) Morphologische und entwicklungsgeschichtliche Untersuchungen zum Phylogenetischen System der Acari: Acariformes Zachv. II. Acaridiae: Acaridae. *Mitteilungen aus dem Zoologischen Museum in Berlin* 35, 347–417.

Knülle, W. (1965) Die Sorption und Transpiration des Wasserdampfes bei der Mehlmilbe (*Acarus siro* L.). *Zeitschrift für Vergleichende Physiologie* 49(6), 586–604.

Knülle, W. (1987) Genetic variability and ecological adaptability of hypopus formation in a stored product mite. *Experimental and Applied Acarology* 3, 21–32.

Knülle, W. (2003) Interaction between genetic and inductive factors controlling the expression of dispersal and dormancy morphs in dimorphic astigmatic mites. *Evolution* 57(4), 828–838.

Knülle, W. and Wharton, G.W. (1964) Equilibrium humidities in arthropods and their ecological significance. Proceedings 1st international Congress of Acarology, Fort Collins, Colorado, U.S.A., 2–7 September 1963, Acarologia 6 (fasc. hors. Ser.), 299–306.

Ko, S.-H., Shin, T.-Y., Lee, J.-Y., Choi, C.-J. and Woo, S.-D. (2021) Screening and Evaluation of Acaropathogenic Fungi against the Bulb Mite *Rhizoglyphus robini*. *Journal of Asia-Pacific Entomology* 24, 991–996.

Koç, K. (1998) A New Record of *Acaropsella volgini* (Acari: Prostigmata, Cheyletidae) for the Fauna of Turkey. *Turkish Journal of Zoology* 22(3), 197–197. Available at: https://journals.tubitak.gov.tr/cgi/viewcontent.cgi?article=2904&context=zoology (accessed 22 April 2023).

Koch, C.L. (1835–1844) Deutschlands Crustaceen, Myriapoden und Arachniden - Ein Beitrag zur Deutschen Fauna, 1/40, (Herrich-Schäffer, G.A.W. editor), Pustet, F., Regensburg, Germany.

Koch, C.L. (1839a) Deutschlands Crustaceen, Myriapoden und Arachniden, fasc. 23, pi. 22. l. III.

Koch, C.L. (1839b) *Deutschlands Crustaceen, Myriapoden und Arachniden. Ein Beitrag zur Deutschen Fauna*. Herrich-Schäffer, G.A.W. (ed.) Pustet, F., Regensburg, Germany, 27(22).

Koch, C.L. (1841) Deutschlands Crustaceen, Myriapoden und Arachniden. *Deutschlands Insecten*, 32 (15), pls. 1-24 + content + explanation pls. 1–24.

Koch, C.L. (1847) System der Myriapoden, mit den Verzeichnissen und Berichtigungen zu Deutschlands Crustaceen, Myriapoden und Arachniden Heft 1–40. Pustet, Regensburg, Germany, 1–270 + plates I–X.

Koeniger, N., Koeniger, G. and Delfinado-Baker, M. (1983) Observations on mites of the Asian honeybee species (*Apis cerana*, *Apis dorsata*, *Apis florea*). *Apidologie* 14, 197–204.

Koenraadt, C.J.M. and Dicke, M. (2010) The role of volatiles in aggregation and host-seeking of the haematophagous poultry red mite *Dermanyssus gallinae* (Acari: Dermanyssidae). *Experimental and Applied Acarology* 50, 191–199.

Kohli, R. and Mathur, S. (1993) Life processes of *Tyrophagus putrescentiae* Schrank (Acarina: Acaridae) as influenced by photoperiod. *Crop Research* 6, 311–316.

Koistinen, T., Ruoppi, P., Putus, T., Pennanen, S., Harju, A. and Nuutinen, J. (2006) Occupational sensitization to storage mites in the personnel of a water-damaged grocery store. *International Archives of Occupational and Environmental Health* 79(7), 602–606.

Komlovsky, J.S. and Jenser, G. (1992) Little known predatory mite species of Hungary (Acari: Stigmaeidae). *Acta Phytopathologica et Entomologica Hungarica* 27(1–4), 361–363.

Kongchuensin, M. (1983) Geographic variation and combining abilities of sex ratio and offspring number in the two-spotted spider mite, *Tetranychus urticae* Koch. PhD thesis, Ohio State University, Columbus, OH.

Konopická, J., Bohatá, A., Nermut', J., Jozová, E., Mráček, Z., Palevsky, E. and Zemek, R. (2021) Efficacy of Soil Isolates of Entomopathogenic Fungi against the Bulb Mite, *Rhizoglyphus robini* (Acari: Acaridae). *Systematic and Applied Acarology* 26, 1149–1167.

Kontschán, J. (2018) *Macrocheles kekensis* sp n., a new macrochelid mite associated with a centoniin beetle from Hungary (Acari, Mesostigmata). *Zookeys* 768, 97–104.

Kontschán, J., Park, S.J., Lim, J., Hwang, J.M. and Seo, H.Y. (2014) New Mesostigmata records and species from the Korean Peninsula. *Opuscula Zoologica* 45, 17–23.

Kopecky, J., Perotti, M.A., Nesvorna, M., Erban, T. and Hubert, J. (2013) *Cardinium* endosymbionts are widespread in synanthropic mite species (Acari: Astigmata). *Journal of Invertebrate Pathology* 112, 20–23.

Kopecky´, J., Nesvorna', M. and Hubert, J. (2014) *Bartonella*-like bacteria carried by domestic mite species. *Experimental and Applied Acarology* 64, 21–32.

Koraiem, M.K. and Fahmy, IA. (1999) Studies on house dust mites in Great Cairo, Egypt. *Journal of the Egyptian Society of Parasitology* 129(1), 131–138.

Kordestani Mahani, B., Ardeshir, F. Sarailu, M.H. and Kamali, H. (2014) Mite fauna of some nuts and dried fruits in Mashhad. Proceedings of 21st Iranian Plant Protection Congress, Urmia, Iran, p. 961.

Korn, W. (1983) Zur Vergesellschaftung der Gamasidenarten *Poecilochirus carabi* G. u. R. Canestrini 1882 (= *P. necrophori* Vitzthum 1930). *P. austroasiaticus* Vitzthum 1930 und *P. subterraneus* Müller 1859 mit Aaskäfern aus der Familie der Silphidae. *Spixiana* 6, 251–280.

Korsgaard, J. (1979) Husstovmider (Pyroglyphidae, Acari) i danskeboliger. *Ugeskrift for Læger* 141, 888–892 (in Danish).

Korsgaard, J. and Hallas T.E. (1979) Tarsonemid mites in Danish house dust. *Allergy* 34(4), 225–232.

Korunić, Z. (2001) Allergenic components of stored agro products. *Arhiv za higijenu rada i toksikologiju* 52(1), 43–48.

Kosik-Bogacka, D.I., Kalisińska, E., Henszel, Ł. and Kuźna-Grygiel, W. (2010) Acarological Characteristics of Dust Originating from Urban and Rural Houses in Northwestern Poland. *Polish Journal of Environmental Studies* 19(6), 1239–1247.

Kosik-Bogacka, D.I., Kalisinska, E., Henszel, L. and Kuzna-Grygie, W. (2012) Seasonal Dynamics of House Dust Mites in Dust Samples Collected From Sleeping Places in North-Western Poland. *Zoonoses Public Health* 59, 8–15.

Kostiainen, T.S. and Hoy, M.A. (1996) The Phytoseiidae as biological control agents of pest mites and insects. A bibliography (1960–1994). Monography 17, University of Florida, Agric. Exp. Stn., 1–355.

Kowal, J., Nosal, P., Niedziółka, R. and Kornas, S. (2014) Presence of blood-sucking mesostigmatic mites in rodents and birds kept in pet stores in the Cracow area, Poland. *Annals of Parasitology* 60, 61–64.

Kramer, P. (1881) *Scirus. Architectura & Natura* 81, 17, t. 3, fig. 9.

Krantz, G.W. (1955) Some mites injurious to farm-stored grain. *Journal of Economic Entomology* 48(6), 754–755.

Krantz, G.W. (1961) The biology and ecology of granary mites of the Pacific Northwest. I. Ecological considerations. *Annals of the Entomological Society of America* 54(2), 169–74.

Krantz, G.W. (1962) A review of the genera of the family Macrochelidae Vitzthum 1930 (Acarina: Macrochelidae). *Acarologia* 4, 143–173.

Krantz, G.W: (1963) A new parholaspid mite from Costa Rica. *Proceedings of the Entomological Society of Washington* 65, 149–152.

Krantz, G.W. (1972) Macrochelidae from Hamburg (Acari Mesostigmata), with descripitions of two new species. *Ent. Mitt. Zool. Mus. Hamburg A*, 263–275.

Krantz, G.W. (1974) *Phaulodinychus mitis* (Leonardi 1899) (Acari: Uropodidae), an intertidal mite exhibiting plastron respiration. *Acarologia* 16, 11–20.

Krantz, G.W. (1978) *A Manual of Acarology*. 2nd edn, Oregon State University Bookstore, Corvallis, OR, pp. 1–509.

Krantz, G.W. (1983) Mites as biological control agents of dung-breeding flies, with special reference to the Macrochelidae. In: Hoy, M.A., Cunningham, G.L. and Knutson, L. (eds) *Biological Control of Pests by Mites'*. University of California Press, Berkeley, CA, pp. 91–98.

Krantz, G.W. (1986) Acari: Rhodacaridae. In: Botosaneanu, L. (ed.) *Stygofauna Mundi*. E.J. Brill, Leiden, The Netherlands, pp. 632–634.

Krantz, G.W. (1998a) A new genus and two new species of hypoaspidine mites (Acari: Laclapidae) associated with old world carpenter bees of the tribe Xylocopini (Hymenoptera: Apidae: Xylocopa). *International Journal of Acarology* 24, 291–300.

Krantz, G.W. (1998b) Reflections on the biology, morphology and ecology of the Macrochelidae. *Experimental and Applied Acarology* 22(3), 125–137.

Krantz, G.W. (2009a) Form and function. In: Krantz, G.W. and Walter, D.E. (eds) *A Manual of Acarology*. Texas Tech University Press, Lubbock, TX, pp. 5–53.

Krantz, G.W. (2009b) Introduction. In: Krantz, G.W. and Walter, D.E. (eds) *A Manual of Acarology*. Texas Tech University Press, Lubbock, TX, pp. 1–2.

Krantz, G.W. (2009c) Origins and phylogenetic relationships. In: Krantz, G.W. and Walter, D.E. (eds) *A Manual of Acarology*. Texas Tech University Press, Lubbock, TX, pp. 3–4.

Krantz, G.W. (2009d) Habits and habitats. In: Krantz, G.W. and Walter, D.E. (eds) *A Manual of Acarology*. Texas Tech University Press, Lubbock, TX, pp. 64–82.

Krantz, G.W. (2018) *Allogynaspis flechtmanni*, a new genus and species of the subfamily Macrochelinae (Acari: Mesostigmata: Macrochelidae) from southeastern Brazil, with comments on cheliceral dentition, reproductive strategies, and postepigynal platelets. *Zootaxa* 455, 150–160.

Krantz, G.W. and Moser, J.C. (2012) A new genus and species of Macrochelidae (Acari: Mesostigmata) associated with the Texas leafcutting ant, *Atta texana* (Buckley) in Louisiana, U.S.A. *International Journal of Acarology* 38, 576–582.

Krantz, G.W. and Royce, L.A. (1994) Observations on the biology and behavior of *Macrocheles mycotrupetes* Krantz and Mellott (Acari: Macrochelidae). *International Journal of Acarology* 2, 115–121.

Krantz, G.W. and Walter, D.E. (2009) *A Manual of Acarology.* Texas Tech University Press, Lubbock, TX, pp. 1–806.

Krantz, G.W. and Wernz, J.G. (1979) Sperm transfer in *Glyptholapis Americana.* In: Rodriguez, J.G. (ed.) *Recent Advances in Acarology.* Vol. 2, Academic Press, New York, San Francisco, London, pp. 441–446.

Krantz, G.W. and Whitaker, J.O. Jr. (1988) Mites of the genus *Macrocheles* (Acari: Macrochelidae) associated with small mammals in North America. *Acarologia* 29(3), 225–259.

Krantz, G.W., OConnor, B.M., Foreyt, W.J. and Fain, A. (2003) A new species of *Aplodontopus* (Acari: Astigmata: Chortoglyphidae) from the yellow-bellied marmot, *Marmota flaviventris* (Rodentia: Sciuridae) in eastern Washington, U.S.A., with observations on its pathology. *International Journal of Acarology* 29, 99–105.

Krezeczkowski, K. (1961) Researches on the occurence of the flour mite (*Tyrophagus noxius* Zachw., *Tyroglyphidae, Acarina*) and its food preferences. *Prace Naukowe Instytutu Ochrony Roślin* 3, 101–127.

Krištofík, J., Šustek, Z. and Mašán, P. (2002) Arthropods (Pseudoscorpionida, Acari, Coleoptera, Siphonaptera) in the nests of red-backed shrike (*Lanius collurio*) and lesser grey shrike (*Lanius minor*). *Biologia* 57(5), 603–613.

Krištofík, J., Mašán, P., Šustek, Z. and Karaska, D. (2009) Arthropods in the nests of lesser spotted eagle (*Aquila pomarina* Brem). *Biologia* 64(5), 974–980.

Krištofík, J., Mašán, P., Šustek, Z. and Nuhličková, S. (2013) Arthropods (Acarina, Coleoptera, Siphonaptera) in nests of hoopoe (*Upupa epops*) in Central Europe. *Biologia* 68(1), 155–161.

Krizkova-Kudlikova, I. and Hubert, J. (2008) Development of polyclonal antibodies for the detection of *Tribolium castaneum* contamination in wheat grain. *Journal of Agricultural and Food Chemistry* 56, 8035–8040.

Krizkova-Kudlikova, I., Stejskal, V. and Hubert, J. (2007) Comparison of Detection Methods for *Acarus siro* (Acari: Acaridida: Acarididae) Contamination in Grain. *Journal of Economic Entomology* 100(6), 1928–1937.

Krombein, K.V. (1967) Trap Nesting Wasps and Bees: Life Histories, Nests and Associates. Smithsonian Press, Washington, DC, pp. 1–496.

Krop, E., Houba, R., Spithoven, J., de Wind, S., Rooyackers, J. and Heederik, D. (2012) Occupational allergy to mites and moths in a biological pesticides company. *Allergy* 67, 63–73.

Kučerová, Z. and Horák, P. (2004) Arthropod Infestation in Samples of Stored Seeds in the Czech Republic. *Czech Journal of Genetics and Plant Breeding* 40(1), 11–16.

Kučerová, Z. and Stejskal, V. (2009) Morphologic diagnosis of the eggs of stored-products mites. *Experimental and Applied Acarology* 49, 173–183.

Kumar, R. (2017) *Insect Pests of Stored Grain.* Apple Academic Press, Oakville, Canada, pp. 1–412.

Kunavong, C., Chanbang, Y., Burana-Phanichphan, S. and Insankh, A. (2014) Control of *Suidasia pontifica* Oudemans mites in chicken feed using formaldehyde incorporated with propionic acid. *Agricultural Journal* 30(3), 243–252.

Kuo, J.S. and Nesbitt, H.H.J. (1970) The internal antomy and histology of adult *Caloglyphus mycophagus* (Mégnin) (Acarina: Acaridae). *Canadian Journal of Zoology* 48, 505–517.

Kurosa, K. (1976) *Nature and Life in Southeast Asia.* Vol. VII, JSPS, Tokio, Japan, pp. 61–81.

Kuwahara, M. (1986) Resistance of the bulb mite, *Rhizoglyphus robini* Claparede (Acarina: Acaridae), to insecticides. I. Resistance patterns to organophosphorus insecticides. *Japanese Journal of Applied Entomology and Zoology* 30, 290–295 (in Japanese).

Kuwahara, Y. (2004) Chemical ecology of astigmatid mites. In: Cardé, R.T. and Millar, J.G. (eds) *Advances in Insect Chemical Ecology.* University Press, Cambridge, UK, pp. 76–109.

Kuwahara, Y. (2010) How astigmatic mites control the emission of two or even three types of pheromones from the same gland. In: Sabelis, M.W. and Bruin, J. (eds) *Trends in Acarology: Proceedings of the 12th International Congress (21–26 August 2006).* Amsterdam, Springer, Dordrecht, The Netherlands, pp. 241–248.

Kuwahara, Y., Ishii, S. and Fukami, H. (1975) Neryl formate: Alarm pheromone of the cheese mite, *Tyrophagus putrescentiae* (Schrank) (Acarina. Acaridae). *Cellular and Molecular Life Sciences* 31(10), 1115–1116.

Kuwahara, Y., Thi My Yen, L., Tominaga, Y., Matsumodo, K. and Wada, Y. (1982) 1,3,5,7-Tetramethyldecyl Formate, Lardolure: Aggregation Pheromone of the Acarid Mite, *Lardoglyphus konoi* (Sasa *et* Asanuma) (Acarina: Acaridae). *Agricultural and Biological Chemistry* 46(9), 2283–2291.

Kuwahara, Y., Akimoto, K., Leal, W.S., Nakao, H. and Suzuki, T. (1987) Isopiperitenone: a new alarm pheromone of the acarid mite, *Tyrophagus similis* (Acarina, Acaridae). *Agricultural and Biological Chemistry* 51, 3441–3442.

Kuwahara, Y., Shibata, C., Akimoto, K., Kuwahara, M. and Suzuki, T. (1988) Pheromone study on acarid mites. XIII. Identification of neryl formate as an alarm pheromone from the bulb mite, *Rhizoglyphus robini* Claparede (Acari: Acaridae). *Applied Entomoloy and Zoology* 23, 76–80.

Kuwahara, Y., Leal, W.S., Nakano, Y., Kaneko, Y., Nakao, H. and Suzuki, T. (1989a) Pheromone study on astigmatid mites XXIII. Identification of the alarm pheromone on the acarid mite, *Tyrophagus neiswanderi* and species specificities of alarm pheromones among four species of the same genus. *Applied Entomoloy and Zoology* 24, 424–429.

Kuwahara, Y., Leal, W.S., Suzuki, T., Maeda, M. and Masutani, T. (1989b) Antifungal activity of *Caloglyphus polyphyllae* sex pheromone and other mite exudates. *Naturwissenschaften* 76, 578–579.

Kuwahara, Y., Koshii, T., Okamoto, M., Matsumoto, K. and Suzuki, T. (1991) Chemical ecology on astigmatid mites. XXX. Neral as the alarm pheromone of *Glycyphagus domesticus* (De Geer) (Acarina: Glycyphagidae). *Japanese Journal of Sanitary Zoology* 42, 29–32.

Kuwahara, Y., Sato, M., Koshii, T. and Suzuki, T. (1992) Chemical ecology of astigmatid mites XXXII. 2-hydroxy–6-methyl-benzaldehyde, the sex pheromone of the brown-legged grain mite *Aleuroglyphus ovatus* (Troupeau) (Acarina: Acaridae). *Applied Entomoloy and Zoology* 27, 253–260.

Kuznetsov, N.N. (1978) Revision of the genus *Stigmaeus* (Acariformes, Stigmaeidae). *Zoologicheskii Zhurnal* 57, 682–694 (in Russian).

Kuznetsov, N.N. and Livshitz, I.Z. (1973) New and relatively unknown species of the tydeid mites (Acariformes, Tydeidae) of the Crimean fauna. *Naucnye Doklady Vyssei Skoly, Biologiche skie Naucki* 3, 13–18 (in Russian).

Kuznetsov, N.N. and Livshits, I.Z. (1979) Predatory mites of the Nikita Botanical Garden (Acariformes: Bdellidae, Cunaxidae, Camerobiidae). *Proceedings of State Nikitsky Botanical Gardens* 79, 51–104 (in Russian).

Labandeira, C.C., Phillips, T.L. and Norton, R.A. (1997) Oribatid mites and the decomposition of plants tissues in Paleozoic coal-swamp forests. *Palaios* 12, 319–353.

Łabanowski, G.S. and Soika, G. (1992) Occurrence and injurious effects of *Rhizoglyphus echinopus* – a new pest of *Hemerocallis hybrida*. *Acta Horticulturae* 325, 739–742.

Laboulbène, A. and Robin, C. (1862) Description de l'*Acarus (Tyroglyphus) entomophagus* Laboulbène et observations anatomiques sur le genre *Tyroglyphus*. *Annales de la Société entomologique de France* 2, 317–338.

Laboulbène, J.A. (1852) Description de l'*Acarus entomophagus*. *Annales de la Société entomologique de France*, 54.

Laffi, F. and Ponti, I. (1997) *Acari Dannosi alle Piante*. Informatore Agrario, Verona, Italy, pp. 1–118.

Lafosse Marin, S., Iraola, V., Merle, S. and Fernandez-Caldas, E. (2006) Etude de la faune acarologique des matelas de l'ile de la Martinique. *Revue Française d'Allergologie et d'Immunologie Clinique* 46, 62–67.

Lai, N.Q. and Wen, T.-H. (1982) A report of the distribution of dust mites in civil houses in Guang Zhou and some other cities. *Journal of Guangzhou Medical College* 3(Supplement), 20–25 (in Chinese).

Lake, F.R., Ward, L.D., Simpson, R.J., Thompson, P.J. and Stewart, G.A. (1991) House dust mite-derived amylase: allergenicity and physicochemical characterization. *Journal of Allergy and Clinical Immunology* 87 (6), 1035–1042.

Lammers, G.A., Bronneberg, R.G.G., Vernooij, J.C.M. and Stegeman, J.A. (2017) Experimental validation of the AviVet trap, a tool to quantitatively monitor the dynamics of *Dermanyssus gallinae* populations in laying hens. *Poultry Science* 96, 1563–1572.

Lange, W.H. and Bacon, O.G. (1958) Crown mite damage on spinach. *California Agriculture* 12(2), 9, 16.

Lareschi, M. (1996) Estudio preliminar de la comunidad de roedores (Rodentia: Muridae) y sus ectoparásitos (Acari, Phthiraptera y Siphonaptera) en Punta Lara (Buenos Aires). *Revista de la Sociedad Entomológica Argentina* 55, 113–120.

Lareschi, M. and Mauri, R. (1998) Dermanyssoidea. In Morrone, J.J. and Coscarón, S. (eds) *Biodiversidad de artrópodos argentinos*. Una perspectiva biotaxonómica, Ediciones Sur, La Plata, Argentina, pp. 581–590.

Lareschi, M., Gettinger, D., Venzal, J.M., Arzua, M., Nieri-Bastos, F.A., Barros-Battesti, D.M. and Gonzalez, E.M. (2006) First report of mites (Gamasida: Laelapidae) parasitic on wild rodents in Uruguay, with new host records. *Neotropical Entomology* 35(5). DOI: 10.1590/S1519–566X2006000500005.

Larson, D.G. (1969) The critical equilibrium activity of adult females of the house dust mite, *Dermatophagoides farinae* Hughes. Dissertation presented in partial fulfillment of the requirements for the Degree Doctor of Philosophy in the Graduate School at the Ohio State University, USA, pp. 1–35.

Latifi, M., Glida, H., Torabi, A. and Bertrand, M. (2005) *Holaspina* (syn. *Parholaspulus*) *persicum*, a new species of Parholaspididae from Iran (Acari: Mesostigmata). *Acarologia* 46, 189–94.

Latifian, M. and Rad, B. (2022) Population assessment of common storage pests in deiri date palm cultivar using spectrophotometric method. *International Journal of Tropical Journal of Insect Science* 42, 345–354.

Latta, R. (1939) Observations on the nature of bulb mite attack on Easter lilies. *Journal of Economic Entomology* 32, 125–128.

Leal, W.S., Nakano, Y., Kuwahara, Y., Nakao, H., and Suzuki, T. (1988) Pheromone study of acarid mites XVII. Identification of 2-hydroxy–6-methyl-benzaldehyde as the alarm pheromone of the acarid mite, *Tyrophagus perniciosus* (Acarina: Acaridae), and its distribution among related mites. *Applied Entomoloy and Zoology* 23, 422–427.

Leal, W.S., Kuwahara, Y. and Suzuki, T. (1989) β-acaridial, the sex pheromone of the acarid mite *Caloglyphus polyphyllae*. Pheromone study of acarid mites, 21(1). *Naturwissenschaften* 76, 332–333.

Learner, M.A. and Chawner, H.A. (1998) Macroinvertebrate associations in sewage filter-beds and their relationship to operational practice. *Journal of Applied Ecology* 35, 720–747.

LeBeck, L.M. and Leppla, N.C. (2010) Guidelines for purchasing and using commercial natural enemies and biopesticides in North America. UF/IFAS EDIS publication. University of Florida. IN849. Available at: file:///C:/Users/User/Downloads/in84900.pdf (accessed 17 March 2021).

Leblanc, M., Berry, K., Graciano, S., Becker, B. and Reuter, J.D. (2014) Falsepositive results after environmental pinworm PCR testing due to Rhabditid nematodes in Corncob bedding. *Journal of the American Association for Laboratory Animal Science* 53, 717–724.

Leclercq, M. and Watrin, P. (1973) Entomologie et médecine légale: acariens et insectes trouvés sur un cadavre humain en décembre 1971. *Bulletin et annales de la Société royale belge d'entomologie* 109, 195–201.

Lee, D.C. (1974) Rhodacaridae (Acari: Mesostigmata) from near Adelaide, Australia. III, Behaviour and development. *Acarologia* 16, 21–44.

Lee, H.-S. (2015) Chemical Composition and Acaricidal Effects of Essential Oils Extracted from *Ligustrum japonicum* against acaridae and pyroglyphid mites. *Journal of Applied Biological Chemistry* 58, 197–199.

Lee, W.-K. and Cho, W.Y. (1980) Studies on the mites (Order Acarina) in Korea I. Suborder Sarcoptiformes. *Korean Journal of Parasitology* 18(2), 119–144.

Lee, S.Y. and Lee, W.K. (1996) Three Soil Mites of the Genus *Holaspina* (Parholaspisae Mesostigmata) from Korea. *Korean Journal of Soil Zoology* 1(1), 24–27 (in Korean).

Lee, W.-K., Lim, J.W. and Lee, S.-Y. (1998) Soil inhabiting Acaridae and Histiostomidae (Acari: Astigmata) from Korea. *Korean Journal of Soil Zoology* 3(2), 63–71.

Lee, C.P., Sung, B. and Lee, H. (2006) Acaricidal activity of fennel seed oils and their main components against *Tyrophagus putrescentiae*, a stored-food mite. *Journal of Stored Products Research* 42, 8–14.

Lee, C.-H., Park, J.-M., Song, H.-Y., Jeong, E.-Y. and Lee, H.-S. (2009) Acaricidal activities of major constituents of essential oil of *Juniperus chinensis* leaves against house dust and stored food mites. *Journal of Food Protection* 72, 1686–1691.

Lee, S.J., Kim, H.K. and Kim, G.H. (2019) Toxicity and effects of essential oils and their components on *Dermanyssus gallinae* (Acari: Dermanyssidae). *Experimental and Applied Acarology* 78, 65–78.

Legner, E.F. and Bowen, W.R. (1973) Influence of available poultry manure breeding habitat on emergence density of synanthropic flies (Diptera). *Annals of the Entomological Society of America* 66, 533–538.

Lehtinen, P.T. and Oksala, I. (1991) Astigmatic and prostigmatic mites of grain stores, mills and sawmills in Finland. In: Schuster, R. and Murphy, P.W. (eds) The Acari. Reproduction, development and life-history strategies, Springer Science+Business Media, Dordrecht, The Netherlands, pp. 519–520.

Leigh, L.A. Dini and Frean, J.A. (2005) Clinical significance of mites in urine. *Journal of Clinical Microbiology* 43, 6200–6201.

Leite, R.G., Harada, E.S., Salgado, D.D., Neto, M.M. and Bueno, L.G.F. (2021) Behavior of laying hens in cage free system at different ages and egg quality. *Research, Society and Development* 10, e6010413833.

Lehman, R.D. (1982) Mites (Acari) of Pennsylvania conifers. *Transactions of the American Entomological Society* 180, 181–286.

Leśna, I., Sabelis, M.W., Bolland, H.R. and Conijn, C.G.M. (1995) Candidate natural enemies for control of *Rhizoglyphus robini* Claparede (Acari: Astigmata) in lily bulbs: exploration in the field and pre-selection in the laboratory. *Experimental and Applied Acarology* 19, 655–669.

Leśna, I., Sabelis, M. and Conijn, C. (1996) Biological control of the bulb mite, *Rhizoglyphus robini*, by the predatory mite, *Hypoaspis aculeifer*, on lilies: predator-prey interaction at various spatial scales. *Journal of Applied Ecology* 33, 369–376.

Leśna, I., Conijn, C.G.M., Sabelis, M.W. and Straalen, N.M. van (2000) Biological control of the bulb mite, *Rhizoglyphus robini*, by the predatory mite, *Hypoaspis aculeifer*, on lilies: predator-prey dynamics in the soil, under greenhouse and field conditions. *Biocontrol Science and Technology* 10, 179–193.

Leśna, I., Wolfs, P., Faraji, F., Roy, L., Komdeur, J. and Sabelis, M.W. (2009) Candidate predators for biological control of the poultry red mite *Dermanyssus gallinae*. *Experimental and Applied Acarology* 48(1–2), 63–80.

Leśna, I., Sabelis, M.W., Van Niekerk, T.G.C.M. and Komdeur, J. (2012) Laboratory tests for controlling poultry red mites (*Dermanyssus gallinae*) with predatory mites in small 'laying hen' cages. *Experimental and Applied Acarology* 58(4), 371–383.

Levinson, H.Z., Levinson, A.R. and Mueller, K. (1991) Functional adaptation of two nitrogenous waste products in evoking attraction and aggregation of flour mites (*Acarus siro* L.). Anz. *Schädl.kd. Pflanzenschutz Umweltschutz* 64(3), 55–60.

Lewandowski, M., Dmowska, E. and Ignatowicz, S. (1999) Fauna of the oyster mushroom houses. *Progress in Plant Protection* 39, 463–466.

Li, Y. (1987) A new pest mite (*Histiostoma sapromyzarum*) injurious to asparagus. *Journal of Southwest Agricultural University* 9(1), 46–47 (in Chinese).

Li, C.-P. (1999) Preliminary study of the acaroid mites breeding in stored Chinese traditional medicinal materials. *Chinese Journal of Parasitic Disease Control* 12(1), 72–73.

Li, L.-S. and Fan, Q.-H. (1997) A survey of food mites from four provinces of China. *Systematic and Applied Acarology* 2, 247–250.

Li, C. and Li, L. (1990) Human pulmonary acariasis in Anhui Province: an epidemiological survey. *Chinese Journal of Parasitology & Parasitic Diseases* 8, 41–44 (in Chinese).

Li, G. and Meng, Y. (1992) Investigation in sucking mice free blood habit of *Hypoaspis lubrica*. *Chinese Journal of Vector Biology and Control* 3, 9–12 (in Chinese).

Li, L.-S., Xuan, J.-Y. and Fan, Q.-H. (1992a) Taxonomic investigation of food mites in Sichuan Province. *Journal of Southeast University* 14(1), 23–34 (in Chinese).

Li, L.-S., Zhang, X.-W. and Guo, Y.-Q. (1992b) The acute lethal effects of low oxygen and high carbon dioxide on *Tyrophagus putrescentiae* at different temperatures. *Grain Storage* 21(5), 3–7.

Li, C.-P., Cui, Y.-B., Wang, J., Yang, Q.-G. and Tian, Y. (2003) Acaroid mite, intestinal and urinary acariasis. *World Journal of Gastroenterology* 9(4), 874–877.

Li, C.-P., Jiang, J.-J., He, J. and Wang, H.-Y. (2005) Constitution and diversity of Acaridida community in storage Chinese traditional medicines in Huainan region. *Acta Arachnologica Sinica* 14(2), 100–103.

Li, Q.-P., Bai, H.-Q. and Luo, Y.-Z. (2008) Threshold of developmental and effective thermal sum of *Rhizoglyphus robini* Claparede. *Journal of Yangzhou University* 29(2), 91–94.

Li, C., Zhan, X., Sun, E., Zhao, J., Wang, H., He, J. and Wang, J. (2015a) The density and species of mite breeding in the stored products in China. *Nutrición Hospitalaria* 31(2), 798–807.

Li, C., Jiang, Y., Guo, W. and Chen, Q. (2015b) Morphologic features of *Sancassania berlesei* (Acari: Astigmata: Acaridae), a common mite of stored products in China. *Nutrición Hospitalaria* 31(4), 1641–1646.

Li, C., Zhan, X., Zhao, J. and Wei, G. (2015c) *Gohieria fusca* (Acari: Astigmata) found in the filter dusts of air conditioners in China. *Nutrición Hospitalaria* 31(2), 808–812.

Li, Y.-Y., Zhang, Y., Hu, Q., Liu, L., XU, X.-N., Liu, H. and Wang, J.-J. (2015d) Effect of bran moisture content and initial population density on mass production of *Tyrophagus putrescentiae* (Schrank) (Acari: Acaridae). *Systematic and Applied Acarology* 20(5), 497–506.

Li, G.-Y., Pattisonc, N. and Zhang, Z.-Q. (2021) Immature development and survival of *Neoseiulus cucumeris* (Oudemans) (Acari: Phytoseiidae) on eggs of *Tyrophagus curvipenis* (Fain & Fauvel) (Acari: Acaridae). *Acarologia* 61(1), 84–93.

Liana, M.K. (2005) First copulation increases longevity and fecundity of *Histiostoma feroniarum* (Acari: Astigmata: Acaridida) females. *Experimental and Applied Acarology* 35(3), 173–181.

Liang, L. and Zhang, Z-Q. (1997) Key to species of the genus *Molothrognathus* (Prostigmata: Caligonellidae) with description of a new species from Iran. *Systematic and Applied Acarology Special Publications* 1, 19–24.

Liao, C.E., Hsu, E.L., Tsai, J.J. and Ho, C.M. (2009) Immunologic characterization and allergenicity of recombinant Tyr p 3 allergen from the storage mite *Tyrophagus putrescentiae. International Archives of Allergy and Immunology* 150, 15–24.

Liao, C.E., Ho, C.M. and Tsai, J.J. (2010) Prevalence of *Tyrophagus putrescentiae* hypersensitivity in subjects over 70 years of age in a veterans' nursing home in Taiwan. *International Archives of Allergy and Immunology* 152(4), 368–377.

Liao, J.-R. Ho, C.-C. and Ko, C.-C. (2020) Checklist of Mite Families Ascidae and Blattisociidae (Acari: Mesostigmata) of Taiwan. *Forms – Entomology* 40, 122–143.

Lichen, Y., Zhang, Z.-Q. and Limin, H. (2010) Two new species of *Pyemotes* closely related to *P. tritici* (Acari: Pyemotidae). *Zootaxa* 2723, 1–40, 15–27.

Liebisch, G., Hack, R. and Smid, G. (2011) Efficacy of spinosad against the poultry red mite, *Dermanyssus gallinae* (Mesostigmata: Dermanyssidae), in laboratory and field trials. In: Moraes, G.J. de and Proctor, H. (eds) *Acarology XIII: Proceedings of the International Congress. Zoosymposia* 6, 282–287.

Lignières, J. (1893) Etude zoologique et anatomique du *Tyroglyphus malus* et de sa nymphe hypopiale. *Mémoires de la Société zoologique de France* 6, 5–15.

Lima-Barbero, J.F., Contreras, M., Mateos-Hernandez, L., Mata-Lorenzo, F.M., Triguero-Ocana, R. *et al.* (2019a) A vaccinology approach to the identification and characterization of *Dermanyssus gallinae* candidate protective antigens for the control of poultry red mite infestations. *Vaccines* 7, 190.

Lima-Barbero, J.F., Contreras, M., Bartley, K., Price, D.R.G., Nunn, F. *et al.* (2019b) Reduction in oviposition of poultry red mite (*Dermanyssus gallinae*) in hens vaccinated with recombinant Akirin. *Vaccines* 7(3), 121.

Lima-Barbero, J.F., Villar, M., Höfle, U. and de la Fuente, J. (2020) Challenges for the Control of Poultry Red Mite (*Dermanyssus gallinae*). In: Bastidas Pacheco, G.A. and Kamboh, A.A. (eds) *Parasitology and Microbiology Research*. IntechOpen, London. DOI: 10.5772/intechopen.82990.

Lin, K.S. (2003) *A List of Natural Enemies of Insect Pests*. Vol. III. Taiwan Agricultural Research Institute Press, Taichung, pp. 1–282.

Lin, J.-Z. and Zhang, Z.-Q. (2002) Tarsonemidae of the world: Key to genera, geographical distribution, systematic catalogue and annotated bibliograhy. *Systematic and Applied Acarology Society, London*, 1–440.

Lin, J.-F., Lin, W.-J., Lin, X., Ruan, Q.-C. and Yan, W. (1992a) Study of fumigation with aluminium phosphide at low concentration against *Acarus siro* L. (flour mite). *Grain Storage* 21(3), 12–18.

Lin, W., Lin, J. and Zhu, M. (1992b) The life history of *Tyrophagus tropicus* Robertson. *Journal of Zhengzhou Grain College*, 1992–04 (summary). Available at: http://en.cnki.com.cn/Article_en/CJFDTOTAL-ZZLS199204011.htm (accessed 15 May 2020).

Lin, X., Ruan, Q.-C., Lin, J.-F., Lin, W.-J. and Li, S.-C. (2000) Investigation on stored product mites in Fujian provinces of China. *Journal of Grain Storage* 29, 13–17.

Lin, Y.C., Chan, M.L., Ko, C.W. and Hsieh, M.Y. (2004) Nail infestation by *Liposcelis bostrychophila* Badonnel. *Clinical and Experimental Dermatology* 29, 620–621.

Lindquist, E.E. (1969) Review of holarctic tarsonemid mites (Acarina: Prostigmata) parasitizing eggs of pine bark beetles. *Memoirs of the Entomological Society of Canada* 60, 1–111.

Lindquist, E.E. (1970) Review of the genus *Heterotarsonemus* (Acarina: Tarsonemidae). *Canadian Entomologist* 102, 812–829.

Lindquist, E.E. (1972) A new species of *Tarsonemus* from stored grain (Acarina: Tarsonemidae). *Canadian Entomologist* 104, 1699–1708.

Lindquist, E.E. (1975a) Associations between mites and other arthropods in forest floor habitats. *Canadian Entomologist* 107, 425–437.

Lindquist, E.E. (1975b) *Digamasellus* Berlse, 1905, and *Dendrolaelaps* Halbert, 1915, with descriptions of new taxa of Digamasellidae (Acarina: Mesostigmata). *Canadian Entomologist* 107, 1–43.

Lindquist, E.E. (1984) Current theories on the evolution of major groups of Acari and on their relationships with other groups of Arachnida, with consequent implications for their classification. In: Griffiths, D.A. and Bowman, C.E. (eds) *Acarology VI*. Volume 1, Ellis Horwood, Chichester, UK, pp. 28–62.

Lindquist, E.E. (1986) The world genera of Tarsonemidae (Acari: Heterostigmata): a morphological, phylogenetic, and systematic revision, with a reclassification of family-group taxa in the Heterostigmata. *Memoirs of the Entomological Society of Canada* 136, 1–517.

Lindquist, E.E. (1996) External anatomy and notation of structures. In: Lindquist, E.E., Sabelis, M.W. and Bruin, J. (eds) *Eriophyoid Mites: Their Biology, Natural Enemies and Control*. World Crop Pests, Volume 6, Elsevier Science, Amsterdam, The Netherlands, pp. 3–31.

Lindquist, E.E. and Evans, G.O. (1965) Taxonomic concepts in the Ascidae, with a modified setal nomenclature for the idiosoma of the Gamasina (Acarina: Mesostigmata). *Memoirs of the Entomological Society of Canada* 47, 1–64.

Lindquist, E.E. and Hunter, P.E. (1965) Some mites of the genus *Proctolaelaps* Berlese (Acarina: Blattisociidae) associated with forest insect pests. *Canadian Entomologist* 97, 15–32.

Lindquist, E.E., Krantz G.W. and Walter, D.E. (2009a) Classification. In: Krantz, G.W. and Walter, D.E. (eds) *A Manual of Acarology*. 3rd edn, Texas Tech University Press, Lubbock, TX, pp. 97–103.

Lindquist, E.E., Krantz, G.W. and Walter, D.E. (2009b) Order Mesostigmata. In: Krantz, G.W. and Walter, D.E. (eds) *A Manual of Acarology*. 3rd edn, Texas Tech University Press, Lubbock, TX, pp. 124–232.

Linnaeus, C. (1758) Systema Naturae per regna tria naturae, secundum classes, ordines, genera, species cum caracteribus differentiis, synonymis, locis. Editio decima, reformata, Holmiae, pp. 1–821.

Liscombe, E.A.R. and Watters, F.L. (1962) Insect and mites infestation in empty granaries in the Prairie Provinces. *Canadian Entomologist* 94, 433–441.

Liu, T.-S. (1995) Control of bulb mites on gladiolus by soil treatment. *Bull. Taichung Dist. Agric. Improv. Stn* 49, 33–40 (in Chinese).

Liu, T.S. and Tzeng, A.K. (1994) Biological and damaging characteristics of bulb mite, *Rhizoglyphus robini* Claparede. *Plant Protection Bulletin (Taipei)* 36, 177–187.

Liu, T., Jin, D.-C., Guo, J.-J. and Li, L. (2006) Development and growth of *Tyrophagus putrescentiae* (Schrank) (Acarina: Acaridae) bred under different temperatures with different nutriments. *Acta Entomologica Sinica* 49(4), 714–718.

Liu, T., Jin, D.-C. and Guo, J.-J. (2008) Laboratory population life table of *Tyrophagus putrescentiae* at different temperatures. *Plant Protein* 33(1), 68–71.

Liu Z., Zhang, K. and Zhang, Z.-Q. (2024) Enhancing the efficiency of egg collection of the astigmatid mite *Carpoglyphus lactis* (Acari: Carpoglyphidae) as a diet for predatory mites. *Systematic and Applied Acarology* 29(2), 355–358.

Livshitz, V.Z. and Mitrofanov, V.I. (1972) Phytoseiid mites from Crimea (Parasitiformes: Phytoseiidae). In: Kochkin, M.A. *et al.* (eds) *Pests and Diseases of Fruit and Ornamental Plants*. The All-Union V.I. Lenin Academy of Agricultural Sciences, The state Nikita Botanical Garden, Yalta, Proc. LXI, 13–64 (in Russian).

Livshitz, V.Z. and Mitrofanov, V.I. (1974) A new species of parasitic mite (Acariformes, Pyemotidae) from Uzbekistan. *Zoologicheskii Zhurnal* 53, 288–289 (in Russian).

Livshitz, V.Z. and Mitrofanov, V.I. (1975) New species of Ameroseiidae (Parasitiformes) from Crimea. *Zoologicheskii Zhurnal* 54, 462–464 (in Russian).

Lloyd, C.M. (2009) Dust mites' dirty dealings in the lung. *Nature Medicine* 15, 366–367.

Locher, N., Al-Rasheid, K.A., Abdel-Ghaffar, F. and Mehlhorn, H. (2010) In vitro and field studies on the contact and fumigant toxicity of a neem-product (Mite-Stop R) against the developmental stages of the poultry red mite *Dermanyssus gallinae*. *Parasitology Research* 107, 417–423.

Longstaff, B.C., Greenslade, P.J.M., Colloff, M., Reid, I, Hart, P. and Packer, I. (1999) Managing Soils in Agriculture - The impact of soil tillage practices on soil fauna. A report for the Rural Industries Research and Development Corporation, Publication No. 99/18, 1–64.

Longworth, T.J. (1973) Some aspects of the ecology of the acari of virgin and reclaimed blanket bog at Glenamoy, Co. Mayo. Ph.D. Thesis, National University of Ireland, Dublin, pp. 1–76.

López, M.D., Pascual-Villalobos, M.J. (2010) Mode of inhibition of acetylcholinesterase by monoterpenoids and implications for pest control. *Industrial Crops and Products* 31, 284–288.

López-Conde, I., Carballas, C., Carballada, F., Iraola, V., Fernández Caldas, E. and Boquete, M. (2005) Cutaneous Sensitivity to *Chortoglyphus arcuatus* and *Glycyphagus domesticus* in Spain. *Journal of Allergy and Clinical Immunology* Abstracts S163, 115 (2). Available at: https://www.jacionline.org/action/showPdf?pii=S0091-6749%2804%2903918-1 (accessed 5 July 2020).

Louadì, K. and Robaux, P. (1992) Etude de populations d'acariens pulvicoles dans l'Est algerien selon les gradients climatiques propres a cette region. *Acarologia* 33(2), 177–191.

Lozzia, G.C. and Ottoboni, F. (1989) Acarofauna del formaggio "Fontina" in stagionatura. In: Domenichini, G. (ed.), Atti 4° Simposio, La Difesa Antiparassitaria nelle Industrie Alimentari e la Protezione degli Alimenti, 23–25 settembre 1987, Piacenza, Italy, CCIAA, 225–233.

Lucky, A.W., Sayers, C., Argus, J.D. and Lucky, A. (2001) Avian mite bites acquired from a new source-pet gerbils: Report of two cases and review of the literature. *Archives of Dermatological* 137, 167–170.

Lukás, J., Stejskal, V., Jarošík, V., Hubert, J. and Žďárková, E. (2007) Differential natural performance of four *Cheyletus* predatory mite species in Czech grain stores. *Journal of Stored Products Research* 43, 97–102.

Lukoschus, F.S., de Cock, A.W.A.M. and Fain, A. (1971) Life cycle of *Melesodectes auricularis* Fain and Lukoschus (Glycyphagidae, Sarcoptifonnes). *Tijdschrift Voor Entomologie* 114, 173–183.

Lukoschus, F.S., Fain, A. and Driessen, F.M. (1972) Life cycle of *Apodemopus apodemi* (Fain 1965) (Glycyphagidae: Sarcoptifonnes). *Tijdschrift Voor Entomologie* 115, 325–339.

Lukoschus, F.S., Janssen Duijghuijsen, G.H.S. and Fain, A. (1979) Parasites of Western Australia IV. Observations on the genus *Marsupiopus* Fain 1968 (Acarina: Astigmata: Glycyphagidae). *Records of the Western Australian Museum* 7, 37–55.

Lundh, J., Wiktelius, D. and Chirico, J. (2005) Azadirachtin-impregnated traps for the control of *Dermanyssus gallinae*. *Veterinary Parasitology* 130, 337–342.

Lung-Shut, L. (1984) Stored grain mites in China: Their distribution and effects. In: Griffiths D.A. and Bowman, C.A. (eds) *Acarology VI*, 2, Ellis Horwood Ltd., London, pp. 1002–1005.

Luo, L.P., Guo, X.G., Qian, T.J., Wu, D., Men, X.Y. and Dong, W.G. (2007) Distribution of gamasid mites on small mammals in Yunnan Province, China. *Journal of Insect Science* 14, 71–78.

Luong, L.T., Penoni, L.R., Horn, C.J. and Polak, M. (2015) Physical and physiological costs of ectoparasitic mites on host flight endurance. *Ecological Entomology* 40, 518–524.

Lustgraaf, B. (1977) L'ovopositeur des tétraniques. *Acarologia* 18(4), 642–650.

Luxton, M. (1972) Studies on the oribatid mites of a Danish beech wood soil. I, Nutritional biology. *Pedobiologia* 12, 434–463.

Luxton, M. (1981) Studies on the astigmatic mites of a Danish beech wood soil. *Pedobiologia* 22, 29–38.

Luxton, M. (1995) Patterns of food intake by some astigmatic mites of beech woodland soil (Acari: Astigmata). *Pedobiologia* 39, 238–242.

Lynch, C.A. (1989) Two new species of the genus *Tyrophagus* (Acari: Acaridae). *Journal of Zoology* 219, 545–567.

Ma, E.P. and Wang, R.S. (1992) Tarsonemid mites. In: Chen, X.B. and Ma, E.P. (eds) *Researches of Acarology in China*. Chongquing Publishing House, Chongqing, China, pp. 34–37.

Ma, M., Li, S.-C. and Fan, Q.-H. (2015) Mites and ticks (Acari) in Shanxi Province, China: An annotated checklist. *Zootaxa* 4006(1), 1–39.

Maaz, D., Krücken, J., Blümke, J., Richter, D., McKay-Demeler, J. *et al.* (2018) Factors associated with diversity, quantity and zoonotic potential of ectoparasites on urban mice and voles. *PLoS ONE* 13(6), e0199385. DOI: 10.1371/journal.pone.0199385.

Machado, O. (1935) Sobre dois casos de tyroglyphose intestinal. *Biology and Medicine* 2(7,8), 16.

Macan, J., Kanceljak, B., Plavek, D. and Milkovic-Kraus, S. (2003) Differences in mite fauna between the continental and Mediterranean climates of Croatia: Microscopy and Dustscreen™ test findings. *Allergy* 58, 780–783.

Macchioni, F., Cioni, P.L., Flamini, G., Morelli, I., Perrucci, S. *et al.* (2002) Acaricidal activity of pine essential oils and their main components against *Tyrophagus putrescentiae*, a stored food mite. *Journal of Agricultural and Food Chemistry* 50, 4586–4588.

Macfarlane, R.P. (2005) Mites associated with bumble bees (*Bombus*: Apidae) in New Zealand. *Records of the Canterbury Museum* 19, 29–34.

Madanlar, N. and Önder, F. (1996) Mites associated with cultivated mushrooms in Turkey. *XX International Congress of Entomology, 25–31 August 1996*. Firenze, Italy, poster n° pp. 15–138.

Madej, G., Barczyk, G. and Gawenda, I. (2011) Importance of microhabitats for preservation of species diversity, on the basis of mesostigmatid mites (Mesostigmata, Arachnida, Acari). *Polish Journal of Environmental Studies* 20(4), 961–968.

Magowski, W. (1994) Discovery of the first representative of the mite subcohort Heterostigmata (Arachnida: Acari) in the Mesozoic Siberian amber. *Acarologia* 35(3), 229–241.

Mahgoob, A.E.A., Badawy, A.I. and Badoor, I.M. (2006) Survey of mites associated with grain residues and mixed flour in warehouses and mills in great Cairo. *Arab University Journal of Agricultural Science* 14(1), 509–529.

Mahmood, S.H. (1992) Mite fauna of stored grain seeds in central Iraq. *Journal of Stored Products Research* 28, 179–181.

Mahmood, S.H. and Al-Dulaimi, S.I. (1986) Biological studies on the mite *Macrocheles muscaedomesticae* (Acarina: Macrochelidae). *Journal of Biological Science Research* 17(2), 329–338.

Mahmood, S.U., Bashir, M.H., Afzal, M. and Khan, B.S. (2011) Estimation of germination losses caused by mites in wheat drawn from farmer's holdings of Tehsil Toba Tek Singh. *Pakistan Entomology* 33, 143–146.

Mahunka, S. (1961) Contributions to the tyroglyphid fauna of Hungary (Acari). *Annales Universitatis Scientiarum Budapestinensis de Rolando Eotvos Nominatae. Sectio Biologica* 4, 113–118.

Mahunka, S. (1970) Atkák V.–Acari V. *Magyarország Állatvilága (Fauna Hungariae 101)* 18, 1–76.

Mahunka, S. (1971) Investigations on coprophilous and stercoricolous macrochelidae (Acarina: Gamasina) in Hungary, as possible agents in control of synanthropous flies. *Parasitology Hungary* 4, 215–226.

Mahunka, S. (1986) A survey of the family Carabodidae C. L. Koch, 1836 (Acari: Oribatida). *Acta zoologica Hungarica* 21, 73–135.

Mahunka, S. (1987) A survey of the family Carabodidae C. L. Koch, 1836 (Acari: Oribatida), II. *Acta zoologica Hungarica* 33, 399–434.

Mahunka, S. and Zaki, A.M. (1990) *Acarophenax rackae* sp. n., a new mite species from Egypt (Acari, Tarsonemina, Acarophenacidae). *Parasitology Hungary* 23, 121–126.

Majidi, M. and Akrami, M.A. (2013) Mites associated with the date palm (*Phoenix dactylifera* L.) in Larestan (Fars Province), southern Iran. *Persian Journal of Acarology* 2(2), 335–339.

Makaida, M.V., Pakhomov, O.Y. and Brygadyrenko, V. (2021) Effect of increased ambient temperature on seasonal generation number in *Lucilia sericata* (Diptera, Calliphoridae). *Folia Oecologica* 48(2), 191–198.

Makarova, O.L. (2012) Gamasid mites (Parasitiformes, Mesostigmata) of the European Arctic and their distribution patterns. *Zoologicheskii Zhurnal* 91, 907–927.

Makert, G.R., Vorbruggen, S., Krautwald-Junghanns, M.E., Voss, M., Sohn, K., Buschmann, T. and Ulbert, S. (2016) A method to identify protein antigens of *Dermanyssus gallinae* for the protection of birds from poultry mites. *Parasitology Research* 115, 2705–2713.

Mąkol, J. and Wohltmann, A. (2012) An annotated checklist of terrestrial Parasitengona (Actinotrichida: Prostigmata) of the world, excluding Trombiculidae and Walchiidae. *The Annals of Zoology* 62(3), 359–562.

Mąkol, J. and Wohltmann, A. (2013) Corrections and additions to the checklist of terrestrial Parasitengona (Actinotrichida: Prostigmata) of the world, excluding Trombiculidae and Walchiidae. *The Annals of Zoology* 61(1), 15–27.

Makovcová, S., Samsinak, K. and Vobrázková, E. (1982) Biologický aspekt roztocov z domáceho prachu a ich vyznam v alergii na domaci prach. I. Cast. *Fysiatrickya Reumatologicky Vestnik* 60, 69–74.

Malabanan, J.M. and Corpuz-Raros, L.A. (1998) Mites associated with *Apis mellifera* Linnaeus and *Apis cerana* Fabricius in the southern Tagalog region, Philippines. *Philippine Entomologist* 12, 123–135.

Malainual, N., Vichyanond, P. and Phan-Urai, P. (1995) House dust mite fauna in Thailand. *Clinical & Experimental Allergy* 25(6), 554–60.

Maleki, S., Ostovan, H., Baniameri, V. and Joharchi, O. (2016) Biodiversity of mesostigmatic soil mite fauna (Acari: Mesostigmata) of a city park located in Tehran, Iran. *Journal of Entomological Society of Iran* 36(3), 181–194.

Malik, A., Gulati, R., Duhan, K. and Poonia, A. (2018) *Tyrophagus putrescentiae* (Schrank) (Acari: Acaridae) as a pest of grains: A review. *Journal of Entomology and Zoology Studies* 6(2), 2543–2550.

Manning, M.J. and Halliday, R.B. (1994) Biology and reproduction of some Australian species of Macrochelidae (Acarina). *Australian Entomology* 21, 89–94.

Manson, D.C.M. (1972a) A contribution to the study of the genus *Rhizoglyphus* Claparede (Acarina: Acaridae). *Acarologia* 13, 621–650.

Manson, D.C.M. (1972b) Three new species, and a redescription of mites of the genus *Schwiebia* (Acarina: Tyroglyphidae). *Acarologia* 14(1), 71–80.

Manson, D. C.M. (1973). *Suidasia reticulata* (Acarina: Acaridae) a new species of mite from New Zealand. *New Zealand Entamology* 5(2), 192–197.

Manson, D.C.M. (1983) List of new mite/host plant records from New Zealand. *New Zealand Entomology* 7, 41–403.

Manson, D.C.M. and Oldfield, G.N. (1996) Life forms, deuterogyny, diapause and seasonal development. In: Lindquist, E.E., Sabelis, M.W. and Bruin, J. (eds) *Eriophyoid Mites: Their Biology, Natural Enemies and Control. World Crop Pests, Volume 6.* Elsevier Science, Amsterdam, The Netherlands, pp. 173–183.

Manton, S.M. (1977) *The Arthropoda: Habits, Functional Morphology and Evolution.* Carendon Press, Oxford, U.K., pp. 1–527.

Manu, M. and Onete, M. (2016) Comparative ecological characterization of the soil mite populations (Acari: Mesostigmata) from some anthropogenic ecosystems, Romania. Muzeul Olteniei Craiova. Oltenia. Studii şi comunicări. *Ştiinţele Naturii* 32(1), 173–180.

Marchenko, I.I. (2016) A new species of *Neparholaspis* (Acari: Parholaspididae) from Russia, with additional information on *Neparholaspis evansi* Krantz, 1960. *Zootaxa* 4154(4), 453–465.

Marchiori, C., Oliveira, T. and Linhares, X. (2001) Artrópodes associados a massas fecais bovinas no sul do estado de Goiás. *Neotropical Entomology* 30, 19–24.

Market, G.R., Vorbrüggen, S., Krautwald-Junghanns, M.-E., Voss, M., Sohn, K., Buschmann, T. and Ulbert, S. (2016) A method to identify protein antigens of *Dermanyssus gallinae* for the protection of birds from poultry mites. *Parasitology Research* 115(7),2705–2713.

Mariana, A., Ho, T.M. and Heah, S.K. (1996) Life cycle, longevity and fecundity of *Blomia tropicalis* (Acari: Glycyphagidae) in a tropical laboratory. *Southeast Asian Journal of Tropical Medicine Public Health* 27(2), 392–395.

Mariana, A., Ho, T.M., Sofian-Azirun, M. and Wong, A.L. (2000) House dust mite fauna in the Klang Valley, Malaysia. *Southeast Asian Journal of Tropical Medicine Public Health* 31(4), 712–721.

Mariana, A., Heah, S.K., Wong, A.L. and Ho, T.M. (2010) The occurrence of arthropods in processed rice products in Malaysia. *Asian Pacific Journal of Tropical Medicine* 3(7), 552–554.

Mariano, R.M.S., Gonçalves, A.A.M., Oliveira, D.S., Ribeiro, H.S., Pereira, D.F.S. *et al.* (2023) A review of major patents on potential malaria vaccine targets. *Pathogens* 12(2), 247.

Marin, J., Ocete, R., Pedroza, M., Zalacain, A., de Miguel, C., Lopez, M.A. and Salinas, M.R. (2009) Influence of the mite *Carpoglyphus lactis* (L) on the aroma of pale and dry wines aged under flor yeasts. *Journal of Food Composition and Analysis* 22(7–8), 745–750.

Marquardt, T. and Kaczmarek, S. (2019) Pre-ovipositional and ovipositional behaviour of *Trichouropoda ovalis* (C.L. Koch) and *Uroobovella marginata* (C.L. Koch) (Parasitiformes: Uropodina: Trematuridae, Urodinychidae) with notes on egg incubation and hatching behaviour. *Journal of Natural History* 53(15–16), 991–1000.

Marquardt, T., Kaczmarek, S. and Halliday, B. (2015) Ovoviviparity in *Macrocheles glaber* (Müller) (Acari: Macrochelidae), with notes on parental care and egg cannibalism. *International Journal of Acarology* 41(1), 71–76.

Marsella, R. and Saridomichelakis, M.N. (2010) Environmental and oral challenge with storage mites in beagles experimentally sensitized to *Dermatophagoides farinae*. *Veterinary Dermatology* 21, 105–111.

Marshall, V.G. (1964) A new parholaspid mite from eastern Canada with notes on the genus *Neparholaspis* Evans (Acarina: Mesostigmata). *Acarologia* 6, 417–420.

Marshall, V.G. (1970) Tydeid mites (Acarina: Prostigmata) from Canada. I, New and redescribed species of *Lorryia*. *Annales de la Société entomologique de France* 15, 17–52.

Martin, N.A. and Workman, P. (1985) Pest control in boxes of Cymbidium orchid flowers with dichlorvos-impregnated plastic. In: Hartley, M.J. (ed.) Proc. NZ Wed and Pest Control Conf. 38 (1985), New Zealand Weed and Pest Control Society, Henderson, Auckland, New Zealand, 169–171.

Martínez Canzonieri, C.F., Baggio, D. and Lizarralde de Grosso, M. (1996) Nuevos hallazgos de fauna acarologica en polvo de habitacion en la provincial de Tucuman. *Acta Zoological Lilloana* 43(2), 289–291.

Martinez Maranon, R. and Hoffmann, A. (1976) 3 cases of human intestinal mite infestation in the South of Veracruz. *Revista Panamericana de Salud Pública* 36, 187–201.

Martins-Hatano, F., Raíces, D.S., Gazeta, G.S., Serra-Freire, N.M., Gettinger, D. and Bergallo, H.G. (2011) Community composition of laelapine mites (Acari: Laelapidae) associated with the nests and fur of *Cerradomys subflavus* (Wagner, 1842). *Journal of Natural History* 45(27), 1679–1688.

Martins, L.M.L., Ventura, A.C.N., Brazis, P. and Cerrato, S. (2020) An investigation of cross-reactivity between the poultry red mite and house dust and storage mites in dogs with contact with chickens infested with red mites. Vet Dermatol. 31(2), 167–169.

Martynov, V.O., Hladkyi, O.Y., Kolombar, T.M. and Brygadyrenko, V.V. (2019) Impact of essential oil from plants on migratory activity of *Sitophilus granarius* and *Tenebrio molitor*. *Regulatory Mechanisms in Biosystems* 10(4), 359–371.

Maroufpoor, M. and Ostovan. H. (2017) Fauna of mites associated with stored wheat in Kurdistan Province and one new record for Iranian mite fauna. *The Journal of Animal Environment* 9(1), 149- 158

Marutani, M., Muniappan, R. and Denton, G.R.W. (1992) Biological control of red coconut scale in Guam. *The Journal of Plant Protection Tropical* 9, 63–67.

Marx Jr, J.J., Twiggs, J.T., Ault, B.J., Merchant, J.A. and Fernandez-Caldas, E. (1993) Inhaled aeroallergen and storage mite reactivity in a Wisconsin farmer nested case-control study. *American Review of Respiratory Disease* 147(2), 354–358.

Marzouk, A.S., Main, A.J., Soliman, S.S. and Montasser, A.A. (2006) Ectoparasites infesting commensal rodents in a rural area of Egypt, with special reference to newly recorded species. *International Journal of Science and Research* 15, 67–72.

Mašán, P. (1998) *Ameroseius fungicolis* sp. n. and *A. callosus* sp. n., two new ameroseiid species (Acarina, Mesostigmata) associated with wood-destroying fungi. *Biologia* 53, 645–649.

Mašán, P. (1999a) Mites (Acarina) associated with burying beetles and carrion beetles (Coleoptera: Silphidae) and description of *Poecilochirus mrciaki* sp.n. (Mesostigmata, Gamasina). *Biologia* 54, 515–524.

Mašán, P. (1999b) New mite species of the cohort Uropodina (Acarina, Mesostigmata) from Slovakia. *Biologia* 54, 121–133.

Mašán, P. (2001) Mites of the Cohort Uropodina (Acarina, Mesostigmata) in Slovakia. Annot. Zool. bot. 223, 1–320.

Mašán, P. (2003) *Macrochelid mites of Slovakia (Acari, Mesostigmata, Macrochelidae)*. Institute of Zoology, Slovak Academy of Sciences, Bratislava, Slovakia, pp. 1–149.

Mašán, P. (2017) A revision of the family Ameroseiidae (Acari, Mesostigmata), with some data on Slovak fauna. *Zookeys* 704, 1–228.

Mašán, P. and Fenďa, P. (2010) *A Review of the Laelapid Mites Associated with Terrestrial Mammals in Slovakia, with a Key to the European Species (Acari: Mesostigmata: Dermanyssoidea)*. Institute of Zoology, Slovak Academy of Sciences, Bratislava, Slovakia, pp. 1–187.

Mašán, P. and Halliday, B. (2010) Review of the European genera of Eviphididae (Acari: Mesostigmata) and the species occurring in Slovakia. *Zootaxa* 2585, 1–122.

Mašán, P. and and Stanko, M. (2005) Mesostigmatic mites (Acari) and fleas (Siphonaptera) associated with nests of mound-building mouse, *Mus spicilegus* Petényi, 1882 (Mammalia, Rodentia). *Acta Parasitology* 50(3), 228–234.

Mascarenhas, C.S., Coimbra, A., Muller, G. and Brum, W. (2009) Ocorrência de *Ornithonyssus bursa* (Berlese, 1888) (Acari: Macronyssidae) em filhotes de *Megascops choliba* (corujinha-do-mato) e *Pitangus sulphuratus* (bem-te-vi), no Rio Grande do Sul, Brasil. *Revista Brasileira de Parasitologia Veterinária* 18, 69–70.

Masnavipour, M., Kazemi, S., Latifi, M. and Ziaaddini, M. (2011) Edaphic mesostigmatic mites (Acari: Mesostigmata) fauna of north-western Kerman Province, Iran. In: Kazemi, S. and Saboori, A. (eds) *International Centre for Science, High Thechnology and Environmental Sciences*, 22–23 December 2011, Kerman, Iran, 41.

Masoudian, F. and Khanjani, M. (2013) Mites associated with some medicinal plants (Asteraceae) in Hamedan, *Iranian Journal of Crop Protection* 2(2), 209–218.

Masoudian, F., Khanjani, M., Saboori, A. and Irani-Nejad, K.H. (2017) Ontogeny of *Tyrophagus perniciosus* Zakhvatkin (Acari: Acaridae) from Western Iran. *Persian Journal of Acarology* 6(4), 225–243.

Masoudian, F., Khanjani, M., Saboori, A. and Irani-Nejad, K.H. (2018) Redescription of four species of the genus *Tyrophagus* (Acari: Acaridae) from western Iran. *Acarina* 26(1), 111–132.

Materna, J. (2000) Oribatid communities (Acari: Oribatida) inhabiting saxicolous mosses and lichens in the Krkonose Mts. (Czech Republic). *Pedobiologia* 44, 40–62.

Mathur, S. (1979) A new record of *Tyroborus lini* Oudemans (Acaridae: Astigmata) from India. *Acarology Newsletter* 9, 3.

Mathur, S. and Manju Dalal (1989) Influence of food quality on the development and growth of the acarid mite, *Suidasia nesbitti* (Acari: Acaridae). In: Channabasavanna, G.P. and Viraktamath, C.A. (eds) *Progress in Acarology*, E.J. Brill, Leiden, The Netherlands, pp. 231–239.

Mathys, G. and Tencalla, Y. (1959) *Note Preliminaire Sur la Biologie et la Valeur Prédatrice de Proctolaelaps Hypudaei Oudms. (Acarien, Mesostigmata, Aceosejidae)*. Publication Stations Fédérales d'Essais Agricoles de Lausanne, Switzerland, pp.645–654.

Matsumoto, K. (1970) Studies and the environmental factors for the breeding of grain mites.10. Effect of the temperature and relative humidity on the breeding and hypopus formation in *Lardoglyphus konoi*. *Japanese Journal of Sanitary Zoology* 21(4), 213–219.

Matsumoto, K. (1973) Studies and the environmental factors for the breeding of grain mites.11. The effect of nutrient on hypopus formation in *Lardoglyphus konoi*. *Japanese Journal of Sanitary Zoology* 24(1), 1–7.

Matsumoto, T. and Satoh, A. (2004) The occurrence of mite-containing wheat flour. *Pediatric Allergy and Immunology* 15(5), 469–471.

Matsumoto, K., Okamoto, M. and Wada, Y. (1986) Effect of relative humidity on life cycle of the house dust mites, *Dermatophagoides farinae* and *D. pteronyssinus*. *Japanese Journal of Sanitary Zoology* 37, 79–90.

Matsumoto, K., Okamoto, M. and Yamaura, H. (1995) Studies on the environmental factors for the breeding of grain mites: XV. The effect of temperature and relative humidity on the life cycle of individually reared *Glycyphagus domesticus* (De Geer, 1778) (Acari: Glycyphagidae). *Medical Entomology and Zoology* 46(1), 31–39.

Matsumoto, T., Goto, Y. and Miike, T. (2001) Anaphylaxis to mite contamined flour. *Allergy* 56, 247.

Matsumoto, T., Hisano, T., Hamaguchi, M. and Miike, T. (1996) Systemic anaphylaxis after eating storage-mite-contaminated food. *International Archives of Allergy and Immunology* 109, 197–200.

Maunder, J. (1990) House dust mites in the work place. *Environmental Health Magazine*, 304–306.

Maunsell, K., Wraith, D.G. and Cunnington, A.M. (1968) Mites and house-dust allergy in bronchial asthma. *Lancet* 2091(7555), 1267–1270.

Maurer, V. and Baumgartner, J. (1992) Temperature influence on life table statistics of the chicken mite *Dermanyssus gallinae* (Acari: Dermanyssidae). *Experimental and Applied Acarology* 15, 27–40.

Maurer, V. and Perler, E. (2006) Silicas for control of the poultry red mite *Dermanyssus gallinae*. In: *Proceedings of Joint Organic Congress, May 30–31 2006*. Odense, Denmark.

Maurer, V., Bieri, M. and and Fölsch, D.W. (1988) Das Suchverhalten von *Dermanyssus gallinae* Hühnerställen. *Archiv Fur Geflügelkunde* 52, 209–215.

Maurer, V., Perler, E. and Heckendorn, F. (2009) In vitro effects of oils, silicas and plant preparations against the poultry red mite *Dermanyssus gallinae*. *Experimental and Applied Acarology* 48, 31–41.

Mauri, R. (1982) Algunos acaros nuevos para la fauna argentina. *Revista de la Sociedad Entomológica Argentina* 41(14), 93–96.

Mauri, R. and Alzuet A.D.B. De (1980) The genus *dermatophagoides* in Argentine (*Acarina: Pyroglyphidae*). *Obra del Centenario del Museo de La Plata* 6, 227–232.

Mazzuco, R.M., Lebarbenchon, M.A., Moecke, E.H.S., Meyer, E., Spíndola-Batti, M.A., Fröde, T. and Morato, E.F. (2000) Anafilaxia por ingestão de ácaros de estocagem. *Revista Brasileira de Alergia e Imunopatologia* 23, 189.

McCulloch, A.E. (1938) Acariasis of the Urinary Tract Caused by *Histogaster*. *Canadian Medical Association Journal* 38, 351–352.

McDaniel, B. and Theron, P.D (1979) A new distribution record and a morphological variant of *Terpnacarus glebulentus* Theron (Acari: Terpnacaridae) (South Dakota, South Africa). *Entomological News* 90, 143–144.

McDonald, L.G. and Tovey, E. (1992) The role of water temperature and laundry procedures in reducing house dust mite populations and allergen content of bedding. *Journal of Allergy and Clinical Immunology* 90, 599–608.

McDonald, L.G. and Tovey, E. (1993) The effectiveness of benzyl benzoate and some essential plant oils as laundry additives for killing house dust mites. *Journal of Allergy and Clinical Immunology* 92, 771–772.

McEvoy, M. and Southall, R. (2004) Redefinition of the functions of a window to achieve improved air quality and energy performance in European housing. In: de Wit, M. (ed.) *Built Environments and Environmental Buildings. PLEA (Passive and Low Energy Architecture), Conference Proceedings, 19–22 September 2004.* Technische Universiteit Eindhoven, The Netherlands, pp. 837–842.

McGarry, J.W., McCall, P.J. and Welby, S. (2001) Arthropod dermatitis acquired in the UK and overseas. *Lancet* 357, 2105–2106.

McGraw, J.R. and Farrier, M.H. (1969) Mites of the superfamily Parasitoidea (Acarina: Mesostigmata) associated with *Dendroctonus* and *Ips* (Coleoptera: Scolytidae). *North Carolina Agricultural Experiment Station Technical Bulletin* 192, 1–162.

McGregor, E.A. (1950) Mites of the genus *Neophyllobius*. *Bulletin of the Southern California Academy of Sciences* 49(2), 55–70.

McIntyre, D.A. (1992) The control of house dust mites by ventilation: A pilot study. In: *Proceedings of 13th AIVC Conference: Ventilation for Energy Efficiency and Optimum Indoor Air Quality, 15–18 September 1992, Nice.* Air Filtration and Ventilation Centre, France, Coventry, U.K., pp. 497–507.

McMurtry, J.A. and Croft, B.A. (1997) Life-styles of phytoseiid mites and their role in the biological control. *Annual Review of Entomology* 42, 291–321.

McMurtry, J.A., Huffaker, C.B. and van de Vrie, M. (1970) Ecology of tetranychid mites and their natural enemies: a review - I. Tetranychid enemies: their biological characters and the impact of spray practices. *Hilgardia* 40(11), 331–390.

McMurtry, J.A., Moraes, G.J. and Sourassou, N.F. (2013) Revision of the lifestyles of phytoseiid mites (Acari: Phytoseiidae) and implications for biological control strategies. *Systematic and Applied Acarology* 18, 297–320.

McSorely, R. and Walter, D.E. (1991) Comparison of soil extraction methods for nematodes and microarthropods. *Agriculture, Ecosystems & Environment* 34, 201–207.

Mégnin, P. (1894) *La Faune des Cadavres: Application de L'entomologie à la Médecine Légale*. G. Masson: Gauthier-Villars et fils, Paris, France, pp. 1–214.

Mehl, R. (1977) *Haemogamasus pontiger* Berlese (Acari, Mesostigmata) in Norway. *Rhizocrinus* 9, 1–4.

Mehl, R. (1998) Occcurrence of mites in Norway and the rest of Scandinavia. *Allergy* 53(48), 28–35.

Mehlhorn, H. (2001) Mites. In H. Mehlhorn (ed.) *Encyclopedic Reference of Parasitology.* Springer, Berlin and Heidelberg, Germany, pp. 364–373.

Meier, F.A., Scherrer, S. and Honegger, R. (2002) Faecal pellets of lichenivorous mites contain viable cells of the lichen-forming ascomycete *Xanthoria parietina* and its green algal photobiont, *Trebouxia arboricola*. *Biological Journal of the Linnean Society London* 76, 259–268.

Meleney, W.P. (1985) Mange mites and other parasitic mites. In: Gaafar, S.M., Howard, W.E. and Marsh, R.E. (eds.) *Parasites, Pests and Predators, World Animal Science B2*. Elsevier, Amsterdam, The Netherlands, pp. 317–346.

Mellanby, K. (1972) *Scabies*. 2nd ed., E.W. Classey Ltd., Middlesex, U.K., pp. 1–81.

Melnyk, J.P., Scott-Dupree, A., Marcone, M.F., Hill, A. and Hill, A. (2010) Identification of cheese mite species inoculated on Mimolette and Milbenkase cheese through cryogenic scanning electron microscopy. Journal of Dairy Science 93(8), 3461–3468.

Mendel, Z. and Gerson, U. (1982) Is the mite *Lorryia formosa* Cooreman (Prostigmata: Tydeidae) a sanitizing agent in citrus groves? *Acta Oecology* 3, 47–51.

Menezes, C., Coletto-Silva, A., Gazet, G.S. and Kerr, W.E. (2009) Infestation by *Pyemotes tritici* (Acari, Pyemotidae) caused death of stingless bee colonies (Hymenoptera: Meliponina). *Genetics and Molecular Research* 8, 630–634.

Ment, D., Raman, S., Gal, S., Ezra, D. and Palevsky, E. (2020) Interactions of *Metarhizium brunneum*–7 with phytophagous mites following different application strategies. *Insects* 11, 330.

Mentz, M.B., Silva, G.L. and Silva, C.E. (2015) Dermatitis caused by the tropical fowl mite *Ornithonyssus bursa* (Berlese) (Acari: Macronyssidae): a case report in humans. *Revista da Sociedade Brasileira de Medicina Tropical* 48, 786–788.

Mercado, D., Puerta, L. and Caraballo, L. (2001) Life-cycle of *Suidasia medanensis* (= *pontifica*) (Acari: Suidasiidae) under laboratory conditions in a tropical environment. *Experimental and Applied Acarology* 25, 751–755.

Merget, R., Bauer, T., Küpper, H.U., Philippou, S., Bauer, H.D., Breitstadt, R. and Bruening, T. (2002) Health hazards due to the inhalation of amorphous silica. *Archives of Toxicology* 75(11–12), 625–634.

Mertins, J.W. and Hartdegen, R.W. (2003) The ground skink, *Scincella lateralis*, an unusual host for phoretic deutonymphs of a uropodine mite, *Fuscuropoda marginata*, with a review of analogous mite-host interactions. *Texas Journal of Science* 55, 33–42.

Messelink, G. and van Holstein-Saj, R. (2008) Improving thrips control by the soil-dwelling predatory mite *Macrocheles robustulus* (Berlese). *Integrated Control in Protected Crops, Temperate Climate. Bulletin IOBC/WPRS* 32, 135–138.

Metwally, A.M, Al-Azazzy, M.M and Abd El-Hady, M.A.H. (2014) Mites Associated with Coleoptera. *Acarines* 8(1), 55–58.

Metwally, A.M., AbdAllah, A.A., Gamal El- Din, Hala M., El-Bltagy and Hala M. (2016) Mites associated with stored products. *Annals of Agricultural Sciences Moshtohor* 54(3), 649–658.

Meyer, M.K.P. (1969) Some stigmaeid mites from South Africa (Acari: Trombidiformes). *Acarologia* 11(2), 227–271.

Meyer Smith, M.K.P. (1981) Mite pests of crops in Southern Africa, Science bulletin 397, *Department Agriculture Fisher, Pretoria, South Africa*, 1–92.

Meyer, M.K.P. and Ryke, P.A.J. (1960) Mites of the superfamily Eupodoidea (Acarina: Prostigmata) associated with South African plants South Africa. *Journal of Agricultural Science* 3(4), 481–496.

Meyer, M.K.P. and Ueckermann, E.A. (1989) African Raphignathoidea (Acari: Prostigmata), Entomology Memoirs, *Department Agriculture Water Supply, South Africa* 74, 1–57.

Meyer-Kuhling, B., Pfister, K., Müller-Lindloff, J. and Heine, J. (2007) Field efficacy of phoxim 50% (ByeMite) against the poultry red mite *Dermanyssus gallinae* in battery cages stocked with laying hens. *Veterinary Parasitology* 147, 289–296.

Michael, A.D. (1878) On a species of Acarus of the genus *Cheyletus*, believed to be new. *Transmission Research Microscopic Society* 1(3), 133–138.

Michael, A.D. (1884) *British Oribatidae. Vol. I.* Ray Society Publisher, London, U.K., pp. 1–333.

Michael, A.D. (1888) *British Oribatidae. Vol. II.* Ray Society Publisher, London, U.K., 1–657.

Michael, A.D. (1889) Observations on the special internal anatomy of *Uropoda krameri*. J. R. Microsc. Soc., pp. 1–15.

Michael, A.D. (1901), *1903 British Tyroglyphidae I, II.* Ray Society Publisher, London, pp. 1–474.

Michael, A.D. (1903) *British Tyroglyphidae. Vol. 2.* Ray Society Publisher, London, pp. 1–183+plates i–xxxix.

Michalska, K., Skoracka, A., Návia, D. and Amrine, J.W. (2010) Behavioural studies on eriophyoid mites: an overview. *Experimental and Applied Acarology* 51, 31–59.

Midthassel, A., Leather, S.R. and Baxter, I.H. (2013) Life table parameters and capture success ratio studies of *Typhlodromips swirskii* (Acari: Phytoseiidae) to the factitious prey *Suidasia medanensis* (Acari: Suidasidae). *Experimental and Applied Acarology* 61(1), 69–78.

Migeon, A. and Dorkeld, F. (2006–2013) Spider Mites Web: A Comprehensive Database for the Tetranychidae. Available at: http://www.montpellier.inra.fr/CBGP/spmweb (accessed 22 Marcg 2012).

Mignon, B. and Losson, B. (2008) Dermatitis in a horse associated with the poultry mite (*Dermanyssus gallinae*). *Veterinary Parasitology* 19(1), 38–43.

Miguel, R.F. (2019) *Eficiência de Métodos Não Convencionais Para O Controle Do Ácaro Vermelho Dermanyssus gallinae (De Geer, 1778) (Mesostigmata: Dermanyssidae)*, Federal Technological University of Paraná, Santa Helena, Brazil.

Mihm, J.A. and Chiang, H.C. (1976) Laboratory studies on the life cycle and reproduction of some soil- and manure-unhabiting predatory mites (Acarina: Laelapidae). *Pedobiologia* 16, 353–363.

Mills, L.R. (1973) Morphology of glands and ducts in the two-spotted spider mite, *Tetranychus urticae* Koch, 1836. *Acarologia* 15, 218–236.

Mineiro, J.L. de C., Lindquist, E.E. and Moraes, G.J. de (2009) Edaphic ascid mites (Acari: Mesostigmata: Ascidae) from the state of São Paulo, Brazil, with description of five new species. *Zootaxa* 2024, 1–32.

Miranda, R.J., Diomodes, Q.A. and Almanza, A. (2002) House dust mites from urban and rural houses on the lowland Pacific slopes of Panama. *Systematic and Applied Acarology* 7(1), 23–30.

Mirfakhraii, S. (1994) Faunistic Study on House Mites and Biology of Important Species in Urmieh. M.Sc. thesis, Tarbiat Modarres University, Tehran, pp. 1–173.

Misayo, M., Tomohiko, N., Daigo, O. and Toshio, F. (2005) Control of acarid mites *Tyrophagus similis* Volgin by soil sterilization. *Annual Report Kansai Plant Protection Society* 47, 1–8.

Mitchell, R.D. (1972a) The tracheae of water mites. *Journal of Morphology* 136, 327–335.

Mitchell, R. (1972b) The sex ratio of the spider mite *Tetranychus urticae*. *Entomologia Experimentalis et Applicata* 15, 299–304.

Mitchell, R.D. and Nadchatram, M. (1969) Schizekenosy: the substitute for defecation in chigger mites. *Journal of Natural History* 3, 121–124.

Mitchell, M.J. and Parkinson, D. (1976) Fungal feeding of oribatid mites (Acari: Cryptostigmata) in an aspen woodland. *Soil Ecology* 57(2), 302–312.

Miura, A. and M. Sasa (1961) Observations on the life cycle of the common grain mite, *Tyrophagus dimidiatus* (Hermann) by individual rearings. *Japanese Journal of Clinical and Experimental Medicine* 31, 333–339.

Miyamoto, J. (1974) Studies on the biology of the mites of grain and house dust: Part 1. The effect of temperature on the growth and oviposition of *Glycyphagus destructor* (Schrank, 1781). *Medical Entomology and Zoology* 25(3), 267–272.

Miyamoto, T., Oshima, S., Domae, A., Takahashi, K., Izeki, M., Tanaka, T. and Ishizaki, T. (1970) Allergenic potency of different house dusts in relation to contained mites. *Annals Allergy* 28, 405–412.

Mizoguchi, A., Mori, N., Nishida, R. and Kuwahara, Y. (2003) Alpha-acaridial, a female sex pheromone from an alarm pheromone emitting mite, *Rhizoglyphus robini*. *Journal of Chemical Ecology* 29, 1681–1690.

Mizoguchi, A., Murakami, K., Shimizu, N., Mori, N., Nishida, R. and Kuwahara, Y. (2005) S-isorobinal as the female sex pheromone from an alarm pheromone emitting mite, *Rhizoglyphus setosus*. *Experimental and Applied Acarology* 36, 107–117.

Mmen, F.M. and Sinha, R. N. (1991) Tydeid mites associated with stored grain and oil seeds in Canada, with descriptions of a new genus and five new species (Acari: Tydeidae). *Canadian Journal of Zoology* 69, 1226–1253.

Modarres Awal, M. (1994) *List of Agricultural Pests and their Natural Enemies in Iran*. Ferdowsi University of Mashhad, Iran, pp. 1–364.

Modarres Awal, M. (1997) *List of Agricultural Pests and their Natural Enemies in Iran (Revised Edition)*. Ferdowsi University of Mashhad, Iran, pp. 1–429.

Mohamed, A.E.M. (2013) Biological aspects and life table parameters of predator gamasid ascid mite, *Blattisocius dentriticus* (Berlese) (Acari: Gamasida: Ascidae). *Egyptian Academic Journal of Biological Sciences* 6(2), 97–105.

Mohamed, M.M.R. (2019) Taxonomy and Behavior of Some Predacious Gamasid Mites used in Biological Control. PHD Thesis, Agriculture Science, University of Agriculture, Cairo, Egypt.

Mohamed, M.I., Zaher, M.A. and Hassan, M.F. (1982) Observations on *Cheyletus cacahuamilpenis*, a predator of the tenuipalpid mite *Dolichotetranchus floridanus*. *Entomologia* 27, 343–348.

Mohapatra, D., Kumar, S., Kotwaliwale, N. and Singh, K. K. (2017). Critical factors responsible for fungi growth in stored food grains and non-Chemical approaches for their control. *Industrial Crops Production* 108, 162–182.

Molina, C., Aiache, J.M., Tourreau, A. and Jeanneret, A. (1975) Enquête épidémiologique et immunologique chez les fromagers. *Revista Française D'Allergology Et D'Immunology Clinical* 15, 89–91.

Molina, C., Tourreau, A., Aiache, J.M., Brun, J., Jeanneret, A. and Roche, G. (1977) Manifestations allergiques chez les fromagers. *Revista Française D'Allergology Et D'Immunology Clinical* 17, 235–245.

Momen, F.M. and Sinha, R.N. (1991) Tydeid mites associated with stored grain and oilseeds in Canada, with descriptions of a new genus and five new species (Acari: Tydeidae). *Canadian Journal of Zoology* 69(5), 1226–1254.

Moniez, R. (1892) Contribution à l'histoire naturelle du *Tyroglyphus mycophagus* Mégnin. *Mémoires de la Société Zoologique de France* 5, 584–601.

Montanarella, L., Pennock, D.J., McKenzie, N., Badraoui, M., Chude, V. *et al.* (2016) World's soils are under threat. *Soil* 2, 79–82. DOI: 10.5194/soil-2-79-2016.

Montazeri, N., Nemati, A., Ostovan, H. and Gwiazdowicz, D.J. (2011) Fauna of some mesostigmatic mites (Acari: Mesostigmata) in Shiraz. Iran. In: *Proceeding First Persian Congress Acarology*, Kerman, Iran, p. 44.

Montealegre, F., Quiñones, C.M.T., Michelen, V.M.D., Bayona, M.M.D., Fernández-Caldas, E., Vasquez, O., Colón, F.M.D. Chardón, D.M.D. and García, M.R.N. (1997) Prevalence of Skin Reaction to Aeroallergens in Asthmatic in Puerto Rico. PRHSJ 16(4), 35–367.

Monteiro, E.S. and Arruda, G.P. de (1969) Ocorrência de ácaros em rações de aves. Resumos II Reun. *Anais da Sociedade Entomológica do* **Brasil** *Recife Pernambuco*, pp. 1–26.

Moraes, G.J. de, McMurtry, J.A. and Denmark, H.A. (1986) *A Catalog of the Mite Family Phytoseiidae. References to Taxonomy, Synonymy, Distribution and Habitat.* EMBRAPA - DDT, Brasilia, Brazil, pp. 1–353.

Moraes, G.J. de, McMurtry, J.A., Denmark, H.A. and Campos, C.B. (2004) A revised catalog of the mite family Phytoseiidae. *Zootaxa* 434, 1–494.

Moraes, G.J. de, Venancio, R., Santos, V.L.V. dos and Paschoal, A.D. (2015) Potential of Ascidae, Blattisociidae and Melicharidae (Acari: Mesostigmata) as biological control agents of pest organisms. In: Carrillo, D., Moraes, G.J. de and Peña, J.E. (eds) *Prospects for biological Control of Plant Feeding Mites and Other Harmful Organisms, Progress in Biological Control Series*, Springer International Publishing, Cham, Switzerland 19, pp. 33–75.

Moraes, G.J. de, Britto, E.P., Mineiro, J.L.D.C. and Halliday, B. (2016) Catalogue of the mite families Ascidae Voigts & Oudemans, Blattisociidae Garman and Melicharidae Hirschmann (Acari: Mesostigmata). *Zootaxa* 4112(1), 1–299.

Moraes, G.J. de, Moreira, G.F., Freire, R.A.P., Beaulieu, F., Klompen, H. and Halliday, B. (2022) Catalogue of the free-living and arthropod-associated Laelapidae Canestrini (Acari: Mesostigmata), with revised generic concepts and a key to genera. *Zootaxa* 5184 (1), 1–590. DOI: 10.11646/zootaxa.5184.1.1.

Morales, M., Iraola, V., Leonor, J.R. and Carnes, J. (2013) Enzymatic Activity of Allergenic House Dust and Storage Mite Extracts. *Journal of Medical Entomology* 50(1), 147–54.

Moraza-Zorrilla, L.M. and Peña, M.A. (2006) Two new taxa of Parholaspidae Evans, 1956 (Acari: Mesostigmata) from Tenerife (Canary Islands). *Graellsia* 62(1), 75–86.

Moreira, N.S. (1975) Acarinos Pyroglyphidae e outros Sarcoptiformes em amostras de po domicilar em Belo Horizonte, Minas Gerais. MS thesis, Federal University of Minas Gerais, pp. 1–80.

Moreira, N.S. (1978) Redescrição de *Chortoglyphus arcuatus* (Troupeau, 1878) (Sarcoptiformes, Chortoglyphidae). *Revista Brasileira de Biologia* 38(2), 245–249.

Moreira, G.F. and de Moraes, G.J. (2015) The potential of free-living laelapid mites (Mesostigmata: Laelapidae) as biological control agents. In: Carrillo, D., de Moraes, G.J. and Peña, J.E. (eds) *Prospects for Biological Control of Plant Feeding Mites and Other Harmful Organisms.* Springer International Publishing, Cham, Switzerland, pp. 77–102.

Moreno, A. (2015) Clase Arachnida Orden Astigmata, *Revista IDE@ - SEA* 15, 1–19.

Morgan, M.S. and Arlian, L.G. (2006) Enzymatic Activity in Extracts of Allergy-Causing Astigmatid Mites. *Journal of Medical Entomology* 43(6), 1200–1207.

Morgan, M.S., Vyszenski-Moher, D.L. and Arlian L.G. (2015) Population Growth and Allergen Content of Cultured Euroglyphus maynei House Dust Mites. Int. Arch. Allergy Immunol. 166(4), 267–272.

Mori, N., Kuwahara, Y. and Kurosa, K. (1998) Rosefuran: the sex pheromone of an acarid mite. *Journal of Chemical Ecology* 24, 1771–1779.

Moro, C.V., Thioulouse, J., Chauve, C. and Zenner, L. (2011) Diversity, geographic distribution, and habitat-specific variations of microbiota in natural populations of the chicken mite, *Dermanyssus gallinae. Journal of Medical Entomology* 48, 788–796.

Morrone, F. and Mayworm, M.A.S. (2001) Acaricidal action of foliar extracts of *Coffea* species (Rubiaceae) on *Dermanyssus gallinae* (De Gerr, 1778) (Acari, Dermanyssidae). *Archives of Biological Institution* 68, 43–47.

Mossadegh, M.S. (1997) Some mites of the honey bee *Apis mellifera* L. hives in Iran. *Journal of Scientific Agriculture* 19(1–2), 7–16.

Mosbech, H., Gravesen, S., Heinig, J.H., Korsgaard, J., Schou, C. and Ostergaard, P.A. (1991) Diagnostic procedures - exposure and environment. *Allergy* 46(11), 23–25.

Moser, J.C. (1975) Biosystematic of the straw itch mite with special reference to nomenclature and dermatology. *Transactions of the Entomological Society of London* 127, 185–191.

Moser, J.C. and Roton, L.M. (1972) Reproductive compatilibity between two widely separated populations of *Pyemotes scolyti* (Acarina: Pyemotidae). *Pan-Pacific Entomologist* 48, 97–99.

Moser, J.C., Cross, E.A. and Roton, L.M. (1971) Biology of *Pyemotes parviscolyti* (Acarina: Pyemotidae). *Entomophaga* 16, 367–379.

Moser, J. C., Smiley, R.L. and Otvos, I.S. (1987) A new *Pyemotes* (Acari: Pyemotidae) reared from the Douglas fir cone moth. *International Journal of Acarology* 13, 141–147.

Moses, J.A, Digvir, S.J. and Alagusundaram, K. (2015) Climate change and its implications on stored food grains. *Agricultural Research* 4, 21–30.

Moshkin, V.S. and Brygadyrenko, V.V. (2022) Influence of air temperature and humidity on *Stratiolaelaps scimitus* (Mesostigmata, Laelapidae) locomotor activity in a laboratory experiment. *Biosystems Diversity* 30(2), 191–197.

Mosquera, S. (1988) *Eucheyletia hardyi* Baker, 1949 y *Haemogamasus pontiger* (Berlese, 1904), dos acaros nuevos para la fauna Argentina. *Neotropica* 34, 73–76.

Moss, W.W. (1962) Studies on the morphology of the trombidiid mite *Allothrombidium lerouxi* Moss (Acari). *Acarologia* 4, 313–345.

Moss, W.W. (1968) An illustrated key to the species of the acarine genus *Dermanyssus* (Mesostigmata: Laelapoidea: Dermanyssidae). *Journal of Medical Entomology* 5(1), 67–84.

Mossa, A.-T.H., Afia, S.I., Mohafrash, S.M. and Abou-Awad, B.A. (2018) Formulation and characterization of garlic (*Allium sativum* L.) essential oil nanoemulsion and its acaricidal activity on eriophyid olive mites (Acari: Eriophyidae). *Environmental Science and Pollution Research* 25, 10526–10537.

Mostafa, E.M. (2011) Incidence of mites inhabiting stored onion bulbus in Egypt, with description of a new species of the genus *Lasioseius* Berlese (Acari. Gamasina). *Journal of Plant Protection and Pathology* 2(10), 855–863.

Mostafa, A.M., Hanem, H.I.S., Yassin, E.M.A and Abdel-Khalik, A.R. (2013) Effect of different diets on the biology of the astigmatid mite *Tyrophagus putrescentiae*. *Egyptian Journal of Agricultural Research* 91(4), 1439–1446.

Mostafa, A.M., Ibrahim, W.L.F., Abou El-Nour, B.M, Yassin, E.M.A. and Abou El-Einien, N.F.E. (2017) Notes on the biological aspects of the fungivorous mites *Proctolaelaps pygmaeus* (Müller) (Mesostigmata: Ascidae) and *Glycyphagus ornatus* (Astigmata: Glycyphagidae) feeding on different fungi at different temperature degrees. *Egyptian Academic Journal of Biological Sciences* 10(5), 45–52.

Mothes, U. and Seitz, K.A. (1981a) Fine structure and function of the prosomal glands of the two-spotted spider mite, *Tetranychus urticae* (Acari, Tetranychidae). *Cell Tissue Research* 221, 339–349.

Mothes, U. and Seitz, K.A. (1981b) Functional microscopic anatomy of the digestive system of the *Tetranychus urticae* (Acari: Tetranychidae). *Acarologia* 22, 257–270.

Mourier, H. and Poulsen, K.P. (2000) Control of insects and mites in grain using a high temperature/short time (HTST) technique. *Journal of Stored Products Research* 36(3), 309–318.

Moya Borja, G.E. (1981) Effects of *Macrocheles muscadomesticae* (Scopoli) on the sexual behavior and longevity of *Dermatobia hominis*. *Revista Brasileira de Biologia* 41, 237–241.

Mowafi, M.H. (2005) Effect of food type on development and reproduction of mite species *Lasioseius africanus*. *The Egyptian Journal of Biological Pest Control* 15, 93–95.

Mrciak, M. (1979) Contribution to the knowledge of gamasid mites (Acari, Gamasoidea) of small mammals in Hungary. *Parasitology Hungary* 12, 99–104.

Mrciak, M. and Brander, T. (1965) Beitrag zur Kenntnis der Milben (Acarina) einiger Säugetiere in Finnland. *Lounais-Hämeen luonto* 17, 3–13.

Mueller, D.K., Kelley, P.J. and Van Ryckeghem, A.R. (2006) Mold mites *Tyrophagus putrescentiae* (Shrank) in stored products. In: Lorini, I., Bacaltchuk, B., Beckel, H., Deckers, D., Sundfeld, E. *et al.* (eds) *Proceedings of the 9th International Working Conference on Stored Product Protection, 15 to 18 October 2006 Campinas, São Paulo, Brazil.* Brazilian

Post-harvest Association–ABRAPOS, Passo Fundo, RS, Brazil, pp. 1117–1122.

Muesken, H., Franz, J.T., Heutelbeck, A.R.R., Schulz, T., Tuchola· P. *et al.* (2004) *Gohieria fusca* (GF) and *Chortoglyphus arcuatus* (CA): Allergenic properties of two mite specie. *Journal of Allergy and Clinical Immunology* 113(2), 298.

Muir, W.E., Yaciuk, G. and Sinha, R.N. (1977) Effects on temperature and insect and mite populations of turning and transferring farm-stored wheat. *Canadian Agriculture Engineering* 19, 25–28.

Mul, M.F. and Koenraadt, C.J.M. (2009) Preventing introduction and spread of *Dermanyssus gallinae* in poultry facilities using the HACCP method. *Experimental and Applied Acarology* 48, 167–181.

Mul, M., van Niekerk, T., Chirico, J., Maurer, V., Kilpinen, O. *et al.* (2009) Control methods for *Dermanyssus gallinae* in systems for laying hens: Results of an international seminar. *World's Poultry Science Journal* 65(4), 589–600.

Mul, M.F., van Riel, J.W., Meerburg, B.G., Dicke, M., George, D.R. and Groot Koerkamp, P.W.G. (2015) Validation of an automated mite counter for *Dermanyssus gallinae* in experimental laying hen cages. *Experimental and Applied Acarology* 66, 589–603.

Mul, M.F., Ploegaert, J.P., George, D.R., Meerburg, B.G., Dicke, M. and Koerkamp, P.W.G. (2016) Structured design of an automated monitoring tool for pest species. *Biosystem Engineering* 151, 126–140.

Mul, M.F., van Riel, J.W., Roy, L., Zoons, J., André, G. *et al.* (2017) Development of a model forecasting *Dermanyssus gallinae*'s population dynamics for advancing Integrated Pest Management in laying hen facilities. *Veterinary Parasitology* 245, 128–140.

Mul, M.F., Neijenhuis, F. and van Niekerk, T.G.C.M. (2020) *Aanpak Van Vogelmijt bij Pluimvee. Project Report 1226.* Wageningen University and Research, Wageningen. Available at: file:///C:/Users/User/Downloads/Rapport%201226%20-%20MIP-Aanpak%20vogelmijt%20bij%20pluimvee%20in%20kaft3.pdf (accessed 12 April 2022).

Mullen, A.J. (1943) Dried-fruit Industry in Argentina and Chile. *US. Depatment Commerce Foreign Communication Weekly* 11(8), 10–12, 38–39.

Mullen, G.R. and OConnor, B.M. (2009) Mites (Acari). In: Mullen, G.R. and Durden, L.A. (eds) *Medical and Veterinary Entomology*. 2nd edn, Academic Press, Elsevier, Amsterdam, The Netherlands, pp. 433–492.

Mullen, G.R. and OConnor, B.M. (2019) Mites (Acari). In: Mullen, G.R. and Durden, L.A. (eds) *Medical and Veterinary Entomology.* 3rd edn, Academic Press, Elsevier, Amsterdam, The Netherlands, pp. 533–602.

Müller, J. (1859) Beitrag zur Höhlenfauna Mährens. *Lotos* 9, 26–33.

Müller, J. (1860) Insectenepizoen der mährischen Fauna. Jahreshefte der Naturwissenschaftlichen Section der K. K. Mährisch-Schlesische Gesellschaft für Ackerbau, *Natur-und Landeskunde in Brünn für* 1859, 157–184.

Muma, M.H. (1955) Phytoseiidae (Acarina) associated with Citrus in Florida. *Annals of the Entomological Society of America* 48, 262–272.

Muma, M.H. (1975) Mites associated with citrus in Florida. *Bulletin University Florida Agricultural Experiment Station U.S.A.* 640A, 92.

Mumcuoglu, Y. (1977) House dust mites in Switzerland. III. Allergenic properties of the mites. *Acta Allergology* 32, 333–349.

Mumcuoglu, Y. and Gerber-Hobl, E. (1977) The importance of house dust mites and stored food mites as house dust allergens in the rural population. *Schweizerische Medizinische Wochensch rift* 107(51), 1909–1912.

Mumcuoglu, K.Y. and Lutsky, I. (1990) A prevalence survey of poultry house mites in Israel. *Acarologia* 31, 51–56.

Mumcuoglu, K.Y., Abed, Y., Armenios, B., Shaheen, S., Jacobs, J., Bar-Sela, S. and Richter, E. (1994) Asthma in Gaza refugee camp children and its relationship with house dust mites. *Annals Allergy* 72, 163–166.

Mumcuoglu, K.Y., Gat, Z., Horowitz, T., Miller, J., Bar-Tana, R., Ben-Zvi, A. and Naparstek, Y. (1999) Abundance of house dust mites in relation to climate in contrasting agricultural settlements in Israel. *Medical and Veterinary Entomology* 13, 252–258.

Munhbayarlah, S., Park, J.W., Ko, S.H., Ree, H.I. and Hong, C.S. (1998) Identification of *Tyrophagus putrescentiae* allergens and evaluation of cross-reactivity with *Dermatophagoides pteronyssinus. Yonsei Medical Journal* 39(2), 109–115.

Muñoz Cárdenas, K.A. (2017) What lies beneath? Linking litter and canopy food webs to protect ornamental crops, (Thesis, fully internal, University of Amsterdam), pp. 1–74. Available at: https://pure.uva.nl/ws/files/13307391/03.pdf (accessed 10 May 2021).

Murata, S., Taniguchi, A., Isezaki, M., Fujisawa, S., Sakai, E. *et al.* (2021) Characterisation of a cysteine protease from poultry red mites and its potential use as a vaccine for chickens. *Parasite* 28, 9.

Murillo, P., Klimov, P., Hubert, J. and OConnor, B. (2018) Investigating species boundaries using DNA and morphology in the mite *Tyrophagus curvipenis* (Acari: Acaridae), an emerging invasive pest, with a molecular phylogeny of the genus *Tyrophagus*. *Experimental and Applied Acarology* 75, 167–189.

Murillo, A.C., Abdoli, A., Blatchford, R.A., Keogh, E.J. and Gerry, A.C. (2020) Parasitic mites alter chicken behaviour and negatively impact animal welfare. *Scientific Reports* 10, 8236.

Murphy, P.W. and Sardar, M.A. (1991) Resource allocation and utilization contrasts in *Hypoaspis aculeifer* (Can.) and *Alliphis halleri* (G.&R. Can.) (Mesostigmata) with emphasis on food source. In: Schuster, R. and Murphy, P.W. (eds) *The Acari. Reproduction, Development and Life-History Strategies*, Chapman & Hall, London, U.K., pp. 301–311.

Murray, A.B. and Zuk, P. (1979) The seasonal variation in a population of house dust mites in a North American city. *Journal of Allergy and Clinical Immunology* 64, 266–269.

Musken, H., Franz, J.T., Wahl, R., Paap, A., Cromwell, O., Masuch, G. and Bergmann, K.C. (2000) Sensitization to different mite species in German farmers: clinical aspects. *The Journal of Investigational Allergology and Clinical Immunology* 10(6), 346–351.

Mustafa, A.M., Shalaby, F.F. Yassin, E.M.A., Khalil, A.M., Eissa, Y.A.E. and Shahata, F.E. (2016) Biological studies of laelapid predacious mites, *Androlaelaps casalis*, Berlese and *Laelaps astronomicus* Koch in two food types under three temperature degrees. *Menoufia Journal of Plant Protection Research* 1, 111–119.

Na, Y.E., Kim, S.I., Bang, H.S., Kim, B.S. and Ahn, Y.J. (2011) Fumigant toxicity of cassia and cinnamon oils and cinnamaldehyde and structurally related compounds to *Dermanyssus gallinae* (Acari: Dermanyssidae). *Veterinary Parasitology* 78, 324–329.

Nachev, P.D. and Trenchev, G. (1987) The acarofauna associated with some Coccidae in Bulgaria. *Pochvoznanie, Agrokhimiya i Rastitelna Zashchita* 22, 81–86.

Naderi, S., Mirfakhraie, S. and Safaralizadee, M. (2016) Fauna of Mesostigmata (Acari: Trombidiformes) in straw caches of some of the villages in Marivan and Urmia. *Proceedings of the 22nd Iranian Plant Protection Congress, 27–30 August*. University of Tehran, Karaj, Iran, p. 505.

Nagakura, T., Yasueda, H., Obata, T., Kanmuri, M., Masaki, T. and Ihara, N. (1996) Major *Dermatophagoides* mite allergen, Der 1, in soft toys. *Clinical and Experimental Allergy* 26, 585–589.

Naghibinejad, M., Ostovan, H., Kamali, K. and Ahadiyat, A. (2011) Mites of the families Ascidae, Blattisociidae and Ameroseiidae in Estahban region. *Journal of Entomology Research* 4, 63–78.

Nakamae, H., Kishi, S., Fujisaki, K., Oshiro, S. and Furuta, K. (1997) Incidence of the parasitism of chicken mite *Dermanyssus gallinae* parasitizing and propagating on chicken even in the daytime and their life cycle. *Japanese Poultry Science* 34(4), 240–247.

Nakao, H. (1989) Studies on acarid mites injurious to vegetable plants (Acari: Astigmata) I. Occurrence of damage to spinach by acarid mites. *Prefectural Agricultural Experiment Stations, Bulletin* 59, 41–47.

Nakao, H. (1991) Studies on acarid mites (Acari: Astigmata) damaging vegetable plants. II. Damage to vegetable seedlings. *Japanese Journal of Applied Entomology Zoology* 35, 303–309.

Nakao, H. and Kurosa, K. (1988) Description of four species of acarid mites newly recorded from Japan, with reference to the damage caused to crops (Acari: Astigmata). *Japanese Journal of Applied Entomology Zoology* 32, 135–142.

Nangia, N. (1986) Studies on mites associated with stored products, bioecology, and evaluation of different methods of control of acarids (Astigmata: Acaridae), Ph.D. thesis, University Agricultural Sciences, Bangalore, India, pp. 1–253.

Nangia, N. and Channa Basavanna, G.P. (1989a) Influence of temperatue and relative humidity on the development of *Tyrolichus casei* (Astigmata: Acaridae). In: Channa Basavanna, G.P. and Viraktamath, C.A. (eds) *Progress in Acarology, Vol. 2*. E.J. Brill, Leiden, The Netherlands, pp. 259–264.

Nangia, N. and ChannaBasavanna, G.P. (1989b) Acarines associated with stored products in Karnataka, India. In: Channa Basavanna, G.P. and Viraktamath, C.A. (eds) *Progress in acarology, Vol. 2*. E.J. Brill, Leiden, The Netherlands, pp. 241–248.

Narayanan, E.S. and Ghai, S. (1961) A new species of *Melichares* (*Blattisocius*) Keegan (Aceosejinae). *Proceedings of the National Institute of Sciences of India* 27B, 18–20.

Nascimento, M.M. (2019) *Armadilhas Impregnadas Com Beauveria Bassiana (Bals.) Vuill. Visando Ao Controle de Dermanyssus Gallinae (De Geer) (Acari: Dermanyssidae)*. State University of Western Paraná, Cascavel, Brazil.

Nascimento, M.M., Alves, L.F.A., de Oliveira, D.G.P., Lopes, R.B. and Guimarães, A.T.B. (2020) Laboratory and field evaluation of an autoinoculation device as a tool to manage poultry red mite, *Dermanyssus gallinae*, infestations with *Beauveria bassiana*. *Experimental and Applied Acarology* 80, 151–165.

Nasrollahi, S., Khanjani, M. and Mirfakhraei, S. (2018) A new species of the genus *Raphignathus* (Acari: Raphignathidae) from Kurdistan, Iran, with a key to world species. *Systematic and Applied Acarology* 23(10), 2070–2081.

Nath, N.S., Bhattacharya, I., Tuck, A.G., Schlipalius, D.I. and Ebert, P.R. (2011) Review Article: Mechanisms of phosphine toxicity. *Journal of Toxicology* 2011, 1–9.

Navarro, S. (2006) Modified atmospheres for the control of stored product insects and mites. In: Heaps, J.W. (ed.) *Insect Management for Food Storage and Processing*. AACC International, St. Paul, Minnesota, U.S.A., pp. 105–145.

Navarro, S. (2012). Advanced grain storage methods for quality preservation and insect control based on aerated or hermetic storage and IPM. *Journal of Agricultural Engineering* 49(1), 13–20.

Navarro, S., Lider, O. and Gerson, U. (1985) Response of adults of the grain mite, *Acarus siro* L. to modified atmospheres. *Journal of Agricultural and Entomology* 2(1), 61–68.

Navarro, J.M., Mendoza Meza, D.L. and Mercado Bermúdez, D. (2008) Identificación de ácaros del polvo casero en colchones y almohadas de niños alérgicos de Santa Marta, Colombia. Duazary. *Revista Facultad Cientifica Salud* 5(1), 24–31.

Navarro-Campos, C., Wäckers, F.L. and Pekas, A. (2016) Impact of factitious foods and prey on the oviposition of the predatory mites *Gaeolaelaps aculeifer* and *Stratiolaelaps scimitus* (Acari: Laelapidae). *Experimental and Applied Acarology* 70, 69–78.

Navpreet, K. and Sakshi, D. (2018) Seasonal abundance and diversity of astigmatic mites in storage facilities of Punjab, India. *Agricultural Science Digest* 38(1), 32–35.

Nawar, M.S., Rakha, M.A. and Ali, F.S. (1990). Laboratory studies on predaceous mite, *Lasioseius bispinosus* Evans (Acari: Mesostigmata: Ascidae) on various kinds of food substances. *Bulletin of the Entomological Society of Egypt* 69, 247–255.

Nayak, M.K. (2006) Management of mould mite *Tyrophagus putrescentiae* (Schrank) (Acarina: Acaridae): A case study in stored animal feed. *International Pest Control* 48, 128–130.

Nayak, M.K. (2017) *Grain Storage Mite Pests*. Storedgrain, Brisbane, Australia. Available at: https://storedgrain.com.au/grain-storage-mite-pests/ (accessed 18 April 2018).

Nayak, M.K., Collins, P.J., Throne, J.E. and Wang, J.J. (2014) Biology and management of psocids infesting stored products. *Annual Review of Entomology* 59, 279–297.

Ndakidemi, B., Mtei, K. and Ndakidemi, P.A. (2016) Impacts of synthetic and botanical pesticides on beneficial insects. *Agricultural Science* 7, 364.

Ndiaye, G.A. and Marchand, B. (1989) *Lardoglyphus konoi* and *Suidasia pontifica* damaging dried, braised sardinelles in Senegal. A scanning electron microscopy study. *Acarologia* 30(2), 131–137.

Nechita, I.S., Poirel, M.T., Cozma, V. and Zenner, L. (2015) The repellent and persistent toxic effects of essential oils against the poultry red mite, *Dermanyssus gallinae*. *Veterinary Parasitology* 214, 348–352.

Nemati, A., Kamali, K. and Mossadegh, M.S. (2000) A faunistic survey of Laelapidae (Acari: Mesostigmata) mites in the soil of Ahvaz region, Khuzestan Province. *Congress, 5-8 September 2000, University of Technology*, Isfahan, Iran, p. 333.

Nemati, A., Gwiazdowicz, D.J., Riahi, E. and Mohseni, M. (2012). Catalogue of the Iranian Mesostigmatid mites part 1: Family Ascidae. *The International Journal of Agriculture and Crop Sciences* 4, 1005–1011.

Nemati, A., Riahi, E., Khalili-Moghadam, A. and Gwiazdowicz, D.J. (2018) A catalogue of the Iranian Mesostigmata (Acari): Additions and updates of the previous catalogue. *Persian Journal of Acarology* 7(2), 115–191.

Nermut', J., Zemek, R., Mráček, Z., Palevsky, E. and Půža, V. (2019) Entomopathogenic nematodes as natural enemies for control of *Rhizoglyphus robini* (Acari: Acaridae)? *Biological Control* 128, 102–110.

Nermut', J., Konopická, J., Zemek, R., Kopãcka, M., Bohatá, A. and Půža, V. (2020) Dissemination of *Isaria fumosorosea* spores by *Steinernema feltiae* and *heterorhabditis bacteriophora*. *Journal of Fungi* 6, 359.

Nesbitt, H. H.J. (1944) Three new mites of the subfamily Rhizoglyphinae. *Canadian Entomologist* 76(2), 21–27.

Nesbitt, H.H.J (1945) A revision of the family acaridae (Tyroglyphidae), order acari, based on comparative morphological studies. Part I. historical, morphological, and general taxonomic studies. *The Canadian Journal of Research* 23(6), 139–188.

Nesbitt, H.H.J (1951) A taxonomic study of the Phytoseiinae (family Lelaptidae) predaceous upon Tetranychidae of economic importance. *Zoologische Verhandelingen* 12, 1–64.

Nesvorna, M., Gabrielova, L. and Hubert, J. (2012) *Tyrophagus putrescentiae* is able to graze and develop on *Fusarium* fungi of mycotoxins importance under laboratory conditions. *Journal of Stored Products Research* 48(1), 37–45.

Nevin, F.R., 1935. Anatomy of *Cnemidocoptes mutans* (R. and L.), the scaly-leg mite of poultry. *Annals of the Entomological Society of America* 28, 338–367.

Newstead, R. and Duvall, H.M. (1918) Bionomic, morphological and economic report on the Acarids of stored grain and flour. *Royal Society Reports of the Grain Pests (War) Committee* 2, 1–48.

Newstead, R. and Duvall, H.M. (1920) Bionomic, morphological and economic report on the acarids of stored grain and flour - Part I. *Royal Society Reports of the Grain Pests (War) Committee* 8, 1–25.

Newstead, R. and Morris, H.M. (1920) Bionomic, morphological and economic report on the acarids of stored grain and flour - Part II. *Royal Society Reports on Grain Pests (War) Committee* 8, 4–15+Pl. I-V.

Nguyen, D.T., Vangansbeke, D., Lü, X. and De Clercq, P. (2013) Development and reproduction of the predatory mite *Amblyseius swirskii* on artificial diets. *BioControl* 58, 369–377.

Nguyen, D.T., Vangansbeke, D. and de Clercq, P. (2014) Artificial and factitious foods support the development and reproduction of the predatory mite *Amblyseius swirskii*. *Experimental and Applied Acarology* 62, 181–194.

Niedbała, W. (1998) Ptyctimous mites of the Pacific Islands. Recent knowledge, origin, descriptions, redescriptions, diagnoses and zoogeography (Acari: Oribatida). *Genus* 9(4), 431–558.

Niedbała, W. (2004) Protoplophoridae (Acari: Oribatida) of the world. *Annales Zoologici* 54(4), 807–834.

Nielsen, P.S. (1998) The use of *Blattisocius tarsalis* (Acari: Ascidae) for biological control in flour mills. In: Jin, Z., Liang, Q., Liang, Y., Tan, X. and Guan, L. (eds) *Proceedings of the 7th Interternational Working Conference on Stored-product Protection, Vol. 2, 14–19 October 1998, Beijing, China*. Sichuan Publishing House of Science and Technology, Chengdu, China, pp. 1265–1268.

Nielsen, S. (2001) Developmental time of *Blattisocius tarsalis* (Acari: Ascidae) at different temperatures. *Experimental and Applied Acarology* 25, 605–608.

Niklas, K.J., Tiffney, B.H. and Knoll, A.H. (1983) Patterns in vascular land plant diversification. *Nature* 303, 614–616.

Nikolaou, P., Marciniak, P., Adamski, Z. and Ntalli, N. (2021) Controlling stored products' pests with plant secondary metabolites: A review. *Agriculture* 11, 879. DOI: 10.3390/agriculture11090879.

Niogret, J., Lumaret, J.P. and Bertrand, M. (2006) Review of the phoretic association between coprophilous insects and macrochelid mites (Acari: Mesostigmata) in France. *Elytron* 20, 99–121.

Noble, W.E., Poe, S.L. and Stall, R.E. (1978) Surface sterilization of bulb mites for phytopathogen transmission studies. *Florida Entomologist* 61, 120.

Noei, J. (2007) Identification of rice storage mites in Guilan Province under different storage conditions. M.Sc. thesis. University of Guilan, Rasht, Iran, pp. 1–152.

Noei, J. and Ostovan, H. (2012) Introduction and identification key of stored astigmatic mites (Acari: Astigmata) of rice in Guilan Province. *Plant Pests Research* 2(1), 29–38.

Noei, J., Hajizadeh, J., Salehi, L., Ostovan, H. and Faraji, F. (2007) Stigmaeid mites associated with stored rice in northern Iran (Acari: Stigmaeidae). *International Journal of Acarology* 33, 153–156.

Noei, J., Hajizadeh, J., Salehi, L. and Ostovan, H. (2008a) *Prostigmatic Stored Mites of Rice in Guilan Province*. Abstract Book of 18th

Iranian Plant Protection Congress, Hamedan, Iran, p. 278.

Noei, J., Hajizadeh, J., Salehi, L. and Ostovan, H. (2008b) Mesostigmatic stored mites of rice in Guilan Province. In: Manzari, S. (ed.) *Abstract Book of the 18th Iranian plant Protection Congress, 24–27 August 2008*. Bu-Ali Sina University, Hamadan, Iran, p. 277.

Noei, J., Hajizadeh, J., Faraji, F., Ostovan H. and Salehi, L. (2009) First report of two species of predatory mites (Acari: Phytoseiidae) for Iranian fauna. *Journal of Entomological Society of Iran* 29(2), 113–116.

Noei, J., Maroufpoor, M., Faizi, F. and Ostovan, H. (2017) Second record of *Erythraeus* (*Zaracarus*) *coleopterus* (Acari: Erythraeidae) from Iran with new morphological data. *Persian Journal of Acarology* 6(2), 71–80.

Noferi, A., Serino, W., Sacerdoti, G. and Blythe, M.E. (1974) Survey on concentration and type of house dust mites (HDM) in dusts of specifically sensitised people in Naples. Preliminary data. *Folia Allergology Immunology and Clinical* 21, 466–469.

Noguchi, S., Mori, N., Kurosa, K. and Kuwahara, Y. (1998) Chemical ecology of astigmatid mites XLIX. beta-acaridial (2(E)-(4-methyl–3-pentenylidene)-butanedial), the alarm pheromone of *Tyrophagus longior* Gervais (Acarina: Acaridae). *Applied Entomoloy and Zoology* 33, 53–57.

Nordberg, S. (1936) Biologisch-Okologische Untersuchungen über die Vogelnidicolen. *Acta Zoologica Fennica* 21, 1–168.

Nordenfors, H. and Chirico, J. (2001) Evaluation of a sampling trap for *Dermanyssus gallinae* (Acari: Dermanyssidae). *Journal of Economic Entomology* 94, 1617–1621.

Nordenfors, H. and Höglund, J. (2000) Long term dynamics of *Dermanyssus gallinae* in relation to mite control measures in aviary systems for layers. *British Poultry Science* 41(5), 533–540.

Nordenfors, H., Hoglund, J. and Uggla, A. (1999) Effects of temperature and humidity on oviposition, molting, and longevity of *Dermanyssus gallinae* (Acari: Dermanyssidae). *Journal of Medical Entomology* 36, 68–72.

Nordvall, S.L., Eriksson, M., Rylander, E. and Schwartz, B. (1988) Sensitisation of children in the Stockholm area to house dust mite. *Acta paediatrica Scandinavica* 77, 716–720.

Norris, J.D. (1958) Obervations on the control of mite infestations in stored wheat by *Cheyletus* spp (Acarina: Cheyletidae). *Annals of Applied Biology* 46, 411–422.

Norton, R.A. (1977) A review of F. Grandjean's system of leg chaetotaxy in the Oribatei and its application to Damaeidae. In: Dindal, D.L. (ed.) *Biology of Oribatid Mites, Volume 33*. SUNY College of Environmental Science and Forestry, Syracuse, New York, U.S.A., pp. 33–62.

Norton, R.A. (2001) Systematic relationships of Nothrolohmanniidae and the evolutionary plasticity of body form in Enarthronota (Acari: Oribatida). *In*: Halliday, R.B., Walter, D.E., Proctor, H.C., Norton, R.A. and Colloff, M.J. (eds) *Acarology: Proceedings of the 10th International Congress of Acarology, June 1998*. CSIRO Publishing, Camberra, Melbourne, Australia, pp. 58–75.

Norton, R.A. and Alberti, G. (1997) Porose integumental organs of oribatid mites (Acari: Oribatida). 3, Evolutionary and ecological aspects. *Zoologica (Stuttgart)* 146, 115–143.

Norton, R.A. and Behan-Pelletier, V. (2009) Suborder oribatida. In: Krantz, G.W. and Walter, D.E. (eds) *A Manual of Acarology.* Texas Tech University Press, Lubbock, U.S.A., pp. 430–564.

Norton, R.A., OConnor, B.M. and Johnston, D.E. (1983) Systematic relationships of the Pediculochelidae (Acari: Acariformes). *Proceedings of the Entomological Society of Washington* 85(3), 493–512.

Norton, R.A., Bonamo, P.M., Grierson, J.D. and Shear, W.A. (1988) Oribatid mites fossils from a terrestrial Devonian deposit near Gilboa, New York. *Journal of Paleontology* 62, 259–269.

Norton, R.A., Kethley, J.L., Johnston, D.E. and OConnor, B. (1993) Phylogenetic perspectives on genetic systems and reproductive modes of mites. In: Wrensch, D.L. and Ebbert, M.A. (eds) *Evolution and Diversity of Sex Ratio in Insects and Mites*. Chapman and Hall, New York, U.S.A., pp. 8–99.

Norton, R.A., Alberti, G., Weigmann, G. and Woas, S. (1997) Porose integumental organs of oribatid mites (Acari, Oribatida). 1, Overview of types and distribution. *Zoology* 146, 1–31.

Nunn, F., Baganz, J., Bartley, K., Hall, S., Burgess, S. and Nisbet, A.J. (2020) An improved method for in vitro feeding of adult female *Dermanyssus gallinae* (poultry red mite) using Baudruche membrane (goldbeater's skin). *Parasites and Vectors* 13(1), 585.

Nuttall, T.J., Hill, P.B., Bensignor, E. and Willemse, T. (2006) House dust and forage mite allergens and their role in human and canine atopic dermatitis. *Veterinary Dermatology* 17, 223–235.

Nutting, W.B. (ed.) (1984) *Mammalian Diseases and Arachnids, Vol. I, Pathogen Biology and Clinical Management, 1–277. Vol. II Medico-Veterinary Laboratory and Wildlife Diseases and Control.* CRC Press, Boca Raton, U.S.A., pp. 1–280.

Nuzzaci, G. and Alberti, G. (1996) Internal anatomy and physiology. In: Lindquist, E.E., Sabelis, M.W. and Bruin, J. (eds) *Eriophyoid Mites: Their Biology, Natural Enemies and Control. World Crop Pests, Volume 6*. Elsevier Science, Amsterdam, The Netherlands, pp. 101–150.

O'Callaghan, M., Glare, T.R. and Lacey, L.A. (2012) Bioassay of bacterial entomopathogens against insect larvae. In: Lacey, L.A. (ed.) *Manual of Techniques in Invertebrate Pathology.* 2nd edn, Elsevier Ltd., The Netherlands, pp. 101–127.

O'Donell, A.E. and Axtell, R.C. (1965) Predation by *Fuscoropoda vegetans* (Acarina: Uropodidae) on house fly *Musca domestica. Annals of the Entomological Society of America* 58(3), 403–404.

O'Donnell, A.E. and Nelson, E.L. (1967) Predation *Fuscuropoda vegetans* (Acarina. Uropodidae) and *Macrocheles muscaedomesticae* (Acarina: Macrochelidae) on the eggs of the little house fly, *Fannia canicularis. Journal of the Kansas Entomological Society* 40, 441–443.

O'Donovan, W.J. (1922) Dermatitis due to *Carpoglyphus passularum. The British Journal of Dermatology and Syphilis* 34, 297–298.

O'Farrell, A.F. and Butler, P.M. (1948) Insects and Mites associated with the storage and manufacture of foodstuffs in Northern Ireland. *The Economic Proceedings of the Royal Dublin Society* 3(22), 343–407.

Oatman, E.R. (1963) Mite species on apple foliage in Winsconsin. In: Naegele, J.A. (ed.) *Advances in Acarology: Vol. 1.* Comstock Publishing, Ithaca, N.Y., U.S.A., pp. 21–24.

Oboussier, H. (1939) Beitrage zur biologie und Anatomie der Wohnungsmilben. *Zeitschrift für angewandte Entomologie* 26(2), 253–296.

OConnor, B.M. and Eickwort, G.C. (1988) Morphology, ontogeny, biology and systematics of the genus *Vidia* (Acari: Winterschmidtiidae). *Acarologia* 29, 147–174.

OConnor, B.M. (1979) Evolutionary origins of astigmatid mites inhabiting stored products. In: Rodriguez, J.G. (ed.) *Recent Advances in Acarology: Vol. I.* Academic Press, New York, San Francisco, London, pp. 273–278.

OConnor, B.M. (1982) Astigmata. In: Parker, S.P. (ed.) *Synopsis and Classification of Living Organisms.* McGraw-Hill, New York, U.S.A., pp. 146–169.

OConnor, B.M. (1984) Acarine-fungal relationships: The evolution of symbiotic associations. In: Wheeler, Q. and Blackwell, M. (eds) *Fungus-Insect Relationships: Perspectives in Ecology and Evolution.* Columbia University Press, New York, U.S.A., pp. 354–381.

OConnor, B.M. (1988) Cevolution in astigmatid mite-bee associations. In: Needham, G.R., Page, R.E. Jr, Delfinado-Baker, M. and Bowman, C. (eds) *Africanized Honey Bees and Bee Mites.* Ellis-Horwood, Ltd., Chichester, U.K., pp. 339–346.

OConnor, B.M. (1990) Ecology and host associations of *Histiogaster arborsignis* (Acari: Acaridae) in the Great Lakes region, particularly in the Huron Mountains of Northern Michigan. *The Great Lakes Entomologist* 23, 205–209.

OConnor, B.M. (1991) A preliminary report on the arthropod-associated astigmatid mites (Acari: Acariformes) of the Huron Mountains of Northern Michigan. *Michigan Academy* 24, 307–320.

OConnor, B.M. (1994) Life-history modifications in astigmatid mites. In: Houck, M.A. (ed.) *Mites, Ecological and Evolutionary Analyses of Life-History Patterns.* Springer Science+Business Media Dordrecht, The Netherlands, pp. 136–159.

OConnor, B.M. (1998) Keys to North American Genera: Family Acaridae Latreille, p. 1802. Available at: http://insects.ummz.lsa.umich.edu/PEET/nagenera.html (accessed 18 May 2015).

OConnor, B.M. (2008a) Family Acaridae Latreille, 1802 (Species listing from the 'Biology Catalog' of Texas A&M University). Available at: http://insects.tamu.edu/research/collection/hallan/Acari/Family/Acaridae.txt http://bug.tamu.edu/research/collection/hallan/Acari/Family/Acaridae.txt (accessed 18 May 2015).

OConnor, B.M. (2008b) Family pyroglyphidae cunliffe, 1958. In: OConnor, B., Needham, G., Welbourn, C., Kelly, D. and Klompen, H. (eds) *Medical Veterinary Acarology, Mite Week, 58th Annual Acarology Summer Programm, Ohio State University, USA, June 28th to July 4th.*

OConnor, B.M. (2009a) Astigmatid mites (Acari: Sarcoptiformes) of forensic interest. *Experimental and Applied Acarology* 49, 125–133.

OConnor, B.M. (2009b) Cohort astigmatina. In: Krantz, G.W. and Walter, D.E. (eds) *A Manual of Acarology*. Texas Tech University Press, Lubbock, Texas, U.S.A., pp. 565–657.

OConnor, B.M. and Houck, M.A. (1989) Two new genera of Hemisarcoptidae (Acari: Astigmata) from the Huron Mountains of northern Michigan. *The Great Lakes Entomologist* 22, 1–10.

OConnor, B.M. and Klimov, P. (2003) Systematics and ecology of North American bee-associated mites: Potential threats to native and introduced pollinators. Available at: http://insects.ummz.lsa.umich.edu/beemites/Objectives.htm (accessed 18 December 2022).

OConnor, B.M. and Klimov, P. (2012) Family Pyemotidae Oudemans, 1937 (1897). Available at: http://insects.ummz.lsa.umich.edu/beemites/Species_Accounts/Pyemotidae.htm (accessed 4 October 2019).

OConnor, B.M. and Klimov, P. (2016) Species and subspecies of mites associated with bees of the World. Available at: http://insects.ummz.lsa.umich.edu/beemites/Species_Accounts/ (accessed 21 May 2020).

Odaka, M., Matsuo, K., Ogino, K., Kanazawa, T., Baba, R. *et al.* (2019) Efficacy of a novel mixture of substances derived from food and food additives for controlling *Dermanyssus gallinae* (Mesostigmata: Dermanyssidae). *Applied Entomoloy and Zoology* 54(1), 31–38.

Oehlschlaegel, G., Bayer, F., Disko, R., Fechter, H. and Mahunka, S. (1983) *Tarsonemus hominis* in skin connective tissue. *Hautarzt* 34, 632–634.

Ofek, T., Gal, S., Inbar, M., Lebiush-Mordechai, S., Tsror, L. and Palevsky, E. (2014) The role of onion-associated fungi in bulb mite infestation and damage to onion seedlings. *Experimental and Applied Acarology* 62, 437–448.

Oh, M.H., Kim, S.S. and Boo, K.S. (1985) Studies on the Morphology and Biology of a Parasitic Mite, *Pyemotes tritici* L.-F. and M. on the Cigarette Beetle (*Lasioderma serricorne* F.). *Korean Journal of Applied Entomology* 24(1) 62, 15–18.

Oh, H., Ishii, A., Tongu, Y. and Itano, K. (1986) Microorganisms associated with the house dust mite, *Dermatophagoides*. *Japanese Journal of Sanitary Zoology* 37, 229–235.

Oh, S.I., Do, Y.J., Kim, E., Yi, S.W. and Yoo, J.G. (2020) Prevalence of poultry red mite (*Dermanyssus gallinae*) in Korean layer farms and the presence of avian pathogens in the mite. *Experimental and Applied Acarology* 81, 223–238.

Okabe, K. (1999) Morphology and ecology in deutonymphs of non-psoroptid Astigmata. *Journal of the Acarological Society of Japan* 8(2), 89–101.

Okabe, K. and Amano, H. (1990) Attractancy of alcohols isolated from culture filtrates of *Fusarium* fungi for the robine bulb mite, *Rhizoglyphus robini* Claparede (Acari: Acaridae), in sand. *Applied Entomoloy and Zoology* 25, 397–404.

Okabe, K. and Amano, H. (1991) Penetration and population growth of the robine bulb mite, *Rhizoglyphus robini* Claparede (Acari: Acaridae), on healthy and *Fusarium*-infected rakkyo bulbs. *Applied Entomoloy and Zoology* 26, 129–136.

Okabe, K., Nakamura, T. and Gotoh, T. (2000) Preliminary study of mites associated with the Japanese honey bee, *Apis cerana japonica* Radoszkowski (Hymenoptera: Apidae). *Journal of the Acarological Society of Japan* 9(1), 23–30.

Okabe, K, Miyazaki, K and Yamamoto, H. (2001) Population increase in mushroom pest mites on cultivated *Hypsozygus marmoreus* and their vectoring of weed fungi between mushroom cultivation media. *Japanese Journal of Applied Entomology and Zoology* 45, 75–81.

Okamoto, M. (1984) Studies on the environmental factors for the life cycle of *Carpoglyphus lactis*: 1. The effects of relative humidities on individual rearing. *Medical Entomology and Zoology* 35(3), 269–275.

Okamoto, M. (1986) The effects of various temperatures on the life cycle of *Carpoglyphus lactis*. *Japanese Journal of Sanitary Zoology* 37, 221–227.

Oksentiuk, Y.R. (2020) Acarid Mites-Pests of Zhytomyr Polissia Stocks (Species Diversity, Peculiarities of Harmfulness, Methods of its Forecasting, Recommendations for Control and Prevention). Manuscript. Thesis for the Candidate Degree in Biological Sciences: Specialty 03.00.08-Zoology. I. I. Schmalhausen Institute of Zoology of National Academy of Sciences of Ukraine, Kyiv, pp. 1–24.

Oksentiuk, Y.R., Liashevych, A.M. and Lupaina, I.S. (2022) Acarid mites species communities in agricultural and industrial objects of storing and concentration of nutritive substrates in Zhytomyr Polissia. *Zoodiversity* 56(4), 232–330.

Oku, K., Yano, S. and Takafuji, A. (2003) Spider mite's use of a refuge during the quiescent stage in the presence of a predator. *Entomologia Experimentalis et Applicata* 108, 71–74.

Ólafsson, E. and Ingimarsdottir, M. (2009) The land-invertebrate fauna on Surtsey during 2002–2006. *Surtsey Research* 113–128.

Oldfield, G.N. and Michalska, K. (1996) Spermatophore deposition, mating behavior and population mating structure. In: Lindquist, E.E., Sabelis, M.W. and Bruin, J. (eds) *Eriophyoid Mites: Their Biology, Natural Enemies and Control, World Crop Pests, Volume 6*. Elsevier Science, Amsterdam, The Netherlands, pp. 185–198.

Oldfield, G.N., Hobza, R.F. and Wilson, N.S. (1970) Discovery and characterization of spermatophores in the Eriophyoidea (Acari). *Annals of the Entomological Society of America* 63, 520–526.

Oldfield, G.N., Newell, I.M. and Reed, D.K. (1972) Insemination of protogynes of *Aculus cornutus* from spermatophores and description of the sperm cell. *Annals of the Entomological Society of America* 65, 1080–1084.

Oliveira, T.M. (2021) *Bioeconomia Das Infestações de Moscas Sinantrópicas E Ácaros Hematófagos Em Galpão de Granja de Postura Do Estado de Minas Gerais*. Federal University of Minas Gerais, Belo Horizonte, Brazil.

Oliveira, A.J. and Daemon, E. (2003) Qualitative and quantitative assessment of mites (Acari) in domiciliary dust in rural dwellings in the "Zona da Mata" region, Minas Gerais, Brazil. *Revista Brasileira de Zoologia* 20(4). Available at: https://www.scielo.br/scielo.php?script=sci_arttext&pid=S0101-81752003000400020&lng=en&tlng=en (accessed 6 July 2020).

Oliveira, C.M., Navia, D. and Frizzas, M.R. (2007) First record of *Tyrophagus putrescentiae* (Schrank) (Acari: Acaridae) in soybean plants under no tillage in Minas Gerais, Brazil. *Ciência Rural* 37(3), 876–877.

Oliveira, C.R.F., Faroni, L.R.A., Guedes, R.N.C., Goncalves, J.R. and Garcia, F.M. (2007) Biologia de *Acarophenax lacunatus* (Cross and Krantz) (Prostigmata: Acarophenacidae) sobre *Tribolium castaneum* (Herbst) (Coleoptera: Tenebrionidae) e *Cryptolestes ferrugineus* (Stephens) (Coleoptera: Cucujidae). *Neotropical Entomology* 36, 459–464.

Oliveira, D.G.P., Alves, L.F.A. and Sosa-Gomez, D.R. (2014) Advances and perspectives of the use of the entomopathogenic fungi *Beauveria bassiana* and *Metarhizium anisopliae* for the control of arthropod pests in poultry production. *British Poultry Science* 16, 1–12.

Oliveira, D.G.P., Kasburg, C.R. and Alves, L.F.A. (2020) Efficacy of *Beauveria bassiana* against the poultry red mite, *Dermanyssus gallinae* (De Geer, 1778) (Mesostigmata: Dermanyssidae), under laboratory and hen house conditions. *Systematic and Applied Acarology* 25, 895–905.

Oliveira, T.M., Teixeira, C.M., Araújo, I.L., Rezende, L.C., Cunha, L.M., Diniz, S.A. and Silva, M.X. (2020) Epidemiology and risk assessment associated with the presence of hematophagous mites in poultry laying sheds. *Brazilian Journal of Veterinary Research and Animal Science* 72, 2148–2156.

Oliver, J.H. Jr. (1962) A mite parasitic in the cocoons of earthworms. *Journal of Parasitology* 48, 120–123.

Oliver, J.H. Jr. (1971) Parthenogenesis in mites and ticks (Arachnida: Acari). *American Zoologist* 11, 283–299.

Oliver, J.H. Jr. (1977) Cytogenetics of mites and ticks. *Annual Review of Entomology* 22, 407–429.

Olivry, T. and Mueller, R.S. (2017) Critically appraised topic on adverse food reactions of companion animals (3): Prevalence of cutaneous adverse food reactions in dogs and cats. *BMC Veterinary Research* 13, 51. DOI: 10.1186/s12917-017-0973-z.

Olivry, T. and Mueller, R.S. (2019) Critically appraised topic on adverse food reactions of companion animals (8): Storage mites in commercial pet foods. *BMC Veterinary Research* 15, 385.

Olsen, A.R. (1983) Food-contaminating mites from imported foods entering the United States through southern California. *International Journal of Acarology* 9(4), 189–193.

Olsen, A.R. (1988) New record of *Lardoglyphus konoi* (Sasa and Asanuma), on dried lizard carcasses (Acarina: Acaridae). *The Pan-Pacific Entomologist* 64(2), 114.

Olsen, A.R., Sidebottom, T.H. and Knight, S.A. (eds) (1995) *Fundamentals of Microanalytical Entomology: A Practical Guide to Detecting and Identifying Filth in Foods*. CRC Press, Boca Raton, Florida, U.S.A., pp. 1–316.

Olsson, S. and Hage-Hamsten, M. (2000) Allergens from house dust and storage mites: Similarities and differences, with emphasis on the storage mite *Lepidoglyphus destructor*. *Clinical and Experimental Allergy* 30(7), 912–919.

Olzsanowski, Z. (1996) A monograph of the Nothridae and Camisiidae of Poland (Acari: Oribatida: Crotonioidea). *Genus International Journal of Invertebrate Taxonomy*, Polish Taxonomical Society, Wroclaw, Poland. pp. 1–204.

Ong, G.S., Somerville, C.P., Jones, T.W. and Walsh, J.P. (2012) Anaphylaxis triggered by benzyl benzoate in a preparation of depot testosterone undecanoate. *Case Reports in Medicine* 2012, 384054. DOI: 10.1155/2012/384054.

Ordman, D. (1971) The incidence of 'climate asthma' in South Africa, its relation to the distribution of mites. *South African Medical Journal* 45, 739–743.

Oshima, S. (1970) Studies on the mite fauna of the house-dust of Japan and Taiwan with special reference to house-dust allergy. *Medical Entomology and Zoology* 21(1), 1–17.

Oshima, S. (1980) Family acaridae. In: Ehara, S. (ed.) *Illustrations of the Mites and Ticks of Japan*. Zenkoku Noson Kyoiku Kyokai, Tokio, Japan, 380–385.

Osman, S.A.A., Yassin, E.M.A. and Khalil, A.M. (2016) Occurrence of mite species associated with some organic manures at el-menofia governorate. *Egyptian Academic Journal of Biological Sciences* 9(4), 65–70.

Ostovan, H. (1993) Faunistic study of stored product mites in Kazerun and biology of important species. M.Sc. thesis, Tarbiat Modarres University, Tehran, Iran, pp. 1–172.

Ostovan, H. (1997) Mites associated with elm bark beetle (Col: Scolytidae) and biocenotic aspects of *Pyemotes scolyti* (Oud.) (Acari: Pyemotidae) as biocontrol agent for *Scolytus multistriatus* (Marsham). Ph. D. dissertation, Islamic Azad University, Tehran, Iran, pp. 1–130.

Ostovan, H. and Kamali, K. (1994) Mites of family Ascidae (Acari: Mesostigmata) associated with stored products in Kazerun, South of Iran. *Journal of Entomological Society of Iran* 14, 9–18.

Ostovan, H. and Kamali, K. (1995a) Some snout mites (Acari: Bdellidae) from Iran and a key for their identification. *Journal of Agricultural Science* 1, 29–43.

Ostovan, H. and Kamali, K. (1995b) New records of six species of astigmatic mites (Acari: Astigmata) infesting stored products in Iran. *Journal of Agricultural Science* 1(2), 53–66.

Ostovan, H. and Kamali, K. (1996) Biology of two bulb mites: *Rhizoglyphus echinopus* (Fumouze and Robin) and *R. robini* Claparede under two different laboratory conditions. *Journal of Agricultural Science* 2(5/6), 47–59.

Ostovan, H. and Mossadegh, M.S. (1999) Mites associated with honey bee (*Apis mellifera* L.) hives in Iran. In: *Proceedings of the 36th International Congress of Apiculture (Apimondia), 12–17 September*. Vancouver, Canada, p. 254.

Otto, J.C. (1999) Four new species of *Agaue* (Acarina: Halacaridae) from the Great Barrier Reef Marine Prk. *Cahiers de Biologie Marine* 40, 273–281.

Ottoboni, F. (1992) House dust mites prevention in Italy. *Bollettino di Zoologia agraria e di Bachicoltura* 24, 113–120.

Ottoboni, F., Falagiani, P. and Centanni, S. (1983) *Gli Acari Allergenici*. B.I. Sieroter, Milan. pp. 389–419.

Ottoboni, F., Cantoni, A. and Lozzia, G.C. (1984–1985) L'ipopia di *Glycyphagus privatus* Oudemans 1903 (Acarina: Glycyphagidae). *Bollettino di Zoologia Agraria e Bachicoltura* 18, 167–175.

Ottoboni, F., Mazzuccato, S., Rigamonti, I.E. and Lozzia, G.C. (1993) Mites infesting flour and malt (Acari infestanti farine e malto). *Tecnica Molitoria* 44(8), 673–676.

Oudemans, A.C. (1901) Notes on Acari. Third series. *Nederlandsch Tijdschrift voor de Dierkunde* VII(2), 50–88.

Oudemans, A.C. (1902a) Acarologische Aanteekeningen. *Entomologische Berichte* 7, 42–45.

Oudemans, A.C. (1902b) New list of Dutch Acari. Second part. *Tijdschrift voor Entomologie* 45, 1–52.

Oudemans, A.C. (1903a) Notes on Acari. *Tijdschrift voor Entomologie* l, 1–24.

Oudemans, A.S. (1903b) Acarologische aanteekeningen. VI. *Entomologische Berichte* I, 83–88.

Oudemans, A.C. (1904) Entomologische Berichten. *Tijdschrift voor Entomologie* XLVII, 129, 8.

Oudemans, A.C. (1905a) *Aleurobius farris* nov. sp., acarologische 454 aanteekeningen XX. *Entomologische Berichte* 2(26), 20.

Oudemans, A.C. (1905b) Acarologische Aanteekeningen XV. *Entomologische Berichte* 21, 207–210.

Oudemans, A.C. (1905c) Notes on Acari, 12th series (Parasitidae, Oribatidae, Tyroglyphidae). *Nederlandsch Tijdschrift voor de Dierkunde* 8, 202–239.

Oudemans, A.C. (1905d) Acariden von Borkum und Wangeroog (Parasitidae, Thrombidiidae, Oribatidae). IX Serie der "Notes on Acari".

Abhandlungen herausgegeben vom Naturwissenschaftlichen Verein zu Bremen 18, 77–98.

Oudemans, A.C. (1905e) Acarologische Aanteekeningen. XIX. *Entomol. ber.* 2 (25), 4–12.

Oudemans, A.C. (1906) Acarologische aanteekeningen XXV. *Entomologische Berichte* 2, 121–123.

Oudemans, A.C. (1910) Acarologische aanteekeningen XXXII. *Entomologische Berichte* III 53, 67–74.

Oudemans, A.C. (1914a) Beschrijving van een Weinig Bekende en Drie Nieuwe Soorten van *Anoetus*. *Tijdschrift voor Entomologie* 57, 107–119.

Oudemans, A.C. (1914b) *Glycyphagus tibodas* Oudemans. *Entomologische Berichte* 4, 102–103.

Oudemans, A.C. (1916) Myrmekofile Acari uit Salatiga. *Entomol. ber.* 4(88), 266–268.

Oudemans, A.C. (1917) Acarologische Aanteekeningen LXII. *Entomol. ber.* 4(93), 241–260.

Oudemans, A.C. (1924a) Nieuwste onderzoekingen in de groep der Tyroglyphidae, vulgo kaasmijten. *Tijdschrift voor Entomologie* 67, XXII–XXXIX.

Oudemans, A.C. (1924b) Acarologische aanteekeningen LXXVII. *Entomologische Berichte* 140(6), 317–336.

Oudemans, A.C. (1924c) Acarologische aanteekeningen. LXXVI. *Entomologische Berichte* 139(VI), 300–310.

Oudemans, A.C. (1924d) Acarologische aanteekeningen. LXXVII. *Entomologische Berichte* 136(VI), 317–336.

Oudemans, A.C. (1924e) Acarologische aanteekeningen. LXXVII. *Entomologische Berichte* 140(VI), 319–320.

Oudemans, A.C. (1925) Acarologische aanteekeningen. LXXIX. *Entomologische Berichte* 146(VII), 26–34.

Oudemans, A.C. (1929a) Acarologische aanteekeningen. 100. *Entomologische Berichte* 8(170), 28–36.

Oudemans, A.C. (1929b) Acarologische aanteekeningen. XCIX. *Entomologische Berichte* 8(169), 11–20.

Oudemans, A.C. (1930a) Acarologische aanteekeningen. CIV. *Entomologische Berichte* 8(175), 135–140.

Oudemans, A.C. (1930b) Acarologische aanteekeningen. CII. *Entomologische Berichte* 8(175), 69–74.

Oudemans, A.C. (1932) Acarologische aanteekeningen. CXII. *Entomologische Berichte* 8(183), 350–364.

Oudemans, A.C. (1936) Neues über *Pediculoides* Targ. Tozz. 1878. In: E. Strand (ed.) *Festschrift zum 60. Geburtstage von Professor Dr. Embrik Strand. Ordinarius für Zoologie und Direktor des Systematisch-Zoologischen Instituts und der Hydrobiologischen Station der Universität Lettlands, Riga; Dr. rer. nat. h. c., M.A.N., F.R.E.S., F.L.S., F.Z.S., etc. Vol. 1.* Izdevniecība “Latvija”, Riga, pp. 391–404.

Oudemans, A.C. (1937) Verslag. Tijdschrift voor Entomologie 80, IV–XVI.

Oudemans, A.C. (1938) Nieuwe vondsten op het gebied der Systematiek en der nomencaltuur der Acari. *Tijdschrift voor Entomologie* 81, II–X.

Oudemans, A.S. (1904) Acarologische aanteekeningen. XI. *Tijdschrift voor Entomologie* 1, 153–155.

Overmeer, W.P.J. (1972) Notes on mating behaviour and sex ratio control of *Tetranychus urticae* Koch (Acarina: Tetranychidae). *Entomologische Berichte* 32, 240–244.

Overmeer, W.P.J. and Harrison, R.A. (1969) Notes on the control of the sex ratio in populations of the two-spotted spider mite, *Tetranychus urticae* Koch (Acarina: Tetranychidae). *New Zealand Journal of Zoology* 12, 920–928.

Özar, A.Ġ., Önder, P., Sarıbay, A. Özkut, Ş., Gündoğdu, M. *et al.* (1986) Investigations on improving the measures for controlling pests and diseases occurring on figs in the Aegean Region. *Turkish Journal of Agriculture and Forestry* 10, 263–277.

Özbek, H.H. (2017) A review of the macrochelid mites of Turkey (Acari: Macrochelidae), with new records and descriptions of three new species. *Zootaxa* 4317, 559–572.

Özbek, H.H., Bal, D.A. and Doğan, S. (2015) The genus *Macrocheles* Latreille (Acari: Mesostigmata: Macrochelidae) from Kelkit Valley (Turkey), with three newly recorded mite species. *Turkish Journal of Zoology* 39, 1–13. DOI: 10.3906/zoo-1409-14.

Özer, M., Toros, S., Çobanoğlu, S. and Çınarlı, S. (1986) Beneficial mites of stored products in İzmir province. *2-14 Subat 1986, Adana, Turkie*, pp. 280–292.

Özer, M., Toros, S., Çobanoğlu, S., Çınarlı, S. and Emekçi, M. (1989) The description, distribution and habitats of Acarina species harmfull to stored grains and grain products and dried fruits in Izmir Province. *Turkish Journal of Agriculture and Forestry* 13, 1154–1189.

Özkan, M., Ayyıldız, N. and Soysal, Z. (1988) The Acari fauna of Turkey. *Turk Zooloji Dergisi* 12, 75–85.

Özman, S.K. and Žďárková, E. (2000) Mites of stored hazelnuts in Turkey. *Proceedings of the XXI International Congress of Entomology, 20–26 August 2000*. Abstract Book II, Foz Do Iguassu, Brazil, p. 1034.

Özman, S.K. and Çobanoğlu, S. (2001) Current status of hazelnut mites in Turkey. *Acta Horticulturae* 556, 479–487.

Packard, A.S. (1878) *A Guide to the Study of Insects*. Hemy Holt and Co., New York, U.S.A., pp. 1–715.

Pagani, M. (1987) Esperimenti di mezzi fisici di lotta contro gli acari dei salumi stagionati. In: Domenichini, G. (ed.) *Atti del 4° Simposio Sulla Difesa Antiparassitaria Nelle Industrie Alimentari e la Protezione degli Alimenti*, Tipolitografia, Piacenza, Italy, pp. 255–265.

Pagani, M. and Ciampitti, M. (1991) Indagine sull'acarofauna in salamifici del Piacentino. *Atti del XVI Congresso Nazionale Italiano di Entomologia*. Bari Martina Franca (Ta), Italy, pp. 149–154.

Pagani, M. and Zanotti, M.R. (1994) Associazione di muffe e acari su formaggi a breve stagionatura. *Entomologia, 13-18 June 1994, Arti Grafiche Friulane*, Udine, Italy, pp. 671–676.

Pageat, P. (2017) Allomone Repulsive and Kairomone Attractive Compositions for Controlling Arachnds. U.S. Patent 9, 844, 217, 19 December 2017.

Pagliarini, N. (1979) Studies on the mites of stored cereals in Yugoslavia. In: Rodriguez J.G. (ed.) *Recent Advances in Acarology*. Academic Press, New York, U.S.A., pp. 305–309.

Pagliarini, N. and Hrlec, G. (1982) Results of studies on the effectiveness of some pesticides for the control of stored-product mites. *Zastita Bilja* 33, 51–56.

Paik, Y., Takaoka, M., Matsuoka, H. and Ishii, A. (1992) Mite fauna and mite antigen in house dust from houses in Seoul, Korea. *Japanese Journal of Sanitary Zoology* 43(1), 29–35.

Pajak, C. and Solarz, K. (2013) Allergenic mites from selected farming environments in the south Poland. In: Buczek, A. and Błaszak, C. (eds) *Arthropods, The Medical and Veterinary Aspects*. PAN, Koliber, Lublin, Poland, pp. 49–59.

Palevsky, E., Soroker, V., Weintraub, P., Mansour, F., Abo-Moch, F. and Gerson, U. (2001) How species-specific is the phoretic relationship between the broad mite, *Polyphagotarsonemus latus* (Acari: Tarsonemidae), and its insect hosts? *Experimental and Applied Acarology* 25, 217–224.

Palevsky, E., Konopická, J., Rueda-Ramírez, D. and Zemek, R. (2022) A review of prospective biocontrol agents and sustainable soil practices for bulb mite (Acari: Acaridae) management. *Agronomy* 12, 1491. DOI: 10.3390/agronomy12071491.

Palin, S.L., Winstanley, R., Mclntyre, D.A. and Edwards, R.E. (1993) Energy implications of domestic ventilation strategy. *Proceedings of the 14th AIVC Conference: Energy Impact of Ventilation and Air Infiltration, 21–23 September 1993*. Air Filtration and Ventilation Centre, Copenhagen, Denmark. Coventry, United Kingdom, pp. 141–148.

Palyvos, N.E., and Emmanouel, N.G. (2004) A preliminary study on the predation by *Cheyletus malaccensis* Oudemans Acarina: Cheyletidae on *Tyrophagus putrescentiae* Schrank Acarina: Acaridae and eggs of *Ephestia kuehniella* Zeller Lepidoptera: Pyralidae. In: *Proceedings of 4th Meeting of COST Action 842: WG4: Biocontrol of Arthropod Pests in Stored Products, 24–25 May 2004*. Athens, Greece, pp. 46–47.

Palyvos, N.E. and Emmanouel, N.G. (2005) A study of the oviposition of the predator mite *Cheyletus malaccensis* Oudemans (Acari: Cheyletidae), at three constant temperatures. *Phytophaga* 14, 727–732.

Palyvos, N.E. and Emmanouel, N.G. (2009) Temperature dependant development of the predatory mite *Cheyletus malaccensis* (Acari: Cheyletidae). *Experimental and Applied Acarology* 47, 147–158.

Palyvos, N.E. and Emmanouel, N.G. (2011) Reproduction, survival, and life table parameters of the predatory mite *Cheyletus malaccensis* (Acari: Cheyletidae) at various constant temperatures. *Experimental and Applied Acarology* 54, 139–150.

Palyvos, N.E., Emmanouel, N.G. and Saitanis, C.J. (2008) Mites associated with stored products in Greece. *Experimental and Applied Acarology* 44, 213–226.

Pan, J.-W. (1985) Discovery of *Blattisocius dentriticus* and *Blattisocius keegani* in salted meat in stored houses. *Entomological Knowledge* 6, 263–264.

Pan, Z.-W. and Teng, K.-F. (1980) *Economic Insect Fauna of China Fasc. 17 Acarina: Gamasina*. Science Press Ltd., Beijing, pp. 1–155.

Pande, Y.D. and Berthet, P. (1973) Studies on the food and feeding habits of soil oribatei in a black pine plantation. *Oecologia* 12(4), 413–426.

Pankiewicz-Nowicka, D., Boczek, J. and Davis, R. (1984) Food selecion in *Tyrophagus putrescentiae* (Schrank) (Acaria: Acarididae). *Journal of the Georgia Entomological Society* 19(3), 317–321.

Pankiewicz-Nowicka, D., Boczek, J. and Davis, R. (1986) Attraction by selected organic compounds to *Tyrophagus putrescentiae* (Schrank) (Acaria: Acarididae). *Annals of the Entomological Society of America* 79, 293–299.

Paredes-León, R. (2013) *Sistemática de la familia Pterygosomatidae (Acari: Prostigmata).* Tesis de Doctorado. Universidad Nacional Autónoma de México, Mexico City, pp. 1–127.

Paredes-León, R., Klompen, H. and Pérez, T.M. (2012) Systematic revision of the genera *Geckobiella* Hirst, 1917 and *Hirstiella* Berlese, 1920 (Acari: Prostigmata: Pterygosomatidae) with description of a new genus for American species parasites on geckos formerly placed in *Hirstiella*. *Zootaxa* 3510, 1–40.

Paredes-León, R., Cuervo-Pineda, N.and Perez, T.M. (2013) Pterygosomatid mites from Cuba, with the description of a new species of *Bertrandiella* (Acari: Prostigmata: Pterygosomatidae). *Revista Mexicana de Biodiversidad* 84(4), 1142–1152.

Park, J.W., Ko, S.H., Yong, T.S., Ree, H.I., Jeoung, B.J. and Hong, C.S. (1999) Cross-reactivity of *Tyrophagus putrescentiae* with Dermatophagoides farinae and *Dermatophagoides pteronyssinus* in urban areas. *Annals of Allergy, Asthma and Immunology* 83(6 Pt 1), 533–539.

Park, H.S., Jee, Y.K., Kim, Y.K., Lee, S.K., Lee, M.H. and Kim, Y.Y. (2002) Identification of immunoglobulin E binding components of the two-spotted spider mite *Tetranychus urticae*: allergenic relationships with the citrus red mite and house-dust mite. *Allergy and Asthma Proceedings* 23(3), 199–204.

Park, J.-H., Lee, N.-H., Yang, Y.-C. and Lee, H.-S. (2017) Food protective effects of 3-methylbenzaldehyde derived from *Myosotis arvensis* and its analogues against *Tyrophagus putrescentiae*. *Scientific Reports* 7, 1–10.

Park, S.E., Lee, M.R., Lee, S.J., Kim, J.C., Parker, B.L. *et al.* (2021a) Strategic positioning of *Beauveria bassiana* sensu lato JEF-410 in management of poultry red mite, *Dermanyssus gallinae* (Mesostigmata: Dermanyssidae). *Biocontrol* 1(2022). DOI: 10.1007/s10526-021-10110-w.

Park, J., Mostafiz, M.M., Hwang, H.-S., Jung, D.-O. and Lee, K.-Y. (2021b) Comparing the life table and population projection of *Gaeolaelaps aculeifer* and *Stratiolaelaps scimitus* (Acari: Laelapidae) Based on the age-stage, two-sex life table theory. *Agron* 11, 1062. DOI: 10.3390/agronomy11061062.

Parker, R. and Gerson, U. (1994) Dispersal of the broad mite, *Polyphagotarsonemus latus* (Banks) (Heterostigmata: Tarsonemidae), by the greenhouse whitefly, *Trialeurodes vaporariorum* (Westwood) (Homoptera: Aleyrodidae). *Experimental and Applied Acarology* 18, 581–585.

Parkinson, C.L. (1990) Population increase and damage by three species of mites on wheat at 20°C and two humidities. *Experimental and Applied Acarology* 8, 179–193.

Parkinson, C.L., Jamieson, N., Eborall, J. and Armitage, D.M. (1991) Comparison of the fecundity of three species of grain store mites on fungal diets. *Experimental and Applied Acarology* 12(3), 297–302.

Parminder Singh Dehar and Mohd Yousuf Paray (2020) A report on the occurrence, abundance and distribution of cheese Mite, *Tyrolichus casei* Oudemans, 1910 (Acari: Acaridae) from some stored food products in Punjab. *Research Review International Journal of Multidisciplinary* 5, 40–42.

Parveen, R. and Gupta, S. K. (2019) Some new records of mites occurring in mushroom in South Bengal. *International Journal of. Zoology Studies* 4(5), 8–12.

Parveen, R. and Gupta, S.K. (2020a) Diversity of mites (ACARI) on wild mushrooms from West Bengal. *International Journal of Agriculture and Plant Science* 2(1), 1–7.

Parveen, R. and Gupta, S.K. (2020b) Bio-Efficacy of some green pesticides under laboratory condition against *Suidasia nesbitti* Sasa (Acari. Suidasiidae) infesting milky white mushroom, *Calocybe indica* Purkayastha and Chandra. *International Journal of. Zoology Studies* 5(1), 4–5.

Parveen, R. and Gupta, S. K. (2020c) A compendium on mushroom mites in India. *Bionotes* 22 (1), 11–20.

Patil, M.P., Niphadkar, P.V. and Bapat, M.M. (2001) *Psocoptera* spp. (book louse): a new major household allergen in Mumbai. *Annals of Allergy, Asthma and Immunology* 87, 151–155.

Pavlovic, I., Pavlicević, A. and Vasic, A. (2016) House flies *Musca domestica* as transmitter of red poultry mite *Dermanyssus gallinae*-Case report. *Agricultural Science Research Journals* 6(5), 117–119.

Pavlićević, A., Pavlović, I. and Kulić, S. (2021) Farmers' economic interest in *Dermanyssus gallinae* control. *Biotechnology in Animal Husbandry* 37, 171–182.

Pawar, S., Jogdand, S., Jadhav, M. and Deokar, T. (2016) Effect of environment on seasonal dynamics of rat house mites at Pune. *International Journal of Entomology Research* 1(4), 26–32.

Pearson, R.S.B. and Cunnington, A.M. (1973) The importance of mites in house dust sensitivity in Barbadian asthmatics. *Clinical Allergy* 3, 299–306.

Pekar, S. and Hubert, J. (2008) Assessing biological control of *Acarus siro* by *Cheyletus malaccensis* under laboratory conditions: Effect of temperatures and prey density. *Journal of Stored Products Research* 44(4), 335–340.

Pekar S. and Žďárková, E. (2004) A model of the biological control of *Acarus siro* by *Cheyletus eruditus* (Acari: Acaridae, Cheyletidae) on grain. *Journal of Pest Science* 77(1), 1–10.

Peñalver, E., Arillo, A., Delclòs, X., Peris, D., Grimaldi, D.A. *et al.* (2017) Ticks parasitised feathered dinosaurs as revealed by Cretaceous amber assemblages. *Nature Communications* 8, 1924. DOI: 10.1038/s41467-017-01550-z.

Penman, D.R. and Cone, W.W. (1972) Behaviour of male two-spotted spider mites in response to quiescent female deutonymphs and to web. *Annals of the Entomological Society of America* 65, 1289–1293.

Penman, D.R. and Cone, W.W. (1974) Role of web, tactile stimuli and female sex pheromone in attraction of male two-spotted spider mites to quiescent female deutonymphs. *Annals of the Entomological Society of America* 67, 179–182.

Pennanen, S.M. and Harju, A.T. (2003) Mites in facilities for laboratory animals. *Scandinavian Journal of Work, Environment and Health* 29, 314–316.

Pereira, C. and Castro, M.P. (1945) Contribuição para o conhecimento de espécie tipo de *Macrocheles* Latr. (Acarina): *M. muscaedomesticae* (Scopoli, 1772) emend. *Arquivos do Instituto Biológico (Sao Paulo)* 16, 153–189.

Pereira, C. and Castro, M.P. (1947) Forese e partenogenese arrenótoca em *Macrocheles muscaedomesticae* (Scopoli) (Acarina: Macrochelidae) a sua significacao ecólogica. *Arquivos do Instituto Biológico (Sao Paulo)* 18, 71–89.

Pereira, F.L., Silva, D.A., Sopelete, M.C., Sung, S.S. and Taketomi, E.A. (2004) Mite and cat allergen exposure in Brazilian public transport vehicles. *Annals of Allergy, Asthma and Immunology* 93(2), 179–184.

Pérez Lozano, A. (1979) Environmental control in asthmatic homes. The role of cheylatus mites. Preliminary report. *Allergologia et Immunopath ologia* VII, 303–306.

Pérez-Rodríguez, J., Calvo, J., Urbaneja, A. and Tena, A. (2018) The soil mite *Gaeolaelaps (Hypoaspis) aculeifer* (Canestrini) (Acari: Laelapidae) as a predator of the invasive citrus mealybug *Delottococcus aberiae* (De Lotto) (Hemiptera: Pseudococcidae): Implications for biological control. *Biological Control* 127, 64–69.

Pernek, M., Wirth, S. Blomquist, S., Avtzis, D. and Moser, J. (2012) New associations of phoretic mites on *Pityokteines curvidens* (Coleoptera, Curculionidae, Scolytinae). *Open Life Sciences* 7(1), 63–68.

Perotti, M.A. (1998) Moscas sinantrópicas (Diptera: Muscidae y Fanniidae) en producciones avícolas del sud este bonaerense. *Natura Neotropical* 29, 145–154.

Perotti, M.A. (1999) Localization of distant preys by *Glyptholaspis confusa* Foa (Acari: Macrochelidae). *Revista de la Sociedad Entomológica Argentina* 58, 106–108.

Perotti, M.A. (2001) Prey location and predation rates of predatory mites (Acari: Macrochelidae) on immature stages of pest flies (Diptera: Muscidae). *Systematic and Applied Acarology* 6, 27–34.

Perotti, M.A. (2009) Preface. The literature on forensic acarology. *Experimental and Applied Acarology* 49(1–2), 1–2.

Perotti, M.A. and Braig, H.R. (2009) Phoretic mites associated with animal and human decomposition. *Experimental and Applied Acarology* 49, 85–124.

Perotti, M.A. and Brasesco, M.J.A. (1996) Especificidad forética de *Macrocheles muscaedomesticae* (Acari: Macrochelidae). *Austral Ecology* 6, 3–8.

Perotti, M.A. and Brasesco, M.J.A. (1997) Orientación química de *Macrocheles muscaedomesticae* Scopoli (Acari: Macrochelidae) y percepción a distancia de posturas de *Musca domestica* (Diptera: Muscidae) y *Calliphora vicina* (Diptera: Calliphoridae). *Revista de la Sociedad Entomológica Argentina* 56, 67–70.

Perotti, M.A., Goff, M.L., Baker, S.A., Turner, B.D. and Braig, H.R. (2009) Forensic acarology: an introduction. *Experimental and Applied Acarology* 49, 3–13.

Perron, R. (1954) Untersuchungen über Bau, Entwicklung und Physiologie der Milbe *Histiostoma laboratorium* Hughes. *Acta Zoologica* 35, 71–176.

Perrucci, S. (1995) Acaricidal activity of some essential oils and their constituents against *Tyrophagus longior*, a mite of stored food. *Journal of Food Protection* 58, 560–563.

Petanović, R. (2001) Stetne grinje u proizvodnji evetnih kultura u yatvorenom postoru (4): Acaride (Acaridae). *Biljni Lekar* 29, 155–159 (in Serbian).

Petersen, H. (1982) Structure and size of soil animal populations. *Oikos* 39, 306–329.

Petrova, A.D. (1977) Family parholaspidae. In: Ghilarov, M.S. and Bregetovae, N.G. (eds) *Handbook for the Identification of Soil Inhabiting Mites.* Mesostigmata, Nauka, Leningrad Inc., pp. 315–346.

Petrova, A.D. and Koshchanova, R.E. (1986) *Ameroseius gilarovi* sp.n. (Mesostigmata, Ameroseiidae) – novyi vid gamazovykh kleshchei s polivnykh zemelj Uzbekistana. *Nauchnye Doklady Vysshei Shkoly. Biologicheskie Nauki* 12, 30–33.

Petrunov, B., Todorov, D. and Donchev, P. (1977) Sensitization to mites in house dust among children suffering bronchial asthma. *Zhurnal Mikrobiologii, Epidemiologii, i Immunobiologii* 6, 104–108.

Pfliegler, W.P., Schönhofer, A., Niedbała, W., Vella, P., Sciberras, A. and Vella, A. (2017) New records of mites (Acari) and harvestmen (Opiliones) from Malta with a preliminary checklist of Maltese Arachnida. *Soil Organic* 89(2), 85–110.

Philips, J.R. (2009) The mite (Acarina) fauna of trogid beetles (Coleoptera: Trogidae). *International Journal of Acarology* 35, 1–17.

Philips, J.R. and Norton, R.A. (1978) *Lardoglyphus falconidus* n.sp. (Acarina: Acaridae) from the nest of an American kestrel (*Falco sparverius* L.). *Acarologia* 20, 128–137.

Phillips, T.W. and Schilling, M.W. (2013) Fumigation alternatives for the ham mite, *Tyrophagus putrescentiae.* In: *Proceedings, Annual International Research 22 Conference on Methyl Bromide Alternatives and Emission Reduction, 4–6 November 2013.* San Diego, California, U.S.A., p. 38.

Phillips, T.W. and Throne, J.E. (2010) Biorational approaches to managing stored-product insects. *Annual Review of Entomology* 55, 375–397.

Phillips, T.W., Hassan, Md.M., Aikins, M.J. and Schilling, M.W. (2008) Efficacy of sulfuryl fluoride to control ham mites and red-legged ham beetles. In: *Proceedings, Annual International Research Conference on Methyl Bromide Alternatives and Emissions Reductions, 11–14 November 2008.* Orlando, Florida, U.S.A., pp. 89–1 to 89–2.

Phillipsen, W.J. and Coppel, H.C. (1977a) *Acotyledon formosani* sp.n. associated with the Fonnosan subterranean termite, *Coptotermes formosanus* Shiraki (Acarina: Acaridae-Isoptera: Rhinotermitidae). *Journal of the Kansas Entomological Society* 50, 399–409.

Phillipsen, W.J. and Coppel, H.C. (1977b) *Histiostoma formosana* sp.n. associated with the Formosan subterranean tennite, *Coptotermes formosanus* Shiraki (Acarina: Acaridae-Isoptera: Rhinotermitidae). *Journal of the Kansas Entomological Society* 50, 496–502.

Piao, X. (1990) Studies on the mites infesting herb medicine in China and Japan and their derivation. *Japan Society of Medical Entomology and Zoology* 41(1), 1–7.

Pielou, D.P. and Verma, A.N. (1968) The arthropod fauna associated with the birch bracket fungus, *Polyporus betulinus*, in eastern Canada. *Canadian Entomologist* 100, 1179–1199.

Pike, A.J. and Wickens, K. (2008) The house dust mite and storage mite fauna of New Zealand dwellings. *New Zealand Entomologist* 31, 17–22.

Pike, A.J., Cunningham, M.J. and Lester, P.J. (2005) Development of *Dermatophagoides pteronyssinus* (Acari: Pyroglyphidae) at Constant and Simultaneously Fluctuating Temperature and Humidity Conditions. *Journal of Medical Entomology* 42(3), 266–269.

Pilkington, L.J., Messelink, G., van Lenteren, J.C. and Le Mottee, K. (2010) "Protected biological control" – biological pest management in the greenhouse industry. *Biological Control* 52, 216–220.

Pillai, P.R.P. (1957) Pests of stored fish and prawns. *Bulletin Central Research Institute Trivandrum*, V. III ser. c, l–73.

Pillai, P.R.P. and Winston, P.W. (1969) Life history and biology of *Caloglyphus anomalus* Nesbitt (Acarina: Acaridae). *Acarologia* 2, 295–303.

Pillers, A.W. N. (1920) *Glycyphagus domesticus* De Geer, an accidental parasite in the ear of the domesticated rabbit. *The Veterinary Journal* 76(4), 126–128.

Pimental, D., Rumsey, M., W., and Streams, F. A. (1960). Rearing tyroglyphid mites on *Neurospora*. *Annals of the Entomological Society of America* 53, 549.

Pinhão, R.C. and Grácio, A.J.S. (1978) House dust mites in Lisbon – a short notice. *Anais do Instituto de Medicina Tropical* 5, 361.

Pinto da-Rocha, R. (1994) Invertebrados cavernicolas da porção meridional da Provincia Espeleologica do Vale do Ribeira, sul do Brasil. *Revista Brasileira de Zoologia* 10(2), 229–255.

Piotrowska, K. (2009) The effects of pasture plant species composition and diversity on earthworm, mite and soil microbial communities. Ph. D. Thesis, University College Dublin, Dublin, pp. 1–267.

Pitariu, T., Dinulescu, N., Panaitescu, D. and Silard, R. (1978) Cholangiocholecystitis, an acute attack with acarids in B bile. *Revista de Igiena, Bacteriologie, Virusologie, Parazitologie, Epidemiologie, Pneumoftiziologie. Bacteriologia, Virusologia, Parazitologia, Epidemiologia* 23, 189–192.

Pitariu, T., Popescu, I.G. and Banescu, O. (1979) Acarids of pathological significance in urine. *Revista de Igiena, Bacteriologie, Virusologie, Parazitologie, Epidemiologie, pneumoftiziologie. Bacteriologia, Virusologia, Parazitologia, Epidemiologia* 24, 55–59.

Piu, G. and Ballero, M. (1989) Pathology caused by acari. *Giornale Italiano di Medicina del Lavoro ed Ergonomia* 11(3–4), 193–200.

Platts-Mills, T.A.E. and de Weck, A.L. (1989) Dust mite allergens and asthma – a worldwide problem. Report of an International Workshop, Bad Kreuznach, Federal Republic of Germany, September, 1987. *Journal of Allergy and Clinical Immunology* 83, 416–427.

Platts-Mills, T.A.E., Tovey, E.R., Mitchell, E.B., Moszoro, H., Nock, P. and Wilkins, S.R. (1982) Reduction of bronchial hyperactivity during prolonged allergen avoidance. *Lancet* 2, 675–678.

Platts-Mills, T.A.E., Vervloet, D., Thomas, W.R., Aalberse, R.C. and Chapman, M.D. (1997) Indoor allergens and asthma: report of the third international workshop. *Journal of Allergy and Clinical Immunology* 100(Suppl), 1–24.

Podder, S., Biswas, H., Gupta, S.K. and Kumar Saha, G. (2009) Life-cycle of house dust mite *Dermatophagoides pteronyssinus* (Acari: Pyrogylphidae) under laborator y conditions in Kolkata Metropolis. *Acarina* 17(2), 239–242.

Podder, S., Biswas, H. and Kumar Saha, G. (2021) A faunistic survey of house dust mites of Kolkata, West Bengal, India. *Acarological Studies* 3(1), 16–21.

Poe, S.L. (1971) Microfauna populations on gladiolus corms. *Florida Entomologist* 54, 127–133.

Poe, S.L., Noble, W.E. and Stall, R.E. (1979) Acquisition and retention of *Pseudomonas marginata* by *Anoetus feroniarum* and *Rhizoglyphus robini*. In: Rodriguez, J.G. (ed.) *Recent Advances in Acarology, Volume I*. Academic Press, New York, U.S.A., pp. 119–124.

Poinar, G.O. Jr. (1992) *Life in Amber*. Stanford University Press, Redwood City, California, U.S.A., pp. 1–368.

Poinar, G.O. Jr. (1995) First fossil soft ticks, *Ornithodoros antiquus* n. sp. (Acari: Argasidae) in Dominican amber with evidence of their mammalian host. *Experientia* 51, 384–387.

Poinar, G.O. Jr and Brown, A.E. (2003) A new genus of hard ticks in Cretaceous Burmese amber (Acari: Ixodida: Ixodidae). *Systematic Parasitology* 54, 199–205.

Poinar, G.O. Jr, Pike, E.T.M. and Krantz, G.W. (1993) Animal-animal parasitism. *Science* 361, 307–308.

Poinar, G.O. Jr, Krantz, G.W., Boucout, A.J. and Pike, T.M. (1997) A unique Mesozoic parasitic association. *Science Nature* 84, 321–322.

Polezhaev, W.G. (1938) Influence de la faim sur la formation des hypopes enkystes chez *Glycyphagus destructor* (Schrk.). *Journal of Zoological (Moscow)* 17, 617–621.

Polezhaev, W.G. (1940) Effect of atmospheric humidity and temperature on the appearance of hypopial stage in *Glycyphagus destructor* Schrk. and *Tyroglyphus farinae* L. *Uchenya Zapiski Mask, Gosud University Zoologia* 42, 185–196.

Pons, M.J., Castañé, C. and Riudavet, J. (2008) Combination of biological control and CO_2 treatments against *Plodia interpunctella* (Hübner). *Bulletin IOBC/WPRS* 40, 283–287.

Pope, A.M., Patterson, R. and Burge, H. (eds) (1993) *Indoor Allergens. Assessing and Controlling Adverse Health Effects*. National Academy Press, Washington, U.S.A., pp. 1–320.

Popescu, I.G. and Banescu, O. (1975) Presence of *Dermatophagoides pteronyssinus* in the houses of asthmatics sensitised to house dust. A one-year study. *Revue Roumaine de Medecine, Medecine Internet* 13, 293–295.

Porshakov, A.M., Yakovlev, S.A. and Kurnyaeva, A.D. (2017) Gamasid mites of small mammals in the semi-desert territories of the Saratov Trans-Volga region. *Parazitologiia* 51(2), 132–142.

Porto, B.C.Z., Couto, J.C.M., Stainki, D.R. and Monteiro, S.G. (2017) Dermatitis by *Cheyletus eruditus* in man. *Veterinary Parasitology: Regional Studies and Reports* 9, 63–64.

Portus, M. and Gomez, M.S. (1979) *Thyreophagus gallegoi* a new mite from flour and house dust in Spain (Acaridae, Sarcoptiformes). *Acarologia* 21, 477–481.

Posthumus, J. and Borish, L. (2012) A 71-year-old man with anaphylaxis after eating grits. *Allergy and Asthma Proceedings* 33(1), 110–113.

Potts, M.F. and Rodriguez J.G. (1978) Effects of spice oils on *Tyrophagus putrescentiae* (Schrank). *Proceeding of the 57th Annual Conference of Entomology Social of America North Central Branch* 33, 27–28.

Pound, J.M. and Oliver, J.H. Jr. (1976) Reproductive morphology and spermatogenesis in *Dermanysuss gallinae* (DeGeer) (Acari: Dermanyssidae). *J. Morphol.* 150(4), 825–842.

Prasad, V. (1968) Three moth mites (Acarina: Mesostigmata) from Hawaii, with description of a new species. *Annals of the Entomological Society of America* 61(1), 129–132.

Prasse, J. (1968) Untersuchungen über Oogenese, Befruchtung, Eifurchung und Spermatogenese bei *Caloglyphus berlesei* Michael 1903 und *C. michaeli* Oudemans 1924 (Acari, Acaridae). *Biologisches Zentralblatt* 87, 757–775.

Premachandra, W.T.S.D., Borgemeister, C., Berndt, O., Ehlers, R.U. and Poehling, H.M. (2003) Combined releases of entomopathogenic nematodes and the predatory mite *Hypoaspis aculeifer* to control soil-dwelling stages of western flower thrips *Frankliniella occidentalis*. *Biocontrol* 48, 529–541.

Price, L.A. and Mills, K.A. (1988) The toxicity of phosphine to the immature stages of resistant and susceptible strains of some common stored-product beetles, and implications for their control. *Journal of Stored Products Research* 24, 51–59.

Price, D.R.G., Küster, T., Øines, Ø., Oliver, E.M., Bartley, K. *et al.* (2019) Evaluation of vaccine delivery systems for inducing long-lived antibody responses to *Dermanyssus gallinae* antigen in laying hens. *Avian Pathology* 48(1), 60–74.

Prieto Ursua, L., Sienra Monge, J.J., del Rio Navarro, B.E. and Mercado Ortiz, V.M. (1995) Skin tests for different species of acari in asthmatic children from Mexico City. *Revista Alergia Mexico* 42(3), 49–53.

Principato, M., Lisi, F., Moretta, I., Samra, N. and Puccetti, F. (2005) *Sancassania berlesei* (Michael, 1903): an opportunistic mite infesting litters in poultry farms causing dermatitis in humans and animals. *Italian Journal of Animal Science* 4(3), 306–309.

Prinzing, A. and Wirtz, H.-P. (1997) The epiphytic lichen, *Evernai punastri* L., as a habitat for arthropods: Shelter from desiccation, food-limitation and indirect mutualism. In: Stork, N.E., Adis, J. and Didham, R.K. (eds) *Canopy Arthropods*. Chapman and Hall, London, U.K., pp. 477–494.

Prisich, I.I, Dobarskaia, L.I. and Zosimova, A.G. (1986) Acariasis in children with chronic digestive system diseases. *Meditsinskaia Parazitologiia* 4, 50–51.

Pritchard, A.E. and Baker, E.W. (1955) A revision of the spider mite family tetranychidae. In: *Memoirs of the Pacific Coast Entomological Society, Volume 2*. Pacific Coast Entomological Society, San Francisco, California, U.S.A., pp. 1–472.

Pritchard, J., Kuster, T., Sparagano, O. and Tomley, F. (2015) Understanding the biology and control of the poultry red mite *Dermanyssus gallinae*: A review. *Avian Pathology* 44(3), 143–153.

Proctor, H. and Owens, I. (2000) Mites and birds: Diversity, parasitism and coevolution. *Trends in Ecology and Evolution* 15, 358–364.

Pucheu-Haston, C.M., Bizikova, P., Eisenschenk, M.N., Santoro, D., Nuttall, T. and Marsella, R. (2015) Review: The role of antibodies, autoantigens and food allergens in canine atopic dermatitis. *Veterinary Dermatology* 26, 115–e30. doi: 10.1111/vde.12201.

Puerta, L., Fernández-Caldas, E., Lockey, R.F. and Caraballo, L.R. (1993) Sensitization to *Chortoglyphus arcuatus* and *Aleuroglyphus ovatus* in *Dermatophagoides* spp. Allergic Individuals. *Clinical and Experimental Allergy* 23(2), 117–23.

Pugh, P.J.A. (1996) Using artifical substrata to monitor how cryptofaunal acari colonize littoral algae on sub-antarctic South Georgia. *Acarologia* 37, 189–200.

Pugh, P.J.A. and King, P.E. (1985) Feeding in intertidal Acari. *Journal of Experimental Marine Biology and Ecology* 94, 269–280.

Pugliese, N., Circella, E., Marino, M., Devirgilio, C., Cocciolo, G. *et al.* (2019) Circulation dynamics of *Salmonella enterica* subsp. *enterica* ser. *Gallinarum* biovar *Gallinarum* in a poultry farm infested by *Dermanyssus gallinae*. *Medical and Veterinary Entomology* 33, 162–170.

Pulpán, J. and Verner, P.H. (1965) Control of tyroglyphoid mites in stored grain by the predatory mite Cheyletus eruditus (Schrank). *Canadian Journal of Zoology* 43, 417–432.

Puspitarini, R.D., Fernando, I., Widjayanti, T., Ramadhatin, A. and Husna, N.L. (2021) Physicochemical characteristics of stored products affect host preference and biology of *Acarus siro* (Acari: Acaridae). *Journal of Crop Protection* 10(2), 281–293.

Putatunda, B.N. (2002) Mites (Acarina) in post harvest storage of some food grains in Himachal Pradesh, India. *Plant Protection Bulletin (Faridabad)* 54(1/2), 25–27.

Putatunda, B.N. and Abrol, D.P. (2003) Mites associated with bees in Jammu and Kashmir, India. *Zoos' Print Journal* 18(2), 1021–1024.

Pytelkova, J., Lepšik, M., Šanda, M., Talacko, P., Marešova, L. and Mareš, M. (2012) Enzymatic activity and immunoreactivity of Aca s 4, an alpha-amylase allergen from the storage mite *Acarus siro*. *BMC Biochemistry* 13(3), 1–8.

Qayyum, M.A., Khan, B.S., Bashir, M.H., Sahi, S.T. and Ozman-Sullivan, S.K. (2016) First record of *Macrocheles matrius* (Hull, 1925) (Acari: Macrochelidae) from Turkey. *International Journal of Agriculture and Biology* 18, 813–816.

Qayyum, H.A. and Chaudhri, W.M. (1979) Mites of the genus *Hemicheyletia* (Acarina: Cheyletidae) from Pakistan. *Pakistan Journal of Zoology* 11, 167–172.

Qin, T.-K. and Rohitha, B.H. (1996) The astigmatic mite *Histiostoma feroniarum* (Acari: Astigmata: Histiostomatidae) in New Zealand. *New Zealand Entomologist* 19, 65–69.

Qu, S.X., Li, H.P., Ma, L., Hou, L.J., Lin, J.S., Song, J.D. *et al.* (2015) Effects of different edible mushroom hosts on the development, reproduction and bacterial community of *Tyrophagus putrescentiae* (Schrank). *Journal of Stored Products Research* 61, 70–75.

Que, S., Xin, T., Yi, J., Zou, Z., Li, L. and Xia, B. (2016) Complete mitochondrial genome of *Aleuroglyphus ovatus* (Acari: Acaridae). *Mitochondrial DNA, Part A* 27(1), 563–564.

Quintero-Gutiérrez, E.J. and Halliday, B. (2021) Review of the mite family Parholaspididae Evans, 1956 (Acari: Mesostigmata). *Zootaxa* 5005(4), 401–459.

Quintero, M.T. and Acevedo, H.A. (1991) Mites of fermented liquid foods in Mexico. In: Dusbabek, F. and Bukva, V. (eds) *Modern acarology Volume I. Proceedings of the VIII International Congress of Acarology, Ceske Budejovice, Czechoslovakia, 6–11 August 1990*. Academia, Prague, pp. 611–614.

Racewicz, M. (2001) House dust mites (Acari: Pyroglyphidae) in the cities of Gdańsk and Gdynia (Northern Poland). *Annals of Agricultural and Environmental Medicine* 8(1), 33–38.

Rack, G. (1959) *Acarophenax dermestidarum* sp.n. (Acarina, Pyemotidae), ein Eiparasit von *Dermestes*-arten. *Zeitschrift f r Parasitenkunde* 19, 411–431.

Rack, G. (1962) Milben aus Taubennestern mit Beschreibung einer neuen Art, *Acheles gracilis* (Acari: Raphignatidae). *Zoologischer Anzeiger* 168(7–10), 276–292.

Rack, G. (1963) *Klemannia* (Acarina, Ameroseiidae), ein neuer Wohnungslästling. *Entomologische Mitteilungen aus dem Zoologischen Museum Hamburg* 2(44), 407–413.

Rack, G. (1964) Über das Auftreten der Milbe *Kleemannia plumigera* Oudemans, 1930 in Neubauten. Mit besonderer Berücksichtigung des Milbenbefalls im neuen Rathaus zu Flensburg. *Der Praktische Schädlingsbekämpfer* 16(8), 125–127.

Rack, G. (1968a) Milben als Hausungeziefer u. Vorratsschädlinge in Nordwestdeutschland. *Entomologische Mitteilungen aus dem Zoologischen Museum Hamburg* 3(62), 249–267.

Rack, G. (1968b) Milben als Hausungeziefer u. Vorratsschädlinge in Nordwestdeutschland. *Entomologische Mitteilungen aus dem Zoologischen Museum Hamburg* 3(62), 1–20.

Rack, G. (1970) Massenvorkommen von *Neoseiulus bakeri*, Hughes 1948 (Acarina, Phytoseiidae) und sechs weiterer Milbenarten in einem Neubau. *Anzahl Schädl.kd. Pflanzenschutz* 43(12), 184–186.

Rack, G. (1971) Milben in Neubaten. *Der Praktische Schädlingsbekämpfer* 23(11), 149–152.

Rack, G. (1972) Weitere Beobachtungen über Milben in Neubauten mit Beschreibung einer neuen Art, *Ameroseius pseudoplumosus* sp. n. (Acarina, Mesostigmata). *Entomologische Mitteilungen aus dem Zoologischen Staatsinstitut und Zoologischen Museum Hamburg* 4(77), 249–257.

Rack, G. (1980) *Procalvolia zacheri* (Oudemans, 1929) in einem Neubau bei Stade (Acarina, Sarcoptiformes, Saproglyphidae). *Entomologische Mitteilungen aus dem Zoologischen Staatsinstitut und Zoologischen Museum Hamburg* 6(107), 303–307.

Rack, G. (1988) Die Hausmilbe, *Glycyphagus domesticus* (De Geer, 1778) (Acariformes, Astigmata, Glycyphagidae). *Der Praktische Schadlingsbekampfer* 40, 243–248.

Radinovsky, S. (1962) Biological, developmental, and behavioral studies of *Leiodinychus krameri* (Can.) (Acarina: Uropodidae). A thesis submitted to Oregon state University in partial fulfillment of the requirements for the degree of Doctor of Philosophy, pp. 1–124.

Radinovsky, S. (1965a) The biology and ecology of granary mites of the Pacific Northwest. 3. Life history and development of *Leiodinychus krameri* (Acarina: Uropodidae). *Annals of the Entomological Society of America* 58(3), 259–267.

Radinovsky, S. (1965b) The biology and ecology of granary mites of the Pacific Northwest. IV. Various aspects of the reproductive behavior of *Leiodinychus krameri* (Acarina: Uropodidae). *Annals of the Entomological Society of America* 58(3), 267–272.

Radinovsky, S. and Krantz, G.W. (1961) The biology and ecology of granary mites of the Pacific northwest. II. Techniques for laboratory observations and rearing. *Annals of the Entomological Society of America* 51, 512–518.

Radovsky, F.J. (1967) *The Macronyssidae and Laelapidae (Acarina: Mesostigmata) Parasitic on Bats*. University of California Publications in Entomology, Berkeley and Los Angeles, U.S.A., pp. 1–288.

Radovsky, F.J. (1985) Evolution of mammalian mesostigmatid mites. In: Kim, K.C. (ed.) *Coevolution of Parasitic Arthropods and Mammals*. John Wiley, New York, U.S.A., pp. 441–504.

Radwan, J. (1995) Male morph determination in two species of acarid mites. *Heredity (Edinb.)* 74, 669–673.

Raele, D.A., Galante, D., Pugliesi, N., La Salandra, G., Lomuto, M. and Cafiero, M.A. (2018) First report of *Coxiella burnetii* and *Borrelia burgdorferi* sensu lato in poultry red mites, *Dermanyssus gallinae* (Mesostigmata, Acari), related to urban outbreaks of dermatitis in Italy. *New Microbes and New Infections* 23, 103–109.

Rahil, A.A.R. (2005) Mites in marjoram and chamomile cultivations in Fayoum and the relative abundance of twentysix species. Arab Univ. *Journal of Agricultural Science* 13(3), 1033–1046.

Rahimi, H. (1991) Faunistic study of arthropods (Arthropoda) associated with saffron in Gonabad and Ghayen, and biology of important species. M. Sc. thesis. Faculty of Agriculture, Shahid Chamran Universiy, Ahwaz, Iran, pp. 1–141.

Rahiminejad, V. and Hajiqanbar, H. (2015) A new species of the genus *Acarophenax* (Acari: Heterostigmatina: Acarophenacidae) associated with *Sphindus* sp. (Coleoptera: Sphindidae) from Iran. *Persian Journal of Acarology* 4(3), 277–286.

Rai, J.K., Armendt, J., Bernhardt, V., Pasquerault, T., Lindstrom, A. and Perotti, M.A. (2020) Mites (Acari) as a relevant tool in trace evidence and postmortem analyses of buried corpses. *Journal of Forensic Sciences* 65, 2174–2183.

Raina, V., Srivastava, S., Aggarwal, K., Ramesh, S. and Kumar, S. (2001) Essential oil composition of *Cinnamomum zeylanicum* Blume leaves from Little Andaman, India. *Flavour and Fragrance Journal* 16, 374–376.

Ramachandran, D., Lindo, Z. and Meehan, M.L. (2021) Feeding rate and efficiency in an apex soil predator exposed to short-term temperature changes. *Basic and Applied Ecology* 50, 87–96.

Ramakers, P.M.J. (1980) Biological control of *Thrips tabaci* (Thysanoptera: Thripidae) with *Amblyseius* spp. (Acari: Phytoseiidae). *Bulletin IOBC/WPRS* 3, 203–208.

Ramirez-Suarez, A., Zvaleta-Nejia, E., Kawasoe, S.O., Galvez, M.C.S. and Carraso, J.V. (2002) A possible role for *Rhizoglyphus robustus* Nesbitt (Astigmata: Acaridae) in transmission of *Sclerotium cepivorum* Berk. (Deuteromycetes: Mycelia-Sterila). *Applied Entomoloy and Zoology* 34(4), 663–669.

Ramjan, S.F.R., Loo, A.H.B., Lim, Y.P. and Chew, F.T. (2005) Catalogue of the major transcripts of the storage mite, *Aleuroglyphus ovatus*: Revealing the putative allergenic repertoire. *Journal of Allergy and Clinical Immunology* 115(2). Available at: https://www.jacionline.org/action/showPdf?pii=S0091-6749%2804%2903619-X (accessed 21 May 2020).

Ramsay, M., Mason, P.C. and Hunter, A.C. (1975) Chicken mite (*Dermanyssus gallinae*) infesting a dog. *New Zealand Veterinary Journal* 23, 155.

Ranabhat, N.B., Goleva, I. and Zebitz, C.P.W. (2014) Life tables of *Neoseiulus cucumeris* exclusively fed with seven different pollens. *Biocontrol* 59(2) 195–203.

Randall, A.J., Hillier, A., Cole, L.K., Kwochka, K.W., Needham, G. and Wassom, D.L. (2005) Quantitation of house dust mite allergens (Der f 1 and group 2) on the skin and hair of dogs. *American Journal of Veterinary Research* 66(1), 143–149.

Ranganath, H.R., Channabasavanna, G.P. and Krishna Rao, N.S. (1983) *Neosuidasia faini* gen. *et* sp. nov. (Acari: Astigmata: Saproglyphidae) with remarks on the subfamily Suidasiinae. *Inte rnational Journal of Acarology* 7, 63–69.

Rangel, J. and Ward, L. (2018) Evaluation of the predatory mite *stratiolaelaps scimitus* for the biological control of the honey bee ectoparasitic mite *Varroa destructor*. *Journal of Apicultural Research* 57, 425–432.

Rao, J. and Prakash, A. (1985) *Tyrophagus palmarum* (Oudemans) mite in rice seedlings and leaf sheaths. *International Rice Research Newsletter* 10(4), 13–14.

Rao, J. and Prakash, A. (1986) Mass rearing *Tyrophagus palmarum*. *International Rice Research Newsletter* 11(2), 26.

Rasp, G. (1991) Sensitization against storage mites in allergic rhinopathy. *Laryngo-Rhino-Otologie* 70(12), 678–680.

Raspotnig, G., Leutgeb, V., Krisper, G. and Leis, H.J. (2011) Discrimination of *Oribotritia* species by oil gland chemistry (Acari, Oribatida). *Experimental and Applied Acarology* 54, 211–224.

Raut, S.K. (1996) Factors determining the effectivenessof the mite *Fuscuropoda marginata* in the control of the slug pests *Laevicaulis alte*. In: Enderson, I.F. (ed.) *Slug and Snail Pests in Agriculture*. British Crop Protection Council, Farnham, U.K., pp. 247–254.

Raut, S.K. and Bhattacharya, S.S. (1999) Pests and diseases of betelvine (Piper betle) and their natural enemies in India. *Experimental and Applied Acarology* 23, 319–325.

Raut, S.K. and Ruma Sarkar (1991) The influence of temperature on the life-cycle of *Rhizoglyphus robini* Claparede (Acari: Acaridae). *International Journal of Acarology* 17, 145–148.

Rawhat, U.N., Tantray, A.Y., Kouser, N., Allie, K.A., Wani, S.M. *et al.* (2021) Influence of ecological and edaphic factors on biodiversity of soil nematodes. *Saudi Journal of Biological Sciences* 28(5), 3049–3059.

Rawlins, W.A. (1955) *Rhizoglyphus solani*, a pest of onions. *Journal of Economic Entomology* 48, 334.

Recio-García, M.D. and García-Cachán, M.D. (2002) *Identification of Dry Cured Ham Mites, 48" ICoMST, 25–30 August 2002*. Rome, Italy, Università di Parma, pp. 950–951.

Reck, G.F. (1976) *Katalog Akarofauny Gruzinskoy SSR (Catalogue of the acarofauna of the Georgian SSR)*. Tbilisi, Metsniereba, USSR, pp. 1–127.

Ree, H.-I., Jeon, S.-H., Lee, I.-Y., Hong, C.-S. and Lee, D.-K. (1997) Fauna and geographical distribution of house dust mites in Korea. *Korean Journal of Parasitology* 35(1), 9–17.

Reeves, R.M. (1988) Distribution and habitat comparisons for *Carabodes* collected from conifer branches with descriptions of *brevis* Banks and *higginsi* sp.n. (Acari: Oribatida: Carabodidae). *Proceedings of the Entomological Society of Washington* 90, 373–392.

Refaei, G.S. and El-Naggar, M.E. (2008) Mites associated with honeybee, *Apis mellifera* in Egypt. *Egyptian Journal of Agricultural Research* 86(4), 1355–1371.

Reger, J.F. (1977) A fine structure study on vitelline envelope formation in the mite, *Caloglyphus anomalus*. *Journal of Submicroscopic Cytology and Pathology* 9, 115–125.

Regev, S. and Cone, W.W. (1975) Evidence of farnesol as a male sex attractant of the two-spotted spider mite, *Tetranychus urticae* Koch (Acarina: Tetranychidae). *Environmental Entomology* 4, 307–311.

Regev, S. and Cone, W.W. (1976) Evidence of gonadotropic effect of farnesol in the two-spotted spider mite *Tetranychus urticae*. *Environmental Entomology* 5, 517–519.

Reis, A.E. and Paschoal, A.D. (1968) Alguns ácaros de produtos armazenados do estado de São Paulo. *Solo* 60, 73–74.

Rennie, J. (1921) Notes on acarine disease. VI. *Bee World* 13, 115–117.

Rentfrow, G., Hanson, D.J., Schilling, M.W. and Mikel, W.B. (2006) Methyl bromide use to combat mite infestation in dry-cured ham during production. In: *Proc. Annual International Research Conference on Methyl Bromide Alternatives and Emission Reduction, 5–9*

November 2006. Orlando, Florida, U.S.A:, IISD, pp. 108-1–108-2.

Rentfrow, G., Hanson, D.J., Schilling, M.W. and Mikel, W.B. (2008) The use of methyl bromide to control insects in country hams in the Southeastern United States. University of Kentucky Extension/National Country Ham Association, Extension Publication, ASC-171.

Requena-García, F., Cabrero-Sañudo, F., Olmeda-García, S., González, J. And Valcárcel, F. (2017) Influence of environmental temperature and humidity on questing ticks in central Spain. *Experimental and Applied Acarology* 71, 277–290.

Retsch (2022) Milling Sieving Assisting. Available at: https://www.retsch.com/ (accessed 21 May 2023).

Revsbech, P. and Andersen, G. (1987) Storage mite allergy among grain elevator workers. *Allergy* 42, 423–429.

Revsbech, P. and Dueholm, M. (1990) Storage mite allergy among bakers. *Allergy* 45, 204–208.

Rezende, L.C., Cunha, L.M., Teixeira, C.M., Oliveira, P.R. and Martins, N.R.S. (2014) Mites affecting hen egg production: Some considerations for Brazilian farms. *Ciência Rural* 43, 1230–1237.

Rezk, H.A. (2000) Mites associated with stored dried-dates in Egypt and the role of *Blattisocius keegani* Fox as a biological control agent. *Alexandria Journal of Agricultural Research* 45(1), 179–191.

Rezk, H.A. (2004) Influence of temperature on life history parameters of the American house dust mite, *Dermatophagoides farinae* Hughes (Acari: Pyroglyphidae). *Alexandria Journal of Agricultural Research* 49(3), 25–31.

Rezk, H.A., El-Hamid, M.M.A. and El-Latif, M.A.A. (1996) House dust mites in Alexandria region, Egypt. *Alexandria Journal of Agricultural Sciences* 41(2), 209–216.

Ribeiro, H.S.; Pereira, D.F.S., Melo-Junior, O., Mariano, R.M.S., Leite, J.C. *et al.* (2021) Vaccine approaches applied to controlling dog ticks. *Ticks Tick Borne Disease* 12(3), 101631.

Richards, K.W. and Richards, L.A. (1977) A new species of *Macrocheles* (Acarina: Macrochelidae) found in bumble bee nests (Hymenoptera: Apidae). *Canadian Entomologist* 109, 711–719.

Ridgway, C. and Chambers, J. (1996) Detection of external and internal insect infestation in wheat by near-infrared reflectance spectroscopy. *Journal of the Science of Food and Agriculture* 71, 251–264.

Ridgway, C., Chambers, J., Portero-Larragueta, E. and Prosser, O. (1999) Detection of mite infestation in wheat by electronic nose with transient flow sampling. *Journal of the Science of Food and Agriculture* 79(15), 2067–2074.

Ridgway, C., Davies, R. and Chambers, J. (2001) Imaging for the high-speed detection of pest insects and other contaminants in cereal grain in transit. Paper presented at the 2001 ASAE Annual International Meeting, California, July 30–August 1, 2001.

Rietschel, R.L. and Fowler Jr., J.F. (2008) *Fisher's Contact Dermatitis*. 6th edn., Hamilton, BC Decker Inc., Ontario, Canada, pp. 1–428.

Rifà, E.V. and Griffiths, D. (2018) U.S. No. US20180160688A1. Available at: http://patents.google.com/patent/US20180160688A1/en (accessed 11 November 2020).

Rijn, P.C.J. van and Tanigoshi, L.K. (1999) Pollen as food for the predatory mites *Iphiseius degenerans* and *Neoseiulus cucumeris* (Acari: Phytoseiidae): Dietary range and life history. *Experimental and Applied Acarology* 23, 785–802.

Rim I.-S. and Jee, C.-H. (2006) Acaricidal effects of herb essential oils against *Dermatophagoides farinae* and *D. pteronyssinus* (Acari: Pyroglyphidae) and qualitative analysis of a herb *Mentha pulegium* (pennyroyal). *Korean Journal of Parasitology* 44(2), 133–138.

Ripka, G., Fain, A., Kaźmierski, Kreiter, S. and Magowski, W. Ł. (2005) New Data to the Knowledge of the Mite Fauna of Hungary (Acari: Mesostigmata, Prostigmata and Astigmata). *Acta Phytopathologica et Entomologica Hungarica* 40 (1–2), 159–176.

Rivard, I. (1962) Some effects of prey density on survival speed of development and fecundity of the predaceous mite *Melichares dentriticus* (Berl.) (Acarina: Aceosejidae). *Canadian Journal of Zoology* 40, 1233–1236.

Riudavets, J. (2004) Pest problems and their control in stored dried fruits, cocoa and spices, aromatic and medicinal herbs: description of the status in Spain. In: *Proc. 4th meeting of COST Action 842: WG4: Biocontrol of Arthropod Pests in Stored Products*. Athens, Greece, pp. 11–12.

Riudavets, J., Castane, C., Alomar, O., Pons, M.J. and Gabarra, R. (2010) The use of carbon dioxide at high pressure to control nine stored-product pests. *Journal of Stored Products Research* 46, 228–233.

Rivard, I. (1959) Influence of humidity on longevity, fecundity, and rate of increase of the grain-infesting mite *Tyrophagus castellani* (Hirst) (Acarina: Acaridae) reared on mould cultures. *Canadian Entomologist* 91, 31–35.

Rivard, I. (1960) A technique for individual rearing of the predacious mite, *Melichares dentriticus* (Berlese) (Acarina: Aceosejidae) with notes on its life history and behaviour. *Canadian Entomologist* 92, 834–839.

Rivard, I. (1961) Influence of temperature and humidity on mortality and rate of development of immature stages of the mite *Tyrophagus putrescentiae* (Schrank) (Acarina: Acaridae) reared on mold cultures. *Canadian Journal of Zoology* 39, 419–426.

Rizk, G.A., Soliman, Z.R. and Ali, M.A. (1979) Survey on mites associated with citrus and grape wine in Minia region, Egypt. *Bulletin of Entomological Social Egypt* 62, 105–110.

Roberts, M. (1991) The parasites of the Polynesian rat within and beyond New Zealand. *International Journal of Parasitology* 21, 777–783.

Robertson, P.L. (1946) Tyroglyphid Mites in Stored Products in New Zealand. *Transactions and Proceedings of the Royal Society of New Zealand* 76(2), 185–207.

Robertson, P.L. (1959) Revision of the genus *Tyrophagus* with a discussion on its taxonomic position in the Acarina. *Australian Journal of Zoology* 7, 146–181.

Robertson, P.L. (1961) A morphological study of variation in *Tyrophagus* (Acarina), with particular reference to populations infesting cheese. *Bulletin of Entomological Research* 52, 501–529.

Robin, C. (1860) Mémoire zoologique et anatomique sur diverses espèces d'Acariens de la famille des Sarcoptides. *Bulletin de la Société Impériale des Naturalistes de Moscou* 33, 184–293.

Rodriguez, J.G. and Stepien, Z.A. (1973) Biology and population dynamics of *Caloglyphus berlesei* (Michael) (Acarina: Acaridae) in xenic diet. *Journal of the Kansas Entomological Society* 46, 176–183.

Rodriguez, J.G. and Wade, C.F. (1961) The nutrition of *Macrocheles muscaedomesticae* (Acarina: Macrochelidae) in relation to its predatory action on the house fly eggs. *Annals of the Entomological Society of America* 54, 782–788.

Rodriguez, J.G., Wade, C.F. and Wells, C.N. (1962) Nematodes as a natural food for *Macrocheles muscaedomesticae* (Acarina: Macrochelidae) a predator of the house fly egg. *Annals of the Entomological Society of America* 55, 507–511.

Rodriguez, J.G., Singh, P. and Taylor, B. (1970) Manure mites and their role in fly control. *Journal of Medical Entomology* 7, 335–341.

Rodriguez, J.G., Potts, M.F. and Patterson, C.G. (1984) Mycotoxin-producing fungi: effects on stored product mites. In: Griffiths, D.A. and Bowman, C.E. (eds) *Acarology 6, Vol. 2*. Ellis Horwood, Chichiester, U.K., pp. 343–350.

Rodriguez-Reina, J.M., Garcia-Mari, F. and Ferragut, F. (1992) Predatory activity of phytoseiid mites on different developmental stages of theWestern flower thrips *Frankliniella occidentalis*. *Boletín de Sanidad Vegetal. Plagas* 18, 253–263.

Rohde, C.J. (1959) Studies on the biologies of two mite species, predator and prey, including some effects of gamma radiation on selected developmental stages. *Ecology* 40(4), 572–579.

Rohdendorf, B.B. (1940) Mites of the families Cheyletidae and Pediculoidae. *Uchenye Zapiski* 42, 69–98.

Rojas, E.W. and Klimov P.B. (2007) Mites of the genus *Rhizoglyphus* (Acari: Acaridae) infesting cultivated bulbs in central and southern Chile, with taxonomic notes on *Acarus hyacinthi* Boisduval and *Rhizoglyphus frickorum* Nesbitt. *International Journal of Acarology* 33, 87–90.

Romański, B., Dziedziczko, A., Pawlik, K., Wilewska-Kłubo, T., Wojtanowski, I. and Bikowska, M. (1977) Allergy to house dust in bronchial asthma I. Incidence and problem of the character of house dust allergen. *Polskie Archiwum Medycyny Wewnetrznej* 57, 21–26.

Rosa, A.E. and Flechtmann, C.H.W. (1979) Mites in house dust from Brazil. *International Journal of Acarology* 5, 195–198.

Rosas, M.A. and Casanueva, M.E. (1992) The house-dust mite: bionomy of *Glycyphagus destructor* (Schrank 1781) (Acari: Astigmata). *Gayana Zoologia* 56, 27–45.

Rosen, S., Yeruham, I. and Braverman, Y. (2002) Dermatitis in humans associated with the mites *Pyemotes tritici*, *Dermanyssus gallinae*, *Ornithonyssus bacoti* and *Androlaelaps casalis* in Israel. *Medical and Veterinary Entomology* 16(4), 442–444.

Rosický, B., Černý, V., Daniel, M., Dusbábek, F., Palička, P., Samšiňák, K. and Hájková, Z. (1979) *Roztoi a Klišťata Škodici Zdravi Loveka*. Academia, Praha, 1–213.

Rota, P. and Serini Bolchi, G. (1974) Reports of mites in cultures of field mushrooms. *Bollettino di Zoologia Agraria e di Bachicoltura* 12, 211–215.

Roy, L. and Buronfosse, T. (2011) Using mitochondrial and nuclear sequence data for disentangling population structure in complex pest species: A case study with *Dermanyssus gallinae*. *PLoS One* 6(7), e22305. DOI: 10.1371/journal.pone.0022305.

Roy, L. and Chauve, C.M. (2007) Historical review of the genus *Dermanyssus* Dugès, 1834 (Acari: Mesostigmata: Dermanyssidae). *Parasite* 14, 87–100.

Rout, G. and Misera, B.K. (1987) Biology and habits of *Cheyletus malaccensis* Oudemans (Cheyletidae: Prostigmata). *Bulletin Grain Technology* 25, 122–128.

Roy, L., Dowling, A.P.G., Chauve, C.M. and Buronfosse, T. (2009a) Delimiting species boundaries within *Dermanyssus* Dugés, 1834 (Acari: Dermanyssidae) using a total evidence approach. *Molecular Phylogenetics and Evolution* 50, 446–470.

Roy, L., Dowling, A.P.G., Chauve, C.M., Leśna, I., Sabelis, M.W. and Buronfosse, T. (2009b) Molecular phylogenetic assessment of host range in five *Dermanyssus* species. *Experimental and Applied Acarology* 48, 115–142.

Roy, L., Chauve, C.M. and Buronfosse, T. (2010) Contrasted ecological repartition of the Northern Fowl Mite *Ornithonyssus sylviarum* (Mesostigmata: Macronyssidae) and the Chicken Red Mite *Dermanyssus gallinae* (Mesostigmata: Dermanyssidae). *Acarologia* 50(2), 207–219.

Roy, L., Chiron, G., Lubac, S. and Bicout, D.J. (2014) Tape-traps as an easy-to-use tool for monitoring and surveillance of the poultry red mite in cage and free-range layer farms. In: *XIVth European Poultry Conference, 23–27June 2014*. Stavanger, Norway.

Roy, L., El Adouzi, M., Moraza, M.L., Chiron, G., de Janti, E.V., Le Peutrec, G. and Bonato, O. (2017) Arthropod communities of laying hen houses: an integrative pilot study toward conservation biocontrol of the poultry red mite *Dermanyssus gallinae*. *Biological Control* 114, 176–194.

Roy, L., Arriaga-Jimenez, A. and El Adouzi, M. (2018) Composition anti-acariens. Montpellier: Patent no WO2018109417A1; FR3060258A1 filed Dec. 2016 issued May 2018 UPVM3, UCNRS M. SATT AxLR.

Rozman, V. and Kalinović, I. (2001) Pests in livestock factories and their control (Štetnici u tvornicama stočne hrane i njihovo suzbijanje). *Krmiva* 43(2), 69–73.

Ruf, A. (1991) Do females eat males? Laboratory studies on the population development of *Hypoaspis aculeifer* (Acari: Parasitiformes). In: Dusbábek, F. and Bukva, V. (eds) *Modern Acarology, vol. 2*. Academia, Prague and *SPB* Academic, The Hague, The Netherlands, pp. 487–492.

Ruoppi, P., Koistinen, T. and Pennanen, S. (2005) Sensitisation to mites in laboratory animal workers with rhinitis. *Occupational and Environmental Medicine* 62, 612–615.

Russel Shoemake, R. (1970) A Review of the Family Eviphididae (Acarina: Mesostigmata). A thesis submitted to Oregon State University, in partial fulfillment of the requirements for the degree of Doctor of Philosophy, pp. 1–263.

Russel, D.J., Schulz, M.M. and OConnor, B.M. (2004) Mass occurrence of astigmatid mites on human remains. *Abhandlungen und Berichte Naturforschende Gesellschaft zu Gorlitz* 76, 51–56.

Russell, R.C., Otranto, D. and Wall, R.L. (2013) *The Encyclopedia of Medical and Veterinary Entomology*. CABI Publishing, Oxford, U.K., pp. 1–440.

Ryke, P.A.J. (1959) A revision of hypoaspid mites associated with Myriapoda, with descriptions of three new species of the subgenus *Julolaeleps* Berl. (Acarina: Laelaptidae). *Parasitology* 49, 6–22.

Ryke, P.A.J. (1964) Acarina associated with *Protea* flowers in the Cape Province. *Journal of the Entomological Society of Southern Africa* 26,337–354.

Ryono, A., Mori, N., Okabe, K. and Kuwahara, Y. (2001) Chemical ecology of astigmatid mites LVIII. 2-Hydroxy-6-methylbenzaldehyde: Female sex pheromone of *Cosmoglyphus hughesi* Samsinak (Acari: Acaridae). *Applied Entomoloy and Zoology* 36, 77–81.

Ryu, J.S., Ree, H.I., Min, D.Y. and Ahn, M.H. (2003) A human case of house dust mite *Tarsonemus floricolus* collected from sputum. *Korean Journal of Parasitology* 41, 171–173.

Sabelis, M.W. (1996) Phytoseiidae. In: Lindquist, E.E., Sabelis, M.W. and Bruin, J. (eds) *Eriophyoid Mites: Their Biology, Natural Enemies and Control. World Crop Pests, Vol. 6*. Elsevier Science, Amsterdam, The Netherlands, pp. 427–456.

Sabelis, M.W. and Van de Baan, H.E. (1983) Location of distant spider mite colonies by phytoseiid predators: Demonstration of specific kairomones emitted by *Tetranychus urticae* and *Panonychus ulmi*. *Entomologia Experimentalis et Applicata* 33, 303–314.

Sabelis, M.W., Afman, B. and Slim, P.J. (1984) Location of distant spider mite colonies by *Phytoseiulus persimilis* Athias-Henriot (Acarina: Phytoseiidae): localization and extraction of a kairomone. In: Griffiths, D. and Bowman, C. (eds) *Acarology VI, Vol. 1*. Ellis Horwood, Chichester, U.K., pp. 431–440.

Saboori, A. and Cobanoglu, S. (2010) A new species of larval *Erythraeus* and a new record of larval *Grandjeanella* (Acari: Erythraeidae) from Turkey. *International Journal of Acarology* 36(3), 249–253.

Saboori, A., Kamali, K. and Pakdel, A. K.(2009) Mites (Acari) found in flour mills in the Karaj region of Iran. *Systematic and Applied Acarology* 14(3), 191–196.

Sacchi Carmona Rodrigueiro, T. and Pires do Prado, A. (2004) *Macrocheles muscaedomesticae* (Acari, Macrochelidae) and a species of *Uroseius* (Acari, Polyaspididae) phoretic on *Musca domestica* (Diptera, Muscidae): effects on dispersal and colonization of poultry manure. *Iheringia Série Zoologia* 94, 181–185.

Sade, K., Roitman, D. and Kivity, S. (2010) Sensitization to *dermatophagoides and blomia tropicalis*, and other mites in atopic patients. *Journal of Asthma* 47, 849–852.

Sadieva, B.E. (1984) A mite and the pathogen of brown mosaic. *Zasch Rast* 7, 20.

Salarzehi, S., Hajizadeh, J., Hakimitabar, M. and Ueckermann, E. (2018) A contribution to the knowledge of cheyletid mites of Iran with redescription of *Eucheyletia flabellifera* (Michael, 1878) (Prostigmata: Cheyletidae). *Acarologia* 58(2), 457–470.

Salarzehi, S., Hajizadeh, J. and Ueckermann, E.A. (2019) A new species of *Cheletonella* Womersley (Prostigmata: Cheyletidae) from Iran and a key to the species. *Acarologia* 59(2), 188–195.

Saleh, S.M., El-Helaly, M.S. and El-Gayar, F.H. (1985) Survey on stored product mites of Alexandria (Egypt). *Acarologia* 26(1), 87–93.

Saleh, S.M., El-Helaly, M.S. and El-Gayar, F.H. (1986) Life history of the predatory mite *Cheyletus malaccensis* (Oudemans). *Acarologia* 27(1), 37–40.

Saleh, S.M., Keladan, N.L. and Shaker, N. (1991) Control of European house dust mite *Dermatophagoides pteronyssinus* (Trouessart) with *Bacillus* spp. *Acarologia* 32(3), 257–260.

Sallam MN. (2013) Insect Damage postharvest Operations. Postharvest Compendium, Food and Agriculture Organization of the United Nations. International Centre of Insect Physiology and Ecology (ICIPE). Available at: http://www.fao.org/3/a-av013e.pdf (accessed 10 January 2023).

Salmane, I. (1999) Soil free-living predatory Gamasina mites (Acari, Mesostigmata) from the coastal meadows of Riga Gulf, Latvia. *Latvijas Entomologs* 37, 104–114.

Salmane, I. (2005) Addition to the Latvian Mesostigmata (Acari, Parasitiformes) check-list. *Latvijas Entomologs* 42, 58–62.

Saloña, M.I., Moraza, M.L, Carles-Tolrá, M., Iraola, V., Bahillo, P. *et al.* (2010) Searching the soil: Forensic importance of edaphic fauna after the removal of a corpse. *Journal of Forensic Science* 55(6), 1652–1655.

Samšiňák, K. (1966) Die Neuerrichtung der Gattung *Cosmoglyphus* Oudemans, 1932 gleichzeitig ein Beitrag zum Problem der 'Copra itch'. *Zoologischer Anzeiger* 176, 27–42.

Samšiňák, K. and Daniel, M. (1978) Five new species of mesostigmatic mites from *Geotrupes* beetles collected in the Himalayan region of Nepal (Acari: Meostigmata). *International Journal of Acarology* 4, 91–100.

Samšiňák, K., Dusbabek, F. and Vobrazkova, E. (1972) Note on the house dust mites in Czechoslovakia. *Folia Parasitologica* 19, 383–384.

Samšiňák, K., Palička, P., Zítek, K., Mališ, L. and Vobrázková, E. (1976) Are the mites of the genus *Tarsonemus* really parasites of man? *Folia Parasitologica* 23(1), 91–93.

Samšiňák, K., Vobrázková, E. and Spicak, V. (1978) Investigations on the fauna of beds in flats, children's sanatoria and old-age homes. *Folia Parasitologica* 25, 157–163.

Samšiňák, K., Chmela, J. and Vobrázková, E. (1979) *Pyemotes herfsi* (Oudemans, 1936) as causative agent of another mass dermatitis in Europe (Acari, Pyemotidae). *Folia Parasitologica* 26(1), 51–54.

Samung, Y., Apiwathnasorn, C., Wonglakorn, S. and Phayakkaphon, A. (2006) The first reported case of mites, *Suidasia pontifica*, in the external ear canal of a Thai agricultural worker. *Journal of Tropical Medicine and Parasitology* 29, 65–67.

Sanchez Palacios, A., Garcia Marrero, J.A., Liminana Canal, J.M., Schaman, F., Alonso Bordon, J. and Gutierrez Perez, A.T. (1991) Skin cross reactivity between *Dermatophagoides pteronyssinus* and storage mites in atopic children. *Journal of Investigational Allergology and Clinical Immunology* 1(3), 179–184.

Sánchez, A., Sánchez, J., Munera, M. and Cardona, R. (2019) The phylogenetic relationship of Glyciphagidae, Pyroglyphidae, Chortoglyphidae and Acaridae mite families according to the sequence of their main allergens. *Revista Alergia México* 66(2), 163–177.

Sánchez-Borges, M., Suárez-Chacón, R., Capriles-Hulett, A. and Caballero-Fonseca, F. (2005) An update on oral anaphylaxis from mite ingestion. *Annals of Allergy, Asthma and Immunology* 94, 216–221.

Sánchez-Borges, M., Suárez-Chacón, R., Capriles-Hulett, A., Caballero-Fonseca, F. and Fernández Caldas, E. (2013) Anaphylaxis from ingestion of mites: pancake anaphylaxis. *Journal of Allergy and Clinical Immunology* 131, 31–35.

Sanchez-Covista, A., Rodriguez-Rodriguez, J.A., De La Torre, F. and Garcia-Robaina, J.C. (1999) Mite fauna of house dust of the island of Tenerife. *Acarologia* 40, 55–58.

Sánchez-López, J. (2002) Control de ácaros contaminantes del jamón ibérico. Tesis Doctoral. Universidad de Extremadura, pp. 1–276.

Sanchez-Medina, M. and Sanchez-Gutierrez, G. (1973) Acaros en polvo de habitaciones a diferentes Alturas y climas de Colombia. *Alergia* 20, 171–188.

Sanchez-Machin, I., Glez-Paloma, P.R., Iglesias-Souto, J., Iraola, V. and Matheu, V. (2010) Oral mite anaphylaxis. *Allergy* 65, 1345–1347.

Sanchez-Machin, I., Gonzalez-Poza, R., Iglesias-Souto, J., Iraola, V. and Matheu, V. (2011) Asthma and rhinitis by storage mites. *Allergy* 66(12), 1615–1616.

Sánchez-Ramos, I., Álvarez-Alfageme, F. and Castañera, P. (2007a) Development and survival of the cheese mites, *Acarus farris* and *Tyrophagus neiswanderi* (Acari: Acaridae), at constant temperatures and 90% relative humidity. *Journal of Stored Products Research* 43, 64–72.

Sánchez-Ramos I, Álvarez-Alfageme F, Castañera P (2007b) Reproduction, longevity and life table parameters of *Tyrophagus neiswanderi* (Acari: Acaridae) at constant temperatures. *Experimental and Applied Acarology* 43(3), 213–226.

Sánchez-Ramos, I., Alvarez-Alfageme, F. and Castañera, P. (2007c) Effects of relative humidity on development, fecundity and survival of three storage mites. *Experimental and Applied Acarology* 41(1–2), 87–100.

Sánchez-Ramos, I. and Castañera, P. (2000) Acaricidal activity of natural monoterpenes on *Tyrophagus putrescentiae* (Schrank), a mite of stored food. *Journal of Stored Products Research* 37(1), 93–101.

Sánchez-Ramos, I. and Castañera, P. (2003a) Laboratory evaluation of selective pesticides against the storage mite *Tyrophagus putrescentiae* (Acari: Acaridae). *Journal of Medical Entomology* 40, 475–481.

Sánchez-Ramos, I. and Castanera, P. (2003b) Acaricidal activity of natural monoterpenes on *Tyrophagus putrescentae* (Schrank), a mite of stored food. *Journal of Stored Products Research* 37, 93–101.

Sánchez-Ramos, I. and Castañera, P. (2005) Effect of temperature on reproductive parameters and longevity of *Tyrophagus putrescentiae* (Acari: Acaridae). *Experimental and Applied Acarology* 36, 93–105.

Sánchez-Ramos, I. and Castañera, P. (2007) Evaluation of low humidity treatments to control *Acarus farris* (Acari: Acaridae) in Cabrales cheese. *Experimental and Applied Acarology* 41(4), 243–249.

Sánchez-Ramos, S. and Castañera, P. (2008) Effect of temperature on reproductive parameters and the longevity of *Acarus farris* (Acari: Acaridae). *Journal of Stored Products Research* 43(4), 578–586.

Sánchez-Ramos, I. and Castañera, P. (2009) Chemical and physical methods for the control of the mite *Acarus farris* on Cabrales cheese. *Journal of Stored Products Research* 45, 61–66.

Sánchez-Ramos, I., Hernández, C.A., Castanera, P. and Ortego, F. (2004) Proteolytic activities in body and faecal extracts of the storage mite, *Acarus farris*. *Medical and Veterinary Entomology* 18, 378–386.

Sandhu, G.S. (1976) Record of bulb mite *Rhizoglyphus echinopus* (Fumouze and Robin) from stored onions in India. *Science Culture (London)* 42(4), 221–222.

Sandhu, R., Jain, A.K. and Putatunda, B.N. (2005) New records of Mesostigmatid mites of stored food products in Hisar, Haryana. *Agricultural Science Digest* 25(4), 278–280.

Sandner, H. and Wasylik, A. (1973) The mites of the sparrow nests and the danger of infestation of granaries by them. *Ekol Polska A* 21, 323–338.

Sansuwan, S. and Chanbang, Y. (2020) Control of *Suidasia pontifica* Using heat and formaldehyde incorporated with propionic acid in swine feed. *Journal of Agriculture* 33(1), 39–47.

Santos da Silva, E., Asam, C., Lackner, P., Hofer, H., Wallner, M. *et al.* (2017) Allergens of *Blomia tropicalis*: An overview of recombinant molecules. *International Archives of Allergy and Immunology* 172(4), 203–214.

Santos, J.C., Rueda-Ramírez, D., Demite, P.R., De Moraes, G.J. (2018) Ascidae, Blattisociidae and Melicharidae (Acari: Mesostigmata): Zoogeographic analyses based on newly available databases. *Zootaxa* 4377(4), 542–564.

Santos, J.C., Demite, P.R. and Moraes, G.J. de (2020a) Ascidae Database. Available at: http://www.lea.esalq.usp.br/acari/ascidae (accessed 24 April 2020).

Santos, J.C., Demite, P.R. and de Moraes, G.J. (2020b) Blattisociidae Database. Available at: http://www.lea.esalq.usp.br/acari/blattisociidae (accessed 27 April 2020).

Santos, J.C., Demite, P.R. and de Moraes, G.J. (2020c) Melicharidae Database. Available at: http://www.lea.esalq.usp.br/acari/melicharidae (accessed 1 May 2020).

Santos, A.M., Jurkevicz, R.M.B., Romani, I., Simone, M.R.C. and Antonucci, A.M. (2022) Use of *Eborellia* sp. (Dermaptera: Forficulidae) for biological control of Mesostigmatid mites in laying hen farms. *Brazilian Journal of Development* 8, 55412–55429.

Sanyal, A.K. (2004) Diversity of invertebrate Fauna of Schirmacher Oasis, East Antarctica. Nineteenth Indian expedition to Antarctica, scientific report. *Technical Publication* 17, 173–187.

Sarathep, P. and Phonkaew W. (2018) A lost world disease: Copra itch outbreak caused by *Tyrophagus longior* mite. *IDCases* 12, 58–63.

Sardar, M.A. and Murphy, P.W. (1987) Feeding tests of grassland soil-inhabiting gamasine predators. *Acarologia* 28, 117–121.

Sargison, N.D., Jacinavicius, F.C., Fleming, R.H., Chaudhry, U.N. and Costa-Junior, L.M. (2020) Investigation of a gamasid mite infestation in a UK textile mill caused by *Dermanyssus gallinae* (De Geer, 1778) (Mesostigmata: Dermanyssidae) special lineage L1. *Parasitology International* 78, 102146.

Saridomichelakis, M.N., Marsella, R., Lee, K.W., Esch, R.E., Farmaki, R. and Koutinas, A.F. (2008) Assessment of cross-reactivity among five species of house dust and storage mites. *Veterinary Dermatology* 19, 67–76.

Sarraf, M.H.R., Pourasgari, H. and Kavousi, A. (2015) Demographic parameters of dried fruit mite *Carpoglyphus lactis* (Acari: Carpoglyphidae) on baker's yeast in two different temperatures. *Iranian Journal of Plant Protection Science* 45(2), 329–336.

Sarwar, M. (2016) Comparative life history characteristics of the mite predator *Neoseiulus cucumeris* (Oudemans) (Acari: Phytoseiidae) on mite and pollen diets. *International Journal of Pest Management* 62, 140–148.

Sarwar, M. (2020) House dust mites: Ecology, biology, prevalence, epidemiology and elimination. In: Bastidas Pacheco, G.A. and Kamboh, A.A. (eds) *Parasitology and Microbiology Research*, IntechOpen, DOI: 10.5772/intechopen.91891. Available at: https://www.intechopen.com/chapters/71977 (accessed 2 April 2022).

Sasa, M. (1947) Title unknown. *Nisshin Igaku* 34(3), 167–170.

Sasa, M. (1950) Mites of the genus *Dermatophagoides* Bogdanoff, 1864, found from three cases of human acariasis. *Japanese Journal of Experimental Medicine* 20, 519–525.

Sasa, M. (1952) Two new acaroid (tyroglyphoid) mites of possible sanitary or economic importance in Japan. *Japanese Journal of Experimental Medicine* 22, 511–521.

Sasa, M. (1964) Special problems of mites in stored food and drugs in Japan. Proc. 1st. int. Congr. Acarology 1963. *Acarologia* 6, 390–391.

Sasa, M. (1965) Mites, In: *An Introduction to Classification, Bionomics and Control of Acarina*. Univ, Tokyo Press, Tokyo, Japan, pp. 1–486.

Sasa, M. and Asanuma, K. (1951) A new food infesting acarid mite, *Hoshikadania konoi* n.g., n. sp., with notes on the comparative-morphology of its different sexes and stages. *Japanese Journal of Experimental Medicine* 21, 209–222.

Sato, M., Kuwahara, Y., Matsuyama, S., Suzuki, T., Okamoto, M. and Matsumoto, K. (1993) Male and female sex pheromones produced by *Acarus immobilis* Griffiths (Acaridae: Acarina). *Naturwissenschaften* 80, 34–36.

Satyal, P., Craft, J.D., Dosoky, N.S. and Setzer, W.N. (2017) The chemical compositions of the volatile oils of garlic (*Allium sativum*) and wild garlic (*Allium vineale*). *Foods* 6, 63.

Sayadi, Z. (2012) Study on fauna of stored product mites in sistan region, Iran. Master thesis, University of Zabol, Iran.

Sayedi, M., Saboori, A. Kamali, K., Kharazi Pakdel, A. and Tork, M. (2006) Fauna of stored product mites of Karaj, Iran. In: Manzari, S. (ed.) *Abstract Book of the 17th Iranian Plant Protection Congress*. Karaj, Iran, p. 187.

Sayedi, M., Saboori, A., Kamali, K. and Pakdell, A.K. (2009) Mites (Acari) found in flour mills in the Karaj region of Iran. *Systematic and Applied Acarology* 14(3), 191–196.

Sayedi, M., Saboori, A. and Allahyari, H. (2012) Preliminary observations on mites found in domesticated animal food factories in Karaj, Iran. *Persian Journal of Acarology* 1(2), 119–125.

Sayadi, Z., Saboori, A., Khani, A. and Ramrodi, S. (2012) Some stored product mites in Sistan region, Iran. In: *Proc. 20th Iranian Plant Protection Congress*, Shiraz, Iran, p. 492.

Schaarschmidt, L. (1959) Systematik und Okologie der Tarsonemiden. In: Stammer, H.J. (ed.) *Beitrage zur Systematik und Okologie Mitteleuropaischer Acarina, Bund 1, Abschnitt 5*. Akademische Verlagsgesellschaft Geest and Portig, Leipzig, Germany, pp. 713–823.

Schatz, H., Behan-Pelletier, V.M., OConnor, B.M. and Norton, R.A. (2011) Suborder Oribatida van der Hammen, 1968. In: Zhang, Z.-Q. (ed.) *Animal Biodiversity: An Outline of Higher-level Classification and Survey of Taxonomic Richness*, pp. 141–148.

Scheucher, R. (1957) Systematik und Ökologie der deutschen Anoetinen. In: Stammer, H.J. (ed.) *Beiträge zur Systematik und Ökologie mitteleuropäischer Acarina*. Akademische Verlagsgesellschaft, Jena, Germany, pp. 233–384.

Schiavone, A., Pugliese, N., Otranto, D., Samarelli, R., Circella, E., De Virgilio, C. and Camarda, A. (2022) *Dermanyssus gallinae*: The long journey of the poultry red mite to become a vector. *Parasites and Vectors* 15(29). DOI: 10.1186/s13071-021-05142-1.

Schicht, S., Qi, W., Poveda, L. and Strube, C. (2013) The predicted secretome and transmembranome of the poultry red mite *Dermanyssus gallinae*. *Parasites and Vectors* 6, 259.

Schicht, S., Qi, W., Poveda, L. and Strube, C. (2014) Whole transcriptome analysis of the poultry red mite *Dermanyssus gallinae* (De Geer, 1778). *Parasitology* 141, 336–346.

Schliesske, J. (1990) On the gall mite fauna (Acari: Eriophyoidea) of *Cocos nucifera* L. in Costa Rica. *Plant Research Development* 31, 74–81.

Schlipalius, D.I., Valmas, N., Tuck, A.G., Jagadeesan, R., Ma, L. *et al.* (2012). A core metabolic enzyme mediates resistance to phosphine gas. *Science* 338(6108), 807–810.

Schlüter, S., Gil, E., Doniger, T., Applebaum, I. and Steinberger, Y. (2022) Abundance and community composition of free-living nematodes as a function of soil structure under different vineyard managements. *Applied Soil Ecology* 170, 104291.

Schmidt, G. (1976) Der Einfluss der von den Beutetieren hinterlassen Spuren auf Suchverhalten und Sucherfolg von *Phytoseiulus persimilis* A.H. (Acarina: Phytoseiidae). *Zeitschrift fr Angewandte Entomologie* 82, 216–218.

Schmidt, H.U. (2009) Die Aufnahme von Wasserdampf aus der Atmosphäre und die kritische Gleichgewichtsluftfeuchte der in Vorräten lebenden Raubmilbe *Cheyletus malaccensis* Oudemans, 1903 (Acari: Cheyletidae). *Journal of Applied Entomology* 87(1–4), 122–131.

Schmidt, A.R., Jancke, S., Lindquist, E.E., Ragazzi, E., Roghi, G. *et al.* (2012) Arthropods in amber from the Triassic Period. *Proceedings of the National Academic Science U.S.A* 109, 14796–14801.

Schneider, K., Renker, K., Scheu, S. and Maraun, M. (2004) Feeding biology of oribatid mites: A minireview. *Phytophaga* 14, 247–256.

Schneider, K., Renker, K. and Maraun, M. (2005) Oribatid mite (Acari, Oribatida) feeding on ectomycorrhizal fungi. *Mycorrhiza* 16, 67–72.

Schneider, C.A., Rasband, W.S. and Eliceiri, K.W. (2012) NIH image to image J: 25 years of image analysis. *Nature Methods* 9, 671–675.

Schopf, Y.J.M. (1974) Permo-Triassic extinctions: relation to sea-floor spreading. *The Journal of Geology* 82, 129–143.

Schrank, F. Von Paula (1781) Envmeratio insectorvm Austriae indigenorum. Cum Figuris. pp. (1–24), 1–548, (1–4), Tab. I–IV (= 1–4).

Schuhl, J.F. (1977) Asma y alergia a acaros en Montevideo. *Allergologia Immunopathologia (Madras)* 5, 117–122.

Schulz, J. (2014) Massnahmen zur Bekämpfung der roten Vogelmilbe (*Dermanyssus gallinae*) in der ökologischen Legehennenhaltung. (Ph.D. thesis). Freie Universität Berlin, Berlin, Germany, pp. 1–126.

Schulz, S., Fuhlendorff, J., Steidle, J., Collatz, J. and Franz, J.-T. (2004) Identification and biosynthesis of an aggregation pheromone of the storage mite *Chortoglyphus arcuatus*. *Chembiochem* 5(11), 1500–1507.

Schulz, J., Berk, J., Suhl, J., Schrader, L., Kaufhold, S. *et al.* (2014) Characterization, mode of action, and efficacy of twelve silica-based acaricides against poultry red mite (*Dermanyssus gallinae*) *in vitro*. *Parasitology Research* 113(9), 3167–3175.

Schuster, R. (1972) Spinnvermögen der Tydeiden (Milben). *Naturwissenschaften* 59, 275.

Schuster, R. (2012) Sex ratio, fitness and capacity for population increase in *Pyemotes tritici* (L.-F. and M.) (Pyemotidae). In: Schuster, R. (ed.) *The Acari: Reproduction, Development and Life-History Strategies*. Springer Science and Business Media, Germany, pp. 209–219.

Schweizer, J. (1922) Beitrag zur Kenntnis der terrestrischen Milbenfauna der Schweiz. *Verhandlungen der Naturforschenden Gesellschaft in Basel* 33, 23–112.

Schweizer, J. (1949) Die Landmilben des Schweizerischen Nationalparkes. 1. Teil: parasitiformes reuter 1909. *Ergebnisse der Wissenschaftlichen Untersuchung des Schweizerischen Nationalparkes* 2(21), 1–99.

Schweizer, J. (1961) Die Landmilben der Schweiz (Mittelland, Jura und Alpen). Parasitiformes Reuter. *Denkschriften der schweizerischen Naturforschenden Gesellschaft* 84, 1–207.

Schweizer, J. and Bader, C. (1963) Die Landmilben der Schweiz (Mittelland, Jura and Alpen). Trombidiformes Reuter. *Denkschriften der Schweizerischen Naturforschenden Gesellschaft* 84, 209–378.

Scopoli, J.A. (1772) *Annus V. Historico Naturalis*. Sumtib. C.G., Hilscheri, Lipsiae, Germany, pp. 8–128.

Seeman, O.D., Beard, J.J. and Otto, J. (2023) Intraspecific variation in *Storchia pacifica* (Summers) (Trombidiformes: Stigmaeidae), a widespread species intercepted by Australian biosecurity from seventeen countries. *Systematic and Applied Acarology* 28(9), 1516–1526.

Seesaard, T., Goel, N., Kumar, M. and Wongchoosuk, C. (2022) Advances in gas sensors and electronic nose technologies for agricultural cycle applications. *Computers and Electronics in Agriculture* 193, 106673.

Seiedy, M., Saboori, A., Kamali, K. and Pakdel, A.K. (2009) Mites (Acari) found in flour mills in the Karaj region of Iran. *Systematic and Applied Acarology* 14(3), 191–196.

Sellnick, M. (1923) Die mir bekannten Arten der Gattung *Tritia* Berl. *Acari. Blätter fur Milbenkunde* 3, 7–22.

Sellnick, M. (1940) Die Milbenfauna Islands–Göteb. *Kungliga Svenska Vetenskapsakademiens Handlingar Stockholm (B)* 6(14), 1–129.

Sellnick, M. (1958) Milben aus landwirtschaftlichen Betrieben Nord-schwedens. *Medd Vaxtskyddsanst Stockholm* 11(71), 9–59.

Sellnick, M. and Forsslund, K.-H. (1955) Die Camisiidae Schwedens (Acar. Oribat.). *Arkiv för Zoologi Series* 8(4), 473–530.

Sen, D.P. (2005) Advances in fish processing technology. *Sunil Sacdev*. Allied Publishers Private Limited, New Dheli, India, pp. 1–282.

Seniczak, S. (1978) The morphology of juvenile forms of soil mites of the family Achipteriidae (Acari, Oribatei), I. *Annales Zoologici* 34(5), 89–99.

Seniczak, S. and Seniczak, A. (2007) Morphology of juvenile stages of *Parachipteria bella* (Sellnick, 1928) and *P. willmanni* Hammen, 1952 (Acari: Oribatida: Achipteriidae). *Annales Zoologici* 57(3), 533–540.

Seniczak, S. and Seniczak, A. (2010) 'Differentiation of body form of Protoplophoroidea (Acari: Oribatida) in the light of ontogeny of three species', *Journal of Natural History* 44(7), 389– 419.

Seniczak, S and Seniczak, A. (2012) Differentiation of external morphology of Oribatulidae (Acari: Oribatida) in light of the ontogeny of three species. *Zootaxa* 3184, 1–34.

Seniczak, A., Seniczak, S., Starý, J., Kaczmarek, S., Jordal, B.H. *et al.* (2021) High diversity of mites (Acari: Oribatida, Mesostigmata) supports the high conservation value of a broadleaf forest in Eastern Norway. *Forests* 12, 1098. DOI: 10.3390/f12081098.

Sepasgozarian, H. (1971) *Mites of Economic Importance in Iran*. Faculty of Agriculture Publication, University of Tehran-Karaj, pp. 245–247.

Sepasgozarian, H. (1977) The Twenty years of researches in acarology in Iran. *Journal of Iranian Society English* 56, 40–50.

Sepasgozarian, H. (1978) Stored pests of Iran and their control. *Tehran University Publication* 1026, 1–287.

Sepasgozarian, H. (1996) II. Addendum to the world species list of the superfamily Raphignatoidea (Acari). *Entomologische Mitteilungen aus dem Zoologischen Staatsinstitut und Zoologischen Museum Hamburg* 12(153), 31–34.

Sepasgozarian, H. and Mumcuoglu, K.Y. (1979) Faunistische und Okologische studien der Hausstaubmilben in Iran. *International Journal of Acarology* 4, 131–137.

Serpa, L.L.N., Franzolin, M.R., Barros-Battesti, D.M. and Kakitani, I. (2004) *Tyrophagus putrescentiae* predando insetos adultos de *Aedes aegypti* e *Aedes albopictus* em laboratório. *Revista de Saúde Pública* 38, 735–737.

Sesay, H.R. and Dobson, R.M. (1972) Studies on the mite fauna of house dusts in Scotland with special reference to that of bedding. *Acarologia* 14(3), 384–392.

Sevastyanov, V.D. (1975) Family Anoetidae. In: Gilyarov, M.S. (ed.) *Opredelitel' obitayushchikh v pochve kleshchey. Sarcoptiformes (Identification key to soil inhabiting mites. Sarcoptiformes).* Nauka Publishing, Moscow, USSR, pp. 382–416.

Severino, G., Oliver, J.H.and Pound, J.M. (1984) Synganglial and neurosecretory morphology of the chicken mite *Dermanyssus gallinae* (De Geer) (Mesostigmata: Dermanyssidae). *Journal of Morphology* 181, 49–68.

Seyd, E.L. and Seaward, M.R.D. (1984) The association of oribatid mites with lichens. *Zoological Journal of the Linnean Society* 80, 369–420.

Shahhosseini, M.J. and Kamali, K. (1989) Checklist of insect, mites and rodents affecting stored products in Iran. *Journal of Entomological Society of Iran* 5, 1–47.

Shamsi, M.H., Saboori, A. and Faraji, F. (2008) Fauna of ascid mites (Acari: Mesostigmata) in Damghan region, Semnan Province, Iran. In: Bertrand, M., Kreiter, S., McCoy, K.D., Migeon, A., Navajas, M., Tixier, M.-S. and Vial, L. (eds) Montpellier, EURAAC, France, pp. 245–249.

Sharp, J.L. and Haramoto, F.H. (1970) *Dermatophagoides pteronyssinus* (Trouessart) and other acarina in house dust in Hawaii. *Proceedings of the Hawaiian Entomological Society* 20, 583–589.

Shaw, G. (1794) In: Shaw, G. and Nodder F.P. *Vivarium Naturae or The Naturalist's Miscellany.* Nodder & Co., London, p. 187.

Shcherback, G. (1980) Kleshchi semeistva Rhodacaridae palearktiki. Naukova Dumka Kiev (in Russian).

Shchur, L.E. and Golovach, G.P. (1981) Distribution of acaroids mites in group and individual nests of the bank vole (*Chlethrionomys glareolus* Schreb.). *Vestnik Zoologii* 2, 86–88.

Sheals, J. G. (1956) Notas on a collection of soil Acari. *Entomologist's Monthly Magazine* 92, 99–103.

Shehata, M. and Baker, A. (1996) Mites infesting phlebotomine sandflies in southern Sinai, Egypt. *Medical and Veterinary Entomology* 10, 193–196.

Shen, Z.-P. (1979) Studies on the life history of *Carpoglyphus lactis* (L.) (Astigmata: Carpoglyphidae). *Acta Entomologica Sinica* 31(1), 60–66.

Shen Z.-P. (1984) A list and survey of stored-product mites in China. *Proc. 3rd International Working Conference on Stored Product Entomology, October 23–28, 1983.* Kansas State University, Manhattan, Kansas, U.S.A., pp. 680–686.

Shen, Z.-P. (1985) Checklist and brief review of stored products in China. *Journal of Grain Storage* 60, 3–8.

Shen, Z.-P. (1988) Study on the life history of mite *Suidasia nesbitti* Hughes. *Acta Entomologica Sinica* 31(1), 60–66.

Shen, Z.-P. (1989) The studies on life history of three flour mites and their damage to stored grain and food. *Grain Storage* 18(1), 3–7.

Shen, Z.-P. (1993) The life history of *Suidasia nesbitti* Hughes under natural condition. *Journal of Jilin Grain College* 1, 1–7.

Shen, Z.-P. (2005) Checklist of stored products mites in China. *Journal of Heilongjiang Grain 5*, 25–31.

Shen, Z.-P. (2006) Identification and control of important stored food mites in China. *Journal of Heilongjiang Grain* 4, 31–34.

Shen, X.-L., Zhao, Y.-J. and Wang, D.-X. (1992) A recent survey of stored product mites in Henan Province. *Journal of Zhengzhou Grain College* 3, 81–88.

Shepherd, A.W. (2012) Grain storage in Africa: Learning from past experiences. *Food Chain* 2(2), 149–163.

Shereef, G.M., Nawar, M.S. and Ali, F.S. (1990) Morphological and biological studies on *Macrocheles glaber* (Acari: Mesostigmata: Macrochelidae). *Bulletin de la Société Entomologique d'Égypte* 69, 305–313.

Shibata, S., Kuwahara, Y., Sato, M., Matsuyama, S., and Suzuki, T. (1998) Sex pheromone activity of 2-hydroxy-6-methylbenzaldehyde analogs against males of two astigmatid mites, *Aleuroglyphus ovatus* and *Acarus immobilis*. *Journal of Pesticide Science* 23, 34–39.

Shih, C.T. (1979) The influence of age of female *Tetranychus kanzawai* on sex ratio and life cycle of its progeny. In: Rodriguez, J.G. (ed.) *Recent Advances in Acarology, Vol. 1*. Academic Press, New York, U.S.A., pp. 511–517.

Shimp, R.L. Jr., Rowe, C., Reiter, K., Chen, B., Nguyen, V. *et al.* (2013) Development of a Pfs25-EPA malaria transmission blocking vaccine as a chemically conjugated nanoparticle. *Vaccine* 31(28), 2954–2962.

Shinkaji, N., Okabe, K. and Amano, H. (1986) Rhizoglyphine mites (Acari: Acaridae) infesting rakkyo (*Allium chinense* G. Don) and Chinese chive (*A. tuberosum* Rottler), and their susceptibility to dimethoate and disulfoton. *Japanese Journal of Applied Entomoloy and Zoology* 30, 285–289.

Shiralizadeh, S., Esfandiari, M., Shishehbor, P. and Farahi, S. (2021) Effect of temperature on the functional response of the predatory mite *Macrocheles muscaedomesticae* (Acari: Macrochelidae) by feeding on eggs of the house fly, *Musca domestica* (Dip.: Muscidae). *Plant Protection* 44(2), 19–31.

Sklyar V.E. (2003) Two new species of the genus *Iphidozercon* (Mesostigmata, Aceosejidae) from Ukraine. *Acarina* 11, 61–64.

Sidorchuk, E. and Bertrand, M. (2013) New fossil labidostomatids (Acari: Labidostomatidae) from Eocene amber and presence of an apustulate species in Europe. *Acarologia* 53, 25–39.

Sidorchuk, E.A., Schmidt, A.R., Ragazzi, E. Roghi, G. and Lindquist, E.E. (2014) Plant-feeding mite diversity in Triassic amber (Acari: Tetrapodili). *Journal of Systematic Palaeontology* 13(2), 129–151.

Siegert, M. (2016) Perfil nutricional de *Aleuroglyphus ovatus* e *Tyrophaghus putrescentiae* alimentado-se de matérias primas utilizadas na preparação de rações. *Dissertação Apresentada ao Programa de Pós-Graduação em Biotecnologia.* do Centro Universitário UNIVATES, Brazil, pp. 1–80.

Siegert, M.K., Silva G.L., Toldi, M., Johann L. and Ferla, N.J. (2018) Assessment on abiotic factors and the presence of storage mites in an animal feed factory. *Systematic and Applied Acarology* 23(12), 2317–2330.

Siepel, H. (1994) Life-history tactics of soil microarthropods. *Biology and Fertility of Soils* 18, 263–278.

Siepel, H. and de Ruiter-Dijkman, E.M. (1993) Feeding guilds of oribatid mites based on their carbohydrase activities. *Soil Biology and Biochemistry* 25, 1491–1497.

Siepel, H., Cremers, H. and Vierbergen, B. (2016) Provisional checklist of the astigmatic mites of the The Netherlands (Acari: Oribatida: Astigmatina). *Nederlandse Faunistische Mededelingen* 47, 49–88.

Siepel, H., Cremers, H., Dimmers, W., Loomans, A. and Vierbergen, B. (2018) Checklist of the mesostigmatic mites of The Netherlands (Acari: Mesostigmata). *Nederlandse Faunistische Mededelingen* 51, 115–188.

Sierra, S. and Martinez, H. (1987) *Suidasia medanensis Oudemans (Acari: Acaridae) una Plaga de Productos Almacenados.* Informe Técnico. Laboratorio Central de Cuarentena Vegetal (LCCV), Cuba, p. 6.

Sillos, P.P. (2002) *Sazonalidade E Testes de Eficácia Com Drogas Contra O Dermanyssus Gallinae (De Geer, 1778) E Ornithonyssus Sylviarum (Canestrini and Fanzago, 1877) Nas Granjas de Postura Da Região Metropolitana de Curitiba.* Federal University of Paraná, Curitiba, Brazil.

Silva, D.E. (2014) Diversidade da acarofauna associada a ninhos de aves silvestres no município de São Sepé, Rio Grande do Sul, Brazil. Dissertação apresentada ao Programa de Pós Graduação em Ambiente e Desenvolvimento, do Centro Univerisitário Univates, pp. 1–86.

Silva, G.L., Ferla, N.J., Toldi, M. and Faleiro, D.C.C. (2013) Mite fauna (Acari) associated to commercial laying hens and bird nests in Vale do Taquari, RS, Brazil. *Biotemas* 26(4), 255–264.

Silva, C., Paredes- León, R., Casanueva, M., Escobar-Huerta, G. and Moreno Salas, L. (2015) A new genus and species of pterygosomatid mite (Acari: Pterygosomatidae) parasitizing *Callopistes maculatus* (Squamata: Teiidae) from Chile. *Zootaxa* 3972(1), 65–74.

Silva, G.L., Metzelthin, M.H., Silva, O.S. and Ferla, N.J. (2016a) Catalogue of the mite family Tydeidae (Acari: Prostigmata) with the world key to the species. *Zootaxa* 4135(1), 1–68.

Silva, G.L., Radaelli, T.F.S. and Esswein, T.Z., Ferla, N.J. and Silva, O.S.S. (2016b) Comparison of biological development of *Blattisocius dentriticus* (Blattisocidae) fed on *Tyrophagus putrescentiae* (Acaridae) and *Megninia ginglymura* (Analgidae). *International Journal of Acarology* 42(8), 405–411.

Silva, D.E., Silva, G.L., Nascimento, J.M. and Ferla, N.J. (2018) Mite fauna associated with bird nests in southern Brazil. *Systematic and Applied Acarology* 23, 426–440.

Silva, C.A.D., Castilho, R.C., Galvao Filho, A.L.A. and Zanuncio, J.C. (2020) *Proctolaelaps bickleyi* (Acari: Mesostigmata: Melicharidae): First record of its association with cotton boll weevil. *Neotropical Entomology* 49, 311–313.

Silva, G.G.D., Zaldívar, M.F., Oliveira, L.A.R., Mariano, R.M.D.S., Lair, D.F. *et al.* (2023) Advances in non-chemical tools to control poultry hematophagous mites. *Veterinary Science* 10, 589. DOI: 10.3390/vetsci10100589.

Simser, J.A., Macaluso, K.R., Mulenga, A. and Azad, A.F. (2004) Immune-responsive lysozymes from hemocytes of the American dog tick, *Dermacentor variabilis* and an embryonic cell line of the Rocky Mountain wood tick, *D. Andersoni*. *Insect Biochemistry and Molecular Biology* 34, 1235–1246.

Singh, M. (1990) Qualitative and quantitative changes in some food grains due to mite infestation. Ph. D Thesis, H.A.U., Hisar.

Sinha, R.N. (1963) Stored product acarology in Canada. In: Naegele, J.A. (ed.) *Advances in Acarology*. Cornell University Press, Ithaca, New York, U.S.A., pp. 70–88.

Sinha, R.N., (1964a) Ecological relationships of stored-product mites and seed-borne fungi. *Acarologia* 6(2), 372–389.

Sinha, R.N. (1964b) Mites of stored grain in Western Canada: Ecology and methods of survey. *Proceedings of the Entomological Society of Manitoba* 20, 19–33.

Sinha, R.N. (1964c). Effect of low temperature on the survival of some storage mites. *Acarologia* 6, 336–341.

Sinha, R.N. (1966a) Feeding and reproduction of some stored-product mites on seed-borne fungi. *Journal of Economic Entomology* 59(5), 227–232.

Sinha, R.N. (1966b) *Aëroglyphus robustus*, a pest of stored grain. *Journal of Economic Entomology* 59(3), 686–688.

Sinha, R.N. (1968a) Climate and potential range of distribution of stored-product mites in Japan. *Journal of Economic Entomology* 61(1), 70–75.

Sinha, R.N. (1968b) Seasonal changes in Mite Populations in Rural Granaries in Japan. *Annals of the Entomological Society of America* 61, 938–949.

Sinha, R.N. (1968c) Adaptive significance of mycophagy in stored product Arthropoda. *Evolution* 22(4), 785–798.

Sinha, R.N. (1974) Seasonal abundance of insects and mites in small farm granaries. *Environmental Entomology* 3, 854–862.

Sinha, R.N. (1979a) Ecology of microflora in stored grain. *Annai Technology Agriculture* 28, 191–109.

Sinha, R.N. (1979b) Role of Acarina in the stored grain ecosystem. In: Rodriguez, J.G. (ed.) *Recent Advances in Acarology, Vol. 1*. Academic Press, New York, U.S.A., pp. 263–272.

Sinha, R.N. (1984) Acarine community in the stored rapeseed ecosystem, In: Griffiths, D.A. and Bowman, C.E. (eds) *Proceedings of the 6th International Congress of Acarology, Vol. 2, 5–11 September 1982*. Edinburgh University, Scotland, Ellis Horwood Ltd, Publisher, Halsted Press, Chichester, New York, U.S,A, pp. 1017–1025.

Sinha, R.N. and Harasymek, L. (1974) Survival and reproduction of stored-product mites and beetles on fungal and bacterial diets. *Environmental Entomology* 3(2), 243–246.

Sinha, R.N. and Mills, J.T. (1968) Feeding and reproduction of the grain mite and the mushroom mite on some species of *Penicillium*. *Journal of Economic Entomology* 61(6), 1548–1552.

Sinha, R.N. and Wallace, H.A.H. (1966) Association of granary mites and seed-borne fungi in stored grain and outdoor and indoor habitats. *Annals of the Entomological Society of America* 59, 1170–1181.

Sinha, R.N., and Wallace, H.A.H. (1973) Population dynamics of stored-product mites. *Oecologia* 12, 315–327.

Sinha, R.N. and White, N.D.G. (1982) Moisture relations of stored triticale and its susceptibility to infestation by mites and infection by microflora. *Canadian Journal of Plant Science* 62, 351–360.

Sinha, R.N., Wallace, H.A.H. and Chebib, F.S. (1969a) Principal-component analysis of interrelations among fungi, mites and insects in grain bulk ecosystems. *Ecology* 50, 536–547.

Sinha, R.N., Wallace, H.A.H. and Chebib, F.S. (1969b) Canonical correlations between groups of acarine, fungal and environmental variables in bulk grain ecosystems. *Researches on Population Ecology* 11, 92–104.

Sioutas, G., Petridou, E., Minoudi, S., Papadopoulos, K.V., Symeonidou, I. *et al.* (2023) Isolation of *Listeria monocytogenes* from poultry red mite (*Dermanyssus gallinae*) infesting a backyard chicken farm in Greece. *Scientific Reports* 13, 685.

Sitnikova, L.G. (1978) The main evolutionary trends of the Acari and the problem of their monophyletism. *Entomologiceskoe Obozrenie* 57, 431–457.

Skelton, A.C., Cameron, M.M., Pickett, J.A. and Birkett, M.A. (2010) Identification of neryl formate as the airborne aggregation pheromone for the American house dust mite and the European house dust mite (Acari: Epidermoptidae). *Journal of Medical Entomology* 47(5), 798–804.

Skoda-Smith, S., Mullen, G.R., Oi, F., Atkinson, T.P., Birmingham, M.D. and Auburn, A.L.W. (1996) Angioedema following dust mite exposure presenting as suspected food allergy. *Journal of Allergy and Clinical Immunology* 97, 228.

Skubala, P., Madej, Solarza, K. and Klis, G. (2003) Old mine underground galleries as the habitat for mites (Acari). In: Tajovský, K., Schlaghamerský, J. and Pižl, V. (eds) *Contributions to Soil Zoology in Central Europe I: Proceedings 7th Central European Workshop on Soil Zoology, 14–16 April 2003*. Czech Republic, Ústav půdní biologie AV ČR, pp. 141–145.

Skubala, P., Marzec, A. and Sokolowska, M. (2006) Accidental acarophagy: Mites found on fruits, vegetables and mushrooms. *Biology Letters* 43, 249–255.

Skvarla, M.J., Fisher, J.R. and Dowling, A.P.G. (2014) A review of Cunaxidae (Acariformes, Trombidiformes): Histories and diagnoses of subfamilies and genera, keys to world species, and some new locality records. *ZooKeys* 418, 1–103.

Sleeckx, N., Kempen, I. and Zoons, J. (2017) The potential of an integrated PRM strategy in practice. In: Sparagano *et al.* (eds) *Improving Current Understanding and Research for Sustainable Control of the Poultry Red Mite Dermanyssus gallinae*. Abstract book 3rd COST Conference COREMI, 19–22 September, Oeiras, Portugal, p. 31.

Sleeckx, N., Van Gorp, S., Koopman, R., Kempen, I., Van Hoye, K. *et al.* (2019) Production losses in laying hens during infestation with the poultry red mite *Dermanyssus gallinae*. *Avian Pathology* 48(1), 17–21.

Smallman, B.N. (1942) *Entomological Investigations. Grain Research Laboratory*. 15th Annual Report (1941), Winnipeg, Manitoba, pp. 40–43.

Smart, L.E., Martin, J.L., Limpalaër, M., Bruce, T.J. and Pickett, J.A. (2013) Responses of herbivore and predatory mites to tomato plants exposed to jasmonic acid seed treatment. *Journal of Chemical Ecology* 39, 1297–1300.

Smiley, R.L. (1987) Mites (Acari). In: Gorham, J.R. (ed.) *Insect and Mite Pests in Food: an Illustrated Key*. USDA, Agricultural Handbook 655, pp. 3–43.

Smiley, R.L. (1992) *The Predatory Mite Family Cunaxidae (Acari) of the World with a New Classification*. Indira Publishing House, West Bloomfield, Michigan, U.S.A., pp. 1–356.

Smiley, R.L. and Moser, J.C. (1968) New species of mites from pine (Acarine: Tarsocheylidae, Eupalopsellidae, Caligonellidae, Cryptognathidae, Raphignathidae, and Neophyllobiidae). *Proceedings of the Entomological Society of Washington, USA* 70, 307–317.

Smith, H.H. (1923) Curing meat on the farm. *Colorado Agricultural College Extension Service, Fort Collins, Colorado, USA*.

Smith, K.G. (1969) Infestation by mites as a factor in the long term storage of flour. In: Evans, G.O. (ed.) *Proc. 2nd int. Congr. Acarology 1967*. Academia Kaidó, Budapest, Hungary, pp. 249–253.

Smith, T. (2023) *Bulb Mites-UMass Extension Greenhouse Crops and Floriculture Program*. University of Massachsetts Amherst. Available at: https://ag.umass.edu/greenhouse-floriculture/fact-sheets/bulb-mites (accessed 17 April 2023).

Smrž, J. (2003) Microanatomical and biological aspects of bacterial associations in *Tyrophagus putrescentiae* (Acari: Acaridida). *Experimental and Applied Acarology* 31(1–2), 105–113.

Smrž, J. (2006) Types of haemocytes in saprophagous soil mites (Acari: Oribatida, Acaridida), and the correlation between their presence and certain processes within mites. *European Journal of Entomology* 103, 679–686.

Smrž, J. and Catska, V. (1987) Food selection of the field population of *Tyrophagus putrescentiae* (Schank) (Acari, Acarida). *Zeitschrift fr Angewandte Entomologie* 104, 329–335.

Smrž, J. and Catska, V. (1989) The effect of the consumption of some soil fungi on the internal microanatomy of the mite *Tyrophagus putrescentiae* (Schrank) (Acari: Acaridida). *Acta Universitatis Carolinae. Biologica* 33, 81–93.

Smrž, J. and Catska, V. (2010) Mycophagous mites and their internal associated bacteria cooperate to digest chitin in soil. *Symbiosis* 52, 33–40.

Smrž, J. and Kocourkova, J. (1999) Mite communities of two epiphytic lichen species (*Hypogymnia physodes* and *Parmelia sulcata*) in the Czech Republic. *Pedobiologia* 43, 385–390.

Smrž, J., Soukalová, H., Čatská, V. and Hubert, J. (2016) Feeding patterns of *Tyrophagus putrescentiae* (Sarcoptiformes: Acaridae) indicate that mycophagy is not a single and homogeneous category of nutritional biology. *Journal of Insect Science* 16(1), 94.

Soares, K.R. and Ximenes, L.F. (2022) *Egg Production; Northeast Economic Studies Technical Office-ETENE*. Moosic, PA, USA.

Soares, L.W., Kuwahara, Y. Suzuki, T. and Kurosa, K. (1989) The alarm pheromone of the mite *Suidasia medanensis* Oudemans, 1924 (Acariformes, Suidasiidae). *Agricultural and Biological Chemistry* 53(10), 2703–2709.

Šobotník, J., Alberti, G., Weyda, F. and Hubert, J. (2008a) Ultrastructure of the digestive tract in *Acarus siro* (Acari: Acaridida). *Journal of Morphology* 269(1), 54–71.

Šobotník, J., Kudlikova-Krizkova, I., Vancova, M., Munzbergova, Z. and Hubert, J. (2008b) Chitin in the peritrophic membrane of *Acarus siro* (Acari: Acaridae) as a target for novel acaricides. *Journal of Economic Entomology* 101(3), 1028–1033.

Sokół, R., Szkamelski, A. and Barski, D. (2008) Influence of light and darkness on the behaviour of *Dermanyssus gallinae* on layer farms. *Polish Journal of Veterinary Sciences* 11, 71–73.

Solarz, K. (1986) The allergenic fauna of house-dust mites in some Silesian towns. *Wiadomości Parazytologiczne* 32, 431–433.

Solarz, K. (1995) A preliminary survey of allergenic mites in railway trucks transporting stored food products at Strzemieszyce Railway Station (Upper Silesia, Poland). In: Kropczyñska, D., Boczek, and Tomczyk, A. (eds) *The Acari: Physiological and Ecological Aspects of Acari-Host Relationships*. Oficyna Dabor, Warszawa, Poland, EURAAC, pp. 281–287.

Solarz, K. (1997) Seasonal dynamics of house dust mite populations in bed/mattress dust from two dwellings in Sosnowiec (upper Silesia, Poland): an attempt to assess exposure. *Annals of Agricultural and Environmental Medicine* 4, 253–261.

Solarz, K. (2000) Some species of mites (Acari) from house dust in Upper Silesia (Poland). *Acta Zoologica Cracoviensia* 43, 241–259.

Solarz, K. (2001a) Risk of exposure to house dust pyroglyphid mites in Poland. *Annals of Agricultural and Environmental Medicine* 8, 11–24.

Solarz, K. (2001b) Pyroglyphidae (Acari: Astigmata) in Poland. Distribution, biology, population ecology and epidemiology. *Acta Zoologica Cracoviensia* 44, 435–528.

Solarz, K. (2010) Temporal changes in the composition of house-dust-mite fauna in Poland. *Acta Zoologica Cracoviensia* 53B (1–2), 39–64.

Solarz, K. (2011) House Dust Mites, Other Domestic Mites and Forensic Medicine. In: Vieira, D.N. (ed.) *Forensic Medicine–From Old Problems to New Challenges*, InTech, Rijeka, Croatia, pp. 327–358.

Solarz, K. (2012) House dust mites and storage mites (Acari: Oribatida: Astigmatina). Identification keys. *Institut of Systematics and Evolution of Animals*. Polish Academy of Sciences, Cracow, pp. 1–120.

Solarz, K. and Pająk, C. (2019a) Risk of exposure of a selected rural population in South Poland to allergenic mites. Part II: acarofauna of farm buildings. *Experimental and Applied Acarology* 77(1), 387–399.

Solarz, K. and Pająk, C. (2019b) Risk of exposure of a selected rural population in South Poland to allergenic mites. Part I: indoor acarofauna of one-family houses. *Experimental and Applied Acarology* 77, 375–386.

Solarz, K. and Solarz, D. (1996) The allergenic mites in coal-mine dust from coal mines in Upper Silesia (Poland). *Annals of Agricultural and Environmental Medicine* 3, 49–55.

Solarz, K., Madej, G., Źbikowska-Zdun, K. and Dudziak, G. (2002) Mites of orders Acaridida, Gamasida and Oribatida in coal mines of Upper Silesian region (Poland). In: Ignatowicz, S. (ed.) *Postępy Polskiej Akarologii*. Wydawnictwo SGGW, Warszawa, Poland, pp. 179–193.

Solarz, K., Pająk, C., Pawelczyk, O., Bobiński, R., Ciechacka, M., Dutka, M. and Ulman-Wlodarz, I. (2022) Mites occurring in farm buildings as allergic agents and indicators in forensic analyses. *Acarologia* 62 (1), 36–47.

Solarz, K., Szilman, E. and Szilman, P. (1994) Allergenic mites from the other Acaridida of agricultural environments. *Abstracts of 17th Congress of Polish Parasitological Society, 15–17 September 1994*. Gdynia, Poland, p. 81.

Solarz, K., Szilman, E. and Szilman, P. (1995) *Dermatophagoides evansi* Fain, Hughes et Johnston, 1967 - the new for Polish fauna species of mite of the family Pyroglyphidae (Acari: Astigmata: Psoroptidia). *Prz-Zool* 39, 271–277.

Solarz, K., Szilman, P. and Szilman, E. (1997) Preliminary study on the occurrence and species composition of astigmatic mites (Acari: Astigmata) in samples of dust, debris and residues from farming environments in Poland. *Annals of Agricultural and Environmental Medicine* 4, 249–252.

Solarz, K., Szilman, P. and Szilmanm E. (1998) Allergenic acarofauna from nests of selected species of synanthropic birds in Poland. *Bulletin of the Scandinavian Society for Parasitology* 8, 81–82.

Solarz, K., Szilman, P. and Szilmanm E. (1999) Allergenic mites associated with bird nests in Poland (Astigmata: Pyroglyphidae, Acaridae, Glycyphagidae). In: Bruin, J., Van Der Geest, L.P.S. and Sabelis, M.W. (eds) *Ecology and Evolution of the Acari*. Kluwer Academic Publishers, Dordrecht, The Netherlands, pp. 651–656.

Solarz, K., Szilman, P. and Szilmanm E. (2000) Further studies on an occurrence of allergenic mites in nests of synanthropic and semisynanthropic birds in Southern Poland (Acaridida: Pyroglyphidae, Acaridae, Glycyphagidae). *Proc. 26th Symposium of Acarology, Kazimierz Dolny, Poland, TN, September 24–26, 1999*. In: Ignatowicz, S. (ed.) Akarologia polska u progu XXI wieku. *Wydawnictwo SGGW, Warszawa*, Poland, pp. 340–351.

Solarz, K., Szilman, P., Szilman, E., Krzak, M. and Jagła, A. (2004) Some allergenic species of astigmatid mites (Acari, Acaridida) from different synanthropic environments in southern Poland. *Acta Zoologica Cracoviensia* 47(3–4), 125–145.

Solarz, K., Senczuk, L., Maniurka, H., Cichecka, E. and Peszke, M. (2007) Comparisons of the allergenic mite prevalence in dwellings and certain outdoor environments of the Upper Silesia (southwest Poland). *International Journal of Hygiene and Environmental Health* 210(6), 715–724.

Soler Cruz, M.D., Vega Robles, M.C., Jespersen, J.B., Kilpinen, O., Birkett, M., Dewhirst, S. and Pickett, J. (2005) Scanning electron microscopy of foreleg tarsal sense organs of the poultry red mite, *Dermanyssus gallinae* (De Geer) (Acari: Dermanyssidae). *Micron Technology* 36(5), 415–421.

Soliman, Z.R. (1975) The family Bdellidae in Giza (Egypt) with a description of a new species. *Bulletin - Zoological Society of Egypt* 27, 47–50.

Soliman, Z.R. and Zaher, M.A. (1975) Bdellid mites of Lattakia, Syria. *Bulletin de la Société Entomologique* 59, 73–82.

Soliman, Z.R., Zaher, M.A. and El-Bishlawy, S.M. (1975) Studies on the biology of the predaceous mite *Cunaxa capreolus* Berl. (Acari, Prostigmata, Cunaxidae). *Anzeiger für Schädlingskunde,* Pflanz enschutz *und Umweltschutz* 48, 124–126.

Soliman, Z.R., Zaher, M.A. and Mohamed, M.I. (1978) Biology and predaceous efficiency of *Macrocheles matrius* (Hull) (Acari, Mesostigmata). *Zeitschrift für Angewandte Entomologie* 85(1–4), 225–230.

Solomon, M.E. (1946) Tyroglyphid mites in stored products; nature and amount of damage to wheat. *Annals of Applied Biology* 33(3), 280–289.

Solomon, M.E. (1962) Ecology of the flour mite, *Acarus siro* L. (= *Tyroglyphus farinae* De Ge.). *Annals of Applied Biology* 50, 178–184.

Solomon, M.E. (1966) Moisture gains, losses and equilibria of flour mites, *Acarus siro* L., in comparison with larger arthropods. *Entomologia Experimentalis et Applicata* 9, 24–41.

Solomon, M.E. (1969) Establishment, growth and decline of populations of the grain mite *Acarus siro* L., on a handful of wheat. In: Evans, G.O. (ed.) *Proceedings of the 2nd International Congress of Acarology, 19–25 July 1967*, Sutton Bonington, UK, Akademiai Kiado, Budapest, Hungary, pp. 255–260.

Solomon, M.E., Hill, S.T., Cunnington, A.M. and Ayerst, G. (1964) Storage fungi antagonistic to the flour mite (*Acarus siro* L.). *Journal of Applied Ecology* 1, 119–125.

Soltani, A., Azizi, K., Saleh, V. and Dabaghmanesh, T. (2011) The fauna and distribution of house dust mites in residential homes of Bandar Abbas District, Southern Iran. *Experimental and Applied Acarology* 54, 269–276.

Somerfield, K.G. (1980) Insects and mites associated with dried milk product storage areas in New Zealand. *New Zealand Journal of Experimental Agriculture* 8, 83–85.

Sonenshine, D.E. (1985) Pheromones and other semiochemicals of the Acari. *Annual Review of Entomology* 30, 1–28.

Sonenshine, D.E. and Gregson, J.D. (1970) A contribution to the internal anatomy and histology of the bat tick *Ornithodoros kelleyi* Cooley and Kohls, 1941. I, The alimentary system, with notes on the food channel in *Ornithodoros denmarki* Kohls, Sonenshine and Clifford. *Journal of Medical Entomology* 7, 46–64.

Song, H.Y., Yang, J.Y., Suh, J.W. and Lee, H.S. (2011) Acaricidal activities of apiol and its derivatives from *Petroselinum sativum* seeds against *Dermatophagoides pteronyssinus*, *Dermatophagoides farinae*, and *Tyrophagus putrescentiae*. *Journal of Agricultural and Food Chemistry* 59, 7759–7764.

Sorenson, J.T., Kinn, D.N., Doutt, R.L. (1983) Biological observations on *Bdella longicornis*: A small predatory mite in California vineyards (Acari: Bdellidae). *Entomography* 2, 297–305.

Sorokin, S. (1948) Population of grain mites in hay. *Zoologicheskii Zhurnal* 27, 487–491.

Sorokin, S.B. (1953) Eigenarten in der Verbreitung von Getreidemilben in Nestern von Mäusen auf landwirtschaftlich genutzten Flächen. *Zoologicheskii Zhurnal* 32, 60–76.

Sorokin, S.V. (1951) The causes of the formation of the hypopi in cereal mites. *Zoologicheskii Zhurnal* 30, 523–539.

Sorush, M.J. (1994) Identification of pomegranate mites and bioloy of *Tenuipalpus punicae* P. and B. in Saveh. M.Sc. thesis. Shahid Chamran Univerity, Ahvaz, Iran, pp. 1–103.

Sosnina, E.F., Vysotskaya, S.O., Markov, G.N. and Atanasov, L.K. (1965) Predatory mites of the family Bdellidae (Acarina, Prostigmata) from nests of rodents in Bulgaria. *Proceedings Zoological Institute of the Russian Academy of Sciences* 35, 272–287.

Souhaila, M.H. (1992) Mite fauna in stored grain seeds in central Iraq. *Journal of Stored Products Research* 28(3), 179–181.

Sourassou, N., de Moraes, G.J., Delalibera, I. and Corrêa, A.S. (2015) Phylogenetic analysis of Ascidae *sensu lato* and related groups (Acari: Mesostigmata: Gamasina) based on nuclear ribosomal DNA partial sequences. *Systematic and Applied Acarology* 20, 225–240.

Sousa, J.M., Gondim, J.R., Manoel, G.C., Barros, R. and Oliveira, J.V. (2005) Mites in stored foods commercialized in supermarkets and markets in Recife. *Neotropical Entomology* 34(2), 303–309.

Southcott, R.V. (1961) Studies on systematics and biology of Erythraeoidea (Acarina), with critical revision of genera and subfamilies. *Australian Journal of Zoology* 9(3), 367–610.

Southcott, R.V. (1991) A further revision of *Charletonia* (Acarina: Erythraeidae) based on larvae, protonymphs and deutonymphs. *Invertebrate Taxonomy* 5, 61–131.

Southcott, R.V. (1993) Larvae of *Leptus* (Acarina: Erythraeidae) ectoparasitic on higher insects of Australia and New Guinea. *Invertebrate Taxonomy* 7, 1473–1550.

Southcott, R.V. (1996) Description of two new larval smarididae (Acarina) from Australia. *Records of the South Australian Museum* 30(1), 1–12.

Sparagano, O. (2020) A nonexhaustive overview on potential impacts of the poultry red mite (*Dermanyssus gallinae*) on poultry production systems. *Journal of Animal Science* 98(Supplement 1), 58–62.

Sparagano, O.A.E. and Ho, J. (2020) Parasitic mite Fauna in Asian poultry farming systems. *Frontiers in Veterinary Science* 7, 400.

Sparagano, O.A.E., Pavlićević, A., Murano, T., Camarda, A., Sahibi, H. *et al.* (2009) Prevalence and key Figures for the poultry red mite *Dermanyssus gallinae* infections in poultry farm systems. *Experimental and Applied Acarology* 48(1–2), 3–10.

Sparagano, O., Khallaayoune, K., Duvallet, G., Nayak, S. and George, D. (2013) Comparing terpenes from plant essential oils as pesticides for the poultry red mite (*Dermanyssus gallinae*). *Transboundary and Emerging Diseases* 60, 150–153.

Sparagano, O., George, D., Harrington, D. and Giangaspero, A. (2014a) Significance and control of the poultry red mite, *Dermanyssus gallinae*. *Annual Review of Entomology* 59, 447–466.

Sparagano, O., George, D.R., Finn, R.D., Giangaspero, A., Mul, M. F. *et al.* (2014b) *Dermanyssus gallinae* and poultry production: Impact, management, and a predicted compatibility matrix for integrated approaches. In: Birger, S., European Poultry Conference and World's Poultry Science Association (Hg.) (eds) *Proc. XIVth European Poultry Conference, Stavanger, Norway, 23–27 June 2014*. WPSA Publishing, Norway, pp. 131–141.

Spieksma, F.Th.M. (1967) The house-dust mite *Dermatophagoides pteronyssinus* (Trouessart, 1897) producer of the house-dust allergen

(Acari: Psoroptidae). Thesis. Bateljee and Terpistra, Leiden, pp. 1–65.

Spieksma, F.Th.M. (1968) Die huistofmijt *Dermatophagoides pteronyssinus* (Trouessart, 1897) de producent van het huisstofallergen. *Entomologische Berichte (Amst.)* 28, 27–38.

Spieksma, F.Th.M. and Spieksma-Boezeman, M.I.A. (1967) The mite fauna of house dust with particular reference to the house dust mite *Dermatophagoides pteronyssinus* (Trouessart, 1897) (Psoroptidae, Sarcoptiformes). *Acarologia* 9, 226–241.

Spieksma, F.Th.M., Zuidema, P. and Leupen, M.J. (1971) High altitude and house dust mites. *The BJM* 1(5740), 82–84.

Spiewak, R., Gora, A. and Dutkiewicz, J. (2001) Work-related skin symptoms and type I allergy among eastern-polish farmers growing hops and other crops. *Annals of Agricultural and Environmental Medicine* 8(1), 51–56.

Stables, L.M. (1980) The effectiveness of some recently developed pesticides against stored-product mites. *Journal of Stored Products Research* 16, 143–146.

Stafford, K.A., Lewis, P.D. and Coles, G.C. (2006) Preliminary study of intermittent lighting regimens for red mite (*Dermanyssus gallinae*) control in poultry houses. *Veterinary Record* 158, 762–763.

Stanaland, B.E., Fernandez-Caldas, E., Jacinto, C.M., Trudeau, W.L. and Lockey, R.F. (1998) Positive nasal challenge responses to *Blomia tropicalis*. *Journal of Allergy and Clinical Immunology* 97(5) 1045–1049.

Stanley, M.S. (2007) An analysis of the history of marine animal diversity. *Paleobiology* 33(4), 1–55.

Stará, J., Erban, T. and Hubert, J. (2010) The Effect of Chitin Metabolic Effectors on the Population Increase of Stored Product Mites. *Experimental and Applied Acarology* 52(2), 155–167.

Stará, J., Nesvorná, M. and Hubert, J. (2011a) Long-term pre-exposure of the pest mite *Tyrophagus putrescentiae* to sub-lethal residues of bifenthrin on rapeseed did not affect its susceptibility to bifenthrin. *Crop Protection* 30, 1227–1232.

Stará, J., Nesvorna´, M. and Hubert, J. (2011b) The toxicity of selected acaricides against five stored product mites under laboratory assay. *Journal of Pest Science* 84, 387–391.

Stará, J., Stejskal, V., Nesvorna, M., Plachy, J. and Hubert, J. (2011c) Efficacy of selected pesticides against synanthropic mites under laboratory assay. *Pest Management Science* 67, 446– 447.

Steenberg, T. and Kilpinen, O. (2003) Fungus infection of the chicken mite *Dermanyssus gallinae*. *Bulletins IOBC/WPRS* 26, 23–26.

Steenberg, T. and Kilpinen, O. (2014) Synergistic interaction between the fungus *Beauveria bassiana* and desiccant dusts applied against poultry red mites (*Dermanyssus gallinae*). *Experimental and Applied Acarology* 62, 511–524.

Steenberg, T., Kilpinen, O. and Moore, D. (2006) Fungi for control of the poultry red mite, *Dermanyssus gallinae*. In: *Proceeding International Workshop Implement Biocontrol Pract. Temp. Reg. Present and Near Future, Nov. 1–3, 2005*. Aarhus Universitet, Det Jordbrugsvidenskabelige Fakultet Publisher, DIAS Rep., Flakkebjerg, Danmark, pp. 71–74.

Stefaniak, O. and Senaczak, S. (1976) The microflora of alimentary canal of *Achipteria coleoptrata* (Acarina: Oribatei). *Pedobiologia* 16, 185–194.

Stein, W. (1960) Untersuchungen über die Möglichkeit einer Bekämpfung von Raubmilben (*Blattisocius tenivorus*) in Zuchten der Getreidemotte (*Sitrotoga cerrealella* (Oliv.)) durch Anwendung von Akeariziden. *Zeitschrift für Pflanzenkrankheiten und Pflanzenschutz* 67, 77–87.

Steinkraus, D. C. and Cross, E.A. (1993) Description and life history of *Acarophenax mahunkai*, n. sp. (Acari, Tarsonemina: Acarophenacidae), an egg parasite of the lesser mealworm (Coleoptera: Tenebrionidae). *Annals of the Entomological Society of America* 86, 239–249.

Steinman, H.A. and Lee, S. (2003) Concomitant clinical sensitivity (CCS) and Crossreactivity. *Allergy Clinical Immunology International* 15(1), 18–29.

Steinman, H. and Ruden, S. (2007) *Mites, Allergy-Which Allergens? Thermo scientific*. X-O Graf Tryckeri AB, Uppsala, Sweden, pp. 1–90.

Stejskal, V. (2001) A new concept of economic injury level that includes penalization of damage to quality and safety of agricultural products. *Plant Protection Science* 37, 151–156.

Stejskal, V. and Hubert, J. (2008) Risk of occupational allergy to stored grain arthropods and false pest-risk perception in Czech grain stores. *Annals of Agricultural and Environmental Medicine* 15(1), 29–35.

Stejskal, V. and Hubert, J. (2009) Differential levels of mite infestation of wheat and barley in Czech grain stores. *Journal of Insect Science* 16, 255–262.

Stejskal, V., Hubert, J., Kučerová, Z., Munzbergová, Z., Lukáš, J. and Žďárková, E. (2003) The influence of the type of storage on pest infestation of stored grain in the Czech Republic. *Plant, Soil and Environment* 49(2), 55–62.

Stejskal, V., Aulicky, R. and Kučerová, Z. (2014) Pest control strategies and damage potential of seed-infesting pests in the Czech stores-A review. *Plant Protection Science* 50(4), 165–173.

Stejskal, V., Vendl, T., Li, Z. and Aulicky, R. (2019) Minimal thermal requirements for development and activity of stored product and food industry pests (Acari, Coleoptera, Lepidoptera, Psocoptera, Diptera and Blattodea): A review. *Insects* 10(149). Available at: file:///C:/Users/Utente/Downloads/insects-10-00149.pdf (accessed 26 May 2020).

Stejskal, V., Vendl, T., Aulicky, R. and Athanassiou, C. (2021) Synthetic and natural insecticides: Gas, liquid, gel and solid formulations for stored-product and food-industry pest control. *Insects* 12(7), 590.

Stenhouse, D.A. (2015) Phoresy of *Uroobovella marginata* (C.L. Koch) (Acari: Mesostigmata) on *Pterostichus madidus* (Fabricius) (Carabidae). *Coleopt* 24(1), 44.

Stenius, B. and Cunnington, A.M. (1972) House dust mites and respiratory allergy: a qualitative survey of species occurring in Finnish house dust. *Scandinavian Journal of Infectious Diseases* 53, 338–348.

Sternberg, S. and Tahori, A.S. (1962) Control of *Aleuroglyphus ovatus* (Troupeau) in flea cultures. *Bulletin of Entomological Research* 53(2), 257–264.

Sternlicht, M. and Goldenberg, S. (1971) Fertilisation, sex ratio and postembryonic stages of the citrus bud mites *Aceria sheldoni* (Ewing) (Acarina, Eriophyidae). *Bulletin of Entomological Research* 60, 391–397.

Stewart, G.A., Bird, C.H. and Thompson, P.J. (1992a) Do the group II dust mite allergens correspond to lysozyme? *Journal of Allergy and Clinical Immunology* 90 (1), 141–142.

Stewart, G.A., Bird, C.H., Krska, K.D., Colloff, M.J. and Thompson, P.J. (1992b) A comparative study of allergenic and potentially allergenic enzymes from *Dermatophagoides pteronyssinus*, *D. farinae* and *Euroglyphus maynéi*. *Experimental and Applied Acarology* 16(1–2), 165–180.

Stewart, G.A., Ward, L.D., Simpson, R.J. and Thompson, P.J. (1992c) The group III allergen from the house dust mite *Dermatophagoides pteronyssinus* is a trypsin-like enzyme. *Immunology* 75 (1), 29–35.

Stewart, G.A., Kollinger, M.R., King, C.M. and Thompson, P.J. (1994) A comparative study of three serine proteases from *Dermatophagoides pteronyssinus* and *D. farinae*. *Allergy* 49(7), 553–560.

Stewart, G.A., Hage-Hamsten van, M., Krska, K., Thompson, P.J. and Olsson, S. (1998) An enzymatic analysis of the storage mite *Lepidoglyphus destructor*. *Comparative Biochemistry and Physiology* 119B, 341–347.

Stewart, G.A., Lake, F.R. and Thampson P.J. (1991) Faecally derived hydrolytic enzymes from *Dermatophagoides pteronyssinus*: physicochemical characterization of potential allergens. *International Archives of Allergy and Immunology* 95 (2–3), 248–56.

Stingeni, L. Principato, M. and Lisi, P. (1997) Glicifagosi: due casi di dermatite papulo-vescico-pustolosa da *Glycyphagus domesticus* e *Lepidoglyphus destructor* (Astigmata: Glycyphagidae). *Annali Italiani di Dermatologia Clinica e Sperimentale* 51, 91–95.

Stingeni, L., Bianchi, L., Tramontana, M., Moretta, I. and Principato, M. (2015) Indoor dermatitis due to *Aeroglyphus robustus*. *British Journal of Dermatology* 174(2), 454–456.

Stingeni, L., Hansel, K., Casciola, G., Bianchi, L., Tramontana, M. *et al.* (2023) Human ectoparasitosis by mites of the genus *Pyemotes* Amerling 1861 (Acarina: Pyemotidae). *Italian Journal of Dermatology and Venereology* 158(1), 4–14.

Stirmanov, A.V. (2013) Acaro del Pane Integrale-*Gohieria fusca Ouds*. Available at: http://www.pesticidy.ru/%D0%9A%D0%BB%D0%B5%D1%89_%D0%B1%D1%83%D1%80%D1%8B%D0%B9_%D1%85%D0%BB%D0%B5%D0%B1%D0%BD%D1%8B%D0%B9 (accessed 10 August 2020).

Storaas, T., Steinsvag, S.K., Florvaag, E., Irgens, A. and Aasen, T.B. (2005) Occupational rhinitis: diagnostic criteria, relation to lower airway symptoms and IgE sensitization in bakery workers. *Acta Oto-Laryngologica* 125(11), 1211–1217.

Stored Grain Pest (1955) *Farmer's Bulletin, 1260*, United States Department of Agriculture, Washington, DC, U.S.A., pp. 1–46.

Strandtmann, R.W. and Wharton, G.W. (1958) In: Yunker, C.E. (ed.) *Manual of Mesostigmatid Mites Parasitic on Vertebrates*, College Park, Institute of Acarology, University of Maryland, U.S.A., pp. 1–69.

Stratil, H.U. and Knülle, W. (1985) Die induktion des hypopusstadiums bei der im lagergetreide vorkommenden milbe *Glycyphagus destructor* (Schrank, 1781) (Astigmata, Glycyphagidae). *Zeitschrift f r Angewandte Entomologie* 99, 350–365.

Stratil, H.U., Startil, H.H.and Knülle, W. (1980) The innate capacity for increase of the stored product mite *Glycyphagus destructor* (Schrank, 1781) at different temperatures and humidities. *Journal of Applied Entomology* 90(1–5), 209–220.

Stroiński, A., Felska, M. and Mąkol, J. (2013) A review of host-parasite associations between terrestrial parasitengona (Actinotrichida: Prostigmata) and bugs (Hemiptera). *Annales Zoologici* 63(1), 195–221.

Strong K.L. and Halliday, R.B. (1994) Three new species of *Hypoaspis* Canestrini (Acarina: Laelapidae) associated with large Australian cockroaches. *Journal of the Australian Entomological Society* 33, 87–96.

Strong, D.R. Jr, Lawton, J.H. and Southwood, T.R.E. (1984) *Insects on Plants: Community Patterns and Mechanisms*. Blackwell Scientific Publications, Oxford, U.K., pp. 1–313.

Studzinski, A. and Malachowska, D. (1975) The occurrence of Acarina on cruciferous plants (Cruciferae) in Poland in the years 1970–1971. *Prace Naukowe Instytutu Ochrony Roślin* 15(2), 153–166.

Sturhan, D. and Hampel, G. (1977) Pflanzenparasitische Nematoden als Beute der Wurzelmilbe *Rhizoglyphus echinopus* (Acarina, Tyroglyphidae). *Anzeiger für Schädlingskunde, Pflanzenschutz und Umweltschutz* 50, 115–118.

Suárez, A. (2004) Catálogo de ácaros de la provincia de Guantánamo. *Fitosanidad* 8(1), 23–31.

Subías, L.S. (2019) Listado sistemático, sinonímico y biogeográfico de los Ácaros Oribátidos (Acariformes: Oribatida) del mundo (Excepto fosiles), 14ª actualización, pp. 1–536. Available at: http://bba.bioucm.es/cont/docs/RO 1.pdf (accessed 2 September 2019).

Subías, L.S. and Arillo, A. (2002) Oribatid mites fossils from the Upper Devonian of South Mountain, New York and the Lower Carboniferous of Country Antrim, Northern Ireland (Acariform, Oribatida). *Estudios del Museo de Ciencias Naturales de Álava* 17, 93–106.

Sublett, J.W. and Bernstein, D.I. (2011) Occupational rhinitis. *Immunology and Allergy Clinics of North America* 31, 787–796.

Sullivan, P. (2003) Overview of Cover Crops and Green Manures. Available at: https://attra.ncat.org/product/overview-of-cover crops-and-green-manures/ (accessed 17 April 2023).

Sumangala, K. and Haq, M.A. (2001) Survey of the mite fauna associated with *Apis* spp. in Kerala, southern India. In: Halliday, R.B., Walter, D.E., Proctor, H.C., Norton, R.A. and Colloff, M.J. (eds). *Acarology: Proceedings of the 10th International Congress*. Melbourne, Australia, CSIRO Publishing, pp. 565–568.

Summers, F.M. (1962) The genus *Stigmaeus* (Acari: Stigmaeidae). *Hilgardia* 33, 491–537.

Summers, F.M. (1964) Three uncommon genera of the mite family Stigmaeidae. *Proceedings of the Entomological Society of Washington* 66(3), 184–192.

Summers, F.M. and Price, D.W. (1970) Review of the mite family Cheyletidae. Univ. Calif. publ. entomol. 61, 1–153.

Summers, F.M. and Schlinger, E.I. (1955) Mites of the family Caligonellidae (Acari). *Hilgardia* 23, 539–561.

Summers, F.M. and Witt, R.L. (1971) The gnathosoma of *Cheyletus cacahuamilpensis* Baker (Acarina. Cheyletidae). *Proceedings of the Entomological Society of Washington* 73, 158–168.

Summers, F.M., Gonzales-Rodriguez, R.H. and Witt, R.L. (1973) The mouthparts of *Bryobia rubrioculus* (Sch.) (Acarina: Tetranychidae). *Proceedings of the Entomological Society of Washington* 75, 96–111.

Sun, Q.-T., Chen, R-Z. and Meng, Z.-J. (2002) Biological characteristics of *Acarus siro* Linnaeus and its comprehensive prevention and control. *Jilin Agricultural University* 24(3), 30–32.

Sun, W., Xia, L. and Wu, Y. (2023) Life histories and functional responses of two predatory mites feeding on the stored-grain pest *Liposcelis bostrychophila* badonnel (Psocoptera: Liposcelididae). *Insects* 14, 478. DOI: 10.3390.

Suñer, D., Casadevall, M., Domínguez, M. and Vicens, J. (1985) Estudio de los ácaros, dípteros y coleópteros presentes en la fabricación del jamón curado. *Cárnica* 2000 18, 169–182.

Sung, B., Lim, J. and Lee, H. (2006) Food protective and color alteration effects of acaricidal aldehydes on *Tyrophagus putrescentiae* (Schrank). *Journal of Food Protection* 69, 1728–1731.

Sushma, Gulati, R., Verma, D. and Khushbu (2022) Tactis for management of stored food mites. In: Shukla, M., Wani, M.A., Hajam, M.A. and Rathar, A.M.U. (eds) *Agriculture Science: Research and Review*. Bhumi Publishing, India VII, pp. 102–108.

Suski, Z.W. (1972) Tarsonemid mites on apple trees in Poland. X. Laboratory studies on the biology of certain mite species of the family Tarsonemidae (Acarina, Heterostigmata). *Zeszyty Problemowe Postępów Nauk Rolniczych* 129, 111–137.

Swift, S.F. (1996) Hawaiian Raphignathoidea: family Caligonellidae (Acari: Prostigmata), with descriptions of five new species and key to genera and species. *Annals of the Entomological Society of America* 89, 313–327.

Swift, S.F. and Goff, M.L. (1987) The family Bdellidae (Acari: Prostigmata) in the Hawaiian Islands. *International Journal of Acarology* 13(1), 29–49.

Sylwestrowicz, Z.A. and Kostecka, D.B. (1991) Oribatei and Mesostigmata of pine forests from Poland. In: Dusbábek, F. and Bukva, V. (eds) *Modern Acarology, Vol.1, Proc. VIII Int. Congress Acarology, Budlojvice, Czechoslovacia, 6–11 August 1990*. Academic, Prague and SPB academic Publishing, The Hague, The Netherlands, pp. 365–371.

Szabo, A., Korodi, I., Tempfli, B. and Penzes, B. (2010) Phytoseiid mites in the Hungarian Vineyards (Acari: Phytoseiidae). *Acta Phytopathology Entomology Hungarici* 45(2), 337–347.

Szafranek, P. and Lewandowski, M. (2017) Mite community on Polish mushroom farms. *International Journal of Acarology* 43(3), 217–222.

Szakacs, T.A., Wood, H., Russel, C., Nelder, M.P. and Patel, S.N. (2020) An apparent, locally acquired case of rickettsialpox (*Rickettsia akari*) in Ontario, Canada. *Journal of the Association of Medical Microbiology and Infectious Disease* 5(2), 115–119.

Szalay, Von L. (1931) Beiträge zur Kenntnis der Arachnoideenfauna der Aggteleker Höle. *Annales Historico-Naturales Musei Nationalis Hungarici* XXVII, 351–370.

Szelecz, I., Losch, S., Seppey, C.V.W., Lara, E., Singer, D., Sorge, F., Tschui, J., Perotti, M.A. and Mitchell, E.A.D. (2018) Comparative analysis of bones, mites, soil, chemistry, nematodes and soil microeukaryotes from a suspected homicide to estimate the post-mortem interval. *Scientific Reports* 8, 25.

Szilman, P., Szilman, E., Szilman, M., Meszyńska, E., Maniurka, H., Solarz, K. and Sieroń, A.L. (2006) Occupational exposure to allergenic mites among workers of the Silesian *Zoological Biology Letter* 43(2), 375–380.

Szlendak, E. and Kraszpulski, P. (1991) Energy budget of the grain mite, *Acarus siro* (Acari: Acaridae). *Experimental and Applied Acarology* 10, 221–230.

Szlendak, E., Boczek, J., Bruce, W. and Davis, R. (1985) Effect of gamma-radiated males on egg production in *Acarus siro*. *Florida Entomologist* 68(2), 286–290.

Szlendak, E., Conyers, C., Muggleton, J. and Thind, B.B. (2000) Pirimiphos-methyl resistance in two stored product mites, *Acarus siro* and *Acarus farris*, as detected by impregnated paper bioassay and esterase activity assays. *Experimental and Applied Acarology* 24, 45–54.

Szwabowicz, A. and Miedzobrodski, K. (1957) Toxicity of the meal mite *Tyrophagus farinae*. I. Experiments in mice, guinea pigs, pigeons and fowls. *Vetinary Bull* 28, 475–478.

Szwabowicz, A., Miedzobrodski, K. and Donigiewick, K. (1957) Toxicity of *Tyrophagus farinae* for animals. II Experimental research in horses and sheep. Med. vet. (Varsovie) 13, 722–724.

Szwabowicz, A., Miedzobrodski, K. and Donigiewick, K. (1958a) Toxicity of *Tyrophagus farinae* for animals. III Experiments in swine. Vet. Bull. 29, 344–346.

Szwabowicz, A., Miedzobrodski, K. Pankowa, J. and Holnicka, B. (1958b) Toxicity of *Tyrophagus farinae* to animals. IV. Experiments on chickens and ducks. *MedVet (Varsovie)* 14, 556–558.

Tabari, M.A., Jafari, A., Jafari, M. and Youssefi, M.R. (2023) Laboratory and field efficacy of terpene combinations (carvacrol, thymol and menthol) against the poultry red mite (*Dermanyssus gallinae*). *Veterinary Parasitology* 313, 109842.

Taboada, M.D.F. (1954). Pulmonary acariasis in Spain; an illustrative case report. *The BMJ* 1(4859), 437–438.

Taborda, J.V.S. (2018) Viabilidade econômica do sistema cage-free para poederias comerciais, Brazil University, Descalvado, Brazil.

Tagami, K. and Halliday, B. (2013) Mites of the family Histiostomatidae associated with the Australian giant earwig, *Titanolabis colossea* (Dohrn) (Dermaptera: Anisolabididae). *Australian Journal of Entomology* 52(3), 218–226.

Taghavi, A. (1996) Faunistic study of mites and biology of *Brevipalpus obovatus* on tea Plant *Camellia sinensis* (L.) in west Mazandaran. M.Sc. thesis. Shahid Chamran University, Ahvaz, 1–151.

Taghavi, A., Kamali, K. and Sahragard, A. (1998) Mites associated with tea plant in western regions of Mazandaran province. *Proc. 13th Iranian Plant Protection Congress*, Karaj, Iran, p. 100.

Taha, H.A., El-Naggar, M.E.E., Ibrahim, G.A. and Soliman, S.M. (1988) Biological studies on two acarine predators of the grain mite, *Caloglyphus rhizoglyphoides* Zachvatkin. *Agricultural Research Review* 66, 137–144.

Taha, H.A., Mahmoud, H.I., Hassan, M.I., Omar, N.R. *et al.* (2010) Effect of different food types on the biology, fecundity and life table parameters of the stored grain mite *Gohieria fusca*, Oud., Acari: Astigmata: Lapidophoridae. *Egyptian Journal of Agricultural Research* 88(1), 133–142.

Taha, H.A., Metwally, A.M., Atwa, W. and El-Sanady, M. (2007) Biological and prediction studies on the two acarine predators *Lasioseius sewai* Nasr and Abou-Awad and *Blattisocius keegani* Fox fed on the grain mite *Tyrophagus putrescentiae* (Acari: Acaridae). *Egyptian Journal of Agricultural Research* 5, 1659–1668.

Takaku, G., Hartini, S., Dwibadra, D. and Corpus-Raros, L.A. (2012) Macrochelid mites (Acari: Gamasina: Macrochelidae) in the Philippines. *Journal of the Acarological Society of Japan* 21, 94–124.

Taketomi, E.A., Justino, C.M., Pereira, F.L., Segundo, G.R., Sopelete, M.C., Sung, S.J. and Silva, DA. (2006) Taxis but not private cars are mite allergen reservoirs in Brazil. *Journal of Investigational Allergology and Clinical Immunology* 16(1), 34–36.

Takken, W. and Knols, B.G.J. (eds) (2010) *Olfaction in Vector-Host Interactions. Ecology and Control of Vector-Borne Diseases, vol. 2.* Wageningen Academic Publishers, Wageningen, The Netherlands, pp. 1–437.

Tang, J.C., Wong, S.F., Mak, J.W. and Ho, T.M. (2011) Antigenic profile of *Blomia tropicalis*, *Aleuroglyphus ovatus* and *Glycycometus malaysiensis*. *Tropical Biomedicine* 28(2), 223–36.

Tareev, V.N. (1970) Akaroidnye kleshchi (Acaroidea) Primorskogo Kraya. Abstract of Candidate of Science Thesis [Acaroid mites (Acaroidea) of Primorskiy Kray]. Avtoreferat dissertatsii kandidata biologicheskikh nauk., Vladivostok (in Russian).

Tareev, V.N. and Dubinina H.V. (1985) On the fauna of dust inhabiting mites from Primorje. *Parazitologiya* 19, 27–31.

Tavassoli, M., Allymehr, M., Pourseyed, S.H., Ownag, A., Bernousi, I. *et al.* (2011) Field bioassay of *Metarhizium anisopliae* strains to control the poultry red mite *Dermanyssus gallinae*. *Veterinary Parasitology* 178(3–4), 374–378.

Taylor, D. (2006) Innate immunity in ticks: a review. *Journal of the Acarological Society of Japan* 15, 109–127.

Taylor, C. (2018) Variety of Life. A Quick Guide to the Diversity of Living. Available at: http://taxondiversity.fieldofscience.com/2018/04/digamasellidae.html (accessed 20 February 2020).

Tee, R.D., Gordon, D.J., van Hage-Hamsten, M., Gordon, S., Nunn, A.J., Johansson, S.G. and Taylor, A.J. (1992) Comparison of allergic responses to dust mites in U.K. bakery workers and Swedish farmers. *Clinical and Experimental Allergy* 22(2), 233–239.

Telford, M.J. and Thomas, R.H. (1998) Expression of homeobox genes show chelicerate arthropods retain their deuterocerebral segment. *Proceedings of the National Academic Science U.S.A.* 95, 10671–10675.

Teng, B., Qiu, Z.X. and Wanh, H. (1988) Investigation of mite infestation of respiratory tract in Chinese herb workers. *Chinese Journal of Parasitic Disease Control (*now *Journal of Pathogen Biology)* 1, 38–39, 77 (in Chinese).

Tenquist, J.D. and Charleston, W.A.G. (2001) A revision of the annotated checklist of ectoparasites of terrestrial mammals in New Zealand. *Journal of the Royal Society of New Zealand* 31(3) 481–542.

Terho, E.O., Leskinen, L., Husman, K. and Kärenlampi, L. (1982) Occurrence of storage mites in Finnish farming environments. *Allergy* 37(1), 15–19.

Terho, E.O., Husman, K., Vohlonen, I., Rautalahti, M. and Tukinainen, H. (1985) Allergy to storage mites or cow dander as a cause of rhinitis among Finnish farmers. *Allergy* 40, 23–26.

Terra, W.R. (2001) The origin and functions of the insect peritrophic membrane and peritrophic gel. *Archives of Insect Biochemistry and Physiology* 47 (2), 47–61.

Terra, W.R. and Ferreira, C. (1994) Insect digestive enzymes: properies, compartmentalizaion and function. *Comparative Biochemistry and Physiology* 109B (1), 1–62.

Terra, W.R., Ferreira, C. and Baker, J. (1996) Compatmentalization of digestion. In: Lehane, M.J. and Billingsley, P.F. (eds) *Biology of the Insect Midgut.* Chapman and Hall, NY, U.S.A., pp. 206–235.

Théodoridès, J. (1955) Contribution à l'étude des parasites et phorétiques de coléoptères terrestres. Actualités scientifiques et industrielles 1217. *Vie et Milieu* 4, 1–310.

Theron, P.D. (1976) New species of the genus *Terpnacarus* Grandjean (Acari: Terpnacaridae) with notes on the biology of one species. *Journal of the Entomological Society of Southern Africa* 39(1), 131–141.

Thewke, S.E. and Enns, W.R. (1979) Three new species of *Hemicheyletia* Volgin (Acarina: Cheyletidae) with a key to known world species. *Journal of the Kansas Entomological Society* 52, 218–225.

Thind, B.B. (2000) Determination of low levels of mite and insect contaminants in food and feed stuffs by a modified flotation method. *Journal of the Association of the Official Analytical Chemists* 83(1), 113–119.

Thind, B.B. (2005) A new versatile and robust trap for detection and monitoring of stored product pests in the cereal and allied industries. *Experimental and Applied Acarology* 35, 1–15.

Thind, B.B. and Clarke, P.G. (2001) The occurrence of mites in cereal-based food destined for human consumption and possible consequences of infestation. *Experimental and Applied Acarology* 25(3), 203–215.

Thind, B.B. and Dunn, J.A. (2002) A laboratory evaluation of a regulated airflow through wheat at four combinations of temperature and humidity on the productivity of three species of stored product mites. *Experimental and Applied Acarology* 27(1–2), 89–102.

Thind, B.B. and Ford, H.L. (2004) Management of storage mites in production premises through early detection and monitoring using BT mite traps. *Phytopaga* 14, 733–744.

Thind, B.B. and Ford, H.L. (2006) Laboratory studies on the use of two new arenas to evaluate the impact of the predatory mites *Blattisocius tarsalis* and *Cheyletus eruditus* on residual populations of the stored product mite *Acarus siro*. *Experimental and Applied Acarology* 38, 167–180.

Thomas, W.R., Smith, W.A., Hales, B.J., Mills, K.L. and O'Brien, R.M. (2002) Characterization and immunobiology of house dust mite allergens. *International Archives of Allergy and Immunology* 129(1), 1–18.

Thomas, H.Q., Zalom, F.G. and Nicola, N.L. (2011) Laboratory studies of *Blattisocius keegani* (Fox) (Acari: Ascidae) reared on eggs of navel orangeworm: potential for biological control. *Bulletin of Entomological Research* 101(5), 499–504.

Thomas, E., Chiquet, M., Sander, B., Zschiesche, E. and Flochlay, A.S. (2017) Field efficacy and safety of fluralaner solution for administration in drinking water for the treatment of poultry red mite (*Dermanyssus gallinae*) infestations in commercial flocks in Europe. *Parasites and Vectors* 10, 457.

Thomas, E., Zoller, H., Liebisch, G., Alves, L.F.A., Vettorato, L., Chiummo, R.M. and Sigognault-Flochay, A. (2018) In vitro activity of fluralaner and commonly used acaricides against *Dermanyssus gallinae* isolates from Europe and Brazil. *Parasites and Vectors* 11, 361.

Thompson, P.J., Gillon, R.L., Bird, C., Krska, K. and Stewart, G.A. (1991) The effect of a combined acaricide/cleaning agent on house dust mite allergen load in carpet and mattress. *Australian and New Zealand Journal of Medicine* 21, 660.

Thor, S. (1931) Bdellidae, Nicoletiellidae, Cryptognathidae. *Tierreich* 56, 1–87.

Tian, Q.Y. (1984) Gamasid mites from Shanxi. *Journal of Shanxi Medical University* 1, 15–18.

Tian, J.-G. and Sun, L.-P. (1991) Toxicity determination of chlorpyrifos-methyl and pirimiphos-methyl to *Acarus siro* and *Cheyletus aversor*. *Grain Storage* 20(4), 31–36.

Timm, R.M. (1972) Mites (Acari Laelapidae) parasitic on the meadow vole, *Microtus pennsylvanicus*. *Acarologia* 14(1), 18–20.

Timms, S., Ferro, D.N. and Emberson, R. M. (1981) General biology and nomenclature of *Sancassania berlesei* (Michael). *Acarologia* 22, 385–390.

Tirry, L. and Van Leeuwen, T. (2014) DERGAL: 'Geïntegreerde plaagbestrijding van *Dermanyssus gallinae* bij leghennen. *Project of the Federal Agency for the Safety of the Food Chain*, Belgium.

Tixier, M.-S., Allam, L., Douin, M. and Kreiter, S. (2016) Phytoseiidae (Acari: Mesostigmata) of Morocco: new records, descriptions of five new species, re-descriptions of two species, and key for identification. *Zootaxa* 4067(5), 501–551.

Tjying, I.S. (1970) Survey and study on the mites of stored grain in Taiwan. *Plant Protection Bulletin* 12(3), 113–123.

Tjying, I.S., Tseng Y.-H. (1971) A field survey of the mites infesting stored sugar in Taiwan. *Plant Protection Bulletin* 13(4), 147–155.

Todorov, D.A. (1979) (Mites of the family Pyroglyphidae (Sarcoptiformes) in house dust and methods for their isolation. *Acta Zoologica Bulgarica* 13, 64–71.

Toldrá, F., Flores, M. and Sanz, Y. (1997) Dry-cured ham flavor: enzymatic generation and process influence. *Food Chemistry* 59(4), 523–530.

Thomas, H.Q., Zalom, F.G. and Nicola, N.L. (2011) Laboratory studies of *Blattisocius keegani* (Fox) (Acari: Ascidae) reared on eggs of navel orangeworm: potential for biological control. *Bulletin of Entomological Research* 101, 499–504.

Tomer, H., Blum, T., Arye, I., Faigenboim, A., Gottlieb, Y. and Ment, D. (2018) Activity of native and commercial strains of *Metarhizium* spp. against the poultry red mite *Dermanyssus gallinae* under different environmental conditions. *Veterinary Parasitology* 262, 20–25.

Tomita, A., Shimuzu, N. and Mori, N. *et al.* (2003) Chemical ecology of astigmatid mites. LXXI. Neril formate (Z)-3,7-dimethyl-2,6-octadienyl formate as the alarm pheromone of *Tyroborus lini* Oudemans 1924, and its recovery after forced discharge. *Journal of the Acarological Society of Japan* 12, 11–19.

Torres-Cabra, E. and Lagos-Lopez, M. I. (2019) Evaluation of essential coriander (*Coriandrum sativum* L.) oil against the red poultry mite (*Dermanyssus gallinae*) (De Geer, 1778) (Acari: Dermanyssidae) under laboratory conditions. *Ciencia y Tecnología Agropecuaria* 20(1), 53–68.

Tovey, E.R. (1992) Allergen exposure and control. *Experimental and Applied Acarology* 16, 181–202.

Tovey, E.R. and Woolcock, A.J. (1994) Direct exposure of carpets to sunlight can kill all mites. The *Journal of Allergy and Clinical Immunology* 93, 1072–1075.

Tovey, E.R., Chapman, M.D. and Platts-Mills, T.A.E. (1981) Mite faeces are a major source of house dust allergens. *Nature* 289, 592–593.

Tovey, E.R., McDonald, L., Peat, J. and Marks, G. (2000) Domestic mite species and Der p 1 allergen levels in nine locations in Australia. *Allergy Clinic Immunology* 12, 226–231.

Trach, V.A. and Joharchi, O. (2018) Mites of the family Laelapidae (Acari, Mesostigmata) associated with scarab beetles in Ukraine. *Vestnik Zoologii* 52(3), 217–228.

Trach, V.A. and Khaustov, A.A. (2017) Mites of the genus *Proctolaelaps* Berlese 1923 (Acari: Mesostigmata: Melicharidae) associated with bark beetles in Asian Russia. *Acarina* 25(2), 151–163.

Tragardh, I. (1905) Acariden aus Aegypten und dem Sudan. Results of the Swedish zoological expedition to Egypt and the White Nile, 1901, part II, pp. 1–124.

Travé, J. (1963) Écologie et biologie des Oribates (Acariens) saxicoles et arboricoles. *Vie Milieu Suppliment* 14, 1–267.

Traver, J. (1951) Unusual Scalp Dermatitis in Humans Caused by the Mite, *Dermatophagoides* (Acarina, Epidermoptidae). *Proceedings of the Entomological Society of Washington* 53(1), 1–12.

Treat, A.E. (1967) Mites from noctuid moths. *Journal of the Lepidopterists' Society* 21, 169–179.

Treat, A.E. (1969) Association of the mite *Blattisocius tarsalis* with the moth *Epizeuxis aemula*. *Journal of the New York Entomological Society* 77, 171–175.

Treat, A.E. (1970) Two tydeid mites from the ears of noctuid moths. *American Museum Novitates* 2426, 1–14.

Treat, A.E. (1975) *Mites of Moths and Butterflies*. Comstock Publ. Assoc., Ithaca, London, U.K., pp. 1–362.

Trematerra, P. and Fleurat-Lessard, F. (2015) Food industry practices affecting pest management. *Stewart Postharvest Review* 11(1), 1–7.

Trinca, J.C., Stringer, K.C., Drummond, F.H. and Bristow, V.G. (1969) Allergenicity of the house-dust mite *Dermatophagoides pteronyssinus*. *Medical Journal of Australia* 1, 1177–1179.

Trivedi, T.P., Rajagopal, D. and Tandon, P.L. (1994) Life table for establishment of potato tuber moth, *Phthorimaea operculella*. *Journal of Indian Potato Association* 1–2, 97–105.

Trouessart, E.L. (1887) Diagnose d'especes nouvelles de Sarcoptides plumicoles (Analgesinae). *Bulletin de la Société d'Études Scientifiques* 16, 85–156.

Trouessart, E.L. (1901) Sur deux espèces, formant un genre nouveau de Sarcoptides détriticoles parasites des fourrures. *Bulletin de la Société zoologique de France* 26, 82–84.

Tryjanowski, P., Baraniak, E., Bajaczyk, R., Gwiazdowicz, D.J., Konwerski, S., Olszanowski, Z. and Szymkowiak, P. (2001) Arthropods in nests of the red-backed shrike (*Lanius collurio*) in Poland. *Belgian Journal of Zoology* 131(1), 69–74.

Tseng, Y.-H. (1971a) A field survey of the mites infesting stored sugar in Taiwan. *Plant Protection Bulletin* 13(4), 147–155.

Tseng, Y.-H. (1971b) A new ascid mite (*Lasioseius martini* n. sp.), recovered from shallots in Taiwan (Acarna: Ascidae). Plant Prot. Bull. 13(1), 1–5.

Tseng, Y.-H. (1972) Bureau of Commodity Inspection and Quarantine, Ministry of Economic Affairs, Republic of China, pp. 1–55.

Tseng, Y.-H. (1973) *A review of the mite family Acaridae (Acarina: Sarcoptiformes) from Taiwan*. Tainan Branch Office, Bureau of Commodity

Inspection and Quarantine, Ministry of Economic Affairs, Republic of China 1, pp. 1–32.

Tseng, Y.-H. (1977) A contribution to the knowledge of Formosan cheyletid mites (Acarina: Prostigmata). *Proceedings of the National Science Council* 10, 213–263.

Tseng, Y.-H. (1978) Mites of the family Bdellidae from Taiwan (Acarina: Prostigmata). *Journal of Agricultural Research of China* 104, 25–51.

Tseng, Y.-H. (1979) Studies on the mites infesting stored food products on Taiwan. In: Rodriguez J.G. (ed.), *Recent Advances in Acarology*, Academic Press, New York, U.S.A., pp. 311–316.

Tseng, Y.-H. (1980) Taxonomical study of the mite family Cunaxidae from Taiwan (Acarina: Tromoidiformes). Quart. J. Taiwan Mus. 33(3–4), 253–277.

Tseng, Y.-H. (1982) A catalogue and bibliography of the Acari of Taiwan (Arachnida: Acarina). Bureau of Commodity Inspection and Quarantine, Ministry of Economic Affairs, Taiwan Branch Office, pp. 1–164.

Tseng, Y.-H. (1984) Mites associated with weeds, paddy rice, and upland rice fields in Taiwan. In: Griffiths, D.A. and Bowman, C.E. (eds) *Acarology VI. Vol. 2.* Ellis Horwood Limited, London, U.K., pp. 770–780.

Tseng, Y.-H. (1989) Redescription of Tyrophagus kentinus Tseng, Tyrophagus bambusae Tseng and Aleuroglyphagus formosanus Tseng and describes a new record species Rhizoglyphus canadii Manson. *Chinese Journal of Entomology* 3, 37–50.

Tseng, Y.-H. and Chang, A.-F. (1973) Mites of stored food from Taiwan. *Agriculture Science* 21(5, 6), 172–178.

Tu, D., Zhu, Z.-M., Xia, B., Wen, C.-G. and Yu, L.-P. (2001) Description of species of genus *Glycyphagus* (Acarina: Glycyphagidae). *Journal of Nanchang University (Natural Science)* 25(4), 356–364.

Tucci, E.C. and Guimarães, J.H. (1998) Biologia de *Dermanyssus gallinae* (De Geer, 1778) (Acari, Dermanyssidae). *Revista Brasileira de Parasitologia Veterinária* 7(1), 27–30.

Tucci, E.C., do Prado, A.P. and Pires de Araújo, R. (2008) Thermal requirements of *Dermanyssus gallinae* (De Geer, 1778) (Acari: Dermanyssidae). *Revista Brasileira de Parasitologia Veterinária* 17(2), Doi: 10.1590/S1984-29612008000200002.

Tuma, D., Sinha, R.N., Muir, W.E. and Abramson, D. (1990) Odor volatiles associated with mite-infested bin-stored wheat. *Journal of Chemical Ecology* 16, 713–724.

Tuovinen, T., Heikkilä, P., Juvonen, S., Lindqvist, B. and Tuovinen, T. (2010) Kanapunkki Hallintaan Munintakanaloissa. Control of Red Poultry Mite in Laying Hen Houses. Research report MMM12-3/312/2010. Joikioinen: MTT Kasvintuotannon tutkimus.

Turgu, V. and Kumral, N.A. (2019) Alternative control agents of the dried fruit mite, *Carpoglyphus lactis* (L.) (Acari: Carpoglyphidae) on dried apricots. *Acarological Studies* 1(2), 59–64.

Türk, E. and Türk, F. (1957) Systematic und ökologie der tyrogliphiden mitteleuropas. In: Stammer, H.J. (ed.) *Beiträge zur Systematik und Ökologie Mitteleuropäischer Acarina, Bund 1, Abschnitt 1.* Akademische Verlagsgesellschaft Geest and Portig, Leipzig, Germany, pp. 1–231.

Turk, F.A. (1953) A synonymic catalogue of British Acari. Parts I-II. *The Annals and Magazine of Natural History* 12, 81–99.

Turk, F.A. and Turk, S.M. (1952) Records and descriptions of mites new to the British fauna together with short notes on the biology of sundry species. *The Annals and Magazine of Natural History* 11(12), 475–506.

Turner, B. and Bishop, J. (1998) What's bugging your kitchen? *Environmental Health* 106, 310–314.

Turner, B.D. (1994) *Liposcelis-Bostrychophila (Psocoptera, Liposcelididae)*, a stored food pest in the UK. *International Journal of Pest Management* 40, 179–190.

Tytar, V.M. and Oksentiuk, Y.R. (2019) Modelling the bioclimatic niche of a cohort of selected mite species (Acari, Acariformes) associated with the infestation of stored products. *Vestnik Zoologii* 53(5), 399–416.

Uchikawa, K. and Rack, G. (1978) *Eulaelaps stabularis* (Koch, 1839) and *Eulaelaps oudemansi* Turk, 1945 (Mesostigmata: Haemogamasidae). *Acarologia* 20(2), 163–172.

Ueckermann, E.A. and Khanjani, M. (2003) Iranian caligonellidae (Acari: Prostigmata), with description of two new species and re-description of *Molothrognathus fulgidus* Summers and Schlinger, with a key to genera and species. *Acarologia* 43, 291–298.

Ueckermann, E.A. and Smith Meyer, M.K.P. (1987) Afrotropical stigmaeidae (Acari: Prostigmata). *Phytophylactica* 19, 371–397.

Ueckermann, A., Rastegar, J., Saboori, A. and Ostovan, H. (2007) Some mites of the superfamily

Bdelloidea (Acari: Prostigmata) of Karaj (Iran), with descriptions of two new species and redescription of *Bdelloides kazeruni*. *Acarologia* 61, 127–138.

Ujvary, I. (1999) Nicotine and other alkaloids. In: Yamomoto, I. and Casida, J.E. (eds) *Nicotinoid Insecticides and the Nicotinic Acetylcholine Receptor*. Springer Verlag, Tokyo, Japan, 29–70.

Ulrichs, C., Han, Y.J., Abdelhamid, M.T. and Mewis, I. (2020) Management of the poultry red mite, *Dermanyssus gallinae*, using silica-based acaricides. *Experimental and Applied Acarology* 82, 243–254.

Uluçay, İ. and Koç, K. (2014) A new species of *Neophyllobius* and description of male of *Neophyllobius yunusi* (Acari: Camerobiidae) from Turkey. *International Journal of Acarology* 40(1), 15–22.

Unc, A., Eshel, G., Unc, G.A., Doniger, T., Sherman, C., Leikin, M. and Steinberger, Y. (2021) Vineyard soil microbial community under conventional, sustainable and organic management practices in a mediterranean climate. *Soil Research* 59, 253.

Upton, M.S. and Mantle, B.L. (2010) Methods for Collecting, Preserving and Studying Insects and Other Terrestrial Arthropods, 5th edn., Australian Entomological Society Inc., Miscelaneous Pubbications 3, Paragon Printers, Australasia, Canberra, Australia, pp. 1–83.

USDA (1981) New North American Record. *Plant Pest News* 1(2), 2.

Usher, M.B. and Davis P.R. (1983) The biology of *Hypoaspis aculeifer* (Canestrini) (Mesostigmata): Is there a tendency towards social behaviour? *Acarologia* 24(3), 243–250.

Vacante, V. (1985) Prima raccolta di acari Tetranichidi in Sicilia. *Phytophaga* 1, 41–114.

Vacante, V. (1988) Note su *Tyrophagus neiswanderi* Johnston and Bruce (Acarina, Acaridae). I. Descrizione morfologica degli adulti. In: Atti XV, *Congresso Nazionale Italiano di Entomologia, 13–17 Giugno 1988*. L'Aquila, Italy, 213–222.

Vacante, V. (1989a) Gli acari degli agrumi in Italia. 3. Gli acarididi (Hemisarcoptidae Oudemans, Acaridae Ewing and Nesbitt, Saproglyphidae Oudemans e Glycyphagidae Berlese). *Bollettino di Zoologia Agraria e di Bachicoltura* 21, 45–84.

Vacante, V. (1989b) I Coleotteri Curculionidi della fragola in Sicilia. I Contributo. *La Difesa delle Piante* 12, 107–118.

Vacante, V. (1997a) Acarofauna delle colture protette mediterranee. *Notiziario delle Piante* 7, 43–52.

Vacante, V. (1997b) Note sugli acari della polvere delle case in Italia. I Pyroglyphidae Cunliffe. *Rivista di parassitologia* XIV(LVIII), 347–366.

Vacante, V. (2016) *The Handbook of Mites of Economic Plants*. CABI Publishing, Wallingford, U.K., pp. 1–872.

Vacante, V. and Benuzzi, M. (2007) *Difesa delle Colture in Serra*. Edagricole, Bologna (now Milan), Italy, pp. 1–437.

Vacante, V. and Calabretta, C. (1982) L'acaro delle abitazioni, *Glycyphagus domesticus* (De Geer) (Acarina, Tyroglyphidae): cenni di morfologia, aspetti bio-etologici e suggerimenti per la lotta. *Agriculture Technology* 34(2), 121–129.

Vacante, V. and Gerson, U. (2012) *Integrated Control of Citrus Pests in the Mediterranean Region*. Bentham Sciences Publishers, United Arab Emirates, pp. 1–286.

Vacante, V. and Nucifora, A. (1984–85) Gli Acari degli agrumi in Italia. I. Specie rinvenute e chiave per il riconoscimento degli ordini, dei sottordini e delle famiglie. *Bollettino di Zoologia Agraria e di Bachicoltura* 18, 115–166.

Vacante, V. and Nucifora, A. (1987) Gli Acari degli agrumi in Italia. II. I Gamasidi (Phytoseiidae Berlese e Ascidae Voigts and Oudemans): chiave per il riconoscimento dei generi e delle specie e note di biologia. *Bollettino di Zoologia Agraria e di Bachicoltura* 19, 145–184.

Vafei, R., Rostovan, H. and Attaran, M.R. (2002) Biology of the straw Itch Mite, *Pyemotes tritici* (Acari: Pyemotidae). In: Hojat Jalali, A.A. (ed.) *Proc. 15th Iranian Plant Protection Congress, 7–11 September 2002, Vol. I.* Razi University, Kermanshah, Iran, p. 145.

Valdivieso, R., Iraola, V., Estupinan, M. and Fernandez-Caldas, E. (2006) Sensitization and exposure to house dust and storage mites in high-altitude areas of Ecuador. *Annals of Allergy, Asthma and Immunology* 97, 532–538.

Valentine, J.W. (1970) How many marine invertebrate fossil species? *Journal of Paleontology* 44, 410–415.

Valerio, C.R., Murray, P., Arlian, L.G. and Slater, J.E. (2005) Bacterial 16S ribosomal DNA in house dust mite cultures. *Journal of Allergy and Clinical Immunology* 116(6), 1296–1300.

Valiente Moro, C., Chauve, C. and Zenner, L. (2005) Vectorial role of some Dermanyssoid mites (Acari, Mesostigmata, Dermanyssoidea). *Parasite* 12(2), 99–109.

Valiente Moro, C., De Luna, C.J., Tod, A., Guy, J.H., Sparagano, O. and and Zenner, L. (2009) The poultry red mite (*Dermanyssus gallinae*): A potential vector of pathogenic agents. *Experimental and Applied Acarology* 48, 93–104.

Vanhaelen, M.A., Vanhaelen-Fastre and Geeraets, J. (1980) Occurrence in mushrooms (Homobasidiomycetes) of cis-and trans-octa-1, 5 dien-3-ol attractants to the cheese mite *Tyrophagus putrescentiae* (Schrank) (Acarina: Acaridae). *Experientia* 36, 406–407.

Vantornhout, I., Minnaert, H., Tirry, L., and De Clercq, P. (2005) Influence of diet on life table parameters of *Iphiseius degenerans* (Acari: Phytoseiidae). *Experimental and Applied Acarology*, 35(3), 183–195.

Várfalvyová, D., Stanko, M. and Miklisová, D. (2011) Composition and seasonal changes of mesostigmatic mites (Acari) and fleas fauna (Siphonaptera) in the nests of *Mus spicilegus* (Mammalia: Rodentia). *Biologia* 66(3), 528–534.

Vargas, M.V. and Mairena, H.A. (1991) House dust mites from the metropolitan area of San José, Costa Rica. *International Journal of Acarology* 17, 141–144.

Veerman, A. (1992) Diapause in phytoseiid mites: A review. *Experimental and Applied Acarology* 14, 1–60.

Veerman, C., Pinto Correia, T., Bastioli, C., Biro, B., Bouma, J. *et al.* (2020) *European Commission; Directorate-General for Research and Innovation, Caring for Soil is Caring for life: Ensure 75% of Soils are Healthy by 2030 for Food, People, Nature and Climate: Report of the Mission Board for Soil Health and Food.* Publications Office: Brussels, Belgium, p. 2020. Available at: https://data.europa.eu/doi/10.2777/821504 (accessed 17 April 2023).

Vercammen-Grandjean, P.H. (1973) Sur les statuts de la famille des Trombidiidae Leach, 1815 (Acarina: Prostigmata). *Acarologia* 15, 102–114.

Vercammen-Grandjean, P.H. (1975) Les organes de Claparède et les papilles genitales de certains acariens sont-ils des organes respiratoires? *Acarologia* 17, 624–630.

Verner, P.H., Ždárková, E. and Reška, M. (1982) Výskyt a migrace skladištních roztočů a brouků na některých přírodních stanovištích. *Bohemian Century* 11, 95–116.

Verts, B.J. and Carraway, L.N. (2002) *Neotoma lepida.* Mamm. *Species* 699, 1–12.

Vervloet, D., Penaud, A., Razzouk, H., Senff, M., Arnaud, A., Boutin, C. and Charpin, J. (1982) Altitude and house dust mites. *Journal of Allergy and Clinical Immunology* 69, 290–296.

Vidal, C., Boquete, O. and Gude, F. (2004) High prevalence of storage mite sensitization in a general adult population. *Allergy* 59, 401–405.

Vijayambika, V. and John, P.A. (1974) Obesrvations on the environmental regulation of hypopial formation in the fish-mite *Lardoglyphus konoi. Acarologia* 16(1), 160–165.

Vijayambika, V. and John, P.A. (1975) Internal morphology and histology of the fish mite *Lardoglyphus konoi* (Sasa and Asanuma) (Acarina: Acaridae). 2. The reproductive system. *Acarologia* 17, 106–113.

Vinarski, M.V. and Korallo-Vinarskaya, N.P. (2016) An annotated catalogue of the gamasid mites associated with small mammals in Asiatic Russia. The family Laelapidae s.str. (Acari: Mesostigmata: Gamasina). *Zootaxa* 4111 (3), 223–245.

Vincent, C., Hallman, G., Panneton, B. and Fleurat-Lessard, F. (2003) Management of agricultural insects with physical control methods. *Annual Review of Entomology* 48, 261–281.

Vitzthum, H.G. (1941) Acarina. In: Bronns, H.G. (ed.) *Klassen und Ordnungen des Tierreichs, Band 5, Abt. 4, Buch 5, Lief 4*, Becker and Erler, Leipzig, German, pp. 429–448.

Vitzthum, H.G. von (1919) Acarologische Beobachtungen. 3 Reihe. *Archiv für Naturgeschichte* 85A, 1–62.

Vitzthum, H.G. von (1923) Acarologische Beobachtungen. 7 Reihe. *Archiv für Naturgeschichte* 89A(2), 97–181.

Vitzthum, H.G. von (1926) Acarapis-Studien. *Archiv für Bienenkönigin* 7, 293–313.

Vitzthum, H.G. von (1928) Die Milben-seuche unserer Höonigbiene. *Wiener Tierärztliche Monatsschrift* 15, 861–862.

Vitzthum, H.G. von (1929) Spinnentiere. 5. Ordnung: Milben, Acari. In: Brohmer, P., Ehrmann, E. and Ulmer, G. (eds) *Die Tierwelt Mitteleuropas. III. Band. 3.* Lieferung, Quelle and Meyer, Leipzig, Germany, pp. 1–155.

Vitzthum, H.G. von (1936) Über das Vorkommen von saprohytischen und parasitaren Milben im Wabenwerk der Honigbiene. *Deutscher Imkerführer Mitteilungen*10, 460–461.

Vitzthum, H.G. von (1940) Die Deutonympha von *Carpoglyphus lactis* (L. 1758) (Acari: Tyroglyphidae). *Zoologischer Anzeiger* 129, 197–201.

Vitzthum, H.G. von (1943) Acarina. Bronns Tierreich 5, Abt 4, Buch 5, Lf. 5, 11011.

Vogel, P., Morelo Dal Bosco, S. and Juarez Ferla, N. (2015) Mites and the implications on human health. *Nutrient Hospital* 31(2), 944–951.

Voigt, P. (1990) Damage caused by mould mites in cucumbers. *Nachrichten Pflanzenschutz* 44, 246–247.

Voigts, H, and Oudemans, A. C. (1904) Neue Milben aus der Umgegend von Bremen. *Zoologischer Anzeiger* 27, 651–656.

Volgin, V.I. (1948) New Species of Mites of the Genus *Tyrophagus* Ouds., 1923 (Tyroglyphidae, Acarina). Doklady Akademii Nauk USSR, Zoology. *Zoology* 3, 509–511.

Volgin, V.I. (1949) Materials on systematics of mites of the genus *Tyrophagus* Ouds., 1923 (Tyroglyphidae, Acarina). Doklady Akademii Nauk USSR. *Zoology* 65(3), 385–388.

Volgin, V.I. (1963) Two new genera of predatory mites of the family Cheyletidae. *Entomological Review* 42, 504–508.

Volgin, V.I. (1965) Two new species of acaroid mites (Acarina, Acaroidea) of the fauna of the USSR (Rus.). *Entomologiceskoe obozrenie* 44(1), 203–211.

Volgin, V.I. (1969) Acarina of the family Cheyletidae of the world. Akademia Nauk, Leningrad, USSR, 1–432 (in Russian). In: *English, 1987*. Amerind Publishing Co., New Delhi, India, pp. 1–532.

Volgin, V.D. (1975) Family acaridae, glycyphagidae and saproglyphidae. In: Gilyarov, M.S. and Krivolutsky, D.A. (eds) *Handbook for the Identification of Soil-inhabiting Mites, Sarcoptiformes*. Zoological Institute of the Academy of Sciences, Petrograd, USSR, pp. 416–476.

Volgin, V.D. (1980) Novie vidi kleshchei sem. Saproglyphidae (Acariformes, Acaroidea). *Parazitology Sborn* 29, 159–168.

Volgin, V.I. and Mironov, S.V. (1979) Novie vidi i novii rod kleshchei semeistva Saproglyphidae (Acarina, Acaroidea). *Trudy Zoologicheskogo Instituta, Akademiya Nauk SSSR* 85, 91–98.

Vollset, I., Larsen, H.J. and Mehl, R. (1986) Immediate type hypersensitivity in dogs induced by storage mites. *Research in Veterinary Science* 40, 123–127.

Voorhorst, R.M.I.A., Spieksma-Boezeman, M.I.A. and Spieksma, F.Th.M. (1964) Is a mite (*Dermatophagoides* sp.) the producer of the house-dust allergen. *Allergie Asthma* 10(6), 329–334.

Voorhorst, R., Spieksma, F.Th.M., Varekamp, H., Leupen, M.J. and Lykema, A.W. (1967) The housedust mite (*Dermatophagoides pteronyssinus*) and the allergens it produces. Identity with the housedust allergen. *Journal of Allergy* 39, 325–339.

Voorhorst, R., Spieksma, F.T.M. and Varekamp, H. (1969) *House-dust Atopy and the House-dust Mite*. Stafleu's Scientific Publication Co., Leiden, The Netherlands, pp. 1–159.

Vysotzkaja, S.O. (1961) The tyroglyphoid mites (Sarcoptiformes) from the nests of rodents and insectivorous mammals of the Leningrad region. *Parazitology Sborn* 20, 267–282.

Waap, H., Aguin-Pombo, D. and Maia, M. (2020) Case report: Human dermatitis linked to *Ornithonyssus bursa* (Dermanyssoidea: Macronyssidae) infestation in Portugal. *Frontiers in Veterinary Science* 7, 567902.

Wade, C.F. and Rodriguez, J.G. (1961) Life history of *Macrocheles muscaedomesticae* (Acarina: Macrochelidae), a predator of the house fly. *Annals of the Entomological Society of America* 54(6), 776–781.

Wafa, A. K., Zaher, M.A. and Soliman., Z.R. (1970) Life-history of the predator mite, *Acaropsis aegyptiaca* Wafa and Soliman. *Bulletin de la Société entomologique d'Égypte* 54, 155–158.

Wainstein, B.A. and Kuznetsov, N.N. (1978) Identification key of soil inhabiting mites, Trombidiformes. In: Gilyarov, M.S. (ed.) *Raphignathoidea*. Nauka, Moscow, pp. 147–171.

Waite, G.K., and Gerson, U. (1994) The predator guild associated with *Aceria litchii* (Acari: Eriophyidae) in Australia and China. *Entomophaga* 39(4), 275–280.

Wakefield, M.E. (2006) Storage arthropod pest detection-current status and future directions. In: Lorini, I., Bacaltchuk, B., Beckel, H., Deckers, D., Sundfeld, E. *et al.* (eds) *Proc. 9th International Working Conference on Stored Product Protection, 15 to 18 October 2006, Campinas, São Paulo, Brazil*. Brazilian Post-harvest Association-ABRAPOS, Passo Fundo, RS, Brazil, pp. 371–384.

Wakefield, M.E. and Dunn, J.A. (2005) Effectiveness of the BT mite trap for detecting the storage mite pests, *Acarus siro*, *Lepidoglyphus destructor* and *Tyrophagus longior*. *Experimental and Applied Acarology* 35, 17.

Walker, N.A. (1964) Euphthiracaroidea of California sequoia litter: with a reclassification of the families and genera of the world (Acarina:

Oribatei). *Fort Hays Studies NS Science* 3, 1–154.

Wall, R.L. and Shearer, D. (2001) *Veterinary Ectoparasites: Biology, Pathology and Control.* 2nd edn, Oxford, Blackwell Science Ltd., U.K., pp. 1–262.

Wallace, D.R.J. (1960) Observations on hypopus development in the Acarina. *Journal of Insect Physiology* 5, 212–229.

Wallace, M.M.H. and Mahon, J.A. (1972) The taxonomy and biology of Australian Bdellidae (Acari). I. Subfamilies Bdellinae, Spinibdellinae and Cytinae. *Acarologia* 14(4), 544–580.

Wallwork, J.A. (1958) Notes on the feeding-behaviour of some forest soil Acarina. *Oikos* 9, 260–271.

Wallwork, J.A. (1959) The distribution and dynamics of some forest soil mites. *Ecology* 40, 557–563.

Walshaw, M.J. and Evans, C.C. (1987) The effect of seasonal and domestic factors on the distribution of *Euroglyphus maynéi* in the homes of *Dermatophagoides pteronyssinus* allergic patients. *Clinical and Experimental Allergy* 17, 7–14.

Walter, D.E. (1988) Predation and mycophagy by endeostigmatid mites (Acariformes: Prostigmata). *Experimental and Applied Acarology* 4, 159–166.

Walter, D.E. (1996) Living on leaves: Mites tomentia and leaf domatia. *Annual Review of Entomology* 41, 101–104.

Walter, D.E. (2009a) Reproduction and embryogenesis. In: Krantz, G.W. and Walter, D.E. (eds) *A Manual of Acarology.* Texas Tech University Press, Lubbock, Texas, U.S.A., pp. 54–56.

Walter, D.E. (2009b) Suborder Endeostigmata. In: Krantz, G.W. and Walter, D.E. (eds) *A Manual of Acarology.* Texas Tech University Press, Lubbock, Texas, U.S.A., pp. 421–429.

Walter, D.E. and Campbell, N.J.H. (2003) Exotic vs. endemic biocontrol agents: Would the real *Stratiolaelaps miles* (Berlese) (Acari: Mesostigmata: Laelapidae), Please stand up? *Biological Control* 26, 253–269.

Walter, D.E. and Kaplan, D.T. (1991) Observations on *Coleoscirus simplex* (Acarina: Prostigmata), a predatory mite that colonizes greenhouse cultures of rootknot nematode (*Meloidogyne* spp.), and a review of feeding behavior in the Cunaxidae. *Experimental and Applied Acarology* 12, 47–59.

Walter, D.E. and Krantz, G.W. (2009) Collecting, rearing, and preparing specimens. In: Krantz, G.W. and Walter, D.E. (eds) *A Manual of Acarology.* Texas Tech University Press, Lubbock, Texas, U.S.A., pp. 83–96.

Walter, D.E., and Lindquist, E.E. (1989). Life history and behavior of mites in the genus *Lasioseius* (Acari: Mesostigmata: Ascidae) from grassland soils in Colorado, with taxonomic notes and description of a new species. *Canadian Journal of Zoology* 67, 2797–2813.

Walter, D.E. and Lindquist, E.E. (1995) The distribution of parthenogenetic ascid mites (Acari: Parasitiformes) do not support the biotic uncertainty hypothesis. *Experimental and Applied Acarology* 19, 423–442.

Walter, D.E. and Lindquist, E.E. (1997) Australian species of *Lasioseius* (Acari: Mesostigmata: Ascidae): the porulosus group and other species from rainforest canopies. *Invertebrate Taxonomy* 11, 525–547.

Walter, D.E. and O'Dowd, D.J. (1995) Life on the forest phyllophane, hairs, little house, and myriad mites. In: Lowman, M.D. and Nadkarni, N. (eds) *Forest canopies.* Academic Press, New York, U.S.A., 325–351.

Walter, D.E. and Oliver, J.H. (1990) *Geolaelaps oreithyiae*, n. sp. (Acari, Laelapidae), a thelytokous predator of arthropods and nematodes, and a discussion of clonal reproduction in the mesostigmata. *Acarologia* 30, 293–303.

Walter, D.E. and Proctor, H.C. (1999) *Mites: Ecology, Evolution and Behaviour.* University of New South Wales Press, Sydney, Australia and CAB International, Wallingford, U.K., pp. 1–322.

Walter, D.E. and Proctor, H.C. (2001) Mites in Soil. In: *An Interactive Key to Mites and Other Soil Microarthropods.* Australian Biological Resources Study, Canberra, Australia (CD-ROM).

Walter, D.E., Kethley, J. and Moore, J.C. (1987) A heptane flotation method for recovering microarthropods from arid grassland soils, with comparisons to the Merchant-Crossley high-gradient extraction method and estimates of microarthropods biomass. *Pedobiologia* 30, 221–232.

Walter, D.E., Lindquist, E.E., Smith, I.M., Cook, D.R. and Krantz, G.W. (2009) Order Trombidiformes. In: Krantz, G.W. and Walter, D.E. (eds) *A Manual of Acarology.* Texas Tech University Press, Lubbock, Texas, U.S.A., pp. 233–420.

Walzl, M.G. (1992) Ultrastructure of the reproductive system of the house dust mites, *Dermatophagoides farinae* and *D. pteronyssinus* (Acari: Pyroglyphidae) with remarks on spermatogenesis and oogenesis. *Experimental and Applied Acarology* 16, 85–116.

Wang, S.-C. (1964) Five new records to Chinese Acaridae. *Acta Entomologica Sinica* 13, 900–902.

Wang, C.-L. (1983) The infestation and control of bulb mite (*Rhizoglyphus robini*) on gladiolus. *Journal of Agricultural Research of China* 32, 75–82.

Wang, X.-Z. (1989) Acarina. In: Deng, G.F., Wang, H.F., Xin, J.L., Wang, D.Q., Wu, W.N. and Wang, X.-Z. (eds) *Synopsis of the China Acarology*. Science Press, Beijing, China, pp. 1–240.

Wang, C.-L. and Lin, R.-T. (1986) Studies on soil treatments for the control of the bulb mite (*Rhizoglyphus robini* Claparede) on *Gladiolus*. *Journal of Agricultural Research of China* 35, 230–234.

Wang, H.F., Cui, Y.Q. and Liu, Y.H. (2000) Acari: Oribatida. In: Huang, B.K. (ed.) *Fauna of Insects in Fujian Province of China, Vol. 9*. Fujian Science and Technology Publishing House, Fuzhou, China, pp. 296–323.

Wang, B.-M., Wang, Z.-Q., Wu, Z.-Y., Wang, X.-W. and Fan, Q.-H. (2008) The occurrence and control of *Carpoglyphus lactis* (Linnaeus) (Acari: Carpoglyphidae). *Entomology Journal of East China*. Available at: http://en.cnki.com.cn/Article_en/CJFDTotal-HDKC200802015.htm (accessed 1 July 2020).

Wang, C., Ma, Y., Huang, Y., Su, S., Wang, L. *et al.* (2019) Darkness increases the population growth rate of the poultry red mite *Dermanyssus gallinae*. *Parasites and Vectors* 12, 213. DOI: 10.1186/s13071-019-3456-1.

Wang, C., Xu, X., Yu, H., Huang, Y., Li, H., Wan, Q. and Pan, B. (2020) Low-temperature storage of the poultry red mite, *Dermanyssus gallinae*, facilitates laboratory colony maintenance and population growth. *Parasitology* 14(7), 740–746.

Waqas, M.S., Saad Elabasy, A.S., Zaky Shoaib, A.A., Cheng, X., Zhang, Q. and Shi, Z. (2020) Lethal and sublethal effect of heat shock on *Phenacoccus solenopsis* Tinsley (Hemiptera: Pseudococcidae). *Journal of Thermal Biology* 92, 102679.

Warner, C.E. and Bohane, S.J. (1978) Preliminary study of a diet infested with *Acarus siro* (L.) on adult food consumption and foetal status in mice. *Journal of Stored Products Research* 14, 13–17.

Warner, J.A., Frederick, J.M., Bryant, T.N., Weich, C., Raw, G.J. *et al.* (1992) Mechanical ventilation and high-efficiency vacuum cleaning: A combined strategy of mite and mite allergen reduction in the control of mite-sensitive asthma. *Journal of Allergy and Clinical Immunology* 105(1), 75–82.

Wasylik, A. (1959) Mite fauna (Tyroglyphoidea) in the nests of the common sparrow (*Passer domesticus* L.). *Ekologia Polska* B5, 187–190.

Wasylik, A. (1963) A method of continue analyse of the Acarina of birds' nests. *Ekologia Polska* B9, 219–224.

Wasylik, A. (1964) Remarks on the dispersion of certain mites. *Ekologia Polska* B10, 189–193.

Wasylik, A. (1971) Nest types and the abundance of mites. *Ekologia Polska* 19, 689–699.

Wasylik, A. (1973) The mites (Acaroidea) inhabiting the nests of the tree sparrow *(Passer montanus* L.). *Ekologia Polska* A21, 869–899.

Watson, J. (2008) New building, old parasite: mesostigmatid mites-an ever-present threat to barrier rodent facilities. *Journal of Lab Animal Research* 49, 303–309.

Weaver, D.K. and Petroff, A.R. (2004) Stored grain, Grain mite. High Plains IPM Guide. Available at: https://wiki.bugwood.org/uploads/GrainMite-StoredGrain.pdf (accessed 21 April 2020).

Webster, L.M.I., Thomas, R.H. and McCormack, G.P. (2004) Molecular systematics of *Acarus siro* s. lat., a complex of stored food pests. *Molecular Phylogenetics and Evolution* 32(3), 817–822.

Wegner, Z. (1980) Study on the occurrence of allergogenic mites on sea-going ships. *Materia y Sesji Naukowej "Zagadnienia zdrowotne w gospodarce morskiej", Szczecin, Poland, TN, September 16–18, 1976*. Szczecińskie Towarzystwo Naukowe, Szczecin, pp. 193–200.

Weigmann, G. (1991) Oribatid communities in transects from bogs to forest in Berlin indicating the isotope qualities. In: Dusbábek, F. and Bukva, V. (eds) *Modern Acarology, Vol.1 Proc. VIII International Congres of Acarology, ceske Budlojvice, Czechoslovacia, 6–11 August 1990*. Academic, Prague and SPB academic Publishing, The Hague, The Netherlands, pp. 359–364.

Wen, T., Hong, S., Shen, S., Cal, L., Xu, W. *et al.* (1988) Preliminary survey of mite allergy in China. In: de Weck, A.L. and Todt, A. (eds) *Mite Allergy: A Worldwide Problem*. UCB Institute of Allergy, Brussels, pp. 23–24.

Wen, D.C., Shyur, S.D., Ho, C.M., Chiang, Y.C., Huang, L.H., Lin, M.T. *et al.* (2005) Systemic anaphylaxis after the ingestion of pancake contaminated with the storage mite *Blomia freemani*. *Annals of Allergy, Asthma and Immunology* 95(6), 612–614.

Wen M.-F., Chi, H., Lian, Y.-X., Zheng, Y.-H., Fan, Q.-H. and You, M.-S. (2017) Population characteristics of *Macrocheles glaber* (Acari: Macrochelidae) and *Stratiolaelaps scimitus* (Acari: Laelapidae) reared on a mushroom fly *Coboldia fuscipes* (Diptera: Scatopsidae). *Insect Science* 26(2), 322–332.

Weng, Z.-K., Zhang, Y.-X., Ye, S.-Z., Lin, J.-Z. and Zhang, J.-H. (1995) Studies on using predacious mite to control pest mite in feed products. *Fujian Journal of Agricultural Sciences* 10(2), 44–47.

Wergin, W.P., Ochoa, R., Erbe, E.F. and Joy, D.C. (2000) Oscillating trichobothria in the low temperature SEM: biological capacitors or charging artifacts. *Scanning* 22, 140–141.

Westerboer, I. (1963) Die Familie Podocinidae Berlese 1916. In: Stammer, H.J. (ed.) *Beiträge zur Systematik und Ökologie Mitteleuropäischer Acarina. Band II. Mesostigmata 1*., Akademische Verlagsgesellschaft Geest and Portig K.G., Leipzig, Germany, pp. 179–450.

Westerboer, I. and Bernhard, F. (1963) Die familie phytoseiidae berlese 1916. In: Stammer, H.-J. (eds) *Beiträge zur Systematik und Ökologie Mitteleuropäischer Acarina, Band II, Mesostigmata 1*. Akademische Verlagsgesellschaft Geest and Portig K.-G., Leipzig, Germany, pp. 778–804.

Weygoldt, P. (1998) Evolution and systematics of the Chelicerata. Proceedings EURAAC III Symposium, Amsterdam. *Experimental and Applied Acarology* 22, 63–79.

Weygoldt, P. and Paulus, H.F. (1979a) Untersuchungen zur morphologie, taxonomie and phylogenie der Chelicerata. I, morphologische untersuchungen. *Journal of Zoological Systematics and Evolutionary Research* 17, 85–116.

Weygoldt, P. and Paulus, H.F. (1979b) Untersuchungen zur morphologie, taxonomie and phylogenie der Chelicerata. II, Cladogramme und die Entfaltung der Chelicerata. *Zeitschrift Zoology System Evolutionsforschung* 17, 177–200.

Wharton, G.W. (1970) Mites and commercial extracts of house dust. *Science* 167(1), 382–383.

Wharton. G.W. (1976) House dust mites. 1. *Journal of Medical Entomology* 12, 577–621.

Wharton, G.W. and Brody, A.R. (1972) A peritrophic membrane of the mite *Dermatophagoides farinae*: Acariformes. *Journal of Parasitology* 58, 801–804.

Wharton, G.W. and Devine, T.L. (1969) Exchange of water between a mite *Laelaps echidnina* and the surrounding air under equilibrium conditions. *Journal of Insect Physiology* 14, 1303–1318.

Wharton, G.W. and Furumizo, R.T. (1977) Supracoxal gland secretions as a source of fresh water for Acarida. *Acarologia* 19, 112–116.

Whitaker, J.O. and Wilson, N. (1974) Host and distribution lists of mites (Acari), parasitic and phoretic, in the hair of wild mammals of North America, north of Mexico. *The American Midland Naturalist* 91, 1–67.

White, N.D.G. (1995) Insects, mites, and insecticides in stored grain ecosystems. In: Jayas, D.S., White, N.D.G. and Muir, W.E. (eds) *Stored-Grain Ecosystems*. Marcel Dekker Inc., New York, Basel, Hong Kong, pp. 123–168.

White, N.D.G. and Jayas, D.S. (1991) Control of insects and mites with carbon dioxide in wheat stored at cool temperatures in Nonairtight Bins. *Journal of Economic Entomology* 84(6), 1933–1942.

White, N.D.G. and Jayas, D.S. (1993) Effectiveness of carbon dioxide in compressed gas or solid formulation for the control of insects and mites in stored wheat and barley. *Phytoprotection* 74(2), 101–111.

White, N.D.G. and Sinha, R.N. (1979) Natural regulation of *Tarsonemus granarius* numbers in stored wheat ecosystems-A multivariate assessment. In: Rodriguez, J.G. (ed.) *Recent Advances in Acarology, Vol. 1*. Academic Press, New York, U.S.A., pp. 291–298.

White, N.D.G. and Sinha, R.N. (1981) Life history and population dynamics of the mycophagous mite *Tarsonemus granarius* Lindquist (Acarina: Tarsonemidae). *Acarologia* 22(4), 353–360.

White, N.D.G. and Sihna, R.N. (1990) Effect of chlorpyrifos-methyl on oat ecosystems in farm granaries. *Journal of Economic Entomology* 83(3), 1128–1134.

White, N.D.G., Henderson, L.P. and Sinha, R.N. (1979) Effects of infestations by three stored product mites on fat acidity, seed germination, microflora of stored wheat. *Journal of Economic Entomology* 72(5), 763–766.

White, N.D.G., Jayas, D.S. and Sinha, R.N. (1994) Impact of pirimiphos-methyl and cold temperatures on arthropod populations in stored wheat. *Phytoprotection* 75(2), 79–90.

White, N.D.G., Jayas, D.S. and Demianyk, C.J. (1997a) Degradation and biological impact of chlorpyrifos-methyl on stored wheat and pirimiphos-methyl on stored maize in western Canada. *Journal of Stored Products Research* 33(2), 125–135.

White, N.D.G., Jayas, D.S. and Demianyk, C.J. (1997b) Movement of grain to control stored-product insects and mites. *Phytoprotection* 78(2), 75–84.

White, N.D.G., Fields, P.G., Demianyk, C.J., Timlick, B. and Jayas, D.S. (2011) Arthropods of stored cereals, oilseeds, and their products in Canada: Artificial ecosystems on grasslands. In: Floate, K.D. (ed.) *Arthropods of Canadian Grasslands, Vol. 2, Inhabitants of a Changing Landscape*. Biological Survey of Canada, pp. 267–289.

Wickman, M., Nordvall, S.L., Pershagen, G., Korsgaard, J. and Johansen, N. (1993) Sensitisation to domestic mites in a cold temperate region. *American Review of Respiratory Disease* 148, 58–62.

Wiĕcko, E. (1986) Acarina znalezione w kurzu obiektów portowych Gdyni. *Wiadomości Parazytologiczne* 32, 445–447.

Wildies, T. (2000) On the morphology of *Tyrophagus tropicus* Robertson, 1959 (Acari: Acaridae). *Entomologische Mitteilungen aus dem Zoologischen. Museum Hamburg* 13(161), 147–150.

Wilkin, D.R. (1975) The effects of mechanical handling and the admixture of acaricides on mites in farm-stored barley. *Journal of Stored Products Research* 11, 87–95.

Wilkin, D.R. (1979) The control of mites in cheese stores. Proc. V International Congress of Acarology, 6–12 August 1978, East Lansing, Michigan, Michigan State University, *Recent Advances in Acarology* 1, 221–229.

Wilkin, D.R. and Stables, L.M. (1985) The effects of dusts containing etrimfos, ethacrifos or pirimiphosmethyl on mites in the surface layers of stored barley. *Experimental and Applied Acarology* 1, 203–211.

Williams, R.W. (1958) An infestation of a human habitation by *Dermanyssus gallinae* (De Geer, 1778) (Acarina: Dermanyssidae) in New York City resulting in sanguisugent attacks upon the occupants. *American Journal of Tropical Medicine and Hygiene* 7(6), 627–629.

Williams, G.C. (1961) Infestation in feed mills. *Agriculture Merch* 41(4), 72–75, 80.

Williams, P. and Norris, K. (2001) Near-infrared technology in the agricultural and food industries. American Association of Cereal Chemists, St. Paul, Minnesota, U.S.A., pp. 1–312.

Williams, G.L., Smiley, R.L. and Redington, B.C. (1978) A taxonomic study of the genus *Haemogamasus* in North America, with descriptions of two new species (Acari: Mesostigmata, Laelaptidae). *International Journal of Acarology* 4, 235–273.

Willis, R.R. and Axtell, R.C. (1968) Mite predators of the house fly: A comparison of *Fuscuropoda vegetans* and *Macrocheles muscaedomesticae*. *Journal of Economic Entomology* 61, 1669–1674.

Willmann, C. (1931) Moosmilben oder Oribatiden (Criptostigmata) in Dahal. *Die Tierwelt Deutschlands* 22, 79–200.

Willmann, C. (1939) *Terrestrische Acari de Nord- und Ostseeküste*. Abh. Naturwiss. Ver. Bremen 31, pp. 521–550.

Willmann, C. (1952) Parasitisch Milben an Kleinsäugern. *Zeitschrift fur Parasitenkunde* 15, 392–428.

Wilson, H.T. (1947) Tropical eosinophilia in East Africa. *The BMJ* 1, 801–804.

Wilson, E.O. (1971) *The Insect Societies*. The Belknap Press of Harvard University Press, Cambridge, Massachusetts, pp. 1–562.

Wirth, S. (2004) Phylogeny, biology and character transformations of the Histiostomatidae (Acari, Astigmata). Ph.D. thesis. Freie Universität, Berlin. Available at: https://refubium.fu-berlin.de/handle/fub188/1 (accessed 17 February 2022).

Wise, G.U., Henessey, M.K. and Axtell, R.C. (1988) A new species of manure inhabiting mite in the genus Poecilochirus (Acari: Mesostigmata: Parasitidae) predacious on house fly eggs and larvae. *Ann. Entomol. Soc. Am.* 81, 209–224.

Wiśniewski, J. and Hirschmann, W. (1991) *Protogamasellopsis posnaniensis* nov. spec. (Acarina: Mesostigmata) aus Palmenhaus in Polan. *Bulletin of the Polish Academy of Sciences. Biological Science* 39(2), 189–194.

Wiśniewski, J. and Hirschmann, W. (1993) Gangsystematik der Parasitiformes. Teil 548. Katalog der Ganggattungen, Untergattungen, Gruppen und Arten der Uropodiden der Erde (Taxonomie, Literatur, Grösse, Verbreitung, Vorkommen). Acarologie. Schriftenr. Vgl. Milbenkunde 40, 1–220.

Witaliński, W. (2000) *Aclerogamasus stenocornis* sp.n., a fossil mite from the Baltic amber (Acari: Gamasida: parasitidae). *Genus (Warsaw)* 11, 619–626.

Witaliński, W., Joflczy J. and Godula, J. (1986) Spermatogenesis and sperm structure before and after insemination in two acarid mites, *Acarus siro* L. and *Tyrophagus putrescentiae* (Schrank) (Acari: Astigmata). *Acarologia* 27, 41–51.

Witaliński, W., Szlendak, E. and Boczek, J. (1990) Anatomy and ultrastructure of the reproductive systems of *Acarus siro* (Acari: Acaridae). *Experimental and Applied Acarology* 10, 1–31.

Witaliński, W., Liana, M. and Alberti, G. (2002) Fine structure and probable function of ring organs in the mite *Histiostoma feroniarum* (Acari: Actinotrichida: Acaridida: Histiostomatidae). *Journal of Morphology* 253, 255–263.

Woas, S. (2002) Acari: Oribatida. In: Adis, J. (ed.) *Amazonian Arachnida and Myriapoda*. Pensoft Publisher, Sofia, Moscow, pp. 21–291.

Woertz, J., Kinney, K., Kraakman, N., van Heiningen, W.N.M., van Eekert, M.H.A. and van Groenestijn, J.W. (2002) Mite growth on fungus under various environmental conditions and its potential application to biofilters. *Experimental and Applied Acarology* 27, 265–276.

Wohltmann, A. (2010) Notes on the taxonomy and biology of Smarididae (Acari: Prostigmata: Parasitengona). *Annales Zoologici* 60(3), 355–381.

Womersley, H. (1941) Studies in Australian Acarina. (2). Tyroglyphidae (s. l.). *Records of the South Australian Museum* 6(4), 451–488.

Womersley, H. (1956) On some new Acarina-Mesostigmata from Australia, New Zealand and New Guinea. *Zoological Journal of the Linnean Society* 42, 505–599.

Wong, S.F., Chong. A.L., Mak, J.W., Tan, J., Ling, S.J. and Ho, M.T. (2011) Molecular identification of house dust mites and storage mites. *Experimental and Applied Acarology* 55, 123–133.

Wongsathuaythong, S. and Lakshana, P. (1972) House-dust mite survey in Bangkok and other provinces of Thailand. *Journal of the Medical Association of Thailand* 55, 272–286.

Wood, T.G. (1967) New Zealand mites of the family Stigmaeidae (Acari, Prostigmata). *Transactions and Proceedings of the Royal Society of New Zealand* 9, 93–139.

Wood, T. G. (1971) Stigmaeidae (Acari, Prostigmata) from the British Solomon Islands. *Acarologia* 13(1), 65–67.

Wood, T.G. (1973) Revision of Stigmaeidae (Acari: Prostigmata) in the Berlese collection. *Acarologia* 15(1), 76–95.

Woodcock, A.A. and Cunnington, A.M. (1980) The allergenic importance of house dust and storage mites in asthmatics in Brunei, S.E. Asia. *Clinical and Experimental Allergy* 10(5), 609–615.

Wooddy, M.W. and Fashing N.J. (1993) The ability of *Rhizoglyphus robini* Claparede (Astigmata: Acaridae) to subsist solely on a diet of filter paper. *International Journal of Acarology* 19, 345–348.

Woodring, J.P. (1963) The nutrition and biology of saprophytic Sarcoptiformes. In: Naegele, J.A. (ed.) *Advances in Acarology, vol. 1*. Comstock Publishing Associates, Ithaca, NY, U.S.A., pp. 89–111.

Woodring, J.P. (1969a) Environmental regulation of andropolymorphism in tyroglyphids (Acari). In: Evans, G.O. (ed.) In: *Proc 2nd international Congress of Acarology, Sutton Bonington, U.K., 19–25 July 1967*. Akademiai Kiado, Budapest, Hungary, pp. 433–440.

Woodring, J.P. (1969b) Observations on the biology of six species of acarid mites. *Annals of the Entomological Society of America* 62, 102–108.

Woodring, J.P. (1973) Comparative morphology, functions and homologies of the coxal glands in oribatid mites (Arachnida: Acari). *Journal of Morphology* 139, 407–429.

Woodring, J.P. and Carter, S.C. (1974) Internal and external morphology of the deutonymph of *Caloglyphus boharti* (Arachnida: Acari). *Journal of Morphology* 144, 275–295.

Woodring, J.P. and Galbraith, C.A. (1976) The anatomy of the adult uropodid *Fuscuropoda agitans* (Arachnida: Acari), with comparative observations on other Acari. *Journal of Morphology* 150, 19–58.

Woodring, J.P. and Moser, J.C. (1970) Six new species of anoetid mites associated with North America Scolytidae. *Canadian Entomologist* 102, 1237–1257.

Woodroffe, G.E. (1953) An ecological study of the insects and mites in the nests of certain birds in Britain. *Bulletin of Entomological Research* 44, 739–772.

Woodroffe, G.E. (1954) An additional note on the fauna of the birds'nests in Britain. *Bulletin of Entomological Research* 45, 135–136.

Woodroffe, G.E. (1956) Some insects and mites associated with bat-roosts, with a discussion of the feeding habits of the cheyletids (Acarina). *Entomologist's Monthly Magazine* 92, 138–141.

Woodrow, J.E., LePage, J.T., Miller, G.C. and Hebert, V.R. (2014) Determination of methyl isocyanate in outdoor residential air near metam-sodium soil fumigations. *Journal of Agricultural and Food Chemistry* 62, 8921–8927.

Woolley, T.A. (1961) A review of the phylogeny of mites. *Annual Review of Entomology* 6, 263–284.

Woolley, T.A. (1988) Acarology: Mites and human welfare. John Wiley & Sons, Inc., New York, U.S.A., pp. 1–484.

Workman, P.J. and Martin, N.A. (2002) Effect of pesticides on *Cymbidium* orchid pollen-cap mite and its predator *Hypoaspis* sp. *New Zealand Plant Protection* 55, 380–384.

Wraith, D.G., Cunnington, A.M. and Seymour, W.M. (1979) The role and allergenic importance of storage mites in house dust and other environments. *Clinical and Experimental Allergy* 9(6), 545–561.

Wrensch, D.L. (1979) Components of reproductive success in spider mites. In: Rodriguez, J.G. (ed.) *Recent Advances in Acarology, Vol. 1*. Academic Press, New York, U.S.A., pp. 154–164

Wrensch, D.L. and Young, S.S.Y. (1978) Effects of density and host quality on rate of development, survivorship and sex ratio in the carmine spider mite. *Environmental Entomology* 7, 499–501.

Wrensch, D.L. and Young, S.S.Y. (1983) Relationships between primary and tertiary sex ratio in the two-spotted spider mite (Acarina: Tetranychidae). *Annals of the Entomological Society of America* 76, 786–789.

Wright, E.M. and Chambers, R.J. (1994) The biology of the predatory mite *Hypoaspis miles* (Acari: Laelapidae), a potential biological control agent of *Bradysia paupera* (Dipt.: Sciaridae). *Entomophaga* 39, 225–235.

Wright, H.W., Bartley, K., Nisbet, A.J., McDevitt, R.M., Sparks, N.H., Brocklehurst, S. and Huntley, J.F. (2009) The testing of antibodies raised against poultry red mite antigens in an in vitro feeding assay; preliminary screen for vaccine candidates. *Experimental and Applied Acarology* 48(1–2), 81–91.

Wright, H.W., Bartley, K., Huntley, J.F. and Nisbet, A.J. (2016) Characterisation of tropomyosin and paramyosin as vaccine candidate molecules for the poultry red mite, *Dermanyssus gallinae*. *Parasites and Vectors* 9, 544.

Wu, H.-H. (1999) The most common house dust mites of house in Taiwan. *Chinese Journal of Entomology, Special Publication* 12, 179–191.

Wu, J. Liu, Z.G., Ran, P.X. and B. Wang (2009a) Influence of environmental characteristics and climatic factors on mites in the dust of air-conditioner filters. *Indoor Air* 19 474–481. DOI: 10.1111/j.1600-0668.2009.00611.x.

Wu, J., Yang, F., Liu, Z., Wu, G. and Ran, P. (2009b) The alimentary canal of *Blomia tropicalis* (Acari: Astigmata: Echymyopodidae): the application of three-dimensional reconstruction technology. *Experimental and Applied Acarology* 47(3), 215–24.

Wu, Z., Zhang, X., Luo, J. and Fan, Q. (2009c) Effects of five preys on growth and reproduction of *Lasioseius* Sp. *Journal of Fujian Agriculture and Forestry University (National Science of Education)* 38, 581–584.

Wunderle, I. (1992) Die Oribatiden-Gemeinschaften (Acari) der verschiedenen Habitate eines Buchenwaldes. *Carolinea* 50, 79–144.

Wurst, E. (1993) Investigations on the anatomy and the behaviour of the fur mite *Listrophorus leuckarti*. *Stuttgarter Beiträge zur Naturkunde* 503A, 1–68.

Xia, B., Luo, D.L., Zou, Z., and Zhu, Z. (2009) Effect of temperature on the life cycle of *Aleuroglyphus ovatus* (Acari: Acaridae) at four constant temperatures. *Journal of Stored Products Research* 45, 190–194.

Xia, B., Zhang, T. and Zou, Z.-W. (2007) The predation of *Cheletomorpha lepidopterorum* on *Tyrophagus putrescentiae*. *Journal of Nanchang University (Natura. Science)* 31, 579–582.

Xie, L., Yan, Y. and Zhang, Z.-Q. (2018) Development, survival and reproduction of *Stratiolaelaps scimitus* (Acari: Laelapidae) on four diets. *Systematic and Applied Acarology* 23(4), 779–794.

Xin, J.-L. (1965) Recent progress of the studies on storage mites. In: Xin, J.-L. and Xu, Y.-Q. (eds) *The Progress of Acarology*. Shanghai Science and Technology Press, Shanghai, China, pp. 83–117.

Xu, X., Wang, C., Zhang, S., Huang, Y., Pan, T., Wang, B. and Pan, B. (2019) Acaricidal efcacy of orally administered macrocyclic lactones against poultry red mites (*Dermanyssus gallinae*) on chicks and their impacts on mite reproduction and blood-meal digestion. *Parasites Vectors* 12, 345.

Xu, X., Wang, C., Zhang, S., Huang, Y., Pan, T., Wang, B. and Pan, B. (2020) Evaluation of the vaccine efficacy of three digestive protease antigens from *Dermanyssus gallinae* using an in vivo rearing system. *Vaccine* 17, 7842–7849.

Xua, Y., Zhua, Y.-Z., Wua, J.-Q. and Zhang, F.-P. (2020) A new species of the genus *Paralycus* from Fujian, China. *Acarologia* 60(2), 481–487.

Xuan, J.-Y., Wang, Z.-S., Wang, C.-B. and Yu, S.-D. (1993) Preliminary research of relationship between *Tyrophagus putrescentiae* and two parasitic fungi. *Grain Storage* 22(3), 20–26.

Yadav, A. (1989) Functional and numerical responses of *Cheyletus malaccensis* to its prey in a bulk of stored wheat. Ph.D. Thesis, University (H.A.U.), Hisar, India.

Yadav, A.E., Morgan, B.L., Vyzenski-Moher, D.L. and Arlian, D.L. (2006) Prevalance of IgE serum to storage mites in Southwestern Ohio population. *Annals of Allergy, Asthma and Immunology* 96, 356–362.

Yagofarov, F.F. and Galikeev, Kh.L. (1987) Some data on the ecology of *Dermatophagoides* mites in the town of Semipalatinsk. *Parazitologia* 21(2), 148–151.

Yan, J., Qin, Z.-L. and Li, G.-C. (1989) A report on the prevention and control of *Tyrophagus putrescentiae* using natural plant essential oil. *Technology Communication Grain Oil Storage Science Technology* 3, 23–24.

Yan, X.-Y., Yang, N.-Z., Yuan, D.-Z., Wu, M.-S. and Chen, J.-H. (1992) Studies on the life-history of *Aleuroglyphus ovatus* Troupeau (Acarina, Acaridae). *Technology Communication Grain Oil Storage Science Technology* 6, 53–55.

Yang, B., Cai, J. and Cheng, X. (2011) Identification of astigmatid mites using ITS2 and COI regions. *Parasitology Research* 108, 497–503.

Yang, X., Liu, Y.-B., Singh, R. and Phillips, T.W. (2022) Nitric oxide fumigation for control of ham mite, *Tyrophagus putrescentiae* (Sarcoptiformes: Acaridae). *Journal of Economic Entomology* 115(2), 501–507.

Yasemin, A.Y. and Ayyildiz, N. (2017) Oribatulid mites (Acari, Oribatida, Oribatulidae) from the southwestern region of the Amanos Mountains. *Plant Protection Bulletin* 57(4), 473–484.

Yassin, E.M.A., Kassem, E.M.K. and Mahmoud, R.H. (2018a) Effect of food kinds on developmental stages and fecundity of the astigmatid mites, *Tyrophagus putrescentiae* (Schrank) and *Rhizoglyphus robini* Claparede (Astigmata: Acaridae). *Journal of Plant Protection and Pathology* 9(10), 633–637.

Yassin, E.A., Osman, S.A. and Rahouma, A.K.A. (2018b) Occurrence of different mites associated with different cereals and legumes crops in different locations of Egypt. *Egyptian Academic Journal of Biological Sciences* 11(4), 51–58.

Yassin, E.M.A., Khalil, A.M. and Osman, S.A.A. (2016) Survey of mites associated with stored onion bulbs at El-Menofia governatorate. *Egyptian Journal of Agricultural Research* 94(4), 821–828.

Yassin, E.M.A., Abdel-Khalik, A.R. Abdul-Aziz, S.A. and Osman, S.A. (2017) Studies on mites associated with some stored hay in different regions of Egypy. *Menoufia Journal of Plant Protection* 2, 191–201.

Yasueda, H., Mita, H., Yui, Y. and Shida, T. (1986) Isolation and characterization of two allergens from *Dermatophagoides farinae. International Archives of Allergy and Immunology* 81(3), 214–23.

Ye, S.-S. and Zhang, Z.-Q. (2014) Age and size at maturity in *Tyrophagus curvipenis* (Acari: Acaridae) when fed on three different diets. *Systematic and Applied Acarology* 19(4), 506–512.

Yi, F.C., Chen, J. Y., Chee, K. K., Chua, K. Y. and Lee, B. W. (2009) Dust mite infestation of flour samples. *Allergy* 64, 1788–1789.

Yi, Y.J., Cheng, X., Yang, Z.F. and Zhang, S.H. (2016) MaxEnt modeling for predicting the potential distribution of endangered medicinal plant (*H. riparia* Lour) in Yunnan, China. *Ecological Engineering* 92, 260–269.

Yildirim Kumral, A., Turgu, V., Yildiz, E., Kumral, N.A., Çevik, R.N. *et al.* (2019) Determination of the effect of ferric oxide and ozone gas applications on the fruit quality parameters and microbial contamination load caused by dried fruit mite, *Carpoglyphus lactis* (L.) (Acari: Carpoglyphidae) infestation. *Bursa Uludağ Üniversitesi Ziraat Fakültesi Dergisi* 33(1), 67–82.

Yong, H., Feng-Xia, D., Zhao-Dan, Shao-Sheng, W. and Chao-Pin, L. (2016a) Preliminary investigation of Suidasia nesbitti breeding status in ground dust flour in Qiqihaer City. *Zhongguo Xue Xi Chong Bing Fang Zhi Za Zhi* 29(2), 225–227.

Yong, H., Jin-Hong, H. and Chao-Pin, L. (2016b) Investigation of *Lardoglyphus konoi* (Astigmata) breeding in stored *Solenognathus. Chinese Journal of Schistosomiasis Control* 28(2), 202–205.

Yoshikawa, M. and Bennett, P.H. (1979) House dust mites in Columbus, Ohio. *The Ohio Journal of Science* 79, 280–282.

Yoshizawa, T., Yamamoto, T. and Yamamoto R. (1971) Synergestic attractancy of cheese components for cheese mite *Tyrophagus putrescentiae* (Schrank). *Batyu-Kagaku Tokyo* 36(1), 1–6.

Yoshizawa, T., Tamamoto, T. and Yamamoto, R. (1972) Isolation and structural elucidation of cheese components which attract the cheese mite, *Tyrophagus putrescentiae*. *Memoirs of the Tokyo University of Agriculture* 15, 1–29.

Young, S.S.Y., Wrensch, D.L. and Kongschuensin, M. (1986) Control of sex ratio by female spider mites. *Entomologia Experimentalis et Applicata* 40, 53–60.

Yousef, A.A., El-Badry, E.A. and Metwally, S.H. (1979) Life history of the anoetid mite *Histiostoma cataglyphi* Yousef et Metwally, with a description of the immature stages (Acari, Astigmata, Anoetoidea). *Zeitschrift für angewandte Entomology* 87, 225–229.

Yousef, E.A., Zaher, M.A. and Kandil, M.M. (1982) Effect of prey and temperature on the development and biology of *Cheyletus malaccensis* Oudemans (Acari: Cheyletidae). *Acarologia* 25, 39–42.

Yu, L. and Zhang, Z.-Q. (2019) New Zealand pyemotes (Trombidiformes: Pyemotidae). *Systematic and Applied Acarology* 24(6), 1014–1047. Available at: https://biotaxa.org/saa/article/view/saa.24.6.7 (accessed 8 Ocober 2019).

Yu, X., Fan, Q.H. and Xu, J.L. (2002) Effective cumulative temperature of *Tyrophagus putrescentiae* (Schrank) (Acari: Acaridae). *Entomology Journal of East China* 11(1), 55–58.

Yu, X., Zeng, X.-R., Shi, Y. and Fan, Q.-H. (2006) Toxicity measurements of fourteen insecticides/acaricides to *Tyrophagus putrescentiae* (Acari: Acaridae). *Entomology Journal of East China* 15(2), 158–160.

Yunker, C.E. (1961) The genera *Bak*, new genus, and *Cheletomimus* Oudemans, with descriptions of three new species (Acarina: Cheyletidae). *Canadian Entomologist* 93(11), 1023–1035.

Zachvatkin, A. (1935) A short key to the granary mites. In: *Izdanie Komiteta Zagotovitelnix s.-x.* Moskwa productov pri SNK, SSSR, Russian, pp. 1–31.

Zachvatkin, A. (1936) Notes systématiques sur les Acariens, habitant les greniers I-II, Résumé. *Bulletin de la Société Impériale des Naturalistes de Moscou* 45(4), 268–270.

Zachvatkin, A.A. (1941) Tiroglifoidnye kleshchi tyroglyphoidea. In: Zernov, S.A. (ed.) *Fauna SSSR: Paukoobraznuye*. Akademiya Nauk SSSR, Moskva-Leningrad, pp. 1–475 (in Russian).

Zachvatkin, A.A. (1949) Novye vidy Am" arnykh kleshcheï iz Azerbaïdzhanskoï SSR. *Entomologiceskoe Obozrenie* 30(3/4), 287–290.

Zachvatkin, A.A. (1953) Einige ergebnisse des studiums der fauna der getreidemilben in der UdSSR. In: Rodendorf, B.B. (ed.) *Sammelwerk wissenschaftlicher Arbeiten von A.A. Zachvatkin*. Verlag der Moskauer Universitat, Moscow, pp. 169–176.

Zachvatkin, A.A. (1959) *Fauna of the U.R.S.S., Arachnoidea, Vol. VI, 1, Tyroglyphoidea (Acari)*. American Institute of Biological Sciences, Washington, DC., U.S.A., pp. 1–573.

Zachvatkin, A.A. and Volgin, V.I. (1955) *Getreide und Speichermilben. Milben der Nagetierfauna der USSR. Bestimmung der Fauna der USSR*. Verlag Akademie der Wissenschaften, Moskau, Leningrad, pp. 86–110.

Zaher, M.A. (1986) Survey and Ecological Studies on Phytophagous, Predaceous and Soil Mites in Egypt. II-A: Predaceous and Nonphytophagous Mites (Nile Valley and Delta). Text. PL 480 Programme U.S.A., Project No. EG-ARS-30, Grant No. FG-EG-139, pp. 1–567.

Zaher, M.A. and Shehata, K.K. (1963) Biological studies of *Tydeus californicus* (Banks) in Egypt (A.R.E.) (Acarina, Tydeidae). *Bulletin of the Entomological Society of Egypt* 67, 297–300.

Zaher, M.A. and Soliman, Z.R. (1967) The family Cheyletidae in the U.A.R. with a description of four new species. *Bulletin of the Entomological Society of Egypt* 51, 21–26.

Zaher, M.A. and Soliman, Z.R. (1971a) Life history of the predatory mite, *Cheletogenes ornatus* (Canestrini and Fanzago) (Acarina: Cheyletidae). *Bulletin of the Entomological Society of Egypt* 55, 85–89.

Zaher, M.A. and Soliman, Z.R. (1971b) Life-history of the predatory mite *Cheyletus malaccensis* Oudemans. *Bulletin of the Entomological Society of Egypt* 55, 49–53.

Zaher, M.A., Soliman, Z.R. and El-Bishlawy, S.M. (1975) Feeding habits of the predaceous mite, *Cunaxa capreolus* (Acarina: Cunaxidae). *Entomophaga* 20, 209–213.

Zaher, M.A., Yousef, A.A. and Kandil, M.M. (1981) Effect of food on the biology of *Cheletogenes ornatus* (C. & F.) (Acari: Prostigmata: Cheyletidae). *Acarologia* 22, 361–366.

Zaher, M.A., Mohamed, M.I. and Abdel-Halim, S.M. (1986) Incidence of mites associated with stored seeds and food products in upper Egypt. *Experimental and Applied Acarology* 2, 19–24.

Zakhvatkin, A.A. (1952) Division of the Acarina into orders and their position in the system of the Chelicerata. *Parazitology Sborn* 14, 5–46.

Zander, E. (1947) Handbuch der Bienenkunde in Einzeldarstellungen. In: *Krankheiten und Schädlinge der erwachsenen Bienen*. Stuttgart, Verlag Eugen Ulmer, pp. 1–147.

Zannou, I.D., Adebo, H.O., Zannou, E. and Hell, K. (2013) Mites associated with stored grain commodities in Benin, West Africa. *Experimental and Applied Acarology* 61, 449–470.

Zare, M., Rahmani, H., Faraji, F. and Akrami, M.A. (2012) Ascoidea (Acari: Mesostigmata) of Zanjan County of Iran with record of two new species. *Abstract Book of the 20th Iranian Plant Protection Congress*. Shiraz University, Shiraz, Iran, p. 459.

Ždárková, E. (1967) Stored food mites in Czechoslovakia. *Journal of Stored Products Research* 3(2), 155–175.

Ždárková, E. (1971) Orientation of *Tyrophagus putrescentiae* (Schrank) towards olfactory stimuli. In: *Proceedings of the 3rd International Congress of Acarology, Prague, Czechoslovakia, August 31–September 6, 1971*. Springer, Netherlands, pp. 241–246.

Ždárková, E. (1979) Cheyletid fauna associated with stored products in Czchoslovakia. *Journal of Stored Products Research* 15, 11–16.

Ždárková, E. (1991a) Stored product acarology. In: Dusbák, F. and Bukva, V. (eds) *Modern acarology, Proc. VIII International Congress of Acarology, České Budejovice, Czechoslovakia, 6–11 August 1990, Vol. 1*. SPB Academic Publishing bv, The Hague, The Netherlands, pp. 211–218.

Ždárková, E. (1991b) Application of the biopreparation "Cheyletin" in empty stores. In: Dusbábek, F. and Bukva, V. (eds) *Modern acarology, Proc. VIII International Congress of Acarology, České Budejovice, Czechoslovakia, 6–11 August 1990, Vol. 1*. SPB Academic Publishing bv, The Hague, The Netherlands, pp. 607–610.

Ždárková, E. (1994) The effectiveness of organophosphate acaricides on stored product mites interacting in biological control. *Experimental and Applied Acarology* 18, 747–751.

Ždárková, E. (1998) Biological control of storage mites by *Cheyletus eruditus*. *Integrated Pest Management Reviews* 3(2), 111–116.

Ždárková, E. and E. Horák (1990) Preventive biological control of stored food mites in empty stores using *Cheyletus eruditus* (Schrank). *Crop Protection* 9(5), 378–382.

Ždárková, E. and Horak, E. (1993) The effects of diets with mites (*Acarus siro* L.: Acaridae) on the performance and carcass quality of pigs. *Zeitschrift für angewandte Zoologie* 79(1), 117–126.

Ždárková, E. and Horák, P. (1997) Bionomics of *Eucheyletia taurica* volgin, a species new to Central Europe (Acarina: Cheyletidae). *Acarologia* 38, 123–125.

Ždárková, E. and Pekár, S. (2007) Is *Lepidoglyphus pilosus* a true species or a form of *L. s destructor* (Acari: Glycyphagidae)? Morphometrical analysis. *Acta Societatis Zoologicae Bohemicae* 71, 61–66.

Ždárková, E. and Reska, M. (1976) Weight losses of groundnuts (*Arachis hypogae* A. L.) from infestation by the mites *Acarus siro* L. and *Tyrophagus putrescentiae* (Schrank). *Journal of Stored Products Research* 12, 101–104.

Ždárková, E. and Voracek, V. (1993) The effect of physical factors on survival of stored food mites. *Experimental and Applied Acarology* 17, 197–204.

Zedan, M.A.-A. (1988) Studies on Predator-Prey Interactions between *Hypoaspis aculeifer* Canestrini (Acarina: Laelapidae) and *Rhizoglyphus echinopus* (Fum. & Rob.) (Acarina: Acaridae) under Laboratory Conditions. *Revue Zoologique Africaine* 102, 381–387.

Zeity, M. and Gowda, C.C. (2013) A new species of *Neophyllobius* (Acari: Camerobiidae) from India. *International Journal of Acarology* 39 (7), 547–550.

Zeman, P. (1988) Surface skin lipids of birds-a proper host kairomone and feeding inducer in the poultry red mite, *Dermanyssus gallinae*. *Experimental and Applied Acarology* 5, 163–173.

Zeman, P. and Jurík, M. (1981) A contribution to the knowledge of fauna and ecology of gamasoid mites in cavity nests of birds in Czechoslovakia. *Folia Parasitologica* 28, 265–271.

Zemek, R. and Prenerová, E. (1997) Powdery mildew (Acomycotina: Erisiphales)–an alternative food for the predatory mite *Typhlodromus pyri* Scheuten (Acari: Phytoseiidae). *Experimental and Applied Acarology* 21, 405–414.

Zeytun E., Doğan, S., Ünver, E. and Özçüçek F. (2018) Evaluation of dermatophagoides *pteronyssinus* (Trouessart) and *D. farinae* Hughes (Acari: Pyroglyphidae) sensitivity in patients with allergic rhinitis: A comparative study. *Systematic and Applied Acarology* 23(2), 206–215.

Zhan X.D., Li, C.P. and Chen, Q. (2017a) *Carpoglyphus lactis* (Carpoglyphidae) infestation in the stored medicinal Fructus Jujubae. *Nutricion Hospitalaria* 34(1), 171–174.

Zhan, X., Xi, Y., Li, C., Zhao, J. and Wei, G. (2017b) Composition and diversity of acaroids mites (Acari: Astigmata) community in the stored rhizomatic traditional Chinese medicinal materials. *Nutricion Hospitalaria* 34(2), 454–459.

Zhang, Z.-Q. (1995) Variance and covariance of ovipositional rates and developmental rates in the phytoseiidae (Acari: Mesostigmata): A philognetic consideration. *Experimental and Applied Acarology* 19, 139–146.

Zhang, Z.-Q. (2003) *Mites of Greenhouses: Identification, Biology and Control.* CAB International, Wallingford, U.K., pp. 1–244.

Zhang, Z.-Q. (2011) Animal biodiversity: An outline of higher-level classification and survey of taxonomic richness. *Zootaxa*, 3148 (1), 1–237.

Zhang, Z.-Q. (2012) New Zealand records of *Carpoglyphus lactis* (Acari: Carpoglyphidae). *Systematic and Applied Acarology* 17(2), 239–240.

Zhang, Z.-Q. and Fan, Q.-H. (2010) Blattisociidae of China: a review, with a checklist. *Zoosymposia* 4, 280–287.

Zhang, A.-H. and Li, Y.-R. (2002) A new species of the genus *Sinosuidasia* from China (Acari: Acaridae). *Acta Zootaxonomica Sinica* 27(3), 483–485.

Zhang, Y. and Lin, J. (1991) Studies of *Gnorimus chaudhrii* (Acari: Phytoseiidae), an important predator of tarsonemid mites in rice fields. *Chinese Journal of Biological Control* 7, 163–165.

Zhang, G.-L. and Jin, L.-C. (1963) Studies on the relative toxicity of some fumigants to the grain-infesting mite *Tyrophagus castellanii* Hirst. *Acta Phytophylactica Sinica* 2(2), 173–178.

Zhang, N.A. and Xie, L. (2021) The lifespans of the potential biological control agents in the family Blattisociidae (Acari: Mesostigmata). *Zoosymposia* 20, 91–103.

Zhang, K. and Zhang, Z.-Q. (2022) *Amblyseius herbicolus* mothers prefer to oviposit near eggs of non-kin in the absence of prey *Carpoglyphus lactis* (Acari: Phytoseiidae, Carpoglyphidae). *Systematic and Applied Acarology* 27, 2347–2354.

Zhang, Y., Lin, J., Zhang, Z.-Q., Saito, Y. and Ji, J. (2003) Studies on the life history of *Amblyseius cucumeris* (Acari: Phytoseiidae) feeding on *Aponychus corpuzae* (Acari: Tetranychidae). *Systematic and Applied Acarology* 8, 67–74.

Zhang, Z.-Q., Fan, Q.-H., Pesic, V., Smith, H., Bochkov, A.V. *et al.* (2011) Order trombidiformes reuter, 1909. In: Zhang, Z.-Q. (ed.) *Animal Biodiversity: An Outline of Higher-level Classification and Survey of Taxonomic Richness.* Zootaxa, 3148 (1), pp. 129–138.

Zhang, Y.-K., Chen, Y.-T., Yang, K. and Hong, X.-Y. (2016) A review of prevalence and phylogeny of the bacterial symbiont *Cardinium* in mites (subclass: Acari). *Systematic and Applied Acarology* 21(7), 978–990.

Zhao, J.-X. (2011) Localization of cross-reactivity allergens in the body of *Tarsonemus granarius* (Acari: Tarsonemidae) by immunohistochemistry. Available at: https://www.semanticscholar.org/paper/Localization-of-cross-reactivity-allergens-in-the-Xue/8aeb8d2390fec0c87d17b0e300b870806adb9592 (accessed 21 July 2019).

Zhao, Y., Phillips, T.W., Aikins, M.J., Abbar, S., Amoah, B., Mikel, W.B. and Schilling, M.W. (2012) Sensory evaluation and pest mortality of dry cured ham fumigation with phosphine in a simulated aging house. In: *Proceedings, Annual International Research Conference on Methyl Bromide Alternatives and Emission Reduction. 6–8 November 2012.* Orlando, Florida, U.S.A., IISD, p. 78.

Zhao, Y., Abbar, S., Phillips, T.W. and Schilling, M. W. (2015) Phosphine fumigation and residues in dry-cured ham in commercial applications. *Meat Science* 107, 57–63.

Zhao, J.-H., Zhan, X.-D., Sun, E.-T. and Li, C.-P. (2015) Investigation on *Tarsonemus granarius* breeding in the stored Chinese herb safflower. *Chinese Journal of Vector Biology and Control.* Available at: http://en.cnki.com.cn/Article_en/CJFDTotal-ZMSK201506014.htm (accessed 21 July 2019).

Zhao, Y., Abbar, S., Phillips, T.W., Williams, J.B., Smith, B.S. and Schilling, M.W. (2016) Developing food-grade coatings for dry-cured hams to protect against ham mite infestation. *Meat Science* 113, 73–79.

Zheleva, M. and Angelkova, E. (1964) Contribution to the study of the tyroglyphid fauna in Bulgaria. *Godishnik na Sofiiskiya Universitet, Biologo-Geologo-Geografski Fakultet* 56, 141–149.

Zhu, Y.-X. and Yang, Q.-G. (2003) Preliminary investigation on acaroid mites in stored foods. *Journal of Tropical Diseases and Parasitology* 1(2), 96–97. Available at: http://en.cnki.com.cn/Article_en/CJFDTotal-YXDZ200307023.htm (accessed 26 June 2020).

Zhu, Z.-M., Zhou, K., Xia, B. Yu, L.-P. and Tu, D. (2000) Effect of temperature on the development period of *Cheyletus trouessarti* Oudemans. *Journal of Nanchang University (Natural Science)* 24, 307–309.

Ziaee, M., Ebadollahi, A. and Wakil, W. (2021) Integrating inert dusts with other technologies in stored products protection. *Toxin Reviews* 40(4), 404–419.

Zoons, J. (2004) The effect of light programs on red mite (*Dermanyssus gallinae*) in battery cage housing. In: Perry, G.C. (ed.) *Welfare of the Laying Hen. CABI Poultry Science Symposium Series Number 27*. CAB International, Wallingford, UK, pp. 1–416.

Zoons, J. and Deschuytere, H. (2018) DERMANYSSUS: Potentiële bestrijdingsmethoden van *Dermanyssus gallinae* bij pluimvee in de praktijk. In: *Project of the Federal Agency for the Safety of the Food Chain*. Brussels

Zou, P. and Wang, X.-Z. (1989) A new species and two new records of Acaridae associated with edible fungi from China (Acarina: Acaroidea). *Acta Agriculturae Shanghai* 5(2), 21–24.

Zou, Z.-W., Xia, B., Gong, Z.-Q., Yu, L.-P. and Zhu, Z.-M. (2003) A study on population dynamic and spatial pattern of *Suidasia nesbitti* Hughes (Acari: Acaridae). *Science Technology and Engineering* 3(6), 565–567.

Zriki, G., Blatrix, R. and Roy, L. (2020) Predation interactions among henhouse-dwelling arthropods, with a focus on the poultry red mite *Dermanyssus gallinae*. *Pest Management Science* 76, 3711–3719.

Zuryn, S., Kuang, J. and Ebert, P. (2008) Mitochondrial modulation of phosphine toxicity and resistance *in Caenorhabditis elegans*. *Toxicological Sciences* 102(1), 179–186.

Zyromska-Rudska, A. (1964) Abundance, dynamics and age structure of populations of *Carpoglyphus lactis* (L.) (Acarina, Glycyphagidae) in closed cultures. *Ekologia Polska Series A* 12, 355–367.

Taxonomic Index

Page numbers in **bold** type refer to illustrations and tables.

A

Absidia spinosa 318
Acacia baileyana 159
Acalvolia squamata 375
Acalypha marginata (copperleaf) 403
Acaridae 275, 313
acarid mite *(Tyrophagus similis)* 54, 56
Acarophenax spp.
  A. assanovi 130
  A. dermestidarum 130
  A. lacunatus 130
  A. rackae 130
  A. tribolii 130, **131**
Acaropsellina spp.
  A. docta 137
  A. sollers 137
  A. tyrophagus 137
  A. volgini 137
Acarus spp. 42, 48, 185, 187
  A. chaetoxysilos 187
  A. entomophagus 274
  A. eruditus 140
  A. farris 46, 57, 65, 75, 80, 187–191, **188**, 295
  A. gracilis 34, 187, 196
  A. immobilis 55, 57, 80, 187, 191–192, 196
  A. mirabilis 187
  A. putrescentiae 209, 227
  A. siro (flour mite) 37, 48, 63, 67, 80, 86, 95, 96, 187, 190, 192–206, **193**, 216, 217, 222, 231, 232, 236, 244, 248, 283–285, 288, 290, 295, 296, 303, 318, 319, 357, 397, 402, 411, 413, 452, 456, 462, 484
  A. tribolii 130
Aceria spp.
  A. litchii (lychee erinose mite) 408
  A. oleae (olive gall mite) 485
Acer spp.
  A. pseudoplatanus 227
  A. saccharum (sugar maple trees) 458
Acotyledon spp. 50
  A. formosani 61
Acrocephalus arundinaceus (great reed warbler) 398
Aculops pelekassi (citrus rust mite) 403
Aegopodium podagraria (bishop's goutweed) 380
Aeroglyphus spp. 50, 292–293, 298, 410
  A. peregrinans 292, 293
  A. robustus (warty grain mite) 61, 174, 292–297, **294**, 318, 410, 452
African snipe (*Gallinago nigripennis*) 360
Afrocalvolia spp. 52
  A. nataliae 52
Agaricus spp.
  A. bisporus (button mushroom) 266, 269, 270, 340, 343
  A. bitorquis (pavement mushroom) 340, 343
Agrilus spp.
  A. auroguttatus 167
  A. coxalis 167
Agrimonia pilosa (hairy agrimony) 385
Ahasverus advena (stored-product beetle) 297, 411
Akodon azarae (Azara's grass mouse) 453
alarm pheromone emitting mite *(Rhizoglyphus robini)* 32, 55, 79
Aleurobius farris 189, 192, 196, 204, 205, 217
Aleuroglyphus spp. 185, 245, 248
  A. beklemishevi 245
  A. formosana 245
  A. ovatus (brown-legged grain mite) 61, 78, 89, 203, 205, 237, 244–250, **246**, 283–286, 291, 321, 483
  A. siculus 245
Alliea spp. 69, 137
  A. prasadi 137
Alliphis spp.
  A. halleri 447–449, **448**
  A. siculus 447–448, **448**
Allium spp.
  A. chinense 261
  A. fistulosum 262, 264
  A. sativum 236
  A. tuberosum 261
almond moth *(Ephestia cautella)* 404, 413
Alnion glutinosae (wet alder forest) 398
Alternaria alternatae (brown spot) 67, 173, 232, 295, 330
Amblydromalus limonicus 482
Amblyseiinae 480

Amblyseius spp. 480, 483
A. alpinus 484
A. arutanjani 198
A. barkeri 198, 484
A. eharai 284
A. herbarius 198
A. herbicolus 482, 483
A. largoensis 482
A. obtusus **481**, 483–484
A. swirskii (Swirski mite) 284, 483
A. tamatavensis 482, 483
American bird mite *(Dermanyssus americanus)* 416
American house dust mite *(Dermatophagoides farinae)* 35, 38, 108, 349–358, **353**, 366
Ameroseiidae 399
Ameroseius spp. 392
A. aegyptiacus 394
A. dendrovagans 394
A. gilarovi 393
A. lanceosetis 392
A. marginalis 392
A. plumigera 392–393
A. plumigerus 394
A. plumosus 392, 394
A. sichanensis 392
Ampezzoa spp. 5
A. triassica 5
Amyelois transitella (navel orangeworm) 410
Anagasta kuehniella 412
Anas platyrhynchos (mallard) 398
Ancistrocerus antilope 294, 295
Androlaelaps spp. 451
A. casalis 70, 198, 269, 295, 296, 331, 332, 431, 436, **451**, 451–453
A. fahrenholzi 453
A. fenilis 453
A. glasgowi 453
A. shealsi 453
A. zaheri 245
Angelicae sinensis (female ginseng) 360
Anobium punctatum 164
ant *(Camponotus ligniperda)* 252
Anthonomus spp. 165
A. grandis 165
Anthophora spp. 163
A. abrupta 163
Apatura iris 279
Aphelenchoides ritzemabosi 264
Aphelenia 368
Aphiochaeta rufipes 493
Apis spp. 247, 298, 466
A. cerana (Asian honey bee) 155, 164, 165, 207, 215, 241, 282, 298, 374
A. dorsata 215, 241
A. florea (red dwarf honey bee) 374
A. mellifera (European honeybee) 207, 241, 274, 292–295, 298, 343, 399, 408, 412
Aplodontopus spp. 286
A. marmotae 287
Apodemus spp. 398
A. flavicollis (yellow-necked mouse) 463
A. sylvaticus (wood mouse) 463
Aponychus corpuzae (bamboo spider mite) 486
Apostigmaeus pacificus 156, **158**
apple tree *(Pyrus malus)* 411
Aquila spp.
A. heliaca (eastern imperial eagle) 398
A. pomarina (lesser spotted eagle) 466, 469
Arachis spp. 83
A. hypogaea 83
Arachnia spp. 240
Araucaria angustifolia (Paraná pine) 410
Arctoseius spp.
A. butleri **396**, 397
A. cetratus 397
A. insularis 70, 397
Artemia spp. 463–464
Asca spp. 397
Asian honey bee *(Apis cerana)* 374
Aspergillus spp. 173, 196, 231, 269, 349
A. amstelodami 349
A. candidus 196, 295
A. flavus 86, 196
A. fumigates 229
A. glaucus 174, 295
A. niger 86, 258, 330
A. ochraceus 196, 258, 318
A. penicillioides 349, 355–356, 361
A. repens 198, 216, 330, 349
A. restrictus 68, 197
A. ruber 198, 216, 330
Aspidiotus nerii (oleander scale) 375
Attagenus piceus 412
Auricularia spp.
A. auricula-judae (jelly ear fungus) 244, 392
A. polytricha 229
Australian bush fly *(Musca vetustissima)* 468
Australian giant earwig *(Titanolabis colossea)* 343
Austrian pine *(Pinus nigra)* 412
Austroglycyphagus spp. 292
A. geniculatus 34
A. lukoschusi 482
A. malaysiensis 89
Avena sativa 211, 296
Azadirachta indica 106, 236
Azara's grass mouse *(Akodon azarae)* 453

B

Bacillus spp. 240, 279, 295
B. cereus 232
B. megaterium 318
B. subtilis (hay bacillus) 318
backyard hens *(Gallus gallus)* 427

baker's yeast *(Saccharomyces cerevisiae)* 355
bamboo spider mite *(Aponychus corpuzae)* 486
bank vole *(Myodes (Clethrionomys) glareolus)* 347, 471
barber's pole worm *(Haemonchus contortus)* 415
barn swallow *(Hirundo rustica)* 493
Bartonella 199
Bartonellaceae 199
Bdella spp. 132
  B. muscorum 132
bearded hedgehog *(Hericium erinaceus)* 343, 345
Beauveria spp. 260
  B. bassiana (white muscardine fungus) 68, 432
Bemisia tabaci (silverleaf whitefly) 482, 483
Berlesella spp. 95
  B. alata 95
bishop's goutweed *(Aegopodium podagraria)* 380
black locust *(Robinia pseudoacacia)* 398
black pine bark beetle *(Hylastes ater)* 402
Blattisociidae 395
Blattisocius spp. 69, 404–405
  B. dentriticus 69, 105, 202, 203, 404–405, **405**, 408–409
  B. keegani 69, 296, 303, 404, **405**, 409, 414, 452
  B. mali 69, 284, 404–405, **406**, 410–411
  B. patagiorum 405
  B. quadridentatus 69, 404
  B. tarsalis 284, 404–405, **407**, 411–414, 475–476
Blomia spp. 31, 297–298, 300–301
  B. freemani 94, 300–303, **302**
  B. gracilipes 300
  B. keegani 303
  B. khalilovae 300
  B. kulagini 303
  B. thori 300
  B. tjibodas 89, 303
  B. tropicalis 36, 90, 249, 295, 300, 301, 303, 482
Bombus spp. 466
  B. ionellus 239
  B. lapidarium 274
  B. lapidarius 306
  B. terrestris (buff-tailed bumblebee) 323
Borrelia spp.
  B. afzelii 421
  B. burgdorferi 421
Bothrocheyla spp. 69
  B. pavlovskii 137
Bothynus spp. 97
  B. gibbosus 97
Brachionus quadridentatus 411
Brachybacterium 295
Brachytydeus spp. 175
  B. australensis 176, **176**
  B. bedfordiensis 176
  B. brusti 176
  B. formosa 175
  B. hughesae 176
  B. manitobensis 176
  B. oregonensis 176
bracket fungi *(Polyporus)* 297
Bradysia spp. 462
  B. matogrossensis (darkwinged fungus gnat) 473
Brevibacterium 295
Brevipalpus reticulatus 176
Brevundimonas spp. 88
broad mite *(Polyphagotarsonemus latus)* 168, 484
Bromelia margarethae 343
brown flour mite *(Gohieria fusca)* 308–312, **309**
brown-legged mite *(Aleuroglyphus ovatus)* 245–250, **246**, 321, 483
brown rat *(Rattus norvegicus)* 409
brown spot *(Alternaria alternata)* 67, 173, 232, 295, 330
Bubas bubalus (domestic water buffalo) 347
Bubo bubo (eagle owl) 493
buff-tailed bumblebee *(Bombus terrestris)* 323
bulb mite *(Rhizoglyphus echinopus)* **262**, 262–265, 473
button mushroom *(Agaricus bisporus)* 340, 343
Buxus spp. 211

C

Cadra cautella 285
Caenorhabditis elegans (roundworm) 415
Calliphoridae 471
Calluna vulgaris (Scotch heather) 408
Calocybe indica (milky mushroom) 372
Caloglyphus spp. 54, 265, 266, 484
  C. anomalus 32, 43
  C. berlesei 32, 34, 43, 283, 284, 321, 403, 456
  C. boharti 32, 43, 58
  C. hughesi 82, 266, 284
  C. michaeli 32
  C. mycophagus 32, 284, 492
  C. polyphyllae 55
  C. redickorzevi 85, 232
  C. rodriguezi 252
Calolampra spp. 227
Calvolia spp. 41
  C. fraxini 52
  C. hebeclinii 375, **376**
Calvoliella spp. 49
Caminella spp. 29
  C. peraphora 29, **30**
Camisia spp.
  C. carrolli 378
  C. segnis 378
Camponotus ligniperda (ant) 252
Cantharids 274
cape jasmine *(Gardenia jasminoides)* 403
Carabodes spp. 379
  C. brevis 379
  C. femoralis 379
  C. granulatus (granulated ground beetle) 379
  C. labyrinthicus 379
  C. niger 379

Carabodes spp. (*Continued*)
C. radiatus 379
C. willmanni 379
Caradrina morpheus 408
Carcinops pumilio (poultry house pill beetle) 471
Cardinium 270, 295
Carpoglyphus spp. 31, 276
adult 277
C. berlesei 231
C. hemipterus 285
C. lactis (dried fruit mite) 55, 60, 67, 88, 244, 276–285, **278**, 303, 321, 402, 403, 411
C. munroi 276, 277
C. wardleorum 276, 277
dichotomous key 277
hypopus 277
pesticide industry 277
Carpophilus hemipterus (dried-fruit beetle) 403
Cartodere constricta 296
Catenotaenia spp. 96, 314
C. pusilla (tapeworm) 97, 314
cattle tick *(Rhipicephalus microplus)* 429, 430
Caudacheles spp. 69
C. lieni 137
Caudiferidae 50
Cerambycidae 443
Ceratitis (coffee fruit fly) 298
Ceratozetes spp. 380
C. gracilis 380
Ceroplastes destructor 211
Cetonia aurata (rose chafer) 466
Chaetodactylidae 313
Chaetomium spp. 173
Chamobates spp.
C. cuspidatus (fiery costus) 381, **381**
C. incisus 381
cheese mite *(Tyrolichus casei)* **206**, 206–208
Chelachecaropsis spp. 69
C. bakeri 137
Chelacheles spp.
C. hellenicus 137
C. robustus 137
C. strabismus 137
Cheletogenes spp. 69
C. ornatus 137, **138**
Cheletomimus spp. 69
C. gracilis 137
Cheletomorpha lepidopterorum 137, **139**
Cheletonella spp.
C. caucasica 136
C. pilosa 138
C. vespertillionis 136
Cheylabididae 50
Cheyletia spp. 69, 138
C. papillifera 138
Cheyletus spp. 203, 249
C. aversor 203, 204
C. bidentatus 139
C. cacahuamilpensis 139
C. eruditus (storage mite predator) 70, 77, 139, **140**, 198, 199, 202–204, 284, 296, 321, 331, 332, 357, 413, 431, 453, 456
C. malaccensis 38, 39, 69, 70, 87, 105, 141, **142**, 198, 202–204, 234, 237, 249, 303, 431
C. malayensis 144
C. ornatus 137
C. rhizoglyphoides 87
C. trouessarti 69, 142, **143**, 357
Chibidania 368
Chinese herb medicine *(Equisetum hiemale)* 385
Chloris chloris (European greenfinch) 398
Chlorophyllum hortense 347
Choloepus didactylus 360
Chortoglyphus spp. 286–287
C. arcuatus 56, 80, 287–291, **289**
Chrysanthemum 219, 460
Cicer arietinum 244
Ciconia ciconia (white stork) 460, 493
cigarette beetle *(Lasioderma serricorne)* 410
Cinclus aquaticus 215
Cinnamomum spp. 107
C. zeylanicum 107–108
Circus aeruginosus (western marsh harrier) 398
Cissus antarctica 227
Citrullus spp. 106
C. colocynthis 106
Citrus aurantium 212
citrus root weevil *(Diaprepes abbreviatus)* 404
citrus rust mite *(Aculops pelekassi)* 403
citrus yellow mite *(L. formosa)* 175
Clacipitaceae 260
Cladophialophora spp. 64
Cladosporium spp. 269
C. cladosporiodes 198, 216, 330
Clerodendrum splendens (flaming glory bower) 403
Coboldia fuscipes 469
coconut rhinoceros beetle *(Oryctes nasicornis)* 394
Coendu spp. 156
C. rothschildi 156
Coffea arabica (Arabian coffee) 227
coffee fruit fly *(Ceratitis)* 298
Coleopters 274
Coleoscirus spp. 146
C. simplex 147
collection mite *(Thyreophagus entomophagus)* 272–275, **273**
Colletidae 52
Collocalidectes spp. 95
Colocasia esculenta 227
Columba oenas (Stock dove) 493
Columbina picui (Picui ground dove) 416
common hairy mite *(Glycyphagus (Lepidoglyphus) destructor)* 314, 320, 325–332, **326**, 456
common idia moth *(Epizeuxis aemula)* 412

common noctule *(Nyctalus noctula)* 468
common orange lichen *(Xanthoria parietina)* 380
controlled atmospheres (CA) 217
copperleaf *(Acalypha marginata)* 403
copra itch mite *(Tyrophagus putrescentiae)* 224–237, **225**
Copris hispanus (Spanish dung beetle) 448, 475
Cordyceps fumosorosea 260
Cordycipitaceae 260
coriander *(Coriandrum sativum)* 427
Coriandrum sativum (coriander) 427
Coritex lycii 365
Cormocephalus aurantiipes (orange-footed centipede) 343
Cornupalpatum spp. 5
C. burmanicum 5
Corvus spp.
C. cornix (hooded crow) 493
C. monedula (jackdaw) 493
Corylus avellana (hazelnut) 402, 412
Corynebacterium 240, 295
Cosmochthonioidea 390
Cosmochthonius spp. 382
C. foveolatus 382
C. lanatus 382
C. ponticus 382
C. reticulatus 382
Cosmoglyphus spp. 265
C. hughesi 55, 304
Cosmolaelaps spp. 450
C. barbatus 261
C. jaboticabalensis 261
Coxiella burnetii 421
cribrate weevil *(Otiorhynchus cribricollis)* 343
crocodile *(Crocodylus johnsoni)* 471
Crocodylus johnsoni (crocodile) 471
Cryptes baccarum 227
Cryptolestes spp. 404, 410
C. ferrugineus (rusty grain beetle) 72, 130, 296, 297, 410
C. turcicus (flour mill beetle) 85, 410
Cryptotermes spp. 165
C. brevis 165
Crytoplophora abscondita 390
Ctenocolletacarus 265
Ctenoglyphus, genera 304–305
C. canestrinii 304
C. palmifer 304
C. plumiger 304–307, **306**
Ctenophthalmidae 263
Cucumis melo (musk melon) 485
Cunaxa spp.
C. capreolus 9, **10**, 146, **147**
C. setirostris 146, **147**
C. sinensis 146
Curcuma spp. 236
C. onga 236
Curvularia spp.
C. geniculata 318
C. tetramera 173
Cylindrocarpon 258
Cymbaeremaeus cymba 383
Cymbidium spp. 216
C. clauboda 216
Czenspinskia spp. 49
C. transversostriata 60

D

Dacus spp.
D. cucurbitae (melon fly) 411
D. dorsalis (oriental fruit fly) 411
Dahlia variabilis 343
darkwinged fungus gnat *(Bradysia matogrossensis)* 473
date-plum *(Diospyros lotus)* 317
Datura spp. 106
D. stramonium 106
Deinocroton spp. 5
D. draculin 5
Delichon urbica 224
Delottococcus aberiae (mealybug) 459
Dendrocopos major (great spotted woodpecker) 493
Dendroctonus micans (great spruce bark beetle) 402
Dermacentor spp. 31
D. variabilis 31
Dermanyssidae 451, 456
Dermanyssus spp. 416–417
D. americanus (American bird mite) 416
D. chelidonis 351
D. gallinae (poultry red mite) 29, 31, 33, 46, 55, 78, 416–436, **417**, 452, 457
D. hirundinis 416
Dermatobia spp. 474–475
D. hominis (human bot fly) 474
Dermatophagoides spp. 60, 93, 303, 319, 348–351
D. africanus 351
D. farinae (American house dust mite) 35, 38, 108, 199, 249, 283, 284, 295, 349–358, **353**, 366
D. longior 351
D. microceras (house dust mite) 351
D. pteronyssinus (European house dust mite) 38, 61, 89, 90, 108, 173, 199, 283, 284, 291, 312, 319, 320, 349–351, 354, 357–363, **359**, 366
D. scheremetewskyi 351
Dermestes spp. 130, 182
D. carnivorus 182, 183
D. maculatus 182, 183
Dermoglyphidae 50
Diamesoglyphus spp. 304
D. intermedius 304
Diaprepes abbreviatus (citrus root weevil) 404
Digamasellus presepum 437, **438**

Dinychidae 490
Diospyros lotus (date-plum) 317
Diplogaster spp. 471, 473
Disonycha 211
Dithinozercoidae 490
Ditylenchus spp.
    D. destructor 264
    D. dipsaci 264
Dityophora 227
domestic water buffalo *(Bubas bubalus)* 347
dor beetle *(Geotrupes stercorarius)* 448
Douglas fir *(Pseudotsuga menziesii)* 378
downy woodpecker *(Dryobates pubescens)* 471
dried fish mite *(Lardoglyphus konoi)* 179–183, **180**
dried fruit beetle *(Carpophilus hemipterus)* 403
dried fruit mite *(Carpoglyphus lactis)* 276, 277–285, **278**, 321, 402, 403, 411
Drosophila spp.
    D. bipectinata 338
    D. hydei 472, 473
    D. nigrospiracula 473
Drosophilidae 471
drosophilid fly *(Phortica semivirgo)* 411
Dryobates pubescens (downy woodpecker) 471
dung beetles *(Geotrupes silvaticus)* 469
Duponchelia fovealis (European pepper moth) 476

E

eagle owl *(Bubo bubo)* 493
eared dove *(Zenaida auriculata)* 452
Earliella scabrosa 343, 345
earth-boring dung beetle *(Geotrupes laevistriatus)* 448
eastern imperial eagle *(Aquila heliaca)* 398
Echimyopodidae 48
Echinolaelaps 450
eelgrass *(Zostera marina)* 393
Egretta alba modesta 239
Elizabethkingia 279
elongate-bodied springtail *(Folsomia fimetaria)* 457
Enchytraeus albidus (white worms) 475
English oak (*Quercus robur*) 380
Enicmus minutus (fungus beetle) 222
Entonyssidae 51
Ephestia spp. 165, 408, 410
    E. cautella (almond moth) 285, 404, 413
    E. kuehniella (Mediterranean flour moth) 205, 410, 413, 463
Epidermoptidae 47, 50
Epizeuxis aemula (common idia moth) 412
Eptesicus serotinus (serotine bat) 336
Equisetum hiemale (Chinese herb medicine) 385
Ereynetes spp. 178
    E. ueckermanni 178, **178**
Erwinia 279, 295
Erysipelothrix spp. 240
    E. rhusiopathiae 421
Erythraeus spp. 155
    E. garmsaricus 155
Escherichia coli 421
Euandrolaelaps spp. 451
    E. sardous 463, **463**
Eucaliptus globulus 108
Eucalyptus spp. 106
Eucheyletia spp.
    E. flabellifera 144
    E. taurica **144**, 145
Eucomis montana (pineapple flowers) 458
Eugamasus spp. 431
    E. butleri 70, 431, 478, **479**
Eugenia spp. 236
    E. caryophyllata 107
    E. uniflora 107
Eulaelaps spp. 451
    E. stabularis 453, **454**
Euoniticellus pallens 493
Euonthophagus gibbosus 493
Eupodes spp. 148
    E. variegatus 148, **148**
Eurasian coot *(Fulica atra)* 397, 398
Eurasian harvest mouse *(Micromys minutus)* 394
Eurasian tree sparrow *(Passer montanus)* 393, 398
Euroglyphus spp. 349, 363–364
    E. longior 351, 363
    E. maynéi (Small house dust mite) 38, 78, 183, 284, 349–350, **363**, 363–366
European cranberry bush *(Viburnum opulus)* 403
European earwig *(Forficula auricularia)* 343
European greenfinch *(Chloris chloris)* 398
European honeybee *(Apis mellifera)* 343, 399, 408, 412
European house dust mite *(Dermatophagoides pteronyssinus)* 319, 320, 349–351, 354, 357–363, **359**, 366
European pepper moth *(Duponchelia fovealis)* 476
Eurotium spp. 196, 232, 349
    E. amstelodami 196, 349, 361
    E. repens 196
    E. restrictus 67
    E. ruber 196
Euseius concordis 482
Eustathia spp. 95
    E. cultrifer 95
Eustathiidae 50
Eutermes exitiosis 270
Euzetes spp. 30
Evernia punastri (oakmoss) 379
Ewingella 279
Exoneura concinnula 211

F

Fa1culiferidae 50
Fagus grandifolia 176
Fannia canicularis (lesser housefly) 493

feather mite *(Megninia ginglymura)* 409, 431
female ginseng *(Angelicae sinensis)* 360
Fessonia 154
F. papillosa 154
Fibulanoetus 50
fiery costus *(Chamobates cuspidatus)* 381, **381**
fir bark beetle *(Pityokteines curvidens)* 387
Flagrosuctobelba 390
flaming glory bower *(Clerodendrum splendens)* 403
Flammulina 343, 345
flour mill beetle *(Cryptolestes turcicus)* 85, 410
flour mite *(Acarus siro)* 192–206, **193**, 318, 319, 357, 397, 402, 411, 413, 452, 456, 462, 484
Foeniculum vulgare 236–237
Folsomia fimetaria (elongate-bodied springtail) 457
Fomitopsis betulina 274
Forficula auricularia (European earwig) 343
Francisella tularensis (rabbit fever) 453
Frankliniella occidentalis (western flower thrips) 403, 460, 484
free-living predatory mite *(Parasitus fimetorum)* 474
Freesia 227, 253, 264
The French fly *(Tyrophagus similis)* 237–240, **238**
Freyanidae 50
Frieseomelitta 165, 167
F. varia 165, 167
Fringilla coelebs 256
Fulica atra (Eurasian coot) 397, 398
fungus beetle *(Enicmus minutus)* 222
Furnarius rufus (red ovenbird) 416
Fusarium spp. 232, 257, 258, 264, 269
F. graminearum 344–345
F. moniliformae 173
F. moniliforme 408–409
F. oxysporum 324, 330
Fuscozetes spp. 380
Fusohericia 50

G

Gabucinia spp. 60
G. delibata 60
Gabuciniidae 50
Gaeolaelaps spp. 450, 451, 457
G. aculeifer 261, **457**, 457–459
Galagoides demidovii (Prince Demidoff's bushbaby) 298
Galendromus 480
Galleria spp. 293
G. mellonella 165, 293, 294, 295
Gallinago spp.
G. nigripennis (African snipe) 360
G. paraguaiae (Pantanal snipe) 470
Gallus gallus (backyard hens) 427
Gamasellodes spp. 240
G. bicolor 397
Gamasholaspis 477
Gamasiphis lanceolatus 439, **439**
Gamasus spp.
G. pygmaeus 402
G. stabularis 453
Ganoderma lucidum 229, 347
Gardenia jasminoides (cape jasmine) 403
Garmania spp.
G. domestica 69, 399
G. pomorum 402
Gastronyssidae 51
Gaudoglyphidae 50
Geniculifera paucispora 261
Geolaelaps aculeifer 450–451
Geotrichum 232
Geotrupes spp. 448
G. laevistriatus (earth-boring dung beetle) 448
G. silvaticus (dung beetles) 469
G. stercorarius (dor beetle) 448
Gladiolus 212, 264
Gliocladium roseum 318
glossy grain mite *(Tarsonemus granarius)* **169**, 169–171
Glycycometus 50, 292, 297–298
G. geniculatus 292, 298–299, **299**
G. kualalumpurensis 292, 297, 298
G. malaysiensis 297, 298
Glycyphagidae 47, 50, 313
Glycyphagoidea 47–48, 91
Glycyphagus spp. 31, 42, 50, 85, 97, 200, 313–315
G. aegyptiacus 315
G. caudatus 315
G. (Lepidoglyphus) destructor (common hairy mite) 48, 67, 72, 141, 198, 204, 216, 222, 236, 314, 320, 325–332, **326**, 456
G. domesticus (house mite) 42, 53, 56, 62, 80, 96, 97, 187, 190, 192, 200, 283, 284, 291, 295, 298, 313–321, **316**, 324, 410, 452, 456, 458, 484
G. (Lepidoglyphus) fustifer 314, **332**, 332–334
G. michaeli 314–315
G. ornatus 33, 34, 283, 314, 321–324, **322**
G. privatus 314, 323–325, **324**
Glycyrrhiza glabra 236
Glyphanoetus spp.
G. fulmeki 339
G. phyllotrichus 339
Glyptholaspis confusa 473
Gnorimus spp. 276
Gohieria spp. 308
G. fusca (brown flour mite) 61, 67, 288, 290, 308–312, **309**
Gossypium spp. 394
grain mite *(Acarus siro)* 192–206, **193**
granary mite *(Tarsonemus granarius)* **169**, 171
granulated ground beetle *(Carabodes granulatus)* 379
Graphognathus leucoloma (white-fringed beetle) 345
Grapholita spp. 164
G. molesta 164
greater mouse-eared bat *(Myotis myotis)* 445

great reed warbler *(Acrocephalus arundinaceus)* 398
great spotted woodpecker *(Dendrocopos major)* 493
great spruce bark beetle *(Dendroctonus micans)* 402
green muscardine fungus *(Metarhizium anisopliae)* 432
Guanolichus 50

H

Haematobia irritans (horn fly) 468, 473
Haematonectria haematococca 258
Haemogamasus spp. 451
  H. glasgowi 456
  H. pontiger 319, 408, 453, **455,** 456–457
Haemolaelaps spp. 450
  H. casalis 451
  H. glasgowi 453
Haemonchus contortus (barber's pole worm) 415
hairy agrimony *(Agrimonia pilosa)* 385
Halarachnidae 51
Haliaeetus albicilla (white-tailed sea eagle) 460
Halictidae 52
Haplochthoniidae 390
Haplochthonius simplex 385, **385**
hay bacillus *(Bacillus subtilis)* 318
hazelnut *(Corylus avellana)* 402, 412
hedgehog tick *(Ixodes hexagonus)* 343
Hemisarcoptes spp. 46, 47, 52, 95, 336
  H. cooremani 90
  H. malus **337**
Hemisarcoptidae 47, 49, 51, 52
Hericia spp. 50
  H. hericia 58
Hericium erinaceus (bearded hedgehog) 343, 345
Heterocoptidae 47, 51
Heterodera avenae 239, 264
Heterorhabditis spp. 260
  H. amazonensis 261
  H. bacteriophora 261
Hieracium pilosella (mouse-ear hawkweed) 458
Hirundo rustica (barn swallow) 493
hispid cotton rat *(Sigmodon hispidus)* 471
Hister spp.
  H. merdarius 343
  H. stercorarius 343
  H. unicolor 343
Histiogaster spp. 16, 49, 52, 185, 275
  H. arborsignis 46
  H. carpio 275
  H. rotundus 55, 185
Histiostoma spp. 49–50, 52, 216, 339–340
  H. feroniarum 58, 339–346, **341**
  H. formosana 61
  H. heinemanni 61, 339–340
  H. humidiatus 339
  H. laboratorium 34, 42, 58, 342
  H. onioni 339
  H. papillata 339
  H. pulchrum 338
  H. sapromyzarum 58, 339–340, **346,** 346–347
Histiostomatidae 46–52
Holaspina spp. 477
  H. alstoni 477
honeybees (*Apis* spp.) 298
hooded crow *(Corvus cornix)* 493
Hoogstraalacarus spp. 51
hoopoe *(Upupa epops)* 468, 469
Hormodendrum cladosporoides 173
Hormosianoetus spp. 50
  H. laboratorium 338
  H. unidentatum 338
horn fly *(Haematobia irritans)* 468, 473
Horstia 52
Horstiella 52
Hoshikadania konoi 182
house-dust mite *(Dermatophagoides microceras)* 351
housefly *(Musca domestica)* 408–409, 466, 468–471, 473–475
house mite *(Glycyphagus domesticus)* 313, 314–321, **316,** 324, 410, 452, 456, 458, 484
house mouse *(Mus musculus)* 306, 394, 471
house mouse mite *(Liponyssoides sanguineus)* 416
hulless oats *(Avena sativa)* 174
human bot fly *(Dermatobia hominis)* 474
Hyacinthus orientalis 412
Hylastes ater (black pine bark beetle) 402
Hymenoptera 375
Hypoaspidinae 450
Hypoaspisella spp. 451
  H. linteyini 460
  H. lubrica 459–460, **460**
Hypoaspis spp. 220, 450
  H. aculeifer (killer mite) 198, 240, 257, 258, 259, 260, 431, 475
  H. freemani 451
  H. miles 198, 431, 460, 475
Hypodectes propus 42, 47
Hypodectidae 42
Hypoderatidae 48
Hypogastrura denticulata 458

I

Incertae sedis 186
Indianmeal moth *(Plodia interpunctella)* 414
Inocybe umbonata 347
Iphidozercon gibbus 397, 398
Iphis casalis 451
Iphiseiodes zuluagai 482
Iphiseius degenerans 483
Iris spp. 460
Isaria spp. 68, 432
  I. fumosorosea 68, 432
Isoglyphus 265
Ixodes hexagonus (hedgehog tick) 343

J

jackdaw *(Corvus monedula)* 493
Jasminum spp. 107
  J. officinale 107
jelly ear fungus *(Auricularia auricula-judae)* 392
Justicia adhatoda (Malabar nut) 403

K

Kanzawa spider mite *(Tetranychus kanzawai)* 40, 41
keys 117–129
  Acaridae, genera 186–187
  fumigants 296
killer mite *(Hypoaspis aculeifer)* 198, 240, 257, 258, 259, 260, 431, 475
Kleemannia spp. 392
  K. insignis 392–393
  K. pavida 392, 484
  K. plumigera **392**, 394, 484
  K. plumosa **393**, 394
  K. plumosus 70, 431, 456, 484
  K. pseudoplumosa 394
Kocuria 295
Kramerellidae 50
Krantzoseius 404
Krivolutskiella pubescens 382

L

Labidophorinae 42
Labidostoma (Pseudocornutella) 5
Laccaria laccata 347
Lackerbaueria 52
Lactobacillus spp. 240
  L. fermentum 355
Laelaps spp.
  L. aculeifer 457
  L. fahrenholzi 453
  L. glasgowi 453
  L. miles 460
  L. pontiger 453
Laevicaulis alte (tropical leatherleaf slug) 493
Laminalloptes spp. 95
Laminosioptidae 50
Lanius spp.
  L. collurio (redbacked shrike) 398, 412
  L. minor (lesser grey shrike) 398
Lardoglyphus spp. 43, 50, 179, 183
  L. angelinae 179
  L. konoi (dried fish mite) 32, 179–183, **180**, 284
  L. radovskyi 179
  L. robustisetosus 179
  L. zackeri 179, 181
large flying fox *(Pteropus vampyrus)* 298
Larvacarus 42
Lasiobelba spp. 386
Lasioderma serricorne (cigarette beetle) 410
Lasioseius spp. 261, 404, 414
  L. africanus 261
  L. allii 261, 404
  L. berlesei 69, 404, 414
  L. bispinosus 257
  L. dentriticus 405
  L. hystrix 402
  L. penicilliger 69, 404, **408**, 414–415
  L. sugawarai 404, 415
  L. tarsalis 411
Lavandula spp. 108
  L. angustifolia 108
Leiodinychus spp.
  L. krameri 231, 443, 445, 456, 490
  L. orbicularis 446
Leioproctus spp. 155
Leioseius bicolor 398
Lemurnyssidae 51
Lentinula edodes (shiitake mushroom) 248, 340, 343, 345
Lentinus edodes 185
Lepidoglyphus spp. 313–314, 334
  L. destructor 56, 63–65, 73, 75, 80, 86, 107, 174, 187, 190, 192, 199–202, 205, 217, 244, 284–286, 295–297, 303
  L. michaeli 334–335, **335**
  L. pilosus 315
lesser grey shrike *(Lanius minor)* 398
lesser housefly *(Fannia canicularis)* 493
lesser spotted eagle *(Aquila pomarina)* 466, 469
Leuconostoc 279
Levant voles *(Microtus guentheri)* 323
Linobia 51
Liponyssoides spp. 416–417
  L. sanguineus (house mouse mite) 416
Listeria monocytogenes 421
Listrophoridae 51
Lithobius spp. 490
  L. forficatus 490
Lobalgidae 51
Locusta migratoria 227, 274
Lolium perenne (perennial ryegrass) 458
Lomelacarus faini 313–314
long-nosed hocicudo *(Oxymycterus nasutus)* 453
Lophioglyphus liciosus 47
Lophuromyopodinae 48
Lorryia spp. 175, 176
  L. formosa (citrus yellow mite) 175
Lucilia spp. 468
  L. arborea (woodlark) 398
lychee erinose mite *(Aceria litchii)* 408
Lycoriella solani (mushroom sciarid larvae) 462
Lyprauta spp. 468, 476

M

Maccullochella peelii peelii (Murray cod) 339
Macrocheles spp. 465–466, 468
M. autumnalis 466
M. dimidiatus 466
M. embersoni 261, 468, 473–474
M. glaber 70, 431, 465, 466, 468–469
M. matrius 70, 431, **467**, 468–470
M. merdarius 70, 431, 468, 470
M. muscaedomesticae 70, 261, 431, 465, 466, **468**, 468–474, 493
M. mycotrupetes 465
M. praedafimetorum 466
M. punctillatus 475
M. robustulus 261, 468, 473, 475–476
M. rotundiscutis 466
M. similis 465
M. subbadius 466, 473
Macronyssidae 456
Macroseius biscutatus 480
Madaglyphus spp. 186, 275
M. javensis 275
Malabar nut *(Justicia adhatoda)* 403
Malayoglyphus 348
mallard *(Anas platyrhynchos)* 398
Mallota spp. 46
Manihot spp. 156
M. esculenta 156
Mealia maynéi 365
mealybug *(Delottococcus aberiae)* 459
Medicago spp. 210
Mediolata spp. 156
M. granaria 156, **157**
Mediterranean flour moth *(Ephestia kuehniella)* 410, 413, 463
Megachilidae 52
Megninia ginglymura (feather mite) 409, 431
Melaleuca spp. 107
M. leucadendra 107
Melanotus rufipes 347
Melesodectinae 48
Melia spp. 106
M. azedarach 106
Melichares spp. 399
M. agilis 69, 399, **400**, 402
M. dentriticus 208, 405
M. keegani 409
Melicharidae 395
Melipona spp.
M. asilvai 165
M. subnitida 165
Meloidogyne spp. 270, 415
M. javanica (sugarcane eelworm) 474
melon fly *(Dacus cucurbitae)* 411
Mentha pulegium (pennyroyal) 358
Metalabidophorinae 48
Metapronematus spp. 83
Metarhizium spp. 105, 260
M. anisopliae (green muscardine fungus) 68, 260, 32
M. brunneum 260
M. indigoticum 260
M. pemphigi 260
M. pinghaense 260
Mexecheles spp. 69
M. votandinii 145
Mexican bean weevil *(Zabrotes subfasciatus)* 410
Microcerotermes diversus 266
Micrococcus 240
Micromys minutus (Eurasian harvest mouse) 394
Micronychites 47
Micronychitoides 47
Microtus guentheri (Levant voles) 323
Middle East blind mole-rat *(Spalax ehrenbergi)* 343, 458
milky mushroom *(Calocybe indica)* 372
mink *(Mustela vison)* 469
Miomantis caffra 211
Molothrognathus spp.
M. brasiliensis 134
M. mehrnejadi 134
Monacrosporium eudermatum 261
Monobiacarus 52
Morchella spp.
M. conica 224
M. esculenta 224
mould mite *(Tyrophagus putrescentiae)* 224–237, **225**, 318, 320, 321, 357, 371, 374, 399, 408–410, 411, 452–453, 458–459, 462, 484
mouse-ear hawkweed *(Hieracium pilosella)* 458
Mucor spp. 67, 232
Mucroseius 404
Murray cod *(Maccullochella peelii peelii)* 339
Musca spp.
M. domestica (housefly) 189, 408–409, 466, 468–471, 473–475
M. vetustissima (Australian bush fly) 468
mushroom sciarid larvae *(Lycoriella solani)* 462
musk melon *(Cucumis melo)* 485
Mus spp.
M. musculus (house mouse) 306, 394, 471
M. spicilegus (steppe mouse) 470
Mustela vison (mink) 469
Mycetoglyphus spp. 48, 185, 186
M. fungivorus 58, 185, 250–252, **251**
M. sevastianovi 250
M. ungivorus 46
Mycetosancassania 265
Mycoplasma spp.
M. gallisepticum 421
M. synoviae 421
Myialges 47
Myocoptes musculinus 95
Myocoptidae 51
Myodes (Clethrionomys) glareolus (bank vole) 347, 471

Myotis spp.
M. dasycneme (pond bat) 394
M. myotis (greater mouse-eared bat) 445
Myroxylon balsamum 107

N
Naiacus 49
Naiadacarus arboricola 46, 60, 61
Nanacaroides 336
Nanacarus spp. 336
N. minutus 336
navel orangeworm *(Amyelois transitella)* 410
Necrobia spp.
N. ruficollis (red-shouldered ham beetle) 410
N. rufipes (red-legged ham beetle) 182, 410
Neoeucheyla spp. 69, 137
N. iranica 145
Neomys fodiens (water shrews) 469
Neophyllobius spp. 149
N. saxatilis 149, **149**
Neoregelia margaretae 343
Neoseiulus spp. 480
N. barkeri **481**, 482–485
N. californicus 482, 483
N. cucumeris 198, 284, **482**, 483–487
N. tunus 482–483
Neosuidasia faini 367
Neotoma spp. 61
Nepenthaceae 49
Neurospora crassa 269
Nipponiella simplex 382
Nocardia 240
Nocardioides alpinus 483, 484
Nodele calamondin 145
northern fowl mite *(Ornithonyssus sylviarum)* 416, 418, 420
northern hopping mouse *(Notomys alexis)* 471
Notaspis marginatus 490
Notomys alexis (northern hopping mouse) 471
Nyctalus noctula (common noctule) 468
Nycteriglyphus spp. 51
N. asiaticus 47
Nysius spp. 88, 199
N. plebeius 199

O
oakmoss *(Evernia punastri)* 379
Ochrolichidae 50
Ocimum spp. 236
O. sanctum 236
Oconnoriidae 50
Odontoscirus spp.
O. kazeruni 132
O. longirostris 132
oleander scale *(Aspidiotus nerii)* 375
Oligoryzomys 453
olive gall mite *(Aceria oleae)* 485
olive rust mite *(Tegolophus hassani)* 485
Ololaelaps 450
Ommatoiulus moreleti (Portuguese millipedes) 343
onion thrips *(Thrips tabaci)* 483, 484
Onthophagus spp. 475
Oodinychus ovalis 490
Ophiocordycipitaceae 260
Oplitis spp. 441
O. arboricavi 441, **442**
Oppia spp.
O. clavipectinata 386
O. nitens 386
Oppiella spp. 386
orange-footed centipede *(Cormocephalus aurantiipes)* 343
Orbilia oligospora 261
Oribatula spp. 387
O. exilis 387
O. tibialis 387
Oribatulidae 387
oriental fruit fly *(Dacus dorsalis)* 411
Ornithagalum oligophyllum 402
Ornithodoros moubata 213
Ornithonyssus spp.
O. bacoti (tropical rat mite) 452
O. bursa (tropical fowl mite) 416, 420
O. sylviarum (northern fowl mite) 416, 418, 420
Oryctes nasicornis (coconut rhinoceros beetle) 270, 394
Oryctolagus cuniculus 239
Oryzaephilus spp. 404, 410
O. mercator 85, 319
O. surinamensis 72, 165, 201, 285, 410, 413
Oryza sativa 263
Osmoderma spp. 97
O. eremicola 97
Otanoetus spp. 47, 51
Otiorhynchus spp. 58
O. cribricollis (cribrate weevil) 58, 343
Otodectes 34
Oudemansicheyla denmarki 145
Oxymycterus spp.
O. nasutus (long-nosed hocicudo) 453
O. rufus (red hocicudo) 453
oyster mushroom *(Pleurotus ostreatus)* 343, 345

P
Paecilomyces spp. 68
P. fumosoroseus 68, 432
Palaeopsylla minor 263
Panaeolus cyanescens 343, 345
Panagrellus spp. 468, 475
P. redivivus 415
Panagrolaimus spp. 471, 473
Pantanal snipe *(Gallinago paraguaiae)* 470

Parachipteria spp.
P. punctata 377
P. willmanni 377
Paragamasus spp.
P. subtruncus 478
P. truncus 478
Paralychus spp. 390
P. chongqingensis 390
Paraná pine *(Araucaria angustifolia)* 410
Paraneognathus spp.
P. oblongus 134
P. wangae 134, **135**
Parasitus spp.
P. fimetorum (free-living predatory mite) 257, 261, 474
P. paraconsanguineus 70, 431
Paratriophtydeus coineaui 174, 296, 297
partridges *(Perdix perdix)* 471
Parus major 398, 460
Passaloglyphus 51–52
Passer spp.
P. domesticus 306
P. montanus (Eurasian tree sparrow) 256, 393, 398
pavement mushroom *(Agaricus bitorquis)* 340, 343
Pedetes capensis 48
Pedetopodidae 48
Pelzneria spp. 50
Penicillium spp. 86, 196, 231, 232, 269, 295, 318
P. aurantiogriseum 86
P. chrysogenum 86, 258
P. claviforme 67
P. crustosum 86
P. cyclopium 198, 216, 330
P. frequentans 229
P. griseofulvum 67
pennyroyal *(Mentha pulegium)* 358
Perafrotydeus meyerae 177
Perdix perdix (partridges) 471
perennial ryegrass *(Lolium perenne)* 458
Pergamasus diversus 198
Peromyscus leucopus (white-footed mouse) 471
Petasites spp. 398
Phaseolus spp. 252
P. vulgaris 109, 203, 205
Phauloppia spp. 387
P. lucorum 387, **388**
Phoeniculum spp. 107
P. vulgare 107
Phoenicurus phoenicurus 463
Phoracantha spp. 167
P. semipunctata 167
Phortica semivirgo (drosophilid fly) 411
Phragmition spp. 398
Phyllophaga spp. 97
P. operculella (potato tuber moth) 411, 475–476
Phyllozetes 382
Phytoseiidae 276
Phytoseiinae 480
Phytoseiulus spp. 480
P. persimilis 39, 56, 482, 483
Pica pica 370
Picui ground dove *(Columbina picui)* 416
Pimeliaphilus podapolipophagus 150, **150**
pineapple flowers *(Eucomis montana)* 458
Pinus spp.
P. nigra (Austrian pine) 108, 412
P. pinea 108, 236–237
P. silvestris (Scots pine) 398, 402
Piper marginatum 107
Pipistrellus pipistrellus 336
Pityokteines curvidens (fir bark beetle) 387
Pityophthorus bisulcatus 161
Plasmodium spp. 421
Plecotus auritus 452
Pleurotus spp.
P. ostreatus (oyster mushroom) 266, 269, 343, 345
P. sajur-caju 343, 345
Plodia interpunctella (Indianmeal moth) 285, 414
Plumeria spp. 227
Pneumocoptidae 51
Poecilochirus 478
Polianthes tuberosa 156
pollen-cap mite *(Tyrophagus neiswanderi)* 217–220, **218**
Polyphagotarsonemus latus (broad mite) 484
Polyphylla fullo 270
Polyporus (bracket fungi) 297
pond bat *(Myotis dasycneme)* 394
Portuguese millipedes *(Ommatoiulus moreleti)* 343
Potamonidae 51
potato *(Solanum tuberosum)* 343, 475
potato tuber moth *(Phthorimaea operculella)* 411, 475–476
poultry house pill beetle *(Carcinops pumilio)* 471
poultry red mite *(Dermanyssus gallinae)* 416–436, **417**, 452, 457
Pretydeus, genera 175
Prince Demidoff's bushbaby *(Galagoides demidovii)* 298
Prionoplus reticularis 215, 227
Procalvolia zacheri 375
Proctolaclaps hypudaei 402
Proctolaelaps spp. 404
P. bickleyi 399, 402
P. hystrix 399, 402
P. pomorum 399, **400**, 402
P. pygmaeus 399, **401**, 402, 403, 484
P. regalis 399, 403
P. scolyti 399, **401**, 403
P. ventrianalis 399, 403
Proctophyllodidae 17–18, 50
Proctotydaeus spp. 176
P. sinhai 176
Proprioseiopsis spp. 480
P. regetovae 483, 487
Protea mellifera 403

Protogamasellopsis posnaniensis 440
Protogamasellus spp. 240, 397
P. denticus 252
P. hibernicus 397, 398
P. minutus 257, 261
P. zaheri 261
Pseudaulacaspis pentagona (white peach scale) 403
Pseudolorryia 175
Pseudomonas spp. 88, 258
Pseudoparasitus 450
Pseudotsuga menziesii (Douglas fir) 378
Psocoptera spp. 227
Psoptidae 17–18
Psoroptidae 51
Psoroptidia 51
Psoroptidid mites 50
Psoroptoididae 50
Pterodectes amaurochalinus 95
Pteroherpus garrulacis 95
Pterolichidae 50
Pteropus vampyrus (large flying fox) 298
Pterostichus madidus 493
Ptiloxenidae 50
Ptyssalgidae 50
Puffinus tenuirostris 212
Pycnonotus leucotis (white-eared bulbul) 328
Pyemotes spp. 163
P. herfsi 70, 161, **162**, 163–165, 431
P. islandicus 163
P. parviscolyti 161
P. tritici 163, 165–167, 452
P. ventricosus 161, **162**, 163
Pygmephorus islandicus 163
Pyroglyphidae 50, 91
Pyroglyphus spp. 348
P. africanus 351
Pyropia yezoensis (seaweed) 427
Pyrus malus (apple tree) 411

Q

Quercus robur (English oak) 380

R

rabbit fever (*Francisella tularensis*) 453
Radix spp.
R. Asteris 354
R. codonopsis 370
R. gentianae macrophyllae 365
R. morindae officinalis 365
R. ophiopogonis 365
R. sanguisorbae 370
Ramusella spp. 386
Ranunculus spp. 212, 227
R. damascena 402
Raphignathus spp.
R. conspicuus 151
R. giselae 151, **152**
R. gracilis 151, **152**
R. hecmatanaensis 151, **153**
R. orientalis 151
R. zhaoi 151, **153**
Rattus spp.
R. exulans 211
R. norvegicus (brown rat) 211, 239, 409
Rectijanuidae 50
redbacked shrike (*Lanius collurio*) 398, 412
red date (*Zizyphus jujuba*) 323
red dwarf honey bee (*Apis florea*) 374
red hocicudo (*Oxymycterus rufus*) 453
red-legged ham beetle (*Necrobia rufipes*) 410
red ovenbird (*Furnarius rufus*) 416
red-shouldered ham beetle (*Necrobia ruficollis*) 410
Rhabditella spp.
R. axei 473–474
R. leptura 471, 473
R. muscicola 452
Rhabditidae 468
Rhabditis spp. 415
R. sccanica 411
Rhinonyssidae 51
Rhipicephalus microplus (cattle tick) 429, 430
Rhizoglyphoides 265, 266
Rhizoglyphus spp. 42, 43, 46, 48–50, 55, 185, 186, 216, 252–253, 259, 261, 402, 451, 457
R. callae 253
R. costarricensis 253
R. echinopus (bulb mite) 43, 185, 249, 253, 261–265, **262**, 473
R. robini (robine bulb mite) 185, 253–262, **254**, 261, 264, 408–409, 411, 459
R. setosus 55, 185, 253
R. stolonifer 258
Rhizoma spp.
R. acori tatarinowii 370, 373
R. chuanxiong 370, 373
R. cimicifugae 354
R. dioscoreae 354
R. homalomenae 365
Rhyncoptidae 51
Rhysotritia ardua 384, **384**
Rickettsia akari (rickettsialpox) 416
rickettsialpox *(Rickettsia akari)* 416
Riparia riparia 458, 459–460, 469
robine bulb mite *(Rhizoglyphus robini)* 253–262, **254**, 408–409, 411, 459
Robinia pseudoacacia (black locust) 398
Rodentopus 304
Rosa spp.
R. canina 403
R. damascena 403
rose chafer (*Cetonia aurata*) 466
Rosensteinia 47

Rosensteiniidae 47, 50, 51
Rosmarinus officinalis 107
roundworm (*Caenorhabditis elegans*) 415
rusty grain beetle (*Cryptolestes ferrugineus*) 410
Rutelinae 52

S
Saccharomyces cerevisiae (baker's yeast) 211, 355
Saintpaulia 227
Salmonella spp. 421
S. enterica 423
Sancassania spp. 48–50, 52, 185, 186, 265–267
S. anomalus 266
S. berlesei (wheat grain mite) 58, 185, 266–271, **268**
S. boharti 58
S. krameri 266
S. mycophagus 266
S. oudemansi 266
S. redikorzevi 266
S. rhizoglyphoides 266
S. rodionovi 269
Sandersonia aurantiaca 343, 345
Sapracarus tuberculatus 48
Saproglyphus spp. 375
Sarcocystis nesbitti 236
Sarcoptes spp. 351
S. scabiei 60, 351
Sarcoptidae 51
Sarraceniaceae 49
Satureja hortensis 106
Sayornis spp. 471
Scalopus aquaticus 256
scaly grain mite *(Suidasia nesbitti)* 367–373, **369**
scaly mite *(Suidasia pontifica)* 367, 368, **372**, 372–374
Scapheremaeus palustris 383, **383**
Scapteromys spp.
S. aquaticus 453
S. tumidus 453
Scheloribates laevigatus 387
Schinus terebinthifolia 107
Schistocerca gregaria 242
Schizotrachelus bakeri 347
Schulzea 52
Schwiebea spp. 43, 48, 49, 50, 275, 402
Scirpus tuberosus 227
Sciurus carolinensis 471
Sclerotinia narcissicola 264
Scolitus spp.
S. multistriatus 161
S. ventralis 161
Scolytidae 443
Scotch heather *(Calluna vulgaris)* 408
Scots pine *(Pinus silvestris)* 398, 402
seaweed *(Pyropia yezoensis)* 427
Secale cereale 256
seed mite *(Tyrophagus longior)* 213–217, **214**
Sejus spp. 2
S. temperaticus 70, 431
Sellea 50
Sellnickiella spp. 5
S. balticae 5
serotine bat *(Eptesicus serotinus)* 336
Serratia 279
shiitake mushroom *(Lentinula edodes)* 340, 343, 345
Sigmodon hispidus (hispid cotton rat) 471
silverleaf whitefly *(Bemisia tabaci)* 482, 483
Sinella coeca 458
Sinosuidasia spp.
S. jinyunensis 367
S. orientalis 367
Sitophilus spp. 201
S. granarius 72, 201, 205
Sitotroga spp. 165
S. cerealella 165, 412
small house dust mite *(Euroglyphus maynéi)* 349–350, **363**, 363–366
Smaris spp. 154
S. mamillatus 154, **154**
Solanum tuberosum (potato) 343, 475
Solenognathus 183
Solenopsis invicta 167
Sorbus spp. 408
Sorex spp.
S. araneus 336
S. vulgaris 336
Spalax ehrenbergi (Middle East blind mole-rat) 343, 458
Spanish dung beetle *(Copris hispanus)* 448, 475
Sphaeroceridae 471
Sphexicozela spp. 52
Spinanoetus spp. 50
Spinibdella spp. 132, **133**
S. lignicola 132
Sporendonema sebi 67, 197
Sporidesmium mucosum 224
stable fly *(Stomoxys calcitrans)* 466, 468, 473–474
Staphylococcus 240, 295
starling *(Sturnus vulgaris)* 493
Stathmopoda 211
Steinernema spp. 260
S. huense 261
Steneotarsonemus spp. 168
Stenotrophomonas spp. 88
steppe mouse *(Mus spicilegus)* 470
Stigmaeus elongatus 157
stock dove *(Columba oenas)* 493
Stomoxys calcitrans (stable fly) 466, 468, 473–474
storage mite *(Lepidoglyphus destructor)* 35, 56, 63–65, 73, 75, 80, 86, 107
storage mite predator *(Cheyletus eruditus)* 321, 331, 332, 357, 413, 431, 453, 456

Storchia spp.
S. pacifica 156
S. robusta 156
stored-product beetle *(Ahasverus advena)* 411
Stratiolaelaps spp. 450, 451
S. gurabensis 460, **461**
S. miles 261, 450, 460–462
S. scimitus 431, 460, 462–464
Streetacarus spp. 95
Strelitzia spp. 403
Stromatinia 258
Sturnophagoides spp. 348
Sturnus spp. 398
S. vulgaris (starling) 211, 215, 224, 239, 493
Suctobelbidae 390
sugarcane eelworm *(Meloidogyne javanica)* 474
sugar maple trees *(Acer saccharum)* 458
Suidasia spp. 85
S. medanensis 367, 368, 374
S. nesbitti (scaly grain mite) 284, 367–373, **369**
S. pontifica (scaly mite) 91, 92, 367, 368, **372**, 372–374
Swirski mite *(Amblyseius swirskii)* 483
Sylvia spp. 398
Syncephalastrum racemosum 318
Syringobiidae 50
Syringophilidae 50
Syringophilinae 50

T
Tadarida brasiliensis 452
tapeworm *(Catenotaenia pusilla)* 314
Tarsonemidae 168–169
Tarsonemus spp. 169–171, 319, 357
T. bilobatus 169, **170**
T. confusus 169, **171**
T. floricolus 169, **171**
T. fusarii 169, 172
T. granarius (glossy grain mite) **169**, 169–173, 283, 284, 296, 297
T. hominis 92, 170
T. rakoviensis 170
T. waitei 9, **11**, 170, **171**
Tectocepheus velatus 379, **379**
Tegolophus hassani (olive rust mite) 485
Teladorsagia circumcincta 415
Tenebrio molitor 270
Tenebrionidae 52
Tenuipalpidae 3, 437
Terpnacarus glebulentus 159, **160**
Terrapene carolina (turtle) 471
Tetragonisca angustula 165
Tetranychidae 33, 437
Tetranychus spp. 483
T. kanzawai (Kanzawa spider mite) 40, 41
T. urticae (two-spotted spider mite) 31, 34, 40, 54, 56, 271, 403, 482–485
Theobroma cacao 227, 241
Thoracosathidae 50
Thrips tabaci (onion thrips) 483, 484
Thyreophagus spp. 16, 49, 52, 186, 410
T. cracentiseta 272
T. entomophagus (collection mite) 89, 91, 272–275, **273**
T. gallegoi 271
T. macfarlanei 272
Tineola spp. 227
T. bisselliella 164, 412
Titanolabis colossea (Australian giant earwig) 343
Trebouxia arboricola 380
Trematura jacksonia 490
Trematuridae 490
Tribolium spp. 227, 404, 410
T. castaneum 130, 174, 296, 297, 370, 410, 412, 413
T. confusum 130, 295, 319, 410, 412, 452, 456
Trichoderma spp. 432
T. album 68
Trichomonas hominis 284
Trichophyton 232
Trichoribates spp.
T. novus 380
T. trimaculatus 380
Trichouropoda spp. 443
T. elegans 443
T. jacksonia 443
T. orbicularis 443–444, **445**
T. patavina 443, **444**
Triticum aestivum 109, 211, 296
Troglodytes musculus 134
Trogoderma spp. 137, 404, 410
T. parabile 410
tropical fowl mite *(Ornithonyssus bursa)* 416, 420
tropical leatherleaf slug *(Laevicaulis alte)* 493
tropical rat mite *(Ornithonyssus bacoti)* 452
Tropilaelaps spp. 450
Tsukamurella spp. 421
Tullbergia spp.
T. kraussbaueri 457
T. putrescentiae 457
Turbinoptidae 50
Turdus spp. 398
T. leucomelas 470
T. merula 398
T. philomelos 256
turtle *(Terrapene carolina)* 471
two-spotted spider mite *(Tetranychus urticae)* 31, 34, 40, 54, 56, 403, 482–485
Tydaeolus sphaeroclaviger 177
Tydeidae 175, **176**
tydeid mite *(Tydeus californicus)* 175
Tydeus spp. 175, 177

T. australensis 15, **17**
T. californicus (tydeid mite) 175
T. interruptus **176**, 177, 408, 456
Typha orientalis 463–464
Typhlodrominae 480
Typhlodromus spp.
T. bakeri 484
T. cucumeris 485
T. domesticus 69, 399
T. mali 410
T. pomorum 402
T. (Anthoseius) transvaalensis 483, 487
Tyreophagus entomophagus 15, **18**
Tyroborus spp. 185, 242
T. entomophagus 274
T. farinae (grain mite) 196
T. lini 55, 242–245, **243**, 285, 286, 295
T. ueckermanni 242
Tyrolichus spp. 185, 206, 244
T. casei (cheese mite) **206**, 206–208
Tyrophagus spp. 48, 49, 75, 77, 91, 92, 95, 97, 185, 200, 208–209, 220, 241, 321, 411, 451, 484
T. amboinensis 213
T. americanus 213
T. australasiae 213
T. bambusae 209
T. breviceps 213
T. brevicrinatus 209
T. castellani 213
T. cocciphilus 213
T. communis 185, 212, 213, 227, 228
T. curvipenis 185, 209–212, **210**, 463–464, 486
T. dimidiatus 209
T. fanetzhangorum 209, 211, 212–213, 228
T. formicetorum 97
T. longior 46, 48, 55, 60, 73, 183, 185, 187, 192, 198, 200, 201, 204, 205, 208, 209, 211, 213–217, **214**, 217, 222, 236, 283, 330, 331, 407–408
T. miripes 209, 242
T. nadinus 213
T. neiswanderi (pollen-cap mite) 7, **9**, 54, 65, 185, 189, 190, 209, 211, 217–220, **218**
T. neotropicus 213
T. noxius 227
T. oudemansi 209
T. palmarum 46, 48, 209, 220–222, **221**
T. perniciosus 185, 209, 211, 222–224, **223**
T. putrescentiae (mould mite) 55, 56, 61, 63, 64, 65, 67, 73, 74, 83, 86, 87, 96, 104–105, 183, 185, 187, 190, 192, 196, 199, 202–206, 208–213, 222, 224–237, **225**, 242, 248, 256–258, 283, 284, 295, 303, 318, 320, 321, 357, 371, 374, 399, 408–411, 452–453, 458–459, 462, 484
T. savasi 209
T. similis (The French fly) 54, 56, 185, 209, 220, 237–240, **238**, 250, 257
T. tropicus 240–242, **241**
T. vanheurni 221, 222
T. zachvatkini 209

U
Uloma spp. 52
Upupa epops (hoopoe) 468, 469
Urodinychidae 490
Uroobovella spp.
U. marginata 490–493, **491**
U. obovata 490, 493
U. pulchella 490
Uropoda spp.
U. acuminata 493
U. krameri 443
Uropodella spp. 488
U. laciniata 488, **489**
Uroseius acuminatus 490, **492**

V
Varroa destructor (varroa mite) 457
varroa mite *(Varroa destructor)* 457
Verticillium spp. 67
V. albo-atrum 264
Vespacarus 47, 52
Vespula germanica 211
Vexillariidae 50
Viburnum opulus (European cranberry bush) 403
Virgilia divaricata 227

W
Wallemia spp. 349
W. sebi 198, 361
warty grain mite *(Aëroglyphus robustus)* 293–297, **294**, 318, 410, 452
water shrews *(Neomys fodiens)* 469
western flower thrips *(Frankliniella occidentalis)* 403, 460, 484
western marsh harrier *(Circus aeruginosus)* 398
wet alder forest *(Alnion glutinosae)* 398
wheat grain mite *(Sancassania berlesei)* 267–271, **268**
white-eared bulbul *(Pycnonotus leucotis)* 328
white-footed mouse *(Peromyscus leucopus)* 471
white-fringed beetle *(Graphognathus leucoloma)* 345
white muscardine fungus *(Beauveria bassiana)* 432
white peach scale *(Pseudaulacaspis pentagona)* 403
white stork *(Ciconia ciconia)* 460, 493
white-tailed sea eagle *(Haliaeetus albicilla)* 460
white worms *(Enchytraeus albidus)* 475
Winterschmidtiidae 47, 52
woodlark *(Lullula arborea)* 398
wood mouse *(Apodemus sylvaticus)* 463

X

Xanthoria parietina (common orange lichen) 380
Xenopsylla cheopis 249
Xenorhabdus 295
Xolalgidae 50
Xylocopa spp. 298, 373–374
  X. nigrita 298
  X. (Koptorthosoma) tenuiscapa 336
Xylocopinae 52

Y

yellow-necked mouse *(Apodemus flavicollis)* 463
ylocopinae 52

Z

Zabrotes subfasciatus (Mexican bean weevil) 410
Zachvatkin 139
Zachvatkiniola spp.
  Z. reticulata 145
Zanthoxylum bungeanum 244
Zea mays 296
Zenaida auriculata (eared dove) 452
Zenarge turneri 239
Zerconopsis remiger 397, 398
Zethacarus 52
Zethovidia 52
Zizyphus jujuba (red date) 298, 323
Zostera marina (eelgrass) 393
Zwickia 49

Subject Index

Page numbers in **bold** type refer to figures.

Abbasi, H. 132
Acari
    circulatory system 29
    digestive system 35–37
    early tick evolution and ecology 5
    ecological role 3–4
    excretory system 37
    glandular system 33
    hydrostatic pressure 15
    indigestible solid products 37
    internal morphology 29–39
    nervous system 29–30
    origins 4–6
    phylogenesis 4–6
    reproductive systems
        female 31–33
        male 33
    salivary glands 33
    systematic position 3–4
Acaridae 45, 47, 49, 54, 97, 275
    adaptation 185
    homotype and bimotype 184
    life cycle 185
acarid mites 92
acariformes 3, 15, 31, 159–167
    classification 115–129
    morphological feature 115
Acarina
    anemoreceptors 27
    body division 7–9
    circumcapitular furrow 8
    external morphology 7–28
    genital papillae 8
    integument 7
    legs 18–21, **20**
    mechanosensory and chemosensory setae 28
    osmoregulatory organs 28
    primary segmentation 7
    reproductive organs 23–26
    respiratory organs 21–23
    secretory organs 28
    sensory receptors 26–28
    setae of 26, **27**
    vibroreceptors 27
acarine grain pests 456
Acarophenacidae 130–131
accessory glands 33
accidental acarophagy 345
Achipteriidae 377
ACO-PCA-ANN 77
Actinochitinosi 5
adaptation
    andropolymorphism 185
    feeding 52
    human-made habitats 295
    semiaquatic 34
adult instar 42–43
adult pre-oviposition period (APOP) 459, 463–464, 469
adults
    and deutonymphs 119
    longevity 459, 469
    and nymph mortality 104
aeration 99, 100, 110, 199, 201, 260, 331
aerial mites 45
Aëroglyphidae 50, 59, 292–299
aggregation pheromone 55, 182, 291, 436
agro-system 259
airborne fungi 307
air-conditioning systems 311, 356, 422
Akiyama, M. 55
Alberti, G. 35
Algophagidae 49, 50
Algophagopsis 49
Algophagus 49, 50
allergenic diseases 56, 302–303, 308, 311, 312, 314
allergenicity of mites 89
allergic contact dermatitis 89
allergic hypersensitivity 81
allergic rhinitis 92
allergological importance 89
Allocalvolia 50
allomones 54, 55
Alloptidae 50
α-amylases 36
Amblypygi 3
Ameranoetus 50

Ameroseiidae
distribution 394
heterogeneous substrates and habitats 391
microhabitat requirements 391–392
synanthropic habitats 391
Ameroseius 52
Amoah, B. 74, 103, 105, 106, 111
Ampezzoa 5
amphitoky 44
Anactinochitinosi 5
Analgidae 50
anaphylactic reactions 89
anaphylaxis 94
André, H.M. 40
andropolymorphism 185
angiosperms 108
animal health 95–96
Anthophorinae 52
anti-dust mite protection 356
anti-resistance strategy 424
Aphodanoetus 50
Apidae 52
Apiol 107
apodemes 15
Arachnida 5
Arachnids 3
Araneae 3
Arlian, L.G. 38
Armitage, D. 75
Armitage, D.M. 72
Arthropoda 344
arthropod-vertebrate interactions 5
artificial neural networks (ANN) 77
Artu-vetrin Skin Test 96
Ascidae 15
artificial habitats 395
microclimatic conditions 397–398
paraphyletic structure groups 395
stage of decomposition 398
ascid mites 69
Askinasia 51
Aspaly, G. 63, 65
Association of Natural Biocontrol Producers (ANBP) 487
asthma and allergic rhinitis 92
Astigmata 46, 50, 53, 312
astigmatid mites 51, 53, 456
habitats shifts in nest-inhabiting 53–54
life-history patterns 49
Astigmatina mites 7, 10, 15, 21, 23, 25, 33, 34, 35, 37, 40, 41, 482
Athias-Binche, F. 57
atopic eczema 350
Atopomelidae 51
attract-and-kill technique 436
Audycoptidae 51
Auricanoetus 51
Avena sativa 156
avian pathogenic *Escherichia coli* (APEC) 423
AviVet trap 78, 425
Azadirachtin (neem) 107

bacteria
enzymatic activity 35
isolation and cultivation methods 87
symbiotic/parasitic 295
and viral pathogens 29
bacterial 16S rRNA gene copies 88
Baker's itch 199
Baloghella 48
Barbosa, M.F.C. 161
Barker, P.S. 140
Bartonella-like bacteria 88
basilar sclerite 21
Bdellidae 132–133
beehive debris 275, 282, 307, 314
Behan-Pelletier, V.M. 116, 117
beneficial insects 109
Berlese-Tullgren funnel 72
Beron, P. 276
Bingham, G.V. 98
bioassay procedures 54
bio-ecology of parasitic mites 60
bioinformatic alignment analyses 90
biological control 105–106. *see also* natural enemies
CO_2-comprising modified atmospheres 414
greenhouse pests 485
phytoseiid mites 482
biological cycle 197
biological instars 40
bird hosts 53
bird mite disease 423
bird nests 48, 53, 179, 290, 317, 398, 487
bisexual reproduction 472
Blattisociidae
bioassays under microcosm conditions 411
development time of lifespan 411
distribution 405–407, 409–412, 414
food density levels 413–414
immature stages and adult lifespans 404
infective larval stages 415
life cycle 408–409
podonotal and opisthonotal shields 404
residual populations of storage mites 413
Blattisocius 15
Blattný, C. 293
Blomia 303
Boczek, J. 42, 140
Boletacarus 49
Boletoglyphus (Acaridae) 50
Bollaerts, D. 290
Boner, A. 312
Bonomoia 49

botanical extracts 109
bothridial organs 26
Boulanova, E. 311
bran moisture content (BMC) 66
Breny, R. 290
British Pest Control Association (BPCA) 110, 320
Bromeliaglyphus 49
bronchial provocation tests 320
Brooks, D.R. 47
Bruce, W.A. 164
Bruststiele 41
Brygadyrenko, V.V. 63
BT Trap 73
bursa copulatrix 25, 31, 206, 293, 301, **302**, 308, 326
Butler, P.M. 312
ByeMite® 434

cagefree systems 423
calcofluor/difluorbenzuron 37, 245
Caligonellidae 134–135
Calvin, D. 82, 99
calyptostasis 40
Camerik, A.M. 57
Camerobiidae 149
Camerobiids
    geographic distribution 149
    morphological characters 149
Camisiidae 378
Canestrini, G. 42
Canestriniidae 47, 51
cannibalism 466, 471, 484
capitulum 9–11
Carabodidae 379
carbamate 434
carbon dioxide 105
carbon sulphide 105
Carpoglyphidae
    life cycle of 276
    morphological characteristics 276
Carpoglyphus (Carpoglyphidae) 49
Castanera, P. 108, 190, 191
cation exchange capacity (CEC) 428
Celaenopsoidea 13
Ceratozetidae 380
Cerophagus 52
Chaetodactylus 42, 52
chaetotaxy 15, 41, 172, 192–195, 338, 448
Chambers, J. 75, 83, 99, 101, 110
Chamobatidae 381
Charleston, W.A.G. 310
Chatterjee, K. 310
cheese mites 199
chelicerae 11–14, **13**, 120–121
Chelicerata 3, 6
chemical communication 54
chemical control methods 102
chemical methods 106
chemoperception 415
Cheyletidae 46, 134–145
cheyletid mites 69, 134–145
chicken tick rash 423
Chiloanoetus 52
Chilocorus 46
Chirodiscidae 51
Chiroptoglyphus 47, 51
Chirorhynchobiidae 51
chitin 37
chitinase 37
chitin digestion and utilization 87
chlorpyrifosmethyl 173–174, 204, 236
Chmielewski, W. 59, 187, 281, 304, 305
Chortoglyphidae 48, 80, 287–291
chortolure 55, 291
chromolipids 37
circulatory system 29
circumgastric 'notogaster' 15
cladistic analysis 389
Claparèdè's organ 41
Clarke, P.G. 82, 94
Clemmons, E.A. 80, 91, 95
Codex standard 77
Coenobitidae 51
cold treatments 100
Coleoptera 51
collection methods 71
Collins, D. 110
colony system 422
concealed antigens 429
condylophores 276
Cone, W.W. 54
Congovidia 49, 52
conservation biological control 468
control procedures 98–111
Cook, D.A. 101
cooling grain and products 111
Copronomoia 50
copulation 60, 161, 164, 228, 307
copulatory organs 33
Cornere, B.M. 311
corrosion to metals 103–104
Cosmochthoniidae 382
cosmopolitan mites 66, 228, 290, 311, 384
coxal glands 33–34
critical control point method 425
critical equilibrium activity (CEA) 38, 39
Crooker, A.R. 35
Crustacea 3
Crypturoptidae 50
Ctenocolletacarus 52
Ctenoglyphid mites 304–305
Ctenoglyphinae 48, 313
Cui, Y. 91
Cunaxidae 146–147

Cunaxid mites 146
Cunnington, A.M. 197
curative control measures in cereals 111
Cureulanoetus 52
Cutcher, J.J. 58
Cymbaeremaeidae 383
cysteine 36
Cyta latirostris 132
cytochrome oxidase subunit I (COI) region 79
Cytoditidae 50

damage produced by storage mites 80–97
defensins in astigmatid mites 29
DEFRA (Department for Environment Food and Rural Affairs) 101
Deinocrotonidae 5
delayed hypersensitivity 89
deposition process 40
Dermanyssidae 13
  acaricidal efficacy of products 428
  action threshold 425
  antagonistic effects 436
  automated mite counter 425
  biocides/veterinary medicines 434
  circular testis development 419
  compatibility matrix of control strategies 435
  control methods 423–424
  dermatological problems of humans 422
  diagnostic characteristics 416–417
  economic importance 417–418
  environmental disequilibrium 432
  genitoventral/epigynal shield 418
  hematophagous ectoparasites of vertebrates 416
  host-searching behaviour 420
  hygiene regulations 424
  median lethal concentration and time 427
  PRM *in vitro* feeding assay 429
  reservoirs/vectors of zoonotic pathogens 421–422
  vectorial competence 421
dermatitis 83, 92–94, 136, 199, 206, 453, 456
Dermatophagoides 96, 141
Desmonomatides 116
detection methods 71
deutonymphal instars 40
deutonymphs 41–42
  non-phoretic dispersal 59
  phoretic dispersal 58–59
deutovum 40
Diasecticide™ 205
Diaspididae 52
diatomaceous earth (DE) 101, 102, 427–428
dichloro-diphenyltrichloroethane (DDT) 434
diflubenzuron 37, 245
Digamasellidae 437–438
digestive system 35–37
Dinabandhoo, C.L. 164
diplodiploidy 44
Diptilomiopidae 46
direct damage to stored food 80–84
direct sperm transfer 25
dispersal (deutonymphal) 56–57
  and environment, adverse changes 60
  morph, loss of 59
  (deutonymphal) morph, loss of 59
  of permanent parasites 59–60
Dithinozercoidae 490
DNA extraction and rearing 78
Dogra, G.S. 164
dorsal and ventral shields 15
Douglas, K.T. 80, 91, 95
Dowell, F. E. 75, 76
Dryacide 201, 205–206
dry dog food (DDF) 232
Dubinin, V.B. 47, 59
dung-inhabiting beetle 178, 466
Dunn, J.A. 73, 75, 76
dynamic adaptive model (DAP) 426
Dynastinae 52

early tick evolution 5
eastern equine encephalitis 423
Echimyopodidae 48, 53, 300–303
ecology 5, 45–70
  allomones 55
  bird hosts 53
  chemical communication 54
  dispersal 56–57
  dispersal and adverse changes in the environment 60
  dispersal of permanent parasites 59–60
  factors influencing population dynamics 60–63
  feeding habits 45
  food resource 66–68
  free-living species 45–46
  habitats 45, 48–49
    by animals through non-interactive processes 50–51
    by host interaction with environment 51
    by insects 51–53
    by plants and fungi 49–50
    shifts in nest-inhabiting Astigmatid mites 53–54
  host associations 46–47
  kairomones 56
  loss of dispersal (deutonymphal) morph 59
  mammal hosts 53
  natural enemies of storage mites 68–70
  non-phoretic dispersal by deutonymphs 59
  parasitic species 46
  parasitism, evolution of 47–48
  pheromones 54–55
  phoresy 57–58
  phoretic dispersal by deutonymphs 58–59
  synomones 56
  temperature and relative humidity 63–66

ectomycorrhizal fungi 379, 387
ectoparasitic mites 473
edible bird nests (EBNs) 94–95
eggs, sterilized horse dung substrate 469
elattostasis 40
electronic nose 77
elevated temperatures 100
embryonal somites 15
Endeostigmata 159–160
endosymbionts 432
energy-consuming mechanism 38
Ensliniella 47, 52
enteric acariasis 92
ENTOMITE-A 459
entomopathogenic fungi 431–432
enzyme-linked immunosorbent assay (ELISA) 71, 76, 200, 351
ephemeral 49
    micro-ecosystems 466
epimeral plates 15
Erban, T. 73
Ereynetidae 178
Eriophyidae 25, 32, 46
eriophyids 485
Eriophyoidea 5, 41, 45
eriophyoid pest mites, life cycle 40
Eriophyoids 3
erythematous papules 452
Erythraeidae 155
essential oils 426–427
Esser, J.R. 182
eupathidia 19
EU pesticides database 205
Euphthiracaridae 384
Eupodidae 148
Eurofins Food & Feed Testing laboratory network 104
European breeding models 421
European Chemical Agency 434
European Medicine Agency 434
Eviphididae
    dorsal idiosomal chaetotaxy 448–449
    podospermal with holodorsal shield 447
evolutionary adaptations 14–15
Ewingia 51
excretory system 37
exposed antigens 429
extrusion movements 12–13
Exzolt® 435
Ezequiel, S. 136

faeces-associated or environmental bacteria 88
Fagacarus 50
Fain, A. 46
famuli 19
Fan, Q.-H. 185, 224
Faroni, L.R.D. 130
Fashing, N.J. 46, 50
fecundity 264, 459, 469–470, 473
feeding 293, 305, 311–313, 340, 367, 387, 409
    habits 45
    mould mites 231
    nematodes 270
feed-to-gain ratio 199
female sperm access system 478
filamentous bacteria 37
flotation technique 76, 82, 95
flotation test 74–75
fluralaner 435
Food & Environmental Research Agency (FERA) 321
food-grade oils and preservatives 106
food preservation techniques 98–99
food production systems 434
food resource 66–68
forensic importance, storage mites 96–97
formadin 434
fowl mite dermatitis 423
Franzolin, M.R. 83
free-living predatory mites 134
free-living species 45–46
fruit fly-mite systems 473
fumigants 98, 103, 284
fumigation 106
    chamber 111
fungal mycelia 173, 257, 349
fungivory 85
fungus
    feeding habits 176
    fungus-friendly 68
    robine bulb mite 258

gamasoidosis 423
g-aminobutyric acid (GABA) 265
gas chromatography-mass spectrometry (GC-MS) 54, 84, 101, 358
Generally Regarded as Safe (GRAS) 106
genetic control 109
genetic dimorphism 325
genital capsule 168
genital opening 287
Gill, N.K. 311
glandular system 33
Glycyphagidae 53
    damp dwellings 328–329
    genetic heterogeneity and ecological plasticity 324
    habitats 313
    human allergic sensitivities 331
    human intestinal and urinary acariasis 319
    infestation rates 314
    life cycle 318, 323–325
    mite trap and flotation test 320
    modified atmospheres on eggs 332
    packaging test 320–321

Glycyphagidae (*Continued*)
  reproductive rates 330
  statistical models 330
  systematic structure 313
  temperate climate 329
glycyphagid mites. *see* Glycyphagidae
Glyphanoetus 50, 52
gnathosoma 5, 9–11, **11**, **12**, 223, 368
  subcheliceral plate 9
gnatosoma 119
Gohieria 7
gonotrophic cycles 5
grain mites
  and beetles 93
  infection 95
  natural enemies 105
grain preparation 111
grain protectants 204
grain storage 196
Grandjean, F. 40, 41
Grandjean's organ **27**, 34, 184, 194, 209, 210, 213, 222–226, 263
Griffiths, D.A. 32, 48, 57, 67, 187
grocer's itch 82, 199, 315
gross reproductive rate (GRR) 344, 409, 459
Guimaraes, J.M. 98, 100–101, 103, 105, 109, 279
Gupta, S.K. 310

habitats 45, 48–49
  by animals through non-interactive processes 50–51
  by host interaction with environment 51
  by insects 51–53
  by plants and fungi 49–50
  shifts in nest-inhabiting Astigmatid mites 53–54
haemocoel 29
haemocytes 29
Hagstrum, D.W. 98, 290
Halictacarus 52
Halliday, R.B. 92
Hammen, L. van der 308
hand-operated vacuum apparatus 72
Haplochthoniidae 385
haplodiploidy 44
Hassan, M.F. 104
hazard analysis 425
heat 100–101
Hemisarcoptidae 46, 336–337
Henriot, Athias 13
herb storehouses 284
heteromorphic male 255, 267
Heterostigmata 19
Heterostigmatina 116
Heterozerconina 13
HGCA-funded project 101
Hiroyuki, H. 55
Histiostomatidae
  aquatic environment 338
  bursa copulatrix 342
  decaying fungi 347
  life cycle 344
  necrotic tissue 339
  occupational exposure to allergenic mites 343
  opisthonotal glands 338, 342
  phoretic deutonymph (hypopus) 339–340
  propodosomatic and hysterosomatic sclerites 338
holonotal shield 15
home environment 356
homeomorphic male 255, 263
honeycombs 276, 307
honeydew 175, 480
Horn, T.B. 134
host associations 46–47
host-microbe interactions 88
host plants 215, 228, 260, 269, 403
  chemical compounds 54
  and substrates 269
Houck, M.A. 46, 57, 59
house-dust mites 35, 36, 55, 75, 90, 96, 295, 349–350
HSI system with chemometric algorithms 77
Hubert, J. 35, 65, 80, 81`, 83, 85, 86, 87, 88, 109
Hughes, A.M. 46, 51, 57, 63, 66, 67, 140, 186, 191, 206, 304, 310
Hussey, N.W. 57
hydrostatic pressure 15
hygiene 74, 78, 99, 110, 232, 282, 366, 424, 432
hyperspectral imaging (HSI) system 76–77
hypopial deutonymphs 293
hypopus 41–42, **188**, 188–189, 192–195
hysterosoma 34, 287

identification
  keys (*see Taxonomic Index*)
  pheromone 291
  storage mites 78–79
idiosoma 15–19, **16**, 288, 293, 305, 308, 310, 391
  of Astigmatina 16
  external structures 16
  locomotion 16
  osmoregulation 16
  respiration 16
  secretion 16
  sensory receptors 16
    types 26
  sperm transfer 16
idiosomatic shields 15
imagochrysalis 40
immunodiagnostic method 75–76, 200
immunological control strategy 429
immunotherapeutic extracts 355

indigestible solid products 37
induction pores 31
inert dusts 101–102
inhalational allergy 89
initial population density (IPD) 66
insecticides 100, 106, 111, 172–173
integrated control system 109
integrated pest management (IPM) programmes 111, 205–206, 233, 237, 435–436
  interventions 74
  for stored products 102
internal seminal receptacle 25
internal transcribed spacer region (ITS-2) 79, 196
intradermal skin test 141
isoxazolines 434
Ixodides 5

Jacobs, S. 82, 99
jasmonic acid (JA) 108
Jayas, D.S. 296
Joharchi, O. 69, 450
John, P.A. 182

kairomones 54, 56
Kanakobia 51
Kanoetus 51–52
Kaszabanoetus 52
Kennethiella 47, 52
Kethley, J.B. 161
Kew, P.E. 94
keys
  Acaridae 126
  Acarophenacidae 122
  Achipteriidae 124
  Aeroglyphidae 125
  Ameroseiidae 128
  Anystides 119
  Arborichthoniidae 123
  Ascidae 128
  Bdellidae 120
  Bdelloidea 120
  Caligonellidae 121
  Camerobiidae 121
  Camisiidae 123
  Carabodidae 124
  Carpoglyphidae 126
  Ceratozetidae 125
  Chamobatidae 124
  chelicerae 120–121
  Cheyletidae 120
  Chortoglyphidae 126
  Cosmochthoniidae 123
  Cryptognatidae 120
  Cunaxidae 120
  Cymbaeremaeidae 124
  Dermanyssidae 129
  Digamasellidae 127
  Echimyopodidae 126
  Ereynetidae 120
  Erythraeidae 119
  Erythraeoidea 119
  Euphthiracaridae 122
  Eupodidae 120
  Eupodides 119
  Eviphididae 129
  Glycyphagidae 125
  Haplochthoniidae 123
  Hemisarcoptidae 126
  Heterocheylidae 121
  Histiostomatidae 125
  identification (*see Taxonomic Index*)
  Laelapidae 129
  Lardoglyphidae 125
  Macrochelidae 128
  Monogynaspida 118
  Ologamasidae 127
  Oplitidae 129
  Oppiidae 124
  Oribatulidae 124
  Parasitengonina 119
  Parasitidae 127
  Parholaspididae 128
  Pediculochelidae 123
  Phytoseiidae 128
  Protoplophoroidae 122
  Pterygosomatidae 120
  Pyemotidae 122
  Pyemotoidea 121
  Pyroglyphidae 125
  Raphignathidae 121
  Raphignathina 119
  Raphignathoidea 120
  Rhodacaridae 128
  Smarididae 119
  Stigmaeidae 121
  Suidasiidae 126
  Tarsocheylidae 121
  Tarsonemidae 121
  Tarsonemoidea 121
  Terpnacaridae 118
  Trematuridae 129
  Trombidiae 119
  Tydeidae 120
  Tydeoidea 120
  Uropodellidae 119
  Uropodidae 129
  Winterschmidtiidae 126
Khaing, T.M. 79
Khanjani, M. 134
Khaustov, A.A. 148
Kiliç, T. 102
Klimov, P.B. 48, 212–213

Kopecký, J. 199
Kosik-Bogacka, D.I. 290–291
Kuwahara, Y. 55

Labidophoridae
  morphological characteristics 308
  sexual dimorphism 308
Labidophorinae Zachvatkin 313
Labidostomatidae 5
Labidostomatides 115
Laelapidae
  decaying oatmeal spillage 461
  distribution 451, 453–454, 457–458
  ectoparasites 450
  habitats 450, 452, 456, 458
  laboratory conditions 459
  life history 452–453
  life stages 459
  pre-oviposition and post-oviposition periods 462–463
  rotting substrates and nests of rodents 460
  sex-determining mechanism 450
  soil organisms 457
Lardoglyphidae 179–183
Larson, D.G. 38
larva 41
Larvacarus (Tenuipalpidae) 3
larval shield characteristics, retention of 15
Latifian, M. 76
Leal, W.S. 55
Lee, H.-S. 107
Lepidoptera 100, 165, 276, 293, 410, 476
less intensive farming systems 422
Levinson, H.Z. 85
Li, C. 66, 90
life cycle 264, 290, 311
  of Acari 40
  Carpoglyphidae 276
  male and female duration 229
  oviposition 228
Lindquist, E.E. 161
linear model 485
logistic regression 474
Louadi, K. 310
low air pressure 100
low-temperature scanning electron microscopy (LTSEM) 78
Loxanoetus 47, 51
Lukoschus, F.S. 46
lung allergy 350
lure-and-kill technique 436
lysozyme-like enzymes 36
lysozymes in astigmatid mites 29

Macrochelidae 13
  biology and ecology 465
  deutonymphs 471
  distribution 468–470
  ectoparasites 473
  habitats 465
  hypothesis 466
  insecticides 474
  life cycle 469, 471–472
  oedipal reproduction 472
  organic substrates 466
  phoretic associations 465
  predation and oviposition rates 473–474
  sciarid density 475
  survival strategy 466–467
macrochelid mites. *see* Macrochelidae
macrocyclic lactones 434
Macroharpa 52
macro-mite packing 476
Magowski, W. 130
malpighian tubules 37
mammal hosts 53
Megachilopus 52
Melicharidae
  characterization 399
  extensive papular dermatosis 402–403
  spermathecal apparatus 399
  systematic history 402
Meliponocoptes 52–53
Meliponoecius 52–53
mesonotal shield 15
Mesostigmata 5, 15, 31, 32, 33
  stages 15
metals corrosion 103–104
metapodosoma 15
methyl bromide 103, 235, 262, 284, 285
microarthropods 45
microbial profiles 88
microflora-producing exogenous enzymes 35
micromycetes 85
  in stored products, vectors 84–87
microsetae 19
microvilli 22
Mills, L.R. 34, 35
Mimolette cheese 199
Minyacarus 5
mite alarm 276
mite cheese 208
mite fitness 88
mite-infested grains 89
mite monitoring 72–74
Mite Monitoring Score (MMS) 424
mites
  micromycetes in stored products, vectors 84–87
  proliferation 110
mites in the dust of air-conditioner filters (MACF) 173
mitogenome 248
Mizoguchi, A. 55
modified atmospheres (MA) 205
Mohamed, A.E.M. 139

moisture content-relative humidity (MC-RH) 295
moisture treatments (mite control) 191
molecular diagnosis 79
molecular methods 78
Molothrognathus 134
monitoring and threshold 77–78
Monogynaspida 116
Morales, M. 91
mortality 197, 244–245, 260, 284
Moshkin, V.S. 63
Mullen, A.J. 66, 91
Mullen, G.R. 186
mushroom cultivation 232, 371, 457
mycotoxins 81

Naczi, R. 59
Naiadacarus (Acaridae) 50
Nanacaroides 49
Nanacarus 49
natural aldehydes 108
natural enemies
  contamination 234
  pathogens and predators 198
  predatory mite 107
  species for specific mites
    grain mite *(Acarus siro)* 192–206, **193**
    haemogamasid mite *(Haemogamasus pontiger)* 319
    poultry red mite *(Dermanyssus gallinae)* 430–431
    Tyrophagus putrescentiae 224–237, **225**
  of storage mites 68–70
Nayak, M.K. 83, 99, 103
near-infrared reflectance spectroscopy (NIRS) method 76
nectarivory 46
neighbour-joining method 90
Neoguanolichus 50
Neohorstia 52
Neohyadesia 49
neopodospermy 419
Neotropacarus 49
neotype fixation 212–213, 228
nervous system 29–30
nests of birds 53
neural networks 76
neuronal cell bodies 30
Nikolaou, P. 106
nitric oxide (NO) fumigation 105
Noei, J. 146
Noguchi, S. 55
non-copulatory sperm transfer 25
non-feeding deutonymphs 41
non-phoretic dispersal by deutonymphs 59
non-woven polypropylene systems 358
Norton, R.A. 116
Notostigmata 5
Nycterilichus 47
nymphochrysalis 40, 41

obligatory phoresy 57
occupational allergy, occurrence 92
O'Connor, B.M. 45–49, 46, 52, 57, 58, 59, *59*, 61, 62, 84, 88, 186, 212–213, 304
octotaxic system 377
Okabe, K. 58
Oksentiuk, Y.R. 66, 189, 196
Olivry, T. 96
Ologamasidae 439
ontogenesis 7, 264, 307
oogenesis 344
oophagy 458
operational taxonomic units (OTUs) 199
Opilioacaridida 41
Opiliones 3
opisthosmatic ganglion 31
opisthosoma 15, 29, 34, 115, 119, 161, 292–293, 308, 364
opisthosomatic shield 15
Oplitidae 441–442
Oppiidae 386
organochlorides 204
organophosphates 204, 265, 433–434
organophosphorus compounds 102–103
Oribatidae 5, 15, 34, 45
oribatid fossils 5
oribatid, microanatomy of 29
oribatid mites 390
Oribatulidae 387–388
orthoquinones 7
Oryza sativa 156
osmoregulatory organs, acarina 28
Ostovan, H. 165
Ottoboni, F. 107
Oulenzia 49
ovicidal activity 108
oviporus 31, 118
oviposition period 25, 61, 197–198, 219, 456, 469
ovoviviparity 469
ovulation process 348, 354
oxygen levels 297
ozone gas 104, 285

packaging tests 77
Packard, A.S. 140
Pająk, C. 93
palpi **14**, 14–15
Palpigradi 3
pancake syndrome 374
Pantopoda 3
parahaploidy 44

paranal suckers 277
Paraneognathus 134
parasitic species 46
Parasitidae 465–466, 478–479
 of Mesostigmata 25–26
Parasitiformes 3, 32, 37, 116
parasitism 473
 evolution of 47–48
Parholaspididae 477
Park, H.S. 68
Parkinson, C.L. 86–87
parthenogenesis 44, 166, 173, 414
partial least squares analysis 76
PC™ trap 74
Pediculochelidae 389
pelargonic acid 191
Penalver, E. 5
pennyroyal fumigation test 358
Penthaleidae 45
Perron, R. 57
pest species 76
pheromones 54–55
Phillips, T.W. 103
phoresy 57–58
phoretic deutonymphs 49, 265, 276, 293, 298, 339
phoretic dispersal by deutonymphs 58–59
phosphine 103–104, 111, 204, 235, 284–285
photoperiods 280
phylogenesis 4–6
phylogenetic analysis 21, 79, 90, 211, 248, 467, 477
physical agents 109
physical control measures 99–100
phytophages 46
phytophagy 46
Phytoptidae 46
Phytoseidae. *see* phytoseiid mites
phytoseiid mites (Phytoseidae) 46
 classification of lifestyles 483
 corpuzae populations 486
 developmental rate 485
 distribution 484–485, 487
 efficacy 484
 feasibility 486
 feeding habits 482
 inhabitants 480
 life cycle 480, 485
 mating effects 486
 phytophagous arthropods 480
 preys 482–485
pirimiphos-methyl residues 72, 174, 204, 236, 459
pirimiphos-methyl sprays 217
Pitariu, T. 92
plant alkaloids 106
plant-derived extracts 106, 107, 236, 423, 426
plant extracts and oils 106–109
plant mites 93
Podapolipidae 3
podocephalic glands 34
podonotal shield 15
podosoma 31
podospermy 13
pollard itch 371
pollenophagy 46
polymerase chain reaction 78
population dynamics 60–63
post-imaginal moulting 40
postoral digestive system 35–37
Prasse, J. 32
predacious mites 474
predation
 biological mode 202
 predator-prey model 203
 predatory mites 260, 261
predatory species 68
prelarva 40
preparing technique 78
preserving technique 78
pretarsal claws 119, 120
preventive control
 curative control measures 111
 storage 199–200
prey availability 109
Pritchard, A.E. 31, 105
prodorsal peritremes 119
prolarva 40
propodosoma 15, 163, 287
Prostigmata 25, 40, 115, 161–167
protease activity 232
protein hydrolysis 36
proterosomatic shield 15
protocol of Fera 74
protonymph 41
protonymphal instars 40
Protoplophoridae 390
pseudogale 423
Pseudoscorpionida 3
pseudostigmatic organs 26
psora dermanyssica 423
Pterygosomatidae 150
pulmonary acariasis 92
push-pull technique 436
Puspitarini, R.D. 67
putative gut symbionts 88
Pyemotidae 161–167
pygidial shield 15
pyrethrin 433
pyrethroids 433–434
Pyroglyphidae
 allergic rhinitis and asthmatic symptoms 351
 asthmatic attacks 350
 burning itching sensation 348
 diagnostic tests 350–351
 drought-resistant quiescent protonymph stage 355
 epidermal dandruff from human skin 360

human squamal cells and skin debris 349
hydrating and dehydrating regimes 355
life cycle 348, 361
maximum acaricidal effect 358
modelling techniques 357
nutritional value 349
population density 363
seasonal dynamics and age structures 362
systematic organization 348

Q-perch® 433
quill mites 50

Rad, B. 76
radioallergosorbent tests 351
Rahil, A.A.R. 305
ratio-response models 203
receptaculum seminis 25, 32, 179, 193, 206, 267, **309**, 352
reproduction
egg production 202
immature stages 190
population doubling time 190
robine bulb mite 256
sex ratio and genetic system 43
reproductive days 459
respiratory syndromes 345
Rezk, H.A. 310
Rhodacaridae 440
rickettsialpox 416
Rickstick method 424, 425
Rogers' equation 474

salivary glands 10, 33, 421
Samung, Y. 91
Sánchez-Ramos, I. 91, 190, 191
sanitation 102
Santos da Silva, E. 303
Saproglyphus 50
Sarcoptiformes 5, 116, 159–160
Sato, M. 55
Sayadi, Z. 134
scanning electron microscopy (SEM) 78
scavengers 51, 96, 97, 175, 185, 492
schizeckenosy 37
Schizomida 3
Schmidt, G. 38, 56
Schrank, F. 140
Schulz, J. 101
sclerotized shields 35
Scorpiones 3
secretory organs, acarine 28
Seeman, O.D. 156
Seesaard, T. 77
Semi Attractive Traps (SAT) 424, 425
seminal vesicle 33, 348
Sennertia 52
Sennertionyx 52
sense organs 31
Seolianoetus 51–52
Sepasgozarian, H. 149
serine proteases 36
Severino, G. 31
sex ratio 280, 469
sexual dimorphism 30
Sharpe Schoolfield Ikemoto (SSI) model 485
Shibata, S. 55
sieves 71–72
silica-based products 427–428
Silico-sec® 205
silk glands 34–35
silk production 34–35
Simplified Passive Tape (SPT) 424–426
Sinha, R.N. 62, 85, 295, 296
Skelton, A.C. 55
skin tests 91, 141, 291, 312, 350, 351
Smarididae 154
Smrž, J. 87
Šobotník, J. 36
soil mites 45
Solarz, K. 93, 187, 189, 214–215, 304, 305, 312
solenidion 19, 163, 178, 185, 195, 210, 212, 287, 301, 305
solenostomes 31
Solifugae 3
Solomon, M.E. 82
somatic adult archetypical acariform 7, 8
sorting 102
soybean trypsin protease inhibitor (STI) 37, 245
species distribution modelling (SDM) 66
spent growth medium (SGM) 36
sperm access system 344
spermatodactyl 13
spermatogenesis 419
spermatophores 13, 25
spermatozoa 446, 493
Sphaerolichida 115
spinneret 15
Spinosad (Elector®) 434
sporocarps 48, 175, 266, 269, 270, 391, 392
spraying with organophosphorus 106
Steinman, H.A. 90
Stejskal, V. 83, 93
Stigmaeidae 46, 156–158
stigmata 22
storage mites 80–84, 92
allergy 88–92
and animal health 95–96
and bacteria 87–88
and citrus mites 94
environmental factors 91
forensic importance 96–97

storage mites (*Continued*)
- good sanitation 99
- house dust and 91
- identification 78–79
- management 109–111
- prevention 99
- primary habitat 90

stored-product mites 37, 48, 58, 69, 80, 81, 102, 199, 205, 245, 266, 284–286, 296, 411
stored-product pests 80
stylophore 12, 115, 168
suboesophageal ganglion 29, 31
sucker-like setae 277
Suidasiidae
- bio-efficacy of green pesticides 371
- diagnostic characteristics 367–368
- economic importance 368
- propodosoma of lamellar setae 367

sulfuryl fluoride 104
Summers, F.M. 156
supercooling point (SCP) 259
supracoxal complex 34
supraoesophageal ganglion 30–31
synganglion tissue 31
synomones 54, 56
systemic anaphylaxis 94–95, 303
Szwabowicz, A. 81

Taha, H.A. 311
Tarsonemidae 32, 46
- bio-ecology 168–169
- morphological characteristics 168–169
- systematics 168–169
- websites 171

T cell-type reaction 89
temperature 280
- adult and nymph survival 433
- egg mortality and hatching 43, 61, 63–66, 190–191
- fecundity and lifespans 201, 264
- and moisture 295
- and relative humidity 63–66
- total developmental time of females 485
- treatments used in mite control 191

temporary internal acariasis 91
tendons 278, 313
Tenquist, J.D. 310
Tenuipalpidae 46
Terpnacaridae 159–160
Terpnacarid mites 159–160
testes 33
Tetranychidae 25, 46
tetranychid mites 27, 483, 487
Tetranychoidea 12, 45
Tetrapodili 5
Thermokill method 433
Thind, B.B. 73, 75, 82, 94
Thomas, W.R. 69
Thyreophagus (Acaridae) 52
tibio-tarsal articulation 300
tocospermy 13
Tomita, A. 55
Tortonia 52
total preadult duration 459
total pre-oviposition period (TPOP) 459, 469
tracheal system 117
Trach, V.A. 69, 450
transmission-blocking vaccines (TBVs) 429–430
transmission electron microscopy (TEM) 78
traps for mites 72–74
Trematuridae
- distribution 443–444
- spermatophore 446
- uncertainty and in-depth studies 443
- unstable microhabitats 445

Triasacaroidea 5
Triasacarus 5
trichobothria 19
tridecane 84, 295, 296
triticale 295
tritonymph 42
tritonymphal instars 40
tritrophic system 482
Trombidiformes 5, 32, 33, 46, 115, 161–167
Trouessartiidae 50
T-type thermocouples 285
Tullgren-Berlese funnel 72
Türk, E. 43, 46
Tydeidae 116, **176**
- bio-ecology 175–177
- morphological characteristics 175–177
- systematics 175–177

Tydeoidea 116
tydeoids 483
Tyrophagus (Acaridae) 35, 59, 96, 97
Tytar, V.M. 66

Ueckermann, E.A. 132, 134
United Nation's Food and Agriculture Organization (FAO) 421
United States Centers for Disease Control and Prevention 164
urinary acariasis 92
Uropodellidae 488–489
Uropodidae
- characterization 490
- deutonymphs 493
- distribution 490–491, 493
- life history 492–493
- reproduction 490

Uropygi 3

Vacante, V. 45
vaccination 429–430
Van Impe, G. 40
vasa deferentia 33
Velcro trap 425
ventral idiosoma 15
Verhulst's mathematical model 425
verticils 19–20, **21**
Vidia 52
Vijayambika, V. 182
vitamin A deficiencies 81
vitamin D deficiencies 81
vitamins 58
Vitzthum, H.G. von 279
Vogel, P. 89
volatile organic compounds (VOCs) 428

Wafa, A. K 137
Wakefield, M.E. 73
Walter, D.E. 132, 148
Wandernymphe 493
washing 101
Wasylik, A. 53, 290
water absorption capacity (WAC) 428
water balance 37–39
water-filled treeholes 50
web nest-producing mites 483
Wen, D.C. 303
western equine encephalitis (WEE) 423
White, N.D.G. 83, 295, 296
Winterschmidtia 52
Winterschmidtiidae 375–376
Witaliński, W. 32, 33
Woertz, J. 64
Woodring, J.P. 58
Wu, H.-H. 36

xerophilic fungi 361
X-ray fluorescence 428

Yang, B. 105

Ždárková, E. 67, 83, 310, 312
zeolite formulations 205
Zhang, Z.-Q. 161, 185, 224
Zootaxa bibliographic analysis, acarological papers 3

Printed and bound by CPI Group (UK) Ltd, Croydon, CR0 4YY
16/09/2026
14963198-0003